Lectures on Algebra

Volume I

S S Abhyankar

Purdue University, USA

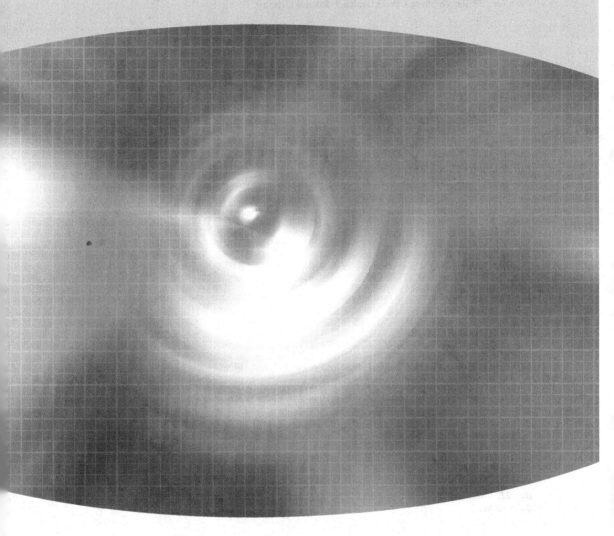

Lectures on Algebra

Volume I

 **World Scientific**

NEW JERSEY · LONDON · SINGAPORE · BEIJING · SHANGHAI · HONG KONG · TAIPEI · CHENNAI

Published by

World Scientific Publishing Co. Pte. Ltd.

5 Toh Tuck Link, Singapore 596224

USA office: 27 Warren Street, Suite 401-402, Hackensack, NJ 07601

UK office: 57 Shelton Street, Covent Garden, London WC2H 9HE

British Library Cataloguing-in-Publication Data
A catalogue record for this book is available from the British Library.

LECTURES ON ALGEBRA

Volume I

ISBN-13   978-981-256-826-7
ISBN-10   981-256-826-3

Printed in Singapore

PREFACE

In universities and colleges it has become customary to give two algebra courses, the first being called abstract algebra and the second linear algebra. The present volume, Lectures on Algebra I, is meant as a text-book for an abstract algebra course, while the forthcoming sequel, Lectures on Algebra II, should serve as a text-book for a linear algebra course. The author's fondness for algebraic geometry shows up in both volumes, and his recent preoccupation with the application of group theory to the calculation of Galois groups is evident in the second volume which contains more local rings and more algebraic geometry. Both volumes are based on the author's lectures at Purdue University during the last several years. An attempt has been made to make these volumes self-contained. The reader may prefer to start with the sixth lecture which gives a rapid summary of the first five lectures. He may also find it helpful to look at the detailed contents printed at the end of the volume just before the index; this is particularly significant for the enormous (about 300 pages) Section §5 of Lecture L5.

When in a certain lecture we are referring to an item from another lecture, the citation of the other lecture precedes the citation of the item. Thus, for instance, in the proof of Theorem (Q4)(T13) in Lecture L5, the reference L4§5(O11) is to Observation (O11) of §5 of Lecture L4, whereas the reference (T11) is to Theorem (T11) of Lecture L5.

Frequently, assertions made in one place are proved or expanded later on. For this purpose, forward reference is indicated by [cf.]. For instance, on page 4, at the end of the sentence "This is unique up to isomorphism, i.e., between two copies of it there is a one-to-one onto map preserving sums and products;" the phrase "[cf. L5§5(Q32)(T138.2)]" means that the proof of the preceding claim that there is a unique field GF(q) of q elements, will be given in Theorem (T138.2) of Quest (Q32) of Section §5 of Lecture L5.

Like a Russian Petrushka doll, there are many books within this book. For instance, the first three lectures, L1, L2, L3 constitute a booklet on a basic abstract algebra course. The sixth lecture L6 by itself constitutes another such course. These two alternatives togther make up a larger such booklet. The fourth lecture L4 is a booklet on commutative algebra. It is continued in the fifth lecture L5 which may be viewed as a treatise on commutative algebra. Within it, Quest (Q31) is a pamphlet on Suslin's work on projective modules and special linear groups over multivariable polynomial rings. Finally, Sections §§2-5 of L3, §§8-9 of L4, and Quests (Q33)-(Q35) of L5§5 form a short course on algebraic geometry.

Thanks are due to Sudhir Ghorpade, Nan Gu, Nick Inglis, Valeria Grant Perez, Avinash Sathaye, David Shannon, Balwant Singh, Umud Yalcin, and Ikkwon Yie for much help. Thanks are also due to NSF Grant DMS 99-88166 and NSA Grant MSP H98230-05-1-0040 for financial support.

Shreeram S. Abhyankar,
Mathematics Department,
Purdue University,
West Lafayette, IN 47907, USA,
e-mail: ram@cs.purdue.edu

CONTENTS

Lecture L1: Quadratic Equations

§1: WORD PROBLEMS

Consider the following word problem. One morning I went to the garden and plucked some roses. Seeing that there were not enough I went to as many flower shops as I had roses and from each of them I purchased as many roses as I had plucked. Thus armed with sufficient flowers I went to the Ganesh Temple, Ganesh being the God of Learning and especially Mathematics. I put at his feet four times as many roses as I had plucked. Then I went to the Shiva Temple and deposited ten roses. The remaining eight roses I took to my spouse. How many roses did I pluck? Now in algebra what we do not know we call x. So having plucked x roses, from the various shops I bought $x \times x = x^2$ roses. Thus armed with $x^2 + x$ roses I offered to Ganesh $4x$ roses and then having deposited ten to Shiva the eight for my spouse makes $x^2 + x = 4x + 10 + 8$. Bringing everything to one side gives us the quadratic equation

$$x^2 - 3x - 18 = 0.$$

We can solve this quadratic equation by completing the square. In other words we want to add a quantity to $x^2 - 3x$ to make it a complete square. By the binomial theorem we have $(x + y)^2 = x^2 + 2xy + y^2$. So we want $2y = -3$, i.e., $y = \frac{-3}{2}$ and hence $y^2 = \frac{9}{4}$. Thus transferring 18 to the right hand side and then adding $\frac{9}{4}$ to both sides we get $x^2 - 3x + \frac{9}{4} = \frac{9}{4} + 18$, i.e., $(x - \frac{3}{2})^2 = \frac{9}{4} + 18 = \frac{9+72}{4} = \frac{81}{4} = (\frac{9}{2})^2$. Therefore $x - \frac{3}{2} = \pm\frac{9}{2}$. Thus $x = \frac{3}{2} + \frac{9}{2} = 6$ or $x = \frac{3}{2} - \frac{9}{2} = -3$. Discarding the negative solution -3 as not applicable in this case we conclude that I plucked 6 roses.

This completing the square method of solving a quadratic equation was conceived around 500 A.D. by the Indian mathematician Shreedharacharya. It was put in verse form in 1150 A.D. by another Indian mathematician Bhaskaracharya in his book [Bha] on Algebra called Beejganit. Thus, writing Y for the unknown, we have the quadratic equation

$$aY^2 + bY + c$$

where a, b, c are its coefficients with $a \neq 0$. To make it monic, i.e., to arrange the coefficient of the highest degree term to be 1, we divide by a to get

$$Y^2 + BY + C$$

with $B = \frac{b}{a}$ and $C = \frac{c}{a}$. Now we complete the square by writing

$$Y^2 + BY + C = \left(Y + \frac{B}{2}\right)^2 + C - \frac{B^2}{4}.$$

Putting back the values of B and C, we get the solutions

$$Y = \frac{-b \pm \sqrt{b^2 - 4ac}}{2a}.$$

In the above word problem, first I thought of bringing 5 flowers to my spouse. That would make $a = 1, b = -3, c = -15$ and putting this in the above formula we would get the solutions to be $\frac{3 \pm \sqrt{69}}{2}$. Since 69 has no integer square root I would be faced with the dilemma of picking an irrational number of flowers. That is why I decided to bring 8 flowers to my spouse since I knew this would lead to a constant term in my equation which could be factored into two factors whose negative sum equals the middle term. What I am referring to is the identity

$$(Y - \alpha)(Y - \beta) = Y^2 - (\alpha + \beta)Y + \alpha\beta = Y^2 + BY + C$$

which tells us that $C = \alpha\beta$ and $B = -(\alpha + \beta)$, i.e., the constant term is the product of the roots and the coefficient of Y is the negative sum of the roots.

Around 1500 A.D., similar but much more complicated formulas for solving cubic and quartic equations by radicals were given by Cardano and Ferrari in Italy. Here by radicals we mean successively extracting square roots, cube roots, and so on. Then for the next three hundred years people tried to solve general quintic equations, where by solving they meant solving by radicals. This was proved to be impossible, first by the Norwegian mathematician Abel in 1820 and then by the young French mathematician Galois in 1830 [Gal]. But Galois went much further and gave a criterion of solvability for any polynomial equation of any degree. What he did was to associate to the equation a certain finite permutation group, which is now called the Galois group, and to prove that the equation can be solved by radicals if and only if its Galois group is solvable in a technical sense which he introduced.

To explain this let

$$f(Y) = a_0 Y^n + a_1 Y^{n-1} + \cdots + a_n \quad \text{with} \quad a_0 \neq 0$$

be a polynomial of any degree n. The coefficients $a_0, a_1, \ldots, a_n$ could be rational numbers, or real numbers, or complex numbers. All these are examples of fields, i.e., collections of objects in which we can carry out the operations of addition, subtraction, multiplication and division; in the last case the quantity we are dividing by is required to be nonzero. Letting the coefficients $a_0, a_1, \ldots, a_n$ of f belong to any field K, we find its roots $\alpha_1, \ldots, \alpha_n$ in some bigger field giving us

$$f(Y) = a_0(Y - \alpha_1) \ldots (Y - \alpha_n).$$

We assume the polynomial f to be separable by which we mean that the roots $\alpha_1, \ldots, \alpha_n$ are all distinct. The Galois group of f is going to be a certain group of permutations of the roots $\alpha_1, \ldots, \alpha_n$. A permutation on a set S is a bijection, i.e., a one-to-one onto map, of S onto itself. Under composition the set of all bijections of

S forms a group which we denote by $\mathrm{Sym}(S)$. In case the set S is finite and its size $|S|$ (= the number of elements in it) is n, it is customary to write S_n for $\mathrm{Sym}(S)$. For the size of an infinite set S we put $|S| = \infty$. Let us now formally introduce the terms set, map, composition, bijection, group, and field.

§2: SETS AND MAPS

A map $\phi : S \to T$ from a set S (= a collection of objects called its elements) to a set T is a rule which to every $x \in S$, i.e., element x of S, associates $\phi(x) \in T$, which is called the image of x under ϕ; this is sometimes indicated by writing $x \mapsto \phi(x)$ which is read as x maps to $\phi(x)$. The map ϕ is injective (= one-to-one) if for all $x \neq y$ in S we have $\phi(x) \neq \phi(y)$. The map ϕ is surjective (= onto) if for every $y \in T$ there is some $x \in S$ with $y = \phi(x)$. The map ϕ is bijective if it is injective as well as surjective. A bijection (resp: an injection, a surjection) is a bijective (resp: injective, surjective) map. The composition of maps $\phi : S \to T$ and $\psi : T \to U$ is the map $\psi\phi : S \to U$ given by $(\psi\phi)(x) = \psi(\phi(x))$. For a bijection $\phi : S \to T$ the inverse $\phi^{-1} : T \to S$ is defined by $\phi^{-1}(y) = x \Leftrightarrow \phi(x) = y$. The logical symbols $A \Rightarrow B$, $A \Leftarrow B$, and $A \Leftrightarrow B$ stand for A implies B, A is implied by B, and A is true if and only if B is true respectively.

§3: GROUPS AND FIELDS

A group is a set G in which: any two elements x, y can be multiplied to produce a third element xy; this multiplication is associative, i.e., for all x, y, z in G we have $(xy)z = x(yz)$, and so we may omit the parenthesis; there is an identity element 1 in G with $1x = x1 = x$ for all $x \in G$; and every $x \in G$ has an inverse $x^{-1} \in G$ with $xx^{-1} = x^{-1}x = 1$. Clearly $\mathrm{Sym}(S)$ is a group under composition, with the identity map $x \mapsto x$ as 1. In particular S_n is a group with $|S_n| = n!$, where $n!$ is read as n factorial and stands for the product $1 \times 2 \times \cdots \times n$. If S is infinite then clearly $\mathrm{Sym}(S)$ is infinite.

The group G is commutative or abelian if $xy = yx$ for all x, y in G. If G is abelian we may write $x + y$ and 0 and $-x$ in place of xy and 1 and x^{-1} respectively, and when we do so, we may call G an additive abelian group. In the general case when we are writing xy, we may call G a multiplicative group; this time G may or may not be commutative.

A field K is an additive abelian group with an associative commutative multiplication defined on it such that the nonzero elements form a commutative multiplicative group and the two operations are connected by the distributive laws saying that for all x, y, z in K we have $x(y + z) = xy + xz$ and $(y + z)x = yx + zx$. The rational numbers, the real numbers, and the complex numbers clearly form infinite fields; they are usually denoted by $\mathbb{Q}$, $\mathbb{R}$, and $\mathbb{C}$ respectively. To get a finite field, in the eventful year 1830, the 19 year old Galois

[Gal] noticed that for any prime p, the integers $0, 1, \ldots, p-1$ form a field when they are added and multiplied modulo p, i.e., after adding or multiplying the answer is to be divided by p and replaced by the remainder. This gives the Galois Field GF(p). At the Chicago World's Fair of 1893, E. H. Moore [Mo1; Mo2] extended Galois' thought by showing that for every power q of p, there is a field GF(q) of q elements. This is unique up to isomorphism, i.e., between any two copies of it there is a one-to-one onto map preserving sums and products; [cf. L5§5(Q32)(T138.2)]. If $1+1+\ldots$ is never zero in a field K then K is said to be of characteristic 0. Otherwise $1+1+\cdots+1 = 0$ when 1 is repeated a certain number of times, and the smallest such number is easily seen to be a prime number, which is then called the characteristic of K. For instance the real and complex fields are of characteristic zero, and the Galois Field GF(q) is of characteristic p.

Just as a set S is a collection of objects, a subcollection R is called a subset and denoted by $R \subset S$; we may also write $S \supset R$ and call S an overset of R; this is characterized by the implication: $x \in R \Rightarrow x \in S$. For $R_1 \subset S$ and $R_2 \subset S$, their intersection $R_1 \cap R_2$ is defined to be the set of all $x \in S$ such that $x \in R_1$ and $x \in R_2$, and their union $R_1 \cup R_2$ is defined to be the set of all $x \in S$ such that $x \in R_1$ or $x \in R_2$. Similarly for the intersection and union of more than two subsets. For $R_1 \subset S$ and $R_2 \subset S$ we define $R_1 \setminus R_2$ to be the set of all $x \in S$ such that $x \in R_1$ but $x \notin R_2$; we may call this the complement of R_2 in R_1. Note that, similar to $x \neq y$, the symbols $x \notin R_2, R_1 \not\subset R_2$, and so on, denote the negations of $x \in R_2, R_1 \subset R_2$, and so on. The symbol $\emptyset$ denotes the empty set. Given any set-theoretic map $\phi : S \to T$, i.e., given any map ϕ of a set S into a set T, for any $R \subset S$ we put $\phi(R) = \{\phi(x) : x \in R\}$, i.e., $\phi(R) =$ the set of all elements of T which can be expressed in the form $\phi(x)$ with $x \in R$; we call $\phi(R)$ the image of R under ϕ; we also put $\operatorname{im}(\phi) = \phi(S)$ and call this the image of ϕ; finally, for any $U \subset T$ we put $\phi^{-1}(U) = \{x \in S : \phi(x) \in U\}$. The set consisting of elements $x_1, \ldots, x_e$ (which may or may not be distinct) is denoted by $\{x_1, \ldots, x_e\}$. However, the group whose only element is 1 may be denoted by 1 instead of the more cumbersome $\{1\}$.

If $H \subset G$ are groups under the same operations (and the same identity elements) then we say that H is a subgroup of G or G is an overgroup of H, and we denote this by $H \leq G$ or $G \geq H$; also we write $H < G$ or $G > H$ to mean $H \leq G$ and $H \neq G$. If $H \leq G$ is such that for all $h \in H$ and $g \in G$ we have $ghg^{-1} \in H$ then we say that H is a normal subgroup of G and indicate this by writing $H \triangleleft G$. If $H \triangleleft G$ with $H \neq G$, then H is called a proper normal subgroup of G. A group G is simple if $G \neq 1$ and G has no proper normal subgroup $\neq 1$, where 1 denotes the identity group having only one element. A group G is cyclic if it consists of powers of a single element; we denote such a group by Z_r where r is the order of G, i.e., the number of elements in G which may be infinite; thus $r = \infty$ or $r = $ a positive integer. Note that a finite simple group is abelian if and only if it is cyclic of prime order.

For $H \leq G$ and $x \in G$, we put $xH = \{xy : y \in H\}$, i.e., $xH = $ the set of all elements of G which can be expressed in the form xy with $y \in H$, and we call xH a left coset of H in G; similarly $Hx = \{yx : y \in H\}$ is called a right coset of H in G; the left cosets of H in G form a partition of G, and so do the right cosets of H in G. Note that a partition of a set S is a collection of nonempty subsets of S such that S is their union and any two of them have an empty intersection. For $H \leq G$, by G/H we denote the set of all left cosets of H in G; if $H \lhd G$ then G/H becomes a group by defining $(xH)(yH) = (xy)H$; we now call G/H the factor group of G by H. For $H \leq G$ we put $[G : H] = |G/H|$ = the number of left cosets of H in G and we call this the index of H in G; this notation is used especially when the index is finite; if G is finite then the index is clearly finite; if G is infinite then the index may be finite or infinite.

A finite group G is said to be solvable if there is a chain $1 = G_0 \lhd G_1 \lhd \cdots \lhd G_r = G$ (of subgroups of G) such that the factor group G_i/G_{i-1} is abelian for $1 \leq i \leq r$; by interposing a few more subgroups and deleting some, we can then arrange that $G_i/G_{i-1} = Z_{p_i}$ = a cyclic group of prime order p_i for $1 \leq i \leq r$. Actually, for any finite group G there is a chain $1 = G_0 \lhd G_1 \lhd \cdots \lhd G_r = G$ such that the factor group G_i/G_{i-1} is simple for $1 \leq i \leq r$. A version of the famous Jordan-Hölder Theorem says that then the simple factors G_{i+1}/G_i are uniquely determined by G except for the order in which they occur. This should remind us of the UFD (= unique factorization domain) property of integers according to which every positive integer n can be uniquely written as a product $n = \prod_{i=1}^{r} p_i$ of prime numbers p_i; here the uniqueness is up to order. Thus simple groups may be compared to prime numbers. So just as we would like to know all prime numbers, we would also like to know all finite simple groups. The latter desire was fulfilled in 1980 when a complete listing of all finite simple groups was achieved; this is called CT = the Classification Theorem of finite simple groups. Eventually we shall include a proof of the Jordan-Hölder Theorem, but not of CT. Indeed the proof of CT is about ten thousand pages. However, eventually we may be able to "describe" it. When invoking CT it is always good to make it clear that this is being done.

The first step towards CT was taken by Galois when he showed that the Alternating Group A_n is simple for $n \geq 5$. To introduce A_n, we note that any $\sigma \in S_n$, i.e., any permutation on n letters, can be written as a product of transpositions, where by a transposition we mean a permutation which permutes only two letters while leaving all the other letters unchanged. Moreover, the parity, i.e., the evenness or oddness, of the number of transpositions occurring in such a product depends only on σ. The permutation σ is called even or odd according as the parity of the product is even or odd. Now A_n is defined to be the group of all even permutations. Clearly $|A_n| = |S_n|/2 = n!/2$ and hence the index $[S_n : A_n] = 2$. Therefore $A_n \lhd S_n$. Since A_n is simple for $n \geq 5$, it follows that S_n is unsolvable for $n \geq 5$. It only remains to note that the Galois group of the "general" quintic is S_5. Therefore the general quintic cannot be solved by radicals. This is Galois' proof. It applies to the general equation of degree n for any $n \geq 5$.

Thus we have the hierarchy of groups in increasing order of complexity: cyclic, abelian, solvable, simple, unsolvable.

§4: RINGS AND IDEALS

Generalizing slightly, if in the definition of a group we do not require the existence of the inverse x^{-1} then what we get is called a monoid, and what we get by further dropping the existence of identity is called a semigroup. The terms commutative or abelian semigroup, additive abelian semigroup, multiplicative semigroup, subsemigroup, and oversemigroup are obvious generalizations of the corresponding terms for groups; similarly for a monoid; note that the identity element of a submonoid is required to coincide with the identity element of the monoid. A semigroup H is cancellative means for any x, y, z in it we have the implications which say that: $xy = xz$ or $yx = zx \Rightarrow y = z$. If H is a subsemigroup of a group G then clearly H is cancellative. As an example of an additive abelian group we have the set of all integers, usually denoted by $\mathbb{Z}$, as an example of a submonoid of $\mathbb{Z}$ we have the set of all nonnegative integers, usually denoted by $\mathbb{N}$, and as an example of a subsemigroup of $\mathbb{Z}$ we have the set of all positive integers, usually denoted by $\mathbb{N}_+$.

Turning to fields, if $K \subset L$ are fields under the same operations (and the same zero and identity elements) then we say that K is a subfield of L or L is an overfield of K. As examples, the rational numbers are a subfield of the real numbers which themselves are a subfield of the complex numbers. Again generalizing slightly, a ring R is an additive abelian group which is also a commutative multiplicative monoid such that the two operations are connected by the distributive laws saying that for all x, y, z in R we have $x(y + z) = xy + xz$ and $(y + z)x = yx + zx$; we call R a null ring if in it we have $1 = 0$, i.e., equivalently if $|R| = 1$. The concepts of a subring and an overring are defined in an obvious manner; in particular they have the same 0 and 1. A domain is a nonnull ring whose nonzero elements form a cancellative multiplicative monoid. If R is a subring of a field K then R is clearly a domain. In particular the ring of integers, being a subring of the field of rational numbers, is a domain which, as said above, is usually denoted by $\mathbb{Z}$. Dropping the commutativity of multiplication in a field (resp: ring, domain) we get the notion of a skew-field (resp: skew-ring, skew-domain). In the definitions of fields and rings we have spelled out two distributive laws, although they follow from each other, exactly because in case of skew-fields and skew-rings they do not. Moreover, in the definition of a skew-field, by requiring the left-distributive law $x(y + z) = xy + xz$, but not requiring the right-distributive law $(y + z)x = yx + zx$, we get the notion of a near-field. Finite near-fields were exhaustively dealt with by L. E. Dickson in 1905.

Dickson was the first Ph.D. student of E. H. Moore in the newly opened University of Chicago which had become a vigorous center of mathematical research

in the westward expansion of the New World. At that time another disciple of E.
H. Moore was the Scottish mathematician J. H. M. Wedderburn. Having classified
finite fields in 1893, Moore asked his two young followers to study near-fields and
skew-fields respectively. Dickson gave a complete list of finite near-fields which was
later shown to be exhaustive by the German mathematician H. Zassenhaus as part
of his 1935 Hamburg Ph.D. thesis written under the guidance of another famous
German mathematician E. Artin. On his part Wedderburn showed that a finite
skew-field is necessarily a field.

As an additive abelian group, every subgroup I of a ring R is a normal subgroup
and so we can form the factor group R/I; note that a typical member of R/I is a
residue class (= additive coset) $a + I$ with $a \in R$. By an ideal in a ring R we mean
an additive subgroup I of R such that for all $a \in R$ and $x \in I$ we have $ax \in I$,
and when that is so we make R/I into a ring by defining $(a + I)(b + I) = (ab) + I$
for all a, b in R. We call R/I the residue class ring of R modulo I. Also we call
I a maximal ideal or a prime ideal in R according as R/I is a field or a domain.
Likewise, we call I a nonzero ideal or a nonunit ideal in R according as $I \neq \{0\}$ or
$I \neq R$; note that every ideal contains the zero ideal $I = \{0\}$ and is contained in
the unit ideal $I = R$; moreover, the zero ideal is prime $\Leftrightarrow R$ is a domain; likewise,
I is the unit ideal $\Leftrightarrow R/I$ is the null ring. For any $a \in R$ we put $aR =$ the set of
all multiples az of a with $z \in R$, and for any $W \subset R$ we put $WR =$ the set of all
finite linear combinations $a_1 z_1 + \cdots + a_e z_e$ with $a_1, \ldots, a_e$ in W and $z_1, \ldots, z_e$ in
R, and we note that these are ideals in R (by convention an empty sum is zero and
hence $0 \in WR$); they are called ideals generated by a and W respectively; an ideal
of the form aR is called a principal ideal in R and a is called a generator of it; in
case $W = \{w_1, w_2, \ldots\}$, we may denote WR by $(w_1, w_2, \ldots)R$ and call $w_1, w_2, \ldots$
its generators. It is easy to see that every ideal in $\mathbb{Z}$ is of the form $p\mathbb{Z}$ with $p \in \mathbb{N}$;
moreover: $p\mathbb{Z}$ is a maximal ideal in $\mathbb{Z} \Leftrightarrow p\mathbb{Z}$ is a nonzero prime ideal in $\mathbb{Z} \Leftrightarrow p$ is a
prime number. If p is a prime number then we may identify $\mathbb{Z}/p\mathbb{Z}$ with the Galois
field $GF(p)$.

The concepts of normal subgroup and ideal can be motivated in terms of homomorphisms thus. For groups G and J, a homomorphism $\phi : G \to J$ is a map
such that $\phi(1) = 1$ and $\phi(xy) = \phi(x)\phi(y)$ for all x, y in G; the kernel of ϕ, denoted
by $\ker(\phi)$, is the set of all $x \in G$ with $\phi(x) = 1$; note that ϕ is injective iff (=
if and only if) $\ker(\phi) = 1$ and when that is so we call ϕ a monomorphism; if ϕ
is surjective, i.e., if $\operatorname{im}(\phi) = J$, then we call ϕ an epimorphism; if ϕ is bijective
then we call it an isomorphism; if $J = G$ and ϕ is an isomorphism then we call ϕ
an automorphism of G; the set of all automorphisms of G is clearly a subgroup of
$\operatorname{Sym}(G)$ and we denote it by $\operatorname{Aut}(G)$. If $G \to J$ is a group homomorphism then
clearly $\operatorname{im}(\phi) \leq J$ and $\ker(\phi) \triangleleft G$. Conversely, for any $H \triangleleft G$ we get a group epimorphism $G \to G/H$ with $\ker(\phi) = H$ by taking $\phi(x) = xH$ for all $x \in G$; we
call ϕ the canonical epimorphism of G onto G/H. Note that for additive abelian
groups G and J the definition of a homomorphism $\phi : G \to J$ becomes $\phi(0) = 0$ and

$\phi(x+y) = \phi(x) + \phi(y)$ for all x, y in G. For rings S and T, a (ring) homomorphism $\phi : S \to T$ is a homomorphism of additive abelian groups such that $\phi(1) = 1$ and $\phi(xy) = \phi(x)\phi(y)$ for all x, y in S; now im(S) is a subring of T and ker(ϕ) is an ideal in S, and the definitions of monomorphism, epimorphism, isomorphism, and automorphism carry over from the group case. Conversely, given any ideal I in a ring S, for the canonical ring epimorphism $\phi : S \to S/I$ we have ker(ϕ) $= I$. The set of all ring automorphisms of a ring S is clearly a subgroup of Sym(S) and we denote it by Aut(S); momentarily letting this Aut be written as Ring-Aut(S) and writing Group-Aut(S) for the automorphism group of the additive abelian group S we clearly have Ring-Aut(S) $\leq$ Group-Aut(S) $\leq$ Sym(S); it is usually clear from the context which automorphism group is being considered and so it is not necessary to use such a cumbersome notation. Finally for a subring R of a ring S, by an R-automorphism of S we mean an automorphism ϕ of S which leaves R elementwise fixed, i.e., $\phi(x) = x$ for all $x \in R$; these form a subgroup of Aut(S) and we denote it by $\mathrm{Aut}_R(S)$; this is particularly relevant when R is a subfield of a field S.

The above definitions of homomorphism, monomorphism, and so on, have obvious generalizations to semigroups, monoids, and so on. Also the concepts of subskew-field, overskew-field, and so on, are defined in an obvious manner.

§5: MODULES AND VECTOR SPACES

Any overring S of a ring R is a module over R in the following sense. A module over R is an additive abelian group V together with a scalar multiplication which to every $r \in R$ and $v \in V$ associates $rv \in V$ so that for all r, s in R and v, w in V we have $(r + s)v = rv + sv$, $r(v + w) = rv + rw$, $(rs)v = r(sv)$, and $1v = v$. Elements $v_1, \ldots, v_e$ in V are linearly dependent (over R) if $r_1 v_1 + \cdots + r_e v_e = 0$ for some $r_1, \ldots, r_e$ in R such that $r_i \neq 0$ for some i; otherwise they are linearly independent (over R), and then clearly they must be distinct, i.e., $v_i \neq v_j$ whenever $i \neq j$; an element $v \in V$ is linearly dependent (over R) on $v_1, \ldots, v_e$ if $v = r_1 v_1 + \cdots + r_e v_e$ for some $r_1, \ldots, r_e$ in R; by $Rv_1 + \cdots + Rv_e$ we denote the set of all $v \in R$ which are linearly dependent on $v_1, \ldots, v_e$; finally, the elements $v_1, \ldots, v_e$ form a basis of V (over R) if they are linearly independent and every element of V is linearly dependent on them; in all this e is a nonnegative integer; note that if $e = 0$ then by convention $Rv_1 + \cdots + Rv_e = \{0\} = 0$. For any $W \subset V$, we put $RW = \cup(Rv_1 + \cdots + Rv_e)$ where the union is over all finite sequences $v_1, \ldots, v_e$ in W, and we say that W is linearly independent to mean that every finite set of distinct elements in W is linearly independent. By a submodule U of V we mean $U \subset V$ such that $RU = U$. A submodule U of V is generated by (= spanned by) $W \subset V$ if $U = RW$, and U is finitely generated if $U = RW$ for a finite $W \subset V$. By a basis of a submodule U of V we mean a linearly independent subset W of V such that W generates U. Clearly 0 and V are submodules of V, and every submodule of V is a module over R in an obvious sense. For submodules $U_1, \ldots, U_e$, the submodule generated by $U_1 \cup \cdots \cup U_e$ is denoted by $U_1 + \cdots + U_e$.

When in the above definitions we want to stress the reference to the ring R, we may say R-module, R-submodule, R-linearly independent, R-basis, and so on.

For modules V and V^* over a ring R, by an R-homomorphism (or R-linear map) $\phi : V \to V^*$ we mean a homomorphism ϕ of the underlying additive abelian groups such that for all $r \in R$ and $v \in V$ we have $\phi(rv) = r\phi(v)$; clearly $\ker(\phi)$ and $\operatorname{im}(\phi)$ are submodules of V and V^* respectively. Now the meaning of R-monomorphism, R-isomorphism, and so on, is obvious.

For a skew-ring R, the above description defines a left R-module V, and an obvious change defines a right R-module V with a scalar multiplication which to every $r \in R$ and $v \in V$ associates $vr \in V$. The rest of the above definitions now apply with obvious modifications.

By a vector space over a field K we mean a K-module V, and by a subspace of V we mean a K-submodule of V, and so on. Elements of a vector space V, are frequently called vectors. It can be shown that a vector space V always has a basis, and there is a bijection between any two bases. The number of elements in a basis of V is denoted by $\dim_K V$, and we call this the dimension of V over K or the K-dimension of V. Note that then either $\dim_K V \in \mathbb{N}$ or $\dim_K V = \infty$. It can be shown that any set of generators of V contains a basis of V and hence if V is generated by a finite number of vectors $v_1, \ldots, v_e$ then $\dim_K V \leq e$.

As said above, any overring of a ring R is an R-module in an obvious sense. In particular R is an R-module, and an ideal in R is exactly an R-submodule of R.

In particular, any overring L of a field K is a K-vector space, and we put $[L : K] = \dim_K L$. In particular this applies to an overfield L of K, and then we call $[L : K]$ the field degree of L/K, i.e., of L over K. For a ring or skew-ring R, by R^+ we denote the underlying additive abelian group of R, and by $R^\times$ we denote the set of all nonzero elements of R; note that $K^\times$ is a multiplicative group in case of a field or skew-field K. We extend the last notation to any additive abelian monoid (such as an additive abelian group, or a module, or a vector space) V and denote the set of all nonzero elements in it by $V^\times$. For an overfield L of a field K, $[L : K]$ usually denotes the field degree rather than the index of K^+ in L^+.

For a homomorphism $\phi : V \to V^*$ of monoids (resp: additive abelian monoids; in particular, modules) we write $\phi = 1$ (resp: $\phi = 0$) to mean $\operatorname{im}(\phi) = 1$ (resp: $\operatorname{im}(\phi) = 0$), where we note that the submonoid $\{1\}$ (resp: $\{0\}$) is denoted by 1 (resp: 0).

§6: POLYNOMIALS AND RATIONAL FUNCTIONS

Given any ring R, we can consider the polynomial ring $R[X]$ in an "indeterminate" X over R, and regard this as an overring of R. A typical polynomial in X

over R, i.e., a typical member of $R[X]$, has a unique (finite) expression

$$A(X) = \sum A_i X^i \quad \text{with} \quad A_i \in R$$

and i varying over a finite set of nonnegative integers. The degree of A is the largest i with $A_i \neq 0$; if $A = 0$, i.e., if $A_i = 0$ for all i, then the degree is taken to be $-\infty$; A is nonconstant means it does not belong to R, i.e., its degree is a positive integer; A is monic means $A \neq 0$ and $A_e = 1$ where e is the degree of A. For any other polynomial

$$B(X) = \sum B_i X^i \quad \text{with} \quad B_i \in R$$

the sum

$$C(X) = A(X) + B(X) = \sum C_i X^i$$

is given by componentwise addition

$$C_i = A_i + B_i$$

and the product

$$D(X) = A(X)B(X) = \sum D_i X^i$$

is given by the "Cauchy multiplication rule"

$$D_k = \sum_{i+j=k} A_i B_j.$$

The "formal" X-derivative $A_X(X) = \frac{dA(X)}{dX}$ of $A(X)$ is defined by putting

$$A_X(X) = \sum i A_i X^{i-1}.$$

Clearly $A \mapsto A_X$ gives an R-derivation of $R[X]$, i.e.,

$$(A + B)_X = A_X + B_X \quad \text{and} \quad (AB)_X = A_X B + A B_X$$

with $A_X = 0$ for all $A \in R$. If R is a domain then the degree of the product of two nonzero polynomials equals the sum of their degrees, and hence in particular the product is nonzero. It follows that if R is a domain then so is the polynomial ring $R[X]$. For an indeterminate X over a field K, by $K(X)$ we denote the field of rational functions in X over K. Members of $K(X)$ are "quotients" of polynomials $A(X)/B(X)$ with nonzero B. The above K-derivation of $K[X]$ can be extended to a K-derivation of $K(X)$ by the "quotient rule"

$$(u/v)_X = (u_X v - u v_X)/v^2.$$

Similarly, given any ring R, we can consider the polynomial ring $R[X_1, \ldots, X_m]$ in a finite number of "indeterminates" $X_1, \ldots, X_m$ over R, and regard this as an overring of R. A typical polynomial in $X_1, \ldots, X_m$ over R, i.e., a typical member of $R[X_1, \ldots, X_m]$ has a unique (finite) expression

$$A(X_1, \ldots, X_m) = \sum A_{i_1 \ldots i_m} X_1^{i_1} \ldots X_m^{i_m} \quad \text{with} \quad A_{i_1 \ldots i_m} \in R$$

and $(i_1, \ldots, i_m)$ varying over a finite set of m-tuples of nonnegative integers. The degree of A is the maximum of $i_1 + \cdots + i_m$ with $A_{i_1 \ldots i_m} \neq 0$; if $A = 0$, i.e., if $A_{i_1 \ldots i_m} = 0$ for all $i_1, \ldots, i_m$, then the degree is taken to be $-\infty$; A is nonconstant means it does not belong to R, i.e., its degree is a positive integer. For any other polynomial

$$B(X_1, \ldots, X_m) = \sum B_{i_1 \ldots i_m} X_1^{i_1} \ldots X_m^{i_m} \quad \text{with} \quad B_{i_1 \ldots i_m} \in R$$

the sum

$$C(X_1, \ldots, X_m) = A(X_1, \ldots, X_m) + B(X_1, \ldots, X_m) = \sum C_{i_1 \ldots i_m} X_1^{i_1} \ldots X_m^{i_m}$$

is given by componentwise addition

$$C_{i_1 \ldots i_m} = A_{i_1 \ldots i_m} + B_{i_1 \ldots i_m}$$

and the product

$$D(X_1, \ldots, X_m) = A(X_1, \ldots, X_m) B(X_1, \ldots, X_m) = \sum D_{i_1 \ldots i_m} X_1^{i_1} \ldots X_m^{i_m}$$

is given by the "Cauchy multiplication rule"

$$D_{k_1 \ldots k_m} = \sum_{i_1 + j_1 = k_1, \ldots, i_m + j_m = k_m} A_{i_1 \ldots i_m} B_{j_1 \ldots j_m}.$$

The "formal" partial derivative $A_{X_j}(X_1, \ldots, X_m) = \frac{\partial A(X_1, \ldots, X_m)}{\partial X_j}$ of $A(X_1, \ldots, X_m)$ is defined by putting

$$A_{X_j}(X_1, \ldots, X_m) = \sum i_j A_{i_1 \ldots i_m} X_1^{i_1} \ldots X_{j-1}^{i_{j-1}} X_j^{i_j - 1} X_{j+1}^{i_{j+1}} \ldots X_m^{i_m}$$

and noting that $A \mapsto A_{X_j}$ gives an $R[X_1, \ldots, X_{j-1}, X_{j+1}, \ldots, X_m]$-derivation of $R[X_1, \ldots, X_m]$. If R is a domain then the degree of the product of two nonzero polynomials equals the sum of their degrees, and hence in particular the product is nonzero. It follows that if R is a domain then so is the polynomial ring $R[X_1, \ldots, X_m]$. For a finite number of indeterminates $X_1, \ldots, X_m$ over a field K, by $K(X_1, \ldots, X_m)$ we denote the field of rational functions in $X_1, \ldots, X_m$ over K. Note that members of $K(X_1, \ldots, X_m)$ are "quotients" of polynomials $A(X_1, \ldots, X_m)/B(X_1, \ldots, X_m)$ with nonzero B. Again by the "quotient rule" we can extend the above

$$K[X_1, \ldots, X_{j-1}, X_{j+1}, \ldots, X_m]\text{-derivation of } K[X_1, \ldots, X_m]$$

to a

$$K(X_1, \ldots, X_{j-1}, X_{j+1}, \ldots, X_m)\text{-derivation of } K(X_1, \ldots, X_m).$$

Given any overring S of the ring R, any finite number of elements $x_1, \ldots, x_m$ in S can be "substituted" for the indeterminates $X_1, \ldots, X_m$ in any $A(X_1, \ldots, X_m) \in R[X_1, \ldots, X_m]$ to get

$$A(x_1, \ldots, x_m) = \sum a_{i_1 \ldots i_m} x_1^{i_1} \ldots x_m^{i_m} \in S$$

and we define

$$R[x_1, \ldots, x_m] = \{A(x_1, \ldots, x_m) : A(X_1, \ldots, X_m) \in R[X_1, \ldots, X_m]\}$$

and we note that then $R[x_1, \ldots, x_m]$ is a subring of S. We call $R[x_1, \ldots, x_m]$ the subring generated by $x_1, \ldots, x_m$ over R. The R-epimorphism $R[X_1, \ldots, X_m] \to R[x_1, \ldots, x_m]$ and the R-homomorphism $R[X_1, \ldots, X_m] \to S$ which send every polynomial $A(X_1, \ldots, X_m)$ to $A(x_1, \ldots, x_m)$ are called natural maps or substitution maps. An element $x \in S$ is algebraic over R if it satisfies a polynomial equation $A(x) = 0$ with $0 \neq A(X) \in R[X]$; otherwise it is transcendental over R; we may express this by saying that x/R is algebraic or transcendental respectively. For $W \subset S$, we say that W is algebraic over R if every $x \in W$ is algebraic over R; again we may express this by saying that W/R is algebraic. Finite number of elements (i.e., a finite sequence) $x_1, \ldots, x_m$ are algebraically dependent over R if they satisfy a polynomial equation $A(x_1, \ldots, x_m) = 0$ with $0 \neq A(X_1, \ldots, X_m) \in R[X_1, \ldots, X_m]$; otherwise they are algebraically independent over R; we may again express this by saying that $(x_1, \ldots, x_m)/R$ are algebraically dependent or algebraically independent respectively. Finite number of elements $x_1, \ldots, x_m$ in S form a transcendence basis of S/R ($= S$ over R) if they are algebraically independent over R, and S is algebraic over $R[x_1, \ldots, x_m]$. For $W \subset S$ we let $R[W] = \cup R[x_{j_1}, \ldots, x_{j_e}]$ with union over all finite sequences $x_{j_1}, \ldots, x_{j_e}$ in W, and note that this is also a subring of S which we call the subring generated by W over R; note that $R[\emptyset] = R$; we say that W is a set of algebraically independent elements over R, or W/R is algebraically independent, if every finite subset of W is algebraically independent over R, and we say that W is a transcendence basis of S/R ($= S$ over R) if W/R is algebraically independent and $S/R[W]$ is algebraic.

Given any overfield L of a field K, for any finite number of elements $x_1, \ldots, x_m$ in L, by $K(x_1, \ldots, x_m)$ we denote the set of all u/v with $u \in K[x_1, \ldots, x_m]$ and $0 \neq v \in K[x_1, \ldots, x_m]$, and we note that this is a subfield of L which we call the subfield generated by $x_1, \ldots, x_m$ over K; similarly for $W \subset L$ we have the subfield $K(W) = \{u/v : u \in K[W], 0 \neq v \in K[W]\}$, and $K(\emptyset) = K$. It can be shown that any overfield L of a field K has a transcendence basis over K, and there is a bijection between any two transcendence bases of L/K. The number of elements in a transcendence basis of L/K is denoted by $\mathrm{trdeg}_K L$, and we call this the transcendence degree of L/K. Note that then either $\mathrm{trdeg}_K L \in \mathbb{N}$ or $\mathrm{trdeg}_K L = \infty$. It can be shown that any set of field generators of L/K (i.e., any $W \subset L$ with $L = K(W)$) contains a transcendence basis of L/K, and hence if $L = K(x_1, \ldots, x_e)$ for a finite number of elements $x_1, \ldots, x_e$ in L then $\mathrm{trdeg}_K L \leq e$.

§7: EUCLIDEAN DOMAINS AND PRINCIPAL IDEAL DOMAINS

For any field K, it can be shown that the one variable (= indeterminate) polynomial ring $K[X]$ as well as the many variable polynomial ring $K[X_1, \ldots, X_m]$ are UFDs in the following sense. A unit in a ring R is an element u in R such that $uu' = 1$ for some $u' \in R$; note that then u' is unique and, in accordance with the field case, we may denote it by u^{-1}, and for any $v \in R$ we may write v/u for vu^{-1}; the set of all units in R forms a multiplicative group which we denote by $U(R)$; note that for a field K we have $U(K) = K^\times$. We extend the notation $U(R)$ to denote the multiplicative group of all (two-sided) units in a skew-ring R; similarly, the (two-sided) inverse of a unit u in a skew-ring R may be denoted by u^{-1}; again note that for a skew-field K we have $U(K) = K^\times$. Elements z, z' in a ring R are associates of each other (in R) if the ideals zR and $z'R$ coincide; note that in case of a domain R, this is equivalent to saying that $z' = uz$ for some unit u in R. An irreducible element in a ring R is a nonzero nonunit element z in R such that z cannot be expressed as $z = z'z''$ where z' and z'' are nonzero nonunit elements in R. A ring R is a UFD (= unique factorization domain) if R is a domain and every nonzero element z in R can uniquely be expressed as $z = uz_1 \ldots z_r$ where u is a unit in R and $z_1, \ldots, z_r$ are irreducible elements in R; here uniqueness means that if $z = u'z_1' \ldots z_{r'}'$ is any other such factorization then $r' = r$ and there is a permutation $\sigma(1), \ldots, \sigma(r)$ of $1, \ldots, r$ such that $z_{\sigma(i)}'$ is an associate of z_i for $1 \leq i \leq r$.

In case of $K[X]$ or $K[X_1, \ldots, X_m]$, the nonzero elements of K are exactly all the units in $K[X]$ or $K[X_1, \ldots, X_m]$ respectively. In case of $K[X]$, every nonzero nonunit in $K[X]$ is the associate of a unique nonconstant monic polynomial. Analogous to $\mathbb{Z}$, every nonzero nonunit ideal in $K[X]$ is of the form $A(X)K[X]$ where $A(X)$ is a nonconstant monic polynomial; moreover this ideal is prime $\Leftrightarrow$ it is maximal $\Leftrightarrow A(X)$ is irreducible. In particular the rings $K[X]$ and $\mathbb{Z}$ are PIDs (= Principal Ideal Domains), where by a PID we mean a domain R in which every ideal is principal, i.e., it is of the form xR for some $x \in R$. Moreover both these rings are EDs (= Euclidean Domains), where by an ED we mean a domain R together with a Euclidean Function, i.e., a map $\phi : R^\times \to \mathbb{N}$ such that for all x, y in $R^\times$ we have: $y \in xR \Rightarrow \phi(x) \leq \phi(y)$, and $y \notin xR \Rightarrow y = qx + r$ for some $q \in R$ and $r \in R^\times$ with $\phi(r) < \phi(x)$. If this r (and hence also q) is unique (depending on x and y) then we may call R a special ED, and if there is a subset $\widetilde{R}$ of R such that r becomes unique when required to belong to $\widetilde{R}$ then we may call R a quasispecial ED relative to the quasispecial subset $\widetilde{R}$. In case of $K[X]$ we take $\phi(x) =$ the X-degree of x, and we note that then $K[X]$ becomes a special ED. In case of $\mathbb{Z}$ we take $\phi(x) = |x|$, i.e., $\phi(x) = x$ or $-x$ according as $x \geq 0$ or $x < 0$, and we note that then $\mathbb{Z}$ becomes a quasispecial ED relative to the quasispecial subset $\mathbb{N}$. It can be shown that: ED $\Rightarrow$ PID $\Rightarrow$ UFD, and: R is a UFD $\Rightarrow R[X]$ is a UFD. Now by induction on m we see that $K[X_1, \ldots, X_m]$ is a UFD.

§8: ROOT FIELDS AND SPLITTING FIELDS

In view of what we have said above, in case of the univariate (= one variable) polynomial ring $K[Y]$ over a field K, for any irreducible $g(Y) \in K[Y]$ the residue class ring $K[Y]/g(Y)K[Y]$ is a field L' which we may regard as an overfield of K, and calling the image of Y under the canonical epimorphism $K[Y] \to L'$ to be α_1 we have $L' = K[\alpha_1] = K(\alpha_1)$ with $g(\alpha_1) = 0$. We call L' a root field of g over K. Given any nonconstant $f(Y) \in K[Y]$ of some degree n, by taking $g(Y)$ to be an irreducible factor of $f(Y)$ in $K[Y]$ we get $f(\alpha_1) = 0$ and $0 \neq f_1(Y) = f(Y)/(Y - \alpha_1)$ is of degree $n - 1$. If $n - 1 > 0$ then repeating this process we find an element α_2 in an overfield of $K(\alpha_1)$ such that $f_1(\alpha_2) = 0$ and $0 \neq f_2(Y) = f_1(Y)/(Y - \alpha_2)$ is of degree $n - 2$. Continuing this process n times we find elements $\alpha_1, \ldots, \alpha_n$ in an overfield of K such that

$$f(Y) = a_0(Y - \alpha_1) \ldots (Y - \alpha_n)$$

where a_0 is the coefficient of Y^n in $f(Y)$. The overfield $L^* = K[\alpha_1, \ldots, \alpha_n] = K(\alpha_1, \ldots, \alpha_n)$ of K is called a splitting field of f over K and as notation we put

$$\mathrm{SF}(f, K) = L^* = K(\alpha_1, \ldots, \alpha_n).$$

For any overfield $\overline{L}$ of L^*, we call L^* the splitting field of f over K in $\overline{L}$; it can be shown that then for any K-isomorphism $\phi : L^* \to L^{**}$ where L^{**} is any overfield of K in $\overline{L}$ we must have $L^{**} = L^*$, i.e., ϕ must be a K-automorphism of L^*. To define the concept of a K-isomorphism, given any overrings S and T of a ring R, a homomorphism (resp: monomorphism, epimorphism, isomorphism) $\phi : S \to T$ is called an R-homomorphism (resp: R-monomorphism, R-epimorphism, R-isomorphism) if $\phi(x) = x$ for all $x \in R$; note that an R-automorphism of S is an R-isomorphism $S \to S$. With this definition at hand, it can be shown that any two splitting fields of f over K are K-isomorphic, i.e., there is a K-isomorphism of one onto the other. Thus the notation $\mathrm{SF}(f, K)$ is unique up to isomorphism. Upon letting

$$f(Y) = a_0 Y^n + a_1 Y^{n-1} + \cdots + a_n$$

with $a_0, a_1, \ldots, a_n$ in K, for the Y-derivative $f_Y(Y)$ of $f(Y)$ we have

$$f_Y(Y) = na_0 Y^{n-1} + (n-1)a_1 Y^{n-2} + \cdots + a_{n-1}.$$

It is easily seen that f is separable (i.e., the roots $\alpha_1, \ldots, \alpha_n$ are distinct) iff $f(Y)$ and $f_Y(Y)$ have no nonconstant common factor in $K[Y]$. By a Galois extension of K we mean an overfield of K which is the splitting field of some nonconstant separable univariate polynomial over K. Assuming f to be separable, by the Galois group $\mathrm{Gal}(L^*, K)$ of L^* over K we mean the group $\mathrm{Aut}_K(L^*)$. For any $\sigma \in \mathrm{Gal}(L^*, K)$ we have $f(Y) = a_0(Y - \sigma(\alpha_1)) \ldots (Y - \sigma(\alpha_n))$ and hence $(\alpha_1, \ldots, \alpha_n) \mapsto (\sigma(\alpha_1), \ldots, \sigma(\alpha_n))$ gives a permutation σ^* of $\{\alpha_1, \ldots, \alpha_n\}$. Clearly

$\sigma \mapsto \sigma^*$ gives a monomorphism $\mathrm{Gal}(L^*, K) \to S_n =$ the permutation group on $\{\alpha_1, \ldots, \alpha_n\}$. We define the image of this monomorphism to be the Galois group of f over K and denote it by $\mathrm{Gal}(f, K)$. Thus we have a natural isomorphism $\mathrm{Gal}(L^*, K) \to \mathrm{Gal}(f, K)$ and hence in particular $\mathrm{Gal}(L^*, K)$ is a finite group.

Now Galois himself [cf. L6§6(E6)] characterized $\mathrm{Gal}(f, K)$ as the set of all relations preserving permutations of the roots, i.e., the set of all $\tau \in S_n$ such that for every $P(X_1, \ldots, X_n) \in K[X_1, \ldots X_n]$ with $P(\alpha_1, \ldots, \alpha_n) = 0$ we have $P(\tau(\alpha_1), \ldots, \tau(\alpha_n)) = 0$. For any intermediate field L, i.e., a field L with $K \subset L \subset L^*$, it is clear that $L^* = \mathrm{SF}(f, L)$ and $\mathrm{Gal}(L^*, L) \leq \mathrm{Gal}(L^*, K)$. Likewise, for any $H \leq \mathrm{Gal}(L^*, K)$, upon letting $\mathrm{fix}_{L^*}(H) = \{z \in L^* : h(z) = z \text{ for all } h \in H\}$ it is clear that this is an intermediate field; we call this the fixed field of H. Eventually, following Galois, we shall prove the following [cf. L6§6(E1)]:

FUNDAMENTAL THEOREM OF GALOIS THEORY (T1). The mapping $L \mapsto \mathrm{Gal}(L^*, L) = H$ gives an inclusion reversing bijection of the set Λ of all fields L between K and L^* onto the set Γ of all subgroups H of $\mathrm{Gal}(L^*, K)$, and the inverse bijection is given by $H \mapsto \mathrm{fix}_{L^*}(H) = L$, where inclusion reversing means: $L_1 \subset L_2 \Leftrightarrow \mathrm{Gal}(L^*, L_1) \supset \mathrm{Gal}(L^*, L_2)$. Moreover $H \triangleleft \mathrm{Gal}(L^*, K) \Leftrightarrow L$ is a Galois extension of K, and when that is so $\mathrm{Gal}(L^*, K)/H$ is naturally isomorphic to $\mathrm{Gal}(L, K)$; in greater detail, by restricting $\sigma \in \mathrm{Gal}(L^*, K)$ to L we get $\overline{\sigma} \in \mathrm{Gal}(L, K)$, and $\sigma \mapsto \overline{\sigma}$ gives an epimorphism $\mathrm{Gal}(L^*, K) \to \mathrm{Gal}(L, K)$ with kernel H. Finally for every $L \in \Lambda$ we have $|\mathrm{Gal}(L^*, L)| = [L^* : L]$.

By an r-cyclic extension of K, where r is a positive integer, we mean a Galois extension L of K such that $\mathrm{Gal}(L, K)$ is a cyclic group of order r. The following theorem dates back to 1775 [cf. L6§6(E3)]:

LAGRANGE RESOLVENT THEOREM (T2). Assume that K has r distinct r-th roots of 1. Then any r-cyclic extension of K is obtained by adjoining an r-th root of some nonzero element a of K, i.e., it is of the form $K(\alpha)$ with $0 \neq \alpha^r = a \in K$. Conversely, for any $0 \neq a \in K$, the splitting field of $Y^r - a$ over K is an s-cyclic extension of K for some divisor s of r.

The above two theorems have the following consequence [cf. L6§6(E3)]:

SOLVABILITY THEOREM (T3). Assume that K has $n!$ distinct $n!$-th roots of 1. Then $\mathrm{Gal}(f, K)$ is solvable if and only if the equation $f = 0$ can be solved by radicals, i.e., if and only if there is a chain of fields $K = K_0 \subset K_1 \subset \cdots \subset K_t = K(\alpha_1, \ldots, \alpha_n)$ such that for $1 \leq i \leq t$ we have $K_i = K_{i-1}(u_i^{1/r_i})$ with $0 \neq u_i \in K_{i-1}$ and $r_i \in \mathbb{N}_+$.

The following theorem dates back to 1665 [cf. L6§6(E4)]:

NEWTON'S SYMMETRIC FUNCTION THEOREM (T4). For the generic n-th degree polynomial

$$F(Y) = Y^n + X_1 Y^{n-1} + \cdots + X_n$$

over $K = k(X_1, \ldots, X_n)$, where $X_1, \ldots, X_n$ are indeterminates over a field k, we have $\mathrm{Gal}(F, K) = S_n$.

We have already noted the following theorem dating back to 1830 [cf. L6§6(E5)]:

GALOIS' SYMMETRIC GROUP THEOREM (T5). The group S_n is unsolvable for $n \geq 5$.

Finally, as a consequence of (T3), (T4), (T5) we get the [cf. L6§6(E5)]:

UNSOLVABILITY THEOREM (T6). Assume that the field k contains $n!$ distinct $n!$-th roots of 1. Then, for $n \geq 5$, the generic n-th degree polynomial $F(Y)$ over $K = k(X_1, \ldots, X_n)$ cannot be solved by radicals.

§9: ADVICE TO THE READER

We shall now fill in some details of argument which were left out in the above discourse so as not to break the smooth flow of thought. This will be done in a series of Definitions (D1), (D2), ..., Remarks (R1), (R2), ..., Examples (X1), (X2), ..., Exercises (E1), (E2), ..., and Notes (N1), (N2), ... We shall end up with a Concluding Note in the style of Aesop's Fables. The same pattern may be followed in the succeeding lectures. In any lecture, assertions such as Theorems, Corollaries, Claims, and Propositions, will be labelled (T1), (T2), ..., and formulas and other items will be labelled (1), (2), ..., (A1), (A2), ..., (B1), (B2), ..., and so on. Sometimes there may be Observations (O1), (O2), ... In a lecture, say in Lecture L3, items from another lecture, say Lecture L2, will be cited as L2(D1), L2(T4), and so on. Some details left out in a lecture may very well be covered in a later lecture. The reader may need a certain amount of patience to follow this method of concentric circles, whereby the ideas expounded at a time are explained and expanded in an outer circle. The patience will be rewarded by entertainment and ease of absorption. Also the discerning reader should have no difficulty in separating out the heuristic geometric discourse from the official logical dogma.

§10: DEFINITIONS AND REMARKS

DEFINITION (D1). [**Divisibility and Prime Ideals**]. Let R be a ring. Given elements x, y in R, we say that x divides y (in R) or y is divisible by x (in R) to mean that $y \in xR$, i.e., $y = xz$ for some $z \in R$; we may express this by writing $x|y$. Recall that $z \in R$ is irreducible (in R) means z is a nonzero nonunit in R which is not the product of any two nonzero nonunits in R. Similarly $z \in R$ is said to be prime (in R) if z is a nonunit and for all x, y in R we have: $z|(xy) \Rightarrow z|x$ or $z|y$. Recall that an ideal I in R is a prime ideal or a maximal ideal according as R/I is a domain or a field. This is easily seen to be equivalent to the more standard definition, i.e.,

(A1) $\begin{cases} \text{an ideal } I \text{ in a ring } R \text{ is maximal} \\ \Leftrightarrow I \neq R \text{ and there is no ideal strictly between } I \text{ and } R \end{cases}$

and

(A2) $\begin{cases} \text{an ideal } I \text{ in a ring } R \text{ is prime} \\ \Leftrightarrow I \neq R \text{ and for all } x, y \text{ in } R \setminus I \text{ we have } xy \notin I. \end{cases}$

By (A2) it follows that

(A3) $\begin{cases} \text{an element } z \text{ in a ring } R \text{ is prime} \\ \Leftrightarrow \text{ the principal ideal } zR \text{ is prime.} \end{cases}$

Note that a maximal ideal is always prime but not conversely; for instance for a field K, the zero ideal in the univariate polynomial ring $K[Y]$ is prime but not maximal; more interestingly, in the bivariate (two variable) polynomial ring $K[X, Y]$, the ideal generated by X is prime but not maximal since the residue class ring is obviously isomorphic to the univariate polynomial ring $K[Y]$.

REMARK (R1). [**Division Algorithm**]. Given any $A = A(X)$ in the univariate polynomial ring $R[X]$ over a ring R, we call A monic or submonic (in X) if $A \neq 0$ and, upon letting a to be the coefficient of X^d in A where d is the degree of A, we have $a = 1$ or $a \in U(R)$ respectively. Assuming $A(X)$ to be submonic, the division algorithm in $R[X]$ says that given any $B(X) \in R[X]$ there exist unique $q(X)$ and $r(X)$ in $R[X]$ such that $B(X) = q(X)A(X) + r(X)$ and the degree of $r(X)$ is $< d$. It follows that the univariate polynomial ring $K[X]$ over a field K is a special ED. Note that the division algorithm in $\mathbb{Z}$ says that given any x, y in $\mathbb{Z}$ with $x \neq 0$ there exist unique q, r in $\mathbb{Z}$ such that $y = qx + r$ and $0 \leq r < |x|$. This makes $\mathbb{Z}$ a quasispecial ED with $\mathbb{N}$ as the quasispecial subset.

REMARK (R2). [**Long Division or Euclidean Algorithm**]. In the ancient Indian literature the following procedure for finding a gcd (= greatest common divisor) of two nonzero elements x_1, x_2 in a euclidean domain R is called the method of Long Division, and in the European literature it is called the Euclidean Algorithm. Namely, upon letting ϕ to be the euclidean function, let us inductively define q_{i+2}, x_{i+2} for $i = 1, 2, \ldots$ by saying that if $x_i \notin x_{i+1}R$ then $q_{i+2} \in R$ and $x_{i+2} \in R^\times$

are such that $x_i = q_{i+2}x_{i+1} + x_{i+2}$ and $\phi(x_{i+2}) < \phi(x_{i+1})$. This stops after t steps, with $\phi(x_2) \geq t \in \mathbb{N}$, when $x_{t+1} \in x_{t+2}R$. Note that if R is a special ED then the elements q_{i+2}, x_{i+2} are unique, and if R is a quasispecial ED then they can be chosen uniquely. At any rate, without assuming special or quasispecial, this algorithm enables us to express x_{t+2} as a linear combination of the elements x_1, x_2, and to express the elements x_1, x_2 as multiples of x_{t+2}, i.e., to write $x_{t+2} = ax_1 + bx_2$, $x_1 = cx_{t+2}$, $x_2 = dx_{t+2}$, with a, b, c, d in R. Thus the element x_{t+2} is a generator of the ideal in R generated by x_1, x_2, and hence x_{t+2} is a gcd of x_1, x_2 in the following sense.

For a subset S of a ring R (which need not be an ED), by a cd (= common divisor) of S (in R) we mean $x \in R$ such that $x|y$ for all $y \in S$, and by a gcd of S (in R) we mean a cd x of S such that x is divisible by every cd of S; any gcd of S may be denoted by $\gcd(S)$; note that if the ideal in R generated by S is principal then a gcd of S is nothing but a generator of that ideal; moreover, without assuming the ideal SR to be principal, if $x \in R$ is a gcd of S then: $y \in R$ is a gcd of $S \Leftrightarrow x$ and y are associates of each other. By a cd of a finite or infinite sequence of elements $y_1, y_2, \ldots$ in a ring R we mean a cd of $S = \{y_1, y_2, \ldots\}$, and by a gcd of $y_1, y_2, \ldots$ we mean a gcd of $S = \{y_1, y_2, \ldots\}$; again any gcd of $y_1, y_2, \ldots$ may be denoted by $\gcd(y_1, y_2, \ldots)$. Although a gcd is determined only up to associates, for euphony we may sometimes call it the gcd.

Again assuming R to be an ED, we extend the above procedure of finding a gcd to a finite sequence $y_1, \ldots, y_s$ in $R^{\times}$, with $s \in \mathbb{N}$, by putting $z_0 = 0, z_1 = y_1, z_2 = \gcd(z_1, y_2), z_3 = \gcd(z_2, y_3), \ldots, z_s = \gcd(z_{s-1}, y_s)$, where the gcd is as found in the above two element procedure, and noting that now clearly z_s is a generator of the ideal in R generated by $y_1, \ldots, y_s$, and hence z_s is a gcd of $y_1, \ldots, y_s$. For a finite sequence $y_1, \ldots, y_s$ in R we apply this procedure to the sequence obtained by deleting all the zeros from the sequence $y_1, \ldots, y_s$, and thereby obtain a generator of the ideal in R generated by $y_1, \ldots, y_s$, which is hence a gcd of $y_1, \ldots, y_s$.

REMARK (R3). [**ED $\Rightarrow$ PID**]. By extending the above method to infinite sequences we can show that ED $\Rightarrow$ PID. But this can be shown more easily, although nonconstructively, thus. Given any nonzero ideal I in a euclidean domain R with euclidean function ϕ, take $0 \neq x \in I$ having the smallest ϕ-value; if there were $y \in I \setminus xR$ then we could find $q \in R$ and $r \in R^{\times}$ such that $y = qx + r$ and $\phi(r) < \phi(x)$, which would be a contradiction because $0 \neq r \in I$. In case R is the univariate polynomial ring $K[X]$ over a field K, for any $S \subset R$ we put $\mathrm{GCD}(S) = 0$ if $S \subset \{0\}$ and otherwise we put $\mathrm{GCD}(S) =$ the unique monic generator of the ideal in R generated by S, and for any finite or infinite sequence $y_1, y_2, \ldots$ in R we put $\mathrm{GCD}(y_1, y_2, \ldots) = \mathrm{GCD}(\{y_1, y_2, \ldots\})$. In case R is the ring of integers $\mathbb{Z}$, for any $S \subset R$ we put $\mathrm{GCD}(S) =$ the unique nonnegative generator of the ideal in R generated by S, and for any finite or infinite sequence $y_1, y_2, \ldots$ in R we put $\mathrm{GCD}(y_1, y_2, \ldots) = \mathrm{GCD}(\{y_1, y_2, \ldots\})$. Here GCD stands for the Greatest Common

Divisor (in R). Note that in the above two situations, the GCD is a gcd. To make the reference to the ring R explicit we may write $\gcd_R$ and GCD_R instead of gcd and GCD respectively.

REMARK (R4). [**ED** $\Rightarrow$ **UFD**]. Later on we shall prove PID $\Rightarrow$ UFD [cf. L5§5(Q18.4)]. Here we shall give a direct proof of ED $\Rightarrow$ UFD. First let us establish some generalities about UFDs. So let R be a domain. Recall that R is a UFD means

(F1) every nonzero element in R is a product of a unit and irreducible elements

and

(F2) the irreducible factors in (F1) are unique up to order and associates.

It can be shown that if R is a UFD then

(F3) every irreducible element in R is prime.

It can also be shown that if (F1) and (F3) hold then so does (F2). In other words

(B1) in a domain R with (F1) we have: (F2) $\Leftrightarrow$ (F3).

It is clear that any subset S of a UFD R, and hence any sequence $y_1, y_2, \ldots$ in R, has a gcd; namely if $S \subset \{0\}$ then 0 is the only gcd of S, and if S has a nonzero element z then we can write $z = uz_1^{e_1} \ldots z_t^{e_t}$ where u is a unit in R, $e_1, \ldots, e_t$ are positive integers, and $z_1, \ldots, z_t$ are irreducibles in R such that no two of them are associates of each other; now upon letting d_i to be the largest integer such that $z_i^{d_i}$ divides every element of S, we see that $x \in R$ is a gcd of S iff $x = vz_1^{d_1} \ldots z_t^{d_t}$ for some unit v. Elements $y_1, y_2, \ldots$ in a UFD are coprime (or relatively prime) means 1 is a gcd of them.

To establish some generalities about EDs, recall that R is an ED means it has a euclidean function, i.e., a map $\phi : R^\times \to \mathbb{N}$ such that for all x, y in $R^\times$ we have:

(G1) $\qquad\qquad\qquad y \in xR \Rightarrow \phi(x) \le \phi(y)$

and

(G2) $y \notin xR \Rightarrow y = qx + r$ for some $q \in R$ and $r \in R^\times$ with $\phi(r) < \phi(x)$.

It is easily seen that given any euclidean function $\phi : R^\times \to \mathbb{N}$, for all x, y in $R^\times$ we have

(G3) $\qquad\qquad x$ and y are associates $\Rightarrow \phi(x) = \phi(y)$

and

(G4) $\qquad\qquad y \in xR$ and $\phi(x) = \phi(y) \Rightarrow x$ and y are associates.

With ϕ as above, for all $z \in R^{\times}$ by (G1) we know that $\phi(z) \geq \phi(1)$; by induction on $\phi(z)$ let us show that z can be expressed as a product of a unit and irreducible elements; for $\phi(z) = \phi(1)$ this follows from (G4); so let $\phi(z) > \phi(1)$ and assume for all values of $\phi(z)$ smaller than the given one; if z is irreducible then we have nothing to show; otherwise $z = z'z''$ where z' and z'' are nonzero nonunits, and by (G4) we get $\phi(z') < \phi(z)$ and $\phi(z'') < \phi(z)$ and hence by the induction hypothesis z' and z'' are products of units and irreducibles, and putting these factorizations together we get a desired factorization of z. Thus

(B2)   every ED satisfies (F1).

Again with ϕ as above, let x, y, z in $R^{\times}$ be such that z is irreducible with $xy \in zR$ and $x \notin zR$; then 1 is a gcd of x, z and hence by (R2) we can write $1 = ax + bz$ with a, b in R; multiplying this equation by y we get $y = axy + byz$ and hence $y \in zR$. Thus

(B3)   every ED satisfies (F3).

By (B1), (B2), (B3) we see that ED $\Rightarrow$ UFD.

REMARK (R5). [**Factorization of Integers**]. Applying the above Remarks (R1) to (R4) to the case $R = \mathbb{Z}$ we see that: R is an ED, PID, and UFD, with $U(R) = \{\pm 1\}$. Every nonzero nonunit element in R is an associate of a unique integer > 1. Every irreducible element in R is an associate of a unique positive irreducible element, i.e., a prime number. Every nonzero element z in the quotient field $\mathbb{Q}$ of R has a unique factorization $z = \pm p_1^{e_1} \ldots p_t^{e_t}$ where $e_1, \ldots, e_t$ are nonzero integers and $p_1, \ldots, p_t$ are pairwise distinct prime numbers (by convention an empty product is 1); moreover $z \in R \Leftrightarrow e_i > 0$ for $1 \leq i \leq t$. All the nonzero prime ideals in R are maximal, and they are of the form pR where p is a prime number. For any positive integer n the residue class ring $S = R/(nR)$ is clearly a ring of size n; moreover S is a field $\Leftrightarrow n$ is a prime p, and then we call S the Galois Field of p elements and denote it by $\mathrm{GF}(p)$.

REMARK (R6). [**Factorization of Univariate Polynomials**]. Applying the above Remarks (R1) to (R4) to the univariate polynomial ring $R = K[Y]$ over a field K we see that: R is an ED, PID, and UFD, with $U(R) = K^{\times}$. Every nonzero nonunit element in R is an associate of a unique nonconstant monic polynomial. Every irreducible element in R is an associate of a unique nonconstant monic irreducible polynomial. Every nonzero element $h = h(Y)$ in the quotient field $K(Y)$ of R has a unique factorization $h = ag_1^{e_1} \ldots g_t^{e_t}$ where $a \in K^{\times}$, $e_1, \ldots, e_t$ are nonzero integers, and $g_1 = g_1(Y), \ldots, g_t = g_t(Y)$ are pairwise distinct nonconstant monic irreducible polynomials; moreover $h \in R \Leftrightarrow e_i > 0$ for $1 \leq i \leq t$. All the nonzero prime ideals in R are maximal, and they are of the form gR where $g = g(Y)$ is a nonconstant irreducible polynomial, which may be chosen to be monic. For any nonconstant $f = f(Y) \in R$ of some degree $n > 0$, clearly $K \cap (fR) = \{0\}$ and hence we may identify K with a subring of the residue class ring $L = R/(fR)$, and then upon letting y to be the image of Y under the residue class map $R \to L$ we see that $1, y, \ldots, y^{n-1}$ is vector space basis of L over K and hence $[L : K] = n$; moreover L is a field $\Leftrightarrow f$ is irreducible.

DEFINITION (D2). [**Characteristic**]. For any ring R there is unique homomorphism $\mathbb{Z} \to R$. The unique nonnegative generator of its kernel is called the characteristic of R and is denoted by $\mathrm{ch}(R)$. Clearly if R is a domain then $\mathrm{ch}(R)$ is either 0 or a prime number. The image of the homomorphism $\mathbb{Z} \to R$ is called the prime subring of R. Clearly $\mathrm{ch}(R) = 0 \Leftrightarrow$ the prime subring of R is isomorphic to $\mathbb{Z}$; in this case we may identify the prime subring with $\mathbb{Z}$. Also clearly $\mathrm{ch}(R)$ is a prime number $p \Leftrightarrow$ the prime subring is a field; in this case the prime subring is obviously isomorphic to the Galois field $\mathrm{GF}(p)$, and we identify the former with the latter. Needless to say that $\mathrm{ch}(R) = 1 \Leftrightarrow$ the prime subring is the null ring $\Leftrightarrow R$ is the null ring.

DEFINITION (D3). [**Minimal Polynomial**]. Let L be an overfield of a field K. Then $y \in L$ is transcendental or algebraic over K according as the kernel of the K-homomorphism $K[Y] \to L$ which sends Y to y is zero or nonzero, i.e., according as $[K[y] : K] = \infty$ or $[K[y] : K] = $ a positive integer n. In the algebraic case n is the field degree of the field $K[y] = K(y)$ over K, and the unique monic generator of the said kernel is a monic irreducible polynomial of degree n in Y over K; we call this polynomial the minimal polynomial of y/K, i.e., of y over K. Note that for any $y \in L$ we always have $K[y] \subset K(y)$ and hence:

$$y \text{ is algebraic over } K \Leftrightarrow K[y] = K(y) \Leftrightarrow [K[y] : K] < \infty \Leftrightarrow [K(y) : K] < \infty.$$

DEFINITION (D4). [**Least Common Multiple**]. In grade school when we learn of cd and gcd we also learn of cm (= common multiple) and lcm (= least common multiple). To introduce these let R be a ring and let $y_1, \ldots, y_s$ be a finite sequence of elements in R with $s > 0$. By a cm of $y_1, \ldots, y_s$ (in R) we mean $x \in R$ such that x is divisible by y_i for $1 \leq i \leq s$. By an lcm of $y_1, \ldots, y_s$ (in R) we mean a cm of $y_1, \ldots, y_s$ which divides every cm of $y_1, \ldots, y_s$, and we denote it by $\mathrm{lcm}(y_1, \ldots, y_s)$. Clearly if $\mathrm{lcm}(y_1, \ldots, y_s)$ exists then it is unique up to associates. For euphony we may sometimes speak of the lcm instead of an lcm.

Now assume that R is a UFD. Then we can find a finite number of irreducible elements $z_1, \ldots, z_t$ in R, no two of which are associates of each other, such that for $1 \leq i \leq s$ we have $y_i = u_i z_1^{e_{i1}} \ldots z_t^{e_{it}}$ where u_i is a unit in R and $e_{i1}, \ldots, e_{it}$ are nonnegative integers. For $1 \leq j \leq t$ let $l_j = \max\{e_{ij} : 1 \leq i \leq s\}$ (with $l_j = 0$ if $e_{ij} = 0$ for all i), and let $y = z_1^{l_1} \ldots z_t^{l_t}$. Then clearly y is an lcm of $y_1, \ldots, y_s$. In case R is the univariate polynomial ring $K[Y]$ over a field K (resp: the ring of integers $\mathbb{Z}$) we define the LCM (= Least Common Multiple) of $y_1, \ldots, y_s$ (in R) to be the unique monic (resp: positive) lcm of $y_1, \ldots, y_s$, and we denote it by $\mathrm{LCM}(y_1, \ldots, y_s)$; note that by taking $z_1, \ldots, z_t$ to be monic (resp: positive) we get $\mathrm{LCM}(y_1, \ldots, y_s) = y$. To make the reference to the ring R explicit we may write lcm_R and LCM_R instead of lcm and LCM respectively.

Continuing with the assumption that R is a UFD, any element ζ in the quotient field of R can be expressed as $\zeta = \xi/\eta$ with $\xi \in R$ and $\eta \in R^\times$ such that the elements ξ and η are coprime, and we call ξ/η a reduced form of ζ. In case R is the univariate polynomial ring $K[Y]$ over a field K (resp: the ring of integers $\mathbb{Z}$) this becomes unique by requiring η to be monic (resp: positive), and so we may call ξ/η the reduced form of ζ.

REMARK (R7). [**Relative Algebraic Closure**]. To continue the discussion of (R6) and (D3), let L be an overfield of a field K, and let K' and K'' be subfields of L with $K \subset K' \subset K''$. It can be shown that then

(C1) $\quad \begin{cases} [K'' : K] < \infty \Leftrightarrow [K'' : K'] < \infty \text{ and } [K' : K] < \infty, \\ \text{and in that case we have } [K'' : K] = [K'' : K'][K' : K]. \end{cases}$

(Hint: if $(x_i)_{1 \leq i \leq m}$ and $(y_j)_{1 \leq j \leq n}$ are bases of K'/K and K''/K' respectively then $(x_i y_j)_{1 \leq i \leq m, 1 \leq j \leq n}$ is a basis of K''/K). It follows that

(C2) $\quad \begin{cases} \text{if } [K'' : K] < \infty \text{ then every } y \in K'' \text{ is algebraic over } K \text{ and} \\ \text{the degree of the minimal polynomial of } y \text{ over } K \text{ divides } [K'' : K]. \end{cases}$

It also follows that

(C3) $\quad \begin{cases} \text{if } K'' \text{ is algebraic over } K', \text{ and } K' \text{ is algebraic over } K, \\ \text{then } K'' \text{ is algebraic over } K. \end{cases}$

Moreover it follows that

(C4) $\quad \begin{cases} \text{if } K'' = K(W) \text{ where } W \subset K'' \text{ is such that} \\ \text{every } y \in W \text{ is algebraic over } K \\ \text{then } K'' = K[W] \text{ and } K'' \text{ is algebraic over } K. \end{cases}$

Finally it follows that

(C5) $\quad \begin{cases} \text{the set } \widetilde{K} \text{ of all elements of } L \text{ which are algebraic over } K \\ \text{is a subfield of } L; \end{cases}$

we call $\widetilde{K}$ the (relative) algebraic closure of K in L; the adjective relative is sometimes used to distinguish this from an (absolute) algebraic closure of K by which we mean an overfield $\overline{K}$ of K such that $\overline{K}$ is algebraic over K and every nonconstant univariate polynomial over K has a root in $\overline{K}$.

REMARK (R8). [**Uniqueness of Splitting Field**]. Let L be an overfield of a field K, and let $f(Y)$ be a nonconstant polynomial of some degree $n > 0$ in Y over K such that f splits completely in L, i.e., $f(Y) = a_0(Y - \alpha_1) \ldots (Y - \alpha_n)$ with $0 \neq a_0 \in K$ and $\alpha_1, \ldots, \alpha_n$ in L; we call $L^* = K(\alpha_1, \ldots, \alpha_n)$ the splitting field of f over K in L. Let L' be an overfield of a field K' such that there is an isomorphism $\phi : K \to K'$ and f' splits completely in L' where $f'(Y)$ is obtained by applying ϕ to the coefficients of $f(Y)$. Now upon letting $a_0' = \phi(a_0)$ we clearly have $f'(Y) = a_0'(Y - \alpha_1') \ldots (Y - \alpha_n')$ with $\alpha_1', \ldots, \alpha_n'$ in L'. We want to show that there is an isomorphism $\phi^* : L^* \to L'^* = K'(\alpha_1', \ldots, \alpha_n')$ such that $\phi^*(x) = \phi(x)$ for all $x \in K$. We do this by induction on n. For $n = 1$ the assertion is obvious because then $L^* = K$ and $L'^* = K'$. So let $n > 1$ and assume for $n - 1$. Let $g(Y)$ be the minimal polynomial of α_1 over K, and let $g'(Y)$ be obtained by applying ϕ to the coefficients of $g(Y)$. Then clearly $g'(Y)$ is the minimal polynomial of α_i' over K' for some i, and upon relabelling $\alpha_1', \ldots, \alpha_n'$ suitably we may assume that $i = 1$. Now $K_1 = K(\alpha_1)$ and $K_1' = K'(\alpha_1')$ are root fields of $g(Y)$ and $g'(Y)$ over K and K' respectively, and they are naturally isomorphic to $K[Y]/(g(Y)K[Y])$ and $K'[Y]/(g'(Y)K'[Y])$ respectively, and therefore there is an isomorphism $\phi_1 : K_1 \to K_1'$ such that $\phi_1(\alpha_1) = \alpha_1'$ and $\phi_1(x) = \phi(x)$ for all $x \in K$. Let $f_1(Y) = (Y - \alpha_2) \ldots (Y - \alpha_n)$ and $f_1'(Y) = (Y - \alpha_2') \ldots (Y - \alpha_n')$. Then $f_1(Y)$ is a polynomial of degree $n - 1$ over K_1, and $f_1'(Y)$ is obtained by applying ϕ_1 to its coefficients. Also clearly L^* and L'^* are the splitting fields of f_1 and f_1' over K_1 and K_1' in L and L' respectively. Therefore by the induction hypothesis there exists an isomorphism $\phi^* : L^* \to L'^*$ such that $\phi^*(x) = \phi_1(x)$ for all $x \in K_1$.

§11: EXAMPLES AND EXERCISES

EXAMPLE (X1). [**Special Permutations**]. A permutation, say of the integers $\{1, 2, \ldots, 6\}$, is a rearrangement like

$$\begin{pmatrix} 1 \ 2 \ 3 \ 4 \ 5 \ 6 \\ 2 \ 4 \ 6 \ 1 \ 5 \ 3 \end{pmatrix}$$

where we send 1 to 2, 2 to 4, 3 to 6, 4 to 1, 5 to 5 and 6 to 3. Let us call the above permutation σ. Let us consider another permutation, say τ, given by

$$\begin{pmatrix} 1 \ 2 \ 3 \ 4 \ 5 \ 6 \\ 4 \ 3 \ 6 \ 1 \ 2 \ 5 \end{pmatrix}.$$

To multiply τ by σ, we first apply σ and then follow it by τ. Thus

$$\tau\sigma = \begin{pmatrix} 1 \ 2 \ 3 \ 4 \ 5 \ 6 \\ 4 \ 3 \ 6 \ 1 \ 2 \ 5 \end{pmatrix} \begin{pmatrix} 1 \ 2 \ 3 \ 4 \ 5 \ 6 \\ 2 \ 4 \ 6 \ 1 \ 5 \ 3 \end{pmatrix} = \begin{pmatrix} 1 \ 2 \ 3 \ 4 \ 5 \ 6 \\ 3 \ 1 \ 5 \ 4 \ 2 \ 6 \end{pmatrix}$$

because: σ sends 1 to 2 and τ sends 2 to 3, σ sends 2 to 4 and τ sends 4 to 1, σ

sends 3 to 6 and τ sends 6 to 5, σ sends 4 to 1 and τ sends 1 to 4, σ sends 5 to 5 and τ sends 5 to 2, σ sends 6 to 3 and τ sends 3 to 6. Doing it the other way round we get

$$\sigma\tau = \begin{pmatrix} 1\ 2\ 3\ 4\ 5\ 6 \\ 2\ 4\ 6\ 1\ 5\ 3 \end{pmatrix} \begin{pmatrix} 1\ 2\ 3\ 4\ 5\ 6 \\ 4\ 3\ 6\ 1\ 2\ 5 \end{pmatrix} = \begin{pmatrix} 1\ 2\ 3\ 4\ 5\ 6 \\ 1\ 6\ 3\ 2\ 4\ 5 \end{pmatrix}$$

which illustrates that in general multiplication is not commutative. The above permutation σ can be written as a product of disjoint cycles

$$\begin{pmatrix} 1\ 2\ 3\ 4\ 5\ 6 \\ 2\ 4\ 6\ 1\ 5\ 3 \end{pmatrix} = (124)(36)(5)$$

where the cyclic permutation (124) is the permutation which sends 1 to 2, 2 to 4, and 4 to 1 (and leaves all the other numbers unchanged) , i.e., in "long form"

$$(124) = \begin{pmatrix} 1\ 2\ 3\ 4\ 5\ 6 \\ 2\ 4\ 3\ 1\ 5\ 6 \end{pmatrix}$$

and similarly, the cyclic permutation (36) sends 3 to 6 and 6 to 3, and, finally, the cyclic permutation (5) sends 5 to 5. Since (5) leaves all the other numbers unchanged, so indeed (5) is simply the identity permutation

$$(5) = \begin{pmatrix} 1\ 2\ 3\ 4\ 5\ 6 \\ 1\ 2\ 3\ 4\ 5\ 6 \end{pmatrix}.$$

The cyclic permutation (124) is a cycle of length 3. A cycle of length 2, such as (36), is called a transposition. The cycles (124), (36) and (5) are disjoint means no two of them have any number in common.

EXAMPLE (X2). [**General Permutations**]. A permutation σ of a finite set, say the set of integers $\{1, 2, \ldots, n\}$, i.e., $\sigma \in S_n$, is written as

$$\begin{pmatrix} 1 & 2 & \cdots & n \\ \sigma(1) & \sigma(2) & \cdots & \sigma(n) \end{pmatrix}$$

or more generally as

$$\begin{pmatrix} i_1 & i_2 & \cdots & i_n \\ \sigma(i_1) & \sigma(i_2) & \cdots & \sigma(i_n) \end{pmatrix}$$

where $i_1, i_2, \ldots, i_n$ is any labelling of $1, 2, \ldots, n$. Thus, by flipping the representation of σ we get

$$\sigma^{-1} = \begin{pmatrix} \sigma(1) & \sigma(2) & \cdots & \sigma(n) \\ 1 & 2 & \cdots & n \end{pmatrix}.$$

We can choose $\sigma(1)$ in n different ways, and then we can choose $\sigma(2)$ in $n-1$ different ways, and then we can choose $\sigma(3)$ in $n - 2$ different ways, and so on. Therefore

$|S_n| = n!$, i.e., the symmetric group of degree n has size $n!$. A permutation on n letters is called a permutation of degree n, to remind us of the fact that an element of the Galois group of a polynomial of degree n is a permutation of the n roots. For a positive integer $e \le n$, an e-cycle or a cycle of length e in S_n is a permutation τ for which there is a rearrangement $i_1, \ldots, i_n$ of $1, \ldots, n$ such that $\tau(i_1) = i_2, \tau(i_2) = i_3, \ldots, \tau(i_{e-1}) = i_e, \tau(i_e) = i_1, \tau(i_{e+1}) = i_{e+1}, \ldots, \tau(i_n) = i_n$; we denote this by $(i_1 i_2 \ldots i_e)$, or for clarity by $(i_1, i_2, \ldots, i_e)$. As in the above example, any permutation σ of $\{1, 2, \ldots, n\}$ can be written as a product of disjoint cycles, say of lengths $e_1, \ldots, e_h$; note that we include cycles of length 1; also note that every $i \in \{1, \ldots, n\}$ occurs in exactly one of the h cycles, and we have $e_1 + \cdots + e_h = n$. Now, a cycle of length e can be expressed as a product of $e - 1$ transpositions, where by a transposition we mean a cycle of length 2; for example

$$(12345) = (12)(23)(34)(45).$$

Therefore, σ can be written as a product of $e_1 + e_2 + \cdots + e_h - h$ transpositions; the permutation σ is called even or odd according as the integer $e_1 + e_2 + \cdots + e_h - h$ is even or odd. To justify this terminology we have to show that if the integer $e_1 + e_2 + \cdots + e_h - h$ is even (resp: odd) then any way of writing σ as a product of transposition must contain an even (resp: odd) number of them. One way of proving this is described in the following Exercise (E1). From the definition of evenness, it follows that the set of all even permutations of $1, 2, \ldots, n$ forms a subgroup of S_n. This subgroup of S_n is called the alternating group of degree n and is denoted by A_n. It can easily be seen that $A_n \triangleleft S_n$ and $|S_n/A_n| = 2$ or 1 according as $n > 1$ or $n = 1$.

EXERCISE (E1). Let S_n be the group of all permutations of $\{1, \ldots, n\}$ where n is a positive integer. Show that every $\sigma \in S_n$ can be expressed as a product of disjoint cycles, and investigate how far this expression is unique; in particular show that the lengths of the cycles are determined by σ. Show that a cycle of length e is a product of $e - 1$ transpositions. Consider the polynomial of degree $n(n-1)$ in n variables over $\mathbb{Z}$ given by

$$D(X_1, \ldots, X_n) = \prod_{i < j \text{ in } \{1,\ldots,n\}} (X_i - X_j)$$

and for any $\sigma \in S_n$ put

$$D(X_1, \ldots, X_n)^\sigma = \prod_{i < j \text{ in } \{1,\ldots,n\}} (X_{\sigma(i)} - X_{\sigma(j)}).$$

Show that for all σ, τ in S_n we have

$$D(X_1, \ldots, X_n)^{(\tau\sigma)} = (D(X_1, \ldots, X_n)^\sigma)^\tau$$

and for all transpositions θ we have

$$D(X_1, \ldots, X_n)^\theta = -D(X_1, \ldots, X_n)$$

and from this deduce that the parity (i.e., the evenness or oddness) of the number of transpositions whose product is σ depends only on σ and not on the particular product representation. Moreover by defining the signature $\mathrm{sgn}(\sigma)$ to be 1 or -1 according as the said number of transpositions is even or odd we have

$$D(X_1, \ldots, X_n)^\sigma = \mathrm{sgn}(\sigma)D(X_1, \ldots, X_n).$$

As we shall discuss later [cf. L5§5(Q32)], the square of the above mentioned polynomial $D(X_1, \ldots, X_n)$ is the modified Y-discriminant of the n-th degree monic polynomial in Y whose roots are $X_1, \ldots, X_n$, i.e., of the polynomial $\prod_{1 \le i \le n}(Y - X_i)$.

EXERCISE (E2). Show that $[S_n : A_n] = 1$ or 2 according as $n = 1$ or $n > 1$. Deduce that $A_n \triangleleft S_n$. Generalize by showing that a subgroup of index 2 is always normal.

EXERCISE (E3). Show that the symmetric group S_n is generated by the transpositions $(12), (13), \ldots, (1n)$; hint: if $1, r, s$ are distinct then $(rs) = (1r)(1s)(1r)$. Show that A_n is generated by 3-cycles $(123), (124), \ldots, (12n)$; hint: if $1, r, s$ are distinct then $(1s)(1r) = (1rs)$, and if $1, 2, r, s$ are distinct then $(12s)(12s)(12r)(12s) = (1rs)$.

EXERCISE (E4). Show that if $H \triangleleft G$ (i.e., if H is a normal subgroup of a group G) then: G/H is abelian $\Leftrightarrow$ for all x, y in G we have $xyx^{-1}y^{-1} \in H$. The element $xyx^{-1}y^{-1}$ is called the commutator of x, y and is denoted by $[x, y]$. The subgroup of a group G generated by the commutators of various pairs of elements of G is called the commutator subgroup of G or the (first) derived subgroup of G, and is denoted by $[G, G]$ or G'. We may rephrase this exercise by saying that for any subgroup H of a group G we have: $G' \le H \Leftrightarrow H \triangleleft G$ and G/H is abelian.

EXERCISE (E5). As a consequence of (E3) and (E4) prove that if $n \ge 5$ then A_n is simple and S_n is unsolvable. Hint: for any distinct integers a, b, c, d, e upon letting $x = (adc)$ and $y = (cbe)$ we have $xyx^{-1}y^{-1} = (abc)$, and hence if $H \triangleleft G \le S_n$ with G/H abelian are groups such that G contains all 3-cycles then so does H; therefore S_n is unsolvable; for simplicity of A_n [cf. L6§6(E43)].

EXERCISE (E6). For $1 \le n \le 4$, find all subgroups of S_n and show that they are all solvable. Simplify the verification by showing that if $H \le G$ are finite groups with G solvable then H is solvable.

EXERCISE (E7). Prove the existence and uniqueness of the division algorithm equations $y = qx + r$ and $B(X) = q(X)A(X) + r(X)$ asserted in (R1). Hint: division is repeated subtraction, or equivalently, subtraction is slow division. Thus for the first equation, if $y \geq x > 0$ then for $r = y - x$ we have $0 \leq r < y$. Similarly for the second equation, if $\deg_X B(X) = e \geq d = \deg_X A(X) \geq 0$ and the coefficients of X^e and X^d in $B(X)$ and $A(X)$ are b and a respectively then for $r(X) = B(X) - (b/a)X^{e-d}A(X)$ we have $\deg_X r(X) < e$.

EXERCISE (E8). Give details of the equations $x_{t+2} = ax_1 + bx_2$, $x_1 = cx_{t+2}$, $x_2 = dx_{t+2}$ asserted in the Euclidean Algorithm described in (R2). Hint: the first step $x_1 = q_3 x_2 + x_3$ of the algorithm gives us $x_3 = x_1 - q_3 x_2$ and combining this with the second step $x_2 = q_4 x_3 + x_4$ gives us $x_4 = x_2 - q_4 x_3 = -q_4 x_1 + (1 + q_3 q_4)x_2$, and so on.

EXERCISE (E9). Prove claim (B1) concerning a UFD made in (R4). Prove that an ED has properties (G3) and (G4) stated in (R4), and prove claims (B2) and (B3) concerning an ED made in (R4).

EXERCISE (E10). Concerning algebraic field extensions, fill in the details of argument in (R6) and (D3), and prove assertions (C1) to (C5) of (R7). Hint: recall the "rationalization of surds" method from school math according to which

$$\frac{3 + \sqrt{3}}{2 + \sqrt{3}} = \frac{(3 + \sqrt{3})(2 - \sqrt{3})}{(2 + \sqrt{3})(2 - \sqrt{3})} = \frac{3 - \sqrt{3}}{1} = 3 - \sqrt{3}.$$

Generalizing this, given an algebraic element y over a field K with $[K(y) : K] = n$, we let $G(Y)$ be the minimal polynomial of y over K, and then for any $A(Y) \in K[Y]$ and $B(Y) \in K[Y] \setminus (G(Y)K[Y])$, by division algorithm and long division we find $C(Y), H(Y)$ in $K[Y]$ with $\deg_Y C(Y) < n$ such that $A(Y) = B(Y)C(Y) + G(Y)H(Y)$, and this gives

$$\frac{A(y)}{B(y)} = C(y)$$

proving that $1, y, \ldots, y^{n-1}$ is a K-vector-space basis of $K(y) = K[y]$.

§12: NOTES

NOTE (N1). [**Derivations**]. In calculus derivatives are defined in terms of limits. In algebra we define derivatives in terms of formal rules of manipulation. In greater detail, by a derivation of a ring S with values in an S-module V we mean a homomorphism $D : S \to V$ of the underlying additive groups such that for all x, y in S we have $D(xy) = xD(y) + yD(x)$. For any $s \in S$ we put $(sD)(x) = sD(x)$ for all $x \in S$, and for any other derivation of D' of S with values in V we put $(D + D')(x) = D(x) + D'(x)$ for all $x \in S$. This makes the set of all derivations of S

with values in V into an S-module which we denote by $\mathrm{Der}(S, V)$. If $D(x) = 0$ for all x in a subring R of S then we say that D is an R-derivation of S or a derivation of S/R. The submodule of all derivations of S/R with values in V is denoted by $\mathrm{Der}_R(S, V)$. By a derivation of S we mean a member of $\mathrm{Der}(S, S)$, and by an R-derivation of S we mean a member of $\mathrm{Der}_R(S, S)$. Now, referring to the section on Polynomials and Rational Functions, for any ring R the formal derivative $A \mapsto A_X$ is an R-derivation of $R[X]$, and if R is a field K then this induces a K-derivation of $K(X)$; likewise, for $1 \leq j \leq m$, the formal partial derivative $A \mapsto A_{X_j}$ is an $R[X_1, \ldots, X_{j-1}, X_{j+1}, \ldots, X_m]$-derivation of $R[X_1, \ldots, X_m]$, and if R is a field K then this induces a $K(X_1, \ldots, X_{j-1}, X_{j+1}, \ldots, X_m)$-derivation of $K(X_1, \ldots, X_m)$. Note that in the said section, in the defining equation $A_X(X) = \sum i A_i X^{i-1}$ for $i = 0$ we take $i A_i X^{i-1} = 0$, and in the defining equation of $A_{X_j}(X_1, \ldots, X_m)$ for $i_j = 0$ we take the corresponding term to be zero. This avoids negative powers.

EXERCISE (E11). Let D be a derivation of a ring S. Show that for all x in the prime subring of S we have $D(x) = 0$. Show that if S is a field then for any u, v in S with $v \neq 0$ we have (the quotient rule): $D(u/v) = (vD(u) - uD(v))/v^2$. Deduce that if S is a domain then D has a unique extension to the quotient field K of S, i.e., there is a unique derivation D' of K such that $D'(z) = D(z)$ for all $z \in S$.

EXERCISE (E12). Let $A(X_1, \ldots, X_m) \in R[X_1, \ldots, X_m]$ where $X_1, \ldots, X_m$ be indeterminates over a ring R, and let $x_1, \ldots, x_m$ be elements in an overring S of R. Show that for any D in $\mathrm{Der}_R(S, S)$ we have (the chain rule):

$$D(A(x_1, \ldots, x_m)) = \sum_{1 \leq j \leq m} A_{X_j}(x_1, \ldots, x_m) D(x_j).$$

More generally show that for any D in $\mathrm{Der}(S, S)$ we have (the more general chain rule):

$$D(A(x_1, \ldots, x_m)) = A_D(x_1, \ldots, x_m) + \sum_{1 \leq j \leq m} A_{X_j}(x_1, \ldots, x_m) D(x_j)$$

where $A_D(X_1, \ldots, X_m) \in R[X_1, \ldots, X_m]$ is obtained by applying D to the coefficients of $A(X_1, \ldots, X_m)$.

NOTE (N2). [**Very Long Proofs**]. As said before, CT = the Classification Theorem of finite simple groups has a proof of about ten thousand pages. Actually, the 1980 proof had some gaps which were repaired (?) only in 2004. Now a proof which is that long is not very digestible. In doing mathematics it is good to make clear what your assumptions are, and to base your arguments on results whose proofs you have checked. But results like CT can be useful guides in the search for truth. For applications of CT see Kleidman-Liebeck's book [KLi].

§13: CONCLUDING NOTE

Algebra means what you do not know, call it x. The experimental data will produce an equation in x. Solve it. The value of x so obtained will tell you what you wanted to know.

Lecture L2: Curves and Surfaces

§1: MULTIVARIABLE WORD PROBLEMS

In the last lecture I switched from calling the unknown quantity x to calling it Y. I did this to make room for bivariate (= two variable) equations because a word problem may simultaneously have two unknown quantities.

For instance, let us again go to the garden and pluck some roses and some gardenias. Then let us go to two more gardens and pluck the same number of roses as well as gardenias as we did from the first garden. From yet one more garden, we pluck only roses, as many as from the first garden, but do not pluck any gardenias. Counting up all the flowers we see that we have collected a total of fifty which is a nice number to make a bouquet. Needing also a garland, we go to four more gardens and from each pluck the same number of roses as well as gardenias as from the first garden, and make a garland of the sixty flowers so collected. How many roses and how many gardenias did we pluck from the first garden? Calling X the number of roses we plucked from the first garden and Y the number of gardenias which were plucked there, we get the two simultaneous equations:

$$4X + 3Y - 50 = 0 \quad \text{and} \quad 4X + 4Y - 60 = 0.$$

Subtracting the first equation from the second we get $Y = 10$ and substituting this in the first or the second equation we obtain $X = 5$.

Now the word problem may be more complicated, such as going to as many shops as the number of roses first picked and from each of them buy the same number of roses, and so on. This could lead to equations such as:

$$Y^2 + 7XY + 3Y + 2X - 9 = 0 \quad \text{and} \quad 4Y^3 + 2XY^2 + 2Y - 11 = 0$$

or even three variable equations such as

$$Y^3 + XZ - 5 = 0 \quad \text{and} \quad Y^2 + 4X^2 - 4 = 0.$$

Of course, we need not always pick flowers. We could shoot arrows instead or count the number of elementary particles as they collide in a cyclotron. Or we may compare distances between stars, or what have you. At any rate, such word problems frequently lead to several polynomial equations in several variables. We solve them, or try to solve them, and interpret the found values of the variables as measurements of the physical quantities we started with.

To solve these equations we could take recourse to the geometric method initiated by Descartes around 1630. For instance the first two linear equations represent two lines in the (X, Y)-plane, and they intersect in the point $(5, 10)$. Turning to the second set of equations and plotting them as curves in the (X, Y)-plane, the first equation, which is of degree 2, is a hyperbola and the second equation, which is of degree 3, is a cubic curve. By Bézout's theorem they intersect in $2 \times 3 = 6$ points, giving us six solutions of the two simultaneous equations. The third set of trivariate (= three variable) equations represent surfaces in the (X, Y, Z)-space. Their degrees being 3 and 2, again by Bézout's theorem they intersect in a space curve of degree $3 \times 2 = 6$. Note that the second equation is a vertical elliptical cylinder.

So what is this theorem of Bézout? Well, Descartes' analytic geometry eventually evolved into algebraic geometry and Bézout's theorem, proved around 1770, was the first substantial theorem of that new discipline. It says that in the plane, a curve of degree m and a curve of degree n always meet in mn points. Likewise, in three-space, a surface of degree m and a surface of degree n always meet in a curve of degree mn. Moreover, in three-space 3 surfaces of degrees d_1, d_2, d_3 will meet in $d_1 d_2 d_3$ points. All this, as we shall see, provided things are counted properly! But learning to count properly may take a while.

Quite generally, our word problem may lead to several polynomial equations in several variables. Generalizing the idea of a curve in the plane and a surface in 3-space, for any positive integer N, the points in N-space satisfying an N-variate (= N variable) polynomial equation $F(X_1, \ldots, X_N) = 0$ are said to form a hypersurface, and the degree of F is called the degree of that hypersurface. Again as a curve in the plane is 1-dimensional and a surface in 3-space is 2-dimensional, so a hypersurface in N-space is $(N - 1)$-dimensional, the intersection of two of them is $(N - 2)$-dimensional, the intersection of three of them is $(N - 3)$-dimensional, and so on; all this only "in general." If our word problem leads to N hypersurfaces $F_i(X_1, \ldots, X_N) = 0$ for $1 \leq i \leq N$, then Bézout says that they meet in $d_1 \ldots d_N$ points where d_i is the degree of F_i. If for some $M < N$ we have only the first M hypersurfaces $F_1 = 0, \ldots, F_M = 0$ then Bézout says that they meet in an $(N - M)$-dimensional variety of degree $d_1 \ldots d_M$. Again provided everything is counted properly. We also have to ensure that the hypersurfaces are "mutually independent"; for instance the equation $F_3 = 0$ should not be a consequence of the equations $F_1 = 0$ and $F_2 = 0$, as it would be say if $F_3 = F_1^2 + F_2$. Also we must attach precise meanings to a variety, its dimension, and its degree. Very briefly, a variety is the geometric locus consisting of the common solutions of a finite number of polynomial equations in a finite number of variables, its dimension is the "degree of freedom" of a point moving on it and is definable in terms of suitable transcendence degrees, its degree is the number of points in which it intersects a complementary dimensional linear subspace and is definable in terms of suitable field degrees, and so on.

To get ready for this proper counting, let us start by considering an algebraic curve C in the (X, Y)-plane over a field k. Now C is given by a bivariate polynomial equation $f(X, Y) = 0$ where

$$f(X, Y) = a_0(X)Y^n + \sum_{1 \leq i \leq n} a_i(X)Y^{n-i}$$

with $a_i(X) \in k[X]$ and $0 \neq a_0(X) \in k[X]$.

To discuss points in the (X, Y)-plane let us introduce cartesian products. The cartesian product $S \times T$ of sets S and T consists of all pairs (α, β) with $\alpha \in S$ and $\beta \in T$. Similarly the cartesian product $S_1 \times \cdots \times S_m$ of sets $S_1, \ldots, S_m$ consists of all m-tuples $(\gamma_1, \ldots, \gamma_m)$ with $\gamma_i \in S_i$. If $S_1 = \cdots = S_m = S$ then $S_1 \times \cdots \times S_m$ is denoted by S^m. Now for a typical point (α, β) in the (X, Y)-plane k^2 we have $\alpha \in k$ and $\beta \in k$; this point is on C if $f(\alpha, \beta) = 0$. Assuming f to be irreducible in $k[X, Y]$, and projecting vertically onto the X-axis, above each point $X = \alpha$, in general there lie n points $(\alpha, \beta_1), \ldots, (\alpha, \beta_n)$ on the curve C. The degree of C, i.e., the (X, Y)-degree of f, could be greater than the covering degree n, i.e., the Y-degree of f.

To ensure that in general $f(\alpha, Y)$ has n roots $\beta_1, \ldots, \beta_n$ in k, we need to assume that the field k is algebraically closed, which means every nonconstant univariate polynomial over k has a root in k. For instance the complex field $\mathbb{C}$ is algebraically closed. This used to be called the fundamental theorem of algebra, though it may be argued that it is neither fundamental nor a theorem of algebra. At any rate, most of its proofs involve analysis and topology. Many of them were given by Gauss between 1800 and 1840. Then around 1880, Kronecker constructed an algebraic closure of any field in a purely algebraic manner. Here by an algebraic closure of a field K we mean an algebraically closed overfield $\overline{K}$ of K which is algebraic over K. Sometimes we may call $\overline{K}$ an absolute algebraic closure of K to distinguish it from the (relative) algebraic closure of K in an overfield L of K by which we mean the subfield of L consisting of all elements of L which are algebraic over K.

Indeed, over any field K the method of first constructing a root field of a univariate irreducible polynomial and then using it to construct a splitting field $\text{SF}(F, K)$ of any nonconstant univariate $F(Y) \in K[Y]$ described in Lecture L1 is due to Kronecker. In turn this may be used to construct an algebraic closure $\overline{K}$ of K. Namely, let us label all the distinct nonconstant monic polynomials in $K[Y]$ as $F_1(Y), F_2(Y), \ldots$, and then after letting $K_{i+1} = \text{SF}(F_i, K_i)$ with $K_1 = K$, let us put $\overline{K} = \cup_{i=1}^{\infty} K_i$.

This is ok if K is countable, i.e., if K is either finite or there is a bijection of $\mathbb{N}_+$ onto K, because in both these cases the nonconstant monic polynomials in $K[Y]$ can be labelled as $F_1, F_2, \ldots$. But if K is uncountable then what to do? Well, let us try making "longer" sequences. At the end of all positive integers let us put a symbol ω, thus getting the "sequence" $1, 2, \ldots, \omega$, where $\omega > i$ for all $i \in \mathbb{N}_+$. Then start again with $\omega + 1, \omega + 2, \ldots, \omega^2 + \omega, \ldots, \omega^2 + 2\omega, \ldots, 2\omega^2, \ldots, 3\omega^2, \ldots$ and then $\omega^3, \ldots, \omega^4, \ldots, \omega^\omega, \omega^\omega + 1, \omega^\omega + 2, \ldots, \omega^\omega + \omega, \ldots, \omega^{\omega^\omega}$, and so on to kingdom come! But will the kingdom ever come? Nobody knows! In other words, it is not known if in this "constructive" manner we will ever get beyond a countable set. So we take recourse to the axiom of choice or the well ordering principle or Zorn's Lemma. Let us proceed to describe these three postulates and show their mutual equivalence.

The axiom of choice says that given any set of nonempty subsets of a set, it is possible to pick one element from each subset. To reformulate this in terms of maps, for any set T let $\widehat{\mathcal{P}}(T) =$ the power set of T, i.e., $\widehat{\mathcal{P}}(T) =$ the set of all subsets of T, and let $\widehat{\mathcal{P}}^\times(T) =$ the restricted power set of T, i.e., $\widehat{\mathcal{P}}^\times(T) = \widehat{\mathcal{P}}(T) \setminus \{\emptyset\}$. Then the axiom of choice says that given any map $\phi : S \to \widehat{\mathcal{P}}^\times(T)$ from a set S to the restricted power set $\widehat{\mathcal{P}}^\times(T)$ of a set T, there exists a map $\psi : S \to T$ such that for all $x \in S$ we have $\psi(x) \in \phi(x)$. This being a very reasonable postulate, we take it for granted. Before introducing the other two postulates, let us discuss various orders.

A po (= partial order) on a set is a binary relation $\leq$ which holds between some pairs $(x, y) \in S^2$ and is such that for all x, y, z in S we have: $x \leq y$ and $y \leq z \Rightarrow x \leq z$, and for all x, y in S we have $x \leq y$ and $y \leq x \Leftrightarrow x = y$; we write $x < y$ to mean $x \leq y$ and $x \neq y$. A poset (= partially ordered set) is a set with a partial order on it. A lo (= linear order = order) on a set S is a partial order on S such that for all x, y in S we have either $x \leq y$ or $y \leq x$. A loset (= linearly ordered set) is a set with a linear order on it. A chain in a poset S is a nonempty subset of S which is a loset under the induced partial order. A maximal element of a poset S is an element y of S such that for all $x \in S$ we have: $y \leq x \Rightarrow y = x$. An upper bound of a nonempty subset R of a poset S is an element y of S such that $x \leq y$ for all $x \in R$, and a smallest element of R is an element z of R such that $z \leq x$ for all $x \in R$. A wo (= well order) on a set S is a po on S such that every nonempty subset R of S has a smallest element; the said smallest element is clearly unique, i.e., it is determined by R; we may denote it by $\min(R)$. A woset (= well ordered set) is a set with a well order on it. For any $x \neq y$ in a woset S, we have $x < y$ or $y < x$ according as the smallest element of $\{x, y\}$ is x or y. Therefore every woset is a loset. The well ordering principle says that every set has a well order. A poset has the Zorn property if every chain in it has an upper bound. Zorn's Lemma says that if a nonempty poset has the Zorn property then it has a maximal element.

As said above, out of the three postulates, the axiom of choice is self-evident. Out of the remaining two, the well ordering principle is surprising but Zorn's Lemma is more versatile. At any rate, as we shall see in a moment, the three are equivalent. After establishing the equivalence, we shall complete the above proof of the existence of an algebraic closure first by using the well ordering principle and then by Zorn's Lemma. We shall also show that any two algebraic closures $\overline{K}$ and $\overline{K}'$ of a field K are K-isomorphic, i.e., there is a K-isomorphism $\overline{K} \to \overline{K}'$. Consequently we may sometimes speak of the algebraic closure of K, rather than an algebraic closure of K.

Now bijections define cardinals or cardinality. Two sets have the same cardinality if there is a bijection between them. For instance, all triads, i.e., sets of size three, define the cardinal 3, or as they used to say in school "new math," the threeness of three. The cardinality of a set S is denoted by $|S|$. It can be shown that cardinals are linearly ordered by defining $|S| \leq |T|$ to mean that there is an injection from S to T. The assertion that cardinals are partially ordered under this binary relation, is known as the Schroeder-Bernstein Theorem; we can reformulate it by saying that if there are injections $S \to T$ and $T \to S$ then there is a bijection $S \to T$. To see that this partial order is a linear order, what we need to prove is that given any sets S and T, either there exists an injection from S to T, or there exists an injection from T to S.

As bijections between (unordered) sets define cardinals, order preserving maps between well ordered sets define ordinals in the following manner. A bijection $\phi : S \to T$ between two posets S and T is order preserving means: $x \le y$ in $S \Leftrightarrow \phi(x) \le \phi(y)$ in T. An ordinal is an equivalence class of wosets under the equivalence of order preserving bijections. For instance, the sequence $1, 2, 3$ defines the ordinal 3. In the above illustrations, ω is the first infinite ordinal. Some other ordinals are "polynomials" in ω of degree $d \in \mathbb{N}_+$ such as $u_0 \omega^d + \sum_{1 \le i \le d} u_i \omega^{d-i}$ with $u_i \in \mathbb{N}$ and $u_0 \in \mathbb{N}_+$. Still bigger ordinals are the exponentials ω^ω, ω^{ω^ω}, and so on. A lower segment in a poset S is a subset R of S such that: $x \le y$ in S with $y \in R \Rightarrow x \in R$. Let us denote the ordinal of a woset S by $||S||$, and for any wosets S and T let us define $||S|| \le ||T||$ to mean that there is an order preserving bijection from S onto a lower segment of T. By using Zorn's Lemma it can be shown that under this relationship, any set of ordinals is well ordered. An ordinal is a limiting or nonlimiting ordinal according as it does not or does have an immediate predecessor, i.e., according as the set of all smaller ordinals does not or does have a maximal element. For instance $\omega, 2\omega, \ldots, \omega^\omega$ are limiting ordinals, whereas $1, 2, \ldots, \omega + 1, \omega + 2, \ldots, \omega^\omega + 1$ are nonlimiting ordinals.

The idea of polynomials in ω is relevant for constructing the algebraic closure of the meromorphic series field $k((X))$ in characteristic $p \ne 0$. So let us proceed to introduce meromorphic series.

§2: POWER SERIES AND MEROMORPHIC SERIES

Writing a polynomial in an indeterminate X over any ring R in ascending powers of X and letting them go to infinity on the right hand side we get a power series in X, and letting them have a finite number of terms on the left, i.e., terms with negative exponents, we get a meromorphic series. We denote the power series ring by $R[[X]]$, and we denote the meromorphic series ring by $R((X))$. We regard $R[X]$ as a subring of $R[[X]]$, and we regard $R[[X]]$ as a subring of $R((X))$. More precisely, a meromorphic series $A(X)$ in X over R, i.e., a member of $R((X))$, has a unique expression

$$A(X) = \sum_{i \in \mathbb{Z}} A_i X^i \quad \text{with} \quad A_i \in R$$

such that $\text{Supp}(A)$ is a well ordered subset of $\mathbb{Z}$ where the support of $A = A(X)$, or the X-support of $A = A(X)$, is defined by putting

$$\text{Supp}(A) = \text{Supp}_X A = \{i \in \mathbb{Z} : A_i \ne 0\}$$

and we note that: $\mathrm{Supp}(A) = \emptyset \Leftrightarrow A = 0$. We define the order of A, or the X-order of A, by putting

$$\mathrm{ord}(A) = \mathrm{ord}_X A = \min \mathrm{Supp}(A)$$

i.e.,

$$\mathrm{ord}(A) = \begin{cases} \text{the smallest element of } \mathrm{Supp}(A) & \text{if } A \neq 0 \\ \infty & \text{if } A = 0. \end{cases}$$

As in the case of polynomials, for any other meromorphic series

$$B(X) = \sum_{i \in \mathbb{Z}} B_i X^i \quad \text{with} \quad B_i \in R$$

the sum

$$C(X) = A(X) + B(X) = \sum_{i \in \mathbb{Z}} C_i X^i$$

is given by componentwise addition

$$C_i = A_i + B_i$$

and the product

$$D(X) = A(X)B(X) = \sum_{i \in \mathbb{Z}} D_i X^i$$

is given by the "Cauchy multiplication rule"

$$D_k = \sum_{i+j=k} A_i B_j.$$

It is this rule which makes us require $\mathrm{Supp}(A)$ to be well ordered, because otherwise there may be infinitely many pairs (i, j) with $A_i \neq 0 \neq B_j$ and $i + j = k$. At any rate, in an infinite summation, by convention we disregard all the zero terms, and so the summation makes sense provided there are only finitely many nonzero terms. The power series ring is defined by putting

$$R[[X]] = \{A = A(X) \in R((X)) : \mathrm{ord}(A) \geq 0\}.$$

As in the polynomial case, the "formal" X-derivative $A_X(X) = \frac{dA(X)}{dX}$ of $A(X)$ in $R((X))$ is defined by putting

$$A_X(X) = \sum_{i \in \mathbb{Z}} i A_i X^{i-1}$$

and again this gives an R-derivation of $R((X))$. If R is a domain then the order of the product of two nonzero meromorphic series equals the sum of their orders, and hence in particular the product is nonzero. It follows that if R is a domain then so is the meromorphic series ring $R((X))$ as well as the power series ring $R[[X]]$.

For finitely many variables (= indeterminates) $X_1, \ldots, X_m$ over a ring R, we consider only the power series ring $R[[X_1, \ldots, X_m]]$, which we regard as an overring of R. A power series in $X_1, \ldots, X_m$ over R, i.e., a member of $R[[X_1, \ldots, X_m]]$, has a unique expression

$$A(X_1, \ldots, X_m) = \sum_{(i_1, \ldots, i_m) \in \mathbb{N}^m} A_{i_1 \ldots i_m} X_1^{i_1} \ldots X_m^{i_m} \quad \text{with} \quad A_{i_1 \ldots i_m} \in R.$$

The order of A, or the $(X_1, \ldots, X_m)$-order of A, is the minimum of $i_1 + \cdots + i_m$ with $A_{i_1 \ldots i_m} \neq 0$; if $A = 0$, i.e., if $A_{i_1 \ldots i_m} = 0$ for all $i_1, \ldots, i_m$, then the order is taken to be ∞; we denote the order of A by $\mathrm{ord}(A)$ or $\mathrm{ord}_{(X_1, \ldots, X_m)} A$. For any other power series

$$B(X_1, \ldots, X_m) = \sum_{(i_1, \ldots, i_m) \in \mathbb{N}^m} B_{i_1 \ldots i_m} X_1^{i_1} \ldots X_m^{i_m} \quad \text{with} \quad B_{i_1 \ldots i_m} \in R$$

the sum

$$C(X_1, \ldots, X_m) = A(X_1, \ldots, X_m) + B(X_1, \ldots, X_m)$$
$$= \sum_{(i_1, \ldots, i_m) \in \mathbb{N}^m} C_{i_1 \ldots i_m} X_1^{i_1} \ldots X_m^{i_m}$$

is given by componentwise addition

$$C_{i_1 \ldots i_m} = A_{i_1 \ldots i_m} + B_{i_1 \ldots i_m}$$

and the product

$$D(X_1, \ldots, X_m) = A(X_1, \ldots, X_m) B(X_1, \ldots, X_m)$$
$$= \sum_{(i_1, \ldots, i_m) \in \mathbb{N}^m} D_{i_1 \ldots i_m} X_1^{i_1} \ldots X_m^{i_m}$$

is given by the "Cauchy multiplication rule"

$$D_{k_1 \ldots k_m} = \sum_{i_1 + j_1 = k_1, \ldots, i_m + j_m = k_m} A_{i_1 \ldots i_m} B_{j_1 \ldots j_m}.$$

The "formal" partial derivative $A_{X_j}(X_1, \ldots, X_m) = \frac{\partial A(X_1, \ldots, X_m)}{\partial X_j}$ of $A(X_1, \ldots, X_m)$ in $R[[X_1, \ldots, X_m]]$ is defined by putting

$$A_{X_j}(X_1, \ldots, X_m) = \sum_{(i_1, \ldots, i_m) \in \mathbb{N}^m} i_j A_{i_1 \ldots i_m} X_1^{i_1} \ldots X_{j-1}^{i_{j-1}} X_j^{i_j - 1} X_{j+1}^{i_{j+1}} \ldots X_m^{i_m}$$

with the understanding that for $i_j = 0$ the corresponding term in the above summation is zero, and we note that $A \mapsto A_{X_j}$ gives an $R[[X_1, \ldots, X_{j-1}, X_{j+1}, \ldots, X_m]]$-derivation of $R[[X_1, \ldots, X_m]]$. If R is a domain then the order of the product of two nonzero power series equals the sum of their orders, and hence in particular the product is nonzero. It follows that if R is a domain then so is the power series ring $R[[X_1, \ldots, X_m]]$. For a finite number of variables $X_1, \ldots, X_m$ over a field K, by

$K((X_1, \ldots, X_m))$ we denote the field of meromorphic functions in $X_1, \ldots, X_m$ over K. Members of the meromorphic function field $K((X_1, \ldots, X_m))$ are "quotients" of power series $A(X_1, \ldots, X_m)/B(X_1, \ldots, X_m)$ with nonzero B. By the "quotient rule" we extend the above $K[[X_1, \ldots, X_{j-1}, X_{j+1}, \ldots, X_m]]$-derivation of the domain $K[[X_1, \ldots, X_m]]$ to a $K((X_1, \ldots, X_{j-1}, X_{j+1}, \ldots, X_m))$-derivation of the field $K((X_1, \ldots, X_m))$.

Formally speaking, for any ring R, we may regard $R((X))$ as the set of all functions (= maps) from $\mathbb{Z}$ to R with well ordered support. At any rate, $R((X)) \subset R^{\mathbb{Z}}$, where for any sets S and T, by S^T we denote the set of all functions $\phi : T \to S$. In case S has a special element e, like the zero element in a ring, by the support of ϕ we mean the set of all $t \in T$ such that $\phi(t) \neq e$, and we denote it by $\mathrm{Supp}(\phi)$. By $(S^T)_{\mathrm{finite}}$ let us denote the set of all $\phi \in S^T$ such that $\mathrm{Supp}(\phi)$ is finite, and in case T is a poset, by $(S^T)_{\mathrm{wellord}}$ let us denote the set of all $\phi \in S^T$ such that $\mathrm{Supp}(\phi)$ is well ordered. Note that then $R((X)) = (R^{\mathbb{Z}})_{\mathrm{wellord}}$, $R[[X]] = R^{\mathbb{N}}$, and $R[X] = (R^{\mathbb{N}})_{\mathrm{finite}}$. In case of a finite number of variables $X_1, \ldots, X_m$ we have $R[[X_1, \ldots, X_m]] = R^{\mathbb{N}^m}$ and $R[X_1, \ldots, X_m] = (R^{\mathbb{N}^m})_{\mathrm{finite}}$. In case of $R((X))$, the function which sends $1 \in \mathbb{Z}$ to $1 \in R$ and every other element of $\mathbb{Z}$ to $0 \in R$ corresponds to X. In case of $R[[X_1, \ldots, X_m]]$, the function which sends $(0, \ldots, 0, 1, 0 \ldots, 0) \in \mathbb{N}^m$ with 1 in the i-th spot to $1 \in R$ and every other element of $\mathbb{N}^m$ to $0 \in R$ corresponds to X_i. Similarly for $R[[X]]$ and $R[X]$.

To speak formally about quotient fields, we first introduce the concept of an equivalence relation. An equivalence relation on a set S is a binary relation $x \sim y$ which holds between certain pairs $(x, y) \in S^2$ and which is reflexive: for all $x \in S$ we have $x \sim x$, symmetric: for all x, y in S we have $x \sim y \Rightarrow y \sim x$, and transitive; for all x, y, z in S we have $x \sim y$ and $y \sim z \Rightarrow x \sim z$. By putting all the elements which are equivalent to each other in one box, the set S gets divided into a disjoint partition, i.e., a family (= set) of nonempty subsets any two of which have an empty intersection. This family of boxes = equivalence classes is denoted by $S/\sim$ and is called the quotient of S by $\sim$; the map which sends every $x \in S$ to the box containing it is called the natural surjection $S \to S/\sim$.

Now the quotient field F of a domain E is defined to be the set of all equivalence classes of pairs $(u, v) \in E^2$ with $v \neq 0$ under the equivalence relation: $(u, v) \sim (u', v') \Leftrightarrow uv' = vu'$. The equivalence class containing (u, v) is denoted by u/v or $\frac{u}{v}$, and we add and multiply the equivalence classes by the rules $\frac{u_1}{v_1} + \frac{u_2}{v_2} = \frac{u_1 v_2 + u_2 v_1}{v_1 v_2}$ and $\frac{u_1}{v_1} \times \frac{u_2}{v_2} = \frac{u_1 u_2}{v_1 v_2}$. Also we "identify" any $u \in E$ with $\frac{u}{1} \in F$. This makes F a field, and E a subring of F. We may write $\mathrm{QF}(E)$ for F.

Given two finite sets of variables $X_1, \ldots, X_m$ and $Z_1, \ldots, Z_d$ over a ring R, and given any power series $P^{(l)}(Z_1, \ldots, Z_d)$ "devoid" of constant term, i.e.,

$$P^{(l)}(Z_1, \ldots, Z_d) = \sum_{(j_1, \ldots, j_d) \in \mathbb{N}^d} P^{(l)}_{j_1 \ldots j_d} Z_1^{j_1} \ldots Z_d^{j_d} \quad \text{with} \quad P^{(l)}_{j_1 \ldots j_d} \in R$$

such that $P^{(l)}_{0\ldots0} = 0$, in any power series

$$A(X_1,\ldots,X_m) = \sum_{(i_1,\ldots,i_m)\in\mathbb{N}^m} A_{i_1\ldots i_m} X_1^{i_1}\ldots X_m^{i_m} \quad\text{with}\quad A_{i_1\ldots i_m}\in R$$

we can "substitute" $P^{(l)}$ for X_l for $1 \le l \le m$ to get

$$A^{(P)}(Z_1,\ldots,Z_d) = A(P^{(1)}(Z_1,\ldots,Z_d),\ldots,P^{(m)}(Z_1,\ldots,Z_d))$$
$$= \sum_{(j_1,\ldots,j_d)\in\mathbb{N}^d} A^{(P)}_{j_1\ldots j_d} Z_1^{j_1}\ldots Z_d^{j_d} \quad\text{with}\quad A^{(P)}_{j_1\ldots j_d}\in R.$$

This makes sense because it does not involve infinite sums of coefficients. Indeed, for any $i = (i_1,\ldots,i_m)\in\mathbb{N}^m$, upon writing

$$P^{(1)}(Z_1,\ldots,Z_d)^{i_1}\ldots P^{(m)}(Z_1,\ldots,Z_d)^{i_m} = \sum_{(j_1,\ldots,j_d)\in\mathbb{N}^d} P^{(i)}_{j_1\ldots j_d} Z_1^{j_1}\ldots Z_d^{j_d}$$

with $P^{(i)}_{j_1\ldots j_d}\in R$, clearly we have

$$P^{(i)}_{j_1\ldots j_d} = 0 \quad\text{whenever}\quad j_1 + \cdots + j_d < i_1 + \cdots + i_m.$$

Thus $A \mapsto A^{(P)}$ gives the R-homomorphism $\phi : R[[X_1,\ldots,X_m]] \to R[[Z_1,\ldots,Z_d]]$, which we call the substitution homomorphism. Note that by taking $d = 0$ we get $R[[Z_1,\ldots,Z_d]] = R$ and by taking $P^{(l)} = 0$ for $1 \le l \le m$ we get the R-epimorphism $R[[X_1,\ldots,X_m]] \to R$ given by $A(X_1,\ldots,X_m) \mapsto A(0,\ldots,0) = A_{0\ldots0}$. The power series $P^{(1)}(Z_1,\ldots,Z_d),\ldots,P^{(m)}(Z_1,\ldots,Z_d)$ (which are assumed devoid of constant terms) are said to be analytically independent or analytically dependent over R according as $\ker(\phi) = 0$ or $\ker(\phi) \ne 0$. If $P^{(1)}(Z_1,\ldots,Z_d),\ldots,P^{(m)}(Z_1,\ldots,Z_d)$ are power series (devoid of constant terms) over a field K such that the substitution map $\phi : K[[X_1,\ldots,X_m]] \to K[[Z_1,\ldots,Z_d]]$ is injective, then ϕ uniquely extends to an injective K-homomorphism $K((X_1,\ldots,X_m)) \to K((Z_1,\ldots,Z_d))$ which we again call the substitution map.

Considering the geometric series identity

(i) $$(1 - X)(1 + X + X^2 + \ldots) = 1$$

in the univariate power series ring $R[[X]]$ over a ring R, and taking any power series $P(Z_1,\ldots,Z_d)$ in a finite number of variables $Z_1,\ldots,Z_d$ over R with $P(0,\ldots,0) = 0$ and substituting it for X we get

$$(1 - P(Z_1,\ldots,Z_d))P'(Z_1,\ldots,Z_d) = 1$$

where

$$P'(Z_1,\ldots,Z_d) = \sum_{0\le i<\infty} P(Z_1,\ldots,Z_d)^i \in R[[Z_1,\ldots,Z_d]].$$

If the element $Q(Z_1, \ldots, Z_d) \in R[[Z_1, \ldots, Z_d]]$ is such that $Q(0, \ldots, 0)$ is a unit in R then letting $P(Z_1, \ldots, Z_d) = 1 - Q(0, \ldots, 0)^{-1} Q(Z_1, \ldots, Z_d)$ we get $P(0, \ldots, 0) = 0$ and $Q(Z_1, \ldots, Z_d) = Q(0, \ldots, 0)(1 - P(Z_1, \ldots, Z_d))$ and therefore by letting $Q'(Z_1, \ldots, Z_d) = Q(0, \ldots, 0)^{-1} P'(Z_1, \ldots, Z_d)$ we get $Q(Z_1, \ldots, Z_d) Q'(Z_1, \ldots, Z_d) = 1$. Thus we have shown that if $Q(0, \ldots, 0)$ is a unit in R then $Q(Z_1, \ldots, Z_d)$ is a unit in $R[[Z_1, \ldots, Z_d]]$. Conversely if $Q(Z_1, \ldots, Z_d)$ is a unit in $R[[Z_1, \ldots, Z_d]]$ then for some $Q^*(Z_1, \ldots, Z_d)$ in $R[[Z_1, \ldots, Z_d]]$ we have $Q(Z_1, \ldots, Z_d) Q^*(Z_1, \ldots, Z_d) = 1$, and putting $Z_1 = \cdots = Z_d = 0$ we get $Q(0, \ldots, 0) Q^*(0, \ldots, 0) = 1$, and hence $Q(0, \ldots, 0)$ is a unit in R. Recalling that $U(R)$ is the group of all units in R, we conclude that

(ii)
$$\begin{cases} U(R[[Z_1, \ldots, Z_d]]) = \{Q(Z_1, \ldots, Z_d) \in R[[Z_1, \ldots, Z_d]] : \\ \qquad\qquad\qquad\qquad\qquad Q(0, \ldots, 0) \in U(R)\}. \end{cases}$$

Applying the above consideration to the power series ring $K[[X_1, \ldots, X_m]]$ in a finite number of variables $X_1, \ldots, X_m$ over a field K, we see that the units in $K[[X_1, \ldots, X_m]]$ are exactly those power series whose order is zero. It follows that in the univariate case, the meromorphic series field $K((X))$ is a concrete representation of the quotient field of $K[[X]]$, i.e., there is no discrepancy in the two meanings of $K((X))$.

§3: VALUATIONS

Again considering the power series ring $K[[X_1, \ldots, X_m]]$ in a finite number of variables over a field K, we can uniquely extend the order function from the domain $K[[X_1, \ldots, X_m]]$ to its quotient field $K((X_1, \ldots, X_m))$ by defining the order, or the $(X_1, \ldots, X_m)$-order, of u/v for any u, v in $K[[X_1, \ldots, X_m]]$ with $v \neq 0$ by putting

$$\mathrm{ord}(u/v) = \mathrm{ord}_{(X_1, \ldots, X_m)} u/v = \mathrm{ord}(u) - \mathrm{ord}(v)$$

with the convention that $\mathrm{ord}(0) = \infty$. Now for all x, y in $K((X_1, \ldots, X_m))$ we have

(V1)
$$\mathrm{ord}(xy) = \mathrm{ord}(x) + \mathrm{ord}(y)$$

and

(V2)
$$\mathrm{ord}(x + y) \geq \min(\mathrm{ord}(x), \mathrm{ord}(y))$$

with obvious conventions about ∞. In particular, for all x in $K((X_1, \ldots, X_m))$ we have

(V3)
$$\mathrm{ord}(x) = \infty \Leftrightarrow x = 0.$$

The second inequality also holds for all x, y in $R[[X_1, \ldots, X_m]]$ for any ring R. To make this transparent it is best to express every power series as a sum of homogeneous polynomials by "collecting terms of like degree." In greater detail, a polynomial

$$H(X_1, \ldots, X_m) = \sum H_{i_1 \ldots i_m} X_1^{i_1} \ldots X_m^{i_m} \quad \text{with} \quad H_{i_1 \ldots i_m} \in R$$

is homogeneous of degree j if for all $(i_1, \ldots, i_m) \in \operatorname{Supp}(H)$ we have $i_1 + \cdots + i_m = j$; note that the zero polynomial is regarded as homogeneous of every degree. Now a power series

$$A(X_1, \ldots, X_m) = \sum_{(i_1, \ldots, i_m) \in \mathbb{N}^m} A_{i_1 \ldots i_m} X_1^{i_1} \ldots X_m^{i_m} \quad \text{with} \quad A_{i_1 \ldots i_m} \in R$$

can uniquely be written as

$$A(X_1, \ldots, X_m) = \sum_{0 \leq j < \infty} A_j(X_1, \ldots, X_m)$$

where the polynomial

$$A_j(X_1, \ldots, X_m) = \sum_{i_1 + \cdots + i_m = j} A_{i_1 \ldots i_m} X_1^{i_1} \ldots X_m^{i_m}$$

is homogeneous of degree j. Given any other power series

$$B(X_1, \ldots, X_m) = \sum_{0 \leq j < \infty} B_j(X_1, \ldots, X_m)$$

where $B_j(X_1, \ldots, X_m) \in R[X_1, \ldots, X_m]$ is homogeneous of degree j, assume $A \neq 0 \neq B$ and let $\operatorname{ord}(A) = a$ and $\operatorname{ord}(B) = b$. Then

$$A(X_1, \ldots, X_m) = A_a(X_1, \ldots, X_m) + \text{terms of degree} > a$$

and

$$B(X_1, \ldots, X_m) = B_b(X_1, \ldots, X_m) + \text{terms of degree} > b$$

with

$$A_a(X_1, \ldots, X_m) \neq 0 \neq B_b(X_1, \ldots, X_m).$$

It follows that

$$A(X_1, \ldots, X_m) + B(X_1, \ldots, X_m)$$
$$= \begin{cases} A_a(X_1, \ldots, X_m) + \text{terms of degree} > a & \text{if } a < b \\ B_b(X_1, \ldots, X_m) + \text{terms of degree} > b & \text{if } a > b \\ A_a(X_1, \ldots, X_m) + B_a(X_1, \ldots, X_m) + \text{terms of degree} > a & \text{if } a = b \end{cases}$$

and hence $\operatorname{ord}(A + B) \geq \min(a, b)$ with equality in case of $a \neq b$. Also note that if R is a domain then

$$A(X_1, \ldots, X_m)B(X_1, \ldots, X_m)$$
$$= A_a(X_1, \ldots, X_m)B_b(X_1, \ldots, X_m) + \text{terms of degree} > a + b$$

and $0 \neq A_a(X_1, \ldots, X_m)B_b(X_1, \ldots, X_m) \in R[X_1, \ldots, X_m]$ is homogeneous of degree $a + b$, and hence $\operatorname{ord}(AB) = a + b$.

To generalize the idea of meromorphic series, let G be an ordered abelian group, i.e., G is an additive abelian group which is also an ordered set such that for all x, y, x', y' in G we have: $x \leq y$ and $x' \leq y' \Rightarrow x + x' \leq y + y'$. For instance $G = \mathbb{Z}$ or $\mathbb{Q}$ or $\mathbb{R}$. Or G could be the set $\mathbb{R}^{[d]}$ of lexicographically ordered d-tuples of real numbers $r = (r_1, \ldots, r_d), s = (s_1, \ldots, s_d), \ldots$, where lexicographic order means: $r \leq s \Leftrightarrow$ either $r_i = s_i$ for $1 \leq i \leq d$, or for some j with $1 \leq j \leq d$ we have $r_i = s_i$ for $1 \leq i < j$ and $r_j < s_j$. Or G could be a subgroup of any of these.

Inspired by properties (V1) to (V3), we define a valuation of a field K to be a map $v : K \to G \cup \{\infty\}$ such that for all x, y in K we have (V1) to (V3) with ord replaced by v, with the conventions about ∞ that: for all $g \in G$ we have $g < \infty$ and $g + \infty = \infty$, and also $\infty + \infty = \infty$. We call G the assigned value group of v, and by the value group of v we mean the subgroup of G given by

$$G_v = \{v(x) : x \in K^\times\}.$$

By the valuation ring of v we mean the ring

$$R_v = \{x \in K : v(x) \geq 0\}$$

and we note that this is clearly a subdomain (= subring which is a domain) of K which is its quotient field. We say that v is trivial over a subfield k of K, or that v is a valuation of K/k, to mean that

$$x \in k^\times \Rightarrow v(x) = 0$$

and we note that this is equivalent to saying that $k \subset R_v$, i.e., k is a subfield of R_v. Now the ord function is obviously a valuation v of $K((X_1, \ldots, X_m))/K$ with

$$G_v = G = \mathbb{Z} \quad \text{and} \quad K[[X_1, \ldots, X_m]] \subset R_v.$$

Note that in case of $m = 1$ the above displayed inclusion is an equality.

To construct a valuation having a given ordered abelian group G as its value group, and at the same time to generalize the idea of the univariate meromorphic series field $K((X))$ over any given field K, let

$$K((X))_G = \{A \in K^G : \operatorname{Supp}(A) \text{ is well ordered}\}$$

where the support of $A \in K^G$, denoted by $\operatorname{Supp}(A)$, is defined by putting

$$\operatorname{Supp}(A) = \{g \in G : A(g) \neq 0\}.$$

For A, B in $K((X))_G$, the sum is componentwise, i.e., for any $g \in G$ we have

$$(A + B)(g) = A(g) + B(g)$$

and the product is by the "Cauchy rule" which says that for all $g \in G$ we have

$$(AB)(g) = \sum_{i+j=g} A(i)B(j)$$

and the well ordering enters in showing that

(V4) $\begin{cases} \{(i,j) \in G^2 : i + j = g \text{ with } i \in \mathrm{Supp}(A) \text{ and } j \in \mathrm{Supp}(B)\} \\ \text{is a finite set.} \end{cases}$

We also have

(V5) $\mathrm{Supp}(A)$ and $\mathrm{Supp}(B)$ are well ordered $\Rightarrow \mathrm{Supp}(AB)$ is well ordered.

Note that $A = 0$ means $A(g) = 0$ for all $g \in G$. Also $A = 1$ means $A(0) = 1$ and $A(g) = 0$ for all other $g \in G$. For any $A \in K((X))_G$ we define

$$\mathrm{ord}(A) = \begin{cases} \min \mathrm{Supp}(A) & \text{if } A \neq 0 \\ \infty & \text{if } A = 0 \end{cases}$$

and we note that then for all x, y in $K((X))_G$ we have (V1) to (V3). Mimicking the proof of (ii) using (i) in §2, we can show that

(V6) $0 \neq A \in K((X))_G \Rightarrow AA' = 1$ for some $A' \in K((X))_G$.

Thus $K((X))_G$ is a field, and ord is a valuation of $K((X))_G/K$ with value group G. Concretely speaking, a typical member of $K((X))_G$ can be written as

$$A(X) = \sum_{i \in G} A_i X^i$$

where we are writing A_i for the previous $A(i)$. Assuming $G \neq 0$ and designating a suitable nonzero element of G as 1, the function which sends $1 \in G$ to $1 \in K$ and sends every other element of G to $0 \in R$ now corresponds to X. For instance, if $G = \mathbb{Z}^{[d]}$ or $\mathbb{Q}^{[d]}$ or $\mathbb{R}^{[d]}$, then designate $(1, 0, \ldots, 0)$ to be $1 \in G$. We also put

$$K[[X]]_G = \{A(X) \in K((X))_G : \mathrm{ord}(A) \geq 0\}$$

and we note that $K[[X]]_G$ is the valuation ring of the ord valuation of $K((X))_G$.

Taking $\mathbb{Q}$ to be the exponent group, i.e., taking $\mathbb{Q}$ for G, for any $n \in N_+$ we put

$$\mathbb{Q}_n = \{j \in \mathbb{Q} : nj \in \mathbb{Z}\}$$

and for any prime p we put

$$\mathbb{Q}_{n,p} = \{j \in \mathbb{Q} : njp^r \in \mathbb{Z} \text{ for some } r \in \mathbb{N}_+\}$$

where r may depend on j, and we let

$$K((X))_{\text{newt}} = \cup_{n \in \mathbb{N}_+} \{A(X) \in K((X))_{\mathbb{Q}} : \text{Supp}(A) \subset \mathbb{Q}_n\}$$

and in case K is of characteristic p we let

$$K((X))_{\text{gnewt}} = \cup_{n \in \mathbb{N}_+} \{A(X) \in K((X))_{\mathbb{Q}} : \text{Supp}(A) \subset \mathbb{Q}_{n,p}\}.$$

These are clearly subfields of $K((X))_{\mathbb{Q}}$, and they are the respective quotient fields of the domains

$$K[[X]]_{\text{newt}} = \{A(X) \in K((X))_{\text{newt}} : \text{ord}(A) \geq 0\}$$

and

$$K[[X]]_{\text{gnewt}} = \{A(X) \in K((X))_{\text{gnewt}} : \text{ord}(A) \geq 0\}.$$

We shall give several proofs of the following theorem which dates back to 1660.

NEWTON'S THEOREM (T1). Let k be an algebraically closed field of characteristic 0. Then $k((X))_{\text{newt}}$ is an algebraic closure of $k((X))$.

We shall also prove the following generalization of Newton's Theorem.

GENERALIZED NEWTON'S THEOREM (T2). Let k be an algebraically closed field. Then $k((X))_{\mathbb{Q}}$ is an algebraically closed overfield of $k((X))$. If k is of characteristic $p > 0$, then $k((X))_{\text{gnewt}}$ is an algebraically closed overfield of $k((X))$, which is strictly bigger than the algebraic closure of $k((X))$; more precisely, $k((X))_{\text{gnewt}}$ contains elements whose supports are such that their ordinals are not polynomials in ω, but the ordinal of the support of any element which is algebraic over $k((X))$ is a polynomial in ω.

§4: ADVICE TO THE READER

The next section on Zorn's Lemma and Well Ordering, spread over seven pages, is rather heavy logical stuff. This inspires me to verbatim reproduce the following lines of advice from the preface of the Second Part of the great and famous Algebra Text-Book of George Chrystal written in 1889 [Chr]:

"I may here give a word of advice to young students reading my second volume. The matter is arranged to facilitate reference and to secure brevity and logical sequence; but it by no means follows that the volume should be read straight through at a first reading. Such an attempt would probably sicken the reader both of the author and the subject. Every mathematical book that is worth anything must be read "backwards and forwards," if I may use the expression. I would modify the advice of a great French mathematician and say, "Go on, but often return to strengthen your faith." When you come on a hard or dreary passage, pass it over;

and come back to it after you have seen its importance or found the need for it further on."

So, in a first reading, without loss of understanding, the reader may prefer to immediately proceed to the section after next where I shall give a Utilitarian Summary of the logical stuff.

§5: ZORN'S LEMMA AND WELL ORDERING

We want to show the equivalence of the three postulates: Axiom of Choice, Well Ordering Principle, and Zorn's Lemma. Let us abbreviate these as (AC), (WO), and (ZL), and restate them. Let us also include an Alternate Version of Zorn's Lemma which we abbreviate as (ZL*).

(AC). Given any map $\phi : S \to \widehat{\mathcal{P}}^\times(T)$ from a set S to the restricted power set $\widehat{\mathcal{P}}^\times(T)$ of a set T, there exists a map $\psi : S \to T$ such that for all $x \in S$ we have $\psi(x) \in \phi(x)$.

(WO). Every set has a well order.

(ZL). If a nonempty poset has the Zorn property then it has a maximal element. Recall that a poset has the Zorn property means every chain in it has an upper bound.

(ZL*). If a nonempty poset has the weak Zorn property then it has a maximal element, where a poset has the weak Zorn property means every woset in it (= well ordered subset of it) has an upper bound.

First let us prove the following:

LEMMA (T3). Let P be a set of subsets of a poset S. Assume that every member of P is well ordered (in the partial order induced from S). Also assume that P is a loset (in the "lower segment" partial order on P according to which $A \leq B$ in P means A is a lower segment of B). Let $Q = \cup_{R \in P} R$. Then Q is well ordered (in the partial order induced from S).

PROOF. Given any nonempty subset Q' of Q, we can take $R \in P$ such that $R \cap Q'$ is nonempty. Let $x = \min(R \cap Q')$. We claim that then $x = \min(Q')$. So let there be given any $y \in Q'$. If $y \in R$ then obviously $x \leq y$ (in the partial order of S). If $y \notin R$ then $y \in R_1$ for some $R_1 \in P$; since $y \in R_1 \setminus R$, R_1 cannot be a lower segment of R; since P is a loset, R must be a lower segment of the woset R_1; therefore, since $y \in R_1 \setminus R$ and $x \in R$, we must have $x \leq y$.

Now we are ready to prove the:

EQUIVALENCE THEOREM (T4). $(AC) \Rightarrow (ZL^*) \Rightarrow (ZL) \Rightarrow (WO) \Rightarrow (AC)$.

PROOF OF (AC) $\Rightarrow$ (ZL*). Assume (AC), and let S be any nonempty poset in which every woset has an upper bound. Suppose if possible that S has no maximal element. By taking ϕ to be the identity map $\widehat{\mathcal{P}}^{\times}(S) \to \widehat{\mathcal{P}}^{\times}(S)$ in (AC), we get a map $\psi : \widehat{\mathcal{P}}^{\times}(S) \to S$ such that for all $\emptyset \neq R \subset S$ we have $\psi(R) \in R$. Let $W(S)$ be the set of all nonempty well ordered subsets of S, and for every $R \in W(S)$ let R^* be the set of all upper bounds of R which do not belong to R; by assumption R has an upper bound and S has no maximal element, and hence we can take an element in S which is strictly bigger than that upper bound thereby getting an upper bound not in R; thus R^* is nonempty and so $\psi(R^*) \in R^*$. For every $y \in S$ let $L(y) = \{x \in S : x < y\}$, and for every $R \subset S$ let $L(y, R) = L(y) \cap R$. Let $z = \psi(S)$. Let $V(S)$ be the set of all $R \in W(S)$ such that: (i) $\min(R) = z$ and (ii) for all $z \neq y \in R$ we have $\psi(L(y, R)^*) = y$. Clearly $\{z\} \in V(S)$ and hence $V(S) \neq \emptyset$. We claim that $V(S)$ is a loset (in the "lower segment" partial order). To see this, given any R_1 and R_2 in $V(S)$ let $R_0 = \{y \in R_1 \cap R_2 : L(y, R_1) = L(y, R_2)\}$. Then $z \in R_0$ and hence $R_0 \neq \emptyset$. Clearly R_0 is a lower segment of R_1 as well as R_2. Suppose if possible that $R_0 \neq R_i$ for $1 \leq i \leq 2$; then upon letting $y_i = \min(R_i \setminus R_0)$ we get $R_0 = L(y_i, R_i)$ for $1 \leq i \leq 2$; consequently $y_1 = \psi(L(y_1, R_1)^*) = \psi((R_0)^*) = \psi(L(y_2, R_2)^*) = y_2$, and hence $y_1 = y_2 \in R_0$ which is a contradiction. Therefore either $R_0 = R_1$ or $R_0 = R_2$, and hence either R_1 is lower segment of R_2 or R_2 is lower segment of R_1. This proves the claim. Now, upon letting $T = \cup_{R \in V(S)} R$, by the above lemma we get $T \in W(S)$ and from this it follows that $T \in V(S)$. Consequently, upon letting $\widehat{T} = T \cup \{\psi(T^*)\}$, we see that $\widehat{T} \in V(S)$ with $\widehat{T} \not\subset T$ which is a contradiction. Therefore S must have a maximal element. Thus (AC) $\Rightarrow$ (ZL*).

PROOF OF (ZL*) $\Rightarrow$ (ZL). Obvious.

PROOF OF (ZL) $\Rightarrow$ (WO). Assume (ZL), and given any set S let $W(S)$ be the set of all well ordered subsets of S. Note that $W(S)$ contains a subset of S as many times as the ways it can be well ordered. Put the "lower segment" partial order on $W(S)$, i.e., $R_1 \leq R_2$ in $S \Leftrightarrow R_1$ is order isomorphic to a lower segment of R_2. Given any chain P in $W(S)$, upon letting $T = \cup_{R \in P} R$ and upon putting the partial order on T induced by the partial orders of the various $R \in P$, we see that $T \in W(S)$ and $R \leq T$ for all $R \in T$, i.e., T is an upper bound of P. Therefore by (ZL), the set $W(S)$ has a maximal element $\widehat{T}$. If $\widehat{T} \neq S$ then by taking $y \in S \setminus \widehat{T}$ and letting $\widetilde{T} = \widehat{T} \cup \{y\}$ and putting the same partial order on $\widetilde{T}$ as $\widehat{T}$ with the understanding that $x \leq y$ for all $x \in \widehat{T}$, we get $\widetilde{T} \in W(S)$ with $\widehat{T} < \widetilde{T}$ which is a contradiction. Therefore $\widehat{T} = S$. Thus (ZL) $\Rightarrow$ (WO).

PROOF OF (WO) $\Rightarrow$ (AC). Assume (WO), and put some well order on T; now given any $\phi : S \to \widehat{\mathcal{P}}^{\times}(T)$ as in the statement of (AC), for every $x \in S$ let $\psi(x) = \min(\phi(x))$. Thus (WO) $\Rightarrow$ (AC).

REMARK (R1). The above theorem indirectly tells us that (ZL) $\Leftrightarrow$ (ZL*). Of this the implication (ZL*) $\Rightarrow$ (ZL) is obvious. Here is a direct proof of the implication (ZL) $\Rightarrow$ (ZL*). Given a poset S, by a weakly cofinal subset S we mean a subset R such that there is no $y \in S$ for which we have $x < y$ for all $x \in R$. Let $W(S)$ be the set of all subsets of S which are well ordered (in the partial order induced from S). In the "lower segment" partial order, clearly $W(S)$ has the Zorn property because the union of any chain is clearly an upper bound for it. Therefore, assuming S to be nonempty and noting that any singleton subset $\{y\}$ with $y \in S$ obviously belongs to $W(S)$, by (ZL) we get a maximal element T in $W(S)$. If T is not weakly cofinal in S then by taking $y \in S \setminus T$ such that $x < y$ for all $s \in T$ and letting $\widehat{T} = T \cup \{y\}$ we get $\widehat{T} \in W(S)$ with $T < \widehat{T}$ which is a contradiction. Therefore T is weakly cofinal in S. It follows that if S has no maximal element then T has no upper bound in S. Thus (ZL) $\Rightarrow$ (ZL*).

REMARK (R2). In (R1) we have shown that: (ZL) $\Rightarrow$ every poset has a weakly cofinal well ordered subset.

REMARK (R3). Given a poset S, by a cofinal subset of S we mean a subset R such that for every $x \in S$ we have $x \leq y$ for some $y \in R$. As a variation of $(R1)$ and $(R2)$, we can show that: (ZL) $\Rightarrow$ every loset has a cofinal well ordered subset, and we can use this implication to prove the implication (ZL) $\Rightarrow$ (ZL*).

REMARK (R4). [**Existence of Algebraic Closure by Well Ordering**]. The usual mathematical induction is based on the fact that the positive integers are well ordered. Similarly transfinite induction is based on well ordered indexing sets, and so in effect on the well ordering principle (WO). For instance we can prove the existence of an algebraic closure of a field K, which need not be countable, thus. Let the set S of all nonconstant monic polynomials in $K[Y]$ be indexed by a well ordered indexing set I and write them as $F_i(Y)$ with $i \in I$; for instance we could take $I = S$ and well order it by the well ordering principle. Let $K_1 = K$ with $1 = \min(I)$. For any $1 \neq j \in I$, having defined K_i for all $i < j$ in I, put $K_j = \mathrm{SF}(F_j, \cup_{i \in I \text{ with } i < j} K_i)$. Now $\overline{K} = \cup_{i \in I} K_i$ is an algebraic closure of K. If this sketch of a proof is not very convincing, because of the problems such as where to take the two unions, the reader is referred to the original source of this proof which is the 1910 Crelle Journal paper of Steinitz reprinted as a book [Ste] by Chelsea. In Remark (R6) we shall give a (perhaps more convincing) proof using Zorn's Lemma. But first in Remark (R5) we shall use Zorn's Lemma to show the existence of prime ideals and maximal ideals.

REMARK (R5). [**Existence of Maximal Ideals by Zorn's Lemma**]. Let I be an ideal in a ring R. Recall that I is a maximal ideal if the residue class ring R/I

is a field, or equivalently if $I \neq R$ and there is no ideal between I and R other than these two. Assuming $I \neq R$, let S be the set of all ideals in R which contain I but are different from R; in the partial order given by inclusion, the union of any chain is an upper bound for it, and hence by Zorn's Lemma S has a maximal element M. Obviously M is a maximal ideal in R.

REMARK (R6). [**Existence of Algebraic Closure by Zorn's Lemma**]. Alternatively, we can use Zorn's Lemma (ZL) to prove the existence of an algebraic closure of a field K thus. Let I be the set of all nonconstant monic polynomials in $K[Y]$. For any $F = F(Y) \in I$, let $D(F)$ be the Y-degree of F, let $\alpha_{F,1}, \ldots, \alpha_{F,D(F)}$ be the roots of F in a splitting field K_F of F over K, let P_F be the polynomial ring in indeterminates $X_{F,1}, \ldots, X_{F,D(F)}$ over K, and note that then $F(Y) = (Y - \alpha_{F,1}) \ldots (Y - \alpha_{F,D(F)})$ and we have a unique K-epimorphism $\phi_F : P_F \to K_F$ given by $\phi_F(X_{F,j}) = \alpha_{F,j}$ for $1 \leq j \leq D(F)$; since the image of ϕ_F is a field, its kernel M_F is a maximal ideal in P_F. Now consider the polynomial ring Q_I in the infinitely many variables $\{X_{F,j} : F \in I$ and $1 \leq j \leq D(F)\}$ over K. Note that the polynomial ring $R[\{X_l : l \in L\}]$ in an infinite number of variables $\{X_l : l \in L\}$ over a ring R is constructed in an obvious manner, and it equals the union of the polynomial rings $\cup R[\{X_h : h \in H\}]$ taken over all finite subsets H of L; thus each polynomial involves only a finite number of variables; any two polynomials together involve only a finite number of variables and so they can be added and multiplied; finally, if $W = \{x_l \in S : l \in L\}$ is any subset of any overring S of R then $X_l \mapsto x_l$ gives us an R-homomorphism $R[\{X_l : l \in L\}] \to S$ (called the substitution map) whose image is the subring $R[W]$ of S. In our case, for every $F \in I$, the ring P_F is a subring of the ring Q_I. Moreover, we have the partition $\{(F,j) : F \in I$ and $1 \leq j \leq D(F)\} = \coprod_{F \in I}\{(F,j) : 1 \leq j \leq D(F)\}$. The following Lemma can be proved easily:

LEMMA (T5). If $L = \coprod_{H \in \Omega} H$ is a partition of a nonempty set L (where Ω is a set of pairwise disjoint nonempty subsets of L), if $R_L = R[\{X_l : l \in L\}]$ is the polynomial ring in indeterminates $\{X_l : l \in L\}$ over a field R, if for each $H \in \Omega$ we are given a nonunit ideal J_H in the polynomial ring $R_H = R[\{X_h : h \in H\}]$, and if J_L is the ideal in R_L generated by all the ideals J_H as H varies over Ω, then J_L is a nonunit ideal in R_L; [cf. L6§6(E44)].

In our situation, by (T5) we see that the ideal in Q_I generated by all the ideals M_F as F varies over I is a nonunit ideal in Q_I, and hence by (R5) there exists a maximal ideal M_I in Q_I such that for every $F \in I$ we have $M_F \subset M_I$. Clearly $M_I \cap K = 0$ and hence we may identify K with a subfield of the field $\overline{K} = Q_I/M_I$. For every $F \in I$ we must have $P_F \cap M_I = M_F$ and hence we may identify K_F with a subfield of $\overline{K}$. Now $\overline{K}$ contains a splitting field of F over K for every $F \in I$. Also $\overline{K} = K[\{\alpha_{F,j} : F \in I$ and $1 \leq j \leq D(F)\}]$ and hence $\overline{K}$ is algebraic over K.

Therefore $\overline{K}$ is an algebraic closure of K.

REMARK (R7). [**Uniqueness of Algebraic Closure by Zorn's Lemma**]. Let L be an overfield of a field K, and let I be a set of nonconstant polynomials in Y with coefficients in K such that every $f = f(Y) \in I$ splits completely in L, i.e., upon letting $n(f)$ to be the Y-degree of f and $F = F(Y)$ to be the monic associate of f, we have

(W1) $$F(Y) = (Y - \alpha_{f,1})\ldots(Y - \alpha_{f,n(f)})$$

with $\alpha_{f,1}, \ldots, \alpha_{f,n(f)}$ in L. We call

(W2) $$L^* = K(\{\alpha_{f,j} : f \in I \text{ and } 1 \le j \le n(f)\})$$

the splitting field of I over K in L; note that if $I = \emptyset$ then $L^* = K$; also note that if I is the set of all nonconstant polynomials in Y with coefficients in K then L^* is an (absolute) algebraic closure of K as well as the (relative) algebraic closure of K in L. Let L' be an overfield of a field K' such that there is an isomorphism $\phi : K \to K'$ and every $f' = f'(Y) \in I'$ splits completely in L' where $I' \subset K'[Y]$ is obtained by applying ϕ to the coefficients of the various members of I. Let L'^* be the splitting field of I' over K' in L'. We want to show that there is an isomorphism $\phi^* : L^* \to L'^*$ such that $\phi^*(x) = \phi(x)$ for all $x \in K$. For any $J \subset I$ let $J' \subset I'$ be obtained by applying ϕ to the coefficients of the various members of J, and let L_J and $L'_{J'}$ be the splitting fields of J and J' over K and K' in L and L' respectively. Let A be the set of all pairs (J, ϕ_J) where $J \subset I$ and $\phi_J : L_J \to L'_{J'}$ is an isomorphism such that $\phi_J(x) = \phi(x)$ for all $x \in K$. Now A becomes a poset by declaring $(J, \phi_J) \le (J_0, \phi_{J_0})$ to mean that $J \subset J_0$ and $\phi_{J_0}(x) = \phi_J(x)$ for all $x \in L_J$. By taking $J = \emptyset$ and $\phi_\emptyset = \phi$ we see that A is nonempty. Moreover, given any chain B in A, by taking

(W3) $$J_B = \bigcup_{(J,\phi_J)\in B} J$$

it can be shown that

(W4) $$\begin{cases} \text{there is a unique isomorphism } \phi_{J_B} : L_{J_B} \to L'_{J'_B} \text{ such that} \\ \text{for all } (J,\phi_J) \in B \text{ and } x \in L_J \text{ we have } \phi_{J_B}(x) = \phi_J(x) \end{cases}$$

and moreover

(W5) $$(J_B, \phi_{J_B}) \text{ is an upper bound of } B \text{ in } A.$$

Therefore by Zorn's Lemma, A has maximal elements. Finally by L1(R8) we see that

(W6) $$(J_1, \phi_{J_1}) \text{ is maximal in } A \Rightarrow J_1 = I.$$

REMARK (R8). [**Existence of Vector Space Bases and Transcendence Bases by Zorn's Lemma**]. We shall now show that any vector space has a basis and any two of its bases have the same cardinality (which is called the dimension of the vector space). At the same time we shall show that any overfield has a transcendence basis and any two of its transcendence bases have the same cardinality (which is called the transcendence degree of the overfield). So let K be a field and let L be either a vector space over K or an overfield of K. Let W be a subset of L. Just for a few minutes, in the vector space case (resp: overfield case): let us call W independent if every finite subset of W is linearly (resp: algebraically) independent over K, and let us call W generating if L coincides with KW (resp: if L is algebraic over $K(W)$); in both cases call W a basis if it is independent and generating; given any other subset U of L, let us say that U is dependent on W if every $u \in U$ belongs to KW (resp: is algebraic over $K(W)$). It is easy to show that

(W7) $\qquad \begin{cases} W \text{ is a basis} & \Leftrightarrow \text{ it is a minimal generating set} \\ & \Leftrightarrow \text{ it is a maximal independent set} \end{cases}$

where the minimality means that W is a generating set but there is no generating set W' with $W' \subset W$ and $W' \neq W$, and the maximality means that W is an independent set but there is no independent set W' with $W \subset W'$ and $W \neq W'$. It is also easy to see that

(W8) $\qquad W$ is generating $\Rightarrow$ there exists a basis W' with $W' \subset W$.

Obviously L is generating and hence by taking $W = L$ in (W8) we get the existence of a basis. To prove equicardinality, assuming W to be a basis and given any other basis U, let A be the set of all triples (U', W', ϕ') where $U' \subset U$, $W' \subset W$, and bijection $\phi' : U' \to W'$, are such that $(W \setminus W') \cup U'$ is a basis. Now A becomes a poset by declaring $(U', W', \phi') \leq (U'', W'', \phi'')$ to mean that $U' \subset U''$, $W' \subset W''$, $\phi''(U') = W'$, and $\phi''(u) = \phi'(u)$ for all $u \in U'$. By taking $U' = \emptyset = W'$ and ϕ' to be the obvious map, we see that A is nonempty. Moreover, given any chain B in A, by taking

$$U_B = \bigcup_{(U',W',\phi') \in B} U' \quad \text{and} \quad W_B = \bigcup_{(U',W',\phi') \in B} W'$$

it can be shown that

(W9) $\qquad \begin{cases} \text{there is a unique bijection } \phi_B : U_B \to W_B \text{ such that} \\ \text{for all } (U', W', \phi') \in B \text{ and } u \in U' \text{ we have } \phi_B(u) = \phi'(u) \end{cases}$

and moreover

(W10) $\qquad (U_B, W_B, \phi_B)$ is an upper bound of B in A.

Therefore by Zorn's Lemma, A has maximal elements. Finally it can be shown that

(W11) (U^*, W^*, ϕ^*) is maximal in $A \Rightarrow U^* = U$ and $W^* = W$.

This completes the proof.

REMARK (R9). [**Linear Order on Cardinals by Zorn's Lemma**]. For the cardinals $|S|$ and $|T|$ of sets S and T, we recall that $|S| = |T|$ means there is a bijection $S \to T$, and $|S| \le |T|$ means there is an injection $S \to T$. To prove that this is a partial order, i.e., to prove the Schroeder-Bernstein Theorem, given any injections $\phi : S \to T$ and $\psi : T \to S$ we have to find a bijection $S \to T$. Let $f = \psi\phi$ and $R = \psi(T)$. Then clearly: (*) $f : S \to S$ is an injection and $f(S) \subset R \subset S$; we claim that (*) implies the existence of a bijection $g : S \to R$. Obviously ψ induces a bijection $h : R \to T$, and hence the claim yields the bijection $hg : S \to T$. To prove the claim, for any $x \in S$ let $f^0(x) = x$ and $f^{n+1}(x) = f(f^n(x))$ for all $n \in \mathbb{N}$; also let $f^\infty(x) = \{f^n(x) : n \in \mathbb{N}\}$, and for any $V \subset S$ let $f^\infty(V) = \cup_{x \in V} f^\infty(x)$. Let $R' = f^\infty(R \setminus f(S))$, and note that then $R' \subset R$. Since $f(S) \subset R$ and $R' \subset R$, we get a map $g : S \to R$ by putting $g(x) = x$ or $f(x)$ according as $x \in R'$ or $x \in S \setminus R'$. It can be shown that then

(W12) g is a bijection.

To prove that $\le$ is a linear order, given any sets U and W, we want to show that either there exists a bijection of U onto a subset of W or there exists a bijection of W onto a subset of U. To do this let A be the set of all triples (U', W', ϕ') where $U' \subset U$, $W' \subset W$, and $\phi' : U' \to W'$ is a bijection. Now A becomes a poset by declaring $(U', W', \phi') \le (U'', W'', \phi'')$ to mean that $U' \subset U''$, $W' \subset W''$, $\phi''(U') = W'$, and $\phi''(u) = \phi'(u)$ for all $u \in U'$. By taking $U' = \emptyset = W'$ and ϕ' to be the obvious map, we see that A is nonempty. Moreover, given any chain B in A, by taking

$$U_B = \bigcup_{(U', W', \phi') \in B} U' \quad \text{and} \quad W_B = \bigcup_{(U', W', \phi') \in B} W'$$

it can be shown that

(W13) $\begin{cases} \text{there is a unique bijection } \phi_B : U_B \to W_B \text{ such that} \\ \text{for all } (U', W', \phi') \in B \text{ and } u \in U' \text{ we have } \phi_B(u) = \phi'(u) \end{cases}$

and moreover

(W14) (U_B, W_B, ϕ_B) is an upper bound of B in A.

Therefore by Zorn's Lemma, A has maximal elements. Finally it can be shown that

(W15) (U^*, W^*, ϕ^*) is maximal in $A \Rightarrow$ either $U^* = U$ or $W^* = W$.

This completes the proof.

REMARK (R10). [**Well Order on Ordinals by Zorn's Lemma**]. For the ordinals $||S||$ and $||T||$ of well ordered sets S and T, we recall that $||S|| = ||T||$ means there is an order preserving bijection $S \to T$, and $||S|| \le ||T||$ means there is an order preserving bijection of S onto a lower segment of T. To prove that this is a partial order, given any order preserving bijections $\phi : S \to T'$ and $\psi : T \to S'$, where T' and S' are lower segments of T and S respectively, we have to find an order preserving bijection $S \to T$. Let $R = \psi(T')$ and for every $x \in S$ let $f(x) = \psi(\phi(x))$. Then clearly: (*) R is a lower segment of S and $f : S \to R$ is an order preserving bijection; we claim that (*) implies $R = S$ and $f : S \to S$ is the identity map of S. Since ϕ and ψ are injections, the claim yields the conclusion that $T' = T$ and hence $\phi : S \to T$ is an order preserving bijection. The above claim is clearly subsumed under the following stronger claim which can be proved easily:

(W16) $\begin{cases} \text{if } f' : S \to T' \text{ and } f'' : S \to T'' \text{ are order preserving bijections} \\ \text{of a well ordered set } S \text{ onto lower segments } T' \text{ and } T'' \\ \text{of a well ordered set } T \text{ then we must have } T'' = T' \text{ and } f'' = f'. \end{cases}$

To prove that $\le$ is a linear order, given any well ordered sets U and W, we want to show that either there exists an order preserving bijection of U onto a lower segment of W or there exists an order preserving bijection of W onto a lower segment of U. To do this let A be the set of all triples (U', W', ϕ') where U' is a lower segment of U, W' is a lower segment of W, and $\phi' : U' \to W'$ is an order preserving bijection. Now A becomes a poset by declaring $(U', W', \phi') \le (U'', W'', \phi'')$ to mean that U' is a lower segment of U'', W' is a lower segment of W'', $\phi''(U') = W'$, and $\phi''(u) = \phi'(u)$ for all $u \in U'$. By taking $U' = \emptyset = W'$ and ϕ' to be the obvious map, we see that A is nonempty. Moreover, given any chain B in A, by taking

$$U_B = \bigcup_{(U', W', \phi') \in B} U' \quad \text{and} \quad W_B = \bigcup_{(U', W', \phi') \in B} W'$$

it can be shown that

(W17) $\begin{cases} \text{there is a unique order preserving bijection } \phi_B : U_B \to W_B \\ \text{such that for all } (U', W', \phi') \in B \text{ and } u \in U' \text{ we have } \phi_B(u) = \phi'(u) \end{cases}$

and moreover

(W18) $\qquad (U_B, W_B, \phi_B)$ is an upper bound of B in A.

Therefore by Zorn's Lemma, A has maximal elements. Finally it can be shown that

(W19) $\qquad (U^*, W^*, \phi^*)$ is maximal in $A \Rightarrow$ either $U^* = U$ or $W^* = W$.

This completes the proof that $\le$ is a linear order. Let us now set up an order preserving bijection between a well ordered set and its lower segments. Namely, given any element y in a well ordered set C, clearly $\{x \in C : x < y\}$ is a lower segment of C different from C; conversely given any lower segment L of C different from C, by taking y to be the smallest element of $C \setminus L$, we get $L = \{x \in C : x < y\}$; it follows that under the above defined partial order $\leq$ on well ordered sets,

(W20)
$$\begin{cases} y \mapsto \{x \in C : x < y\} \text{ gives an order preserving bijection} \\ \text{of any well ordered set } C \text{ onto} \\ \text{the set of all lower segments of } C \text{ other than } C. \end{cases}$$

Finally let us show that under the above defined partial order $\leq$ on well ordered sets, any set D of well ordered sets is well ordered. Namely, given any nonempty subset E of D, take $C \in E$, and let $E' = \{C' \in E : C' \leq C\}$. Then by (W20) we see that E' has a smallest element C'', and clearly C'' is a smallest element of E.

§6: UTILITARIAN SUMMARY

In the material of the previous section through (R3) we have shown the equivalence of the axiom of choice, Zorn's Lemma, and the well ordering principle. In (R4) we have shown that the well ordering principle implies the existence of algebraic closure. In (R5) we have shown that Zorn's Lemma implies that, given any ideal I of any ring R with $I \neq R$, there exists a maximal ideal M of R with $I \subset M$. In (R6), (R7), and (R8) we have shown that Zorn's Lemma implies the existence and (up to isomorphism) uniqueness of algebraic closure, where uniqueness means: given any isomorphism $\phi : K \to K'$ of fields and given any algebraic closures L^* and L'^* of K and K' respectively, ϕ can be extended to an isomorphism of L^* onto L'^*, i.e., there exists an isomorphism $\phi^* : L^* \to L'^*$ such that $\phi^*(x) = \phi(x)$ for all $x \in K$; note that if J' is the set of all members of $K'[Y]$ obtained by applying ϕ to the coefficients of the members of a set J of nonconstant members of $K[Y]$, and if L_J and $L'_{J'}$ are the splitting fields of J and J' over K and K' in L^* and L'^* respectively, then we automatically have $\phi^*(L_J) = L'_{J'}$; in particular this shows that ϕ can be extended to an isomorphism of L_J onto $L'_{J'}$. In (R8) we have shown that Zorn's lemma implies the existence of vector spaces bases and transcendence bases. In (R9) we have shown that Zorn's Lemma implies that any set of cardinals is linearly ordered. Finally in (R10) we have shown that Zorn's Lemma implies that any set of ordinals is well ordered.

Without loss of understanding, all this may be taken for granted. As an exception, the material of (R8) through display (W8), i.e., the first 21 lines in the discussion of the existence of vector space bases and transcendence bases, should be read.

§7: DEFINITIONS AND EXERCISES

DEFINITION (D1). [**Real and Complex Numbers**]. Real numbers may be defined as equivalence classes of Cauchy sequences of rational numbers, and the complex number field $\mathbb{C}$ may then be defined as the splitting field of the quadratic polynomial $Y^2 + 1$ over the real number field $\mathbb{R}$. By analogy with $\mathbb{N}_+$, by $\mathbb{Q}_+$ we denote the set of all positive rationals; moreover, by $\mathbb{Q}_{0+}$, $\mathbb{Q}_-$, and $\mathbb{Q}_{0-}$ we denote the set of all nonnegative rationals, negative rationals, and nonpositive rationals respectively; as usual, the absolute value of any $r \in \mathbb{Q}$ is denoted by $|r|$, i.e., $|r| = r$ or $-r$ according as $r \in \mathbb{Q}_{0+}$ or $r \in \mathbb{Q}_-$; by context, this will not be confused with the size $|S|$ of a set S. A sequence $x = (x_i)_{1 \leq i < \infty}$ in $\mathbb{Q}$ is Cauchy means for every $\epsilon \in \mathbb{Q}_+$ there exists $N_\epsilon \in \mathbb{N}_+$ such that for all $i > N_\epsilon$ and $j > N_\epsilon$ we have $|x_i - x_j| < \epsilon$. This is equivalent to the Cauchy sequence $x' = (x'_i)_{1 \leq i < \infty}$, in symbols $x \sim x'$, if for every $\epsilon \in \mathbb{Q}_+$ there exists $M_\epsilon \in \mathbb{N}_+$ such that for all $i > M_\epsilon$ we have $|x_i - x'_i| < \epsilon$. Now $\mathbb{R}$ may be defined to be the quotient $C_\mathbb{Q} / \sim$ of the set $C_\mathbb{Q}$ of all Cauchy sequences in $\mathbb{Q}$ by the equivalence relation $\sim$. We "identify" $\mathbb{Q}$ with a subset of $\mathbb{R}$ by sending any $q \in \mathbb{Q}$ to the equivalence class of the Cauchy sequence $(q_i)_{1 \leq i < \infty}$ with $q_i = q$ for all i. If $y = (y_i)_{1 \leq i < \infty}$ is another Cauchy sequence in $\mathbb{Q}$ then the sequences $(x_i + y_i)_{1 \leq i < \infty}$ and $(x_i y_i)_{1 \leq i < \infty}$ are Cauchy sequences whose equivalence classes are unchanged if x and y are replaced by equivalent Cauchy sequences; this makes $\mathbb{R}$ into a ring and in fact an overfield of $\mathbb{Q}$.

DEFINITION (D2). [**Ordered Fields**]. The order relation $\leq$ can be extended from $\mathbb{Q}$ to $\mathbb{R}$ by declaring that the equivalence class of x is $\leq$ the equivalence class of y if the Cauchy sequences x and y in their equivalence classes can be chosen so that $x_i \leq y_i$ for all i; see (E3) below. Like $\mathbb{Z}$ and $\mathbb{Q}$, this makes $\mathbb{R}$ an ordered domain, i.e., a domain whose underlying additive abelian group is an ordered abelian group and in which the product of any positive elements (i.e., elements which are greater than zero) is again positive. Out of these $\mathbb{Q}$ and $\mathbb{R}$ are ordered fields, i.e., fields whose underlying domains are ordered domains. We extend the notation $\mathbb{Q}_+$, $\mathbb{Q}_{0+}$, $\mathbb{Q}_-$, and $\mathbb{Q}_{0-}$ to any ordered abelian group G (and hence to ordered domains and ordered fields) by putting $G_+ = \{g \in G : g > 0\}$, $G_{0+} = G_+ \cup \{0\}$, $G_- = \{g \in G : g < 0\}$, and $G_{0-} = G_- \cup \{0\}$; also we define the absolute value of any $g \in G$ by putting $|g| = g$ or $-g$ according as $g \in G_{0+}$ or $g \in G_-$. In particular this defines the sets R_+, R_{0+}, R_-, R_{0-}, and defines the absolute value $|r|$ of any $r \in \mathbb{R}$. Note that now $\mathbb{Z}_+ = \mathbb{N}_+$ and $\mathbb{Z}_{0+} = \mathbb{N}$. An ordered abelian group G is archimedean means for all x, y in G_+ we have $nx > y$ for some $n \in \mathbb{N}_+$. Clearly $\mathbb{R}$ is an archimedean ordered field, i.e., an ordered field whose underlying additive ordered abelian group is archimedean. A sequence $x = (x_i)_{1 \leq i < \infty}$ in an ordered abelian group G is Cauchy means for every $\epsilon \in G_+$ there exists $N_\epsilon \in \mathbb{N}_+$ such that for all $i > N_\epsilon$ and $j > N_\epsilon$ we have $|x_i - x_j| < \epsilon$; the sequence x is convergent means it converges to a limit $\xi \in G$, i.e., for every $\epsilon \in G_+$ there exists $M_\epsilon \in \mathbb{N}_+$ such that for all $i > M_\epsilon$ we have $|\xi - x_i| < \epsilon$; we indicate this by some standard notation such as $x_i \to \xi$ as $i \to \infty$ or $\lim_{i \to \infty} x_i = \xi$. An ordered abelian group is complete means in it every Cauchy sequence is convergent. Clearly $\mathbb{R}$ is a complete field, i.e., an ordered field whose underlying additive group is complete as an ordered abelian group.

In the next several Definitions and Exercises we extend the above construction of the reals to embed any ordered abelian group in a complete ordered abelian group.

DEFINITION (D3). [**Torsion Subgroups and Divisible groups**]. By the subgroup of a group G generated by elements $x_1, x_2, \ldots$ in G we mean the smallest subgroup of G which contains these elements. The order of an element in a group is the order of the subgroup generated by it. The subgroup of an additive abelian group G generated by all of its elements of finite order is called the torsion subgroup of G; if this is zero then G is said to be torsion free. An additive abelian group G is divisible means for every $g \in G$ and $n \in \mathbb{Z}^{\times}$ there is $h \in G$ with $nh = g$.

EXERCISE (E1). Show that for any elements a, b in a torsion free additive abelian group and any nonzero integer n we have: $na = nb \Rightarrow a = b$. Show that any ordered abelian group is torsion free, and hence any ordered field is of characteristic zero. Show that for all x, y in an ordered field we have $|xy| = |x||y|$. Show that the usual order on $\mathbb{Q}$ is the only order on it which makes it an ordered field.

EXERCISE (E2). Show that for all x, y in an ordered abelian group G we have $|x+y| \leq |x|+|y|$, and from this deduce that any convergent sequence in G is Cauchy and has a unique limit.

HINT: Divide the argument into two cases according as there does or does not exist a sequence of positive elements $(\epsilon_i)_{1 \leq i < \infty}$ in G tending to zero. In both the cases what we have to show is that: $x_i \to \xi$ and $x_i \to \xi'$ in $G \Rightarrow (x_i)$ is Cauchy and $\xi' = \xi$. Upon letting $y_i = \xi - x_i$ and $\eta = \xi - \xi'$, it suffices to show that: $y_i \to 0$ and $y_i \to \eta$ in $G \Rightarrow (y_i)$ is Cauchy and $\eta = 0$.

In the first case, we claim that for any $\epsilon > 0$ in G we have $\epsilon_l + \epsilon_l < \epsilon$ for some $l \in \mathbb{N}_+$; namely, since $\epsilon_i \to 0$, for some $m \in \mathbb{N}_+$ we must have $\epsilon_m < \epsilon$; again since $\epsilon_i \to 0$, for some $n \in \mathbb{N}_+$ we must have $\epsilon_n < \epsilon - \epsilon_m$; it suffices to take $l = m$ or n according as $\epsilon_m \leq \epsilon_n$ or $\epsilon_m > \epsilon_n$. Now $y_i \to 0 \Rightarrow$ there exists $N \in \mathbb{N}_+$ such that for all $i > N$ we have $|y_i| < \epsilon_l$, and hence for all $i > N$ and $j > n$ we have $|y_i - y_j| < \epsilon$, which shows that (y_i) is Cauchy. Since $y_i \to \eta$, by choosing N large enough we can also arrange that for all $i > N$ we have $|\eta - y_i| < \epsilon_l$, and this yields $|\eta| < |y_i| + |\eta - y_i| < \epsilon_l + \epsilon_l < \epsilon$, which shows that $\eta = 0$.

In the second case, $y_i \to 0 \Rightarrow$ there exists $N \in \mathbb{N}_+$ such that for all $i > N$ we have $y_i = 0$, because otherwise we can find an infinite sequence of integers $0 < j_1 < j_2 < \ldots$ such that $y_{j_i} \neq 0$ for all i and upon letting $\epsilon_i = |y_{j_i}|$ this would produce a sequence of positive elements $(\epsilon_i)_{1 \leq i < \infty}$ in G tending to zero. Therefore clearly (y_i) is Cauchy and $\eta = 0$.

DEFINITION (D4). [**Rational and Real Completions**]. Given any torsion free additive abelian group G, let $G^* = G \times \mathbb{Z}^\times / \sim$ where the equivalence relation $\sim$ is given by: $(g, n) \sim (g', n') \Leftrightarrow n'g = ng'$, and embed G in G^* by identifying every $g \in G$ with the equivalence class containing $(g, 1)$. Define addition in G^* by taking equivalence classes in the proposed equation $(g, n) + (h, m) = (mg + nh, nm)$. Note that then G^* is a divisible additive abelian group such that for every $\overline{g} \in G^*$ we have $n\overline{g} \in G$ for some $n \in \mathbb{Z}^\times$. We call G^* the rational completion of G. Note that if G is divisible then $G^* = G$.

Now assume that G is an ordered abelian group. Note that then the order on G uniquely extends to an order on G^* so that G^* becomes an ordered abelian group. G is said to be orderwise complete if every Cauchy sequence in G is convergent and has a limit in G. Moreover, G is said to be an orderwise completion of a subgroup H if G is orderwise complete and every element of G is the limit of a sequence all of whose members belong to H. Let $\overline{G} = C_G / \sim$ where the equivalence relation $\sim$ on the set C_G of all Cauchy sequences in G is defined as in (D1) with G replacing $\mathbb{Q}$. As in (D1) and (D2), $\overline{G}$ is made into an ordered abelian group and G is embedded into it; see (E4) below. Clearly $\overline{G}$ is an orderwise completion of G, and we call it the ordered completion of G. Also we call $\overline{G^*}$ the real completion of G. Note that G^* is archimedean $\Leftrightarrow$ G is archimedean $\Leftrightarrow$ $\overline{G}$ is archimedean. Also note that if G is complete then $\overline{G} = G$. Finally note that if G is divisible and complete then $\overline{G^*} = G$.

In particular if K is an ordered field then its underlying additive group is divisible and hence $\overline{K^*} = \overline{K}$; as in (D1) and (D2) we make this into an ordered field; it is a complete ordered field, i.e., an ordered field whose underlying additive group is orderwise complete.

EXERCISE (E3). In (D2) show that the induced relation $\leq$ on the equivalence classes of Cauchy sequences in $\mathbb{Q}$ is a linear order.

EXERCISE (E4). In (D4) show that the induced relation $\leq$ on the equivalence classes of Cauchy sequences in an ordered abelian group is a linear order.

EXERCISE (E5). Let G be any nonzero archimedean ordered abelian group. Show that given any $g > 0$ in G and $x > 0$ in $\mathbb{R}$, there exists a unique order monomorphism (i.e., a group monomorphism which is order preserving) $\phi : G \to \mathbb{R}$ such that $\phi(g) = x$, and there exists a unique order isomorphism (i.e., a group isomorphism which is order preserving) $\psi : \overline{G^*} \to \mathbb{R}$ such that $\psi(g) = x$. Moreover, for these maps we always have $\phi(h) = \psi(h)$ for all $h \in G$.

EXERCISE (E6). Show that $x \mapsto x^2$ gives a surjection $\mathbb{R} \to \mathbb{R}_{0+}$.

HINT. The usual method of finding the decimal expansion $1.14142\ldots$ of $\sqrt{2}$ can be explained in terms of decimal expansions of integers by saying that $1^2 < 2 < 2^2$,

$14^2 < 2 \times 10^2 < 15^2$, $141^2 < 2 \times 10^4 < 142^2$, $1414^2 < 2 \times 10^6 < 1415^2$, $14142^2 < 2 \times 10^8 < 14143^2$, and so on. More generally let $n > 1$ and $d > 1$ be any integers, and let $y > 0$ and $i > 0$ be any integers. Then clearly there is a unique integer x_i such that $x_i^d \leq yn^{di} < (x_i + 1)^d$; x_i is nothing but the n-adic expansion of the largest integer $\leq y^{1/d}n^i$. Obviously the sequence (x_i/n^i) is Cauchy and for its limit x in $\mathbb{R}_+$ we have $x^d = y$. Any positive rational can be written in the form y/z^d where y and z are positive integers, and then we get $x/z \in \mathbb{R}_+$ with $(x/z)^d = y/z^d$. Finally, any $\eta \in \mathbb{R}_+$ can be written as the limit of a sequence (η_j) in $\mathbb{Q}_+$ and then taking $\xi_j \in \mathbb{R}_+$ with $\xi_j^d = \eta_j$ we get a Cauchy sequence (ξ_j) for whose limit $\xi \in \mathbb{R}_+$ we have $\xi^d = \eta$.

EXERCISE (E7). Show that the identity map is the only field automorphism of $\mathbb{R}$. Hint: by (E6) any field automorphism of $\mathbb{R}$ is order preserving, and since it must send 1 to 1, by (E5) it must be the identity map.

EXERCISE (E8). In contrast with (E7), assuming the fact that $\mathbb{C}$ is an algebraic closure of $\mathbb{R}$, show that $\mathbb{C}$ has uncountably many field automorphisms.

DEFINITION (D5). [**Rational and Real Ranks**]. In view of the first sentence of (E1), any torsion free divisible additive abelian group may clearly be regarded as a $\mathbb{Q}$-vector-space. The $\mathbb{Q}$-vector-space dimension of the rational completion of a torsion free additive abelian group G is called the rational rank of G and is denoted by $r(G)$. Alternatively, $r(G)$ may be characterized as the cardinal of a maximal ($=$ nonenlargeable) $\mathbb{Z}$-linearly independent subset H of G, where independent means that for any finite number of distinct elements $x_1, \ldots, x_d$ in H and any integers $n_1, \ldots, n_d$ we have: $n_1 x_1 + \cdots + n_d x_d = 0 \Rightarrow n_1 = \cdots = n_d = 0$.

Now let G be an ordered abelian group. By a segment of G we mean a nonempty subset H of G such that: $g \in G$ and $h \in H$ with $|g| \leq |h| \Rightarrow g \in H$. By an isolated subgroup of G we mean a subgroup which is a segment. The next Exercise says that the set $S(G)$ of all nonzero isolated subgroups of G is linearly ordered by inclusion. The order-type of $S(G)$ is called the real rank of G and is denoted by $\rho(G)$. Two linearly ordered sets are said to have the same order-type if there is an order preserving bijection between them. An ordinal is the order-type of a well ordered set. Examples of nonordinal order-types are the order-types of $\mathbb{Z}$ or $\mathbb{Q}$ or $\mathbb{Q}_+$. We write $\rho(G) \in \mathbb{N}$ or $\rho(G) = \infty$ according as $S(G)$ is finite or infinite. As a comparison, recall that the dimension $\dim_K V$ of a vector space V over a field K is the common cardinality of a vector space basis of V/K, and we write $\dim_K V \in \mathbb{N}$ or $\dim_K V = \infty$ according as the said basis is finite or infinite. Similarly the transcendence degree $\mathrm{trdeg}_K L$ of an overfield L of a field K is the common cardinality of a transcendence basis of L/K, and we write $\mathrm{trdeg}_K L \in \mathbb{N}$ or $\mathrm{trdeg}_K L = \infty$ according as the said basis is finite or infinite.

In the section on Valuations, for any nonnegative integer d, we introduced the ordered abelian group $\mathbb{R}^{[d]}$ as the lexicographically ordered set of all d-tuples of real numbers. Upon replacing $\mathbb{R}$ by any ordered abelian group G we generalize this to $G^{[d]}$. As a further generalization, given any well ordered set T we introduce the ordered abelian group $G^{[T]}$ as the set G^T of all functions $f : T \to G$ with componentwise addition and the order defined by setting: $f < g \Leftrightarrow f \neq g$ and $f(i) < g(i)$ where i is the smallest $j \in T$ with $f(j) \neq g(j)$.

EXERCISE (E9). Show that for any isolated subgroups H_1, H_2 of an ordered abelian group G we have either $H_1 \subset H_2$ or $H_2 \subset H_1$.

EXERCISE (E10). Let G be a nonzero archimedean ordered abelian group; by (E5) this is equivalent to saying that, up to order isomorphism, G is a nonzero subgroup of $\mathbb{R}$; for instance $G = \mathbb{R}$. For any nonempty well ordered set T, describe the nonzero isolated subgroups of $G^{[T]}$ and show that they are in an order reversing bijective correspondence with T; hint: $i \mapsto \{f \in G^T : f(j) = 0 \text{ for all } j < i\}$ is the desired bijection from T to the nonzero isolated subgroups. In particular, for any positive integer d, the ordered abelian group $\mathbb{G}^{[d]}$ has exactly d nonzero isolated subgroups.

EXERCISE (E11). Let G be any ordered abelian group. Show that $r(G) = 0 \Leftrightarrow \rho(G) = 0 \Leftrightarrow G = 0$. Show that G is archimedean $\Leftrightarrow \rho(G) \leq 1$. Show that $r(G) < \infty \Rightarrow \rho(G) \leq r(G)$. Show that $r(G) = \rho(G) = d \in \mathbb{N}_+ \Rightarrow$ there exists an order monomorphism $G \to \mathbb{Q}^{[d]}$ and an order isomorphism $G \to \mathbb{Q}^{[d]}$. See the Hint in (E12) below.

EXERCISE (E12). Show that, given any ordered abelian group G with $\rho(G) = d \in \mathbb{N}_+$, there exists an order monomorphism $G \to \mathbb{R}^{[d]}$ and there exists an order isomorphism $\overline{G^*} \to \mathbb{R}^{[d]}$. Hint: make induction on d starting with (E5), and use the fact that if H is any nonzero isolated subgroup of G then in a natural manner G/H can be made into an ordered abelian group with $\rho(H) + \rho(G/H) = d$.

DEFINITION (D6). [**Dedekind Cuts**]. Instead of using Cauchy sequences to prove (E5), we can use Dedekind Cuts. So let G be a nonzero divisible archimedean ordered abelian group. A Dedekind cut of G is a pair (L, U) of nonempty subsets of G with $U = G \setminus L$ such that for all $l \in L$ and $u \in U$ we have $l < u$ and there is no $u' \in U$ with $U = \{u \in G : u' \leq u\}$. For any $t \in G$ we get a Dedekind cut (L_t, U_t) with $L_t = \{l \in G : l \leq t\}$ and $U_t = \{u \in G : t < u\}$. Let D_G be the set of all Dedekind cuts of G. It can be shown that G is complete $\Leftrightarrow t \mapsto (L_t, R_t)$ gives a surjection $G \to D_G$. Indeed, D_G may be defined to be the completion of G. At any rate, for proving (E5), given any $h \in G$ let $\theta(h)$ be the real number which corresponds to the Dedekind cut (L, U) of $\mathbb{Q}$ where $L = \{m/n \in \mathbb{Q} \text{ with } m \in \mathbb{Z} \text{ and } n \in \mathbb{N}_+ : nh \leq mg\}$ and $U = \mathbb{Q} \setminus L$, and take $\phi(h) = x\theta(h)$.

EXERCISE (E13). Prove the geometric series identity (i) in the univariate power series ring $R[[X]]$ over any ring R, and from it deduce description (ii) of units in the multivariate power series ring $R[[Z_1, \ldots, Z_d]]$. Prove claims (V4) to (V6) and thereby show that, for any field K and any ordered abelian group G, $K((X))_G$ is a field with a valuation whose value group is G. [cf. L6§6(E45)].

EXERCISE (E14). Complete the existence proof for vector space bases and transcendence bases by establishing claims (W7) to (W11) made in (R8). Complete the proof that cardinals are linearly ordered by establishing claims (W12) to (W15) made in (R9). Complete the proof that ordinals are well ordered by establishing claims (W16) to (W20) made in (R10).

DEFINITION (D7). [**Approximate Roots**]. In the Hint to (E6) we showed how to use n-adic expansions of positive integers to find successive approximations to the d-th root of a positive integer. Mixing a generalization of this with a generalization of the completing the square method of solving quadratic equations leads us to the concept of approximate roots of polynomials. So consider a monic polynomial

$$F = F(Y) = Y^N + \sum_{1 \leq i \leq N} A_i Y^{N-i}$$

of degree $N \geq 0$ in Y with coefficients A_i in a ring R. If N is a unit in R then we can generalize the completing the square idea to completing the N-th power by writing

$$F(Y) = (Y - A_1/N)^N + \sum_{2 \leq i \leq N} A_i'(Y - A_1/N)^{N-i}$$

with $A_i' \in R$, i.e., by killing the coefficient of Y^{N-1}. To generalize this further let $D > 0$ be an integer which divides N. Instead of assuming N to be a unit in R, assume D to be a unit in R; note that in case of a field R this is equivalent to assuming that the characteristic of R does not divide D and so characteristic zero is always ok. Now we look for a monic polynomial

$$G = G(Y) = Y^{N/D} + \sum_{1 \leq i \leq N/D} B_i Y^{(N/D)-i}$$

of degree N/D in Y with coefficients B_i in R such that G^D is as close to being equal to F as possible. As (E15) below shows, if we interpret this as requiring $\deg_Y(F - G^D) < N - (N/D)$ then a unique G exists, and we call it the approximate D-th root of F (relative to Y) and denote it by $\mathrm{App}_D(F)$ or $\mathrm{App}_{D,Y}(F)$. Recall that for any m, n in $\mathbb{N}$ with $n > 1$, the n-adic expansion of m consists of writing $m = \sum_{i \geq 0} m_i n^i$ where integers $0 \leq m_i < n$ are the digits of the expansion. Likewise for any f, g in $R[Y]$ with g monic of positive Y-degree, the g-adic expansion of f

consists of writing $f = \sum_{i \geq 0} f_i g^i$ where $f_i \in R[Y]$ with $\deg_Y f_i < \deg_Y g$ are the digits of the expansion. By (E16) below these expansions exist and are unique. Moreover, if f is monic of Y-degree $N > 0$ and the Y-degree of g is $N/D \in \mathbb{N}_+$ where $D \in \mathbb{N}_+$ is a unit in R, then $f_D = 1$ and $f_i = 0$ for all $i > D$; we try completing the D-th power by putting $\tau_f(g) = \tau_{f,Y}(g) = g + (f_{D-1}/D)$ and calling it the f-Tschirnhausen of g (relative to Y); see Krull's charming little book [Kr1] for the reference to the 1683 work of Tschirnhausen who was a friend of Leibnitz. By (E17) below, starting with any monic g of degree N/D and applying τ_f to it N/D times will produce the approximate D-th root of f.

EXERCISE (E15). Let F be a monic polynomial of degree $N > 0$ in Y over a ring R. Let $D > 0$ be an integer such that D divides N and D is a unit in R. Show that there exists a unique monic polynomial G of degree N/D in Y over R such that $\deg_Y(F - G^D) < N - (N/D)$. Hint: With display as in (D7), the last condition gives the equations $A_i = DB_i + P_i(B_1, \ldots, B_{i-1})$ for $1 \leq i \leq N/D$ where the coefficient of Y^{N-i} in G^D equals $DB_i + P_i(B_1, \ldots, B_{i-1})$ with P_i a polynomial over $\mathbb{Z}$; since D is a unit in R, these can be solved successively (in a unique manner).

EXERCISE (E16). Given integers $m \geq 0$ and $n > 1$, show the unique existence of the n-adic expansion of m. Given univariate polynomials f, g over a ring R with g monic of positive degree, show the unique existence of the g-adic expansion $f = \sum_{i \geq 0} f_i g^i$ of f. Show that if f is monic of degree $N > 0$ and the degree of g is N/D where D is a positive integer factor of N, then $f_D = 1$ and $f_i = 0$ for all $i > D$, and moreover: $\mathrm{App}_D(f) = g \Leftrightarrow f_{D-1} = 0$.

EXERCISE (E17). Let f, g be univariate monic polynomials of positive degrees N and N/D over a ring R where D is a positive integer which divides N, and is a unit in R. Let $\tau_f(g) = \overline{g}$, and let $f = \sum_{0 \leq i \leq D} f_i g^i$ and $f = \sum_{0 \leq i \leq D} \overline{f}_i \overline{g}^i$ be the g-adic and $\overline{g}$-adic expansions of f respectively. Show that if $f_{D-1} \neq 0 \neq \overline{f}_{D-1}$ then $\deg(\overline{f}_{D-1}) < \deg(f_{D-1})$. From this deduce that $\tau_f^{N/D}(g) = \mathrm{App}_D(f)$.

§8: NOTES

NOTE (N1). [Algebraic Closedness of Complex Numbers]. Having constructed the real number field $\mathbb{R}$, and having defined the complex number field $\mathbb{C}$ as the splitting field of the irreducible quadratic $Y^2 + 1$ over $\mathbb{R}$, eventually we shall sketch one or two out of the numerous available proofs of the algebraic closedness of $\mathbb{C}$.

NOTE (N2). [Integers and Rational Numbers]. The construction of the reals may be backgrounded by noting that the rationals are the elements of the quotient field of the integers. Here we stop. To quote Kronecker: God made the integers; all the rest is the work of man.

§9: CONCLUDING NOTE

When there are several things which we do not know, we call them X, Y, Z and so on. This leads to one or more polynomial equations in these variables. One equation in two variables gives rise to a plane curve. One equation in three variables produces a surface in three space. Two equations in two variables give rise to the finite number of points in which the two curves intersect. Two equations in three variables produce one or more space curves in which the two surfaces intersect. Bézout makes a proper counting of the intersections.

Lecture L3: Tangents and Polars

§1: SIMPLE GROUPS

After proving the simplicity of the alternating group A_n for $n \geq 5$, Galois enlarged the list of finite simple groups by showing that the projective special linear group PSL$(2, p)$, which we shall define in a moment, is simple for any prime $p \geq 5$. Jordan extended this by proving the simplicity of PSL(n, p) for all $n \geq 3$; his proof may be found in his 1870 book [Jor] which was the first expansion of Galois' ideas on equation solving. Then Moore proved the simplicity of PSL$(2, q)$ for any power q of p with $q > p$; indeed this was his main aim in the paper [Mo1] which he presented at the Chicago World's Fair of 1893 which was the birthplace of the Galois field GF(q). Finally, in his influential 1901 book [Dic], Dickson established the simplicity of PSL(n, q) for all $n \geq 3$ and $q > p$. Eventually we shall prove all these simplicity statements. [cf. L6§6(E46)].

To define PSL we start by introducing determinants and matrices.

Given any ring R, for any integers $m \geq 0$ and $n \geq 0$, by an $m \times n$ matrix over R we mean a "rectangular array" $A = (A_{ij})$ with $A_{ij} \in R$ for $1 \leq i \leq m$ and $1 \leq j \leq n$. By the (i, j)-th entry of A we mean the element A_{ij}. The $1 \times n$ matrix whose $(1, j)$-th entry is A_{ij} is called the i-th row of A, and the $m \times 1$ matrix whose $(i, 1)$-th entry is A_{ij} is called the j-th column of A. The set of all $m \times n$ matrices over R is denoted by MT$(m \times n, R)$. For any $B = (B_{ij}) \in$ MT$(m \times n, R)$, we define the sum $A + B = ((A + B)_{ij}) \in$ MT$(m \times n, R)$ by putting $(A + B)_{ij} = A_{ij} + B_{ij}$. Clearly MT$(m \times n, R)$ becomes an additive abelian group whose 0 is the matrix with all entries reduced to the 0 of R. Next, for any integer $r \geq 0$ and any $B = (B_{ij}) \in$ MT$(n \times r, R)$, we define the product $AB = ((AB)_{ij}) \in$ MT$(m \times r, R)$ by putting $(AB)_{ij} = \sum_{1 \leq l \leq n} A_{il} B_{lj}$. This is easily seen to be associative, i.e., for any integer $s \geq 0$ and any $C \in$ MT$(r \times s, R)$ we have

(M1) $$(AB)C = A(BC).$$

Now assume $m = n$. Then MT$(n \times n, R)$ becomes a skew-ring whose 1 is the matrix $1_n = ((1_n)_{ij})$ where $(1_n)_{ij} =$ the Kronecker δ_{ij} which equals 1 or 0 according as $i = j$ or $i \neq j$. The determinant of A is defined by putting

(M2) $$\det(A) = \sum_{\sigma \in S_n} \text{sgn}(\sigma) \prod_{1 \leq i \leq n} A_{i\sigma(i)}$$

where S_n is the permutation group (= the symmetric group) on $(1, 2, \ldots, n)$, and $\text{sgn}(\sigma)$ (= the sign of σ) is defined to be $+1$ or -1 according as σ is even or odd, i.e., according as σ is a product of an even or odd number of transpositions. It can be shown that for any A, B in MT$(n \times n, R)$ we have

(M3) $$\det(AB) = \det(A)\det(B).$$

Recalling that $U(R)$ is the group of all units in R, it is easily seen that

(M4) $U(\text{MT}(n \times n, R)) = \{A \in \text{MT}(n \times n, R) : \det(A) \in U(R)\}.$

The group $U(\text{MT}(n \times n, R))$ is called the n-dimensional general linear group over R and it is denoted by $\text{GL}(n, R)$. The kernel of the determinant map $\text{GL}(n, R) \to U(R)$, i.e., the homomorphism given by $A \mapsto \det(A)$, is called the n-dimensional special linear group over R and is denoted by $\text{SL}(n, R)$. In other words $\text{SL}(n, R) = \{A \in \text{GL}(n, R) : \det(A) = 1\}$. Being the kernel of a homomorphism, $\text{SL}(n, R) \lhd \text{GL}(n, R)$, i.e., $\text{SL}(n, R)$ is a normal subgroup of $\text{GL}(n, R)$. Note that the determinant of a 0×0 matrix is 1; namely $|S_0| = 1$ because $|\emptyset^\emptyset| = 1$, and $\prod_{1 \leq i \leq 0} A_{i\sigma(i)} = 1$ because by convention the product over an empty family is 1 (just as the sum over an empty family is 0). Therefore $\text{MT}(0 \times 0, R) = 0 = 1$, i.e., this is the null ring, and $\text{SL}(0, R) = \text{GL}(0, R) = 1$, i.e., these are identity groups.

For a field k and an integer $n > 0$, we have an injective homomorphism $k^\times \to \text{GL}(n, k)$ which sends every $a \in k^\times$ to the matrix whose (i, i)-th entry is a for all i and whose (i, j)-th entry is 0 for all $i \neq j$. The image of this homomorphism is called the n-dimensional homothety group over k and is denoted by $\text{HL}(n, k)$. In other words, $\text{HL}(n, k)$ is the group of all nonzero scalar matrices. As we shall eventually see, an $n \times n$ matrix induces a linear transformation of the n dimensional vector space k^n over k. The transformation induced by a nonzero scalar matrix sends a geometric configuration to a configuration which is similar to it, like two similar triangles. For the use of the word homothety for such a transformation, see the last chapter starting on page 267 of the classic Pure Geometry book [Ask] by Askwith published in 1921 (first edition 1903) by the Cambridge University Press. We can easily see that

(M5) $\text{HL}(n, k) \lhd \text{GL}(n, k)$

and we define $\text{PGL}(n, k) = \text{GL}(n, k)/\text{HL}(n, k)$ and $\text{PSL}(n, k) = \text{SL}(n, k)/(\text{SL}(n, k) \cap \text{HL}(n, k))$, and call these the n-dimensional projective general linear group over k and the n-dimensional projective special linear group over k respectively. In case k is the Galois field $\text{GF}(q)$ where q is a power of a prime p, we may write $\text{GL}(n, q)$, $\text{SL}(n, q)$, $\text{PGL}(n, q)$ and $\text{PSL}(n, q)$ in place of $\text{GL}(n, k)$, $\text{SL}(n, k)$, $\text{PGL}(n, k)$ and $\text{PSL}(n, k)$ respectively.

As said above, eventually we shall prove the simplicity of $\text{PSL}(n, q)$ for $n \geq 2$ with $(n, q) \neq (2, 2), (2, 3)$. Now the list of finite simple groups can be enlarged by considering finite orthogonal groups, i.e., the finite analogs of groups of distance preserving transformations of the ordinary euclidean space. The said orthogonal groups, as well as the related symplectic groups and unitary groups, are based on the geometry of quadrics; later on we shall give a detailed treatment of these groups. Collectively, the finite linear groups $\text{GL}(n, q)$, $\text{SL}(n, q)$, $\text{PGL}(n, q)$ and $\text{PSL}(n, q)$, together with the finite orthogonal, symplectic and unitary groups, are called finite classical groups. If we do not insist the field k to be finite then they are simply called classical groups. This terminology goes back to Hermann Weyl's book [Wey]. Dickson [Dic] called all of them linear groups.

§2: QUADRICS

To motivate the geometry of a quadric, let me start with the following:

HOMEWORK PROBLEM (H1). Draw the two tangent lines to a circle D from a point A, and let Π be the line joining the two points of contact. Call Π the polar of A, and A the pole of Π. Show that if the polar of A passes through a point B then the polar of B passes through A.

If done pedantically, by analytic geometry, this may take five pages. The pure geometry solution given on page 14 of Askwith's book [Ask] cited above takes one page. Soon I shall show you a half-line proof. That proof will be based on homogeneous coordinates which are the mainstay of analytic projective geometry. Pure geometry may be regarded as synthetic projective geometry. Affine, i.e., ordinary, space is converted into projective space by adding points at infinity to it.

In the above problem, the circle may be replaced by any conic such as an ellipse or parabola or hyperbola, or even by a quadric or a hyperquadric. A quadric is a surface in 3-space given by a quadratic equation, such as a sphere or an ellipsoid or a paraboloid or a hyperboloid. In case of a quadric, the polar of a point A is a plane which intersects the quadric in the points of contact of various tangent lines from A to the quadric. More generally a hyperquadric in N-space is given by a quadratic equation in N variables, i.e., it is a hypersurface of degree 2; similarly, a hyperplane is a hypersurface of degree 1. Now the polar of a point A is the hyperplane which intersects the hyperquadric in the points of contact of the tangent lines from A to the hyperquadric. This gives rise to the following:

GENERALIZED HOMEWORK PROBLEM (H2). In case of a hyperquadric, show that if the polar hyperplane of a point A passes through a point B then the polar hyperplane of B passes through A.

The half-line solution I spoke of actually works in the general hyperquadric case!

So what are tangent lines, tangent cones, singular points, singular curves, and so on? Seeing that analytic geometry, like its sister discipline of projective geometry, has mostly been eliminated from college courses, let me first review these concepts from the viewpoint of calculus and then from the viewpoint of analytic geometry.

In calculus we consider a plane curve C given by an explicit equation $Y = \Phi(X)$, and introduce the concept of the tangent T to C at a point $P = (U, V)$ of C with $V = \Phi(U)$ as the limit of chords (i.e., secant lines) of C through P. Then we show that the slope of T is the value of $\Phi'(X) = \frac{dY}{dX}$ at $X = U$. This gives us $Y - V = \Phi'(U)(X - U)$ as the equation of T. If C is given by an implicit equation $F(X, Y) = 0$ then we use implicit differentiation to get $F_X dX + F_Y dY = 0$ where subscripts indicate partial derivatives. This yields $\frac{dY}{dX} = \frac{-F_X}{F_Y}$. If $F_Y(U, V) \neq 0$ then the slope of T is $\frac{-F_X(U,V)}{F_Y(U,V)}$ giving $Y - V = \frac{-F_X(U,V)}{F_Y(U,V)}(X - U)$ as the equation of T, and now by clearing the denominator and bringing things to one side we get

$$F_X(U, V)(X - U) + F_Y(U, V)(Y - V) = 0$$

as the equation for T. Likewise, if $F_X(U, V) \neq 0$ then for the "antislope of T" (= the tan of the angle T makes with the Y-axis) we have $\frac{dX}{dY}\big|_{(U,V)} = \frac{-F_Y(U,V)}{F_X(U,V)}$ giving $X - U = \frac{-F_Y(U,V)}{F_X(U,V)}(Y - V)$ as the equation of T, and again by clearing the denominator and bringing things to one side we get the same equation for T as displayed above. Thus the calculus method works if either F_X or F_Y is nonzero at P. But if F_X and F_Y are both zero at P then it breaks down and the books simply declare that P is then a singular point of C. Note that $F(U, V) = 0 \neq F_Y(U, V)$ is exactly the condition under which, according to the Implicit Function Theorem, $F(X, Y) = 0$ can be solved near P giving $Y = \Phi(X)$ with $V = \Phi(U)$. Similarly, if $F(U, V) = 0 \neq F_X(U, V)$ then we solve $F(X, Y) = 0$ to get $X = \Psi(Y)$ with $U = \Psi(V)$. The point P is singular exactly means that the Implicit Function Theorem does not work at P.

§3: HYPERSURFACES

As I recently became aware, the modern calculus treatment of tangents to a surface $S : F(X, Y, Z) = 0$ in 3-space is even more troublesome. What is done is to introduce the gradient to S at a point P of S as the vector (F_X, F_Y, F_Z) evaluated at P. The gradient is also called the normal to S at P, and the plane perpendicular to it is called the tangent plane to S at P. More generally, given a hypersurface $S : F(X_1, \ldots, X_N) = 0$ in N-space, the gradient to S at a point $P = (U_1, \ldots, U_N)$ of S is the vector of partials $(F_{X_1}, \ldots, F_{X_N})$ evaluated at P; thinking of P as a variable point, the gradient of S at P is the vector $(F_{U_1}, \ldots, F_{U_N})$ where F_{U_i} is the partial derivative of $F(U_1, \ldots, U_N)$ with respect to U_i. Again, the gradient is also called the normal to S at P, and the hyperplane T_P perpendicular to it is called the tangent hyperplane to S at P. Clearly $F_{U_1}(X_1 - U_1) + \cdots + F_{U_N}(X_N - U_N) = 0$ may be taken as the equation of T_P. Alternatively, T_P is defined to be the hyperplane spanned by the tangent lines to the various curves on S through P. Both these definitions break down if all the N partials are zero at P, i.e., if P is a singular point of S. The alternative definition has the further difficulty of having to define tangents to space or hyperspace curves, and of proving that sufficiently many of them can be drawn on the surface or hypersurface S.

All these difficulties vanish if, instead of calculus, we follow the method of analytic geometry.

To explain the method of analytic geometry, let us intersect the hypersurface

$$S : F(X_1, \ldots, X_N) = 0$$

in N-space by a line L through a point

$$P = (U_1, \ldots, U_N)$$

of S. Let the line L be given parametrically by the equations

$$L : X_i = U_i + R_i t \text{ for } 1 \le i \le N$$

where t is a parameter and the components of the vector $R = (R_1, \ldots, R_N)$ are proportional to the "direction cosines" of the line; for the idea of direction cosines see the 1903 book [Bel] of R. J. T. Bell on Three Dimensional Analytic Geometry; in effect, they are the cosines of the angles a line makes with the various coordinate axes. To find the points of intersection, we substitute the equations of L in the equation of S. This gives

$$F(U_1 + R_1 t, \ldots, U_N + R_N t) = h(t) = t^\mu h^*(t) \text{ with } h^*(0) \ne 0$$

where $\mu = \mathrm{ord}_t h(t)$. Now $t = 0$ is a root of $h(t)$ of multiplicity μ, and so we visualize that μ points of S have coalesced at P. We call μ the intersection multiplicity of S and L at P and we denote it by $\mathrm{int}_P(S, L)$. To indicate the dependence of μ on P and R, we may write $\mu_{P,R}$ for μ. Expanding F around P we have

$$F(X_1, \ldots, X_N) = \sum G_{i_1 \ldots i_N} (X_1 - U_1)^{i_1} \ldots (X_N - U_N)^{i_N}.$$

Let

$$\nu = \mathrm{ord}_P F$$

i.e., let ν be the smallest value of $i_1 + \cdots + i_N$ with $G_{i_1 \ldots i_N} \ne 0$. We call ν the multiplicity of S at P and we denote it by $\mathrm{mult}_P S$. Geometrically ν can be characterized by noting that, with P fixed, it is the minimum of $\mu_{P,R}$ taken over all R, i.e., taken over all lines through P. To indicate the dependence of ν on P, we may write ν_P for ν. We call P a simple or double or triple or ... point of S according as $\nu_P = 1$ or 2 or 3 or ... ; by a singular point of S we mean a point of S which is not simple. The line L is said to be tangent to S at P if its intersection multiplicity with S at P is greater than the multiplicity of S at P, i.e., if $\mu_{P,R} > \nu_P$. Let

$$G_\nu(X_1, \ldots, X_N) = \sum_{i_1 + \cdots + i_N = \nu} G_{i_1 \ldots i_N} (X_1 - U_1)^{i_1} \ldots (X_N - U_N)^{i_N}.$$

Now a cone with vertex P is a hypersurface passing through P such that, for every point $P' \ne P$ on it, the entire line PP' is on it. Clearly $G_\nu = 0$ is a cone with vertex P, and a line L through P is on it iff L is tangent to S at P. We call $G_\nu = 0$ the tangent cone of S at P. If P is a simple point of S, i.e., if $\nu = 1$, then for

$1 \leq i \leq N$ we have $F_{U_i} = G_{0\ldots010\ldots0}$ with 1 in the i-th place, and hence the said cone is reduced to the hyperplane

$$F_{U_1}(X_1 - U_1) + \cdots + F_{U_N}(X_N - U_N) = 0$$

which we call the tangent hyperplane of S at P.

Note that so far we did not assume F to be a polynomial. So at least when P is the origin $(0, \ldots, 0)$, the above discussion applies to an analytic hypersurface, i.e., to the case when F is a (formal) power series.

Now suppose that $F(X_1, \ldots, X_N)$ is a polynomial of degree d. Then d is also called the degree of the hypersurface S; if $N = 2$ then S is a curve of degree d, and if $N = 3$ then S is a surface of degree d. To characterize d geometrically, we note that now $h(t)$ is a polynomial of degree d, and hence it has d roots giving d points of intersection of S with L. This is so for most values of U_i and R_i. For special values of U_i and R_i, the equation $h(t) = 0$ may have less than d distinct roots because some of them may be multiple roots. Thus d may be characterized as the maximum number of points in which a line meets S; here and in what follows, the line is assumed not to lie on S. Moreover, if there are c distinct roots and their multiplicities are $e_1, \ldots, e_c$, then for the sum $e = e_1 + \cdots + e_c$ we have $e = d$, and for the corresponding points $P_1, \ldots, P_c$ in which S meets L we have $\mathrm{int}_{P_i}(S, L) = e_i$. Thus the sum e of the intersection multiplicities of S and L at their points of intersection equals the degree of S. This is the first case of Bézout's Theorem, which is the oldest Theorem of algebraic geometry and was enunciated by Bézout in his 1770 book [Bez].

Here is an example which shows that we need to be careful in this. Namely, with $N > 1$ and $d > 1$, let

$$F(X_1, \ldots, X_N) = X_1^{d-1} X_2 + X_3 + \cdots + X_N - 1$$

and $P = (1, 1, 0, \ldots, 0)$. Then

$$\begin{aligned} h(t) &= (1 + R_1 t)^{d-1}(1 + R_2 t) + R_3 t + \cdots + R_N t - 1 \\ &= R_1^{d-1} R_2 t^d + \text{(terms of degree } < d \text{ in } t). \end{aligned}$$

Therefore if either $R_1 = 0$ or $R_2 = 0$ then the sum e is smaller than the degree d of S, because some points of intersection of S and L "have gone to infinity." To take care of this we must go over to projective spaces and homogeneous coordinates.

For an elementary treatment of all these things, you may like to browse in my 1990 AMS book [A04].

§4: HOMOGENEOUS COORDINATES

Again let $F(X_1, \ldots, X_N)$ be a polynomial of degree d and consider the hypersurface S in N-space given by $F(X_1, \ldots, X_N) = 0$. Expanding F we have

$$F(X_1, \ldots, X_N) = \sum F_{i_1 \ldots i_N} X_1^{i_1} \ldots X_N^{i_N}.$$

Collecting terms of like degree we have

$$F(X_1, \ldots, X_N) = \sum_{j=0}^{d} F_j(X_1, \ldots, X_N)$$

where

$$F_j(X_1, \ldots, X_N) = \sum_{i_1 + \cdots + i_N = j} F_{i_1 \ldots i_N} X_1^{i_1} \ldots X_N^{i_N}$$

is a homogeneous polynomial of degree j, i.e., every term in it is of degree j. For some j it may happen that F_j has no term in it, i.e., F_j may be reduced to zero. But F_d is not reduced to zero because F is of degree d.

Now for very large values of the variables, the values of the terms of degree d are much bigger than the values of the terms of degree $< d$, i.e., F_d dominates the rest of F. Therefore, heuristically speaking, we may say that $F_d = 0$ gives the points at infinity of S.

Algebraically, we "homogenize" F by multiplying terms of degree j by the $(d-j)$-th power of a new variable X_{N+1} to get a homogeneous polynomial $f(X_1, \ldots, X_{N+1})$ of degree d. By "dehomogenizing" f, i.e., by substituting $X_{N+1} = 1$ in it, we get back F. Thus

$$f(X_1, \ldots, X_{N+1}) = \sum_{j=0}^{d} X_{N+1}^{d-j} F_j(X_1, \ldots, X_N)$$

and

$$F(X_1, \ldots, X_N) = f(X_1, \ldots, X_N, 1).$$

Alternatively, we substitute

$$X_i = x_i / x_{N+1} \quad \text{for} \quad 1 \le i \le N$$

in F and then we clear the denominators by multiplying by x_{N+1}^d to get

$$f(x_1, \ldots, x_{N+1}) = x_{N+1}^d F(x_1 / x_{N+1}, \ldots, x_N / x_{N+1})$$

$$= \sum_{j=0}^{d} x_{N+1}^{d-j} F_j(x_1, \ldots, x_N).$$

By substituting $x_{N+1} = 0$ in $f(x_1, \ldots, x_{N+1})$ we get $F_d(x_1, \ldots, x_N)$, and so we may say that $F_d = 0$ is the intersection of $f = 0$ with the "hyperplane at infinity" given by $x_{N+1} = 0$.

Geometrically, let

$$\mathbb{A}^N = \mathbb{A}_k^N = k^N = \{P = (U_1, \ldots, U_N) : U_i \in k \text{ for } 1 \leq i \leq N\}$$

be the affine N-space over the field k. (If you do not want to be bothered by "fields" and such, you may simply take k be the set of all real numbers $\mathbb{R}$, and then the affine N-space is your familiar $\mathbb{R}^N$). We obtain the projective N-space $\mathbb{P}^N = \mathbb{P}_k^N$ over k when we augment $\mathbb{A}^N$ by the "hyperplane at infinity" in the following manner. First, instead of representing a point P of $\mathbb{A}^N$ by a single N-tuple $(U_1, \ldots, U_N)$, let us represent it by all the $(N+1)$-tuples $u = (u_1, \ldots, u_{N+1})$ such that $u_{N+1} \neq 0$ and $U_i = u_i/u_{N+1}$ for $1 \leq i \leq N$; these are called homogeneous coordinates of P. Now the remaining $(N+1)$-tuples, i.e., the $(N+1)$-tuples $u = (u_1, \ldots, u_{N+1})$ with $u_{N+1} = 0$, represent points of the hyperplane at infinity, again with the understanding that different $(N+1)$-tuples represent the same point iff they are proportional. In other words, let us say that two $(N+1)$-tuples $(u_1, \ldots, u_{N+1})$ and $(u_1', \ldots, u_{N+1}')$ are equivalent iff they are proportional, i.e., iff $(u_1', \ldots, u_{N+1}') = (cu_1, \ldots, cu_{N+1})$ for some nonzero constant c. Now $\mathbb{P}^N$ may be identified with the set of all equivalence classes of $(N+1)$-tuples, with the understanding that the zero $(N+1)$-tuple $(0, \ldots, 0)$ is excluded from consideration; in other words

$$\mathbb{P}^N = \mathbb{P}_k^N = (k^{N+1} \setminus \{(0, \ldots, 0)\})/ \sim$$

where $\sim$ is the said equivalence relation.

A projective hypersurface S of degree d, i.e., a hypersurface in the projective space $\mathbb{P}^N$, is given by an equation $f(x_1, \ldots, x_{N+1}) = 0$ where $f(x_1, \ldots, x_{N+1})$ is a nonzero homogeneous polynomial of degree d; this is unambiguous because for any projective point P we have $f(u_1, \ldots, u_{N+1}) = 0$ for one homogeneous coordinate tuple $(u_1, \ldots, u_{N+1})$ of P iff $f(u_1', \ldots, u_{N+1}') = 0$ for every homogeneous coordinate tuple $(u_1', \ldots, u_{N+1}')$ of P; when this is so we say that $P \in S$; otherwise $P \notin S$. In particular, a projective hyperplane is given by a homogeneous linear equation $a_1 x_1 + \cdots + a_{N+1} x_{N+1} = 0$ where the coefficients $a_1, \ldots, a_{N+1}$ are determined up to proportionality, and at least one of them is nonzero; if $a_i \neq 0$ for some $i \leq N$ then this corresponds to the hyperplane $a_1 X_1 + \cdots + a_N X_N + a_{N+1} = 0$ in $\mathbb{A}^N$; if $a_i = 0$ for all $i \leq N$ then we get the hyperplane at infinity given by $x_{N+1} = 0$. The portion of S "at finite distance," i.e., $S \cap \mathbb{A}^N$, is the affine hypersurface of degree d given by the equation $F(X_1, \ldots, X_N) = 0$ where $F(X_1, \ldots, X_N) = f(X_1, \ldots, X_N, 1)$; the remaining portion of S consists of the points of S lying on the hyperplane at infinity. Any hyperplane H in $\mathbb{P}^N$ is clearly a copy of $\mathbb{P}^{N-1}$, and its complement $\mathbb{P}^N \setminus H$ is a copy of $\mathbb{A}_k^N$; in particular, this is so for the hyperplane $H_i : x_i = 0$, i.e., $\mathbb{P}^N \setminus H_i$ is a copy $\mathbb{A}_{k,i}^N$ of $\mathbb{A}_k^N$; moreover, $\mathbb{P}^N = \cup_{i=1}^{N+1}(\mathbb{P}^N \setminus H_i)$, i.e., the projective N-space can be covered by $N+1$ copies of the affine N-space; in other words

$$\mathbb{P}_k^N = \cup_{i=1}^{N+1}(\mathbb{P}^N \setminus H_i) = \cup_{i=1}^{N+1}\mathbb{A}_{k,i}^N.$$

Therefore the definitions of simple point, singular point, multiplicity, intersection multiplicity, tangent cone, and tangent hyperplane can be extended to projective hypersurfaces.

Let $u = (u_1, \ldots, u_{N+1})$ be homogeneous coordinates of a point P of S. Let $x = (x_1, \ldots, x_{N+1})$ and $f_x = (f_{x_1}, \ldots, f_{x_{N+1}})$ and $f_u = (f_{u_1}, \ldots, f_{u_{N+1}})$, and as usual let us put

$$x \cdot f_x = x_1 f_{x_1} + \cdots + x_{N+1} f_{x_{N+1}}$$

and

$$x \cdot f_u = x_1 f_{u_1} + \cdots + x_{N+1} f_{u_{N+1}}$$

and so on. By Euler's Theorem on Homogeneous Polynomials we have $x \cdot f_x = df(x)$, and hence (because $P \in S$) we get $u \cdot f_u = 0$. For a moment assume that $u_{N+1} \neq 0$ and let $U_i = u_i/u_{N+1}$ for $1 \leq i \leq N$. Then by the equation $f(x_1, \ldots, x_{N+1}) = x_{N+1}^d F(x_1/x_{N+1}, \ldots, x_N/x_{N+1})$, we see that $f_{u_i} = u_{N+1}^{d-1} F_{U_i}$ for $1 \leq i \leq N$; since P is singular for S iff $F_{U_i} = 0$ for $1 \leq i \leq N$, in view of the equation $u \cdot f_u = 0$ we conclude that: P is singular for S iff $f_{u_i} = 0$ for $1 \leq i \leq N+1$. In case P is a simple point of S, we know that the tangent hyperplane T_P of S at P has the affine equation $F_{U_1}(X_1 - U_1) + \cdots + F_{U_N}(X_N - U_N) = 0$; since $u \cdot f_u = 0$ and $f_{u_i} = u_{N+1}^{d-1} F_{U_i}$ for $1 \leq i \leq N$, we conclude that T_P is given by the homogeneous equation $x \cdot f_u = 0$. By symmetry it follows that both these conclusions remain valid without assuming $u_{N+1} \neq 0$.

Thus, for any point P of S with homogeneous coordinates $u = (u_1, \ldots, u_{N+1})$, we have that:

(I1) P is a singular point of S iff $f_{u_i} = 0$ for $1 \leq i \leq N+1$, and

(I2) if P is a simple point of S then the tangent hyperplane T_P of S at P is given by the homogeneous linear equation $x \cdot f_u = 0$.

For any point A of $\mathbb{P}^N$ with homogeneous coordinates $y = (y_1, \ldots, y_{N+1})$, let us define the polar Π_A of A (relative to S) to be the hypersurface of degree $d - 1$ defined by the homogeneous equation $y \cdot f_x = 0$. By (I1) we see that:

(I3) if P is any singular point of S then $P \in \Pi_A$. By (I2) we see that:

(I4) if P is any simple point of S then: $P \in \Pi_A \Leftrightarrow A \in T_P$. By (I4) we see that:

(I5) if P is any simple point of S with $P \neq A$ then: $P \in \Pi_A \Leftrightarrow$ the line AP is tangent to S at P.

It follows that if S is a nonsingular (= having no singular point) hyperquadric then the present definition of polar coincides with the definition given when we assigned Homework Problems (H1) and (H2).

To do the Homework Problems (H1) and (H2), assume that S is a nonsingular hyperquadric. Let A and B be any points in $\mathbb{P}^N$ with homogeneous coordinates $y = (y_1, \ldots, y_{N+1})$ and $z = (z_1, \ldots, z_{N+1})$ respectively. Then the polars Π_A and Π_B are given by the equations $y \cdot f_x = 0$ and $z \cdot f_x = 0$ respectively. We can write

$$f(x_1, \ldots, x_{N+1}) = \sum_{1 \leq i \leq N+1} a_i x_i^2 + \sum_{1 \leq j < j' \leq N+1} b_{jj'} x_j x_{j'}$$

with constants a_i and $b_{jj'}$, and then we get

$$y \cdot f_z = \sum_{1 \le i \le N+1} 2a_i y_i z_i + \sum_{1 \le j < j' \le N+1} b_{jj'}(y_j z_{j'} + y_{j'} z_j) = z \cdot f_y.$$

Therefore

$$B \in \Pi_A \Leftrightarrow y \cdot f_z = 0 \Leftrightarrow z \cdot f_y = 0 \Leftrightarrow A \in \Pi_B.$$

§5: SINGULARITIES

Reverting to affine equations, let $S : F(X_1, \ldots, X_N) = 0$ be a hypersurface in the affine N-space over a field k. Given a point P of S, by a translation of the coordinate system, we may assume it to be the origin $(0, \ldots, 0)$. This amounts to making the k-automorphism of the polynomial ring $k[X_1, \ldots, X_n]$ given by $X_i \mapsto X_i + U_i$ where $(U_1, \ldots, U_N)$ are the original coordinates of P. Let $\text{mult}_P S = \nu$. Then

$$F(X_1, \ldots, X_N) = F_\nu(X_1, \ldots, X_N) + (\text{terms of degree} > \nu)$$

where $F_\nu(X_1, \ldots, X_N)$ is a nonzero homogeneous polynomial of degree ν.

First let us think of S as an analytic hypersurface at P, i.e., let us assume F to be a power series. Let $k[[X_1, \ldots, X_N]]$ be the ring of all formal power series in $X_1, \ldots, X_N$ with coefficients in k; here formal means we are disregarding questions of convergence, and we are going to do only algebraic manipulations; for instance, we permit the power series $\sum i^i X^i$ whose radius of convergence is zero. We shall discuss the notion of radius of convergence later on. Later on we shall also show that power series rings over a field are UFDs. Using this fact we can write

$$F = E_0 E_1^{r_1} \ldots E_t^{r_t}$$

where $r_1, \ldots, r_t$ are positive integers, $E_1, \ldots, E_t$ are pairwise coprime irreducible power series with $E_1(0, \ldots, 0) = \cdots = E_t(0, \ldots, 0) = 0$, and E_0 is a power series with $E_0(0, \ldots, 0) \ne 0$. We call $S_1 : E_1 = 0, \ldots, S_t : E_t = 0$ the analytic branches of S (at P). Let $\text{mult}_P S_i = \nu_i$ for $1 \le i \le t$. Then $r_1 \nu_1 + \cdots + r_t \nu_t = \nu$, and for $1 \le i \le t$ we have

$$E_i(X_1, \ldots, X_N) = E_{i\nu_i}(X_1, \ldots, X_N) + (\text{terms of degree} > \nu_i)$$

where $E_{i\nu_i}(X_1, \ldots, X_N)$ is a nonzero homogeneous polynomial of degree ν_i. For the "initial form" F_ν of F we have the factorization

$$F_\nu(X_1, \ldots, X_N) = E_0(0, \ldots, 0) \prod_{1 \le i \le t} E_{i\nu_i}(X_1, \ldots, X_N)^{r_i} \text{ with } E_0(0, \ldots, 0) \ne 0.$$

We say that S has a t-fold NC (= Normal Crossing) at P if all the analytic branches have a simple point at P and their tangent hyperplanes are linearly independent, i.e., if $\nu_1 = \cdots = \nu_t = 1$ and upon letting

$$E_{i1} = \sum_{1 \le j \le N} E'_{ij} X_j \text{ with } E'_{ij} \in k$$

we have that the rank of the $t \times N$ matrix (E'_{ij}) is t. Note that the rank of an $m \times n$ matrix (a_{ij}) over k is the largest nonnegative integer r for which there exist sequences $1 \le p_1 < \cdots < p_r \le m$ and $1 \le q_1 < \cdots < q_r \le n$ such that the determinant of the $r \times r$ matrix $(a_{p_i q_j})$ is nonzero. Also note that if S has a t-fold NC at P then obviously $t \le N$. We may say that, in the sense of analytic geometry, S has a simple point at P iff it has a 1-fold NC at P.

Now let us think of S as an algebraic hypersurface, i.e., let us assume F to be a polynomial of some degree d. Since the polynomial ring $k[X_1, \ldots, X_N]$ is a UFD, we can write

$$F = \Gamma_0 \Gamma_1^{\rho_1} \ldots \Gamma_\tau^{\rho_\tau}$$

where $\rho_1, \ldots, \rho_\tau$ are positive integers, $\Gamma_1, \ldots, \Gamma_\tau$ are pairwise coprime irreducible polynomials with $\Gamma_1(0, \ldots, 0) = \cdots = \Gamma_\tau(0, \ldots, 0) = 0$, and Γ_0 is a polynomial with $\Gamma_0(0, \ldots, 0) \ne 0$. We call $\Sigma_1 : \Gamma_1 = 0, \ldots, \Sigma_\tau : \Gamma_\tau = 0$ the algebraic branches of S at P. Concerning these two factorizations, you may do the following Homework Problem, where we recall that a partition of a set W is a collection of nonempty subsets of W whose union is W and which are pairwise disjoint, i.e., have pairwise empty intersection; we may express this by writing $W = \coprod_{i \in I} W_i$ where W_i are the subsets indexed by an indexing set I.

HOMEWORK PROBLEM (H3). Show that the analytic factorization is a refinement of the algebraic factorization, i.e., there is a partition $\{1, \ldots, t\} = \coprod_{1 \le i \le \tau} W_i$ such that for $1 \le i \le \tau$ we have $\rho_i = r_j$ for all $j \in W_i$ and $\Gamma_i = E_i^* \prod_{j \in W_i} E_j$ where E_i^* is a power series with $E_i^*(0, \ldots, 0) \ne 0$. Equivalently, an irreducible polynomial is never divisible by the square of a power series which is zero at the origin, and any two coprime polynomials remain coprime as power series.

Now let us study in greater detail the case of an algebraic plane curve C : $F(X, Y) = 0$ of degree d. Suppose C has a double point at $P = (0, 0)$. Here are some examples of this.

Node: $Y^2 - X^2 - X^3 = 0$. Two tangents $Y - X = 0$ and $Y + X = 0$ at P.

Tacnode: $Y^2 - X^4 - X^5 = 0$. Only one tangent $Y = 0$ at P.

Cusp: $Y^2 - X^3 = 0$. Again only one tangent $Y = 0$ at P.

All these are of the form $Y^2 = B(X)$ where $B(X)$ is a polynomial of odd degree. Therefore they are algebraically irreducible because a polynomial of odd degree cannot have a polynomial square root. The last (cusp) is analytically irreducible because a power series of odd order cannot have a power series as square root. But the first two are analytically reducible (provided k is not of characteristic 2) because a power series of even order with constant term 1 has a power series square root. In case of characteristic zero, this follows from Newton's Binomial Theorem for fractional exponents thus. By the said Theorem we have the power series identity

$$(\text{T1}) \qquad (1+X)^r = 1 + \sum_{i>0} \frac{r(r-1)\ldots(r-i+1)}{i!} X^i$$

where r is any rational number [cf. §12(E16)]. For X we may substitute any power series in X of positive order. Therefore by taking $r = 1/n$ we see that any power series $A(X)$ whose constant term is 1 has an n-th root for every positive integer n. If the order of the power series $B(X)$ is en for some integer $e \geq 0$ and the coefficient of X^{en} in $B(X)$ is 1 then we can write it as $B(X) = (X^e)^n A(X)$ with $A(0) = 1$. Therefore $B(X)$ has an n-th root. It follows that

$$Y^2 - X^2 - X^3$$
$$= [Y - X + (\text{terms of degree} > 1)] \times [Y + X + (\text{terms of degree} > 1)]$$

and

$$Y^2 - X^4 - X^5$$
$$= [Y - X^2 + (\text{terms of degree} > 2)] \times [Y + X^2 + (\text{terms of degree} > 2)].$$

Consequently, at a node we have a 2-fold NC, but at a tacnode we do not have an NC. At a cusp we of course do not have an NC.

§6: HENSEL'S LEMMA AND NEWTON'S THEOREM

The fact that by L2§3(T1) we can find an n-th root of any $A(X) \in k[[X]]$ with $A(0) = 1$, and for the said n-th root $\alpha(X) \in k[[X]]$ we have $\alpha(0) = 1$, can be restated by saying that the polynomial $Y^n - A(X)$ is divisible by $Y - \alpha(X)$ in $k[[X]][Y]$. Moreover by carrying out the division we get the factorization

$$Y^n - A(X) = [Y - \alpha(X)] \times [Y^{n-1} + \beta_1(X)Y^{n-2} + \cdots + \beta_{n-1}(X)]$$

where $\beta_i(X) \in K[[X]]$ with $\beta_i(0) = 1$ for $1 \leq i \leq n-1$. Around 1900 this was generalized by Hensel into his celebrated lemma which has many incarnations. Let us now state and prove the following simplest version of it, where we note that two univariate polynomials over a field are coprime means they have no nonconstant common factor.

BASIC HENSEL'S LEMMA (T2). Let

$$F(X,Y) = Y^n + A_1(X)Y^{n-1} + \cdots + A_n(X)$$

be a monic polynomial of degree $n > 0$ in Y with coefficients $A_i(X)$ in $k[[X]]$ for $1 \leq i \leq n$ where k is any field. Assume that

$$F(0, Y) = \overline{G}(Y)\overline{H}(Y)$$

where $\overline{G}(Y)$ and $\overline{H}(Y)$ are coprime monic polynomials of positive degrees r and s in Y with coefficients in k. Then there exist unique monic polynomials $G(X, Y)$ and $H(X, Y)$ of degrees r and s in Y with coefficients in $k[[X]]$ such that

$$F(X, Y) = G(X, Y)H(X, Y)$$

and

$$G(0, Y) = \overline{G}(Y) \quad \text{and} \quad H(0, Y) = \overline{H}(Y).$$

PROOF. Rewriting F, G, H as power series in X with coefficients polynomials in Y we have

$$\begin{cases} F(X, Y) = F_0(Y) + F_1(Y)X + \cdots + F_i(Y)X^i + \ldots \\ G(X, Y) = G_0(Y) + G_1(Y)X + \cdots + G_i(Y)X^i + \ldots \\ H(X, Y) = H_0(Y) + H_1(Y)X + \cdots + H_i(Y)X^i + \ldots \end{cases}$$

where

$$\begin{cases} F_0(Y) = F(0, Y) = \text{ a monic polynomial of degree } n \text{ over } k \\ G_0(Y) = \overline{G}(Y) = \text{ a monic polynomial of degree } r \text{ over } k \\ H_0(Y) = \overline{H}(Y) = \text{ a monic polynomial of degree } s \text{ over } k \\ F_i(Y) \in k[Y] \text{ of degree } < n \text{ for all } i > 0 \\ G_i(Y) \in k[Y] \text{ of degree } < r \text{ to be found for all } i > 0 \\ H_i(Y) \in k[Y] \text{ of degree } < s \text{ to be found for all } i > 0 \end{cases}$$

and where by the coprimeness we can find $P(Y)$ and $Q(Y)$ in $k[Y]$ such that

$$(1) \qquad\qquad P(Y)G_0(Y) + Q(Y)H_0(Y) = 1.$$

The requirement $F = GH$ is equivalent to saying that for every $l \geq 0$ we want to satisfy the equation

$$(2) \qquad\qquad \sum_{i+j=l} G_i(Y)H_j(Y) = F_l(Y).$$

For $l = 0$ this is given. Now let $l > 0$ and suppose we have found G_i and H_j for all $i < l$ and $j < l$ such that the above equation is satisfied for every value of l smaller than the given value. By sending the known terms to the right we see that (2) is equivalent to requiring that

$$(2') \qquad\qquad H_l(Y)G_0(Y) + G_l(Y)H_0(Y) = F_l'(Y)$$

where

$$F_l'(Y) = F_l(Y) - \sum_{i+j=l \text{ with } i < l \text{ and } j<l} G_i(Y)H_j(Y) \in k[Y] \text{ is of degree } < n.$$

Upon letting $H_l'(Y) = P(Y)F_l'(Y)$ and $G_l'(Y) = Q(Y)F_l'(Y)$, by (1) we get

$$(1') \qquad\qquad H_l'(Y)G_0(Y) + G_l'(Y)H_0(Y) = F_l'(Y).$$

Dividing $H_l'(Y)$ by $H_0(Y)$ we find $G_l''(Y)$ and $H_l(Y)$ in $k[Y]$ such that $H_l'(Y) = G_l''(Y)H_0(Y) + H_l(Y)$ and the degree of $H_l(Y)$ is $<$ than the degree of $H_0(Y)$ which is s. Now upon letting $G_l(Y) = G_l'(Y) + G_l''(Y)G_0(Y)$, by $(1')$ we get

$$(1'') \qquad\qquad H_l(Y)G_0(Y) + G_l(Y)H_0(Y) = F_l'(Y).$$

In the above equation, the degrees of the first and the last terms are $< n$ and so is the degree of the middle term $G_l(Y)H_0(Y)$, but in it the degree of $H_0(Y)$ is $s = n - r$ and hence the degree of the other factor $G_l(Y)$ must be $< r$. Thus the degree conditions are satisfied and hence $(1'')$ yields $(2')$, which completes the induction on l.

  The uniqueness can also be proved by induction thus. Given $l > 0$, suppose $G_1(Y),\ldots,G_{l-1}(Y)$ and $H_1(Y),\ldots,H_{l-1}(Y)$ are unique. Let $G_l^*(Y)$ and $H_l^*(Y)$ be any polynomials of degrees $< r$ and $< s$ respectively such that

$$(2^*) \qquad\qquad H_l^*(Y)G_0(Y) + G_l^*(Y)H_0(Y) = F_l'(Y).$$

Subtracting (2^*) from $(2')$ we get

$$(2^{**}) \qquad\qquad H_l^{**}(Y)G_0(Y) = -G_l^{**}(Y)H_0(Y)$$

where $G_l^{**}(Y) = G_l(Y) - G_l^*(Y) \in k[Y]$ is of degree $< r$ and $H_l^{**}(Y) = H_l(Y) - H_l^*(Y) \in k[Y]$ is of degree $< s$. Since $G_0(Y)$ and $H_0(Y)$ are coprime and by the above equation $G_0(Y)$ divides $G_l^{**}(Y)H_0(Y)$, it must divide $G_l^{**}(Y)$; but the degree of $G_l^{**}(Y)$ is smaller than the degree of $G_0(Y)$ and hence we must have $G_l^{**}(Y) = 0$, i.e., $G_l^*(Y) = G_l(Y)$. Similarly $H_l^*(Y) = H_l(Y)$. This proves uniqueness.

  COMPLETING THE POWER. Now we shall use Hensel's Lemma to prove Newton's Theorem stated in L2(T1), i.e., in (T1) of Lecture L2. An important ingredient of the proof will be Shreedharacharya's completing the square method in its incarnation of "completing the n-th power." By this we mean, given an n-th degree monic polynomial

$$F(Y) = Y^n + A_1Y^{n-1} + \cdots + A_n$$

with coefficients $A_1,\ldots,A_n$ in a ring R in which n is a unit (for instance R could be a field whose characteristic does not divide n), "by completing the n-th power" we can write

$$F(Y) = \left(Y + \frac{A_1}{n}\right)^n + \tilde{A}_2\left(Y + \frac{A_1}{n}\right)^{n-2} + \cdots + \tilde{A}_n$$

with $\tilde{A}_2, \ldots, \tilde{A}_n$ in R, and this gives us

$$\tilde{F}(Y) = F(Y - (A_1/n)) = Y^n + \tilde{A}_2 Y^{n-2} + \cdots + \tilde{A}_n.$$

The advantage of "killing" the coefficient of Y^{n-1} is to detect multiple roots. So for a moment suppose that the coefficients $A_1, \ldots, A_n$ belong to an algebraically closed field K of characteristic 0, and let $\alpha_1, \ldots, \alpha_n$ be the roots of F in K, i.e.,

$$F(Y) = (Y - \alpha_1) \ldots (Y - \alpha_n).$$

Assume that $A_1 = 0$ but $A_j \neq 0$ for some j with $2 \leq j \leq n$. Then, since $A_j \neq 0$ for some j we must have $\alpha_i \neq 0$ for some i. If $\alpha_1 = \cdots = \alpha_n$ then we would get

$$F(Y) = (Y - \alpha_1)^n = Y^n - n\alpha_1 Y^{n-1} + \cdots$$

which would give us the contradiction $A_1 = -n\alpha_1 \neq 0$. Therefore upon relabelling $\alpha_1, \ldots, \alpha_n$ we can arrange matters so that for some positive integer $r < n$ we have $\alpha_1 = \cdots = \alpha_r \neq \alpha_{r+j}$ for $1 \leq j \leq s = n - r$. Now upon letting

$$G(Y) = (Y - \alpha_1) \ldots (Y - \alpha_r) \quad \text{and} \quad H(Y) = (Y - \alpha_{r+1}) \ldots (Y - \alpha_{r+s})$$

we see that $F(Y) = G(Y)H(Y)$ where $G(Y)$ and $H(Y)$ are coprime monic polynomials of positive degrees in Y with coefficients in K. Thus we have proved the following:

LEMMA (T3). If $F(Y) = Y^n + \sum_{1 \leq i \leq n} A_i Y^{n-i}$ is a monic polynomial of degree $n > 0$ with coefficients A_i in an algebraically closed field K of characteristic 0, and if $A_1 = 0$ but $A_j \neq 0$ for some j with $2 \leq j \leq n$, then $F(Y) = G(Y)H(Y)$ where $G(Y)$ and $H(Y)$ are coprime monic polynomials of positive degrees in Y with coefficients in K.

PROCESSES ON ROOTS. To put the completing the power method in proper perspective, again consider a monic polynomial

$$F(Y) = Y^n + \sum_{1 \leq i \leq n} A_i Y^{n-i}$$

of degree $n > 0$ in Y with coefficients A_i in a field K, and let $\alpha_1, \ldots, \alpha_n$ be the roots of F in an overfield L of K, i.e.,

$$F(Y) = \prod_{1 \leq i \leq n} (Y - \alpha_i).$$

To "increase the roots" by $B \in L$, by putting

$$F^*(Y) = F(Y - B) = Y^n + \sum_{1 \leq i \leq n} A_i^* Y^{n-i}$$

with $A_i^* \in L$ we get

$$F^*(Y) = \prod_{1 \leq i \leq n} (Y - (\alpha_i + B)).$$

Likewise to "multiply the roots" by $0 \neq C \in L$, by putting

$$F^{**}(Y) = C^n F(YC^{-1}) = Y^n + \sum_{1 \leq i \leq n} A_i^{**} Y^{n-i}$$

with $A_i^{**} \in L$ we get

$$F^{**}(Y) = \prod_{1 \leq i \leq n} (Y - C\alpha_i).$$

As the third "standard process" on roots, to "reciprocate the roots" we assume $A_n \neq 0$ and then by putting

$$F^{***}(Y) = Y^n F(Y^{-1}) = 1 + A_1 Y + \cdots + A_n Y^n$$

i.e., by "reading F backwards" we get

$$F^{***}(Y) = A_n \prod_{1 \leq i \leq n} (Y - (1/\alpha_i)).$$

With this preparation let us now prove a slight variation of Newton's Theorem which we call:

BASIC NEWTON'S THEOREM (T4). Let there be given any monic polynomial

$$F(X, Y) = Y^n + A_1(X)Y^{n-1} + \cdots + A_n(X)$$

of degree $n > 0$ in Y with coefficients $A_1(X), \ldots, A_n(X)$ in $k((X))$ where k is an algebraically closed field of characteristic 0. Then for some integer $m > 0$ we have a factorization

$$F(T^m, Y) = \prod_{1 \leq i \leq n} (Y - y_i(T)) \quad \text{with} \quad y_i(T) \in k((T)).$$

Moreover, if $A_i(X) \in k[[X]]$ for $1 \leq i \leq n$ then $y_i(T) \in k[[T]]$ for $1 \leq i \leq n$.

For the reason given above we call our proof:

SHREEDHARACHARYA'S PROOF OF NEWTON'S THEOREM. We make induction on n. For $n = 1$ we have nothing to show. So let $n > 1$ and assume true for all values of n smaller than the given value. By completing the n-th power we have

$$F^*(X,Y) = F(X, Y - (A_1(X)/n)) = Y^n + \sum_{1 \leq i \leq n} A_i^*(X)Y^{n-i}$$

where $A_i^*(X) \in k((X))$ with $A_1^*(X) = 0$. If $A_j^*(X) = 0$ for $2 \leq j \leq n$ then we have nothing to show. So assume that $A_j^*(X) \neq 0$ for some j with $2 \leq j \leq n$. Let

$$u = \min_{1 \leq i \leq n} \frac{\mathrm{ord}_X A_i^*(X)}{i}.$$

Then $u = v/w$ where v and w are integers with $w > 0$. Note that if $A_i(X) \in k[[X]]$ for $1 \leq i \leq n$ then clearly $v \geq 0$. Let

$$F^{**}(Z,Y) = Z^{-nv} F^*(Z^w, YZ^v) = Y^n + \sum_{1 \leq i \leq n} A_i^{**}(Z)Y^{n-i}$$

where $A_i^{**}(Z) = A_i^*(Z^w)Z^{-iv} \in k[[Z]]$ for $1 \leq i \leq n$ with $A_1^{**}(Z) = 0$ and

$$A_j^{**}(0) \neq 0 \text{ for some } j \text{ with } 2 \leq j \leq n.$$

Now first applying (T3) and then (T2) we get

$$F^{**}(Z,Y) = F'(Z,Y)F''(Z,Y)$$

where $F'(Z,Y)$ and $F''(Z,Y)$ are monic polynomials of degrees $n' > 0$ and $n'' > 0$ in Y with coefficients in $k[[Z]]$. By the induction hypothesis we can find integers $m' > 0$ and $m'' > 0$ such that

$$F'(T^{m'}, Y) = \prod_{1 \leq i \leq n'} (Y - y_i'(T)) \quad \text{and} \quad F''(T^{m''}, Y) = \prod_{1 \leq i \leq n''} (Y - y_i''(T))$$

with $y_i'(T)$ and $y_i''(T)$ in $k[[T]]$. Upon letting $m = wm'm''$ we see that m is a positive integer and

$$F^*(T^m, Y) = \prod_{1 \leq i \leq n} (Y - y_i^*(T)) \quad \text{with} \quad y_i^*(T) \in k((T))$$

and if $v \geq 0$ then $y_i^*(T) \in k[[T]]$ for $1 \leq i \leq n$. It follows that

$$F(T^m, Y) = \prod_{1 \leq i \leq n} (Y - y_i(T)) \quad \text{with} \quad y_i(T) \in k((T))$$

and if $A_i(X) \in k[[X]]$ for $1 \leq i \leq n$ then $y_i(T) \in k[[T]]$ for $1 \leq i \leq n$.

§7: INTEGRAL DEPENDENCE

The last sentence of Theorem (T4) is a bit informal. It does not make completely clear whether the roots $y_i(T)$ asserted in the factorization display are automatically in $k[[T]]$ when the coefficients $A_i(X)$ are assumed to be in $k[[X]]$, or whether they can be so chosen. Of course the uniqueness of roots does show the two statements to be equivalent. To prove the apparently stronger statement, i.e., the implication $A_i(X) \in k[[X]]$ for all $i \Rightarrow y_i(T) \in k[[T]]$ for all i, let us introduce the ideas of integral dependence and integral closure.

To go back in history, what is $\sqrt{2}$ and why is it not rational? At any rate, the need to know $\sqrt{2}$ arose since, by the Pythagoras Theorem, it is the length of the hypotenuse of an isosceles right angle triangle whose two equal sides have length one. Now $1^2 = 1, 2^2 = 4, \ldots$, and hence there is no integer whose square is two. Suppose there is a rational number t whose square is two, i.e., t satisfies the equation $y^2 - 2 = 0$. We can write $t = u/v$ where u and v are integers with $v \neq 0$; we may assume that this is a reduced form of t, i.e., u and v have no common prime factor; we may also assume that $v > 0$. Since $(u/v)^2 - 2 = 0$ we get $u^2 = 2v^2$ which shows that every prime factor of v divides u. Therefore we must have $v = 1$. Thus t must be an integer which we have shown to be impossible. The belief that although $\sqrt{2}$ is not rational, it is some sort of a number, makes it desirable to "complete" the field $\mathbb{Q}$ to get the field $\mathbb{R}$; we can then accommodate $\sqrt{2} = 1.4142\ldots, \pi = 3.1415\ldots$, and so on.

The above argument can be generalized to show that $\mathbb{Z}$ is a normal domain, i.e., any element t in its quotient field $\mathbb{Q}$ which is integral over it belongs to it. An element t in an overring S of a ring R is said to be integral over R if it satisfies a monic polynomial equation over R, i.e., if $F(t) = 0$ for some monic polynomial $F(Y) = Y^n + A_1 Y^{n-1} + \cdots + A_n$ of some degree $n > 0$ with coefficients $A_1, \ldots, A_n$ in R. We may say that t/R is integral to mean that t is integral over R. A subset T of S is integral over R means every $t \in T$ is integral over R; again we may indicate this by saying that T/R is integral. By the integral closure of R in S we mean the set of all elements in S which are integral over R. It can be shown that sums and products of integral elements are integral; moreover, any $t \in R$ is integral over R since it satisfies the equation $Y - t = 0$; consequently [cf. §10(E2)]:

(J1) $\quad \begin{cases} \text{the integral closure of a ring } R \text{ in an overring } S \text{ is a subring of } S, \\ \text{and } R \text{ is a subring of the said integral closure.} \end{cases}$

It can also be shown that [cf. §10(E2)]

(J2) $\quad \begin{cases} \text{if an overring } S \text{ of a ring } R \text{ is integral over } R, \text{ and} \\ \text{an overring } T \text{ of } S \text{ is integral over } S, \text{ then } T \text{ is integral over } R. \end{cases}$

A ring R is integrally closed in an overring S means R coincides with its integral closure in S. Thus a domain is normal means it is integrally closed in its quotient field.

Since $\mathbb{Z}$ is a UFD, its normality will follow by showing that

(J3)   every UFD is normal.

To put this in proper perspective, let us first show that

(J4) $\begin{cases} \text{for any valuation } v : K \to G \cup \{\infty\} \text{ of any field } K, \\ \text{the valuation ring } R_v = \{x \in K : v(x) \geq 0\} \text{ of } v \text{ is normal.} \end{cases}$

To prove this let us extend v to a valuation w of $K(Y)$ thus. For any

$$A = A(Y) = \sum A_i Y^i \in K[Y] \quad \text{with} \quad A_i \in K$$

let

$$w(A) = \min v(A_i) \text{ taken over all } i$$

and note that if $A = 0$ then $w(A) = \infty$ and if $A \neq 0$ then $w(A) \in G_v =$ the value group of v. For verifying that w satisfies the valuation axioms, we observe that the valuation axioms imply that

(J5) $$v(1) = 0$$

and for any finite number of elements $x_1, \ldots, x_m$ in K we have

(J6) $$v(x_1 + \cdots + x_m) \geq \min(v(x_1), \ldots, v(x_m))$$

and

(J7) if $v(x_1) < v(x_i)$ for $2 \leq i \leq m$ then $v(x_1 + \cdots + x_m) = v(x_1)$.

Now given any $B = \sum B_j Y^j \in K[Y]$ with $B_j \in K$ we clearly have

$$w(A + B) \geq \min(w(A), w(B)).$$

Moreover, if $A \neq 0 \neq B$ then, upon letting $i' = \min\{i : w(A) = v(A_i)\}$ and $j' = \min\{j : w(B) = v(B_j)\}$, and upon letting $C = AB = \sum C_l Y^l$ with $C_l \in K$, for all l we have

$$C_l = \sum_{i+j=l} A_i B_j$$

and hence $v(C_l) \geq w(A) + w(B)$, and we also have

$$C_{i'+j'} = A_{i'} B_{j'} + \sum_{i+j=i'+j' \text{ and either } i < i' \text{ or } j < j'} A_i B_j$$

and hence $v(C_{i'+j'}) = w(A) + w(B)$. Thus we have

(J8) $$w(AB) = w(A) + w(B) \text{ for all } A \neq 0 \neq B \text{ in } K[Y].$$

Therefore we may unambiguously put $w(A/B) = w(A) - w(B)$, and then it follows that

(J9)
$$\begin{cases} w : K(Y) \to G \cup \{\infty\} \text{ is a valuation of } K(Y) \\ \text{such that } w(x) = v(x) \text{ for all } x \in K, \text{ and} \\ \text{the value group of } w \text{ coincides with the value group of } v. \end{cases}$$

Given a domain R, let us say that R is overnormal to mean that:

(*)
$$\begin{cases} \text{if } A(Y), B(Y), C(Y) \text{ are monic polynomials in } Y \text{ with coefficients} \\ \text{in the quotient field } K \text{ of } R \text{ such that } A(Y)B(Y) = C(Y) \in R[Y], \\ \text{then } A(Y) \in R[Y] \text{ and } B(Y) \in R[Y]. \end{cases}$$

Using (J8) let us show that

(J10)
$$\begin{cases} \text{for any valuation } v : K \to G \cup \{\infty\} \text{ of any field } K, \\ \text{the valuation ring } R_v = \{x \in K : v(x) \geq 0\} \text{ of } v \text{ is overnormal.} \end{cases}$$

Namely, since $v(1) = 0$, we see that if $A = A(Y) \in K[Y]$ and $B = B(Y) \in K[Y]$ are monic in Y then $w(A) \leq 0$ and $w(B) \leq 0$, and if $C = C(Y) \in R_v[Y]$ is monic in Y then $w(C) = 0$, and hence if also $AB = C$ then by (J8) we get $w(A) = 0 = w(B)$, and therefore $A(Y) \in R_v[Y]$ and $B(Y) \in R_v[Y]$. Obviously

(J11)   every overnormal domain is normal.

In greater detail, let y in the quotient field K of a domain R satisfy $C(y) = 0$ with nonconstant monic $C(Y) \in R[Y]$; then $C(Y) = A(Y)B(Y)$ with monic $A(Y) = Y - y \in K[Y]$ and monic $B(Y) = C(Y)/A(Y) \in K[Y]$, and hence assuming R overnormal we get $A(Y) \in R[Y]$, i.e., $y \in R$. Now (J4) follows from (J10) and (J11). Given a ring R, let us say that R is a valuation ring to mean that R is a domain which is the valuation ring of some valuation of its quotient field. Given a domain R, let us say that R is an intval domain to mean that R is the intersection of a nonempty family of valuation rings of valuations of its quotient field. Now obviously

(J12)
$$\begin{cases} \text{if a domain } R \text{ with quotient field } K \text{ is the intersection of a nonempty} \\ \text{family of normal (resp: overnormal) domains with quotient field } K, \\ \text{then } R \text{ is normal (resp: overnormal).} \end{cases}$$

By (J10) and the overnormal part of (J12) we see that

(J13)   every intval domain is overnormal.

Likewise by (J4) and the normal part of (J12) we see that

(J14)   every intval domain is normal.

§8: UNIQUE FACTORIZATION DOMAINS

Let R be a UFD with quotient field K. Recall that $U(R)$ is the set of all units in R. Let W be a set of irreducible elements in R such that no two elements of W are associates of each other and every irreducible element in R is an associate of some element of W. Then every $x \in K^\times$ can be uniquely expressed as

$$x = u_x \prod_{z \in W} z^{v_z(x)} \quad \text{with} \quad u_x \in U(R) \quad \text{and} \quad v_z(x) \in \mathbb{Z}$$

such that $v_z(x) = 0$ for all except a finite number of z, with the understanding that in the product the terms which are reduced to 1, i.e., those for which $v_z(x) = 0$ should be disregarded; also put $v_z(0) = \infty$. Then $v_z : K \mapsto \mathbb{Z} \cup \{\infty\}$ is a valuation of K with value group $\mathbb{Z}$, and clearly we have

(J15) $$R = \bigcap_{z \in W} R_{v_z}$$

and hence

(J16) every UFD is an intval domain

and now (J3) follows from (J14). As above, we get a valuation $w_z : K(Y) \mapsto \mathbb{Z} \cup \{\infty\}$ when for every

$$A = A(Y) = \sum A_i Y^i \in K[Y] \quad \text{with} \quad A_i \in K$$

we put

$$w_z(A) = \min v_z(A_i) \text{ taken over all } i$$

and so on. In particular, as in (J8), we have

$$w_z(AB) = w_z(A) + w_z(B) \text{ for all } A \neq 0 \neq B \text{ in } R[Y].$$

This being so for all $z \in W$, we get the GAUSS LEMMA which says that

(J17) $$c(AB) = c(A)c(B) \text{ for all } A \neq 0 \neq B \text{ in } R[Y]$$

where we define the content $c(A)$ of $0 \neq A = A(Y) \in R[Y]$ by putting $c(A) = \gcd(A_0, A_1, \dots)$ with gcd as in L1(R4). Note that this defines $c(A)$ up to associates in R, and (J17) means that for any values of $c(A)$ and $c(B)$, $c(A)c(B)$ is a value of $c(AB)$. As a consequence of (J17) it can be shown that

(J18) $$R \text{ is a UFD} \Rightarrow R[Y] \text{ is a UFD}.$$

For any positive integer r, by induction and (J18) we see that

(J19) $$R \text{ is a UFD} \Rightarrow R[X_1, \dots, X_r] \text{ is a UFD}$$

and hence in particular

(J20) K is a field $\Rightarrow K[X_1,\ldots,X_r]$ is a UFD.

As a consequence of (J1) and (J20) it can be shown that:

(J21)
$$
\begin{cases}
\text{for any field } K \text{ and positive integers } r, m_1,\ldots,m_r, \\
\text{the } K\text{-homomorphism } \alpha : K[X_1,\ldots,X_r] \to K[T_1,\ldots,T_r] \\
\text{with } X_i \mapsto T_i^{m_i} \text{ is injective,} \\
\text{and by identifying } K(X_1,\ldots,X_r) \\
\text{with its image under the monomorphism} \\
\beta : K(X_1,\ldots,X_r) \to K(T_1,\ldots,T_m) \text{ induced by } \alpha, \\
\text{we have that } [K(T_1,\ldots,T_r) : K(X_1,\ldots,X_r)] = m_1\ldots m_r, \text{ and} \\
K[T_1,\ldots,T_r] \\
= \text{ the integral closure of } K[X_1,\ldots,X_r] \text{ in } K(T_1,\ldots,T_r).
\end{cases}
$$

§9: REMARKS

REMARK (R1). [**Factorization of Univariate Power Series**]. Recall that an ED is a domain R with a euclidean function, i.e., a function $\phi : R^\times \to \mathbb{N}$ such that for all x, y in $R^\times$ we have: (G1) $y \in xR \Rightarrow \phi(x) \le \phi(y)$, and (G2) $y \notin xR \Rightarrow y = qx + r$ for some $q \in R$ and $r \in R^\times$ with $\phi(r) < \phi(x)$; moreover, if the r (and hence also the q) is unique then we call R a special ED. We have noted that the univariate polynomial ring over a field is a special ED. Now we observe that the univariate power series ring $R = K[[X]]$ over a field K is even a simpler example of a special ED. To see this, it suffices to take $\phi(A) = \operatorname{ord}_X A$ for all $A = A(X) \in R^\times$.

It follows that: R is a PID and UFD with $U(R) = $ the set of all power series of zero order, and $X^i R$ with i varying in $\mathbb{N}$ are exactly all the distinct nonzero ideals in R. Every nonzero nonunit element in R is an associate of a unique power X^i of X with integer $i > 0$. Every irreducible element in R is an associate of X. Every nonzero element $h = h(X)$ in the quotient field $K((X))$ of R has a unique factorization $h = aX^e$ where $a \in U(R)$ and e is an integer; moreover $h \in R \Leftrightarrow e \ge 0$. The only nonzero prime ideal in R is XR and it is maximal. For any nonzero $f = f(X) \in R$ of some order $n > 0$, clearly $K \cap (fR) = \{0\}$ and hence we may identify K with a subring of the residue class ring $L = R/(fR)$, and then upon letting x to be the image of X under the residue class map $R \to L$ we see that $1, x, \ldots, x^{n-1}$ is vector space basis of L over K and hence $[L : K] = n$; moreover L is a field $\Leftrightarrow n = 1$.

For any $S \subset R$ we put $\operatorname{GCD}(S) = 0$ if $S \subset \{0\}$ and otherwise we put $\operatorname{GCD}(S) = $ the unique power X^i of X which is a generator of the ideal in R generated by S, and for any finite or infinite sequence $y_1, y_2, \ldots$ in R we put $\operatorname{GCD}(y_1, y_2, \ldots) = $

GCD($\{y_1, y_2, \dots\}$); note that then the GCD is obviously a gcd. Likewise, for any finite sequence $y_1, \dots, y_s$ of elements in $R^\times$ with $s > 0$, by LCM($y_1, \dots, y_s$) we denote the unique lcm of $y_1, \dots, y_s$ which is a power X^i of X. To make the reference to the ring R explicit we may write GCD$_R$ and LCM$_R$ instead of GCD and LCM respectively. Finally we note any $0 \neq \zeta \in K((X))$ has a unique reduced form $\zeta = \xi/X^i$ with $\xi \in R$ and $i \in \mathbb{N}$, and we call this the reduced form of ζ.

It follows that

(J22) $\qquad\qquad\qquad$ K is a field $\Rightarrow$ $K[[X]]$ is a UFD.

To justify the last sentence of (T4), as a consequence of (J1) and (J22) it can be shown that:

(J23) $\qquad \begin{cases} \text{for any field } K \text{ and positive integer } m, \text{ the } K\text{-homomorphism} \\ \alpha : K[[X]] \to K[[T]] \text{ with } X \mapsto T^m \text{ is injective and,} \\ \text{by identifying } K((X)) \text{ with its image under the monomorphism} \\ \beta : K((X)) \to K((T)) \text{ induced by } \alpha, \\ \text{we have that } [K((T)) : K((X))] = m \\ \text{and } K[[T]] = \text{the integral closure of } K[[X]] \text{ in } K((T)). \end{cases}$

REMARK (R2). [**Kronecker's Theorem**]. Later on we shall prove the converse of (J14) saying that [cf. L4§12(R7)]

(J24) $\qquad\qquad\qquad$ every normal domain is intval.

By (J13) and (J24) we get KRONECKER'S THEOREM which says that

(J25) $\qquad\qquad\qquad$ every normal domain is overnormal.

Note that by (J13) and (J16) we already get the special case of (J25) saying that

(J26) $\qquad\qquad\qquad$ every UFD is overnormal.

§10: ADVICE TO THE READER

As I have said before, an elementary treatment of many of the things said above, can be found in my 1990 book [A04] entitled Algebraic Geometry for Scientists and Engineers. Although this Engineering Book is not a logical part of the present book, I recommend it as collateral reading.

§11: HENSEL AND WEIERSTRASS

We now turn to Hensel's Lemma for more variables, and at the same time we shall deal with the very versatile WPT = Weierstrass Preparation Theorem discovered by Weierstrass around 1870. As a companion to WPT we shall also prove the equally versatile WDT= Weierstrass Division Theorem. Before stating these we note that, following the treatment given in my Local Analytic Geometry book [A06], here our scheme of proof will be WDT $\Rightarrow$ WPT $\Rightarrow$ Hensel, all for any number of variables. For a comparison between Hensel's Lemma and WPT, and for a proof of bivariate WPT similar to the proof of Basic Hensel's Lemma (T2) given above, see pages 119-124 of my Engineering Book [A04].

To get ready for stating WDT and WPT, let $X_1, \ldots, X_m, Y$ be $m+1$ variables over a field k where m is any nonnegative integer, and consider the power series ring

$$S = k[[X_1, \ldots, X_m, Y]] = R[[Y]] \quad \text{with} \quad R = k[[X_1, \ldots, X_m]].$$

As we have seen before, $\delta = \delta(X_1, \ldots, X_m, Y) \in S$ is a unit in S iff $\delta(0, \ldots, 0) \neq 0$. Likewise $\epsilon = \epsilon(X_1, \ldots, X_m) \in R$ is a unit in R iff $\epsilon(0, \ldots, 0) \neq 0$. Equivalently $h = h(X_1, \ldots, X_m) \in R$ is a nonunit in R iff $h(0, \ldots, 0) = 0$, i.e., iff $h \in M(R)$ where $M(R)$ is the ideal in R generated by $X_1, \ldots, X_m$. Note that $M(R)$ is the kernel of the epimorphism $R \to k$ given by $h \mapsto h(0, \ldots, 0)$ and hence $M(R)$ is a maximal ideal in R. Since $M(R)$ is the ideal of all nonunits, it is the unique maximal ideal in R; see observation (1$^\bullet$) below. We define the Weierstrass degree of any $f = f(X_1, \ldots, X_m, Y) \in S$ relative to (Y, R) by putting

$$\text{wideg}(f) = \text{wideg}_{Y,R} f = \text{ord}_Y f(0, \ldots, 0, Y).$$

If $f^* \in S$ is of the form

$$f^* = Y^d + \sum_{1 \leq j \leq d} f_j^* Y^{d-j} \text{ with } f_j^* \in M(R) \text{ for } 1 \leq j \leq d \in \mathbb{N}$$

then we say that f^* is distinguished or, in greater detail, we say that f^* is a distinguished polynomial (of degree d) in Y over R. Note that then

$$d = \deg_Y f^* = \text{wideg}_{Y,R} f^*$$

with the understanding that for any ring R and any power series $A = \sum_{i \in \mathbb{N}} A_i Y^i \in R[[Y]]$ with $A_i \in R$ we put

$$\deg_Y A = \max \text{Supp}_Y A$$

where by convention the said max equals $-\infty$ iff the Supp is empty, i.e., iff $A = 0$, and similarly the said max equals ∞ iff the Supp is an infinite set, i.e., iff $A \notin R[Y]$. Recall that elements f, g in a ring R are said to be associates in R if for some elements f', g' in R we have $f = gg'$ and $g = ff'$.

Now let us state WDT and WPT together with some supplements.

SUPPLEMENTED WEIERSTRASS PREPARATION THEOREM (T5). Let $X_1, \ldots, X_m, Y$ be $m+1$ variables over a field k where m is any nonnegative integer, and let $S = k[[X_1, \ldots, X_m, Y]] = R[[Y]]$ with $R = k[[X_1, \ldots, X_m]]$. Then we have the following.

WDT (T5.1). Given any $g \in S$ and any $f \in S$ with $\mathrm{wideg}(f) = d \in \mathbb{N}$ we can uniquely write $g = qf + r$ where $q \in S$ and $r \in R[Y]$ with $\deg_Y r < d$.

WPT (T5.2). Given any $f \in S$ with $\mathrm{wideg}(f) = d \in \mathbb{N}$ we can uniquely write $f = \delta f^*$ where δ is a unit in S and f^* is a distinguished polynomial of degree d in Y over R.

SUPPLEMENT (T5.3). If $g \in R[Y]$ and $f \in S$ is distinguished then the equation $g = qf + r$ in (T5.1) is the division identity in $R[Y]$, i.e., $q \in R[Y]$. Moreover, if g is also distinguished and $r = 0$ then q is distinguished.

SUPPLEMENT (T5.4). If f, g in S are distinguished then: f and g are associates in $S \Leftrightarrow f = g$.

SUPPLEMENT (T5.5). If $f \in S$ is distinguished and $f = F_1 \ldots F_e$ with $F_1, \ldots, F_e$ in S then, for $1 \le i \le e$, F_i is an associate in S of a distinguished $F_i^* \in S$, and we have $f = F_1^* \ldots F_e^*$.

We shall deduce (T5) from a more general version of it. To start this off, we observe that the above rings S and R are complete Hausdorff quasilocal rings in the following sense.

A quasilocal ring is a ring R with a unique maximal ideal, which we denote by $M(R)$. It can easily be seen that

$(1^\bullet)$ $\quad\begin{cases} \text{a ring is quasilocal iff all the nonunits in it form an ideal;} \\ \text{this ideal is then the unique maximal ideal in that ring.} \end{cases}$

As a consequence of $(1^\bullet)$ and the valuation axioms L2(V1) to L2(V3) it can be seen that

$(2^\bullet)$ $\quad\begin{cases} \text{the valuation ring } R_v \text{ of any valuation } v \text{ of a field } K \\ \text{is a quasilocal ring with } M(R_v) = \{x \in K : v(x) > 0\}. \end{cases}$

For ideals B, C in any ring R, the product BC is the ideal in R defined by $BC = $ the set of all finite sums $\sum b_i c_i$ with $b_i \in B$ and $c_i \in C$. Similarly for the product $B_1 \ldots B_n$ of a finite number of ideals $B_1, \ldots, B_n$ in R. If $B_1 = \cdots = B_n$ then we write B^n for $B_1 \ldots B_n$. In particular $B^0 = R$. Note that products and sums of ideals are connected by the distributive law: $(B_1 + \cdots + B_n)C = B_1 C + \cdots + B_n C$.

Let R be a quasilocal ring. R is Hausdorff means $\cap_{i \in \mathbb{N}} M(R)^i = 0$. For any $h \in R$ we define the R-order of h by putting

$$\mathrm{ord}_R h = \max\{i \in \mathbb{N} : h \in M(R)^i\}$$

and we note that then R is Hausdorff means for any $h \in R$ we have: $\operatorname{ord}_R h = \infty \Leftrightarrow h = 0$. A sequence $x = (x_i)_{1 \leq i < \infty}$ in R is Cauchy means for every $E \in \mathbb{N}$ there exists $N_E \in \mathbb{N}$ such that for all $i > N_E$ and $j > N_E$ we have $x_i - x_j \in M(R)^E$. The sequence x is convergent means it converges to a limit $\xi \in R$, i.e., for every $E \in \mathbb{N}$ there exists $N_E \in \mathbb{N}$ such that for all $i > N_E$ we have $\xi - x_i \in M(R)^E$; we indicate this by some standard notation such as $x_i \to \xi$ as $i \to \infty$ or $\lim_{i \to \infty} x_i = \xi$. If R is Hausdorff then a limit when it exists is unique. R is complete means every Cauchy sequence in it has a limit in it. If R is a complete Hausdorff quasilocal ring and $y_j \in M(R)^{u(j)}$ with $u(j) \to \infty$ as $j \to \infty$ then a sequence like $\sum_{i' \leq j \leq i} y_j$ has a unique limit in R which we denote by $\sum_{j \geq i'} y_j$.

Note that if R is a quasilocal ring then the univariate power series ring $S = R[[Y]]$ is a quasilocal ring with $M(S) = M(R)S + YS$, and if R is complete and Hausdorff then so is S.

Clearly the rings S and R of (T5) are complete Hausdorff quasilocal rings. In particular R is Hausdorff because for any $h \in R$ we have $\operatorname{ord}_R h = \operatorname{ord}_{(X_1,\ldots,X_m)} h$. The above power series rings are actually local rings in the following sense.

If all ideals in a ring are finitely generated then we call it a noetherian ring. We shall eventually give a proof of the famous Hilbert Basis Theorem which says that a polynomial ring over a noetherian ring is noetherian, and then from this, by using WPT, we shall deduce that a power series ring is also noetherian. By a local ring we mean a noetherian quasilocal ring. Eventually we shall prove the Krull Intersection Theorem which says that every local ring is Hausdorff.

For $S = R[[Y]]$ with R quasilocal, the above concepts of wideg and distinguished apply except that now for any $f = \sum_{i \geq 0} f_i Y^i \in S$ with $f_i \in R$ we have

$$\operatorname{wideg}(f) = \operatorname{wideg}_{Y,R} f = \operatorname{ord}_Y \psi(f)$$

where $\psi : S \to (R/M(R))[[Y]]$ is the natural map given by $\psi(f) = \sum_{i \geq 0} \phi(f_i) Y^i$ with $\phi : R \to R/M(R)$ being the residue class epimorphism. Note that then

(3•) f is distinguished $\Leftrightarrow \deg(f) = \operatorname{wideg}(f) < \infty$ and f is monic in Y.

Also note that for any f, f' in S we have

(4•) $\deg(ff') = \deg(f) + \deg(f')$

and

(5•) $\operatorname{wideg}(ff') = \operatorname{wideg}(f) + \operatorname{wideg}(f')$.

Finally note that $\delta = \sum_{i \geq 0} \delta_i Y^i \in S$ with $\delta_i \in R$ is a unit in S iff δ_0 is a unit in R.

Now let us state the said generalization of (T5).

ABSTRACT WEIERSTRASS PREPARATION THEOREM (T6). Let R be a complete Hausdorff quasilocal ring and consider the univariate power series ring $S = R[[Y]]$. Then we have the following.

(T6.1). Given any $g \in S$ and any $f \in S$ with wideg$(f) = d \in \mathbb{N}$ we can uniquely write $g = qf + r$ where $q \in S$ and $r \in R[Y]$ with $\deg_Y r < d$.

(T6.2). Given any $f \in S$ with wideg$(f) = d \in \mathbb{N}$ we can uniquely write $f = \delta f^*$ with δ a unit in S and f^* a distinguished polynomial of degree d in Y over R.

(T6.3). If $g \in R[Y]$ and $f \in S$ is distinguished then the equation $g = qf + r$ in (T6.1) is the division identity in $R[Y]$, i.e., $q \in R[Y]$. Moreover, if g is also distinguished and $r = 0$ then q is distinguished.

(T6.4). If f, g in S are distinguished then: f and g are associates in $S \Leftrightarrow f = g$.

(T6.5). If $f \in S$ is distinguished and $f = F_1 \ldots F_e$ with $F_1, \ldots, F_e$ in S then, for $1 \le i \le e$, F_i is an associate in S of a distinguished $F_i^* \in S$, and we have $f = F_1^* \ldots F_e^*$.

Items (T6.1) to (T6.5) are identical with items (T5.1) to (T5.5) respectively, and hence (T5) follows from (T6). To prove (T6.1) we write

$$f = \sum_{i \ge 0} f_i Y^i \quad \text{with} \quad f_i \in R$$

and upon letting

$$\overline{f} = \sum_{i \ge d} f_i Y^{i-d} \quad \text{and} \quad \widetilde{f} = \sum_{0 \le i < d} f_i Y^i$$

we see that $\overline{f}$ is a unit in S and hence upon letting

$$G = g \quad \text{and} \quad F = -\widetilde{f}/\overline{f} \quad \text{and} \quad Q = q\overline{f}$$

we see that

$$qf = (Y^d - F)Q$$

and

$(\bullet)$
$$\begin{cases} G = \sum_{i \ge 0} G_i Y^i \text{ with } G_i \in R \text{ for all } i \\ \text{and} \\ F = \sum_{i \ge 0} F_i Y^i \text{ with } F_i \in M(R) \text{ for all } i \end{cases}$$

and therefore (T6.1) is equivalent to the following:

LEMMA (T7). Let R be a complete Hausdorff quasilocal ring and consider the univariate power series ring $S = R[[Y]]$. Let G, F in S be as in $(\bullet)$, and let d be any nonnegative integer. Then $\deg_Y[G - (Y^d - F)Q] < d$ for a unique $Q \in S$.

Moreover for this Q we have $Q = \sum_{u \geq 0} Q^{(u)}$ where $Q^{(u)} \in M(R)^u[[Y]]$ are given by the recursive formulas

$$(\bullet\bullet) \quad \begin{cases} Q_e^{(0)} = G_{d+e} & \text{for all } e \geq 0; \\ Q_e^{(u+1)} = \sum_{0 \leq j \leq d+e} F_j Q_{d+e-j}^{(u)} & \text{for all } u \geq 0 \text{ and } e \geq 0; \\ Q^{(u)} = \sum_{e \geq 0} Q_e^{(u)} Y^e & \text{for all } u \geq 0. \end{cases}$$

PROOF OF (T7). To prove uniqueness it suffices to show that

$$\deg_Y (Y^d - F)Q < d \text{ for } Q \in S \Rightarrow Q = 0.$$

If for some $u \geq 0$ we have $Q \in M(R)^u[[Y]]$, i.e., upon letting $Q = \sum_{i \geq 0} Q_i Y^i$ with $Q_i \in R$ we have $Q_i \in M(R)^u$ for all i, then for any $e \geq 0$, by the above degree condition, we get

$$0 = \text{coeff of } Y^{d+e} \text{ in } (Y^d - F)Q = Q_e + \text{terms in } M(R)^{u+1}$$

and hence $Q_e \in M(R)^{u+1}$. Thus by induction on u we can conclude that for all $e \geq 0$ we have $Q_e \in M(R)^u$ for all $u \geq 0$ and hence $Q_e = 0$. Therefore $Q = 0$.

To prove existence we try to find $Q^{(u)} \in M(R)^u[[Y]]$ such that for $Q = \sum_{u \geq 0} Q^{(u)}$ we have $\deg_Y[G - (Y^d - F)Q] < d$, i.e.,

$$\deg_Y \left[G - \left(\sum_{u \geq 0} Q^{(u)} \right) Y^d + \left(\sum_{u \geq 0} Q^{(u)} \right) F \right] < d.$$

Solving $\deg_Y[G - Q^{(0)}Y^d] < d$ we get $Q^{(0)} = \sum_{e \geq 0} Q_e^{(0)} Y^e$ with $Q_e^{(0)} = G_{d+e}$ for all $e \geq 0$. It remains to find $Q^{(u)} \in M(R)^u[[Y]]$ for all $u \geq 1$ such that

$$\deg_Y \left[\left(\sum_{u \geq 1} Q^{(u)} \right) Y^d - \left(\sum_{u \geq 0} Q^{(u)} \right) F \right] < d.$$

Since $F \in M(R)[[Y]]$, we can successively find $Q^{(u)}$ by "solving modulo $M(R)^u$." Upon doing this by recursively letting $Q_e^{(u+1)} = \sum_{0 \leq j \leq d+e} F_j Q_{d+e-j}^{(u)}$ for all $u \geq 0$ and $e \geq 0$, we see that the above degree inequality is satisfied by taking $Q^{(u)} = \sum_{e \geq 0} Q_e^{(u)} Y^e$ for all $u \geq 0$.

PROOF OF (T6). As said above, (T6.1) is equivalent to (T7).

The first assertion in (T6.3) follows from the uniqueness part of (T6.1) and the second assertion follows from $(3^\bullet)$ to $(5^\bullet)$.

To prove (T6.4) let f, g in S be distinguished, and assume that they are associates in S. Upon relabelling f, g we may suppose that $\text{wideg}(g) = e \leq d = \text{wideg}(f)$. Let $\psi : S \to (R/M(R))[[Y]]$ be the natural map whose restriction to R is the residue class epimorphism $R \to R/M(R)$. Since f and g are associates we have $g = ff'$ for some $f' \in S$. Applying ψ to the equation $g = ff'$ we get the equation

$Y^e = Y^d \psi(f')$ in $(R/M(R))[[Y]]$ and hence $e = d$. Now $g - f = f(f' - 1)$ with $\deg_Y(g - f) < d$ and hence by the uniqueness part of (T6.1) we get $g - f = 0$.

To prove (T6.2) let there be given $f \in S$ with $\mathrm{wideg}(f) = d \in \mathbb{N}$. Taking Y^d for g in (T6.1) we get $Y^d = qf + r$ where $q \in S$ and $r \in R[Y]$ with $\deg_Y r < d$. By applying the above map ψ to the equation $Y^d = qf + r$ we see that $r \in M(R)[Y]$ and q is a unit in S. Let $f^* = Y^d - r$ and $\delta = 1/q$. Then f^* is distinguished and $f = \delta f^*$. Uniqueness follows from (T6.4).

In view of ($3^{\bullet}$) to ($5^{\bullet}$), (T6.5) follows from (T6.2).

This completes the proof of (T6).

As said above, (T5) follows by taking $R = k[[X_1, \ldots, X_m]]$ in (T6).

Next we come to:

ABSTRACT HENSEL'S LEMMA (T8). Let $R = k[[X_1, \ldots, X_m]]$ where m is any nonnegative integer and k is an algebraically closed field, or more generally let R be any complete Hausdorff quasilocal ring whose residue field $R/M(R)$ is algebraically closed. Let $\phi : R \to R/M(R)$ be the residue class epimorphism and let $\overline{\phi} : R[Y] \to \phi(R)[Y]$ be defined by putting $\overline{\phi}(f) = \sum_{i \geq 0} \phi(f_i) Y^i$ for all $f = \sum_{i \geq 0} f_i Y^i \in R[Y]$ with $f_i \in R$. Let

$$F(Y) = Y^n + \sum_{1 \leq j \leq n} A_j Y^{n-j} \quad \text{with} \quad n \in \mathbb{N}_+ \quad \text{and} \quad A_j \in R.$$

Then we have the following.

(T8.1) Let $a_1, \ldots, a_h$ be the distinct roots of $\overline{\phi}(F(Y))$ in $\phi(R)$ and let $e_1, \ldots, e_h$ be their multiplicities, i.e.,

$$\overline{\phi}(F(Y)) = \prod_{1 \leq i \leq h} (Y - a_i)^{e_i}.$$

Then there exist unique monic polynomials $F_1(Y), \ldots, F_h(Y)$ of degrees $e_1, \ldots, e_n$ in Y over R such that $F(Y) = F_1(Y) \ldots F_h(Y)$ and $\overline{\phi}(F_i(Y)) = (Y - a_i)^{e_i}$ for $1 \leq i \leq h$.

(T8.2) If $\overline{\phi}(F(Y)) = \overline{G}(Y) \overline{H}(Y)$ where $\overline{G}(Y)$ and $\overline{H}(Y)$ are coprime monic polynomials of positive degrees r and s in Y with coefficients in $\phi(R)$, then there exist unique monic polynomials $G(Y)$ and $H(Y)$ of degrees r and s in Y with coefficients in R such that $F(Y) = G(Y)H(Y)$ with $\overline{\phi}(G(Y)) = \overline{G}(Y)$ and $\overline{\phi}(H(Y)) = \overline{H}(Y)$.

PROOF. By (T6.2) and (T6.3) we see that

(T8.3)   (T8.2) is true when $\overline{H}(Y) = Y^s$.

Now it is easy to see that (T8.3) $\Rightarrow$ (T8.1) $\Rightarrow$ (T8.2).

§12: DEFINITIONS AND EXERCISES

DEFINITION (D1). [**Minors and Cofactors**]. Consider an $m \times n$ matrix

$$A = \begin{pmatrix} A_{11} & \cdots & A_{1j} & \cdots & A_{1n} \\ \cdots & \cdots & \cdots & \cdots & \cdots \\ \cdots & \cdots & \cdots & \cdots & \cdots \\ A_{i1} & \cdots & A_{ij} & \cdots & A_{in} \\ \cdots & \cdots & \cdots & \cdots & \cdots \\ \cdots & \cdots & \cdots & \cdots & \cdots \\ A_{m1} & \cdots & A_{mj} & \cdots & A_{mn} \end{pmatrix}$$

over a ring R. By the $r \times s$ submatrix of A located at the $(u_1, \ldots, u_r; v_1, \ldots, v_s)$-th spots, where $r \geq 0$ and $s \geq 0$ with $1 \leq u_1 < \cdots < u_r \leq m$ and $1 \leq v_1 < \cdots < v_s \leq n$ are integers, we mean the $r \times s$ matrix $B = (B_{pq})$ over R given by $B_{pq} = A_{u_p v_q}$; if $r = s$ then we may call B a square submatrix of A of order r. By the rank of A, denoted by $\mathrm{rk}(A)$, we mean the largest order of its square submatrix whose determinant is a unit in R. By the row rank of A, denoted by $\mathrm{rrk}(A)$, we mean max r such that A has an $r \times n$ submatrix of rank r. By the column rank of A, denoted by $\mathrm{crk}(A)$, we mean max s such that A has an $m \times s$ submatrix of rank s. If A is a square matrix, i.e., if $m = n$, then for $1 \leq i \leq n$ and $1 \leq j \leq n$, by the (i,j)-th square submatrix of A we mean the $(n-1) \times (n-1)$ square submatrix of A obtained by deleting the i-th row and j-th column, i.e., located at the $(1, \ldots, i-1, i+1, \ldots, n; 1, \ldots, j-1, j+1, \ldots, n)$-th spots, and $(-1)^{i+j}$ times its determinant is called the (i,j)-th cofactor of A. The term cofactor is motivated by (E1) below. The word minor is used as a synonym for the determinant of a square submatrix. Thus, by a minor of A of order r we mean the determinant of a square submatrix of A of order r. Likewise, if $m = n$ then the (i,j)-th cofactor of A is $(-1)^{i+j}$ times the (i,j)-th minor of A.

EXERCISE (E1). Show that the determinant of an $n \times n$ matrix $A = (A_{ij})$ over a ring R can be calculated by expanding in terms of any row or any column, i.e., show that for $1 \leq i \leq n$ and $1 \leq j \leq n$ we have $\det(A) = \sum_{1 \leq j \leq n} A_{ij} C_{ij} = \sum_{1 \leq i \leq n} A_{ij} C_{ij}$ where C_{ij} is the (i,j)-th cofactor of A.

EXERCISE (E2). Verify the product rules for matrices and determinants, i.e, prove claims (M1) and (M3). Hint: do these things "directly"; a "tricky" proof of (M3) may be given in terms of "block matrices" and "Laplace development."

DEFINITION (D2). [**Transpose and Adjoint**]. The transpose of the $m \times n$ matrix A over a ring R depicted in (D1) is the $n \times m$ matrix obtained by interchanging rows and columns, i.e., if we denote the transpose by $A' = (A'_{ij})$ then $A'_{ij} = A_{ji}$. If $m = n$ then by the adjoint of A we mean the $n \times n$ matrix which is the transpose of the matrix of cofactors of A, i.e., if we denote the adjoint of A by $A^* = (A_{ij}^*)$ and the (i,j)-th cofactor of A by C_{ij} then $A_{ij}^* = C_{ji}$. For elements $a_1, \ldots, a_n$ in R, by $\text{diag}_n(a_1, \ldots, a_n)$ or $\text{diag}(a_1, \ldots, a_n)$ we denote the $n \times n$ diagonal matrix whose (i,j)-th entry is a_i or 0 according as $i = j$ or $i \neq j$. By the $n \times n$ identity matrix we mean the matrix $\text{diag}_n(1, \ldots, 1)$; we may sometimes denote this by 1_n. For the $m \times n$ matrix $A = (A_{ij})$ and any $r \in R$, by rA we denote the $m \times n$ matrix whose (i,j)-th entry is rA_{ij}. This makes the set $\text{MT}(m \times n, R)$ of all $m \times n$ matrices over R into an R-module. For a field k, it converts $\text{MT}(m \times n, k)$ into an mn dimensional vector space over k.

EXERCISE (E3). Show that units in the group of $n \times n$ matrices over a ring R are those matrices whose determinants are units in R, i.e., prove claim (M4). Hint: for an $n \times n$ matrix $A = (A_{ij})$ over R, show that $AA^* = A^*A = \det(A)1_n$ where A^* is the adjoint of A and 1_n is the $n \times n$ identity matrix.

EXERCISE (E4). Show that for $n \times n$ matrices over a field k with positive integer n, the nonzero scalar matrices form a normal subgroup of the group of all matrices of nonzero determinant, i.e., prove claim (M5).

EXERCISE (E5). Prove Euler's Theorem on Homogeneous Polynomials, i.e., show that for any homogeneous polynomial $H = H(Z_1, \ldots, Z_N)$ of degree d in a finite number of variables $Z_1, \ldots, Z_N$ over a field k we have $\sum_{1 \leq i \leq N} Z_i H_{Z_i} = dH$ where H_{Z_i} is the partial derivative of H with respect to Z_i.

EXERCISE (E6). Given any nonzero polynomial $G(Z_1, \ldots, Z_N)$ in variables $Z_1, \ldots, Z_N$ over an infinite field k, where N is a positive integer, show that there exists $(a_1, \ldots, a_N) \in k^N$ such that $G(a_1, \ldots, a_N) \neq 0$.

EXERCISE (E7). In §11, prove characterization $(1^\bullet)$ of quasilocal rings, and from it deduce claim $(2^\bullet)$ about valuation rings.

DEFINITION (D3). [**Matrices and Linear Maps**]. For any nonnegative integer n and any module V over any ring R, by componentwise addition and scalar multiplication, V^n becomes an R-module. In particular R^n is an R-module. An R-module U is finitely generated, or is a finite R-module, means there exists a surjective R-linear map $R^n \to U$ for some nonnegative integer n. Thinking of members of R^n as column vectors, let us identify R^n with $\text{MT}(n \times 1, R)$. So the $m \times n$ matrix $A = (A_{ij})$ depicted in (D1) corresponds to the R-linear map $\phi_A : R^n \to R^m$ given by $\phi_A(y) = Ay$ for all $y \in R^n = \text{MT}(n \times 1, R)$. Now $A \mapsto \phi_A$ gives a bijection of $\text{MT}(m \times n, R)$ onto the set of all R-linear maps $R^n \to R^m$. Moreover, for any $B \in \text{MT}(n \times r, R)$ we have $\phi_A\phi_B = \phi_{AB}$. In case R is a field k, we have $\text{rrk}(A) = \dim_k \text{im}(\phi_A) = \text{crk}(A) = n - \dim_k \ker(\phi_A)$. In particular if R is a field k and $m = n$ then: ϕ_A is a linear k-automorphism of $R^n \Leftrightarrow \det(A) \neq 0$.

DEFINITION (D4). [**Linear and Polynomial Automorphisms**]. Consider the rings $P = k[X_1, \ldots, X_n] \subset k[[X_1, \ldots, X_n]] = S$ where $X_1, \ldots, X_n$ are a finite number of variables over a field k. Let $A = (A_{ij})$ be the $m \times n$ matrix depicted in (D1) where we take $R = k$ and $m = n$. Let $N = n^2$ and let $(a_1, \ldots, a_N) \in k^N$ be given by $a_{(i-1)n+j} = A_{ij}$. Let $F'(Z_1, \ldots, Z_N) \in T = k[Z_1, \ldots, Z_N]$ be equal to $\det(Z)$ where $Z = (Z_{ij})$ is an $n \times n$ matrix of indeterminates and $Z_{(i-1)n+j} = Z_{ij}$.

Assuming $\det(A) \neq 0$, i.e., $F'(a_1, \ldots, a_N) \neq 0$, we get a unique k-automorphism $\sigma_A : P \to P$ such that $\sigma_A(X_i) = \sum_{1 \leq j \leq n} A_{ij} X_j$ for $1 \leq i \leq n$. Clearly $\sigma_A(P_+) = P_+$ where $P_+ = \{f \in P : \mathrm{ord}_S f > 0\}$. Hence by (E8) below, σ_A uniquely extends to an automorphism $\tau_A : S \to S$. We call τ_A a k-linear automorphism of S. If A can be chosen so that for a given nonzero $F^*(Z_1, \ldots, Z_N) \in T$ we have $F^*(a_1, \ldots, a_N) \neq 0$ then we say that τ_A is a generic k-linear automorphism of S.

Given any $0 \neq F \in S$, upon letting $\mathrm{ord}_S F = \nu$, we have that ν is a nonnegative integer and we can uniquely write

$$F = F(X_1, \ldots, X_n) = F_\nu(X_1, \ldots, X_n) + (\text{terms of degree} > \nu)$$

where $F_\nu = F_\nu(X_1, \ldots, X_n) \in P$ is a nonzero homogeneous polynomial of degree ν. We call F_ν the initial form of F relative to S or relative to $(X_1, \ldots, X_n)$, and denote it by $\mathrm{info}_S F$ or $\mathrm{info}_{(X_1, \ldots, X_n)} F$; we also put $\mathrm{info}_S 0 = 0$. Geometrically speaking $F_\nu = 0$ is the tangent cone of the analytic hypersurface $F = 0$ at $(0, \ldots, 0)$. Now suppose $n > 0$. If for $F''(Z_1, \ldots, Z_N) = F_\nu(Z_{1n}, \ldots, Z_{nn})$ we have $F''(a_1, \ldots, a_N) \neq 0$ then clearly

$$\tau_A(F)(X_1, \ldots, X_n) = \sigma_A(F_\nu)(X_1, \ldots, X_n) + (\text{terms of degree} > \nu)$$

where $\sigma_A(F_\nu)(X_1, \ldots, X_n) \in P$ is a nonzero homogeneous polynomial of degree ν with $\sigma_A(F_\nu)(0, \ldots, 0, 1) \neq 0$, and hence $\mathrm{ord}_S \tau_A(F) = \nu$ with $\mathrm{info}_S \tau_A(F) = \sigma_A(F_\nu)$ and upon writing $(X_1, \ldots, X_m, Y)$ for $(X_1, \ldots, X_n)$ we get $\mathrm{wideg}(\tau_A(F)) = \nu$. If k is infinite then by taking $G = F^* F' F''$ and applying (E6) we will have found a generic k-linear automorphism $\tau_A : S \to S$ such that $\mathrm{wideg}(\tau_A(F)) = \nu$.

Without assuming k to be infinite, we claim that we can always find a polynomial k-automorphism $\tau : S \to S$ such that $\mathrm{wideg}(\tau(F)) < \infty$, where an automorphism $\tau : S \to S$ is called a polynomial k-automorphism if it is a k-automorphism with $\tau(P) = P$; note that then we must have $\tau(P_+) = P_+$. To see this, for any $E = (e_1, \ldots, e_n) \in \mathbb{N}_+^n$ with $e_n = 1$ let $\sigma_E : P \to P$ be the unique k-automorphism such that $\sigma_E(X_i) = X_i + X_n^{e_i}$ for $1 \leq i \leq n-1$ and $\sigma_E(X_n) = X_n$. By (E8) this uniquely extends to an automorphism $\tau_E : S \to S$. Writing

$$F = F(X_1, \ldots, X_n) = \sum_{i \in \mathbb{N}^n} F_i X^i$$

with $F_i \in k$ and $X^i = X_1^{i_1} \ldots X_n^{i_n}$, we put $\mathrm{Supp}(F) = \{i \in \mathbb{N}^n : F_i \neq 0\}$ and call this the support of F. Note that $\mathrm{Supp}(F)$ is nonempty because we assumed $F \neq 0$; also note that if $F \in P$ then $\mathrm{Supp}(F)$ is finite. Let us lexicographically order $\mathbb{N}^n$ by saying that for $i \neq j$ in $\mathbb{N}^n$, upon letting t to be the smallest subscript with $i_t \neq j_t$ we have: $i < j \Leftrightarrow i_t < j_t$. In this ordering any nonempty set has a smallest element and any nonempty finite set has a largest element. So let r be the smallest element of $\mathrm{Supp}(F)$, and in case of $F \in P$ let s be the largest element of $\mathrm{Supp}(F)$. Upon letting deg and $\deg_{X_n}$ be the $(X_1, \ldots, X_n)$-degree and the X_n-degree respectively, by (E9) below we see that if

$$(*) \qquad \begin{cases} e_{n-u} > \sum_{0 \leq v < u} e_{n-v} r_{n-v} \\ \text{for } 1 \leq u \leq n-1 \end{cases}$$

then

$$(**) \qquad \begin{cases} \deg(\sigma_E(X^r)) = \deg_{X_n}(\sigma_E(X^r)) < \deg_{X_n}(\sigma_E(X^{r^*})) \\ \text{for all } r \neq r^* \in \mathrm{Supp}\ (F) \end{cases}$$

and hence $\mathrm{wideg}(\tau_E(F)) = \deg_{X_n}(\sigma_E(X^r)) = e_1 r_1 + \cdots + e_n r_n < \infty$, and in case of $F \subset P$, if

$$(') \qquad \begin{cases} e_{n-u} > \max_{s' \in \mathrm{Supp}(F)} \sum_{0 \leq v < u} e_{n-v}(s'_{n-v} - s_{n-v}) \\ \text{for } 1 \leq u \leq n-1 \end{cases}$$

then

$$('') \qquad \begin{cases} \deg(\sigma_E(X^s)) = \deg_{X_n}(\sigma_E(X^s)) > \deg_{X_n}(\sigma_E(X^{s'})) \\ \text{for all } s \neq s' \in \mathrm{Supp}\ (F) \end{cases}$$

and hence $\deg(\sigma_E(F)) = \deg_{X_n}(\sigma_E(F)) = \deg_{X_n}(\sigma_E(X^s)) = e_1 s_1 + \cdots + e_n s_n$.

Thus, in addition to finding E so that $\mathrm{wideg}(\tau_E(F)) < \infty$, in case of $F \in P$ we have also found E so that $\deg(\sigma_E(F)) = \deg_{X_n}(\sigma_E(F))$, i.e., $\sigma_E(F)$ is regular in X_n in the following sense. Given any $0 \neq F \in P$, upon letting $\deg(F) = \mu$, we have that μ is a nonnegative integer and we can uniquely write

$$F = F(X_1, \ldots, X_n) = F_\mu(X_1, \ldots, X_n) + (\text{terms of degree} < \mu)$$

where $F_\mu = F_\mu(X_1, \ldots, X_n) \in P$ is a nonzero homogeneous polynomial of degree μ. We call F_μ the degree form of F relative to P or relative to $(X_1, \ldots, X_n)$, and denote it by $\mathrm{defo}_P F$ or $\mathrm{defo}_{(X_1, \ldots, X_n)} F$; we also put $\mathrm{defo}_P 0 = 0$. F is regular in X_n means $\deg(F) = \deg_{X_n}(F)$, i.e., equivalently $F_\mu(0, \ldots, 0, 1) \neq 0$. Geometrically speaking $F_\mu = 0$ is the infinite portion of the hypersurface $F = 0$, and F is regular in X_n means the X_n-axis, i.e., the line $X_1 = \cdots = X_{n-1} = 0$ does not meet this portion. If k is infinite then by taking F_μ for F'' in the above argument, we can arrange that $\sigma_A(F)$ is regular in X_n and has the same degree as F.

All this can be applied to any finite family of polynomials or power series by letting F be the product of that family.

EXERCISE (E8). In (D4) show that any k-linear automorphism $\sigma : P \to P$ with $\sigma(P_+) = P_+$ can be uniquely extended to an automorphism $\tau : S \to S$.

EXERCISE (E9). In (D4) show that $(*) \Rightarrow (**)$, and show that $(') \Rightarrow ('')$.

EXERCISE (E10). Let $Y_1, \ldots, Y_n$ be a finite number of indeterminates over a ring R, and let I be the ideal in $S = R[[Y_1, \ldots, Y_n]]$ generated by $Y_1, \ldots, Y_n$. Show that then S is I-complete and I-Hausdorff according to (D5) below. Also show that if R is quasilocal then S is quasilocal with $M(S) = M(R)S + I$. Finally show that if R is complete Hausdorff quasilocal then so is S.

EXERCISE (E11). Give details for the proof of Abstract Hensel's Lemma (T8).

DEFINITION (D5). [**Hausdorff Modules and Bilinear Maps**]. Let us now generalize the notions of Hausdorff and complete from the case of quasilocal rings to more general situations. So let I be an ideal in a ring R and let V be an R-module. V is I-Hausdorff, or Hausdorff relative to I, means $\cap_{i \in \mathbb{N}} I^i V = 0$, where for any ideal J in R, by JV we denote the submodule of V given by $JV = \{\sum j_i v_i : j_i \in J \text{ and } v_i \in V \text{ (finite sums)}\}$. For any $h \in V$ we define the (R, I)-order of h by putting $\mathrm{ord}_{(R,I)} h = \max\{i \in \mathbb{N} : h \in I^i V\}$ and we note that then V is I-Hausdorff means for any $h \in V$ we have: $\mathrm{ord}_{(R,I)} h = \infty \Leftrightarrow h = 0$. A sequence $x = (x_i)_{1 \leq i < \infty}$ in V is I-Cauchy, or Cauchy relative to I, means for every $E \in \mathbb{N}$ there exists $N_E \in \mathbb{N}$ such that for all $i > N_E$ and $j > N_E$ we have $x_i - x_j \in I^E V$. The sequence x is I-convergent, or convergent relative to I, means it converges to an I-limit $\xi \in V$, i.e., for every $E \in \mathbb{N}$ there exists $N_E \in \mathbb{N}$ such that for all $i > N_E$ we have $\xi - x_i \in I^E V$; we indicate this by some standard notation such as $x_i \to \xi$ as $i \to \infty$ or $\lim_{i \to \infty} x_i = \xi$. If V is I-Hausdorff then a limit when it exists is unique. V is I-complete means every I-Cauchy sequence in it has a limit in it. If V is I-complete and I-Hausdorff and $y_j \in I^{u(j)} V$ with $u(j) \to \infty$ as $j \to \infty$ then a sequence like $\sum_{i' \leq j \leq i} y_j$ has a unique limit in V which we denote by $\sum_{j \geq i'} y_j$. If R is quasilocal with $I = M(R)$ and $V = R$ then, according to the previous definitions of the above notions, the adjective "relative to I" may be dropped.

Let U, V, W be modules over R. A map $f : V \times W \to U$ is R-bilinear means for every $v \in V$ the map $W \to U$ given by $w \mapsto f(v, w)$ is R-linear and for every $w \in W$ the map $V \to U$ given by $v \mapsto f(v, w)$ is R-linear. In a natural manner we can regard $U/(IU), V/(IV), W/(IW)$ as modules over the ring $R/(IR)$, and then in a natural manner f induces a (R/IR)-bilinear map $(V/(IV)) \times (W/(IW)) \to U/(IU)$.

The following three Exercises deal with the above set-up. In particular, in (E14) we remove the hypothesis of $R/M(R)$ being algebraically closed from (T8.2).

EXERCISE (E12). [**Completeness Lemma**]. Let I be an ideal in a ring R, let U, V, W be any R-modules, and let n be any positive integer.

(E12.1) Show that if V is I-Hausdorff then V^n is I-Hausdorff.

(E12.2) Show that if V is I-complete then V^n is I-complete.

(E12.3) Show that if (x_i) is an I-Cauchy sequence in V such that some subsequence $(x_{r(j)})$ has an I-limit x in V, then x is an I-limit of the sequence (x_i).

(E12.4) Show that for any R-linear map $\phi : W \to V$ we have $\phi(I^n W) \subset I^n V$ and if ϕ is surjective then we have $\phi(I^n W) = I^n V$.

(E12.5) Show that for any R-bilinear map $\theta : V \times W \to U$ we have $\theta(I^n V \times I^n W) \subset I^n U$.

(E12.6) Show that if W is I-complete and there exists a surjective R-linear map $W \to V$ then V is I-complete.

(E12.6*) Show that (E12.6) is not true with Hausdorff replacing complete.

(E12.7) Show that if V is a finite R-module and R is I-complete then V is I-complete.

HINT. (E12.1) to (E12.5) are straightforward. For (E12.6*) take $W = R = k[X]$ where X is an indeterminate over a field k, let $I = XR$, and let ϕ be the residue class epimorphism $R \to V = R/(X-1)R$. (E12.7) follows from (E12.2) and (E12.6). To prove (E12.6), let (x_i) be any Cauchy sequence in V. Then we can find positive integers $r(1) < r(2) < \ldots$ such that for all j, m, n in $\mathbb{N}_+$ with $m \geq r(j)$ and $n \geq r(j)$ we have $x_m - x_n \in I^j V$. Pick any $y_1 \in \phi^{-1}(x_1)$. For all $j > 1$, by (E12.4) there exists $y_j \in \phi^{-1}(\{x_{r(j)}\}) \cap (y_1 + I^j W)$. Clearly (y_j) is I-Cauchy in W and hence it has an I-limit η in W. Let $\xi = \phi(\eta)$. Then by (E12.4), ξ is an I-limit of $(x_{r(j)})$ in V, and hence by (E12.3), ξ is an I-limit of (x_i).

EXERCISE (E13). [**Hensel's Sublemma**]. Let I be an ideal in a ring R such that R is I-complete. Let U, V, W be finite R-modules such that U is I-Hausdorff. Let $\phi : R \to \overline{R} = R/I$, $\overline{\phi}_U : U \to \overline{U} = U/(IU)$, $\overline{\phi}_V : V \to \overline{V} = V/(IV)$, $\overline{\phi}_W : W \to \overline{W} = W/(IW)$ be the canonical epimorphisms. Let $\theta : V \times W \to U$ be an R-bilinear map and let $\overline{\theta} : \overline{V} \times \overline{W} \to \overline{U}$ be the $\overline{R}$-bilinear map induced by θ. Let there be given $F \in U$ with $\overline{G} \in \overline{V}$ and $\overline{H} \in \overline{W}$ such that (i) $\overline{\phi}_U(F) = \overline{\theta}(\overline{G}, \overline{H})$ and (ii) $\overline{U} = \overline{\theta}(\overline{G}, \overline{W}) + \overline{\theta}(\overline{V}, \overline{H})$, where as usual $\overline{\theta}(\overline{G}, \overline{W}) = \{\overline{\theta}(\overline{G}, \widetilde{W}) : \widetilde{W} \in \overline{W}\}$ and $\overline{\theta}(\overline{V}, \overline{H}) = \{\overline{\theta}(\widetilde{V}, \overline{H}) : \widetilde{V} \in \overline{V}\}$. Show that then there exist $G \in (\overline{\phi}_V)^{-1}(\{\overline{G}\})$ and $H \in (\overline{\phi}_W)^{-1}(\{\overline{H}\})$ such that $F = \theta(G, H)$.

HINT. Let $G_0 = H_0 = 0$. By induction on $l \geq 1$ we find (i*) $G_l \in (\overline{\phi}_V)^{-1}(\{\overline{G}\})$ and $H_l \in (\overline{\phi}_W)^{-1}(\{\overline{H}\})$ such that: (ii*) $G_l - G_{l-1} \in I^{l-1} V$ and $H_l - H_{l-1} \in I^{l-1} W$ with: (iii*) $F - \theta(G_l, H_l) \in I^l U$. By (i) we can do it for $l = 1$. So let $l > 1$ and assume we have done it for $l - 1$. Then $F - \theta(G_{l-1}, H_{l-1}) \in I^{l-1} U$ and hence $F - \theta(G_{l-1}, H_{l-1}) = \sum \xi_j x_j$ with $\xi_j \in I^{l-1}$ and $x_j \in U$ where the sum is finite. By (ii) we find $y_j \in V$ and $z_j \in W$ such that $x_j - \theta(G, z_j) - \theta(y_j, H) \in IU$. Now it suffices to take $G_l = G_{l-1} + \sum \xi_j y_j$ and $H_l = H_{l-1} + \sum \xi_j z_j$. In view of (ii*), by

(E12.7), (G_l) and (H_l) have I-limits G and H in V and W respectively. In view of (i*), by (E12.4) we see that $G \in (\overline{\phi}_V)^{-1}(\{\overline{G}\})$ and $H \in (\overline{\phi}_W)^{-1}(\{\overline{H}\})$. In view of (iii*), by (E12.5) we have $F = \theta(G, H)$.

EXERCISE (E14). [**Hensel's Lemma**]. Let R be a complete Hausdorff quasilocal ring. Let $\phi : R \to \overline{R} = R/M(R)$ be the residue class epimorphism and let $\overline{\phi} : R[Y] \to \overline{R}[Y]$ be the obvious extension of ϕ. Let

$$F(Y) = Y^n + \sum_{1 \le j \le n} A_j Y^{n-j} \quad \text{with} \quad n \in \mathbb{N}_+ \quad \text{and} \quad A_j \in R.$$

Assume that $\overline{\phi}(F(Y)) = \overline{G}(Y)\overline{H}(Y)$ where $\overline{G}(Y)$ and $\overline{H}(Y)$ are coprime monic polynomials of positive degrees r and s in Y with coefficients in $\overline{R}$. Show that then there exist monic polynomials $G(Y)$ and $H(Y)$ of degrees r and s in Y with coefficients in R such that $F(Y) = G(Y)H(Y)$ with $\overline{\phi}(G(Y)) = \overline{G}(Y)$ and $\overline{\phi}(H(Y)) = \overline{H}(Y)$.

HINT. Let U, V, W be the R-modules consisting of all members of $R[Y]$ of degrees $\le n, r, s$ respectively. Let $I = M(R)$. Let $\overline{\phi}_U : U \to \overline{U} = U/(IU)$, $\overline{\phi}_V : V \to \overline{V} = V/(IV)$, $\overline{\phi}_W : W \to \overline{W} = W/(IW)$ be the epimorphisms induced by $\overline{\phi}$. Let $\theta : V \times W \to U$ be the R-bilinear map given by multiplication, i.e., given by $(\widetilde{V}, \widetilde{W}) \mapsto \widetilde{V}\widetilde{W}$. Let $\overline{\theta} : \overline{V} \times \overline{W} \to \overline{U}$ be the $\overline{R}$-bilinear map induced by θ. In view of (E12.1), by (E13) we find $\widehat{G}(Y), \widehat{H}(Y)$ in $R[Y]$ of degree $\le r$ and $\le s$ respectively such that $\overline{\phi}(\widehat{G}(Y)) = \overline{G}(Y)$ and $\overline{\phi}(\widehat{H}(Y)) = \overline{H}(Y)$ with $F(Y) = \widehat{G}(Y)\widehat{H}(Y)$. It follows that the degree of $\widehat{G}(Y)$ is r and therefore for the coefficient G' of Y^r in it we have $\phi(G') = 1$, and the degree of $\widehat{H}(Y)$ is s and for the coefficient H' of Y^s in it we have $\phi(H') = 1$. Therefore G', H' are units in R with $G'H' = 1$. Now it suffices it to take $G(Y) = \widehat{G}(Y)/G'$ and $H(Y) = \widehat{H}(Y)/H'$.

EXERCISE (E15). Let D be a derivation of a field K. Let y, z be nonzero elements of K such that $z^r = y$ where r is a rational number. Show that then $D(y) = rz^{r-1}D(z)$.

OBSERVATION (O1). I am not saying that given y and r this defines z uniquely. The matter is made precise by saying that for some way of writing $r = m/n$ with $m \in \mathbb{Z}$ and $n \in \mathbb{N}_+$, and for some $x \in K$ with $y = x^m$, we have $z = x^n$, and then interpreting the equation $D(y) = rz^{r-1}D(z)$ as the equation $zD(y) = ryD(z)$.

EXERCISE (E16). Prove Binomial Theorem (T1) for fractional exponents. In other words, for any rational number $r = m/n$ with $m \in \mathbb{Z}$ and $n \in \mathbb{N}_+$, upon letting

$$y(X) = 1 + \sum_{i>0} \frac{r(r-1)\dots(r-i+1)}{i!} X^i \in k[[X]]$$

where X is an indeterminate over a field k of characteristic zero, show that $y(X)$ is the unique element in $k[[X]]$ with $y(0) = 1$ such that

$$(1 + X)^r = y(X) \quad \text{or equivalently} \quad (1 + X)^m = y(X)^n.$$

HINT. By taking $F(X, Y) = Y^n - (1 + X)^m$ in Basic Hensel (T2) we see that there exists a unique

$$y(X) = 1 + \sum_{i>0} y_i X^i \in k[[X]] \quad \text{with} \quad y_i \in k$$

such that

$$(1 + X)^m = y(X)^n$$

i.e.,

$$(1 + X)^r = 1 + \sum_{i>0} y_i X^i \in k[[X]].$$

For any $i > 0$, in view of (E15) and (O1), differentiating both sides i times with respect to X and then putting $X = 0$ we get

$$r(r - 1) \ldots (r - i + 1) = i! y_i.$$

EXERCISE (E17) Prove Binomial Theorem for integral exponents. In other words, show that for any $n \in \mathbb{N}$ and any elements x, y in any ring R we have

$$(x + y)^n = \sum_{0 \le i \le n} \binom{n}{i} x^i y^{n-i}$$

where $\binom{n}{i}$ is the number of ways of choosing i things out of n things, i.e., the size of the set of all i size subsets of a set of size n. Also show that

$$\binom{n}{i} = \frac{n!}{i!(n - i)!} = \frac{n(n - 1) \ldots (n - i + 1)}{i!}.$$

HINT. In addition to the usual high-school proof, give a second proof using (E16).

OBSERVATION (O2). Newton first proved (E17), then he generalized it to (E16), and finally he generalized it to (T4).

EXERCISE (E18) Show that a univariate polynomial ring over a UFD is a UFD, i.e., complete the proof of (J18). Also complete the proofs of claims (J21) and (J23) about power series rings.

OBSERVATION (O3). From this lecture, it only remains to do Homework (H3) and to complete the proofs of claims (J1), (J2), and (J24).

§13: NOTES

NOTE (N1). [**Preparation for WPT**]. To take care of the finiteness of wideg in WPT, consider the power series ring $S = k[[X_1, \ldots, X_n]]$ in a finite number of variables $X_1, \ldots, X_n$ over a field k. Given $0 \neq F = F(X_1, \ldots, X_n) \in S$, we want to find an automorphism $\tau : S \to S$ such that after taking $(X_1, \ldots, X_m, Y)$ for $(X_1, \ldots, X_n)$ we would have wideg$(\tau(F)) < \infty$. If k is infinite then by (D4) and (E8) we can achieve this by means of a generic k-linear automorphism $\tau = \tau_A$ of S; geometrically speaking what we do is to rotate coordinates so that a generic line not on the tangent cone of $F = 0$ becomes the X_n-axis, i.e., becomes the line $X_1 = \cdots = X_{n-1} = 0$. Without assuming k to be infinite, by (D4) and (E9) we can always achieve it by means of a polynomial k-automorphism $\tau = \tau_E$ of S; what we do is that instead of varying the coefficients of the equations of the automorphism over the field, we suitably vary their exponents over $\mathbb{N}_+$ which is an infinite set.

NOTE (N2). [**Abhyankar's Proof of Newton's Theorem**]. What we called Shreedharacharya's Proof of Newton's Theorem, is also known as Abhyankar's Proof of Newton's Theorem; see pages 575 and 659 of volume II of Jacobson's Basic Algebra Book [Jac].

§14: CONCLUDING NOTE

Analytic geometry was initiated by Descartes in 1637. Descartes' work was pushed further, first in France (1760-1790) by Bézout, then in England (1840-1890) by Cayley, Sylvester, and Salmon, then in Germany (1860-1910) by Plücker, Klein, and Noether, and simultaneously in Italy (1860-1920) by Cremona, Castelnuovo, Enriques, and Severi, and finally in the United States (1930 onwards) by Zariski, Weil, Chevalley, and others. Somewhere along the line, in this long march, analytic geometry changed its name to algebraic geometry. By stressing synthetic arguments and augmenting ordinary space by inserting points at infinity, analytic geometry is metamorphosed into projective geometry. As said above, analytic geometry and projective geometry have mostly been eliminated from the college curriculum in the last forty years or so. But I was lucky. In my time I had four years of analytic geometry. Also I had two years of projective geometry. In the one year course in Bombay College, my text for projective geometry was the classic Pure Geometry book [Ask] of Askwith. This was followed up by the one year projective geometry course I had at Harvard with Zariski where our text was the equally famous book [VYo] of Veblen and Young. Out of all these illustrious names, whom shall we elect as the father of algebraic geometry? By reading the relevant material in Klein's famous Entwicklung der Mathematik book [Kle] of 1910 which is reconfirmed in Karen Parshall's recent book [PaR] on Sylvester's Correspondence, I am convinced that George Salmon of Dublin, Ireland, is the true father of algebraic geometry and his 1852 book [Sal] on Higher Plane Curves is the cradle of that subject. In any case, on page 17 of Askwith's book, you will find Salmon's generalization of the pole-polar property of a circle.

Lecture L4: Varieties and Models

§1: RESULTANTS AND DISCRIMINANTS

One way of proving Bézout's Theorem, and of doing Homework L3(H3), is by using resultants and discriminants.

Assuming n, m to be nonnegative integers, the Y-Resultant of two polynomials

$$f(Y) = a_0 Y^n + a_1 Y^{n-1} + \cdots + a_n$$
$$g(Y) = b_0 Y^m + b_1 Y^{m-1} + \cdots + b_m$$

is the determinant

$$\text{Res}_Y(f, g) = \det(\text{Resmat}_Y(f, g))$$

of the $n + m$ by $n + m$ matrix

$$\text{Resmat}_Y(f, g) = \begin{pmatrix} a_0 & a_1 & \cdot & \cdot & \cdot & \cdot & a_n & 0 & \cdot & \cdot & \cdot & \cdot & 0 \\ 0 & a_0 & a_1 & \cdot & \cdot & \cdot & & a_n & 0 & \cdot & \cdot & \cdot & 0 \\ \cdot & \cdot & \cdot & \cdot & \cdot & \cdot & \cdot & \cdot & \cdot & \cdot & \cdot & \cdot & \cdot \\ \cdot & \cdot & \cdot & \cdot & \cdot & \cdot & \cdot & \cdot & \cdot & \cdot & \cdot & \cdot & \cdot \\ \cdot & \cdot & \cdot & \cdot & \cdot & \cdot & \cdot & \cdot & \cdot & \cdot & \cdot & \cdot & \cdot \\ 0 & 0 & \cdot & \cdot & a_0 & a_1 & \cdot & \cdot & \cdot & \cdot & \cdot & a_n \\ b_0 & b_1 & \cdot & \cdot & \cdot & \cdot & b_m & 0 & \cdot & \cdot & \cdot & \cdot & 0 \\ 0 & b_0 & b_1 & \cdot & \cdot & \cdot & & b_m & 0 & \cdot & \cdot & \cdot & 0 \\ \cdot & \cdot & \cdot & \cdot & \cdot & \cdot & \cdot & \cdot & \cdot & \cdot & \cdot & \cdot & \cdot \\ \cdot & \cdot & \cdot & \cdot & \cdot & \cdot & \cdot & \cdot & \cdot & \cdot & \cdot & \cdot & \cdot \\ \cdot & \cdot & \cdot & \cdot & \cdot & \cdot & \cdot & \cdot & \cdot & \cdot & \cdot & \cdot & \cdot \\ 0 & 0 & \cdot & \cdot & \cdot & b_0 & b_1 & \cdot & \cdot & \cdot & \cdot & b_m \end{pmatrix}$$

where the first m rows consist of the coefficients of f and the last n rows consist of the coefficients of g. More precisely, the first row starts with the coefficients of f, these are shifted one step to the right to get the second row, shifted two steps to the right to get the third row, and so on for the first m rows, then the $(m + 1)$-st row starts with the coefficients of g, these are shifted one step to the right to get the $(m + 2)$-nd row, and so on for the next n rows. The matrix is completed by stuffing zeroes elsewhere. The determinant $\text{Res}_Y(f, g)$ is sometimes called the Sylvester resultant of f and g because it was introduced by Sylvester in his 1840 paper [Syl] where he enunciated the following:

BASIC FACT (T1). If the coefficients a_i, b_j belong to a domain R then we have: $\text{Res}_Y(f, g) = 0 \Leftrightarrow n + m \neq 0$ and either $a_0 = 0 = b_0$ or f and g have a common root in some overfield of R.

In case $n > 0$, the Y-Discriminant of f is defined to be the Y-Resultant of f and f_Y, i.e.,

$$\text{Disc}_Y(f) = \text{Res}_Y(f, f_Y).$$

where we view f_Y to be the polynomial

$$f_Y(Y) = na_0 Y^{n-1} + (n-1)a_1 Y^{n-2} + \cdots + a_{n-1}$$

i.e., we let the discriminant to be the determinant of the appropriate $2n-1$ by $2n-1$ matrix without considering whether na_0 equals zero or not.

From the Basic Fact (T1) we deduce the following:

COROLLARY (T2). If $n > 1$ and the coefficients a_i belong to a domain R then: $\text{Disc}_Y(f) = 0 \Leftrightarrow$ either $a_0 = 0$ or f has a multiple root in some overfield of R.

OBSERVATION (O1). [**Resultant and Projection**]. If $X_1, \ldots, X_N$ are indeterminates over a field k with $N \in \mathbb{N}_+$ and R is either the polynomial ring $k[X_1, \ldots, X_N]$ or the power series ring $k[[X_1, \ldots, X_N]]$, then $\text{Res}_Y(f, g)$ equals a polynomial or power series $\Phi = \Phi(X_1, \ldots, X_N)$. If a_0 and b_0 are in $k^\times$ with $nm \neq 0$ and k is algebraically closed then, in the polynomial case, by the Basic Fact it follows that the hypersurface $\Phi = 0$ in the N-space of $(X_1, \ldots, X_N)$ is the projection of the intersection of the hypersurfaces $f = 0$ and $g = 0$ in the $(N+1)$-space of $(X_1, \ldots, X_N, Y)$. Moreover, without assuming k to be algebraically closed but assuming that a_0 and b_0 are nonzero constants, in the polynomial case as well as the power series case, by the Basic Fact it follows that: Φ is identically zero (i.e., Φ is the zero element of R) $\Leftrightarrow$ f and g have a nonconstant common factor in $R[Y]$.

OBSERVATION (O2). [**Discriminant and Projection**]. Again if $X_1, \ldots, X_N$ are indeterminates over a field k with $N \in \mathbb{N}_+$ and R is either the polynomial ring $k[X_1, \ldots, X_N]$ or the power series ring $k[[X_1, \ldots, X_N]]$, then $\text{Disc}_Y(f)$ equals a polynomial or power series $\Delta = \Delta(X_1, \ldots, X_N)$. Now if a_0 is in $k^\times$ with $n > 1$ and k is algebraically closed then, in the polynomial case, for all values $(U_1, \ldots, U_N)$ of $(X_1, \ldots, X_N)$ in k, the equation $f = 0$ has n roots which may or may not be distinct, and by the Corollary it follows that these roots are not distinct iff $\Delta(U_1, \ldots, U_N) = 0$. In other words, when we project the hypersurface $f = 0$ in $(N+1)$-space onto the N-space, above most points there lie n points, and $\Delta = 0$ is the locus of those points above which there lie less than n points. Moreover, without assuming k to be algebraically closed but assuming that a_0 is a nonzero constant, in the polynomial case as well as the power series case, by the Corollary it follows that: Δ is identically zero $\Leftrightarrow$ f has a nonconstant multiple factor in $R[Y]$.

OBSERVATION (O3). [**Isobaric Property**]. View the coefficients a_i, b_j as indeterminates over $\mathbb{Z}$. Give weight i to a_i, and j to b_j. Then $0 \neq \text{Res}_Y(f, g) \in$

$\mathbb{Z}[a_0, \ldots, a_n, b_0, \ldots, b_m]$ is isobaric of weight mn, i.e., for the weight of any monomial $a_0^{i_0} \ldots a_n^{i_n} b_0^{j_0} \ldots b_m^{j_m}$ occurring in $\mathrm{Res}_Y(f,g)$ we have $(\sum_{0 \leq r \leq n} r i_r) + (\sum_{0 \leq s \leq m} s j_s) = mn$. In particular, the principal diagonal $a_0^m b_m^n$ has weight mn, and it does not cancel out because there is no other term of b_m-degree n in the resultant; the principal diagonal of an $N \times N$ matrix (A_{ij}) is the term $A_{11} A_{22} \ldots A_{NN}$; [cf. §12(R6)(14) and its proof following §12(R6)(15)]. The resultant being isobaric of weight mn is the fundamental fact behind various cases of Bézout's Theorem. The following two Observations, where we use the set-up of Observation (O1), illustrate this for plane curves and general hypersurfaces respectively.

OBSERVATION (O4). [**Plane Bézout**]. Let $N = 1$ with $X = X_1$, and assume that a_0, b_0 are nonzero elements in k, and f, g are polynomials of total (X, Y)-degrees n and m respectively. By the isobaric property we see that then always $\deg_X \Phi \leq mn$ and "in general" $\deg_X \Phi = mn$. Hence the n-degree plane curve $f = 0$ meets the m-degree plane curve $g = 0$ in mn points "counted properly." The possibility of $\deg_X \Phi < mn$ is explained by saying that some intersections have "gone to infinity."

OBSERVATION (O5). [**Hyperspatial Bézout**]. Let N be general and assume that a_0, b_0 are nonzero elements in k, and f, g are polynomials of total $(X_1, \ldots, X_N, Y)$-degrees n and m respectively. By the isobaric property we see that then always $\deg_{(X_1, \ldots, X_N)} \Phi \leq mn$ and "in general" $\deg_{(X_1, \ldots, X_N)} \Phi = mn$. Hence, in the $(N + 1)$-dimensional space, the n-degree hypersurface $f = 0$ and the m-degree hypersurface $g = 0$ meet along a "secundum" (= a subvariety of dimension two less than dimension of the ambient space) which projects onto the (mn)-degree hypersurface $\Phi = 0$ in N-dimensional space. Again the possibility of $\deg_{(X_1, \ldots, X_N)} \Phi < mn$ says that some intersections have "gone to infinity."

EXAMPLE (X1). [**Resultant and Discriminant in terms of Roots**]. If the coefficients a_i, b_j belong to a domain R and $a_0 \neq 0 \neq b_0$ then, upon writing

$$f(Y) = a_0 \prod_{1 \leq i \leq n} (Y - \alpha_i) \quad \text{and} \quad g(Y) = b_0 \prod_{1 \leq j \leq m} (Y - \beta_j)$$

with $\alpha_1, \ldots, \alpha_n$ and $\beta_1, \ldots, \beta_m$ in an overfield of R, we have

$$\mathrm{Res}_Y(f,g) = a_0^m b_0^n \prod_{1 \leq i \leq n} \prod_{1 \leq j \leq m} (\alpha_i - \beta_j)$$

$$= a_0^m \prod_{1 \leq i \leq n} g(\alpha_i)$$

$$= (-1)^{mn} b_0^n \prod_{1 \leq j \leq m} f(\beta_j)$$

and

$$\text{Disc}_Y(f) = (-1)^{n(n-1)/2} a_0^n \prod_{1 \leq i < j \leq n} (\alpha_i - \alpha_j)^2.$$

EXAMPLE (X2). [**Quadratic Resultant**]. Considering the quadratic polynomials

$$f(Y) = aY^2 + bY + c \quad \text{and} \quad g(Y) = a'Y^2 + b'Y + c'$$

and calculating the 4×4 determinant

$$\begin{pmatrix} a & b & c & 0 \\ 0 & a & b & c \\ a' & b' & c' & 0 \\ 0 & a' & b' & c' \end{pmatrix}$$

we get $\text{Res}_Y(f,g) = (a^2 c'^2 + a'^2 c^2) + (b^2 a'c' + b'^2 ac) - (abb'c' + a'b'bc) - 2aca'c'$.

EXAMPLE (X3). [**Quadratic Discriminant**]. Considering the quadratic

$$f(Y) = aY^2 + bY + c$$

and calculating the 3×3 determinant

$$\begin{pmatrix} a & b & c \\ 2a & b & 0 \\ 0 & 2a & b \end{pmatrix}$$

we get $\text{Disc}_Y(f) = -a(b^2 - 4ac)$.

EXAMPLE (X4). [**Cubic Discriminant**]. Considering the cubic

$$f(Y) = a_0 Y^3 + a_1 Y^2 + a_2 Y + a_3$$

and calculating an appropriate 5×5 determinant we get

$$\text{Disc}_Y(f) = -a_0(a_1^2 a_2^2 - 4a_0 a_2^3 - 4a_1^3 a_3 - 27a_0^2 a_3^2 + 18a_0 a_1 a_2 a_3).$$

EXAMPLE (X5). [**Special Quartic Discriminant**]. Considering the quartic

$$f(Y) = Y^4 + pY^2 + qY + r$$

and calculating an appropriate 7×7 determinant we get

$$\text{Disc}_Y(f) = 16p^4 r - 4p^3 q^2 - 128p^2 r^2 + 144pq^2 r - 27q^4 + 256r^3.$$

EXAMPLE (X6). [**General Quartic Discriminant**]. Considering the quartic

$$f(Y) = a_0 Y^4 + a_1 Y^3 + a_2 Y^2 + a_3 Y + a_4$$

and calculating an appropriate 7×7 determinant we get

$$\begin{cases} \text{Disc}_Y(f) = a_0(256a_0^3 a_4^3 - 192a_0^2 a_1 a_3 a_4^2 - 128a_0^2 a_2^2 a_4^2 + 144a_0^2 a_2 a_3^2 a_4 - 27a_0^2 a_3^4 \\ \quad +144a_0 a_1^2 a_2 a_4^2 - 6a_0 a_1^2 a_3^2 a_4 - 80a_0 a_1 a_2^2 a_3 a_4 + 18a_0 a_1 a_2 a_3^3 + 16a_0 a_2^4 a_4 \\ \quad -4a_0 a_2^3 a_3^2 - 27a_1^4 a_4^2 + 18a_1^3 a_2 a_3 a_4 - 4a_1^3 a_3^3 - 4a_1^2 a_2^3 a_4 + a_1^2 a_2^2 a_3^2). \end{cases}$$

OBSERVATION (O6). [**Future Plans**]. Further details of the concrete explicit approach to equation solving expounded in this section, including proofs of the various claims, will be given by and by. First let us push forward the abstract approach via ideals and varieties.

§2: VARIETIES

The concepts of curves, surfaces, and hypersurfaces give rise to the idea of a variety. A variety consists of the common solutions of a finite number of polynomial equations in a finite number of variables. To study these in any detail we need to build up some facts about ideals in a ring. For instance, why we do not get anything extra by allowing an infinite number of equations is explained by proving that every ideal in a multivariate (finitely many variables) polynomial ring over a field is finitely generated. This is called the Hilbert Basis Theorem. It is one of the numerous versatile theorems of commutative algebra obtained by Hilbert [Hi1] in the period 1890-1920. As said before, a ring in which every ideal is finitely generated is called a noetherian ring. This is in honor of Emmy Noether who, in 1925-1935, stressed the importance of that property. In a moment we shall prove the Hilbert Basis Theorem, and at the same time we shall prove that the multivariate power series ring over a field is also noetherian. A variety can be decomposed into its (finitely many) "irreducible components." The ideal theoretic counterpart of this will be the "primary decomposition" of ideals in a noetherian ring which we shall take up next. This will also set up an inclusion reversing correspondence between irreducible varieties and prime ideals. Modding out the polynomial ring by the prime ideal corresponding to an irreducible variety will produce the "affine coordinate ring" of the variety; its quotient field is the "function field" of the variety. Global properties of a variety, such as its dimension and degree, are reflected in the affine coordinate ring and the function field.

Local properties of a variety "near a point," such as whether the point is simple or singular, are reflected in the properties of the "local ring" of the point on the variety. The construction of the local ring generalizes the formation of the quotient field of a domain. By taking the local ring of an irreducible subvariety of an irreducible variety we can detect whether most points of the subvariety are simple points of the original variety; in technical terms, this will be so iff the said local ring is "regular." After talking about "localization" we shall resume the discussion of varieties, first in affine space, and then in projective space. The idea of localization will also enable us to reinterpret varieties as "models" which are suitable collections of local rings. In turn, the language of models will facilitate the discussion of various operations on varieties, such as the blowing-up of subvarieties of a variety for resolving its singularities.

§3: NOETHERIAN RINGS

Let us now prove the:

HILBERT BASIS THEOREM (T3). For any noetherian ring R and any positive integer m, the multivariate polynomial ring $R[X_1, \ldots, X_m]$ is noetherian.

PROOF. Clearly $R[X_1, \ldots, X_m] = R[X_1, \ldots, X_{m-1}][X_m]$ and hence by induction the general case follows from the case of $m = 1$. So assume $m = 1$ and let $X = X_1$. Given any ideal J in $S = R[X]$ we want to show that it is finitely generated. Let d denote the X-degree. For every $i \in \mathbb{N}$ let $J_i = \{r \in R : d(s - rX^i) < i$ for some $s \in J\}$. For all r, r' in R and s, s' in J we have:

(i') $d(s - rX^i) < i$ and $d(s' - r'X^i) < i \Rightarrow d((s + s') - (r + r')X^i) < i$ with $s + s' \in J$;

(ii') $d(s - rX^i) < i \Rightarrow d((r's) - (r'r)X^i) < i$ with $r's \in J$; and

(iii') $d(s - rX^i) < i \Rightarrow d(sX - rX^{i+1}) < i + 1$ with $sX \in J$.

Therefore J_i is an ideal in R with $J_i \subset J_{i+1}$. Since R is noetherian, we can find

($'$) $\begin{cases} \text{a finite number of generators } (r_{ij})_{1 \le j \le e(i)} \text{ of } J_i \\ \text{together with elements } (s_{ij})_{1 \le j \le e(i)} \text{ in } J \\ \text{such that } d(s_{ij} - r_{ij}X^i) < i \text{ for } 1 \le j \le e(i). \end{cases}$

Let $J_\infty = \cup_{i \in \mathbb{N}} J_i$. Then J_∞ is an ideal in R. Again since R is noetherian, we can find a finite number of generators $(r_j)_{1 \le j \le e}$ of J_∞ together with $l(j) \in \mathbb{N}$ and $(s_{\infty j}) \in J$ such that $d(s_{\infty j} - r_j X^{l(j)}) < l(j)$ for $1 \le j \le e$. Take $l \in \mathbb{N}$ such that $l \ge l(j)$ for $1 \le j \le e$ and let $s_j = s_{\infty j} X^{l - l(j)}$. Then for $1 \le j \le e$ we have $s_j \in J$ and $d(s_j - r_j X^l) < l$. Now

($''$) $\begin{cases} \text{for every } i \ge l \text{ in } \mathbb{N} \text{ we may take } e(i) = e \\ \text{and for } 1 \le j \le e \text{ we may take } r_{ij} = r_j \text{ and } s_{ij} = s_j X^{i-l}. \end{cases}$

For every $i \in \mathbb{N}$, by ($'$) we see that

(1') $\begin{cases} \text{given any } s \in J \text{ with } d(s) = i \text{ there exists } a_j \in R \text{ such that} \\ d(s - \sum_{1 \le j \le e(i)} a_j r_{ij} X^i) < i \text{ and hence } d(s - \sum_{1 \le j \le e(i)} a_j s_{ij}) < i. \end{cases}$

In view of $(1')$ and $('')$, by induction we see that

$(2')$ $\begin{cases} \text{given any } s \in J \text{ with } d(s) \geq l \text{ there exists } a_{jc} \in R \text{ such that} \\ d(s - \sum_{1 \leq j \leq e} A_j s_{lj}) < l \text{ where } A_j = \sum_{l \leq c \leq d(s)} a_{jc} X^{c-l} \in S. \end{cases}$

In view of $(1')$, by induction we also see that

$(3')$ $\begin{cases} \text{given any } s \in J \text{ with } d(s) < l \text{ there exists } b_{ij} \in R \text{ such that} \\ s = \sum_{0 \leq i < l} \sum_{1 \leq j \leq e(i)} b_{ij} s_{ij}. \end{cases}$

By first applying $(2')$ and then $(3')$, we see that the elements $(s_{ij})_{1 \leq j \leq e(i)}^{0 \leq i \leq l}$ constitute a finite number of generators of J.

Next, by changing degree to order in the above proof, let us prove the:

POWER SERIES VERSION OF HILBERT BASIS THEOREM (T4). For any noetherian ring R and any positive integer m, the multivariate power series ring $R[[X_1, \ldots, X_m]]$ is noetherian.

PROOF. Clearly $R[[X_1, \ldots, X_m]] = R[[X_1, \ldots, X_{m-1}]][[X_m]]$ and hence by induction the general case follows from the case of $m = 1$. So assume $m = 1$ and let $X = X_1$. Given any ideal J in $S = R[[X]]$ we want to show that it is finitely generated. Let d denote the X-order. For every $i \in \mathbb{N}$ let $J_i = \{r \in R : d(s - rX^i) > i$ for some $s \in J\}$. For all r, r' in R and s, s' in J we have:

(i*) $d(s - rX^i) > i$ and $d(s' - r'X^i) > i \Rightarrow d((s + s') - (r + r')X^i) > i$ with $s + s' \in J$;

(ii*) $d(s - rX^i) > i \Rightarrow d((r's) - (r'r)X^i) > i$ with $r's \in J$; and

(iii*) $d(s - rX^i) > i \Rightarrow d(sX - rX^{i+1}) > i + 1$ with $sX \in J$.

Therefore J_i is an ideal in R with $J_i \subset J_{i+1}$. Since R is noetherian, we can find

$(*)$ $\begin{cases} \text{a finite number of generators } (r_{ij})_{1 \leq j \leq e(i)} \text{ of } J_i \\ \text{together with elements } (s_{ij})_{1 \leq j \leq e(i)} \text{ in } J \\ \text{such that } d(s_{ij} - r_{ij} X^i) > i \text{ for } 1 \leq j \leq e(i). \end{cases}$

Let $J_\infty = \cup_{i \in \mathbb{N}} J_i$. Then J_∞ is an ideal in R. Again since R is noetherian, we can find a finite number of generators $(r_j)_{1 \leq j \leq e}$ of J_∞ together with $l(j) \in \mathbb{N}$ and $(s_{\infty j}) \in J$ such that $d(s_{\infty j} - r_j X^{l(j)}) > l(j)$ for $1 \leq j \leq e$. Take $l \in \mathbb{N}$ such that $l \geq l(j)$ for $1 \leq j \leq e$ and let $s_j = s_{\infty j} X^{l-l(j)}$. Then for $1 \leq j \leq e$ we have $s_j \in J$ and $d(s_j - r_j X^l) > l$. Now

$(**)$ $\begin{cases} \text{for every } i \geq l \text{ in } \mathbb{N} \text{ we may take } e(i) = e \\ \text{and for } 1 \leq j \leq e \text{ we may take } r_{ij} = r_j \text{ and } s_{ij} = s_j X^{i-l}. \end{cases}$

For every $i \in \mathbb{N}$, by (*) we see that

$$(1^*) \quad \begin{cases} \text{given any } s \in J \text{ with } d(s) = i \text{ there exists } a_j \in R \text{ such that} \\ d(s - \sum_{1 \le j \le e(i)} a_j r_{ij} X^i) > i \text{ and hence } d(s - \sum_{1 \le j \le e(i)} a_j s_{ij}) > i. \end{cases}$$

In view of (1*) and (**), by induction we see that

$$(2^*) \quad \begin{cases} \text{given any } s \in J \text{ with } d(s) \ge l \text{ there exists } a_{jc} \in R \text{ such that} \\ s = \sum_{1 \le j \le e} A_j s_{lj} \text{ where } A_j = \sum_{l \le c < \infty} a_{jc} X^{c-l} \in S. \end{cases}$$

In view of (1*), by induction we also see that

$$(3^*) \quad \begin{cases} \text{given any } s \in J \text{ with } d(s) < l \text{ there exists } b_{ij} \in R \text{ such that} \\ d(s - \sum_{0 \le i < l} \sum_{1 \le j \le e(i)} b_{ij} s_{ij}) \ge l. \end{cases}$$

By first applying (3*) and then (2*), we see that the elements $(s_{ij})_{1 \le j \le e(i)}^{0 \le i \le l}$ constitute a finite number of generators of J.

§4: ADVICE TO THE READER

The next long section on Ideals and Modules is rather heavy abstract dry stuff. You may prefer to skip it in a first reading and return to it as necessary. At any rate, to soften the blow here is a brief summary.

The totality of common solutions of a bunch of polynomial equations $f_l = f_l(X_1, \ldots, X_N) = 0$, with l varying over an indexing set L, in a finite number of variables $X_1, \ldots, X_N$ over a field k, is unchanged by replacing the polynomial family $(f_l)_{l \in L}$ by another family which generates the same ideal I in the polynomial ring $R = k[X_1, \ldots, X_N]$. By Hilbert's Basis Theorem (T3) we can choose the other family to be finite. Thus there is no loss of generality to let the variety V of common solutions be defined by a finite number of equations. The variety V may be decomposed into its irreducible components; for instance the variety in the (X, Y) plane given by the equation $(X^2 + Y^2 - 1)(Y^2 - X^2) = 0$ has the circle $X^2 + Y^2 = 1$ and the pair of lines $Y = \pm X$ as its irreducible components. Algebraically this corresponds to making a primary decomposition of I in R which generalizes the fact that an integer is a product of powers of prime numbers. The said decomposition amounts to writing $I = Q_1 \cap \cdots \cap Q_n$ with "primary" ideals $Q_1, \ldots, Q_n$. The "radical" of Q_i is a prime ideal P_i. The pair prime-primary ideal generalizes the pair prime number and its power. The primary decomposition has certain uniqueness properties. The operations of sums, products, and quotients of integers are generalized to sums, intersections (as well as products), and colons (or quotients) of ideals. Most of this further generalizes from ideals to modules. At once dealing with modules has the extra advantage that in various arguments the roles of the first and the second elements in a product rs are separately stressed as r belonging to the ring and s belonging to the module.

Two circles meet at a point of tangency with "intersection multiplicity" two, a tangent meets a cubic curve at a point of inflexion with intersection multiplicity three, and so on. To algebracize intersection multiplicity we introduce the concept of the length of a module, which itself generalizes the idea of the dimension of a vector space. To formalize the idea of the dimension of a variety we introduce the ideas of the height and the depth of an ideal as well as the idea of the dimension of a ring.

§5: IDEALS AND MODULES

To continue the general discussion of ideals and modules started in L1§4 and L1§5, let V be a module over a ring R.

The sums and intersections of any (indexed) family of submodules $(U_l)_{l \in L}$ of V are again clearly submodules of V, where the latter is set-theoretic while the former is defined to be the union $\cup(U_{l_1} + \cdots + U_{l_n})$ taken over all finite sequences $l_1, \ldots, l_n$ in L. Note that the members of an "indexed family" need not be distinct, i.e., we may have $U_l = U_{l^*}$ for $l \neq l^*$ in the "indexing set" L.

For any $B \subset R$ and any submodule C of V we put

$$BC = \left\{ \sum b_i c_i : b_i \in B \text{ and } c_i \in C \text{ (finite sums)} \right\}$$

and we note that this is a submodule of C. For any elements $a, a_1, a_2, \ldots$ in R and any submodule C of V we put

$$aC = \{a\}C \quad \text{and} \quad (a_1, a_2, \ldots)C = \{a_1, a_2, \ldots\}C$$

and we note that these are again submodules of C.

For any submodule U of V we put

$$\text{rad}_V U = \{r \in R : r^e V \subset U \text{ for some } e \in \mathbb{N}_+\}$$

and we call this the radical of U in V, and we note that it is an ideal in R.

By taking $V = R$, all the above applies to ideals in R. In particular, sums and intersections of any families of ideals in R are again ideals in R, and the radical

$$\text{rad}_R I = \{r \in R : r^e \in I \text{ for some } e \in \mathbb{N}_+\}$$

of any ideal I in R is an ideal in R. The product of any finite number of ideals in R was already introduced in the paragraph following item $(2^\bullet)$ of L3§11, and in case of two ideals B and C it agrees with the above definition.

For any $B \subset R$ or $a \in R$, and any $C \subset V$, we put

$$(C : B)_V = \{v \in V : B(Rv) \subset C\} \quad \text{and} \quad (C : a)_V = \{v \in V : a(Rv) \subset C\}$$

and we call these the colons (or quotients) of C by B and C by a in V respectively, and we note that if C is a submodule of V then these are submodules of V which contain C. For any $B \subset V$ or $a \in V$, and any $C \subset V$, we put

$$(C : B)_R = \{r \in R : r(RB) \subset C\} \quad \text{and} \quad (C : a)_R = \{r \in R : r(Ra) \subset C\}$$

and we call these the colons (or quotients) of C by B and C by a in R respectively, and we note that if C is a submodule of V then these are ideals in R. Note that for $V = R$ the two meanings of colon coincide.

An element x in R is a zerodivisor in R means $xy = 0$ for some $y \in R^{\times} = R \setminus \{0\}$. Now nonP means the negation of a property P, and so nonzerodivisor means an element which is not a zerodivisor. Note that a domain is a nonnull ring having no nonzero zerodivisor. An element x in R is a nilpotent means $x^e = 0$ for some positive integer e.

Just as in the ring of integers, a prime number generates a prime ideal, a positive power of a prime number generates an ideal which is primary as well as irreducible in the following sense.

An ideal I in the ring R is said to be primary if $I \neq R$ and for any elements r, s in R with $rs \in I$ and $s \notin I$ we have $r^e \in I$ for some positive integer e. Just as the primeness (resp: maximalness) of I is equivalent to the residue class ring R/I being a domain (resp: field), so the primaryness of I is clearly equivalent to R/I being a nonnull ring in which every zerodivisor is nilpotent.

An ideal I in the ring R is said to be irreducible if $I \neq R$ and I cannot be expressed in the form $I = J_1 \cap \cdots \cap J_e$ where e is a positive integer and $J_1, \ldots, J_e$ ideals in R with $I \subset J_i$ and $I \neq J_i$ for $1 \leq i \leq e$.

More generally: a submodule U of V is said to be primary if $U \neq V$ and for any $r \in R$ and $s \in V$ with $rs \in U$ and $s \notin U$ we have $r \in \mathrm{rad}_V U$; likewise, U is said to be irreducible if $U \neq V$ and U cannot be expressed in the from $U = W_1 \cap \cdots \cap W_e$ where e is a positive integer and $W_1, \ldots, W_e$ submodules of V with $U \subset W_i$ and $U \neq W_i$ for $1 \leq i \leq e$.

OBSERVATION (O7). [**Primary for a Prime Ideal**]. For ideals P, Q in a ring R, consider the conditions:

(i$^{\bullet}$) Q is primary with $\mathrm{rad}_R Q = P$.
(ii$^{\bullet}$) $P \subset \mathrm{rad}_R Q \neq R$ and: $r \in R$ and $s \in R \setminus Q$ with $rs \in Q \Rightarrow r \in P$.
(iii$^{\bullet}$) Q is primary and P is prime with $\mathrm{rad}_R Q = P$.

By a cyclical proof (i$^{\bullet}$) $\Rightarrow$ (ii$^{\bullet}$) $\Rightarrow$ (iii$^{\bullet}$) $\Rightarrow$ (i$^{\bullet}$) we see that these three conditions are equivalent. When they are satisfied, we say that Q is primary for P, or that Q is P-primary.

OBSERVATION (O8). [**Colons and Intersections of Primary Ideals**]. As easy consequences of (O7) we see that for ideals P, Q in a ring R we have the following:

($1^\bullet$) Q is P-primary and $B \not\subset Q$ for ideal B in $R \Rightarrow (Q : B)_R$ is P-primary.

($2^\bullet$) $Q = Q_1 \cap \cdots \cap Q_h$ where Q_i is P-primary for $1 \le i \le h \in \mathbb{N}_+ \Rightarrow Q$ is P-primary.

($3^\bullet$) P is a maximal ideal in R with $P \subset \mathrm{rad}_R Q \ne R \Rightarrow Q$ is P-primary.

($4^\bullet$) P is a maximal ideal in R and $P^e \subset Q \ne R$ (for instance $Q = P^e$) with $e \in \mathbb{N}_+ \Rightarrow Q$ is P-primary.

OBSERVATION (O9). [**Primaries and Powers of Primes**]. In connection with the above claim ($4^\bullet$), let us construct some relevant examples.

EXAMPLE (X7). Here is an example of

($\bullet'$) a maximal ideal P and a P-primary Q which is not a power of P.

Let $e > 1$ and $m > 1$ be integers, and let $R = k[[X_1, \ldots, X_m]]$ where $X_1, \ldots, X_m$ are indeterminates over a field k. Let $P = M(R) = (X_1, \ldots, X_m)R$, let W be the set of all monomials of degree e in $X_1, \ldots, X_m$, i.e., set of all expressions $X_1^{i_1} \ldots X_m^{i_m}$ with nonnegative integers $i_1, \ldots, i_m$ for which $i_1 + \cdots + i_m = e$. Let W' be a subset of W which contains $X_1^e, \ldots, X_m^e$ but is different from W, and let Q be the ideal in R generated by W'.

EXAMPLE (X8). Also here is an example of

($\bullet''$) a nonmaximal prime $\overline{P}$ with $\overline{P}^e$ which is not $\overline{P}$-primary but $\mathrm{rad}_{\overline{R}} \overline{P}^e = \overline{P}$.

In the notation of the Example (X7), let e', e'' be positive integers with $e' + e'' = e$, let $\phi : R \to \overline{R} = R/(X_1^e - X_2^{e'} X_3^{e''})R$ be the residue class map, and let $\overline{P}$ be the ideal in $\overline{R}$ generated by $\phi(X_1)$ and $\phi(X_2)$.

OBSERVATION (O10). [**Factor Modules and Isomorphism Theorems**]. For a module V over ring R, and a submodule U of V, the additive abelian factor group V/U is made into an R-module in a unique manner so that the residue class map (of additive abelian groups) $V \to V/U$ becomes an R-epimorphism; the resulting R-module V/U is called the factor module; note that U is the kernel of the residue class epimorphism $V \to V/U$. In L1§4 we have already dealt with the special case of this when $V = R$ and $U =$ an ideal I in R, and then we made R/I into a ring; now R/I also becomes an R-module.

For any R-epimorphisms $\phi : V \to V'$ and $\phi^* : V \to V^*$ where V, V', V^* are R-modules, if $\ker(\phi^*) \subset \ker(\phi)$ then clearly there exists a unique R-epimorphism $\phi' : V^* \to V'$ such that $\phi = \phi'\phi^*$; conversely if there exists an R-epimorphism $\phi' : V^* \to V'$ such that $\phi = \phi'\phi^*$ then clearly we must have $\ker(\phi^*) \subset \ker(\phi)$.

(1^{**}) For any R-epimorphism $\phi : V \to V'$ (of R-modules) and any R-submodule T of V, let $\ker(\phi) = U$ with $\phi(T) = T'$, and let $\alpha : T \to T'$ be the map induced by ϕ, i.e., given by $\alpha(x) = \phi(x)$ for all $x \in T$; then clearly α is an R-epimorphism with

$\phi^{-1}(T') = U + T$ and $\ker(\alpha) = U \cap T$, and hence ϕ induces a unique R-isomorphism $\alpha' : T/(U \cap T) \to T'$ such that $\alpha = \alpha'\alpha^*$ where $\alpha^* : T \to T/(U \cap T)$ is the residue class epimorphism; also clearly $\phi(U + T) = T'$ and for the map $\beta : U + T \to T'$ induced by ϕ we have that β is an R-epimorphism with $\ker(\beta) = U$ and hence ϕ induces a unique R-isomorphism $\beta' : (U + T)/U \to T'$ such that $\beta = \beta'\beta^*$ where $\beta^* : U + T \to (U + T)/U$ is the residue class epimorphism; thus ϕ induces the R-isomorphism $(\alpha')^{-1}\beta' : (U+T)/U \to T/(U \cap T)$. Note that in all this, by taking ϕ to be the residue class epimorphism $V \to V/U$ we can let U be any preassigned R-submodule of V.

$(2^{*\bullet})$ For any R-epimorphism $\phi : V \to V'$ (of R-modules) with $\ker(\phi) = U$, clearly $W \mapsto \phi(W)$ gives a bijection of the set of all R-submodules of V containing U onto the set of all R-submodules of V'; this bijection is inclusion preserving, i.e., for any submodules $W \subset W^*$ of V we have $\phi(W) \subset \phi(W^*)$. Moreover, for any R-submodule W of V with $U \subset W$, upon letting $\phi(W) = W'$, we have $\phi^{-1}(W') = W$, and ϕ induces an R-epimorphism $W \to W'$ with kernel U, and it also induces R-isomorphisms $W/U \to W'$ and $V/W \to V'/W'$.

Assertions $(1^{*\bullet})$ and $(2^{*\bullet})$ are sometimes called the First and the Second Isomorphism Theorems. We may use them tacitly, i.e., without explicit mention.

OBSERVATION (O11). [**Abelian Groups and Isomorphism Theorems**]. Any additive abelian group G may be viewed as a $\mathbb{Z}$-module by declaring $ng = g + g + \cdots + g$ taken n times; this for all $n \in \mathbb{N}$ and $g \in G$; the definition is completed by putting $(-n)g = -(ng)$ for all $n \in \mathbb{N}_+$ and $g \in G$. This makes all considerations of R-modules over a ring R, for instance the above two theorems of (O10), applicable to additive abelian groups. In particular an ideal I in a ring R and a subring H of R are both additive subgroups of R and hence their sum $I + H$ is defined as an additive subgroup of R; actually $I + H$ is a subring of R and I is an ideal in $I + H$; likewise $I \cap H$ is an ideal in H. The said two theorems may be supplemented by the following Third, Fourth, and Fifth Isomorphism Theorems which may also be used tacitly.

$(3^{*\bullet})$ Let $\phi : R \to R'$ be a ring epimorphism with $\ker(\phi) = I$. Then for any subring H of R and the additive group epimorphism $\alpha : H \to H' = \phi(H)$ induced by ϕ we have that H' is a subring of R' with $\phi^{-1}(H') = I + H$ and $\ker(\alpha) = I \cap H$, and hence ϕ induces a unique ring isomorphism $\alpha' : H/(I \cap H) \to H'$ such that $\alpha = \alpha'\alpha^*$ where $\alpha^* : H \to H/(I \cap H)$ is the residue class epimorphism; also clearly $\phi(I + H) = H'$ and for the map $\beta : I + H \to H'$ induced by ϕ we have that β is a ring epimorphism with $\ker(\beta) = I$ and hence ϕ induces a unique ring isomorphism $\beta' : (I+H)/I \to H'$ such that $\beta = \beta'\beta^*$ where $\beta^* : I+H \to (I+H)/I$ is the residue class epimorphism; thus ϕ induces the ring isomorphism $(\alpha')^{-1}\beta' : (I + H)/I \to$

$H/(I \cap H)$. Also $J \mapsto \phi(J)$ gives an inclusion preserving bijection of the set of all ideals in R containing I onto the set of all ideals in R'. Moreover, for any ideal J in R with $I \subset J$, upon letting $\phi(J) = J'$, we have $\phi^{-1}(J') = J$, and ϕ induces an additive group epimorphism $J \to J'$ with kernel I, and it also induces an additive group isomorphism $J/I \to J'$ and a ring isomorphism $R/J \to R'/J'$.

(4**) Let $\phi : R \to R'$ be a ring epimorphism. Then for any ideal J' in R' upon letting $J = \phi^{-1}(J')$ we have that J is primary (resp: prime, maximal) iff J' is primary (resp: prime, maximal), and for any ideal $\overline{J}'$ in R' upon letting $\overline{J} = \phi^{-1}(\overline{J}')$ we have that $\mathrm{rad}_R J = \overline{J}$ iff $\mathrm{rad}_{R'} J' = \overline{J}'$, and for any family of ideals $(J_l')_{l \in L}$ in R' upon letting $J_l = \phi^{-1}(J_l')$ we have that $J = \sum_{l \in L} J_l$ iff $J' = \sum_{l \in L} J_l'$, and we have that $J = \cap_{l \in L} J_l$ iff $J' = \cap_{l \in L} J_l'$.

(5**) Let $\phi : V \to V'$ be an R-epimorphism of R-modules where R is a ring. Then for any R-submodule W' of V' upon letting $W = \phi^{-1}(W')$ we have that W is primary iff W' is primary, and for any family of R-submodules $(W_l')_{l \in L}$ of V' upon letting $W_l = \phi^{-1}(W_l')$ we have that $W = \sum_{l \in L} W_l$ iff $W' = \sum_{l \in L} W_l'$, and we have that $W = \cap_{l \in L} W_l$ iff $W' = \cap_{l \in L} W_l'$.

OBSERVATION (O12). [**Annihilators and Colons**]. For a module V over a ring R, colons in R can be expressed in terms of annihilators thus. For any $B \subset V$ or $a \in V$, we put

$$\mathrm{ann}_R B = \{r \in R : r(RB) = 0\} \quad \text{and} \quad \mathrm{ann}_R a = \{r \in R : r(Ra) = 0\}$$

and we call these the annihilators of B and a in R respectively, and we note that they are clearly ideals in R. Obviously

$$\mathrm{ann}_R B = (0 : B)_R \quad \text{and} \quad \mathrm{ann}_R a = (0 : a)_R$$

and more generally for any submodule C of V we have

$$\mathrm{ann}_R((C + RB)/C) = (C : B)_R \quad \text{and} \quad \mathrm{ann}_R((C + Ra)/C) = (C : a)_R.$$

If I is any ideal in R with $I \subset \mathrm{ann}_R V$ then the R-module structure of V induces a unique (R/I)-module structure on V such that, upon letting $\phi : R \to R/I$ be the residue class epimorphism, for all $r \in R$ and $v \in V$ we have $\phi(r)v = rv$. Note that for any ideal J in R we clearly have

$$\mathrm{ann}_R(R/J) = J.$$

Also note that for any R-submodule U of V we clearly have

$$\mathrm{rad}_V U = \mathrm{rad}_R \mathrm{ann}_R(V/U).$$

OBSERVATION (O13). [**Primary Submodules**]. To generalize the material of (O7) and (O8) to modules, let P be an ideal in a ring R and let Q be an R-submodule of an R-module V. Again by a cyclical proof we see that the following three conditions are mutually equivalent, and when they are satisfied, we say that Q is primary for P, or that Q is P-primary; [cf. §13(E11)].

(i'$^{\bullet}$) Q is primary with $\mathrm{rad}_V Q = P$.

(ii'$^{\bullet}$) $P \subset \mathrm{rad}_V Q \neq R$ and: $r \in R$ and $s \in V \setminus Q$ with $rs \in Q \Rightarrow r \in P$.

(iii'$^{\bullet}$) Q is primary, $\mathrm{ann}_R(V/Q)$ is primary (as ideal), and P is prime with $\mathrm{rad}_V Q = P$.

As easy consequences of the above equivalence we have the following [cf. §13(E11)]:

(1'$^{\bullet}$) Q is P-primary and $B \not\subset Q$ for submodule B of $V \Rightarrow (Q : B)_R$ is P-primary.

(2'$^{\bullet}$) $Q = Q_1 \cap \cdots \cap Q_h$ where Q_i is P-primary for $1 \leq i \leq h \in \mathbb{N}_+ \Rightarrow Q$ is P-primary.

(3'$^{\bullet}$) P is a maximal ideal in R with $P \subset \mathrm{rad}_V Q \neq R \Rightarrow Q$ is P-primary.

(4'$^{\bullet}$) P is a maximal ideal in R and $P^e V \subset Q \neq V$ (for instance $Q = P^e V$) with $e \in \mathbb{N}_+ \Rightarrow Q$ is P-primary.

OBSERVATION (O14). [**Noetherian Modules with ACC and MXC**]. Generalizing the noetherian condition on a ring R, an R-module V is said to be noetherian if every submodule U of it is finitely generated, i.e., $U = R\overline{U}$ for some finite subset $\overline{U}$ of V; let us call this the NNC = noetherian condition on V. In the proofs of the Hilbert Basis Theorems (T3) and (T4), we implicitly used the fact that the NNC is equivalent to the ACC = ascending chain condition which says that for any ascending chain $U_1 \subset U_2 \subset \ldots$ of submodules of V we have $U_e = U_{e+1} = U_{e+2} = \ldots$ for some $e \in \mathbb{N}_+$. Yet another equivalent condition is the MXC = maximal condition which says that every nonempty family $(U_l)_{l \in L}$ of submodules of V has a maximal element, i.e., there exists $l' \in L$ for which there is no $l'' \in L$ with $U_{l'} \subset U_{l''}$ and $U_{l'} \neq U_{l''}$. To prove the equivalence of the three conditions we go in the following cyclical manner:

(5$^{\bullet}$) NNC $\Rightarrow$ ACC $\Rightarrow$ MXC $\Rightarrow$ NNC.

To see that NNC $\Rightarrow$ ACC, given any ascending chain $U_1 \subset U_2 \subset \ldots$ of submodules of V let $U = \cup_{i \in \mathbb{N}_+} U_i$; then U is a submodule of V and hence by NNC we can find a finite set of generators $u_1, \ldots, u_d$ of U; now $u_j \in U_{e(j)}$ for some $e(j) \in \mathbb{N}_+$, and it suffices to take $e \in \mathbb{N}_+$ with $e \geq e(j)$ for $1 \leq j \leq d$. To see that ACC $\Rightarrow$ MXC, let $(U_l)_{l \in L}$ be any nonempty family of submodules of V; take any $l(1) \in L$; if $U_{l(1)}$ is not maximal then for some $l(2) \in L$ we have $U_{l(1)} \subset U_{l(2)}$ with $U_{l(1)} \neq U_{l(2)}$; if $U_{l(2)}$ is not maximal then for some $l(3) \in L$ we have $U_{l(2)} \subset U_{l(3)}$ with $U_{l(2)} \neq U_{l(3)}$;

and so on; by ACC this must stop and we get $l' \in L$ with $U_{l'}$ maximal. To see that MXC $\Rightarrow$ NNC, let U be any submodule of V; the family (= the set)$\widehat{U}$ of all finitely generated submodules of U is nonempty since it contains the zero module; by MXC this family contains a maximal member U'; for any $x \in U \setminus U'$, the module $U'' = U' + Rx$ is clearly in the family and for it we have $U' \subset U''$; therefore by the maximality of U' we must have $U' = U''$, i.e., $x \in U'$; thus $U = U'$ and hence U is finitely generated.

OBSERVATION (O15). [**Irreducible and Primary Submodules**]. Let us show that for a module V over a ring R we have the following:

(6$^\bullet$) V satisfies MXC $\Rightarrow$ every submodule U of V can be expressed as an intersection $U = \cap_{1 \leq i \leq h} U_i$ of irreducible submodules $U_1, \ldots, U_h$ of V with $h \in \mathbb{N}$. CONVENTION: $\cap_{1 \leq i \leq 0} U_i = V$.

(7$^\bullet$) V satisfies ACC $\Rightarrow$ every irreducible submodule U of V is primary.

To prove (6$^\bullet$), consider the family (= the set) of all submodules $U \neq V$ of V which are not finite intersections of irreducible submodules; if this family were nonempty then by MXC we could take a maximal member U of it; clearly U is not irreducible and hence $U = W_1 \cap \cdots \cap W_e$ where e is a positive integer and $W_1, \ldots, W_e$ submodules of V with $U \subset W_i$ and $U \neq W_i$ for $1 \leq i \leq e$; by the maximality of U, each W_i can be expressed as finite intersections of irreducible submodules of V and this provides an expression of U as a finite intersection of irreducible submodules of V; this is a contradiction and hence the said family must be empty. To prove (7$^\bullet$), let $U \neq V$ be a nonprimary submodule of V; then for some $r \in R$ and $s \in V \setminus U$ we have $rs \in U$ and $r^e V \not\subset U$ for all $e \in \mathbb{N}_+$; now $(U : r^e)_V$ gives an increasing sequence of submodules of V and hence, by ACC, for some $e \in \mathbb{N}_+$ we must have $(U : r^e)_V = (U : r^{e+1})_V$; it follows that $U + Rs$ and $U + r^e V$ are submodules of V which contain U but are different from U; conversely, for any element t which belongs to both these submodules we have $u + \rho s = t = u' + r^e v$ with u, u' in U, $\rho \in R$ and $v \in V$; this implies $r(u + \rho s) = rt = ru' + r^{e+1} v$; consequently $r^{e+1} v \in U$ and hence $r^e v \in U$ and therefore $t \in U$; thus U equals the intersection of the said two submodules and hence U is nonirreducible.

Now every prime ideal is obviously irreducible. However, in connection with the above assertion (7$^\bullet$), let us construct a relevant example.

EXAMPLE (X9). Here is an example of

($\bullet'''$) a primary nonirreducible ideal in a noetherian ring.

Let $R = k[[X_1, \ldots, X_m]]$ where $X_1, \ldots, X_m$ are indeterminates over a field k with integer $m > 1$, let $e > 0$ be an integer, let $P = M(R) = (X_1, \ldots, X_m)R$, let W be the set of all monomials of degree e in $X_1, \ldots, X_m$, let $W_1, \ldots, W_n$ be a finite number of nonempty subsets of W such that $W_1 \cap \cdots \cap W_n = \emptyset$. By (4$^\bullet$) we know that P^{e+1} is P-primary and it can easily be seen that $P^{e+1} \neq (P^{e+1} + W_i R)$ for $1 \leq i \leq n$ but $P^{e+1} = \cap_{1 \leq i \leq n}(P^{e+1} + W_i R)$ and hence P^{e+1} is not irreducible.

OBSERVATION (O16). [**Spectrum and Irredundant Decomposition**]. For a ring R we put

$$\text{spec}(R) = \text{the set of all prime ideals in } R$$

and we call this the spectrum of R, and we put

$$\text{mspec}(R) = \text{the set of all maximal ideals in } R$$

and we call this the maximal spectrum of R. For any $J \subset R$ we put

$$\text{vspec}_R J = \{P \in \text{spec}(R) : J \subset P\}$$

and we call this the spectral variety of J in R, and we put

$$\text{mvspec}_R J = \text{mspec}(R) \cap \text{vspec}_R J$$

and we call this the maximal spectral variety of J in R, and we put

$$\text{nvspec}_R J = \text{the set of all minimal members of } \text{vspec}_R J$$

i.e., the set of all P in $\text{vspec}_R J$ such that for all $P' \neq P$ in $\text{vspec}_R J$ we have $P' \not\subset P$, and we call this the minimal spectral variety of J in R.

We put

$$\text{svt}(R) = \text{the set of all spectral varieties in } R$$

i.e., the set of all subsets of $\text{spec}(R)$ of the form $\text{vspec}_R J$ with J varying over the set of all subsets of R. For any $W \subset \text{spec}(R)$ we put

$$\text{ispec}_R W = \cap_{P \in W} P$$

and we call this the spectral ideal of W in R; note that by convention: $\text{ispec}_R W = R \Leftrightarrow W = \emptyset$. We put

$$\text{rd}(R) = \text{the set of all radical ideals in } R$$

i.e., the set of all ideals J in R such that $\text{rad}_R J = J$. Later on we shall show that, when R is a multivariate polynomial ring over a field, ispec_R gives a bijection $\text{svt}(R) \rightarrow \text{rd}(R)$ whose inverse is given by vspec_R. First we want to set up the machinery for proving partial uniqueness of the primary decomposition hinted at in (O15).

For any R-module V we put

$$\text{ass}_R V = \{P \in \text{spec}(R) : \text{ann}_R a \text{ is } P\text{-primary for some } a \in V\}$$

and we call this the associator of V in R; note that we automatically have $a \neq 0$ because the annihilator of 0 is R. We also put

$$\text{nass}_R V = \text{the set of all minimal members of } \text{ass}_R V$$

i.e., the set of all P in $\text{ass}_R V$ such that for all $P' \neq P$ in $\text{ass}_R V$ we have $P' \not\subset P$, and we call this the minimal associator of V in R. Note that for any R-submodule U of V we clearly have

$$\text{ass}_R(V/U) \subset \text{vspec}_R(\text{ann}_R(V/U)).$$

For any R-submodule U of V, by an associated (resp: associated minimal, associated embedded) prime of U in V we mean a member of $\text{ass}_R(V/U)$ (resp: a member of $\text{nass}_R(V/U)$, a member of $\text{ass}_R(V/U) \setminus \text{nass}_R(V/U)$). These terms are motivated by an assertion about an irredundant primary decomposition of an R-submodule U of V, i.e., an expression of the form:

$$(\bullet) \quad \begin{cases} U = \cap_{1 \leq i \leq n} Q_i \quad (\text{CONVENTION: } \cap_{1 \leq i \leq 0} Q_i = V) \\ \text{where } n \in \mathbb{N} \text{ and } Q_i \text{ is a primary submodule of } V \text{ for } 1 \leq i \leq n \\ \text{such that } \cap_{j \in \{1,\ldots,n\} \setminus \{i\}} Q_j \not\subset Q_i \text{ for all } i \in \{1,\ldots,n\}, \text{ and} \\ \text{upon letting } \text{rad}_V Q_i = P_i \text{ we have } P_i \neq P_j \text{ for all } i \neq j \text{ in } \{1,\ldots,n\}. \end{cases}$$

The said assertion says that:

(8$^\bullet$) In ($\bullet$) we have

$$\text{ass}_R(V/U) = \{P_1, \ldots, P_n\} \text{ and } \text{nass}_R(V/U) = \text{nvspec}_R(\text{ann}_R(V/U)).$$

Clearly:

(8.1$^\bullet$) In ($\bullet$) we have $\text{vspec}_R(\text{ann}_R(V/U)) = \cup_{1 \leq i \leq n} \text{vspec}_R P_i$.

Therefore, assertion (8$^\bullet$) is equivalent to the following assertion:

(8.2$^\bullet$) In ($\bullet$) we have

$$\{P \in \text{spec}(R) : (U : a)_R \text{ is } P\text{-primary for some } a \in V\} = \{P_1, \ldots, P_n\}.$$

As a supplement to (8.2$^\bullet$) we shall also prove:

(8.3$^\bullet$) If R is noetherian then in ($\bullet$) we have

$$\{P \in \text{spec}(R) : (U : a)_R = P \text{ for some } a \in V\} = \{P_1, \ldots, P_n\}.$$

Before proving (8.2$^\bullet$) and (8.3$^\bullet$), let us prove the following claim:

(8.4$^\bullet$) In ($\bullet$), for any $i \in \{1, \ldots, n\}$ and $a \in \cap_{j \in \{1,\ldots,n\}\setminus\{i\}}Q_j$ with $a \notin Q_i$ we have that $(U : a)_R$ is P_i-primary. Moreover, if $(U : a)_R \neq P_i$ then for some $x \in R$ with $xa \notin Q_i$ we have $(U : a)_R \neq (U : xa)_R$. Finally, if R is noetherian then for some $z \in R$ with $za \notin Q_i$ we have $(U : za)_R = P_i$.

Namely, in view of (ii'$^\bullet$) we see that $\mathrm{ann}_R(V/Q_i) \subset (U : a)_R \subset P_i$ and hence $P_i \subset \mathrm{rad}_R(U : a)_R \neq R$. Moreover:

$$\begin{cases} r \in R \setminus P_i \text{ and } s \in R \text{ with } rs \in (U : a)_R \\ \Rightarrow r(sa) \in U \subset Q_i \\ \Rightarrow sa \in Q_i \text{ [by (ii'$^\bullet$) because } r \notin P_i] \\ \Rightarrow sa \in U \text{ [because } U = \cap_{1 \leq i \leq n}Q_i \text{ and } a \in \cap_{j \in \{1,\ldots,n\}\setminus\{i\}}Q_j] \\ \Rightarrow s \in (U : a)_R \end{cases}$$

and hence by (ii$^\bullet$) we see that $(U : a)_R$ is P_i-primary. Moreover, if $(U : a)_R \neq P_i$ then for some x, y in $R \setminus (U : a)_R$ we have $yx \in (U : a)_R$; now: $yx \in (U : a)_R \Rightarrow y(xa) \in U \Rightarrow y \in (U : xa)_R \Rightarrow (U : a)_R \neq (U : xa)_R$; also: $x \notin (U : a)_R$ and $a \in \cap_{j \in \{1,\ldots,n\}\setminus\{i\}}Q_j \Rightarrow xa \notin Q_i$. If $(U : x_1a)_R \neq P_i$ with $x_1 = x$ then replacing a by x_1a we find $x_2 \in R$ such that $x_2x_1a \notin Q_i$ and $(U : x_1a)_R \neq (U : x_2x_1a)_R$. And so on. Now clearly $(U : a)_R \subset (U : x_1a)_R \subset (U : x_2x_1a)_R \subset \ldots$. It follows that if R satisfies ACC then this must stop and we find $z = x_ex_{e-1}\ldots x_1 \in R$ such that $za \notin Q_i$ and $(U : za)_R = P_i$.

To prove (8.2$^\bullet$) and (8.3$^\bullet$), given $i \in \{1, \ldots, n\}$, by the irredundancy we can find a as in (8.4$^\bullet$) and then by that claim we see that $(U : a)_R$ is P_i-primary, and if R is noetherian then after replacing a by za with suitable $z \in R$ we have $(U : a)_R = P_i$. Conversely let $P \in \mathrm{spec}(R)$ be such that $(U : a)_R$ is P-primary for some $a \in V$. Now

$$P = \mathrm{rad}_R(U : a)_R = \cap_{1 \leq i \leq n}\mathrm{rad}_R(Q_i : a)_R$$

and by (8.4$^\bullet$) we have $\mathrm{rad}_R(Q_i : a)_R = P_i$ or R according as $a \notin Q_i$ or $a \in Q_i$. Therefore P is the intersection of some of the P_i and is hence equal to one of them.

OBSERVATION (O17). [**Associator and Tight Associator**]. For a module V over a ring R we tighten the definition of ass given above by putting

$$\mathrm{tass}_R V = \{P \in \mathrm{spec}(R) : P = \mathrm{ann}_R a \text{ for some } a \in V\}$$

and we call this the tight associator of V in R; again note that we automatically have $a \neq 0$. The following assertion says that in the noetherian case of (8$^\bullet$), ass and tass coincide.

(9$^\bullet$) If R is noetherian then in ($\bullet$) we have $\mathrm{tass}_R(V/U) = \mathrm{ass}_R(V/U)$.

This follows from (8.2$^\bullet$) and (8.3$^\bullet$).

OBSERVATION (O18). [**Tacit Use of PIC and PMC**]. In the proof of (8.4$^\bullet$) in (O16) we cited (ii$^\bullet$) and (ii$'^\bullet$) when we should have really said by the equivalence of conditions (i$^\bullet$) to (iii$^\bullet$) asserted in (O7) and by the equivalence of conditions (i$'^\bullet$) to (iii$'^\bullet$) asserted in (O13). These are conditions for an ideal to be P-primary and for a module to be P-primary. We may call these PIC (= Primary Ideal Conditions) and PMC (= Primary Module Conditions) respectively and we may use them tacitly.

OBSERVATION (O19). [**Multiplicative Sets and Isolated Components**]. By a multiplicative set in a ring R we mean a subset S of R with $1 \in S$ such that: x, y in $S \Rightarrow xy \in S$. For a submodule U of a module V over R we put

$$[U : S]_V = \cup_{s \in S}(U : s)_V$$

and we call this the isolated S-component of U in V; note that $[U : S]_V$ is a submodule of V with $U \subset [U : S]_V$. To distinguish this from $(U : B)_V$ for $B \subset R$ we observe that

$$(U : B)_V = \cap_{b \in B}(U : b)_V.$$

For a prime ideal P in R we put

$$[U : P]_V = [U : R \setminus P]_V$$

and we call this the isolated P-component of U in V; there should be no confusion between these notations because always $1 \in S$ and $1 \notin P$. More generally, for any prime ideals $P_1, \ldots, P_n$ in R with $n \in \mathbb{N}_+$ we put

$$[U : (P_1, \ldots, P_n)]_V = [U : \cap_{1 \leq i \leq n}(R \setminus P_i)]_V$$

and we call this the isolated $(P_1, \ldots, P_n)$-component of U in V. In (O16), we used the parenthetical colon operation (:) to prove the uniqueness of the primes P_i occurring in ($\bullet$), and now we shall use the bracketed colon operation [:] to prove a partial uniqueness of the primaries Q_i. The proof applies to a primary decomposition which need not be irredundant, i.e., when for a submodule U of V we have:

($\bullet\bullet$) $\begin{cases} U = \cap_{1 \leq i \leq n} Q_i \quad \text{(CONVENTION: } \cap_{1 \leq i \leq 0} Q_i = V) \\ \text{where } n \in \mathbb{N} \text{ and } Q_i \text{ is a primary submodule of } V \text{ for } 1 \leq i \leq n \\ \text{and } \mathrm{rad}_R Q_i = P_i \text{ for } 1 \leq i \leq n. \end{cases}$

Now the uniqueness assertion is the following:

(10•) In (••) let $1 \leq i(1) < i(2) < \cdots < i(h) \leq n$ be any sequence of integers with $h \in \mathbb{N}_+$ such that: $j \in \{1, \ldots, n\}$ with $P_j \subset P_{i(l)}$ for some $l \in \{1, \ldots, h\} \Rightarrow j \in \{i(1), \ldots, i(h)\}$. Then we have $[U : (P_{i(1)}, \ldots, P_{i(h)})]_V = \cap_{1 \leq l \leq h} Q_{i(l)}$.

In view of a property of finite unions of prime ideals which we shall prove in a moment (see (22•) of Observation (O24) below), assertion (10•) follows by taking $S = \cap_{1 \leq l \leq h}(R \setminus P_{i(l)})$ in the following more general assertion:

(11•) In (••) let S be any multiplicative set in R and let $1 \leq i(1) < i(2) < \cdots < i(h) \leq n$ be the unique sequence of integers such that for any $m \in \{1, \ldots, n\}$ we have: $m \in \{i(1), \ldots, i(h)\} \Leftrightarrow S \cap P_m = \emptyset$. Then we have $[U : S]_V = \cap_{1 \leq l \leq h} Q_{i(l)}$.

To prove (11•) let $Q' = \cap_{1 \leq l \leq h} Q_{i(l)}$ and $Q'' = \cap_{1 \leq l \leq n-h} Q_{j(l)}$ where $1 \leq j(1) < j(2) < \cdots < j(n-h) \leq n$ is the unique sequence of integers such that for any $m \in \{1, \ldots, n\}$ we have: $m \in \{j(1), \ldots, j(n-h)\} \Leftrightarrow S \cap P_m \neq \emptyset$. Then $U = Q' \cap Q''$. We want to show that $[U : S]_V = Q'$. Now

$$\begin{cases} v \in [U : S]_V & \Rightarrow \quad xv \in U \text{ for some } x \in S \subset \cap_{1 \leq l \leq h}(R \setminus P_{i(l)}) \\ & \Rightarrow \quad v \in \cap_{1 \leq l \leq h} Q_{i(l)} = Q' \text{ [by PMC]}. \end{cases}$$

To prove the other implication, for $1 \leq l \leq n-h$ we can first take $y_l \in S \cap P_{j(l)}$ and then we can find $e(l) \in \mathbb{N}_+$ such that $y_l^{e(l)} V \subset Q_{j(l)}$. Now $y = \prod_{1 \leq l \leq n-h} y_l^{e(l)} \in S$ and

$$\begin{cases} v \in Q' & \Rightarrow \quad yv \in Q' \cap Q'' = U \\ & \Rightarrow \quad v \in [U : S]_V. \end{cases}$$

OBSERVATION (O20). [**Zerodivisors and Quasiprimary Decomposition**]. To generalize the contents of (O16) to (O19), PIC and PMC suggest the following weakenings of primaries in a ring R and in an R-module V. Given ideals P, Q in R we say that Q is P-quasiprimary if P is prime with $Q \neq R$ and: $r \in R$ and $s \in R \setminus Q$ with $rs \in Q \Rightarrow r \in P$. More generally, given an ideal P in R and a submodule Q of V we say that Q is P-quasiprimary if P is prime with $Q \neq V$ and: $r \in R$ and $s \in V \setminus Q$ with $rs \in Q \Rightarrow r \in P$.

By fixing $s \in R \setminus Q$ in the above implication about ideals we see that $r \in Q \Rightarrow rs \in Q \Rightarrow r \in P$ and hence: Q is a P-quasiprimary ideal in $R \Rightarrow Q \subset P$. Similarly, by fixing $s \in V \setminus Q$ in the above implication about modules we see that $r \in \mathrm{ann}_R(V/Q) \Rightarrow rs \in Q \Rightarrow r \in P$ and hence: Q is a P-quasiprimary submodule of $V \Rightarrow \mathrm{ann}_R(V/Q) \subset P$. Moreover in the ideal case $\mathrm{rad}_R Q = R$ iff $Q = R$, and in the module case $\mathrm{rad}_R Q = R$ iff $Q = V$. Therefore in (O7) we see that (ii•) $\Rightarrow Q \subset P$, and in (O13) we see that (ii'•) $\Rightarrow \mathrm{ann}_R(V/Q) \subset P$. This remark was implicitly used in the cyclical proofs asserted in (O7) and (O13).

By a zerodivisor of V we mean $r \in R$ such that $rv = 0$ for some $0 \neq v \in V$, and we put

$$Z_R(V) = \text{the set of all zerodivisors of } V.$$

We also put

$$S_R(V) = R \setminus Z_R(V)$$

and we note that, by the nonP convention, a nonzerodivisor of V means an element of $S_R(V)$; we also note that $S_R(V)$ is a multiplicative set in R. For a submodule U of V, members of $Z_R(V/U)$ or $S_R(V/U)$ may respectively be called zerodivisors or nonzerodivisors mod U. In particular, for an ideal I in R, members of $Z_R(R/I)$ or $S_R(R/I)$ may respectively be called zerodivisors or nonzerodivisors mod I.

[The following six terms: loass, lmass, lpass, zoass, zmass, and zpass, are used only in this Observation (O20), with the single exception of lmass to be used in the proof of (O24)(20•). So the readers NEED NOT MEMORIZE them].

For any ideal I in R and multiplicative set S in R with $I \cap S = \emptyset$, we put

$$\text{loass}_R(I, S) = \text{the set of all ideals } J \text{ in } R \text{ with } I \subset J \text{ and } J \cap S = \emptyset$$

and we call this the local overideal associator of I at S in R, and we put

$$\text{lmass}_R(I, S) = \text{the set of all maximal members of loass}_R(I)$$

i.e., the set of all $J \in \text{loass}_R(I, S)$ such that for all $J \subset J' \in \text{loass}_R(I, S)$ we have $J = J'$, and we call this the local maximal associator of I at S in R, and we put

$$\text{lpass}_R(I, S) = \text{vspec}_R(I) \cap \text{loass}_R(I, S)$$

i.e., the set of all $J \in \text{loass}_R(I, S)$ such that J is a prime ideal in R, and we call this the local prime associator of I at S in R.

For any nonunit ideal I in R we clearly have $I \cap S_R(R/I) = \emptyset$ and we put

$$\text{zoass}_R(I) = \text{loass}_R(I, S_R(R/I))$$

and we call this the zerodivisor overideal associator of I in R, and we put

$$\text{zmass}_R(I) = \text{lmass}_R(I, S_R(R/I))$$

and we call this the zerodivisor maximal associator of I in R, and we put

$$\text{zpass}_R(I) = \text{lpass}_R(I, S_R(R/I))$$

and we call this the zerodivisor prime associator of I in R. Note that an ideal J in R consists only of zerodivisors mod I iff $J \cap S_R(R/I) = \emptyset$; in this case we may say that J is a zerodivisor ideal mod I. Similarly, a member of $\text{zoass}_R(I)$ (resp: $\text{zmass}_R(I)$,

$\mathrm{zpass}_R(I))$ may be called a zerodivisor overideal (resp: zerodivisor maximal ideal, zerodivisor prime ideal) mod I.

In $(12^\bullet)$ below we shall show that $\mathrm{lmass}_R(I, S) \subset \mathrm{lpass}_R(I, S)$ and hence in particular $\mathrm{zmass}_R(I) \subset \mathrm{zpass}_R(I)$. In $(13^\bullet)$ below we shall give a quasiprimary decomposition for any nonunit ideal I in R, and $(14^\bullet)$ below we shall generalize it to a quasiprimary decomposition for any submodule U of V with $U \neq V$.

To achieve the said generalization, we note that for any submodule U of V with $U \neq V$ we clearly have $\mathrm{ann}_R(V/U) \cap S_R(V/U) = \emptyset$ and we put

$$\mathrm{zoass}_V(U) = \mathrm{loass}_R(\mathrm{ann}_R(V/U), S_R(V/U))$$

and we call this the zerodivisor overideal associator of U in V, and we put

$$\mathrm{zmass}_V(U) = \mathrm{lmass}_R(\mathrm{ann}_R(V/U), S_R(V/U))$$

and we call this the zerodivisor maximal associator of U in V, and we put

$$\mathrm{zpass}_V(U) = \mathrm{lpass}_R(\mathrm{ann}_R(V/U), S_R(V/U))$$

and we call this the zerodivisor prime associator of U in V. Note that an ideal J in R consists only of zerodivisors mod U iff $J \cap S_R(V/U) = \emptyset$; in this case we may say that J is a zerodivisor ideal mod U. Similarly, a member of $\mathrm{zoass}_V(U)$ (resp: $\mathrm{zmass}_V(U)$, $\mathrm{zpass}_V(U)$) may be called a zerodivisor overideal (resp: zerodivisor maximal ideal, zerodivisor prime ideal) mod U.

By a principal component of a nonunit ideal I in R we mean an ideal in R of the form $[I : P]_R$ for some $P \in \mathrm{zmass}_R(I)$; in $(13^\bullet)$ we shall show that $[I : P]_R$ is P-quasiprimary. Likewise, by a principal component of a submodule U of V with $U \neq V$ we mean a submodule of V of the form $[U : P]_V$ for some $P \in \mathrm{zmass}_V(U)$; in $(14^\bullet)$ we shall show that $[U : P]_V$ is P-quasiprimary. Note that since 1 belongs to every multiplicative set S in R, for any ideal I in R we clearly have $I \subset [I : S]_R$ and hence in particular for every prime ideal P in R we have $I \subset [I : P]_R$. Similarly, for any submodule U of V we have $U \subset [U : S]_V$ and $U \subset [U : P]_V$.

$(12^\bullet)$ For any ideal I in R and multiplicative set S in R with $I \cap S = \emptyset$ we have

$$\emptyset \neq \mathrm{lmass}_R(I, S) \subset \mathrm{lpass}_R(I, S) \subset \mathrm{vspec}_R(I) \subset \mathrm{spec}(R)$$

and for any ideal I_1 in R with $I \subset I_1$ and $I_1 \cap S = \emptyset$ we have

$$\mathrm{lmass}_R(I_1, S) \subset \mathrm{lmass}_R(I, S).$$

In particular, for any nonunit ideal I in R we have

$$\emptyset \neq \mathrm{zmass}_R(I) \subset \mathrm{zpass}_R(I) \subset \mathrm{vspec}_R(I) \subset \mathrm{spec}(R)$$

and for any ideal H in R we have

$$H \cap S_R(R/I) = \emptyset \Rightarrow H \subset P \text{ for some } P \in \mathrm{zmass}_R(I).$$

More generally, for any submodule U of V with $U \neq V$ we have

$$\emptyset \neq \mathrm{zmass}_V(U) \subset \mathrm{zpass}_V(U) \subset \mathrm{vspec}_R(\mathrm{ann}_R(V/U)) \subset \mathrm{spec}(R)$$

and for any ideal H in R we have

$$H \cap S_R(V/U) = \emptyset \Rightarrow H \subset P \text{ for some } P \in \mathrm{zmass}_V(U).$$

(13•) Given any nonunit ideal I in R, let $W = \mathrm{zmass}_R(I)$ with $W' = \mathrm{vspec}_R(I)$, and for every $P \in \mathrm{spec}(R)$ let $I_{[P]} = [I : P]_R$. Then for every $P \in \mathrm{spec}(R) \setminus W'$ we have $I_{[P]} = R$, for every $P \in W'$ the ideal $I_{[P]}$ is P-quasiprimary, and we have the quasiprimary decompositions $I = \cap_{P \in W} I_{[P]} = \cap_{P \in W'} I_{[P]} = \cap_{P \in \mathrm{spec}(R)} I_{[P]}$ where the first equation may be paraphrased by saying that I is the intersection of its principal components. Moreover, for every $P \in \mathrm{nvspec}_R(I)$ the ideal $I_{[P]}$ is P-primary.

(14•) Given any submodule U of V with $U \neq V$, let $W = \mathrm{zmass}_V(U)$ with $W' = \{P \in \mathrm{spec}(R) : U_{[P]} \neq V\}$ where for every $P \in \mathrm{spec}(R)$ we are putting $U_{[P]} = [U : P]_V$. Then for every $P \in W'$ the submodule $U_{[P]}$ is P-quasiprimary, and we have the quasiprimary decompositions $U = \cap_{P \in W} U_{[P]} = \cap_{P \in W'} U_{[P]} = \cap_{P \in \mathrm{spec}(R)} U_{[P]}$ where the first equation may be paraphrased by saying that U is the intersection of its principal components. Also $W' \subset \mathrm{vspec}_R(\mathrm{ann}_R(V/U))$, and if V is a finitely generated R-module then $W' = \mathrm{vspec}_R(\mathrm{ann}_R(V/U))$. Moreover, for every $P \in \mathrm{nvspec}_R(\mathrm{ann}_R(V/U))$ the submodule $U_{[P]}$ is P-primary.

PROOF OF (12•). To prove the first display, let us start by noting that $\mathrm{lmass}_R(I, S)$ is nonempty because of Zorn's Lemma. To show that any P in $\mathrm{lmass}_R(I, S)$ is prime, given any $x_i \in R \setminus P$ for $1 \leq i \leq 2$, by the maximality of P we can find $y_i \in (x_i R + P) \cap S$, and then $y_i = r_i x_i + p_i$ with $r_i \in R$ and $p_i \in P$, and now $y_1 y_2 \in S \subset R \setminus P$ and $y_1 y_2 - r_1 r_2 x_1 x_2 = r_1 x_1 p_2 + r_2 x_2 p_1 + p_1 p_2 \in P$, and hence $x_1 x_2 \notin P$. The rest of the first display is straightforward and so is the second display. The third display follows by taking $S_R(R/I)$ for S in the first display, and the fourth display follows from the second display by also taking $H + I$ for I_1. The fifth display follows by taking $\mathrm{ann}_R(V/U)$ for I and $S_R(V/U)$ for S in the first display, and the sixth display follows from the second display by also taking $H + I$ for I_1.

PROOF OF (13•). Follows by taking $(U, V) = (I, R)$ in (14•).

PROOF OF (14•). Let us put $I = \mathrm{ann}_R(V/U)$. Since $U \subset U_{[P]}$ for every P in $\mathrm{spec}(R)$, we have $U \subset \cap_{P \in W'} U_{[P]} = \cap_{P \in \mathrm{spec}(R)} U_{[P]} \subset \cap_{P \in W} U_{[P]}$. Hence to prove the "quasiprimary decompositions" it suffices to show that $u \in \cap_{P \in W} U_{[P]} \Rightarrow u \in U$.

Now given any $u \in \cap_{P \in W} U_{[P]}$, upon letting $H = (U : u)_R$, we see that:

$$\begin{cases} P \in W \Rightarrow u \in U_{[P]} \\ \Rightarrow \text{ for some } r \in R \setminus P \text{ we have } ru \in U \text{ and hence } r \in H \\ \Rightarrow H \not\subset P. \end{cases}$$

Consequently by the last display in $(12^\bullet)$ we conclude that $H \cap S_R(V/U) \neq \emptyset$ and hence $u \in U$.

Let there be given any prime ideal P in R. Now

$$\begin{cases} r \in R \setminus P \text{ and } s \in V \text{ with } rs \in U_{[P]} \\ \Rightarrow xrs \in U \text{ for some } x \in R \setminus P \\ \Rightarrow (xr)s \in U \text{ with } xr \in R \setminus P \\ \Rightarrow s \in U_{[P]}. \end{cases}$$

This shows that $U_{[P]}$ is P-quasiprimary for every $P \in W'$. In view of PMC, it only remains to establish the following three implications.

(a) $I \not\subset P \Rightarrow U_{[P]} = V$.

(b) $U_{[P]} = V$ and $V = Rv_1 + \cdots + Rv_h$ with h in $\mathbb{N}_+$ and $v_1, \ldots, v_h$ in $V \Rightarrow I \not\subset P$.

(c) $P \in \mathrm{nvspec}_R(I) \Rightarrow P \subset \mathrm{rad}_V U_{[P]}$.

To prove (a) note that:

$$\begin{cases} I \not\subset P \\ \Rightarrow sV \subset U \text{ for some } s \in R \setminus P \\ \Rightarrow V \subset (U : s)_V \text{ for some } s \in R \setminus P \\ \Rightarrow V \subset [U : P]_V \\ \Rightarrow U_{[P]} = V. \end{cases}$$

To prove (b) note that:

$$\begin{cases} U_{[P]} = V = Rv_1 + \cdots + Rv_h \text{ for some } v_1, \ldots, v_h \text{ in } V \text{ with } h \in \mathbb{N}_+ \\ \Rightarrow x_i v_i \in U \text{ with } x_i \in R \setminus P \text{ for } 1 \leq i \leq h \\ \Rightarrow xV \subset U \text{ where } x = x_1 \ldots x_h \in R \setminus P \\ \Rightarrow x \in I \text{ with } x \in R \setminus P \\ \Rightarrow I \not\subset P. \end{cases}$$

To prove (c) assume that $P \in \mathrm{nvspec}_R(I)$. Given any $y \in P$ let S be the smallest multiplicative set in R containing y and $R \setminus P$. If I were disjoint from S (i.e., if $I \cap S = \emptyset$) then, by the first display in $(12^\bullet)$, there exists $P' \in \mathrm{spec}(R)$ with $I \subset P'$ and $P' \cap S = \emptyset$. But $P' \cap S = \emptyset$ implies that $P' \cap (R \setminus P) = \emptyset$, i.e., $P' \subset P$ which in view of the minimality of P implies that $P' = P$ and hence $y \in P'$ which is a contradiction. Therefore I is not disjoint from S, i.e., $y^n z \in I$ for some

$n \in \mathbb{N}$ and $z \in R \setminus P$. It follows that $y^n z V \subset U$ and hence $y^n V \subset U_{[P]}$. Therefore $y \in \mathrm{rad}_V U_{[P]}$.

OBSERVATION (O21). [**Length of a Module**]. The dimension of a vector space V over a field R can be characterized as the maximum length of strictly increasing finite chains of subspaces $V_0 \subset V_1 \subset \cdots \subset V_n$.

Generalizing this we define the length $\ell_R(V)$ of a module V over a ring R to be the maximum length $n \in \mathbb{N}$ of a finite sequence

(α) $$V_0 \subset V_1 \subset \cdots \subset V_n$$

of submodules of V with

$$0 = V_0 \neq V_1 \neq \cdots \neq V_{n-1} \neq V_n = V.$$

We call such a sequence a normal series of V of length n. When there does not exist a bound for the lengths of such sequences then we put $\ell_R(V) = \infty$. Note that $\ell_R(V) = 0 \Leftrightarrow V = 0$. By a simple module we mean a module of length 1, i.e., a nonzero module which has no nonzero submodule different from the whole module. By a composition series of V we mean a normal series (α) such that the factor module V_i/V_{i-1} is simple for $1 \leq i \leq n$. Two normal series (α) and

(α') $$V_0' \subset V_1' \subset \cdots \subset V_{n'}'$$

of V are equivalent means $n = n'$ and there is a permutation σ of $\{1, \ldots, n\}$ such that the factor modules $V_{\sigma(i)}/V_{\sigma(i)-1}$ and V_i'/V_{i-1}' are isomorphic for $1 \leq i \leq n$. The normal series (α') is a refinement of the normal series (α) means $n' \geq n$ and there is a subsequence $0 < j_1 < \cdots < j_{n-1} < n'$ such that $V_{j_i}' = V_i$ for $0 < i < n$.

To slightly generalize the above concepts, let W be any submodule of V. By a normal series between W and V of length $n \in \mathbb{N}$ we mean a sequence (α) of submodules of V with

$$W = V_0 \neq V_1 \neq \cdots \neq V_{n-1} \neq V_n = V.$$

This is called a composition series between W and V if the factor module V_i/V_{i-1} is simple for $1 \leq i \leq n$. Two normal series (α) and (α') between W and V are equivalent means $n = n'$ and there is a permutation σ of $\{1, \ldots, n\}$ such that the factor modules $V_{\sigma(i)}/V_{\sigma(i)-1}$ and V_i'/V_{i-1}' are isomorphic for $1 \leq i \leq n$. The normal series (α') between W and V is a refinement of the normal series (α) between W and V means $n' \geq n$ and there is a subsequence $0 < j_1 < \cdots < j_{n-1} < n'$ such that $V_{j_i}' = V_i$ for $0 < i < n$.

To put some meat behind these definitions let us prove the following Special Jordan-Hölder Theorem $(15^\bullet)$ to $(18^\bullet)$ whose general form, which we shall prove later on, deals with groups which are not necessarily abelian. As a remark to be used tacitly in the proof of $(15^\bullet)$ below, let us note that if n and n' are nonnegative integers with $n \leq n'$ such that the length of any normal series of V is at most n', and if (α) is a normal series of V of length n, then (α) can be refined to a composition series of V of length at most n'. Namely, if (α) is not a composition series then V_i/V_{i-1} is not simple for some $i \in \{1, \ldots, n\}$ and we can insert a submodule between V_{i-1} and V_i which is different from both. Now we must have $n + 1 \leq n'$, and so on.

$(15^\bullet)$ If the module V has a composition series then any two composition series of V are equivalent (and hence in particular have the same length which equals $\ell_R(V)$), and every normal series of V can be refined into a composition series of V.

PROOF. By induction on the nonnegative integer n let us show that if V has a composition series (α) of length n then for the length n' of any normal series (α') of V we have $n' \leq n$. This being obvious for $n \leq 1$, let $n > 1$ and assume true for all smaller values of n. If $n' \leq 1$ then we have nothing to show. So assume $n' > 1$. If $V'_{n'-1} = V_{n-1}$ then $V'_0 \subset V'_1 \subset \cdots \subset V'_{n'-1}$ is a normal series of V_{n-1} of length $n' - 1$ and $V_0 \subset V_1 \subset \cdots \subset V_{n-1}$ is a composition series of V_{n-1} of length $n - 1$, and hence by the induction hypothesis $n' - 1 \leq n - 1$ and therefore $n' \leq n$. If $V'_{n'-1} \neq V_{n-1}$ with $V'_{n'-1} \subset V_{n-1}$ then $V'_0 \subset V'_1 \subset \cdots \subset V'_{n'-1} \subset V_{n-1}$ is a normal series of V_{n-1} of length n' and $V_0 \subset V_1 \subset \cdots \subset V_{n-1}$ is a composition series of V_{n-1} of length $n-1$, and hence again by the induction hypothesis $n' \leq n-1$ and therefore $n' \leq n$. Finally suppose that $V'_{n'-1} \not\subset V_{n-1}$. Then, since there is no submodule between V_{n-1} and V other than these two, we must have $V_{n-1} + V'_{n'-1} = V$, and hence by the First Isomorphism Theorem $(1^{\bullet\bullet})$, the factor modules V/V_{n-1} and $V'_{n'-1}/(V_{n-1} \cap V'_{n'-1})$ are isomorphic, and therefore $V'_{n'-1}/(V_{n-1} \cap V'_{n'-1})$ is simple. Since $V_0 \subset V_1 \subset \cdots \subset V_{n-1}$ is a composition series of V_{n-1} of length $n - 1$ and $V_{n-1} \cap V'_{n'-1} \subset V_{n-1}$ with $V_{n-1} \cap V'_{n'-1} \neq V_{n-1}$, by the induction hypothesis it follows that the length of any normal series of $V_{n-1} \cap V'_{n'-1}$ is at most $n - 2$, and hence $V_{n-1} \cap V'_{n'-1}$ has a composition series of length at most $n - 2$; since $V'_{n'-1}/(V_{n-1} \cap V'_{n'-1})$ is simple, it follows that $V'_{n'-1}$ has a composition series of length at most $n - 1$. Since $V'_0 \subset V'_1 \subset \cdots \subset V'_{n'-1}$ is a normal series of $V'_{n'-1}$ of length $n' - 1$, by the induction hypothesis we get $n' - 1 \leq n - 1$ and hence $n' \leq n$.

This completes the induction. It follows that every composition series of V has length n, and any normal series of V can be refined into a composition series of V. It only remains to show that, assuming (α) and (α') to be composition series of V of lengths n and n' with $n = n'$, there is a permutation σ of $\{1, \ldots, n\}$ such that the factor modules $V_{\sigma(i)}/V_{\sigma(i)-1}$ and V'_i/V'_{i-1} are isomorphic for $1 \leq i \leq n$. Again we make induction on n. For $n \leq 1$ we have nothing to show. So let $n > 1$ and assume true for all values of n smaller than the given one. If $V'_{n-1} = V_{n-1}$ then deleting the last terms from (α) and (α') we get two composition series of V_{n-1} of length $n - 1$, which are equivalent by the induction hypothesis, and hence (α) and (α') are equivalent. Now assume that $V'_{n-1} \neq V_{n-1}$. Then $V'_{n-1} + V_{n-1} = V$ and hence, upon letting $V^*_{n^*} = V'_{n-1} \cap V_{n-1}$, by the First Isomorphism Theorem $(1^{\bullet\bullet})$, the two modules $V_{n-1}/V^*_{n^*}$ and V/V'_{n-1} are isomorphic, and so are $V'_{n-1}/V^*_{n^*}$ and

V/V_{n-1}. Therefore by taking a fixed composition series

$$(\beta) \qquad\qquad V_0^* \subset V_1^* \subset \cdots \subset V_{n^*}^*$$

of $V_{n^*}^*$ we get the two equivalent composition series

$$(\gamma) \qquad\qquad V_0^* \subset V_1^* \subset \cdots \subset V_{n^*}^* \subset V_{n-1} \subset V$$

and

$$(\gamma') \qquad\qquad V_0^* \subset V_1^* \subset \cdots \subset V_{n^*}^* \subset V_{n-1}' \subset V$$

of V. By deleting the last terms in (α) and (γ) we get two composition series of V_{n-1} which are equivalent by the induction hypothesis, and hence (α) and (γ) are equivalent; clearly we must have $n^* = n - 2$. Similarly (α') and (γ') are equivalent. Therefore (α) and (α') are equivalent.

(16•) If W is a submodule of V such that there is a composition series between W and V then any two composition series between W and V are equivalent (and hence in particular have the same length which equals $\ell_R(V/W)$), and every normal series between W and V can be refined into a composition series between W and V. If W is a submodule of V with residue class epimorphism $\phi : V \to V/W$, then, for any normal series (α) between W and V, the series $\phi(V_0) \subset \phi(V_1) \subset \cdots \subset \phi(V_n)$ is a normal series of V/W and for $1 \le i \le n$ the factor modules $\phi(V_i)/\phi(V_{i-1})$ and V_i/V_{i-1} are isomorphic; in particular: (α) is a composition series between W and $V \Leftrightarrow \phi(V_0) \subset \phi(V_1) \subset \cdots \subset \phi(V_n)$ is a composition series of V/W.

PROOF. The first sentence follows by taking V/W for V in (15•). The second sentence follows by the Second Isomorphism Theorem (2••).

(17•) For any submodule W of V we have $\ell_R(V) = \ell_R(W) + \ell_R(V/W)$, where by convention $\infty = \infty + \infty = \infty + $ integer. In greater detail, if (α) is a composition series between W and V, and $W_0 \subset W_1 \subset \cdots \subset W_m$ is a composition series of W, then $W_0 \subset W_1 \subset \cdots \subset W_m \subset V_1 \subset \cdots \subset V_n$ is a composition series of V.

PROOF. Follows from (16•).

(18•) For any submodules U and W of the module V we have $\ell_R(U) + \ell_R(W) = \ell_R(U + W) + \ell_R(U \cap W)$, again with the understanding that $\infty + \infty = \infty + $ integer.

PROOF. Taking $(U + W, U)$ for (V, W) in (17•) we get

$$\ell_R(U + W) = \ell_R(U) + \ell_R((U + W)/U).$$

The factor modules $(U + W)/U$ and $W/(U \cap W)$ are isomorphic by the First Isomorphism Theorem (1••), and hence by taking $(W, U \cap W)$ for (V, W) in (17•) we get

$$\ell_R(W) = \ell_R(U \cap W) + \ell_R((U + W)/U).$$

We are done by adding the two equations, and taking care of infinities.

OBSERVATION (O22). [**Heights and Depths of Ideals, and Dimensions of Rings**]. Replacing chains of submodules by chains of prime ideals we get the definitions of heights and depths of ideals and the definition of the dimension of a ring.

By the dimension $\dim(R)$ of a ring R we mean the maximum length $n \in \mathbb{N}$ of a finite sequence

(δ) $$P_0 \subset P_1 \subset \cdots \subset P_n$$

of prime ideals in R with

$$P_0 \neq P_1 \neq \cdots \neq P_{n-1} \neq P_n.$$

We call such a sequence a prime sequence in R of length n. When there does not exist a bound for such sequences then we put $\dim(R) = \infty$ provided R is not the null ring; if R is the null ring then we put $\dim(R) = -\infty$.

By the height $\mathrm{ht}_R P$ (resp: depth $\mathrm{dpt}_R P$) of a prime ideal P in R we mean the maximum length of a prime chain (δ) in R with $P = P_n$ (resp: $P = P_0$), again with the understanding that when there does not exist a bound for such sequences then we put $\mathrm{ht}_R P = \infty$ (resp: $\mathrm{dpt}_R P = \infty$).

We define the height $\mathrm{ht}_R J$ and the depth $\mathrm{dpt}_R J$ of a nonunit ideal J in R by putting

$$\mathrm{ht}_R J = \min\{\mathrm{ht}_R P : P \in \mathrm{vspec}_R J\} \in \mathbb{N} \cup \{\infty\}$$

and

$$\mathrm{dpt}_R J = \max\{\mathrm{dpt}_R P : P \in \mathrm{vspec}_R J\} \in \mathbb{N} \cup \{\infty\}$$

and we note that for $J \in \mathrm{spec}(R)$ these reduce to the previous definitions. Also note that for a nonunit ideal J in R, $\mathrm{dpt}_R J$ is the maximum length of a prime chain (δ) in R with $J \subset P_0$. Finally note that for an ideal J in R, $\mathrm{vspec}_R J$ is empty iff $J = R$, and let us complete the definitions of height and depth by putting $\mathrm{ht}_R R = -\infty = \mathrm{dpt}_R R$ and let us observe that now

$$\mathrm{dpt}_R J = \dim(R/J).$$

EXAMPLE (X10). Taking R to be the n-variable polynomial ring $k[X_1, \ldots, X_n]$ over a field k and P_i to be the ideal in R generated by $X_1, \ldots, X_i$, and putting $X_1 = \cdots = X_i = 0$ we get an epimorphism $R \to k[X_{i+1}, \ldots, X_n]$ with kernel P_i for $0 \leq i \leq n$ which shows that (δ) is a prime sequence in R. It follows that $\dim(R) \geq n$

with $\mathrm{ht}_R P_i \geq i$ and $\mathrm{dpt}_R P_i \geq n - i$ for $0 \leq i \leq n$. Later on we shall show that these three inequalities are actually equalities. [cf. L5§5(Q10)(T47)].

OBSERVATION (O23). [**Modular Law**]. The following Modular Law gives a useful identity concerning a module V over a ring R:

(19$^\bullet$) For any submodules T, U, W of V with $U \subset T$ we have

$$T \cap (U + W) = U + (T \cap W).$$

To see this note that the RHS is obviously contained in the LHS. Conversely, if v is any element in the LHS then, because $v \in U + W$, we get $v = u + w$ with $u \in U$ and $w \in W$. Now $w = v - u \in T$ because $v \in T$ and $u \in U \subset T$. Thus $w \in T \cap W$ and hence $v \in U + (T \cap W)$.

OBSERVATION (O24). [**Generalities on Prime Ideals**]. In the following assertions (20$^\bullet$) to (23$^\bullet$) we give some useful properties of prime ideals.

(20$^\bullet$) In any ring R, the set of all nilpotent elements, i.e., the radical of the zero ideal, equals the intersection of all the prime ideals, with the understanding that if there are no prime ideals (i.e., if the ring is the null ring) then the said intersection is the unit ideal. [Note the analogy with L2(R5) about the existence of maximal ideals].

PROOF. Since any prime ideal contains 0, it must contain every nilpotent element. Conversely, given any nonnilpotent element v in R we want to find a prime ideal P not containing it. We can do this by letting $I = 0$ and $S = \{v^n : n \in \mathbb{N}\}$ in (12$^\bullet$), and taking $P \in \mathrm{lmass}_R(I, S)$.

(21$^\bullet$) Let P be a prime ideal in a ring R, and let $J_1, \ldots, J_n$ be ideals in R with $n \in \mathbb{N}_+$ such that $J_1 \cap \cdots \cap J_n \subset P$. Then for some $i \in \{1, \ldots, n\}$ we have $J_i \subset \mathrm{rad}_R J_i \subset P$. Moreover, if $J_1 \cap \cdots \cap J_n = P$ then for some $i \in \{1, \ldots, n\}$ we have $J_i = P$.

PROOF. If $J_i \not\subset P$ for $1 \leq i \leq n$ then by taking $x_i \in J_i \backslash P$ we get $x_1 \ldots x_n \in J_1 \cap \cdots \cap J_n \subset P$ which contradicts the primeness of P. Therefore for some $i \in \{1, \ldots, n\}$ we must have $J_i \subset P$ and by taking radicals we get $\mathrm{rad}_R J_i \subset P$. The rest is obvious.

(22$^\bullet$) Let J be an ideal in a ring R, and let $P_1, \ldots, P_n$ be prime ideals in R with $n \in \mathbb{N}_+$ such that $J \subset P_1 \cup \cdots \cup P_n$. Then for some $i \in \{1, \ldots, n\}$ we have $J \subset P_i$.

PROOF. By discarding some of the primes $P_1, \ldots, P_n$ we may assume that

$P_j \not\subset P_i$ whenever $j \neq i$, and then we can take $b_{i,j} \in P_j \setminus P_i$ whenever $j \neq i$. Suppose if possible that $J \not\subset P_i$ for $1 \leq i \leq n$. For $1 \leq i \leq n$ fix $c_i \in J \setminus P_i$, and let $b_i = c_i \prod_{j \neq i} b_{i,j}$. Now $b_i \in J \setminus P_i$, and $b_i \in P_j$ whenever $j \neq i$. Let $b = b_1 + \cdots + b_n$. Then $b \in J$ with $b \notin P_i$ for $1 \leq i \leq n$, which is a contradiction.

(23$^\bullet$) Any nonunit ideal J in a noetherian ring R contains a nonempty finite product of prime ideals, i.e., $P_1 \ldots P_n \subset J$ for some prime ideals $P_1, \ldots, P_n$ with $n \in \mathbb{N}_+$.

PROOF. Suppose that the family of nonunit ideals in R which do not contain any nonempty finite product of prime ideals is nonempty. Since R is noetherian, the said family contains a maximal element J. Since J belongs to the said family, it cannot be prime and hence we can find elements a, b in $R \setminus J$ with ab in J. Let $A = J + aR$ and $B = J + bR$. Then A and B are nonunit ideals in R such that $J \subset A$ and $J \subset B$ with $AB \subset J$ and $A \neq J \neq B$. By the maximality of J, the ideals A and B contain nonempty finite products of prime ideals and, since $AB \subset J$, so does J. This is a contradiction.

OBSERVATION (O25). [**Artinian Modules with DCC and MNC**]. For a module V over a ring R, analogous to the ACC and the MXC, we can consider the DCC = descending chain condition which says that for any descending chain $U_1 \supset U_2 \supset \ldots$ of submodules of V we have $U_e = U_{e+1} = U_{e+2} = \ldots$ for some $e \in \mathbb{N}_+$, and the MNC = minimal condition which says that every nonempty family $(U_l)_{l \in L}$ of submodules of V has a minimal element, i.e., there exists $l' \in L$ for which there is no $l'' \in L$ with $U_{l'} \supset U_{l''}$ and $U_{l'} \neq U_{l''}$. We claim that:

(24$^\bullet$)   DCC $\Leftrightarrow$ MNC.

Namely, by reversing inclusions and changing maximal to minimal in the proof of ACC $\Rightarrow$ MXC given in (O14) we get DCC $\Rightarrow$ MNC. Conversely, if V does not satisfy DCC then we can find an infinite sequence $U_1 \supset U_2 \supset \ldots$ of submodules of V with $U_1 \neq U_2 \neq \ldots$ and clearly the family $(U_l)_{l \in \mathbb{N}_+}$ has no minimal element.

We say that V is an artinian R-module to mean that V satisfies the DCC and hence the MNC. We say that R is an artinian ring to mean that R is an artinian R-module.

In assertions (25$^\bullet$) to (33$^\bullet$) below, we shall prove some general properties of artinian and noetherian modules.

(25$^\bullet$) Let W be any submodule of V. Then V is an artinian module $\Leftrightarrow$ W and V/W are artinian modules. Similarly V is a noetherian module $\Leftrightarrow$ W and V/W are noetherian modules.

PROOF. If DCC holds in V then it obviously holds in W and by the Second

Isomorphism Theorem (2^{**}) it also holds in V/W. Conversely, assume DCC holds in W and V/W, and let $U_1 \supset U_2 \supset \dots$ be any decreasing sequence of submodules of V. Then $U_1 \cap W \supset U_2 \cap W \supset \dots$ and $(U_1 + W)/W \supset (U_2 + W)/W \supset \dots$ are decreasing sequences of submodules of W and V/W respectively, and hence for some $e \in \mathbb{N}_+$ we have $U_e \cap W = U_{e+1} \cap W = U_{e+2} \cap W = \dots$ and $U_e + W = U_{e+1} + W = U_{e+2} + W = \dots$. For every integer $d \geq e$ we get

$$
\begin{cases}
U_d = U_d \cap (U_d + W) & \text{because } U_d \subset U_d + W \\
\quad = U_d \cap (U_{d+1} + W) & \text{because } U_d + W = U_{d+1} + W \\
\quad = U_{d+1} + (U_d \cap W) & \text{by Modular Law } (19^{\bullet}) \\
\quad = U_{d+1} + (U_{d+1} \cap W) & \text{because } U_d \cap W = U_{d+1} \cap W \\
\quad = U_{d+1} & \text{because } U_{d+1} \cap W \subset U_{d+1}.
\end{cases}
$$

Therefore $U_e = U_{e+1} = U_{e+2} = \dots$. For the noetherian case, in the undisplayed material change DCC to ACC and $\supset$ to $\subset$, and in the displayed material change $(d, d+1)$ to $(d+1, d)$.

$(26^{\bullet})$ Assuming $V = V_1 + \dots + V_s$, where $V_1, \dots, V_s$ are a finite number of submodules of V, we have the following. If the modules $V_1, \dots, V_s$ are artinian then so is V. Similarly, if the modules $V_1, \dots, V_s$ are noetherian then so is V.

PROOF. Since, $V_1 + \dots + V_s = (V_1 + \dots + V_{s-1}) + V_s$, by induction the general case follows from the $s = 2$ case. This case follows from $(25^{\bullet})$ by noting that, in view of the First Isomorphism Theorem (1^{**}), the modules $(V_1 + V_2)/V_1$ and $V_2/(V_1 \cap V_2)$ are isomorphic.

$(27^{\bullet})$ If R is an artinian ring and V is a finitely generated R-module then V is an artinian module. Similarly, if R is a noetherian ring and V is a finitely generated R-module then V is a noetherian module.

PROOF. In view of $(26^{\bullet})$ it is sufficient to show that, assuming $V = Rx$ with $x \in V$, if DCC (resp: ACC) holds for ideals in R then DCC (resp: ACC) holds for submodules of V. So assume DCC (resp: ACC) for ideals in R, and given any descending (resp: ascending) sequence $(U_i)_{i \in \mathbb{N}_+}$ of submodules of $V = xR$ let $J_i = \{a \in R : ax \in U_i\}$. Then $(J_i)_{i \in \mathbb{N}_+}$ is a descending (resp: ascending) sequence of ideals in R, and for all $i \in \mathbb{N}_+$ we have $U_i = \{ax : a \in J_i\}$. By DCC (resp: ACC) for ideals we can find $e \in \mathbb{N}_+$ such that $J_e = J_{e+1} = J_{e+2} = \dots$. In both the cases it follows that $U_e = U_{e+1} = U_{e+2} = \dots$.

$(28^{\bullet})$ $\ell_R(V) < \infty \Leftrightarrow$ the module V is artinian as well as noetherian.

PROOF. If $\ell_R(V) = n < \infty$ then every descending or ascending sequence of submodules of V has at most n distinct members, and hence V is artinian as well as noetherian. Conversely suppose that V is artinian as well as noetherian. Let $U_0 = V$. If $U_0 = 0$ then $\ell_R(V) = 0$. If $U_0 \neq 0$ then, because V is noetherian, we can take a maximal member U_1 in the family of submodules of U_0 different from U_0. If $U_1 = 0$ then $\ell_R(V) = 1$. If $U_1 \neq 0$ then, again because V is noetherian, we can take a maximal member U_2 in the family of submodules of U_1 different from U_1. And so on. Since V is artinian, this must stop. Thus we get $n \in \mathbb{N}$ together with a decreasing sequence $U_0 \supset U_1 \supset \cdots \supset U_n$ of submodules of V with $V = U_0 \neq U_1 \neq \cdots \neq U_{n-1} \neq U_n = 0$ such that U_i/U_{i+1} is simple for $0 \leq i < n$. It follows that $\ell_R(V) = n < \infty$.

(29$^\bullet$) If R is a field then $\dim_R V = \ell_R V$ and we have that: V is an artinian module $\Leftrightarrow \dim_R V < \infty \Leftrightarrow V$ is a noetherian module.

PROOF. Obvious.

(30$^\bullet$) If there are maximal ideals $P_1, \ldots, P_n$ in R with $n \in \mathbb{N}_+$ such that $P_1 \ldots P_n = 0$ then: R is an artinian ring $\leftrightarrow R$ is a noetherian ring.

PROOF. Let $V_i = P_1 \ldots P_{n-i}$. Then $0 = V_0 \subset V_1 \subset \cdots \subset V_n = R$ is an increasing sequence of ideals in R. For $1 \leq i \leq n$ we have $P_{n-i+1} \subset \mathrm{ann}_R(V_i/V_{i-1})$ and hence the R-module V_i/V_{i-1} may be regarded as a vector space over the field R/P_{n-i+1}. Therefore our assertion follows from (29$^\bullet$).

(31$^\bullet$) If R is any ring and J is an ideal in R with $\ell_R J = 1$ then $\mathrm{ann}_R J$ is a prime ideal in R.

PROOF. Now $J \neq 0$ and hence $\mathrm{ann}_R J \neq R$. Given any a, b in $R \setminus \mathrm{ann}_R J$, let $A = aR$ and $B = bR$. Then AJ and BJ are nonzero ideals contained in J. Since $\ell_R J = 1$, we get $AJ = J = BJ$. Consequently $ABJ = J$ and hence ab belongs to $R \setminus \mathrm{ann}_R J$.

(32$^\bullet$) If R is a domain then: R is an artinian ring $\Leftrightarrow R$ is a field. If R is an artinian ring then every prime ideal in R is maximal.

PROOF. Assume R to be an artinian domain, and let $0 \neq x \in R$. Then $xR \supset x^2R \supset \ldots$ is a decreasing sequence of ideals in R, and hence for some $e \in \mathbb{N}_+$ and $y \in R$ we must have $x^e = yx^{e+1}$ which gives $x^e(1 - yx) = 0$. Since $x^e \neq 0$ we get $yx = 1$. Thus R is a field. The rest is now obvious.

(33$^\bullet$) R is an artinian ring iff it is either the null ring or a zero-dimensional noetherian ring.

PROOF. The case of a null ring being obvious let us suppose that R is nonnull. First assume R to be a zero-dimensional noetherian ring. Then by $(23^{\bullet})$ we can find a positive integer n and maximal ideals $P_1, \ldots, P_n$ in R such that $P_1 \ldots P_n \subset 0$ and hence $P_1 \ldots P_n = 0$. By $(30^{\bullet})$ we see that R is artinian. Conversely assume R to be artinian. Then we can take a minimal element I in the set of all ideals in R which can be expressed as nonempty finite products of prime ideals. We shall prove that $I = 0$ and, in view of $(30^{\bullet})$ and $(32^{\bullet})$, this will show that R is a zero-dimensional noetherian ring. Suppose if possible that $I \neq 0$. Let $A = \operatorname{ann}_R I$. Then A is a nonunit ideal in R, and hence we can take a minimal element B in the nonempty set of all ideals in R which contain A but are different from A. Let $P = (A : B)_R$ and let $\phi : R \to R/A$ be the residue class epimorphism. Now $\phi(R)$ is an artinian ring and $\ell_{\phi(R)} \phi(B) = 1$. Therefore by $(31^{\bullet})$ we see that $\operatorname{ann}_{\phi(R)} \phi(B)$ is a prime ideal in $\phi(R)$. Clearly $\operatorname{ann}_{\phi(R)} \phi(B) = \phi(P)$ and hence P is a prime ideal in R. Now $PB \subset A$ and hence $IPB = 0$; also $(0 : I)_R = A \subset B \subset (0 : IP)_R$ with $A \neq B$; consequently $IP \subset I$ with $IP \neq I$ in contradiction with the minimality of I.

OBSERVATION (O26). [**Comaximal Ideals**]. To give an example of a situation when intersections and products of ideals coincide we introduce the following definition. Ideals A and B in a ring R are said to be comaximal if $A + B = R$. For any ideals A, B we always have $AB \subset A \cap B$, and if $A + B = R$ then $a + b = 1$ for some $a \in A$ and $b \in B$ and hence for any $c \in A \cap B$ we get $c = ac + bc \in AB$. Thus for comaximal ideals A, B we have $AB = A \cap B$.

Instances of comaximality are enlarged by noting that ideals A, B are comaximal if their radicals are comaximal; namely, if $A + B \neq R$ then, by Zorn's Lemma, $A + B \subset M$ for some maximal ideal, and now $A \subset M$ and $B \subset M$, and hence $\operatorname{rad}_R A \subset M$ and $\operatorname{rad}_R B \subset M$, and therefore $\operatorname{rad}_R A + \operatorname{rad}_R B \subset M$ in contradiction to the comaximality of $\operatorname{rad}_R A$ and $\operatorname{rad}_R B$. Clearly A, B are comaximal $\Leftrightarrow$ there are elements a, b in R such that $\phi(a) = 0 = \psi(b)$ and $\phi(b) = 1 = \psi(a)$ where $\phi : R \to R/A$ and $\psi : R \to R/B$ are the residue class epimorphisms (for $\Rightarrow$ take a, b as in the above paragraph; for $\Leftarrow$ write $(1 - b) + b = 1$). Clearly A, B are comaximal $\Leftrightarrow$ no maximal ideal in R contains both A and B; hence if $A_1, \ldots, A_n, B_1, \ldots, B_m$ are ideals in R with n, m in $\mathbb{N}_+$ such that A_i, B_j are comaximal for each i, j then $A_1 \ldots A_n, B_1 \ldots B_m$ are comaximal and so are $A_1 \cap \cdots \cap A_n, B_1 \cap \cdots \cap B_m$.

The above equality of the product and intersection can be generalized by showing that for any finite number of pairwise comaximal ideals $A_1, \ldots, A_n$ we have $\prod_{1 \le i \le n} A_i = \cap_{1 \le i \le n} A_i$. We do this by induction on n, without using the above special case of $n = 2$. For $i = 1$ it is obvious. So let $n > 1$ and assume for all smaller values of n. Then by the induction hypothesis

$$\cap_{1 \le i \le n} A_i \subset \cap_{i \neq 1} A_i = \prod_{i \neq 1} A_i \quad \text{and} \quad \cap_{1 \le i \le n} A_i \subset \cap_{i \neq 2} A_i = \prod_{i \neq 2} A_i$$

and by assumption $A_1 + A_2 = R$ and hence

$$
\begin{cases}
\cap_{1 \le i \le n} A_i & = (A_1 + A_2)\,(\cap_{1 \le i \le n} A_i) \\
& = A_1\,(\cap_{1 \le i \le n} A_i) + A_2\,(\cap_{1 \le i \le n} A_i) \\
& \subset A_1\left(\prod_{i \ne 1} A_i\right) + A_2\left(\prod_{i \ne 2} A_i\right) \\
& = \prod_{1 \le i \le n} A_i
\end{cases}
$$

and obviously $\prod_{1 \le i \le n} A_i \subset \cap_{1 \le i \le n} A_i$ and therefore $\prod_{1 \le i \le n} A_i = \cap_{1 \le i \le n} A_i$.

Continuing to assume $A_1, \ldots, A_n$ to be pairwise comaximal, for any $i \ne j$, since $A_i + A_j = R$, we can write $a_{i,j} + a_{j,i} = 1$ with $a_{i,j} \in A_j$ and $a_{j,i} \in A_i$. Let $a_i = \prod_{j \ne i} a_{i,j}$. Then $a_i \equiv 1 \bmod A_i$ and $a_i \equiv 0 \bmod A_j$ for $j \ne i$, i.e., $a_i - 1 \in A_i$ and $a_i \in A_j$ for $j \ne i$. Given any $x_i \in R$ for $1 \le i \le n$, upon letting $x = a_1 x_1 + \cdots + a_n x_n$ we see that $x \equiv x_i \bmod A_i$ for $1 \le i \le n$.

What we have proved above may be summarized in assertions (34$^\bullet$) and (35$^\bullet$).

(34$^\bullet$) Ideals A, B in a ring R are comaximal $\Leftrightarrow$ $\mathrm{rad}_R A$ and $\mathrm{rad}_R B$ are comaximal $\Leftrightarrow$ no maximal ideals in R contains both A and B $\Leftrightarrow$ there are elements a, b in R such that $\phi(a) = 0 = \psi(b)$ and $\phi(b) = 1 = \psi(a)$ where $\phi : R \to R/A$ and $\psi : R \to R/B$ are the residue class epimorphisms. If $A_1, \ldots, A_n, B_1, \ldots, B_m$ are ideals in R with n, m in $\mathbb{N}_+$ such that A_i and B_j are comaximal for each i, j then $A_1 \ldots A_n$ and $B_1 \ldots B_m$ are comaximal and so are $A_1 \cap \cdots \cap A_n$ and $B_1 \cap \cdots \cap B_m$.

(35$^\bullet$) If $A_1, \ldots, A_n$ are a finite number of pairwise comaximal ideals in a ring R then $\prod_{1 \le i \le n} A_i = \cap_{1 \le i \le n} A_i$, and given any $x_i \in R$ for $1 \le i \le n$ there exists $x \in R$ with $x \equiv x_i \bmod A_i$ for $1 \le i \le n$.

OBSERVATION (O27). [**Affine Rings and Hilbert Basis Theorem**]. A ring S is said to be an affine ring over a ring R if there is an R-epimorphism of a finite variable polynomial ring over R onto S, i.e., if S is a finitely generated ring extension of R.

Recall that a ring is said to be noetherian if ideals in it are finitely generated. Obviously, a ring is a field iff in it the zero ideal and the unit ideal (which are generated by 0 and 1 respectively) are distinct and there are no other ideals in it. So in particular a field is noetherian. Also the ring of integers is noetherian because in it every ideal is actually principal. By (25$^\bullet$) of (O25), a homomorphic image of a noetherian ring is noetherian. Consequently the Hilbert Basis Theorems (T3) and (T4) may be restated thus.

(36$^\bullet$) Finite variable polynomial rings and power series rings over a noetherian ring are noetherian. Also homomorphic images of noetherian rings are noetherian. Therefore affine rings over noetherian rings are noetherian. In particular affine rings over a field, and affine rings over the ring of integers, are noetherian.

§6: PRIMARY DECOMPOSITION

The fact that every positive integer can be written as a product of powers of prime numbers can be reformulated by saying that every nonzero ideal in the ring of integers can be expressed as a finite product, and hence finite intersection, of ideals generated by powers of prime numbers. In (O15) to (O20) we have generalized this to obtain primary and quasiprimary decompositions of ideals in a ring R. Let us restate these generalizations by first recalling some relevant definitions.

The radical of an ideal I in R is defined by $\text{rad}_R I = \{r \in R : r^e \in I \text{ for some } e \in \mathbb{N}_+\}$. By $\text{spec}(R)$ and $\text{mspec}(R)$ we denote the set of all prime and maximal ideals in R respectively. We put $\text{vspec}_R I = \{P \in \text{spec}(R) : I \subset P\}$, and $\text{nvspec}_R I =$ set of all minimal members of $\text{vspec}_R I$. I is primary means $I \neq R$ and: $r \in R$ and $s \in R \setminus I$ with $rs \in I \Rightarrow r^e \in I$ for some $e \in \mathbb{N}_+$. If I is primary then $\text{rad}_R I$ is prime and we say that I is $(\text{rad}_R I)$-primary. We say that I is P-quasiprimary if $P \in \text{spec}(R)$ with $I \neq R$ and: $r \in R$ and $s \in R \setminus I$ with $rs \in I \Rightarrow r \in P$; then I is P-primary iff I is P-quasiprimary and $P \subset \text{rad}_R I$; see (O7).

For any $a \in R$ we put $(I : a)_R = \{r \in R : ra \in I\}$. We put $\text{ass}_R(R/I) = \{P \in \text{spec}(R) : (I : a)_R \text{ is } P\text{-primary for some } a \in R\}$ and $\text{tass}_R(R/I) = \{P \in \text{spec}(R) : (I : a)_R = P \text{ for some } a \in R\}$; we also put $\text{nass}_R(R/I) =$ set of all minimal members of $\text{ass}_R(R/I)$; members of $\text{ass}_R(R/I)$ and $\text{nass}_R(R/I)$ are called associated primes and associated minimal primes of I in R respectively. By a multiplicative set in R we mean a subset S of R with $1 \in S$ such that: x, y in $S \Rightarrow xy \in S$. For any multiplicative set S in R we put $[I : S]_R = \cup_{a \in S}(I : a)_R =$ the isolated S-component of I in R. For any $P \in \text{spec}(R)$ we put $[I : P]_R = [I : R \setminus P]_R =$ the isolated P-component of I in R. More generally for any $P_1, \ldots, P_n$ in $\text{spec}(R)$ with $n \in \mathbb{N}_+$ we put $[I : (P_1, \ldots, P_n)]_R = [I : \cap_{1 \leq i \leq n}(R \setminus P_i)]_R =$ the isolated $(P_1, \ldots, P_n)$-component of I in R.

An irredundant primary decomposition of I is an expression of the form

(†) $\begin{cases} I = \cap_{1 \leq i \leq n} Q_i \quad \text{(CONVENTION: } \cap_{1 \leq i \leq 0} Q_i = R) \\ \text{where } n \in \mathbb{N} \text{ and } Q_i \text{ is a primary ideal in } R \text{ for } 1 \leq i \leq n \\ \text{such that } \cap_{j \in \{1, \ldots, n\} \setminus \{i\}} Q_j \not\subset Q_i \text{ for all } i \in \{1, \ldots, n\}, \text{ and} \\ \text{upon letting } \text{rad}_R Q_i = P_i \text{ we have } P_i \neq P_j \text{ for all } i \neq j \text{ in } \{1, \ldots, n\}. \end{cases}$

A (not necessarily irredundant) primary decomposition of I is an expression of the form

(††) $\begin{cases} I = \cap_{1 \leq i \leq n} Q_i \quad \text{(CONVENTION: } \cap_{1 \leq i \leq 0} Q_i = R) \\ \text{where } n \in \mathbb{N} \text{ and } Q_i \text{ is a primary ideal in } R \text{ for } 1 \leq i \leq n \\ \text{and } \text{rad}_R Q_i = P_i \text{ for } 1 \leq i \leq n. \end{cases}$

Given (††), first by (2•) of (O8) we arrange the fourth line of (†) to hold, and then by discarding some of the Q_i we arrange the third line of (†) to hold. Thus from a primary decomposition we extract an irredundant primary decomposition. Therefore by (6•) and (7•) of (O15) we get the first sentence of the following Decomposition Theorem (T5) concerning the existence of (†); by (12•) and (13•) of (O20) and (35•) of (O26) we get the second sentence of (T5) and also Decomposition Theorem (T8). For primes occurring in (†), by (8.1•) to (8.3•) of (O16) we get the following Uniqueness Theorem (T6), and for the primaries occurring in (†), by (10•) and (11•) of (O19) we get the following Partial Uniqueness Theorem (T7).

PRIMARY DECOMPOSITION THEOREM FOR IDEALS (T5). If R is noetherian then every ideal I in R has an irredundant primary decomposition. Without assuming R to be noetherian, if for an ideal I in R we have that

(† † †) $\qquad\qquad$ $\mathrm{vspec}_R I$ is a finite subset of $\mathrm{mspec}(R)$

then I has an irredundant primary decomposition (†) for which

$$Q_1 \cap \cdots \cap Q_n = Q_1 \ldots Q_n.$$

PRIME UNIQUENESS THEOREM FOR IDEALS (T6). In decomposition (†), $\mathrm{nvspec}_R I = \mathrm{nass}_R(R/I)$ and $\mathrm{ass}_R(R/I) = \{P_1, \ldots, P_n\}$. If R is noetherian then in decomposition (†), $\mathrm{ass}_R(R/I) = \mathrm{tass}_R(R/I)$. If a nonunit ideal I in R satisfies († † †) then in its decomposition (†), $\mathrm{nvspec}_R I = \mathrm{nass}_R(R/I) = \mathrm{ass}_R(R/I)$.

PRIMARY UNIQUENESS THEOREM FOR IDEALS (T7). In (††) (and hence in (†)) let $1 \le i(1) < i(2) < \cdots < i(h) \le n$ be any sequence of integers with $h \in \mathbb{N}_+$ such that: $j \in \{1, \ldots, n\}$ with $P_j \subset P_{i(l)}$ for some $l \in \{1, \ldots, h\} \Rightarrow j \in \{i(1), \ldots, i(h)\}$; then we have $[I : (P_{i(1)}, \ldots, P_{i(h)})]_R = \cap_{1 \le l \le h} Q_{i(l)}$. More generally in (††) (and hence in (†)) let S be any multiplicative set in R and let $1 \le i(1) < i(2) < \cdots < i(h) \le n$ be the unique sequence of integers such that for any $m \in \{1, \ldots, n\}$ we have: $m \in \{i(1), \ldots, i(h)\} \Leftrightarrow S \cap P_m = \emptyset$; then we have $[I : S]_R = \cap_{1 \le l \le h} Q_{i(l)}$. Hence in particular, by (T6) we see that, in (†), for every $P_i \in \mathrm{nass}_R(R/I)$ we have $[I : P_i]_R = Q_i$.

QUASIPRIMARY DECOMPOSITION THEOREM FOR IDEALS (T8). For any nonunit ideal I in R, let W be the set of all maximal members of the set of all ideals J with $I \subset J$ and $J \cap \{x \in R : xy \in R \setminus I \text{ for all } y \in R \setminus I\} = \emptyset$, let $W' = \mathrm{vspec}_R(I)$, and for every $P \in \mathrm{spec}(R)$ let $I_{[P]} = [I : P]_R$. Then W is a nonempty subset of $\mathrm{vspec}_R I$, for every $P \in \mathrm{spec}(R) \setminus W'$ we have $I_{[P]} = R$, for every $P \in W'$ the ideal $I_{[P]}$ is P-quasiprimary, and we have the quasiprimary decompositions $I = \cap_{P \in W} I_{[P]} = \cap_{P \in W'} I_{[P]} = \cap_{P \in \mathrm{spec}(R)} I_{[P]}$. Moreover, for every $P \in \mathrm{nvspec}_R(I)$ the ideal $I_{[P]}$ is P-primary.

§6.1: PRIMARY DECOMPOSITION FOR MODULES

Turning to a submodule U of an R-module V, the radical of U in V is defined by $\mathrm{rad}_V U = \{r \in R : r^e V \subset U$ for some $e \in \mathbb{N}_+\}$. U is primary means $U \neq V$ and: $r \in R$ and $s \in V \setminus U$ with $rs \in U \Rightarrow r \in \mathrm{rad}_V U$. If U is primary then $\mathrm{rad}_V U \in \mathrm{spec}(R)$ and we say that U is $(\mathrm{rad}_V U)$-primary. We say that U is P-quasiprimary if $P \in \mathrm{spec}(R)$ with $U \neq V$ and: $r \in R$ and $s \in V \setminus U$ with $rs \in U \Rightarrow r \in P$; then U is P-primary iff U is P-quasiprimary and $P \subset \mathrm{rad}_V U$; see (O13).

For any $a \in R$ we put $(U : a)_V = \{v \in V : av \in U\}$. For any $a \in V$ we put $(U : a)_R = \{r \in R : ra \in U\}$. We put $\mathrm{ass}_R(V/U) = \{P \in \mathrm{spec}(R) : (U : a)_R$ is P-primary for some $a \in V\}$ and $\mathrm{tass}_R(V/U) = \{P \in \mathrm{spec}(R) : (U : a)_R = P$ for some $a \in V\}$; we also put $\mathrm{nass}_R(V/U) =$ set of all minimal members of $\mathrm{ass}_R(V/U)$; members of $\mathrm{ass}_R(V/U)$ and $\mathrm{nass}_R(V/U)$ are called associated primes and associated minimal primes of U in V respectively. Note that $\mathrm{ann}_R(V/U) = \{r \in R : rV \subset U\}$. For any multiplicative set S in R we put $[U : S]_V = \cup_{a \in S}(U : a)_V =$ the isolated S-component of U in V. For any $P \in \mathrm{spec}(R)$ we put $[U : P]_V = [U : R \setminus P]_V =$ the isolated P-component of U in V. More generally for any $P_1, \ldots, P_n$ in $\mathrm{spec}(R)$ with $n \in \mathbb{N}_+$ we put $[U : (P_1, \ldots, P_n)]_V = [U : \cap_{1 \leq i \leq n}(R \setminus P_i)]_V =$ the isolated $(P_1, \ldots, P_n)$-component of U in V.

An irredundant primary decomposition of U is an expression of the form

$$(\dagger') \quad \begin{cases} U = \cap_{1 \leq i \leq n} Q_i \quad (\text{CONVENTION: } \cap_{1 \leq i \leq 0} Q_i = V) \\ \text{where } n \in \mathbb{N} \text{ and } Q_i \text{ is a primary submodule of } V \text{ for } 1 \leq i \leq n \\ \text{such that } \cap_{j \in \{1, \ldots, n\} \setminus \{i\}} Q_j \not\subset Q_i \text{ for all } i \in \{1, \ldots, n\}, \text{ and} \\ \text{upon letting } \mathrm{rad}_V Q_i = P_i \text{ we have } P_i \neq P_j \text{ for all } i \neq j \text{ in } \{1, \ldots, n\}. \end{cases}$$

A (not necessarily irredundant) primary decomposition of U is an expression of the form

$$(\dagger\dagger') \quad \begin{cases} U = \cap_{1 \leq i \leq n} Q_i \quad (\text{CONVENTION: } \cap_{1 \leq i \leq 0} Q_i = V) \\ \text{where } n \in \mathbb{N} \text{ and } Q_i \text{ is a primary submodule of } V \text{ for } 1 \leq i \leq n \\ \text{and } \mathrm{rad}_V Q_i = P_i \text{ for } 1 \leq i \leq n. \end{cases}$$

Given $(\dagger\dagger')$, first by $(2'^\bullet)$ of (O13) we arrange the fourth line of $(\dagger')$ to hold, and then by discarding some of the Q_i we arrange the third line of $(\dagger')$ to hold. Thus from a primary decomposition we extract an irredundant primary decomposition. Therefore by $(6^\bullet)$ and $(7^\bullet)$ of (O15) we get the first sentence of the following Decomposition Theorem (T5$'$) concerning the existence of $(\dagger')$; by $(12^\bullet)$ and $(14^\bullet)$ of (O20) we get its second sentence and also the Decomposition Theorem (T8$'$). For primes occurring in $(\dagger')$, by $(8.1^\bullet)$ to $(8.3^\bullet)$ of (O16) we get the following Uniqueness Theorem (T6$'$), and for the primaries occurring in $(\dagger')$, by $(10^\bullet)$ and $(11^\bullet)$ of (O19) we get the following Partial Uniqueness Theorem (T7$'$).

PRIMARY DECOMPOSITION THEOREM FOR MODULES (T5'). If V is noetherian (i.e., every submodule of V is finitely generated) then every submodule U of V has an irredundant primary decomposition. Without assuming V to be noetherian, if for a submodule U of V we have that

$$(\dagger \; \dagger \; \dagger') \qquad \text{vspec}_R(\text{ann}_R(V/U)) \text{ is a finite subset of mspec}(R)$$

then U has an irredundant primary decomposition $(\dagger')$.

PRIME UNIQUENESS THEOREM FOR MODULES (T6'). In decomposition $(\dagger')$, $\text{nvspec}_R(\text{ann}_R(V/U)) = \text{nass}_R(V/U)$ and $\text{ass}_R(V/U) = \{P_1, \ldots, P_n\}$. If R is noetherian then in decomposition $(\dagger')$, $\text{ass}_R(V/U) = \text{tass}_R(V/U)$. If a submodule U of V with $U \neq V$ satisfies $(\dagger \; \dagger \; \dagger')$ then in its decomposition $(\dagger')$, $\text{nvspec}_R(\text{ann}_R(V/U)) = \text{nass}_R(V/U) = \text{ass}_R(V/U)$.

PRIMARY UNIQUENESS THEOREM FOR MODULES (T7'). In $(\dagger\dagger')$ (and hence in $(\dagger')$) let $1 \leq i(1) < i(2) < \cdots < i(h) \leq n$ be any sequence of integers with $h \in \mathbb{N}_+$ such that: $j \in \{1, \ldots, n\}$ with $P_j \subset P_{i(l)}$ for some $l \in \{1, \ldots, h\} \Rightarrow j \in \{i(1), \ldots, i(h)\}$; then we have $[U : (P_{i(1)}, \ldots, P_{i(h)})]_V = \cap_{1 \leq l \leq h} Q_{i(l)}$. More generally in $(\dagger\dagger')$ (and hence in $(\dagger')$) let S be any multiplicative set in R and let $1 \leq i(1) < i(2) < \cdots < i(h) \leq n$ be the unique sequence of integers such that for any $m \in \{1, \ldots, n\}$ we have: $m \in \{i(1), \ldots, i(h)\} \Leftrightarrow S \cap P_m = \emptyset$; then we have $[U : S]_V = \cap_{1 \leq l \leq h} Q_{i(l)}$. Hence in particular, by (T6') we see that, in $(\dagger')$, for every $P_i \in \text{nass}_R(V/U)$ we have $[U : P_i]_V = Q_i$.

QUASIPRIMARY DECOMPOSITION THEOREM FOR MODULES (T8'). For any submodule U of V with $U \neq V$, let $I = \text{ann}_R(V/U)$, let W be the set of all maximal members of the set of all ideals J with $I \subset J$ and $J \cap \{x \in R : xy \in V \setminus U \text{ for all } y \in V \setminus U\} = \emptyset$, and let $W' = \{P \in \text{spec}(R) : U_{[P]} \neq V\}$ where for every $P \in \text{spec}(R)$ we are putting $U_{[P]} = [U : P]_V$. Then W is a nonempty subset of $\text{vspec}_R I$, for every $P \in W'$ the submodule $U_{[P]}$ is P-quasiprimary, and we have the quasiprimary decompositions $U = \cap_{P \in W} U_{[P]} = \cap_{P \in W'} U_{[P]} = \cap_{P \in \text{spec}(R)} U_{[P]}$. Also $W' \subset \text{vspec}_R(\text{ann}_R(V/U))$, and if V is a finitely generated R-module then $W' = \text{vspec}_R(\text{ann}_R(V/U))$. Moreover, for every $P \in \text{nvspec}_R I$ the submodule $U_{[P]}$ is P-primary.

§7: LOCALIZATION

The quotient field $\text{QF}(E)$ of a domain E introduced in L2§2 gets generalized to the total quotient ring $\text{QR}(R)$ of a ring R when we replace the set $E^\times$ of all nonzero elements of E as denominators by the set $S_R(R)$ of all nonzerodivisors of R, i.e., those $r \in R$ for which: $rr' = 0$ with $r' \in R \Rightarrow r' = 0$. Note that the set $U(R)$ of all units in R is a subset of $S_R(R)$ which itself is a multiplicative set in R, where we recall that a multiplicative set in R is a subset S of R with $1 \in S$ such that: $s \in S$ and $s' \in S \Rightarrow ss' \in S$.

Still more generally we may let any multiplicative set S in R play the role of denominators and thereby obtain the localization R_S of R at S thus. We define R_S to be the set of all equivalence classes of pairs $(u, v) \in R \times S$ under the equivalence relation: $(u, v) \sim (u', v') \Leftrightarrow v''(uv' - u'v) = 0$ for some $v'' \in S$. The equivalence class containing (u, v) is denoted by u/v or $\frac{u}{v}$, and we add and multiply the equivalence classes by the rules $\frac{u_1}{v_1} + \frac{u_2}{v_2} = \frac{u_1 v_2 + u_2 v_1}{v_1 v_2}$ and $\frac{u_1}{v_1} \times \frac{u_2}{v_2} = \frac{u_1 u_2}{v_1 v_2}$. Also we "send" any $u \in R$ to $u/1 \in R_S$. This makes R_S into a ring and $u \mapsto u/1$ gives the "canonical" ring homomorphism

$$\phi : R \to R_S.$$

Clearly

$$\phi(S) \subset U(R_S) \quad \text{and} \quad \ker(\phi) = [0 : S]_R$$

where we recall that for any ideal I in R we have defined

$$\begin{cases} [I : S]_R &= \{r \in R : rs \in I \text{ for some } s \in S\} \\ &= \text{the isolated } S\text{-component of } I \text{ in } R. \end{cases}$$

The localization of R at $S_R(R)$ is called the total quotient ring of R and denoted by $\mathrm{QR}(R)$. Clearly $[0 : S_R(R)]_R = 0$ and hence the canonical map of R into $\mathrm{QR}(R)$ is injective. By "identifying" every $u \in R$ with $u/1 \in \mathrm{QR}(R)$, we may and we shall regard $\mathrm{QR}(R)$ to be an overring of R. Note that then $S_R(R) \subset U(\mathrm{QR}(R))$ and every element of $\mathrm{QR}(R)$ is of the form u/v with $u \in R$ and $v \in S_R(R)$; also note that $\mathrm{QR}(R)$ is its own total quotient ring. Conversely, if T is an overring of R with $S_R(R) \subset U(T)$ then clearly there is a unique R-injection of $\mathrm{QR}(R)$ into T whose image is the set of all u/v with $u \in R$ and $v \in S_R(R)$; we may call this image the total quotient ring of R in T and again denote it by $\mathrm{QR}(R)$. Note that if R is a domain then $\mathrm{QF}(R) = \mathrm{QR}(R)$, and in this case the said image may be called the quotient field of R in T. In particular we may talk of the quotient field of a domain in an overfield.

Let us revert to the general multiplicative set S in R and note that: $\ker(\phi) = 0 \Leftrightarrow S \subset S_R(R)$. If $S \subset S_R(R)$ then clearly there is a unique R-injection of R_S into $\mathrm{QR}(R)$; we call its image the localization of R at S in $\mathrm{QR}(R)$ and continue to denote it by R_S; let us call this the GOOD case; thus in the good case R_S is an overring of R. In the general case the passage from R to R_S can be achieved in two steps. First we take the residue class map $\alpha : R \to R/[0 : S]_R$. Now $\alpha(S)$ is a multiplicative set in $\alpha(R)$ with $\alpha(S) \subset S_{\alpha(R)}(\alpha(R))$ and we take $\alpha(R) \subset \overline{Q} = \alpha(R)_{\alpha(S)}$ as in the good case, and we consider the map $\overline{\gamma} = \beta\alpha : R \to \overline{Q}$ where $\beta : \alpha(R) \to \overline{Q}$ is the natural injection. Now we may identify R_S with $\overline{Q}$ via the unique isomorphism $\overline{\psi} : R_S \to \overline{Q}$ such that $\overline{\psi}\phi = \overline{\gamma}$.

The isomorphism $\overline{\psi}$ is generalized in the following Characterization Theorem (T9) for the localization R_S. In Theorems (T10) to (T12) we give some basic properties of localization.

CHARACTERIZATION THEOREM FOR LOCALIZATION (T9). Let $\gamma : R \to Q$ be a ring homomorphism. Then: there is an isomorphism $\psi : R_S \to Q$ with $\psi\phi = \gamma \Leftrightarrow \ker(\gamma) = [0 : S]_R$ with $\gamma(S) \subset U(Q)$ and every element in Q can be written as $\gamma(u)/\gamma(v)$ with $u \in R$ and $v \in S$. Moreover, if $\gamma(S) \subset U(Q)$ then there is a unique homomorphism $\psi : R_S \to Q$ with $\psi\phi = \gamma$.

PROOF. Straightforward.

TRANSITIVITY OF LOCALIZATION (T10). Let $\gamma : R \to R_{S'}$ and $\phi' : R_S \to (R_S)_{\phi(S')}$ be the canonical maps where S' is a multiplicative set in R with $S \subset S'$. Then there is a unique isomorphism $\psi : (R_S)_{\phi(S')} \to R_{S'}$ with $\psi\phi'\phi = \gamma$. Moreover, if every element of S' is of the form ss' with $s \in S$ and $s' \in U(R)$, then ϕ' is an isomorphism.

PROOF. Follows from (T9).

PERMUTABILITY OF LOCALIZATION AND SURJECTION (T11). Let $\delta : R \to R'$ be a ring epimorphism with $\ker(\delta) \cap S = \emptyset$, and consider the map $\gamma = \phi'\delta : R \to \delta(R)_{\delta(S)}$ where $\phi' : \delta(R) \to \delta(R)_{\delta(S)}$ is the canonical homomorphism. Then there exists a unique epimorphism $\psi : R_S \to \delta(R)_{\delta(S)}$ with $\psi\phi = \gamma$.

PROOF. Follows from (T9).

IDEAL CORRESPONDENCE FOR LOCALIZATION (T12). For any ideal I in R let $I_S = \phi(I)R_S$, and note that the two meanings of R_S do coincide. For any ideal I in R let $I_{[S]} = [I : S]_R$, and let C be the set of all ideals I in R for which $I_{[S]} = I$. Finally let E be the set of all ideals in R_S. Then we have the following.

(a) For any ideal I in R we have $I_S = \{\phi(t)/\phi(s) : (t, s) \in I \times S\}$. Moreover: $I_S \ne R_S \Leftrightarrow I \cap S = \emptyset$.

(b) For any ideal I in R we have $I_{[S]} = \phi^{-1}(I_S)$.

(c) An ideal I in R is a contracted ideal (i.e., for some ideal J in R_S we have $I = \phi^{-1}(J)$) iff $I \in C$.

(d) Every ideal J in R_S is an extended ideal (i.e., for some ideal I in R we have $J = \phi(I)R_S$).

(e) $I \mapsto I_S$ gives a bijection of C onto E, and its inverse is given by $J \mapsto \phi^{-1}(J)$. The sets C and E are closed with respect to the ideal theoretic operations of radicals, intersections, and quotients, i.e., for instance: $I \in C \Rightarrow \mathrm{rad}_R I \in C$; recall that quotients = parenthetical colons, where for ideals I, H in R we have

$(I : H)_R = \{r \in R : rh \in I \text{ for all } h \in H\}$. Moreover, the said bijection commutes with these operations, i.e., for instance: $I \in C \Rightarrow \mathrm{rad}_{R_S} I_S = (\mathrm{rad}_R I)_S$.

(f) If $\overline{P}$ and $\overline{Q}$ are ideals in R such that $\overline{P}$ is prime and $\overline{Q}$ is $\overline{P}$-primary with $\overline{Q} \cap S = \emptyset$ then: $\overline{P} \cap S = \emptyset$, $\overline{P}$ and $\overline{Q}$ belong to C and both contain $\ker(\phi)$, $\overline{P}_S$ is prime, and $\overline{Q}_S$ is $(\overline{P}_S)$-primary.

(g) A primary ideal, and hence in particular a prime ideal, in R is a contracted ideal iff it belongs to C iff it is disjoint from S. Moreover, $\overline{P} \mapsto \overline{P}_S$ gives a bijection of $C \cap \mathrm{spec}(R)$ onto $\mathrm{spec}(R_S)$. Likewise, given any $\overline{P} \in C \cap \mathrm{spec}(R)$, $\overline{Q} \mapsto \overline{Q}_S$ gives a bijection of all $\overline{P}$-primaries in R onto the set of all $(\overline{P}_S)$-primaries in R_S, and this bijection preserves intersections and quotients.

(h) Given an irredundant primary decomposition $I = \cap_{1 \le i \le n} \overline{Q}_i$ for an ideal I in R, let $1 \le i(1) < i(2) < \cdots < i(h) \le n$ be the unique sequence of integers such that for any $m \in \{1, \ldots, n\}$ we have: $m \in \{i(1), \ldots, i(h)\} \Leftrightarrow S \cap \overline{P}_m = \emptyset$. Then $I_S = \cap_{1 \le j \le h}(\overline{Q}_{i(j)})_S$ is an irredundant primary decomposition in R_S, and $I_{[S]} = \cap_{1 \le j \le h} \overline{Q}_{i(j)}$ is an irredundant primary decomposition in R. In particular: $I = I_{[S]} \Leftrightarrow S \cap P_i = \emptyset$ for $1 \le i \le n$.

(i) If R is a noetherian ring then so is R_S.

(j) If R is a noetherian ring then $\ker(\phi) =$ the intersection of all primary ideals in R which are disjoint from S, and also $\ker(\phi) =$ the intersection of those primary ideals which occur in an irredundant primary decomposition of 0 in R and which are disjoint from S.

PROOF OF (a). By the definition of R_S, or by (T9), every element of R_S can be written as $\phi(r)/\phi(s)$ with $(r, s) \in R \times S$. Therefore, since $I_S = \phi(I)R_S$, every element of I_S can be written as a finite sum $\sum_i \phi(t_i)\phi(r_i)/\phi(s_i)$. Now this sum equals $\phi(t)/\phi(s)$ where $t = \sum_i t_i r_i \prod_{j \ne i} s_j \in I$ and $s = \prod_i s_i \in S$. Thus $I_S = \{\phi(t)/\phi(s) : (t, s) \in I \times S\}$. It follows that

$$
\begin{cases}
1 \in I_S \\
\Leftrightarrow 1 = \phi(t)/\phi(s) \text{ for some } (t, s) \in I \times S \\
\Leftrightarrow \phi(t - s) = 0 \text{ for some } (t, s) \in I \times S \\
\Leftrightarrow s'(t - s) = 0 \text{ for some } (t, s, s') \in I \times S \times S \quad \text{because } \ker(\phi) = [0 : S]_R \\
\Leftrightarrow I \cap S \ne \emptyset.
\end{cases}
$$

Consequently: $I_S \ne R_S \Leftrightarrow I \cap S = \emptyset$.

PROOF OF (b). For any $r \in R$ we have

$$
\begin{cases}
r \in \phi^{-1}(I_S) \\
\Leftrightarrow \phi(r) \in I_S \\
\Leftrightarrow \phi(r) = \phi(t)/\phi(s) \text{ for some } (t, s) \in I \times S \qquad \text{by (a)} \\
\Leftrightarrow \phi(rs - t) = 0 \text{ for some } (t, s) \in I \times S \\
\Leftrightarrow s'(rs - t) = 0 \text{ for some } (t, s, s') \in I \times S \times S \quad \text{by (T9)} \\
\Leftrightarrow r \in I_{[S]}.
\end{cases}
$$

PROOF OF (c). If $I \in C$ then, upon letting $J = I_S$, by (b) we see that I is contracted. Conversely suppose $I = \phi^{-1}(J)$ for an ideal J in R_S. Then $I_S \subset J$ and hence $\phi^{-1}(I_S) \subset \phi^{-1}(J) = I$. Therefore by (b) we get $I_{[S]} \subset I$ and hence $I_{[S]} = I$.

PROOF OF (d). Given an ideal J in R_S let $I = \phi^{-1}(J)$. By (T9), any $u \in J$ can be expressed as $u = \phi(r)/\phi(s)$ with $(r, s) \in R \times S$, and multiplying this by $\phi(s)$ we get $\phi(r) \in J$. Consequently $r \in I$ and hence $u \in I_S$. Thus $J \subset I_S$ and obviously $I_S \subset J$. Therefore $J = I_S$.

PROOF OF (e). In view of (c) and (d), this follows from the following Lemma (T13.3).

PROOF OF (f). Let $\overline{P}$ and $\overline{Q}$ be ideals in R such that $\overline{P}$ is prime and $\overline{Q}$ is $\overline{P}$-primary with $\overline{Q} \cap S = \emptyset$. Then $s \in \overline{P} \cap S \Rightarrow s^e \in \overline{Q} \cap S$ for some $e \in \mathbb{N}_+$ which contradicts the assumption that $\overline{Q} \cap S = \emptyset$. Consequently $\overline{P} \cap S = \emptyset$, and hence $\overline{P}$ and $\overline{Q}$ belong to C, and therefore by (c) they are contracted ideals and hence they contain $\ker(\phi)$. Now by (a) and (e) we see that $\overline{P}_S \subset \mathrm{rad}_{R_S}\overline{Q}_S \neq R_S$ and hence, in view of PIC = (O7), for proving that $\overline{P}_S$ is prime and $\overline{Q}_S$ is $(\overline{P}_S)$–primary, it suffices to show that: $x \in R_S$ and $x' \in R_S \setminus \overline{Q}_S$ with $xx' \in \overline{Q}_S \Rightarrow x \in \overline{P}_S$. Given x, x' as postulated, by (a) we can write $x = \phi(r)/\phi(s), x' = \phi(r')/\phi(s'), xx' = \phi(r'')/\phi(s'')$ with r, r', r'' in R and s, s', s'' in S such that $r' \notin \overline{Q}$ and $r'' \in \overline{Q}$. Equating two values of xx' we obtain $\phi(rr')/\phi(ss') = \phi(r'')/\phi(s'')$ and hence by (T9) we get $s^*(rr's'' - r''ss') = 0$ for some $s^* \in S$. Consequently $r(r's''s^*) = r''ss's^* \in \overline{Q}$ with $r's''s^* \in R \setminus \overline{Q}$, and hence $r \in \overline{P}$. Therefore $x = \phi(r)/\phi(s) \in \overline{P}_S$.

PROOF OF (g). In view of (c), (e) and (f), this follows from the following Lemmas (T13.1) and (T13.2) together with the additional observation that, say by (O8), if $\overline{Q}, \widehat{Q}$ are $\overline{P}$-primaries then so are $\overline{Q} \cap \widehat{Q}$ and $(\overline{Q} : \widehat{Q})_R$, except that: if $\widehat{Q} \subset \overline{Q}$ then $(\overline{Q} : \widehat{Q})_R = R$ and $\widehat{Q}_S \subset \overline{Q}_S$ with $(\overline{Q}_S : \widehat{Q}_S)_{R_S} = R_S = R^e$.

PROOF OF (h). Let $\overline{I} = \cap_{1 \leq j \leq h}\overline{Q}_{i(j)}$ and $\overline{J} = \cap_{1 \leq j \leq h}(\overline{Q}_{i(j)})_S$. The first equation is obviously an irredundant primary decomposition in R. By (e) to (g) the second equation is an irredundant primary decomposition in R_S and we have $\overline{J} = \overline{I}_S$. Also clearly: $I = \overline{I} \Leftrightarrow S \cap P_i = \emptyset$ for $1 \le i \le n$. By (T7) we have $\overline{I} = I_{[S]}$. It only remains to show that $I_S = \overline{J}$. By the following Lemma (T13.5) we have $I_S \subset \overline{J}$. Conversely, by (a), any $x \in \overline{J}$ can be expressed as $x = \phi(t)/\phi(s)$ with $(t, s) \in \overline{I} \times S$. We can take $s' = \prod_{l \in \{1,\dots,n\} \setminus \{i(1),\dots,i(h)\}} s'_l$ with $s'_l \in S \cap \overline{Q}_l$ and then $s't \in I$ and hence by (a) we get $x = \phi(ts')/\phi(ss') \in I_S$.

PROOF OF (i). Follows from (d).

PROOF OF (j). Assuming R to be noetherian, by (T9) we know that $\ker(\phi) = 0_S$ and, in view of (T5), by taking $I = 0$ in (h) we see that $\ker(\phi) =$ the intersection of those primary ideals which occur in an irredundant primary decomposition of 0 in R and which are disjoint from S, and hence by (g) we see that $\ker(\phi) =$ the intersection of all primary ideals in R which are disjoint from S.

LEMMA ON CONTRACTED AND EXTENDED IDEALS (T13). Let $\theta : R \to T$ be a ring homomorphism. For any ideal I in R let $I^e = \phi(I)T$, and for any ideal J in T let $J^c = \phi^{-1}(J)$; note that if θ is the natural injection of R into an overring T then $I^e = IT$ and $J^c = J \cap R$. Let C be the set of all contracted ideals in R (relative to θ), i.e., ideals which can be expressed as J^c with ideal J in T. Let E be the set of all extended ideals in T (relative to θ), i.e., ideals which can be expressed as I^e with ideal I in R. Then we have the following.

(T13.1) For any ideals $I \subset \overline{I}$ in R we have $I^e \subset \overline{I}^e$. For any ideals $J \subset \overline{J}$ in T we have $J^c \subset \overline{J}^c$. For any ideal I in R we have $I \subset (I^e)^c$ and $I^e = ((I^e)^c)^e$. For any ideal J in T we have $(J^c)^e \subset J$ and $J^c = ((J^c)^e)^c$.

(T13.2) If P, Q are ideals in T such that P is prime and Q is P-primary, then P^c is prime and Q^c is (P^c)-primary. If $J = \cap_{1 \le i \le n} Q_i$ is a primary decomposition in T then $J^c = \cap_{1 \le i \le n} Q_i^c$ is a primary decomposition in R; however the decomposition of J could be irredundant without the decomposition of J^c being irredundant.

(T13.3) $I \mapsto I^e$ gives a bijection of C onto E, and its inverse is given by $J \mapsto J^c$. The set C is closed under the ideal theoretic operations of radicals, intersections, and quotients. Moreover, if the set E is also closed under one of these three operations then the said bijection commutes with it. Finally the set E is always closed under the ideal theoretic operations of sums and finite products.

PROOF. The proofs of (T13.1) and (T13.2) being straightforward, we shall prove (T13.3). Clearly $I \in C \Rightarrow I = (I^e)^c$. Also clearly $J \in E \Rightarrow J = (J^c)^e$. Therefore $I \mapsto I^e$ gives a bijection $C \to E$ whose inverse is given by $J \mapsto J^c$. For any ideals I and J in R and T respectively, we clearly have

(T13.4) $(\mathrm{rad}_R I)^e \subset \mathrm{rad}_T(I^e)$ and $(\mathrm{rad}_T J)^c = \mathrm{rad}_R(J^c)$.

From the second part of the above display it follows that C is closed under the operation of radicals, and if E is also closed under it then the said bijection commutes with it. For any families of ideals $(I_l)_{l \in L}$ and $(J_l)_{l \in L}$ in R and T respectively, we clearly have

(T13.5) $\qquad (\cap_{l \in L} I_j)^e \subset \cap_{l \in L} I_l^e \quad$ and $\quad (\cap_{l \in L} J_j)^c = \cap_{l \in L} J_l^c$

and

(T13.6) $\qquad (\sum_{l \in L} I_l)^e = \sum_{l \in L} I_l^e \quad$ and $\quad \sum_{l \in L} J_l^c \subset (\sum_{l \in L} J_l)^c$

and if the indexing set L is finite then we also have

(T13.7) $\qquad (\prod_{l \in L} I_l)^e = \prod_{l \in L} I_l^e \quad$ and $\quad \prod_{l \in L} J_l^c \subset (\prod_{l \in L} J_l)^c$

From the second part of (T13.5) it follows that C is closed under the operation of intersections, and if E is also closed under it then the said bijection commutes with it. From the first parts of (T13.6) and (T13.7) it follows that E is closed under the operation of sums and finte products respectively.

To turn to quotients let $I, \overline{I}$ be any ideals in R and let $J, \overline{J}$ be any ideals in T. Then clearly

(T13.8) $\qquad (I : \overline{I})_R^e \subset (I^e : \overline{I}^e)_T \quad$ and $\quad (J : \overline{J})_T^c \subset (J^c : \overline{J}^c)_R.$

We claim that

(T13.9) $\qquad \overline{J} \in E \Rightarrow (J : \overline{J})_T^c = (J^c : \overline{J}^c)_R.$

Namely, assuming $\overline{J} \in E$, we have

$$(J^c : \overline{J}^c)_R^e \overline{J} = (J^c : \overline{J}^c)_R^e (\overline{J}^c)^e = ((J^c : \overline{J}^c)_R (\overline{J}^c))^e \subset (J^c)^e$$

(where the first equation is because of the assumption $\overline{J} \in E$ and the second equation is because of the first part of (T13.7)) and hence we get

$$(J^c : \overline{J}^c)_R^e \subset (J : \overline{J})_T$$

and therefore we have

$$((J^c : \overline{J}^c)_R^e)^c \subset ((J : \overline{J})_T)^c$$

and hence, because obviously $(J^c : \overline{J}^c)_R \subset ((J^c : \overline{J}^c)_R^e)^c$, we get

$$(J^c : \overline{J}^c)_R \subset ((J : \overline{J})_T)^c$$

and therefore, because of the second part of (T13.8), we conclude that

$$(J : \overline{J})_T^c = (J^c : \overline{J}^c)_R.$$

This proves (T13.9). By taking $J = I^e$ and $\bar{J} = \bar{I}^e$ in (T13.9) we see that

(T13.10) $\qquad I \in C$ and $\bar{I} \in C \Rightarrow \begin{cases} (I : \bar{I})_R \in C \text{ and} \\ (I^e : \bar{I}^e)_T^c = (I : \bar{I})_R. \end{cases}$

By (T13.10) we conclude that C is closed under quotients, and if E is also closed under it then the said bijection commutes with it.

§7.1: LOCALIZATION AT A PRIME IDEAL

Given any prime ideal P in R we get a particularly interesting case of localization by taking $S = R \setminus P$. We put

$$R_P = R_{R \setminus P}$$

and call this the localization of R at P. Let

$$\Phi : R \to R_P$$

be the canonical map and note that now

$$\Phi(R \setminus P) \subset U(R_P) \quad \text{and} \quad \ker(\Phi) = [0 : P]_R$$

where, for any ideal I in R, by definition

$$\begin{cases} [I : P]_R &= [I : R \setminus P]_R \\ &= \{r \in R : rs \in I \text{ for some } s \in R \setminus P\} \\ &= \text{the isolated } P\text{-component of } I \text{ in } R. \end{cases}$$

Note that if $Z_R(R) \subset P$, where $Z_R(R)$ is the set of all zerodivisors in R, then we are in the good case, and R_P may be regarded as a subring of $QR(R)$.

The following Theorems (T14) to (T17) follow by taking $S = R \setminus P$ in the above Theorems (T9) to (T12) respectively.

CHARACTERIZATION THEOREM FOR PRIME LOCALIZATION (T14). Let $\gamma : R \to Q$ be a ring homomorphism. Then: there is an isomorphism $\psi : R_P \to Q$ with $\psi\Phi = \gamma \Leftrightarrow \ker(\gamma) = [0 : P]_R$ with $\gamma(R \setminus P) \subset U(Q)$ and every element in Q can be written as $\gamma(u)/\gamma(v)$ with $u \in R$ and $v \in R \setminus P$. Moreover, if $\gamma(R \setminus P) \subset U(Q)$ then there is a unique homomorphism $\psi : R_P \to Q$ with $\psi\Phi = \gamma$.

TRANSITIVITY OF PRIME LOCALIZATION (T15). Let $\gamma : R \to R_{P'}$ and $\Phi' : R_P \to (R_P)_{\Phi(P')}$ be the canonical maps where P' is a prime ideal in R with $P' \subset P$. Then there is a unique isomorphism $\psi : (R_P)_{\Phi(P')} \to R_{P'}$ with $\psi\Phi'\Phi = \gamma$.

PERMUTABILITY OF PRIME LOCALIZATION AND SURJECTION (T16). Let $\delta : R \to R'$ be a ring epimorphism with $\ker(\delta) \subset P$, and consider the map $\gamma = \Phi'\delta : R \to \delta(R)_{\delta(P)}$ where $\Phi' : \delta(R) \to \delta(R)_{\delta(P)}$ is the canonical homomorphism. Then there exists a unique epimorphism $\psi : R_P \to \delta(R)_{\delta(P)}$ with $\psi\Phi = \gamma$.

IDEAL CORRESPONDENCE FOR PRIME LOCALIZATION (T17). For any ideal I in R let $I_P = \Phi(I)R_P$, and note that the two meanings of R_P do coincide. For any ideal I in R let $I_{[P]} = [I : P]_R$, and let C be the set of all ideals I in R for which $I_{[P]} = I$. Finally let E be the set of all ideals in R_P. Then we have the following.

(a) For any ideal I in R we have $I_P = \{\Phi(t)/\Phi(s) : (t, s) \in I \times (R \setminus P)\}$. Moreover: $I_P \neq R_P \Leftrightarrow I \subset P$.

(b) For any ideal I in R we have $I_{[P]} = \Phi^{-1}(I_P)$.

(c) An ideal I in R is a contracted ideal (i.e., for some ideal J in R_P we have $I = \Phi^{-1}(J)$) iff $I \in C$.

(d) Every ideal J in R_P is an extended ideal (i.e., for some ideal I in R we have $J = \Phi(I)R_P$).

(e) $I \mapsto I_P$ gives a bijection of C onto E, and its inverse is given by $J \mapsto \Phi^{-1}(J)$. The sets C and E are closed with respect to the ideal theoretic operations of radicals, intersections, and quotients, i.e., for instance: $I \in C \Rightarrow \mathrm{rad}_R I \in C$; recall that quotients = parenthetical colons, where for ideals I, H in R we have $(I : H)_R = \{r \in R : rh \in I \text{ for all } h \in H\}$. Moreover, the said bijection commutes with these operations, i.e., for instance: $I \in C \Rightarrow \mathrm{rad}_{R_P} I_P = (\mathrm{rad}_R I)_P$.

(f) If $\overline{P}$ and $\overline{Q}$ are ideals in R such that $\overline{P}$ is prime and $\overline{Q}$ is $\overline{P}$-primary with $\overline{Q} \subset P$ then: $\overline{P} \subset P$, $\overline{P}$ and $\overline{Q}$ belong to C and both contain $\ker(\Phi)$, $\overline{P}_P$ is prime, and $\overline{Q}_P$ is $(\overline{P}_P)$-primary.

(g) A primary ideal, and hence in particular a prime ideal, in R is a contracted ideal iff it belongs to C iff it is contained in P. Moreover, $\overline{P} \mapsto \overline{P}_P$ gives a bijection of $C \cap \mathrm{spec}(R)$ onto $\mathrm{spec}(R_P)$. Likewise, given any $\overline{P} \in C \cap \mathrm{spec}(R)$, $\overline{Q} \mapsto \overline{Q}_P$ gives a bijection of all $\overline{P}$-primaries in R onto the set of all $(\overline{P}_P)$-primaries in R_P, and this bijection preserves intersections and quotients.

(h) Given an irredundant primary decomposition $I = \cap_{1 \leq i \leq n} \overline{Q}_i$ for an ideal I in R, let $1 \leq i(1) < i(2) < \cdots < i(h) \leq n$ be the unique sequence of integers such that for any $m \in \{1, \ldots, n\}$ we have: $m \in \{i(1), \ldots, i(h)\} \Leftrightarrow \overline{P}_m \subset P$. Then $I_P = \cap_{1 \leq j \leq h}(\overline{Q}_{i(j)})_P$ is an irredundant primary decomposition in R_P, and $I_{[P]} = \cap_{1 \leq j \leq h}\overline{Q}_{i(j)}$ is an irredundant primary decomposition in R. In particular: $I = I_{[P]} \Leftrightarrow P_i \subset P$ for $1 \leq i \leq n$.

(i) If R is a noetherian ring then so is R_P.

(j) If R is a noetherian ring then $\ker(\Phi)$ = the intersection of all primary ideals in R which are contained in P, and also $\ker(\Phi)$ = the intersection of those primary ideals which occur in an irredundant primary decomposition of 0 in R and which are contained in P.

Some obvious consequences of the above Theorem (T17) are worth repeating as:

LOCAL RING CONSTRUCTION THEOREM (T18). R_P is a quasilocal ring, i.e., a ring with a unique maximal ideal $M(R_P)$. We have

$$M(R_P) = \Phi(P)R_P \quad \text{with} \quad \Phi^{-1}(M(R_P)) = P.$$

The mapping $\overline{P} \mapsto \Phi(\overline{P})R_P$ gives a bijection of $\{\overline{P} \in \operatorname{spec}(R) : \overline{P} \subset P\}$ onto $\operatorname{spec}(R_P)$ whose inverse is given by Φ^{-1}, and hence in particular, referring to (O22) for the definitions of dimension and height, we have $\dim(R_P) = \operatorname{ht}_R P$. If R is noetherian then R_P is a local ring, i.e., a noetherian quasilocal ring. If $Z_R(0) \subset P$ (resp: if R is a domain) then regarding R_P to be a subring of $\operatorname{QR}(R)$ (resp: of $\operatorname{QF}(R)$), for any ideals I and J in R and R_P we have

$$\Phi(I)R_P = IR_P \quad \text{and} \quad \Phi^{-1}(J) = J \cap R.$$

§8: AFFINE VARIETIES

Postponing the discussion of varieties in projective space to a later opportunity, we shall now discuss varieties in affine space. Let

$$k \subset \kappa \subset \lambda$$

be fields and let N be a nonnegative integer. Consider the affine N-space

$$\mathbb{A}_\kappa^N = \kappa^N = \{\alpha = (\alpha_1, \ldots, \alpha_N) : \alpha_i \in \kappa \text{ for } 1 \le i \le N\}$$

over κ, and the N-variable polynomial ring

$$B_{N,k} = k[X_1, \ldots, X_N]$$

over k. For any $J \subset B_{N,\lambda}$ let

$$\mathbb{V}_\kappa(J) = \{\alpha \in \mathbb{A}_\kappa^N : f(\alpha) = 0 \text{ for all } f \in J\}$$

where for $f = f(X_1, \ldots, X_N) \in B_{N,\lambda}$ and $\alpha = (\alpha_1, \ldots, \alpha_N) \in \mathbb{A}_\kappa^N$ we are writing $f(\alpha) = f(\alpha_1, \ldots, \alpha_N)$. We call this the affine variety defined by J. For polynomials $f, g, \ldots$ in $B_{N,\lambda}$ we may write $\mathbb{V}_\kappa(f, g, \ldots)$ in place of $\mathbb{V}_\kappa(\{f, g, \ldots\})$ and we may informally call this the affine variety $f = g = \cdots = 0$. By a variety in $\mathbb{A}_\kappa^N$ defined over k we mean a subset of $\mathbb{A}_\kappa^N$ which can be expressed in the form $\mathbb{V}_\kappa(J)$ for some $J \subset B_{N,k}$. We put

$$\operatorname{avt}_k(\mathbb{A}_\kappa^N) = \text{ the set of all varieties in } \mathbb{A}_\kappa^N \text{ defined over } k.$$

Members of $\operatorname{avt}_k(\mathbb{A}_k^N)$ may be called varieties in $\mathbb{A}_k^N$. For any $U \subset \mathbb{A}_\kappa^N$ we put

$$\mathbb{I}_k(U) = \{f \in B_{N,k} : f(\alpha) = 0 \text{ for all } \alpha \in U\}$$

and we note that this is an ideal in $B_{N,k}$; we call it the ideal of U in $B_{N,k}$. For points $\alpha, \beta, \ldots$ in $\mathbb{A}_\kappa^N$ we may write $\mathbb{I}_k(\alpha, \beta, \ldots)$ in place of $\mathbb{I}_k(\{\alpha, \beta, \ldots\})$.

Given U, U' in $\mathrm{avt}_k(\mathbb{A}_\kappa^N)$, U' is a subvariety of U means $U' \subset U$, and U' is a proper subvariety of U means $U' \subset U$ with $U' \neq U$. Given U in $\mathrm{avt}_k(\mathbb{A}_\kappa^N)$, U is reducible means U is the union of two proper subvarieties; U is irreducible means it is nonempty and nonreducible. We put

$$\mathrm{iavt}_k(\mathbb{A}_\kappa^N) = \text{the set of all irreducible members of } \mathrm{avt}_k(\mathbb{A}_\kappa^N).$$

Referring to (O22) for the definition of the dimension dim of a ring, for any U in $\mathrm{avt}_k(\mathbb{A}_\kappa^N)$ we define the dimension of U by putting

$$\dim(U) = \dim(B_{N,k}/\mathbb{I}_k(U)).$$

Given $U \in \mathrm{avt}_k(\mathbb{A}_\kappa^N)$, for any $f \in B_{N,k}$ we get a "polynomial map" $\mathbb{A}_\kappa^N \to \kappa$ given by $\alpha \mapsto f(\alpha)$ and, letting α vary only over U, this gives a polynomial map $U \to \kappa$. Since two such maps on U coincide iff the difference of the corresponding members of $B_{N,k}$ belongs to the ideal $\mathbb{I}_k(U)$, the ring of all these κ-valued polynomial maps on U may be identified with the residue class ring $B_{N,k}/\mathbb{I}_k(U)$. Indeed this is the basic motivation behind the concept of residue class rings. We denote $B_{N,k}/\mathbb{I}_k(U)$ by $k[U]^*$ and call it the affine coordinate ring of U. In case $\mathbb{I}_k(U)$ is not the unit ideal in $B_{N,k}$, we may and we do identify k with a subfield of $k[U]^*$ and we note that then, according to (O27), $k[U]^*$ becomes an affine ring over k, and hence in particular it is a noetherian ring.

In case $U \in \mathrm{avt}_k(\mathbb{A}_\kappa^N)$ is such that $\mathbb{I}_k(U)$ is a prime ideal in $B_{N,k}$, we denote the quotient field of $k[U]^*$ by $k(U)^*$, and we call $k(U)^*$ the function field of U (over k). We denote the localization of $B_{N,k}$ at $\mathbb{I}_k(U)$ by $R_k(U)^*$, and call it the local ring of U (over k); note that by (T18) this is indeed a local ring. We denote the residue field $R_k(U)^*/M(R_k(U)^*)$ of $R_k(U)^*$ by $k(U)^\sharp$, and we call this field $k(U)^\sharp$ the alternative function field of U (over k); again we may and we do identify k with a subfield of $k(U)^\sharp$. Let $\Phi_U : B_{N,k} \to k[U]^*$ and $\Psi_U : R_k(U)^* \to k(U)^\sharp$ be the residue class epimorphisms. Now clearly there is a unique homomorphism $\psi_U : R_k(U)^* \to k(U)^*$ such that for all $f \in B_{N,k}$ and $g \in B_{N,k} \setminus \mathbb{I}_k(U)$ we have $\psi_U(f/g) = \Phi_U(f)/\Phi_U(g)$. Obviously $\mathrm{im}(\psi_U) = k(U)^*$ and $\ker(\psi_U) = M(R_k(U)^*)$ and hence there is a unique isomorphism $\phi_U : k(U)^\sharp \to k(U)^*$ for which we have $\phi_U \Psi_U = \psi_U$. We call ψ_U and ϕ_U the natural epimorphism and the natural isomorphism respectively. Note that $\mathbb{A}_k^N \subset \mathbb{A}_\kappa^N$, and the ideal $\mathbb{I}_k(\alpha)$ of any point α in $\mathbb{A}_k^N$ is the maximal ideal in $B_{N,k}$ generated by $X_1 - \alpha_1, \ldots, X_N - \alpha_N$; also $\mathbb{V}_\kappa(\mathbb{I}_k(\alpha)) = \{\alpha\}$, and writing α in place of $\{\alpha\}$ we get $k = k[\alpha]^* = k(\alpha)^* \subset R_k(\alpha)^* = $ the local ring of α.

The above notions are particularly relevant when k or κ is algebraically closed, or more precisely, when κ contains an algebraic closure of k. To take care of fields which may not be algebraically closed, we recall that analogously, for any subset J of any ring A, in (O16) we have defined the spectral variety, the maximal spectral variety, and the minimal spectral variety of J in A by putting

$$\text{vspec}_A J = \{P \in \text{spec}(A) : J \subset P\}$$

and

$$\text{mvspec}_A J = \text{mspec}(A) \cap \text{vspec}_A J$$

and

$$\text{nvspec}_A J = \text{the set of all minimal members of } \text{vspec}_A J$$

where $\text{spec}(A)$ and $\text{mspec}(A)$ are the sets of all prime and maximal ideals in A respectively. Moreover, for any $U \subset \text{spec}(A)$ we have defined the spectral ideal of U in A by putting

$$\text{ispec}_A U = \cap_{P \in U} P$$

and we have put

$$\text{rd}(A) = \text{the set of all radical ideals in } A$$

i.e., ideals which are their own radicals. We have also put

$$\text{svt}(A) = \text{the set of all spectral varieties in } A$$

i.e., the set of all subsets of $\text{spec}(A)$ of the form $\text{vspec}_A J$ with J varying over the set of all subsets of A; members of $\text{svt}(A)$ may be called varieties in $\text{spec}(A)$. Now we put

$$\text{msvt}(A) = \text{the set of all maximal spectral varieties in } A$$

i.e., the set of all subsets of $\text{mspec}(A)$ of the form $\text{mvspec}_A J$ with J varying over the set of all subsets of A; members of $\text{msvt}(A)$ may be called varieties in $\text{mspec}(A)$.

Given U, U' in $\text{msvt}(A)$ or $\text{svt}(A)$, U' is a subvariety of U means $U' \subset U$, and U' is a proper subvariety of U means $U' \subset U$ with $U' \neq U$. Given U in $\text{msvt}(A)$ or $\text{svt}(A)$, U is reducible means U is the union of two proper subvarieties; U is irreducible means it is nonempty and nonreducible. We put

$$\text{imsvt}(A) = \text{the set of all irreducible members of } \text{msvt}(A)$$

and

$$\text{isvt}(A) = \text{the set of all irreducible members of } \text{svt}(A).$$

For any U in $\text{msvt}(A)$ or $\text{svt}(A)$ we define the dimension of U by putting

$$\dim(U) = \dim(A/\text{ispec}_A U).$$

If A is an affine ring over k then for any U in $\text{msvt}(A)$ or $\text{svt}(A)$ we denote $A/\text{ispec}_A(U)$ by $k[U]^*$ and call it the affine coordinate ring of U. In case $\text{ispec}_A U$

is not the unit ideal in A, we may identify k with a subfield of $k[U]^*$ and we note that then $k[U]^*$ becomes an affine ring over k, and hence it is a noetherian ring.

In case $\text{ispec}_A U$ is a prime ideal in A, we denote the quotient field of $k[U]^*$ by $k(U)^*$, and we call $k(U)^*$ the function field of U (over k). We denote the localization of A at $\text{ispec}_A U$ by $R_k(U)^*$ and call it the local ring of U (over k); we note that by (T18) this is indeed a local ring. We denote the residue field $R_k(U)^*/M(R_k(U)^*)$ by $k(U)^\sharp$ and we call it the alternative function field of U (over k); we may and we do identify k with a subfield of $k(U)^\sharp$. Let $\Phi_U : A \to k[U]^*$ and $\Psi_U : R_k(U)^* \to k(U)^\sharp$ be the residue class epimorphisms. Clearly there is a unique homomorphism $\psi_U : R_k(U)^* \to k(U)^*$ such that for all $f \in A$ and $g \in A \setminus \text{ispec}_A U$ we have $\psi_U(f/g) = \Phi_U(f)/\Phi_U(g)$. Obviously $\text{im}(\psi_U) = k(U)^*$ and $\ker(\psi_U) = M(R_k(U)^*)$ and hence it follows that there is a unique isomorphism $\phi_U : k(U)^\sharp \to k(U)^*$ for which we have $\phi_U \Psi_U = \psi_U$. We call ψ_U and ϕ_U the natural epimorphism and the natural isomorphism respectively.

As GEOMETRIC MOTIVATION for the definitions, consider the variety $f = 0$ in $\mathbb{A}_\kappa^N$ where f is a nonconstant polynomial in $B_{N,k}$. For $N = 1$ this is a point-set (= a finite set of points) on the line, for $N = 2$ it is a plane curve, for $N = 3$ it is a surface, for $N = 4$ it is a solid, and in general it is a hypersurface, i.e., something defined by a single equation. For instance, by taking $f = X_1^2 + \cdots + X_N^2 - 1$ we get a point-pair, a circle, a sphere, a solid ball, and so on. Intuitively, a point-set is zero-dimensional, a curve is one-dimensional, a surface is two-dimensional, a solid is three-dimensional, ..., a hypersurface is $(N-1)$-dimensional. The hypersurface is irreducible if the polynomial f is irreducible.

This gives rise to the INTUITIVE DEFINITION of an r-dimensional irreducible algebraic variety U in the affine N-space $\mathbb{A}_\kappa^N$ as a geometric object which can be parametrized by an irreducible hypersurface in $(r+1)$-space. Here by a geometric object we mean something defined by a bunch of polynomial equations. Let P be the ideal generated by these polynomials in $B_{N,k}$. The irreducibility of U suggests that we require P to be a prime ideal. The dimensionality r of U suggests that we require it to equal $\text{trdeg}_k k(U)^*$ where trdeg_k denotes transcendence degree over k and $k(U)^*$ is the quotient field $\text{QF}(k[U]^*)$ of $k[U]^* = B_{N,k}/P$. Referring to (O22) for the definitions of the depth dpt and the height ht of an ideal, eventually we shall prove the following [cf. L5§5(Q10)(T47)]:

DIMENSION THEOREM (T19). For any prime ideals $P \subset P'$ in $B = B_{N,k}$, upon letting $\phi : B \to B/P$ be the residue class epimorphism, we have

$$\text{trdeg}_k \text{QF}(B/P) = \dim(B/P) = \text{dpt}_B P = N - \text{ht}_B P$$

and

$$\text{trdeg}_k \text{QF}(B/P) = \text{ht}_{\phi(B)}\phi(P') + \text{dpt}_{\phi(B)}\phi(P').$$

Eventually we shall also prove the following [cf. L5§5(Q32)(T144)]:

PRIMITIVE ELEMENT THEOREM (T20). If k is algebraically closed then for any prime ideal P in $B = B_{N,k}$, upon letting $\phi : B \to B/P$ be the residue class epimorphism, and after changing the variables by a linear transformation, it can be arranged that $\phi(X_1), \ldots, \phi(X_r)$ is a transcendence basis of $\mathrm{QF}(B/P)$ over k and either $N = r$ or $\phi(X_{r+1})$ is a primitive element of $\mathrm{QF}(B/P)$ over $\mathrm{QF}(\phi(B_{r,k}))$, i.e., $\mathrm{QF}(B/P) = \mathrm{QF}(\phi(B_{r,k}))(\phi(X_{r+1}))$. [In the latter case we can find an irreducible polynomial $F(X_1, \ldots, X_{r+1})$ in $P \cap B_{r+1,k}$ and then the irreducible variety U in $\mathbb{A}_\kappa^N$ defined by P is "parametrized" by the irreducible hypersurface $F = 0$ in $\mathbb{A}_\kappa^{r+1}$; in the former case U is "parametrized" by $\mathbb{A}_\kappa^r$].

Now (T20) provides the rationale behind calling the transcendence degree as dimension. In turn (T19) connects it to the ring-theoretic definition of dimension which we have chosen for U in avt or msvt or svt. At any rate, as suggested by the above GEOMETRIC MOTIVATION, in avt or msvt or svt, we call U a point-set or a curve or a surface or a solid according as $\dim(U) = 0$ or 1 or 2 or 3 respectively. Another rationale for choosing the ring-theoretic definition of dimension can be found in the following Theorems (T21) and (T22) according to which: a curve has points on it, a surface has curves on it, a solid has surfaces on it, and so on.

Most parts of the following Inclusion Relations Theorem (T21) for the operators $\mathbb{V}$, $\mathbb{I}$, vspec, and ispec are straightforward. We shall soon complete the proof by establishing the remaining parts [see the cf. between (T21) and (T22)].

INCLUSION RELATIONS THEOREM (T21). Recall that A is any ring. Let

$$\begin{cases} (R, S, V, V', V'', I) \text{ stand for} \\ (B_{N,k}, \mathbb{A}_\kappa^N, \mathbb{V}_\kappa, \mathrm{avt}_k(\mathbb{A}_\kappa^N), \mathrm{iavt}_k(\mathbb{A}_\kappa^N), \mathbb{I}_k) \\ \text{or } (A, \mathrm{mspec}(A), \mathrm{mvspec}_A, \mathrm{msvt}(A), \mathrm{imsvt}(A), \mathrm{ispec}_A) \\ \text{or } (A, \mathrm{spec}(A), \mathrm{vspec}_A, \mathrm{svt}(A), \mathrm{isvt}(A), \mathrm{ispec}_A) \end{cases}$$

and let I' be the set of all ideals J in R such that $J = I(U)$ for some $U \subset S$. Then we have the following.

(T21.1) For any subsets J and J' of R we have: $J \subset J' \Rightarrow V(J') \subset V(J)$, and we have: $\mathrm{rad}_R(JR) = \mathrm{rad}_R(J'R) \Rightarrow V(J) = V(J')$.

(T21.2) For any subsets U and U' of S we have: $U \subset U' \Rightarrow I(U') \subset I(U)$.

(T21.3) For any family of ideals $(J_l)_{l \in L}$ in R we have $V(\sum_{l \in L} J_l) = \cap_{l \in L} V(J_l)$, and if the family is finite then we have $V(\cap_{l \in L} J_l) = V(\prod_{l \in L} J_l) = \cup_{l \in L} V(J_l)$.

(T21.4) For any family of subsets $(U_l)_{l \in L}$ of S we have $I(\cup_{l \in L} U_l) = \cap_{l \in L} I(U_l)$.

(T21.5) For any $J \subset R$ we have $J \subset I(V(J))$ and: $J = I(V(J)) \Leftrightarrow J \in I' \Rightarrow J = \mathrm{rad}_R(JR)$.

(T21.6) For any $U \subset S$ we have $U \subset V(I(U))$ and: $U = V(I(U)) \Leftrightarrow U \in V'$.

(T21.7) $V'' = \{U \in V' : I(U) \in \operatorname{spec}(R)\}$.

(T21.8) Assuming R is noetherian (this is certainly so in case $R = B_{N,k}$), every $U \in V'$ can be expressed as a finite union $U = \cup_{1 \le i \le h} U_i$ with $U_i \in V''$, and this decomposition is unique up to order if it is irredundant, i.e., if no U_i can be deleted from it. [The irredundancy can obviously be achieved by deleting some of the U_i, and after it is achieved, the remaining U_i are called the IRREDUCIBLE COMPONENTS of U].

It was said in (O16) that later on we shall show that $U \mapsto \operatorname{ispec}_{B_{N,k}} U$ gives a bijection $\operatorname{svt}(B_{N,k}) \to \operatorname{rd}(B_{N,k})$ whose inverse is given by $J \mapsto \operatorname{vspec}_{B_{N,k}} J$. Now this assertion is the eighth amongst the following eight equivalent versions (T22.1) to (T22.8) of the famous Hilbert Nullstellensatz which we shall soon prove [cf. (T49) and (T50) of L5§5(Q11), and the material between (T47) and (C15) of L5§5(Q10].

HILBERT NULLSTELLENSATZ (T22).

(T22.1) A field which is an affine ring over k is algebraic over k.

(T22.2) For any maximal ideal J in $B_{N,k}$, the field $B_{N,k}/J$ is algebraic over k, i.e., over the image of k under the residue class map $B_{N,k} \to B_{N,k}/J$.

(T22.3) For any maximal ideal J in $B_{N,k}$ there is a unique set of N generators $(f_i(X_1, \ldots, X_i) \in B_{i,k})_{1 \le i \le N}$ such that for $1 \le i \le N$ we have

$$f_i(X_1, \ldots, X_i) = X_i^{n_i} + g_i(X_1, \ldots, X_i)$$

where n_i is a positive integer and $g_i(X_1, \ldots, X_i) \in B_{i,k}$ with

$$\deg_{X_j} g_i(X_1, \ldots, X_i) < n_j \quad \text{for} \quad 1 \le j \le i.$$

(T22.4) If k is algebraically closed then the mapping $\alpha \mapsto \mathbb{I}_k(\alpha)$ gives a bijection $\mathbb{A}_k^N \to \operatorname{mspec}(B_{N,k})$.

(T22.5) If κ contains an algebraic closure of k then for any ideal J in $B_{N,k}$ we have that $\mathbb{I}_k(\mathbb{V}_\kappa(J)) = \operatorname{rad}_{B_{N,k}} J$.

(T22.6) If κ contains an algebraic closure of k then the mapping $U \mapsto \mathbb{I}_k(U)$ gives inclusion reversing bijections

$$\operatorname{avt}_k(\mathbb{A}_\kappa^N) \to \operatorname{rd}(B_{N,k}) \quad \text{and} \quad \operatorname{iavt}_k(\mathbb{A}_\kappa^N) \to \operatorname{spec}(B_{N,k})$$

whose inverses are given by $J \mapsto \mathbb{V}_\kappa(J)$.

(T22.7) The mapping $U \mapsto \operatorname{ispec}_{B_{N,k}} U$ gives inclusion reversing bijections

$$\operatorname{msvt}(B_{N,k}) \to \operatorname{rd}(B_{N,k}) \quad \text{and} \quad \operatorname{imsvt}(B_{N,k}) \to \operatorname{spec}(B_{N,k})$$

whose inverses are given by $J \mapsto \operatorname{mvspec}_{B_{N,k}} J$.

(T22.8) The mapping $U \mapsto \operatorname{ispec}_{B_{N,k}} U$ gives inclusion reversing bijections

$$\operatorname{svt}(B_{N,k}) \to \operatorname{rd}(B_{N,k}) \quad \text{and} \quad \operatorname{isvt}(B_{N,k}) \to \operatorname{spec}(B_{N,k})$$

whose inverses are given by $J \mapsto \mathrm{vspec}_{B_{N,k}} J$.

§8.1: SPECTRAL AFFINE SPACE

Henceforth we shall consider points and varieties in $\mathbb{A}_k^N$ rather than in $\mathbb{A}_\kappa^N$.

As we have already noted, the ideal $\mathbb{I}_k(\alpha)$ of any point α in $\mathbb{A}_k^N$ is the maximal ideal in $B_{N,k}$ generated by $X_1 - \alpha_1, \ldots, X_N - \alpha_N$. In other words, α is the intersection of the N hyperplanes $X_1 - \alpha_1 = 0, \ldots, X_N - \alpha_N = 0$. Also clearly $\alpha \neq \alpha'$ in $\mathbb{A}_k^N \Rightarrow \mathbb{I}_k(\alpha) \neq \mathbb{I}_k(\alpha')$. So via $\mathbb{I}_k$ we may ENLARGE $\mathbb{A}_k^N$ into $\mathrm{spec}(B_{N,k})$ which we may denote by $(\mathbb{A}_k^N)^\sigma$ and call it the spectral affine N-space over k. The mapping $\alpha \mapsto \mathbb{I}_k(\alpha)$ gives an injection $\mathbb{A}_k^N \to (\mathbb{A}_k^N)^\sigma$. The image of $\mathbb{A}_k^N$ under this injection may be denoted by $(\mathbb{A}_k^N)^{\rho\sigma}$ and called the rational spectral affine N-space over k. Also we may denote $\mathrm{mspec}(B_{N,k})$ by $(\mathbb{A}_k^N)^{\mu\sigma}$ and call it the maximal spectral affine N-space over k. Thus

$$\mathbb{A}_k^N \approx (\mathbb{A}_k^N)^{\rho\sigma} \subset (\mathbb{A}_k^N)^{\mu\sigma} \subset (\mathbb{A}_k^N)^\sigma$$

where $\approx$ stands for isomorphism; here it is set-theoretic isomorphism, i.e., bijection. Now (T22.3) says that a point of $(\mathbb{A}_k^N)^{\mu\sigma}$ is the intersection of N hypersurfaces; moreover, these hypersurfaces are hyperplanes, i.e., they are of degree one, iff it is a point of $(\mathbb{A}_k^N)^{\rho\sigma}$.

§8.2: MODELIC SPEC AND MODELIC AFFINE SPACE

To describe another way of enlarging $\mathbb{A}_k^N$, for any domain A we put

$$\mathfrak{V}(A) = \text{the set of all localizations } A_P \text{ with } P \text{ varying over } \mathrm{spec}(A)$$

and we call this the modelic spec of A. By (T18) we see that A_P is a quasilocal ring for which $M(A_P) = PA_P$ and $M(A_P) \cap A = P$ with $\dim(A_P) = \mathrm{ht}_A P$, and if A is noetherian then A_P is a local ring R. Thus $P \mapsto A_P$ gives an inclusion reversing bijection of $\mathrm{spec}(A)$ onto $\mathfrak{V}(A)$, and the reverse bijection is given by $R \mapsto M(R) \cap A$. We denote $\mathfrak{V}(B_{N,k})$ by $(\mathbb{A}_k^N)^\delta$ and call it the modelic affine N-space over k. Now $P \mapsto (B_{N,k})_P$ gives an inclusion reversing bijection $(\mathbb{A}_k^N)^\sigma \to (\mathbb{A}_k^N)^\delta$ whose inverse is given by $R \mapsto M(R) \cap B_{N,k}$. The images of $(\mathbb{A}_k^N)^{\rho\sigma}$ and $(\mathbb{A}_k^N)^{\mu\sigma}$ under this map may be denoted by $(\mathbb{A}_k^N)^{\rho\delta}$ and $(\mathbb{A}_k^N)^{\mu\delta}$ and called the rational modelic affine N-space over k and the minimal modelic affine N-space over k respectively; note that because of the inclusion reversing property, $(\mathbb{A}_k^N)^{\mu\delta}$ is the set of all minimal members of $(\mathbb{A}_k^N)^\delta$. Now the isomorphisms

$$(\mathbb{A}_k^N)^{\rho\sigma} \approx (\mathbb{A}_k^N)^{\rho\delta} \quad \text{and} \quad (\mathbb{A}_k^N)^{\mu\sigma} \approx (\mathbb{A}_k^N)^{\mu\delta} \quad \text{and} \quad (\mathbb{A}_k^N)^\sigma \approx (\mathbb{A}_k^N)^\delta$$

give rise to the enlargements

$$\mathbb{A}_k^N \approx (\mathbb{A}_k^N)^{\rho\delta} \subset (\mathbb{A}_k^N)^{\mu\delta} \subset (\mathbb{A}_k^N)^\delta.$$

Given any $U \in (\mathbb{A}_k^N)^\sigma$, by (T22) we see that $k(U)^* = k \Leftrightarrow U \in (\mathbb{A}_k^N)^{\rho\sigma}$ whereas $k(U)^*/k$ is algebraic $\Leftrightarrow U \in (\mathbb{A}_k^N)^{\mu\sigma}$; consequently, points in $(\mathbb{A}_k^N)^{\rho\sigma}$ may be called rational and points in $(\mathbb{A}_k^N)^{\mu\sigma}$ may be called algebraic. Similarly, for any $R \in (\mathbb{A}_k^N)^\delta$, after identifying k with a subfield of $R/M(R)$, we have that $R/M(R) = k \Leftrightarrow R \in (\mathbb{A}_k^N)^{\rho\delta}$ whereas $(R/M(R))/k$ is algebraic $\Leftrightarrow R \in (\mathbb{A}_k^N)^{\mu\delta}$; consequently, points in $(\mathbb{A}_k^N)^{\rho\delta}$ may be called rational and points in $(\mathbb{A}_k^N)^{\mu\delta}$ may be called algebraic. By (T22) we also see that if κ contains an algebraic closure of k then "points" of $(\mathbb{A}_k^N)^\sigma$ are the prime ideals of "irreducible varieties in $\mathbb{A}_\kappa^N$ defined over k," and the corresponding "points" of $(\mathbb{A}_k^N)^\delta$ are their local rings.

§8.3: SIMPLE POINTS AND REGULAR LOCAL RINGS

The embedding dimension of a local ring R is defined by putting

emdim(R) = the smallest number of elements which generate $M(R)$.

We shall soon prove the [cf. L5§5(Q7)(T29)]:

DIM–EMDIM THEOREM (T23). For any local ring R we have

$$\text{emdim}(R) \geq \dim(R)$$

and hence in particular (since R noetherian $\Rightarrow M(P)$ is finitely generated)

emdim(R) and dim(R) are nonnegative integers.

A local ring R is said to be regular if emdim(R) = dim(R).

Concerning regular local rings we shall soon prove the following theorem, where we recall that a domain is normal means it is integrally closed in its quotient field (see L3§7), and for any element x in any quasilocal ring R we have put ord$_R x = \max\{i \in \mathbb{N} : x \in M(R)^i\}$ where the max is taken to be ∞ if the set of i is unbounded (see L3§11). [cf. L5§5(Q4)(T7), L5§5(Q14)(T64), L5§5(Q15)(T71), L5§5(Q17)(T81.7)].

ORD VALUATION THEOREM (T24). Let R be any local ring. Then:
(T24.1) For all $0 \neq x \in R$ we have ord$_R x \in \mathbb{N}$, i.e., $\cap_{i\in\mathbb{N}} M(R)^i = 0$.
(T24.2) If R is regular then for all nonzero elements x and y in R we have ord$_R(xy)$ = ord$_R x$ + ord$_R y$; consequently R is a normal domain and we get a valuation ord$_R$: QF$(R) \to \mathbb{Z}$ by putting ord$_R(x/y)$ = ord$_R x$ − ord$_R y$. [We shall continue to use this extended meaning of ord$_R$].
(T24.3) For any $x \in M(R) \setminus Z_R(R)$ we have that: (i) $R/(xR)$ is a local ring whose dimension is one less than the dimension of R, (ii) if $R/(xR)$ is regular then ord$_R x = 1$, and (iii) if ord$_R x = 1$ and R is regular then $R/(xR)$ is regular.

[Note that if R is regular, or more generally if R is a domain, then the condition $x \in M(R) \setminus Z_R(R)$ is equivalent to the condition $0 \neq x \in M(R)$].

(T24.4) If R is regular then its localization R_P at any prime ideal P in it is regular.

Theorems (T19) and (T22.3) tell us that, the localization of $B_{N,k}$ at any $U \in (\mathbb{A}_k^N)^{\mu\sigma}$ is an N-dimensional regular local domain, i.e., equivalently, every $R \in (\mathbb{A}_k^N)^{\mu\delta}$ is an N-dimensional regular local domain. Geometrically this says that every point of $(\mathbb{A}_k^N)^{\mu\sigma}$ is simple, and so is every point of $(\mathbb{A}_k^N)^{\mu\delta}$. By (T24.4) it follows that the localization of $B_{N,k}$ at any $U \in (\mathbb{A}_k^N)^\sigma$ is a regular local domain, i.e., equivalently, every $R \in (\mathbb{A}_k^N)^\delta$ is a regular local domain.

EXAMPLE (X11). Consider a hypersurface $S : F = 0$ in $\mathbb{A}_k^N$ with $F = F(X_1, \dots, X_N) \in B_{N,k} \setminus k$ and let $\alpha = (\alpha_1, \dots, \alpha_N)$ be a point of S, i.e., $\alpha_1, \dots, \alpha_N$ are elements in k with $F(\alpha_1, \dots, \alpha_N) = 0$. Expanding F around α we get

$$F(X_1, \dots, X_N) = \sum G_{i_1 \dots i_N}(X_1 - \alpha_1)^{i_1} \dots (X_N - \alpha_N)^{i_N}$$

with $G_{i_1 \dots i_N} \in k$. Let ν be the smallest value of $i_1 + \dots + i_N$ for which $G_{i_1 \dots i_N} \neq 0$. In L3§3 we have put $\mathrm{ord}_\alpha F = \nu$, called ν the multiplicity $\mathrm{mult}_\alpha S$ of S at α, called α a ν-fold point of S, and said that α is a simple or singular point of S according as $\nu = 1$ or $\nu > 1$. In terms of (formal) partial derivatives we have

$$\nu > 1 \text{ (i.e., } \alpha \text{ is a singular point of } S) \iff \left.\frac{\partial F}{\partial X_i}\right|_\alpha = 0 \text{ for } 1 \leq i \leq N.$$

In the present notation $\mathrm{ord}_{R(\alpha)^*} F = \nu$, and by (T24) the local ring $R(\alpha)^*/(FR(\alpha)^*)$ is regular iff $\nu = 1$, i.e., iff α is a simple point of S.

§9: MODELS

As described in L3§4, we can construct the projective N-space $\mathbb{P}_k^N$ over a field k by patching up $N+1$ copies of the affine N-space $\mathbb{A}_k^N$ over k. Similarly, by patching up several modelic specs we can construct a model. More precisely, we introduce the following definitions.

Let A be a subring of a field K. Referring to L2§3 for the definition of a valuation v of K and the definition of the valuation ring R_v of v, recall that, according to L3§7 and L3§11, R_v is a quasilocal domain with quotient field K, and by a valuation ring we mean the valuation ring of some valuation of its quotient field. By a valuation of K/A (K over A) we mean a valuation v of K such that $A \subset R_v$. By a valuation ring of K we mean a valuation ring with quotient field K, and by a valuation ring of K/A we mean a valuation ring of K which contains A. We put

$$\mathfrak{R}(K) = \text{the set of all valuation rings of } K$$

and

$$\mathfrak{R}(K/A) = \text{the set of all valuation rings of } K/A$$

and we call these the Riemann-Zariski space of K and the Riemann-Zariski space of K/A respectively. We also put

$$\mathfrak{R}'(K) = \text{the set of all quasilocal domains with quotient field } K$$

and

$$\mathfrak{R}'(K/A) = \text{the set of all members of } \mathfrak{R}'(K) \text{ which contain } A$$

and we call these the quasitotal Riemann-Zariski space of K and the quasitotal Riemann-Zariski space of K/A respectively. Finally we put

$$\mathfrak{R}''(K) = \text{the set of all quasilocal domains which are subrings of } K$$

and

$$\mathfrak{R}''(K/A) = \text{the set of all members of } \mathfrak{R}''(K) \text{ which contain } A$$

and we call these the total Riemann-Zariski space of K and the total Riemann-Zariski space of K/A respectively. A quasilocal ring R is dominated by a quasilocal ring S means R is a subring of S and $M(R) \subset M(S)$ or equivalently $M(R) = R \cap M(S)$; alternatively we may say that S dominates R; we may indicate this by writing

$$R < S \quad \text{or} \quad S > R.$$

This converts $\mathfrak{R}(K), \mathfrak{R}(K/A), \mathfrak{R}'(K), \mathfrak{R}'(K/A), \mathfrak{R}''(K), \mathfrak{R}''(K/A)$ into posets ($=$ partially ordered sets). A valuation v dominates a quasilocal ring R means R_v dominates R. We shall soon prove the following Theorems (T25) to (T28) about valuations; [cf. §12(R7)].

VALUATION MAXIMALITY THEOREM (T25). $\mathfrak{R}(K)$ (resp: $\mathfrak{R}(K/A)$) is the set of all maximal members of $\mathfrak{R}''(K)$ (resp: $\mathfrak{R}''(K/A)$), i.e., R in $\mathfrak{R}''(K)$ (resp: $\mathfrak{R}''(K/A)$) belongs to $\mathfrak{R}(K)$ (resp: $\mathfrak{R}(K/A)$) iff R is not dominated by any member of $\mathfrak{R}''(K)$ (resp: $\mathfrak{R}''(K/A)$) other than itself.

VALUATION EXISTENCE THEOREM (T26). Every member of $\mathfrak{R}''(K)$ is dominated by some member of $\mathfrak{R}(K)$.

VALUATION EXTENSION THEOREM (T27). Let L be any overfield of K. Then every member V of $\mathfrak{R}(K)$ is dominated by some member W of $\mathfrak{R}(L)$, and for any such W we have $V = W \cap K$. Conversely, for any $W \in \mathfrak{R}(L)$, upon letting $V = W \cap K$, we have that W dominates V and $V \in \mathfrak{R}(K)$.

VALUATION CHARACTERIZATION THEOREM (T28). A domain R is a valuation ring iff for all x, y in R we have either $x \in yR$ or $y \in xR$. Equivalently, a domain R with quotient field K is a valuation ring of K iff for every $x \in K^{\times}$ we have either $x \in R$ or $1/x \in R$.

To proceed with the definition of models, by a premodel of K (resp: K/A) we mean a nonempty subset E of $\mathfrak{R}'(K)$ (resp: $\mathfrak{R}'(K/A)$). The premodel E is irredundant means any member of $\mathfrak{R}(K)$ (resp: $\mathfrak{R}(K/A)$) dominates at most one member of E. Note that a quasilocal domain S can dominate at most one member R of an irredundant premodel E, and if R exists then we call it the center of S on E. To see the uniqueness of R, let S dominate another member R' of E; identifying K with a subfield of the quotient field L of S, by (T26) we can find a valuation ring W of L dominating S and then by (T27) $W \cap K$ is a valuation ring of K which dominates R as well as R' and hence $R = R'$.

By a semimodel (resp: model) of K/A we mean an irredundant premodel E of K/A which can be expressed as a union $E = \cup_{l \in \Lambda} \mathfrak{V}(B_l)$ for some family (resp: finite family) $(B_l)_{l \in \Lambda}$ of subrings B_l of K with quotient field K such that B_l is an overring of (resp: affine ring over) A. Note that if B is any subring of K with quotient field K such that B is an overring of (resp: affine ring over) A then $\mathfrak{V}(B)$ is a semimodel (resp: model) of K/A; we call it an affine semimodel (resp: affine model) of K/A. Also note that if E is any irredundant premodel of K then: E is a semimodel of $K/A \Leftrightarrow$ for every $R \in E$ we have $A \subset R$ and $\mathfrak{V}(R) = \{R' \in E : R \subset R'\}$. Finally note that every model of K/A is a semimodel of K/A, and if A is noetherian then every member of E is a local ring.

A quasilocal ring S dominates a set E of quasilocal rings (or E is dominated by S) means S dominates some member of E; we may indicate this by writing $E < S$ (or $S > E$). A set E' of quasilocal rings dominates a set E of quasilocal rings (or E is dominated by E') means every member of E' dominates E; we may indicate this by writing $E < E'$ (or $E' > E$). A set E' of quasilocal rings properly dominates a set E of quasilocal rings (or E is properly dominated by E') means E' dominates E and every member of E is dominated by some member of E'. A semimodel or model of K/A is complete means it is dominated by $\mathfrak{R}(K/A)$. Note that if K is the quotient field of A then $\mathfrak{V}(A)$ is a complete model of K/A. Also note that if E is a semimodel (resp: complete semimodel) of K/A then E dominates (resp: properly dominates) $\mathfrak{V}(A)$.

A model E (of K/A) is said to be normal (resp: noetherian) if every $R \in E$ is normal (resp: noetherian). A model E is said to be nonsingular if every $R \in E$ is a regular local ring. Note that by (T24), every nonsingular model is a normal noetherian model. By the dimension $\dim(E)$ of a model E we mean $\max\{\dim(R) : R \in E\}$; note that then

$$\dim(E) \in \mathbb{N} \cup \{\infty\}.$$

§9.1: MODELIC PROJ AND MODELIC PROJECTIVE SPACE

As we have seen in §8.1, for the N-variable polynomial ring $B_{N,k}$ over a field k, the modelic spec $\mathfrak{V}(B_{N,k})$ is bijective with $\mathrm{spec}(B_{N,k})$ (i.e., there is a bijection between the two), and hence if k is algebraically closed then its portion consisting of N-dimensional members is bijective with the affine space $\mathbb{A}_k^N$. We proceed to show that similarly a certain portion of a certain model is bijective with the projective space $\mathbb{P}_k^N$. So for any family $(x_l)_{l \in \Lambda}$ of elements in K, with $x_j \neq 0$ for some $j \in \Lambda$, we put

$$\mathfrak{W}(A; (x_l)_{l \in \Lambda}) = \bigcup_{j \in \Lambda \text{ with } x_j \neq 0} \mathfrak{V}(A[(x_l/x_j)_{l \in \Lambda}])$$

and we call this the modelic proj of $(x_l)_{l \in \Lambda}$ over A. Here $A[(x_l/x_j)_{l \in \Lambda}]$ denotes the smallest subring of K which contains A and which contains x_l/x_j for all $l \in \Lambda$. If Λ is a finite set, say $\Lambda = \{1, \ldots, n\}$, then we may write $\mathfrak{W}(A; x_1, \ldots, x_n)$ instead of $\mathfrak{W}(A; (x_l)_{l \in \Lambda})$ and call it the modelic proj of $(x_1, \ldots, x_n)$ over A. Clearly for all i, i' in Λ with $x_i \neq 0 \neq x_{i'}$ we have $\mathrm{QF}(A[(x_l/x_i)_{l \in \Lambda}]) = \mathrm{QF}(A[(x_l/x_{i'})_{l \in \Lambda}])$ and, upon letting K' be this common quotient field, $\mathfrak{W}(A; (x_l)_{l \in \Lambda})$ is obviously a premodel of K'/A. Moreover, for any $i' \in \Lambda$ with $x_{i'} \neq 0$, upon letting $y_l = x_l/x_{i'}$, we see that $(y_l)_{l \in \Lambda}$ is a family of elements in K' such that $y_j \neq 0$ for some $j \in \Lambda$ and $\mathfrak{W}(A; (x_l)_{l \in \Lambda}) = \mathfrak{W}(A; (y_l)_{l \in \Lambda})$. According to Theorem (T29), which is stated below and which we shall soon prove, $\mathfrak{W}(A; (x_l)_{l \in \Lambda})$ is actually a semimodel of K'/A and if Λ is finite then it is in fact a complete model of K'/A; [cf. §12(R8)]. By a projective model of K/A we mean a premodel E of K/A such that $E = \mathfrak{W}(A; x_1, \ldots, x_n)$ for some finite number of elements $x_1, \ldots, x_n$ in an overfield of K with $x_j \neq 0$ for some $j \in \{1, \ldots, n\}$; by what we have just said, E is then indeed a complete model of K/A and we have $E = \mathfrak{W}(A; y_1, \ldots, y_n)$ for a finite number of elements $y_1, \ldots, y_n$ in K at least one of which is nonzero.

MODELIC PROJ THEOREM (T29).   Given any family $(x_l)_{l \in \Lambda}$ of elements in K, with $x_j \neq 0$ for some $j \in \Lambda$, let $E = \mathfrak{W}(A; (x_l)_{l \in \Lambda})$ and let $K' = \mathrm{QF}(A[(x_l/x_i)_{l \in \Lambda}])$ where $i \in \Lambda$ with $x_i \neq 0$; as noted above the QF is independent of the choice of i. Also let $y_l = x_l/x_{i'}$ for any fixed $i' \in \Lambda$ with $x_{i'} \neq 0$. Then we have the following.

(T29.1) E is a premodel of K'/A and $E = \mathfrak{W}(A; (x_l/x)_{l \in \Lambda})$ for all $0 \neq x \in K$. In particular $(y_l)_{l \in \Lambda}$ is a family of elements in K' such that $y_j \neq 0$ for some $j \in \Lambda$, and we have $E = \mathfrak{W}(A; (y_l)_{l \in \Lambda})$.

(T29.2) Given any $R \in E$ and any subring S of K such that S is a quasilocal ring dominating R, there exists $j \in \Lambda$ with $x_j \neq 0$ such that $x_l/x_j \in S$ for all $l \in \Lambda$. Moreover, for any such j we have $R = B_Q$ where $B = A[(x_l/x_j)_{l \in \Lambda}]$ and $Q = B \cap M(S)$.

(T29.3) E is a semimodel of K'/A, and if Λ is finite then E is in fact a complete model of K'/A.

To justify the terms modelic proj and projective model, we proceed to show that there is a natural injection of the projective space $\mathbb{P}_k^N$ into a projective model. Referring to L3§4 for details, recall that

$$\mathbb{P}_k^N = (k^{N+1} \setminus \{(0,\ldots,0)\})/\sim$$

where for $u = (u_1,\ldots,u_{N+1})$ and $u' = (u'_1,\ldots,u'_{N+1})$ in $k^{N+1}\setminus\{(0,\ldots,0)\}$ we have: $u \sim u' \Leftrightarrow (u'_1,\ldots,u'_{N+1}) = (cu_1,\ldots,cu_{N+1})$ for some $0 \neq c \in k$. Thus $\mathbb{P}_k^N$ is the set of all equivalence classes under this equivalence relation. Let H_i be the hyperplane in $\mathbb{P}_k^N$ consisting of all equivalence classes $\beta \in \mathbb{P}_k^N$ of those $u \in k^{N+1}$ for which $u_i = 0$, and let $\mathbb{A}_{k,i}^N$ be the copy of $\mathbb{A}_k^N$ consisting of those $v = (v_1,\ldots,v_{N+1}) \in k^{N+1}$ for which $v_i = 1$. Now $\beta \mapsto (u_1/u_i,\ldots,u_{N+1}/u_i)$ gives a bijection $\phi_i : \mathbb{P}_k^N \setminus H_i \to \mathbb{A}_{k,i}^N$, and identifying $\mathbb{P}_k^N \setminus H_i$ with $\mathbb{A}_{k,i}^N$ via this bijection we have

$$\mathbb{P}_k^N = \cup_{i=1}^{N+1}(\mathbb{P}_k^N \setminus H_i) = \cup_{i=1}^{N+1}\mathbb{A}_{k,i}^N.$$

Now $B_{N+1,k} = k[X_1,\ldots,X_{N+1}]$ and we take its quotient field $k(X_1,\ldots,X_{N+1})$ to be K; also we take k to be A. For $1 \leq i \leq N+1$, upon letting $Y_{li} = X_l/X_i$ (with $Y_{ii} = 1$), we get a copy $B_{N,k,i} = k[Y_{1i},\ldots,Y_{N+1,i}]$ of $B_{N,k}$ whose modelic spec $\mathfrak{V}(B_{N,k,i})$ may be identified with $(\mathbb{A}_{k,i}^N)^\delta$; let $\theta_i : \mathbb{A}_{k,i}^N \to (\mathbb{A}_{k,i}^N)^\delta$ be the natural injection whose image is $(\mathbb{A}_{k,i}^N)^{\rho\delta}$. For all i,j in $\{1,\ldots,N+1\}$ we clearly have $\mathrm{QF}(B_{N,k,i}) = \mathrm{QF}(B_{N,k,j})$ and we let K' stand for this common subfield of K. Then $\mathfrak{V}(k;X_1,\ldots,X_{N+1})$ is a projective model of K'/k. We denote $\mathfrak{V}(k;X_1,\ldots,X_{N+1})$ by $(\mathbb{P}_k^N)^\delta$ and call it the modelic projective N-space over k. By the definition of $\mathfrak{V}$ we get

$$(\mathbb{P}_k^N)^\delta = \bigcup_{1 \leq i \leq N+1} (\mathbb{A}_{k,i}^N)^\delta.$$

We put

$$(\mathbb{P}_k^N)^{\rho\delta} = \bigcup_{1 \leq i \leq N+1} (\mathbb{A}_{k,i}^N)^{\rho\delta} \quad \text{and} \quad (\mathbb{P}_k^N)^{\mu\delta} = \bigcup_{1 \leq i \leq N+1} (\mathbb{A}_{k,i}^N)^{\mu\delta}$$

and we respectively call these the rational and minimal modelic projective N-spaces over k. By §8.2 and §8.3 it follows that $(\mathbb{P}_k^N)^\delta$ is a nonsingular N-dimensional complete model, and for any $R \in (\mathbb{P}_k^N)^\delta$, after identifying k with a subfield of $R/M(R)$, we have that: $R \in (\mathbb{P}_k^N)^{\rho\delta} \Leftrightarrow R/M(R) = k$, and we have that: $R \in (\mathbb{P}_k^N)^{\mu\delta} \Leftrightarrow \dim(R) = N \Leftrightarrow R$ is a minimal member of $(\mathbb{P}_k^N)^\delta \Leftrightarrow R/M(R)$ is algebraic over k.

We claim that there exists a unique injection $\theta : \mathbb{P}_k^N \to (\mathbb{P}_k^N)^\delta$, such that for $1 \leq i \leq N+1$ we have $\theta(\alpha) = \theta_i(\alpha)$ for all $\alpha \in \mathbb{A}_{k,i}^N$; note that then the image of θ

must be $(\mathbb{P}_k^N)^{\rho\delta}$. After proving this claim we will have

$$\mathbb{P}_k^N \approx (\mathbb{P}_k^N)^{\rho\delta} \subset (\mathbb{P}_k^N)^{\mu\delta} \subset (\mathbb{P}_k^N)^{\delta}$$

extending the enlargements of §8.2 from the affine case to the projective case. The uniqueness in the claim follows from the existence. To prove the existence, let β be the equivalence class of $u = (u_1, \ldots, u_{N+1}) \in k^{N+1} \setminus \{(0, \ldots, 0)\}$. For any $i \in \{1, \ldots, N+1\}$ with $u_i \neq 0$, let $R = \theta_i(v)$ where $v = (u_1/u_i, \ldots, u_{N+1}/u_i) \in \mathbb{A}_{k,i}^N$, and define $\theta(\beta) = R$. To show that θ is well-defined, for any $i' \in \{1, \ldots, N+1\}$ with $u_{i'} \neq 0$, let $R' = \theta_{i'}(v')$ where $v' = (u_1/u_{i'}, \ldots, u_{N+1}/u_{i'}) \in \mathbb{A}_{k,i'}^N$. What we must prove is that $R = R'$. Since R and R' both belong to the irredundant premodel $(\mathbb{P}_k^N)^{\delta}$, it suffices to find a local domain S which dominates both R and R'; take S to be the localization of $B_{N+1,k}$ at the maximal ideal generated by $X_1 - u_1, \ldots, X_{N+1} - u_{N+1}$. This defines θ as a map from $\mathbb{P}_k^N$ to $(\mathbb{P}_k^N)^{\delta}$. To see that it is injective, let β^* be the equivalence class of $u^* = (u_1^*, \ldots, u_{N+1}^*) \in k^{N+1} \setminus \{(0, \ldots, 0)\}$, such that $\theta(\beta) = \theta(\beta^*)$. We have to show that $\beta = \beta^*$. This is obvious if $u_i^* \neq 0$, because $\theta_i : \mathbb{A}_{k,i}^N \to (\mathbb{A}_{k,i}^N)^{\delta}$ is injective. Moreover, $u_i \neq 0$ with $\theta(\beta) = \theta_i(v) = R \Rightarrow X_j/X_i \in R$ for all j but $X_j/X_i \in M(R)$ for exactly those j for which $u_j = 0$; therefore, since $\theta(\beta^*) = \theta(\beta) = R$, we must have $u_i^* \neq 0$.

§9.2: MODELIC BLOWUP

The following Theorem (T30), which we shall soon prove, says that the modelic blowup, which we shall now introduce, is nothing but the coordinate free incarnation of the modelic proj; [cf. §12(R9)]. By and by we shall see that it is also the main key to the desingularization of curves and surfaces in particular and varieties in general. For any nonzero A-submodule P of K we put

$$\mathfrak{W}(A, P) = \bigcup_{0 \neq x \in P} \mathfrak{V}(A[Px^{-1}])$$

and we call this the modelic blowup of A at P. Here $A[Px^{-1}]$ denotes the smallest subring of K which contains A and which contains y/x for all $y \in P$.

MODELIC BLOWUP THEOREM (T30). Recalling that A is a subring of a field K, and P is a nonzero A-submodule of K, we have the following.

(T30.1) For any $0 \neq x \in P$ we have $(A[Px^{-1}])P = (A[Px^{-1}])x$ and hence $RP = Rx$ for every $R \in \mathfrak{V}(A[Px^{-1}])$. In particular, if P is a nonzero ideal in A then for every $R \in \mathfrak{V}(A[Px^{-1}])$ we have that PR is a nonzero principal ideal in R.

(T30.2) Given any family $(x_l)_{l \in \Lambda}$ of generators of P, for all $0 \neq x \in P$ we have $A[Px^{-1}] = A[(x_l/x)_{l \in \Lambda}]$, and we have $\mathfrak{W}(A; (x_l)_{l \in \Lambda}) \subset \mathfrak{W}(A, P)$.

(T30.3) For any family $(x_l)_{l \in \Lambda}$ of generators of P we have

$$\mathfrak{W}(A, P) = \mathfrak{W}(A; (x_l)_{l \in \Lambda}).$$

(T30.4) For all $x \neq 0 \neq y$ in P we have $\mathrm{QF}(A[Px^{-1}]) = \mathrm{QF}(A[Py^{-1}])$ and letting K' denote this common QF we have that $\mathfrak{W}(A, P)$ is a semimodel of K'/A, and if P is a finitely generated A-module then $\mathfrak{W}(A, P)$ is a projective model of K'/A. In particular, if P is a finitely generated ideal in A then $\mathfrak{W}(A, P)$ is a projective model of $\mathrm{QF}(A)/A$. If $P = Ax$ for some $0 \neq x \in K$ then $\mathfrak{W}(A, P) = \mathfrak{W}(A; x) = \mathfrak{V}(A)$.

(T30.5) If P is an ideal in A then

$$\{R \in \mathfrak{V}(A) : PR = R\} = \{R \in \mathfrak{W}(A, P) : PR = R\}.$$

(T30.6) If A is quasilocal and P is an ideal in A then:

$$P \text{ is a principal ideal in } A \Leftrightarrow \mathfrak{W}(A, P) = \mathfrak{V}(A) \Leftrightarrow A \in \mathfrak{W}(A, P).$$

§9.3: BLOWUP OF SINGULARITIES

We shall now make some EXAMPLES to indicate how modelic blowup helps out in desingularizing plane curves and surfaces in 3-space.

EXAMPLE (X12). [**Nodal and Cuspidal Cubics**]. Consider the nodal cubic $f = 0$ with $f = Y^2 - X^2 - X^3$ and the cuspidal cubic $g = 0$ with $g = Y^2 - X^3$ discussed in L3§5. For A take the bivariate polynomial ring $k[X, Y]$ over a field k. For P take the ideal of the origin in the (X, Y)-plane, i.e., the maximal ideal in A generated by (X, Y). Now $\mathfrak{W}(A, P) = \mathfrak{V}(A') \cup \mathfrak{V}(A'')$ where A' and A'' are the bivariate polynomial rings $k[X', Y']$ and $k[X'', Y'']$ where $(X' = X, Y' = Y/X)$ and $(X'' = X/Y, Y'' = Y)$ respectively. Substituting the first set of equations in f we get $f = X'^2(Y'^2 - 1 - X')$ and hence, in the (X', Y')-plane, the "total transform" of the nodal cubic consists of the "exceptional line" $X' = 0$ together with the "proper transform" $Y'^2 - 1 - X' = 0$ which is a parabola meeting the exceptional line in two distinct points. Likewise, substituting the first set of equations in g we get $g = X'^2(Y'^2 - X')$ and hence, in the (X', Y')-plane, the "total transform" of the cuspidal cubic consists of the "exceptional line" $X' = 0$ together with the "proper transform" $Y'^2 - X' = 0$ which is a parabola meeting the exceptional line at a single point where it is tangent. At any rate, in both the cases, a single blowup has resolved the singularity. You may reconfirm this by taking a similar look in the (X'', Y'')-plane. We shall talk more about exceptional lines and proper transforms later. In the meantime you may like to glance at the relevant pictures on pages 2, 35, 131-134, and 154-155 of my Engineering Book [A04].

EXAMPLE (X13). [**Higher Cusps**]. Consider the plane curve $g_s = 0$ with $g_s = Y^2 - X^{2s+1}$ where $s > 1$ is an integer; at the origin this has a double point which is a higher cusp, say a cusp of height s (pictured on pages 155-156 of [A04]). With notation as in the above Example (X12), substituting the first set of equations in g_s we get $g_s = X'^2(Y'^2 - X'^{2(s-1)+1})$ and hence, in the (X', Y')-plane, the total transform of our curve consists of the exceptional line $X' = 0$ together with the proper transform $Y'^2 - X'^{2(s-1)+1} = 0$ having a cusp of height $s - 1$. Repeating the process we get a cusp of height $s - 2$, a cusp of height $s - 3$, ..., a cusp of height 1 (like $Y^2 - X^3 = 0$), and finally a simple point. In Max Noether's picturesque terminology, we say that the plane curve $g_s = 0$, in addition to having a double point at the origin, has one double point in its first neighborhood, one double in the second neighborhood, ..., and one double point in the $(s-1)$-th neighborhood, thus making a total of $s - 1$ double points infinitely near to the origin. So (including the origin) we have a totality of s double points. By and by we shall give precision to the idea of infinitely near points. In the meantime you may enjoy reading about them, from a somewhat heuristic viewpoint, in Lecture 19 on Infinitely Near Singularities (pages 145-158) of the Engineering Book [A04].

EXAMPLE (X14). [**Circular Cones and Fermat Cones**]. Consider the horizontal circular cone $X^2 - Y^2 - Z^2 = 0$ (description on pages 197-198 of [A04]), or more generally the Fermat Cone $f = 0$ with $f = X^n \pm Y^n \pm Z^n$, where $n > 1$ is an integer which is nondivisible by the characteristic (discussed on page 202 of [A04]). For A take the trivariate polynomial ring $k[X, Y, Z]$ over a field k. For P take the ideal of the origin in the (X, Y, Z)-space, i.e., the maximal ideal in A generated by (X, Y, Z). Now $\mathfrak{W}(A, P) = \mathfrak{V}(A') \cup \mathfrak{V}(A'') \cup \mathfrak{V}(A''')$ where A', A'', and A''' are the trivariate polynomial rings $k[X', Y', Z']$, $k[X'', Y'', Z'']$, and $k[X''', Y''', Z''']$ where $(X' = X, Y' = Y/X, Z' = Z/X)$, $(X'' = X/Y, Y'' = Y, Z'' = Z/Y)$, and $(X''' = X/Z, Y''' = Y/Z, Z''' = Z)$ respectively. Substituting the first set of equations in f we get $f = X'^n(1 \pm Y'^n \pm Z'^n)$ and hence, in the (X', Y', Z')-space, the total transform of the Fermat Cone consists of the exceptional plane $X' = 0$ together with the proper transform $1 \pm Y'^n \pm Z'^n = 0$ which is a nonsingular "cylinder."

§10: EXAMPLES AND EXERCISES

EXERCISE (E1). [**Algebraic Closure**]. In L2(R6), while establishing the existence of an algebraic closure $\overline{K}$ of any given field K, we found an overfield $\overline{K}$ of K such that: ($\flat$) $\overline{K}$ is algebraic over K and contains a splitting field of every nonconstant monic polynomial F over K. The implication ($\flat$) $\Rightarrow$ ($\flat\flat$), where ($\flat\flat$) says that $\overline{K}$ is algebraically closed, follows from the fact that an algebraic extension of an algebraic extension is an algebraic extension, which was proved in L1(R7)(C3). To have another transparent proof of the implication ($\flat$) $\Rightarrow$ ($\flat\flat$), show that: given any algebraic field extension $\overline{K}/K$ and given any nonconstant monic G in $\overline{K}[Y]$, there exists a nonconstant monic F in $K[Y]$ such that G divides F in $\overline{K}[Y]$.

EXERCISE (E2). [**Integral Closure**]. Recall that: an element t in an overring S of a ring R is said to be integral over R if it satisfies a monic polynomial equation over R, i.e., if $F(t) = 0$ for some monic polynomial

$$(\sharp) \qquad\qquad F(Y) = Y^n + A_1 Y^{n-1} + \cdots + A_n$$

of some degree $n > 0$ with coefficients $A_1, \ldots, A_n$ in R; this is indicated by saying that t/R is integral; a subset T of S is integral over R means every $t \in T$ is integral over R; this is indicated by saying that T/R is integral; finally, by the integral closure of R in S we mean the set of all elements in S which are integral over R. In L3§7 we asserted without proof that

(J1) $\begin{cases} \text{the integral closure of a ring } R \text{ in an overring } S \text{ is a subring of } S, \\ \text{and } R \text{ is a subring of the said integral closure} \end{cases}$

and

(J2) $\begin{cases} \text{if an overring } S \text{ of a ring } R \text{ is integral over } R, \text{ and} \\ \text{an overring } T \text{ of } S \text{ is integral over } S, \text{ then } T \text{ is integral over } R. \end{cases}$

Note that assertions (J1) and (J2) are the integral extension analogues of the algebraic extension assertions L1(R7)(C5) and L1(R7)(C3) out of which the last was cited in (E1) above. Also note that to prove (J1) it suffices to show that

(J0) $\begin{cases} \text{if } t_1, \ldots, t_m \text{ are any finite number of elements in a ring } S \\ \text{which are integral over a subring } R \text{ of } S \text{ then the ring} \\ R[t_1, \ldots, t_m] \text{ is a finitely generated } R\text{-module and is integral over } R. \end{cases}$

As an exercise, prove (J0) and (J2) in the following five STEPS (E2.1) to (E2.5), where S is an overring of a ring R.

STEP (E2.1). For $t \in S$ let V be an $R[t]$-submodule of S with $\mathrm{ann}_{R[t]} V = 0$ such that V is finitely generated as an R-module, and let J be an ideal in R such that $tV \subset JV$. Then $F(t) = 0$ for some $F(Y)$ with $A_i \in J^i$ for $1 \leq i \leq n$.

HINT. Let $X_1, \ldots, X_n$ be R-generators of V with $n \in \mathbb{N}_+$. Then for $1 \leq i \leq n$ we have $tX_i = \sum_{1 \leq j \leq n} C_{ij} X_j$ with $C_{ij} \in J$. Let $A_{ij} = \delta_{ij} t - C_{ij}$ with Kronecker δ_{ij}. Let $Y_i = 0$ for $1 \leq i \leq n$, and let $B^{(j)} = (B_{il}^{(j)})$ be the $n \times n$ matrix obtained by replacing the j-th column of the $n \times n$ matrix $A = (A_{ij})$ by the column vector $(Y_1, \ldots, Y_n)$, i.e., the $n \times 1$ matrix whose $(i, 1)$-th entry is Y_i for $1 \leq i \leq n$. Then $\sum_{1 \leq j \leq n} A_{ij} X_j = Y_i$ for $1 \leq i \leq n$, and hence by Cramer's Rule $\det(A) X_j = \det(B^{(j)}) = 0$ for $1 \leq j \leq n$; see (E4) below. Consequently $\det(A) \in \mathrm{ann}_{R[t]} V$ and hence $\det(A) = 0$. Let $F(Y) = \det(Y 1_n - C)$ where C is the $n \times n$ matrix (C_{ij}) and where we recall that 1_n is the $n \times n$ identity matrix. Then clearly $F(Y) = Y^n + \sum_{1 \leq i \leq n} A_i Y^{n-i}$ with $A_i \in J^i$ for $1 \leq i \leq n$, and $F(t) = \det(A) = 0$.

STEP (E2.2). For any $t \in S$ the following four conditions are equivalent.

(1) t/R is integral.

(2) The ring $R[t]$ is finitely generated as an R-module.

(3) $R[t] \subset T$ for some subring T of S such that T is finitely generated as an R-module.

(4) There exists an $R[t]$-submodule V of S with $\operatorname{ann}_{R[t]} V = 0$ such that V is finitely generated as an R-module

HINT. (1) $\Rightarrow$ (2): from $F(t) = 0$, by induction we get $t^{n+i} \in R + Rt + \cdots + Rt^{n-1}$ for all $i \in \mathbb{N}$, and hence $(1, t, \ldots, t^{n-1})$ are R-generators of $R[t]$. (2) $\Rightarrow$ (3): take $T = R[t]$. (3) $\Rightarrow$ (4): take $V = T$ and note that then $1 \in V$ and hence for any r in $\operatorname{ann}_{R[t]} V$ we have $r = r1 = 0$. (4) $\Rightarrow$ (1): take $J = R$ in STEP (E2.1).

STEP (E2.3). If S is a finitely generated R-module, and an overring T of S is a finitely generated S-module, then T is a finitely generated R-module.

HINT. If $(x_i)_{1 \le i \le n}$ are R-generators of S, and $(y_j)_{1 \le j \le m}$ are S-generators of T, then $(x_i y_j)_{1 \le i \le n, 1 \le j \le m}$ are R-generators of T.

STEP (E2.4). (J0).

HINT. In view of (3) $\Rightarrow$ (1) of (E2.2) it suffices to show that if elements $t_1, \ldots, t_m$ of S are integral over R then $R[t_1, \ldots, t_m]$ is a finitely generated R-module. Make induction on m. For $m = 1$ this follows from (1) $\Rightarrow$ (2) of (E2.2). Also $m - 1 \Rightarrow m$ follows from the $m = 1$ case and (E2.3).

STEP (E2.5). (J2).

HINT. Assume S/R and T/S are integral. Then for any $t \in T$ we have $F(t) = 0$ for some $F(Y)$ as in ($\natural$) above with $A_1, \ldots, A_n$ in S. Now by (E2.2) and (E2.4) we see that $R[A_1, \ldots, A_n]$ is a finitely generated R-module and $R[t, A_1, \ldots, A_n]$ is a finitely generated $R[A_1, \ldots, A_n]$-module. Therefore $R[t, A_1, \ldots, A_n]$ is a finitely generated R-module by (E2.3), and hence t/R is integral by (E2.2).

EXERCISE (E3). [**Conditions of Integral Dependence**].

(i) Note that in (E2.1) as well as (E2.2)(4), if either $V \ne 0$ and S is a domain, or if $1 \in V$ (for instance if V is an overring of R), then the condition $\operatorname{ann}_{R[t]} V = 0$ is automatically satisfied.

(ii) Show that if R is noetherian then (2) of (E2.2) is equivalent to saying that: ($2'$) $R[t]$ is contained in a finitely generated R-submodule of S.

(iii) By an example show that if R is nonnoetherian then ($2'$) $\nRightarrow$ (2).

EXAMPLE (X15). [**Nonarchimedean Valuations**]. Let v be a valuation of a field K such that the value group G_v of v is nonarchimedian, let α, β be elements in G_v such that $\alpha > n\beta > 0$ for all $n \in \mathbb{N}_+$, let x, y be elements in the valuation ring $R = R_v$ of v with $v(x) = \alpha$ and $v(y) = \beta$; see L2§7(D2), L2§3, L3§7. Let $t = 1/y$ and $V = Ru$ with $u = 1/x$. Then t/R is nonintegral but $R[t]$ is contained in the finitely generated R-module V.

EXERCISE (E4). [**Cramer's Rule for Linear Equations**]. Establish the Rule cited in (E2.1) above, i.e., given a finite number of simultaneous linear equations

$$\sum_{1 \leq j \leq n} A_{ij} X_j = Y_i \quad \text{for} \quad 1 \leq i \leq n$$

where the entries of the $n \times n$ matrix $A = (A_{ij})$ as well as the elements X_j, Y_i belong to a ring R, show that

(E4.1) $\det(A) X_j = \det(B^{(j)}) \quad \text{for} \quad 1 \leq j \leq n$

where the $n \times n$ matrix $B^{(j)} = (B_{il}^{(j)})$ is given by

$$B_{il}^{(j)} = \begin{cases} A_{il} & \text{if } l \neq j \\ Y_i & \text{if } l = j \end{cases}$$

i.e., $B^{(j)}$ is obtained by replacing the j-th column of A by the column vector $(Y_1, \ldots, Y_n)$. Hence in particular, if $\det(A)$ is a nonzerodivisor in R then, in the total quotient ring $\text{QR}(R)$, we have

(E4.2) $X_j = \det(B^{(j)})/\det(A) \quad \text{for} \quad 1 \leq j \leq n.$

HINT. Let $A^{(j)}$ be the matrix obtained by multiplying the j-th column of A by X_j. Then $B^{(j)}$ is obtained by adding to the j-th column of $A^{(j)}$ linear combinations of the remaining columns. Hence by L3(E1) we get

$$\det(A^{(j)}) = \det(A) X_j \quad \text{and} \quad \det(B^{(j)}) = \det(A^{(j)}).$$

EXERCISE (E5). [**Solutions of Homogeneous Linear Equations**]. Consider the system of homogeneous linear equations

$$\sum_{1 \leq j \leq N} a_{ij} X_j = 0 \quad \text{for} \quad 1 \leq i \leq M$$

where $a = (a_{ij})$ is an $M \times N$ matrix over a field k with M, N in $\mathbb{N}_+$. Let S be the solution space of the system, i.e., the set of all $X = (X_1, \ldots, X_N) \in \mathbb{A}_k^N$ satisfying the system; clearly S is a k-linear subspace of $\mathbb{A}_k^N$. Let n be the rank of a, and note that then $0 \leq n \leq \min(M, N)$. Show that if $M = N$ then: $S = \{(0, \ldots, 0)\} \Leftrightarrow n = N$, i.e., $\Leftrightarrow \det(a) \neq 0$. More generally, without any assumption, show that $\dim_k S = N - n$ by finding a k-linear injection $\phi : \mathbb{A}_k^{N-n} \to \mathbb{A}_k^N$ with $\operatorname{im}(\phi) = S$. Moreover, do this by establishing the explicit formulas which are described below. By relabelling a_{ij} we may assume that $\det(A) \neq 0$ where $A = (A_{ij})$ is the $n \times n$ matrix with $A_{ij} = a_{ij}$ for $1 \leq i \leq n$ and $1 \leq j \leq n$, i.e., A is the left upper corner submatrix of a. Let

$$Y_i = - \sum_{n < \lambda \leq N} a_{i\lambda} X_\lambda \quad \text{for} \quad 1 \leq i \leq M.$$

For $1 \leq j \leq n$ let $B^{(j)}$ be the $n \times n$ matrix obtained by replacing the j-th column of A by the column vector $(Y_1, \ldots, Y_n)$. Now for every $Z = (Z_1, \ldots, Z_{N-n}) \in \mathbb{A}_k^{N-n}$ we put $\phi(Z) = x = (x_1 \ldots, x_N) \in \mathbb{A}_k^N$ where

$$x_j = \begin{cases} \det(B^{(j)})/\det(A) & \text{if } 1 \leq j \leq n \\ Z_{j-n} & \text{if } n < j \leq N. \end{cases}$$

HINT. By the above displayed definition of x_j, and by looking at the expansion of $\det(B^{(j)})$ by its j-th column, we see that ϕ is an injective k-linear map. So it only remains to show that its image is S. Now the given homogeneous system is clearly equivalent to the nonhomogeneous system

$$\sum_{1 \leq j \leq n} A_{ij} X_j = Y_i \quad \text{for} \quad 1 \leq i \leq n$$

together with the extra equations $\sum_{1 \leq j \leq N} a_{ij} X_j = 0$ for $n < i \leq M$. Therefore, by solving the above nonhomogeneous system by Cramer's Rule (E4) and then feeding the solution (E4.2) in the extra equations, we are reduced to showing that for $n < i \leq M$ we have

$$Y_i - \sum_{1 \leq j \leq n} a_{ij} \det(B^{(j)})/\det(A) = 0.$$

So given any such i, let C be the $(n+1) \times (n+1)$ matrix obtained by bordering A by the last column $(Y_1, \ldots, Y_n, Y_i)$ and the last row $(a_{i1}, \ldots, a_{in}, Y_i)$, and for $n < l \leq N$ let $C^{(l)}$ be the $(n+1) \times (n+1)$ matrix obtained by bordering A by the last column $(a_{1l}, \ldots, a_{nl}, a_{il})$ and the last row $(a_{i1}, \ldots, a_{in}, a_{il})$. Expanding by the last column we see that $\det(C) = - \sum_{n < l \leq N} \det(C^{(l)}) X_l$; but $\det(C^{(l)}) = 0$ because $C^{(l)}$ is an $(n+1) \times (n+1)$ submatrix of a; therefore $\det(C) = 0$. Expanding by the last row we also see that

$$\det(C) = Y_i \det(A) - \sum_{1 \leq j \leq n} a_{ij} \det(B^{(j)})$$

and this completes the proof.

EXERCISE (E6). [**Conditions for a Common Factor**]. Let n and m be positive integers, and let f and g be polynomials of degrees $\leq n$ and $\leq m$ in Y with coefficients in a field k, respectively. Show that

$$
\begin{cases}
ef + gh = 0 \text{ for some } e \text{ and } h \text{ in } k[Y] \\
\text{with } \deg_Y e \leq m - 1 \text{ and } \deg_Y h \leq n - 1 \\
\text{such that at least one of } e \text{ and } h \text{ is nonzero} \\
\Leftrightarrow \text{ either } \deg_Y f < n \text{ and } \deg_Y g < m \\
\quad \text{or } f \text{ and } g \text{ have a common root in some overfield of } k.
\end{cases}
$$

HINT. $k[Y]$ is a UFD. Moreover, if f, g have a nonconstant common irreducible factor then in its splitting field we get a common root for them.

EXERCISE (E7). [**Sylvester Resultant**]. For polynomials

$$f = f(Y) = a_0 Y^n + a_1 Y^{n-1} + \cdots + a_n$$
$$g = g(Y) = b_0 Y^m + b_1 Y^{m-1} + \cdots + b_m$$

with coefficients in a domain R, where n, m are nonnegative integers, establish the Basic Fact (T1), i.e., show that: $\mathrm{Res}_Y(f, g) = 0 \Leftrightarrow n + m \neq 0$ and either $a_0 = 0 = b_0$ or f and g have a common root in some overfield of R.

HINT. The case when $mn = 0$ being straightforward, assume $mn \neq 0$. Let $A = (A_{ij})$ be the $(m + n) \times (m + n)$ matrix over $k = \mathrm{QF}(R)$ whose transpose is $\mathrm{Resmat}_Y(f, g)$. In view of (E6) it suffices to show that $\det(A) = 0$ iff $ef + gh = 0$ for some e and h in $k[Y]$ with $\deg_Y e \leq m - 1$ and $\deg_Y h \leq n - 1$ such that at least one of e and h is nonzero. Now

$$e = e(Y) = X_1 Y^{m-1} + X_2 Y^{m-2} + \cdots + X_m$$
$$h = h(Y) = X_{m+1} Y^{n-1} + X_{m+2} Y^{n-2} + \cdots + X_{m+n}$$

where elements $X_1, \ldots, X_{m+n}$, at least one of which is nonzero, are to be found in k, such that $ef + gh = 0$. Upon letting

$$ef + gh = \sum_{1 \leq i \leq m+n} D_i Y^{m+n-i}$$

we see that for $1 \leq i \leq m + n$ we have

$$D_i = \sum_{1 \leq j \leq m+n} A_{ij} X_j.$$

Hence by (E5), $\det(A) = 0$ iff there exists $(X_1, \ldots, X_{m+n}) \in \mathbb{A}_k^{m+n} \setminus \{(0, \ldots, 0)\}$ for which $D_1 = \cdots = D_{m+n} = 0$. This completes the proof.

EXERCISE (E8). [**Elementary Properties of Determinants**]. Prove the following properties of $\det(A)$ where $A = (A_{ij})$ is an $n \times n$ matrix over a ring R which were used in (E4) to (E7) above. [(E8.1) to (E8.3) are vacuous for $n = 0$].

(E8.1) If a row (or a column) of A is multiplied by an element of R then $\det(A)$ gets multiplied by that element.

(E8.2) If to a row (or column) of A we add an R-linear combination of the remaining rows (or columns) of A then $\det(A)$ is unchanged.

(E8.3) If the rows (or columns) of A are permuted according to a permutation σ then $\det(A)$ gets multiplied by $\text{sgn}(\sigma)$. In particular if we interchange two rows (or columns) then $\det(A)$ gets multiplied by -1.

(E8.4) If A is replaced by its transpose then $\det(A)$ is unchanged.

EXERCISE (E9). [**Blowup of a Point in the Plane and Three Space**]. In Examples (X12) and (X13) of §9.3 check what happens in $\mathfrak{V}(A'')$, and in Example (X14) of §9.3 check what happens in $\mathfrak{V}(A'')$ and $\mathfrak{V}(A''')$.

EXAMPLE (X16). [**Blowup of a Line in Three Space**]. In the Examples cited above we blew up a point and studied the effect on singularities of curves and surfaces. Now let us blowup the line

$$L : X = Y = 0$$

in the (X, Y, Z)-space over a field k, and study what effect it has on a surface singularity at the origin $Q : X = Y = Z = 0$. So consider the surface

$$S : f = Y^3 - X^2 Z^{n-2} = 0$$

of degree $n > 5$ with double line L through the triple point Q, and the nonconical cubic surface

$$T : g = Y^2 - X^2 Z = 0$$

with double line L through the double point Q, discussed on pages 204-205 of my Engineering Book [A04]. Let P be the ideal generated by (X, Y) in $A = k[X, Y, Z]$. Then

$$\mathfrak{W}(A, P) = \mathfrak{V}(A') \cup \mathfrak{V}(A'')$$

where A' and A'' are the trivariate polynomial rings $k[X', Y', Z]$ and $k[X'', Y'', Z]$ where $(X', Y') = (X, Y/X)$ and $(X'', Y'') = (X/Y, Y)$ respectively. Substituting the first set of equations in f we get

$$f = X'^2(X'Y'^3 - Z^{n-2})$$

and hence, in the (X', Y', Z)-space, the total transform of S consists of the exceptional plane $X' = 0$ together with the proper transform $X'Y'^3 - Z^{n-1} = 0$ which has a quadruple point at the origin. Likewise, substituting the first set of equations in g we get

$$g = X'^2(Y'^2 - Z)$$

and hence, in the (X', Y', Z)-space, the total transform of T consists of the exceptional plane $X' = 0$ together with the proper transform $Y'^2 - Z = 0$ which is a nonsingular parabolic cylinder. The reason why the singularity of T was resolved but the singularity of S was worsened (a triple point becoming a quadruple point) is that for T we were in the equimultiple case because the multiplicity of $Q = 2 =$ the multiplicity of L, whereas for S we were in the nonequimultiple case because the multiplicity of $Q = 3 > 2 =$ the multiplicity of L.

EXAMPLE (X17). [**Tight Associators and Monomial Ideals**]. In §5(O17) we showed that for a submodule U of a module V over a noetherian ring R, the associator $\mathrm{ass}_R(V/U)$ coincides with the tight associator $\mathrm{tass}_R(V/U)$, i.e., if P is a prime ideal in R such that $(U : a)_R$ is P-primary for some $a \in V$ then $(U : b)_R = P$ for some $b \in V$. Here is an example that in case R is nonnoetherian, this need not be so even when V is R and U is an ideal Q in R.

So let R be the polynomial ring $k[X_1, X_2, X_3, \ldots]$ in an infinite number of indeterminates $(X_i)_{i \in \mathbb{N}_+}$ over a field k, and let P and Q be the ideals in R generated by $X_1, X_2, X_3, \ldots$, and $X_1^2, X_2^2, X_3^2 \ldots$ respectively, i.e., by $(X_i)_{i \in \mathbb{N}_+}$ and $(X_i^2)_{i \in \mathbb{N}_+}$ respectively. In a moment we shall show that P belongs to $\mathrm{ass}_R(R/Q)$ but not to $\mathrm{tass}_R(R/Q)$. Actually we shall do this in a somewhat more general set-up.

We start by proving some claims about monomial ideals. So let $(X_i)_{i \in I}$ be any family of indeterminates over the field k and consider [cf. L6§6(D10)] the polynomial ring $R_I = k[(X_i)_{i \in I}]$. Let W be the set of all maps $w : I \to \mathbb{N}$ whose support $\mathrm{supp}(w) = \{i \in I : w(i) \neq 0\}$ is finite, and for any such w write $X^w = \prod_{i \in I} X_i^{w(i)}$. Note that then the set of all monomials $M = \{X^w : w \in W\}$ is a (free) vector space basis of R_I over k, and for any $f \in R_I$, after writing $f = \sum_{w \in W} a_w X^w$ with $a_w \in k$ we have that its support $\mathrm{supp}(f) = \{X^w : w \in W \text{ with } a_w \neq 0\}$ is finite. For any u and t in W we write $u \geq t$ to mean that $u(i) \geq t(i)$ for all $i \in I$, and we observe that this is so iff $X^u = X^t X^w = X^{t+w}$ for some $w \in W$, where we define $t + w \in W$ by putting $(t + w)(i) = t(i) + w(i)$ for all $i \in I$. For any $T \subset W$, let

$$L(T) = \{X^t : t \in T\}$$

be the corresponding set of monomials, let

$$\widehat{T} = \{u \in W : u \geq t \text{ for some } t \in T\}$$

be the corresponding "saturated" subset of W, and let

$$N(T) = \{f \in R_I : \mathrm{supp}(f) \subset L(T)\}.$$

Note that then $N(T)$ is the k-vector-subspace of R_I generated by $L(T)$, and $L(T)$ is a k-vector-space-basis of $N(T)$. Let us consider a monomial ideal in R_I, i.e., an ideal generated by $L(T)$ for some $T \subset W$; let us denote this ideal by $J(T)$, i.e.,

$$J(T) = L(T)R_I.$$

We claim that for any $T \subset W$ we have

(1) $J(T) = N(\widehat{T})$ and $J(T) \cap M = L(\widehat{T}).$

To prove this, note that obviously $N(\widehat{T}) \subset J(T)$. Conversely, any $f \in J(T)$ can be written as $f = \sum_{t \in T'} X^t f_t$ where T' is a finite subset of T and $f_t \in R_I$. Now we can find a finite subset W' of W such that for all $t \in T'$ we have $f_t = \sum_{w \in W'} a_{tw} X^w$ with $a_{tw} \in k$. Now $T^* = \{t + w : (t, w) \in T' \times W'\}$ is a finite subset of $\widehat{T}$ and we have $f = \sum_{u \in T^*} b_u X^u$ where $b_u = \sum_{\{(t,w) \in T' \times W' : t+w=u\}} a_{tw} \in k$ for all $u \in T^*$. Therefore $\operatorname{supp}(f) \subset \widehat{T}$. This proves the first part of (1), and the second part follows from it.

Consider the ideal in R_I generated by $(X_i)_{i \in I'}$ where I' is any subset of I, i.e., the ideal $J(c(I'))$ where

$$c(I') = \{c(j) : j \in I'\}$$

and where $c(j)$ is the characteristic function of $j \in I$, i.e., $c(j) \in W$ is defined by putting

$$c(j)(i) = \begin{cases} 1 & \text{if } i = j \\ 0 & \text{if } i \neq j. \end{cases}$$

For any $T \subset W$ consider the condition:

($\sharp$)   for all j in some infinite subset I' of I we have $t(j) \neq 1$ for all $t \in T$.

We claim that then:

(2) $(\sharp) \Rightarrow (J(T) : J(c(I)))_{R_I} = J(T).$

To prove this, assuming ($\sharp$), given any $f \in R_I \setminus J(T)$ we want to find $g \in J(c(I))$ with $fg \notin J(T)$. Now $\operatorname{supp}(f)$ is a nonempty finite set and hence we may label its elements as $t_1, \ldots, t_n$ where $n \in \mathbb{N}_+$ and $t_1, \ldots, t_n$ are pairwise distinct members of W. Since $f \notin J(T)$, by (1) we know that some t_i does not belong to $\widehat{T}$ and hence by relabelling we may assume that $t_1 \notin \widehat{T}$. Since I' is an infinite set, we can find $j \in I'$ such that $t_1(j) = 0$; then $(t_1 + c(j))(j) = 1$; since $t_1 \notin \widehat{T}$, by ($\sharp$) it follows that $t_1 + c(j) \notin \widehat{T}$. Let $g = X_j$. Then $t_1 + c(j), \ldots, t_n + c(j)$ are exactly all the distinct elements of $\operatorname{supp}(fg)$, and hence by (1) we see that $fg \notin J(T)$. Also obviously $g \in J(c(I))$.

For any $T \subset W$ consider the conditions:

($\sharp'$)   for every $t \in T$ we have $t(i) \neq 0$ for some $i \in I$

and

($\sharp''$)   for every $j \in I$ there exists $d_j \in T$ such that: $d_j(i) \neq 0$ iff $j = i$.

By (1) we see that:

(3) $(\sharp) + (\sharp') \Rightarrow J(T) \subset J(c(I)) \neq J(T)$.

Now $J(c(I))$ is the kernel of the k-epimorphism $R_I \to k$ with $X_i \mapsto 0$ for all $i \in I$, and hence $J(c(I))$ is a maximal ideal in R_I and therefore by §5(O8)(3*) we see that:

(4) $(\sharp') + (\sharp'') \Rightarrow \begin{cases} J(T) \text{ is } J(c(I))\text{-primary with} \\ \mathrm{vspec}_{R_I} J(T) = \mathrm{mvspec}_{R_I} J(T) = \{J(c(I))\}. \end{cases}$

In view of Theorems (T5) and (T6) of §6, by (2) to (4) we conclude that:

(5) $(\sharp) + (\sharp') + (\sharp'') \Rightarrow \mathrm{ass}_{R_I}(R_I/J(T)) = \{J(c(I))\}$ but $\mathrm{tass}_{R_I}(R_I/J(T)) = \emptyset$.

As a special case of (5), assuming I to be infinite, upon letting P to be the ideal in R_I generated by $(X_i)_{i \in I}$ and Q to be ideal in R_I generated by $(X_i^{n_i})_{i \in I}$ where $n_i \in \mathbb{N}_+$ with $n_i > 1$ for infinitely many i, we get $\mathrm{ass}_{R_I}(R_I/Q) = \{P\}$ but $\mathrm{tass}_{R_I}(R_I/Q) = \emptyset$. In particular we have the beginning illustration when $I = \mathbb{N}_+$ with P and Q generated by $(X_i)_{i \in \mathbb{N}_+}$ and $(X_i^2)_{i \in \mathbb{N}_+}$ respectively.

EXAMPLE (X18). **[Isolated Components of a Module]**. As we noted in §5(O20)(14*), given any submodule U of a module V over a ring R with $U \neq V$, if V is finitely generated then: $P \in \mathrm{vspec}(\mathrm{ann}_R(V/U)) \Rightarrow [U : P]_V \neq V$. We shall now show by an example that this is not true without finite generation.

Let V be the bivariate polynomial ring $k[X, Y]$ over a field k. Then $U = k[X, XY]$ is a subring of V, and $R = k[X]$ is a subring of U. So we may regard V as a module over R, and U a submodule. Note that by putting primes on X and Y, the pair $U \subset V$ may be identified with the pair $A \subset A'$ of Examples (X12) and (X13) of §9.3. For any $f = \sum a_{ij} X^i Y^j \in V$ with $a_{ij} \in k$ let $\mathrm{supp}(f) = \{(i, j) \in \mathbb{N} \times \mathbb{N} : a_{ij} \neq 0\}$. Clearly $U = \{f \in V : i \geq j$ for all $(i, j) \in \mathrm{supp}(f)\}$. Any $0 \neq g \in R$ can be written as a finite sum $g = \sum b_i X^i$ where $b_i \in k$ with $b_e \neq 0$ for some $e \in \mathbb{N}$; obviously $(e, e + 1)$ belongs to $\mathrm{supp}(gY^{e+1})$ and hence $gY^{e+1} \notin U$ with $Y^{e+1} \in V$; therefore $\mathrm{ann}_R(V/U) = 0$. Thus for the zero ideal P in R we have $P \in \mathrm{vspec}(\mathrm{ann}_R(V/U))$. For any $f \in V$ we can find a nonnegative integer d which exceeds the degree of f and then by the above criterion we get $X^d f \in U$ with $X^d \in R \setminus P$. Therefore $[U : P]_V = V$.

EXAMPLE (X19). **[Annihilator of a Primary Module]**. As we noted in §5(O13), given any submodule U of a module V over a ring R: U is primary $\Rightarrow$ $\mathrm{ann}_R(V/U)$ is primary. Here is an example showing that the converse is not true. Let R be the bivariate polynomial ring $k[X, Y]$ over a field k, and let U and V be the ideals in R generated by (X^2, XY) and (X, Y) respectively. Then clearly

$\text{ann}_R(V/U)$ is the ideal in R generated by X which is prime and hence primary. But $Y \in R \setminus \text{rad}_R(\text{ann}_R(V/U)) = R \setminus (XR)$ and $X \in V \setminus U$ with $YX \in U$. Therefore U is not primary.

§11: PROBLEMS

To initiate a new column, occasionally I shall list Problems (P1), (P2), ... A Problem is an Exercise which has not been completely worked out. By solving one of these, sometimes the student may get a mild satisfaction, sometimes a Ph.D. thesis, and sometimes fame. If there is some imprecision in the statement of a problem, a part of the exercise is to make it precise.

PROBLEM (P1). [**Explicit Equations of Integral Dependence**]. Redo items (J1), (J2) mentioned in (E2) together with the analogous items (C3), (C5) of L1(R7) cited there, by finding explicit formulas. In other words, let

$$F_i(Y) = Y^{n_i} + A_{i1}Y^{n_i-1} + \cdots + A_{in_i}$$

with A_{ij} in a ring R be the equations of integral dependence satisfied by a finite number of elements t_i in an overring S of R for $1 \le i \le m$. Find explicit formulas for $B_1, \ldots, B_{n'}$ and $C_1, \ldots, C_{n''}$ so that $t_1 + \cdots + t_m$ and $t_1 \ldots t_m$ satisfy the equations

$$G(Y) = Y^{n'} + B_1 Y^{n'-1} + \cdots + B_{n'}$$

and

$$H(Y) = Y^{n''} + C_1 Y^{n''-1} + \cdots + C_{n''}$$

of integral dependence over R. Also let $s \in S$ satisfy

$$F^*(Y) = Y^m + t_1 Y^{m-1} + \cdots + t_m$$

and find formulas for $D_1, \ldots, D_{n^*}$ so that s satisfies the equation

$$E(Y) = Y^{n^*} + D_1 Y^{n^*-1} + \cdots + D_{n^*}$$

of integral dependence over R. [SATISFACTION].

HINT. Think of the A_{ij} as indeterminates over $\mathbb{Z}$. Let $G(Y)$, $H(Y)$, and $E(Y)$ be the minimal polynomials of $t_1 + \cdots + t_m$, $t_1 \ldots t_m$, and s over $\mathbb{Q}((A_{ij})_{1 \le i \le m, 1 \le j \le n_i})$ respectively. Then their coefficients B_i, C_i, and D_i belong to $\mathbb{Z}[(A_{ij})_{1 \le i \le m, 1 \le j \le n_i}]$. These should give you universal explicit formulas.

PROBLEM (P2). [**Uniformization**]. Let K/A be a function field, i.e., let K be the quotient field of an affine domain over a domain A. Assume that A is either a field or the ring of integers $\mathbb{Z}$. Show that, given any valuation ring V of K/A, there exists a nonsingular affine model of K/A which is dominated by V. [THESIS].

PROBLEM (P3). [**Resolution**]. Let K/A be as in the above (P2). Show that there exists a nonsingular projective model of K/A. [FAME].

§12: REMARKS

REMARK (R1). [**Laplace Development**]. To describe the Laplace development of a determinant cited in L3§12(E2), which generalizes the development according to a row or column described in L3§12(E1), let $A = (A_{ij})$ be an $n \times n$ matrix over a ring R, say as depicted in L3§12(D1) with $m = n$. Let $r = (r_1, \ldots, r_b)$ be a sequence of positive integers with $r_1 + \cdots + r_b = n$. Recall that S_n is the set of all bijections σ of $\{1, \ldots, n\}$. Let $S_n(r)$ be the set of all $\sigma \in S_n$ such that $\sigma(r_1 + \cdots + r_{v-1} + w) < \sigma(r_1 + \cdots + r_{v-1} + w + 1)$ for $1 \leq v \leq b$ and $1 \leq w < r_v$. For any σ, τ in $S_n(r)$ and $1 \leq v \leq b$ let $A_{\sigma\tau}^{(v)}$ be the square submatrix of A of order r_v located at the $(\sigma(r_1 + \cdots + r_{v-1} + 1), \ldots, \sigma(r_1 + \cdots + r_v); \tau(r_1 + \cdots + r_{v-1} + 1), \ldots, \tau(r_1 + \cdots + r_v))$-th spots; see L3§12(D1). By a straightforward verification we can check that for any σ in $S_n(r)$ we have

$$\det(A) = \sum_{\tau \in S_n(r)} \mathrm{sgn}(\sigma\tau^{-1}) \prod_{1 \leq v \leq b} \det(A_{\sigma\tau}^{(v)})$$

and for any τ in $S_n(r)$ we have

$$\det(A) = \sum_{\sigma \in S_n(r)} \mathrm{sgn}(\sigma\tau^{-1}) \prod_{1 \leq v \leq b} \det(A_{\sigma\tau}^{(v)}).$$

These may respectively be called the Laplace development of $\det(A)$ according to the row partition $(r_1, \ldots, r_b)$ relative to the permutation σ, and the Laplace development of $\det(A)$ according to the column partition $(r_1, \ldots, r_b)$ relative to the permutation (τ, s); when σ or τ is the identity permutation we may drop the reference to it.

REMARK (R2). [**Block Matrices**]. Let $A = (A_{ij})$ be an $m \times n$ matrix over a ring R, say as depicted in L3§12(D1). Let $(q_1, \ldots, q_a)$ and $(r_1, \ldots, r_b)$ be positive integers with $q_1 + \cdots + q_a = m$ and $r_1 + \cdots + r_b = n$. The definition of a matrix over a ring introduced in L3§1 can be generalized in an obvious manner to define a matrix over (having entries in) any set. In particular we get an $a \times b$ matrix $\widehat{A} = (\widehat{A}_{uv})$ where $\widehat{A}_{uv} = ((\widehat{A}_{uv})_{ij})$ is the $q_u \times r_v$ matrix over R given by putting

$$(\widehat{A}_{uv})_{ij} = A_{q_1 + \cdots + q_{u-1} + i, \, r_1 + \cdots + r_{v-1} + j}$$

i.e., the (i, j)-th entry of $\widehat{A}_{uv}$ equals the $(q_1 + \cdots + q_{u-1} + i, r_1 + \cdots + r_{v-1} + j)$-th entry of A. We call $\widehat{A}$ a block matrix over R, or in greater detail a $(q_1, \ldots, q_a) \times (r_1, \ldots, r_b)$

block matrix over R, and in an obvious sense we have the equation

$$A = \widehat{A}$$

which, for making clear that the LHS is an ordinary matrix while the RHS is a block matrix, may be written as

$$A =_{block} \widehat{A}.$$

Note that the "blocks" $\widehat{A}_{uv}$ are submatrices of A.

We define sums and products of block matrices by an obvious generalization of the definitions for the sums and products of matrices over a ring given in L3§1. For instance, if $B =_{block} \widehat{B}$ where $B = (B_{ij})$ is an $m \times n$ matrix over R and $\widehat{B} = (\widehat{B}_{uv})$ is an $(q_1, \ldots, q_a) \times (r_1, \ldots, r_b)$ block matrix over R then $\widehat{A} + \widehat{B} = ((\widehat{A} + \widehat{B})_{uv})$ is the $(q_1, \ldots, q_a) \times (r_1, \ldots, r_b)$ block matrix over R defined by putting $(\widehat{A} + \widehat{B})_{uv} = \widehat{A}_{uv} + \widehat{B}_{uv}$, and we clearly have $A + B =_{block} \widehat{A} + \widehat{B}$. Likewise, if $B =_{block} \widehat{B}$ where $B = (B_{ij})$ is an $n \times o$ matrix over R and $\widehat{B} = (\widehat{B}_{uv})$ is an $(r_1, \ldots, r_b) \times (s_1, \ldots, s_c)$ block matrix over R with $s_1 + \cdots + s_c = o$ then $\widehat{A}\widehat{B} = ((\widehat{A}\widehat{B})_{uv})$ is the $(q_1, \ldots, q_a) \times (s_1, \ldots, s_b)$ block matrix over R defined by putting $(\widehat{A}\widehat{B})_{uv} = \sum_{1 \leq w \leq n} \widehat{A}_{uw} \widehat{B}_{wv}$, and we clearly have $AB =_{block} \widehat{A}\widehat{B}$.

Sometimes we may even use a "quasiblock matrix," i.e., a block matrix like object in which the sizes of the submatrices ($=$ quasiblocks) are not so systematically arranged. We shall try to indicate this by drawing dots or line segments between the submatrices. For instance, in (R6) below, we will have equations of the type

$$A = \left(\begin{array}{c} A' \\ \hline A^* \mid A^{**} \end{array} \right)$$

where $A' = (A'_{ij})$, $A^* = (A^*_{ij})$, $A^{**} = (A^{**}_{ij})$, are $m' \times n$, $m^* \times n^*$, $m^* \times n^{**}$ matrices over R with $m' + m^* = m$, $n^* + n^{**} = n$, and

$$A_{ij} = \begin{cases} A'_{ij} & \text{if } 1 \leq i \leq m' \text{ and } 1 \leq j \leq n \\ A^*_{i-m',j} & \text{if } m' < i \leq m \text{ and } 1 \leq j \leq n^* \\ A^{**}_{i-m',j-n^*} & \text{if } m' < i \leq m \text{ and } n^* < j \leq n \end{cases}$$

or equations of the type

$$A = \left(\begin{array}{c|c} A' & A'' \\ \hline A^* & A^{**} \end{array} \right)$$

where $A' = (A'_{ij})$, $A'' = (A''_{ij})$, $A^* = (A^*_{ij})$, $A^{**} = (A^{**}_{ij})$, are $m' \times n'$, $m' \times n''$, $m^* \times n^*$, $m^* \times n^{**}$ matrices over R with $m' + m^* = m$, $n' + n'' = n$, $n^* + n^{**} = n$,

and

$$A_{ij} = \begin{cases} A'_{ij} & \text{if } 1 \le i \le m' \text{ and } 1 \le j \le n' \\ A''_{i,j-n'} & \text{if } 1 \le i \le m' \text{ and } n' < j \le n \\ A^*_{i-m',j} & \text{if } m' < i \le m \text{ and } 1 \le j \le n^* \\ A^{**}_{i-m',j-n^*} & \text{if } m' < i \le m \text{ and } n^* < j \le n. \end{cases}$$

For the sake of clarity, in cases as above, we may write $A =_{qblock} (\text{---})$ in place of $A = (\text{---})$.

REMARK (R3). [**Products of Determinants**]. Referring to L3§12(E2), here is the proof of the product formula for determinants using the Laplace development. Given $n \times n$ matrices A, B over a ring R, let $C = (C_{ij})$ be the $(2n) \times (2n)$ matrix over R which is given in terms of an $(n,n) \times (n,n)$ block matrix as $C = \begin{pmatrix} A & 0 \\ -1 & B \end{pmatrix}$ where 1 and 0 are the $n \times n$ identity and zero matrices. For $n < j \le 2n$, we add to the j-th column of C the linear combination $\sum_{1 \le l \le n} C_{il} B_{lj}$ of the first n columns of C, we get the $(2n) \times (2n)$ matrix D given in block form as $D = \begin{pmatrix} A & AB \\ -1 & 0 \end{pmatrix}$ and hence by (E8.2) we get $\det(C) = \det(D)$. By applying the Laplace row development according to the row partition (n, n) we get $\det(C) = \det(A)\det(B)$ and $\det(D) = \det(AB)$. Therefore $\det(AB) = \det(A)\det(B)$.

REMARK (R4). [**Solutions of Nonhomogeneous Linear Equations**]. To generalize (E5), let

$$\sum_{1 \le j \le N} a_{ij} \overline{X}_j = b_i \quad \text{for} \quad 1 \le i \le M$$

be a system of nonhomogeneous linear equations where $a = (a_{ij})$ is an $M \times N$ matrix over a field k with M, N in $\mathbb{N}$, and $\overline{a} = (\overline{a}_{ij})$ is the $M \times (N+1)$ matrix over k obtained by augmenting a by the last column $(b_1, \ldots, b_M)$, i.e., for $1 \le i \le M$ we have $\overline{a}_{ij} = a_{ij}$ or b_i according as $1 \le j \le N$ or $j = N + 1$. Let $\overline{S}$ be the solution space of the system, i.e., the set of all $\overline{X} = (\overline{X}_1, \ldots, \overline{X}_N) \in A_k^N$ satisfying the system. Note that if S is the solution space of the corresponding homogeneous system

$$\sum_{1 \le j \le N} a_{ij} X_j = 0 \quad \text{for} \quad 1 \le i \le M$$

and if $\overline{S}$ is nonempty then it is clearly an additive coset of S in $\mathbb{A}_k^N$, i.e., for any $\overline{X}$ in $\overline{S}$ we have $\overline{S} = \{X + \overline{X} : X \in S\}$. Let n and $\overline{n}$ be the ranks of a and $\overline{a}$ respectively. Then clearly $n \le \overline{n}$. In a moment we shall show that $\overline{S}$ is nonempty iff $n = \overline{n}$, and when that is so, we shall modify the formulas of (E5) to get explicit formulas for the members of $\overline{S}$. As in (E5), by relabelling a_{ij} (and the corresponding b_i) we may assume that $\det(A) \ne 0$ where $A = (A_{ij})$ is the $n \times n$ matrix with $A_{ij} = a_{ij}$ for

$1 \leq i \leq n$ and $1 \leq j \leq n$, i.e., A is the left upper corner submatrix of a. Let

$$\overline{Y}_i = b_i - \sum_{n < \lambda \leq N} a_{i\lambda} \overline{X}_\lambda \quad \text{for} \quad 1 \leq i \leq M.$$

For $1 \leq j \leq n$ let $\overline{B}^{(j)}$ be the $n \times n$ matrix obtained by replacing the j-th column of A by the column vector $(\overline{Y}_1, \ldots, \overline{Y}_n)$. Now for every $Z = (Z_1, \ldots, Z_{N-n}) \in \mathbb{A}_k^{N-n}$ we put $\overline{\phi}(Z) = \overline{x} = (\overline{x}_1 \ldots, \overline{x}_N) \in \mathbb{A}_k^N$ where

$$\overline{x}_j = \begin{cases} \det(\overline{B}^{(j)})/\det(A) & \text{if } 1 \leq j \leq n \\ Z_{j-n} & \text{if } n < j \leq N. \end{cases}$$

We CLAIM that $\overline{S}$ is nonempty iff $n = \overline{n}$, and when that is so, $\overline{\phi}$ is an injective map whose image is $\overline{S}$.

PROOF OF THE CLAIM. By the above displayed definition of $\overline{x}_j$, and by looking at the expansion of $\det(\overline{B}^{(j)})$ by its j-th column, we see that $\overline{\phi}$ is an injective map. So it only remains to show that $\overline{S}$ is nonempty iff $n = \overline{n}$, and when that is so, $\overline{S}$ is the image of $\overline{\phi}$. Now the given system is clearly equivalent to the abbreviated system

$$\sum_{1 \leq j \leq n} A_{ij} \overline{X}_j = \overline{Y}_i \quad \text{for} \quad 1 \leq i \leq n$$

together with the extra equations $\sum_{1 \leq j \leq N} a_{ij} \overline{X}_j = b_i$ for $n < i \leq M$. Therefore, by solving the abbreviated system by Cramer's Rule (E4) and then feeding the solution (E4.2), with bars on X and B, in the extra equations, we are reduced to showing that

$$n = \overline{n} \Leftrightarrow \overline{Y}_i - \sum_{1 \leq j \leq n} a_{ij} \det(\overline{B}^{(j)})/\det(A) = 0 \text{ for } n < i \leq M.$$

So given any such i, let $\overline{C}$ be the $(n+1) \times (n+1)$ matrix obtained by bordering A by the last column $(\overline{Y}_1, \ldots, \overline{Y}_n, \overline{Y}_i)$ and the last row $(a_{i1}, \ldots, a_{in}, \overline{Y}_i)$, and for $n < l \leq N+1$ let $\overline{C}^{(l)}$ be the $(n+1) \times (n+1)$ matrix obtained by bordering A by the last column $(a_{1l}, \ldots, a_{nl}, \overline{a}_{il})$ and the last row $(a_{i1}, \ldots, a_{in}, \overline{a}_{il})$. Expanding by the last column we see that $\det(\overline{C}) = \det(\overline{C}^{(N+1)}) - \sum_{n < l \leq N} \det(\overline{C}^{(l)}) X_j$; but $\det(\overline{C}^{(l)}) = 0$ for $n < l \leq N+1$ because $\overline{C}^{(l)}$ is an $(n+1) \times (n+1)$ submatrix of a; therefore $\det(\overline{C}) = 0$. Expanding by the last row we also see that

$$\det(\overline{C}) = \overline{Y}_i \det(A) - \sum_{1 \leq j \leq n} a_{ij} \det(\overline{B}^{(j)})$$

and this completes the proof.

REMARK (R5). [**The How and Why of Cramer and Sylvester**]. To solve
the simultaneous linear equations

$$\sum_{1\le j\le n} A_{ij}X_j = Y_i \quad \text{for} \quad 1\le i\le n$$

we subtract A_{i1} times the first equation from A_{11} times the i-th equation for $2\le$
$i\le n$. This "eliminates" X_1 and produces $n-1$ equations in the $n-1$ variables
$X_2,\ldots,X_n$. Iterating this procedure n times, the last step produces a value of X_n.
Substituting this in the first $n-1$ equations gives us $n-1$ equations in the $n-1$
variables $X_1,\ldots,X_{n-1}$. Iterating this entire procedure produces a value of X_{n-1},
and so on down the line until we get values of all the variables $X_1,\ldots,X_n$. By
condensing the whole hullabaloo in one giant step we end up with Cramer's Rule.

This "explains" (E4). Similar "explanations" apply to the systems of linear
equations dealt with in (E5) and (R4).

Turning to the higher degree simultaneous equations

$$f(Y) = a_0Y^n + a_1Y^{n-1} + \cdots + a_n$$
$$g(Y) = b_0Y^m + b_1Y^{m-1} + \cdots + b_m$$

dealt with in (E7), since there is only one variable Y, how shall we eliminate it
by itself? Well, we do it by letting the degree play the role of a variable. To wit,
assuming $n\ge m$, we multiply the first equation by b_0 and the second by a_0Y^{n-m},
and then subtract the second from the first. This replaces the equations of degrees
n,m by equations of degrees $n-1,m$ thereby decreasing the sum $n+m$. Iterating
this procedure several times we make the sum of the degrees zero, i.e., we eliminate
Y and get hold of an expression involving only the coefficients $a_0,\ldots,a_n,b_0,\ldots,b_m$,
which, if everything is nice and dandy, ought to produce something resembling the
Sylvester Resultant.

In (E7) we proved the Basic Fact (T1) about the Sylvester Resultant. Here
is an easier proof of that part of (T1) which says that if n,m are positive integers and f,g have a common root in some overfield of R then $\text{Res}_Y(f,g) = 0$.
Namely, call that root Y. Multiplying f by $Y^{m-1},Y^{m-2},\ldots,Y,1$ and g by
$Y^{n-1},Y^{n-2},\ldots,Y,1$ we get $m+n$ homogeneous linear equations in the $m+n$
"variables" $Y^{m+n-1},Y^{m+n-2},\ldots,Y,1$. This "solution vector" is a nonzero vector
because it ends with 1. Therefore by (E5) the determinant of these equations, which
is nothing but $\text{Res}_Y(f,g)$, is zero. Q.E.D.

In greater detail, the matrix $T = (T_{ij})_{1\le j\le m+n}^{1\le i\le m+n}$ of the said equations is nothing
but $\text{Resmat}_Y(f,g)$, and the equations are

$$\sum_{1\le j\le m+n} T_{ij}Y^{m+n-j} = \begin{cases} Y^{m-i}f(Y) = 0 & \text{if } 1\le i\le m \\ Y^{n-i}g(Y) = 0 & \text{if } 1\le i-m\le n. \end{cases}$$

Note that T is the transpose of the matrix A of the HINT to (E7). Letting Y revert
to its role of an indeterminate, and without assuming f,g to have a common root,

but continuing to assume n, m to be positive integers and disregarding the $= 0$ (twice) of the above display, the said display says that by adding to the last column of T certain $R[Y]$-linear combinations of the remaining columns we get the column vector $(Y^{m-1}f(Y), \ldots, Yf(Y), f(Y), Y^{n-1}g(Y), \ldots, Yg(Y), g(Y))$. Expanding the determinant of this new matrix by its last column we get an identity of the form

(5.1) $\quad \begin{cases} \operatorname{Res}_Y(f, g) = u(Y)f(Y) + v(Y)g(Y) \text{ with } u(Y), v(Y) \text{ in } R[Y] \\ \text{of } Y\text{-degrees} \le m - 1 \text{ and } \le n - 1 \text{ respectively.} \end{cases}$

By letting Y be a common root of f, g, this gives yet another proof of the above mentioned part of (T1).

Having proved the Basic Fact (T1), let us briefly comment on the remaining items of §1. Corollary (T2) follows by noting that a nonconstant univariate polynomial over a field has a multiple root iff the polynomial and its derivative have a common root. Observations (O1) and (O2) follow by noting that the operations of forming the resultant or the discriminant commute with the operation of giving values to the auxiliary variables $X_1, \ldots, X_N$. The Isobaric Property (O3) and the Product Formula (X1) will be proved in the next Remark (R6), where we shall also prove a product formula for the resultant matrix. Examples (X2) to (X6) are straightforward computations. Bézout Observations (O4) and (O5) will be dealt with later on. This takes care of the Future Plan Observation (O6).

REMARK (R6). [**Product Formula for the Resultant Matrix**]. To formulate the last identity (5.1) of the above Remark (R5) more precisely, and to prove the Product Formula (X1) of §1 together with its generalizations, for nonnegative integers n, m, in addition to the polynomials

$$f(Y) = a_0 Y^n + a_1 Y^{n-1} + \cdots + a_n$$
$$g(Y) = b_0 Y^m + b_1 Y^{m-1} + \cdots + b_m$$

with coefficients $a_0, \ldots, a_n, b_0, \ldots, b_m$ in a ring R, which is assumed to be nonnull, let us consider polynomials

$$F(Y) = A_0 Y^n + A_1 Y^{n-1} + \cdots + A_n$$
$$G(Y) = B_0 Y^m + B_1 Y^{m-1} + \cdots + B_m$$

whose coefficients $A_0, \ldots, A_n, B_0, \ldots, B_m$ are indeterminates over R. Let

$$\overline{R} = R[A_0, \ldots, A_n, B_0, \ldots, B_m].$$

Since $\operatorname{Resmat}_Y(f, g)$ and $\operatorname{Res}_Y(f, g)$ depend not only on f, g but also on n, m, strictly speaking we should write $\operatorname{Resmat}_Y^{(n,m)}(f, g)$ and $\operatorname{Res}_Y^{(n,m)}(f, g)$ for them and call them the Y-Resultant Matrix and the Y-Resultant of f, g relative to n, m respectively, with the understanding that the reference to n, m may be dropped when it is clear from the context. As a more precise version of (5.1) we have the identity:

$$(6.1) \quad \begin{cases} \mathrm{Res}_Y^{(n,m)}(F,G) = U(Y)F(Y) + V(Y)G(Y) \text{ with} \\ U(Y) = \sum_{1 \le j \le m} Y^{m-j}C_{ij} \text{ and } V(Y) = \sum_{m+1 \le j \le m+n} Y^{m+n-j}C_{ij} \\ \text{in } \overline{R}[Y] \text{ of } Y\text{-degrees} \le m-1 \text{ and } \le n-1 \text{ respectively,} \\ \text{where } C_{ij} \in \overline{R} \text{ is the } (i,j)\text{-th cofactor of } \mathrm{Resmat}_Y^{(n,m)}(F,G) \end{cases}$$

In the rest of this Remark, as an abbreviation, we denote $\mathrm{Resmat}_Y^{(n,m)}(f,g)$ by the $(m+n) \times (m+n)$ matrix

$$T^{(n,m)}(f,g) = (T^{(n,m)}(f,g)_{ij}).$$

and we note that then

$$T^{(n,m)}(f,g)_{ij} \in R \quad \text{and} \quad T^{(n,m)}(F,G)_{ij} \in \overline{R}.$$

We also put

$$P^{(n,m)}(f,g) = \mathrm{Res}_Y^{(n,m)}(f,g)$$

and

$$Q(A_0, \ldots, A_n, B_0, \ldots, B_m) = P^{(n,m)}(F,G)$$

and we note that then

$$P^{(n,m)}(f,g) \in R \quad \text{and} \quad Q(A_0, \ldots, A_n, B_0, \ldots, B_m) \in \overline{R}.$$

Finally, we let

$$\tilde{a}_i = \begin{cases} a_i & \text{if } i \in \{0, \ldots, n\} \\ 0 & \text{if } i \in \mathbb{Z} \setminus \{0, \ldots, n\} \end{cases}$$

and, for any nonnegative integers u, v, we let $S^{(n,u,v)}(f) = (S^{(n,u,v)}(f)_{ij})$ be the $u \times v$ matrix obtained by putting

$$S^{(n,u,v)}(f)_{ij} = \tilde{a}_{j-i}$$

and we call $S^{(n,u,v)}(f)$ the (n, u, v)-spread of f. The resultant matrix can be expressed in terms of the spreads by the block matrix identity:

$$(0) \quad T^{(n,m)}(f,g) = \begin{pmatrix} S^{(n,m,m+n)}(f) \\ S^{(m,n,m+n)}(g) \end{pmatrix}.$$

As direct consequences of the spread notation we have the following two relations.

Relation between Polynomial Addition and Matrix Addition:

$$(1) \qquad \text{if } n = m \text{ then} \quad S^{(n,u,v)}(f+g) = S^{(n,u,v)}(f) + S^{(m,u,v)}(g).$$

Relation between Polynomial Multiplication and Matrix Multiplication:

(2)   if integer $q \geq n + u$ then $S^{(m+n,u,v)}(fg) = S^{(n,u,q)}(f)S^{(m,q,v)}(g)$.

Before using (0) to (2) for obtaining a product rule for T, let us observe some easy properties of P. Since a 0×0 determinant is 1, we see that:

(3)   if $n = 0 = m$ then $P^{(n,m)}(f,g) = 1$.

The determinant of a triangular matrix, i.e., a matrix having zeroes above or under the principal diagonal, equals the product of the diagonal entries. Also the zeroth power of anything (including zero) equals 1. Hence:

(4)   if $n = 0$ then $P^{(n,m)}(f,g) = a_0^m$

whereas

(5)   if $m = 0$ then $P^{(n,m)}(f,g) = b_0^n$.

By respectively expanding according to the first or the last row and then calculating with triangular determinants, i.e., determinants of triangular matrices, we get:

(6) $P^{(n,m)}(f,g) = \begin{cases} \sum_{0 \leq i \leq n}(-1)^i a_i b_0^i b_1^{n-i} & \text{if } m = 1 \\ \sum_{0 \leq i \leq m}(-1)^{m+i} b_i a_0^i a_1^{m-i} & \text{if } n = 1. \end{cases}$

By (E8.1) we see that:

(7)   for all c, d in R we have $P^{(n,m)}(cf, dg) = c^m d^n P^{(n,m)}(f,g)$.

Expanding by a row and invoking (E8.2) we see that the determinant of an $N \times N$ matrix (with positive integer N) is zero if either a row consists only of zeroes or if two rows are identical, and hence:

(8)   if either $mn \neq 0 = f - g$ or $m \neq 0 = f$ or $n \neq 0 = g$ then $P^{(n,m)}(f,g) = 0$.

If $mn = 0$ then by (3) to (5), and if $mn \neq 0$ then by (E8.3), we get the

Permuting Rule for Resultant:

(9) $P^{(m,n)}(g,f) = (-1)^{mn} P^{(n,m)}(f,g)$.

Also we clearly have the

Universality Property of Resultant:

(10) $P^{(n,m)}(f,g) = Q(a_0, \ldots, a_n, b_0, \ldots, b_m)$.

Next we record the following five properties (11) to (15) of the polynomial Q. As notation, $\text{coeff}(U, V)$ denotes the coefficient of a monomial U in a polynomial $V = V(X_1, \ldots, X_r, Y_1, \ldots, Y_s)$ in indeterminates $X_1, \ldots, X_r, Y_1, \ldots, Y_s$ over the ring R. Recall that V is homogeneous of degree t in $(X_1, \ldots, X_r)$ means that for every monomial $U = X_1^{i_1} \ldots X_r^{i_r} Y_1^{j_1} \ldots Y_s^{j_s}$ with $\text{coeff}(U, V) \neq 0$ we have $i_1 + \cdots + i_r = t$. More generally, V is isobaric of weight τ in $(X_1, \ldots, X_r)$ when we give weight u_i to X_i for $1 \leq i \leq r$ means that for every monomial $U = X_1^{i_1} \ldots X_r^{i_r} Y_1^{j_1} \ldots Y_s^{j_s}$ with $\text{coeff}(U, V) \neq 0$ we have $u_1 i_1 + \cdots + u_r i_r = \tau$; clearly this is so iff as a polynomial in an extra indeterminate Z we have $V(Z^{u_1} X_1, \ldots, Z^{u_r} X_r, Y_1, \ldots, Y_s) = Z^\tau V(X_1, \ldots, X_r, Y_1, \ldots, Y_s)$.

Diagonal Property of Resultant:

(11) $\text{coeff}(A_0^m B_m^n, Q) = 1$ and $\text{coeff}(A_n^m B_0^n, Q) = (-1)^{mn}$.

First Homogeneity Property of Resultant:

(12) Q is homogeneous of degree m in $(A_0, \ldots, A_n)$.

Second Homogeneity Property of Resultant:

(13) Q is homogeneous of degree n in $(B_0, \ldots, B_m)$.

Weight Property of Resultant:

(14) $\begin{cases} Q \text{ is isobaric of weight } mn \text{ in } (A_0, \ldots, A_n, B_0, \ldots, B_m) \\ \text{when } A_i \text{ and } B_j \text{ are given weights } i \text{ and } j \\ \text{for } 0 \leq i \leq n \text{ and } 0 \leq j \leq m \text{ respectively.} \end{cases}$

Special Property of Resultant:

(15) $\begin{cases} \text{if } \text{coeff}(U, Q) \neq 0 \text{ for a monomial} \\ U = A_0^{i_0} \ldots A_n^{i_n} B_0^{j_0} \ldots B_m^{j_m} \notin \{A_0^m B_m^n, A_n^m B_0^n\} \\ \text{then } i_n < m \text{ and } j_m < n \text{ and } i_n + j_m \leq \max(m, n) \\ \text{and if also } m \neq n \text{ then } i_n + j_m < \max(m, n). \end{cases}$

To prove these, by looking at the resultant matrix depicted at the beginning of §1 we easily get (11) to (13), whereas (15) follows from (12) to (14). So it only remains to establish (14). To do this, first we note that

$$\begin{cases} Q(Z^0 A_0, \ldots, Z^n A_n, Z^0 B_0, \ldots, Z^m B_m) \\ = \det \left(T^{n,m} (Z^0 A_0 Y^n + \cdots + Z^n A_n, Z^0 B_0 Y^m + \cdots + Z^m B_m) \right) \end{cases}$$

and now for $1 \le i \le m$ we multiply the i-th row of this matrix by Z^i and for $1 \le j \le n$ we multiply its $(m+j)$-th row by Z^j and then upon denoting the resulting matrix by $\widetilde{T}$ we get

$$\det(\widetilde{T}) = Z^\delta Q(Z^0 A_0, \ldots, Z^n A_n, Z^0 B_0, \ldots, Z^m B_m)$$

where $\delta = \sum_{1 \le i \le m} i + \sum_{1 \le j \le n} j$; however, $\widetilde{T}$ is the same as the matrix we obtain when for $1 \le l \le m+n$ we multiply the l-th column of $T^{n,m}(F, G)$ by Z^l and hence

$$\det(\widetilde{T}) = Z^\epsilon Q(A_0, \ldots, A_n, B_0, \ldots, B_m)$$

where $\epsilon = \sum_{1 \le l \le m+n} l$; by summing the three arithmetic series we conclude that $\epsilon - \delta = \frac{(m+n)(m+n+1)}{2} - [\frac{m(m+1)}{2} + \frac{n(n+1)}{2}] = mn$, and hence by the above two displayed equations for $\det(\widetilde{T})$ we get (14).

To start towards the product rule, for a polynomial

$$h(Y) = c_0 Y^{m+n} + c_1 Y^{m+n-1} + \cdots + c_{m+n}$$

with $c_0, c_1, \ldots, c_{m+n}$ in R, we record the following two sum rules.

First Sum Rule for Resultant:

(16) $$P^{(n,m+n)}(f, h + fg) = P^{(n,m+n)}(f, h).$$

Second Sum Rule for Resultant:

(17) $$P^{(m+n,m)}(h + fg, g) = P^{(m+n,m)}(h, g).$$

To prove these, by (0) we have

$$T^{(m+n,m)}(h, g) = \begin{pmatrix} S^{(m+n,m,2m+n)}(h) \\ S^{(m,m+n,2m+n)}(g) \end{pmatrix}$$

and by (1) and (2) we have

$$T^{(m+n,m)}(h + fg, g) = \begin{pmatrix} S^{(m+n,m,2m+n)}(h) + S^{(n,m,m+n)}(f) S^{(m,m+n,2m+n)}(g) \\ S^{(m+n,m+n,2m+n)}(g) \end{pmatrix}$$

and hence by adding certain linear combinations of the last $m+n$ rows of the matrix $T^{(m+n,m)}(h, g)$ to its first m rows we obtain the matrix $T^{(m+n,m)}(h + fg, g)$ and from this we conclude that these two matrices have the same determinant which establishes (17); now by (9) and (17) we also get (16).

Recalling that u, v are any nonnegative integers, let also w be any nonnegative integer, and consider the following two shift rules.

First Shift Rule for Resultant:

$$(18) \qquad P^{(n+w,m)}(f, g) = (-1)^{wm} b_0^w P^{(n,m)}(f, g).$$

Second Shift Rule for Resultant:

$$(19) \qquad P^{(n,m+w)}(f, g) = a_0^w P^{(n,m)}(f, g).$$

The proof of (19) follows by noting that $T^{(n,m+w)}(f, g) = \left(\frac{S^{(n,w,m+n+w)}(f)}{0_{m+n,w} \mid T^{(n,m)}(f,g)} \right)$ where $0_{u,v}$ denotes the $u \times v$ matrix with all entries zero; moreover, by (9) and (19) we get (18). Recall that 1_u is the $u \times u$ identity matrix, and note that for $1 \in R[Y]$ we have $\deg_Y 1 = 0 \le w$. About these special matrices we have the obvious

Relations:

$$(2_0) \qquad \begin{cases} 0_{u,v} + J = J + 0_{u,v} \text{ for every } u \times v \text{ matrix } J, \\ 0_{u,v}J = 0_{u,w} \text{ for every } v \times w \text{ matrix } J, \\ J0_{u,v} = 0_{w,v} \text{ for every } w \times u \text{ matrix } J, \\ 1_u J = J \text{ for every } u \times v \text{ matrix } J, \\ J1_u = J \text{ for every } v \times u \text{ matrix } J, \\ S^{(w,u,u+v+w)}(1) = (0_{u,w}, 1_u, 0_{u,v}) \text{ as block matrix.} \end{cases}$$

The ENLARGED RESULTANT MATRIX is the $(u + m + n + v) \times (u + m + n + v)$ matrix defined as a block matrix by putting

$$(0_1) \qquad T^{(u,n,m,v)}(f, g) = \begin{pmatrix} 1_u & 0_{u,m+n} & 0_{u,v} \\ 0_{m+n,u} & T^{(n,m)}(f, g) & 0_{m+n,v} \\ 0_{v,u} & 0_{v,m+n} & 1_v \end{pmatrix}$$

and we note that clearly (say by Laplace development)

$$(20) \qquad \det\left(T^{(u,n,m,v)}(f, g) \right) = P^{(n,m)}(f, g)$$

and

$$(21) \qquad \det\left(T^{(u,n,m,v)}(f, 1) \right) = a_0^m.$$

By analogy with the equation

$$T^{(n,m+w)}(f,g) = \begin{pmatrix} S^{(n,m+w,m+n+w)}(f) \\ S^{(m+w,n,m+n+w)}(g) \end{pmatrix}$$

we introduce the SKEWSHIFTED RESULTANT MATRIX by putting

(0') $$T^{[n,w,m]}(f,g) = \begin{pmatrix} S^{(n,m,m+n+w)}(f) \\ S^{(m,n+w,m+n+w)}(g) \end{pmatrix}$$

and we note that these two $(m+n+w) \times (m+n+w)$ matrices are clearly related to the $(m+n) \times (m+n)$ matrix $T^{(n,m)}(f,g)$ by the formulas

(22) $$T^{(n,m+w)}(f,g) = \begin{pmatrix} S^{(n,w,m+n+w)}(f) \\ \hline 0_{m+n,w} \mid T^{(n,m)}(f,g) \end{pmatrix}$$

and

(22') $$T^{[n,w,m]}(f,g) = \begin{pmatrix} T^{(n,m)}(f,g) \mid 0_{m+n,w} \\ \hline 0_{w,n} \mid S^{(m,w,m+w)}(g) \end{pmatrix}.$$

As an immediate consequence of (22') we have the

Skewshift Rule for Resultant:

(19') $$\det\left(T^{[n,w,m]}(f,g)\right) = b_m^w P^{(n,m)}(f,g).$$

Finally we introduce the ENLARGED SKEWSHIFTED RESULTANT MATRIX which is the $(u+m+n+w+v) \times (u+m+n+w+v)$ matrix defined as a block matrix by putting

(0'₁) $$T^{[u,n,w,m,v]}(f,g) = \begin{pmatrix} 1_u & 0_{u,m+n+w} & 0_{u,v} \\ 0_{m+n+w,u} & T^{[n,w,m]}(f,g) & 0_{m+n+w,v} \\ 0_{v,u} & 0_{v,m+n+w} & 1_v \end{pmatrix}$$

and we note that by (19') we have

(20') $$\det\left(T^{[u,n,w,m,v]}(f,g)\right) = b_m^w P^{(n,m)}(f,g)$$

and

(21') $$\det\left(T^{[u,0,n,m,v]}(1,g)\right) = b_m^n.$$

Now upon letting n', n^*, m', m^* be any nonnegative integers such that

$$n = n' + n^* \quad \text{and} \quad m = m' + m^*$$

and upon considering any polynomials

$$f'(Y) = a'_0 Y^{n'} + a'_1 Y^{n'-1} + \cdots + a'_{n'}$$
$$g'(Y) = b'_0 Y^{m'} + b'_1 Y^{m'-1} + \cdots + b'_{m'}$$

and

$$f^*(Y) = a^*_0 Y^{n^*} + a^*_1 Y^{n^*-1} + \cdots + a^*_{n^*}$$
$$g^*(Y) = b^*_0 Y^{m^*} + b^*_1 Y^{m^*-1} + \cdots + b^*_{m^*}$$

with coefficients $a'_0, \ldots, a'_{n'}, b'_0, \ldots, b'_{m'}, a^*_0, \ldots, a^*_{n^*}, b^*_0, \ldots, b^*_{m^*}$ in R such that

$$f = f'f^* \quad \text{and} \quad g = g'g^*$$

we can state the following two product rules.

Product Rule for Resultant matrix:

(23) $\qquad T^{(m,n^*,n',0)}(f^*,1)T^{(n,m)}(f,g) = T^{(0,n',m,n^*)}(f',g)T^{(n^*,m+n')}(f^*,g).$

Skew Product Rule for Resultant matrix:

(23') $\qquad T^{[0,0,m',m^*,n]}(1,g^*)T^{(n,m)}(f,g) = T^{(m^*,n,m',0)}(f,g')T^{[n,m',m^*]}(f,g^*).$

To prove (23), we have the following equations where the first and the last equations are deduced by items (0) and (0_1) whereas the middle two equations are deduced by block multiplication followed by items (2) and (2_0).

$$\text{LHS of (23)} = \begin{pmatrix} 1_m & 0_{m,n} \\ 0_{n',m} & S^{(n^*,n',n)}(f^*) \\ 0_{n^*,m} & S^{(n',n^*,n)}(1) \end{pmatrix} \begin{pmatrix} S^{(n,m,m+n)}(f) \\ S^{(m,n,m+n)}(g) \end{pmatrix}$$

$$= \begin{pmatrix} S^{(n,m,m+n)}(f) \\ S^{(m+n^*,n',m+n)}(f^*g) \\ S^{(m+n',n^*,m+n)}(g) \end{pmatrix}$$

$$= \begin{pmatrix} S^{(n',m,m+n')}(f') & 0_{m,n^*} \\ S^{(m,n',m+n')}(g) & 0_{n',n^*} \\ 0_{n^*,m+n'} & 1_{n^*} \end{pmatrix} \begin{pmatrix} S^{(n^*,m+n',m+n)}(f^*) \\ S^{(m+n',n^*,m+n)}(g) \end{pmatrix}$$

$$= \text{RHS of (23)}.$$

To prove (23'), we have the following equations where the first and the last equations are deduced by items (0') and $(0'_1)$ whereas the middle two equations are deduced by block multiplication followed by items (2) and (2_0).

$$\text{LHS of } (23') = \begin{pmatrix} S^{(0,m^*,m)}(1) & 0_{m^*,n} \\ S^{(m^*,m',m)}(g^*) & 0_{m',n} \\ 0_{n,m} & 1_n \end{pmatrix} \begin{pmatrix} S^{(n,m,m+n)}(f) \\ S^{(m,n,m+n)}(g) \end{pmatrix}$$

$$= \begin{pmatrix} S^{(n,m^*,m+n)}(f) \\ S^{(m^*+n,m',m+n)}(fg^*) \\ S^{(m,n,m+n)}(g) \end{pmatrix}$$

$$= \begin{pmatrix} 1_{m^*} & 0_{m^*,m'+n} \\ 0_{m',m^*} & S^{(n,m',m'+n)}(f) \\ 0_{n,m^*} & S^{(m',n,m'+n)}(g') \end{pmatrix} \begin{pmatrix} S^{(n,m^*,m+n)}(f) \\ S^{(m^*,m'+n,m+n)}(g^*) \end{pmatrix}$$

$$= \text{RHS of } (23').$$

Taking determinants of all the matrices in (23), by (19), (20) and (21) we get

(24) $$a_0^{*n'} P^{(n,m)}(f,g) = a_0^{*n'} P^{(n',m)}(f',g) P^{(n^*,m)}(f^*,g)$$

and taking determinants of all the matrices in (23'), by (19'), (20') and (21') we also get

(24') $$b_{m^*}^{*m'} P^{(n,m)}(f,g) = b_{m^*}^{*m'} P^{(n,m')}(f,g') P^{(n,m^*)}(f,g^*).$$

By the "principal of universal identities" let us show that the factors $a_0^{*n'}$ and $b_{m^*}^{*m'}$ can be cancelled out from the above two equations. So consider the polynomials

$$F'(Y) = A_0' Y^{n'} + A_1' Y^{n'-1} + \cdots + A_{n'}'$$
$$G'(Y) = B_0' Y^{m'} + B_1' Y^{m'-1} + \cdots + B_{m'}'$$

and

$$F^*(Y) = A_0^* Y^{n^*} + A_1^* Y^{n^*-1} + \cdots + A_{n^*}^*$$
$$G^*(Y) = B_0^* Y^{m^*} + B_1^* Y^{m^*-1} + \cdots + B_{m^*}^*$$

whose coefficients $A' = (A_0', \ldots, A_{n'}')$, $B' = (B_0', \ldots, B_{m'}')$, $A^* = (A_0^*, \ldots, A_{n^*}^*)$ and $B^* = (B_0^*, \ldots, B_{m^*}^*)$ are indeterminates over R which are assumed to be "independent" of the previous indeterminates $A = (A_0, \ldots, A_n)$ and $B = (B_0, \ldots, B_m)$. Then

$$F'(Y)F^*(Y) = \sum_{0 \le i \le n} \widehat{F}_i(A', A^*, B) Y^{n-i}$$
$$G'(Y)G^*(Y) = \sum_{0 \le j \le m} \widehat{G}_j(A, B', B^*) Y^{m-j}$$

where $\widehat{F}_i$ and $\widehat{G}_j$ are polynomials over R in the exhibited indeterminates. Let

$$\overline{Q}(A', A^*, B) = P^{(n',m)}(F', G)P^{(n^*,m)}(F^*, G)$$
$$\overline{Q}'(A, B', B^*) = P^{(n,m')}(F, G')P^{(n,m^*)}(F, G^*)$$

and

$$\widehat{Q}(A', A^*, B) = Q(\widehat{F}_0(A', A^*, B), \ldots, \widehat{F}_n(A', A^*, B), B_0, \ldots, B_m)$$
$$\widehat{Q}'(A, B', B^*) = Q(A_0, \ldots, A_n, \widehat{G}_0(A, B', B^*), \ldots, \widehat{G}_m(A, B', B^*))$$

where $\overline{Q}, \overline{Q}', \widehat{Q}, \widehat{Q}'$ are polynomials over R in the exhibited indeterminates. Now clearly

$$\widehat{Q}(A', A^*, B) = P^{(n,m)}(F'F^*, G)$$
$$\widehat{Q}'(A, B', B^*) = P^{(n,m)}(F, G'G^*)$$

and $A_0^{*n'}$ and $B_m^{*m'}$ are nonzerodivisors in $R[A', A^*, B]$ and $R[A, B', B^*]$ respectively, and hence by (24) and (24′) we get

$$(25) \qquad \qquad \widehat{Q}(A', A^*, B) = \overline{Q}(A', A^*, B)$$

and

$$(25') \qquad \qquad \widehat{Q}'(A, B', B^*) = \overline{Q}'(A, B', B^*).$$

Upon letting $a = (a_0, \ldots, a_n)$, $b = (b_0, \ldots, b_m)$, $a' = (a'_0, \ldots, a'_{n'})$, $b' = (b'_0, \ldots, b'_{m'})$, $a^* = (a_0^*, \ldots, a_{n^*}^*)$, $b^* = (b_0^*, \ldots, b_{m^*}^*)$, we clearly have

$$P^{(n,m)}(f, g) = \widehat{Q}(a', a^*, b) = \widehat{Q}'(a, b', b^*)$$
$$P^{(n',m)}(f', g)P^{(n^*,m)}(f^*, g) = \overline{Q}(a', a^*, b)$$
$$P^{(n,m')}(f, g')P^{(n,m^*)}(f, g^*) = \overline{Q}'(a, b', b^*)$$

and therefore upon substituting small letters for the corresponding big indeterminates in the polynomial identities (25) and (25′) we get

$$(26) \qquad \qquad P^{(n,m)}(f, g) = P^{(n',m)}(f', g)P^{(n^*,m)}(f^*, g)$$

and

$$(26') \qquad \qquad P^{(n,m)}(f, g) = P^{(n,m')}(f, g')P^{(n,m^*)}(f, g^*).$$

By repeated applications of (26) and (26′) we get the

Product Rule for Resultant:

$$(27) \quad \begin{cases} \text{if } u, v, n(i), m(j) \text{ are in } \mathbb{N} \\ \text{with } n = \sum_{1 \leq i \leq u} n(i) \text{ and } m = \sum_{1 \leq j \leq u} m(j) \\ \text{and } f^{(i)}(Y), g^{(j)}(Y) \text{ are in } R[Y] \\ \text{with } f = \prod_{1 \leq i \leq u} f^{(i)} \text{ and } g = \prod_{1 \leq j \leq v} g^{(j)} \\ \text{then } P^{(n,m)}(f, g) = \prod_{1 \leq i \leq u} \prod_{1 \leq j \leq v} P^{(n(i), m(j))}(f^{(i)}, g^{(j)}). \end{cases}$$

By (3), (4), (5), (6) and (27) we get the following three special product rules.

First Special Product Rule for Resultant:

$$(28) \quad \begin{cases} \text{if } f(Y) = a_0 \prod_{1 \leq i \leq n} (Y - \alpha_i) \text{ with } \alpha_i \text{ in } R \\ \text{then } P^{(n,m)}(f, g) = a_0^m \prod_{1 \leq i \leq n} g(\alpha_i). \end{cases}$$

Second Special Product Rule for Resultant:

$$(29) \quad \begin{cases} \text{if } g(Y) = b_0 \prod_{1 \leq j \leq m} (Y - \beta_j) \text{ with } \beta_j \text{ in } R \\ \text{then } P^{(n,m)}(f, g) = (-1)^{mn} b_0^n \prod_{1 \leq j \leq m} f(\beta_j). \end{cases}$$

Third Special Product Rule for Resultant:

$$(30) \quad \begin{cases} \text{if } f(Y) = a_0 \prod_{1 \leq i \leq n} (Y - \alpha_j) \text{ and } g(Y) = b_0 \prod_{1 \leq j \leq m} (Y - \beta_j) \\ \text{with } \alpha_i \text{ and } \beta_j \text{ in } R \\ \text{then } P^{(n,m)}(f, g) = a_0^m b_0^n \prod_{1 \leq i \leq n} \prod_{1 \leq j \leq m} (\alpha_i - \beta_j). \end{cases}$$

By (8) and (27) we get the

Tiny Resultant theorem:

$$(31) \quad \begin{cases} \text{if } q, n', m' \text{ are in } \mathbb{N} \text{ with } q > 0 \text{ and } n = q + n' \text{ and } m = q + m' \\ \text{and } f = hf' \text{ and } g = hg' \text{ where } h, f', g' \text{ are in } R[Y] \\ \text{with } \deg_Y h \leq q \text{ and } \deg_Y f' \leq n' \text{ and } \deg_Y g' \leq m' \\ \text{then } P^{(n,m)}(f, g) = 0. \end{cases}$$

By (8) and (31) we get the

Small Resultant theorem:

$$(32) \quad \begin{cases} \text{if } R \text{ is a domain with } m + n \neq 0 \text{ and } f \text{ and } g \text{ have a} \\ \text{nonconstant (i.e., of positive } Y\text{-degree) common factor in } R[Y] \\ \text{then } P^{(n,m)}(f, g) = 0. \end{cases}$$

By (32) we get the

Little Resultant theorem:

(33)
$$\begin{cases} \text{if } R \text{ is a domain with } m+n \neq 0 \text{ and } f(\gamma) = 0 = g(\gamma) \\ \text{for some } \gamma \text{ in an overfield of } R \\ \text{then } P^{(n,m)}(f,g) = 0. \end{cases}$$

In (E7) of §10 we proved the Basic Fact (T1) stated in §1. Now, in view of (3) to (5), by (28) or (29) we get another proof of (T1) which we restate as (35). The version (34) is clearly equivalent to (35).

First Version of Resultant theorem:

(34)
$$\begin{cases} \text{if } R \text{ is a field then:} \\ P^{(n,m)} = 0 \Leftrightarrow m+n \neq 0 \text{ and either } a_0 = 0 = b_0 \\ \text{or } f \text{ and } g \text{ have a nonconstant common factor in } R[Y]. \end{cases}$$

Second Version of Resultant theorem:

(35)
$$\begin{cases} \text{if } R \text{ is an algebraically closed field then:} \\ P^{(n,m)} = 0 \Leftrightarrow m+n \neq 0 \text{ and either } a_0 = 0 = b_0 \\ \text{or } f \text{ and } g \text{ have a common root in } R. \end{cases}$$

REMARK (R7). [**Intval Domains, Divisibility Rings, Maximality Rings, and Equivalent Valuations**]. From L2§3, L3§7, and L3§8, recall various terms related to valuations, and note that out of assertions (J1) to (J26) about valuations made in L3§§7-9, we still have to prove that:

(J24)   every normal domain is intval

where a domain is intval means it is an intersection of a nonempty family of valuation rings of its quotient field. We shall now prove this as well as the Valuation Theorems (T25) to (T28) stated in §9. First let us introduce some more concepts.

By a divisibility ring we mean a domain in which, out of any two nonzero elements, at least one divides the other.

In other words, a domain R with quotient field K is a divisibility ring means it satisfies one and hence all of the following five equivalent conditions:

(1) For any nonzero element x in K we have either $x \in R$ or $1/x \in R$.
(2) For any nonzero elements x, y in K we have either $x/y \in R$ or $y/x \in R$.
(3) For any nonzero elements x, y in R we have either $x/y \in R$ or $y/x \in R$.
(4) For any principal ideals I, J in R we have either $I \subset J$ or $J \subset I$.
(5) The set of all principal ideals in R is linearly ordered by inclusion.

To see the equivalence of (1), (2), (3), in (1) change x to z and then write $z = x/y$. The equivalence of (3), (4), (5) is obvious.

Given any divisibility ring R with quotient field K, we introduce an ordered abelian group $\Gamma(R)$ together with a valuation $\gamma_R : K \to \Gamma(R) \cup \{\infty\}$ in the following manner. We call these the divisibility group of R and the divisibility valuation of K induced by R respectively. It will turn out that the value group of γ_R is $\Gamma(R)$, and the valuation ring of γ_R is R.

As a group $\Gamma(R)$ equals the factor group $U(K)/U(R)$ written additively, where we recall that $U(R)$ denotes the group of all units in R. The restriction of γ_R to $U(K) = K^\times$ is the residue class epimorphism and we stipulate that $\gamma_R(0) = \infty$. For any x, y in $K^\times$ we define: $\gamma_R(x) \leq \gamma_R(y) \Leftrightarrow y/x \in R$. Since R is a divisibility ring, it follows that the relation $\leq$ on $\Gamma(R)$ is well-defined (i.e., is independent of the representatives in $K^\times$) and converts it into an ordered abelian group. Using the assumption that R is a divisibility ring, we also see that γ_R is a valuation of K with value group $\Gamma(R)$ and valuation ring R.

By a divisibility ring of a field K we mean a divisibility ring with quotient field K. We may restate the above construction by saying that:

(7.1)
$$\begin{cases} \text{if } R \text{ is a divisibility ring of a field } K \text{ then} \\ \text{the divisibility group } \Gamma(R) \text{ is an ordered abelian group} \\ \text{and the divisibility valuation } \gamma_R \text{ is a valuation of } K \\ \text{with value group } \Gamma(R) \text{ and valuation ring } R. \end{cases}$$

Note that for any valuation v and any nonzero x, y in R_v we have either $v(x) \leq v(y)$ or $v(y) \leq v(x)$ and hence either $y/x \in R_v$ or $x/y \in R_v$, and therefore R_v is a divisibility ring. This may be taken as a motivation for the above definition of a divisibility ring. Also note that for any valuation v of a field K, the restriction of v to $K^\times$ gives an epimorphism $K^\times \to G_v$ with kernel $U(R_v)$ and for any x, y in $K^\times$ we have: $v(x) \leq v(y) \Leftrightarrow y/x \in R_v$. This motivates the above definitions of the divisibility group and the divisibility valuation of a divisibility ring. Thus we have shown that:

(7.2)
$$\begin{cases} \text{a domain is a valuation ring} \\ \text{iff it is a divisibility ring.} \end{cases}$$

Valuations v and w of a field K are equivalent means there is an order (preserving) isomorphism θ of the value group G_v onto the value group G_w such that for all $x \in K^\times$ we have $\theta(v(x)) = w(x)$. By what we have said above, it follows that:

(7.3)
$$\begin{cases} \text{every valuation is equivalent to the} \\ \text{divisibility valuation of its valuation ring} \end{cases}$$

and

(7.4) $\begin{cases} \text{valuations of a field are equivalent} \\ \text{iff their valuation rings coincide.} \end{cases}$

We claim that for any valuation v and for any ideals I and J in R_v we have: (*) either $I \subset J$ or $J \subset I$. To see this, first note that: (**) $x \in I$ and $y \in R_v$ with $v(x) \leq v(y) \Rightarrow y \in I$ (because $y = x(y/x)$ with $y/x \in R_v$). Now suppose $J \not\subset I$ and take $y \in J \setminus I$. Then for every $x \in I$, by (**) we get $v(x) > v(y)$ and hence by applying (**) to J we get $x \in J$. This proves (*). Now in view of the equivalence of (1) to (5), by (7.2) and (*) we see that:

(7.5) $\begin{cases} \text{a domain is a valuation ring} \\ \text{iff the set of all ideals in it is linearly ordered.} \end{cases}$

By a maximality ring of a field K we mean a subring R of K such that R is a quasilocal domain which is not dominated by any quasilocal domain S which is a subring of K with $S \neq R$. We claim that:

(7.6) $\begin{cases} \text{a subring } R \text{ of a field } K \text{ is a valuation ring of } K \\ \text{iff it is a maximality ring of } K. \end{cases}$

Namely, if $R = R_v$ for some valuation v of K and S is a quasilocal domain dominating R such that S is a subring of K with $S \neq R$ then we can find $x \in S \setminus R$; clearly $x \in K \setminus R \Rightarrow v(x) < 0 \Rightarrow (1/x) \in M(R_v) \Rightarrow (1/x) \in M(S)$ (because S dominates R) $\Rightarrow x \notin S$ which is a contradiction.

Conversely assume that R is a quasilocal domain which is not dominated by any quasilocal domain S which is a subring of K with $S \neq R$. Given any $u \in K \setminus R$ let ϕ be the unique R-epimorphism of the univariate polynomial ring $R[X]$ onto $R[u]$ which sends X to u. Let J be the ideal in $R[X]$ generated by $\ker(\phi)$ and $M(R)$. If $J \neq R[X]$ then by Zorn's Lemma we can find a maximal ideal N in $R[X]$ with $J \subset N$, and this would give us a maximal ideal $\phi(N)$ in $R[u]$ whose intersection with R equals $M(R)$; localizing $R[u]$ at $\phi(N)$ gives us a quasilocal domain $R[u]_{\phi(N)}$ which dominates R and is a subring of K different from R; this is a contradiction. Therefore $J = R[X]$, and hence we can find a positive integer n, elements $a_0, \ldots, a_n$ in R, and elements $p_0 \ldots, p_n$ in $M(R)$ such that $a_0 + a_1 u + \cdots + a_n u^n = 0$ and $(a_0 + a_1 X + \cdots + a_n X^n) + (p_0 + p_1 X + \cdots + p_n X^n) = 1$. The last equation implies that $a_0 = 1 - p_0 \in R \setminus M(R)$ and $a_i = -p_i \in M(R)$ for $1 \leq i \leq n$. Consequently a_0 is a unit in R, and upon letting $b_i = a_i/a_0$ we get $b_i \in M(R)$ for $1 \leq i \leq n$ with

$$(1/u)^n + \sum_{1 \leq i \leq n} b_i (1/u)^{n-i} = 0.$$

By minimizing n we may assume that $1/u$ does not satisfy any equation of this type of degree smaller than n. Suppose if possible that $1/u \notin R$. Then in a similar manner we can find a positive integer m together with elements $c_i \in M(R)$ for $1 \leq i \leq m$ such that

$$u^m + \sum_{1 \leq i \leq m} c_i u^{m-i} = 0$$

and again we can assume m to be minimal. Without loss of generality we may assume that $m \geq n$. Multiplying the above displayed equation for $1/u$ by $c_m u^n$ and then subtracting it from the above displayed equation for u we get

$$d_0 u^m + \sum_{1 \leq i \leq m-1} d_i u^{m-i} = 0$$

with $d_0 \in R \setminus M(R)$ and $d_i \in M(R)$ for $1 \leq i \leq m-1$. Dividing throughout by the unit d_0 we obtain

$$u^m + \sum_{1 \leq i \leq m-1} e_i u^{m-i} = 0$$

with $e_i \in M(R)$ for $1 \leq i \leq m-1$. Since $u \neq 0$, by the above equation we must have $m > 1$. Now dividing by u we get the equation

$$u^{m-1} + \sum_{1 \leq i \leq m-1} e_i u^{m-1-i} = 0$$

which contradicts the minimality of m. This shows that R is a divisibility ring with quotient field K and hence we are done by (7.2).

Given any chain (under domination) of quasilocal domains which are subrings of a field K, their union is clearly a quasilocal domain (whose maximal ideal is the union of the maximal ideals of the members of the chain), which is a subring of K and which dominates every member of the chain. Thus the said union is an upper bound of the chain. Therefore, in view of (7.6), by Zorn's Lemma (see L2§5) we conclude that:

(7.7) $\begin{cases} \text{if } R \text{ is a subring of a field } K \text{ such that } R \text{ is quasilocal} \\ \text{then } R \text{ is dominated by some valuation ring of } K. \end{cases}$

We claim that

(7.8) $\begin{cases} \text{if } P \text{ is a prime ideal in a subring } A \text{ of a field } K \text{ then} \\ A \text{ is contained in a valuation ring } T \text{ of } K \text{ with } P = M(T) \cap A. \end{cases}$

This follows by taking $R = A_P$ in (7.7). Next we claim that:

(7.9) $\begin{cases} \text{if } y \text{ is a nonunit in a subring } A \text{ of a field } K \text{ then} \\ A \text{ is contained in a valuation ring } T \text{ of } K \text{ with } y \in M(T). \end{cases}$

This follows from (7.8) by noting that by Zorn's Lemma there exists a prime ideal (in fact a maximal ideal) P in R with $y \in P$.

If w is a valuation of a field L and K is a subfield of L then we get a valuation v of K by putting $v(x) = w(x)$ for all $x \in K$, and clearly $R_v = R_w \cap K$ and R_w dominates R_v. Thus we have shown that:

(7.10)
$$\begin{cases} \text{if } S \text{ is a valuation ring of a field } L \text{ and } K \text{ is a subfield of } L \\ \text{then } S \cap K \text{ is a valuation ring of } K \text{ dominated by } S. \end{cases}$$

If R is a valuation ring of a field K and L is an overfield of K then by (7.7) R is dominated by some valuation ring S of L and by (7.6) and (7.10) we must have $R = S \cap K$. Thus we have shown that:

(7.11)
$$\begin{cases} \text{if } R \text{ is a valuation ring of a field } K \text{ and } L \text{ is an overfield of } K \\ \text{then } R \text{ is dominated by some valuation ring } S \text{ of } L \\ \text{and for any such pair we have } R = S \cap K. \end{cases}$$

This completes the proof of Theorems (T25) to (T28) of §9.

If a subring R of a field K is integrally closed in K then for any $x \in K \setminus R$ we must have $x \notin R[1/x]$ (because otherwise we can write $x = q_0 + q_1(1/x) + \cdots + q_n(1/x)^n$ with n in $\mathbb{N}$ and $q_0, \ldots, q_n$ in R, and multiplying by x^n this would give an equation $x^{n+1} - q_0 x^n - q_1 x^{n-1} - \cdots - q_n = 0$ for x over R), and therefore by taking $A = R[1/x]$ and $y = 1/x$ in (7.9) we can find a valuation ring T of K with $R[1/x] \subset T$ such that $1/x \in M(T)$ and hence such that $x \notin T$. Since every valuation ring is normal, it follows that:

(7.12)
$$\begin{cases} \text{if a subring } R \text{ of a field } K \text{ is integrally closed in } K \text{ then} \\ R \text{ is the intersection of all valuation rings of } K \text{ which contain it.} \end{cases}$$

This completes the proof of (J24).

REMARK (R8). [**Modelic Proj**]. Let A be a subring of a field K. Given any family $(x_l)_{l \in \Lambda}$ of elements in K, with $x_j \neq 0$ for some $j \in \Lambda$, let $E = \mathfrak{W}(A; (x_l)_{l \in \Lambda})$ and let $K' = \mathrm{QF}(A[(x_l/x_i)_{l \in \Lambda}])$ where $i \in \Lambda$ with $x_i \neq 0$, and note that K' is clearly independent of the choice of i. Also let $y_l = x_l/x_{i'}$ for any fixed $i' \in \Lambda$ with $x_{i'} \neq 0$. Theorem (T29) of §9.1 says that then we have the following.

(T29.1) E is a premodel of K'/A and $E = \mathfrak{W}(A; (x_l/x)_{l \in \Lambda})$ for all $0 \neq x \in K$. In particular $(y_l)_{l \in \Lambda}$ is a family of elements in K' such that $y_j \neq 0$ for some $j \in \Lambda$, and we have $E = \mathfrak{W}(A; (y_l)_{l \in \Lambda})$.

(T29.2) Given any $R \in E$ and any subring S of K such that S is a quasilocal ring dominating R, there exists $j \in \Lambda$ with $x_j \neq 0$ such that $x_l/x_j \in S$ for all $l \in \Lambda$. Moreover, for any such j we have $R = B_Q$ where $B = A[(x_l/x_j)_{l \in \Lambda}]$ and $Q = B \cap M(S)$.

(T29.3) E is a semimodel of K'/A, and if Λ is finite then E is in fact a complete model of K'/A.

PROOF OF (T29.1). It suffices to note that if $x_i \neq 0$ then for all $l \in \Lambda$ we have $\frac{x_l/x}{x_i/x} = x_l/x_i$ and hence $A[(\frac{x_l/x}{x_i/x})_{l \in \Lambda}] = A[(x_l/x_i)_{l \in \Lambda}]$.

PROOF OF (T29.2). Since $R \in \mathfrak{W}(A; (x_l)_{l \in \Lambda})$, there exists $j' \in \Lambda$ with $x_{j'} \neq 0$ such that $R \in \mathfrak{V}(B')$ where $B' = A[(x_l/x_{j'})_{l \in \Lambda}]$, and then we have $x_l/x_{j'} \in R$ for all $l \in \Lambda$, and $R = B'_{Q'}$ where $Q' = B' \cap M(R)$. Since S dominates R, we get $x_l/x_{j'} \in S$ for all $l \in \Lambda$, and $Q' = B' \cap M(S)$. Therefore the first assertion follows by taking $j = j'$. To prove the second assertion, given any $j \in \Lambda$ such that $x_j \neq 0$ and $x_l/x_j \in S$ for all $l \in \Lambda$, let $B = A[(x_l/x_j)_{l \in \Lambda}]$ and $Q = B \cap M(S)$. Then $x_j/x_{j'}$ and $x_{j'}/x_j$ are both in S and hence they are units in S. Since $x_j/x_{j'} \in R$, S dominates R, and $x_j/x_{j'}$ is a unit in S, we get that $x_j/x_{j'}$ is a unit in R and hence $x_{j'}/x_j \in R$; consequently $x_l/x_j = (x_l/x_{j'})(x_{j'}/x_j) \in R$ for all $l \in \Lambda$ and hence $B \subset R$; since S dominates R and $Q = B \cap M(S)$, we get that $Q = B \cap M(R)$ and hence $B_Q \subset R$. Again since $x_{j'}/x_j \in B_Q$, S dominates B_Q, and $x_{j'}/x_j$ is a unit in S, we get that $x_{j'}/x_j$ is a unit in B_Q and hence $x_j/x_{j'} \in B_Q$; consequently $x_l/x_{j'} = (x_l/x_j)(x_j/x_{j'}) \in B_Q$ for all $l \in \Lambda$ and hence $B' \subset B_Q$; since S dominates B_Q and $Q' = B' \cap M(S)$, we get that $Q' = B' \cap M(B_Q)$ and hence $B'_{Q'} \subset B_Q$. Since $R = B'_{Q'}$, we conclude that $R = B_Q$.

PROOF OF (T29.3). The first assertion follows from (T29.1) and (T29.2). To prove the second assertion assume that Λ is a finite set. In view of the first assertion it suffices to show that if V is any valuation ring of K' containing A then V dominates $\mathfrak{W}(A; (x_l)_{l \in \Lambda})$. Since Λ is a finite set, there exists $j \in \Lambda$ with $x_j \neq 0$ such that $x_l/x_j \in V$ for all $l \in \Lambda$. Let $B = A[(x_l/x_j)_{l \in \Lambda}]$ and $Q = B \cap M(V)$. Then V dominates B_Q and $B_Q \in \mathfrak{W}(A; (x_l)_{l \in \Lambda})$.

REMARK (R9). [**Modelic Blowup**]. Let A be a subring of a field K, and let P be a nonzero A-submodule of K. Theorem (T30) of §9.2 says that then we have the following.

(T30.1) For any $0 \neq x \in P$ we have $(A[Px^{-1}])P = (A[Px^{-1}])x$ and hence $RP = Rx$ for every $R \in \mathfrak{V}(A[Px^{-1}])$. In particular, if P is a nonzero ideal in A then for every $R \in \mathfrak{V}(A[Px^{-1}])$ we have that PR is a nonzero principal ideal in R.

(T30.2) Given any family $(x_l)_{l \in \Lambda}$ of generators of P, for all $0 \neq x \in P$ we have $A[Px^{-1}] = A[(x_l/x)_{l \in \Lambda}]$, and we have $\mathfrak{W}(A; (x_l)_{l \in \Lambda}) \subset \mathfrak{W}(A, P)$.

(T30.3) For any family $(x_l)_{l \in \Lambda}$ of generators of P we have

$$\mathfrak{W}(A, P) = \mathfrak{W}(A; (x_l)_{l \in \Lambda}).$$

(T30.4) For all $x \neq 0 \neq y$ in P we have $\mathrm{QF}(A[Px^{-1}]) = \mathrm{QF}(A[Py^{-1}])$ and letting K' denote this common QF we have that $\mathfrak{W}(A, P)$ is a semimodel of K'/A, and if P is a finitely generated A-module then $\mathfrak{W}(A, P)$ is a projective model of K'/A. In particular, if P is a finitely generated ideal in A then $\mathfrak{W}(A, P)$ is a projective model of $\mathrm{QF}(A)/A$. If $P = Ax$ for some $0 \neq x \in K$ then $\mathfrak{W}(A, P) = \mathfrak{W}(A; x) = \mathfrak{V}(A)$.

(T30.5) If P is an ideal in A then

$$\{R \in \mathfrak{V}(A) : PR = R\} = \{R \in \mathfrak{W}(A, P) : PR = R\}.$$

(T30.6) If A is quasilocal and P is an ideal in A then:

$$P \text{ is a principal ideal in } A \Leftrightarrow \mathfrak{W}(A, P) = \mathfrak{V}(A) \Leftrightarrow A \in \mathfrak{W}(A, P).$$

PROOF OF (T30.1). Obvious.

PROOF OF (T30.2). The first assertion is obvious. The second follows from it.

PROOF OF (T30.3). In view of (T30.2) it suffices to show that $\mathfrak{W}(A, P) \subset \mathfrak{W}(A; (x_l)_{l \in \Lambda})$. So let any $R \in \mathfrak{W}(A, P)$ be given. Then there exists $0 \neq x \in P$ such that $R = B_Q$ where $B = A[Px^{-1}] = A[(x_l/x)_{l \in \Lambda}] \subset R$ and $Q = B \cap M(R)$. Since $0 \neq x \in P$, there exists a nonempty finite subset Λ' of Λ such that $x = \sum_{l \in \Lambda'} r_l x_l$ with $r_l \in A$. Then $1 = \sum_{l \in \Lambda'} r_l (x_l/x)$ and $r_l \in R$ and $(x_l/x) \in R$ for all $l \in \Lambda'$, and hence there exists $j \in \Lambda'$ such that $x_j/x \notin M(R)$. Consequently $x_j \neq 0$, and x_j/x and x/x_j are units in R. In particular $x_l/x_j = (x_l/x)(x/x_j) \in R$ for all $l \in \Lambda$ and hence $B' \subset R$ where $B' = A[(x_l/x_j)_{l \in \Lambda}]$. Upon letting $Q' = B' \cap M(R)$ we get that $B'_{Q'} \in \mathfrak{W}(A; (x_l)_{l \in \Lambda})$ and R dominates $B'_{Q'}$. Since $x/x_j \in B'_{Q'}$, R dominates $B'_{Q'}$, and x/x_j is a unit in R, we get that x/x_j is a unit in $B'_{Q'}$ and hence $x_j/x \in B'_{Q'}$. Consequently $x_l/x = (x_l/x_j)(x_j/x) \in B'_{Q'}$ for all $l \in \Lambda$ and hence $B \subset B'_{Q'}$; since R dominates $B'_{Q'}$ and $Q = B \cap M(R)$, we get $Q = B \cap M(B'_{Q'})$ and hence $B_Q \subset B'_{Q'}$, i.e., $R \subset B'_{Q'}$. Therefore $R = B'_{Q'}$, and hence $R \in \mathfrak{W}(A; (x_l)_{l \in \Lambda})$. Thus $\mathfrak{W}(A, P) \subset \mathfrak{W}(A; (x_l)_{l \in \Lambda})$.

PROOF OF (T30.4). Follows from (T30.3) and (T29).

PROOF OF (T30.5). Assume P is an ideal in A. First let $R \in \mathfrak{V}(A)$ such that $PR = R$; then $P \not\subset M(R)$ and hence there exists $0 \neq x \in P$ such that $x \notin M(R)$; now $A[Px^{-1}] \subset R$ and hence, upon letting $Q = A[Px^{-1}] \cap M(R)$, we get $R = (A[Px^{-1}])_Q \in \mathfrak{W}(A, P)$. Conversely let $R \in \mathfrak{W}(A, P)$ such that $PR = R$; now $\mathfrak{W}(A, P)$ dominates $\mathfrak{V}(A)$ and hence there exists $R' \in \mathfrak{V}(A)$ such that R dominates R'; since R dominates R' and $PR = R$, it follows that $PR' = R'$; therefore $R' \in \mathfrak{W}(A, P)$ by what we have already proved; since R dominates R as well as R', and by (T30.4) $\mathfrak{W}(A, P)$ is an irredundant premodel of the quotient field of A, we must have $R = R'$; consequently $R \in \mathfrak{V}(A)$.

PROOF OF (T30.6). Assume A is quasilocal and P is an ideal in A. Now: P is a principal ideal in $A \Rightarrow P = xA$ for some $0 \neq x \in A \Rightarrow \mathfrak{W}(A, P) = \mathfrak{W}(A; x) = \mathfrak{V}(A)$.

Also clearly: $\mathfrak{W}(A, P) = \mathfrak{V}(A) \Rightarrow A \in \mathfrak{W}(A, P)$. Finally: $A \in \mathfrak{W}(A, P) \Rightarrow A \in$ $\mathfrak{V}(A[Px^{-1}])$ for some $0 \neq x \in P \Rightarrow A[Px^{-1}] \subset A \Rightarrow A[Px^{-1}] = A \Rightarrow PA =$ $P(A[Px^{-1}]) = x(A[Px^{-1}]) = xA \Rightarrow P$ is a principal ideal in A.

§13: DEFINITIONS AND EXERCISES

EXERCISE (E10). [**Hilbert Basis Theorem**]. Show that an alternative proof of the polynomial case of the Hilbert Basis Theorem, i.e., Theorem (T3) of §3, can be obtained by slightly modifying the proof of the power series case, i.e., the proof of Theorem (T4) of §3.

HINT. In the proof of (T4), everywhere change $[[]]$ to $[]$, and in the second summation of display (2^*) change $< \infty$ to $\leq \deg(s)$.

EXERCISE (E11). [**Primary Modules**]. Let P be an ideal in a ring R and let Q be an R-submodule of an R-module V. Recall that Q is primary if $Q \neq V$ and for any $r \in R$ and $s \in V$ with $rs \in Q$ and $s \notin Q$ we have $r \in \mathrm{rad}_V Q$ where $\mathrm{rad}_V Q = \{r \in R : r^e V \subset Q \text{ for some } e \in \mathbb{N}_+\}$. In §5(O13) (which obviously generalizes §5(O7) and §5(O8) from ideals to modules) we said that the following three conditions are equivalent and when they are satisfied Q is called P-primary.

(i$'^{\bullet}$) Q is primary with $\mathrm{rad}_V Q = P$.
(ii$'^{\bullet}$) $P \subset \mathrm{rad}_V Q \neq R$ and: $r \in R$ and $s \in V \setminus Q$ with $rs \in Q \Rightarrow r \in P$.
(iii$'^{\bullet}$) Q is primary, $\mathrm{ann}_R(V/Q)$ is primary (as ideal), and P is prime with $\mathrm{rad}_V Q = P$.

We also stated the following four consequences of the said equivalence:

(1$'^{\bullet}$) Q is P-primary and $B \not\subset Q$ for submodule B of $V \Rightarrow (Q : B)_R$ is P-primary.
(2$'^{\bullet}$) $Q = Q_1 \cap \cdots \cap Q_h$ where Q_i is P-primary for $1 \leq i \leq h \in \mathbb{N}_+ \Rightarrow Q$ is P-primary.
(3$'^{\bullet}$) P is a maximal ideal in R with $P \subset \mathrm{rad}_V Q \neq R \Rightarrow Q$ is P-primary.
(4$'^{\bullet}$) P is a maximal ideal in R and $P^e V \subset Q \neq V$ (for instance $Q = P^e V$) with $e \in \mathbb{N}_+ \Rightarrow Q$ is P-primary.

The exercise is to give the details of the proof.

HINT FOR THE EQUIVALENCE. The definition of primary can be paraphrased in terms of zerodivisors by saying that

(11.1$'$) Q is primary $\Leftrightarrow Q \neq V$ and $Z_R(V/Q) \subset \mathrm{rad}_V Q$

and similarly for an ideal J in R

$(11.2')$ $\qquad\qquad$ J is primary $\Leftrightarrow J \neq R$ and $Z_R(R/J) \subset \mathrm{rad}_R J$.

Likewise

$(11.3')$ $\qquad\qquad$ $(\mathrm{ii'}^\bullet) \Leftrightarrow P \subset \mathrm{rad}_V Q \neq R$ and $Z_R(V/Q) \subset P$.

Also clearly

$(11.4')$ $\qquad\qquad\qquad$ $Q \neq V \Leftrightarrow \mathrm{rad}_V Q \neq R$

and

$(11.5')$ $\qquad\qquad\qquad$ $\mathrm{rad}_V Q = \mathrm{rad}_R \mathrm{ann}_R(V/Q)$

and

$(11.6')$ $\qquad\qquad\qquad$ $Z_R(R/\mathrm{ann}_R(V/Q)) \subset Z_R(V/Q)$.

Putting $\mathrm{ann}_R(V/Q) = J$, by $(11.1')$ to $(11.6')$ we see that $(\mathrm{i'}^\bullet) \Rightarrow (\mathrm{ii'}^\bullet) \Rightarrow Q$ is primary and $\mathrm{ann}_R(V/Q)$ is primary, and obviously $(\mathrm{iii'}^\bullet) \Rightarrow (\mathrm{i'}^\bullet)$. To complete the equivalence proof, we need to show that $(\mathrm{ii'}^\bullet) \Rightarrow P$ is prime with $\mathrm{rad}_V Q = P$. For any r in $\mathrm{rad}_V Q$, by $(11.5')$ we have $r^e \in \mathrm{ann}_R(V/Q)$ for some positive integer e; letting e to be the smallest such, and assuming $\mathrm{rad}_V Q \neq R$, by $(11.4')$ we get $r^{e-1} \notin \mathrm{ann}_R(V/Q)$; it follows that $r^e V \subset Q$ and $r^{e-1}V \not\subset Q$; since $r^{e-1}V \not\subset Q$, we can find $t \in V$ with $r^{e-1}t \notin Q$, and since $r^e V \subset Q$, we get $r^e t \in Q$; thus:

$(11.7')$ $\qquad$ $r \in \mathrm{rad}_V Q \neq R \Rightarrow \begin{cases} \text{for some } e \in \mathbb{N}_+ \text{ and } t \in R \text{ we have} \\ r^e t \in Q \text{ with } r^{e-1}t \notin Q. \end{cases}$

Assuming $(\mathrm{ii'}^\bullet)$, in the above set-up, by taking $s = r^{e-1}t$ we have $s \in V \setminus Q$ with $rs \in Q$ and hence we get $r \in P$. Thus

$(11.8')$ $\qquad\qquad\qquad$ $(\mathrm{ii'}^\bullet) \Rightarrow P = \mathrm{rad}_V Q$.

Again assuming $(\mathrm{ii'}^\bullet)$, given any $r_1 \in R \setminus P$ and $r_2 \in R$ with $r \in P$ where $r = r_2 r_1$, in view of $(11.8')$, by taking $s_1 = (r_2 r_1)^{e-1}t$ in $(11.7')$ we get $s_1 \in V \setminus Q$ with $r_2 r_1 s_1 \in Q$; since $r_1 \in R \setminus P$ and $s_1 \in V \setminus Q$, by $(\mathrm{ii'}^\bullet)$ we get $s_2 \in V \setminus Q$ where $s_2 = r_1 s_1$; since $r_2 \in R$ and $s_2 \in V \setminus Q$ with $r_2 s_2 \in Q$, by $(\mathrm{ii'}^\bullet)$ we get $r_2 \in P$. Thus $(\mathrm{ii'}^\bullet) \Rightarrow P$ is prime.

HINT FOR THE FOUR CONSEQUENCES. To prove $(1'^\bullet)$, assume that Q is P-primary and $B \not\subset Q$ for submodule B of V. We want to prove that $(Q : B)_R$ is P-primary. By the definition of $(Q : B)_R$ we get $(Q : B)_R B \subset Q$; in view of $(\mathrm{ii'}^\bullet)$ and because Q is P-primary, the last inclusion and the noninclusion $B \not\subset Q$ tell us that $(Q : B)_R \subset P$, and hence: (a) $\mathrm{rad}_R(Q : B)_R \subset P$. Now: $\rho \in \mathrm{rad}_V Q \Rightarrow \rho^e V \subset Q$ for some $e \in \mathbb{N}_+ \Rightarrow \rho^e B \subset Q$ (because $B \subset V$) $\Rightarrow \rho^e \in (Q : B)_R$; thus:

$\text{rad}_V Q \subset \text{rad}_R(Q : B)_R$; since Q is P-primary, we also have $P \subset \text{rad}_V Q$, and hence: (b) $P \subset \text{rad}_R(Q : B)_R$. By (a) and (b) we get: (c) $\text{rad}_R(Q : B)_R = P$. Now: $r \in R$ and $s \in R \setminus (Q : B)_R$ with $rs \in (Q : B)_R \Rightarrow r \in R$ and $s \in R$ such that for some $v \in B$ we have $sv \notin Q$ but $rsv \in Q \Rightarrow r \in \text{rad}_v Q \Rightarrow r \in P$ by (c); thus: (d) $r \in R$ and $s \in R \setminus (Q : B)_R$ with $rs \in (Q : B)_R \Rightarrow r \in P$. in view of (ii'$^\bullet$), by (c) and (d) we see that $(Q : B)_R$ is P-primary.

To prove (2'$^\bullet$), assume that $Q = Q_1 \cap \cdots \cap Q_h$ where Q_i is P-primary for $1 \le i \le h \in \mathbb{N}_+$. Then $\text{rad}_V Q = (\text{rad}_V Q_1) \cap \cdots \cap (\text{rad}_V Q_h)$ and hence $\text{rad}_V Q = P$. Now: $r \in R$ and $s \in V \setminus Q$ with $rs \in Q \Rightarrow r \in R$ and for some $i \in \{1, \ldots, h\}$ we have $s \in V \setminus Q_i$ with $rs \in Q_i \Rightarrow r \in \text{rad}_V Q_i = P$ by (ii'$^\bullet$). Therefore again by (ii'$^\bullet$) we conclude that Q is P-primary.

To prove (3'$^\bullet$), assume P is a maximal ideal in R with $P \subset \text{rad}_V Q \ne R$. Let $r \in R \setminus P$ and $s \in V$ be such that $rs \in Q$. In view of (ii'$^\bullet$), it suffices to show that then $s \in Q$. Since P is maximal, we must have $P + Rr = R$ and hence we can write $1 = c + dr$ with $c \in P$ and $d \in R$. Since $c \in P \subset \text{rad}_V Q$, we can find $m \in \mathbb{N}_+$ with $c^m V \subset Q$. Raising the equation $1 = c + dr$ to the m-th power we get $1 = 1^m = c^m + d'r$ with $d' \in R$ and multiplying throughout by s we obtain $s = c^m s + d'rs \in Q$.

To prove (4'$^\bullet$), in view of (3'$^\bullet$), it suffices to note that $P \subset \text{rad}_V P^e V$.

DEFINITION (D1). [**Support of a Module**]. The support of a module V over a ring R is defined by putting

$$\text{supp}_R V = \{P \in \text{spec}(R) : [0 : P]_V \ne V\}.$$

EXERCISE (E12). [**Support and Annihilator**]. Show that for the support of a module V over a ring R we have the following.

(E12.1) If $V = RB$ with $B \subset V$ then $\text{supp}_R V = \bigcup_{b \in B} \text{vspec}_R(\text{ann}_R b)$.

(E12.2) If V is finitely generated then $\text{supp}_R V = \text{vspec}_R(\text{ann}_R V)$.

HINT FOR (E12.1). Assuming $V = RB$, for any $P \in \text{spec}(R)$ we have

$$\begin{cases} P \in \text{supp}_R V \\ \Leftrightarrow [0 : P]_V \ne V \\ \Leftrightarrow \text{there exists } b \in B \text{ such that } b \notin [0 : P]_V \\ \Leftrightarrow \text{there exists } b \in B \text{ such that } b \notin (0 : s)_V \text{ for all } s \in R \setminus P \\ \Leftrightarrow \text{there exists } b \in B \text{ such that } bs \ne 0 \text{ for all } s \in R \setminus P \\ \Leftrightarrow \text{there exists } b \in B \text{ such that } \text{ann}_R b \subset P \\ \Leftrightarrow P \in \bigcup_{b \in B} \text{vspec}_R(\text{ann}_R b). \end{cases}$$

HINT FOR (E12.2). Assuming $V = RB$ with finite set $B = \{b_1, \ldots, b_n\}$ where $n \in \mathbb{N}_+$, for any $P \in \mathrm{spec}(R)$ we have

$$\begin{cases} P \in \mathrm{supp}_R V \\ \Leftrightarrow P \in \mathrm{vspec}_R(\mathrm{ann}_R b_i) \text{ for some } i \quad [\text{by (E12.1)}] \\ \Leftrightarrow \mathrm{ann}_R b_i \subset P \text{ for some } i \end{cases}$$

and obviously

$$P \in \mathrm{vspec}_R(\mathrm{ann}_R V) \Leftrightarrow \mathrm{ann}_R V \subset P$$

and hence it suffices to show that

$$\mathrm{ann}_R b_i \subset P \text{ for some } i \Leftrightarrow \mathrm{ann}_R V \subset P.$$

The $\Rightarrow$ part of the above is obvious because for all i we have $\mathrm{ann}_R V \subset \mathrm{ann}_R b_i$. Moreover: the negation of the LHS $\Rightarrow$ for all i we have $\mathrm{ann}_R b_i \not\subset P \Rightarrow$ for each i there is $s_i \in R \setminus P$ with $s_i b_i = 0 \Rightarrow s_1 \ldots s_n \in R \setminus P$ with $s_1 \ldots s_n V = 0 \Rightarrow$ the negation of the RHS.

EXERCISE (E13). [**Support and Quasiprimary Decomposition**]. Let U be a submodule of a module V over a ring R. Show that:

(E13.1) If $U \neq V$ and V is finitely generated then

$$\mathrm{supp}_R(V/U) = \{P \in \mathrm{spec}(R) : [U : P]_V \neq V\}$$

and

$$\mathrm{supp}_R(V/U) = \{P \in \mathrm{spec}(R) : [U : P]_V \text{ is } P\text{-quasiprimary}\}$$

(E13.2) If P is a minimal element of $\mathrm{supp}_R(V/U)$ then $[U : P]_V$ is P-primary.

HINT. Use the above Exercise (E12) and Theorem (T8$'$) of §6.1.

DEFINITION (D2). [**Homogenization and Dehomogenization**]. Given a polynomial $f(Y_1, \ldots, Y_N)$ in indeterminates $Y_1, \ldots, Y_N$ over a field k, taking a nonnegative integer $d \geq \deg(f)$, we can "homogenize" f to get a homogenous polynomial $F = F(X_1, \ldots, X_{N+1})$ of degree d in indeterminates $X_1, \ldots, X_{N+1}$ over k by the substitution $Y_i = X_i/X_{N+1}$, i.e., by putting

$$F(X_1, \ldots, X_{N+1}) = X_{N+1}^d f(X_1/X_{N+1}, \ldots, X_N/X_{N+1}).$$

Conversely, given any homogeneous $F = F(X_1, \ldots, X_{N+1})$ of degree d we can "dehomogenize" it by putting

$$f(Y_1, \ldots, Y_N) = F(Y_1, \ldots, Y_N, 1) = X_{N+1}^{-d} F(Y_1 X_{N+1}, \ldots, Y_N X_{N+1}, X_{N+1}).$$

EXERCISE (E14).   [**Embedding Projective Space Into Projective Model**]. In §9.1 we gave a natural injection θ from the projective N-space $\mathbb{P}_k^N$ over a field k into the modelic projective N-space $(\mathbb{P}_k^N)^\delta = \mathfrak{W}(k; X_1, \ldots, X_{N+1})$. There we used the irredundancy of $\mathfrak{W}$. Without using the said irredundancy, give a more elementary construction of the said injection on the following lines by using homogenization–dehomogenization.

Given $u = (u_1, \ldots, u_{N+1}) \in \mathbb{A}_k^{N+1} \setminus \{(0, \ldots, 0)\}$ we want to find its image in

$$\mathfrak{W}(k; X_1, \ldots, X_{N+1}) = \cup_{1 \leq i \leq N+1} \mathfrak{V}(B_{N,k,i})$$

with

$$B_{N,k,i} = k[Y_{1i}, \ldots, Y_{N+1,i}] \quad \text{where} \quad Y_{li} = X_l / X_i.$$

For i, i' in $\{1, \ldots, N+1\}$ with $u_i \neq 0 \neq u_{i'}$ consider the local rings $R_k(v)^*$ and $R_k(v')^*$ where $v = (u_1/u_i, \ldots, u_{N+1}/u_i) \in \mathbb{A}_{k,i}^N$ and $v' = (u_1/u_{i'}, \ldots, u_{N+1}/u_{i'}) \in \mathbb{A}_{k,i'}^N$. If we show that these local rings coincide then we can send the equivalence class of u in $\mathbb{P}_k^N$ to that common local ring. So the exercise is to show that $R_k(v)^* = R_k(v')^*$.

HINT. The case of $i = i'$ being trivial, suppose $i \neq i'$. Relabelling suitably we can arrange that $i = 1$ and $i' = 2$. Disregarding the coordinate Y_{ii} which identically equals 1, we have

$$B_{N,k} = B_{N,k,i} = k[Y_1, \ldots, Y_N]$$

where

$$(Y_1, Y_2, \ldots, Y_N) = (X_2/X_1, X_3/X_1, \ldots, X_{N+1}/X_1)$$

and

$$v = (u_2/u_1, u_3/u_1, \ldots, u_{N+1}/u_1) \in \mathbb{A}_{k,i}^N.$$

Likewise, disregarding the coordinate $Y_{ii'}$ which identically equals 1, we have

$$B'_{N,k} = B_{N,k,i'} = k[Y'_1, \ldots, Y'_N]$$

where

$$(Y'_1, Y'_2, \ldots, Y'_N) = (X_1/X_2, X_3/X_2, \ldots, X_{N+1}/X_2)$$

and

$$v' = (u_1/u_2, u_3/u_2, \ldots, u_{N+1}/u_2) \in \mathbb{A}^N_{k,i'}.$$

Now let us note that the local ring $R_k(u)^*$ is the set of all quotients of polynomials $F(X_1, \ldots, X_{N+1})/G(X_1, \ldots, X_{N+1})$ with $G(u_1, \ldots, u_{N+1}) \neq 0$, the local ring $R_k(v)^*$ is the set of all quotients of polynomials $f(Y_1, \ldots, Y_N)/g(Y_1, \ldots, Y_N)$ with $g(v_1, \ldots, v_N) \neq 0$, and the local ring $R_k(v')^*$ is the set of all quotients of polynomials $f'(Y'_1, \ldots, Y'_N)/g'(Y'_1, \ldots, Y'_N)$ with $g'(v'_1, \ldots, v'_N) \neq 0$. For f/g in $R_k(v)^*$, homogenizing, i.e., multiplying f and g by X_1^d for large enough positive integer d we get $f/g = F(X_1, \ldots, X_{N+1})/G(X_1, \ldots, X_{N+1})$ where F, G are homogeneous of degree d with $G(u_1, \ldots, u_{N+1}) \neq 0$, and hence $F/G \in R_k(u)^*$. Dehomogenizing, i.e., dividing F and G by X_2^d we get $F/G = f'(Y'_1, \ldots, Y'_N)/g'(Y'_1, \ldots, Y'_N)$ with $g'(v'_1, \ldots, v'_N) \neq 0$, and hence $f'/g' \in R_k(v')^*$. Thus $R_k(v)^* \subset R_k(v')^*$, and by symmetry $R_k(v')^* \subset R_k(v)^*$. Therefore $R_k(v)^* = R_k(v')^*$.

§14: NOTES

NOTE (N1). [**Princeton Book**]. My Princeton Book [A01] provides a rapid introduction to parts of this and some of the following Lectures. In particular, Observation §5(O20)(13•) and Theorem §6(T8) stem from Chapter I Section 2 of that book, and Observation §5(O20)(14•) and Theorem §6.1(T8') are their modulifications. In [A01] these items were motivated by the desire of giving a unified treatment of ramification of local rings which are always noetherian and valuation rings which are not (unless the value group is a subgroup of $\mathbb{Z}$).

NOTE (N2). [**Resolution Book**]. The language of models was introduced at the end of my Princeton Book [A01] and was further developed in my Resolution Book [A05]. In particular the material of §8 and §9 is mostly taken from [A05]. More material from [A05] will be included in later Lectures. As a motivation for the idea of an irredundant premodel introduced in §9, the proof of the existence of a natural injection of a projective space into a projective model, which is given in §9.1 and which uses the irredundancy, should be compared with the longer but more elementary proof of the same which is given in §13(E14) and which uses the technique of homogenization–dehomogenization.

NOTE (N3). [**Kyoto Paper**]. The treatment of resultants given in §12(R6) is taken from my Kyoto Paper [A03]. The same is true of the material on Approximate Roots given in (D7) and (E15) to (E17) of L2§7. Later Lectures will contain further material from that paper. A significant feature of the treatment of resultants given in §12(R6) is the establishment of product rules for the resultant matrix and the ensuing deduction of the product formula for the resultant itself. This makes the whole theory applicable to polynomials over a ring with zerodivisors. The methods used are longer but more elementary since they depend only on the grade-school operations of addition and multiplication and do not involve root-extractions.

NOTE (N4). [**Valuations and L'Hospital's Rule**]. The L'Hospital Rule from calculus says that if $x(t), y(t)$ are infinitely differentiable functions of a real variable t which are not identically zero then the limit of $x(t)/y(t)$ as t tends to zero always exists although it may be infinite. Motivated by this, what we have called a divisibility ring in §12(R7) could have been called a L'Hospital ring, and then assertion (7.2) of §12(R7) would have said that a L'Hospital ring is nothing but a valuation ring.

The situation for two or more variable functions is quite different, as shown by the quotient X/Y which takes all possible values as X and Y both tend to zero. This fundamental indeterminacy is mitigated by the introduction of valuations which may be viewed as an algebraization of the idea of limits. As we shall see in later lectures, the said indeterminacies can be "removed" by means of the modelic blowups introduced in §9.2

NOTE (N5). [**What Remains To Be Done**]. Referring to §1(O7) and the last para of §12(R5), the only items from §1 which remain to be dealt with are Bézout Observations (O4) and (O5). The only other assertions from the present Lecture which remain to be proved are Theorems (T19) to (T24) of §8; Theorems (T25) to (T30) of §9 were proved in (R7) to (R9) of §12. Finally, referring to L3§12(O3), we proved (J1) and (J2) in §10(E2) and we proved (J24) in §12(R7), and so the only item from Lecture L3 which remains to be dealt with is Homework (H3).

§15: CONCLUDING NOTE

It is good to keep a balance between algebra and geometry. Let logical formality and narrative discourse follow each other.

Lecture L5: Projective Varieties

§1: DIRECT SUMS OF MODULES

Before studying projective varieties which are defined in terms of homogeneous polynomials, we prepare some algebraic background.

Given any polynomial in a finite number of variables $X_1, \ldots, X_m$ over a ring R, by collecting terms of like degree in it we can uniquely write it as a finite sum of homogeneous polynomials. This is formalized by saying that, as an R-module, the polynomial ring $S = R[X_1, \ldots, X_n]$ is a direct sum

$$S = \sum_{0 \leq i < \infty} S_i$$

where S_i is the set of all homogeneous polynomials of degree i, including the zero polynomial. Moreover, for all $f_i \in S_i$ and $f_j \in S_j$ we have $f_i f_j \in S_{i+j}$. This is expressed by saying that S is a graded ring over R with gradation $(S_i)_{i \in \mathbb{N}}$.

Quite generally, a module V over a ring R is a direct sum of a family $(V_i)_{l \in I}$ of submodules if it is the sum

$$V = \sum_{i \in I} V_i$$

and the sum is direct, i.e., for every finite subset I' of the indexing set I we have: $\sum_{i \in I'} v_i = 0$ with $v_i \in V_i$ for all $i \in I' \Rightarrow v_i = 0$ for all $i \in I'$. Note that $V = \sum_{i \in I} V_i$ means every $v \in V$ can be written as $v = \sum_{i \in I'} v_i$ for some finite subset I' of I and some $v_i \in V_i$ for all $i \in I'$.

The fact that the sum is direct is expressed by writing something like

$$V = \sum_{i \in I} V_i \quad \text{(direct sum)}$$

or something like

$$V = \bigoplus_{i \in I} V_i.$$

In case I is a finite set, say $I = \{1, 2, \ldots, n\}$, we may write something like

$$V = V_1 \oplus V_2 \oplus \cdots \oplus V_n.$$

These are sometimes called internal direct sums to distinguish them from the external direct sums which we shall now introduce and which we shall again denote by $\oplus$. To stress the fact of being internal, the above displays may be written as

$$V = \sum_{i \in I} V_i \quad \text{(internal direct sum)}$$

or

$$V = \bigoplus_{i \in I} V_i \quad \text{(internal)}$$

or

$$V = V_1 \oplus V_2 \oplus \cdots \oplus V_n \quad \text{(internal)}.$$

Recall that $W = U_1 \times \cdots \times U_n$ is the cartesian product (see L2§1) of a finite number of sets $U_1, \ldots, U_n$ and it consists of all n-tuples $u = (u_1, \ldots, u_n)$ with $u_i \in U_i$. Assuming $U_1, \ldots, U_n$ to be R-modules, we convert W into an R-module by componentwise addition and scalar multiplication, i.e., by defining $u + u' = (u_1 + u'_1, \ldots, u_n + u'_n)$ for all $u' = (u'_1, \ldots, u'_n)$, and $ru = (ru_1, \ldots, ru_n)$ for all $r \in R$. With this module structure we call W the (external) direct sum of $U_1, \ldots, U_n$ and indicate this by writing

$$W = U_1 \oplus U_2 \oplus \cdots \oplus U_n \quad \text{(external)}$$

where we shall usually omit the adjective external. If $U_1 = \cdots = U_n = U$ then, as in the case of cartesian products, we put $W = U^n$ and call it the (module theoretic) direct sum of n copies of U.

More generally, the cartesian product of a family $(U_i)_{i \in I}$ of sets U_i, where I is any indexing set (which need not be finite), is defined by putting

$$\prod_{i \in I} U_i = \text{the set of all maps } \phi : I \to \bigcup_{i \in I} U_i \text{ such that } \phi(i) \in U_i \text{ for all } i \in I.$$

In case the U_i are R-modules, this is converted into an R-module again by defining componentwise addition and scalar multiplication, i.e., for any $\psi \in \prod_{i \in I} U_i$ we put $(\phi + \psi)(i) = \phi(i) + \psi(i)$ for all $i \in I$, and for any $r \in R$ we put $(r\phi)(i) = r\phi(i)$ for all $i \in I$; this module is called the (external) direct product of the family, where the adjective external is usually omitted. The set of all $\phi \in \prod_{i \in I} U_i$ whose support $\text{supp}(\phi) = \{i \in I : \phi(i) \neq 0\}$ is finite is called the (external) direct sum of the family and is denoted by

$$\bigoplus_{i \in I} U_i \quad \text{(external)}$$

where the adjective external will usually be omitted. Note that this is a submodule of the direct product $\prod_{i \in I} U_i$ and is hence an R-module.

For any $j \in I$ we get an injective map (called the natural injection) $\mu_j : U_j \to \prod_{i \in I} U_i$ by saying that for every $x \in U_j$ we have $\mu_j(x)(i) = x$ or 0 according as $i = j$ or $i \neq j$, and we get a surjective map (called the natural projection) $\nu_j : \prod_{i \in I} U_i \to U_j$ by saying that for every $y \in \prod_{i \in I} U_i$ we have $\nu_j(y) = y(j)$. Clearly $\nu_j \mu_j$ is the identity map of U_i. All this works for sets U_i. In the module case these maps are R-homomorphisms, and they give rise to an obvious R-monomorphism (again called the natural injection) $\mu_j : U_j \to \oplus_{i \in I} U_i$ and an obvious R-epimorphism (again called the natural projection) $\nu_j : \oplus_{i \in I} U_i \to U_j$. Note that upon letting $\overline{U}_i$ to be the image of μ_i we have

$$\bigoplus_{i \in I} U_i \quad \text{(external)} \ = \bigoplus_{i \in I} \overline{U}_i \quad \text{(internal)}.$$

If the set $U_i = $ a set U for all $i \in I$ then the cartesian product $\prod_{i \in I} U_i$ is denoted by U^I and is called the set-theoretic I-th power of U, and if U is an R-module then the module U^I is called the module theoretic I-th power of U. Finally, if the module $U_i = $ a module U for all $i \in I$ then the module $\oplus_{i \in I} U_i$ is denoted by $U_{\oplus}^I$ and is called the module theoretic restricted I-th power of U.

If $(\alpha_i)_{i \in I}$ is a family of R-homomorphisms $\alpha_i : U_i \to D_i$ where $(D_i)_{i \in I}$ is a family of R-modules, then we get an R-homomorphism $\prod_{i \in I} U_i \to \prod_{i \in I} D_i$ or $\oplus_{i \in I} U_i \to \oplus_{i \in I} D_i$ which sends any ϕ in $\prod_{i \in I} U_i$ or $\oplus_{i \in I} U_i$ to ψ in $\prod_{i \in I} D_i$ or $\oplus_{i \in I} D_i$ given by $\psi(i) = \alpha_i(\phi(i))$ for all $i \in I$; we call this the direct product or the direct sum of the family $(\alpha_i)_{i \in I}$ and denote it by $\prod_{i \in I} \alpha_i$ or $\oplus_{i \in I} \alpha_i$ respectively. Note that the kernels of these maps are $\prod_{i \in I} \ker(\alpha_i)$ and $\oplus_{i \in I} \ker(\alpha_i)$, and their images are $\prod_{i \in I} \mathrm{im}(\alpha_i)$ and $\oplus_{i \in I} \mathrm{im}(\alpha_i)$, respectively. Also note that for the definition of the map $\prod_{i \in I} \alpha_i : \prod_{i \in I} U_i \to \prod_{i \in I} D_i$, it suffices that the $\alpha_i : U_i \to D_i$ be set-theoretic maps; in case $U_i = U$ for all $i \in I$, the diagonal map $U \to U^I$ sends any $x \in U$ to $\phi \in U^I$ with $\phi(i) = x$ for all $i \in I$, and multiplying it from the left by $\prod_{i \in I} \alpha_i$ we get a map $U \to \prod_{i \in I} D_i$ which we call the diagonal product of the family $(\alpha_i)_{i \in I}$; in case of modules, the diagonal map is an R-monomorphism and the diagonal product is an R-homomorphism.

Note that if the indexing set I is finite then the direct product coincides with the direct sum. In particular, if I is finite then the diagonal product $U \to \prod_{i \in I} D_i = \oplus_{i \in I} D_i$ may be called the diagonal direct sum of the family $(\alpha_i)_{i \in I}$.

If T_i is a submodule of U_i for all $i \in I$, then clearly $\prod_{i \in I} T_i$ and $\oplus_{i \in I} T_i$ are submodules of $\prod_{i \in I} U_i$ and $\oplus_{i \in I} U_i$ respectively, and upon letting $\beta_i : U_i \to U_i/T_i$ be the residue class epimorphism we see that $\prod_{i \in I} \beta_i : \prod_{i \in I} U_i \to \prod_{i \in I} (U_i/T_i)$ and $\oplus_{i \in I} \beta_i : \oplus_{i \in I} U_i \to \oplus_{i \in I} (U_i/T_i)$ are R-epimorphisms with kernels $\prod_{i \in I} T_i$ and $\oplus_{i \in I} T_i$ respectively.

For finite cartesian products $U_1 \times \cdots \times U_n$, and for the corresponding external direct sums, we used the tuple notation. In case of indexed families $(U_i)_{i \in I}$, for cartesian products $\prod_{i \in I} U_i$ and the corresponding external direct sums, we used the functional notation. The two are related by saying that the n-tuple $u = (u_1, \ldots, u_n)$ corresponds to the function (= map) $u : \{1, \ldots, n\} \to \cup_{i \in I} U_i$ with $u(i)$ written as u_i. The concepts of the above several paragraphs carry over to the tuple notation. For instance the natural projection $U_1 \oplus \cdots \oplus U_n \to U_j$ picks out the j-th component, and the natural injection $U_j \to U_1 \oplus \cdots \oplus U_n$ puts zeros around it. Given maps $\alpha_i : U_i \to D_i$, their product $\alpha_1 \times \cdots \times \alpha_n : U_1 \times \cdots \times U_n \to D_1 \times \cdots \times D_n$ as well as their direct sum $\alpha_1 \oplus \cdots \oplus \alpha_n : U_1 \oplus \cdots \oplus U_n \to D_1 \oplus \cdots \oplus D_n$ is given by $(u_1, \ldots, u_n) \mapsto (\alpha_1(u_1), \ldots, \alpha_n(u_n))$. Moreover, in case of $U_1 = \cdots = U_n = U$, the diagonal maps $U \to U_1 \times \cdots \times U_n$ and $U \to U_1 \oplus \cdots \oplus U_n$ are given by $u \mapsto (u, \ldots, u)$, and the diagonal product $U \to D_1 \times \cdots \times D_n$ and the diagonal direct sum $U \to D_1 \oplus \cdots \oplus D_n$ are given by $u \mapsto (\alpha_1(u), \ldots, \alpha_n(u))$. We use the notation $\alpha_1 \oplus \cdots \oplus \alpha_n$ for the last map, i.e., we write

$$\alpha_1 \oplus \cdots \oplus \alpha_n : U \to D_1 \oplus \cdots \oplus D_n \quad \text{(diagonal direct sum)}$$

where, as said above, $\alpha_1 \oplus \cdots \oplus \alpha_n(u) = (\alpha_1(u), \ldots, \alpha_n(u))$; we may drop the parenthetical phrase "diagonal direct sum" when clear from the context; note that now the maps $\alpha_i : U \to D_i$ are from a common set U, and hence the notation $\alpha_1 \oplus \cdots \oplus \alpha_n : U \to D_1 \oplus \cdots \oplus D_n$ will not be confused with the similar notation $\alpha_1 \oplus \cdots \oplus \alpha_n : U_1 \oplus \cdots \oplus U_n \to D_1 \oplus \cdots \oplus D_n$ where the maps $\alpha_i : U_i \to D_i$ are from differently labelled sets U_i. For the above diagonal direct sum we clearly have

$$(1) \qquad \ker(\alpha_1 \oplus \cdots \oplus \alpha_n) = \ker(\alpha_1) \cap \cdots \cap \ker(\alpha_n).$$

Most of the discussion of the above several paragraphs applies with internal direct sums replacing external direct sums. For instance the inclusion map $V_j \to \oplus_{i \in I} V_i$ is the natural injection, and the natural projection $\oplus_{i \in I} V_i \to V_j$ is the obvious epimorphism whose kernel is $\oplus_{i \in I \setminus \{j\}} V_i$.

It is usually clear from the context whether we are dealing with internal or external direct sums. For instance when the constituents V_i are submodules of a module V then it must be internal, and when the modules U_i are not given to be submodules of some module then it must be external. For the lengths of finite (internal or external) direct sums of R-modules $U_1, \ldots, U_n$, by induction applied to L4§5(O21)(18•) we see that

$$(2) \qquad \ell_R(U_1 \oplus \cdots \oplus U_n) = \ell_R(U_1) + \cdots + \ell_R(U_n)$$

with the usual convention about sums involving ∞.

All the above matter applies to additive abelian groups by viewing them as modules over $\mathbb{Z}$.

EXAMPLE (X1). [**Annihilators and Direct Sums**]. In L4§10(X18) we gave an example of a submodule T of a non-finitely-generated module U over a ring R together with a prime P in R such that $\text{ann}_R(U/T) \subset P$ but $[T : P]_U = U$, and in L4§10(X19) we gave an example of a nonprimary submodule W of a module V over a ring R such that $\text{ann}_R(V/W)$ is a primary ideal in R. Now using (external) direct sums we give other such examples.

First let x be a nonzero element in a domain R such that $\cap_{i \in \mathbb{N}}(x^i R) = 0$; for instance $R =$ the polynomial ring in one or more indeterminates $X, Y, \ldots$ over a field k, and $x = X$. Let $U = \oplus_{i \in \mathbb{N}} U_i$ and $T = \oplus_{i \in \mathbb{N}} T_i$ with $U_i = R$ and $T_i = x^i R$. Let P be the zero ideal in R. Then P is a prime ideal in R. Given any $0 \neq r \in R$ we have $r \notin x^n R$ for some $n \in \mathbb{N}$; we can take $\phi \in U$ with $\phi(i) = 1$ or 0 according as $i = n$ or $i \neq n$, and then $r\phi \in U$ with $(r\phi)(n) \notin T_n$ and hence $r\phi \notin T$; consequently

$r \notin \operatorname{ann}_R(U/T)$. Thus $\operatorname{ann}_R(U/T) = P$. Given any $\psi \in U$ we can find $m \in \mathbb{N}$ such that $\psi(i) = 0$ for all $i > m$; now $x^m \psi \in T$ with $x^m \in R \setminus P$ and hence $\psi \in [T : P]_U$. Thus $[T : P]_U = U$.

Next let R be any domain having a nonzero prime ideal P; in other words, let R be any domain which is not a field. Let 0_E and 1_E be the images of $0 \in R$ and $1 \in R$ in $E = R/P$. Let $V = R \oplus E$ and let W be the zero submodule of V, i.e., $W = 0 \oplus 0_E$. Then the zero element in V is $(0, 0_E)$ and for any $r \in R$ we have: $r(1, 0_E) = (0, 0_E) \Rightarrow r = 0$; therefore $\operatorname{ann}_R(V/W)$ is the zero ideal in R which is prime and hence primary. But we can take $0 \neq t \in P$ and then $(0, 1_E) \in V \setminus W$ with $t(0, 1_E) = (0, 0_E) \in W$ but no power of t belongs to $\operatorname{ann}_R(V/W)$. Therefore W is not a primary submodule of V.

§2: GRADED RINGS AND HOMOGENEOUS IDEALS

We are now ready to formalize the concept of a graded ring. Recall (see L1§4) that an additive abelian monoid is a nonempty set I together with an element 0 and a binary operation $+$ which to every pair of elements i, j in I associates a third element $i + j$ such that for all i, j, k in I we have: $i + 0 = i$, $i + j = j + i$, and $(i + j) + k = i + (j + k)$. For instance, $\mathbb{Z}$ and $\mathbb{N}$ are additive abelian monoids, and for any positive integer d, so are $\mathbb{Z}^d$ and $\mathbb{N}^d$.

By a graded ring S over a ring R we mean an overring S of R together with a family $(S_i)_{i \in I}$ of R-submodules of S, where I is an additive abelian monoid (see L1§4), such that

$$S = \sum_{i \in I} S_i \quad \text{(internal direct sum of R-modules)}$$

with

$$S_0 = R$$

and such that for all $f_i \in S_i$ and $f_j \in S_j$ with i, j in I we have $f_i f_j \in S_{i+j}$. The family $(S_i)_{i \in I}$ is called the gradation (or grading) of S, and I is called the type of S. Elements of S_i are called homogeneous elements of degree i. Collectively, elements of $\cup_{i \in I} S_i$ are called homogeneous elements.

The direct sum assumption tells us that every $s \in S$ has a unique expression

$$s = \sum_{i \in I} s_i$$

with $s_i \in S_i$; the element s_i is called the homogeneous component of s of degree i, and collectively the elements s_i are called the homogeneous components of s. Note that $s_i = 0$ for almost all (i.e., all except a finite number of) i. Also note that we have Cauchy Multiplication, i.e., if t_j are the homogeneous components of $t \in S$, and $(st)_k$ are the homogeneous components of st, then

$$(st)_k = \sum_{i+j=k} s_i t_j.$$

An ideal J in S is said to be a homogeneous ideal if it satisfies one and hence all of the following three mutually equivalent conditions:

(i) $J = \sum_{i \in I} (J \cap S_i)$ (as additive groups).
(ii) The homogeneous components of every element of J belong to J.
(iii) J is generated (as an ideal in S) by homogeneous elements.

The unique decomposition $s = \sum_{i \in I} s_i$ yields (i) $\Leftrightarrow$ (ii). Obviously (ii) $\Rightarrow$ (iii). To prove (iii) $\Rightarrow$ (ii), let $(x_l)_{l \in L}$ be a family of generators of J with $x_l \in S_{d(l)}$. Any $y \in J$ can be expressed as $y = \sum_{l \in L} z_l x_l$ where $z_l \in S$ with $z_l = 0$ for almost all l. Writing $y = \sum_{i \in I} y_i$ where $y_i \in S_i$ with $y_i = 0$ for almost all i, and $z_l = \sum_{j \in I} z_{lj}$ where $z_{lj} \in S_j$ and $z_{lj} = 0$ for almost all j, we get

$$\sum_{i \in I} y_i = \sum_{l \in L} \left(\sum_{j \in I} z_{lj} x_l \right) = \sum_{i \in I} \left(\sum_{j+d(l)=i} z_{lj} x_l \right)$$

and hence

$$y_i = \sum_{j+d(l)=i} z_{lj} x_l \in J \quad \text{for all} \quad i \in I.$$

Thus we have shown that

(1)   conditions (i), (ii), (iii) are mutually equivalent.

As in the case of an element, we call S_i the homogeneous component of S of degree i, and more generally, for any homogeneous ideal J in S, we call $J \cap S_i$ the homogeneous component of J of degree i. We use i-th homogeneous component as a synonym for homogeneous component of degree i. Also we may say component instead of homogeneous component.

By a graded subring of S we mean a subring T of S such that T is a graded ring of type I over $T \cap R$ with gradation $(T \cap S_i)_{i \in I}$.

Given any other graded ring S' of type I over a ring R' with gradation $(S_i')_{i \in I}$, by a graded ring homomorphism $\phi : S \to S'$ we mean a ring homomorphism such that $\phi(S_i) \subset S_i'$ for all $i \in I$.

Clearly:

(2) $\begin{cases} \text{if } \phi : S \to S' \text{ is any graded ring homomorphism} \\ \text{then im}(\phi) \text{ is a graded subring of } S' \\ \text{and ker}(\phi) \text{ is a homogeneous ideal in } S. \end{cases}$

Conversely, given any homogeneous ideal J in S, upon letting $\psi : S \to S/J$ be the canonical epimorphism and $(S/J)_i = \psi(S_i)$, we can convert S/J into a graded ring of type I over $\psi(R)$ with grading $(S/J)_{i \in I}$, and then clearly $\psi : S \to S/J$ becomes a graded ring epimorphism.

Thus:

(3) $\quad \begin{cases} \text{if } J \text{ is any homogeneous ideal in any graded ring } S \\ \text{then the canonical epimorphism } \psi : S \to S/J \\ \text{is a graded ring epimorphism.} \end{cases}$

In view of (1) we see that:

(4) $\quad \begin{cases} \text{if } (J_l)_{l \in L} \text{ is any family of homogeneous ideals in } S, \\ \text{then the ideals } \sum_{l \in L} J_l \text{ and } \cap_{l \in L} J_l \text{ in } S \text{ are homogeneous,} \\ \text{and if the family is finite} \\ \text{then the ideal } \prod_{l \in L} J_l \text{ in } S \text{ is also homogeneous.} \end{cases}$

In view of (1) and (2) we see that:

(5) $\quad \begin{cases} \text{if } \phi : S \to S' \text{ is any graded ring epimorphism} \\ \text{then } J \mapsto \phi(J) \text{ gives a bijection of} \\ \text{the set of all homogeneous ideals in } S \text{ containing } \ker(\phi) \\ \text{onto the set of all homogeneous ideals in } S'. \end{cases}$

COMMENT (C1). [**Alternative Definition of Graded Rings**]. The assumption $S_0 = R$ implies that in the definition of a graded ring S it is not necessary to mention the subring R. At any rate, alternatively, a graded ring S of type I may be defined to be a ring S together with a family of additive subgroups $(S_i)_{i \in I}$, where the type I is an additive abelian monoid, such that we have the internal direct sum of additive abelian groups $S = \sum_{i \in I} S_i$ with $1 \in S_0$ (where 1 is the identity element of S) and such that for all $f_i \in S_i$ and $f_j \in S_j$ with i, j in I we have $f_i f_j \in S_{i+j}$. Note that in the alternative definition S_0 is a subring of S, and hence the alternative definition is equivalent to saying that, in the previous sense, S is a graded ring of type I over S_0. Also note that if the monoid I is cancellative (i.e., $i + j = i + k$ in $I \Rightarrow j = k$) then the condition that $1 \in S_0$ is automatically satisfied. To see this let 1_j be the homogeneous component of 1 of degree j, and for any $y \in S$ let y_i be the homogeneous component of y of degree i. Then $y_i = y_i 1 = \sum_{j \in I} y_i 1_j$ and hence by comparing homogeneous components of degree i we get $y_i = y_i 1_0$ because by cancellativeness $i + j = i \Rightarrow j = 0$. By summing over $i \in I$ we get $y = y 1_0$. Therefore $1 = 1_0 \in S_0$.

COMMENT (C2). [**Cancellative Monoids and Ordered Monoids**]. In a moment we shall show that if I is cancellative then for any homogeneous ideals J and H in S, the ideal $(J : H)_S$ is also homogeneous. To introduce another useful property of monoids, recall that (see L2§3) an ordered abelian group is an additive abelian group G with a linear order $\leq$ such that: $i \leq i'$ and $j \leq j'$ in $G \Rightarrow i+j \leq i'+ j'$; it follows that then: $i_1 = i'_1, \ldots, i_a = i'_a, j_1 < j'_1, \ldots, j_b < j'_b$ in G with $a \in \mathbb{N}$ and $b \in \mathbb{N}_+ \Rightarrow i_1 + \cdots + i_a + j_1 + \cdots + j_b < i'_1 + \cdots + i'_a + j'_1 + \cdots + j'_b$; note that $j < j'$ means $j \leq j'$ and $j \neq j'$. Analogously, the (additive) abelian monoid I is said to be ordered if there is a linear order $\leq$ on it such that: $i_1 = i'_1, \ldots, i_a = i'_a, j_1 < j'_1, \ldots, j_b < j'_b$ in I with $a \in \mathbb{N}$ and $b \in \mathbb{N}_+ \Rightarrow i_1 + \cdots + i_a + j_1 + \cdots + j_b < i'_1 + \cdots + i'_a + j'_1 + \cdots + j'_b$. In a moment we shall show that if I is ordered then the radical of any homogeneous ideal in S is again homogeneous. Note that if I is ordered then it is obviously cancellative.

COMMENT (C3). [**Integrally or Nonnegatively or Naturally Graded Rings**]. If $I \subset \mathbb{Z}$, i.e., if I is a submonoid of $\mathbb{Z}$, then we say that S is a subintegrally graded ring. If $I \subset \mathbb{Z}$ and $S_i = 0$ for all $i < 0$ in I, then we say that S is a nonnegatively graded ring. If $I = \mathbb{Z}$, then we say that S is an integrally graded ring. Finally, if $I = \mathbb{N}$, then we say that S is a naturally graded ring. Note that in all these four cases, I is an ordered abelian monoid.

§3 IDEAL THEORY IN GRADED RINGS

Again let I be an additive abelian monoid, and let S be a graded ring of type I over a ring R with gradation $(S_i)_{i \in I}$.

Given any $y \in S$ let y_i be the i-th homogeneous component of y, and let the support of y be defined by $\mathrm{supp}(y) = \{i \in I : y_i \neq 0\}$. Note that then

$$y = \sum_{i \in \mathrm{supp}(y)} y_i \quad \text{with} \quad 0 \neq y_i \in S_i.$$

We claim that:

$$(1) \qquad \begin{cases} \text{assuming } I \text{ to be cancellative,} \\ \text{for any homogeneous ideals } J \text{ and } H \text{ in } S, \\ \text{the ideal } (J : H)_S \text{ is homogeneous.} \end{cases}$$

Namely, for any generating set L of H we clearly have $(J : H)_S = \cap_{x \in L}(J : x)_S$ and hence, in view of (iii) and (4) §2, it suffices to show that, for any $j \in I$ and $x \in S_j$, the ideal $(J : x)_S$ is homogeneous. Given any $y \in S$ we get

$$\sum_{i \in \mathrm{supp}(y)} xy_i = xy \in J \quad \text{with} \quad xy_i \in S_{j+i}$$

and, by cancellativeness, for all $i \neq i'$ in $\mathrm{supp}(y)$ we have $j+i \neq j+i'$, and therefore the homogenity of J yields $xy_i \in J$ for all $i \in \mathrm{supp}(y)$; consequently, $y_i \in (J:x)_S$ for all $i \in \mathrm{supp}(y)$. Thus $(J:x)_S$ is homogeneous.

Next we claim that:

(2)
$$\begin{cases} \text{assuming } I \text{ to be ordered,} \\ \text{for any homogeneous ideal } J \text{ in } S, \\ \text{the ideal } \mathrm{rad}_S J \text{ is homogeneous.} \end{cases}$$

Namely, by induction on the size of $\mathrm{supp}(y)$ we shall show that $y \in \mathrm{rad}_S J \Rightarrow y_i \in \mathrm{rad}_S J$ for all $i \in \mathrm{supp}(y)$. If the size is zero then $y = 0$ and we have nothing to show. So let the size be positive and assume true for all smaller values of the size. Let d be the smallest element of $\mathrm{supp}(y)$. Since $y \in \mathrm{rad}_S J$, we have $y^n \in J$ for some $n \in \mathbb{N}_+$, and raising both sides of the displayed expansion of y to the n-th power we see that

$$y^n = y_d^n + \sum_{i \in A} u_i \quad \text{with} \quad y_d^n \in S_{nd} \quad \text{and} \quad u_i \in S_i$$

where A is a finite subset of $\{i \in I : i > nd\}$. Consequently by the homogenity of J we get $y_d^n \in J$ and hence $y_d \in \mathrm{rad}_S J$ and therefore $y - y_d \in \mathrm{rad}_S J$. Clearly $\mathrm{supp}(y - y_d) = \mathrm{supp}(y) \setminus \{d\}$ and $(y - y_d)_i = y_i$ for all $i > d$ in I, and hence we are done by the induction hypothesis.

To prepare for the next two claims, given any $z \in S$ we can write

$$z = \sum_{i \in \mathrm{supp}(z)} z_i \quad \text{with} \quad 0 \neq z_i \in S_i.$$

The usual criterion for a prime ideal may be sharpened by claiming that:

(3)
$$\begin{cases} \text{assuming } I \text{ to be ordered,} \\ \text{for any nonunit homogeneous ideal } J \text{ in } S, \text{ we have that:} \\ J \text{ is prime } \Leftrightarrow \text{ for all homogeneous elements } r, s \text{ in } S \\ \qquad \text{with } s \notin J \text{ and } rs \in J \text{ we have } r \in J. \end{cases}$$

To prove this, assume that for all homogeneous elements r, s in $S \setminus J$ we have $rs \notin J$. It suffices to show that then for all nonzero elements y, z in $S \setminus J$ we have $yz \notin J$. Let d and e be the smallest elements of $\mathrm{supp}(y)$ and $\mathrm{supp}(z)$ such that $y_d \notin J$ and $z_e \notin J$ respectively. Then $y_d z_e \in S_{d+e} \setminus J$ and $yz = x + y_d z_e + \sum_{i \in B} v_i$ with $x \in J$ and $v_i \in S_i$ for all $i \in B$ where B is a finite subset of $\{i \in I : i > d + e\}$. Therefore $yz \notin J$.

Likewise, the usual criterion for a primary ideal may be sharpened by claiming that:

(4) $\begin{cases} \text{assuming } I \text{ to be ordered,} \\ \text{for any nonunit homogeneous ideal } J \text{ in } S, \text{ we have that:} \\ J \text{ is primary } \Leftrightarrow \text{ for all homogeneous elements } r, s \text{ in } S \\ \qquad \text{with } s \notin J \text{ and } rs \in J \text{ we have } r \in \mathrm{rad}_S J. \end{cases}$

To prove this, assume that for all homogeneous elements r, s in S with $s \notin J$ and $rs \in J$ we have $r \in \mathrm{rad}_S J$. It suffices to show that then for all nonzero elements y, z in S with $z \notin J$ and $yz \in J$ we have $y \in \mathrm{rad}_S J$. Let e be the smallest element of $\mathrm{supp}(z)$ such that $z_e \notin J$. Upon replacing z by $z - \sum_{\{i \in \mathrm{supp}(z) : i < e\}} z_i$, without loss of generality we may assume that for the smallest element e of $\mathrm{supp}(z)$ we have $z_e \notin J$. Let the size of $\mathrm{supp}(y)$ be c, and label the elements of $\mathrm{supp}(y)$ as $d(1) < \cdots < d(c)$. By induction on b with $0 \le b \le c$ we shall show that $y_{d(i)} \in \mathrm{rad}_S J$ for $1 \le i \le b$. For $b = 0$ we have nothing to show. So let b be positive and assume true for all values of b smaller than the given one. Now $\sum_{1 \le i < b} y_{d(i)} \in \mathrm{rad}_S J$ and hence

$$\left(\sum_{1 \le i < b} y_{d(i)} \right)^m \in J$$

for some $m \in \mathbb{N}_+$. Let

$$x = \left(y - \sum_{1 \le i < b} y_{d(i)} \right)^m.$$

Then $xz \in J$ and

$$xz = y_{d(b)}^m z_e + \sum_{i \in C} w_i \quad \text{with} \quad y_{d(b)}^m z_e \in S_{md(b)+e} \quad \text{and} \quad w_i \in S_i$$

where C is a finite subset of $\{i \in I : i > md(b) + e\}$. Therefore $y_{d(b)}^m z_e \in J$ and hence $y_{d(b)} \in \mathrm{rad}_S J$. This completes the induction and establishes the claim.

We claim that;

(5) $\begin{cases} \text{assuming } I \text{ to be ordered,} \\ \text{and letting } (H)^* \text{ denote the ideal in } S \\ \text{generated by all homogeneous elements in } H \subset S, \\ \text{for any prime (resp: primary) ideal } J \text{ in } S, \\ \text{the ideal } (J)^* \text{ is also prime (resp: primary).} \end{cases}$

To see this, let J be any prime (resp: primary) ideal in S, and let there be given any homogeneous elements in S with $rs \in (J)^*$ and $s \notin (J)^*$. Now clearly r, s are elements in S with $rs \in J$ and $s \notin J$. Therefore, since J is prime (resp:primary), we get $r \in J$ (resp: $r^n \in J$ for some $n \in \mathbb{N}_+$). Since r is homogeneous, we must have $r \in (J)^*$ (resp: $r^n \in (J)^*$). By (3) (resp: (4)) It follows that $(J)^*$ is prime (resp: primary).

Recall that a primary decomposition of an ideal J in S is an expression of the form

$$J = Q_1 \cap \cdots \cap Q_n$$

where $Q_1, \ldots, Q_n$ are primary ideals in S with $n \in \mathbb{N}$, and this is an irredundant primary decomposition means the prime ideals $\mathrm{rad}_S Q_1, \ldots, \mathrm{rad}_S Q_n$ are pairwise distinct and for $1 \leq i \leq n$ we have

$$J \neq Q_1 \cap \cdots \cap Q_{i-1} \cap Q_{i+1} \cap \cdots \cap Q_n.$$

If the primary ideals $Q_1, \ldots, Q_n$ are all homogeneous then we call this a homogeneous primary decomposition or homogeneous irredundant primary decomposition respectively. We claim that:

(6) $\begin{cases} \text{assuming } I \text{ to be ordered,} \\ \text{if a homogeneous ideal } J \text{ in } S \text{ has a primary decomposition in } S \\ \text{then } J \text{ has a homogeneous irredundant primary decomposition in } S \\ \text{and all the associated primes of } J \text{ in } S \text{ are homogeneous.} \end{cases}$

To see this, letting $J = Q_1 \cap \cdots \cap Q_n$ be a primary decomposition in S, since an intersection of homogeneous ideals is homogeneous, by (5) we get a homogeneous primary decomposition

$$J = (Q_1)^* \cap \cdots \cap (Q_n)^*$$

in S. By L4§5(O8)(2$^\bullet$), in the last decomposition we can arrange that the prime ideals $\mathrm{rad}_S(Q_1)^*, \ldots, \mathrm{rad}_S(Q_n)^*$ are pairwise distinct, and then by discarding some of the $(Q_i)^*$ we can arrange that for $1 \leq i \leq n$ we have

$$J \neq (Q_1)^* \cap \cdots \cap (Q_{i-1})^* \cap (Q_{i+1})^* \cap \cdots \cap (Q_n)^*.$$

Now, in view of L4§6(T6) by (2) we see that all the associated primes of J in S are homogeneous.

COMMENT (C4). [**Positive Portion and Irrelevant Ideals**]. Assuming I to be ordered, we define the positive portion $\Omega(S)$ of S by putting

$$\Omega(S) = \sum_{\{i \in I : i > 0\}} S_i$$

and we note that this is an R-submodule of S, and if S is nonnegatively graded then it is actually a homogeneous ideal in S. Assuming I to be ordered, an ideal J in S is said to be irrelevant or relevant according as $\Omega(S) \subset \mathrm{rad}_S J$ or $\Omega(S) \not\subset \mathrm{rad}_S J$.

COMMENT (C5). [**Relevant Portion and Isolated System of Prime Ideals**]. Assuming S to be nonnegatively graded and noetherian, given any ideal J in S, let $J = Q_1 \cap \cdots \cap Q_n$ be an irredundant primary decomposition of J in S with $\text{rad}_S Q_i = P_i$. By a suitable labelling we may assume that $Q_1, \ldots, Q_m$ are relevant and $Q_{m+1}, \ldots, Q_n$ are irrelevant. It follows that $P_1, \ldots, P_m$ is an "isolated system of prime ideals" of J in S, i.e., for $1 \leq i \leq m$ and $m + 1 \leq j \leq n$ we have $P_j \not\subset P_i$, and hence by L4§6(T7) we see that the intersection $Q_1 \cap \cdots \cap Q_m$ depends only on J and not on the particular irredundant primary decomposition [cf. L5§6(E1)]. We define the relevant portion $(J)^\sharp$ of J by putting

$$(J)^\sharp = Q_1 \cap \cdots \cap Q_m.$$

If J is homogeneous then in view of (6) we can arrange $Q_1, \ldots, Q_n$ to be homogeneous and hence $(J)^\sharp$ is homogeneous.

COMMENT (C6). [**Homogeneous Rings and Semihomogeneous Rings**]. Examples of nonnegatively graded noetherian rings are provided by homogeneous rings over noetherian rings. By a semihomogeneous ring S over a ring R we mean a naturally graded ring S over R such that for its grading $(S_i)_{i \in I}$ with $I = \mathbb{N}$ we have $S = R[S_1]$ as a ring; this is clearly equivalent to saying that there is an R-epimorphism $R[(X_l)_{l \in L}] \to S$ of graded rings where $R[(X_l)_{l \in L}]$ is the polynomial ring in a family of variables over R which is naturally graded by putting $R[(X_l)_{l \in L}]_i = $ the set of all homogeneous polynomials of degree i (including the zero polynomial). By a homogeneous ring S over a ring R we mean a semihomogeneous ring S over R such that S_1 is a finitely generated R-module; this is clearly equivalent to saying that there is an R-epimorphism $R[X_1, \ldots, X_N] \to S$ of graded rings where $R[X_1, \ldots, X_N]$ is a finite variable polynomial ring over R which is naturally graded by putting $R[X_1, \ldots, X_N]_i = $ the set of all homogeneous polynomials of degree i (including the zero polynomial). Needless to say that a homogeneous ring over a noetherian ring is noetherian. Also needless to say that a homogeneous (resp: semihomogeneous) domain is a homogeneous (resp: semihomogeneous) ring which is a domain.

For a moment suppose that S is a noetherian semihomogeneous ring over R, let J be a homogeneous ideal in S, and let the notation be as in Comment (C5) above. By the noetherianness we can find $d_0 \in \mathbb{N}_+$ such that for all $d \geq d_0$ in $\mathbb{N}_+$ we have the inclusion $\Omega(S)^d \subset Q_i$ of homogeneous ideals for $m + 1 \leq i \leq n$, and hence we have the equality $S_d = Q_i \cap S_d$ for $m + 1 \leq i \leq n$, and therefore we have the equality $J \cap S_d = (J)^\sharp \cap S_d$. Instead of saying that there is $d_0 \in \mathbb{N}_+$ such that for all $d \geq d_0$ in $\mathbb{N}_+$ we have $J \cap S_d = (J)^\sharp \cap S_d$, for brevity we may say that $J \cap S_d = (J)^\sharp \cap S_d$ for ALL LARGE ENOUGH $d \in \mathbb{N}_+$. Now let H be a homogeneous ideal in S such that $H \cap S_d = (J)^\sharp \cap S_d$ for all large enough $d \in \mathbb{N}_+$. Then we can take $e \in \mathbb{N}_+$ such that $H \cap S_d = (J)^\sharp \cap S_d = J \cap S_d$ for all $d \geq e$ in $\mathbb{N}_+$. Given any $i \in \{1, \ldots, m\}$

we have $\Omega(S) \not\subset P_i$ and hence there exists $t(i) \in S_1 \setminus P_i$; for any $y \in H \cap S_l$ with $l \in \mathbb{N}$, we have $yt(i)^e \in H \cap S_{e+l} = (J)^\sharp \cap S_{e+l} = J \cap S_{e+l}$ with $t(i)^e \notin P_i$ and hence $y \in Q_i$; this being so for $1 \le i \le m$ we get $y \in (J)^\sharp$; therefore by the homogenity of H we get $H \subset (J)^\sharp$. Applying the same result to $(H)^\sharp$ we get $(H)^\sharp \subset (J)^\sharp$. By symmetry we obtain $(J)^\sharp \subset (H)^\sharp$. Therefore $(H)^\sharp = (J)^\sharp$.

Thus we have shown that:

(7)
$$\begin{cases} \text{assuming } S \text{ to be a noetherian semihomogeneous ring over } R, \\ \text{for any homogeneous ideal } J \text{ in } S \text{ we have} \\ J \cap S_d = (J)^\sharp \cap S_d \text{ for all large enough } d \in \mathbb{N}_+, \\ \text{and moreover } (J)^\sharp \text{ is the largest such homogeneous ideal,} \\ \text{i.e., if } H \text{ is any homogeneous ideal in } S \text{ such that} \\ H \cap S_d = (J)^\sharp \cap S_d \text{ for all large enough } d \in \mathbb{N}_+ \\ \text{then } H \subset (J)^\sharp \text{ and } (H)^\sharp = (J)^\sharp. \end{cases}$$

Again suppose that S is a noetherian semihomogeneous ring over R, let J be a homogeneous ideal in S, and let the notation be as in Comment (C5) above. Now $(J : \Omega(S)^d)_S$ is an increasing sequence of ideals in S and hence by the noetherianness of S we can find $e \in \mathbb{N}_+$ such that $(J : \Omega(S)^d)_S = (J : \Omega(S)^e)_S$ for all $d \ge e$ in $\mathbb{N}_+$. Moreover, by the said noetherianness, for all large enough $d \in \mathbb{N}_+$ we have $\Omega(S)^d \subset Q_i$ for $m + 1 \le i \le n$ and hence $(Q_i : \Omega(S)^d)_S = S$ for $m + 1 \le i \le n$. For $i \in \{1, \ldots, m\}$ we have $\Omega(S) \not\subset P_i$ and hence there exists $t(i) \in S_1 \setminus P_i$ and therefore for all $d \in \mathbb{N}_+$ we get $Q_i \subset (Q_i : \Omega(S)^d)_S \subset (Q_i : t(i)^d)_S \subset Q_i$ and hence $(Q_i : \Omega(S)^d)_S = Q_i$. Since taking colon clearly commutes with finite intersections, for all large enough $d \in \mathbb{N}_+$ we have $(J : \Omega(S)^d)_S = (Q_1 \cap \cdots \cap Q_n : \Omega(S)^d)_S = (Q_1 : \Omega(S)^d)_S \cap \cdots \cap (Q_n : \Omega(S)^d)_S = Q_1 \cap \cdots \cap Q_m = (J)^\sharp$.

Thus we have shown that:

(8)
$$\begin{cases} \text{assuming } S \text{ to be a noetherian semihomogeneous ring over } R, \\ \text{for any homogeneous ideal } J \text{ in } S \text{ we have} \\ (J)^\sharp = (J : \Omega(S)^d)_S \text{ for all large enough } d \in \mathbb{N}_+. \end{cases}$$

In the following two claims (9) and (10), for any F in the polynomial ring $R[X_1, \ldots, X_N]$ in a finite number of indeterminates $X_1, \ldots, X_N$, we shall write $F = \sum F_i$ where F_i is a homogeneous polynomial of degree i (or the zero polynomial).

We claim that:

(9) $\begin{cases} \text{assuming } S \text{ to be a semihomogeneous ring over } R, \\ \text{and } x_1, \ldots, x_N \text{ to be a finite number of elements in } S_1, \\ \text{for any homogeneous ideal } J \text{ in } S \text{ we have that:} \\ F \in R[X_1, \ldots, X_N] \text{ with } F(x_1, \ldots, x_N) \in J \\ \Rightarrow F(tx_1, \ldots, tx_N) \in J \text{ for all } t \in R. \end{cases}$

Namely,

(') $\begin{cases} F(x_1, \ldots, x_N) = \sum F_i(x_1, \ldots, x_N) \text{ with } F_i(x_1, \ldots, x_N) \in S_i \\ \text{and} \\ F(tx_1, \ldots, tx_N) = \sum F_i(x_1, \ldots, x_N) t^i \end{cases}$

and so the homogenity of J tells us that: $F(x_1, \ldots, x_N) \in J \Rightarrow F(tx_1, \ldots, tx_N) \in J$ for all $t \in R$.

As a partial converse to (9) we claim that:

(10) $\begin{cases} \text{assuming } S \text{ to be a homogeneous ring over an infinite field } R, \\ \text{and } x_1, \ldots, x_N \text{ to be a finite number of } R\text{-generators of } S_1, \\ \text{if } J \text{ is an ideal in } S \text{ such that:} \\ F \in R[X_1, \ldots, X_N] \text{ with } F(x_1, \ldots, x_N) \in J \\ \Rightarrow F(tx_1, \ldots, tx_N) \in J \text{ for all } t \in R, \\ \text{then } J \text{ is homogeneous.} \end{cases}$

Namely, in view of (') it suffices to show that, for any $F \in R[X_1, \ldots, X_N]$, the (finite-dimensional) R-vector-subspace U of S generated by $F(tx_1, \ldots, tx_N)_{t \in R}$ coincides with the R-vector-subspace V of S generated by $F_i(x_1, \ldots, x_N)_{i \in \mathbb{N}}$. Clearly V is of some finite dimension ν and by (') we have $U \subset V$. If $U \neq V$ then we can take a basis $z_1, \ldots, z_\nu$ of V such that $z_1, \ldots, z_\mu$ is a basis of U for some $\mu < \nu$, and writing any $v \in V$ as $v = a_1 z_1 + \cdots + a_\nu z_\nu$ with $a_1, \ldots, a_\nu$ in R, we would get a linear map $V \to R$ given by $v \mapsto a_\nu$ which is zero on U but not on V. Thus it suffices to show that if $\phi : V \to R$ is a linear map which is zero on U, it must also be zero on V. So given any linear map $\phi : V \to R$ which is zero on U, let $c_i = \phi(F_i(x_1, \ldots, x_N))$. Suppose if possible that $c_i \neq 0$ for some i and let δ be maximum with $c_\delta \neq 0$. Then $\delta \in \mathbb{N}$ and

$$G(Y) = \sum c_i Y^i$$

is a univariate (nonzero) polynomial of degree δ. Since $|R| > \delta$, we can find $t \in R$ with $G(t) \neq 0$. But by (') we have $G(t) = \phi(F(tx_1, \ldots, tx_N))$ and by the zeroness of ϕ on U we get $\phi(F(tx_1, \ldots, tx_N)) = 0$ which is a contradiction. Therefore $c_i = 0$ for all i and hence ϕ is zero on V.

§4: ADVICE TO THE READER

In L4 we started a new column called Problems (P1), (P2), ... In the next section we shall start a new column called Quests (Q1), (Q2), ... In the last two sections we have started labelling Comments as (C1), (C2), ... The only place where we previously used the letter C as a designation was when in L1(R7) we labelled assertions involving algebraic closure as (C1) to (C5), and later on while referring to those assertions. At any rate, having used up the name Remarks almost to mean subsections, we are calling as Comments what are usually called Remarks. The present "Advice" is an addendum to the "Advice" given in Lecture L1. Also everything said in the "Advice" given in Lecture 4 is applicable here with double force to the next very very very long section.

§5: MORE ABOUT IDEALS AND MODULES

In L4§5, where the items were labelled as Observations (O7) to (O30), we gave the definitions of several concepts and the proofs of several results about Ideals and Modules. The matter was enhanced in L4§6 and L4§7 and in §1 to §3 of the present Lecture. We shall now continue this project, but this time we shall label the items as Quests (Q1), (Q2), ... In particular we shall prove Theorems (T19) to (T30) stated in §§8-9 of L4. For a motivating summary of this very very long section see the next Lecture L6 where we shall summarize Lectures L1 to L5. In a first reading you may prefer to look through the appropriate part of L6 rather than ploughing through the details of this enormous section.

QUEST (Q1) Nilpotents and Zerodivisors in Noetherian Rings

As shown in L4§6, any ideal I in a noetherian ring R has an irredundant primary decomposition $I = Q_1 \cap \cdots \cap Q_e$ where $Q_1, \ldots, Q_e$ are primary ideals whose radicals $P_1, \ldots, P_e$ are distinct prime ideals which depend only on I and not on the particular decomposition. After suitable labelling we may assume that $P_1, \ldots, P_d$ are the minimal primes of I, and $P_{d+1}, \ldots, P_e$ are its embedded primes, i.e., for $1 \leq i \leq e$ we have: $P_j \subset P_i$ for some $j \neq i \Leftrightarrow i > d$. Then, as shown in L4§6,

(1)
$$\begin{cases} \operatorname{ass}_R(R/I) = \operatorname{tass}_R(R/I) = \{P_1, \ldots, P_e\} \\ \text{and} \\ \operatorname{nvspec}_R I = \operatorname{nass}_R(R/I) = \{P_1, \ldots, P_d\} \end{cases}$$

where for the definitions of the terms ass, tass, nass, and nvspec we refer to (O16) and (O17) of L4§5. Let $\phi : R \to R/I$ be the canonical epimorphism, and let $N_R(R/I)$ be the set of all $x \in R$ such that $\phi(x)$ is nilpotent. Then clearly

(2) $$N_R(R/I) = \mathrm{rad}_R I = P_1 \cap \cdots \cap P_d.$$

As a companion to this result, in a moment we shall show that

(3) $$Z_R(R/I) = P_1 \cup \cdots \cup P_e$$

where $Z_R(R/I)$ is the set of all zerodivisors mod I as defined in L4§5(O20). Actually, we shall prove the following more general result:

ZERODIVISOR THEOREM (T1). For any submodule U of any module V over a noetherian ring R, we have

(1') $$\begin{cases} Z_R(V/U) = \cup_{P \in \mathrm{tass}_R(V/U)} P \\ \text{with} \\ \mathrm{tass}_R(V/U) = \emptyset \Leftrightarrow U = V. \end{cases}$$

Moreover, assuming the module V to be finitely generated, there is an irredundant primary decomposition $U = Q_1 \cap \cdots \cap Q_e$ where $Q_1, \ldots, Q_e$ are primary submodules whose radicals $P_1, \ldots, P_e$ are distinct prime ideals in R, which are labelled so that $P_1, \ldots, P_d$ are minimal and $P_{d+1}, \ldots, P_e$ are embedded; with such labelling we have

(2') $$\begin{cases} \mathrm{ass}_R(V/U) = \mathrm{tass}_R(V/U) = \{P_1, \ldots, P_e\} \\ \text{and} \\ \mathrm{nvspec}_R(\mathrm{ann}_R(V/U)) = \mathrm{nass}_R(V/U) = \{P_1, \ldots, P_d\}. \end{cases}$$

PROOF. It suffices to prove the first assertion; the second assertion starting with "Moreover" follows from L4§6.1. Clearly: $U = V \Rightarrow Z_R(V/U) = \emptyset = \mathrm{tass}_R(V/U)$. So assume that $U \neq V$. Then the set of all ideals in R of the form $(U : a)_R$ with $a \in V \setminus U$ is a nonempty set of nonunit ideals. Since R is noetherian, the said set contains a maximal member P. Now $P = (U : a)_R$ for some $a \in V \setminus U$. Given any $r \in R$ and $s \in R \setminus P$ with $rs \in P$, we have $sa \in V \setminus U$ and $P \subset (U : sa)_R$, and hence by the maximality of P we must have $P = (U : sa)_R$; now by the equation $r(sa) \in U$ we get $r \in P$. Thus P is prime and hence P belongs to $\mathrm{tass}_R(V/U)$ which shows that the latter is nonempty.

Clearly for every P in $\mathrm{tass}_R(V/U)$ we have $P = (U : a)_R$ for some $a \in V \setminus U$, and hence $P \subset Z_R(V/U)$. Conversely, given any $s \in Z_R(V/U)$ we have $sb \in U$ for some $b \in V \setminus U$, and by what we have just proved we have $\mathrm{tass}_R((U + Rb)/U) \neq \emptyset$, and hence we can find $r \in R$ together with $P \in \mathrm{spec}(R)$ such that $P = (U : c)_R$ where $c = rb$; now $c \in V \setminus U$ and hence $P \in \mathrm{tass}_R(V/U)$; also $sc \in U$ and hence $s \in P$.

COROLLARY (T2). In the situation of (T1), assuming V to be finitely generated, for any ideal J in R we have that;

$$J \not\subset Z_R(V/U) \Leftrightarrow (U : J)_V = U \Leftrightarrow J \not\subset P_i \text{ for } 1 \leq i \leq e.$$

PROOF. In view of L4§5(O24)(22•), this follows from (T1).

COMMENT (C7). [**Maximality and Conditions for Prime Ideals**]. From the above proof of (T1) as well as from several other instances such as L2§5(R5) and L4§5(O20)(12•), a common message is that a maximal element in a set of ideals satisfying certain conditions tends to be prime. In checking the primeness of a nonunit ideal P in a ring R, we can use one of the following three conditions, each of which is easily seen to be equivalent to the primeness of P: (i) $x \in R$ and $y \in R \setminus P$ with $xy \in P \Rightarrow x \in P$; (ii) I and J are ideals in R with $IJ \subset P$ and $J \not\subset P \Rightarrow I \subset P$; (iii) I and J are ideals in R which contain P but are different from $P \Rightarrow IJ$ is not contained in P. Out of this (i) was stated as (A2) in L1§10. As an exercise, you may shorten the proof of L4§5(O20)(12•) by using (ii) or (iii) instead of (i). Note that L2§5(R5) corresponds to the $S = \{1\}$ case of L4§5(O20)(12•).

COMMENT (C8). [**Embedded Prime and Primary Components**]. As said above, in L4§6 we have shown that every ideal I in a noetherian ring R has an irredundant (finite) primary decomposition $I = Q_1 \cap \cdots \cap Q_e$ where $Q_1, \ldots, Q_e$ are primary ideals whose radicals $P_1, \ldots, P_e$ are distinct prime ideals which depend only on I and not on the particular decomposition. Moreover, after suitable labelling we may assume that $P_1, \ldots, P_d$ are the minimal primes of I, and $P_{d+1}, \ldots, P_e$ are its embedded primes, i.e., for $1 \leq i \leq e$ we have $P_j \subset P_i$ for some $j \neq i \Leftrightarrow i > d$, and then, as shown in L4§6, the primaries $Q_1, \ldots, Q_d$ are uniquely determined by I. As an application of the Krull Intersection Theorem, in Theorem (T13) of (Q4) we shall show that the primaries $Q_{d+1}, \ldots, Q_e$ are NOT determined by I. Actually, from the Krull Intersection Theorem we shall first deduce Theorem (T11) on the intersection of all primaries belonging to the same prime, and then deduce (T13) from (T11). Moreover, from (T11) we shall also deduce Theorem (T12) which says that the members of $\mathrm{ass}_R(R/I)$ can be any preassigned finite set of primes which are not minimal primes of 0, and hence in a domain they can be the members of any finite set of nonzero primes. This shows the existence of ideals with embedded primes in any noetherian ring of dimension at least two.

QUEST (Q2) Faithful Modules and Noetherian Conditions

Let D be a module over a ring R. We say that D is a faithful module to mean that $\mathrm{ann}_R D = 0$. As noted in L4§5(O12), D may be regarded as a module over the residue class ring $R/(0 : D)_R$, and then it is clearly faithful. In a moment we shall show that if D is a noetherian module over R then the ring $R/(0 : D)_R$ must be noetherian. First note that by the Isomorphism Theorems:

(1)
$$\begin{cases} \text{if } C \text{ is any submodule of } D \text{ then:} \\ D \text{ is noetherian} \Leftrightarrow C \text{ and } D/C \text{ are noetherian.} \end{cases}$$

In view of (1), by induction on n we see that:

(2)
$$\begin{cases} \text{if } D \text{ is the (internal or external) direct sum } D_1 \oplus \cdots \oplus D_n \\ \text{of a finite number of modules } D_1, \ldots, D_n \text{ then:} \\ D \text{ is noetherian} \Leftrightarrow D_1, \ldots, D_n \text{ are noetherian.} \end{cases}$$

Obviously:

(3)
$$\begin{cases} \text{if } v_1, \ldots, v_n \text{ are any finite number of generators of } D \\ \text{then upon letting } \alpha_i : R \to D_i = Rv_i \\ \text{be the } R\text{-epimorphism given by } 1 \mapsto v_i \text{ we have that:} \\ \text{the diagonal direct sum } \alpha_1 \oplus \cdots \oplus \alpha_n : R \to D_1 \oplus \cdots \oplus D_n \\ \text{is an } R\text{-homomorphism with kernel } (0 : D)_R. \end{cases}$$

If D is noetherian then it is generated by a finite number of elements $v_1, \ldots, v_n$, and hence by (2) and (3) (and because submodules and homomorphic images of noetherian modules are noetherian) we see that:

(4)   if the module D is noetherian then so is the module $R/(0 : D)_R$.

If J is an ideal in R then clearly: the R-module R/J is noetherian $\Leftrightarrow$ the ring R/J is noetherian; consequently by (4) we conclude that:

(5)   if the module D is noetherian then the ring $R/(0 : D)_R$ is noetherian.

QUEST (Q3) Jacobson Radical, Zariski Ring, and Nakayama Lemma

The Jacobson radical of a ring R is defined by putting

$$\mathrm{jrad}(R) = \bigcap_{P \in \mathrm{mspec}(R)} P$$

and we observe that clearly:

(1) $\mathrm{jrad}(R) = R \Leftrightarrow \mathrm{mspec}(R) = \emptyset \Leftrightarrow \mathrm{spec}(R) = \emptyset \Leftrightarrow R = 0.$

By a Zariski ring we mean a pair (R, J) where R is a noetherian ring and J is an ideal in R with $J \subset \mathrm{jrad}(R)$. This generalizes the idea of a local ring, since for any local ring R, the pair $(R, M(R))$ is clearly a Zariski Ring. After proving the Krull

Intersection Theorem in (Q4), in (Q5) we shall use it to give a Characterization Theorem for Zariski rings.

As a preliminary to the said characterization, for any ideal J in any ring R we put

$$1 - J = \{1 - x : x \in J\}$$

and we note that

(2)   if $J \subset \mathrm{jrad}(R)$ then every element of $1 - J$ is a unit in R

and

(3) $\begin{cases} \text{if } V \text{ is a finitely generated ideal in } R \text{ with } JV = V \\ \text{then } (1 - J) \cap (\mathrm{ann}_R V) \neq \emptyset \end{cases}$

and more generally

(4) $\begin{cases} \text{if } V \text{ is a finitely generated module over } R \text{ with } JV = V \\ \text{then } (1 - J) \cap (\mathrm{ann}_R V) \neq \emptyset. \end{cases}$

To prove Claim (2), if $x \in R$ is such that $1 - x$ in a nonunit in R then we can find a maximal ideal P in R with $1 - x \in P$, and now we must have $x \notin \mathrm{jrad}(R)$ because otherwise we would get $x \in P$ which would yield the contradiction $1 \in P$. Claim (3) follows by taking $t = 1$ in the HINT to L4§10(E2.1), and noting that for the $n \times n$ matrix A of the said HINT we have $\det(A) \in \mathrm{ann}_R V$ and $\det(A) = 1 - x$ with $x \in J$. Claim (4) also follows from the same HINT, in view of the following:

COMMENT (C9). [**Cramer's Rule for Modules**]. Cramer's Rule given in Exercise (E4) of L4§10 remains verbatim valid by letting the elements X_j, Y_i of that Exercise belong to an R-module.

In view of (2), by taking V/U for V in (4) we get the very useful:

NAKAYAMA LEMMA (T3). If U is a submodule of a finitely generated module V over a ring R such that $V = U + JV$ for an ideal J in R with $J \subset \mathrm{jrad}(R)$, then $V = U$.

QUEST (Q4) Krull Intersection Theorem and Artin-Rees Lemma

In Theorem (T24.1) of L4§8.3 we asserted that if R is local then $\cap_{n \in \mathbb{N}} M(R)^n = 0$. This is the famous Intersection Theorem of Krull which we now proceed to prove. Heuristically it says that, in a good situation, a function (such as the one defined by a power series) which is not identically zero has a definite order. Actually we shall give a modification of Krull's original proof by means of the so called:

ARTIN-REES LEMMA (T4). Given any ideals V, W, J in a noetherian ring R, there exists $a \in \mathbb{N}$ such that for all $c \geq b \geq a$ in $\mathbb{N}$ we have

$$J^c W \cap V = J^{c-b}(J^b W \cap V).$$

PROOF. Take a finite set of generators $x_1, \ldots, x_N$ of J. Let $S = R[X_1, \ldots, X_N]$ be the N variable polynomial ring regarded as naturally graded with grading $(S_i)_{i \in \mathbb{N}}$ where S_i is the set of all homogeneous polynomials of degree i (including the zero polynomial). For every $i \in \mathbb{N}$ we have an R-epimorphism $\phi_i : S_i \to J^i$ given by $f(X_1, \ldots, X_N) \mapsto f(x_1, \ldots, x_N)$. Since $J^i W \cap V \subset J^i$, upon letting $H_i = \phi_i^{-1}(J^i W \cap V)$ we get $H_i \subset S_i$ with $\phi_i(H_i) = J^i W \cap V$. By Hilbert Basis Theorem (see L4§3(T3)), S is noetherian and hence we can find $a \in \mathbb{N}$ and a finite number of elements $(y_{i,j})_{1 \leq j \leq d_i}$ in H_i for $0 \leq i \leq a$ with $d_i \in \mathbb{N}$ such that every element in $\cup_{e \in \mathbb{N}} H_e$ is an S-linear combination of the elements $(y_{i,j})_{0 \leq i \leq a, 1 \leq j \leq d_i}$ [cf. L5§6(E2)]. Given any $c \geq b \geq a$ in $\mathbb{N}$ and any $z \in H_c$ we can write

$$z = \sum_{0 \leq i \leq a, 1 \leq j \leq d_i, 0 \leq l \leq e} t_{i,j,l} y_{i,j} \quad \text{with} \quad t_{i,j,l} \in S_l$$

for some $e \geq c$ in $\mathbb{N}$. Equating like degree terms we get

$$z = \sum_{0 \leq i \leq a, 1 \leq j \leq d_i} t_{i,j,c-i} y_{i,j}$$

and hence

$$\phi_c(z) \in \left(\sum_{0 \leq i \leq a} J^{c-i}(J^i W \cap V) \right) \subset J^{c-b}(J^b W \cap V).$$

It follows that $J^c W \cap V \subset J^{c-b}(J^b W \cap V)$. Since the other inclusion is obvious, we get $J^c W \cap V = J^{c-b}(J^b W \cap V)$.

ARTIN-REES COROLLARY (T5). Given any ideals W and J in a noetherian ring R, upon letting

$$V = \bigcap_{n \in \mathbb{N}} J^n W$$

we have

$$J^m V = V \quad \text{for all} \quad m \in \mathbb{N}.$$

PROOF. In (T4) take $m = c - b$.

KRULL INTERSECTION LEMMA (T6). Given any ideals W and J in a noetherian ring R, upon letting

$$V = \bigcap_{n \in \mathbb{N}} J^n W$$

we have that:

$$V = 0 \Leftrightarrow (1 - J) \cap Z_R(W) = \emptyset.$$

PROOF. By (Q3)(3) and (T5) we get $V \neq 0 \Rightarrow (1 - J) \cap Z_R(V) \neq \emptyset$ and obviously $Z_R(V) \subset Z_R(W)$. Conversely suppose there exists $x \in J$ and $0 \neq y \in W$ with $(1 - x)y = 0$. Then $xy = y$ and hence by induction on n we see that $x^n y = y$ for all $n \in \mathbb{N}$. Therefore $y \in \cap_{n \in \mathbb{N}} J^n W = V$ and hence $V \neq 0$.

KRULL INTERSECTION THEOREM (T7). For any local ring R we have

$$\bigcap_{n \in \mathbb{N}} M(R)^n = 0.$$

PROOF. In view of (Q3)(2), this follows from (T6) by taking $W = R$ and $J = M(R)$.

CLOSEDNESS COROLLARY (T8). For any ideals I and J in a noetherian ring R, in the notation of (1) and (3) of (Q1) we have that:

$$\bigcap_{n \in \mathbb{N}} (I + J^n) = I \Leftrightarrow P_i + J \neq R \text{ for } 1 \leq i \leq e.$$

PROOF. Taking $W = R$ in (T6) we see that:

$(*)$ $$\bigcap_{n \in \mathbb{N}} J^n = 0 \Leftrightarrow (1 - J) \cap Z_R(R) = \emptyset,$$

Taking R/I in $(*)$ we see that:

$(**)$ $$\bigcap_{n \in \mathbb{N}} (I + J^n) = I \Leftrightarrow (1 - J) \cap Z_R(R/I) = \emptyset.$$

Now in view of $(**)$ our assertion follows from (1) and (3) of (Q1).

CLOSURE COROLLARY (T9). Let J be any ideal in any noetherian ring R. For any ideal I in R let us put $\overline{I} = \cap_{n \in \mathbb{N}} (I + J^n)$. Then we have the following.

(T9.1) Let $I = Q_1 \cap \cdots \cap Q_e$ be any primary decomposition (which need not be irredundant and which exists by L4§6) where $Q_1, \ldots, Q_e$ are a finite number of primary ideals whose radicals $P_1, \ldots, P_e$ in R are prime ideals labelled so that $P_i + J \neq R$ for $1 \leq i \leq d$ and $P_i + J = R$ for $d + 1 \leq i \leq e$. Then upon letting

$A = Q_1 \cap \cdots \cap Q_d$ and $B = Q_{d+1} \cap \cdots \cap Q_e$ we have $I = A \cap B$ with $\overline{I} = \overline{A} = A$ and $\overline{B} = R$.

(T9.2) Let $I_1, \ldots, I_h$ be any finite number of ideals in R such that $I = I_1 \cap \cdots \cap I_h$. Then $\overline{I} = \overline{I}_1 \cap \cdots \cap \overline{I}_h$.

PROOF. To see that (T9.2) follows from (T9.1), for $1 \leq j \leq h$ we can take a primary decomposition $I_j = Q_{j1} \cap \cdots \cap Q_{je(j)}$ where $Q_{j1}, \ldots, Q_{je(j)}$ are a finite number of primary submodules of V whose radicals are prime ideals $P_{j1}, \ldots, P_{je(j)}$ in R labelled so that $P_{ji} + J \neq R$ for $1 \leq i \leq d(j)$ and $P_{ji} + J = R$ for $d(j) + 1 \leq i \leq e(j)$. Now applying (T9.1) to I as well as I_j we get

$$\overline{I} = \bigcap_{1 \leq j \leq h} \bigcap_{1 \leq i \leq d(j)} Q_{ji} \quad \text{and} \quad \overline{I}_j = \bigcap_{1 \leq i \leq d(j)} Q_{ji}$$

and hence $\overline{I} = \overline{I}_1 \cap \cdots \cap \overline{I}_h$.

To prove (T9.1), first note that clearly $I = A \cap B$. By (T8) we have $\overline{Q}_i = Q_i$ for $1 \leq i \leq d$, and hence

$$\overline{A} = \bigcap_{n \in \mathbb{N}} \left(\left(\bigcap_{1 \leq i \leq d} Q_i \right) + J^n \right) \subset \bigcap_{n \in \mathbb{N}} \left(\bigcap_{1 \leq i \leq d} (Q_i + J^n) \right) = A.$$

But obviously $A \subset \overline{A}$, and hence $\overline{A} = A$.

By L4§5(O26) we see that $B + J^n = R$ for all $n \in \mathbb{N}$, and hence $\overline{B} = R$. For all $n \in \mathbb{N}$ we have

$$A + J^n = AR + J^n = A(B + J^n) + J^n = AB + J^n \subset I + J^n$$

and therefore by intersecting as n varies over $\mathbb{N}$ we get $\overline{A} \subset \overline{I}$. But clearly $\overline{I} \subset \overline{A}$, and hence $\overline{I} = \overline{A}$.

DOMAINIZED KRULL INTERSECTION THEOREM (T10). For any nonunit ideal J in any noetherian domain R we have

$$\bigcap_{n \in \mathbb{N}} J^n = 0.$$

Moreover, if $J \neq 0$ then for all $a \neq b$ in $\mathbb{N}$ we have $J^a \neq J^b$.

PROOF. In view of (Q3)(2), the first assertion follows from (T6) by taking $W = R$. To prove the second assertion, assume that $J \neq 0$. Now if $J^a = J^b$ for some $a \neq b$ in $\mathbb{N}$, say with $a < b$, then by induction on m we would get $J^{a+m(b-a)} = J^a$ for all $m \in \mathbb{N}$, and by the first assertion this would tell us that $J^a = 0$, which would be a contradiction.

INTERSECTION OF PRIMARIES THEOREM (T11). For any prime ideal P in a noetherian ring R, the intersection of all P-primary ideals in R equals $[0 : P]_R$.

COMMENT (C10). [**Power of Krull Intersection**]. Note that, upon letting $0 = Q_1^* \cap \cdots \cap Q_\epsilon^*$ be an irredundant primary decomposition of 0 in R labelled so that for the radicals $P_1^*, \ldots, P_\epsilon^*$ of $Q_1^*, \ldots, Q_\epsilon^*$ we have $P_i^* \subset P$ for $1 \le i \le \delta$ and $P_i^* \not\subset P$ for $\delta + 1 \le i \le \epsilon$, by L4§5(O19) we have $[0 : P]_R = P^*$ where $P^* = Q_1^* \cap \cdots \cap Q_\delta^*$. Also note that, for the canonical homomorphism $\Phi : R \to R_P$ = the localization of R at P, by L4§7.1 we have $[0 : P]_R = \ker(\Phi)$. Finally note that a weaker version of (T11) was proved in L4§7.1(T17)(j) where, upon letting $\widetilde{P}$ = the intersection of all primary ideals in R which are contained in P, it was shown that $\widetilde{P} = P^*$. On the other hand, (T11) says that, letting $\widehat{P}$ = the intersection of all P-primary ideals in R, we have $\widehat{P} = P^*$. The said result is weaker than (T11) because obviously $\widetilde{P} \subset \widehat{P}$. This shows the power of the Krull Intersection Theorem.

PROOF OF (T11). Let H be the set of all P-primary ideals in R, let H' be the set of all $M(R_P)$-primary ideals in R_P, and let $\Phi : R \to R_P$ be the canonical homomorphism. Then by L4§7.1(T16)(g) we see that $Q' \mapsto \Phi^{-1}(Q')$ gives a bijection $H' \to H$, and $\Phi^{-1}(\cap_{Q' \in H'} Q') = \cap_{Q \in H} Q$. By L4§7.1(T16)(i) we know that R_P is a local ring and hence by (T7) we get $\cap_{n \in N} M(R_P)^n = 0$. By L4§5(O8), every positive power of $M(R_P)$ belongs to H', and by noetherianness every member of H' contains a positive power of $M(R_P)$. Therefore $\cap_{Q' \in H'} Q' = 0$. Consequently $\ker(\Phi) = \cap_{Q \in H} Q$, and as noted above $[0 : P]_R = \ker(\Phi)$.

EXISTENCE OF EMBEDDED PRIMES THEOREM (T12). Let $P_1, \ldots, P_e$ with $e > 0$ be any finite number of distinct prime ideals in a noetherian ring R which are all of positive height, i.e., none of which is a minimal prime of 0 in R. Then we have the following.

(T12.1) There exists an ideal I in R whose associated primes are exactly the given ideals $P_1, \ldots, P_e$.

(T12.2) Assume that $e > 1$ with $P_e \not\subset P_i$ for $1 \le i \le e - 1$, and let $J = Q_1 \cap \cdots \cap Q_{e-1}$ be an irredundant primary decomposition in R such that the radicals of $Q_1, \ldots, Q_{e-1}$ are $P_1, \ldots, P_{e-1}$ respectively. Then there exists a (P_e)-primary ideal Q_e in R with $J \not\subset Q_e$. Moreover, for any such Q_e, upon letting $I = J \cap Q_e$, we have that the primary decomposition $I = Q_1 \cap \cdots \cap Q_e$ is irredundant. [In (T12.2) the positive height assumption for P_e is not needed]. [cf. L5§6(E3)].

PROOF. (T12.1) is obvious for $e = 1$ because then we can take $I = P_1$. Therefore (T12.1) follows from (T12.2) by induction on e. To prove (T12.2), first note that because $P_1, \ldots, P_{e-1}$ are of positive height, by (T11) there exists a (P_e)-primary ideal Q_e in R with $J \not\subset Q_e$. It only remains to show that $Q_1 \cap \cdots \cap Q_{i-1} \cap Q_{i+1} \cap \cdots \cap Q_e \not\subset Q_i$ for $1 \le i \le e - 1$. By relabelling we may suppose $i = 1$. By the irredundancy of $J = Q_1 \cap \cdots \cap Q_{e-1}$ there exists $r \in (Q_2 \cap \cdots \cap Q_{e-1}) \setminus Q_1$. Since $P_e \not\subset P_1$, there exists $s \in Q_e \setminus P_1$. Now clearly $rs \in (Q_2 \cap \cdots \cap Q_e) \setminus Q_1$.

NONUNIQUENESS OF EMBEDDED PRIMARIES THEOREM (T13). In any noetherian ring R we have the following.

(T13.1) Let $I = Q_1 \cap \cdots \cap Q_e$ be any irredundant primary decomposition of an ideal I in R with $e > 0$ such that, upon letting $P_1, \ldots, P_e$ be the radicals of $Q_1, \ldots, Q_e$ respectively, P_e is an embedded prime of I. Then there exists an infinite sequence $Q_e = Q_{e0} \supset Q_{e1} \supset Q_{e2} \supset \ldots$ of (P_e)-primary ideals in R with $Q_{e0} \neq Q_{e1} \neq Q_{e2} \neq \ldots$ such that $I = Q_1 \cap \cdots \cap Q_{e-1} \cap Q_{en}$ is an irredundant primary decomposition of I for all $n \in \mathbb{N}$.

(T13.2) Let Q and $\overline{Q}$ be any primary ideals in R whose radicals P and $\overline{P}$ are such that $\overline{P} \subset P$ with $\overline{P} \neq P$. Then there exists a P-primary ideal Q' with $Q' \subset Q$ such that $Q' \neq Q$ and $\overline{Q} \cap Q' = \overline{Q} \cap Q$.

PROOF. (T13.1) obviously follows from (T13.2). To prove (T13.2), in view of the Isomorphism Theorems of L4§5(O11), by passing to $R/(\overline{Q} \cap Q)$ we may suppose that $\overline{Q} \cap Q = 0$. In view of this equation, by (T11) the intersection of all P-primary ideals is 0. Since $\overline{P} \subset P$ with $\overline{P} \neq P$, we also have $Q \neq 0$. Therefore $Q \not\subset Q''$ for some P-primary ideal Q'', and we are done by taking $Q' = Q \cap Q''$.

QUEST (Q5) Nagata's Principle of Idealization

We shall now modulize most of the material of the previous Quest (Q3) by means of the above mentioned principle which converts the study of submodules of a module D over a ring R into the study of those ideals of a ring $\overline{R}$ which are contained in a certain ideal $\widetilde{D}$ of $\overline{R}$. To do this we consider the (external) R-module-theoretic direct sum

$$\overline{R} = R \oplus D$$

which we convert into a ring by declaring that for all (r, v) and (r', v') we have

$$(r, v)(r', v') = (rr', rv' + r'v).$$

We let

$$\widetilde{R} = \alpha(R) \quad \text{with natural injection} \quad \alpha : R \to \overline{R}$$

and we note that $\widetilde{R}$ is a subring of $\overline{R}$ isomorphic to R, and α is an injective ring homomorphism. We also let

$$\widetilde{D} = \beta(D) \quad \text{with natural injection} \quad \beta : D \to \overline{R}$$

and we note that $\widetilde{D}$ is an ideal in $\overline{R}$, and $C \mapsto \beta(C)$ gives a bijection of the set of all R-submodules of D onto the set of all ideals in $\overline{R}$ contained in $\widetilde{D}$; moreover, if $B \subset D$ is any set of R-module generators of C, then the two subrings $\widetilde{R}[\beta(B)]$ and $\widetilde{R}[\beta(C)]$ of $\overline{R}$ coincide. It follows that

(1)   if D is a finitely generated module then $\overline{R}$ is an affine ring over $\widetilde{R}$

and hence

(2) $\begin{cases} \text{if } D \text{ is a noetherian module and } R \text{ is a noetherian ring} \\ \text{then } \overline{R} \text{ is a noetherian ring.} \end{cases}$

We are now ready to prove:

MODULIZED ARTIN-REES LEMMA (T14). Given any submodules V and W of a noetherian module D over a ring R and any ideal J in R, there exists $a \in \mathbb{N}$ such that for all $c \geq b \geq a$ in $\mathbb{N}$ we have

$$J^c W \cap V = J^{c-b}(J^b W \cap V).$$

PROOF. By (Q2), via the canonical epimorphism $\phi : R \to \widehat{R} = R/(0 : D)_R$ we may regard D as a module over the noetherian ring $\widehat{R}$ and then $\widehat{J} = \phi(J)$ is an ideal in $\widehat{R}$ with $\widehat{J}^i(\widehat{J}^j W \cap V) = J^i(J^j W \cap V)$ for all i, j in $\mathbb{N}$. Therefore, replacing R by $\widehat{R}$ we may assume that R is noetherian. Now in the above notation let $\widetilde{V} = \beta(V)$, $\widetilde{W} = \beta(W)$, and $\widetilde{J} = \alpha(J)\overline{R}$. Then $\widetilde{V}, \widetilde{W}, \widetilde{J}$ are ideals in the noetherian ring $\overline{R}$, and for all i, j in $\mathbb{N}$ we have $\widetilde{J}^i(\widetilde{J}^j \widetilde{W} \cap \widetilde{V}) = \beta(J^i(J^j W \cap V))$. Thus we are done by (T4).

MODULIZED ARTIN-REES COROLLARY (T15). Given any submodule W of a noetherian module D over a ring R and any ideal J in R, upon letting

$$U = \bigcap_{n \in \mathbb{N}} J^n W$$

we have

$$J^m U = U \quad \text{for all} \quad m \in \mathbb{N}.$$

PROOF. In (T14) take $m = c - b$.

MODULIZED KRULL INTERSECTION LEMMA (T16). Given any submodule W of a noetherian module D over a ring R and any ideal J in R, upon letting

$$V = \bigcap_{n \in \mathbb{N}} J^n W$$

we have that:

$$V = 0 \Leftrightarrow (1 - J) \cap Z_R(W) = \emptyset.$$

PROOF. By (Q3)(4) and (T15) we get $V \neq 0 \Rightarrow (1 - J) \cap Z_R(V) \neq \emptyset$ and obviously $Z_R(V) \subset Z_R(W)$. Conversely suppose there exists $x \in J$ and $0 \neq y \in W$ with $(1 - x)y = 0$. Then $xy = y$ and hence by induction on n we see that $x^n y = y$ for all $n \in \mathbb{N}$. Therefore $y \in \cap_{n \in \mathbb{N}} J^n W = V$ and hence $V \neq 0$.

MODULIZED KRULL INTERSECTION THEOREM (T17). Let J be an ideal in a noetherian ring R with $J \subset \mathrm{jrad}(R)$. Then for any finitely generated R-module D we have

$$\bigcap_{n \in \mathbb{N}} J^n D = 0.$$

In particular we have

$$\bigcap_{n \in \mathbb{N}} J^n = 0.$$

PROOF. In view of (Q3)(2), the first assertion follows by taking $W = D$ in (T16). The second assertion follows by taking $D = R$ in the first assertion.

MODULIZED CLOSEDNESS COROLLARY (T18). Given any submodule U of a finitely generated module V of a noetherian ring R, and any ideal J in R, in the notation of (1') and (2') of (Q1) we have that:

$$\bigcap_{n \in \mathbb{N}} (U + J^n V) = U \Leftrightarrow P_i + J \neq R \text{ for } 1 \leq i \leq e.$$

PROOF. Taking $W = D$ in (T16) we see that for any finitely generated module D over R we have:

(') $$\bigcap_{n \in \mathbb{N}} J^n D = 0 \Leftrightarrow (1 - J) \cap Z_R(D) = \emptyset.$$

Taking $D = V/U$ in (') we see that:

(") $$\bigcap_{n \in \mathbb{N}} (U + J^n V) = U \Leftrightarrow (1 - J) \cap Z_R(V/U) = \emptyset.$$

Now in view of (") our assertion follows from (1') and (2') of (Q1).

MODULIZED CLOSURE COROLLARY (T19). Let J be any ideal in any noetherian ring R. For any submodule U of a finitely generated module V over R let us put $\overline{U} = \cap_{n \in \mathbb{N}} (U + J^n V)$. Then we have the following.

(T19.1) Let $U = Q_1 \cap \cdots \cap Q_e$ be any primary decomposition (which need not be irredundant and which exists by L4§6.1) where $Q_1, \ldots, Q_e$ are a finite number of primary submodules whose radicals $P_1, \ldots, P_e$ in V are prime ideals in R labelled so that $P_i + J \neq R$ for $1 \leq i \leq d$ and $P_i + J = R$ for $d+1 \leq i \leq e$. Then upon letting $A = Q_1 \cap \cdots \cap Q_d$ and $B = Q_{d+1} \cap \cdots \cap Q_e$ we have $U = A \cap B$ with $\overline{U} = \overline{A} = A$ and $\overline{B} = V$.

(T19.2) Let $U_1, \ldots, U_h$ be any finite number of submodules of V such that $U = U_1 \cap \cdots \cap U_h$. Then $\overline{U} = \overline{U}_1 \cap \cdots \cap \overline{U}_h$.

PROOF. To see that (T19.2) follows from (T19.1), for $1 \leq j \leq h$ we can take a primary decomposition $U_j = Q_{j1} \cap \cdots \cap Q_{je(j)}$ where $Q_{j1}, \ldots, Q_{je(j)}$ are a finite number of primary submodules of V whose radicals are prime ideals $P_{j1}, \ldots, P_{je(j)}$ in R labelled so that $P_{ji} + J \neq R$ for $1 \leq i \leq d(j)$ and $P_i + J = R$ for $d(j) + 1 \leq i \leq e(j)$. Now applying (T19.1) to U as well as U_j we get

$$\overline{U} = \bigcap_{1 \leq j \leq h} \bigcap_{1 \leq i \leq d(j)} Q_{ji} \quad \text{and} \quad \overline{U}_j = \bigcap_{1 \leq i \leq d(j)} Q_{ji}$$

and hence $\overline{U} = \overline{U}_1 \cap \cdots \cap \overline{U}_h$.

To prove (T19.1), first note that clearly $U = A \cap B$. By (T18) we have $\overline{Q}_i = Q_i$ for $1 \leq i \leq d$, and hence

$$\overline{A} = \bigcap_{n \in \mathbb{N}} \left(\left(\bigcap_{1 \leq i \leq d} Q_i \right) + J^n V \right) \subset \bigcap_{n \in \mathbb{N}} \left(\bigcap_{1 \leq i \leq d} (Q_i + J^n V) \right) = A.$$

But obviously $A \subset \overline{A}$, and hence $\overline{A} = A$.

Since R is noetherian, we can find $c(i) \in \mathbb{N}$ with $P_i^{c(i)} V \subset Q_i$ for $d+1 \leq i \leq e$. For any $n \in \mathbb{N}$, by L4§5(O26) we have $P_i^{c(i)} + J^n = R$ and hence we can write $p_{in} + j_{in} = 1$ with $p_{in} \in P_i^{c(i)}$ and $j_{in} \in J^n$. Given any $x \in A$, upon letting $r_n = \prod_{d+1 \leq i \leq e} p_{in}$ we see that $r_n x \in U$ with $r_n x - x \in J^n V$. Thus for all $n \in \mathbb{N}$ we have

$$A + J^n V \subset U + J^n V$$

and therefore by intersecting as n varies over $\mathbb{N}$ we get $\overline{A} \subset \overline{U}$. But clearly $\overline{U} \subset \overline{A}$, and hence $\overline{U} = \overline{A}$.

CHARACTERIZATION THEOREM FOR ZARISKI RINGS (T20).

For an ideal J in a noetherian ring R, the following conditions are mutually equivalent:

(i) $J \subset \mathrm{jrad}(R)$.

(ii) Every element of $1 - J$ is a unit in R.

(iii) For any finitely generated R-module V with $V = JV$ we have $V = 0$.

(iv) For any finitely generated R-module V and any submodule U of V we have $\bigcap_{n \in \mathbb{N}} (U + J^n V) = U$.

(v) For any finitely generated R-module V we have $\cap_{n \in \mathbb{N}} J^n V = 0$.

(vi) For any ideal I in R we have $\cap_{n \in \mathbb{N}} (I + J^n) = I$.

PROOF. By (Q3)(2) we get (i) $\Rightarrow$ (ii). By (Q3)(4) we get (ii) $\Rightarrow$ (iii). By (T18) we get (iii) $\Rightarrow$ (iv) by noting that: if I is any ideal in R with $I + J = R$ then upon letting $V = R/I$ we have $JV = (I + JR)/I = (I + J)/I = V$ and hence (iii) tells us that $V = 0$, i.e., $I = R$. By taking $U = 0$ we get (iv) $\Rightarrow$ (v). By taking $V = R/I$ we get (v) $\Rightarrow$ (vi). Finally to show that (vi) $\Rightarrow$ (i), by taking $I \in \mathrm{mspec}(R)$ in (vi) we get $I + J^n \neq R$ for some $n \in \mathbb{N}$ and hence $J^n \subset I$ and therefore $J \subset I$.

QUEST (Q6) Cohen's and Eakin's Noetherian Theorems

After proving Lemma (T21), as an application of the Maximality Principle stated in (C7), in (T22) we shall give a proof of Cohen's Theorem which says that if all the prime ideals in a ring are finitely generated then so are all the ideals. In L4§5(O25)(27•) we showed that if an overring of a ring is a finitely generated module over the ring then the noetherianness of the ring implies the noetherianness of the overring. Eakin's Theorem says that conversely the noetherianness of the overring implies the noetherianness of the ring. As an application of assertion (Q2)(5) about faithful modules, in (T23) we shall give Formanek's proof of Eakin's Theorem.

LEMMA (T21). Let J be an ideal in a ring R such that, for some $a \in R$, the ideals $J + aR$ and $(J : a)_R$ are finitely generated. Then J is finitely generated.

PROOF. We can take a finite number of elements $x_1, \ldots, x_n$ in J and $r_1, \ldots, r_n$ in R such that the elements $x_1 + ar_1, \ldots, x_n + ar_n$ generate $J + aR$. We can also take a finite number of generators $y_1, \ldots, y_m$ of $(J : a)_R$. We claim that elements $x_1, \ldots, x_n, ay_1, \ldots, ay_m$ generate J. They are certainly in J. Conversely, given any $z \in J$, we clearly have $z \in J + aR$ and hence we can write $z = b_1(x_1 + ar_1) + \cdots + b_n(x_n + ar_n)$ with $b_1, \ldots, b_n$ in R. Now $a(b_1 r_1 + \cdots + b_n r_n) = z - b_1 x_1 - \cdots - b_n x_n \in J$ and hence $b_1 r_1 + \cdots + b_n r_n \in (J : a)_R$ and therefore

$$b_1 r_1 + \cdots + b_n r_n = c_1 y_1 + \cdots + c_m y_m \quad \text{with} \quad c_1, \ldots, c_m \text{ in } R.$$

Thus, upon letting $d_1 = ac_1, \ldots, d_m = ac_m$, we have found $d_1, \ldots, d_m$ in R with $z = b_1 x_1 + \cdots + b_n x_n + d_1 y_1 + \cdots + d_m y_m$.

COHEN'S NOETHERIAN THEOREM (T22). A ring R is noetherian iff every prime ideal in it is finitely generated.

PROOF. The only if part being obvious, assume that every prime ideal in R is finitely generated, and let W be the set of all ideals in R which are not finitely generated. Suppose if possible that W is nonempty. Now the union of any linearly ordered (by inclusion) set of members of W clearly belongs to W and hence, by Zorn's lemma, W has a maximal member J. Clearly J is a nonunit ideal which is not prime. Therefore we can find elements a, b in $R \setminus J$ with $ab \in J$. Since $a \in R \setminus J$, the ideal $J + aR$ properly contains J (i.e., contains J but is different from J); since $b \in R \setminus J$ with $ab \in J$ the ideal $(J : a)_R$ also properly contains J. Therefore by the maximality of J, the ideals $J + aR$ and $(J : a)_R$ are finitely generated, and hence by (T21) so is J, which is a contradiction.

THEOREM (T23). For any ring R we have the following.

FORMANEK (T23.1) Assume that there exists a faithful R-module V such that V is finitely generated and the MXC ($=$ Maximal Condition) holds for the set Γ all submodules of the form IV with I varying over the set of all ideals in R (i.e., every nonempty subset of Γ has a maximal member). Then R is noetherian.

EAKIN (T23.2) Assume that there exists a noetherian overring S of R such that S is a finitely generated R-module. Then R is noetherian.

PROOF. Clearly (T23.2) follows by taking $V = S$ in (T23.1). In view of (Q2)(5), to prove (T23.1), it suffices to show that V is a noetherian module. Assume the contrary, and let Ω be the set of all submodules of V of the form IV with ideal I in R such that the R-module $V/(IV)$ is nonnoetherian. Then Ω is nonempty and hence has a maximal member JV with ideal J in R. Let $W = V/(JV)$ and $T = R/(0 : W)_R$. Then W is a faithful finitely generated nonnoetherian T-module such that for every nonzero ideal I in T the T-module $W/(IW)$ is noetherian. Let $x_1, \ldots, x_n$ be a finite set of T-generators of W, and let Δ be the set of all submodules W^* of W for which the set $\{a \in T^\times : ax_1 \in W^*, \ldots, ax_n \in W^*\}$ is empty. Then $0 \in \Delta$ and the union of any nonempty linearly ordered (by inclusion) subset of Δ clearly belongs to Δ. Therefore by Zorn's lemma, Δ has a maximal member W'. If the T-module W/W' were noetherian then T would be a noetherian ring by (Q2)(5) and hence W would be a noetherian T-module by L4§5(O25)(27•). Therefore W/W' is a nonnoetherian T-module and hence W has a submodule U with $W' \subset U$ and $W' \neq U$ such that the T-module U/W' is not finitely generated. By the maximality of W' there exists $a \in T^\times$ such that $aW \subset U$. Now the T-module $W/(aW)$ is noetherian and hence its T-submodule $U/(aU)$ is finitely generated. Therefore the T-module U is finitely generated, and hence so is U/W'. This is a contradiction.

QUEST (Q7) Principal Ideal Theorems

In L4§8 we indicated (although not yet proved) a correspondence between varieties in affine N-space over a field and ideals in the N variable polynomial ring over that field. More generally there is such a correspondence between subvarieties of a variety and ideals in the affine coordinate ring of that variety. Geometrically speaking, a hypersurface, i.e., a subvariety defined by one equation, has codimension one, i.e., dimension one less than the dimension of the ambient space or the ambient variety. Similarly the codimension of a subvariety defined by a certain number of equations is less equal that number, and in good situations equality prevails. The various principal ideal theorems proved by Krull strengthen this experience by showing that there is a parallel situation for ideals in a noetherian ring. Namely, the height of an ideal generated by a finite sequence of elements is less equal the number of elements in that sequence, and if each of those elements is a nonzerodivisor modulo the previous elements then the height of every minimal prime of the ideal equals that number. We start off by proving the case of one element.

PRINCIPAL IDEAL THEOREM (T24). Let x be an element in a noetherian ring R, and let P be a prime in $\operatorname{nvspec}_R x$. Then $\operatorname{ht}_R P \le 1$ with equality in case x is a nonzerodivisor in R.

COMMENT (C11). [**Minimal Primes**]. We are writing $\operatorname{vspec}_R x$ and $\operatorname{nvspec}_R x$ instead of $\operatorname{vspec}_R\{x\}$ and $\operatorname{nvspec}_R\{x\}$ respectively. A prime ideal P in a ring R is said to be minimal over an ideal I in R if it belongs to $\operatorname{nvspec}_R I$; similarly, P is minimal over x means it belongs to $\operatorname{nvspec}_R x$. Clearly a prime ideal in R is of height zero iff it is minimal over 0; let us denote the set of all these by $\operatorname{nspec}(R)$ and call it the minimal spectrum of (R). By (Q1) we see that:
(1) if R is noetherian then every element of such an ideal is a zerodivisor in R.
By (Q1) it also follows that:
(2) if R is noetherian then every $\overline{P}$ in $\operatorname{vspec}_R I$ contains some P in $\operatorname{nvspec}_R I$.
To prove (2), since $I \subset \overline{P}$, in the notation of (Q1)(1) and (Q1)(2) we have $P_1 \ldots P_d \subset P_1 \cap \cdots \cap P_d = \operatorname{rad}_R I \subset \operatorname{rad}_R \overline{P} = \overline{P}$, and hence by (Q1)(C7) we get $P_i \subset \overline{P}$ for some $i \in \{1. \ldots, d\}$.

PROOF OF (T24). Let $\Phi : R \to S = R_P$ be the canonical homomorphism and let $y = \Phi(x)$. Then S is a local with $\dim(S) = \operatorname{ht}_R P$ and $y \in M(S)$ with $M(S) \in \operatorname{vspec}_S y$. In view of item (1) of the above Comment, it only remains to show that $\dim(S) \le 1$. In other words, given any prime ideal J in S different from $M(S)$, we want to show that $\operatorname{ht}_S J = 0$, i.e., upon letting T be the local ring S_J, we want to show that $\dim(T) = 0$. Let $\alpha : S \to T$ be the canonical homomorphism, and for every $n \in \mathbb{N}_+$ let $J_n = \alpha^{-1}(M(T)^n)$; by L4§5(O8) and L4§7.1(T17) we see that J_n is a J-primary ideal in S. In S we have the descending chain of ideals $yS + J_1 \supset yS + J_2 \supset yS + J_3 \supset \ldots$. Since $M(S)$ is minimal over y, the local ring S/yS is artinian and hence for some $n \in \mathbb{N}_+$ we must have $yS + J_n = yS + J_{n+1}$. Now any $z \in J_n$ can be written as $z = ys + t$ with $s \in S$ and $t \in J_{n+1}$; this gives $ys \in J_n$; since $M(S)$ is minimal over y, we must have $y \notin J$; since J_n is J-primary, we get $s \in J_n$. Thus $J_n = yJ_n + J_{n+1}$. Consequently, since $y \in M(S)$, by (T3) we get $J_n = J_{n+1}$. Therefore by J4§7.1(T17) we conclude that $M(T)^n = M(T)^{n+1}$. Consequently, again by (T3) we get $M(T)^n = 0$. Therefore $\dim(T) = 0$.

COROLLARY (T25). Let H and I be ideals in a noetherian ring R, let $x \in R$, and let $P \in \text{nvspec}_R(H + xR)$ with $I \subset P$ be such that for every $P' \in \text{nvspec}_R H$ we have $I \not\subset P'$. Then $P \in \text{nvspec}_R(H + I)$.

PROOF. Apply (T24) to the images of x and P in R/H.

COROLLARY (T26). Let H and P be ideals in a noetherian ring R such that P is prime with $\text{ht}_R P \geq e \in \mathbb{N}_+$ and for every $P' \in \text{nvspec}_R H$ we have $P \not\subset P'$. Then there exists a sequence of prime ideals $P_0 \subset P_1 \subset \cdots \subset P_e = P$ in R with $P_0 \neq P_1 \neq \cdots \neq P_e$ such that for every $P' \in \text{nvspec}_R H$ we have $P_1 \not\subset P'$.

PROOF. We make induction on e. The assertion being obvious for $e = 1$, let $e > 1$ assume for $e - 1$. Since $\text{ht}_R P \geq e$, there exists a sequence of prime ideals $P_0' \subset P_1' \subset \cdots \subset P_e' = P$ in R with $P_0' \neq P_1' \neq \cdots \neq P_e'$. By L4§5(O24)(22$^\bullet$) there exits $x \in P \setminus P_{e-2}'$ such that for every $P' \in \text{nvspec}_R H$ we have $x \notin P'$. By (C11)(2) there exits $P^* \in \text{nvspec}_R(P_{e-2}' + xR)$ with $P^* \subset P$. Applying (T24) to the images of x and P^* in R/P_{e-2}' we see that $P_{e-2}' \neq P^* \neq P$. It follows that $\text{ht}_R P^* \geq e - 1$ and for every $P' \in \text{nvspec}_R H$ we have $P^* \not\subset P'$. Therefore by the induction hypothesis there exists a sequence of prime ideals $P_0 \subset P_1 \subset \cdots \subset P_{e-1} = P^*$ in R with $P_0 \neq P_1 \neq \cdots \neq P_{e-1}$ such that for every $P' \in \text{nvspec}_R H$ we have $P_1 \not\subset P'$. We are done by taking $P_e = P$.

GENERALIZED PRINCIPAL IDEAL THEOREM (T27). Given any elements $x_1, \ldots, x_n$ in a noetherian ring R, with $n \in \mathbb{N}$, let I be the ideal generated by them. Then $\text{ht}_R I \leq n$. Moreover, if x_i is a nonzerodivisor mod $x_1, \ldots, x_{i-1}$ for $1 \leq i \leq n$, i.e., if the image of x_i is a nonzerodivisor in the residue class ring $R/(x_1, \ldots, x_{i-1})R$ for $1 \leq i \leq n$, then for every P in $\text{nvspec}_R I$ we have $\text{ht}_R P = n$.

PROOF. In view of (C11)(2), to prove $\text{ht}_R I \leq n$, it suffices to show that given any P in $\text{nvspec}_R I$ we have $\text{ht}_R P \leq n$. This being obvious for $n = 0$, letting $n > 0$ and assuming true for $n - 1$ we shall show it for n, and then we shall be done by induction. Let H be the ideal in R generated by $x_1, \ldots, x_{n-1}$. If $P \in \text{nvspec}_R H$ then by the induction hypothesis $\text{ht}_R P \leq n - 1$. So assume the contrary. Then clearly for every $P' \in \text{nvspec}_R H$ we have $P \not\subset P'$. Given any $e \in \mathbb{N}_+$ with $\text{ht}_R P \geq e$ we want to show that $e \leq n$. By (T26) there exists a sequence of prime ideals $P_0 \subset P_1 \subset \cdots \subset P_e = P$ in R with $P_0 \neq P_1 \neq \cdots \neq P_e$ such that for every $P' \in \text{nvspec}_R H$ we have $P_1 \not\subset P'$. Taking $I = P_1$ and $x = x_n$ in (T25) we get $P \in \text{nvspec}_R(H + P_1)$. Consequently by applying the induction hypothesis to the image of P in R/P_1 we get $e - 1 \leq n - 1$, and hence $e \leq n$.

This completes the proof of the height inequalities $\text{ht}_R P \leq n$ and $\text{ht}_R I \leq n$. To prove the height equality $\text{ht}_R P = n$ under the nonzerodivisor assumption, we again make induction on n. The $n = 0$ case being obvious, let $n > 0$ and assume for $n - 1$. Given $P \in \text{nvspec}_R I$, upon letting H be the ideal in R generated by $x_1, \ldots, x_{n-1}$, by (C11)(2) there exists $P' \in \text{nvspec}_R H$ with $P' \subset P$. Applying (T24) to the images of x_n and P in R/H we see that $P' \neq P$. By the induction hypothesis $\text{ht}_R P' = n - 1$ and by the already proved height inequality $\text{ht}_R P \leq n$. Therefore $\text{ht}_R P = n$.

As a consequence of (T27) we have the following theorem which includes a partial converse of (T27).

HEIGHT THEOREM (T28). Let R be a noetherian ring. Then for any nonunit ideal I in R we have $\text{ht}_R I \in \mathbb{N}$; recall that the height of the unit ideal is defined to be $-\infty$. In particular, for any prime ideal P in R, upon letting $\text{ht}_R P = n$, there exists a sequence of prime ideals $P_0 \subset P_1 \subset \cdots \subset P_n = P$ in R with $P_0 \neq P_1 \neq \cdots \neq P_n$, and for any such sequence of prime ideals we have $\text{ht}_R P_j = j$ for $0 \leq j \leq n$. Moreover, given any such sequence of prime ideals, there exist elements $x_1, \ldots, x_n$ in R such that upon letting I_j be the ideal generated by $x_1, \ldots, x_j$, for $0 \leq j \leq n$ we have $P_j \in \text{nvspec}_R I_j$ and $\text{ht}_R P_j' = j$ for all $P_j' \in \text{nvspec} I_j$.

PROOF. Everything except the last sentence immediately follows from (T26). The last sentence being obvious for $n = 0$, let $n > 0$ and assume it for $n - 1$. Then there exist elements $x_1, \ldots, x_{n-1}$ in R such that upon letting I_j be the ideal generated by $x_1, \ldots, x_j$, for $0 \leq j \leq n - 1$ we have $P_j \in \text{nvspec}_R I_j$ and $\text{ht}_R P_j' = j$ for all $P_j' \in \text{nvspec}_R I_j$. By L4§5(O24)(22•) there exists $x_n \in P$ such that for all $P_{n-1}' \in \text{nvspec}_R I_{n-1}$ we have $x_n \notin P_{n-1}'$. Let I_n be the ideal in R generated by $x_1, \ldots, x_n$. By (C11)(2) and (T27) we see that $P_n \in \text{nvspec}_R I_n$ and $\text{ht}_R P_n' = n$ for all $P_n' \in \text{nvspec}_R I_n$. So we are done by induction on n.

As an immediate consequence of (T27) and (T28) we get the following theorem which includes L4§8.3(T23).

EXTENDED DIM–EMDIM THEOREM (T29). For any local ring R, $\text{emdim}(R)$ and $\dim(R)$ are nonnegative integers with $\text{emdim}(R) \geq \dim(R)$. Moreover, $\dim(R)$ can be characterized as the smallest number of elements in R which generate an $M(R)$-primary ideal.

As another consequence of (T27) and (T28) we now prove the:

DIMENSION LEMMA (T30). Let S be the polynomial ring $R[X_1, \ldots, X_m]$ or the power series ring $R[[X_1, \ldots, X_m]]$ where $X_1, \ldots, X_m$ are a finite number of indeterminates over a noetherian ring R. Then we have the following.
(T30.1) S is noetherian, and for any ideal I in R, upon letting I_j be the ideal

in S generated by $X_1, \ldots, X_j$ and I we have $I_j \cap R = I$ for $0 \le j \le m$, and hence for any ideal $H \ne I$ in R we have $HS \ne IS$. Moreover, if $I \ne R$ then $I_0 \subset I_1 \subset \cdots \subset I_m \subset S$ with $I_0 \ne I_1 \ne \cdots \ne I_m \ne S$ and $\mathrm{ht}_S I_j = j + \mathrm{ht}_R I$ for $0 \le j \le m$. Furthermore, if $I \in \mathrm{spec}(R)$ then $I_j \in \mathrm{spec}(S)$ for $0 \le j \le m$. Finally, assuming $I \ne R$ but without assuming $I \in \mathrm{spec}(R)$, if $(L_i)_{1 \le i \le d}$ are all the distinct members of $\mathrm{nvspec}_R I$ then $(L_i S + (X_1, \ldots, X_j)S)_{1 \le i \le d}$ are all the distinct members of $\mathrm{nvspec}_S I_j$ for $0 \le j \le m$.

(T30.2) $\dim(S) = m + \dim(R)$, where by convention the dimension of the null ring is $-\infty$, and for any $n \in \mathbb{N}$ we have $n + (-\infty) = -\infty$ and $n + \infty = \infty$.

(T30.3) If R is a local ring and we are in the power series case then S is a local ring with $M(S) = M(R)S + (X_1, \ldots, X_m)S$ and $\mathrm{emdim}(S) = m + \mathrm{emdim}(R)$, and hence: R is regular $\Leftrightarrow$ S is regular.

(T30.4) If R is a field then $R[X_1, \ldots, X_m]$ is an m-dimensional noetherian ring and $R[[X_1, \ldots, X_m]]$ is an m-dimensional regular local ring, where we note that a ring is a field iff it is a zero-dimensional regular local ring.

PROOF. The noetherianness of S follows from L4§3(T3) and L4§3(T4), and in view of an obvious induction on m, in proving the rest of (T30.1) to (T30.3), without loss of generality we may assume $m = 1$ and write X for X_1. (T30.4) immediately follow from (T30.2) and (T30.3).

In (T30.1), any z in I_1 can be written as a finite sum $z = Xf(X) + \sum y_i g_i(X)$ with $f(X), g_i(X)$ in S and y_i in I, and if $z \in R$ then by substituting $X = 0$ on both sides of the equation we get $z = \sum y_i g_i(0) \in I$. Thus $I_1 \cap R = I$ and hence $I_0 \cap R = I$. Now assume $I \ne R$. Again, any $v \in I_0$ can be written as a finite sum $v = \sum u_i h_i(X)$ with $h_i(X) \in S$ and $u_i \in I$, and equating coefficients we see that the coefficient z_e of X^e in z equals $\sum u_i t_{ie}$ where t_{ie} is the coefficient of X^e in $h_i(X)$; since the last sum belongs to I, for all $e \in \mathbb{N}$ we have $X^e \not\in I_0$; so in particular $X \not\in I_0$, and therefore $I_0 \ne I_1$. Let $\phi : R \to \overline{R} = R/I$ be the canonical epimorphism. Let $\overline{S} = \overline{R}[X]$ or $\overline{R}[[X]]$ respectively. Then: I is prime $\Leftrightarrow$ $\overline{R}$ is a domain $\Leftrightarrow$ $\overline{S}$ is a domain. We get an epimorphism $\psi : S \to \overline{S}$ by letting $\psi(\sum a_e X^e) = \sum \phi(a_e)X^e$ for all $\sum a_e x^e \in S$ with $a_e \in R$. Clearly $\ker(\psi) = I_0$ and hence: I is prime iff I_0 is prime. We get an epimorphism $\theta : \overline{S} \to \overline{R}$ by letting $\theta(\sum b_e X^e) = b_0$ for all $\sum b_e X^e \in \overline{S}$ with $b_e \in \overline{R}$. Clearly $\ker(\theta) = X\overline{S} = \psi(I_1)$ and hence: I is prime iff I_1 is prime. In view of (C11)(2) it now follows that: ($\bullet$) if $(L_i)_{1 \le i \le d}$ are all the distinct members of $\mathrm{nvspec}_R I$ then $(L_i S + (X_1, \ldots, X_j)S)_{1 \le i \le d}$ are all the distinct members of $\mathrm{nvspec}_S I_j$ for $0 \le j \le 1$. Therefore, to prove the equations $\mathrm{ht}_S I_j = j + \mathrm{ht}_R I$, without loss of generality we may assume I to be prime.

So assuming I to be prime, we can find a sequence $P_0 \subset P_1 \subset \cdots \subset P_n = I$ of prime ideals in R with $P_0 \ne P_1 \ne \cdots \ne P_n$, where $\mathrm{ht}_R I = n$, and by what we have proved so far we see that $P_0 S \subset P_1 S \cdots \subset P_n S = I_0 \subset I_1$ is a sequence of prime ideals in S with $P_0 S \ne P_1 S \ne \cdots \ne P_n S = I_0 \ne I_1$. Consequently $\mathrm{ht}_S I_j \ge j + n$ for $0 \le j \le 1$. By (T28) we have $I \in \mathrm{nvspec}_R(x_1, \ldots, x_n)R$ for some $x_1, \ldots, x_n$ in R, and then by ($\bullet$) we get $I_0 \in \mathrm{nvspec}_R(x_1, \ldots, x_n)S$ and $I_1 \in \mathrm{nvspec}_R(x_1, \ldots, x_n, X)S$, and hence by (T27) we get $\mathrm{ht}_S I_j \leq j + n$ for $0 \leq j \leq 1$. Therefore $\mathrm{ht}_S I_j = j + n$ for $0 \leq j \leq 1$. This completes the proof of (T30.1). In view of the argument of the first sentence of this paragraph, to prove (T30.2) it suffices to show that $\dim(S) \leq 1 + \dim(R)$. We shall do this first in the polynomial case and then in the power series case.

For a moment suppose we are in the polynomial case. Since the intersection of any prime in S with R is a prime in R, continuing the assumption that I is prime with $\mathrm{ht}_R I = n$, and letting J be any prime in S with $J \cap R = I$ and $J \neq IS$, in view of (T30.1) it suffices to show that $\mathrm{ht}_S J = 1 + n$. In the above notation let $\overline{J} = \psi(J)$. Then $\overline{J}$ is a nonzero prime ideal in $\overline{S}$ with $\overline{J} \cap \overline{R} = 0$. Let $\mathrm{QF}(\overline{R}) = K$. Then the localization of $\overline{S}$ at the multiplicative set $\overline{R}^\times$ is the univariate polynomial ring $K[X]$ over the field K, and hence $\overline{J} = \psi(F(X))K[X]$ with $F(X) \in J$ of degree $E > 0$ in X such that the coefficient of X^E in F belongs to $R \setminus I$. Upon letting $B = F(X)S$ we get $J \in \mathrm{nvspec}_S(IS + B)$. By (T28) we have $I \in \mathrm{nvspec}_R A$ where $A = (x_1, \ldots, x_n)R$ for some $x_1, \ldots, x_n$ in R. It follows that (see HINT in (C13) below): ($\bullet\bullet$) $J \in \mathrm{nvspec}_S(AS + B)$. By ($\bullet\bullet$) and (T27) we get $\mathrm{ht}_S J \leq 1 + n$. By (T30.1) we have $\mathrm{ht}_S IS = n$ and hence $\mathrm{ht}_S J \geq 1 + n$. Therefore $\mathrm{ht}_S J = 1 + n$.

Henceforth suppose we are in the power series case. To complete the proof of (T30.2) it suffices to show that given any prime ideal J in S there exists a prime ideal I in R with $J \subset IS + XS$; to do this, by L2§2(ii) we see that the ideal $\{f(0) : f(X) \in J\}$ in R is not the unit ideal, and hence by Zorn's Lemma it is contained in a maximal ideal I in R. To prove (T30.3), also assume that R is a local ring. Then in view of L2§2(ii) and L3§11(1$^\bullet$) we see that S is a local ring with $M(S) = M(R)S + XS$, and now clearly $J \subset IS = XS$. Upon letting $\mathrm{emdim}(R) = d$, it only remains to show that $\mathrm{emdim}(S) = 1 + d$. We can write $M(R) = (z_1, \ldots, z_d)R$ with $z_1, \ldots, z_d$ in R. Then $M(S) = (z_1, \ldots, z_d, X)S$ and by (T30.1) we get $M(S) \neq (z_1, \ldots, z_d)S$. For $1 \leq i \leq d$ we also have $M(R) \neq (z_1, \ldots, z_{i-1}, z_{i+1}, \ldots, z_d)R$ and hence by (T30.1) we get $M(S) \neq (z_1, \ldots, z_{i-1}, z_{i+1}, \ldots, z_d, X)S$. Therefore by (T3) we conclude that $\mathrm{emdim}(S) = 1 + d$ [cf. (C14) below].

COMMENT (C12). [**Multivariate Ideal Extensions**]. In the situation of (T30), for any $A \subset R$ let $A[X_1, \ldots, X_m]$ or $A[[X_1, \ldots, X_m]]$ respectively denote the set of all $f(X_1, \ldots, X_m) \in S$ such that the coefficient of any monomial $X_1^{e_1} \ldots X_m^{e_m}$ in f belongs to A. Then, as indicated in the proof of (T30.1), for any ideal I in R we have $IS = I[X_1, \ldots, X_m]$ or $I[[X_1, \ldots, X_m]]$ respectively.

COMMENT (C13). [**Univariate Ideal Extensions**]. Let S be the univariate polynomial ring $R[X]$ over a noetherian ring R, let J be a prime ideal in S, and let $I = J \cap R$. Then I is a prime ideal in R, and as indicated in the proof of the polynomial case of(T30.2), we have the following.

(1) If $J = IS$ then $\mathrm{ht}_S J = \mathrm{ht}_R I$.

(2) If $J \neq IS$ then $\mathrm{ht}_S J = 1 + \mathrm{ht}_R I$ and there exists $F(X) \in J$ of degree $E > 0$ such that the coefficient of X^E in F belongs to $R \setminus I$.

HINT. Here is the detailed proof of claim (••) which says that: given respective prime ideals I and J in a noetherian ring R and a noetherian overring S with $J \cap R = I$, and given respective ideals A and B in R and S with $I \in \mathrm{nvspec}_R A$ and $J \in \mathrm{nvspec}_S(IS + B)$, we have $J \in \mathrm{nvspec}_S(AS + B)$. We can prove this by a double application of a criterion which follows from L4§6(T5) to L4§6(T7) and which says that, for ideals A and I in a noetherian ring R with I prime we have: $I \in \mathrm{nvspec}_R A \Leftrightarrow I = \mathrm{rad}_R[A : I]_R$. In our case $AS + B \subset J$ and hence $\mathrm{rad}_S[AS + B]_S \subset J$. Conversely, given any $z \in J$, because of $J \in \mathrm{nvspec}_S(IS + B)$, we can find $n \in \mathbb{N}_+$ and $b \in B$ such that $z^n = b + s_1 c_1 + \cdots + s_e c_e$ with $s_1, \ldots, s_e$ in S and $c_1, \ldots, c_e$ in I. Because of $I \in \mathrm{nvspec}_R A$, we can find $m \in \mathbb{N}_+$ and $t \in R \setminus I$ such that $t c_i^m = r_i a_i$ with $r_i \in R$ and $a_i \in A$ and for $1 \leq i \leq e$. It follows that for all large enough positive integers μ, ν we have $t^\mu z^\nu \in AS + B$. Clearly $t^\mu \in S \setminus J$ and hence $z \in \mathrm{rad}_S[AS + B]_S$.

COMMENT (C14). [**Module Generation For Local Rings**]. In the last sentence in the proof of (T30) when we said by (T3) we meant by the following consequence of (T3). Let V be a finitely generated module over a local ring R, and let $x_1, \ldots, x_m$ and $y_1, \ldots, y_n$ be finite sequences of generators of V, which are both assumed to be irredundant, i.e., no member of either sequence can be omitted for it to remain a system of generators of V. Then $m = n$. This follows by noting that $V/M(R)V$ is a finite dimensional vector space over the field $R/M(R)$, and for any submodule U of V, by (T3), we have: $V = U + M(R)V \Rightarrow U = V$. Consequently, upon letting $\alpha : V \to V/M(R)V$ to be the canonical epimorphism and ν to be the dimension of the vector space $\alpha(V)$, we see that:

$$\begin{cases} x_1, \ldots, x_m \text{ are irredundant generators of } V \\ \Leftrightarrow \alpha(x_1), \ldots, \alpha(x_m) \text{ are irredundant generators of } \alpha(V) \\ \Leftrightarrow \alpha(x_1), \ldots, \alpha(x_m) \text{ is a basis of } \alpha(V) \\ \Leftrightarrow m = \nu. \end{cases}$$

As an illustration that this is not true without the localness hypothesis, take $V = R = \mathbb{Z}$ with $(x_1, \ldots, x_m) = (1)$ and $(y_1, \ldots, y_n) = (2, 3)$.

QUEST (Q8) Relative Independence and Analytic Independence

In continuation of (T26) and (T27), in Theorem (T34) below we shall show that an ideal I generated by a finite number of elements $x_1, \ldots, x_n$ in a noetherian ring R has height equal to that number n iff the elements are independent in a sense to be defined is a moment. The said theorem will be preceded by Lemmas (T31) to (T33). Out of these, Lemma (T33) will relate the independence of the elements $x_1, \ldots, x_n$ to the modelic blowup of the ring R at the ideal I as described in L4§9.2 which is the coordinate free incarnation of the modelic proj of the ring R at the ideal I as described in L4§9.1; unlike in these references, here the ring R will not be assumed to be a domain. In Theorem (T35) we shall show how this concept of independence is related to the notion of analytic independence of power series which too we shall introduce in a moment.

We shall consider the polynomial ring $R[Y_1, \ldots, Y_n]$ in a finite number of variables over an arbitrary ring R which need not be noetherian. For any $A \subset R$ we let

$$\begin{cases} A[Y_1, \ldots, Y_n]_i = \text{the set of all homogeneous members of } R[Y_1, \ldots, Y_n] \\ \qquad \text{of degree } i \text{ all of whose coefficients belong to } A \\ \qquad \text{(including the zero polynomial in case } 0 \in A) \end{cases}$$

and we put

$$A[Y_1, \ldots, Y_n]_\infty = \bigcup_{i \in \mathbb{N}} A[Y_1, \ldots, Y_n]_i.$$

Let J be an ideal in R. For elements $x_1, \ldots, x_n$ in R let

$$I = (x_1, \ldots, x_n)R.$$

The elements $x_1, \ldots, x_n$ are said to be independent over (R, J), or independent relative to (R, J), if $I \subset J$ and

(†)
$$\begin{cases} F(Y_1, \ldots, Y_n) \in R[Y_1, \ldots, Y_n]_\infty \text{ with } F(x_1, \ldots, x_n) = 0 \\ \Rightarrow F(Y_1, \ldots, Y_n) \in J[Y_1, \ldots, Y_n]_\infty. \end{cases}$$

The elements $x_1, \ldots, x_n$ are said to be independent over R, or independent relative to R, if $I \neq R$ and they are independent over $(R, \mathrm{rad}_R I)$. More precisely, we may say that the sequence $x_1, \ldots, x_n$ is independent over (R, J) or R and so on.

In case R is the power series ring $K[[X_1, \ldots, X_m]]$ in a finite number of variables over a field K, elements $x_1, \ldots, x_n$ in $M(R)$ are said to be analytically independent over K if there is no nonzero $f(Y_1, \ldots, Y_n) \in K[[Y_1, \ldots, Y_n]]$ with $f(x_1, \ldots, x_n) = 0$.

In all the above situations, dependent means not independent.

LEMMA (T31). For any elements $x_1, \ldots, x_n$ in R we have the following.

(T31.1) For the R-epimorphism $\alpha : R[Y_1, \ldots, Y_n] \to R$ which sends Y_i to x_i for $1 \leq i \leq n$, we have

(♯)
$$\ker(\alpha) = (Y_1 - x_1, \ldots, Y_n - x_n)R[Y_1, \ldots, Y_n]$$

and condition (†) is equivalent to the condition:

(†′)
$$\ker(\alpha) \cap R[Y_1, \ldots, Y_n]_\infty \subset J[Y_1, \ldots, Y_n]_\infty.$$

(T31.2) Assuming $I \subset J$, condition (†) is equivalent to the condition:

$$(\ddagger) \quad \begin{cases} F(Y_1,\ldots,Y_n) \in R[Y_1,\ldots,Y_n]_i \text{ with } i \in \mathbb{N} \text{ and } F(x_1,\ldots,x_n) \in I^{i+1} \\ \Rightarrow F(Y_1,\ldots,Y_n) \in J[Y_1,\ldots,Y_n]_i. \end{cases}$$

PROOF OF (T31.1). Clearly (†) $\Leftrightarrow$ (†′). To prove ($\sharp$), by induction on n we shall show that any $f(Y_1,\ldots,Y_n) \in R[Y_1,\ldots,Y_n]$ can be written as $f(Y_1,\ldots,Y_n) = r + \sum_{1 \leq i \leq n}(Y_i - x_i)g_i(Y_1,\ldots,Y_i)$ with $r \in R$ and $g_i(Y_1,\ldots,Y_i) \in R[Y_1,\ldots,Y_i]$ for $1 \leq i \leq n$. This is obvious for $n = 0$. So let $n > 0$ and assume for $n - 1$. By division algorithm in Y_n over $R[Y_1,\ldots,Y_{n-1}]$ we can write $f(Y_1,\ldots,Y_n) = h(Y_1,\ldots,Y_{n-1}) + (Y_n - x_n)g_n(Y_1,\ldots,Y_n)$ with $h(Y_1,\ldots,Y_{n-1}) \in R[Y_1,\ldots,Y_{n-1}]$ and $g_n(Y_1,\ldots,Y_n) \in R[Y_1,\ldots,Y_n]$. By the induction hypothesis we can write $h(Y_1,\ldots,Y_{n-1}) = r + \sum_{1 \leq i \leq n-1}(Y_i - x_i)g_i(Y_1,\ldots,Y_i)$ where r is an element of R and for $1 \leq i \leq n - 1$ we have $g_i(Y_1,\ldots,Y_i) \in R[Y_1,\ldots,Y_i]$.

PROOF OF (T31.2). Clearly ($\ddagger$) $\Rightarrow$ (†). Conversely, assuming (†), for any F as in ($\ddagger$), we can write

$$F(x_1,\ldots,x_n) = G(x_1,\ldots,x_n) \text{ with } G(Y_1,\ldots,Y_n) \in I[Y_1,\ldots,Y_n]_i$$

and letting

$$\overline{F}(Y_1,\ldots,Y_n) = F(Y_1,\ldots,Y_n) - G(Y_1,\ldots,Y_n)$$

we get $\overline{F}(x_1,\ldots,x_n) = 0$; hence $\overline{F}(Y_1,\ldots,Y_n) \in J[Y_1,\ldots,Y_n]_\infty$ and therefore $F(Y_1,\ldots,Y_n) \in J[Y_1,\ldots,Y_n]_i$.

LEMMA (T32). Elements $x_1,\ldots,x_n$ in a noetherian ring R are independent over R iff the elements $Z, x_1,\ldots,x_n$ are independent over the univariate polynomial ring $R[Z]$.

PROOF. Upon letting $H = ZR[Z] + IR[Z]$, by (T30.1) we know that if $I \neq R$ then $H \neq R[Z]$, and clearly

$$R \cap \mathrm{rad}_{R[Z]}H = \mathrm{rad}_R I$$

and hence if the elements $Z, x_1,\ldots,x_n$ are independent over $R[Z]$ then the elements $x_1,\ldots,x_n$ are independent over R. Conversely assuming that the elements $x_1,\ldots,x_n$ are independent over R, let $F(Y_0,\ldots,Y_n) \in R[Z][Y_0,\ldots,Y_n]_i$ with $i \in \mathbb{N}$ be such that $F(Z, x_1,\ldots,x_n) = 0$. Expanding F we have

$$F(Y_0,\ldots,Y_n) = \sum_{0 \leq l \leq i} \sum_{j_1+\cdots+j_n=i-l} F_{lj_1\ldots j_n}(Z)Y_0^l Y_1^{j_1} \ldots Y_n^{j_n}$$

with $F_{lj_1\ldots j_n}(Z) \in R[Z]$ and substituting $(Z, x_1, \ldots, x_n)$ for $(Y_0, \ldots, Y_n)$ on both sides we get

$$\sum_{0 \le l \le i} \sum_{j_1 + \cdots + j_n = i - l} F_{lj_1\ldots j_n}(Z) Z^l x_1^{j_1} \ldots x_n^{j_n} = 0$$

and comparing coefficients of Z^l we obtain

$$\sum_{j_1 + \cdots + j_n = i - l} F_{lj_1\ldots j_n}(0) x_1^{j_1} \ldots x_n^{j_n} \in I^{i+1-l} \quad \text{for} \quad 0 \le l \le i$$

and hence by (T31.2) we get

$$F_{lj_1\ldots j_n}(0) \in \mathrm{rad}_R I \quad \text{for all} \quad l, j_1, \ldots, j_n$$

and therefore

$$F_{lj_1\ldots j_n} \in \mathrm{rad}_{R[Z]} H \quad \text{for all} \quad l, j_1, \ldots, j_n.$$

BLOWUP LEMMA (T33). For elements $x_1, \ldots, x_n$ in a nonnull noetherian ring R with $n \ge 1$, assume x_1 is a nonzerodivisor in R. Let L be the subring of $K = \mathrm{QR}(R)$ given by $L = R[x_2/x_1, \ldots, x_n/x_1]$, let S be the polynomial ring $R[T_2, \ldots, T_n]$, let $\beta : S \to L$ be the R-epimorphism which sends T_i to x_i/x_1 for $2 \le i \le n$, let $B = \ker(\beta)$, let $C = (x_1 T_2 - x_2, \ldots, x_1 T_n - x_n)S$, let R' be the subring of K given by $R' = R[x_n/x_1]$. Then we have the following.

(T33.1) $C \subset B$ and $x_1^m B \subset C$ for some $m \in \mathbb{N}_+$.

(T33.2) If $\mathrm{rad}_R I = J$ then condition (†) is equivalent to the condition:

$$(\dagger'') \qquad B \subset PS \text{ for all } P \in \mathrm{vspec}_R I.$$

Hence the elements $x_1, \ldots, x_n$ are independent over R iff $I \ne R$ and (†″) holds.

(T33.3) If the elements $x_1, \ldots, x_n$ are independent over R then the elements $x_1, \ldots, x_{n-1}$ are independent over R'.

(T33.4) If $\mathrm{ht}_R I = n$ then the elements $x_1, \ldots, x_n$ are independent over R.

(T33.5) If the elements $x_1, \ldots, x_n$ are independent over R then $\mathrm{ht}_R I = n$.

PROOF OF (T33.1). Obviously $C \subset B$. Now $L \subset \overline{R} \subset K$ where $\overline{R}$ is the localization of R at the multiplicative set $(x_1^j)_{j \in \mathbb{N}}$. Extend β to an $\overline{R}$-epimorphism $\overline{\alpha} : \overline{R}[T_2, \ldots, T_n] \to \overline{R}$ with $\overline{\alpha}(T_i) = x_i/x_1$ for $2 \le i \le n$. Given any $z \in B$, clearly $z \in \ker(\overline{\alpha})$ and hence by (T31.1)(♯) we can write $z = \sum_{2 \le i \le n}(T_i - x_i x_1^{-1}) g_i(T_2, \ldots, T_n)$ with $g_i(T_2, \ldots, t_n) \in \overline{R}[T_2, \ldots, T_n]$ for $2 \le i \le n$. Therefore $x_1^e z \in C$ for some $e \in \mathbb{N}_+$. It follows that $x_1^m B \subset C$ for some $m \in \mathbb{N}_+$.

PROOF OF (T33.2). For a moment assume that $\mathrm{rad}_R I = J$ and (†); then given any $P \in \mathrm{vspec}_R I$ and $0 \ne f(T_2, \ldots, T_n) \in B$ of degree i, letting $F(Y_1, \ldots, Y_n) =$

$Y_1^i f(Y_2/Y_1, \ldots, Y_n/Y_1) \in R[Y_1, \ldots, Y_n]_i$, by (T31.1)(†′) we get $F(Y_1, \ldots, Y_n) \in J[Y_1, \ldots, Y_n]_i$; hence $F(Y_1, \ldots, Y_n) \in P[Y_1, \ldots, Y_n]_i$; therefore $f(T_2, \ldots, T_n) \in PS$. Conversely assume $\mathrm{rad}_R I = J$ and (†″); then by (Q1)(2) and (C11)(2) we get $J = \cap_{P \in \mathrm{vspec}_R I} P$ and hence, by (C12), all the coefficients of any $f(T_2, \ldots, T_n) \in B$ belong to J; given any $F(Y_1, \ldots, Y_n) \in R[Y_1, \ldots, Y_n]_\infty$ with $F(x_1, \ldots, x_n) = 0$, letting $f(T_2, \ldots, T_n) = F(1, T_2, \ldots, T_n)$ we clearly have $f(T_2, \ldots, T_n) \in B$ and hence $F(Y_1, \ldots, Y_n) \in J[Y_1, \ldots, Y_n]_\infty$.

PROOF OF (T33.3). Assume the elements $x_1, \ldots, x_n$ are independent over R. Let $B' = \ker(\beta')$ where β' is the R'-epimorphism of $S' = R'[T_2, \ldots, T_{n-1}]$ onto L which sends T_i to x_i/x_1 for $2 \le i \le n-1$. Let $\theta : S \to S'$ be the R-epimorphism which sends $(T_2, \ldots, T_n)$ to $(T_2, \ldots, T_{n-1}, x_n/x_1)$. Then clearly $\theta(B) = B'$. Let $I' = (x_1, \ldots, x_{n-1})R'$. Then clearly $I' = IR' \ne R'$. Given any $P' \in \mathrm{vspec}_{R'} I'$, upon letting $P = P' \cap R$, we clearly have $P \in \mathrm{vspec}_R I$ and hence by (T33.2) we get $B \subset PS$ and therefore $B' \subset P'S'$. Consequently, again by (T33.2) we see that the elements $x_1, \ldots, x_{n-1}$ are independent over R'.

PROOF OF (T33.4). Assume $\mathrm{ht}_R I = n$. Given any $P \in \mathrm{vspec}_R I$, in view of (T33.2), it suffices to show that $B \subset PS$. By (T30.1) we have $PS \in \mathrm{spec}(S)$ and clearly $C \subset IS \subset PS$. Consequently by (C11)(2) we can find $Q \in \mathrm{nvspec}_S C$ with $Q \subset PS$. By (T27) we get $\mathrm{ht}_S Q \le n-1$ and by (T30.1) we get $\mathrm{ht}_S(IS) = \mathrm{ht}_R I \ge n$. Therefore $IS \not\subset Q$. Consequently $x_1 \notin Q$ because otherwise $IS \subset C \subset Q$. Therefore by (T33.1) we get $B \subset Q$ and hence $B \subset PS$.

PROOF OF (T33.5). Assuming the elements $x_1, \ldots, x_n$ to be independent over R, by induction on n we shall show that $\mathrm{ht}_R I = n$. For $n = 1$ this follows from (T24). So let $n > 1$ and assume for $n-1$. By (T27) we have $\mathrm{ht}_R I \le n$. Therefore, given any $P \in \mathrm{vspec}_R I$, it suffices to show that $\mathrm{ht}_R P \ge n$. Let $I' = (x_1, \ldots, x_{n-1})R'$. Then clearly $I' = IR' \ne R'$. In view of the induction hypothesis, by (T33.3) we get $\mathrm{ht}_{R'} I' = n-1$. Let $P' = PR'$. We shall show that then $P' \in \mathrm{vspec}_{R'} I'$ with $\mathrm{ht}_R P > \mathrm{ht}_{R'} P'$ and this will complete the proof. Let $D = R[T_n] \subset S$ and let $\delta : D \to R'$ be the R-epimorphism which sends T_n to x_n/x_1. Then clearly $I' \subset \delta(PD) = P'$. Also $\delta(y) = \beta(y)$ for all $y \in D$, and hence $\ker(\delta) = B \cap D$. By (T30.1) we have $PD = (PS) \cap D \in \mathrm{spec}(D)$ with $\mathrm{ht}_D(PD) = \mathrm{ht}_R P$, and by (T33.2) we have $B \subset PS$. Therefore $PD \in \mathrm{vspec}_D \ker(\delta)$ with $P' = \delta(PD) \in \mathrm{vspec}_{R'} I'$, and it suffices to show that $\mathrm{ht}_D(PD) > \mathrm{ht}_{R'} P'$. Let $h = \mathrm{ht}_{R'} P'$. Then we can find a sequence of prime ideals $\ker(\delta) \subset P_1 \subset P_2 \subset \cdots \subset P_h = PD$ in D with $P_1 \ne P_2 \ne \cdots \ne P_h$. Since x_1 is a nonzerodivisor in R, it follows that $x_1 T_n - x_n$ is a nonzerodivisor in D; since $x_1 T_n - x_n$ clearly belongs to $\ker(\delta)$, by (C11)(2) and (T24) there exists a prime ideal P_0 in D with $P_0 \subset P_1$ and $P_0 \ne P_1$. Therefore $\mathrm{ht}_S(PD) \ge h$.

RELATIVE INDEPENDENCE THEOREM (T34). Elements $x_1, \ldots, x_n$ in a noetherian ring R are independent over R iff $\mathrm{ht}_R I = n$.

PROOF. The assertion being vacuous if R is the null ring, we may assume R to be nonnull. Let $n^* = 1 + n$ and let I^* be the ideal in the univariate polynomial ring $R^* = R[Z]$ generated by the sequence $(x_1^*, \ldots, x_{n^*}^*) = (Z, x_1, \ldots, x_n)$. Then $n^* \geq 1$, the element x_1^* is a nonzerodivisor in the nonnull noetherian ring R^*, by (T30.1) we know that $\mathrm{ht}_{R^*} I^* = n^* \Leftrightarrow \mathrm{ht}_R I = n$, and by (T32) we know that

$$\begin{cases} \text{the elements } x_1^*, \ldots, x_{n^*}^* \text{ are independent over } R^* \\ \Leftrightarrow \text{ the elements } x_1, \ldots, x_n \text{ are independent over } R. \end{cases}$$

Thus we are reduced to (T33.4) and (T33.5).

ANALYTIC INDEPENDENCE THEOREM (T35). In case R is the power series ring $K[[X_1, \ldots, X_m]]$ in a finite number of variables over a field K, and the elements $x_1, \ldots, x_n$ belong to $M(R)$, we have the following.

(T35.1) If the elements $x_1, \ldots, x_n$ are independent over $(R, M(R))$ then they are analytically independent over K.

(T35.2) If $m = n$ and the ideal I is $M(R)$-primary then the elements $x_1, \ldots, x_n$ are analytically independent over K.

PROOF OF (T35.1). Assume $x_1, \ldots, x_n$ are analytically dependent over K. Then there exists $0 \neq f(Y_1, \ldots, Y_n) \in K[[Y_1, \ldots, Y_n]]$ with $f(x_1, \ldots, x_n) = 0$. Now let i be the order of f. Then $i \in \mathbb{N}$ and $f(Y_1, \ldots, Y_n) \in M(R)^i \setminus M(R)^{i+1}$. Consequently $f(Y_1, \ldots, Y_n) = F(Y_1, \ldots, Y_n) \in R[Y_1, \ldots, Y_n]_i \setminus M(R)[Y_1, \ldots, Y_n]_i$. Therefore $x_1, \ldots, x_n$ are dependent over $(R, M(R))$.

PROOF OF (T35.2). Follows from (T34) and (T35.1).

QUEST (Q9) Going Up and Going Down Theorems

In (T30.4) we showed that a finite variable polynomial ring over a field has dimension equal to the number of variables. We want to generalize this by showing that the dimension of an affine domain over a field equals the transcendence degree of its quotient field. To do this we need the going up and going down theorems for prime ideals in integral extensions of rings; the domain case of these theorems is due to Krull, and then they were generalized by Cohen-Seidenberg to rings with zerodivisors. These are applied via Emmy Noether's Normalization Theorem which we shall take up in the next Quest and which says that every affine ring is an integral extension of a polynomial ring.

Note that an integral extension of a ring is an overring which is integral over the ring. Also note that a prime ideal Q in an overring S of a ring R lies above a prime ideal P in R means $P = Q \cap R$; we may also express this by saying that P lies below Q in R.

We extend the notion of a normal domain to a normal ring by saying that the ring R is normal if it is integrally closed in its total quotient ring $\mathrm{QR}(R)$. Clearly the quotient field of a domain is the same thing as its total quotient ring, and hence a domain is a normal domain iff it is a normal ring.

Note that if a multiplicative set T in a ring R is such that $T \cap Z_S(S) = \emptyset$ where $Z_S(S)$ is the set of all zerodivisors in an overring S of R, then the localization R_T may be regarded as a subring of $\mathrm{QR}(S)$. In particular this is so for the localization R_P at a prime ideal P in R with $(R \setminus P) \cap Z_S(S) = \emptyset$. See L4§7.

LOCALIZATION OF NORMALITY LEMMA (T36). For a normal ring R, the localization R_T at a multiplicative set T in R with $T \cap Z_R(R) = \emptyset$ (hence in particular the localization R_P at a prime ideal P in R with $Z_R(R) \subset P$) is a normal ring.

PROOF. Given any y in $\mathrm{QR}(R)$ which is integral over R_T, we can write

$$y^n + \sum_{1 \leq i \leq n} x_i y^{n-i} = 0 \quad \text{with} \quad n \in \mathbb{N}_+$$

and $x_i \in R_T$. We can find $t \in T$ with $tx_i \in R$ for $1 \leq i \leq n$. Multiplying the above equation by t^n we get

$$(ty)^n + \sum_{1 \leq i \leq n} (t^i x_i)(ty)^{n-i} = 0$$

with $t^i x_i \in R$ as an equation of integral dependence for $ty \in R$ over R. Consequently by the normality of R we get $ty \in R$ and hence $y \in R_T$.

PRESERVATION OF INTEGRAL DEPENDENCE LEMMA (T37). For any integral extension S of a ring R we have the following.

(T37.1) If $\psi : S \to S'$ is any ring homomorphism then $\psi(S)$ is integral over $\psi(R)$. Equivalently, if J and I are ideals in S and R with $I = J \cap R$ then after identifying R/I with a subring of S/J we have that S/J is integral over R/I.

(T37.2) If $\psi : S \to S_T$ is the natural homomorphism where T is any multiplicative set in R, then S_T is integral over $\psi(R)_{\psi(T)}$. [Referring to L4§7, upon letting $\phi : R \to R_T$ be the natural homomorphism, there is a unique isomorphism $\theta : R_T \to \psi(R)_{\psi(T)}$ such that $\theta(\phi(x)) = \psi(x)$ for all $x \in R$, and if $T \cap Z_S(S) = \emptyset$ then we may regard R_T and S_T as subrings of $\mathrm{QR}(S)$ with $R_T \subset S_T$].

PROOF. Let

$$y^n + \sum_{1 \leq i \leq n} x_i y^{n-i} = 0 \quad \text{with} \quad n \in \mathbb{N}_+$$

and $x_i \in R$ be an equation of integral dependence for $y \in S$ over R. In case of (T37.1) we get

$$\psi(y)^n + \sum_{1 \leq i \leq n} \psi(x_i)\psi(y)^{n-i} = 0$$

with $\psi(x_i) \in \psi(R)$ as an equation of integral dependence for $\psi(y)$ over $\psi(R)$. In case of (T37.2), for any $t \in T$ we get

$$[\psi(y)/\psi(t)]^n + \sum_{1 \leq i \leq n} [\psi(x_i)/\psi(t^i)][\psi(y)/\psi(t)]^{n-i} = 0$$

with $\psi(x_i)/\psi(t^i) \in \psi(R)_{\psi(T)}$ as an equation of integral dependence for $\psi(y)/\psi(t)$ over $\psi(R)_{\psi(T)}$.

PRESERVATION OF FIELDS AND MAXIMAL IDEALS LEMMA (T38). For any integral extension S of a ring R we have the following.

(T38.1) If S is a domain then: R is a field $\Leftrightarrow$ S is a field.

(T38.2) If P is a prime ideal in R lying below a prime ideal Q in S then: P is maximal $\Leftrightarrow$ Q maximal.

PROOF. (T38.2) follows by taking $(R/P, S/Q)$ for (S, R) in (T38.1). To prove (T38.1) first suppose that S is a domain and R is a field. Then for any $0 \neq y \in S$ we have that y/R is integral and hence y/R is algebraic and so $1/y \in R[y] \subset S$. Therefore S is a field.

Conversely suppose that S is a field and let $0 \neq x \in R$. Then $x^{-1} \in S$ and hence x^{-1}/R is integral, i.e.,

$$(x^{-1})^n + \sum_{1 \leq i \leq n} z_i(x^{-1})^{n-i} = 0 \quad \text{with} \quad n \in \mathbb{N}_+$$

and $z_i \in R$, and multiplying by x^{n-1} we get

$$x^{-1} = -\sum_{1 \leq i \leq n} z_i x^{i-1} \in R.$$

Therefore R is a field.

LYING BELOW LEMMA (T39). Given any overring S of a ring R and any prime ideal P in R, we have the following.

(T39.1) P lies below some prime ideal in S iff $(PS) \cap R = P$.

(T39.2) If P does not lie below any prime ideal in S, then there exists a prime ideal Q in S together with elements $y \in S$, $u \in P \setminus Q$, and $v \in R \setminus P$, such that $Q \cap R \subset P$ and $v - uy \in Q$.

PROOF. The only if part of (T39.1) is obvious. To prove its if part let W be the set of all ideals Q in S with $Q \cap R = P$ and assume that $(PS) \cap R = P$, and to prove (T39.2) let W be the set of all ideals Q in S with $Q \cap R \subset P$. Then W is nonempty (in the first case it contains PS and in the second case it contains 0). In both the cases the union of any linearly ordered (by inclusion) subfamily of W clearly belongs to W. Hence by Zorn's Lemma W contains a maximal member Q. Clearly $1 \notin Q$. If Q were not prime then there would exist elements r, t in $S \backslash Q$ with $rt \in Q$; now the ideals $Q + rS$ and $Q + tS$ properly contain Q (i.e., contain Q but are different from it) and hence by maximality they respectively contain elements r', t' belonging to $R \backslash P$; for suitable a, b in S we have $r' - ar \in Q$ and $t' - bt \in Q$ and this implies $r't' - abrt \in Q$ which gives the contradiction $r't' \in (Q \cap R) \backslash P$. Thus Q must be prime. This proves (T39.1). To prove (T39.2), assuming $Q \cap R \neq P$, we need to show the existence of $y \in S$, $u \in P \backslash Q$, and $v \in R \backslash P$, with $v - uy \in Q$. Since $Q \cap R \subset P$ with $Q \cap R \neq P$, there exists $u \in P \backslash Q$. Now $Q + uS$ properly contains Q, and hence by maximality we have $(Q + uS) \cap R \not\subset P$, and therefore we can find $y \in s$ and $v \in R \backslash P$ with $v - uy \in Q$.

LYING ABOVE THEOREM (T40). Given any integral extension S of a ring R and any prime ideal P in R, there exists a prime ideal Q in S lying above P.

PROOF. Suppose if possible that there is no prime in S lying above P. Then by (T39.2) there exits a prime Q in S together with elements $y \in S$, $u \in P \backslash Q$, and $v \in R \backslash P$, such that $Q \cap R \subset P$ and $v - uy \in Q$. Since $y \in S$ and S/R is integral, we have

$$y^n + \sum_{1 \leq i \leq n} x_i y^{n-i} = 0 \quad \text{with} \quad n \in \mathbb{N}_+$$

and $x_i \in R$. Multiplying by u^n we get

$$(uy)^n + \sum_{1 \leq i \leq n} (u^i x_i)(uy)^{n-i} = 0$$

and hence, because $v - uy \in Q$, we get

$$v^n + \sum_{1 \leq i \leq n} (u^i x_i) v^{n-i} \in Q.$$

Since the above LHS is in R, it must be in P; since $u \in P$, we get $v^n \in P$ and hence $v \in P$ which is a contradiction.

GOING UP THEOREM (T41). Given any integral extension S of a ring R, given any prime ideals $P^* \subset P$ in R, and given any prime ideal Q^* in S lying above P^*, there exists a prime ideal Q in S lying above P such that $Q^* \subset Q$.

PROOF. By (T37.1) S/Q^* may be regarded as an integral extension of R/P^*, and P/P^* may be regarded as a prime ideal in R/P^*, and so our assertion follows from (T40).

PROPER CONTAINMENT LEMMA (T42). Given any integral extension S of a ring R we have the following.

(T42.1) Assuming S to be a domain, for any nonzero ideal J in S we have $J \cap R \neq 0$.

(T42.2) If Q is a prime ideal in S lying above a prime ideal P in R and J is any ideal in S properly containing Q (i.e., containing Q but different from it), then $J \cap R$ properly contains P.

PROOF. (T42.2) follows by applying (T42.1) to S/Q and invoking (T37.1). To prove (T42.1), assume S to be a domain, fix $0 \neq y \in J$, and let

$$y^n + \sum_{1 \leq i \leq n} x_i y^{n-i} = 0 \quad \text{with} \quad n \in \mathbb{N}_+ \quad \text{and} \quad x_i \in R$$

be a smallest degree equation of integral dependence for y over R. Then $0 \neq x_n \in J \cap R$.

RADICAL DESCRIPTION LEMMA (T43). Given any integral extension S of a domain R such that S is a domain, and given any ideal H in R, we have $\mathrm{rad}_S(HS) = \{y \in S : y^n + \sum_{1 \leq i \leq n} x_i y^{n-i} = 0 \text{ with } n \in \mathbb{N}_+ \text{ and } x_i \in H\}$.

PROOF. If $y \in S$ satisfies an equation of the above type then $y^n \in HS$ and hence $y \in \mathrm{rad}_S(HS)$. Conversely, given any $y \in \mathrm{rad}_S(HS)$ we have $y^d \in HS$ for some $d \in \mathbb{N}_+$. We shall show that for any $z \in HS$ we have

$$z^m + \sum_{1 \leq i \leq m} t_i z^{m-i} = 0 \quad \text{with} \quad m \in \mathbb{N}_+ \text{ and } t_i \in H$$

and then by taking $z = y^d$ and $n = dm$ we will get $y^n + \sum_{1 \leq i \leq n} x_i y^{n-i} = 0$ with $x_i \in H$. If $z = hv$ with $h \in H$ and $v \in S$ then letting

$$v^m + \sum_{1 \leq i \leq m} u_i v^{m-i} = 0 \quad \text{with} \quad m \in \mathbb{N}_+ \quad \text{and} \quad u_i \in R$$

be an equation of integral dependence for v over R we get $z^m + \sum_{1 \leq i \leq m} h^i u_i z^{m-i} = 0$ with $h^i u_i \in H$ as a desired equation. Since every element of HS is a finite sum of elements of the form hv, it suffices to show that if z, w in S satisfy equations of the above form then so does $z + w$. So suppose that $z \in S$ satisfies the above displayed equation and $w \in S$ satisfies the equation

$$w^q + \sum_{1 \leq j \leq q} r_j w^{q-j} = 0 \quad \text{with} \quad q \in \mathbb{N}_+ \quad \text{and} \quad r_j \in H.$$

Let the products $(z^i w^j)_{0 \leq i \leq m-1, 0 \leq j \leq q-1}$ be designated as $f_1, \ldots, f_{mq}$. Then every power z^i with $i \geq m$ can be written as a linear combination of $1, z, \ldots, z^{m-1}$ with coefficients in H, and every power w^j with $j \geq q$ can be written as a linear combination of $1, w, \ldots, w^{q-1}$ with coefficients in H. Therefore every product $z^i w^j$ with $i + j \geq m + q - 1$ can be written as a linear combination of $f_1, \ldots, f_{mq}$ with coefficients in H. Hence for $p = m + q - 1$ we have

$$(z + w)^p f_i = \sum_{1 \leq j \leq mq} h_{ij} f_j \quad \text{with} \quad h_{ij} \in H \quad \text{and} \quad 1 \leq i \leq mq.$$

Therefore, letting δ_{ij} be the Kronecker symbol and considering the determinant D of the $(mq) \times (mq)$ matrix $((z + w)^p \delta_{ij} - h_{ij})$, by Cramer's Rule we get $D f_i = 0$ for $1 \leq i \leq mq$. Since one of the f_i equals 1, the determinant D is zero and this gives us an equation of degree p for $z + w$ of the desired type.

GOING DOWN THEOREM (T44). Given any integral extension S of a normal domain R such that S is a domain, given any prime ideals $P^* \subset P$ in R, and given any prime ideal Q in S lying above P, there exists a prime ideal Q^* in S lying above P^* such that $Q^* \subset Q$.

PROOF. Let T be the multiplicative set in S consisting of elements y which can be expressed as $y = ab$ with $a \in R \setminus P^*$ and $b \in S \setminus Q$. Suppose if possible that $T \cap (P^*S) \neq \emptyset$, i.e., for any such elements y, a, b we have $y = ab \in P^*S$. Then by (T43) we have $A(y) = 0$ for some univariate polynomial $A(Y)$ over R of the form

$$A(Y) = \sum_{0 \leq k \leq n} A_k Y^k \quad \text{where} \quad n \in \mathbb{N}_+ \quad \text{and} \quad A_k \in P^*$$

with $A_n = 1$. Let

$$B(Y) = \sum_{0 \leq i \leq m} B_i Y^i \quad \text{where} \quad m \in \mathbb{N}_+ \quad \text{and} \quad B_i \in \mathrm{QF}(R)$$

with $B_m = 1$ be the minimal polynomial of y over $\mathrm{QF}(R)$. Then

$$C(Y) = A(Y)/B(Y) = \sum_{0 \leq j \leq q} C_j Y^j \quad \text{where} \quad q \in \mathbb{N} \quad \text{and} \quad C_j \in \mathrm{QF}(R)$$

with $C_q = 1$. By Kronecker's Theorem (see L3§8(J25) and L4§14(N5)) $B(Y)$ and $C(Y)$ belong to $R[Y]$. Clearly there is a unique minimal i with $B_i \notin P^*$ and a unique minimal j with $C_j \notin P^*$. Now $A_{i+j} = B_i C_j + $ terms in P^*, and hence $A_{i+j} \notin P^*$, and so $i + j = n$, i.e., $i = m$ and $j = q$. Therefore $B_l \in P^*$ for $0 \leq l \leq m - 1$. Clearly

$$B'(Y) = \sum_{0 \leq l \leq m} B_l' Y^l \quad \text{with} \quad B_l' = B_l/a^{m-l}$$

is the minimal polynomial of b over $\mathrm{QF}(R)$, and hence again by Kronecker's Theorem (see L3§8(J25) and L4§14(N5)) we get $B_l' \in R$ for $0 \leq l \leq m - 1$. Since $B_l = a^{m-l} B_l'$

with $a \in R \setminus P^*$ and $B_l \in P^*$, we must have $B_l' \in P^*$. Consequently

$$b^m = - \sum_{0 \le l \le m-1} B_l' b^l \in P^* S \subset Q$$

which is a contradiction. Thus we have shown that $T \cap (P^* S) = \emptyset$. Hence by L4§5(O20)(12•) there exists a prime ideal Q^* in S with $T \cap Q^* = \emptyset$ and $P^* S \subset Q^*$. Now

$$T \cap Q^* = \emptyset \Rightarrow Q^* \cap (S \setminus Q) = \emptyset \Rightarrow Q^* \subset Q.$$

Also

$$T \cap Q^* = \emptyset \Rightarrow (Q^* \cap R) \cap (R \setminus P^*) = \emptyset \Rightarrow Q^* \cap R \subset P^*$$

where the first implication follows by noting that if there existed

$$a' \in (Q^* \cap R) \cap (R \setminus P^*)$$

then by letting $b' = 1 \in S \setminus Q$ we would get $y' = a' b' \in T \cap Q^*$. Since

$$P^* \subset (P^* S) \cap R \subset Q^* \cap R,$$

we must have $Q^* \cap R = P^*$.

DIMENSION COROLLARY (T45). For any integral extension S of a ring R we have the following.

(T45.1) Given any finite sequence $P_0 \subset P_1 \subset \cdots \subset P_n$ of prime ideals in R with $P_0 \ne P_1 \ne \cdots \ne P_n$, there exists a sequence $Q_0 \subset Q_1 \subset \cdots \subset Q_n$ of prime ideals in S with $Q_0 \ne Q_1 \ne \cdots \ne Q_n$ such that $P_i = Q_i \cap R$ for $0 \le i \le n$.

(T45.2) Given any finite sequence $Q_0 \subset Q_1 \subset \cdots \subset Q_n$ of prime ideals in S with $Q_0 \ne Q_1 \ne \cdots \ne Q_n$, upon letting $P_i = Q_i \cap R$ for $0 \le i \le n$, we have that $P_0 \subset P_1 \subset \cdots \subset P_n$ is a sequence of prime ideals in R with $P_0 \ne P_1 \ne \cdots \ne P_n$.

(T45.3) $\dim(S) = \dim(R)$.

(T45.4) For any prime ideal Q in S lying above any prime ideal P in R we have $\mathrm{dpt}_S Q = \mathrm{dpt}_R P$ and $\mathrm{ht}_S Q \le \mathrm{ht}_R P$.

(T45.5) If S is a domain and R is a normal domain, then for any prime ideal Q in S lying above any prime ideal P in R we have $\mathrm{ht}_S Q = \mathrm{ht}_R P$.

PROOF. The first assertion follows from (T40) and (T41). The second assertion follows from (T42). The third and the fourth assertions follow from the first two assertions. The fifth assertion follows from (T42) and (T44).

QUEST (Q10) Normalization Theorem and Regular Polynomials

The above theorem, proved by Emmy Noether in 1926, says that any affine ring A over a field k has a normalization basis over k, by which we mean a finite number of elements $Y_1, \ldots, Y_r$ in A which are algebraically independent over k and which are such that the ring A is integral over the subring $k[Y_1, \ldots, Y_r]$. By definition the affine ring A is a homomorphic image $\phi(B)$ of a finite variable polynomial ring $B = k[X_1, \ldots, X_N]$ under a k-epimorphism $\phi : B \to A$. The basic idea behind the Normalization Theorem is the fact that, by applying a k-automorphism θ to B, any given nonzero member of B can be made regular in X_N. Recall that a nonzero polynomial F of degree d in $X_1, \ldots, X_N$ over k is regular in X_N means

$$F = cX_N^d + \sum_{1 \leq i \leq d} F_i X_N^{d-i} \quad \text{where} \quad F_i \in k[X_1, \ldots, X_{N-1}]$$

with $\deg(F_i) \leq i$ and $0 \neq c \in k$. This was proved in L3§12(D4) where it was noted that if k is infinite then θ can be chosen to be k-linear, i.e.,

$$\theta(X_i) = \sum_{1 \leq j \leq N} M_{ij} X_j \quad \text{for} \quad 1 \leq i \leq N$$

where $M = (M_{ij})$ is an $N \times N$ matrix over k with $\det(M) \neq 0$; note that if $N = 0$ then $0 \neq F = c \in k$ with $d = 0$.

In Theorem (T47) we shall show that the length r of any normalization basis of A over k equals the dimension of A, and is hence independent of the particular normalization basis. Actually, in view of (T45), this follows from (T30.4). However we want to make (Q9) and (Q10) independent of (Q1) to (Q8).

NOETHER NORMALIZATION THEOREM (T46). Every affine ring A over a field k has a normalization basis $Y_1, \ldots, Y_r$ over k.

SUPPLEMENT. Given any k-epimorphism $\phi : B \to A$ where B is a finite variable polynomial ring $k[X_1, \ldots, X_N]$, and given any nonunit ideal I in B with $\ker(\phi) \subset I$, we can choose the normalization basis $Y_1, \ldots, Y_r$ of A over k, for some integer $0 \leq r \leq N$, together with a k-automorphism τ of B and elements $Z_1, \ldots, Z_N$ in B which are algebraically independent over k so that: B is integral over the ring $C = k[Z_1, \ldots, Z_N]$ with $\ker(\phi) \cap C = (Z_{r+1}, \ldots, Z_N)C$ and $(\phi(Z_1), \ldots, \phi(Z_r)) = (Y_1, \ldots, Y_r)$, and for some integer $0 \leq s \leq r$ we have $I \cap C = (Z_{s+1}, \ldots, Z_N)C$ with $(Z_1, \ldots, Z_s) = (\tau(X_1), \ldots, \tau(X_s))$ and $\phi(I) \cap E = (Y_{s+1}, \ldots, Y_r)E$ where $E = k[Y_1, \ldots, Y_r]$.

ADDENDUM. If k is infinite then τ can be chosen to be k-linear.

PROOF. We shall first consider the Special Case when ϕ is the identity map of $B = A$. Note that in this case $r = N$ and we need to find a k-automorphism τ of B (which is k-linear if k is infinite) together with elements $Z_1, \ldots, Z_N$ in B which are algebraically independent over k such that B is integral over $C = k[Z_1, \ldots, Z_N]$ and for the given nonunit ideal I in B we have $I \cap C = (Z_{s+1}, \ldots, Z_N)C$ with $(Z_1, \ldots, Z_s) = (\tau(X_1), \ldots, \tau(X_s))$ for some nonnegative integer $s \leq N$. To satisfy the "need" we make induction on N. This being trivial for $N = 0$, let $N > 0$ and assume for $N - 1$. If $I = 0$ then we have nothing to show. So also assume $I \neq 0$ and fix any $0 \neq Z_N \in I$. Since $I \neq B$, by L3§12(D4) there is a k-automorphism θ of B (which is k-linear if k is infinite) such that

$$\theta(Z_N) = cX_N^d + \sum_{1 \leq i \leq d} F_i X_N^{d-i} \quad \text{where} \quad F_i \in B' = k[X_1, \ldots, X_{N-1}]$$

with $d \in \mathbb{N}_+$ and $0 \neq c \in k$. Let $I' = \theta(I) \cap B'$. Then I' is a nonunit ideal in B' and hence by the induction hypothesis we can find a k-automorphism τ' of B' together with elements $Z_1', \ldots, Z_{N-1}'$ in B' which are algebraically independent over k such that B' is integral over $C' = k[Z_1', \ldots, Z_{N-1}']$ and for the nonunit ideal I' in B' we have $I' \cap C' = (Z_{s+1}', \ldots, Z_{N-1}')C'$ with $(Z_1', \ldots, Z_s') = (\tau'(X_1), \ldots, \tau'(X_s))$ for some nonnegative integer $s \leq N - 1$. Let $Z_i = \theta^{-1}(Z_i')$ for $1 \leq i \leq N - 1$. By the above displayed equation we see that $\theta^{-1}(X_N)$ is integral over $\theta^{-1}(B')[Z_N]$, and hence the elements $Z_1, \ldots, Z_N$ in B are algebraically independent over k and the ring B is integral over the subring $C = k[Z_1, \ldots, Z_N]$. Since $I' \cap C' = (Z_{s+1}', \ldots, Z_{N-1}')C'$, by the above displayed equation we also see that $I \cap C = (Z_{s+1}, \ldots, Z_N)C$. Obviously there is a unique k-automorphism τ of B such that $\tau(X_N) = \theta^{-1}(X_N)$, and $\tau(X_i) = \theta^{-1}(\tau'((X_i))$ for $1 \leq i \leq N - 1$. Clearly $(Z_1, \ldots, Z_s) = (\tau(X_1), \ldots, \tau(X_s))$, and if k is infinite then τ is k-linear.

Having completed the proof of the Special Case, let us turn to the General Case. Applying the Special Case to $(B, \ker(\phi))$ we can find a k-automorphism σ of B (which is k-linear if k is infinite) together with elements $T_1, \ldots, T_N$ in B which are algebraically independent over k such that the ring B is integral over the subring $D = k[T_1, \ldots, T_N]$ and we have $\ker(\phi) \cap D = (T_{r+1}, \ldots, T_N)D$ with $(T_1, \ldots, T_r) = (\sigma(X_1), \ldots, \sigma(X_r))$ for some nonnegative integer $r \leq N$. Let $B' = k[T_1, \ldots, T_r]$ and $I' = I \cap B'$. Applying the special case to (B', I') we find a k-automorphism τ' of B' (which is k-linear if k is infinite) together with elements $Z_1, \ldots, Z_r$ in B' which are algebraically independent over k such that B' is integral over $C' = k[Z_1, \ldots, Z_r]$ and we have $I' \cap C' = (Z_{s+1}, \ldots, Z_r)C'$ with $(Z_1, \ldots, Z_s) = (\tau'(T_1), \ldots, \tau'(T_s))$ for some nonnegative integer $s \leq r$. Let $Z_i = T_i$ for $r + 1 \leq i \leq N$. Then the elements $Z_1, \ldots, Z_N$ are algebraically independent over k, the ring B is integral over the subring $C = k[Z_1, \ldots, Z_N]$, and we have $\ker(\phi) \cap C = (Z_{r+1}, \ldots, Z_N)C$ with $I \cap C = (Z_{s+1}, \ldots, Z_N)C$. Let $Y_i = \phi(Z_i)$ for $1 \leq i \leq r$. Then $Y_1, \ldots, Y_r$ is a normalization basis of A over k, and we have $\phi(I) \cap E = (Y_{s+1}, \ldots, Y_r)E$ where $E = k[Y_1, \ldots, Y_r]$. Obviously there is a unique k-automorphism τ of B such that $\tau(X_i) = \tau'(\sigma(X_i))$ or $\sigma(X_i)$ according as $1 \leq i \leq r$ or $r + 1 \leq i \leq N$. Clearly $(Z_1, \ldots, Z_s) = (\tau(X_1), \ldots, \tau(X_s))$, and if k is infinite then τ is k-linear.

As a consequence of (T45) and (T46) we shall now prove Theorem (T47) which includes L4§8(T19). For convenience we introduce some:

TERMINOLOGY. By a saturated chain of prime ideals in a ring R we mean a finite sequence $P_0 \subset P_1 \subset \cdots \subset P_n$ of prime ideals in R with $P_0 \neq P_1 \neq \cdots \neq P_n$ such that there is no prime ideal P^* in R with $P_i \subset P^* \subset P_{i+1}$ and $P_i \neq P^* \neq P_{i+1}$ for some $i \in \{0, \ldots, n-1\}$; we call n the length of the chain; if $P \subset Q$ are prime ideals in R with $P = P_0$ and $Q = P_n$ then we say that $P_0 \subset P_1 \subset \cdots \subset P_n$ is a saturated prime ideal chain between P and Q; if there is no prime P in R with $P \subset P_0$ and $P \neq P_0$ and there is no prime Q in R with $P_n \subset Q$ and $P_n \neq Q$ then we say that $P_0 \subset P_1 \subset \cdots \subset P_n$ is an absolutely saturated prime ideal chain in R. By an infinite chain of prime ideals in a ring R we mean a sequence of prime ideals $(P_i)_{i \in \mathbb{N}}$ in R such that for all $i \in \mathbb{N}$ we have $P_i \subset P_{i+1}$ with $P_i \neq P_{i+1}$; if $P \subset Q$ are prime ideals in R such that $P \subset P_i \subset Q$ for all $i \in \mathbb{N}$ then we say that this is an infinite prime ideal chain between P and Q.

For a subdomain R of a domain R', by $\mathrm{trdeg}_R R'$ we denote the transcendence degree of $\mathrm{QF}(R')$ over $\mathrm{QF}(R)$. More generally, for a prime ideal P' in a ring R' lying above a prime ideal P in a subring R, after identifying R/P with a subdomain of R'/P', by $\mathrm{trdeg}_{R/P} R'/P'$ we denote the transcendence degree of $\mathrm{QF}(R'/P')$ over $\mathrm{QF}(R/P)$; in case R is a subfield k of R', we write $\mathrm{trdeg}_k R'/P'$ in place of $\mathrm{trdeg}_{k/0} R'/P'$; in case P' and P are the respective maximal ideals in quasilocal rings R' and R, we may write $\mathrm{restrdeg}_R R'$ in place of $\mathrm{trdeg}_{R/P} R'/P'$ and call this the residual transcendence degree of R' over R.

EXTENDED DIMENSION THEOREM (T47). Given any affine ring A over a field k, we have the following.

(T47.1) For any nonunit ideal J in A, upon letting $J_1, \ldots, J_m$ be all the distinct minimal primes of J in R we have

$$\dim(A/J) = \mathrm{dpt}_A J = \max_{1 \leq i \leq m} \mathrm{trdeg}_k(A/J_i) \in \mathbb{N}.$$

Hence, for any prime ideal P in A we have

$$\dim(A/P) = \mathrm{dpt}_A P = \mathrm{trdeg}_k(A/P) \in \mathbb{N}.$$

In particular, if A is a domain then $\dim(A) = \mathrm{trdeg}_k A \in \mathbb{N}$.

(T47.2) Every pair of prime ideals $P \subset Q$ in A has the property which says that: $(*)$ there is no infinite chain of prime ideals between P and Q, and any two saturated chains of prime ideals between P and Q have the same length. Moreover: $(**)$ this common length equals $\dim(A/P) - \dim(A/Q)$.

(T47.3) If A is a domain then the length of every absolutely saturated chain of prime ideals in R equals the dimension of R.

(T47.4) If A is a domain then for every nonunit ideal H in A we have

$$\mathrm{ht}_A H + \mathrm{dpt}_A H = \dim(A).$$

(T47.5) If A is a domain then, for any prime ideal P in A, the length of every saturated prime ideal chain in A between 0 and P equals $\mathrm{ht}_A P$, and the length of every saturated prime ideal chain in A between P and any member of $\mathrm{mvspec}_A P$ equals $\mathrm{dpt}_A P$.

(T47.6) If A is a domain and A' is an affine domain over A then for any prime ideal P' in A' lying above any prime ideal P in A, upon considering the localizations $R' = A'_{P'}$ and $R = A_P$, we have

$$\dim(R') + \mathrm{restrdeg}_R R' = \dim(R) + \mathrm{trdeg}_R R'.$$

(T47.7) In the situation of (T46) we have $\dim(A) = r$ and $\mathrm{dpt}_A \phi(I) = s$.

PROOF. Clearly $\dim(A/J) = \mathrm{dpt}_A J = \max_{1 \le i \le m} \dim(A/J_i)$ and hence to prove (T47.1) it suffices to show that if A is a domain then $\dim(A) = N$ where $N = \mathrm{trdeg}_k A$. We shall do this by induction on N. By (T46) we can arrange that A is an integral extension of $B = k[X_1, \ldots, X_N]$ and then by (T45.3) we have $\dim(A) = \dim(B)$. If $N = 0$ then obviously $\dim(B) = 0$ and hence $\dim(A) = 0$. So let $N > 0$ and assume that our assertion is true for all smaller values of N. By L4§5(O22)(X10) we have $\dim(B) \ge N$. Suppose if possible that $\dim(B) > N$, and let $\overline{Q}_0 \subset \overline{Q}_1 \subset \cdots \subset \overline{Q}_{N+1}$ be a sequence of prime ideals in B for which we have $\overline{Q}_0 \ne \overline{Q}_1 \ne \cdots \ne \overline{Q}_{N+1}$. Then $B/\overline{Q}_1$ is an affine domain over k with $\mathrm{trdeg}_k(B/\overline{Q}_1) < N$ and hence by the induction hypothesis $\dim(B/\overline{Q}_1) < N$ which is a contradiction because obviously $\dim(B/\overline{Q}_1) = \mathrm{dpt}_B \overline{Q}_1 \ge N$. This completes the proof of (T47.1).

Continuing to assume A to be a domain, let $0 = P_0 \subset P_1 \subset \cdots \subset P_n$ be any absolutely saturated prime ideal chain in A. Since $\dim(A) = N$, we must have $n \le N$. By induction on N we shall show that $n = N$, which will prove (T47.3). This being obvious for $N = 0$, let $N > 0$ and assume for all smaller values of N. Since $\dim(A) > 0$, the domain A cannot be a field, and hence we must have $n > 0$. For $0 \le i \le n$ let $Q_i = P_i \cap B$. Then $0 = Q_0 \subset Q_1 \subset \cdots \subset Q_n$ is a prime ideal sequence in B, and by (T45.2) we have $Q_0 \ne Q_1 \ne \cdots \ne Q_n$; we CLAIM that this is an absolutely saturated chain in B. First note that P_n is clearly a maximal ideal in A, and hence Q_n is a maximal ideal in B by (T45.4). Also clearly $\mathrm{ht}_A P_1 = 1$ and B is a normal domain because it is a UFD; consequently $\mathrm{ht}_B Q_1 = 1$ by (T45.5). By the Special Case of (T46) we can find a sequence of integers $N > N(1) > \cdots > N(n) = 0$ and, for $1 \le i \le n$, elements $X_{i,1}, \ldots, X_{i,N}$ in B which are algebraically independent over k such that B is integral over the subring $B_i = k[X_{i,1}, \ldots, X_{i,N}]$ with $Q_i \cap B_i = (X_{i,N(i)+1}, \ldots, X_{i,N})B_i$. Again $\mathrm{ht}_{B_1}(Q_1 \cap B_1) = 1$ by (T45.5), and hence we must have $N(1) = N-1$; consequently $B_1/(Q_1 \cap B_1) = B_1/(X_{1,N})B_1$ is isomorphic to the $(N-1)$-variable polynomial ring $k[X_{1,1}, \ldots, X_{1,N-1}]$. Since B/Q_1 is integral over $B_1/(Q_1 \cap B_1)$, it follows that B/Q_1 is an $(N-1)$-dimensional affine domain over k, and hence by the induction hypothesis, every absolutely saturated prime ideal chain in it has length $N-1$.

Given any $i \in \{1, \ldots, n-1\}$, let $\alpha_i : A \to A/P_i$ and $\beta_i : B \to B/Q_i$ be the residue class epimorphisms, and let $C_i = k[X_{i,1}, \ldots, X_{i,N(i)}]$. Then α_i and β_i induce isomorphisms $C_i \to \alpha_i(C_i)$ and $C_i \to \beta_i(C_i)$, and the domains A/P_i and B/Q_i are integral over the polynomial rings $\alpha_i(C_i)$ and $\beta_i(C_i)$ respectively. Moreover, under the induced isomorphisms, the prime ideal $Q_{i+1} \cap C_i$ in C_i is sent to the prime ideals $(P_{i+1}/P_i) \cap \alpha_i(C_i)$ and $(Q_{i+1}/Q_i) \cap \beta_i(C_i)$ respectively. Consequently, by applying (T45.5) twice we get $\mathrm{ht}_{A/P_i}(P_{i+1}/P_i) = \mathrm{ht}_{C_i}(Q_{i+1} \cap C_i) = \mathrm{ht}_{B/Q_i}(Q_{i+1}/Q_i)$. Since the chain $P_0 \subset P_1 \subset \cdots \subset P_n$ is saturated, we must have $\mathrm{ht}_{A/P_i}(P_{i+1}/P_i) = 1$. Therefore $\mathrm{ht}_{B/Q_i}(Q_{i+1}/Q_i) = 1$. This being so for $1 \le i \le n-1$, our CLAIM is proved. Consequently by going mod Q_1 we see that $Q_1/Q_1 \subset \cdots \subset Q_n/Q_1$ is an absolutely saturated prime ideal chain in B/Q_1 of length $n-1$, and hence $n-1 = N-1$. Therefore $n = N$. This completes the proof of (T47.3).

Still continuing to assume A to be a domain, given any prime ideal P in A, we can clearly find saturated prime ideal chains

$$('\,)\qquad\qquad \widehat{P}_0 \subset \cdots \subset \widehat{P}_h \quad \text{and} \quad \widetilde{P}_0 \subset \cdots \subset \widetilde{P}_d$$

in A such that $\widehat{P}_0 = 0$ and $\widehat{P}_h = P$ with $\mathrm{ht}_A P = h$ and such that $\widetilde{P}_0 = P$ and $\widetilde{P}_d \in \mathrm{mspec}(A)$ with $\mathrm{dpt}_A P = d$. This gives the absolutely saturated prime ideal chain

$$(''\,)\qquad\qquad \widehat{P}_0 \subset \cdots \subset \widehat{P}_h \subset \widetilde{P}_1 \subset \cdots \subset \widetilde{P}_d$$

in A of length $h + d$, and hence by (T47.3) we get $h + d = N$. In particular, given any nonunit ideal H in A, for every P in $\mathrm{vspec}_A H$ we have $\mathrm{ht}_A P + \mathrm{dpt}_A P = N$, and hence $\mathrm{ht}_A H + \mathrm{dpt}_A H = N$. This proves (T47.4).

Again assuming A to be a domain, given any prime ideal P in A, if $(')$ are any saturated prime ideal chains in A such that $\widehat{P}_0 = 0$ with $\widehat{P}_h = P$ and such that $\widetilde{P}_0 = P$ with $\widetilde{P}_d \in \mathrm{mspec}(A)$, then clearly $('')$ is an absolutely saturated prime ideal chain in A of length $h + d$, and hence by (T47.3) we get $h + d = N$. But obviously $\mathrm{ht}_A P \ge h$ and $\mathrm{dpt}_A P \ge d$, and by (T47.4) we have $\mathrm{ht}_A P + \mathrm{dpt}_A P = n$. Therefore we must have $\mathrm{ht}_A P = h$ and $\mathrm{dpt}_A P = d$. This proves (T47.5).

In (T47.2), the nonexistence of an infinite chain between P and Q follows because by (T47.1) we have $\dim(A/P) < \infty$. Moreover, given any saturated prime ideal chain $P = P_0 \subset P_1 \subset P_n = Q$ in A between P and Q, we get a saturated prime ideal chain $0 = P_0/P \subset P_1/P \subset P_n/P = Q/P$ in A/P between 0 and Q/P, and hence by (T47.5) we obtain $n = \mathrm{ht}_{A/P}(Q/P)$ and by (T47.4) we have $\mathrm{ht}_{A/P}(Q/P) = \dim(A/P) - \dim(A/Q)$. This proves (T47.2).

Once again assume that A is a domain and let P, R, A', P', R' be as in (T47.6). Then A' is also an affine domain over k, and hence by (T47.1) we get

$$\dim(A') - \dim(A) = \mathrm{trdeg}_R R'$$

and

$$\mathrm{restrdeg}_R R' = \dim(A'/P') - \dim(A/P).$$

By (T47.3) we also have

$$\dim(R') - \dim(R) = [\dim(A') - \dim(A'/P')] - [\dim(A) - \dim(A/P)].$$

Adding the three equations we obtain

$$\dim(R') + \mathrm{restrdeg}_R R' = \dim(R) + \mathrm{trdeg}_R R'$$

which proves (T47.6).

In the situation of (T46), A is integral over the r-variable polynomial ring $E = k[Y_1, \ldots, Y_r]$ and $A/\phi(I)$ is integral over an isomorphic image of the s-variable polynomial ring $D = k[Y_1, \ldots, Y_s]$, and hence by (T45.3) and (T47.1) we get $\dim(A) = \dim(E) = r$ and $\mathrm{dpt}(A/\phi(I)) = \dim(A/\phi(I)) = \dim(D) = s$. This proves (T47.7).

We shall close this Quest by deducing some parts of the Hilbert Nullstellensatz, i.e., L4§8(T22), as consequences of part (T47.1) of the above Dimension Theorem. The rest of the Nullstellensatz will be dealt with in the next Quest.

PROOF OF (T22.1) TO (T22.6) OF L4§8. The first part (T22.1) says that a field which is an affine ring over a field k is algebraic over k; this follows from (T47.1) because the dimension of a field is 0, and algebraic means whose transcendence degree is 0.

The second part (T22.2) says that for any maximal ideal J in the finite variable polynomial ring $B_{N,k} = k[X_1, \ldots, X_N]$, the field $B_{N,k}/J$ is algebraic over k; this is simply (T22.1) in a different guise.

The third part (T22.3) says that any such maximal ideal J has a unique set of N generators $(f_i(X_1, \ldots, X_i) \in B_{i,k})_{1 \le i \le N}$ such that for $1 \le i \le N$ we have

$$f_i(X_1, \ldots, X_i) = X_i^{n_i} + g_i(X_1, \ldots, X_i)$$

where n_i is a positive integer and $g_i(X_1, \ldots, X_i) \in B_{i,k}$ with

$$\deg_{X_j} g_i(X_1, \ldots, X_i) < n_j \quad \text{for} \quad 1 \le j \le i.$$

As a preamble to the proof, let $\phi : B_{N,k} \to B_{N,k}/J$ be the residue class epimorphism, let k_0 be the subfield $\phi(k)$ of the field $\phi(B_{N,k})$, and for $1 \le i \le N$ let $k_i = \phi(B_{i,k})$ and $x_i = \phi(X_i)$ together with

$$\gamma_i = \{g(X_1, \ldots, X_i) \in B_{i,k} : \deg_{X_j} g(X_1, \ldots, X_i) < n_j \text{ for } 1 \le j \le i\}.$$

By (T22.2) we know that k_N is a finite algebraic field extension of k_0 (i.e., k_N is an overfield of the field k_0 with $[k_N : k_0] < \infty$), and hence for $1 \le i \le N$ we see that $k_i = k_{i-1}(x_i)$ is a finite algebraic field extension of the field k_{i-1}. In proving our claim by induction on N, let us include the extra assertion which says that for $1 \le i \le N$ we have that: the polynomial $f_i(x_1, \ldots, x_{i-1}, Y)$ is the minimal polynomial of x_i over k_{i-1} (and hence the field degree $[k_i : k_{i-1}]$ equals n_i), the polynomial $f_i(X_1, \ldots, X_i)$ is irreducible in $B_{i,k}$, and

$$g(X_1, \ldots, X_i) \mapsto g(x_1, \ldots, x_i) \quad \text{gives a bijection} \quad \gamma_i \to k_i.$$

With this added task the $N = 0$ case is obvious and for $N > 0$ the implication $N - 1 \Rightarrow N$ can be proved thus. Let $B = B_{N,k}$ and $A = B_{N-1,k}$ with $I = J \cap A$. Then ϕ maps A onto the field k_{N-1} and the kernel of the induced epimorphism $A \to k_{N-1}$ is I. Consequently I is a maximal ideal in A. Therefore by the induction hypothesis I has a unique set of $N - 1$ generators $(f_i(X_1, \ldots, X_i) \in B_{i,k})_{1 \le i \le N-1}$ such that for $1 \le i \le N - 1$ we have

$$f_i(X_1, \ldots, X_i) = X_i^{n_i} + g_i(X_1, \ldots, X_i)$$

where n_i is a positive integer and $g_i(X_1, \ldots, X_i) \in B_{i,k}$ with

$$\deg_{X_j} g_i(X_1, \ldots, X_i) < n_j \quad \text{for} \quad 1 \le j \le i.$$

Moreover, for $1 \le i \le N - 1$ we have that: the polynomial $f_i(x_1, \ldots, x_{i-1}, Y)$ is the minimal polynomial of x_i over k_{i-1} (and hence the field degree $[k_i : k_{i-1}]$ equals n_i), the polynomial $f_i(X_1, \ldots, X_i)$ is irreducible in $B_{i,k}$, and

$$g(X_1, \ldots, X_i) \mapsto g(x_1, \ldots, x_i) \quad \text{gives a bijection} \quad \gamma_i \to k_i.$$

Upon letting $n_N = [k_N : k_{N-1}]$ we see that n_N is a positive integer and we can find $f_N = f_N(X_1, \ldots, X_N) \in B$ such that $f_N(x_1, \ldots, x_{n-1}, Y)$ is the minimal polynomial of x_n over k_{N-1} and

$$f_N(X_1, \ldots, X_N) = X_N^{n_N} + g_N(X_1, \ldots, X_N)$$

where $g_N(X_1, \ldots, X_N) \in B$ with

$$\deg_{X_j} g_N(X_1, \ldots, X_N) < n_j \quad \text{for} \quad 1 \le j \le N.$$

Clearly $f_N(X_1, \ldots, X_N)$ is irreducible in B and belongs to J. Given any $h \in B$, by the division algorithm we can uniquely write $h = qf_N + r$ with q, r in B such that $\deg_{X_N} r < n_N$. From this it follows that $g(X_1, \ldots, X_N) \mapsto g(x_1, \ldots, x_N)$ gives a bijection $\gamma_N \to k_N$. Moreover $h \in J \Rightarrow r \in I \Rightarrow r \in (f_1, \ldots, f_{N-1})A \Rightarrow h \in (f_1, \ldots, f_N)B$, and hence $(f_1, \ldots, f_N)B = J$. This completes the induction except for the uniqueness of $f_1, \ldots, f_N$.

To prove the uniqueness let $(F_i(X_1, \ldots, X_i) \in B_{i,k})_{1 \le i \le N}$ be any set of N generators of J in B such that for $1 \le i \le N$ we have

$$F_i(X_1, \ldots, X_i) = X_i^{N_i} + G_i(X_1, \ldots, X_i)$$

where N_i is a positive integer and $G_i(X_1, \ldots, X_i) \in B_{i,k}$ with

$$\deg_{X_j} G_i(X_1, \ldots, X_i) < N_j \quad \text{for} \quad 1 \le j \le i.$$

Now $(IB) \cap A = I$ and hence there is a unique epimorphism $\theta : B \to C = k_{N-1}[Y]$ such that $\theta(X_N) = Y$ and $\theta(z) = \phi(z)$ for all $z \in A$. Let

$$\widehat{f}(Y) = f_N(x_1, \ldots, x_{n-1}, Y) \quad \text{and} \quad \widehat{F}(Y) = F_N(x_1, \ldots, x_{n-1}, Y).$$

Then $\widehat{f}(Y)C = \theta(J) = \widehat{F}(Y)C$ and hence $\widehat{f}(Y) = \widehat{F}(Y)$. Therefore $N_N = n_N$, and given any $h \in B$, by the division algorithm we can uniquely write $h = QF_N + R$ with Q, R in B such that $\deg_{X_N} R < n_N$. From all this we see that $(F_1, \ldots, F_{N-1})A = I$. Hence by the uniqueness part of the induction hypothesis $F_i = f_i$ with $N_i = n_i$ for $1 \leq i \leq N - 1$. It is clear that also $F_N = f_N$. Thus (T22.3) has been proved.

The fourth part (T22.4) says that if k is algebraically closed then the mapping $\alpha \mapsto \mathbb{I}_k(\alpha)$ gives a bijection $\mathbb{A}_k^N \to \mathrm{mspec}(B_{N,k})$, where we recall that for every $\alpha = (\alpha_1, \ldots, \alpha_N) \in \mathbb{A}_k^N = k^N$ we have put $\mathbb{I}_k(\alpha) = \{f \in B_{N,k} : f(\alpha) = 0\}$. Note that for

$$f = f(X_1, \ldots, X_N) \in B_{N,k} \quad \text{and} \quad \alpha = (\alpha_1, \ldots, \alpha_N) \in \mathbb{A}_\kappa^N = \kappa^N$$

where $k \subset \kappa$ are any fields we write

$$f(\alpha) = f(\alpha_1, \ldots, \alpha_N).$$

Clearly (T22.4) follows from (T22.3) or even from (T22.2). To see the latter, assuming k algebraically closed, let there be given any J in $\mathrm{mspec}(B_{N,k})$. By (T22.2) the residue class epimorphism $\phi : B_{N,k} \to B_{N,k}/J$ maps k onto the entire image. Hence there exists $\alpha \in \mathbb{A}_k^N$ such that $\phi(X_i) = \phi(\alpha_i)$ for $1 \leq i \leq N$. Now clearly we have $\mathbb{I}_k(\alpha) = J = (X_1 - \alpha_1, \ldots, X_N - \alpha_N)B_{N,k}$.

To prove the rest of (T22) and some relevant portions of L4§8(T21), henceforth let $k \subset \kappa$ be fields, and let

$$R = B_{N,k} = k[X_1, \ldots, X_N] \quad \text{and} \quad S = \mathbb{A}_\kappa^N = \kappa^N$$

where N is a nonnegative integer. For any $U \subset S$ let

$$I(U) = \mathbb{I}_k(U) = \{f \in R : f(\alpha) = 0 \text{ for all } \alpha \in U\}$$

and for any $J \subset R$ let

$$V(J) = \mathbb{V}_\kappa(J) = \{\alpha \in S : f(\alpha) = 0 \text{ for all } f \in J\}.$$

Let

$$I' = \{I(U) : U \subset S\}$$

and

$$\mathrm{rd}(R) = \text{set of all ideals in } R \text{ which are their own radicals.}$$

let

$$V' = \mathrm{avt}_k(\mathbb{A}_\kappa^N) = \{V(J) : J \subset R\}$$

and

$$V'' = \text{set of all irreducible members of } V'$$

where U in V' is irreducible means U is nonempty and cannot be expressed as union of two members of V' which are subsets of U different from U.

Now clearly

(1) $\begin{cases} \text{for any subsets } J \text{ and } J' \text{ of } R \text{ we have: } J \subset J' \Rightarrow V(J') \subset V(J) \\ \text{and we have: } \text{rad}_R(JR) = \text{rad}_R(J'R) \Rightarrow V(J) = V(J') \end{cases}$

and

(2)   for any subsets U and U' of S we have: $U \subset U' \Rightarrow I(U') \subset I(U)$.

We claim that

(3) $\begin{cases} \text{for any family of ideals } (J_l)_{l \in L} \text{ in } R \text{ we have} \\ V(\sum_{l \in L} J_l) = \cap_{l \in L} V(J_l), \\ \text{and if the family is finite then we have} \\ V(\cap_{l \in L} J_l) = V(\prod_{l \in L} J_l) = \cup_{l \in L} V(J_l) \end{cases}$

and

(4) $\begin{cases} \text{for any family of subsets } (U_l)_{l \in L} \text{ of } S \text{ we have} \\ I(\cup_{l \in L} U_l) = \cap_{l \in L} I(U_l). \end{cases}$

Concerning the first equation of (3), by the first implication of (1) we see that LHS $\subset$ RHS; conversely: $\alpha \in$ RHS $\Rightarrow \alpha \in V(\cup_{l \in L} J_l) \Rightarrow \alpha \in$ LHS where the first implication is obvious and the second follows from the second implication of (1). Concerning the last line of (3), by the first implication of (1) we get the inclusions $\cup_{l \in L} V(J_l) \subset V(\cap_{l \in L} J_l) \subset V(\prod_{l \in L} J_l)$; moreover, for any $\alpha \in S$ we have that: $\alpha \notin \cup_{l \in L} V(J_l) \Rightarrow$ for each $l \in L$ there exists $f_l \in J_l$ with $f_l(\alpha) \neq 0 \Rightarrow f(\alpha) \neq 0$ with $f = \prod_{l \in L} f_l \in \prod_{l \in L} J_l$, and hence $V(\prod_{l \in L} J_l) \subset \cup_{l \in L} V(J_l)$. Concerning equation (4), by (2) we get LHS $\subset$ RHS; conversely: $f \in$ RHS $\Rightarrow f(\alpha) = 0$ for every $\alpha \in \cup_{l \in L} U_l \Rightarrow f \in$ LHS.

We also claim that

(5) $\begin{cases} \text{for any } J \subset R \text{ we have } J \subset I(V(J)) \\ \text{and: } J = I(V(J)) \Leftrightarrow J \in I' \Rightarrow J = \text{rad}_R J \end{cases}$

and

(6) $\begin{cases} \text{for any } U \subset S \text{ we have } U \subset V(I(U)) \\ \text{and: } U = V(I(U)) \Leftrightarrow U \in V'. \end{cases}$

The first line of (5) follows by first applying the first line of (1) and after that applying (2). The first line of (6) follows by first applying (2) and after that applying the first line of (1). Concerning the second line of (5), obviously $J = I(V(J)) \Rightarrow J \in I'$; moreover

$$
\begin{cases}
J \in I' \\
\Rightarrow J = I(U) \text{ for some } U \subset S \\
\Rightarrow V(J) = V(I(U)) \text{ by taking } V \text{ of both sides} \\
\Rightarrow U \subset V(J) \text{ by first line of (6)} \\
\Rightarrow I(V(J)) \subset J \text{ by (2) because } I(U) = J \\
\Rightarrow J = I(V(J)) \text{ by the first line of (5);}
\end{cases}
$$

also:

$$
\begin{cases}
f \in \mathrm{rad}_R J \text{ with } J = I(U) \text{ for some } U \subset S \\
\Rightarrow f^n \in J \text{ for some } n \in \mathbb{N}_+ \\
\Rightarrow f^n(\alpha) = 0 \text{ for all } \alpha \in U \\
\Rightarrow f(\alpha) = 0 \text{ for all } \alpha \in U \\
\rightarrow f \in J
\end{cases}
$$

and hence $J = \mathrm{rad}_R J$. Concerning the second line of (6), obviously $U = V(I(U)) \Rightarrow U \in V'$; moreover

$$
\begin{cases}
U \in V' \\
\Rightarrow U = V(J) \text{ for some } J \subset R \\
\Rightarrow I(U) = I(V(J)) \text{ by taking } I \text{ of both sides} \\
\Rightarrow J \subset I(U) \text{ by first line of (5)} \\
\Rightarrow V(I(U)) \subset U \text{ by (1) because } V(J) = U \\
\Rightarrow U = V(I(U)) \text{ by the first line of (6).}
\end{cases}
$$

Next we claim that

(7) $$ V'' = \{U \in V' : I(U) \in \mathrm{spec}(R)\}. $$

To prove (7) let $U \in V'$. Assuming U to be irreducible, we have $U \neq \emptyset$ and hence $I(U) \neq R$; moreover, for $1 \leq i \leq 2$, given any $f_i \in R \setminus I(U)$, upon letting $U_i = \{\alpha \in U; f_i(\alpha) = 0\}$, we have $U_i \in V'$ with $U_i \subset U$ and $U_i \neq U$; since U is irreducible, we must have $U_1 \cup U_2 \neq U$, and hence $f_1 f_2 \notin I(U)$; thus $I(U)$ is prime. Conversely assuming $I(U)$ to be prime, we have $I(U) \neq R$ and hence $U \neq \emptyset$; moreover, given any U_1, U_2 in V' with $U_1 \cup U_2 = U$, by (4) we have $I(U_1) \cap I(U_2) = I(U)$ and clearly $I(U_1) I(U_2) \subset I(U_1) \cap I(U_2)$ and hence $I(U_1) I(U_2) \subset I(U)$; now assuming $U_1 \neq U$, by (2) and the second line of (6) we get $I(U) \subset I(U_1)$ with $I(U) \neq I(U_1)$, and hence the primeness of $I(U)$ tells us that $I(U_2) \subset I(U)$ and therefore by the first line of (1) we get $U_2 = U$; thus U is irreducible.

Finally we claim that

(8)
$$\begin{cases} \text{every } U \in V' \text{ can be expressed as a finite union} \\ U = \cup_{1 \le i \le h} U_i \text{ with } U_i \in V'', \\ \text{and this decomposition is unique up to order if it is irredundant,} \\ \text{i.e., if no } U_i \text{ can be deleted from it,} \end{cases}$$

(where the irredundancy can be achieved by deleting some of the U_i, and after it is achieved, the remaining U_i are called the IRREDUCIBLE COMPONENTS of U).

Before proving (8) we note that the noetherianness of R tells us that it has no strictly increasing infinite sequence of ideals, i.e., there does not exist any infinite sequence ideals $J_0 \subset J_1 \subset \cdots \subset J_n \subset \ldots$ in R with $J_0 \ne J_1 \ne \cdots \ne J_n \ne \ldots$, and hence by (2) and the second line of (6) we see that

(9)
$$\begin{cases} V' \text{ has no strictly decreasing infinite sequence,} \\ \text{i.e., there does not exist any infinite sequence} \\ W_0 \supset W_1 \supset \cdots \supset W_n \supset \ldots \text{ in } V' \\ \text{with } W_0 \ne W_1 \ne \cdots \ne W_n \ne \ldots. \end{cases}$$

To prove the existence part of (8), suppose if possible that some $U \in V'$ has no irreducible decomposition, i.e., cannot be expressed as a finite union of members of V''. Then U itself cannot be irreducible and we can write $U = U_1 \cup U_2$ with U_1, U_2 in V' such that $U_1 \ne U \ne U_2$; note that the empty variety has the empty irreducible decomposition (with $h = 0$), and hence U is nonempty, and therefore U_1, U_2 are also nonempty. Clearly either U_1 or U_2 has no irreducible decomposition (because if both did then putting them together would produce an irreducible decomposition of U). Say U_1 does not have an irreducible decomposition. Let $W_0 = U$ and $W_1 = U_1$. In this manner we get an infinite sequence $W_0 \supset W_1 \supset \cdots \supset W_n \supset \ldots$ of members of V' with $W_0 \ne W_1 \ne \cdots \ne W_n \ne \ldots$ such that none of the W_n has an irreducible decomposition. This contradicts (9) and so proves the existence part.

To prove the uniqueness part of (8), in addition to assuming the decomposition displayed in (8) to be irredundant, let $U = \cup_{1 \le j \le h'} U_j'$ with $U_j' \in V''$ be another irredundant decomposition. For any $i \in \{1, \ldots, h\}$ we have $V_i = \cup_{1 \le j \le h'}(V_j' \cap V_i)$ and hence, by the irreducibility of V_i, for some j we must have $V_i = V_j' \cap V_i$, i.e., $V_i \subset V_j'$; by symmetry for some i^* we must have $V_j' \subset V_{i^*}$; consequently $V_i \subset V_{i^*}$ and hence $V_i \subset V_{i^*}$; therefore by irredundancy we must have $i = i^*$ and so $V_i = V_j'$. Thus each V_i coincides with some V_j', and similarly each V_j' coincides with some V_i. It follows that $h = h'$ and after suitable relabelling $V_i = V_i'$ for $1 \le i \le h$.

The fifth part (T22.5) of (T22) says that if κ contains an algebraic closure of k then for any ideal J in R we have that $I(V(J)) = \mathrm{rad}_R J$. Here the inclusion $\mathrm{rad}_R J \subset I(V(J))$ follows from the first line of (5). To prove the other inclusion, let there be given any $f \in I(V(J))$. We want to show that then $f^n \in J$ for some $n \in \mathbb{N}_+$. By assumption, κ contains an algebraic closure k' of k. Let $T' =$

$B_{N+1,k'} = k'[X_1, \ldots, X_{N+1}]$ and $H' = JT' + (1 - X_{N+1}f)T'$. Then clearly there is no $\alpha' = (\alpha'_1, \ldots, \alpha'_{N+1}) \in \mathbb{A}^{N+1}_{k'}$ such that $f'(\alpha') = 0$ for all $f' \in H'$. Suppose if possible that $H' \neq T'$. Then H' is contained in some maximal ideal H^* in T', and by (T22.4) we have $H^* = (X_1 - \alpha'_1, \ldots, X_{N+1} - \alpha'_{N+1})T'$. But this gives the contradiction that $f'(\alpha') = 0$ for all $f' \in H'$. Therefore we must have $H' = T'$. Hence, upon letting $T = B_{N+1,k} = k[X_1, \ldots, X_{N+1}]$ and $H = JT + (1 - X_{N+1}f)T$, by the following Comment (C15) we get $H = T$. Consequently we can express 1 as a finite sum

$$1 = (1 - X_{N+1}f)g + \sum_{1 \leq i \leq m} f_i g_i \quad \text{with} \quad f_i \in J \quad \text{and} \quad g, g_i \text{ in } T.$$

Substituting $1/X_{N+1}$ for X_{N+1} in the above identity and then multiplying both sides by X^n_{N+1} for a large enough $n \in \mathbb{N}_+$ we get

$$X^n_{N+1} = (X_{N+1} - f)G + \sum_{1 \leq i \leq m} f_i G_i \quad \text{with} \quad f_i \in J \quad \text{and} \quad G, G_i \text{ in } T.$$

Now substituting f for X_{N+1} in the above identity we obtain

$$f^n = \sum_{1 \leq i \leq m} f_i h_i \quad \text{with} \quad f_i \in J \quad \text{and} \quad h_i \in R$$

and hence $f^n \in J$.

The sixth part (T22.6) of (T22) says that if κ contains an algebraic closure of k then the mapping $U \mapsto I(U)$ gives inclusion reversing bijections $V' \to \mathrm{rd}(R)$ and $V'' \to \mathrm{spec}(B)$ whose inverses are given by $J \mapsto V(J)$. In view of (1), (2), (5), (6) and (7), this follows from (T22.5).

COMMENT (C15). [**Ideals Under Field Extensions**]. In the above proof of (T22.5) we used the fact that: if $k \subset k'$ are any fields and $T = k[X_1, \ldots, X_N]$ and $T' = k'[X_1, \ldots, X_N]$ are finite variable polynomial rings, then for any ideal H in T we have $(HT') \cap T = H$. To see this, we can take a vector space basis $(z_i)_{i \in I}$ of k' over k with $z_j = 1$ for some $j \in I$. By expressing the coefficients of any $f \in T'$ in terms of this basis we can uniquely write f as a finite sum

$$f = \sum_{i \in J} z_i f_i \quad \text{with} \quad f_i \in T$$

where J is a finite subset of I with $j \in J$. Note that then

$$f \in T \Leftrightarrow f_i = 0 \text{ for all } i \in J \setminus \{j\}.$$

Assuming $f \in HT'$ we can write it as a finite sum

$$f = \sum_{l \in L} F_l g_l \quad \text{with} \quad F_l \in T' \quad \text{and} \quad g_l \in H.$$

By taking J large enough, for all $l \in L$ we have

$$F_l = \sum_{i \in J} z_i F_{li} \quad \text{with} \quad F_{li} \in T.$$

Now the uniqueness says that for all $i \in J$ we have

$$f_i = \sum_{l \in L} F_{li} g_l \in H.$$

Hence if $f \in (HT') \cap T$ then $f = f_l \in H$. It follows that $(HT') \cap T = H$.

COMMENT (C16). [**Supplement to Hilbert Nullstellensatz**]. In the context of L4§8(T22.3), let $\phi : B_{N,k} \to B_{N,k}/J$ be the residue class epimorphism, let k_0 be the subfield $\phi(k)$ of the field $\phi(B_{N,k})$, and for $1 \le i \le N$ let $k_i = \phi(B_{i,k})$ and $x_i = \phi(X_i)$ together with

$$\gamma_i = \{g(X_1, \ldots, X_i) \in B_{i,k} : \deg_{X_j} g(X_1, \ldots, X_i) < n_j \text{ for } 1 \le j \le i\}.$$

Then, for $1 \le i \le N$, as shown above: $k_i = k_{i-1}(x_i)$ is a finite algebraic field extension of the field k_{i-1} (i.e., k_i is an overfield of the field k_{i-1} for which we have $[k_i : k_{i-1}] < \infty$ and $k_i = k_{i-1}(x_i)$) with $[k_i : k_{i-1}] = n_i$, the polynomial $f_i(x_1, \ldots, x_{i-1}, Y)$ is the minimal polynomial of x_i over k_{i-1}, the polynomial $f_i(X_1, \ldots, X_i)$ is irreducible in $B_{i,k}$, and $g(X_1, \ldots, X_i) \mapsto g(x_1, \ldots, x_i)$ gives a bijection $\gamma_i \to k_i$.

COMMENT (C17). [**Decomposition of Ideals and Varieties**]. In (1) to (8) above we have proved the first out of the three cases of the Inclusion Relations Theorem, i.e., L4§8(T21). The other two cases will be dealt with in the next Quest. In view of the portions of the Inclusion Relations Theorem and the Hilbert Nullstellensatz proved above, the decomposition of affine algebraic varieties into their irreducible components can be found thus. Let $k \subset \kappa$ be fields such that κ contains an algebraic closure of k. For any ideal J in the finite variable polynomial ring $B = k[X_1, \ldots, X_N]$ take a primary decomposition $J = \cap_{1 \le i \le e} Q_i$ where Q_i is primary for the prime ideal P_i in B. Then $\mathbb{V}_\kappa(P_i) = \mathbb{V}_\kappa(Q_i) =$ an irreducible variety in $\mathbb{A}_\kappa^N$ for $1 \le i \le e$ and $\mathbb{V}_\kappa(J) = \cup_{1 \le i \le e} \mathbb{V}_\kappa(P_i)$ is a decomposition of $\mathbb{V}_\kappa(J)$ into irreducible subvarieties. Moreover, if the primary decomposition of J is irredundant and is so labelled that P_i is a minimal or embedded prime of J according as $1 \le i \le d$ or $d + 1 \le i \le e$, then $\mathbb{V}_\kappa(J) = \cup_{1 \le i \le d} \mathbb{V}_\kappa(P_i)$ is the irredundant decomposition of $\mathbb{V}_\kappa(J)$ into its irreducible components. Conversely, for any variety U in $\mathbb{A}_\kappa^N$ let $U = \cup_{1 \le i \le n} U_i$ be its decomposition into its distinct irreducible components. Then $\mathbb{I}_k(U_i)_{1 \le i \le n}$ are all the distinct associated (automatically minimal) primes of the ideal $\mathbb{I}_k(U)$ in B and we have $\text{rad}_B \mathbb{I}_k(U) = \mathbb{I}_k(U) = \cap_{1 \le i \le n} \mathbb{I}_k(U_i)$. Note that an irreducible component of U can be characterized by saying that it is an irreducible subvariety U' of U such that U has no irreducible subvariety $U'' \ne U'$ with $U' \subset U''$.

QUEST (Q11) Nilradical, Jacobson Spectrum, and Jacobson Ring

At the end of this Quest we are going to give Two Supplements to the Dimension Theorem (T47) which slightly strengthen some parts of it. First we shall complete the proofs of the Inclusion Relations Theorem and the Hilbert Nullstellensatz, i.e., L4§8(T21) and L4§8(T22). To put the matter in proper perspective, we start by defining Nilradical, Jacobson Spectrum, and Jacobson Ring. Recall that in (Q9) and (Q10) we avoided using (Q1) to (Q8). In the current Quest we shall avoid using (Q2) to (Q8), except that we may use the incidental comment (C11) of (Q7) but we shall not use Comment (C8) of (Q1). Also note that (Q9) to (Q11) are independent of §1 to §4 of this Lecture L5.

After making some definitions, we shall prove the Nilradical Theorem (T48). Then we shall prove the Spectral Relations Theorem (T49) which includes the remaining two cases of L4§8(T21). After that we shall prove the Spectral Nullstellensatz (T50) which generalizes the remaining two parts (T22.7) and (T22.8) of the Hilbert Nullstellensatz of L4§8(T22), and adds yet another incarnation (T50.3) of it. Still after that we shall prove the Minimal Primes Theorem (T51) which says that (C11)(2) remains valid without the noetherian hypothesis. Finally we shall prove the Supplementary Dimension Theorems (T52) and (T53).

Recall that in (Q3) we defined the Jacobson radical of a ring R by putting

$$\mathrm{jrad}(R) = \bigcap_{P \in \mathrm{mspec}(R)} P.$$

Now we define the nilradical of R by putting

$$\mathrm{nrad}(R) = \bigcap_{P \in \mathrm{spec}(R)} P.$$

We also define the Jacobson Spectrum of R by putting

$$\mathrm{jspec}(R) = \left\{ P \in \mathrm{spec}(R) : P = \bigcap_{Q \in \mathrm{mvspec}_R P} Q \right\}$$

and we note that then

$$\mathrm{mspec}(R) \subset \mathrm{jspec}(R) \subset \mathrm{spec}(R).$$

The ring R is said to be a Jacobson Ring if $\mathrm{jspec}(R) = \mathrm{spec}(R)$.

In analogy with the definitions of $\mathrm{vspec}_R J$, $\mathrm{mvspec}_R J$, $\mathrm{svt}(R)$, $\mathrm{msvt}(R)$, $\mathrm{isvt}(R)$, and $\mathrm{imsvt}(R)$ made in L4§8, for any $J \subset R$ we put

$$\mathrm{jvspec}_R J = \{P \in \mathrm{jspec}(R) : J \subset P\}$$

and call this the Jacobson spectral variety of J in R, and we put

$$\mathrm{jsvt}(R) = \text{the set of Jacobson spectral varieties in } R$$

i.e., subsets of jspec(R) of the form jvspec$_R J$ for some $J \subset R$ (we may call these varieties in jspec(R)), and we put

$$\text{ijsvt}(R) = \text{the set of all irreducible members of jsvt}(R)$$

where U is irreducible means U is nonempty and U is not the union two proper subvarieties, i.e., subsets which belong to jsvt(R) and are different from U. By an IRREDUCIBLE COMPONENT of $U \in$ jsvt(R) we mean an irreducible subvariety U' of U such that U has no irreducible subvariety $U'' \neq U'$ with $U' \subset U''$; similar definitions hold for IRREDUCIBLE COMPONENTS of varieties in msvt(R) or svt(R). Recall that

$$\text{rd}(R) = \text{the set of all radical ideals in } R$$

and for any $U \subset$ spec(R) we defined the spectral ideal of U by putting

$$\text{ispec}_R U = \cap_{P \in U} P.$$

NILRADICAL THEOREM (T48). For any ring R we have nrad(R) = rad$_R 0$. More generally, for any ideal J in any ring R we have rad$_R J$ = ispec$_R$(vspec$_R J$), i.e., rad$_R J$ is the intersection of all the prime ideals in R which contain J.

PROOF. The first assertion is L4§5(O24)(20$^\bullet$) which itself was a repetition of the first line of L4§5(O20)(12$^\bullet$). The second assertion follows by applying the first assertion to the residue class ring R/J.

SPECTRAL RELATIONS THEOREM (T49). Given any ring R, let

$$\begin{cases} (S, V, V', V'', I) \text{ stand for} \\ (\text{mspec}(R), \text{mvspec}_R, \text{msvt}(R), \text{imsvt}(R), \text{ispec}_R) \\ \text{or } (\text{jspec}(R), \text{jvspec}_R, \text{jsvt}(R), \text{ijsvt}(R), \text{ispec}_R) \\ \text{or } (\text{spec}(R), \text{vspec}_R, \text{svt}(R), \text{isvt}(R), \text{ispec}_R). \end{cases}$$

In all the cases let I' stand for the set of all ideals J in R such that $J = I(U)$ for some $U \subset S$. Then we have the following.

(T49.1) For any subsets J and J' of R we have: $J \subset J' \Rightarrow V(J') \subset V(J)$, and we have: rad$_R(JR) = $ rad$_R(J'R) \Rightarrow V(J) = V(J')$.

(T49.2) For any subsets U and U' of S we have $U \subset U' \Rightarrow I(U') \subset I(U)$.

(T49.3) For any family of ideals $(J_l)_{l \in L}$ in R we have $V(\sum_{l \in L} J_l) = \cap_{l \in L} V(J_l)$, and if the family is finite then we have $V(\cap_{l \in L} J_l) = V(\prod_{l \in L} J_l) = \cup_{l \in L} V(J_l)$.

(T49.4) For any family of subsets $(U_l)_{l \in L}$ of S we have $I(\cup_{l \in L} U_l) = \cap_{l \in L} I(U_l)$.

(T49.5) For any $J \subset R$ we have $J \subset I(V(J))$ and: $J = I(V(J)) \Leftrightarrow J \in I' \Rightarrow J = $ rad$_R J$.

(T49.6) For any $U \subset S$ we have $U \subset V(I(U))$ and: $U = V(I(U)) \Leftrightarrow U \in V'$.

(T49.7) $V'' = \{U \in V' : I(U) \in \mathrm{spec}(R)\}$.

(T49.8) Assuming R is noetherian, every $U \in V'$ can be expressed as a finite union $U = \cup_{1 \le i \le h} U_i$ with $U_i \in V''$, and this decomposition is unique up to order if it is irredundant, i.e., if no U_i can be deleted from it. The irredundancy can be achieved by deleting some of the U_i from any decomposition, and after it is achieved, the remaining U_i are exactly all the IRREDUCIBLE COMPONENTS of U.

(T49.9) Assuming R is noetherian, V' has no strictly decreasing infinite sequence, i.e., there does not exist any infinite sequence $W_0 \supset W_1 \supset \cdots \supset W_n \supset \cdots$ in V' with $W_0 \ne W_1 \ne \cdots \ne W_n \ne \cdots$.

(T49.10) Without assuming R to be noetherian, every $U \in V''$ is contained in an irreducible component of S, and every $U \in V'$ is the union of its irreducible components, although the number of these irreducible components may be infinite.

(T49.11) Without assuming R to be noetherian: (i) for any $T \in S$ we have

$$V(T) = V(I(\{T\})) = \bigcap_{\{W \in V' : T \in W\}} W \quad \text{and} \quad I(V(T)) = T.$$

Moreover: (ii) if $U \in V''$ is such that $I(U) \in S$ then upon letting $P = I(U)$ we have that P is the unique member of S with $U = V(P)$.

(T49.12) Without assuming R to be noetherian, in the last two cases, i.e., when S is either $\mathrm{jspec}(R)$ or $\mathrm{spec}(R)$, for any $U \in V''$ we have that $I(U) \in S$ and upon letting $P = I(U)$ we have that P is the unique member of S with $U = V(P)$.

PROOF. The proofs of (T49.1) and (T49.2) are clear.

Concerning the first equation of (49.3), by the first implication of (49.1) we see that LHS $\subset$ RHS; conversely

$$\alpha \in \mathrm{RHS} \Rightarrow \alpha \in V(\cup_{l \in L} J_l) \Rightarrow \alpha \in \mathrm{LHS}$$

where the first implication is obvious and the second follows from the second implication of (49.1). Concerning the last line of (49.3), by the first implication of (49.1) we get the inclusions

$$\cup_{l \in L} V(J_l) \subset V(\cap_{l \in L} J_l) \subset V(\prod_{l \in L} J_l).$$

Moreover, for any $\alpha \in S$ we have that

$$\begin{cases} \alpha \notin \cup_{l \in L} V(J_l) \\ \Rightarrow \text{ for each } l \in L \text{ there exists } f_l \in J_l \text{ with } f_l \notin \alpha \\ \Rightarrow f \notin \alpha \text{ with } f = \prod_{l \in L} f_l \in \prod_{l \in L} J_l \end{cases}$$

and hence

$$V(\prod_{l \in L} J_l) \subset \cup_{l \in L} V(J_l).$$

Concerning equation (49.4), by (49.2) we get LHS $\subset$ RHS; conversely

$$f \in \text{RHS} \Rightarrow f \in \alpha \text{ for every } \alpha \in \cup_{l \in L} U_l \Rightarrow f \in \text{LHS}.$$

The proofs of (T49.5) and (T49.6) are verbatim the same as the proofs of (5) and (6) given in the previous Quest (Q10), except that the references to (1), (2), (5), (6) should be changed to references to (49.1), (49.2), (49.5), (49.6) respectively, and the second braced display starting with "also" should be replaced by saying that, also:

$$\begin{cases} f \in \text{rad}_R J \text{ with } J = I(U) \text{ for some } U \subset S \\ \Rightarrow f^n \in J \text{ for some } n \in \mathbb{N}_+ \\ \Rightarrow f^n \in \alpha \text{ for all } \alpha \in U \\ \Rightarrow f \in \alpha \text{ for all } \alpha \in U \\ \Rightarrow f \in J \end{cases}$$

and hence $J = \text{rad}_R J$.

To prove (49.7) let $U \in V'$. Assuming U to be irreducible, we have $U \neq \emptyset$ and hence $I(U) \neq R$; moreover, for $1 \leq i \leq 2$, given any $f_i \in R \setminus I(U)$, upon letting $U_i = \{\alpha \in U; f_i \in \alpha\}$, we have $U_i \in V'$ with $U_i \subset U$ and $U_i \neq U$; since U is irreducible, we must have $U_1 \cup U_2 \neq U$, and hence $f_1 f_2 \notin I(U)$; thus $I(U)$ is prime. Conversely assuming $I(U)$ to be prime, we have $I(U) \neq R$ and hence $U \neq \emptyset$; moreover, given any U_1, U_2 in V' with $U_1 \cup U_2 = U$, by (49.4) we have $I(U_1) \cap I(U_2) = I(U)$ and clearly $I(U_1)I(U_2) \subset I(U_1) \cap I(U_2)$ and hence $I(U_1)I(U_2) \subset I(U)$; now assuming $U_1 \neq U$, by (49.2) and the implication portion of (49.6) we get $I(U) \subset I(U_1)$ with $I(U) \neq I(U_1)$, and hence the primeness of $I(U)$ tells us that $I(U_2) \subset I(U)$ and therefore by the first implication of (49.1) we get $U_2 = U$; thus U is irreducible.

The proofs of (T49.8) and (T49.9) are verbatim the same as the proofs of (8) and (9) given in the previous Quest (Q10), except that the references to (2), (6), (8), (9) should be changed to references to (49.2), (49.6), (49.8), (49.9) respectively.

To prove (T49.10), by Zorn's Lemma, given any $U \in V''$ we have $U \subset U^*$ for some maximal $U^* \in V''$, and clearly any such U^* is an irreducible component of S. Given any $P \in U \in V'$, replacing (U, S) by $(V(P), U)$ in the previous sentence, we see that $P \in V(P) \subset U^*$ for some irreducible component U^* of U, and hence every U in V' is the union of its irreducible components. [cf. §6(E13)].

The proof of part (i) of (T49.11) is straightforward. To prove part (ii), let $U \in V''$ be such that upon letting $P = I(U)$ we have $P \in S$; then by (T49.6) we get $U = V(P)$; conversely, if $T \in S$ is such that $U = V(T)$ then by the last equation in part (i) we get $I(U) = T$ and hence $T = P$.

To prove (T49.12), given any $U \in V''$, upon letting $P = I(U)$, by (T49.7) we see that P is a prime ideal in R. It follows that if S is spec$R$ then $P \in S$. On the other hand, if S is jspec(R) then every $Q \in S$ is an intersection of members of mspec(R), and hence by the equation $P = I(U) = \cap_{Q \in U} Q$ with $U \subset S$ we see that P is an intersection of members of mspec(R); thus again $P \in S$. The rest follows from part (ii) of (T49.11).

SPECTRAL NULLSTELLENSATZ (T50). Given any ring R, let

$$\begin{cases} (S, V, V', V'', I) \text{ stand for} \\ (\text{mspec}(R), \text{mvspec}_R, \text{msvt}(R), \text{imsvt}(R), \text{ispec}_R) \\ \text{or } (\text{jspec}(R), \text{jvspec}_R, \text{jsvt}(R), \text{ijsvt}(R), \text{ispec}_R) \\ \text{or } (\text{spec}(R), \text{vspec}_R, \text{svt}(R), \text{isvt}(R), \text{ispec}_R). \end{cases}$$

Then we have the following.

(T50.1) If R is Jacobson or we are in the case of $S = \text{spec}(R)$, then for any ideal J in R we have $I(V(J)) = \text{rad}_R J$.

(T50.2) If R is Jacobson or we are in the case of $S = \text{spec}(R)$, then the mapping $U \mapsto I(U)$ gives inclusion reversing bijections $V' \to \text{rd}(R)$ and $V'' \to \text{spec}(R)$ whose inverses are given by $J \mapsto V(J)$.

(T50.3) If R is an affine ring over a field, then R is a Jacobson ring, i.e., every prime ideal in it is an intersection of maximal ideals.

PROOF. The first part (T50.1) follows from (T48) and, in view of (T49.1), (T49.2), (T49.5), (T49.6) and (T49.7), the second part (T50.2) follows from the first part (T50.1). [cf. §6(E14)].

To prove the third part (T50.3), given any affine ring R over a field k and any prime ideal H in R, we want to show that $H = \cap_{P \in \text{mvspec}_R H} P$. We can take a k-epimorphism $\phi : B \to R$ where $B = k[X_1, \ldots, X_N]$ is a polynomial ring over k in a finite number of variables, and we can take an algebraic closure κ of k. Let $G = \phi^{-1}(H)$ and $J = \cap_{P \in \text{mvspec}_B G} P$. Then $G \subset J$ with $G \in \text{spec}(B)$ and $J \in \text{rd}(R)$. What we want to show is that $G = J$. Suppose if possible that $G \neq J$. Then by the sixth part (T22.6) of the Hilbert Nullstellensatz L4§8(T22), which we proved at the end of the last Quest, there exist $\alpha = (\alpha_1, \ldots, \alpha_N) \in \kappa^N$ such that $\alpha \in \mathbb{V}_\kappa(G) \setminus \mathbb{V}_\kappa(J)$, i.e., upon letting Q be the maximal ideal in $C = \kappa[X_1, \ldots, X_N]$ generated by $X_1 - \alpha_1, \ldots, X_N - \alpha_N$ we have $G \subset Q$ with $J \not\subset Q$. Now C is integral over B and hence upon letting $P = Q \cap B$, by (T38.2) we see that P is a maximal ideal in B. Clearly $G \subset P$ but $J \not\subset P$ which is a contradiction.

MINIMAL PRIMES THEOREM (T51). Given any ring R, every prime ideal in R contains some minimal prime ideal in R, and we have $\text{rad}_R 0 = \cap_{P \in \text{nspec}(R)} P$. More generally, given any ideal J in any ring R, every member of $\text{vspec}_R J$ contains some member of $\text{nvspec}_R J$, and we have $\text{rad}_R J = \cap_{P \in \text{nvspec}_R J} P$.

PROOF. The second assertion follows by applying the first assertion to R/J.

To prove the first assertion, by (T49.10) we see that for any $H \in \mathrm{spec}(R)$ there exists an irreducible component U_H of $\mathrm{spec}(R)$ with $H \in U_H$ and, upon letting $P_H = \mathrm{ispec}_R U_H$, by (T49.12) we see that

$$P_H \in \mathrm{nspec}(R) \quad \text{with} \quad P_H \subset H.$$

It follows that

$$\mathrm{nrad}(R) = \cap_{P \in \mathrm{nspec}(R)} P$$

and hence by (T48) we get

$$\mathrm{rad}_R 0 = \cap_{P \in \mathrm{nspec}(R)} P.$$

FIRST SUPPLEMENTARY DIMENSION THEOREM (T52). Given any affine ring A over a field k, let $(J_i)_{1 \leq i \leq m}$ be all the distinct minimal primes of 0 in A, and assume that $\dim(A/J_i)$ is independent of i. Then for every nonunit ideal H in A we have

$$\mathrm{ht}_A H + \mathrm{dpt}_A H = \dim(A).$$

Moreover, for any prime ideal P in A, the length of every saturated prime ideal chain $\widehat{P}_0 \subset \cdots \subset \widehat{P}_h$ in A with $\mathrm{ht}_A \widehat{P}_0 = 0$ and $\widehat{P}_h = P$ equals $\mathrm{ht}_A P$, and the length of every saturated prime ideal chain $\widetilde{P}_0 \subset \cdots \subset \widetilde{P}_d$ in A with $\widetilde{P}_0 = P$ and $\mathrm{dpt}_A \widetilde{P}_d = 0$ equals $\mathrm{dpt}_A P$.

PROOF. Let $\dim(A) = n$. For $1 \leq i \leq m$, upon letting $\phi_i : A \to A/J_i$ be the residue class epimorphism, by (T47.1) we get $\dim(\phi_i(A)) = n$. Let P be any prime ideal in A.

We can take saturated prime ideal chains $\widehat{P}_0 \subset \cdots \subset \widehat{P}_h$ and $\widetilde{P}_0 \subset \cdots \subset \widetilde{P}_d$ in A with $\mathrm{ht}_A \widehat{P}_0 = 0$, $\widehat{P}_h = P$, $\mathrm{ht}_A P = h$, $\widetilde{P}_0 = P$, $\mathrm{dpt}_A \widetilde{P}_d = 0$, and $\mathrm{dpt}_A P = d$; clearly we must have $\widehat{P}_0 = J_i$ for some i and applying (T47.3) to $\phi_i(A)$ we get $h + d = n$, i.e.: ($\bullet$) $\mathrm{ht}_A P + \mathrm{dpt}_A P = n$. In particular, given any nonunit ideal H in A we have $\mathrm{ht}_A \overline{P} + \mathrm{dpt}_A \overline{P} = n$ for every $\overline{P}$ in $\mathrm{vspec}_A H$ and hence $\mathrm{ht}_A H + \mathrm{dpt}_A H = \dim(A)$.

Given any saturated prime ideal chain $\widehat{P}_0 \subset \cdots \subset \widehat{P}_h$ in A with $\mathrm{ht}_A \widehat{P}_0 = 0$ and $\widehat{P}_h = P$, we can take a saturated prime ideal chain $\widetilde{P}_0 \subset \cdots \subset \widetilde{P}_d$ in A with $\widetilde{P}_0 = P$, $\mathrm{dpt}_A \widetilde{P}_d = 0$, and $\mathrm{dpt}_A P = d$; clearly we must have $\widehat{P}_0 = J_i$ for some i and applying (T47.3) and (T47.5) to $\phi_i(A)$ we get $h + d = n$ with $\mathrm{ht}_{\phi_i(A)} \phi_i(P) = h$ and $\mathrm{dpt}_{\phi_i(A)} \phi_i(P) = d$; obviously $\mathrm{dpt}_{\phi_i(A)} \phi_i(P) = \mathrm{dpt}_A P$, and hence by ($\bullet$) we get $\mathrm{ht}_A P = h$.

Given any saturated prime ideal chain $\widetilde{P}_0 \subset \cdots \subset \widetilde{P}_d$ in A with $\widetilde{P}_0 = P$ and $\mathrm{dpt}_A \widetilde{P}_d = 0$, we can take a saturated prime ideal chain $\widehat{P}_0 \subset \cdots \subset \widehat{P}_h$ in A with $\mathrm{ht}_A \widehat{P}_0 = 0$, $\widehat{P}_h = P$, and $\mathrm{ht}_A P = h$; clearly we must have $\widehat{P}_0 = J_i$ for some i and applying (T47.3) and (T47.5) to $\phi_i(A)$ we get $h + d = n$ with $\mathrm{ht}_{\phi_i(A)} \phi_i(P) = h$

and $\text{dpt}_{\phi_i(A)}\phi_i(P) = d$; obviously $\text{dpt}_{\phi_i(A)}\phi_i(P) = \text{dpt}_A P$, and hence by ($\bullet$) we get $\text{dpt}_A P = d$.

SECOND SUPPLEMENTARY DIMENSION THEOREM (T53). Let A be any affine ring over a field k. Then $\dim(A)$ is the maximum number of elements of A which are algebraically independent over k. Moreover if A^* is any subring of A such that A^* is also an affine ring over k, then $\dim(A^*) \le \dim(A)$.

PROOF. The second assertion follows from the first. To prove the first assertion, let $\dim(R) = n$, let $(J_i)_{1 \le i \le m}$ be all the distinct minimal primes of 0 in A, and let $\phi_i : A \to A/J_i$ be the residue class epimorphism. By (T47.1) we know that $\text{trdeg}_{\phi_i(k)}\phi_i(A) \le n$ for $1 \le i \le m$ with equality for some $j = i$. We can take elements $Z_1, \ldots, Z_n$ in A such that the elements $\phi_j(Z_1), \ldots, \phi_j(Z_n)$ are algebraically independent over $\phi_j(k)$; it follows that the elements $Z_1, \ldots, Z_n$ are algebraically independent over k. Conversely, let $Y_1, \ldots, Y_r$ be any finite number of elements in A which are algebraically independent over k. Then the subring $C = k[Y_1, \ldots, Y_r]$ of A, being a polynomial ring, is a domain and hence $\text{rad}_A 0 \cap C = 0$. But $\text{rad}_A 0 = \cap_{1 \le i \le m} J_i$ and hence we must have $J_i \cap C = 0$ for some i. It follows that the elements $\phi_i(Y_1), \ldots, \phi_i(Y_r)$ in $\phi_i(A)$ are algebraically independent over $\phi_i(k)$, and hence $r \le n$.

COMMENT (C18). [**Again Decomposition of Ideals and Varieties**]. In view of (T49) and (T50), the connection between ideal decomposition and variety decomposition described in (C17) for polynomial rings and affine varieties can be reproduced in a more general setup thus. In the notation of (T49) and (T50), assuming the ring R to be noetherian, for any ideal J in R take a primary decomposition $J = \cap_{1 \le i \le e} Q_i$ where Q_i is primary for the prime ideal P_i in R. Then $V(P_i) = V(Q_i) =$ an irreducible variety in S for $1 \le i \le e$ and $V(J) = \cup_{1 \le i \le e} V(P_i) =$ a decomposition of $V(J)$ into irreducible subvarieties. Moreover, if the primary decomposition of J is irredundant and is so labelled that P_i is a minimal or embedded prime of J according as $1 \le i \le d$ or $d + 1 \le i \le e$, then $V(J) = \cup_{1 \le i \le d} V(P_i)$ is the irredundant decomposition of $V(J)$ into its irreducible components. Conversely, for any variety U in S let $U = \cup_{1 \le i \le n} U_i$ be its decomposition into its distinct irreducible components. Then $I(U_i)_{1 \le i \le n}$ are all the distinct associated (automatically minimal) primes of the ideal $I(U)$ in R and we have $\text{rad}_R I(U) = I(U) = \cap_{1 \le i \le n} I(U_i)$. Note that an irreducible component of U can be characterized by saying that it is an irreducible subvariety U' of U such that U has no irreducible subvariety $U'' \ne U'$ with $U' \subset U''$. All this is particularly significant when R contains a field k and there is a k-epimorphism $\phi : B \to R$ where $B = k[X_1, \ldots, X_N]$ is a finite variable polynomial ring over k, and $\ker(\phi)$ is the ideal $\mathbb{I}_k(U)$ of some $U \in \text{avt}_k(A_\kappa^N)$ for some overfield κ of k containing an algebraic closure of k; we may then regard R as the coordinate ring of U.

QUEST (Q12) Catenarian Rings and Dimension Formula

A ring A is said to be catenarian if every pair of prime ideals $P \subset Q$ in it satisfies condition (∗) of (T47.2) which says that: there is no infinite chain of prime ideals between P and Q, and any two saturated chains of prime ideals between P and Q have the same length. Thus the first part of (T47.2) says that an affine ring over a field is catenarian. A ring A is said to be universally catenarian if it is noetherian and the polynomial ring over it in any finite number of variables is catenarian.

Given an overdomain A' of a domain A, we say that the dimension formula (resp: the dimension inequality) holds for A relative to A' if for every prime ideal P' in A', considering the localizations $R' = A'_{P'}$ and $R = A_P$ with $P = P' \cap A$, we have

$$\dim(R') + \mathrm{restrdeg}_R R' = \dim(R) + \mathrm{trdeg}_R R'$$

$$(\text{resp: } \dim(R') + \mathrm{restrdeg}_R R' \leq \dim(R) + \mathrm{trdeg}_R R').$$

We say that the dimension formula (resp: the dimension inequality) holds in a domain A if it holds for A relative to every affine domain over A. Thus (T47.6) says that the dimension formula holds in any affine domain over a field.

As we shall now see, the above concepts enable us to view parts of the Dimension Lemma (T30) and the Dimension Theorem (T47) from a somewhat more general perspective. Let us start off by giving two sharper versions of (C13). In Lemma (T54) we consider the behavior of prime ideals in a simple ring extension, i.e., in a ring extension $A[x]$ obtained by adjoining a single element x to a ring A, and in Lemma (T55) we shall generalize it to a ring extension obtained by adjoining a finite number of elements.

SIMPLE RING EXTENSION LEMMA (T54). Let A and A' be noetherian domains such that $A' = A[x]$ for some $x \in A'$. Let P' be any prime ideal in A' lying above any prime ideal P in A, and consider the localizations $R' = A'_{P'}$ and $R = A_P$. Then

$$\mathrm{restrdeg}_R R' \leq 1 \quad \text{and} \quad \mathrm{trdeg}_R R' \leq 1$$

and

$$\dim(R') + \mathrm{restrdeg}_R R' \leq \dim(R) + \mathrm{trdeg}_R R'.$$

Moreover, if $\mathrm{trdeg}_R R' = 1$ then

$$\dim(R') + \mathrm{restrdeg}_R R' = \dim(R) + \mathrm{trdeg}_R R'.$$

PROOF. The first display is obvious, and the third follows from (C13). So assume x is algebraic over A. Let $B = A[X]$ be the univariate polynomial ring over A, let $\phi : B \to A'$ be the unique A-epimorphism which sends X to x, let $H = \ker(\phi)$, and let $Q = \phi^{-1}(P')$. Now H is a nonzero prime ideal in B lying above 0 in A, and hence by (C13) we have $\operatorname{ht}_B H = 1$; also clearly $\operatorname{ht}_B H + \operatorname{ht}_{A'} P' \leq \operatorname{ht}_B Q$; consequently

$$1 + \dim(R') \leq \dim(B_Q).$$

Obviously $\operatorname{restrdeg}_R B_Q = \operatorname{restrdeg}_R R'$ and Q is a prime ideal in B lying above P in A, and hence by (C13) we get

$$\dim(B_Q) + \operatorname{restrdeg}_R R' = 1 + \dim(R).$$

The above two displays imply $\dim(R') + \operatorname{restrdeg}_R R' \leq \dim(R)$, and this completes the proof.

MULTIPLE RING EXTENSION LEMMA (T55). Let A and A' be noetherian domains such that $A' = A[x_1, \dots, x_m]$ for some $m \in \mathbb{N}$ and $x_1, \dots, x_m$ in A'. Let P' be any prime ideal in A' lying above any prime ideal P in A, and consider the localizations $R' = A'_{P'}$ and $R = A_P$. Then

$$\operatorname{restrdeg}_R R' \leq m \quad \text{and} \quad \operatorname{trdeg}_R R' \leq m$$

and

$$\dim(R') + \operatorname{restrdeg}_R R' \leq \dim(R) + \operatorname{trdeg}_R R'.$$

Moreover, if either $\operatorname{trdeg}_R R' = m$ or the polynomial ring $B = A[X_1, \dots, X_m]$ in m variables is catenarian, then

$$\dim(R') + \operatorname{restrdeg}_R R' = \dim(R) + \operatorname{trdeg}_R R'.$$

PROOF. The first display is obvious. The second display and the first case of the third display are trivial for $m = 0$ and for $m > 0$ their implication $m - 1 \Rightarrow m$ follows from (T54), and so they are done by induction on m. So now assume that B is catenarian. Let $\phi : B \to A'$ be the unique A-epimorphism which sends X_i to x_i for $1 \leq i \leq m$. Let $H = \ker(\phi)$, and let $Q = \phi^{-1}(P')$. Now H and Q are prime ideals in B lying above 0 and P in A respectively, and hence by the first case of the third display, taking $(A', P', A, P) = (B, H, A, 0)$ and $(A', P', A, P) = (B, Q, A, P)$, we respectively get

$$\operatorname{ht}_B H + \operatorname{trdeg}_R R' = m$$

and

$$\operatorname{ht}_B Q + \operatorname{restrdeg}_R R' = m + \dim(R).$$

Since B is catenarian, by Lemma (T56.1) below we have $\mathrm{ht}_B Q - \mathrm{ht}_B H = \dim(R')$, and hence by the above two displays we get

$$\dim(R') + \mathrm{restrdeg}_R R' = \dim(R) + \mathrm{trdeg}_R R'$$

which completes the proof.

CATENARIAN CONDITION LEMMA (T56). For any ring A we have the following.

(T56.1) If A is a noetherian domain then:

$$\begin{cases} A \text{ is catenarian} \\ \Leftrightarrow \text{ for every pair of prime ideals } P \subset Q \text{ in } A \\ \quad \text{we have } \mathrm{ht}_A Q = \mathrm{ht}_A P + \mathrm{ht}_{(A/P)}(Q/P) \\ \Leftrightarrow \text{ for every pair of prime ideals } P \subset Q \text{ in } A \text{ with } \mathrm{ht}_{(A/P)}(Q/P) = 1 \\ \quad \text{we have } \mathrm{ht}_A Q = 1 + \mathrm{ht}_A P. \end{cases}$$

(T56.2) If A is catenarian then every homomorphic image of A is catenarian and so is the localization of A with respect to any multiplicative set in A.

(T56.3) If A is universally catenarian then every homomorphic image of A is universally catenarian and so is the localization of A with respect to any multiplicative set in A.

(T56.4) If A is universally catenarian then every affine ring over A is universally catenarian.

PROOF. In (T56.1), to see that the first condition implies the second, we can take saturated prime ideal chains $\widehat{P}_0 \subset \cdots \subset \widehat{P}_h$ and $\widetilde{P}_0 \subset \cdots \subset \widetilde{P}_d$ in A with $\widehat{P}_0 = 0$, $\widehat{P}_h = P$, $\mathrm{ht}_A P = h$, $\widetilde{P}_0 = P$, $\widetilde{P}_d = Q$, and $\mathrm{ht}_{(A/P)}(Q/P) = d$; clearly $\widehat{P}_0 \subset \cdots \subset \widehat{P}_h \subset \widetilde{P}_1 \subset \cdots \subset \widetilde{P}_d$ is a saturated chain of length $h + d$ in A between 0 and Q and hence by the catenarianness of A we get $\mathrm{ht}_A Q = h + d$. The second condition obviously implies the third. To see that the third condition implies the first, let $P \subset Q$ be any prime ideals in A together with a saturated prime ideal chain $P_0 \subset \cdots \subset P_n$ in A between P and Q; then by the third condition we have $\mathrm{ht}_A P_i = 1 + \mathrm{ht}_A P_{i-1}$ for $1 \le i \le n$ and hence $\mathrm{ht}_A Q - \mathrm{ht}_A P = n$.

The ideal correspondence between a ring and its homomorphic image given in the Third and the Fourth Isomorphism Theorems of L4§5(O11), and the ideal correspondence between a ring and its localization given in L4§7(T12), yield (T56.2).

A ring epimorphism $\phi : A \to C$ uniquely induces an epimorphism of polynomial rings $\psi : A[X_1, \ldots, X_m] \to C[X_1, \ldots, X_m]$ with $\psi(X_i) = X_i$ for $1 \le i \le m$ and $\psi(z) = \phi(z)$ for all $z \in A$, and hence the first part of (T56.2) implies the first part of (T56.3). Similarly, for any multiplicative set S in any ring A, the canonical homomorphisms $\phi : A \to A_S$ and $\theta : A[X_1, \ldots, X_m] \to (A[X_1, \ldots, x_m])_S$ uniquely induce an isomorphism $\psi : (A[X_1, \ldots, X_m])_S \to A_S[X_1, \ldots, X_m]$ with $\psi(\theta(X_i)) =$

X_i for $1 \leq i \leq m$ and $\psi(\theta(z)) = \phi(z)$ for all $z \in A$, and hence the second part of (T56.2) implies the second part of (T56.3).

A finite variable polynomial ring over an affine ring over A can obviously be expressed as a homomorphic image of a finite variable polynomial ring over A. Consequently (T56.4) follows from the first part of (T56.2).

CATENARIAN DOMAIN THEOREM (T57). For any noetherian domain A we have the following.

(T57.1) The dimension inequality holds in A.

(T57.2) The dimension formula holds for A relative to any finite variable polynomial ring over A.

(T57.3) If A is universally catenarian then the dimension formula holds in A.

PROOF. Follows from (T55).

CATENARIAN RING THEOREM (T58). A ring A is universally catenarian iff A is a noetherian catenarian ring and the dimension formula holds in A/H for every prime ideal H in A.

PROOF. The only if part follows from (T57.3) and the first part of (T56.3). So now assume that A is a noetherian catenarian ring and the dimension formula holds in A/H for every prime ideal H in A. Let $B = A[X_1, \ldots, X_m]$ be any finite variable polynomial ring over A, let $P \subset Q$ be any prime ideals in B, let $\phi : B \to C' = B/P$ be the residue class epimorphism, and let $Q' = \phi(Q)$. Now Q' is a prime ideal in the noetherian domain C' and it suffices to show that the length of every saturated prime ideal chain in C' between 0 and Q' equals $\mathrm{ht}_{C'}Q'$. For showing this it is enough to prove that C' is catenarian, and hence in view of (T56.1) it is enough to prove that, for any pair of prime ideals $I' \subset J'$ in C', upon letting $R' = C'_{I'}$ and $S' = C'_{J'}$, and upon letting $T' = D'_{L'}$ where $\phi' : C' \to D' = C'/I'$ is the residue class epimorphism and $L' = \phi'(J')$, we have $\dim(S') = \dim(R') + \dim(T')$.

Let $H = P \cap A$. Then H is a prime ideal in A, the domain C' is an affine domain over the subdomain $C = \phi(A)$ which is isomorphic to A/H, and the prime ideals $I' \subset J'$ in C' lie above the prime ideals $I \subset J$ in C where $I = I' \cap C$ and $J = J' \cap C$ respectively. Therefore, upon letting $R = C_I$ and $S = C_J$, by the dimension formula assumption we have

$$\dim(R') + \mathrm{restrdeg}_R R' = \dim(R) + \mathrm{trdeg}_C C'$$

and

$$\dim(S') + \mathrm{restrdeg}_S S' = \dim(S) + \mathrm{trdeg}_C C'.$$

Let $D = \phi'(C)$ and $T = D_L$ where $L = L' \cap C$. Then the domain D is a homomorphic image of A, and the domain D' is an affine domain over D, and hence again by the dimension formula assumption we have

$$\dim(T') + \operatorname{restrdeg}_T T' = \dim(T) + \operatorname{trdeg}_T T'.$$

Now C is catenarian by (T56.2), and hence by (T56.1) we get

$$\dim(S) = \dim(R) + \dim(T).$$

Also clearly

$$\operatorname{restrdeg}_T T' = \operatorname{restrdeg}_S S'$$

and

$$\operatorname{trdeg}_T T' = \operatorname{restrdeg}_R R'.$$

By the above six displays we get $\dim(S') = \dim(R') + \dim(T')$, and this completes the proof.

QUEST (Q13) Associated Graded Rings and Leading Ideals

Let I be an ideal in a ring R. Recall that for any $h \in R$ we have put $\operatorname{ord}_{(R,I)} h = \max\{n \in \mathbb{N} : h \in I^n\}$, and the ring R is I-Hausdorff means: $\operatorname{ord}_{(R,I)} h = \infty \Leftrightarrow h = 0$; see L3§12(D5). Also recall that in case R is quasilocal with $I = M(R)$, $\operatorname{ord}_{(R,I)}$ and I-Hausdorff are abbreviated to ord_R and Hausdorff respectively; see the first display after (2•) in L3§11.

We define the associated graded ring of (R, I) to be the naturally graded ring $\operatorname{grad}(R, I)$ whose underlying additive abelian group is obtained by putting

$$\operatorname{grad}(R, I) = \bigoplus_{n \in \mathbb{N}} (I^n / I^{n+1}) \quad \text{(external direct sum of $\mathbb{Z}$-modules).}$$

This is made into a ring by requiring that for all $f_m \in I^m$ and $f_n \in I^n$ with m, n in $\mathbb{N}$ we have

$$\operatorname{grad}_{(R,I)}^m(f_m) \cdot \operatorname{grad}_{(R,I)}^n(f_n) = \operatorname{grad}_{(R,I)}^{m+n}(f_m f_n)$$

where the central dot indicates multiplication and where

$$\operatorname{grad}_{(R,I)}^n : I^n \to \operatorname{grad}(R, I)$$

is the composition of

$$\lambda_n : I^n \to (I^n / I^{n+1}) \quad \text{and} \quad \mu_n : (I^n / I^{n+1}) \to \operatorname{grad}(R, I)$$

where λ_n is the natural surjection and μ_n is the natural injection. We also put

$$\operatorname{grad}(R, I)_n = \operatorname{grad}_{(R,I)}^n(I^n)$$

and we observe that then

$$\ker\left(\operatorname{grad}_{(R,I)}^n\right) = I^{n+1}$$

and

$$\operatorname{im}\left(\operatorname{grad}_{(R,I)}^n\right) = \operatorname{grad}(R,I)_n$$

and moreover

$$\operatorname{grad}(R,I) = \sum_{n\in\mathbb{N}} \operatorname{grad}(R,I)_n \quad \text{(internal direct sum)}.$$

We may call $\operatorname{grad}_{(R,I)}^n$ and $\operatorname{grad}(R,I)_n$ the n-th graded map and the n-th graded component of (R,I) respectively, and for any $h \in I^n$ we may call $\operatorname{grad}_{(R,I)}^n(h)$ the n-th graded image of h relative to (R,I).

Note that the ring $\operatorname{grad}(R,I)$ is nonnull iff $I \neq R$. For any $h \in R$ we put

$$\operatorname{lefo}_{(R,I)}(h) = \begin{cases} \operatorname{grad}_{(R,I)}^n(h) & \text{if } \operatorname{ord}_{(R,I)}h = n \in \mathbb{N} \\ 0 & \text{if } \operatorname{ord}_{(R,I)}h = \infty \end{cases}$$

and we call this the leading form of h relative to (R,I). For any ideal J in R we put

$$\operatorname{ledi}_{(R,I)}(J) = \begin{cases} \text{the homogeneous ideal in } \operatorname{grad}(R,I) \\ \text{generated by } \left(\operatorname{lefo}_{(R,I)}(h)\right)_{h\in J} \end{cases}$$

i.e., equivalently

$$\operatorname{ledi}_{(R,I)}(J) = \begin{cases} \text{the homogeneous ideal in } \operatorname{grad}(R,I) \\ \text{generated by } \bigcup_{n\in\mathbb{N}} \operatorname{grad}_{(R,I)}^n(J \cap I^n) \end{cases}$$

and we call this the leading ideal of J relative to (R,I). Taking any family of generators $x = (x_l)_{l\in L}$ of I and letting

$$y_l = \operatorname{grad}_{(R,I)}^1(x_l)$$

we see that $\operatorname{grad}(R,I)$ is a semihomogeneous (homogeneous if L is finite) ring over the ring $\operatorname{grad}(R,I)_0$ with

$$\operatorname{grad}(R,I) = \operatorname{grad}(R,I)_0[\operatorname{grad}(R,I)_1] = \operatorname{grad}(R,I)_0[(y_l)_{l\in L}]$$

and moreover considering the polynomial ring $k[(X_l)_{l\in L}]$ with $k = R/I$ as a semihomogeneous (homogeneous if L is finite) ring over the ring k (see (C6) of §3), we get a unique graded ring epimorphism

$$\Theta_{(R,I)}^x : k[(X_l)_{l\in L}] \to \operatorname{grad}(R,I) \quad \text{with} \quad k = R/I$$

such that

$$\Theta^x_{(R,I)}(z) = \mu_0(z) \text{ for all } z \in k \quad \text{and} \quad \Theta^x_{(R,I)}(X_l) = y_l \text{ for all } l \in L.$$

The homomorphism $\Theta^x_{(R,I)}$ is said to be induced by (R, I) and x.

Given any ring homomorphism $\phi : R \to R'$ and any ideal I' in R' with $\phi(I) \subset I'$, we get a unique graded ring homomorphism

$$\mathrm{grad}_{(R,I)}(\phi) : \mathrm{grad}(R, I) \to \mathrm{grad}(R', I')$$

such that, for every $n \in \mathbb{N}$, the homomorphism

$$\mathrm{grad}^n_{(R,I)}(\phi) : \mathrm{grad}(R, I)_n \to \mathrm{grad}(R', I')_n$$

induced by $\mathrm{grad}_{(R,I)}(\phi)$ has the property that for all $h \in I^n$ we have

$$\mathrm{grad}^n_{(R,I)}(\phi)(\mathrm{grad}^n_{(R,I)}(h)) = \mathrm{grad}^n_{(R',I')}((\phi(h)).$$

The homomorphisms $\mathrm{grad}_{(R,I)}(\phi)$ and $\mathrm{grad}^n_{(R,I)}(\phi)$ are said to be induced by ϕ. Note that upon letting

$$J = \ker(\phi) \quad \text{and} \quad H = \mathrm{ledi}_{(R,I)}(J)$$

for the homogeneous ideal $\mathrm{ledi}_{(R,I)}(J)$ in $\mathrm{grad}(R, I)$ we have that [cf. §6(E15)]

(1) $$\mathrm{ledi}_{(R,I)}(J) \subset \ker(\mathrm{grad}_{(R,I)}(\phi))$$

and

(2) $$\begin{cases} \text{if } \phi \text{ is surjective with } \phi(I) = I' \\ \text{then } \mathrm{grad}_{(R,I)}(\phi) \text{ is surjective} \\ \text{with } \mathrm{ledi}_{(R,I)}(J) = \ker(\mathrm{grad}_{(R,I)}(\phi)) \end{cases}$$

and

(3) $$\begin{cases} \text{if } \phi \text{ is surjective with } \phi(I) = I' \text{ and } J \text{ is a principal ideal in } R \\ \text{generated by an element } h \text{ whose leading form } \mathrm{lefo}_{(R,I)}(h) \\ \text{is a nonzerodivisor in } \mathrm{grad}(R, I) \\ \text{then } \mathrm{grad}_{(R,I)}(\phi) \text{ is surjective} \\ \text{and } \ker(\mathrm{grad}_{(R,I)}(\phi)) \text{ is the principal ideal in } \mathrm{grad}(R, I) \\ \text{generated by } \mathrm{lefo}_{(R,I)}(h). \end{cases}$$

In case R is a quasilocal ring with $I = M(R)$, in the above three paragraphs we write R in place of (R, I), except that we write $\mathrm{grad}(R)$ and $\mathrm{grad}(R)_n$ in place of $\mathrm{grad}(R, I)$ and $\mathrm{grad}(R, I)_n$ respectively.

A valuation function on a ring R is a map $v : R \to \mathbb{Z} \cup \{\infty\}$ such that for all x, y, z in R we have:

(i) $v(xy) = v(x) + v(y)$,

(ii) $v(x + y) \geq \min(v(x), v(y))$, and

(iii) $v(z) = \infty \Leftrightarrow z = 0$,

with the usual understanding about ∞. Thus in the notation of L3§11, a valuation function is like a valuation with R and $\mathbb{Z}$ playing the roles of the field K and the ordered abelian group G respectively. If v is a valuation function on a nonnull ring R then clearly R is a domain and we get a unique valuation $\bar{v} : \mathrm{QF}(R) \to \mathbb{Z} \cup \{\infty\}$ such that for all x, y in $R^{\times}$ we have $\bar{v}(x/y) = v(x) - v(y)$; we say that the valuation $\bar{v}$ is induced by v; usually we denote $\bar{v}$ also by v. Considering the weaker property: (i*) $v(xy) \geq v(x) + v(y)$, we see that $\mathrm{ord}_{(R,I)}$ satisfies (i*) and (ii); it satisfies (iii) iff R is I-Hausdorff. We shall investigate this matter further in (T60) below and then again in the next Quest. Note that in case R is a local ring with $I = M(R)$, (iii) is the Krull Intersection Theorem.

LEMMA (T59). Let $A = \sum_{n \in \mathbb{N}} A_n$ be a naturally graded nonnull ring. Then A is a domain iff the product of any nonzero homogeneous elements in A is nonzero, i.e., iff: $0 \neq f_m \in A_m$ and $0 \neq g_n \in A_n$ with m, n in $\mathbb{N} \Rightarrow f_m g_n \neq 0$.

PROOF. Given $0 \neq f = \sum_{m \in \mathbb{N}} f_m \in A$ and $0 \neq g = \sum_{n \in \mathbb{N}} g_n \in A$ with $f_m \in A_m$ and $g_n \in A_n$, let m and n be minimal with $f_m \neq 0 \neq g_n$. Then $0 \neq f_m g_n \in A_{m+n}$ and $fg = f_m g_n + \sum_{m+n < q \in \mathbb{N}} h_q$ with $h_q \in A_q$, and hence $fg \neq 0$.

THEOREM (T60). Let I be an ideal in a ring R. Then we have the following.

(T60.1) $I \neq R$ and $\mathrm{ord}_{(R,I)}$ is a valuation function on R iff R is I-Hausdorff and $\mathrm{grad}(R, I)$ is a domain.

(T60.2) Upon letting $R^* = R/J^*$ and $I^* = I/J^*$ with $J^* = \cap_{n \in \mathbb{N}} I^n$, we have that R^* is I^*-Hausdorff, and if $\mathrm{grad}(R, I)$ is a domain then R^* is a domain and $\mathrm{ord}_{(R^*, I^*)}$ is a valuation function on R^*.

(T60.3) Let $\phi : R \to R'$ be a ring epimorphism, with $J = \ker(\phi)$, such that the leading ideal $\mathrm{grad}_{(R,I)}(J)$ is a (homogeneous) prime ideal in $\mathrm{grad}(R, I)$. Then $\cap_{n \in \mathbb{N}} (J + I^n)$ is a prime ideal in R.

PROOF. (T60.1) follows from (T59).

To prove (T60.2) first note that R^* is obviously I^*-Hausdorff, and then let $\phi^* : R \to R^*$ be the canonical epimorphism and assume that $\mathrm{grad}(R, I)$ is a domain. Now clearly $\ker(\phi^*) = J^*$ and $I^* = \phi^*(I)$ with $\mathrm{grad}_{(R,I)}(\phi^*)(J^*) = 0$, and hence $\mathrm{grad}(R^*, I^*)$ is a domain, because $\mathrm{grad}_{(R,I)}(\phi^*)(\ker(\phi^*))$ is the kernel of the induced epimorphism of $\mathrm{grad}(R, I)$ onto $\mathrm{grad}(R^*, I^*)$. Thus we are reduced to (T60.1).

To prove (T60.3), upon letting $I' = \phi(I)$, we see that $\mathrm{grad}(R', I')$ is a domain, because $\mathrm{grad}_{(R,I)}(J)$ is the kernel of the induced epimorphism of $\mathrm{grad}(R, I)$ onto $\mathrm{grad}(R', I')$. Therefore by (T60.2) we see that $R'/ \cap_{n \in \mathbb{N}} I'^n$ is a domain, and therefore $\cap_{n \in \mathbb{N}} I'^n$ is a prime ideal in R'. Clearly $\cap_{n \in \mathbb{N}} (J + I^n) = \phi^{-1}(\cap_{n \in \mathbb{N}} I'^n)$, and hence $\cap_{n \in \mathbb{N}}(J + I^n)$ is a prime ideal in R.

THEOREM (T61). Let R be a local ring with $\dim(R) = d$ and $\mathrm{emdim}(R) = e$, let $x = (x_1, \ldots, x_m)$ be a finite number of generators of $M(R)$, let $X = (X_1, \ldots, X_m)$ be indeterminates over the field $k = R/M(R)$, and recall that $\Theta_R^x : k[X] \to \mathrm{grad}(R)$ is a graded ring epimorphism where $k[X] = \sum_{n \in \mathbb{N}} k[X]_n$ is the naturally graded polynomial ring with $k[X]_n$ = the set of all homogeneous polynomials of degree n (including the zero polynomial). Then we have the following.

(T61.1) $\ker(\Theta_R^x) \cap k[X]_1 = 0 \Leftrightarrow m = e$.

(T61.2) $\ker(\Theta_R^x) = 0 \Leftrightarrow m = d$.

(T61.3) If $m = e$ then: R is regular $\Leftrightarrow \ker(\Theta_R^x) = 0$.

(T61.4) R is regular $\Leftrightarrow$ there is a permutation σ of $\{1, \ldots, m\}$ together with elements a_{ij} in k such that the ideal $\ker(\Theta_R^x)$ in $k[X]$ is generated by the $m - e$ elements $X_{\sigma(i)} - \sum_{1 \le j \le e} a_{ij} X_{\sigma(j)}$ for $e + 1 \le i \le m$.

PROOF. Let $I = M(R)$, let $\lambda_n : I^n \to (I^n/I^{n+1})$ be the natural surjection, and let $\mu_n : (I^n/I^{n+1}) \to \mathrm{grad}(R, I)$ be the natural injection. Also let $y_l = \mu_1(\lambda_1(x_l))$ for $1 \le l \le m$.

To prove (T61.1), note that any $z \in k[X]_1$ can be expressed uniquely as $z = a_1 X_1 + \cdots + a_m X_m$ with $a_1, \ldots, a_m$ in k, and for $1 \le l \le m$ we can write $a_l = \lambda_0(b_l)$ with $b_l \in R$. Observe that $z = 0 \Leftrightarrow$ all the a_l are zero $\Leftrightarrow$ all the b_l belong to I. Now $\Theta_R^x(z) = \mu_1(\lambda_1(b_1 x_1 + \cdots + b_m x_m))$. Since μ_1 is injective, we see that:

$$\Theta_R^x(z) = 0 \Leftrightarrow \lambda_1(b_1 x_1 + \cdots + b_m x_m) = 0.$$

By the Nakayama Lemma (T3) we know that: $m > e \Leftrightarrow \lambda_1(b_1 x_1 + \cdots + b_m x_m) = 0$ for some elements $b_1, \ldots, b_m$ in R at least one of which does not belong to I. This completes the proof of (T61.1).

Now clearly $m = d \Leftrightarrow \mathrm{ht}_R I = m$, and hence (T61.2) follows from the Relative Independence theorem (T34). By definition R is regular iff $e = d$, and hence (T61.2) implies (T61.3). So it only remains to prove (T61.4).

First assume that R is regular. Then $e = d$, and by the Nakayama Lemma (T3) we can find a permutation σ of $\{1, \ldots, m\}$ together with elements b_{ij} in R such that $x_{\sigma(i)} = \sum_{1 \le j \le e} b_{ij} x_{\sigma(j)}$ for $e+1 \le i \le m$. Now $x' = (x_{\sigma(1)}, \ldots, x_{\sigma(e)})$ are generators of I and, upon considering the polynomial ring $k[X']$ with $X' = (X_{\sigma(1)}, \ldots, X_{\sigma(e)})$, by (T61.3) we get $\ker(\Theta_R^{x'}) = 0$. Let $a_{ij} = \lambda_0(b_{ij})$, and let H be the ideal in $k[X]$ generated by the $m - e$ elements $X_{\sigma(i)} - \sum_{1 \le j \le e} a_{ij} X_{\sigma(j)}$ for $e + 1 \le i \le m$. Let $\Delta : k[X] \to k[X']$ be the unique k-epimorphism of graded rings which sends $X_{\sigma(i)}$ to $X_{\sigma(i)}$ or $\sum_{1 \le j \le e} a_{ij} X_{\sigma(j)}$ according as $1 \le i \le e$ or $e + 1 \le i \le m$. Then clearly $\Theta_R^x = \Theta_R^{x'} \Delta$, and hence $\ker(\Theta_R^x) = \ker(\Delta) = H$.

Conversely assume that there exists a permutation σ of $\{1, \ldots, m\}$ together with elements a_{ij} in k such that the ideal $\ker(\Theta_R^x)$ in $k[X]$ is generated by the $m - e$

elements

$$X_{\sigma(i)} - \sum_{1 \le j \le e} a_{ij} X_{\sigma(j)} \quad \text{for} \quad e + 1 \le i \le m.$$

It follows that $x' = (x_{\sigma(1)}, \ldots, x_{\sigma(e)})$ are generators of I modulo I^2 and hence by Nakayama they are generators of I. Now with $k[X'], H, \Delta$ as above we have $\Theta_R^x = \Theta_R^{x'} \Delta$ with $\ker(\Delta) = H$. Consequently $\ker(\Theta_R^{x'}) = 0$ and hence R is regular by (T61.3).

COROLLARY (T62). If R is a d-dimensional regular local ring then $\mathrm{grad}(R)$ is isomorphic to a d variable polynomial ring over the field $R/M(R)$, and hence in particular $\mathrm{grad}(R)$ is a noetherian normal domain.

PROOF. Follows from (T61.3).

COMMENT (C19). [**Graded Rings of Polynomial Rings**]. Let $X = (X_l)_{l \in L}$ be a family of variables over a field k, consider the polynomial ring $R = k[X]$, and let I be the maximal ideal in R generated by X. Then we may "identify" k with R/I as well as with $\mathrm{grad}(R, I)_0$. Now upon letting

$$Y = (Y_l)_{l \in L} \quad \text{with} \quad Y_l = \mathrm{grad}^1_{(R,I)}(X_l) = \mathrm{lefo}_{(R,I)}(X_l)$$

we see that

$$\mathrm{grad}(R, I) = \text{the polynomial ring } k[Y]$$

and

$$\Theta^X_{(R,I)} : k[X] \to k[Y]$$

is the k-isomorphism which sends X_l to Y_l for all $l \in L$.

QUEST (Q14) Completely Normal Domains

Having proved L4§8.3(T24.1) as the Krull Intersection Theorem (T7), in (T64) we shall prove L4§8.3(T24.2) which says that the order function of a regular local ring is a valuation and hence the said ring is a normal domain. For dealing with the normality part of this we introduce the concept of complete normality and prove a lemma about it in (T63).

A domain R is said to be completely normal if: $x \in \mathrm{QF}(R)$ with $dx^n \in R$ for some $d \in R^\times$ and all $n \in \mathbb{N} \Rightarrow x \in R$.

COMPLETE NORMALITY LEMMA (T63). For any given ring R we have the following.

(T63.1) If R is a completely normal domain then R is a normal domain. If R is a noetherian normal domain then R is a completely normal domain.

(T63.2) If I is an ideal in R such that for all $z \in R$ we have $\cap_{n \in \mathbb{N}}(zR + I^n) = zR$, and grad$(R, I)$ is a completely normal domain, then R is a completely normal domain.

(T63.3) For any ideal I in R we have that: $I \neq R$ iff grad(R, I) is nonnull.

(T63.4) If R is noetherian and I is any ideal in R then grad(R, I) is noetherian.

(T63.5) If R is noetherian and has an ideal I with $I \subset$ jrad(R) such that grad(R, I) is a normal domain, then R is a completely normal domain.

PROOF. To prove (T63.1), first assume that R is a completely normal domain, let $x \in \mathrm{QF}(R)$ be integral over R, and let

$$x^m + \sum_{1 \leq i \leq m} a_i x^{m-i} = 0$$

with $m \in \mathbb{N}_+$ and $a_i \in R$ be an equation of integral dependence for x. Write

$$x = y/z \quad \text{with} \quad y \in R \quad \text{and} \quad z \in R^\times.$$

Let

$$d = z^m.$$

Then clearly

$$d \in R^\times \quad \text{with} \quad dx^n \in R \quad \text{for} \quad 0 \leq n \leq m.$$

By induction on n we shall show that for all $m \leq n \in \mathbb{N}$ we have $dx^n \in R$. For $n = m$ we have already said this. So let $n > m$ and assume for all smaller values of n. Multiplying the equation of integral dependence by dx^{n-m} we get the equation

$$dx^n = -\sum_{1 \leq i \leq m} a_i dx^{n-i}$$

whose RHS belongs to R by the induction hypothesis and hence so does the LHS. Therefore by complete normality we get

$$x \in R.$$

Thus R is normal.

Now assume that R is a normal noetherian domain, and let $x \in \mathrm{QF}(R)$ be such that

$$dx^n \in R \quad \text{for some} \quad d \in R^\times \text{ and all} \quad n \in \mathbb{N}.$$

Then $R[x]$ is contained in the finitely (by a single element) generated R-module Rd^{-1}, and hence (because of the noetherianness of R) by L4§5(O25)(27•) we conclude that $R[x]$ is a finitely generated R-module. Consequently by L4§10(E2.2) we see that x is integral over R, and hence (by the normality of R) we get

$$x \in R.$$

Thus R is a completely normal domain.

To prove (T63.2), let I be an ideal in R such that for every element z in R we have $\cap_{n \in \mathbb{N}}(zR + I^n) = zR$, and $\mathrm{grad}(R, I)$ is a completely normal domain. By taking $z = 0$ we see that $\cap_{n \in \mathbb{N}} I^n = 0$, and hence by (T60.2) we see that R is a domain and $\mathrm{ord}_{(R,I)}$ is a valuation function on R. To show that R is completely normal, let there be given any $x \in \mathrm{QF}(R)$ such that $dx^n \in R$ for some $d \in R^\times$ and all $n \in \mathbb{N}$. Write

$$x = y/z \quad \text{with} \quad y \in R \quad \text{and} \quad z \in R^\times.$$

We want to show that $x \in R$, i.e., $y \in zR$. By hypothesis

$$\cap_{m \in \mathbb{N}}(zR + I^m) = zR$$

and hence we are reduced to showing that

$$y \in zR + I^m \quad \text{for all} \quad m \in \mathbb{N}.$$

We do this by induction. Since it is trivial for $m = 0$, let $m > 0$ and assume for $m - 1$. Then

$$y = az + b \quad \text{with} \quad a \in R \quad \text{and} \quad b \in I^{m-1}.$$

Since $dx^n \in R$ for all $n \in \mathbb{N}$, we get $d(x - a)^n \in R$ for all $n \in \mathbb{N}$. Consequently, since $x = y/z = a + (b/z)$, we get

$$db^n \in z^n R \quad \text{for all} \quad n \in \mathbb{N}.$$

Thus for all $n \in \mathbb{N}$ we can write

$$db^n = c_n z^n \quad \text{with} \quad c_n \in R.$$

Since $\mathrm{ord}_{(R,I)}$ is a valuation function on R, taking leading forms relative to (R, I) preserves products, and hence for all $n \in \mathbb{N}$ we have

$$\mathrm{lefo}_{(R,I)}(d) \cdot \big(\mathrm{lefo}_{(R,I)}(b)\big)^n = \mathrm{lefo}_{(R,I)}(c_n) \cdot \big(\mathrm{lefo}_{(R,I)}(z)\big)^n.$$

Since $\mathrm{grad}(R, I)$ is completely normal, this implies that

$$\mathrm{lefo}_{(R,I)}(b)/\mathrm{lefo}_{(R,I)}(z) \in \mathrm{grad}(R, I)$$

and hence

$$\mathrm{lefo}_{(R,I)}(b) = \mathrm{lefo}_{(R,I)}(c) \cdot \mathrm{lefo}_{(R,I)}(z) \quad \text{with} \quad c \in R$$

and therefore (because $b \in I^{m-1}$) by the definition of multiplication in $\mathrm{grad}(R, I)$ we get

$$b - cz \in I^m.$$

Thus

$$y - (a + c)z \in I^m$$

and hence

$$y \in zR + I^m.$$

Part (T63.3) is obvious.

Part (T63.4) follows from L4§5(O27)(36$^\bullet$).

In view of (T20) and (T60), by (T63.1) to (T63.4) we get (T63.5).

ORD VALUATION LEMMA (T64). Let R be a regular local ring. Then for all nonzero elements x and y in R we have $\mathrm{ord}_R(xy) = \mathrm{ord}_R x + \mathrm{ord}_R y$. Consequently R is a normal domain and we get a valuation $\mathrm{ord}_R : \mathrm{QF}(R) \to \mathbb{Z}$ of $\mathrm{QF}(R)$ by putting $\mathrm{ord}_R(x/y) = \mathrm{ord}_R x - \mathrm{ord}_R y$.

[We shall continue to use this extended meaning of ord_R].

PROOF. Follows from (T62) and (T63).

QUEST (Q15) Regular Sequences and Cohen-Macaulay Rings

Let R be a ring. Let $x = (x_1, \ldots, x_n)$ be a sequence of elements in R of length $n \in \mathbb{N}$, and for $0 \le m \le n$ let I_m be the ideal in R generated by $(x_1, \ldots, x_m)$. Let V be a module over R. Recall that $S_R(V) = R \setminus Z_R(V)$ where $Z_R(V)$ is the set of all zerodivisors of V, i.e., elements $r \in R$ with $rv = 0$ for some $0 \ne v \in V$.

We say that x is a V-regular sequence, if $I_n V \ne V$ and for $1 \le m \le n$ we have $x_m \in S_R(V/(I_{m-1}V))$. Moreover if J is an ideal in R with $I_n \subset J$ then we say that x is a V-regular sequence in J. Furthermore if for every $z \in J \cap S_R(V/(I_n V))$ we have $I_n V + zV = V$ then we say that x is a maximal V-regular sequence in J. By taking $V = R$ we get the notions of R-regular and maximal R-regular sequences; note that now the condition $x_m \in S_R(R/(I_{m-1}R))$ is equivalent to the condition $\phi_m(x_m) \in S_{\phi_m(R)}(\phi_m(R))$ where $\phi_m : R \to R/I_{m-1}$ is the residue class epimorphism. We define the (R, J)-regularity of V by putting

$$\mathrm{reg}_{(R,J)}V = \text{the maximum length of a } V\text{-regular sequence in } J$$

and we note that: this is in $\mathbb{N} \cup \{-\infty, \infty\}$; it is $-\infty$ means $V = 0$; it is 0 means $V \ne 0$ and for every $z \in J \cap S_R(V)$ we have $zV = V$; and it is ∞ means $V \ne 0$ and there is no bound on the said length. If R is local with $J = M(R)$ then we call this the R-regularity of V and denote it by $\operatorname{reg}_R V$, i.e.,

$$\operatorname{reg}_R V = \text{the maximum length of a } V\text{-regular sequence in } M(R)$$

and in this case we write $\operatorname{reg}(R)$ in place of $\operatorname{reg}_R R$, i.e.,

$$\operatorname{reg}(R) = \text{the maximum length of an } R\text{-regular sequence in } M(R)$$

and we call it the regularity of R; by (T27) this is a nonnegative integer $\leq \dim(R)$.

By parts (T66.4) and (T66.5) of Lemma (T66) below it follows that if R is noetherian, V finitely generated, and J is an ideal in R with $JV \neq V$, then there exist maximal V-regular sequences in J and they all have the same length which equals $\operatorname{reg}_{(R,J)} V$. In particular if R is local then there exist maximal R-regular sequences in $M(R)$ and the integer $\operatorname{reg}(R)$ equals the length of every such sequence.

The Generalized Principal Ideal Theorem (T27) inspires the following definitions of generating numbers, complete intersections, and CM (= Cohen-Macaulay) modules and rings.

We define the generating number of V in R by putting

$$\operatorname{gnb}_R V - \text{the smallest number of generators of } V$$

and we note that this is ∞ iff V is not finitely generated, and it is 0 iff $V = 0$. In particular this defines the generating number $\operatorname{gnb}_R I$ of any ideal I in R by putting

$$\operatorname{gnb}_R I = \text{the smallest number of generators of } I.$$

An ideal I in R is said to be an ideal-theoretic complete intersection in R if $I \neq R$ and $\operatorname{ht}_R I = \operatorname{gnb}_R I$.

A finitely generated R-module is free means it is isomorphic to R^d for some $d \in \mathbb{N}$, i.e., equivalently, it has a basis $y_1, \ldots, y_d$. In case R is local and V is finitely generated, V is said to be a CM module if $\operatorname{reg}_R V = \dim(R/(0 : V)_R)$.

A local ring R is said to be a CM ring if it is a CM module over itself, i.e., if $\operatorname{reg}(R) = \dim(R)$. Note that by (T27), for any local ring R we have the inequality $\operatorname{reg}(R) \leq \dim(R)$, and so the CM property says that equality holds; this may be compared with the fact that for any local ring R we have $\operatorname{emdim}(R) \geq \dim(R)$ and R is regular means equality holds. A noetherian ring R is said to be a CM ring if, for every maximal ideal P in R, the localization R_P is a CM ring.

An ideal I in a noetherian ring R is said to be unmixed if for every prime P in $\operatorname{ass}_R(R/I)$ we have $\operatorname{ht}_R P = \operatorname{ht}_R I$, i.e., equivalently, if I has no embedded primes and all its minimal primes have the same height. The unmixedness theorem is said to hold in a noetherian ring R if every ideal in R, which can be generated by as many elements as its height, is unmixed; note that by (T27) this is equivalent to saying that if I is any ideal in R, which can be generated by e elements where the nonnegative integer e equals $\operatorname{ht}_R I$, then I has no embedded primes.

Amongst the various results about the above concepts which we shall now prove, in (T71) we shall prove the third part L4§8(T24.3) of the Ord Valuation Theorem.

Before that, in (T67) and (T68) we shall relate regularity with the independence of (Q8) and the graded rings of (Q13). We start off with Lemmas (T65) and (T66). Out of this, the very versatile Lemma (T65) is of general interest.

LOCALIZATION LEMMA (T65). For any ring R we have the following.
(T65.1) If R is a domain then for any ideal I in R we have

$$\bigcap_{P \in \mathrm{mspec}(R)} (IR_P) = I$$

and in particular we have

$$\bigcap_{P \in \mathrm{mspec}(R)} R_P = R.$$

In both the above equations the intersection is taken in $\mathrm{QF}(R)$.
(T65.2) Without assuming R to be a domain, for any ideal I in R we have

$$\{r \in R : \phi_P(r) \in \phi_P(I)R_P \text{ for all } P \in \mathrm{mspec}(R)\} = I$$

where $\phi_P : R \to R_P$ is the canonical homomorphism, i.e., equivalently we have

$$\bigcap_{P \in \mathrm{mspec}(R)} [I : P]_R = I$$

where the intersection is taken in R.

PROOF. In the second equation of (T65.1), clearly RHS $\subset$ LHS; to prove the other inclusion, given any $x \in$ LHS, upon letting

$$J = \{z \in R^\times : xz \in R\} \cup \{0\}$$

we see that J is an ideal in R with $J \not\subset P$ for all $P \in \mathrm{mspec}(R)$, and hence $J = R$ by L2§5(R5), and therefore $1 \in J$, i.e., $x \in R$. This proves the second equation of (T65.1), and hence the first equation of (T65.1) is subsumed under the first equation of (T65.2). In view of L4§7.1(T17)(b), the two equations of (T65.2) are equivalent to each other. In the second equation of (T65.2), clearly RHS $\subset$ LHS; to prove the other inclusion, given any $x \in$ LHS, upon letting

$$J = \{z \in R^\times : xz \in I\} \cup \{0\}$$

we see that J is an ideal in R with $J \not\subset P$ for all $P \in \mathrm{mspec}(R)$, and hence $J = R$ by L2§5(R5), and therefore $1 \in J$, i.e., $x \in I$.

REGULAR SEQUENCE LEMMA (T66). Let V be a module over a ring R, let $x = (x_1, \ldots, x_n)$ be a V-regular sequence in R of length $n \in \mathbb{N}$, let I_m be the ideal generated by $x_1, \ldots, x_m$ for $0 \leq m \leq n$, and let $I = I_n$. Then we have the following.

(T66.1) If $y_1, \ldots, y_n$ are any elements in V with $x_1 y_1 + \cdots + x_n y_n = 0$ then $y_i \in IV$ for $1 \leq i \leq n$.

(T66.2) If $d_1, \ldots, d_n$ are any elements in $\mathbb{N}_+$ then the sequence $x_1^{d_1}, \ldots, x_n^{d_n}$ is V-regular.

(T66.3) If $n > 0$ and $x_n \in S_R(V/(I_m V))$ for $0 \leq m < n$ then the sequence $(x_n, x_1, \ldots, x_{n-1})$ is V-regular.

(T66.4) If R is noetherian and J is an ideal in R with $JV \neq V$ such that $I \subset J$, then there exists an integer $n' \geq n$ together with elements $x_{n+1}, \ldots, x_{n'}$ in J such that $(x_1, \ldots, x_{n'})$ is a maximal V-regular sequence in J.

(T66.5) If R is noetherian, V is finitely generated, J is an ideal in R with $JV \neq V$ such that $I \subset J$, and there exists a maximal V-regular sequence in J of length n, then x is a maximal V-regular sequence in J.

[If R is noetherian and J is an ideal in R with $JV \neq V$, then by taking $n = 0$ in (T66.4) we see the existence of maximal V-regular sequences in J, and moreover if V is finitely generated then by (T66.5) it follows that all maximal V-regular sequences in J have the same length and this common integer equals $\mathrm{reg}_{(R,J)} V$].

(T66.6) If R is noetherian, V is finitely generated, and J is an ideal in R with $JV \neq V$ such that $I \subset J$, then

$$\mathrm{reg}_{(R,J)} V = n + \mathrm{reg}_{(R,J)} (V/(IV)).$$

(T66.7) If R is noetherian, V is finitely generated, and J is an ideal in R with $JV \neq V$ such that $I \subset J$, then upon letting $n' = \mathrm{reg}_{(R,J)} V$ and $\nu = \mathrm{gnb}_R J$ we have $\nu \geq n' \geq n$ and J has generators $\xi_1, \ldots, \xi_\nu$ such that $(\xi_1, \ldots, \xi_{n'})$ is a maximal V-regular sequence in J.

(T66.8) Without assuming the sequence x to be V-regular, but assuming R to be noetherian and V to be finitely generated, if positive integer $m \leq n$ is such that $I_{m-1} \subset Z_R(V)$ and $I_m \not\subset Z_R(V)$, then for some elements $r_1, \ldots, r_{m-1}$ in R we have $r_1 x_1 + \cdots + r_{m-1} x_{m-1} + x_m \in S_R(V)$.

(T66.9) Without assigning a V-regular sequence x in R, if R is noetherian, J is an ideal in R generated by ν generators with $\nu \in \mathbb{N}$, and V is finitely generated with $JV \neq V$ and $\mathrm{reg}_{(R,J)} V = \mu$, then $\nu \geq \mu$ and J has generators $\xi_1, \ldots, \xi_\nu$ such that $(\xi_1, \ldots, \xi_\mu)$ is a V-regular sequence.

PROOF. To prove (T66.1) we make induction on n. For $n = 0$ there is nothing to show. So let $n > 0$ and assume for $n-1$. Let $y_1, \ldots, y_n$ in V with $x_1 y_1 + \cdots + x_n y_n = 0$. Since x_n is a nonzerodivisor for $V/(x_1, \ldots, x_{n-1})V$, we get $y_n = \sum_{1 \leq i \leq n-1} x_i z_i$ with $z_i \in V$. This gives $\sum_{1 \leq i \leq n-1} x_i (y_i + x_n z_i) = 0$ and hence, by the induction hypothesis we get $y_i + x_n z_i \in (x_1, \ldots, x_{n-1})V$ for $1 \leq i \leq n - 1$ and therefore $y_i \in IV$ for $1 \leq i \leq n$.

In (T66.2), the $n = 0$ case being trivial, we may suppose $n > 0$, and we may also assume $d_1 = d \in \mathbb{N}_+$ with $d_2 = \cdots = d_n = 1$ because then the sequence $(x_2, \ldots, x_n)$ will be $(V/(x_1^d V))$-regular and we can repeat the argument. Now we make induction on d. For $d = 1$ we have nothing to show. So let $d > 1$ and assume that the sequence $(x_1^{d-1}, x_2, \ldots, x_n)$ is V-regular. Clearly x_1^d is V-regular. So let $1 < i \le n$ and assume that $(x_1^d, x_2, \ldots, x_{i-1})$ is V-regular. Let there be given any elements $s, t_1, \ldots, t_{i-1}$ in V such that $x_i s = x_1^d t_1 + x_2 t_2 + \cdots + x_{i-1} t_{i-1}$. Then by the induction hypothesis we have $s = x_1^{d-1} u_1 + x_2 u_2 + \cdots + x_{i-1} u_{i-1}$ with $u_1, \ldots, u_{i-1}$ in V. Therefore

$$x_1^{d-1}(x_1 t_1 - x_i u_1) + x_2(t_2 - x_i u_2) + \cdots + x_{i-1}(t_{i-1} - x_i u_{i-1}) = 0$$

and hence by (T66.1) we get

$$x_1 t_1 - x_i u_1 \in (x_1^{d-1}, x_2, \ldots, x_{i-1})V.$$

It follows that $x_i u_1 \in (x_1, \ldots, x_{i-1})V$ and hence $u_1 \in (x_1, \ldots, x_{i-1})V$ and therefore $s \in (x_1^d, x_2, \ldots, x_{i-1})V$. This completes the induction on i and proves (T66.2).

To prove (T66.3) suppose $n > 0$ and $x_n \in S_R(V/(I_m V))$ for $0 \le m < n$. We make induction on n. The case of $n = 1$ being obvious, let $n > 1$ and assume for $n - 1$. Applying the induction hypothesis to the $V/(I_1 V)$-regular sequence $(x_2, \ldots, x_n)$ we see that the sequence $(x_1, x_n, x_2, \ldots, x_{n-1})$ is V-regular. Suppose if possible that $x_1 \in Z_R(V/(x_n V))$. However

$$
\begin{cases}
x_1 \in Z_R(V/(x_n V)) \\
\Rightarrow x_1 y = x_n z \text{ for some } y, z \text{ in } V \text{ with } y \notin x_n V \\
\Rightarrow z = x_1 w \text{ for some } w \in V \text{ (because } x_n \in S_R(V/(x_1 V))) \\
\Rightarrow x_1(y - x_n w) = 0 \\
\Rightarrow y - x_n w = 0 \text{ (because } x_1 \in S_R(V)) \\
\Rightarrow \text{contradiction to } y \notin x_n V.
\end{cases}
$$

Therefore $x_1 \in S_R(V/(x_n V))$, and hence $(x_n, x_1, \ldots, x_{n-1})$ is V-regular.

To prove (T66.4) and (T66.5) assume that R is noetherian and let J be an ideal in R with $JV \ne V$ such that $I \subset J$. If $x_{n+1}, x_{n+2}, \ldots$ are any elements in J such that the sequence $(x_1, \ldots, x_{n+i})$ is V-regular for $i = 1, 2, \ldots$ then, upon letting $I_{n+i} = (x_1, \ldots, x_{n+i})R$ we have that $I_n \subset I_{n+1} \subset I_{n+2} \subset \ldots$ are ideals in R with $I_n \ne I_{n+1} \ne I_{n+2} \ne \ldots$ and hence by the noetherianness of R this must stop at some integer $n' \ge n$. This proves (T66.4).

To prove (T66.5) also assume that V is finitely generated and there exists a maximal V-regular sequence $x' = (x_1', \ldots, x_n')$ in J of length n. By induction on n we shall show that then x is a maximal V-regular sequence in J. This being obvious for $n = 0$, for a moment suppose that $n = 1$. Then $J \subset Z_R(V/(x_1' V))$ and hence, by (Q1)(T1), $J \subset \cup P$ where the union is taken over the finite nonempty subset $\mathrm{tass}_R(V/(x_1' V))$ of $\mathrm{spec}(R)$ and therefore, by L4§5(O24)(22•), $J \subset P$ for some $P \in \mathrm{tass}_R(V/(x_1' V))$. Consequently, there exists $t \in V \setminus (x_1' V)$ such that for every $j \in J$ we have $jt = x_1' j^*$ for some $j^* \in V$. In particular $x_1 t = x_1' u$ for some $u \in V$; since $t \notin x_1' V$ and $x_1 \in S_R(V)$, we must have $u \notin x_1 V$. Multiplying the equation $x_1 t = x_1' u$ by j we get $x_1 jt = x_1' ju$, and multiplying the equation $jt = x_1' j^*$ by x_1 we get $x_1 jt = x_1' x_1 j^*$; equating the two RHSs we obtain $x_1' ju = x_1' x_1 j^*$ and hence $x_1'(ju - x_1 j^*) = 0$; since $x_1' \in S_R(V)$, this yields $ju = x_1 j^*$ and therefore $j \in Z_R(V/(x_1 V))$. Thus $J \subset Z_R(V/(x_1 V))$, and hence $x = (x_1)$ is a maximal V-regular sequence in J.

Now let $n > 1$ and assume for $n - 1$. For $0 \le m < n$, upon letting $I_m' = (x_1', \ldots, x_m')R$, we have $J \not\subset Z_R(V/(I_m V))$ and $J \not\subset Z_R(V/(I_m' V))$. Hence by (Q1)(T1) and L4§5(O24)(22•) we can find $x_n^* \in J$ such that for $0 \le m < n$ we have $x_n^* \notin Z_R(V/(I_m V))$ and $x_n^* \notin Z_R(V/(I_m' V))$. It follows that $(x_1, \ldots, x_{n-1}, x_n^*)$ and $(x_1', \ldots, x_{n-1}', x_n^*)$ are V-regular sequences in J. By (T66.3) we see that $(x_n^*, x_1, \ldots, x_{n-1})$ and $(x_n^*, x_1', \ldots, x_{n-1}')$ are V-regular sequences in J; the second is maximal by applying the $n = 1$ case to the $(V/(I_{n-1}' V))$-regular sequences x_n' and x_n^*. Consequently $(x_1, \ldots, x_{n-1})$ and $(x_1', \ldots, x_{n-1}')$ are $(V/(x_n^* V))$-regular sequences in J; the maximality of the second follows by the maximality of the V-regular sequence $(x_n^*, x_1', \ldots, x_{n-1}')$ and hence the maximality of the first follows by the induction hypothesis. This in turn yields the maximality of the V-regular sequence $(x_1, \ldots, x_{n-1}, x_n^*)$. From this, again by the $n = 1$ case, as applied to the $(V/(I_{n-1} V))$-regular sequences x_n and x_n^*, we get the maximality of the V-regular sequence x. This completes the proof of (T66.5).

In view of the bracketed remark following (T66.5), the proofs of (T66.6) and the inequality $n' \ge n$ of (T66.7) follow from (T66.4). The rest of (T66.7) follows from (T66.9) which itself will be deduced from (T66.8).

To prove (T66.8), without assuming the sequence x to be V-regular, but assuming R to be noetherian, V to be finitely generated, and m to be a positive integer with $m \le n$ such that $I_{m-1} \subset Z_R(V)$ and $I_m \not\subset Z_R(V)$, we want to show that for some elements $r_1, \ldots, r_{m-1}$ in R we have $r_1 x_1 + \cdots + r_{m-1} x_{m-1} + x_m \in S_R(V)$. Now for some $s_1, \ldots, s_m$ in R we have $s_1 x_1 + \cdots + s_m x_m \in S_R(V)$. By (Q1)(T1) we know that $Z_R(V) = P_1 \cup \cdots \cup P_c$ where $P_1, \ldots, P_c$ are prime ideals in R with $c \in \mathbb{N}_+$ such that for all $i \ne j$ we have $P_i \not\subset P_j$ (namely, these are the maximal members of $\mathrm{ass}_R(V)$). Since

$$I_{m-1} \subset Z_R(V)$$

and

$$s_1 x_1 + \cdots + s_m x_m \in S_R(V)$$

we must have $x_m \notin P_i$ for some i. Therefore by suitably labelling $P_1, \ldots, P_c$ we can arrange that $x_m \notin P_i$ for $1 \le i \le b$ and $x_m \in P_j$ for $b + 1 \le j \le c$ where b is an integer with $1 \le b \le c$. By L4§5(O24)(22•), for $1 \le i \le b$ we can take $t_i \in P_i$ such that $t_i \notin P_j$ for $b+1 \le j \le c$. Let $t = t_1 \ldots t_b$. Then $t \in P_i$ for $1 \le i \le b$ and $t \notin P_j$ for $b + 1 \le j \le c$. Now if $1 \le i \le b$ then

$$t \in P_i \text{ with } x_m \notin P_i \Rightarrow ts_1 x_1 + \cdots + ts_{m-1} x_{m-1} + x_m \notin P_i$$

whereas if $b + 1 \leq i \leq c$ then

$$\begin{cases} s_1 x_1 + \cdots + s_m x_m \notin Z_R(V) = P_1 \cup \cdots \cup P_c \text{ with } x_m \in P_i \\ \Rightarrow s_1 x_1 + \cdots + s_{m-1} x_{m-1} \notin P_i \\ \Rightarrow t s_1 x_1 + \cdots + t s_{m-1} x_{m-1} \notin P_i \\ \Rightarrow t s_1 x_1 + \cdots + t s_{m-1} x_{m-1} + x_m \notin P_i. \end{cases}$$

Thus $t s_1 x_1 + \cdots + t s_{m-1} x_{m-1} + x_m \notin P_i$ for $1 \leq i \leq c$, and hence it suffices to take $r_l = t s_l$ for $1 \leq l \leq m - 1$.

To prove (T66.9), without assigning a V-regular sequence x in R, assume that R is noetherian, J is an ideal in R generated by ν generators $\eta_1, \ldots, \eta_\nu$ with $\nu \in \mathbb{N}$, and V is finitely generated with $JV \neq V$ and $\mathrm{reg}_{(R,J)} V = \mu$. We want to show that then $\nu \geq \mu$ and J has generators $\xi_1, \ldots, \xi_\nu$ such that $(\xi_1, \ldots, \xi_\mu)$ is a V-regular sequence. We make induction on ν. If $\nu = 0$ then $\mu = 0$ and we have nothing to show. So let $\nu > 0$ and assume for $\nu - 1$. Let e be the largest nonnegative integer $\leq \nu$ such that $(\eta_1, \ldots, \eta_e) R \subset Z_R(V)$. If $e = \nu$ then $\mu = 0$ and again we have nothing to show. So assume that $e < \nu$. Then by (T66.8) we can find elements $r_1, \ldots, r_e$ in R such that for $\xi_1 = r_1 \eta_1 + \cdots + r_e \eta_e + \eta_{e+1}$ we have $\xi_1 \in S_R(V)$. Clearly J is generated by $\xi_1, \eta_2, \ldots, \eta_e, \eta_{e+2}, \ldots, \eta_\nu$. Let $\phi : R \to R' = R/(\xi_1 R)$ be the canonical epimorphism, let $J' = \phi(J)$ and let $V' = V/(\xi_1 V)$. Then V' is a finitely generated module over the noetherian ring R', J' is an ideal in R' with $J' V' \neq V'$ and $\mathrm{reg}_{(R',J')} V' = \mu - 1$, and J' is generated by the $\nu - 1$ elements $\phi(\eta_1), \ldots, \phi(\eta_e), \phi(\eta_{e+2}), \ldots, \phi(\eta_\nu)$. Therefore by the induction hypothesis we see that $\nu - 1 \geq \mu - 1$ and J' has generators $\xi_2', \ldots, \xi_\nu'$ for which the sequence $(\xi_2', \ldots, \xi_\mu')$ is V'-regular. We are done by taking $\xi_i \in \phi^{-1}(\{\xi_i'\})$ for $2 \leq i \leq \nu$.

INDEPENDENCE VERSUS REGULARITY THEOREM (T67). Let there be given a sequence of elements $x = (x_1, \ldots, x_n)$ in a ring R of length $n \in \mathbb{N}$, let I_m be the ideal generated by $x_1, \ldots, x_m$ for $0 \leq m \leq n$, and let $I = I_n$. Then we have the following.

(T67.1) The sequence x is independent over $(R, I) \Leftrightarrow \ker(\Theta_R^x) = 0$, where the epimorphism $\Theta_R^x : (R/I)[X_1, \ldots, X_n] \to \mathrm{grad}(R, I)$, with $\Theta_R^x(X_i) = \mathrm{grad}_{(R,I)}^1(x_i)$ for $1 \leq i \leq n$, is described in (Q13).

(T67.2) If the sequence x is independent over (R, I) then the (R/I)-module I^e/I^{e+1} is free for all $e \in \mathbb{N}$; more specifically, it is isomorphic to $(R/I)^{(e,n)}$ where $(e, n) = \binom{e+n-1}{e}$. See Comment (C20) below.

(T67.3) If the sequence x is independent over (R, I) then for all $d \in \mathbb{N}_+$ we have $S_R(R/I) \subset S_R(R/I^d)$.

(T67.4) If R is noetherian with $\mathrm{rad}_R I = I \neq R$ then: the given sequence x is independent over $(R, I) \Leftrightarrow I$ is an ideal-theoretic complete intersection in R with $\mathrm{ht}_R I = n$.

(T67.5) If the ring R/I_m is nonnull and (I/I_m)-Hausdorff for $0 \leq m \leq n$, and the sequence x is independent over (R, I), then it is R-regular.

(T67.6) If the sequence x is R-regular then it is independent over (R, I).

(T67.7) If the ring R/I_m is nonnull and (I/I_m)-Hausdorff for $0 \leq m \leq n$, the sequence x is independent over (R, I), the ring R is noetherian, and the ideal I is prime, then R is a domain and R_I is a normal domain. See Comment (C21) below.

(T67.8) If the ring R is noetherian with $I \neq R$ and the sequence x is independent over (R, I), then the ideal I is generated by an R-regular sequence of length n.

(T67.9) If R is noetherian with $\mathrm{rad}_R I = I \neq R$ then: x is R-regular $\Rightarrow I$ is an ideal-theoretic complete intersection in R with $\mathrm{ht}_R I = n \Rightarrow I$ is generated by an R-regular sequence of length n.

(T67.10) Without reference to a specific sequence x in R, we may restate (T67.4) and (T67.9) by saying that: a nonunit radical ideal I in a noetherian ring is an ideal-theoretic complete intersection in R with $\mathrm{ht}_R I = n \Leftrightarrow I$ is generated by a sequence of length n in R which is independent over $(R, I) \Leftrightarrow I$ is generated by a sequence of length n in R which is R-regular; moreover, this remains true after dropping the reference to height and length.

PROOF. (T67.1) and (T67.2) follow from (T31.2).

To prove (T67.3), assuming that x is independent over (R, I), by induction on $d \in \mathbb{N}_+$ we shall show that then $S_R(R/I) \subset S_R(R/I^d)$. This being obvious for $d = 1$, let $d > 1$ and assume for $d - 1$. Given any $t \in S_R(R/I)$ and $y \in R$ with $ty \in I^d$, we want to show that $y \in I^d$. By the induction hypothesis we have $y \in I^{d-1}$ and hence $y = F(x_1, \ldots, x_n)$ for some homogeneous polynomial $F(X_1, \ldots, X_n)$ of degree $d - 1$ over R (where F may be zero since the zero polynomial is regarded as homogeneous of any degree). Since $tF(x_1, \ldots, x_n) = ty \in I^d$, by (T31.2) it follows that all the coefficients of F belong to I. Therefore $y \in I^d$.

By (T27) and (T34) we get (T67.4).

To prove (T67.5) assume that x is independent over (R, I), and the ring R/I_m is nonnull and (I/I_m)-Hausdorff for $0 \leq m \leq n$. By induction on n we shall show that then x is R-regular. This being obvious for $n = 0$, let $n > 0$ and assume for $n - 1$. If $y \in R$ is such that $x_1 y = 0$ then by the independence we must have $y \in I$, and hence for some elements $z_1, \ldots, z_n$ in R we have $y = \sum_{1 \leq i \leq n} x_i z_i$ which gives $\sum_{1 \leq i \leq n} x_1 x_i z_i = 0$ and therefore by the independence we get $z_i \in I$ for $1 \leq i \leq n$, and this yields $y \in I^2$. Proceeding like this, we show that $y \in I^q$ for all $q \in \mathbb{N}_+$. Consequently by the Hausdorffness we get $y = 0$. Thus x_1 is R-regular.

We CLAIM that the sequence $(\phi(x_2), \ldots, \phi(x_n))$ is independent over $(\phi(R), \phi(I))$ where $\phi : R \to R/(x_1 R)$ is the residue class epimorphism; clearly by the induction hypothesis, this will complete the proof of (T67.5). To prove the claim, let $F(X_2, \ldots, X_n) \in R[X_2, \ldots, X_n]$ be homogeneous of some degree $d \in \mathbb{N}$ such that $F(x_2, \ldots, x_n) = x_1 s$ with $s \in R$. We want to show that then all the coefficients of F belong to I. Let e be any nonnegative integer with $s \in I^e$. Write $s = G(x_1, \ldots, x_n)$ where $G(X_1, \ldots, X_n) \in R[X_1, \ldots, X_n]$ is homogeneous of degree *e*. Then

(•) $$F(x_2, \ldots, x_n) = x_1 G(x_1, \ldots, x_n).$$

If $e < d - 1$ then, in view of (•) and (T31.2), by the independence of x over (R, I) we see that all the coefficients of G belong to I, and hence $s \in I^{e+1}$. Proceeding in this manner, we can arrange that $e \geq d - 1$. If $e > d - 1$ then, in view of (•) and (T31.2), by the independence of x over (R, I) we see that all the coefficients of F belong to I. If $e = d - 1$ then, in view of (•), by the independence of x over (R, I) we see that all the coefficients of F belong to I. Thus always all the coefficients of F belong to I.

To prove (T67.6) assume that x is R-regular. By induction on n we shall show that then x is independent over (R, I). This being obvious for $n = 0$, let $n > 0$ and assume for $n - 1$. Given any homogeneous $F(X_1, \ldots, X_n) \in R[X_1, \ldots, X_n]$ of some degree $d \in \mathbb{N}$ with $F(x_1, \ldots, x_n) = 0$, we want to show that all the coefficients of F belong to I. We do this by a second induction on d. It being obvious for $d = 0$, let $d > 0$ and assume for $d - 1$. We can write

(1) $$F(X_1, \ldots, X_n) = G(X_1, \ldots, X_{n-1}) + X_n F'(X_1, \ldots, X_n)$$

with homogeneous $G(X_1, \ldots, X_{n-1}) \in R[X_1, \ldots, X_{n-1}]$ of degree d, and homogeneous $F'(X_1, \ldots, X_n) \in R[X_1, \ldots, X_n]$ of degree $d - 1$. Since $F(x_1, \ldots, x_n) = 0$, by (1) we get $x_n F'(x_1, \ldots, x_n) \in ((x_1, \ldots, x_{n-1})R)^d$ and hence by (T67.3) and the first induction hypothesis we see that

(2) $$F'(x_1, \ldots, x_n) \in ((x_1, \ldots, x_{n-1})R)^d$$

and hence we can write

(3) $$F'(x_1, \ldots, x_n) = G'(x_1, \ldots, x_{n-1})$$

with homogeneous $G'(X_1, \ldots, X_{n-1}) \in R[X_1, \ldots, X_{n-1}]$ of degree d. Upon letting

(4) $$G''(X_1, \ldots, X_{n-1}) = G(X_1, \ldots, X_{n-1}) + x_n G'(X_1, \ldots, X_{n-1})$$

we have $G''(x_1, \ldots, x_{n-1}) = 0$ and hence again by the first induction hypothesis we see that all the coefficients of G'' belong to $(x_1, \ldots, x_{n-1})R$. Consequently by (4) we see that all the coefficients of G belong to I. In a moment we shall show that all the coefficients of F' belong to I, and then by (1) it will follow that all the coefficients of F belong to I which will complete the proof. By (2) we can write

$$F'(x_1, \ldots, x_n) = F''(x_1, \ldots, x_n)$$

where homogeneous $F''(X_1, \ldots, X_n) \in R[X_1, \ldots, X_n]$ of degree $d - 1$ is such that all its coefficients belong to I. Let

$$F'''(X_1, \ldots, X_n) = F'(X_1, \ldots, X_n) - F''(X_1, \ldots, X_n).$$

Then $F'''(X_1, \ldots, X_n) \in R[X_1, \ldots, X_n]$ is homogeneous of degree $d - 1$ and we have $F'''(x_1, \ldots, x_n) = 0$ and hence, by the second induction hypothesis, all the coefficients of F''' belong to I. Therefore all the coefficients of F' belong to I.

To prove (T67.7) assume that the ring R/I_m is nonnull and (I/I_m)-Hausdorff for $0 \leq m \leq n$, the sequence x is independent over (R, I), the ring R is noetherian, and the ideal I is prime. By (T67.1) we see that $\operatorname{grad}(R, I)$ is a domain and hence by (T60) it follows that R is a domain. By (T67.5) we know that x is R-regular and hence it is (R_I)-regular by L4§7.1(T17) and therefore by (T67.6) it is independent over (R_I, IR_I). Consequently $\operatorname{grad}(R_I, IR_I)$ is a normal domain because by (T67.1) it is a polynomial ring over a field. Now R_I is a normal domain by (T63).

To prove (T67.8) assume that the ring R is noetherian with $I \neq R$, and the sequence x, is independent over (R, I). We want to show that I is generated by an R-regular sequence of length n. For $n = 0$ there is nothing to show. So let $n > 0$ and assume for $n - 1$. If $I \subset Z_R(R)$ then by (Q1) we have $I \subset P$ for some $P \in \operatorname{tass}_R(R)$ and hence we can find $t \in R^\times$ such that $tu = 0$ for all $u \in I$. In particular $tx_1 = 0$. Let $I^* = \cap_{q \in \mathbb{N}} I^q$. If $t \notin I^*$ then for some $d \in \mathbb{N}$ we have $t \in I^d \setminus I^{d+1}$, and we can write $t = F(x_1, \ldots, x_n)$ where $F(X_1, \ldots, X_n)$ is a homogeneous polynomial of degree d with coefficients in R at least one of which is not in I. Let $G(X_1, \ldots, X_n) = F(X_1, \ldots, X_n)X_1$. Then $G(X_1, \ldots, X_n)$ is a homogeneous polynomial of degree $d + 1$ with coefficients in R at least one of which is not in I. Consequently by (T67.1) we see that $G(x_1, \ldots, x_n) \notin I^{d+2}$. But $G(x_1, \ldots, x_n) = tx_1 = 0$ which is a contradiction. Therefore $t \in I^*$. Consequently, because $t \neq 0$, by taking $(W, J) = (R, I)$ in the Artin-Rees Corollary Lemma (T5) and invoking (Q3)(3), we can find $v \in I$ such that we have $(1 - v)t = 0$, i.e., $t = vt$. Since $tu = 0$ for all $u \in I$, this gives the contradiction $t = 0$. Therefore $I \not\subset Z_R(R)$. Consequently for some positive integer $m \leq n$ we have $I_{m-1} \subset Z_R(R)$ with $I_m \not\subset Z_R(R)$. Now by (T66.8) we can find elements $r_1, \ldots, r_{m-1}$ in R such that for

$$x_1' = r_1 x_1 + \cdots + r_{m-1} x_{m-1} + x_m$$

we have $x_1' \in S_R(R)$. Clearly $(x_1', x_1, \ldots, x_{m-1}, x_{m+1}, \ldots, x_n)$ is a sequence which generates I and is independent over (R, I). Let $\phi : R \to R/(x_1'R)$ be the canonical epimorphism. Then $\phi(I)$ is a nonunit ideal in the noetherian ring $\phi(R)$ and the sequence $(\phi(x_1), \ldots, \phi(x_{m-1}), \phi(x_{m+1}), \ldots, \phi(x_n))$ generates $\phi(I)$ and is independent over $(\phi(R), \phi(I))$. Therefore by the induction hypothesis $\phi(I)$ is generated by a $\phi(R)$-regular sequence $(x_2^*, \ldots, x_n^*)$. Take

$$x_i' \in \phi^{-1}(\{x_i^*\}) \quad \text{for} \quad 2 \leq i \leq n.$$

Then I is generated by the sequence $(x_1', \ldots, x_n')$ which is R-regular. This completes the proof of (T67.8).

By (T67.4), (T67.6) and (T67.8), we get (T67.9). Finally (T67.10) is only a reformulation.

COMMENT (C20). [**Binomial Coefficients**]. Let $B = k[X_1, \ldots, X_n]$ be the finite variable polynomial ring over a ring k with $n \in \mathbb{N}$. We have the direct sum decomposition $B = \sum_{e \in \mathbb{N}} B_e$ where B_e is the set of all homogeneous polynomials of degree e including the zero polynomial. We may regard B_e as a k-module and we CLAIM that then it is isomorphic to $k^{(e,n)}$ where (e,n) is the binomial coefficient

$$(e,n) = \binom{e+n-1}{e}.$$

[In applying this to (T67.2) we take $k = R/I$]. Note that by definition, for any $d \in \mathbb{Z}$ (and $e \in \mathbb{N}$) we have

$$\binom{d}{e} = \frac{d(d-1)\ldots(d-e+1)}{e!}.$$

By the empty product equal 1 convention, for $n = 0$ we have $(e,n) = 1$ or 0 according as $e = 0$ or $e > 0$, and everything is trivially ok. So henceforth assume that $n > 0$. Then clearly $(e,n) \in \mathbb{N}_+$. In case k is the nullring, we have $B_e = 0$ and we can take a basis of it consisting of (e,n) zeros, and again everything is trivially ok. So henceforth also assume that k is nonnull. Now the set of all monomials degree e in $X_1, \ldots, X_n$ is a basis of B_e and we need to show that the number of these monomials is (e,n). We do this by induction on n. For $n = 1$ this is obvious because then for all $e \in \mathbb{N}$ we clearly have $(e,n) = 1$. So let $n > 1$ and assume for $n - 1$. By grouping our monomials according to their degree in X_n, by the induction hypothesis we see that their number equals the sum $\sum_{0 \le \epsilon \le e}(\epsilon, n-1)$. Thus we want to show that

$$\sum_{0 \le \epsilon \le e} (\epsilon, n-1) = (e,n).$$

This we do by induction on e. For $e = 0$ it is obvious because everything is reduced to 1. So let $e > 0$ and assume for $e - 1$. By the induction hypothesis

$$\text{LHS} = (e-1, n) + (e, n-1)$$

and so we want to show that

$$(e-1, n) + (e, n-1) = (e,n).$$

After multiplying both sides of this by $e!$ we get

$$\left\{ \begin{array}{l} \text{NEW LHS} \\ = [e(n+e-2)(n+e-3)\ldots(n)] + [(n+e-2)(n+e-3)\ldots(n-1)] \\ = [e+(n-1)][(n+e-2)(n+e-3)\ldots(n)] \\ = (n+e-1)(n+e-2)\ldots(n) \\ = \text{NEW RHS.} \end{array} \right.$$

COMMENT (C21). [**Localization of Regular Sequences**]. In proving the last part (T67.7) of the above theorem we used the fact that if $x = (x_1, \ldots, x_n)$ is an R-regular sequence in a noetherian ring R, and $\phi : R \to R_S$ is the canonical homomorphism into the localization of R at a multiplicative set S in it with $(xR) \cap S = \emptyset$, then $(\phi(x_1), \ldots, \phi(x_n))$ is an (R_S)-regular sequence in R_S. In view of (Q1), this follows from L4§7(T12) by noting that the R-regularity of x is equivalent to saying that: $xR \neq R$ and for every $m \in \{1, \ldots, n\}$ and every $P \in \operatorname{ass}_R(x_1, \ldots, x_{m-1})R$ we have $x_m \notin P$.

Also note that, by (T20), the condition "nonnull and Hausdorff" required in (T67.5) and (T67.7) is satisfied if R is nonnull with $I \subset \operatorname{jrad}(R)$.

COROLLARY (T68). Let (R, J) be a nonnull Zariski ring [i.e., R is a nonnull noetherian ring and J is an ideal in R with $J \subset \operatorname{jrad}(R)$ where $\operatorname{jrad}(R)$ is the intersection of all maximal ideals in R; for instance R could be a local ring with $J = M(R)$]. Let $x = (x_1, \ldots, x_n)$ be a sequence of elements in R of length $n \in \mathbb{N}$ such that for the ideal $I = xR$ we have $I \subset J$. Then we have the following.

(T68.1) The sequence x is R-regular $\Leftrightarrow$ it is independent over (R, I).

(T68.2) If $\operatorname{rad}_R I = I$ then: the sequence x is R-regular $\Leftrightarrow$ I is an ideal-theoretic complete intersection in R with $\operatorname{ht}_R I = n$.

(T68.3) If the sequence x is R-regular then the sequence $(x_{\sigma(1)}, \ldots, x_{\sigma(n)})$ is R-regular for every permutation σ of $\{1, \ldots, n\}$.

(T68.4) If I is a prime ideal which is an ideal-theoretic complete intersection in R with $\operatorname{ht}_R I = n$ then R is a domain and for any permutation σ of $\{1, \ldots, n\}$ and any nonnegative integer $m \leq n$, upon letting $x' = (x_{\sigma(1)}, \ldots, x_{\sigma(m)})$ and $I' = x'R$, we have that: the sequence x' is R-regular, the ideal I' is prime, its localization $R_{I'}$ is a normal domain, and it is an ideal-theoretic complete intersection in R with $\operatorname{ht}_R I' = m$.

PROOF. The independence of a sequence is clearly invariant under permutations. Therefore, in view of the second paragraph of (C21) above, everything will follow from (T67.4) to (T67.7) if we show that, assuming I to be prime and the sequence x to be R-regular, the ideal $I_m = (x_1, \ldots, x_m)R$ is prime for every nonnegative integer $m \leq n$. Let $\overline{R} = R/I_m$ and $\overline{I} = I/I_m$. Then the images of $x_{m+1}, \ldots, x_n$ in $\overline{R}$ constitute an $\overline{R}$-regular sequence which generates $\overline{I}$, and hence by (T67.1) and (T67.6) we see that $\operatorname{grad}(\overline{R}, \overline{I})$ is a domain because it is isomorphic to a polynomial ring over the domain R/I. Consequently $\overline{R}$ is a domain by (T20) and (T60). Therefore I_m is prime.

COHEN-MACAULAY LEMMA (T69). Given any local ring R, for it we have the following.

(T69.1) If V is any finitely generated R-module then

(1) $$\mathrm{reg}_R V \le \dim(R/(0:V)_R)$$

and in greater detail

(2) $$\begin{cases} \text{for every } P \in \mathrm{ass}_R V \text{ we have} \\ \mathrm{reg}_R V \le \dim(R/P) \le \dim(R/(0:V)_R) \end{cases}$$

and moreover

(3) $$\begin{cases} V \text{ is CM} \\ \Rightarrow \text{for every } P \in \mathrm{ass}_R V \text{ we have} \\ \quad \mathrm{reg}_R V = \dim(R/P) = \dim(R/(0:V)_R) \\ \quad \text{and hence } 0 \text{ has no embedded primes in } V, \\ \quad \text{i.e., } \mathrm{ass}_R V = \mathrm{nass}_R V. \end{cases}$$

(T69.2) If V is any finitely generated R-module and $x = (x_1, \ldots, x_n)$ is any V-regular sequence in $M(R)$ with $n \in \mathbb{N}$ then

(1*) $$\dim(R/(0:V)_R) = n + \dim(R/(xV:V)_R)$$

and

(2*) $$\mathrm{reg}_R V = n + \mathrm{reg}_R(V/(xV))$$

and

(3*) $$V \text{ is CM} \Leftrightarrow V/(xV) \text{ is CM}$$

and

(4*) $$\begin{cases} V \text{ is CM} \\ \Rightarrow \text{for every } P \in \mathrm{ass}_R(V/(xV)) \text{ we have} \\ \quad \mathrm{reg}_R V = n + \dim(R/P) = \dim(R/(0:V)_R) \\ \quad \text{and hence } 0 \text{ has no embedded primes in } V, \\ \quad \text{i.e., } \mathrm{ass}_R V = \mathrm{nass}_R V. \end{cases}$$

(T69.3) If $x = (x_1, \ldots, x_n)$ is any sequence in $M(R)$ with $n \in \mathbb{N}$ then upon letting

$$\dim(R) = d \quad \text{and} \quad I_n = xR$$

we have

(1') $$d \ge \dim(R/I_n) \ge d - n$$

and

(2') $\quad \dim(R/I_n) = d - n \Leftrightarrow \begin{cases} d \geq n \text{ and } (x_1, \ldots, x_d)R \text{ is } M(R)\text{-primary} \\ \text{for some } x_{n+1}, \ldots, x_d \text{ in } M(R) \end{cases}$

and

(3') $\quad \text{ht}_R I_n = n \Rightarrow \dim(R/I_n) = d - n.$

(T69.4) If $x = (x_1, \ldots, x_n)$ is any R-regular sequence in $M(R)$ with $n \in \mathbb{N}$ then upon letting $I_n = xR$ we have

(1'') $\qquad\qquad\qquad\qquad \text{ht}_R I_n = n$

and

(2'') $\qquad\qquad\qquad\qquad \dim(R) = n + \dim(R/I_n)$

and

(3'') $\qquad\qquad\qquad\qquad \text{reg}(R) = n + \text{reg}(R/I_n)$

and

(4'') $\qquad\qquad\qquad\qquad R \text{ is CM} \Leftrightarrow R/I_n \text{ is CM}.$

(T69.5) Upon letting $\dim(R) = d$, the following four conditions are equivalent:

(i) R is CM.

(ii) Every sequence of length d in $M(R)$ which generates an $M(R)$-primary ideal is R-regular.

(iii) There exists an R-regular sequence of length d in $M(R)$ which generates an $M(R)$-primary ideal.

(iv) There exists an R-regular sequence of length d in $M(R)$.

(T69.6) If R is CM then for every sequence $x = (x_1, \ldots, x_n)$ in $M(R)$ with $n \in \mathbb{N}$, upon letting $\dim(R) = d$ and $I_m = (x_1, \ldots, x_m)R$ for $0 \leq m \leq n$, the following four conditions are equivalent:

(i*) x is R-regular,

(ii*) $\text{ht}_R I_m = m$ for $0 \leq m \leq n$,

(iii*) $\text{ht}_R I_n = n$,

(iv*) $d \geq n$ and $(x_1, \ldots, x_d)R$ is $M(R)$-primary for some $x_{n+1}, \ldots, x_d$ in R.

Moreover, the implications (i*) $\Rightarrow$ (ii*) $\Rightarrow$ (iii*) $\Rightarrow$ (iv*) hold without assuming R to be CM.

(T69.7) If R is regular then R is CM, and for every sequence $x = (x_1, \ldots, x_d)$ of generators of $M(R)$ with $\dim(R) = d$ we have that the sequence x is R-regular.

(T69.8) If $\dim(R) = 0$ then R is CM, and hence in particular every field is CM.

PROOF. In the proof of this Lemma (T69) we shall tacitly use (Q1).

To prove (T69.1) let V be any finitely generated R-module. Then

$$\mathrm{reg}_R V = -\infty \Leftrightarrow V = 0 \Leftrightarrow \mathrm{ass}_R V = \emptyset \Leftrightarrow \dim(R/(0:V)_R) = -\infty$$

and in case of $\mathrm{ass}_R V \neq \emptyset$ we always have

$$\dim(R/(0:V)_R) = \max\{\dim(R/P) : P \in \mathrm{ass}_R V\}$$

and (2) is equivalent to saying that

$$\mathrm{reg}_R V \leq \min\{\dim(R/P) : P \in \mathrm{ass}_R V\} \leq \dim(R/(0:V)_R).$$

Therefore (1) follows from (2), and also (3) follows from (1) and (2). To prove (2) it suffices to show that, given any V-regular sequence $x = (x_1, \ldots, x_n)$ in $M(R)$ with $n \in \mathbb{N}$ and any $P \in \mathrm{ass}_R V$ with $\dim(R/P) = e$, we have $n \leq e$. We do this by induction on n. It being obvious for $n = 0$, let $n > 0$ and assume for $n - 1$. We shall find $Q \in \mathrm{ass}_R(V/(x_1 V))$ such that $P \subset Q$ with $P \neq Q$ and then, because $(x_2, \ldots, x_n)$ is a $(V/(x_1 V))$-regular sequence in $M(R)$, by the induction hypothesis we will get $n - 1 \leq \dim(R/Q) \leq e - 1$ which will complete the proof. Let $W = (x_1 V : P)_V$. Then W is a submodule of V with $x_1 V \subset W$. We shall show that $x_1 V \neq W$, and then we can take $Q \in \mathrm{tass}_R(W/(x_1 V))$ and for any such Q we will have $P \subset Q \in \mathrm{tass}_R(V/(x_1 V)) = \mathrm{ass}_R(V/(x_1 V))$, and this will complete the proof. Suppose if possible that $x_1 V = W$ and let $U = (0 : P)_V$. Then U is a submodule of W and every $u \in U$ can be expressed as $u = x_1 v$ with $v \in V$, and because of $PU = 0$ we get $0 = pu = x_1(pv)$ for all $p \in P$, and since x_1 is a nonzerodivisor of V we further get $pv = 0$ for all $p \in P$, and because of $U = (0 : P)_V$ this implies $v \in U$. Thus $U = x_1 U$ and hence by Nakayama (T3) we get $U = 0$. But because of $P \in \mathrm{ass}_R V = \mathrm{tass}_R V$ we must have $(0 : P)_R \neq 0$ which is a contradiction. This completes the proof of (T69.1).

Although most of (T69.4) follows by taking $V = R$ in (T69.2), we shall give independent proofs of both; moreover, since the ring case is more transparent than the module case, before proving (T69.2) we shall prove (T69.3) and (T69.4). So let there be given any sequence $x = (x_1, \ldots, x_n)$ in $M(R)$ with $n \in \mathbb{N}$, and let $\dim(R) = d$ and $I_n = xR$. The first inequality of $(1')$ is obvious and the second follows from the Generalized Principal Ideal Theorem (T27). By the Extended Dim-Emdim Theorem (T29), the RHS of $(2')$ implies $\dim(R/I_n) \leq d - n$ and this together with the second inequality of $(1')$ implies the LHS of $(2')$. By (T29) we also see that the LHS of $(2')$ implies its RHS. To prove $(3')$, assume $\mathrm{ht}_R I_n = n$; then by (T27) we see that the height of every minimal prime of I_n is n, and clearly we can take a minimal prime P of I_n such that $\dim(R/I_n) = \dim(R/P)$, and since obviously $\dim(R/P) \leq \dim(R) - \mathrm{ht}_R P$, this gives us

$$d - n \leq \dim(R/I_n) = \dim(R/P) \leq \dim(R) - \mathrm{ht}_R P = d - n$$

where the first inequality is the second inequality of $(1')$. From the above display we get $\dim(R/I_n) = d - n$ which proves $(3')$. This completes the proof of (T69.3).

If the sequence x is R-regular then by (T27) we get (1″) and, from it, by (3′) we get (2″), and by taking $(V, I, J) = (R, I_n, M(R))$ in (T66.6) we get (3″), and finally by (2″) and (3″) we get (4″). This completes the proof of (T69.4).

To prove (T69.2) let V be a finitely generated R-module and let $x = (x_1, \ldots, x_n)$ be a V-regular sequence in $M(R)$ with $n \in \mathbb{N}$. We shall prove (1*) by induction on n. This being obvious for $n = 0$, let $n > 0$, and assume for $n - 1$. Let $y = x_n$ and $W = V/(x_1 V + \cdots + x_{n-1} V)$. Then W is a finitely generated R-module and $y \in S_R(W)$. By the induction hypothesis we have $\dim(R/(0 : V)_R) = n - 1 + \dim(R/(0 : W)_R)$ and so it only remains to show that $\dim(R/(0 : W)_R) = 1 + \dim(R/(yW : W)_R)$. Since W is finitely generated, by Comment (C23) below, we see that

(†) $$\mathrm{vspec}_R(yW : W)_R = (\mathrm{vspec}_R(0 : W)_R) \bigcap (\mathrm{vspec}_R(yR)).$$

Now

$$\dim(R/(0 : W)_R) = \max\{\dim(R/P) : P \in \mathrm{nvspec}_R(0 : W)_R\}$$

where $\mathrm{nvspec}_R(0 : W)_R$ is a nonempty set of primes none of which contains the element y, and hence by (†) we get $\dim(R/(0 : W)_R) > \dim(R/(yW : W)_R)$. Moreover, we can first take $P \in \mathrm{nvspec}_R(0 : W)_R$ with $\dim(R/(0 : W)_R) = \dim(R/P)$, and then we can take $Q \in \mathrm{nvspec}_R(P + yR)$ with $\dim(R/(P + yR)) = \dim(R/Q)$. Now by (†) we get $Q \in \mathrm{vspec}_R(yW : W)_R$ and by (T24) and (T29) we get $\dim(R/P) = 1 + \dim(R/Q)$, where the inequality $\dim(R/P) \geq 1 + \dim(R/Q)$ is by (T24) and the inequality $\dim(R/P) \leq 1 + \dim(R/Q)$ is by (T29). Therefore $\dim(R/(0 : W)_R) = 1 + \dim(R/(yW : W)_R)$. This proves (1*). Taking $J = M(R)$ in (T66.6) we get (2*). By (1*) and (2*) we get (3*). Finally, in view of (T69.1)(3), by (1*) to (2*) we get (4*). This completes the proof of (T69.2).

To prove (T69.5) let $\dim(R) = d$. By (T29) we know the existence of a sequence of length d in $M(R)$ which generates an $M(R)$-primary ideal, and hence we get the implication (ii) $\Rightarrow$ (iii). The implications (iii) $\Rightarrow$ (iv) and (iv) $\Rightarrow$ (i) are obvious. So we only need to prove (i) $\Rightarrow$ (ii). We shall do this by induction on d. It being trivial for $d = 0$, let $d > 0$ and assume for $d - 1$. Assuming R to be CM and letting $(x_1, \ldots, x_d)$ be any sequence in $M(R)$ which generates an $M(R)$-primary ideal, by taking $V = R$ in (T69.1)(3) we see that $\dim(R/P) = d$ for every associated prime P of the zero ideal in R. Let $\phi : R \to R/(x_1 R)$ be the residue class epimorphism. Since the ideal in $\phi(R)$ generated by $\phi(x_2), \ldots, \phi(x_d)$ is $M(R/(x_1 R))$-primary, by (T29) we know that $\dim(\phi(R)) \leq d - 1$. Therefore for every associated prime P of the zero ideal in R we must have $x_1 \notin P$. Consequently $x_1 \in S_R(R)$. Therefore it only remains to show that the sequence $(\phi(x_2), \ldots, \phi(x_d))$ is $\phi(R)$-regular, but this follows from the induction hypothesis because by (T69.4) we know that the local ring $\phi(R)$ is CM with $\dim(\phi(R)) = d - 1$. This completes the proof of (T69.5).

In (T69.6), without assuming R to be CM, by (T69.4)(1″) we have

$$(i^*) \Rightarrow (ii^*),$$

obviously we have

$$(\text{ii}^*) \Rightarrow (\text{iii}^*),$$

and by (T69.3)(2′) and (T69.3)(3′) we have

$$(\text{iii}^*) \Rightarrow (\text{iv}^*).$$

Moreover, assuming R to be CM, by (T69.5) we get $(\text{iv}^*) \Rightarrow (\text{i}^*)$. This completes the proof of (T69.6). By (T69.5) we immediately get (T69.7). Finally, (T69.8) is obvious because a sequence of length zero in $M(R)$ is clearly R-regular.

COHEN-MACAULAY THEOREM (T70). For any noetherian ring R we have the following.

(T70.1) If R is CM and $x = (x_1, \ldots, x_n)$ is any R-regular sequence in R with $n \in \mathbb{N}$, then upon letting $I_n = xR$ we have $\mathrm{ht}_R I_n = n$ and for every $P \in \mathrm{ass}_R(R/I_n)$ we have $\mathrm{ht}_R P = n$, and hence in particular I_n is an ideal-theoretic complete intersection in R, and I_n is unmixed.

[By taking $n = 0$ we see that if R is CM then every associated prime of 0 in R is minimal].

(T70.2) If R is CM then for any nonunit ideal I in R we have

$$\mathrm{reg}_{(R,I)} R = \mathrm{ht}_R I.$$

(T70.3) If R is CM and I is a nonunit ideal I in R which is an ideal-theoretic complete intersection in R then I is generated by an R-regular sequence.

(T70.4) If R is CM then its localization R_S at any multiplicative set S in it is CM, and hence in particular its localization R_P at any prime ideal P in it is CM.

(T70.5) If R is CM then R is catenarian.

(T70.6) If R is CM and local then the length of any absolutely saturated prime ideal chain in R equals $\dim(R)$ and for any nonunit ideal I in R we have

$$\mathrm{ht}_R I = \dim(R) - \dim(R/I).$$

(T70.7) R is CM $\Leftrightarrow$ the unmixedness theorem holds in R.

(T70.8) R is a finite variable polynomial ring over a field $\Rightarrow R$ is CM and the unmixedness theorem holds in R.

(T70.9) R is a finite variable power series ring over a field $\Rightarrow R$ is CM and the unmixedness theorem holds in R.

(T70.10) R is CM $\Rightarrow$ every finite variable polynomial ring over R is CM.

(T70.11) R is CM $\Rightarrow$ every finite variable power series ring over R is CM.

(T70.12) R is CM $\Rightarrow$ every homomorphic image of R is universally catenarian.

[In view of (T70.7), by (T70.10) and (T70.11) we obtain (T70.8) and (T70.9) respectively, but we shall give more direct proofs of (T70.8) and (T70.9)].

PROOF. In the proof of this Theorem (T70) we shall tacitly use (Q1).

To prove (T70.1) assume that R is CM, let $x = (x_1, \ldots, x_n)$ be an R-regular sequence in R with $n \in \mathbb{N}$, and let $I_n = xR$. We want to show that then $\mathrm{ht}_R I_n = n$ and for every $P \in \mathrm{ass}_R(R/I)$ we have $\mathrm{ht}_R P = n$. First we shall show this in the local case, i.e., when R is a local ring; then by (T69.4)(1'') we have $\mathrm{ht}_R I_n = n$ and hence by (T27) we get $\mathrm{ht}_R P = n$ for all $P \in \mathrm{nass}_R(R/I_n)$; by (T69.4)(4'') we see that R/I_n is CM as a ring and hence also as an R-module; consequently by taking $V = R/I_n$ in (T69.1)(3) we see that $\mathrm{ass}_R(R/I_n) = \mathrm{nass}_R(R/I_n)$, and hence $\mathrm{ht}_R P = n$ for all $P \in \mathrm{ass}_R(R/I_n)$. Now turning to the general case, given any $P \in \mathrm{ass}_R(R/I_n)$, we can take a maximal ideal Q in R with $P \subset Q$. Let $\phi : R \to R' = R_Q$ be the canonical homomorphism into the localization of R at Q, let $x_i' = \phi(x_i)$ for $1 \le i \le n$, consider the sequence $x' = (x_1', \ldots, x_n')$ in the local ring R', and let $I_n' = x'R'$ and $P' = \phi(P)R'$. Then by L4§7.1(T17) we see that x' is an R'-regular sequence in R' and $P' \in \mathrm{ass}_{R'}(R'/I')$ with $\mathrm{ht}_{R'} P' = \mathrm{ht}_R P$. Therefore by the local case we have $\mathrm{ht}_{R'} P' = n$, and hence $\mathrm{ht}_R P = n$. This being so for every $P \in \mathrm{ass}_R(R/I_n)$, by (T27) it follows that $\mathrm{ht}_R I_n = n$.

To prove (T70.2) assume that R is CM, and let I be a nonunit ideal in R. Upon letting $\mathrm{reg}_{(R,I)} R = n$, we can take a maximal R-regular sequence $x = (x_1, \ldots, x_n)$ in I. Let $I_n = xR$. Since x is maximal, we get $I \subset Z_R(R/I_n)$. Therefore by L4§5(O24)(22•) we have $I \subset P$ for some $P \in \mathrm{ass}_R(R/I_n)$; by (T70.1) we get $\mathrm{ht}_R P = n$ and hence we must have $\mathrm{ht}_R I \le n$. By (T70.1) we also have $\mathrm{ht}_R I_n = n$ and hence, because $I_n \subset I \ne R$, we get $\mathrm{ht}_R I \ge n$. Therefore $\mathrm{ht}_R I = n$.

To prove (T70.3) assume that R is CM and let I be a nonunit ideal in R which is an ideal-theoretic complete intersection in R. Let $\mathrm{reg}_{(R,I)} R = n$. Then by (T70.2) we have $\mathrm{ht}_R I = n$, and hence by taking $(V, J, \mu, \nu) = (R, I, n, n)$ in (T66.9) we can find an R-regular sequence $x = (x_1, \ldots, x_n)$ in R such that $I = xR$.

In proving (T70.4), suppose we have shown that: (a) R is CM $\Rightarrow R_P$ is CM for every $P \in \mathrm{spec}(R)$. Then it follows that: (b) R is CM $\Leftrightarrow R_P$ is CM for every $P \in \mathrm{spec}(R)$. For any multiplicative set S in R and any $Q \in \mathrm{spec}(R_S)$, upon letting $\phi : R \to R_S$ to be the canonical homomorphism and upon letting $P = \phi^{-1}(Q)$, we have that $P \in \mathrm{spec}(R)$ and the local ring $(R_S)_Q$ is isomorphic to the local ring R_P. Therefore by (b) it follows that: (c) R is CM $\Rightarrow R_S$ is CM for every multiplicative set S in R. Thus to prove (T70.4) it suffices to prove (a). So assume R is CM, and given any $P \in \mathrm{spec}(R)$ let $\Phi : R \to R^* = R_P$ be the canonical homomorphism. Let $\mathrm{ht}_R P = n$. Then $\dim(R^*) = n$. By (T70.2) we have $\mathrm{reg}_{(R,P)} R = n$ and hence we can find an R regular sequence $x = (x_1, \ldots, x_n)$ in P. Let $x_i^* = \Phi(x_i)$ for $1 \le i \le n$ and consider the sequence $x^* = (x_1^*, \ldots, x_n^*)$. Then by L4§7.1(T17) we see that x^* is an R^*-regular sequence in $M(R^*)$. Therefore R^* is CM.

To prove (T70.5) assume that R is CM, and given any prime ideals $P \subset Q$ in R let $\phi : R \to R' = R_Q$ be the canonical homomorphism. Let $P' = \phi(P)$ and $\mathrm{ht}_R P = n$. Then by (T70.2) we see that $\mathrm{reg}_{(R,P)} R = n$ and hence we can find an R-regular sequence $x = (x_1, \ldots, x_n)$ in P, and by (T70.1) we get $\mathrm{ht}_R(R/(xR)) = n$ and hence $P \in \mathrm{nass}_R(R/(xR))$. Let $x_i' = \phi(x_i)$ for $1 \le i \le n$ and consider the sequence $x' =$

$(x_1', \ldots, x_n')$. Then by L4§7.1(T17) we see that x' is an R'-regular sequence in P' and $P' \in \mathrm{nass}_{R'}(R'/(x'R'))$. Therefore, since R' is CM by (T70.4), by (T69.2)(4*) we get $\dim(R') = \mathrm{ht}_R P + \dim(R'/P')$. Consequently, since $\dim(R') = \mathrm{ht}_R Q$ and $\dim(R'/P') = \mathrm{ht}_{(R/P)}(Q/P)$, we get $\mathrm{ht}_R Q = \mathrm{ht}_R P + \mathrm{ht}_{(R/P)}(Q/P)$. Thus we are done by (T56.1).

To prove (T70.6) assume that R is a CM local ring. We shall show that then for every prime ideal Q in R we have $\mathrm{ht}_R Q = \dim(R) - \dim(R/Q)$ and from this it will follow that for every nonunit ideal I in R we have $\mathrm{ht}_R I = \dim(R) - \dim(R/I)$. Let $\dim(R) = d$. Given any $P \in \mathrm{ass}_R(R)$, by taking $(V, n) = (R, 0)$ in (T69.2)(4*) we get $\dim(R/P) = d$ and hence by (T70.5) we see that every saturated prime ideal chain in R between P and $M(R)$ has length d. It follows that every absolutely saturated prime ideal chain in R has length d, and hence for every prime ideal Q in R we have $\mathrm{ht}_R Q = d - \dim(R/Q)$.

In (T70.7), the implication $\Rightarrow$ follows from (T70.1) and (T70.3). To prove the reverse implication, assume that the unmixedness theorem holds in R. Given any $P \in \mathrm{spec}(R)$, by (T28) there exist elements $x_1, \ldots, x_n$ in P with $\mathrm{ht}_R P = n$ such that for the ideal $I_m = (x_1, \ldots, x_m)R$ we have $\mathrm{ht}_R I_m = m$ for $0 \le m \le n$. By assumption the ideal is unmixed for $0 \le m \le n$; it follows that for $0 \le m < n$ the element x_{m+1} cannot belong to any associated prime of I_m in R, and hence $x_{m+1} \in S_R(R/I_m)$. Therefore $(x_1, \ldots, x_n)$ is an R-regular sequence in P. Consequently, upon letting $\Phi : R \to R_P$ be the canonical homomorphism, by L4§7.1(T17) we see that $(\Phi(x_1), \ldots, \Phi(x_n))$ is an (R_P)-regular sequence in $M(R_P)$. Therefore R_P is CM. This being so for every $P \in \mathrm{spec}(R)$, it follows that R is CM.

If R is a finite variable polynomial ring over a field then, by (T19) and (T22) of L4§8, which were respectively proved at the end of Quest (Q10) of the current Lecture and in (T47) of that Quest, we see that the localization of R at any maximal ideal in it is a regular local ring which is CM by (T69.7), and hence R is CM by definition. If R is a finite variable power series ring over a field then, R is a regular local ring by (T30.4), and hence R is CM by (T69.7). In view of (T70.7), this proves (T70.8) and (T70.9).

In view of an obvious induction, it suffices to prove (T70.10) and (T70.11) in the univariate case. So consider the univariate polynomial ring $T = R[X]$ or power series ring $T = R[[X]]$ over the CM noetherian ring R. Given any maximal ideal Q in T, in Comment (C24) we shall show that $\mathrm{ht}_T Q = n + 1$ with $n \in \mathbb{N}$, and upon letting $P = R \cap Q$ we have $P \in \mathrm{spec}(R)$ with $\mathrm{ht}_R P = n$ and $Q = PT + FT$ where in the power series case $F = X$ and in the polynomial case F is a polynomial of degree $E \in \mathbb{N}_+$ in which the coefficient of X^E belongs to $R \setminus P$. By (T70.2) we can find an R-regular sequence $x = (x_1, \ldots, x_n)$ in P. Let $\Phi : R \to S = R_P$ and $\phi : T \to T' = T_Q$ be the natural homomorphisms. Then upon letting $y_i = \Phi(x_i) \in S$ for $1 \le i \le n$, by L4§7.1(T17) we see that the sequence $y = (y_1, \ldots, y_n)$ in $M(S)$ is S-regular. We shall show that the sequence $(\phi(x_1), \ldots, \phi(x_n), \phi(F))$ in $M(T')$ is T'-regular and, since Q was any maximal ideal in T, this will complete the proof of the CMness of T. Since x is an R-regular sequence in P, we easily see that it is a T-regular sequence in Q.

First consider the polynomial case. Now $S[X]$ may be regarded as a localization of T at the multiplicative set $R \setminus P$ and so we have a natural homomorphism $\Theta : T \to S[X]$ such that $\Theta(X) = X$ and $\Theta(z) = \Phi(z)$ for all $z \in R$. Also T' may be regarded as a localization of $S[X]$ at the maximal ideal $\Theta(Q)S[X]$ and so we have a natural homomorphism $\Delta : S[X] \to T'$. Clearly $\phi = \Delta\Theta$. Consequently, in view of L4§7.1(T17), it suffices to show that the sequence $(y_1, \ldots, y_n, G)$ is $S[X]$-regular where $G = \Theta(F) \in S[X]$ is of degree $E \in \mathbb{N}_+$ in which the coefficient of X^E is a unit in the local ring S. There is an obvious epimorphism $S[X] \to (S/(yS))[X]$ and under it the image of G is clearly a nonzerodivisor; since y in an S-regular sequence in $M(S)$, it follows that the sequence $(y_1, \ldots, y_n, G)$ is $S[X]$-regular.

Next consider the power series case. Now $F = X$ and we want to show that the sequence $(\phi(x_1), \ldots, \phi(x_n), \phi(X))$ is T'-regular. Again, in view of L4§7.1(T17), it suffices to show that the sequence $(x_1, \ldots, x_n, X)$ is $R[[X]]$-regular. There is an obvious epimorphism $R[[X]] \to (R/(xR))[[X]]$ and under it the image of X is clearly a nonzerodivisor; since x in an R-regular sequence in P, it follows that $(x_1, \ldots, x_n, X)$ is an $R[[X]]$-regular sequence in Q.

Finally, in view of (T56), by (T70.5) and (T70.10) we get (T70.12).

ANOTHER ORD VALUATION LEMMA (T71). Given any local ring R, for any $x \in M(R) \setminus Z_R(R)$ we have that:

(i) $R/(xR)$ is a local ring whose dimension is one less than the dimension of R,

(ii) if $R/(xR)$ is regular then $\operatorname{ord}_R x = 1$, and

(iii) if $\operatorname{ord}_R x = 1$ and R is regular then $R/(xR)$ is regular.

[Note that if R is regular, or more generally if R is a domain, then the condition $x \in M(R) \setminus Z_R(R)$ is equivalent to the condition $0 \neq x \in M(R)$].

PROOF. (i) follows from (T69.1)(2''). Upon letting $\operatorname{emdim}(R) = e$, by Nakayama (T3), we see that $\operatorname{emdim}(R/(xR)) = e - 1$ or e according as $\operatorname{ord}_R x = 1$ or $\neq 1$, and hence (ii) and (iii) follow from (i).

COMMENT (C22). [**History**]. (T70.8) was proved by Macaulay in 1916, and (T70.9) was proved by Cohen in 1946. Lemma (T65) which relates global and local properties of rings occurs in Krull's fundamental 1938 paper.

COMMENT (C23). [**Annihilators of Finitely Generated Modules**]. In the proof of (T69.2)(1*) we used the fact that for any finitely generated module W over a ring R and any element y in R we have

(†) $$\operatorname{vspec}_R(yW : W)_R = (\operatorname{vspec}_R(0 : W)_R) \bigcap (\operatorname{vspec}_R(yR)).$$

Let us now establish this fact. First note that, without assuming W to be finitely generated, we obviously have $(0 : W)_R + (yR) \subset (yW : W)_R$ and hence in the above display the LHS $\subset$ the RHS. For the other inclusion, assuming W to be finitely generated, let us write $W = Rz_1 + \cdots + Rz_e$ where $z_1, \ldots, z_e$ are elements in W with $e \in N_+$. Given any $t \in (yW : W)_R$, we can write

$$tz_i = \sum_{1 \leq j \leq e} ya_{ij}z_j \quad \text{for} \quad 1 \leq i \leq e \quad \text{with} \quad a_{ij} \in R.$$

Let D be the determinant of the $e \times e$ matrix $(t\delta_{ij} - ya_{ij})$ with Kronecker's δ_{ij}. Then by Cramer's Rule (see (C9)), $z_iD = 0$ for $1 \leq i \leq e$, and hence $D \in (0 : W)_R$. By expanding the determinant we see that $D = t^e +$ an element in yR. Therefore $t \in \text{rad}_R((0 : W)_R + yR)$, and hence in (†), the RHS $\subset$ the LHS.

COMMENT (C24). [**Univariate Polynomial and Power Series Rings**]. Let R be a noetherian ring, consider the univariate polynomial ring $T = R[X]$ or power series ring $T = R[[X]]$ over R, and let Q be any maximal ideal in T. In the proof of (T70.10) and (T70.11) we used the fact that then $\text{ht}_SQ = n + 1$ with $n \in \mathbb{N}$, and upon letting $P = R \cap Q$ we have $P \in \text{spec}(R)$ with $\text{ht}_RP = n$ and $Q = PT + FT$ where in the power series case $F = X$ and in the polynomial case F is a polynomial of degree $E \in \mathbb{N}_+$ in which the coefficient of X^E belongs to $R \setminus P$. In the polynomial case this follows from (T30.1) and (C13). To prove it in the power series case assume that $T = R[[X]]$. Then by L2§2(ii), for every unit in R the element $u - X$ is a unit in T, and hence by (T20) we get $X \in Q$. Let $\phi : T \to R$ be the unique R-epimorphism which sends X to 0. Then $\ker(\phi) = XT$ and hence $P \in \text{mspec}(R)$ with $Q = PT + XT$. By (T30.1) it now follows that $\text{ht}_TQ = 1 + \text{ht}_RP$.

QUEST (Q16) Complete Intersections and Gorenstein Rings

A local ring R is said to be a complete intersection if there exists a regular local ring T and an epimorphism $\phi : T \to R$ such that $\ker(\phi)$ is an ideal-theoretic complete intersection in T according to the definition introduced at the beginning of (Q15). By parameters for R we mean a sequence of elements in $M(R)$ whose length equals $\dim(R)$ and which generates an $M(R)$-primary ideal; by a parameter ideal in R we mean an ideal generated by parameters; by (T29) we know that parameters and hence parameter ideals always exist. By regular parameters for R we mean parameters which generate $M(R)$; by definition, they exist iff R is regular. The socle of R is defined by putting

$$\text{soc}(R) = (0 : M(R))_R.$$

This is clearly an ideal in R which is annihilated by $M(R)$, i.e., $M(R) \subset (0 : \text{soc}(R))_R$. Consequently it is a finite dimensional vector space over $R/M(R)$ and we define the socle-size of R to be the nonnegative integer obtained by putting

$$\text{socz}(R) = \text{vector space dimension of } \text{soc}(R) \text{ over } R/M(R).$$

Note that if $\dim(R) = 0$ then $M(R)$ is an associated prime of the zero ideal in R, i.e., $M(R) \in \text{tass}_R R$, and therefore $\text{soc}(R) \neq 0$ and hence $\text{socz}(R) > 0$. We define the CM type of R to be the positive integer obtained by putting

$$\text{cmt}(R) = \min \text{socz}(R/Q) \text{ taken over all parameter ideals } Q \text{ in } R.$$

In (T75) we shall show that

$$R \text{ is CM } \Rightarrow \text{cmt}(R) = \text{socz}(R/Q) \text{ for every parameter ideal } Q \text{ in } R.$$

So far the concept of irreducible ideals (defined in the preamble of L4§5) was used only in L4§5(O15) where it was employed for proving primary decomposition. Here we shall use this concept to define GST (= Gorenstein) rings which lie between complete intersections and CM rings.

The local ring R is said to be GST if every parameter ideal in it is irreducible.

In (T76) we shall show that a local ring R is GST iff it is CM and some parameter ideal in it is irreducible. In (T78) we shall give some more criteria for a one dimensional local CM to be GST.

A noetherian ring R is said to be a local complete intersection if its localization R_P is a complete intersection for every maximal ideal P in it. A noetherian ring R is said to be GST if its localization R_P is GST for every maximal ideal P in it. A noetherian ring R is said to be regular if its localization R_P is regular for every maximal ideal P in it.

In (T77) we shall show that if a noetherian ring is a local complete intersection then it is GST.

As notation, by a smallest (resp: largest) element of a set W of subsets of a set R we mean $U \in W$ such that for every $V \in W$ we have $U \subset V$ (resp: $V \subset U$). Given subsets Q and Q' of a set R, we say that Q' properly contains Q, or Q is properly contained in Q', to mean that $Q \subset Q'$ with $Q \neq Q'$. By an overideal (resp: a proper overideal) of an ideal Q in a ring R we mean an ideal Q' in R such that Q' contains (resp: properly contains) Q; by a smallest proper overideal of Q we mean a smallest element of the set of all proper overideals of Q. By a subideal (resp: a proper subideal) of an ideal Q in a ring R we mean an ideal Q' in R such that Q contains (resp: properly contains) Q'; by a largest proper subideal of Q we mean a largest element of the set of all proper subideals of Q.

In (T72) we shall characterize largest proper subideals. In (T73) we shall characterize smallest proper overideals, which amounts to characterizing zero-dimensional GST local rings; note that a zero-dimensional local ring is GST iff the zero ideal in it is irreducible (since that is the unique parameter ideal in it); in particular (T73) says that a zero dimensional local ring R is GST iff $\text{cmt}(R) = 1$, i.e., iff $\text{socz}(R) = 1$. In (T74) we shall paraphrase this to give several criteria for a primary ideal to be irreducible; in particular (T74) says that an $M(R)$-primary ideal Q in a local ring R is irreducible iff $\operatorname{socz}(R/Q) = 1$. For euphony we are speaking of a largest proper subideal and a smallest proper overideal, rather than the largest proper subideal and the smallest proper overideal, although these are unique when they exist.

LARGEST SUBIDEAL CHARACTERIZATION LEMMA (T72). For any ideal Q in a ring R we have the following.

(T72.1) If Q has a largest proper subideal, then Q is a nonzero principal ideal.

(T72.2) If R is quasilocal and Q is a nonzero principal ideal, then $M(R)Q$ is a largest proper subideal of Q.

(T72.3) If J is a subideal of Q with $\ell_R(Q/J) \le 1$, then the R-modules Q/J and $R/(J : Q)_R$ are isomorphic. Hence, in particular, if $\ell_R(Q/J) = 1$, then we have $(J : Q)_R \in \operatorname{mspec}(R)$.

(T72.4) If R is quasilocal with $P = M(R)$ and J is a subideal of Q with $\ell_R(Q/J) = 1$ then, upon letting $I' = (0 : I)_R$ for every ideal I in R, there exists an R-monomorphism $\psi : J'/Q' \to P'$ and hence, in particular, $\ell_R(J'/Q') \le \ell_R(P')$.

PROOF. To prove (T72.1) assume that Q has a largest proper subideal J. Then we can take $x \in Q \setminus J$. Clearly $xR \subset Q$. If $xR \ne Q$ then we must have $xR \subset J$ which would be a contradiction. Therefore $xR = Q$.

To prove (T72.2) assume that R is quasilocal and Q is a nonzero principal ideal. Then $M(R)Q$ is a subideal of Q. To show that $M(R)Q$ is largest, given any subideal J of Q let $B = (J : Q)_R$. Then B is an ideal in R and obviously $BQ \subset J$. Since $J \subset Q = xR$, every $j \in J$ can be written as $j = xr$ with $r \in R$; but the equation $j = xr$ tells us that $r \in B$, and hence $J \subset BQ$. Thus $J = BQ$; it follows that if $J \ne Q$ then $B \ne R$ and hence $B \subset M(R)$ and therefore $J \subset M(R)Q$.

To prove (T72.3) let J be any subideal of Q with $\ell_R(Q/J) \le 1$. Then $Q = J + tR$ for some $t \in Q$. For any such t we get a unique R-homomorphism $\phi_t : Q/J \to R/(J : Q)_R$ such that for all $q \in Q$ we have $\phi_t(\overline{q}) \cdot \overline{t} = \overline{q}$ where $\cdot$ is multiplication in the $(R/(J : Q)_R)$-module Q/J, and where $^{-} : Q \to Q/J$ is the residue class epimorphism. We easily see that ϕ_t is an isomorphism, and hence the R-modules Q/J and $R/(J : Q)_R$ are isomorphic. It follows that if $\ell_R(Q/J) = 1$, then $\ell_R(R/(J : Q)_R) = 1$ and hence we have $(J : Q)_R \in \operatorname{mspec}(R)$.

To prove (T72.4), assume that R is quasilocal with $P = M(R)$ and J is a subideal of Q with $\ell_R(Q/J) = 1$. For every ideal I in R let $I' = (0 : I)_R$. Since $J \subset Q$, we have $Q' \subset J'$. Since $\ell_R(Q/J) = 1$, we have $Q = J + tR$ for some $t \in Q \setminus J$. By (T72.3) we have $P = (J : Q)_R$. Now

$$P = (J : Q)_R \Rightarrow QP \subset J \Rightarrow QPJ' \subset JJ' = 0 \Rightarrow PJ' \subset Q'$$

and hence we may regard J'/Q' as a vector space over R/P. We will be done by getting a (R/P)-linear injection $J'/Q' \to P'$. Since $tP \subset J$, we obtain an R-homomorphism $\phi : J' \to P'$ by putting $\phi(x) = xt$ for all $x \in J'$. Clearly

$\ker(\phi) = Q'$ and hence ϕ induces an R-monomorphism $\psi : J'/Q' \to P'$ which is obviously (R/P)-linear. It follows that $\ell_R(J'/Q') \le \ell_R(P')$.

ZERO DIMENSIONAL GST RING CHARACTERIZATION LEMMA (T73). For any zero-dimensional local ring R we have the following.

(T73.1) The following 7 conditions are equivalent:

(i) R is GST.

(ii) R has a smallest nonzero ideal.

(iii) $\mathrm{soc}(R)$ is a smallest nonzero ideal in R.

(iv) $\mathrm{soc}(R)$ is a nonzero principal ideal in R.

(v) $\mathrm{socz}(R) = 1$.

(vi) For every ideal H in R we have $H = (0 : \overline{H})_R$ for some ideal $\overline{H}$ in R.

(vii) For every ideal H in R we have $H = (0 : (0 : H)_R)_R$.

(T73.2) If R is GST then, upon letting $I' = (0 : I)_R$ for every ideal I in R, for any ideals H and L in R we have

(1)
$$\begin{cases} (H \cap L)' = H' + L' \quad \text{and} \quad (H + L)' = H' \cap L' \\ \text{and} \\ (HL)' = (H' : L)_R = (L' : H)_R \quad \text{and} \quad ((H : L)_R)' = LH' \end{cases}$$

and for any nonunit ideal H in R we have

(2)
$$R/H \text{ is GST} \Leftrightarrow H' \text{ is a nonzero principal ideal in } R.$$

[(vii) may be paraphrased by saying that $H \mapsto (0 : H)_R$ gives an inclusion reversing involution of the set of all ideals in R onto itself (involution = bijection whose inverse is itself). (1) may be paraphrased by saying that the said involution transforms sums of ideals into intersections, and products into quotients].

PROOF. Let $P = M(R)$, and for any ideal I in R let $I' = (0 : I)_R$.

We shall prove the equivalence of (i) to (v) in one breath, and then afterwards deal with (vi) and (vii). Let $\mathrm{socz}(R) = n$. Now n is a positive integer, and the nonzero ideal $\mathrm{soc}(R)$ in R may be viewed as an n-dimensional vector space over the field R/P, and then the nonzero subideals of the ideal $\mathrm{soc}(R)$ are exactly the nonzero subspaces of this vector space, and in particular the nonzero principal subideals are the one-dimensional subspaces. Clearly for a nonzero vector space V over a field K we have that: $\dim_K V > 1 \Leftrightarrow$ there are two nonzero subspaces intersecting in the zero subspace (namely, if $\dim_K V > 1$ then we may take the two subspaces to be the one-dimensional subspaces generated by linearly independent vectors x and y). Therefore, to prove the equivalence of (i) to (v), it suffices to show that for any nonzero ideal H in R we have $H \cap P' \ne 0$. Now $P^d = 0$ for some $d \in \mathbb{N}_+$, and hence there is a unique $e \in \mathbb{N}_+$ such that $HP^e = 0 \ne HP^{e-1}$. It follows that $HP^{e-1} \subset H \cap P'$ and hence $H \cap P' \ne 0$. This completes the proof of the equivalence of (i) to (v).

Obviously (vii) $\Rightarrow$ (vi). Conversely, let H and $\overline{H}$ be any ideals in R such that $H = \overline{H}'$. Then clearly $\overline{H} \subset H'$. For any ideals $A \subset B$ in R we obviously have $B' \subset A'$ and applying this to the previous sentence we get

$$(H')' \subset \overline{H}'.$$

But by assumption the RHS of the above equation equals H and hence H equals the LHS, i.e., $H = (H')'$. Thus (vi) $\Rightarrow$ (vii). This completes the proof of the equivalence of (vi) to (vii).

We CLAIM that, assuming (v), if $A \subset B$ are any ideals in R with $\ell_R(B/A) = 1$ then $\ell_R(A'/B') \leq 1$. This follows by taking $(A, B) = (J, Q)$ in (T72.4) and noting that (v) is equivalent to saying that $\ell_R(P') = 1$.

Next, we ASSERT that, assuming (v), for any ideal I in R we have $\ell_R(R/I) + \ell_R(R/I') = \ell_R(R)$. To see this, referring to L4§5(O21) and upon letting $\ell_R(R) = n$ and $\ell_R(R/I) = j$, we can take a composition series $V_0 \subset V_1 \subset \cdots \subset V_n$ of $V = R$ with $I = V_{n-j}$. This gives us a second series $V_n' \subset V_{n-1}' \subset \cdots \subset V_0'$ of ideals in R, and by the above claim, for $0 \leq i < n$, the R-module V_i'/V_{i+1}' is either simple or zero. Since the length of the second series is n, all its terms must be distinct. Therefore $\ell_R(R/I') = n - j$. This completes the proof of the assertion.

Assuming (v), by the above assertion, for any ideal H in R we have $\ell_R(R/H) = \ell_R(R/(H')')$; but obviously $H \subset (H')'$, and hence we must have $H = (H')'$. Thus (v) $\Rightarrow$ (vi).

Assuming (vii), $I \mapsto I'$ gives an inclusion reversing involution of the set of all ideals in R onto itself; since P is obviously a largest proper subideal of R, it follows that P' is a smallest nonzero ideal in R, i.e., we have (iii). This completes the proof of (T73.1).

Again assuming (vii), in view of the said inclusion reversing involution, for any ideal H in R, by (T72) we see that: H' is a nonzero principal ideal in R $\Leftrightarrow$ H has a smallest proper overideal, and hence by applying (ii) $\Leftrightarrow$ (i) to R/H we see that: R/H is GST $\Leftrightarrow$ H' is a nonzero principal ideal in R. Thus, in view of (i) $\Rightarrow$ (vii), this proves part (2) of (T73.2).

Turning to part (1) of (T73.2), the second equation in the first line and the first equation in the last line are actually true in any ring R without assuming it to be a zero dimensional local ring. Since (i) $\Rightarrow$ (vii), it suffices to note that in view of the said inclusion reversing involution, the first equation in the first line follows by putting H', L' for H, L in the second equation, and the second equation in the last line follows by putting H', L for H, L in the first equation. This proves part (1) of (T73.2), and so completes the proof of (T73).

IRREDUCIBLE IDEAL CHARACTERIZATION LEMMA (T74). Given any primary ideal Q in a noetherian ring R, upon letting $\mathrm{rad}_R Q = P$, we have the following.

(T74.1) Q is irreducible $\Leftrightarrow Q$ is not the intersection of any two P-primary ideals in R different from $Q \Leftrightarrow \phi(Q)R_P$ is irreducible where $\phi : R \to R_P$ is the natural homomorphism.

(T74.2) If R is local with $M(R) = P$ then the following 7 conditions are equivalent:

(i) Q is irreducible.

(ii) Q has a smallest proper overideal.

(iii) $(Q : P)_R$ is a smallest proper overideal of Q.

(iv) $(Q : P)_R/Q$ is a nonzero principal ideal in R/Q.

(v) $\mathrm{socz}(R/Q) = 1$.

(vi) For every overideal H of Q in R we have $H = (Q : \overline{H})_R$ for some overideal $\overline{H}$ of Q in R.

(vii) For every overideal H of Q in R we have $H = (Q : (Q : H)_R)_R$.

(T74.3) If R is local with $M(R) = P$ and Q is irreducible then, upon letting $I' = (Q : I)_R$ for every overideal I of Q in R, for any overideals H and L of Q in R we have

(1)
$$
\begin{cases}
(H \cap L)' = H' + L' \quad \text{and} \quad (H + L)' = H' \cap L' \\
\text{and} \\
(HL)' = (H' : L)_R = (L' : H)_R \quad \text{and} \quad ((H : L)_R)' = LH'
\end{cases}
$$

and for any nonunit overideal of Q in R we have

(2) H is irreducible $\Leftrightarrow H'/Q$ is a nonzero principal ideal in R/Q.

[(vii) may be paraphrased by saying that $H \mapsto (Q : H)_R$ gives an inclusion reversing involution of the set of all overideals of Q in R onto itself (involution = bijection whose inverse is itself). (1) may be paraphrased by saying that the said involution transforms sums of ideals into intersections, and products into quotients].

PROOF. The second $\Leftrightarrow$ in (T74.1) follows from L4§7.1(T17), and the first $\Rightarrow$ in (T74.1) is obvious. Now assume that Q is not the intersection of any two P-primary ideals different from Q, and let H and L be any overideals of Q with $Q = H \cap L$. We want to show that then either $H = Q$ or $L = Q$. We can take irredundant primary decompositions $H = \cap_{1 \leq i \leq d} Q_i'$ and $L = \cap_{1 \leq j \leq e} Q_j''$ with $\mathrm{rad}_R Q_i' = P_i'$ and $\mathrm{rad}_R Q_j'' = P_j''$. Suppose if possible that P is properly contained in every P_i' and every P_j''. Then we can find a prime ideal $\overline{P}$ which properly contains P, and which is such that $\overline{P}$ is not properly contained in any P_i' or P_j'' but coincides with some P_i' or P_j''. In view of L4§5(O19)(10•), by taking $[\cdot : \overline{P}]_R$ of both sides of the equation

$$
Q = (\cap_{1 \leq i \leq d} Q_i') \bigcap (\cap_{1 \leq j \leq e} Q_j'')
$$

we see that $Q =$ either a $\overline{P}$-primary ideal or an intersection of two $\overline{P}$-primary ideals (which is again $\overline{P}$-primary by L4§5(O8)). This is a contradiction. It follows that

P coincides with: (a) some P_i' but no P_j'', or (b) no P_i' but some P_j'', or (c) some P_i' and some P_j''. Again in view of L4§5(O19)(10•), by taking $[\cdot : P]_R$ of both sides of the above equation we respectively get: (a) $Q = Q_i'$, or (b) $Q = Q_j''$, or (c) $Q = Q_i' \cap Q_j''$. It follows that either $Q = H$ or $Q = L$. This completes the proof of part (T74.1). Parts (T74.2) and (T74.3) follow by applying (T73) to R/Q.

SOCLE SIZE LEMMA (T75). For any local ring R we have the following.

(T75.1) If $\dim(R) = d > 0$ and there are parameters $(x_1, \ldots, x_d)$ for R such that upon letting $Q_e = (x_1^e, \ldots, x_d^e)R$ we have $\mathrm{socz}(R/Q_e) = 1$ for all $e \in \mathbb{N}_+$ then $\mathrm{reg}(R) > 0$.

(T75.2) If $\mathrm{socz}(R/Q) = 1$ for every parameter ideal Q in R then R is CM.

(T75.3) If R is CM then for every parameter ideal Q in R we have $\mathrm{cmt}(R) = \mathrm{socz}(R/Q)$.

PROOF. Let $\dim(R) = d$.

To prove (T75.1) assume that $d > 0$ and there are parameters $(x_1, \ldots, x_d)$ for R such that upon letting $Q_e = (x_1^e, \ldots, x_d^e)R$ we have $\mathrm{socz}(R/Q_e) = 1$ for all $e \in \mathbb{N}_+$. Let $Q = Q_1$. Suppose if possible that $\mathrm{reg}(R) \le 0$. Then by (Q1) there exists $0 \ne t \in M(R)$ with $(0 : t)_R = M(R)$. Now by Krull Intersection (T7) we can find an integer $e > 1$ with $t \notin Q^{e-1}$ and since $d > 0$ we get

$$Q_e \subset Q^{e-1} \quad \text{with} \quad Q_e \ne Q^{e-1}.$$

Therefore by the (v) $\Rightarrow$ (iii) part of (T74) we get $(Q_e : M(R))_R \subset Q^{e-1}$ which contradicts the fact that $t \notin Q^{e-1}$ because clearly

$$(0 : t)_R = M(R) \Rightarrow t \in (0 : M(R))_R \subset (Q_e : M(R))_R.$$

We shall prove (T75.2) by induction on d. It being trivial for $d = 0$, let $d > 0$ and assume for $d - 1$. We are supposing that $\mathrm{socz}(R/Q) = 1$ for every parameter ideal Q in R, and we want to show that R is CM. By (T75.1) there exists an R-regular sequence x_1 of length 1 in $M(R)$. By (T69.4)(2″) we have $\dim(R/(x_1R)) = d - 1$ and, in view of (T69.4)(4″), it suffices to show that $R/(x_1R)$ is CM. But this follows by the induction hypothesis.

We shall prove (T75.3) again by induction on d. It being trivial for $d = 0$, let $d > 0$ and assume for $d - 1$. We are supposing R to be CM, and given any parameters $x = (x_1, \ldots, x_d)$ and $x' = (x_1', \ldots, x_d')$ for R we want to show that $\mathrm{socz}(R/I_d) = \mathrm{socz}(R/I_d')$ where for $0 \le m \le d$ we put $I_m = (x_1, \ldots, x_m)R$ and $I_m' = (x_1', \ldots, x_m')R$. By (T69.6) we know that both the sequences x and x' are R-regular. Therefore, for $0 \le m < d$ we have $M(R) \not\subset Z_R(R/I_m)$ and $M(R) \not\subset Z_R(R/I_m')$ and hence by (Q1) and L4§5(O24)(22•) we can find x_d^* in $M(R)$ such that for $0 \le m < d$ we have $x_d^* \notin Z_R(R/I_m)$ and $x_d^* \notin Z_R(R/I_m')$. It follows that $(x_1, \ldots, x_{d-1}, x_d^*)$ and $(x_1', \ldots, x_{d-1}', x_d^*)$ are R-regular sequences in $M(R)$. Hence by (T66.3) we see that $(x_d^*, x_1, \ldots, x_{d-1})$ and $(x_d^*, x_1', \ldots, x_{d-1}')$ are

R-regular sequences in $M(R)$. Upon letting $\phi : R \to R/(x_d^* R)$ be the residue class epimorphism, it follows that $(\phi(x_1), \ldots, \phi(x_{d-1}))$ and $(\phi(x_1'), \ldots, \phi(x_{d-1}'))$ are $\phi(R)$-regular sequences in $M(\phi(R))$. By (T69.4) and (T69.6) we see that $\phi(I_{d-1})$ and $\phi(I_{d-1}')$ are parameter ideals in the $(d-1)$-dimensional CM local ring $\phi(R)$, and hence by the induction hypothesis $\text{socz}(\phi(R)/\phi(I_{d-1})) = \text{socz}(\phi(R)/\phi(I_{d-1}'))$. Therefore $\text{socz}(R/I_d^*) = \text{socz}(R/I_d'^*)$ where $I_d^* = (x_1, \ldots, x_{d-1}x_d^*)R$ and $I_d'^* = (x_1', \ldots, x_{d-1}'x_d'^*)R$. Let $\psi : R \to S = R/I_{d-1}$ be the residue class epimorphism and let $y = \psi(x_d)$ and $z = \psi(x_d^*)$. By (T69.4), yS and zS are parameter ideals in the 1-dimensional CM ring S, and hence $\text{socz}(S/yS) = \text{socz}(S/zS)$ by the easy to prove $d = 1$ case [cf. L6§6(E51)]. Therefore $\text{socz}(R/I_d) = \text{socz}(R/I_d^*)$. Similarly $\text{socz}(R/I_d') = \text{socz}(R/I_d'^*)$. Consequently $\text{socz}(R/I_d) = \text{socz}(R/I_d')$.

GST RING CHARACTERIZATION THEOREM (T76). A local ring is GST iff it is CM and some parameter ideal in it is irreducible.

PROOF. Follows from (T74) and (T75).

COMPLETE INTERSECTION THEOREM (T77). If a noetherian ring is a local complete intersection then it is GST. In particular, if a noetherian ring is regular then it is a complete intersection and hence it is GST.

PROOF. For the second sentence it suffices to note that a regular local ring is a complete intersection because it is a domain by (T64) and hence the zero ideal in it is an ideal-theoretic complete intersection. For the first sentence it suffices to show that, given any epimorphism $\phi : T \to R$ of local rings where T is regular and $\ker(\phi)$ is an ideal-theoretic complete intersection in T, we have that R is GST. To prove that R is GST, in view of (T74) and (T76), all we have to do is to establish that R is CM and $\text{socz}(R/Q) = 1$ for some parameter ideal Q in R.

To do this establishment, let $n = d - e$ where $d = \dim(T)$ and $e = \dim(R)$. Now T is CM by (T69.7) and hence, in view of (T69.6), by (T70.3) and (T70.6) there exists a parameter ideal J in T generated by elements $x_1, \ldots, x_n, y_1, \ldots, y_e$ such that $x = (x_1, \ldots, x_n)$ is a T-regular sequence in $M(T)$ with $\ker(\phi) = xT$ and $Q = (\phi(y_1), \ldots, \phi(y_e))R$ is a parameter ideal in R. Since T is CM and $\ker(\phi) = xT$, by (T69.4) we see that R is CM. Clearly $\text{socz}(R/Q) = \text{socz}(T/J)$ and so it only remains to show that $\text{socz}(T/J) = 1$.

Since T is CM and $M(T)$ is obviously a parameter ideal in it, by (T75) we have $\text{socz}(T/J) = \text{socz}(T/M(T))$. But clearly $\text{socz}(T/M(T)) = 1$. Therefore $\text{socz}(T/J) = 1$.

ONE DIMENSIONAL GST RING CHARACTERIZATION THEOREM (T78). For any one-dimensional CM local ring R the following 7 conditions are equivalent, where $P = M(R)$ and the rest of the notation is explained in Comment (C25).

(i) R is GST.

(ii) $\ell_R(P^{-1}/R) = 1$.

(iii) For every $t \in S_R(R) \cap P$, the ideal tR is irreducible.

(iv) For every $t \in S_R(R) \cap P$ we have $\ell_R((tR:P)_R/(tR)) = 1$.

(v) For every $t \in S_R(R) \cap P$ we have $\mathrm{socz}(R/(tR)) = 1$.

(vi) For every fractional R-ideal J with $J \cap S_R(R) \neq \emptyset$ we have $J = (J^{-1})^{-1}$.

(vii) For every ideal I in R with $I \cap S_R(R) \neq \emptyset$ we have $\ell_R(I^{-1}/R) = \ell_R(R/I)$.

[In view of (C25.1) and (C25.2), (vi) may be paraphrased by saying that $J \mapsto J^{-1}$ gives an inclusion reversing involution of the set of all fractional ideals J with $J \cap S_R(R) \neq \emptyset$ onto itself].

Before proving (T78) we shall make two Comments.

COMMENT (C25). [**Inverse Modules and Fractional Ideals**]. Given any ring R with total quotient ring K, for any R-submodule J of K we define its inverse module J^{-1} by putting $J^{-1} = (R:J)_K$ and we note that J^{-1} is again an R-submodule of K. By a fractional R-ideal J we mean an R-submodule J of K such that $tJ \subset R$ for some $t \in S_R(R)$; note that if R is noetherian then this is equivalent to saying that J is a finitely generated R-submodule of K; by taking $t = 1$ we see that every ideal in R is a fractional R-ideal. Let us make the following observations concerning an R-submodule J of K.

(C25.1) For any $\tau \in J$ we clearly have $\tau J^{-1} \subset R$, and hence if $S_R(R) \cap J \neq \emptyset$ then J^{-1} is a fractional R-ideal. For any $t \in K$ with $tJ \subset R$ we clearly have $t \in J^{-1}$, and hence if J is a fractional R-ideal then $S_R(R) \cap J^{-1} \neq \emptyset$. [It follows that $J \mapsto J^{-1}$ gives a map of the set of all fractional R-ideals with $J \cap S_R(R) \neq \emptyset$ into itself].

(C25.2) If $J \subset L$ for an R-submodule L of K then clearly $L^{-1} \subset J^{-1}$.

(C25.3) Clearly $J \subset (J^{-1})^{-1}$ and $J^{-1} = ((J^{-1})^{-1})^{-1}$.

(C25.4) For any $t \in S_R(R)$ clearly we have $(tR)^{-1} = t^{-1}R$ and also we have $(tJ)^{-1} = t^{-1}J^{-1}$.

(C25.5) For any $t \in S_R(R) \cap J$ we have $(tR:J)_R = tJ^{-1}$. [Namely, for any $t \in S_R(R)$ we have $(tR:J)_K = tJ^{-1}$, and if $t \in J$ then we have $tJ^{-1} \subset R$].

(C25.6) If R is a domain and J is a fractional R-ideal then we have the implications: $S_R(R) \cap J \neq \emptyset \Leftrightarrow J \neq 0$. [Here the implication $\Rightarrow$ is obviously true without assuming J to be a fractional R-ideal. Assuming J to be a fractional R-ideal we have $tJ \subset R$ for some $t \in S_R(R) = R \setminus \{0\}$; now if $J \neq 0$ then taking $0 \neq j \in J$ and letting $\tau = tj$ we get $0 \neq \tau \in R \cap J$ and hence $S_R(R) \cap J \neq \emptyset$].

COMMENT (C26). [**One Dimensional CM Local Rings**]. For any one-dimensional local ring R with $P = M(R)$ we have the following assertions, where (C26.1) and (C26.2) follow from (Q1), and the proofs of (26.3) and (26.4) are sketched in square brackets.

(C26.1) R is CM iff $S_R(R) \cap P \neq \emptyset$.

(C26.2) R is CM $\Rightarrow$ an ideal Q in R is a parameter ideal in R iff $Q = tR$ for some $t \in S_R(R) \cap P$.

(C26.3) $\ell_R(P^{-1}/R) = 1 \Leftrightarrow R$ is CM and for every $t \in S_R(R) \cap P$ we have $\ell_R((tR : P)_R/(tR)) = 1$. [If R were not CM then the total quotient ring K of R would coincide with R; moreover, for any $t \in S_R(R) \cap P$, taking $J = P$ in (C25.5) we have $tP^{-1} = (tR : P)_R$, and hence in view of the R-isomorphism $x \mapsto tx$ of K onto itself we see that P^{-1}/R is isomorphic to $(tR : P)_R/(tR)$ and therefore $\ell_R(P^{-1}/R) = \ell_R((tR : P)_R/(tR))$].

(C26.4) For any nonunit ideal I in R with $S_R(R) \cap I \neq \emptyset$ such that I is generated by two elements, we have $\ell_R(R/I) = \ell_R(I^{-1}/R) \in \mathbb{N}_+$. [Taking $(V, J) = (R, I)$ in (T66.9) we can find t, u in P with $t \in S_R(R)$ such that $I = (t, u)R$; now

$$\begin{cases} \ell_R(R/I) \\ = \ell_R(R/(tR)) - \ell_R(R/(tR : I)_R) \\ \quad \text{since } y \mapsto uy + tR \text{ gives an } R\text{-epimorphism } R \to I/(tR) \\ \quad \text{whose kernel is } (tR : I)_R \\ = \ell_R(R/(tR)) - \ell_R(R/(tI^{-1})) \\ \quad \text{since by (C25.5) we have } (tR : I)_R = tI^{-1} \\ = \ell_R(tI^{-1}/(tR)) \\ \quad \text{since } tR \subset tI^{-1} \subset R; \end{cases}$$

but in view of the R-isomorphism $x \mapsto tx$ of the total quotient of R onto itself we get $\ell_R(tI^{-1}/(tR)) = \ell_R(I^{-1}/R)$, and hence $\ell_R(R/I) = \ell_R(I^{-1}/R) \in \mathbb{N}_+$].

PROOF OF (T78). Briefly, (T78) follows from (T73) by noting that if $t \in S_R(R)$ then $x \mapsto tx$ gives an R-isomorphism $\theta : K \to K$ of the total quotient ring K of R. To explain this in greater detail, keeping (C26.1) and (C26.2) in mind, we may proceed thus.

By definition of GST we get (i) $\Leftrightarrow$ (iii). For any $t \in S_R(R)$, obviously we have $\ell_R((tR : P)_R/(tR)) = 1 \Leftrightarrow \mathrm{socz}(R/(tR)) = 1$, and by applying the part (i) $\Leftrightarrow$ (v) of (T73.1) to $R/(tR)$ we see that the ideal tR is irreducible $\Leftrightarrow \mathrm{socz}(R/(tR)) = 1$. This proves (iii) $\Leftrightarrow$ (iv) $\Leftrightarrow$ (v). By (C26.3) we get (ii) $\Leftrightarrow$ (iv). Thus we have shown the equivalence of (i) to (v).

Since for any fractional R-ideal J we have $tJ \subset R$ for some $t \in S_R(R)$, upon letting $H = tJ$ we get an ideal H in R, and by applying (C25.4) twice we obtain the equation $(H^{-1})^{-1} = t(J^{-1})^{-1}$. Therefore in view of (C25.1) we see that (vi) is equivalent to saying that:

(vi*) for every ideal H in R with $S_R(R) \cap H \neq \emptyset$ we have $H = (H^{-1})^{-1}$.

Now for a moment assume (v), and let H be any ideal in R with $S_R(R) \cap H \neq \emptyset$. Then we can take some $t \in S_R(R) \cap H \neq \emptyset$. Applying part (v) $\Rightarrow$ (vii) of (T73.1) to $R/(tR)$ we see that $H = (tR : (tR : H)_R)_R$. By (C25.5) we get $(tR : H)_R = tH^{-1}$

and $(tR : (tR : H)_R)_R = t((tR : H)_R)^{-1}$ and therefore by (C25.4) we see that $(tR : (tR : H)_R)_R = (H^{-1})^{-1}$. Consequently $H = (H^{-1})^{-1}$. Thus (v) $\Rightarrow$ (vi*) and hence (v) $\Rightarrow$ (vi).

Next for a moment assume (vi), and let I be any ideal in R with $S_R(R) \cap I \neq \emptyset$. Referring to L4§5(O21) and upon letting $\ell_R(R/I) = n$, we can take a composition series $V_0 \subset V_1 \subset \cdots \subset V_n$ of ideals in R between $I = V_0$ and $R = V_n$. By (C25.2) this gives us a second series $V_n^{-1} \subset V_{n-1}^{-1} \subset \cdots \subset V_0^{-1}$ of R-modules with $R = V_n^{-1}$ and $I^{-1} = V_0^{-1}$. Since the members of the first series are distinct, by (vi) the same is true of the second series. It follows that $\ell_R(I^{-1}/R) \geq n = \ell_R(R/I)$. Starting with a normal series between R and I^{-1}, by taking inverses and invoking (vi), we similarly see that $\ell_R(I^{-1}/R) \leq \ell_R(R/I)$. Consequently $\ell_R(I^{-1}/R) = \ell_R(R/I)$. Thus we have shown that (vi) $\Rightarrow$ (vii).

In view of (C26), by taking $I = P$ in (vii) we get $\ell_R(P^{-1}/R) = \ell_R(R/P) = 1$. Thus we have shown that (vii) $\Rightarrow$ (ii). This completes the proof of (T78).

COMMENT (C27). [**One Dimensional Special GST Rings**]. If R is a local ring with $\dim(R) = 1$ and $\mathrm{emdim}(R) \leq 2$ then by taking $I = M(R)$ in (C26.4) and invoking the (ii) $\Rightarrow$ (i) part of (T78) we see that R is GST. We get a special case of this by letting $\phi : T \to R$ be an epimorphism where T is a two-dimensional regular local ring and $\ker(\phi)$ is a nonzero nonunit principal ideal in T. The special case also follows from (T77).

COMMENT (C28). [**Conductor**]. Let R be a ring. By the conductor of R in an overring $\widetilde{R}$ of R we mean the ideal in R given by putting

$$\mathrm{cond}(R, \widetilde{R}) = (R : \widetilde{R})_R.$$

Note that this may be characterized as the largest ideal in R which remains an ideal in $\widetilde{R}$. Also note that

$$\begin{cases} \text{if } R \subset L \text{ where } L \text{ is a subring of an overring of } \widetilde{R} \\ \text{then } \mathrm{cond}(R, \widetilde{R}) = (R : \widetilde{R})_L \end{cases}$$

which follows by observing that, since $1 \in \widetilde{R}$, for every x in the RHS we have $x = x1 \in R$.

The most common usage of this concept is when we take the integral closure R^* of R in the total quotient ring K of R and consider the conductor C of R in R^*, i.e.,

$$C = (R : R^*)_R.$$

Now referring to (C25), by the above braced display we have

$$C = (R^*)^{-1}$$

and hence by taking $J = R^*$ in the part (i) $\Rightarrow$ (vi) of (T78) we see that

$$\begin{cases} \text{if } R \text{ is a one-dimensional GST local ring} \\ \text{and } R^* \text{ is a finitely generated } R\text{-module} \\ \text{then } S_R(R) \cap C \neq \emptyset \text{ and } R^* = C^{-1}. \end{cases}$$

SOMETHING IS TWICE SOMETHING THEOREM (T79). Let R be a GST local ring of dimension one such that the integral closure R^* of R in the total quotient ring K of R is a finitely generated R-module, and let C be the conductor of R in R^*, i.e., C is the ideal in R given by $C = (R : R^*)_R$. Then

$$\ell_R(R^*/R) = \ell_R(R/C)$$

and hence

$$\ell_R(R^*/C) = 2\ell_R(R/C)$$

where all the lengths are nonnegative integers.

PROOF. In view of (C28) our assertion follows by taking $I = C$ in the part (i) $\Rightarrow$ (vii) of (T78).

QUEST (Q17) Projective Resolutions of Finite Modules

Part (T24.4) of Theorem (T24) of L4§8 says that the localization of a regular local ring at any prime ideal in it is again regular. As a primary aim of the present Quest, we shall prove this in (T81.7) which will complete the task of providing the proofs of Theorems (T19) to (T30) of L4§§8-9, except (T20), as announced at the beginning of the current Section. Our main tools are the concepts of projective resolution and projective dimension (abbreviated as pdim) which we proceed to define. As preparation for proving the Dim–Pdim Theorem (T81), which is the main result of this Quest, first we shall prove the Projective Resolution Lemma (T80).

Let R be a ring. By an exact sequence of R-modules we mean a system (f, W, n) consisting of its length $n \in \mathbb{N}$, its sequence $W = (W_i)_{0 \le i \le n+1}$ of R-modules, and its sequence $f = (f_i : W_{i+1} \to W_i)_{0 \le i \le n}$ of R-homomorphisms, such that for $1 \le i \le n$ we have $\operatorname{im}(f_i) = \ker(f_{i-1})$ and moreover we also have $\ker(f_n) = 0$ and $\operatorname{im}(f_0) = W_0$.

The three properties after "such that" constitute the exactness of the sequence (f, W, n). Individually they are called the exactness of (f, W, n) at W_i or W_{n+1} or W_0 respectively, where the last two are reduced to the first by augmenting the sequence by the unique R-homomorphisms $0 \to W_{n+1}$ and $W_0 \to 0$ respectively.

The entire sequence may also be depicted by the arrowed directed flow

$$0 \to W_{n+1} \to W_n \to \cdots \to W_1 \to W_0 \to 0.$$

The names of the maps may be written on the top or the bottom of the respective arrows. The sequence may also be depicted as a vertical flowing from top to bottom, and then the names of the maps may be written to the left or the right of the respective arrows.

Let V be an R-module and let the notation be as introduced in §1 on Direct Sums of Modules. We say that V is free (or a free R-module) if V is isomorphic to the restricted I-th power $R_{\oplus}^I$ for some indexing set I, and we say that V is finite free (or a finite free R-module) if V is isomorphic to R^n for some $n \in \mathbb{N}$, i.e., to R^I with $|I| = n$. We say that V is a direct summand of a free module (resp: a finite free module) if there exists an R-module U such that the (external) direct sum $V \oplus U$ is free (resp: finite free). By a short exact sequence of R-modules we mean an exact sequence $(f, W, 1)$ of R-modules of length 1; this sequence splits means there exists an R-homomorphism $g_0 : W_0 \to W_1$ such that $f_0 g_0$ is the identity map $W_0 \to W_0$.

The concept of a projective module is obtained by generalizing a crucial property of free modules. The said property says that if $\alpha : E \to F$ is any R-epimorphism of R-modules and $\beta : R^d \to F$ is any R-homomorphism with $d \in \mathbb{N}$ then $\beta = \alpha \gamma$ for some R-homomorphism $\gamma : R^d \to E$; to see this we let $\gamma(e_i) \in \alpha^{-1}(\{\beta(e_i)\})$ for $1 \leq i \leq d$ where $e_i = (0, \ldots, 0, 1, 0, \ldots, 0) \in R^d$ with 1 in the i-th spot.

By a projective R-module we mean an R-module D having the property which says that if $\alpha : E \to F$ is any R-epimorphism of R-modules and $\beta : D \to F$ is any R-homomorphism then $\beta = \alpha \gamma$ for some R-homomorphism $\gamma : D \to E$. We are mostly interested in projective modules which are also finitely generated; we shall call them finite projective modules.

In accordance with the terms finite free and finite projective, a finitely generated module may be called a finite module.

Recall that $\approx$ stands for isomorphism; here it mostly indicates isomorphism of R-modules.

The R-module V is said to be a direct summand of a projective module (resp: a finite projective module) if there exists an R-module U such that the (external) direct sum $V \oplus U$ is projective (resp: finite projective).

By an R-resolution of V we mean an exact sequence (f, W, n) of R-modules with $V = W_0$. The R-resolution (f, W, n) is said to be a projective (resp: finite projective, free, finite free, minimal free) R-resolution if for $1 \leq i \leq n+1$ the module W_i is projective (resp: finite projective, free, finite free, isomorphic to $R^{d(i)}$ with $d(i) = \mathrm{gnb}_R \mathrm{im}(f_{i-1}) \in \mathbb{N}$). The R-resolution (f, W, n) is said to be a preprojective (resp: finite preprojective, prefree, finite prefree, minimal prefree) R-resolution if for $1 \leq i \leq n$ the module W_i is projective (resp: finite projective, free, finite free, isomorphic to $R^{d(i)}$ with $d(i) = \mathrm{gnb}_R \mathrm{im}(f_{i-1}) \in \mathbb{N}$).

By the projective R-dimension of V, denoted by $\mathrm{pdim}_R V$, we mean the smallest length of a projective R-resolution of V if such exists; if there does not exist any projective R-resolution of V then we put $\mathrm{pdim}_R V = \infty$. Note that V is projective iff $\mathrm{pdim}_R V = 0$.

Next, by the global dimension of R, denoted by $\mathrm{gdim}(R)$, we mean the smallest nonnegative integer n, if it exists, such that $\mathrm{pdim}_R V \le n$ for every finite R-module V; otherwise we put $\mathrm{gdim}(R) = \infty$. Thus, $\mathrm{gdim}(R) = n \in \mathbb{N}$ means $\mathrm{pdim}_R V \le n$ for every finite R-module V, and $\mathrm{pdim}_R V_n = n$ for some finite R-module V_n; likewise, $\mathrm{gdim}(R) = \infty$ means for every $n \in \mathbb{N}$ we have $\mathrm{pdim}_R U_n > n$ for some finite R-module U_n.

PROJECTIVE RESOLUTION LEMMA (T80). For any ring R we have the following.

(T80.1) If a short exact sequence $(f, W, 1)$ of R-modules splits then there exists an isomorphism $W_1 \approx W_2 \oplus W_0$. If an R-module V is a direct summand of a finite free R-module then V is a finite module.

(T80.2) An R-module is projective iff every length-one R-resolution of it splits iff it is a direct summand of a free module. A finite R-module is projective iff it is a direct summand of a finite free module.

(T80.3) Every free module is projective. A direct sum of any finite number of projective modules is projective. A direct summand of a projective module is projective.

(T80.4) In case R is quasilocal, for any R-module V we have that:

$$V \text{ is a finite projective } R\text{-module } \Leftrightarrow V \text{ is a finite free } R\text{-module.}$$

(T80.5) Let there be given any two short exact sequences $(f, W, 1)$ and $(f', W', 1)$ of R-modules such that the modules W_1 and W_1' are projective, and assume that there exists an R-isomorphism $g : W_0 \to W_0'$. Then there exists an R-isomorphism $h : W_1 \oplus W_1' \to W_1 \oplus W_1'$ such that $gf_0 l = f_0' l' h$ where $l : W_1 \oplus W_1' \to W_1$ and $l' : W_1 \oplus W_1' \to W_1'$ are the natural projections. Moreover, for any such h we have $h(W_2 \oplus W_1') = W_1 \oplus W_2'$ where we regard $W_2 \oplus W_1'$ and $W_1 \oplus W_2'$ as submodules of $W_1 \oplus W_1'$ via the injective maps f_1 and f_1'.

(T80.6) Conversely, let there be given any two short exact sequences $(f, W, 1)$ and $(f', W', 1)$ of R-modules such that the modules W_1 and W_1' are projective, and assume that there exists an R-isomorphism $h : W_1 \oplus W_1' \to W_1 \oplus W_1'$ with $h(W_2 \oplus W_1') = W_1 \oplus W_2'$ where we regard $W_2 \oplus W_1'$ and $W_1 \oplus W_2'$ as submodules of $W_1 \oplus W_1'$ via the injective maps f_1 and f_1'. Then there exists an R-isomorphism $g : W_0 \to W_0'$ such that $gf_0 l = f_0' l' h$ where $l : W_1 \oplus W_1' \to W_1$ and $l' : W_1 \oplus W_1' \to W_1'$ are the natural projections.

(T80.7) Let there be given any two exact sequences (f, W, n) and (f', W', n) of R-modules of equal length such that the modules W_i and W_i' are projective for $1 \le i \le n$, and assume that there exists an R-isomorphism $g : W_0 \to W_0'$. Then

$$W_1 \oplus W_2' \oplus W_3 \oplus W_4' \oplus \cdots \oplus W_{n+1} \approx W_1' \oplus W_2 \oplus W_3' \oplus W_4 \oplus \cdots \oplus W_{n+1}'$$

or

$$W_1' \oplus W_2 \oplus W_3' \oplus W_4 \oplus \cdots \oplus W_{n+1} \approx W_1 \oplus W_2' \oplus W_3 \oplus W_4' \oplus \cdots \oplus W_{n+1}'$$

according as n is even or odd. Moreover: W_{n+1} is projective iff W_{n+1}' is projective.

(T80.8) (THE COMPARISON LEMMA) Let there be given any exact sequence (f, W, n) of R-modules such that the module W_i is projective for $1 \leq i \leq n$. Then $\mathrm{pdim}_R W_0 \leq n$ iff W_{n+1} is projective. More precisely, if the R-module $\mathrm{im}(f_m)$ is projective for some integer $m \in \{0, 1, \ldots, n\}$ then for the smallest such integer m we have $\mathrm{pdim}_R W_0 = m$, and the R-module $\mathrm{im}(f_i)$ is projective or nonprojective according as $m \leq i \leq n$ or $0 \leq i < m$. Moreover, if $\mathrm{pdim}_R W_0 \geq n$ then we have $n + \mathrm{pdim}_R W_{n+1} = \mathrm{pdim}_R W_0$ with the understanding that $n + \infty = \infty$.

(T80.9) (THE SNAKE LEMMA) Let there be given two short exact sequences $(f', W', 1)$ and $(f, W, 1)$ of R-modules and for $0 \leq j \leq 2$ let there be given a length-two exact sequence $(f^{(j)}, W^{(j)}, 2)$ of R-modules such that for $0 \leq i \leq 2$ we have $W_i' = W_2^{(i)}$ and $W_i = W_1^{(i)}$, and for $0 \leq i \leq 1$ we have $f_1^{(i)} f_i' = f_i f_1^{(i+1)}$. Let $W^* = (W_i^*)_{0 \leq i \leq 5}$ be the sequence of R-modules which is given by putting $W_i^* = W_0^{(i)}$ and $W_{i+3}^* = W_3^{(i)}$ for $0 \leq i \leq 2$. Then there exists a unique sequence $f^* = (f_i^* : W_{i+1}^* \to W_i^*)_{0 \leq i \leq 4}$ of R-homomorphisms such that for $0 \leq i \leq 1$ we have $f_i^* f_0^{(i+1)} = f_0^{(i)} f_i$ and $f_2^{(i)} f_{i+3}^* = f_i' f_2^{(i+1)}$, and for all $(x, y, z) \in W_3^{(0)} \times W_2^{(1)} \times W_1^{(2)}$ with $f_2^{(0)}(x) = f_0'(y)$ and $f_1^{(1)}(y) = f_1(z)$ we have $f_2^*(x) = f_0^{(2)}(z)$. Moreover, putting together the sequences W^* and f^*, we have that $(f^*, W^*, 4)$ is an exact sequence of R-modules. [DIAGRAM: To clarify the statement and the proof of this lemma, the reader may like to draw a diagram in which the short exact sequences are depicted horizontally (flowing from left to right) with the primed sequence on top, and the length-two exact sequences are depicted vertically (flowing from top to bottom) in the left to right order $f^{(2)}, f^{(1)}, f^{(0)}$].

(T80.10) Let there be given any short exact sequence $(f, W, 1)$ of R-modules such that two of the three modules W_0, W_1, W_2 have finite projective R-dimension. Then the third module has finite projective R-dimension and we have

$$\mathrm{pdim}_R W_1 \leq \max(\mathrm{pdim}_R W_0, \mathrm{pdim}_R W_2).$$

Moreover, if

$$\mathrm{pdim}_R W_1 < \max(\mathrm{pdim}_R W_0, \mathrm{pdim}_R W_2)$$

then we have

$$\mathrm{pdim}_R W_0 = 1 + \mathrm{pdim}_R W_2.$$

[In proving (T80.10), without assuming that two of the three modules W_0, W_1, W_2 have finite projective R-dimension, we shall be establishing the following assertion:

(T80.10*) Given any short exact sequence $(f, W, 1)$ of R-modules, there exists a short exact sequence $(f'', W'', 1)$ of R-modules such that for $0 \le i \le 2$ we have that: (i) $\mathrm{pdim}_R W_i = 0 \Rightarrow \mathrm{pdim}_R W_i'' = 0$, (ii) $\mathrm{pdim}_R W_i = \infty \Rightarrow \mathrm{pdim}_R W_i'' = \infty$, and (iii) $0 < \mathrm{pdim}_R W_i < \infty \Rightarrow \mathrm{pdim}_R W_i = 1 + \mathrm{pdim}_R W_i''$].

(T80.11) For any two R-modules V and U we have

$$\mathrm{pdim}_R(V \oplus U) = \max(\mathrm{pdim}_R(V), \mathrm{pdim}_R(U)).$$

(T80.12) In case R is local, for any R-module V and any $x \in M(R) \cap S_R(V)$ we have that:

V is a finite free R-module $\Leftrightarrow V/xV$ is a finite free (R/xR)-module.

(T80.13) In case R is local, for any finite R-module V and any element x in $M(R) \cap S_R(V) \cap S_R(R)$, we have $\mathrm{pdim}_R V = \mathrm{pdim}_{(R/xR)}(V/xV)$.

(T80.14) In case R is local, for any short exact sequence $(f, W, 1)$ of R-modules with W_1 finite free and $\mathrm{reg}_R W_0 = 0 < \mathrm{reg}(R)$, we have $\mathrm{reg}_R W_2 = 1$.

(T80.15) If R is local and V is any finite R-module with $\mathrm{pdim}_R V = n \in \mathbb{N}$ then V has a minimal free R-resolution (f, W, n) of length n. Moreover, if $\mathrm{reg}(R) = 0$ then $n = 0 = \mathrm{reg}_R V$.

(T80.16) If R is noetherian and V is any finite R-module then there exists a family $(f^{(n)}, W^{(n)}, n)_{n \in \mathbb{N}}$ of minimal prefree R-resolutions of V such that for all $n < m$ in $\mathbb{N}$ we have: $W_i^{(n)} = W_i^{(m)}$ for $0 \le i \le n$, $W_{n+1}^{(n)} = \mathrm{im}(f_n^{(m)})$, $f_i^{(n)} = f_i^{(m)}$ for $0 \le i < n$, and $f_n^{(n)} : W_{n+1}^{(n)} \to W_n^{(n)}$ is the natural injection.

(T80.17) If R is noetherian and V is any finite R-module then given any $n \in \mathbb{N}$ there exists a minimal prefree R-resolution (f, W, n) of V. Moreover, for any such R-resolution we have that it is always finite preprojective, and furthermore we also have that: it is finite projective $\Leftrightarrow \mathrm{pdim}_R V \le n \Leftrightarrow W_{n+1}$ is a projective R-module.

PROOF. To prove (T80.1), first suppose that $(f, W, 1)$ is a short exact sequence of R-modules which splits, i.e., there exists an R-homomorphism $g_0 : W_0 \to W_1$ such that $f_0 g_0$ is the identity map $W_0 \to W_0$. Then clearly $(a, b) \mapsto f_1(a) + g_0(b)$ gives an isomorphism of the (external) direct sum $W_2 \oplus W_0$ onto W_1.

Next let V and U be R-modules together with an isomorphism $g : R^n \to V \oplus U$ with $n \in \mathbb{N}$. Then upon letting $p : V \oplus U \to V$ be the natural projection, we get an R-epimorphism $pg : R^n \to V$ and hence V is a finite R-module. This completes the proof of (T80.1).

To prove (T80.2), let V be an R-module. We shall cyclically prove that: V is a direct summand of a free module $\Rightarrow V$ is projective $\Rightarrow$ every length-one R-resolution of V splits $\Rightarrow V$ is a direct summand of a free module; moreover, in the proof it will turn out that for a finite module V, in the last implication, free may be replaced by finite free. First, assuming that for some R-module U and some indexing set I there is an isomorphism $g : R_\oplus^I \to V \oplus U$ where $R_\oplus^I$ is the restricted I-th power, let us show that V is projective. So given any R-epimorphism $\alpha : E \to F$ of R-modules and any R-homomorphism $\beta : V \to F$, we want to find an R-homomorphism $\gamma : V \to E$ such that $\beta = \alpha\gamma$. Let $p : V \oplus U \to V$ be the natural projection, let $q : V \to V \oplus U$ be the natural injection, and let $(e_i)_{i \in I}$ be the natural basis of $R_\oplus^I$, i.e., $e_i : I \to R$ is the map such that $e_i(i) = 1$ and $e_i(j) = 0$ for all $j \neq i$ in I. Since α is surjective, for each $i \in I$ we can take $x_i \in E$ such that $\alpha(x_i) = \beta pg(e_i)$. Since $(e_i)_{i \in I}$ is a basis of $R_\oplus^I$, there is a unique R-homomorphism $\delta : R_\oplus^I \to E$ such that $\delta(e_i) = x_i$ for all $i \in I$. Let $\gamma = \delta g^{-1} q$. Then γ is an R-homomorphism $V \to E$ with $\beta = \alpha\gamma$.

Next, assuming V is projective, given any length-one R-resolution $(f, W, 1)$ of V, let us show that it splits. Let $i : W_0 \to W_0$ be the identity map. Since $f_0 : W_1 \to W_0$ is an R-epimorphism and W_0 is a projective module, we can find an R-homomorphism $g_0 : W_0 \to W_1$ with $f_0 g_0 = i$.

Now, assuming every length-one R-resolution of V splits, let us show that V is a direct summand of a free module. Let $(y_i)_{i \in I}$ be a set of generators of $W_0 = V$ where I is a finite set if V is a finite R-module. Let e_i be the above basis of $W_1 = R_\oplus^I$, and let $f_0 : W_1 \to W_0$ be the unique R-epimorphism with $f_0(e_i) = y_i$ for all $i \in I$. Let $\ker(f_0) = W_2$ and let $f_1 : W_2 \to W_1$ be the natural injection. Then $(f, W, 1)$ is a length-one R-resolution of V. Hence it splits and therefore by (T80.1) there is an isomorphism $W_1 \approx W_2 \oplus W_0$. This completes the proof of (T80.2).

To prove (T80.3), first note that by the preamble to the definition of a projective module, after changing R^d to $R_\oplus^I$ with any indexing set I and letting e_i be as in the above proof of (T80.2), we see that every free module is projective. Next, if $V_1, \ldots, V_m$ are any finite number of projective R-modules then by (T80.2) we can find R-modules $U_1, \ldots, U_m$ such that the modules $V_1 \oplus U_1, \ldots, V_m \oplus U_m$ are free, and now obviously the module $(V_1 \oplus \cdots \oplus V_m) \oplus (U_1 \oplus \cdots \oplus U_m)$ is free and hence again by (T80.2) we conclude that the module $(V_1 \oplus \cdots \oplus V_m)$ is projective. Finally, if V and U are R-modules such that $V \oplus U$ is projective then by (T80.2) we can find an R-module T such that the module $(V \oplus U) \oplus T$ is free, which is obviously isomorphic to the module $V \oplus (U \oplus T)$ and hence again by (T80.2) the module V is projective. This completes the proof of (T80.3).

To prove (T80.4), assume that R is quasilocal and let V be any R-module. First note that by the preamble to the definition of a projective module we see that if V is a finite free R-module then V is a finite projective R-module.

Next, supposing V to be a finite projective R-module, we want to show that V is isomorphic to R^n for some $n \in \mathbb{N}$. So let $\mathrm{gnb}_R V = n$. Then $n \in \mathbb{N}$ and we can find an R-epimorphism $R^n \to V$. Upon letting V' be the kernel of this epimorphism with $\mathrm{gnb}_R V' = n'$ by (T80.1) and (T80.2) we see that $n' \in \mathbb{N}$ and $R^n \approx V \oplus V'$. By Comment (C29)(2) below we have $\mathrm{gnb}_R(V \oplus V') = n + n'$ and $\mathrm{gnb}_R R^n = n$. Consequently $n = n + n'$ and hence $n' = 0$. Therefore $V' = 0$ and hence $R^n \approx V$. This completes the proof of (T80.4).

To prove (T80.5) and (T80.6), given any two short exact sequences $(f, W, 1)$ and $(f', W', 1)$ of R-modules where the modules W_1 and W_1' are projective, let $l : W_1 \oplus W_1' \to W_1$ and $l' : W_1 \oplus W_1' \to W_1'$ be the natural projections, and let us regard $W_2 \oplus W_1'$ and $W_1 \oplus W_2'$ as submodules of $W_1 \oplus W_1'$ via the injective maps f_1 and f_1'. First, assuming that there exists an R-isomorphism $g : W_0 \to W_0'$, we want to show that there exists an R-isomorphism $h : W_1 \oplus W_1' \to W_1 \oplus W_1'$ such that $g f_0 l = f_0' l' h$, and for any such h we have $h(W_2 \oplus W_1') = W_1 \oplus W_2'$. Since the modules W_1 and W_1' are projective, we can find R-homomorphisms $\gamma : W_1 \to W_1'$ and $\gamma' : W_1' \to W_1$ such that $f_0' \gamma = g f_0$ and $g f_0 \gamma' = f_0'$ respectively. Now $(a, a') \mapsto (a, a' - \gamma(a))$ and $(a, a') \mapsto (a - \gamma'(a'), a')$ give R-isomorphisms $t : W_1 \oplus W_1' \to W_1 \oplus W_1'$ and $t' : W_1 \oplus W_1' \to W_1 \oplus W_1'$. Let $h = t^{-1} t'$. Then $h : W_1 \oplus W_1' \to W_1 \oplus W_1'$ is an R-isomorphism with $g f_0 l = f_0' l' h$. Moreover, $f_0 l$ and $f_0' l'$ are R-epimorphisms of $W_1 \oplus W_1'$ onto W_0 and W_0' with kernels $W_2 \oplus W_1'$ and $W_1 \oplus W_2'$ respectively, and hence if $h : W_1 \oplus W_1' \to W_1 \oplus W_1'$ is any R-isomorphism with $g f_0 l = f_0' l' h$ then $h(W_2 \oplus W_1') = W_1 \oplus W_2'$.

Next, assuming that there exists an R-isomorphism $h : W_1 \oplus W_1' \to W_1 \oplus W_1'$ with $h(W_2 \oplus W_1') = W_1 \oplus W_2'$, we want to show that there exists an R-isomorphism $g : W_0 \to W_0'$ with $g f_0 l = f_0' l' h$. This follows from the fact that $f_0 l$ and $f_0' l'$ are R-epimorphisms of $W_1 \oplus W_1'$ onto W_0 and W_0' with kernels $W_2 \oplus W_1'$ and $W_1 \oplus W_2'$ respectively. This completes the proof of (T80.5) and (T80.6).

To prove (T80.7), let there be given any two exact sequences (f, W, n) and (f', W', n) of R-modules of equal length such that the modules W_i and W_i' are projective for $1 \leq i \leq n$ and there exists an R-isomorphism $g : W_0 \to W_0'$. Suppose we have proved the existence of the displayed isomorphism. Then by (T80.3) it follows that W_{n+1} is projective iff W_{n+1}' is projective. To prove the existence of the displayed isomorphism we make induction on n. This is obvious for $n = 0$, and for $n = 1$ we are reduced to (T80.5). So let $n > 1$ and assume for $n - 1$. Let $\overline{W}_i = W_i$ or $\mathrm{im}(f_{n-1})$ according as $0 \leq i \leq n - 1$ or $i = n$. Let $\overline{f}_i = f_i$ or the natural injection $\overline{W}_n \to \overline{W}_{n-1}$ according as $0 \leq i \leq n - 2$ or $i = n - 1$. Let $\overline{W}_i' = W_i'$ or $\mathrm{im}(f_{n-1}')$ according as $0 \leq i \leq n - 1$ or $i = n$. Let $\overline{f}_i' = f_i'$ or the natural injection $\overline{W}_n' \to \overline{W}_{n-1}'$ according as $0 \leq i \leq n - 2$ or $i = n - 1$. Then by applying the induction hypothesis to the exact sequences $(\overline{f}, \overline{W}, n - 1)$ and $(\overline{f}', \overline{W}', n - 1)$ of R-modules we find an R-isomorphism $g : \widehat{W}_0 \to \widehat{W}_0'$ where

$$
\begin{cases}
\widehat{W}_0 = \overline{W}_1' \oplus \overline{W}_2 \oplus \overline{W}_3' \oplus \overline{W}_4 \oplus \cdots \oplus \overline{W}_n \\
\text{and} \\
\widehat{W}_0' = \overline{W}_1 \oplus \overline{W}_2' \oplus \overline{W}_3 \oplus \overline{W}_4' \oplus \cdots \oplus \overline{W}_n'
\end{cases}
$$

or

$$\begin{cases} \widehat{W}_0 = \overline{W}_1 \oplus \overline{W}'_2 \oplus \overline{W}_3 \oplus \overline{W}'_4 \oplus \cdots \oplus \overline{W}_n \\ \text{and} \\ \widehat{W}'_0 = \overline{W}'_1 \oplus \overline{W}_2 \oplus \overline{W}'_3 \oplus \overline{W}_4 \oplus \cdots \oplus \overline{W}'_n \end{cases}$$

according as n is even or odd. Let $\widehat{W}_1$ and $\widehat{W}'_1$ be obtained by removing the overlines from the components in the above definitions of $\widehat{W}_0$ and $\widehat{W}'_0$ respectively. Let $\widehat{f}_0 : \widehat{W}_1 \to \widehat{W}_0$ and $\widehat{f}'_0 : \widehat{W}'_1 \to \widehat{W}'_0$ be the maps which leave the first $n-1$ components unchanged and apply the maps f_{n-1} and f'_{n-1} to the respective last components. Let $\widehat{W}_2 = W_{n+1}$ and $\widehat{W}'_2 = W'_{n+1}$. Let $\widehat{W}_2 \to \widehat{W}_1$ and $\widehat{W}'_2 \to \widehat{W}'_1$ be the obvious injective maps induced by f_n and f'_n respectively. Then $(\widehat{f}, \widehat{W}, 1)$ and $(\widehat{f}', \widehat{W}', 1)$ are short exact sequences of R-modules, and by (T80.3) we see that the modules $\widehat{W}_1$ and $\widehat{W}'_1$ are projective. Therefore by (T80.5) we get the desired isomorphism $\widehat{W}'_1 \oplus \widehat{W}_2 \approx \widehat{W}_1 \oplus \widehat{W}'_2$. This completes the proof of (T80.7).

To prove (T80.8), let there be given any exact sequence (f, W, n) of R-modules such that the module W_i is projective for $1 \le i \le n$. First suppose that we have $\mathrm{pdim}_R V \le n$. Then by definition we have a projective R-resolution (f', W', n') of W_0 with $n' \le n$. For $n' + 1 < i \le n + 1$ let $W'_i = 0$, and for $n' + 1 \le i \le n$ let $f'_i : W'_{i+1} \to W'_i$ be the unique R-homomorphism (with $f'_i(0) = 0$). Then (f', W', n) is a projective R-resolution of W_0, and hence W_{n+1} is a projective R-module by (T80.7).

Next suppose that W_{n+1} is projective; then by definition we have $\mathrm{pdim}_R V \le n$. Thus we have proved that $\mathrm{pdim}_R V \le n$ iff W_{n+1} is projective.

Now suppose that $\mathrm{im}(f_m)$ is projective for some $m \in \{0, 1, \ldots, n\}$ and let m be the smallest such. Given any $j \in \{0, 1, \ldots, n\}$, we define an exact sequence $(f^{(j)}, W^{(j)}, j)$ of R-modules as follows. Let $W_i^{(j)} = W_i$ for $0 \le i \le j$ and let $W_{j+1}^{(j)} = \mathrm{im}(f_j)$. Let $f_i^{(j)} = f_i$ for $0 \le i < j$ and let $f_j^{(j)} : W_{j+1}^{(j)} \to W_j^{(j)}$ be the natural injection. Then $(f^{(j)}, W^{(j)}, j)$ is a preprojective R-resolution of W_0 for $0 \le j \le n$. Therefore, by what we have proved above, we see that $\mathrm{pdim}_R W_0 = m$. Hence, by what we have proved above, we also see that the R-module $\mathrm{im}(f_i)$ is projective or nonprojective according as $m \le i \le n$ or $0 \le i < m$.

Next suppose that $\mathrm{pdim}_R W_0 \ge n$. Given any $j \ge n$ and any preprojective R-resolution $(\overline{f}^{(j)}, \overline{W}^{(j)}, j - n)$ of W_{n+1}, we define an exact sequence $(f^{(j)}, W^{(j)}, j)$ of R-modules thus. Let $W_i^{(j)} = W_i$ or $\overline{W}_{i-n}^{(j)}$ according as $0 \le i \le n$ or $n + 1 \le i \le j + 1$. Let $f_n^{(j)} = f_i$ or $f_n \overline{f}_0^{(j)}$ or $\overline{f}_{i-n}^{(j)}$ according as $0 \le i \le n$ or $i = n$ or $n + 1 \le i \le j$. Then $(f^{(j)}, W^{(j)}, j)$ is a preprojective R-resolution of W_0 for $j \ge n$. If $\mathrm{pdim}_R W_{n+1} = m < \infty$ then by taking $j = n + m$ and $(\overline{f}^{(j)}, \overline{W}^{(j)}, j - n)$ to be a projective R-resolution of W_{n+1}, we see that $(f^{(j)}, W^{(j)}, j)$ is a projective R-resolution of W_0, and hence by what we have proved above we get $\mathrm{pdim}_R W_0 = n + m$.

Likewise, if $\operatorname{pdim}_R W_{n+1} = \infty$ then for every $j \geq n$ we can find $j' \geq j$ together with a preprojective but not projective R-resolution $(\overline{f}^{(j)}, \overline{W}^{(j)}, j-n)$ of W_{n+1}, and then we see that $(f^{(j)}, W^{(j)}, j)$ is a preprojective but not projective R-resolution of W_0, and hence by what we have proved above we get $\operatorname{pdim}_R W_0 = \infty$. This completes the proof of (T80.8).

To prove (T80.9), let there be given two short exact sequences $(f', W', 1)$ and $(f, W, 1)$ of R-modules and for $0 \leq j \leq 2$ let there be given a length-two exact sequence $(f^{(j)}, W^{(j)}, 2)$ of R-modules such that for $0 \leq i \leq 2$ we have $W_i' = W_2^{(i)}$ and $W_i = W_1^{(i)}$, and for $0 \leq i \leq 1$ we have $f_1^{(i)} f_i' = f_i f_1^{(i+1)}$; furthermore let $W^* = (W_i^*)_{0 \leq i \leq 5}$ be the sequence of R-modules which is given by putting $W_i^* = W_0^{(i)}$ and $W_{i+3}^* = W_3^{(i)}$ for $0 \leq i \leq 2$. We want to show that then there exists a unique sequence $f^* = (f_i^* : W_{i+1}^* \to W_i^*)_{0 \leq i \leq 4}$ of R-homomorphisms such that for $0 \leq i \leq 1$ we have $f_i^* f_0^{(i+1)} = f_0^{(i)} f_i$ and $f_2^{(i)} f_{i+3}^* = f_i' f_2^{(i+1)}$, and for all $(x, y, z) \in W_3^{(0)} \times W_2^{(1)} \times W_1^{(2)}$ with $f_2^{(0)}(x) = f_0'(y)$ and $f_1^{(1)}(y) = f_1(z)$ we have $f_2^*(x) = f_0^{(2)}(z)$; also we want to show that, for these sequences W^* and f^*, we have that $(f^*, W^*, 4)$ is an exact sequence of R-modules. Let us break up the proof in several parts.

Part One which says that: for $0 \leq i \leq 1$ there exists a unique R-homomorphism $f_i^* : W_{i+1}^* \to W_i^*$ with $f_i^* f_0^{(i+1)} = f_0^{(i)} f_i$; moreover, for this homomorphism we have $\operatorname{im}(f_i^*) = \operatorname{im}(f_0^{(i)} f_i)$ and $\ker(f_i^*) = f_0^{(i+1)}(\ker(f_0^{(i)} f_i))$.

To see this fix $i \in \{0, 1\}$. Then $f_0^{(i)} f_i : W_1^{(i+1)} \to W_i^*$ is an R-homomorphism and (since $(f^{(i+1)}, W^{(i+1)}, 2)$ is exact) $f_0^{(i+1)} : W_1^{(i+1)} \to W_{i+1}^*$ is an R-epimorphism, and hence the unique existence of the R-homomorphism $f_i^* : W_{i+1}^* \to W_i^*$ with $f_i^* f_0^{(i+1)} = f_0^{(i)} f_i$ is equivalent to the inclusion $\ker(f_0^{(i+1)}) \subset \ker(f_0^{(i)} f_i)$. The exactness of $(f^{(i+1)}, W^{(i+1)}, 2)$ tells us that $\ker(f_0^{(i+1)}) = \operatorname{im}(f_1^{(i+1)})$, and hence the desired inclusion is equivalent to the equation $f_0^{(i)} f_i f_1^{(i+1)}(W_2^{(i+1)}) = 0$ which, in view of the assumed relation $f_1^{(i)} f_i' = f_i f_1^{(i+1)}$, is equivalent to the equation $f_0^{(i)} f_1^{(i)} f_i'(W_2^{(i+1)}) = 0$. In turn, this last equation is a consequence of the fact that $\ker(f_0^{(i)} f_1^{(i)}) = W_2^{(i)}$ which itself follows from the exactness of $(f^{(i)}, W^{(i)}, 2)$. Having proved the unique existence of f_i^*, the "moreover" claim follows from the relation $f_i^* f_0^{(i+1)} = f_0^{(i)} f_i$ and the surjectivity of $f_0^{(i+1)} : W_1^{(i+1)} \to W_{i+1}^*$.

Part Two which says that: for $0 \leq i \leq 1$ there exists a unique R-homomorphism $f_{i+3}^* : W_{i+4}^* \to W_{i+3}^*$ with $f_2^{(i)} f_{i+3}^* = f_i' f_2^{(i+1)}$; moreover, for this homomorphism we have $\operatorname{im}(f_{i+3}^*) = (f_2^{(i)})^{-1}(\operatorname{im}(f_i' f_2^{(i+1)}))$ and $\ker(f_{i+3}^*) = \ker(f_i' f_2^{(i+1)})$.

To see this fix $i \in \{0, 1\}$. Then $f_i' : W_{i+1}' \to W_i'$ is an R-homomorphism and (since $(f^{(i+1)}, W^{(i+1)}, 2)$ and $(f^{(i)}, W^{(i)}, 2)$ are exact) $f_2^{(i+1)} : W_{i+4}^* \to W_{i+1}'$ and $f_2^{(i)} : W_{i+3}^* \to W_i'$ are R-monomorphisms, and hence the unique existence of the R-homomorphism $f_{i+3}^* : W_{i+4}^* \to W_{i+3}^*$ with $f_2^{(i)} f_{i+3}^* = f_i' f_2^{(i+1)}$ is equivalent to the inclusion $\operatorname{im}(f_i' f_2^{(i+1)}) \subset \operatorname{im}(f_2^{(i)})$. The exactness of $(f^{(i)}, W^{(i)}, 2)$ tells us that

$\ker(f_1^{(i)}) = \mathrm{im}(f_2^{(i)})$, and hence the desired inclusion is equivalent to the equation $f_1^{(i)} f_i' f_2^{(i+1)}(W_3^{(i+1)}) = 0$ which, in view of the assumed relation $f_1^{(i)} f_i' = f_i f_1^{(i+1)}$, is equivalent to the equation $f_i f_1^{(i+1)} f_2^{(i+1)}(W_3^{(i+1)}) = 0$. In turn, this last equation is a consequence of the fact that $\ker(f_1^{(i+1)} f_2^{(i+1)}) = W_3^{(i+1)}$ which itself follows from the exactness of $(f^{(i+1)}, W^{(i+1)}, 2)$. Having proved the unique existence of f_{i+3}^*, the "moreover" claim follows from the relation $f_2^{(i)} f_{i+3}^* = f_i' f_2^{(i+1)}$ and the injectivity of $f_2^{(i)} : W_3^{(i)} \to W_2^{(i)}$.

Part Three saying that: there is a unique map $f_2^* : W_3^* \to W_2^*$ such that for all $(x, y, z) \in W_3^{(0)} \times W_2^{(1)} \times W_1^{(2)}$ with $f_2^{(0)}(x) = f_0'(y)$ and $f_1^{(1)}(y) = f_1(z)$ we have $f_2^*(x) = f_0^{(2)}(z)$.

To see this, for any $x \in W_3^{(0)}$ let

$$A_x = \{ y \in W_2^{(1)} : f_2^{(0)}(x) = f_0'(y) \}$$

and for any $y \in W_2^{(1)}$ let

$$B_y = \{ z \in W_1^{(2)} : f_1^{(1)}(y) = f_1(z) \}.$$

In the R-homomorphisms $f_0' : W_2^{(1)} = W_1' \to W_0' = W_2^{(0)}$ and $f_2^{(0)} : W_3^{(0)} \to W_2^{(0)}$, the first is surjective by the exactness of $(f', W', 1)$, and hence:

(1) $A_x \neq \emptyset$.

By the said exactness we know that $\mathrm{im}(f_1') = \ker(f_0')$; consequently:

(2) for any y, y' in A_x we have $y - y' \in \mathrm{im}(f_1')$.

The exactness of $(f^{(0)}, W^{(0)}, 2)$ yields $f_0^{(0)} f_1^{(0)}(x) = 0$ and hence $f_1^{(0)} f_0'(A_x) = 0$; consequently by the assumed relation $f_1^{(0)} f_0' = f_0 f_1^{(1)}$ we see that $f_0 f_1^{(1)}(A_x) = 0$; therefore by the exactness of $(f, W, 1)$ we conclude that:

(3) $f_1^{(1)}(A_x) \subset f_1(W_1^{(2)})$.

The exactness of $(f', W', 1)$ and $(f, W, 1)$ tells us that the two R-homomorphisms $f_1' : W_2' \to W_1'$ and $f_1 : W_2 \to W_1$ are injective and hence by the assumed relation $f_1^{(1)} f_1' = f_1 f_1^{(2)}$ we see that: if $z \in B_y$ and $z' \in B_{y'}$ where y, y' are elements in $W_2^{(1)}$ with $y - y' \in \mathrm{im}(f_1')$, then $z - z' \in \mathrm{im}(f_1^{(2)})$; consequently, since the exactness $(f^{(2)}, W^{(2)}, 2)$ tells us that $\ker(f_0^{(2)} f_1^{(2)}) = W_2^{(2)}$, in view of (2) we conclude that:

(4) $z \in B_y$ and $z' \in B_{y'}$ with y, y' in $A_x \Rightarrow f_0^{(2)}(z) = f_0^{(2)}(z')$.

By (1) and (3) we see that for every $x \in W_3^* = W_3^{(0)}$ there exists $(y, z) \in W_2^{(1)} \times W_1^{(2)}$ with $f_2^{(0)}(x) = f_0'(y)$ and $f_1^{(1)}(y) = f_1(z)$, and by (4) we see that the element $f_0^{(2)}(z) \in W_2^* = W_0^{(2)}$ depends only on x and not on the particular choice of (y, z). This establishes the unique existence of the required map $f_2^* : W_3^* \to W_2^*$.

Part Four which says that: the map f_2^* of Part Three is an R-homomorphism and for it we have $\mathrm{im}(f_2^*) = f_0^{(2)}(\ker(f_0^{(1)} f_1))$ and $\ker(f_2^*) = (f_2^{(0)})^{-1}(\mathrm{im}(f_0' f_2^{(1)}))$.

The claim that the map f_2^* of Part Three is an R-homomorphism follows by noting that if for $1 \le i \le 2$ we have

$$(r_i, x_i, y_i, z_i, w_i) \in R \times W_3^{(0)} \times W_2^{(1)} \times W_1^{(2)} \times W_0^{(2)}$$

with $f_2^{(0)}(x_i) = f_0'(y_i)$ and $f_1^{(1)}(y_i) = f_1(z_i)$ and $f_0^{(2)}(z_i) = w_i$ then, because the various W are R-modules and the various f are R-homomorphisms, upon letting

$$(x, y, z, w) = (r_1 x_1 + r_2 x_2, r_1 y_1 + r_2 y_2, r_1 z_1 + r_2 z_2, r_1 w_1 + r_2 w_2)$$

we have

$$(x, y, z, w) \in W_3^{(0)} \times W_2^{(1)} \times W_1^{(2)} \times W_2^{(0)}$$

with $f_2^{(0)}(x) = f_0'(y)$ and $f_1^{(1)}(y) = f_1(z)$ and $f_0^{(2)}(z) = w$.

To prove the inclusion $\mathrm{im}(f_2^*) \subset f_0^{(2)}(\ker(f_0^{(1)} f_1))$, it suffices to show that if $(x, y, z) \in W_3^{(0)} \times W_2^{(1)} \times W_1^{(2)}$ with $f_2^{(0)}(x) = f_0'(y)$ and $f_1^{(1)}(y) = f_1(z)$ then $z \in \ker(f_0^{(1)} f_1)$. So let $(x, y, z) \in W_3^{(0)} \times W_2^{(1)} \times W_1^{(2)}$ with $f_2^{(0)}(x) = f_0'(y)$ and $f_1^{(1)}(y) = f_1(z)$. The last equation tells us that: $z \in \ker(f_0^{(1)} f_1) \Leftrightarrow f_0^{(1)} f_1^{(1)}(y) = 0$. Moreover, by the exactness of $(f^{(1)}, W^{(1)}, 2)$ we get $f_0^{(1)} f_1^{(1)}(y) = 0$.

To prove the reverse inclusion $\mathrm{im}(f_2^*) \supset f_0^{(2)}(\ker(f_0^{(1)} f_1))$, it suffices to show that, given any element $z \in \ker(f_0^{(1)} f_1)$ we can find $(x, y) \in W_3^{(0)} \times W_2^{(1)}$ such that $f_2^{(0)}(x) = f_0'(y)$ and $f_1^{(1)}(y) = f_1(z)$. Now $f_1(z) \in \ker(f_0^{(1)})$ and by the exactness of $(f^{(1)}, W^{(1)}, 2)$ we have $\mathrm{im}(f_1^{(1)}) = \ker(f_0^{(1)})$. Therefore $f_1^{(1)}(y) = f_1(z)$ for some element $y \in W_2^{(1)}$. In view of the assumed relation $f_1^{(0)} f_0' = f_0 f_1^{(1)}$ we have $f_1^{(0)} f_0'(y) = f_0 f_1^{(1)}(y)$ and hence, because of the equation $f_1^{(1)}(y) = f_1(z)$ we get $f_1^{(0)} f_0'(y) = f_0 f_1(z)$. But by the exactness of $(f, W, 1)$ we have $f_0 f_1(z) = 0$ and hence $f_1^{(0)} f_0'(y) = 0$. Therefore $f_0'(y) \in \ker(f_1^{(0)})$; but by the exactness of $(f^{(0)}, W^{(0)}, 2)$ we have $\mathrm{im}(f_2^{(0)}) = \ker(f_1^{(0)})$, and hence $f_2^{(0)}(x) = f_0'(y)$ for some element $x \in W_0^{(3)}$.

To prove the inclusion $\ker(f_2^*) \subset (f_2^{(0)})^{-1}(\mathrm{im}(f_0' f_2^{(1)}))$, it suffices to show that if $(x, y, z) \in W_3^{(0)} \times W_2^{(1)} \times W_1^{(2)}$ with $f_2^{(0)}(x) = f_0'(y)$ and $f_1^{(1)}(y) = f_1(z)$ and $f_0^{(2)}(z) = 0$ then $f_2^{(0)}(x) \in \mathrm{im}(f_0' f_2^{(1)})$. So let $(x, y, z) \in W_3^{(0)} \times W_2^{(1)} \times W_1^{(2)}$ with $f_2^{(0)}(x) = f_0'(y)$ and $f_1^{(1)}(y) = f_1(z)$ and $f_0^{(2)}(z) = 0$. By the exactness of $(f^{(2)}, W^{(2)}, 2)$ we have $\mathrm{im}(f_1^{(2)}) = \ker(f_0^{(2)})$; hence by the equation $f_0^{(2)}(z) = 0$ we get $z \in \mathrm{im}(f_1^{(2)})$; consequently, in view of the assumed relation $f_1^{(1)} f_1' = f_1 f_1^{(2)}$, we have $f_1(z) \in \mathrm{im}(f_1^{(1)} f_1')$; therefore by the equation $f_1^{(1)}(y) = f_1(z)$ we see that $f_1^{(1)}(y) \in \mathrm{im}(f_1^{(1)} f_1')$, i.e., $f_1^{(1)}(y) = f_1^{(1)}(y')$ for some $y' \in \mathrm{im}(f_1')$; but by the exactness of $(f', W', 1)$ we have $\mathrm{im}(f_1') = \ker(f_0')$, and hence $f_0'(y') = 0$, and therefore by the equation $f_2^{(0)}(x) = f_0'(y)$ we get $f_2^{(0)}(x) = f_0'(y - y')$. By the exactness of $(f^{(1)}, W^{(1)}, 2)$ we have $\mathrm{im}(f_2^{(1)}) = \ker(f_1^{(1)})$, and hence by the equation $f_1^{(1)}(y) = f_1^{(1)}(y')$ we get $y - y' \in \mathrm{im}(f_2^{(1)})$, and therefore by the previous equation $f_2^{(0)}(x) = f_0'(y - y')$ we conclude that $f_2^{(0)}(x) \in \mathrm{im}(f_0' f_2^{(1)})$.

To prove the reverse inclusion $\ker(f_2^*) \supset (f_2^{(0)})^{-1}(\mathrm{im}(f_0' f_2^{(1)}))$, let there be given any $x \in (f_2^{(0)})^{-1}(\mathrm{im}(f_0' f_2^{(1)}))$. Then, in the notation of Part Three, we can find $y \in A_x \cap \mathrm{im}(f_2^{(1)})$. By the exactness of $(f^{(1)}, W^{(1)}, 2)$ we have $\mathrm{im}(f_2^{(1)}) = \ker(f_1^{(1)})$, and hence $f_1^{(1)}(y) = 0$. Consequently by taking $z = 0$ we get $z \in B_y$, and obviously $f_0^{(2)}(z) = 0$. But by the definition of f_2^* we have $f_2^*(x) = f_0^{(2)}(z)$, and hence $x \in \ker(f_2^*)$.

Part Five which says that the sequence $(f^*, W^*, 4)$ is exact.

Namely, by Parts One, Two, and Four we have $\mathrm{im}(f_i^*) = \ker(f_{i-1}^*)$ for $i = 2, 3$. By the exactness of $(f^{(0)}, W^{(0)}, 2)$ and $(f, W, 1)$ we respectively see that $f_0^{(0)}$ and f_0 are surjective, and hence by Part One we deduce the surjectiveness of f_0^*. By the exactness of $(f^{(1)}, W^{(1)}, 2)$ and $(f', W', 1)$ we respectively see that $f_2^{(2)}$ and f_1' are injective, and hence by Part Two we deduce the injectiveness of f_4^*. Moreover, by Part One we see that

$$\mathrm{im}(f_1^*) = \ker(f_0^*) \Leftrightarrow \mathrm{im}(f_0^{(1)} f_1) = f_0^{(1)}(\ker(f_0^{(0)} f_0))$$

and by Part Two we see that

$$\mathrm{im}(f_4^*) = \ker(f_3^*) \Leftrightarrow (f_2^{(1)})^{-1}(\mathrm{im}(f_1' f_2^{(2)})) = \ker(f_0' f_2^{(1)}).$$

To prove the inclusion $\mathrm{im}(f_0^{(1)} f_1) \subset f_0^{(1)}(\ker(f_0^{(0)} f_0))$, it suffices to show that $\mathrm{im}(f_1) \subset \ker(f_0^{(0)} f_0)$. But by the exactness of $(f, W, 1)$ we have $\mathrm{im}(f_1) = \ker(f_0)$, and hence $\mathrm{im}(f_1) \subset \ker(f_0^{(0)} f_0)$.

To prove the reverse inclusion $\mathrm{im}(f_0^{(1)} f_1) \supset f_0^{(1)}(\ker(f_0^{(0)} f_0))$, it suffices to show that, given any element $v \in W_1^{(1)}$ with $f_0(v) \in \ker(f_0^{(0)})$ we can find an element $u \in \mathrm{im}(f_1)$ with $f_0^{(1)}(u) = f_0^{(1)}(v)$. By the exactness of $(f^{(0)}, W^{(0)}, 2)$ we have $\mathrm{im}(f_1^{(0)}) = \ker(f_0^{(0)})$ and hence $f_0(v) \in \mathrm{im}(f_1^{(0)})$; by the exactness of $(f', W', 1)$ we have $\mathrm{im}(f_0') = W_2^{(1)}$ and hence by the assumed relation $f_1^{(0)} f_0' = f_0 f_1^{(1)}$ we can find $w \in \mathrm{im}(f_1^{(1)})$ such that $f_0(w) = f_0(v)$; by the exactness of $(f, W, 1)$ we have $\mathrm{im}(f_1) = \ker(f_0)$ and hence upon letting $u = v - w$ we get $u \in \mathrm{im}(f_1)$; by the exactness of $(f^{(1)}, W^{(1)}, 2)$ we have $\mathrm{im}(f_1^{(1)}) = \ker(f_0^{(1)})$ and hence $f_0^{(1)}(w) = 0$; therefore $f_0^{(1)}(u) = f_0^{(1)}(v)$.

To prove the inclusion $(f_2^{(1)})^{-1}(\mathrm{im}(f_1' f_2^{(2)})) \subset \ker(f_0' f_2^{(1)})$, it suffices to show that for any $x \in W_3^{(1)}$ with $f_2^{(1)}(x) \in \mathrm{im}(f_1' f_2^{(2)})$ we have $f_2^{(1)}(x) \in \ker(f_0')$. By the exactness of $(f', W', 1)$ we have $\mathrm{im}(f_1') = \ker(f_0')$ and hence $f_0'(\mathrm{im}(f_1' f_2^{(2)})) = 0$ and therefore $f_2^{(1)}(x) \in \ker(f_0')$.

To prove the reverse inclusion $(f_2^{(1)})^{-1}(\mathrm{im}(f_1' f_2^{(2)})) \supset \ker(f_0' f_2^{(1)})$, it suffices to show that, given any $x \in W_3^{(1)}$ with $f_2^{(1)}(x) \in \ker(f_0')$ we can find $z \in \mathrm{im}(f_2^{(2)})$ with $f_2^{(1)}(x) = f_1'(z)$. By the exactness of $(f', W', 1)$ we have $\mathrm{im}(f_1') = \ker(f_0')$ and hence $f_2^{(1)}(x) = f_1'(z)$ for some $z \in W_2^{(2)}$; by the exactness of $(f^{(1)}, W^{(1)}, 2)$ we have $\mathrm{im}(f_2^{(1)}) = \ker(f_1^{(1)})$ and hence $f_1^{(1)}(f_2^{(1)}(x)) = 0$ and therefore by the assumed relation $f_1^{(1)} f_1' = f_1 f_1^{(2)}$ we get $f_1 f_1^{(2)}(z) = 0$; by the exactness of $(f, W, 1)$ we know that f_1 is injective and hence $f_1^{(2)}(z) = 0$; by the exactness of $(f^{(2)}, W^{(2)}, 2)$ we have $\text{im}(f_2^{(2)}) = \ker(f_1^{(2)})$ and hence $z \in \text{im}(f_2^{(2)})$.

This completes the proof of Part Five, and hence also the proof of the entire SNAKE LEMMA (T80.9).

To prove (T80.10), let $(f, W, 1)$ be any short exact sequence of R-modules with $\text{pdim}_R W_i = p_i$ for $0 \leq i \leq 2$, and assume that two out of the three quantities p_0, p_1, p_2 are finite. Let $p \in \mathbb{N}$ be the sum of the two smallest quantities. More precisely let p be defined in the following six exhaustive but mutually exclusive cases thus.

Case One, when $p_0 \leq p_1 \leq p_2$, we put $p = p_0 + p_1$.

Case Two, when $p_0 \leq p_2 < p_1$, we put $p = p_0 + p_2$.

Case Three, when $p_1 < p_0 \leq p_2$, we put $p = p_1 + p_0$.

Case Four, when $p_1 \leq p_2 < p_0$, we put $p = p_1 + p_2$.

Case Five, when $p_2 < p_0 \leq p_1$, we put $p = p_2 + p_0$.

Case Six, when $p_2 < p_1 \leq p_0$, we put $p = p_2 + p_1$.

[In other words, if $\min(p_0, p_1, p_2) = p_0$ then we are in Case One or Two according as $\min(p_1, p_2) = p_1$ or $\min(p_1, p_2) \neq p_1$; if $\min(p_0, p_1, p_2) \neq p_0$ and $\min(p_0, p_1, p_2) = p_1$ then we are in Case Three or Four according as $\min(p_0, p_2) = p_0$ or $\min(p_0, p_2) \neq p_0$; if $\min(p_0, p_1, p_2) \neq p_0$ and $\min(p_0, p_1, p_2) \neq p_1$ then we are in Case Five or Six according as $\min(p_0, p_1) = p_0$ or $\min(p_0, p_1) \neq p_0$]. By induction on p we shall prove the CLAIM which says that: $\{p_0, p_1, p_2\} \subset \mathbb{N}$ with $p_1 \leq \max(p_0, p_2)$, and if $p_1 < \max(p_0, p_2)$ and then $p_0 = 1 + p_2$. By (T80.1) to (T80.3) we see that if $p_0 = 0$ then $p_1 = p_2 = 0$. By (T80.8) we see that if $p_1 = p_2 = 0$ then $p_0 = 0$. This settles the Claim for $p = 0$.

Now we shall apply the Snake Lemma (T80.9). To prepare the background for this, first for $0 \leq j \leq 2$ we put $W_1^{(j)} = W_j$ with $W_0^{(j)} = W_j^* = 0$ and we let $f_0^{(j)}$ be the unique R-epimorphism $f_0^{(j)} : W_1^{(j)} \to W_0^{(j)}$. Next we construct a short exact sequence $(f', W, 1)$ of projective R-modules $W_2^{(j)} = W_j'$ together with R-epimorphisms $f_1^{(j)} : W_2^{(j)} \to W_1^{(j)}$ for $0 \leq j \leq 2$ thus. To exhibit such an epimorphism for $j = 2$ (with similar construction for $j = 0$), let $(u_i)_{i \in I}$ be generators of W_2, let $(e_i)_{i \in I}$ be the natural basis of $W_2^{(j)} = R_\oplus^I$ as in the proof of (T80.2), and let $f_1^{(2)} : W_2^{(2)} \to W_1^{(2)} = W_2$ be the unique R-epimorphism which sends e_i to u_i for all $i \in I$. We put $W_2^{(1)} = W_2^{(2)} \oplus W_2^{(0)}$ with external direct sum, and we let $f_1^{(1)} : W_2^{(1)} \to W_1^{(1)} = W_1$ be the R-epimorphism which sends (x, y) to $f_1^{(2)}(x) + f_1^{(0)}(y)$. Also we let $f_1' : W_2' \to W_1'$ and $f_0' : W_1' \to W_0'$ be the natural injection and the natural projection respectively. The modules W_2', W_1', W_0' are projective by (T80.3). It only remains to put $W_{j+3}^* = W_3^{(j)} = \ker(f_1^{(j)})$ and let $f_2^{(j)} : W_3^{(j)} \to W_2^{(j)}$ be the natural injection for $0 \leq j \leq 2$. The conditions of the Snake Lemma are now satisfied and hence by that Lemma there exists an R-homomorphism $f_i^* : W_{i+1}^* \to W_i^*$ for $0 \leq i \leq 4$ such that $(f^*, W^*, 4)$ becomes an exact sequence of R-modules. Let $W_i'' = W_3^{(i)}$ for $0 \le i \le 2$ and $f_i'' = f_{i+3}^*$ for $0 \le i \le 1$. Since $(f^*, W^*, 4)$ is exact with $W_2^* = 0$, we see that $(f'', W'', 1)$ is a short exact sequence of R-modules. Moreover, for $0 \le i \le 2$, since $(f^{(i)}, W^{(i)}, 2)$ is exact with $W_0^{(i)} = 0$, upon letting $g_i'' : W_i'' \to W_i'$ and $g_i' : W_i' \to W_i$ to be the maps $f_2^{(i)}$ and $f_1^{(i)}$ respectively, we get a short exact sequence of R-modules $0 \to W_i'' \to W_i' \to W_i \to 0$; therefore, upon letting $\mathrm{pdim}_R W_i'' = p_i''$, by (T80.8) we see that if $p_i \in \{0, \infty\}$ then $p_i'' = p_i$, and if $0 < p_i < \infty$ then $p_i = 1 + p_i''$. Consequently, if $p > 0$ then two of the three quantities p_0'', p_1'', p_2'' are finite and $p'' < p$ where p'' is defined in terms of p_0'', p_1'', p_2'' exactly as p was defined in terms of p_0, p_1, p_2, and hence the Claim follows by induction. This completes the induction and hence proves (T80.10).

In the above paragraph, without using the assumption that two of the three modules W_0, W_1, W_2 have finite projective R-dimension, we proved the following:

(T80.10*) Given any short exact sequence $(f, W, 1)$ of R-modules, there exists a short exact sequence $(f'', W'', 1)$ of R-modules such that for $0 \le i \le 2$ we have that: (i) $\mathrm{pdim}_R W_i = 0 \Rightarrow \mathrm{pdim}_R W_i'' = 0$, (ii) $\mathrm{pdim}_R W_i = \infty \Rightarrow \mathrm{pdim}_R W_i'' = \infty$, and (iii) $0 < \mathrm{pdim}_R W_i < \infty \Rightarrow \mathrm{pdim}_R W_i = 1 + \mathrm{pdim}_R W_i''$.

To prove (T80.11), given any two R-modules V and U, let $(f, W, 1)$ be the short exact sequence of R-modules obtained by taking $W_2 = V, W_1 = V \oplus U, W_0 = U$, and letting $f_1 : W_2 \to W_1$ and $f_0 : W_1 \to W_0$ be the natural injection and natural projection respectively. Let $\mathrm{pdim}_R W_i = p_i$ for $0 \le i \le 2$. We want to show that then $\max(p_0, p_2) = p_1$. If two of the three quantities p_0, p_1, p_2 are finite then this follows from (T80.10); if none of them is finite then we have nothing to show; if exactly one of them is finite but it is not p_1 then again we have nothing to show. So supposing that $p_1 < \infty = p_0 = p_2$, by induction on p_1 we shall prove our assertion (i.e., deduce a contradiction). If $p_1 = 0$ then by (T80.3) we get $p_0 = p_2 = 0$. So let $p_1 > 0$ and assume for all smaller values of p_1. Let $(f'', W'', 1)$ be the short exact sequence of R-modules as in (T80.10*) and let $\mathrm{pdim}_R W_i''$ for $0 \le i \le 2$. Then by (i) to (iii) above we see that $p_1'' < p_1$ with $p_0'' = p_2'' = \infty$, and hence we are done by the induction hypothesis. This completes the proof of (T80.11).

To prove (T80.12) to (T80.14), assume that R is local. For any R-module V and any $x \in M(R)$ we regard V/xV to be an (R/xR)-module. By (T80.4) we see that, for I= 12, 13, 14, part (T80.I) is equivalent to the following assertion (I).

(12) If V is a finite R-module and $x \in M(R) \cap S_R(V)$ then: V is projective iff V/xV is projective.

(13) If V is a finite R-module and $x \in M(R) \cap S_R(V) \cap S_R(R)$ then we have: $\mathrm{pdim}_R V = \mathrm{pdim}_{(R/xR)}(V/xV)$.

(14) If $(f, W, 1)$ is any short exact sequence of R-modules with finite free W_1 and $\mathrm{reg}_R W_0 = 0 < \mathrm{reg}(R)$, then $\mathrm{reg}_R W_2 = 1$.

Given any finite R-module, upon letting $\mathrm{gnb}_R V = n$ and taking generators $u_1, \ldots, u_n$ of V, let $f_0 : R^n = W_1 \to W_0 = V$ be the R-epimorphism which sends $e_i = (0, \ldots, 0, 1, 0 \ldots, 0)$ (with 1 in the i-th spot) to u_i for $1 \le i \le n$. Also let $f_1 : \ker(f_0) = W_2 \to W_1$ be the natural injection. Thus in all the three situations we have a short exact sequence $(f, W, 1)$ of R-modules with finite free W_1. Also we can take $x \in M(R)$ where $x \in S_R(V)$ or $x \in S_R(V) \cap S_R(R)$ or $x \in S_R(R)$ according as we are in (12) or (13) or (14).

Only assuming that $x \in M(R)$, let us prepare the groundwork for applying the Snake Lemma (T80.9). Let $(f', W', 1) = (f, W, 1)$. For $0 \le j \le 2$ let $(f^{(j)}, W^{(j)}, 2)$ be the length-two exact sequence of R-modules obtained thus. Let $f_0^{(j)} : W_1^{(j)} = W_j \to W_j / x W_j = W_0^{(j)}$ be the residue class epimorphism, let $f_1^{(j)} : W_2^{(j)} = W_j \to W_j = W_1^{(j)}$ be given by $v \mapsto xv$, and let $f_2^{(j)} : W_3^{(j)} = \ker(f_1^{(j)}) \to W_2^{(j)}$ be the natural injection. Let $W_i^* = W_0^{(i)}$ for $0 \le i \le 2$, and $W_{i+3}^* = W_3^{(i)}$ for $0 \le i \le 2$. Now the conditions of the Snake Lemma are satisfied and hence by that Lemma there exists an R-homomorphism $f_i^* : W_{i+1}^* \to W_i^*$ such that $(f^*, W^*, 4)$ becomes an exact sequence of R-modules.

Note that if $x \in S_R(V)$ then $\ker(f_1^{(2)}) = 0$, i.e., $W_3^* = 0$. Also note that if $x \in S_R(R)$ then clearly $x \in S_R(R^d)$ for all $d \in \mathbb{N}$ and hence $\ker(f_1^{(1)}) = 0$, i.e., $W_4^* = 0$ and therefore by the exactness of $(f^*, W^*, 4)$ at W_5^* and W_4^* we also have $W_5^* = 0$.

In case of (12) we are assuming $x \in S_R(V)$, and we want to show that: V is projective iff V/xV is projective. If V is projective then V/xV is projective by (T80.4) and (C29)(6) below. If V/xV is projective then by (C29)(7) below we have $\ker(f_0^*) = 0$ and hence by the exactness of $(f^*, W^*, 4)$ at W_1^* we get $\mathrm{im}(f_1^*) = 0$; since $W_3^* = 0$, by the exactness of $(f^*, W^*, 4)$ at W_2^* we also get $\ker(f_1^*) = 0$. Therefore if V/xV is projective then $W_2^* = 0$, i.e., $W_2/xW_2 = 0$, and hence $W_2 = 0$ by Nakayama (T3). Since $\ker(f_0) = W_2$, it follows that if V/xV is projective then V is finite free and hence it is projective by (T80.3).

In case of (13) we are assuming $x \in S_R(V) \cap S_R(R)$, and we want to show that $\mathrm{pdim}_R V = \mathrm{pdim}_{(R/xR)}(V/xV)$. If both sides of the last equation are ∞ then we have nothing to show. So assuming that their minimum m is a nonnegative integer, we make induction on it. If $m = 0$ then we are reduced to (12). So let $m > 0$ and assume true for all smaller values of it. Since $W_3^* = 0$, the exactness of $(f^*, W^*, 4)$ yields the exactness of $(\overline{f}, \overline{W}, 1)$ where $\overline{W}_i = W_i / x W_i$ for $0 \le i \le 2$, and $\overline{f}_i = f_i^*$ for $0 \le i \le 1$. Since $W_5^* = 0$, we get $x \in S_R(W_2)$. By (T80.4) we know that W_1 is a projective R-module, and likewise by (T80.4) and (C29)(6) below W_1/xW_1 is a projective (R/xR)-module. Applying Lemma (T80.8) to $(f, W, 1)$ we see that $\mathrm{pdim}_R W_2 = \mathrm{pdim}_R V$ or $(\mathrm{pdim}_R V) - 1$ according as $\mathrm{pdim}_R V = \infty$ or $\ne \infty$, and similarly applying the same Lemma (T80.8) to $(\overline{f}, \overline{W}, 1)$ we see that $\mathrm{pdim}_{(R/xR)}(W_2/xW_2) = \mathrm{pdim}_{(R/xR)}(V/xV)$ or $(\mathrm{pdim}_{(R/xR)}(V/xV)) - 1$ according as $\mathrm{pdim}_{(R/xR)}(V/xV) = \infty$ or $\ne \infty$. Therefore we are done by the induction hypothesis.

In case of (14) we are assuming $x \in S_R(R)$ with $\operatorname{reg}_R W_0 = 0$, and we want to show that $\operatorname{reg}_R W_2 = 1$. Since $\operatorname{reg}_R W_0 = 0$, by (Q1) we have $M(R) \in \operatorname{tass}_R W_0$ and hence there exits $0 \neq t \in W_0$ with $M(R) = (0 : t)_R$. Since $W_3^* = (0 : x)_{W_0}$ with $x \in M(R)$, we get $t \in W_3^*$ and hence $M(R) \in \operatorname{tass}_R W_3^*$. Since $W_4^* = 0$, by the exactness of $(f^*, W^*, 4)$ at W_3^* we see that $f_2^* : W_3^* \to W_2^* = W_2/xW_2$ is injective. Therefore, because of $M(R) \in \operatorname{tass}_R W_3^*$, we get $M(R) \in \operatorname{tass}_R(W_2/xW_2)$ and hence $\operatorname{reg}_R(W_2/xW_2) = 0$. Since $W_5^* = 0$, we also have $x \in S_R(W_2)$. Consequently $\operatorname{reg}_R W_2 = 1$. This completes the proof of (T80.12) to (T80.14).

To prove (T80.15) to (T80.17), assuming R to be noetherian, let there be given any finite R-module V. For every $d \in \mathbb{N}$ let $(e_i^{(d)})_{1 \leq i \leq d}$ be the basis of R^d which is given by $e_i^{(d)} = (0, \ldots, 0, 1, 0, \ldots, 0)$ with 1 in the i-th spot. Let $u_1^{(0)}, \ldots, u_{d(0)}^{(0)}$ with $\operatorname{gnb}_R V = d(0)$ be generators of $W_0 = V$ and let $f_0 : W_1 = R^{(d(0))} \to W_0$ be the R-homomorphism given by $e_i^{(d(0))} \mapsto u_i^{(0)}$ for $1 \leq i \leq d(0)$, let $u_1^{(1)}, \ldots, u_{d(1)}^{(1)}$ with $\operatorname{gnb}_R \ker(f_0) = d(1)$ be generators of $\ker(f_0)$ and let $f_1 : W_2 = R^{(d(1))} \to W_1$ be the R-homomorphism given by $e_i^{(d(1))} \mapsto u_i^{(1)}$ for $1 \leq i \leq d(1)$, and in the same manner let $u_1^{(2)}, \ldots, u_{d(2)}^{(2)}$ with $\operatorname{gnb}_R \ker(f_1) = d(2)$ be generators of $\ker(f_1)$ and let $f_2 : W_3 = R^{(d(2))} \to W_2$ be the R-homomorphism given by $e_i^{(d(2))} \mapsto u_i^{(2)}$ for $1 \leq i \leq d(2)$, and so on.

To prove (T80.15), suppose that R is local with $\operatorname{pdim}_R V = n \in \mathbb{N}$. Then, in view of (T80.4) and (T80.8), by (1) and (2) of (C29) below we see that $d(n + 2) = d(n + 3) = \cdots = 0$ and (f, W, n) is a minimal free R-resolution of V where $f = (f_i : W_{i+1} \to W_i)_{0 \leq i \leq n}$; note that if $n > 0$ then $d(i) \neq 0$ for $1 \leq i \leq n + 1$.

Now also suppose that $\operatorname{reg}(R) = 0$. Then by (Q1) we have $M(R) = (0 : x)_R$ for some $0 \neq x \in R$. If $n > 0$ then by (C29)(4) below we have $f_n(W_{n+1}) \subset M(R)W_n$ and hence $f_n(xW_{n+1}) \subset xM(R)W_n = 0$; however, since $x \neq 0 \neq d(n + 1)$, we have $xW_{n+1} \neq 0$ and hence by the injectivity of f_n we get $f_n(xW_{n+1}) \neq 0$, which is a contradiction. Therefore $n = 0$. It follows that $V \approx W_1 = R^{(d(0))}$ and hence the zeroness of $\operatorname{reg}(R)$ implies the zeroness of $\operatorname{reg}_R V$.

To prove (T80.16), dropping the assumption that R is local with $\operatorname{pdim}_R V = n \in \mathbb{N}$, given any $n \in \mathbb{N}$, we get a minimal prefree R-resolution $(f^{(n)}, W^{(n)}, n)$ of V where: $W_i^{(n)} = W_i$ for $0 \leq i \leq n$, $W_{n+1}^{(n)} = \operatorname{im}(f_n)$, $f_i^{(n)} = f_i$ for $0 \leq i < n$, and $f_n^{(n)} : W_{n+1}^{(n)} \to W_n^{(n)}$ is the natural injection. For any $n < m$ in $\mathbb{N}$ we clearly have: $W_i^{(n)} = W_i^{(m)}$ for $0 \leq i \leq n$, $W_{n+1}^{(n)} = \operatorname{im}(f_n^{(m)})$, $f_i^{(n)} = f_i^{(m)}$ for $0 \leq i < n$, and $f_n^{(n)} : W_{n+1}^{(n)} \to W_n^{(n)}$ is the natural injection.

To prove (T80.17), given any $n \in \mathbb{N}$, let $(f, W, n) = (f^{(n)}, W^{(n)}, n)$. Then (f, W, n) is minimal prefree R-resolution of V. Moreover, by (T80.3) we see that any such R-resolution is always finite preprojective, and by (T80.8) we also see that: it is finite projective $\Leftrightarrow \operatorname{pdim}_R V \leq n \Leftrightarrow W_{n+1}$ is a projective R-module. This completes the proof of Lemma (T80).

COMMENT (C29). [**Residual dimension and gnb over Quasilocal Rings**].

$$(\bullet) \qquad \begin{cases} \text{Let } V \text{ be a finite module over a quasilocal ring } R \\ \text{with } \mathrm{gnb}_R V = n \text{ and } \dim_{R/M(R)}(V/(M(R)V)) = \nu \end{cases}$$

i.e., the nonnegative integer n is the smallest number of generators of V, and the nonnegative integer ν is the dimension of $V/(M(R)V)$ as a vector space over the field $R/M(R)$. By Nakayama Lemma (T3), if $U = z_1 R + \cdots + z_m R$ where $z_1, \ldots, z_m$ in V with $m \in \mathbb{N}$ are such that $V = U + M(R)V$ then $V = U$, and from this it follows that

$$(1) \qquad\qquad\qquad n = \nu.$$

As a consequence of the equation $n = \nu$ we can see that

$$(2) \qquad \begin{cases} \text{if } V' \text{ is another finite } R\text{-module with } \mathrm{gnb}_R V' = n' \\ \text{then } \mathrm{gnb}_R(V \oplus V') = n + n', \\ \text{and hence, in particular, for all } d \in \mathbb{N} \text{ we have } \mathrm{gnb}_R R^d = d \\ \text{(because clearly } \mathrm{gnb}_R R = 1). \end{cases}$$

To prove the above consequence, it is sufficient to show that the two modules $(V \oplus V')/(M(R)(V \oplus V'))$ and $(V/(M(R)V)) \oplus (V'/(M(R)V'))$ are isomorphic as modules over R and hence as modules over $(R/M(R))$. More generally, without assuming R to be quasilocal, given any two short exact sequences $(f, W, 1)$ and $(f', W', 1)$ of R-modules, for the external direct sums $W_i'' = W_i \oplus W_i'$, consider the R-homomorphisms $f_1'' : W_2 \oplus W_2' \to W_1 \oplus W_1'$ and $f_0'' : W_1 \oplus W_1' \to W_0 \oplus W_0'$ given by $(a, a') \mapsto (f_1(a), f_1'(a'))$ and $(a, a') \mapsto (f_0(a), f_0'(a'))$ respectively. Then obviously $(f'', W'', 1)$ is a short exact sequence of R-modules. It only remains to note that, in our case, when f_0 and f_0' are the residue class epimorphisms $V \to V/(M(R)V)$ and $V' \to V'/(M(R)V')$, we clearly have $\mathrm{im}(f_1'') = M(R)(V \oplus V')$.

As another consequence of (1) we see that:

$$(3) \qquad \begin{cases} \text{if } T \text{ is any submodule of } V \text{ then} \\ \mathrm{gnb}_R(V/T) = \mathrm{gnb}_R V \Leftrightarrow T \subset M(R)V \end{cases}$$

and hence by (2) it follows that:

$$(4) \qquad \begin{cases} \text{if } (f, W, n) \text{ is any minimal free resolution of } V \text{ then} \\ \text{for } 1 \leq i \leq n \text{ we have } f_i(W_{i+1}) \subset M(R)W_i. \end{cases}$$

Recall that for any $J \subset M(R)$ we may regard V/JV as a module over R/JR and clearly it is a finite module over R/JR. Moreover, if U is any submodule of V

with $V = U + JV$ then $V = U$ by Nakayama (T3). It follows that:

(5) $$J \subset M(R) \Rightarrow \text{gnb}_{(R/JR)}(V/JV) = \text{gnb}_R V.$$

Let $\phi : R \to R/JR$ be the canonical epimorphism. For any $n \in \mathbb{N}$, by sending the element $e_i = (0, \ldots, 0, 1, 0 \ldots, 0) \in R^n$ (with $1 \in R$ in the i-th spot) to the element $\bar{e}_i = (0, \ldots, 0, 1, 0 \ldots, 0) \in (R/JR)^n$ (with $1 \in R/JR$ in the i-th spot) we get an epimorphism $\psi : R^n \to (R/JR)^n$ with kernel JR^n and hence, upon letting $f_0^{(1)} : R^n \to R^n/JR^n$ be the residue class epimorphism, we get a unique isomorphism $\epsilon : (R/JR)^n \to R^n/JR^n$ such that $\epsilon \psi = f_0^{(1)}$. It follows that:

(6) $$\begin{cases} V \text{ is finite free } R\text{-module with } J \subset M(R) \\ \Rightarrow V/JV \text{ is finite free } (R/JR)\text{-module.} \\ [\text{Thus the } \Rightarrow \text{ part of (T80.12) holds} \\ \text{for any } x \in M(R) \text{ without assuming } x \in S_R(V)]. \end{cases}$$

Reverting to the assumption that $\text{gnb}_R V = n$, take generators $u_1, \ldots, u_n$ of V, and let $f_0 : R^n \to V$ be the epimorphism which sends e_i to u_i for $1 \le i \le n$. Also let $f_0^{(0)} : V \to V/JV$ be the residue class epimorphism, and let $\bar{f}_0 : (R/JR)^n \to V$ be the epimorphism which sends $\bar{e}_i$ to $f_0^{(0)}(u_i)$ for $1 \le i \le n$. Then clearly there is a unique epimorphism $f_0^* : R^n/JR^n \to V/JV$ with $f_0^* \epsilon = \bar{f}_0$; note that obviously $f_0^* f_0^{(1)} = f_0^{(0)}$. By (T80.2) we see that if V/JV is projective (as an (R/JR)-module) then $R^n/JR^n \approx \ker(f_0^*) \oplus (V/JV)$ (external direct sum of (R/JR)-modules), and by (2), (5) and (6) we have $\text{gnb}_{(R/JR)}(R^n/JR^n) = n = \text{gnb}_{(R/JR)}(V/JV)$. It follows that:

(7) V/JV is projective (as (R/JR)-module) with $J \subset M(V) \Rightarrow \ker(f_0^*) = 0$.

DIM–PDIM THEOREM (T81). For any ring R we have the following.

(T81.1) [AUSLANDER-BUCHSBAUM] If R is local and V is a nonzero finite R-module with $\text{pdim}_R V < \infty$ then: $\text{pdim}_R V + \text{reg}_R V = \text{reg}(R)$.

(T81.2) If R is noetherian with $R \ne 0$ and (f, W, n) is a finite free R-resolution of a finite R-module V where $W_i \approx R^{d(i)}$ with $d(i) \in \mathbb{N}$ for $1 \le i \le n+1$ then:

$$(0 : V)_R \ne 0 \quad \Leftrightarrow \quad \sum_{1 \le i \le n+1} (-1)^i d(i) = 0 \quad \Leftrightarrow \quad (0 : V)_R \cap S_R(R) \ne \emptyset.$$

(T81.3) If R is noetherian and $I = (x_1, \ldots, x_n)R$ with $n \in \mathbb{N}$ where the sequence $x_1, \ldots, x_n$ is R-regular then: the (R/I)-module I/I^2 is isomorphic to $(R/I)^n$ and we have $\text{pdim}_R(R/I) = n$.

(T81.4) If R is local and I is a nonunit ideal in R such that the (R/I)-module I/I^2 is free and $\text{pdim}_R(R/I) < \infty$ then: I is generated by an R-regular sequence of length n where $n = \text{gnb}_{(R/I)}(I/I^2)$.

(T81.5) If R is local with $\dim(R) = d$ then:

$$\begin{cases} R \text{ is regular} \\ \Leftrightarrow \text{pdim}_R(R/M(R)) < \infty \\ \Leftrightarrow \text{pdim}_R(R/M(R)) = d. \end{cases}$$

(T81.6) [SERRE] If R is noetherian with $\dim(R) = d \in \mathbb{N}$ then:

$$\begin{cases} R \text{ is regular} \\ \Leftrightarrow \text{gdim}(R) = d \\ \Leftrightarrow \text{gdim}(R) < \infty \\ \Leftrightarrow \text{pdim}_R V < \infty \text{ for every finite } R\text{-module } V. \end{cases}$$

(T81.7) If R is a noetherian regular ring then its localization R_P at any prime ideal P in it is regular, and more generally its localization R_S at any multiplicative set S in it is regular.

PROOF. To prove (T81.1), assume that R is a local ring and let V be a finite R-module with $\text{pdim}_R V < \infty$. By induction on $\text{reg}(R)$ we shall show that then $\text{pdim}_R V + \text{reg}_R V = \text{reg}(R)$. In case of $\text{reg}(R) = 0$, this follows from (T80.15). So let $\text{reg}(R) > 0$ and assume for all smaller values of $\text{reg}(R)$.

First suppose that $\text{reg}_R V > 0$. Now the inequalities $\text{reg}(R) > 0$ and $\text{reg}_R V > 0$ respectively imply that $M(R) \not\subset Z_R(R)$ and $M(R) \not\subset Z_R(V)$. Consequently, in view of L4§5(O24)(22•), by (Q1) we see that $M(R) \not\subset Z_R(R) \cup Z_R(V)$. Hence there exists $x \in M(R) \cap S_R(R) \cap S_R(V)$, and therefore by (T80.13) we see that $\text{pdim}_R V = \text{pdim}_{(R/xR)}(V/xV)$. Also it is clear that $\text{reg}(R) = 1 + \text{reg}(R/xR)$ and $\text{reg}_R V = 1 + \text{reg}_{(R/xR)}(V/xV)$. Since $\text{reg}(R/xR) < \text{reg}(R)$, we are done by applying the induction hypothesis to the (R/xR)-module V/xV.

Next suppose that $\text{reg}_R V = 0$. If $\text{pdim}_R V = 0$ then by (T80.4) we would have $V \approx R^d$ for some $d \in \mathbb{N}$; since $\text{reg}(R) > 0$, there exits $x \in S_R(R)$, and for any such x we clearly have $x \in S_R R^d$, and hence $x \in S_R V$, and therefore $\text{reg}_R V > 0$, which is a contradiction. Thus we must have $\text{pdim}_R V > 0$. Since V is a finite module, we can find $e \in \mathbb{N}$ together with an R-epimorphism $f_0 : R^e \to V$. We get a short exact sequence $(f, W, 1)$ of R-modules by taking $W_0 = V$, $W_1 = R^e$, $W_2 = \ker(f_0)$, and $f_1 : W_2 \to W_1$ to be the map given by $f_1(z) = z$ for all $z \in W_2$. Consequently by (T80.4) we see that $\text{pdim}_R W_1 = 0$, and hence by (T80.10) we get $\text{pdim}_R W_2 < \infty$ with $\text{pdim}_R W_0 = 1 + \text{pdim}_R W_2$. By (T80.14) we also have $\text{reg}_R W_2 = 1$. Finally, since $\text{reg}_R W_2 > 0$, by the case considered in the above paragraph we get $\text{pdim}_R W_2 + \text{reg}_R W_2 = \text{reg}(R)$. Putting all these equations together we conclude that $\text{pdim}_R V + \text{reg}_R V = \text{reg}(R)$. This completes the proof of (T81.1). Parts (T81.2) to (T81.7) are respectively proved in items (12) to (17) of Comment (C30) below.

COMMENT (C30). [**Localization of a Module**]. In L4§7 we defined the localization R_S of a ring R at a multiplicative set S in R. Given any R-module V, we shall now generalize this to the localization V_S of V at S.

Recall that R_S is defined to be the set of all equivalence classes of pairs $(u, v) \in R \times S$ under the equivalence relation: $(u, v) \sim (u', v') \Leftrightarrow v''(uv' - u'v) = 0$ for some $v'' \in S$. The equivalence class containing (u, v) is denoted by u/v or $\frac{u}{v}$, and we add and multiply the equivalence classes by the rules $\frac{u_1}{v_1} + \frac{u_2}{v_2} = \frac{u_1v_2+u_2v_1}{v_1v_2}$ and $\frac{u_1}{v_1} \times \frac{u_2}{v_2} = \frac{u_1u_2}{v_1v_2}$. Also we "send" any $u \in R$ to $u/1 \in R_S$. This makes R_S into a ring and $u \mapsto u/1$ gives the

$(\bullet)$ $\quad \begin{cases} \text{canonical ring homomorphism } \phi : R \to R_S \\ \text{for which we have} \\ \ker(\phi) = [0 : S]_R. \end{cases}$

We define V_S to be the set of all equivalence classes of pairs $(x, s) \in V \times S$ under the equivalence relation: $(x, s) \sim (x', s') \Leftrightarrow s''(s'x - sx') = 0$ for some $s'' \in S$. The equivalence class containing (x, s) is denoted by x/s or $\frac{x}{s}$, and we add the equivalence classes by the rule $\frac{x_1}{s_1} + \frac{x_2}{s_2} = \frac{s_2x_1+s_1x_2}{s_1s_2}$ and we scalar-multiply them by the rule $\frac{u}{v} \times \frac{x}{s} = \frac{ux}{vs}$. Also we "send" any $x \in V$ to $x/1 \in V_S$. This makes V_S into a module over R_S and hence also over R, and $x \mapsto x/1$ gives the

$(\bullet\bullet)$ $\quad \begin{cases} \text{canonical } R\text{-homomorphism } \phi_V : V \to V_S \\ \text{for which we have} \\ \ker(\phi_V) = [0 : S]_V \text{ and } V_S = R_S\phi_V(V). \end{cases}$

Given any other R-module U and any R-homomorphism $F : U \to V$, clearly there exists a unique (R_S)-homomorphism $F_S : U_S \to V_S$ such that $\phi_V F = F_S\phi_U$; we call F_S the localization of F at S. If $G : T \to U$ is any other R-homomorphism then clearly $(FG)_S = F_SG_S$. For any prime ideal P in R, we may write V_P and F_P instead of $V_{(R \setminus P)}$ and $F_{(R \setminus P)}$, and call these the localizations of V and F at P, respectively.

Clearly

(1) $\quad \begin{cases} \text{if a homomorphism } F : U \to V \text{ of } R\text{-modules} \\ \text{is injective (resp: surjective, bijective) then} \\ \text{the homomorphism } F_S : U_S \to V_S \text{ of } (R_S)\text{-modules} \\ \text{is injective (resp: surjective, bijective).} \end{cases}$

Also clearly

$$(2) \quad \begin{cases} \text{if for homomorphisms } G : T \to U \text{ and } F : U \to V \\ \text{of } R\text{-modules we have im}(G) = \ker(F) \text{ then} \\ \text{for the homomorphisms } G_S : T_S \to U_S \text{ and } F_S : U_S \to V_S \\ \text{of } (R_S)\text{-modules we have im}(G_S) = \ker(F_S). \end{cases}$$

Again clearly

$$(3) \quad \begin{cases} \text{if } (V_i)_{i \in I} \text{ is any family of } R\text{-modules then there exists} \\ \text{a unique } (R_S)\text{-isomorphism } \psi : (\oplus_{i \in I} V_i)_S \to \oplus_{i \in I} (V_i)_S \\ \text{such that } \psi \phi_T = \oplus_{i \in I} \phi_{V_i} \text{ where } T = \oplus_{i \in I} V_i. \end{cases}$$

In view of (3), by (T80.2) we see that

(4)   if V is a projective R-module then V_S is a projective (R_S)-module.

We claim that $\mathrm{pdim}_R V \geq \mathrm{pdim}_{R_S} V_S$; namely, if $\mathrm{pdim}_R V = \infty$ then we have nothing to show; on the other hand, if $\mathrm{pdim}_R V = n \in \mathbb{N}$ then we can find a length-n projective R-resolution (f, W, n) of V, and by (1), (2) and (4) we see that (f_S, W_S, n) is a length-n projective (R_S)-resolution of V_S, where $(W_S)_i = (W_i)_S$ for $0 \leq i \leq n + 1$ and $(f_S)_i = (f_i)_S$ for $0 \leq i \leq n$, and hence $\mathrm{pdim}_{R_S} V_S \leq n$. Thus we have obtained a generalization of (4) saying that

(5)   if V is any R-module then $\mathrm{pdim}_R V \geq \mathrm{pdim}_{R_S} V_S$.

In (7′) we shall include a sort of converse of (4). In proving (7′) we shall use the claim which says that

$$(5') \quad \begin{cases} \text{if there is an } R\text{-epimorphism } \alpha : R^n \to V \\ \text{where } n \in \mathbb{N} \text{ and } \ker(\alpha) \text{ is a finite } R\text{-module,} \\ \text{and if } H : V_S \to U_S \text{ is an } (R_S)\text{-homomorphism} \\ \text{where } U \text{ is an } R\text{-module,} \\ \text{then for some } \sigma \in S \text{ and some } R\text{-homomorphism } G : V \to U \\ \text{we have } G_S(x) = \sigma H(x) \text{ for all } x \in V_S. \end{cases}$$

To see this let $e_i = (0, \ldots, 0, 1, 0, \ldots, 0) \in R^n$ (with 1 in the i-th spot) for $1 \leq i \leq n$ be the natural basis of R^n. Now $(H \phi_V \alpha)(e_i) \in U_S$ for $1 \leq i \leq n$, and hence we can write

$$(H \phi_V \alpha)(e_i) = u_i / \mu \in U_S \text{ with } \mu \in S \text{ and } u_i \in U \text{ for } 1 \leq i \leq n.$$

We get a unique

$$R\text{-homomorphism } F : R^n \to U \text{ with } F(e_i) = u_i \text{ for } 1 \leq i \leq n$$

and clearly, as an equation in U_S, we have

$$F(y)/s = \mu H(\alpha(y)/s) \text{ for all } y \in R^n \text{ and } s \in S.$$

Now for every $t \in \ker(\alpha)$ we have

$$F(t) \in U \text{ with } \phi_U(F(t)) \in U_S \text{ and } \phi_U(F(t)) = (H\phi_V\alpha)(t) = 0$$

and hence

$$\nu_t F(t) = 0 \text{ for some } \nu_t \in S.$$

Therefore, since $\ker(\alpha)$ is a finite R-module, we can find

$$\nu \in S \text{ with } \nu F(t) = 0 \text{ for all } t \in \ker(\alpha).$$

Since $y \mapsto \nu F(y)$ gives as R-homomorphism $R^n \to U$ and, because of the above display, its kernel contains the kernel of α, it follows that there exists a unique R-homomorphism $G : V \to U$ such that for all $y \in R^n$ we have $G(\alpha(y)) = \nu F(y)$. Let $\sigma = \nu\mu$. Then $\sigma \in S$ and for all $x = \alpha(y)/s \in V_S$ with $y \in R^n$ and $s \in S$ we have

$$G_S(x) = G_S(\alpha(y)/s) = G(\alpha(y))/s = \nu F(y)/s = \sigma H(\alpha(y)/s) = \sigma H(x).$$

A module over a ring can be regarded as a module over a subring in an obvious manner (by restricting the scalars to the subring). In particular, any module B over R_S may be regarded as a module over R, and then $\phi_B : B \to B_S$ is clearly an (R_S)-isomorphism. If A is an R-submodule of B with $R_S A = B$ then, upon letting $I : A \to B$ be the natural injection, $I_S : A_S \to B_S$ is clearly an (R_S)-isomorphism. It follows that if B is a finite (R_S)-module then, by letting A be R-generated by a finite number of (R_S)-generators of B, we get a finite R-module A such that $A_S \approx B$ as (R_S)-modules. Thus

(6)
$$\begin{cases} \text{every } (R_S)\text{-module } B \text{ is } (R_S)\text{-isomorphic to } A_S \\ \text{for some } R\text{-module } A \\ \text{and if } B \text{ is a finite } (R_S)\text{-module} \\ \text{then } A \text{ can be chosen to be a finite } R\text{-module.} \end{cases}$$

We claim that $\gdim(R) \geq \gdim(R_S)$; clearly it suffices to show that for given any finite (R_S)-module B we can find a finite R-module A with $\pdim_R A \geq \pdim_{R_S} B$; so choose A as in (6) and invoke (5). Thus we always have

(7)
$$\gdim(R) \geq \gdim(R_S).$$

As a consequence of $(5')$ we shall now prove the converse of (4) which says that

$(7')$ $\begin{cases} \text{if } R \text{ is noetherian and } V \text{ is a finite } R\text{-module then:} \\ V \text{ is a projective } R\text{-module} \\ \Leftrightarrow V_P \text{ is a projective } (R_P)\text{-module for every } P \in \mathrm{mspec}(R) \end{cases}$

and from this we shall deduce the claim, which is a generalization of $(7')$ and a refinement of (7), saying that

$(7'')$ $\begin{cases} \text{if } R \text{ is noetherian and } V \text{ is a finite } R\text{-module then} \\ \mathrm{pdim}_R V = \max\{\mathrm{pdim}_{R_P} V_P : P \in \mathrm{mspec}(R)\}. \end{cases}$

In view of (4), to prove $(7')$, assuming R to be noetherian and V to be a finite R-module such that V_P is a projective (R_P)-module for every $P \in \mathrm{mspec}(R)$, we want to show that V is a projective R-module. Since V is finitely generated, we can find $n \in \mathbb{N}$ together with an R-epimorphism $\alpha R^n \to V$. Let I be the set of all $r \in R$ such that for some R-homomorphism $G : V \to R^n$ we have $\alpha G = r_V$ where the R-homomorphism $r_V : V \to V$ is given by $z \mapsto rz$. Adding two values of G adds the corresponding values of r, taking $G = 0$ gives $r = 0$, and for any $\lambda \in R$, replacing G by the R homomorphism $V \to V$ given by $z \mapsto \lambda G(y)$ has the effect of replacing r by λr. Therefore I is an ideal in R. Given any $P \in \mathrm{mspec}(R)$, by (1) we know that $\alpha_P : (R^n)_P \to V_P$ is an (R_P)-epimorphism, and hence by (T80.2) there exists an (R_P)-homomorphism $H : V_P \to (R^n)_P$ such that $\alpha_P H$ is the identity map $V \to V$. By (3) we have $(R^n)_P \approx (R_P)^n$ and by noetherianness we know that $\ker(\alpha_P)$ is a finite (R_P)-module; therefore by $(5')$ we can find $\sigma \in R \setminus P$ together with an R-homomorphism $G : V \to R^n$ such that for all $x \in V_S$ we have $G_S(x) = \sigma H(x)$. It follows that $\alpha G = \sigma_V$. Consequently $\sigma \in I$ and hence $I \not\subset P$. This being so for all $P \in \mathrm{mspec}(R)$, by Zorn's Lemma we get $I = R$. Therefore $1 \in I$ and hence, for some R-homomorphism $G : V \to R^n$, αG is the identity map of V. Consequently by (T80.1) and (T80.2) we conclude that V is a projective R-module.

In view of (5) and $(7')$, to prove $(7'')$ it suffices to show that given any $n \in \mathbb{N}$ and any finite R-module V with $\mathrm{pdim}_R V > n$, for some $P \in \mathrm{mspec}(R)$ we have $\mathrm{pdim}_{R_P} V_P > n$. By (T80.17) we can find a finite preprojective R-resolution (f, W, n) of V such that W_{n+1} is not projective. For every $P \in \mathrm{mspec}(R)$ we get a finite preprojective (R_P)-resolution (f_P, W_P, n) of V_P where $(W_P)_i = (W_i)_P$ for $0 \leq i \leq n + 1$, and where $(f_P)_i = (f_i)_P$ for $0 \leq i \leq n$. By $(7')$ we see that $(W_P)_{n+1}$ is not projective, and hence by (T80.17) we get $\mathrm{pdim}_{R_P} V_P > n$.

Recalling that V is any R-module, and recalling the definition of the support $\mathrm{supp}_R V$, by $(\bullet\bullet)$ we see that

(8) $\qquad\qquad\qquad \mathrm{supp}_R V = \{\mathrm{spec}(R) : V_P \neq 0\}$

and by L4§13(E12) we see that

(9)   if $V = RB$ with $B \subset V$ then $\operatorname{supp}_R V = \bigcup_{b \in B} \operatorname{vspec}_R(\operatorname{ann}_R b)$.

and

(10)   if V is a finite R-module then $\operatorname{supp}_R V = \operatorname{vspec}_R(\operatorname{ann}_R V)$.

We claim that

(11) $\begin{cases} \text{assuming } R \text{ to be local and letting} \\ (f, W, n) \text{ be any exact sequence of } R\text{-modules} \\ \text{where } W_i \text{ is a finite projective } R\text{-module for } 0 \leq i \leq n+1, \\ \text{we have } \sum_{0 \leq i \leq n+1}(-1)^i \operatorname{gnb}_R W_i = 0. \end{cases}$

We prove this by induction on n. For $n = 0$ we have $W_1 \approx W_0$ and hence we have nothing to show. For $n = 1$, by (T80.1) and (T80.2) we have $W_1 \approx W_2 \oplus W_0$ and hence we are done by (C29)(2). So let $n > 1$ and assume for $n - 1$. Let $W_i' = W_i$ or $\operatorname{im}(f_{n-1})$ according as $0 \leq i \leq n - 1$ or $i = n$. Let $f_i' = f_i$ for $0 \leq i \leq n - 2$. Let $f_{n-1}' : W_n' \to W_{n-1}'$ be the natural injection. Let $W_i'' = W_n'$ or W_n or W_{n+1} according as $i = 0$ or 1 or 2. Let $f_1'' = f_n$, and let $f_0'' : W_1'' \to W_0''$ be the surjection given by putting $f_0''(x) = f_{n-1}(x)$ for all $x \in W_1''$. By (T80.8) we see that W_n' is a finite projective R-module. It follows that $(f', W', n - 1)$ and $(f'', W'', 1)$ are exact sequences of R-modules. Therefore by the induction hypothesis we have $\sum_{0 \leq i \leq n}(-1)^i \operatorname{gnb}_R W_i' = 0$ and by the $n = 1$ case we have $\sum_{0 \leq i \leq 2}(-1)^i \operatorname{gnb}_R W_i'' = 0$. By adding or subtracting the second equation from the first according as n is odd or even we get $\sum_{0 \leq i \leq n+1}(-1)^i \operatorname{gnb}_R W_i = 0$ which completes the induction.

Next we claim that

(12) $\begin{cases} \text{assuming } R \text{ to be a nonnull noetherian ring, letting} \\ (f, W, n) \text{ be any finite free } R\text{-resolution of a finite } R\text{-module } V \\ \text{where } W_i \approx R^{d(i)} \text{ with } d(i) \in \mathbb{N} \text{ for } 1 \leq i \leq n+1, \\ \text{and letting } D(V) = \sum_{1 \leq i \leq n+1}(-1)^{i+1}d(i), \\ \text{we have: } (0 : V)_R \neq 0 \Leftrightarrow D(V) = 0 \Leftrightarrow (0 : V)_R \cap S_R(R) \neq \emptyset. \end{cases}$

To see this, for any $P \in \operatorname{ass}_R(R)$, by (Q1) and L4§7.1 we have $\operatorname{reg}(R_P) = 0$. Moreover, by (1) to (3) we see that (f_P, W_P, n) is a finite free (R_P)-resolution of the finite (R_P)-module $V_P = (W_P)_0 = (W_0)_P$ where $(W_P)_i = (W_i)_P \approx (R_P)^{d(i)}$ with $d(i) \in \mathbb{N}$ for $1 \leq i \leq n+1$, and $(f_P)_i = (f_i)_P$ for $0 \leq i \leq n$. By (T80.3) this is a projective (R_P)-resolution of V_P and hence $\operatorname{pdim}_{R_P} V_P < \infty$. Therefore by (T81.1) we must have $\operatorname{pdim}_{R_P} V_P = 0$, i.e., V_P is projective. Consequently, in view of (C29)(2), by (11) we get $\mathrm{gnb}_{R_P} V_P = D(V)$. Thus

$$(12') \qquad\qquad P \in \mathrm{ass}_R(R) \Rightarrow \mathrm{gnb}_{R_P} V_P = D(V).$$

Since V_P is projective, by (T80.4) it is finite free, and hence by ($\bullet$) and ($\bullet\bullet$) we see that if $V_P \neq 0$ then $(0 : V)_R \subset [0 : P]_R$. Thus

$$(12'') \qquad\qquad P \in \mathrm{ass}_R(R) \text{ with } V_P \neq 0 \Rightarrow (0 : V)_R \subset [0 : P]_R.$$

If $D(V) = 0$ then, by (8), (10) and (12') we see that $(0 : V)_R \not\subset P$ for every $P \in \mathrm{ass}_R(R)$, and hence by L4§5(O24)(22$^\bullet$) and (Q1) we get $(0 : V)_R \cap S_R(R) \neq \emptyset$. Obviously $(0 : V)_R \cap S_R(R) \neq \emptyset \Rightarrow (0 : V)_R \neq 0$. Thus it only remains to show that $(0 : V)_R \neq 0 \Rightarrow D(V) = 0$, and hence in view of (12') it only remains to show that $(0 : V)_R \neq 0 \Rightarrow V_P = 0$ for some $P \in \mathrm{ass}_R(R)$. So suppose if possible that $(0 : V)_R \neq 0$ and $V_P \neq 0$ for all $P \in \mathrm{ass}_R(R)$. Then by (12'') we get

$$(0 : V)_R \subset \bigcap_{P \in \mathrm{ass}_R(R)} [0 : P]_R$$

but the latter intersection is 0 because by L4§6 we see that

$$(\bullet \bullet \bullet) \qquad \text{for any ideal } I \text{ in a noetherian ring } R \text{ we have } I = \bigcap_{P \in \mathrm{ass}_R(R/I)} [I : P]_R.$$

Thus we have reached a contradiction. This completes the proof of (12).

Now we claim that

$$(13) \qquad \begin{cases} \text{if } R \text{ is noetherian and } I = (x_1, \ldots, x_n)R \text{ with } n \in \mathbb{N} \\ \text{where the sequence } x_1, \ldots, x_n \text{ is } R\text{-regular,} \\ \text{then the } (R/I)\text{-module } I/I^2 \text{ is isomorphic to } (R/I)^n \\ \text{and we have } \mathrm{pdim}_R(R/I) = n. \end{cases}$$

Namely, the fact that $I/I^2 \approx (R/I)^n$ is proved in (T67.2). We prove the equation $\mathrm{pdim}_R(R/I) = n$ by induction on n. For $n = 0$ we are reduced to (T80.3). So let $n > 0$ and assume for $n - 1$. Let $W_0 = R/I$, let $W_2 = W_1 = R/(x_1, \ldots, x_{n-1})R$, let $f_1 : W_2 \to W_1$ be given by $v \mapsto x_n v$, and let $f_0 : W_1 \to W_0$ be the unique epimorphism such that $f_0 g_1 = g_0$ where $g_i : R \to W_i$ is the residue class map for $i = 0, 1$. Then $(f, W, 1)$ is a short exact sequence of R-modules and by the induction hypothesis we have $\mathrm{pdim}_R W_j = n - 1$ for $j = 1, 2$, and hence by (T80.10) we get $\mathrm{pdim}_R W_0 \leq n$, i.e., $\mathrm{pdim}_R(R/I) \leq n$. Therefore, taking $S = R \setminus P$ with $P \in \mathrm{vspec}_R I$, by (5) we get $\mathrm{pdim}_{R_P}(R/I)_P < \infty$. By L4§7.1 we see that $(R/I)_P \approx R_P/(\phi(x_1), \ldots, \phi(x_n))R_P$ where $(\phi(x_1), \ldots, \phi(x_n))$ is an (R_P)-regular sequence, and hence $\mathrm{reg}(R_P) - \mathrm{reg}_{R_P}(R/I)_P = n$. Therefore by (T81.1) we get $\mathrm{pdim}_{R_P}(R/I)_P = n$. Consequently by (5) we get $\mathrm{pdim}_R(R/I) \geq n$ and hence $\mathrm{pdim}_R(R/I) = n$. This completes the proof of (13).

Next we claim that

(14) $\begin{cases} \text{if } R \text{ to be local and } I \text{ is a nonunit ideal in } R \text{ such that} \\ \text{the } (R/I)\text{-module } I/I^2 \text{ is free and } \mathrm{pdim}_R(R/I) < \infty, \\ \text{then } I \text{ is generated by an } R\text{-regular sequence of length } n \\ \text{where } n = \mathrm{gnb}_{(R/I)}(I/I^2). \end{cases}$

We prove this by induction on n. If $n = 0$ then $I = I^2$ and hence by Nakayama (T3) we get $I = 0$. So let $n > 0$ and assume for $n - 1$. We can take elements $\overline{x}_1, x_2, \ldots, x_n$ in I such that $I = (\overline{x}_1, x_2, \ldots, x_n)R + I^2$. Let

$$J = (x_2, \ldots, x_n)R + I^2.$$

Then $J \subset I$ are ideals in R with $\mathrm{vspec}_R I = \mathrm{vspec}_R J$ and $\mathrm{gnb}_R(I/J) = 1$. Let $P_1, \ldots, P_s$ be all the distinct maximal (relative to inclusion) members of $\mathrm{ass}_R(R)$. Then $P_1, \ldots, P_s$ are prime ideals in R and by (Q1) we have $S_R(R) = R \setminus \cup_{1 \le i \le s} P_i$; moreover, in view of (T80.4), $V = R/I$ has a finite free R-resolution; therefore by (12) we get $I \not\subset \cup_{1 \le i \le s} P_i$. Consequently by the $m = 1$ case of (C31.2) below we can find $x_1 \in I \setminus \cup_{1 \le i \le s} P_i$ with $I = x_1 R + J$, and by (Q1) we get $x_1 \in S_R(R)$. Thus we have

$$I = (x_1, \ldots, x_n)R + I^2 \quad \text{with} \quad x_1 \in S_R(R).$$

Let $R' = R/(x_1 R)$ and $I' = I/(x_1 R)$. We shall show that

(i) $$\mathrm{pdim}_{R'}(R'/I') < \infty$$

and

(ii) $\quad$ the (R'/I')-module I'/I'^2 is free with $\mathrm{gnb}_{(R'/I')}(I'/I'^2) = n - 1$

and then by the induction hypothesis we will be able to conclude that I' is generated by an R'-regular sequence of length $n - 1$ and hence I is generated by an R-regular sequence of length n, and this will complete the proof of (14).

To prove (i) note that, since $\mathrm{pdim}_R(R/I) < \infty$, by (T80.10) we get $\mathrm{pdim}_R I < \infty$. Moreover, since $x_1 \in S_R(R)$, we also have $x_1 \in S_R(I)$, and hence by (T80.13) we get $\mathrm{pdim}_{R'}(I/x_1 I) < \infty$. We shall now show that as R-modules, and consequently as R'-modules, we have $I/x_1 I \approx (R/I) \oplus I'$ and therefore we get (i) by (T80.11). For establishing the said direct sum decomposition, first we note that the equation $I = (x_1 R) + J$ gives the equation $I/x_1 I = (x_1 R/x_1 I) + (J/x_1 I)$. Since $x_1 \in S_R(R)$, multiplication by x_1 gives us an isomorphism $R/I \to x_1 R/x_1 I$. Suppose we have shown that

(iii) $$(x_1 R) \cap J = x_1 I.$$

In view of the Isomorphism Theorems L4§5(O11), by (iii) we see that $J/x_1 R \approx J/((x_1 R) \cap J) \approx ((x_1 R) + J)/(x_1 R) \approx I/(x_1 R) = I'$ and $(x_1 R/x_1 I) \cap (J/x_1 I) = 0$, and hence $I/x_1 I \approx (R/I) \oplus I'$.

We shall show that (ii) and (iii) hold without assuming $x_1 \in S_R(R)$. In other words

$(14')$
$$\begin{cases} \text{if } R \text{ is local and } I \text{ is a nonunit ideal in } R \text{ such that} \\ \text{the } (R/I)\text{-module } I/I^2 \text{ is free with } \mathrm{gnb}_{(R/I)}(I/I^2) = n \in N_+ \\ \text{and } x_1, \ldots, x_n \text{ are elements in } I \text{ with } I = (x_1, \ldots, x_n)R + I^2 \\ \text{then, upon letting } R' = R/(x_1 R) \text{ and } I' = I/(x_1 R) \text{ we have that} \\ \text{the } (R'/I')\text{-module } I'/I'^2 \text{ is free with } \mathrm{gnb}_{(R'/I')}(I'/I'^2) = n - 1, \\ \text{and upon letting } J = (x_2, \ldots, x_n)R + I^2 \text{ we have } (x_1 R) \cap J = x_1 I. \end{cases}$$

In (T80.4) and (C29)(7) we have proved that

$(14'')$
$$\begin{cases} \text{the } (R/I)\text{-homomorphism } h : (R/I)^n \to I/I^2, \text{ which sends} \\ e_i = (0, \ldots, 1, 0 \ldots, 0) \in (R/I)^n \text{ (with 1 in the } i\text{-th spot) to } x_i, \\ \text{is an isomorphism.} \end{cases}$$

In view of $(14'')$, for proving $(14')$ we may take an R-epimorphism $g : I \to (R/I)^n$ whose kernel is I^2 and which sends x_i to e_i for $1 \le i \le n$. Now clearly

$$(x_1 R) \cap J = x_1 I.$$

Also clearly $g(J) = (R/I)e_2 + \cdots + (R/I)e_n$ and hence $g(J)$ is a free (R/I)-module and by (C29)(2) we have $\mathrm{gnb}_{(R/I)}g(J) = n - 1$. Moreover, there are certain obvious (R/I)-isomorphisms

$$g(J) \approx J/I^2 \approx I/((x_1 R) + I^2) \approx I'/I'^2.$$

It only remains to note that the rings R/I and R'/I' are naturally isomorphic and the structure of I'/I'^2 as an (R/I)-module coincides with its structure as an (R'/I')-module.

Now we claim that

(15)
$$\begin{cases} \text{if } R \text{ is local with } \dim(R) = d \text{ then:} \\ R \text{ is regular } \Leftrightarrow \mathrm{pdim}_R(R/M(R)) < \infty \Leftrightarrow \mathrm{pdim}_R(R/M(R)) = d. \end{cases}$$

Namely, by (T69.4) and (T69.7) we know that: R is regular $\Leftrightarrow M(R)$ is generated by an R-regular sequence $\Leftrightarrow M(R)$ is generated by an R-regular sequence of length $\dim(R)$. Therefore (15) follows by taking $I = M(R)$ in (13) and (14).

As an application of (15) let us show that

$(15')$ if R is regular local then $\mathrm{gdim}(R) = \dim(R)$.

We prove this by induction on $\dim(R)$. If $\dim(R) = 0$ then R is a field, and hence every finite R-module is free and therefore projective by (T80.3), and this gives $\text{gdim}(R) = 0$. So let $\dim(R) > 0$ and assume for all smaller values of $\dim(R)$. Given any finite R-module V, upon letting $\text{gnb}_R V = n$, we can take a short exact sequence $0 \to V' \to R^n \to V \to 0$ of R-modules. We can take an element x in $M(R)/M(R)^2$, and then by (T71) we see that $x \in S_R(R)$ and R/xR is a regular local ring whose dimension equals $\dim(R) - 1$. Since $V' \to R^n$ is injective, it follows that V' is a finite R-module with $x \in S_R(V')$. Therefore by (T80.13) we have $\text{pdim}_R V' = \text{pdim}_{(R/xR)}(V'/xV')$. Consequently by the induction hypothesis we get $\text{pdim}_R V' \le \dim(R) - 1$, and hence by (T80.3) and (T80.10) we conclude that $\text{pdim}_R V \le \dim(R)$. Therefore $\text{gdim}(R) = \dim(R)$ by (15).

Next we claim that

$$(16) \quad \begin{cases} \text{if } R \text{ is noetherian with } \dim(R) = d \in \mathbb{N} \text{ then:} \\ R \text{ is regular } \Leftrightarrow \text{ gdim}(R) = d \Leftrightarrow \text{ gdim}(R) < \infty \\ \Leftrightarrow \text{ pdim}_R V < \infty \text{ for every finite } R\text{-module } V. \end{cases}$$

We give a cyclical proof of this. Clearly

$$\dim(R) = \max\{\dim(R_P) : P \in \text{mspec}(R)\}$$

and hence, in view of (6), by (7″) and (15′) we see that

$$R \text{ is regular } \Rightarrow \text{ gdim}(R) = d.$$

Obviously

$$\text{gdim}(R) = d \Rightarrow \text{ gdim}(R) < \infty$$

and

$$\text{gdim}(R) < \infty \Rightarrow \text{ pdim}_R V < \infty \text{ for every finite } R\text{-module } V.$$

In view of (5) and (6) [or in view of (5) and the (R_P)-isomorphism $(R/P)_P \approx (R_P/M(R_P))$ for every maximal ideal P in R], by (15) we get

$$\text{pdim}_R V < \infty \text{ for every finite } R\text{-module } V \Rightarrow R \text{ is regular.}$$

Now we claim that

$$(17) \quad \begin{cases} \text{if } R \text{ is a noetherian regular ring then} \\ \text{its localization } R_P \text{ at any prime ideal } P \text{ in it is regular,} \\ \text{and more generally} \\ \text{its localization } R_S \text{ at any multiplicative set } S \text{ in it is regular.} \end{cases}$$

The second assertion follows from the first. To prove the first, given any $P \in \text{spec}(R)$ we can take $Q \in \text{mspec}(R)$ with $P \subset Q$, and then $R_P \approx (R_Q)_{P'}$ with $P' \in \text{spec}(R_Q)$ given by $P' = \phi(P)R_Q$ where $\phi : R \to R_Q$ is the natural homomorphism. Now the local ring R_Q is regular by definition, and hence the local ring R_P is regular by (7) and (16).

COMMENT (C31). [**Prime Avoidance**]. L4§5(O24)(22•) may be restated as:

(C31.1) PRIME AVOIDANCE LEMMA. Let J be an ideal in a ring R, and let $P_1, \ldots, P_n$ be prime ideals in R with $n \in \mathbb{N}_+$ such that $J \subset P_1 \cup \cdots \cup P_n$. Then for some $i \in \{1, \ldots, n\}$ we have $J \subset P_i$.

Now let us prove:

(C31.2) PRIME AVOIDANCE COROLLARY. Let R be a noetherian ring, let $J \subset I$ be ideals in R with $\mathrm{vspec}_R I = \mathrm{vspec}_R J$, let $\mathrm{gnb}_R(I/J) = m$, and let $P_1, \ldots, P_s$ be prime ideals in R with $s \in \mathbb{N}$ such that $I \not\subset P_1 \cup \cdots \cup P_s$. Then there exist elements $a_1, \ldots, a_m$ in I with $I = (a_1, \ldots, a_m)R + J$ and $a_i \notin P_1 \cup \cdots \cup P_s$ for $1 \le i \le m$ such that for every $P \in \mathrm{vspec}_R(a_1, \ldots, a_m)R \setminus \mathrm{vspec}_R I$ we have $\mathrm{ht}_R P \ge m$.

PROOF. By induction on $\mu \in \{0, 1, \ldots, m\}$ we shall find elements $a_1, \ldots, a_m$ in I with $I = (a_1, \ldots, a_m)R + J$ and $a_i \notin P_1 \cup \cdots \cup P_s$ for $1 \le i \le \mu$ such that for every $P \in \mathrm{vspec}_R(a_1, \ldots, a_\mu)R \setminus \mathrm{vspec}_R I$ we have $\mathrm{ht}_R P \ge \mu$. For $\mu = 0$ we have nothing to show. So let $\mu > 0$ and suppose we have found elements $b_1, \ldots, b_m$ in I with $I = (b_1, \ldots, b_m)R + J$ and $b_i \notin P_1 \cup \cdots \cup P_s$ for $1 \le i \le \mu - 1$ such that for every $P \in \mathrm{vspec}_R(b_1, \ldots, b_{\mu-1})R \setminus \mathrm{vspec}_R I$ we have $\mathrm{ht}_R P \ge \mu - 1$. Let $P'_1, \ldots, P'_t$ be all the distinct members of $\mathrm{nvspec}_R(b_1, \ldots, b_{\mu-1})R \setminus \mathrm{vspec}_R I$, let A be the set of maximal (relative to inclusion) members of the set $\{P_1, \ldots, P_s, P'_1, \ldots, P'_t\}$, let $B = \{P \in A : b_\mu \in P\}$, and let $C = \{P \in A : b_\mu \notin P\}$. By assumption we have $\mathrm{vspec}_R I = \mathrm{vspec}_R J$, and clearly $I \not\subset P$ for all $P \in A$; consequently $J \not\subset P$ for all $P \in A$, and hence by (C31.1) we can find $a \in J$ such that $a \notin P$ for all $P \in A$. By L4§5(O24)(21•) we see that $\cap_{P \in C} P \not\subset P$ for all $P \in B$, and hence by (C31.1) we can find $b \in \cap_{P \in C} P$ such that $b \notin P$ for all $P \in B$. Let $a_\mu = b_\mu + ab$. Then $a_\mu \notin P$ for all $P \in A$, and hence $a_\mu \notin P_i$ for $1 \le i \le s$. Also $a_\mu - b_\mu \in J$ and hence, upon letting $a_j = b_j$ for all $j \in \{1, \ldots, \mu, \mu + 1, \ldots, m\}$, we get $I = (a_1, \ldots, a_m)R + J$ with $a_i \notin P_1 \cup \cdots \cup P_s$ for $1 \le i \le \mu$. Clearly $\mathrm{vspec}_R(a_1, \ldots, a_\mu)R \setminus \mathrm{vspec}_R I \subset \mathrm{vspec}_R(b_1, \ldots, b_{\mu-1})R \setminus \mathrm{vspec}_R I$ and hence for any $P \in \mathrm{vspec}_R(a_1, \ldots, a_\mu)R \setminus \mathrm{vspec}_R I$ we have we have $P'_i \subset P$ for some $i \in \{1, \ldots, t\}$ and by hypothesis $\mathrm{ht}_R P'_i \ge \mu - 1$; but $a_\mu \notin P'_i$ and hence $\mathrm{ht}_R P \ge \mu$. This completes the induction, and proves the Corollary.

COMMENT (C32). [**History**]. As said before, the primary aim of (Q17) was to prove (T81.7) which says that a localization of a noetherian regular ring is regular. This is due to Auslander-Buchsbaum (see their 1957 paper [AuB]) and Serre (see his

1958 book [Se1]). In Comment (C45) of Quest (Q29), we shall give an alternative proof for the case of a polynomial ring over a field. This proof will be based on the Noether Normalization Theorem (T46). In my Local Analytic Geometry Book [A06], yet another proof for the case of a power series ring over a field is given which is based on the Weierstrass Preparation Theorem L3§11(T5).

QUEST (Q18) Direct Sums of Algebras, Reduced Rings, and PIRs

In §1 we studied direct sums of modules over a ring R. We shall now extend this to direct sums of R-algebras, where by an R-algebra, or an algebra over R, we mean a ring S together with a ring homomorphism $\phi : R \to S$; we call ϕ the underlying homomorphism of the said algebra, and we regard S as an R-module by taking $rs = \phi(r)s$ for all $r \in R$ and $s \in S$. Instead of saying that S is an algebra over a ring R with underlying homomorphism ϕ, we may simply say that $\phi : R \to S$ is an algebra. The underlying homomorphism need not be named explicitly.

The concept of direct sums of algebras, which will be introduced shortly, will be particularly useful in describing the total quotient rings of reduced noetherian rings, where by a reduced ring we mean a ring whose nilradical is zero. Since the nilradical coincides with the radical of the zero ideal, it follows that a ring is reduced iff it has no nonzero nilpotent element.

Direct sums will also enable us to put the notion of comaximal ideals in a proper perspective. This will be used in studying PIRs = Principal Ideal Rings which are a natural generalization of PIDs = Principal Ideal Domains.

An R-subalgebra of an R-algebra $\phi : R \to S$ is a subring S^* of S which is also an R-submodule of S; note that then automatically $\phi(R)$ is a subring of S^*. Given another R-algebra $\phi' : R \to S'$, an R-algebra-homomorphism $f : S \to S'$ is a ring homomorphism which is also an R-homomorphism. An overring R' of R can be considered as an R-algebra by taking the canonical injection $R \to R'$ to be the underlying homomorphism; note that an R-subalgebra of R' is the same thing as a subring of R' containing R. Also note that if $\phi : R \to S$ is any R-algebra then ϕ is obviously an R-algebra-homomorphism.

Let $\phi_1 : R \to R_1, \ldots, \phi_n : R \to R_n$ be R-algebras with $n \in \mathbb{N}$. Regarding $R_1, \ldots, R_n$ as R-modules, in §1 of the current Lecture we learnt how to form the (external) direct sum $R_1 \oplus \cdots \oplus R_n$. Defining componentwise multiplication, i.e., taking $(x_1, \ldots, x_n)(y_1, \ldots, y_n) = (x_1 y_1, \ldots, x_n y_n)$ for all elements x_i, y_i in R_i, the module $R_1 \oplus \cdots \oplus R_n$ becomes a ring in such a manner that (the diagonal direct sum) $\phi_1 \oplus \cdots \oplus \phi_n : R \to R_1 \oplus \cdots \oplus R_n$ is a ring homomorphism; the R-algebra with this underlying homomorphism is called the R-algebra-theoretic direct sum of $R_1, \ldots, R_n$. Let $\epsilon_1, \ldots, \epsilon_n$ be the identity elements of the rings $R_1, \ldots, R_n$ respectively. Then $(\epsilon_1, \ldots, \epsilon_n)$ is the identity element of the ring $R_1 \oplus \cdots \oplus R_n$. The natural projection $R_1 \oplus \cdots \oplus R_n \to R_i$ is an R-algebra-homomorphism. However the natural injection $R_i \to R_1 \oplus \cdots \oplus R_n$ is not a ring homomorphism if $n > 1$ and R_j is nonnull for some $j \neq i$, because then $(0, \ldots, 0, \epsilon_i, 0 \ldots, 0)$ (with ϵ_i in the i-th spot) is not the identity element of the ring $R_1 \oplus \cdots \oplus R_n$. For any subrings $S_1, \ldots, S_n$ of $R_1, \ldots, R_n$ respectively, $S_1 \oplus \cdots \oplus S_n$ is a subring of $R_1 \oplus \cdots \oplus R_n$; if $S_1, \ldots, S_n$ are R-subalgebras of $R_1, \ldots, R_n$ respectively, then $S_1 \oplus \cdots \oplus S_n$ is an R-subalgebra of $R_1 \oplus \cdots \oplus R_n$. If S is another R-algebra and $\psi_1 : S \to R_1, \ldots, \psi_n : S \to R_n$ are R-algebra-homomorphisms then $\psi_1 \oplus \cdots \oplus \psi_n : S \to R_1 \oplus \cdots \oplus R_n$ is automatically an R-algebra-homomorphism.

Every ring can uniquely be considered to be a $\mathbb{Z}$-algebra, and hence for any rings $R_1, \ldots, R_n$ we get the notion of the ring-theoretic (i.e., $\mathbb{Z}$-algebra-theoretic) direct sum of $R_1, \ldots, R_n$.

When the symbol $\oplus$ is meant to stand for ring theoretic or algebra theoretic direct sum, and not merely module theoretic direct sum, this will be clear from the context. To make it more explicit, in the former cases we may say "multiplicative direct sum" instead of "direct sum". Note that the module theoretic direct sum of an infinite family of nonnull rings $(R_i)_{i \in I}$ cannot be made into a ring by componentwise multiplication because then the element ϵ given by $\epsilon(i) = \epsilon_i$ for all $i \in I$ (which would correspond to 1) in the module theoretic direct product does not belong to the module theoretic direct sum. To stress this point we may speak of "a finite direct sum of rings" instead of "ring-isomorphic to a direct sum of rings" and then it would hardly be necessary to also say multiplicative.

QUEST (Q18.1). [**Orthogonal Idempotents and Ideals in a Direct Sum**]. Consider a direct sum of rings

$$T = R_1 \oplus \cdots \oplus R_n$$

with $n \in \mathbb{N}_+$. For $1 \leq i \leq n$ let

$$f_i : T \to R_i$$

be the natural projection. Also let

$$W = \text{set of all ideals in } T \quad \text{and} \quad V_i = \text{set of all ideals in } R_i.$$

Note that for any subset $J_i \subset R_i$ for $1 \leq i \leq n$, the cartesian product $J_1 \times \cdots \times J_n$ is the subset of T given by $\{(x_1, \ldots, x_n) \in R_1 \times \cdots \times R_n : x_i \in J_i \text{ for } 1 \leq i \leq n\}$. Now clearly

(1) $\begin{cases} J \mapsto f_i(J) \text{ gives a surjection } W \to V_i \text{ for } 1 \leq i \leq n, \\ (J_1, \ldots, J_n) \mapsto J_1 \times \cdots \times J_n \text{ gives a bijection } V_1 \times \cdots \times V_n \to W \\ \text{whose inverse is given by } J \mapsto (f_1(J), \ldots, f_n(J)), \\ \text{and for this bijection the ideal } J_1 \times \cdots \times J_n \text{ in } T \text{ is} \\ \text{prime (resp: primary, maximal) iff } J_i = R_i \text{ for all except one } i \\ \text{and for that } i \text{ the ideal } J_i \text{ is prime (resp: primary, maximal).} \end{cases}$

By way of proof, upon letting $\alpha_i : R_i \to R_i/J_i$ be the residue class epimorphism, we note that $\oplus_{1 \le i \le n} \alpha_i : T \to (R_1/J_1) \oplus \cdots \oplus (R_n/J_n)$ is an epimorphism whose kernel is $J_1 \times \cdots \times J_n$.

Now for $1 \le i \le n$ let

$$K_i = \text{QR}(R_i) \quad \text{and} \quad \overline{R}_i = \text{integral closure of } R_i \text{ in } K_i$$

and let

$$L = K_1 \oplus \cdots \oplus K_n \quad \text{and} \quad \overline{T} = \overline{R}_1 \oplus \cdots \oplus \overline{R}_n.$$

Regarding L as an overring of T, and $\overline{T}$ as a subring of L, we clearly have

$$(2) \quad \begin{cases} L = \text{QR}(T) \text{ with } \overline{T} = \text{integral closure of } T \text{ in } L \\ \text{and hence: } T \text{ is normal iff } R_i \text{ is normal for } 1 \le i \le n. \end{cases}$$

By way of proof, let $x = (x_1, \ldots, x_n) \in K_1 \times \cdots \times K_n$. If x is integral over T then an equation of integral dependence

$$x^d + \sum_{1 \le j \le d} (a_{j1}, \ldots, a_{jn}) x^{d-j} = 0$$

over T with $d \in \mathbb{N}_+$ gives an equation of integral dependence

$$x_i^d + \sum_{1 \le j \le d} a_{ji} x_i^{d-j} = 0$$

over R_i for $1 \le i \le n$. Conversely if x_i is integral over R_i for $1 \le i \le n$ then upon multiplying their equations of integral dependence by their suitable powers we may suppose the equations are of the same positive degree d, i.e.,

$$x_i^d + \sum_{1 \le j \le d} a_{ji} x_i^{d-j} = 0$$

with $a_{ji} \in R_i$ for $1 \le i \le n$. This gives

$$x^d + \sum_{1 \le j \le d} (a_{j1}, \ldots, a_{jn}) x^{d-j} = 0$$

as an equation of integral dependence for x over T.

Let ϵ_i be the identity element of R_i and let $e_i = (0, \ldots, 0, \epsilon_i, 0, \ldots, 0)$ with ϵ_i in the i-th spot. Then

$$f_i(e_i) = \epsilon_i \quad \text{and} \quad g_i(\epsilon_i) = e_i$$

where

$$g_i : R_i \to T$$

is the natural injection. Now e_i is an idempotent in the ring T, i.e., an element whose square is itself; note that an idempotent in T is integral over every subring of T; it follows that if T is contained in the total quotient ring of a normal subring then the subring contains every idempotent of T; note that by a normal subring we mean a subring which is normal. The idempotents $e_1, \ldots, e_n$ are orthogonal (to each other), i.e., $e_i e_j = 0$ whenever $i \neq j$. Also $1 = e_1 + \cdots + e_n$ where 1 is the identity element of T. The three conditions

$$1 = \sum_{1 \leq i \leq n} e_i \quad \text{with} \quad e_i^2 = e_i \text{ for all } i \quad \text{and} \quad e_i e_j = 0 \text{ for all } i \neq j$$

are collectively codified by saying that the elements $e_1, \ldots, e_n$ form a complete system of orthogonal idempotents of T. Multiplying the first equation by $t \in T$ we see that

$$t \in T \Rightarrow t = \sum_{1 \leq i \leq n} r_i e_i \quad \text{with} \quad r_i = t e_i = g_i(f_i(t e_i)).$$

It follows that if R is a subring of T with $e_i \in R$ and $f_i(R) = R_i$ for all i, then for any i we can take $s_i \in R$ with $f_i(s_i) = f_i(t)$ and this gives $r_i = g_i(f_i(t e_i)) = g_i(f_i(s_i e_i)) = s_i e_i \in R$, and hence $T = R$. Therefore, in view of (2), we see that:

(3) $\quad \begin{cases} \text{if } R \text{ is a normal subring of } T \text{ such that} \\ T \text{ is a subring of QR}(R) \text{ with } f_i(R) = R_i \text{ for } 1 \leq i \leq n, \\ \text{then } T = R \text{ and } R_i \text{ is normal for } 1 \leq i \leq n. \end{cases}$

Next we claim that:

(4) $\quad \begin{cases} \text{if } R \text{ is a nonnull reduced noetherian ring with} \\ \text{height-zero primes } Q_1, \ldots, Q_n \text{ such that for } 1 \leq i \leq n \\ \text{there is an epimorphism } \phi_i : R \to R_i \text{ with } \ker(\phi_i) = Q_i, \\ \text{then } \Phi = \phi_1 \oplus \cdots \oplus \phi_n : R \to T \text{ is a ring monomorphism} \\ \text{with } f_i(\Phi(R)) = R_i \text{ for } 1 \leq i \leq n \text{ and in a natural manner} \\ \text{QR}(T) \text{ may be identified with QR}(\Phi(R)). \end{cases}$

Namely, it is clear that Φ a monomorphism with $S_{\Phi(R)}(\Phi(R)) \subset S_T(T)$ and $f_i(\Phi(R)) = R_i$ for $1 \leq i \leq n$, and so it only remains to show that for every $t \in T$ we have $st \in \Phi(R)$ for some $s \in S_{\Phi(R)}(\Phi(R))$. Now for $1 \leq i \leq n$, by $(21^{\bullet})$ and $(22^{\bullet})$ of L4§5(O24) we can find $y_i \in (\cap_{j \in \{1, \ldots, i-1, i+1, \ldots, n\}} Q_j) \setminus Q_i$ and $z_i \in Q_i \setminus (\cup_{j \in \{1, \ldots, i-1, i+1, \ldots, n\}} Q_j)$. Let $s_i = \Phi(y_i + z_i)$. Then by (Q1) we get $s_i \in S_{\Phi(R)}(\Phi(R))$, and clearly $s_i e_i = \Phi(y_i) \in \Phi(R)$. Let $s = s_1 \ldots s_n$. Then $s \in S_{\Phi(R)}(\Phi(R))$ with $s e_i \in \Phi(R)$ for $1 \leq i \leq n$, and hence by what we have shown in the paragraph before (3) we get $st \in \Phi(R)$ for all $t \in T$.

QUEST (Q18.2). [**Localizations of Direct Sums**]. We shall now prove two claims about localizations of direct sums saying that:

(1)
$$\begin{cases} \text{if a noetherian ring } R \text{ is reduced and normal} \\ \text{then } R \text{ is a finite direct sum of normal domains} \\ \text{and for every } Q \in \mathrm{nspec}(R) \text{ the domain } R/Q \text{ is normal} \end{cases}$$

and

(2)
$$\begin{cases} \text{if a ring } R \text{ is a finite direct sum of normal domains} \\ \text{then for every } P \in \mathrm{spec}(R) \text{ the ring } R_P \text{ is a normal domain.} \end{cases}$$

Moreover, as a sort of converse of (1) and (2) we shall prove that:

(3)
$$\begin{cases} \text{if a noetherian ring } R \text{ is such that} \\ \text{for every } P \in \mathrm{mspec}(R) \text{ the ring } R_P \text{ is a normal domain} \\ \text{then } R \text{ is reduced and normal.} \end{cases}$$

Actually, the proof of (3) will be done in two parts saying that:

(4)
$$\begin{cases} \text{if a noetherian ring } R \text{ is such that} \\ \text{for every } P \in \mathrm{mspec}(R) \text{ the local ring } R_P \text{ is a domain} \\ \text{then } R \text{ is reduced} \end{cases}$$

and

(5)
$$\begin{cases} \text{if a noetherian ring } R \text{ is such that} \\ \text{for every } P \in \mathrm{mspec}(R) \text{ the local ring } R_P \text{ is normal} \\ \text{then } R \text{ is normal.} \end{cases}$$

Since the null ring is the direct sum of an empty family of normal domains, and its nspec is empty, (1) follows from (Q18.1)(3) and (Q18.1)(4).

To prove (2) assume that $R = R_1 \oplus \cdots \oplus R_n$ where $R_1, \ldots, R_n$ are normal domains with $n \in \mathbb{N}$. If $n = 0$ then R is the null ring and hence its spec is empty. So assume that $n > 0$. Given any $P \in \mathrm{spec}(R)$, by (Q18.1)(1) there exists $i \in \{1, \ldots, n\}$ together with $P_i \in \mathrm{spec}(R_i)$ such that $P = \{(x_1, \ldots, x_n) \in R_1 \times \cdots \times R_n : x_i \in P_i\}$. Let $Q_i = \{(x_1, \ldots, x_n) \in R_1 \times \cdots \times R_n : x_i = 0\}$. By (Q18.1)(1), Q_i is a height-zero prime in R, and clearly $[0 : P]_R = Q_i$. Therefore by L4§7.1, R_P is isomorphic to $(R_i)_{P_i}$ which is a normal domain by the Localization of Normality Lemma (T36).

We need not prove (3) since it follows from (4) and (5). To prove (4) and (5) assume that R is noetherian. Since the null ring is a reduced normal ring we may assume that R in nonnull. The proof of (4) and (5) will be based on the consequence of L4§7.1 which says that: ($\bullet$) if I is any ideal in R and W is any subset of spec(R), such that every $P \in \text{ass}(R/I)$ we have $P \subset W_P$ for some $W_P \in W$, then $I = \cap_{P \in W}[I : P]_R$.

Taking $I = 0$ with $W = \text{mspec}(R)$ in ($\bullet$), and noting that for every prime Q in R the ideal $[0 : Q]_R$ is the kernel of the natural homomorphism $R \to R_Q$, we see that if R_P is a domain for every $P \in \text{mspec}(R)$ then the zero ideal in R is an intersection of prime ideals and hence R is reduced, which proves (4).

To prove (5) assume that for every $P \in \text{mspec}(R)$ the local ring R_P is normal, and let there be given any $x \in R$ and $y \in S_R(R)$ such that the element x/y in QR(R) is integral over R. We shall show that then $xR \subset yR$ and this will complete the proof of (5). Let

$$(x/y)^m + a_1(x/y)^{m-1} + \cdots + a_m = 0$$

with m in $\mathbb{N}_+$ and $a_1, \ldots, a_m$ in R be an equation of integral dependence for x/y. Given any P in $\text{mspec}(R)$, upon letting $\phi : R \to R_P$ be the natural homomorphism, by L4§7.1 we see that $\phi(y) \in S_{R_P}(R_P)$. Multiplying the above equation by y^m and then applying ϕ we get

$$\phi(x)^m + \phi(a_1)\phi(x)^{m-1}\phi(y) + \cdots + \phi(a_m)\phi(y)^m = 0.$$

Dividing this by $\phi(y)^m$ we get the equation

$$(\phi(x)/\phi(y))^m + \phi(a_1)(\phi(x)/\phi(y))^{m-1} + \cdots + \phi(a_m) = 0$$

which is an equation of integral dependence over R_P for the element $\phi(x)/\phi(y)$ of QR(R_P). Therefore the normality of R_P tells us that $\phi(x) \in \phi(y)R_P$. By L4§7.1 we have

$$\phi^{-1}(\phi(y)R_P) = [yR : P]_R$$

and hence $x \in [yR : P]_R$. Now intersecting over $\text{mspec}(R)$ and invoking ($\bullet$) we get $x \in yR$.

QUEST (Q18.3). [**Comaximal Ideals and Ideal Theoretic Direct Sums**]. We shall now relate the concept of comaximal ideals in a ring R introduced in L4§5(O26) to the concept of direct sums of rings.

In continuation of (Q18.1), given any ring homomorphisms

$$\phi_i : R \to R_i \quad \text{with} \quad \ker(\phi_i) = Q_i$$

for $1 \le i \le n \in \mathbb{N}_+$, consider the ring homomorphism

$$\Phi = \phi_1 \oplus \cdots \oplus \phi_n : R \to T = R_1 \oplus \cdots \oplus R_n.$$

Then by (34•) and (35•) of L4§5(O26) we see that:

(1) $\left\{\begin{array}{l} \Phi \text{ is surjective} \\ \Leftrightarrow \phi_i \text{ is surjective for } 1 \le i \le n \text{ and} \\ \quad \text{the ideals } Q_1, \ldots, Q_n \text{ in } R \text{ are pairwise comaximal} \\ \Leftrightarrow \phi_i \text{ is surjective for } 1 \le i \le n \text{ and} \\ \quad \text{for all } i \ne j \text{ there exists } e_{ij} \in R \text{ with } \phi_i(e_{ij}) = 1 \text{ and } \phi_j(e_{ij}) = 0. \end{array}\right.$

Clearly Φ is injective iff $Q_1 \cap \cdots \cap Q_n = 0$ and hence by (1) we see that:

(2) $\left\{\begin{array}{l} \Phi \text{ is an isomorphism} \\ \Leftrightarrow \phi_i \text{ is surjective for } 1 \le i \le n \text{ and} \\ \quad \text{the ideals } Q_1, \ldots, Q_n \text{ in } R \text{ are pairwise comaximal} \\ \quad \text{with } Q_1 \cap \cdots \cap Q_n = 0. \end{array}\right.$

More generally, given any subring R^* of R let

$$\phi_i^* : R^* \to R_i^* = \phi_i(R^*) \quad \text{with} \quad \ker(\phi_i^*) = Q_i^*$$

be the ring epimorphism induced by ϕ_i, and consider the ring homomorphism

$$\Phi^* = \phi_1^* \oplus \cdots \oplus \phi_n^* : R^* \to T^* = R_1^* \oplus \cdots \oplus R_n^*$$

where we regard T^* to be a subring of T. Then (1) and (2) clearly imply their respective generalizations saying that:

(3) $\left\{\begin{array}{l} \Phi^* \text{ is surjective} \\ \Leftrightarrow \Phi(R^*) = T^* \\ \Leftrightarrow \text{the ideals } Q_1^*, \ldots, Q_n^* \text{ in } R^* \text{ are pairwise comaximal} \\ \Leftrightarrow \text{for all } i \ne j \text{ there exists } e_{ij} \in R^* \text{ with } \phi_i(e_{ij}) = 1 \text{ and } \phi_j(e_{ij}) = 0 \end{array}\right.$

and:

(4) $\left\{\begin{array}{l} \Phi^* \text{ is an isomorphism} \\ \Leftrightarrow \text{the ideals } Q_1^*, \ldots, Q_n^* \text{ in } R^* \text{ are pairwise comaximal} \\ \quad \text{with } Q_1^* \cap \cdots \cap Q_n^* = 0. \end{array}\right.$

In the above paragraph we may take $Q_1^*, \ldots, Q_n^*$ to be any ideals in any ring R^* with $n \in \mathbb{N}_+$ and we may take

$$\phi_i^* : R^* \to R_i^* = R^*/Q_i^*$$

to be the residue class epimorphism.

Let ϵ_i be the identity element of R_i^* and let $e_i = (0, \ldots, 0, \epsilon_i, 0, \ldots, 0)$ with ϵ_i in the i-th spot. Then, as noted in (Q18.1),

$$f_i(e_i) = \epsilon_i \quad \text{and} \quad g_i(\epsilon_i) = e_i$$

where

$$f_i : T^* \to R_i^* \quad \text{and} \quad g_i : R_i^* \to T^*$$

are respectively the natural projection and the natural injection. Moreover, as noted in (Q18.1), the elements $e_1, \ldots, e_n$ form a complete system of orthogonal idempotents of T^*, i.e., upon letting 1 to be the identity element of T^*, we have

$$1 = \sum_{1 \le i \le n} e_i \quad \text{with} \quad e_i^2 = e_i \text{ for all } i \quad \text{and} \quad e_i e_j = 0 \text{ for all } i \ne j.$$

Likewise, as noted in (Q18.1), if T' is any subring of T^* with $e_i \in T'$ and $f_i(T') = R_i^*$ for all i, then $T^* = T'$. For $1 \le i \le n$ let

$$A_i^* = (e_1, \ldots, e_{i-1}, e_{i+1}, \ldots, e_n)T^* \quad \text{and} \quad A_i' = e_i T^*$$

i.e., the ideals in T^* generated by the exhibited elements. Then clearly

$$(5) \quad \begin{cases} \text{for } 1 \le i \le n \text{ we have} \\ \ker(f_i) = A_i^* = A_1' + \cdots + A_{i-1}' + A_{i+1}' + \cdots + A_n' \text{ and} \\ \operatorname{im}(g_i) = A_i' = A_1^* \cap \cdots \cap A_{i-1}^* \cap A_{i+1}^* \cap \cdots \cap A_n^* \text{ with } A_i^* \cap A_i' = 0, \\ \text{the ideals } A_1^*, \ldots, A_n^* \text{ are pairwise comaximal with } A_1^* \cap \cdots \cap A_n^* = 0, \\ \text{and we have } T^* = A_1' + \cdots + A_n' \text{ with } A_i' \cap A_j' = 0 \text{ for all } i \ne j. \end{cases}$$

As a supplement to (4), upon letting

$$Q_i' = Q_1^* \cap \cdots \cap Q_{i-1}^* \cap Q_{i+1}^* \cap \cdots \cap Q_n^*$$

by (4) and (5) we see that

$$(6) \quad \begin{cases} \text{if } \Phi^* \text{ is an isomorphism then:} \\ Q_i^* = Q_1' + \cdots + Q_{i-1}' + Q_{i+1}' + \cdots + Q_n' \text{ with } Q_i^* \cap Q_i' = 0 \text{ for all } i, \\ R^* = Q_1' + \cdots + Q_n' \text{ with } Q_i' \cap Q_j' = 0 \text{ for all } i \ne j, \\ \text{and the elements } \bar{e}_1 = (\Phi^*)^{-1}(e_1), \ldots, \bar{e}_n = (\Phi^*)^{-1}(e_1) \\ \text{form a complete system of orthogonal idempotents of } R^* \\ \text{such that for } 1 \le i \le n \text{ we have} \\ Q_i^* = (\bar{e}_1, \ldots, \bar{e}_{i-1}, \bar{e}_{i+1}, \ldots, \bar{e}_n)R^* \text{ and } Q_i' = \bar{e}_i R^*. \end{cases}$$

Conversely, given any ideals $P_1', \ldots, P_n'$ in R^*, upon letting

$$P_i^* = P_1' + \cdots + P_{i-1}' + P_{i+1}' + \cdots + P_n'$$

we claim that

$$
(7) \quad
\begin{cases}
\text{if } R^* = P_1' + \cdots + P_n' \text{ with } P_i'P_j' = 0 \text{ for all } i \neq j \text{ then:} \\
P_i' = P_1^* \cap \cdots \cap P_{i-1}^* \cap P_{i+1}^* \cap \cdots \cap P_n^* \text{ with } P_i^* \cap P_i' = 0 \text{ for all } i \\
\text{and } P_1^*, \ldots, P_n^* \text{ are pairwise comaximal with } P_1^* \cap \cdots \cap P_n^* = 0.
\end{cases}
$$

Namely, by the condition

$$
R^* = P_1' + \cdots + P_n'
$$

we see that every $t \in R^*$ can be expressed as $t = \sum t_i$ with $t_i \in P_i'$ and summation over $1 \leq i \leq n$. In particular $1 = \sum e_i'$ with $e_i' \in P_i'$. Multiplying both sides of the last equation by any $s \in P_1'$ and using the condition $P_i'P_j' = 0$ we get $s = e_1's$, and hence if also $s \in P_1^*$, i.e., if $s = \sum_{i \neq 1} s_i$ with $s_i \in P_i'$ then we get $s = e_1's = \sum_{i \neq 1} e_1's_i = 0$ by the condition $P_i'P_j' = 0$. Thus $P_1^* \cap P_1' = 0$ and similarly $P_i^* \cap P_i' = 0$ for all i. Therefore by subtracting the equation $t = \sum t_i$ from any other equation $t = \sum t_i'$ with $t_i' \in P_i'$, we see that the elements t_i are uniquely determined by the element t. In particular the elements $e_1', \ldots, e_n'$ are uniquely determined and they form a complete system of orthogonal idempotents of R^*. It follows that P_i' is a ring with identity element e_i', and $t \mapsto (t_1, \ldots, t_n)$ gives a ring isomorphism $R^* \to P_1' \oplus \cdots \oplus P_n'$. Now by (4) to (6) we see that

$$
P_i' = P_1^* \cap \cdots \cap P_{i-1}^* \cap P_{i+1}^* \cap \cdots \cap P_n^*
$$

for all i, and the ideals $P_1^*, \ldots, P_n^*$ are pairwise comaximal with

$$
P_1^* \cap \cdots \cap P_n^* = 0.
$$

Referring to (4) to (7), if

$$
R^* = P_1' + \cdots + P_n' \quad \text{with} \quad P_i'P_j' = 0 \text{ for all } i \neq j
$$

then we may say that R^* is the ideal-theoretic direct sum of $P_1', \ldots, P_n'$ and indicate this by writing something like

$$
R^* = P_1' \oplus \cdots \oplus P_n' \quad \text{(ideal-theoretic direct sum)}.
$$

As another converse of (6) we see that

$$
(8) \quad
\begin{cases}
\text{if } \widehat{e}_1, \ldots, \widehat{e}_n \text{ is a complete system of orthogonal idempotents of } R^* \\
\text{then } R^* \text{ equals the ideal-theoretic direct sum } \widehat{e}_1 R^* \oplus \cdots \oplus \widehat{e}_n R^*.
\end{cases}
$$

In connection with (5) and (6) we note that

$$
(9) \qquad A_i' \cap A_j' = 0 \Rightarrow A_i'A_j' = 0 \quad \text{and} \quad Q_i' \cap Q_j' = 0 \Rightarrow Q_i'Q_j' = 0.
$$

QUEST (Q18.4). [SPIRs = Special Principal Ideal Rings]. In (R4) of L1§10 we announced a future proof of PID ⇒ UFD which we shall now give. Let us start by generalizing the notion of a PID = Principal Ideal Domain to the notion of a PIR = Principal Ideal Ring.

By a PIR we mean a ring R in which every ideal is principal, i.e., of the form xR for some $x \in R$. Every PIR is obviously noetherian. Every homomorphic image of a PIR is obviously a PIR. By an SPIR = Special PIR we mean a zero-dimensional local ring R which is a PIR. In view of L4§5(O25)(33•), by (C7) and (T7) we see that

(1)   a local ring is zero-dimensional iff its maximal ideal is nilpotent

where an ideal P in a ring R is nilpotent means $P^n = 0$ for some $n \in \mathbb{N}_+$; when it exists, the smallest such n may be called the nilpotency index of P.

In CLAIMS (2) to (14) below we shall prove some properties of PIDs, UFDs, and PIRs.

(2) Any two nonzero elements in a PID have a gcd (which by L1§10(R2) is a generator of the ideal they generate).

(3) If t is a nonzero element in a domain R such that the ideal tR is prime then the element t is irreducible. Conversely if t is an irreducible element in a domain R which is a PID or a UFD then the ideal tR is prime and we respectively have $\mathrm{dpt}_R(tR) = 0$ or $\mathrm{ht}_R(tR) = 1$.

In particular, a nonzero ideal in a PID is prime iff it is generated by an irreducible element; moreover, every such ideal is maximal.

[According to L1§7, an irreducible element in a ring is a nonzero nonunit element which is not the product of any two nonzero nonunit elements; also note that a prime ideal is maximal iff its depth is zero].

(4) A nonzero principal ideal in a UFD is prime iff it is generated by an irreducible element. A nonzero prime ideal in a UFD is principal iff it is of height one.

(5) Every nonzero nonunit element in a noetherian domain is a nonempty finite product of irreducible elements.

(6) For a noetherian domain R the following four conditions are equivalent:
(a) R is a UFD,
(b) every irreducible element in R generates a prime ideal,
(c) every height one prime ideal in R is principal, and
(d) the intersection of every pair of principal ideals in R is again principal.

(7) A domain is a PID iff it is a UFD of dimension ≤ 1; moreover, if I is any nonzero ideal in such a domain R then the ring R/I has only a finite number of ideals and hence in particular it is artinian. (Clearly a domain is of dimension ≤ 1 iff every nonzero prime ideal in it is maximal).

(8) A domain R is a PID iff there exists a map $f : R^\times \to \mathbb{N}$ such that for all elements x, y in $R^\times$:

(i) $xR \supset yR \Rightarrow f(x) \le f(y)$,

(ii) $xR = yR \Leftrightarrow xR \supset yR$ with $f(x) = f(y)$, and

(iii) $xR \not\supset yR$ and $yR \not\supset xR \Rightarrow \begin{cases} \text{for some } z \in R^\times \text{ and } a, b \text{ in } R \text{ we have} \\ z = ax + by \text{ with } f(z) < \min(f(x), f(y)). \end{cases}$

[This gives another proof of the implication ED $\Rightarrow$ PID established in L1§10(E3)].

(9) If R is an SPIR and n the index of nilpotency of $M(R)$, then $M(R)^i$ with $i = 0, 1, \ldots, n$ are exactly all the distinct ideals in R, and for any $t \in R$ with $tR = M(R)$ and any $0 \ne x \in R$ we can uniquely write $x = at^m$ with $a \in R \setminus M(R)$ and $m \in \mathbb{N}$ (clearly $\mathrm{ord}_R x = m$). Moreover, a ring R is a field iff it is an SPIR whose maximal ideal has nilpotency index 1.

(10) If R is a PIR and $P' \subset P$ are any prime ideals in R with $P' \ne P$ then P' and P are the only prime ideals in R which are contained in P, and any primary ideal in R contained in P contains P'. If R is a PIR and P' is any nonmaximal prime ideal in R then P' is the only ideal in R which is primary for P'.

(11) If R is a PIR and P' and P'' are any prime ideals in R then either $P' \subset P''$ or $P'' \subset P'$ or $P' + P'' = R$.

(12) A finite direct sum of PIRs is itself a PIR. Every PIR is isomorphic to a finite direct sum of PIDs and SPIRs. A PIR whose zero ideal is primary is an SPIR or a PID according as its dimension is zero or positive. A nonnull PIR is an SPIR or a PID $\Leftrightarrow$ its zero ideal is primary $\Leftrightarrow$ it has exactly one height-zero prime ideal.

(13) If R is a PIR and an R-module V is generated by n elements with $n \in \mathbb{N}$ then every submodule U of V is generated by n elements.

(14) If R is a PID and an R-module V has a basis of n elements with $n \in \mathbb{N}$ then every submodule U of V has a basis of m elements for some $m \le n$.

PROOF. To prove (2) and (3) let R be a PID. Given any two nonzero elements u, v in in $R^\times$, we can write $(u, v)R = wR$ for some $w \in R^\times$, and for any such w we have $u \in wR$, $v \in wR$, and $w = ru + sv$ with r, s in R, and hence w is a gcd of u, v.

To prove (3) let t be a nonzero element in a domain R.

Assuming tR to be prime, let t^*, t^{**} in $R^\times$ be such that $t = t^* t^{**}$; then the primeness of tR tells us that $t^* \in tR$ or $t^{**} \in tR$ and this respectively implies that $t^{**} \in U(R)$ or $t^* \in U(R)$ where $U(R)$ is the set of all units in R; therefore t is irreducible.

Conversely assuming R to be a PID and t to be irreducible, let t', t'' in $R^\times$ be such that $t't'' \in tR$ with $t' \notin tR$; we shall show that then $t'' \in tR$ and this will establish the primeness of tR. Since irreducible t divides the product $t't''$ without dividing t', and since by (2) the elements t, t' have a gcd, such a gcd must belong to $U(R)$. Consequently 1 is gcd of t, t' and hence again by (2) we can write $1 = r't + s't'$ with r', s' in R. Multiplying both sides by t'' we get $t'' = r'tt'' + s't't'' \in tR$. Thus tR is prime. It is also maximal, because every ideal properly containing it (i.e, containing it but different from it) is of the form dR where d divides t but t does not divide d, and hence the irreducibility of t tells us that $d \in U(R)$. Thus $\mathrm{dpt}_R(tR) = 0$.

Now conversely assuming R to be a UFD and t to be irreducible, by (D1)(A1) and (R4)(F3) of L1§10 we see that tR is prime. To prove that tR has height one, given any nonzero prime ideal P in R with $P \subset tR$ we want to show that $P = tR$. We can take a nonzero element in P and, in view of the UFD property, factor it into irreducible factors, and then by the primeness of P, one of these factors, say s, must belong to P. Now the irreducibility of s with $s \in P \subset tR$ yields $P = tR$.

This completes the proof of (3).

In view of (3), to prove (4) it suffices to show that if R is UFD and P is a height one prime ideal in R then P is principal. To see this we can take a nonzero element in P and, in view of the UFD property, factor it into irreducible factors, and then by the primeness of P, one of these factors, say t, must belong to P. By (3) the ideal tR is a nonzero prime ideal, and since $tR \subset P$ with height-one prime P we must have $tR = P$.

To prove (5) let u be a nonzero nonunit in a noetherian domain R. Let $w_1 = u$. If w_1 is not irreducible then we can find a nonzero nonunit w_2 in R such that $w_1 R \subset w_2 R$ with $w_1 R \neq w_2 R$. If w_2 is not irreducible then we can find a nonzero nonunit w_3 in R such that $w_2 R \subset w_3 R$ with $w_2 R \neq w_3 R$. And so on. Since R is noetherian this will stop after n steps for some $n \in N_+$, producing a sequence of nonzero nonunits $w_1, \ldots, w_n$ in R with w_n irreducible and $w_i R \subset w_{i+1} R$ with $w_i R \neq w_{i+1} R$ for $1 \leq i \leq n - 1$. Thus, upon letting $v_1 = w_n$ and $u_1 = u/w_n$ we have proved the FACT that $u = v_1 u_1$ with irreducible v_1 and nonzero u_1 in R such that either $u_1 = 1$ or u_1 is a nonunit (according as $n = 1$ or $n > 1$). If u_1 is a nonunit then by applying the FACT to it we can write $u = v_1 v_2 u_2$ with irreducible v_2 and nonzero u_2 in R such that either $u_2 = 1$ or u_2 is a nonunit. If u_2 is a nonunit then applying the FACT to it, and so on. Upon letting $u_0 = u$, this produces a sequence $u_0 R \subset u_1 R \subset \ldots$ with $u_0 R \neq u_1 R \neq \ldots$. Since R is noetherian this will stop after m steps for some $m \in N_+$, producing a sequence of irreducible elements $v_1, \ldots, v_m$ in R with $u = v_1 \ldots v_m$.

To prove (6) let R be a noetherian domain. Then by (4) we get (a) $\Rightarrow$ (b). In view of (D1)(A1) and (R4)(F3) of L1§10, by (5) we get (b) $\Rightarrow$ (a). By (4) we get (a) $\Rightarrow$ (c). To show that (c) $\Rightarrow$ (b), assuming (c), given any irreducible element t in R we want to prove the ideal tR is prime; by (T24) we can take a height one prime ideal Q in R with $t \in Q$, and by (c) we can write $Q = sR$ with $s \in R^\times$; now $t \in sR = Q$ with t irreducible implies $tR = Q$ and hence tR is prime. To show that (a) $\Rightarrow$ (d), assuming R is a UFD, given any nonzero elements u, v in R we want to find a nonzero element w in R such that $(uR) \cap (vR) = wR$; in view of L1§10(D4) we may take w to be an lcm of u, v. To show that (d) $\Rightarrow$ (b), assuming (d), given any irreducible element x in R we want to prove the ideal xR is prime. By (Q1) we can find a prime ideal P in R with $x \in P$ such that $(xR : y)_R = P$ for some $0 \neq y \in R$; namely, take P to be any associated prime of xR in R. The equation $(xR : y)_R = P$ yields the equation $yP = (yR) \cap (xR)$, and hence by (d) we get

$yP = zR$ for some $0 \neq z \in yR$. It follows that $z/y \in R \setminus U(R)$ with $P = (z/y)R$. Since x is irreducible with $x \in (z/y)R$, we must have $xR = (z/y)R$. Therefore the ideal xR is prime.

To prove the if part of (7) let R be a UFD of dimension ≤ 1. By (4) we see that every nonzero prime ideal in R is a principal ideal generated by an irreducible element. Given any nonzero nonunit ideal Q_0 in R we shall show that Q_0 is principal and this will complete the proof. By Zorn's Lemma Q_0 is contained in a maximal ideal which being prime, is a principal ideal generated by an irreducible element t_1. Since R is a UFD, we can write $Q_0 = t_1 Q_1$ with nonzero ideal Q_1. If $Q_1 \neq R$ then $Q_1 = t_2 Q_2$ with irreducible element t_2 and nonzero ideal Q_2, and so on. Since R is a UFD, this must stop because any fixed $s \in Q_0$ is divisible by $t_1, t_1 t_2, \ldots$. Thus $Q_n = R$ for some $n \in \mathbb{N}_+$. Thus $Q_0 = tR$ with $t = t_1 \ldots t_n$.

To prove the only if part of (7) let R be a PID. By (3), (4) and (6) we see that R is UFD of dimension ≤ 1. Since R is a PID, any nonzero ideal I in R is of the form $I = tR$ with $t \in R^\times$, and the ideals in R which contain I correspond to equivalence classes of divisors of t where two divisors are equivalent iff they are associates. Since R is UFD, these classes are finite in number. Consequently the ring R/I has only a finite number of ideals and hence in particular it is artinian.

To prove the only if part of (8) let R be a PID. By (7) R is a UFD, and for any $x \in R^\times$ we can take $f(x)$ to be the number of irreducible factors in a factorization of x, i.e., if $x = x_0 x_1^{e_1} \ldots x_n^{e_n}$ with unit x_0, pairwise nonassociate irreducibles $x_1, \ldots, x_n$, and positive integers $e_1, \ldots, e_n$, then $f(x) = e_1 + \cdots + e_n$. Now (i) and (ii) are obvious and, in view of L1§10(D4), (iii) follows by taking z to be a gcd of x, y.

To prove the if part of (8) let R be a domain with a map $f : R^\times \to \mathbb{N}$ satisfying conditions (i) to (iii). We want to show that any nonzero ideal I in R is principal. Choose $0 \neq x \in I$ such that $f(x) \leq f(y)$ for all $0 \neq y \in I$. For any $0 \neq y \in I \setminus (xR)$, by (i) and (ii) we get $x \notin yR$ and hence by (iii) we can find $0 \neq z \in I$ with $f(z) < \min(f(x), f(y))$ which is contradiction. Therefore $I = xR$.

To prove (9) let R be an SPIR and let n be the index of nilpotency of $M(R)$. Since R is a PIR we have $M(R) = tR$ for some $t \in R$. Since $M(R)^n = 0$, for any such t and any $0 \neq x \in R$, upon letting $m = \mathrm{ord}_R x$ we have $m \in \mathbb{N}$ and we can write $x = at^m$ for some $a \in R \setminus M(R)$. By (T2) this expression of x with $a \in R \setminus M(R)$ and $m \in \mathbb{N}$ is unique. Since R is a local PIR with $M(R)^n = 0 \neq M(R)^{n-1}$, by (T2), $M(R)^i = t^i R$ with $i = 0, 1 \ldots, n$ are exactly all the distinct ideals in R. It follows that a ring R is a field iff it is an SPIR whose maximal ideal has nilpotency index 1.

To prove (10) let R be a PIR, let $P' \subset P$ be prime ideals in R with $P' \neq P$, and let Q be a primary ideal in R with $Q \subset P$. We shall show that then $P' \subset Q$ and this will complete the proof of (10). Since R is a PIR, we get $P = tR$ and $P' = t'R$ with t, t' in R. Since $P' \subset P$, we have $t' = rt$ with $r \in R$. Since P' is prime with $t \notin P'$, we get $r \in P'$, i.e., $r = st'$ with $s \in R$. Thus $t' = stt'$. Since

$(1 - st)t' = 0 \in Q$ and $1 - st \notin P$, we must have $t' \in Q$, i.e., $P' \subset Q$.

To prove (11) let R be a PIR and let P' and P'' be prime ideals in R with $P' + P'' \neq R$. We want to show that then either $P' \subset P''$ or $P'' \subset P'$. By Zorn's Lemma $P' + P'' \subset P$ for some ideal P in R which is maximal and hence prime. If $P' = P$ then obviously $P'' \subset P'$. If $P' \neq P$ then by taking $Q = P''$ in (10) we see that $P' \subset P''$.

To prove the first part of (12) let $R_1, \ldots, R_n$ be PIRs with $n \in \mathbb{N}_+$ and consider their finite direct sum

$$T = R_1 \oplus \cdots \oplus R_n.$$

By (Q18.1)(1) every ideal J in T is of the form $J = J_1 \times \cdots \times J_n$ with ideal J_i in R_i for $1 \leq i \leq n$. Since R_i is a PIR we can write $J_i = s_i R_i$ with $s_i \in R_i$ and now clearly $J = (s_1, \ldots, s_n)T$. Therefore T is a PIR.

To prove the second part of (12), let R be a nonnull PIR. Let $0 = Q_1 \cap \cdots \cap Q_n$ with $n \in \mathbb{N}_+$ be an irredundant primary decomposition of the zero ideal in R and let $\mathrm{rad}_R Q_i = P_i$. If $n = 1$ then by (1) and (10) we see that R is an SPIR or a PID according as its dimension is zero or positive; conversely, if R is an SPIR or a PID then obviously $n = 1$. If $P_i \subset P_j$ for some $i \neq j$ then by taking

$$(P', P) = (P_i, P_j)$$

in (10) we would get a contradiction to the irredundancy of the decomposition. Therefore $P_1, \ldots, P_n$ are exactly all the distinct height-zero prime ideals in R. In particular, for any $i \neq j$ we have $P_i \not\subset P_j$ with $P_j \not\subset P_i$ and hence by taking

$$(P', P'') = (P_i, P_j)$$

in (11) we see that P_i and P_j are comaximal and therefore by L4§5(O26)(34$^\bullet$) so are Q_i and Q_j. For $1 \leq i \leq n$, upon letting $\phi_i : R \to R_i = R/Q_i$ be the residue class epimorphism, by the $n = 1$ case we see that R_i is either an SPIR or a PID. Considering the ring homomorphism

$$\Phi = \phi_1 \oplus \cdots \oplus \phi_n : R \to R_1 \oplus \cdots \oplus R_n$$

by (Q18.3)(2) we see that Φ is an isomorphism.

To prove (13) and (14) let R be a PIR, let V be an R-module generated by n elements $x_1, \ldots, x_n$ with $n \in \mathbb{N}$, and let U be a submodule of V. By induction on n we shall show that U is generated by n elements and if R is a domain and $x_1, \ldots, x_n$ is a basis of V then U has a basis of m elements for some $m \leq n$. If $n = 0$ then we have nothing to show. Now let $n > 0$ and assume for $n - 1$. Let V' be the submodule of V generated by $x_2, \ldots, x_n$. Then by the induction hypothesis $U \cap V'$ is generated by $n - 1$ elements $y_2, \ldots, y_n$ and if R is a domain and $x_1, \ldots, x_n$ is a basis of V then the elements $y_2, \ldots, y_n$ can be chosen so that $y_2, \ldots, y_m$ is a basis of $U \cap V'$ for some $m \leq n$ with $y_{m+1}, \ldots, y_n = 0$. Let $f : V \to V/V'$ be the residue class epimorphism and let

$$I = \{r \in R : f(rx_1) \in f(U)\}.$$

Then I is an ideal in R with

$$f(U) = If(V)$$

and, since R is a PIR, we can write $I = tR$ for some $t \in R$. Let $y_1 = tx_1$. Then $f(U) = Rf(y_1)$ and hence U is generated by $y_1, \ldots, y_n$ which completes the proof of (13). Now suppose that R is a domain, $x_1, \ldots, x_n$ is a basis of V, and the elements $y_2, \ldots, y_n$ have been chosen so that $y_2, \ldots, y_m$ is a basis of $U \cap V'$ for some $m \leq n$ with $y_{m+1}, \ldots, y_n = 0$. If $f(U) = 0$ then clearly $y_2, \ldots, y_m$ is a basis of U. So assume that $f(U) \neq 0$. Then $f(y_1) \neq 0$ and hence $t \neq 0$ and therefore the domainness of R and the basisness of $x_1, \ldots, x_n$ tells us that for any $s \in R$ we have

$$f(sy_1) = 0 \Rightarrow f(stx_1) = 0 \Rightarrow st = 0 \Rightarrow s = 0.$$

It follows that $y_1, \ldots, y_m$ is a basis of U.

QUEST (Q19) Invertible Ideals, Conditions for Normality, and DVRs

In (T65) we showed that every domain is the intersection of its localizations at its maximal ideals. In (T83) we shall supplement this by showing that a normal noetherian domain is the intersection of its localizations at its height-one primes; recall that a ring is normal means it is integrally closed in its total quotient ring. This result, which is due to Krull, will be deduced from another result (T82) which is also due to Krull and which says that in a normal noetherian domain every nonzero nonunit principal ideal is unmixed and hence, by Krull's Principal Ideal Theorem (T24), all its associated primes have height one. Recall that an ideal in a noetherian ring is unmixed means all its associated primes have the same height, i.e., equivalently, it has no embedded primes and all its minimal primes have the same height. Krull's work until 1935 is reported in his book [Kr2] of that date.

Later on Serre (see page 125 of Matsumura's book [Mat]) generalized (T82) to the case of rings with zerodivisors. We shall prove this result of Serre in (T88), which is stated in terms of the following conditions on a noetherian ring R, and which says that R is a reduced normal ring iff it satisfies conditions (R_1) and (S_2). These conditions deal with the local ring R_P for various prime ideals P in R. In (T86) and (T87) we shall discuss some variations of these conditions. For the definitions of reduced rings and direct sums of rings see (Q18). In (T86) CMR refers to a Cohen-Macaulay Ring, and in (T87) RNR refers to a Reduced Normal Ring.

SERRE CONDITION (R_n). $P \in \mathrm{spec}(R)$ with $\mathrm{ht}_R P \leq n \Rightarrow R_P$ is regular.

SERRE CONDITION (S_n). $P \in \mathrm{spec}(R)$ with $\mathrm{reg}(R_P) < n \Rightarrow R_P$ is CM.

The proof of (T82), more precisely the proof of (T82.4), is based on the three previous assertions (T82.1) to (T82.3) which deal with the notions of inverse and invertibility of ideals in a ring considered as submodules of the total quotient ring. An R-submodule I of the total quotient ring K of a ring R is said to be invertible if $IJ = R$ for some R-submodule J of K, where the product IJ of R-submodules I and J of K is defined to be the subset of K consisting of all finite sums $\sum a_i b_i$ with $a_i \in I$ and $b_i \in J$; clearly IJ is an R-submodule of K. We claim that I is invertible iff $II^{-1} = R$ where I^{-1} is defined in (C25). To see this, the if part being obvious, suppose that $IJ = R$ for some R-submodule J of K; then $y \in J \Rightarrow xy \in R$ for all $x \in I \Rightarrow y \in I^{-1}$; thus $J \subset I^{-1}$; but $J \subset I^{-1}$ and $R \subset IJ \Rightarrow R \subset II^{-1}$; therefore $II^{-1} = R$. Note that we always have $II^{-1} \subset R$, and if $I \subset R$ then $R \subset I^{-1}$ and hence $I \subset II^{-1}$.

As another application of (T82), in (T85) we shall give equivalent conditions for a ring to be a DVR (= Discrete Valuation Ring), i.e., the valuation ring of a real discrete valuation. Note that a valuation is real discrete means its value group is order isomorphic to $\mathbb{Z}$. Also note that a valuation is trivial means its value group is 0, and observe that: a valuation is trivial iff its valuation ring is a field. Finally note that a valuation is real means its value group is order isomorphic to a subgroup of $\mathbb{R}$, i.e., according L2§7(E5), it is archimedean. It follows that a valuation is real discrete iff it is a nontrivial real valuation and its value group has a smallest positive element. In the proof of part (T85.2) of (T85) we shall use the REAL VALUATION CHARACTERIZATION which says that a valuation is real iff its valuation ring has no nonzero nonmaximal prime ideal; this characterization will be proved in a subsequent Quest [cf. (Q21)(4)]. In (T84) we shall discuss generalities about normal local rings.

KRULL NORMALITY LEMMA (T82). For any noetherian domain R we have the following.

(T82.1) If P is an associated prime of a nonzero principal ideal in R then we have $P^{-1} \neq R$.

(T82.2) If R is normal and P is a nonzero maximal ideal in R with $P^{-1} \neq R$ then P is invertible.

(T82.3) If R is local and $M(R)$ is an invertible ideal then $M(R)$ is a nonzero principal ideal and hence R is a one-dimensional regular local ring.

(T82.4) If R is normal then every nonzero nonunit principal ideal in it is unmixed of height one.

(T82.5) If every nonzero nonunit principal ideal in R is unmixed then

$$R = \bigcap_{\{P \in \mathrm{spec}(R) : \mathrm{ht}_R P = 1\}} R_P$$

(where we note that if R has no height-one prime then LHS = QF(R) by (T24) and

RHS $= \mathrm{QF}(R)$ by the convention about empty intersections).

(T82.6) If R is local of dimension 0 then R is regular. If R is local of dimension ≤ 1 then R is CM. If R is normal local of dimension ≤ 1 then R is regular. If R is normal local of dimension ≥ 2 then $\mathrm{reg}(R) \geq 2$. If R is normal local of dimension ≤ 2 then R is CM.

(T82.7) If R is of dimension 0 then R is regular. If R is of dimension ≤ 1 then R is CM. If R is normal of dimension ≤ 1 then R is regular. If R is normal of dimension ≤ 2 then R is CM.

PROOF. To prove (T82.1), let $P \in \mathrm{ass}_R(R/xR)$ with $0 \neq x \in R$. Then by (Q1) we have $P \in \mathrm{tass}_R(R/xR)$, i.e., $P = (xR : y)_R$ for some $y \in R$. Now clearly $y/x \in P^{-1}$ but $y/x \notin R$.

To prove (T82.2), assume that R is normal and let P be a nonzero maximal ideal in R with $P^{-1} \neq R$. Suppose if possible that P is not invertible, i.e., $PP^{-1} \neq R$. Then, since $P \subset PP^{-1} \subset R$ and P is maximal, we must have $P = PP^{-1}$. Since R is noetherian and $P \neq 0$, for any $t \in P^{-1}$, because $PP^{-1} \subset P$, we see that P is a finite $R[t]$-module with $\mathrm{ann}_{R[t]}P = 0$, and hence t/R is integral by L4§10(E2.2)(4), and therefore $t \in R$ by the normality of R. Thus $P^{-1} \subset R$. But, since P is an ideal in R, we also have $R \subset P^{-1}$. Therefore $P^{-1} = R$ which is a contradiction.

To prove (T82.3), assume R is local and $M(R)$ is an invertible ideal. Then $M(R)$ is nonzero (indeed, in any ring R with total quotient ring K, we have

$$(R : 0)_K = K$$

and hence if 0 is invertible then $R = 0K = 0$). If $M(R) = M(R)^2$ then multiplying both sides by $M(R)^n$ we would get

$$M(R)^{n+1} = M(R)^{n+2}$$

for all $n \in \mathbb{N}$ and hence by the Krull Intersection Theorem (T7) we would get $M(R) = 0$, which would be a contradiction. Therefore

$$M(R) \neq M(R)^2$$

and hence we can take $x \in M(R) \setminus M(R)^2$. Since $x \in M(R)$, we get $xM(R)^{-1} \subset R$. If $xM(R)^{-1} \subset M(R)$ then multiplying both sides by $M(R)$, because of the invertibility of $M(R)$ we would get $x \in M(R)^2$, which would be a contradiction. Therefore we must have $xM(R)^{-1} = R$. Now multiplying both sides by $M(R)$, by the invertibility of $M(R)$ we get $xR = M(R)$, and hence by (T29) we see that R is a one-dimensional regular local ring.

To prove (T82.4), assuming R to be normal, let I be any nonzero nonunit principal ideal in R. Then for any associated prime P of I, by L4§7.1 we see that $\dim(R_P) = \mathrm{ht}_R P$ and the maximal ideal of R_P is an associated prime of the nonzero principal ideal IR_P, and hence by (T82.1) to (T82.3) we get $\mathrm{ht}_R P = 1$.

To prove (T82.5), assume that every nonzero nonunit principal ideal in R is unmixed. Clearly

$$R \subset \bigcap_{\{P \in \mathrm{spec}(R):\, \mathrm{ht}_R P = 1\}} R_P.$$

Conversely, any nonzero element in the RHS can be written as x/y with $x \neq 0 \neq y$ in R such that for every height-one prime P in R we have $x/y \in R_P$, i.e., $xR_P \subset yR_P$, and hence by L4§7.1 we get

$$[xR : P]_R = (xR_P) \cap R \subset (yR_P) \cap R = [yR : P]_R.$$

By L4§7.1, for any ideal I in R and any $W \subset \mathrm{spec}(R)$ with $\mathrm{ass}_R(R/I) \subset W$ we have $I = \cap_{P \in W}[I : P]_R$ and, in view of the Krull Principal Ideal Theorem (T24), our assumption tells us that every prime P in $\mathrm{ass}_R(R/xR) \cup \mathrm{ass}R(R/yR)$ has height one. Therefore by intersecting the above display over this set of primes we get $xR \subset yR$, i.e., $x/y \in R$.

To prove (T82.6), assume that R is local. If $\dim(R) = 0$ then R is a field and hence R is regular. If $\dim(R) = 0$ then R is CM because the empty sequence is R-regular. If $\dim(R) = 1$ then R is CM because the sequence x_1 is R-regular for every $0 \neq x_1 \in M(R)$. Assuming $\dim(R) \geq 1$ we can take $0 \neq x_1 \in M(R)$. If R is normal with $\dim(R) = 1$ then $M(R)$ is an associated prime of x_1R and hence R is regular by (T82.1) to (T82.3). If R is normal with $\dim(R) \geq 2$ then, in view of the Prime Avoidance Lemma (C31.1), by (T82.4) we can find an element $x_2 \in M(R)$ such that x_2 does not belong to any associated prime of x_1R, and now the sequence (x_1, x_2) is R-regular by (Q1), and hence $\mathrm{reg}(R) \geq 2$. It follows that if R is normal with $\dim(R) = 2$ then R is CM. This completes the proof of (T82.6).

(T82.7) follows from (T82.6) by noting that: for every $P \in \mathrm{spec}(R)$ we have $\dim(R_P) \leq \dim(R)$; by definition R is CM (resp: regular) iff R_P is CM (resp: regular) for every $P \in \mathrm{mspec}(R)$; and if R is normal then by (T36) we know that R_P is normal for every $P \in \mathrm{spec}(R)$.

KRULL NORMALITY THEOREM (T83). A noetherian domain R is normal iff:

(1)   the local ring R_P is normal for every height-one prime P in R

and

(2)
$$R = \bigcap_{\{P \in \mathrm{spec}(R):\, \mathrm{ht}_R P = 1\}} R_P$$

(where we note that if R has no height-one prime then LHS $= \mathrm{QF}(R)$ by (T24) and RHS $= \mathrm{QF}(R)$ by the convention about empty intersections).

PROOF. In view of (T36), the only if part follows from (T82.4) and (T82.5). The if part follows from the obvious fact that if, in a family of domains with a common quotient field, a particular member equals the intersection of other members which are normal, then it too is normal. To see this take an equation of integral dependence of an element of the quotient field with coefficients in the particular member. All the coefficients then belong to each of the other members. Since each of them is normal, the element must belong to them and hence to their intersection. [Recall that an equation of integral dependence of an element of an overring over the ring is a univariate polynomial with coefficients in that ring which becomes zero when evaluated at that element].

NORMAL LOCAL RING LEMMA (T84). For any local ring R we have the following.

(T84.1) If $M(R)$ is an associated prime of the zero ideal in R then $\mathrm{QR}(R) = R$ and hence R is normal.

(T84.2) If R is reduced and normal then R is a domain.

(T84.3) If R is normal and $M(R)$ is not an associated prime of the zero ideal in R then R is a domain.

PROOF. By (Q1) we see that if $M(R)$ is an associated prime of the zero ideal in R then $M(R) = Z_R(R)$ and hence $\mathrm{QR}(R) = R$ and therefore R is normal. This proves (T84.1).

Next assume that R is reduced and normal. Since R is reduced, in R we have $0 = P_1 \cap \cdots \cap P_u$ where $P_1, \ldots, P_u$ are all the distinct height zero prime ideals. Suppose if possible that $u > 1$. Then by (21$^\bullet$) and (22$^\bullet$) of L4§5(O24) we can find $b \in P_2 \cap \cdots \cap P_u$ such that $b \notin P_1$, and $a \in P_1$ such that $a \notin P_i$ for $2 \leq i \leq u$. Now $ab \in P_1 \cap \cdots \cap P_u$, i.e., $ab = 0$. Also $a + b \notin P_1 \cup \cdots \cup P_u$, and hence $a + b \in S_R(R)$ by (Q1). Therefore in $\mathrm{QR}(R)$ we can put $c = a/(a + b)$. Obviously $c^2 - c = 0$ and hence the normality of R gives $c \in R$. Now

$$a(1 - c) = bc \in P_2 \text{ and } a \notin P_2 \Rightarrow 1 - c \in P_2 \subset M(R)$$

and

$$bc = a(1 - c) \in P_1 \text{ and } b \notin P_1 \Rightarrow c \in P_1 \subset M(R)$$

and these two implications tell us that $1 \in M(R)$ which is a contradiction. This proves (T84.2).

Now assume that R is normal but, instead of assuming that R is reduced, assume that $M(R)$ is not an associated prime of 0 in R, and let b be any nilpotent element in R such that $b \in R$ with $b^u = 0$ for some $u \in \mathbb{N}_+$. We shall show that then $b = 0$ and, in view of (T84.2), this will complete the proof of (T84.3). Since $M(R)$ is not an associated prime of 0 in R, by (Q1) there exists $a \in M(R) \cap S_R(R)$. Now for all $n \in \mathbb{N}_+$ we have $b/a^n \in \mathrm{QR}(R)$ with $(b/a^n)^u = 0$, and hence $b/a^n \in R$ by the normality of R. Thus

$$b \in \bigcap_{n \in \mathbb{N}_+} a^n R \subset \bigcap_{n \in \mathbb{N}_+} M(R)^n = 0$$

where the last equality is by Krull Intersection (T7).

CONDITIONS FOR DVR (T85). For any domain R we have the following.

(T85.1) R is a DVR $\Leftrightarrow$ R is a one dimensional normal local domain $\Leftrightarrow$ R is a one dimensional regular local domain $\Leftrightarrow$ R is a positive dimensional local PID (where a local PID means a local ring which is also a principal ideal domain) $\Leftrightarrow$ R is a local PID in which $M(R)^e$ with $e = 0, 1, 2, \ldots$ are exactly all the distinct nonzero ideals.

(T85.2) If R is the valuation ring of a nontrivial valuation v then: R is noetherian $\Leftrightarrow$ v is real discrete.

(T85.3) If R is local then: R is the valuation ring of a nontrivial valuation $\Leftrightarrow$ R is regular of dimension one.

PROOF. We shall show that

$$\text{DVR} \Rightarrow \text{SLD} \Rightarrow \text{PLD} \Rightarrow \text{RLD} \Rightarrow \text{NLD} \Rightarrow \text{RLD} \Rightarrow \text{DVR}$$

where SLD stands for a local PID in which $M(R)^e$ with $e = 0, 1, 2, \ldots$ are exactly all the distinct nonzero ideals, PLD stands for positive dimensional local PID, RLD stands for one dimensional regular local domain, and NLD stands for one dimensional normal local domain, and this will prove (T85.1). Now DVR $\Rightarrow$ SLD, because given any valuation v of a field with value group $\mathbb{Z}$ and valuation ring R_v, upon taking $x \in R_v$ with $v(x) = 1$, for any nonzero ideal I in R_v we clearly have $I = (xR_v)^{v(I)}$ where $v(I) = \min\{v(y) : 0 \neq y \in I\}$ and $I \mapsto v(I)$ gives a bijection of the set of all nonzero ideals in R_v onto $\mathbb{N}$. Obviously SLD $\Rightarrow$ PLD. By the Extended Dim-Emdim Theorem (T29) we get PLD $\Rightarrow$ RLD. By the Ord Valuation Lemma (T64), we get RLD $\Rightarrow$ NLD. By the Krull Normality Lemma (T82.6) we get NLD $\Rightarrow$ RLD. By the Ord Valuation Lemma (T64), we also get RLD $\Rightarrow$ DVR, because upon letting v be the valuation ord_R we have $R_v = R$; to see this we can write $M(R) = xR$ where $x \in R$ with $v(x) = 1$, and then for any $0 \neq y \in R$ we have $y \in M(R)^{v(y)} \setminus M(R)^{1+v(y)}$ and hence $y/x^{v(y)} \in R \setminus M(R)$, and therefore for any $0 \neq z \in R$ with $v(z) \geq v(y)$ we have $z/y \in R$.

The $\Leftarrow$ part of (T85.2) follows from (T85.1). To prove the $\Rightarrow$ part, assume that R is noetherian and R is the valuation ring of a nontrivial valuation v. Then R is local. Let P be any nonmaximal prime ideal in R. By L4§12(R7)(7.5) ideals in R_v are linearly ordered and hence by (C7) we get $P \subset M(R)^n$ for all $n \in \mathbb{N}_+$, and hence $P = 0$ by Krull Intersection Theorem (T7). Therefore v is real by the REAL VALUATION CHARACTERIZATION stated in the preamble of this Quest. Suppose if possible that v is not real discrete. Then we can find an infinite sequence of nonzero elements $x_1, x_2, \ldots$ in $M(R)$ such that

$$v(x_1) > v(x_2) > \ldots .$$

This gives rise to a sequence of ideals

$$x_1 R \subset x_2 R \subset \ldots$$

in R with

$$x_1 R \neq x_2 R \neq \ldots$$

which contradicts the noetherianness of R.

Clearly (T85.3) follows from (T85.1) and (T85.2).

CONDITIONS FOR CMR (T86). For any noetherian ring R and any $n \in \mathbb{N}$ we have: $(S_n) \Leftrightarrow (S'_n) \Leftrightarrow (S''_n) \Leftrightarrow (S'''_n) \Rightarrow (S^*_n)$ where (S'_n), (S''_n), (S'''_n), and (S^*_n) refer to the following conditions.

(S'_n). $x_1, \ldots, x_m$ are elements in R with $n > m \in \mathbb{N}$ such that the ideal I_j in R generated by $x_1, \ldots, x_j$ is a nonunit ideal with $\mathrm{ht}_R I_j = j$ for $0 \leq j \leq m \Rightarrow$ the ideal I_m is unmixed.

(S''_n). $x_1, \ldots, x_m$ are elements in R with $n > m \in \mathbb{N}$ such that the ideal I_j in R generated by $x_1, \ldots, x_j$ is a nonunit ideal with $\mathrm{ht}_R I_j = j$ for $0 \leq j \leq m \Rightarrow$ the ideal I_m is unmixed and the sequence $x_1, \ldots, x_m$ is R-regular.

(S'''_n). $P \in \mathrm{spec}(R) \Rightarrow \mathrm{reg}(R_P) \geq \min(n, \mathrm{ht}_R P)$.

(S^*_n). $P \in \mathrm{spec}(R)$ with $\mathrm{ht}_R P \leq n \Rightarrow R_P$ is CM.

PROOF. The proof of $(S'''_n) \Leftrightarrow (S_n) \Rightarrow (S^*_n)$ follows by first noting that reg and ht are nonnegative integers such that $\mathrm{reg}(R_P) \leq \mathrm{ht}_R P$ with equality iff R_P is CM, and then by using some simple semantics without regard to any further meaning of these three terms. To see this, given any $P \in \mathrm{spec}(R)$, let us abbreviate $\mathrm{reg}(R_P)$ and $\mathrm{ht}_R P$ to r and h respectively. Then: $(\bullet)$ $r \leq h$. Moreover

$$\begin{cases} (S'''_n) : r \geq \min(n, h), \\ (S_n) : r < n \Rightarrow r = h, \\ (S^*_n) : h \leq n \Rightarrow r = h. \end{cases}$$

To prove $(S_n) \Rightarrow (S^*_n)$, it suffices to note that if $r < n$ then $r = h$ by (S_n), whereas if $h \leq n$ with $n \leq r$ then $h \leq r$ and hence $r = h$ by $(\bullet)$. Next, assuming (S'''_n), to prove (S_n) suppose that $r < n$; then by (S'''_n) we must have $r \geq h$ and hence by $(\bullet)$ we get $r = h$. Now, assuming (S_n), to prove (S'''_n) note that if $r \geq n$ then obviously $r \geq \min(n, h)$, whereas if $r < n$ then by (S_n) we get $r = h = \min(n, h)$.

In the rest of the proof we shall use the following consequence of (Q1):

$$(\dagger) \quad \begin{cases} \text{if } x_1, \ldots, x_m \text{ are elements in a noetherian ring } R \text{ with } m \in \mathbb{N}_+ \\ \text{such that, upon letting } I_j = (x_1, \ldots, x_j), \text{ we have that} \\ I_m \text{ is a nonunit ideal of height } m, \\ I_{m-1} \text{ is an unmixed ideal of height } m-1, \\ \text{and the sequence } x_1, \ldots, x_{m-1} \text{ is } R\text{-regular,} \\ \text{then the sequence } x_1, \ldots, x_m \text{ is } R\text{-regular.} \end{cases}$$

To see this let $P_1, \ldots, P_r$ be the associated primes of of I_{m-1}. By (Q1) we know that $x_m \in S_R(R/I_{m-1}) \Leftrightarrow x_m \notin P_i$ for $1 \leq i \leq r$. Since I_{m-1} is unmixed of height $m-1$ we get $\mathrm{ht}_R P_i = m-1$ for $1 \leq i \leq r$ and hence, because $\mathrm{ht}_R I_m = m$, we must have $x_n \notin P_i$ for $1 \leq i \leq r$.

Obviously $(S_n'') \Rightarrow (S_n')$. To prove $(S_n') \Rightarrow (S_n'')$ we make induction on n. Since there is nothing to prove for $n = 0$, suppose $n > 0$ and assume for $n-1$. Let $x_1, \ldots, x_m$ be any elements in R with $0 \leq m < n$ such that the ideal I_j in R generated by $x_1, \ldots, x_j$ is a nonunit ideal with $\mathrm{ht}_R I_j = j$ for $0 \leq j \leq m$. We are assuming that I_m is unmixed of height m, and we want to establish that that the sequence $x_1, \ldots, x_m$ is R-regular. If $m = 0$ then we have nothing to show. So let $m > 0$. Clearly $(S_n') \Rightarrow (S_{n-1}')$ and hence by applying the $n-1$ case to the elements $x_1, \ldots, x_{m-1}$ we see that I_{m-1} is an unmixed ideal of height $m-1$, and the sequence $x_1, \ldots, x_{m-1}$ is R-regular. Hence we are done by $(\dagger)$.

To prove $(S_n'') \Rightarrow (S_n''')$, assuming (S_n''), given any prime P in R with $\mathrm{ht}_R P = m$, we want to show that $\mathrm{reg}(R_P) \geq m$ or $\mathrm{reg}(R_P) \geq n$ according as $m < n$ or $n \leq m$. By Height Theorem (T28) we can find elements $x_1, \ldots, x_m$ in P such that upon letting $I_j = (x_1, \ldots, x_j)$ we have $P \in \mathrm{nass}_R(R/I_m)$ and: $(\bullet\bullet)$ for $0 \leq j \leq m$ we have $\mathrm{ht}_R I_j = j = \mathrm{ht}_R P_j$ for every $P_j \in \mathrm{nass}_R(R/I_j)$. Let $\phi : R \to R_P$ be the canonical homomorphism. If $m < n$ then by (S_n'') we see that the sequence $x_1, \ldots, x_m$ is R-regular, and hence by L4§7.1 the sequence $\phi(x_1), \ldots, \phi(x_m)$ is (R_P)-regular, and therefore $\mathrm{reg}(R_P) \geq m$. Now assume that $n \leq m$. Then by (S_n'') we see that I_{n-1} is an unmixed ideal of height $n-1$ and the sequence $x_1, \ldots, x_{n-1}$ is R-regular. Hence the sequence $x_1, \ldots, x_n$ is R-regular by $(\dagger)$. Consequently by L4§7.1 the sequence $\phi(x_1), \ldots, \phi(x_n)$ is (R_P)-regular, and therefore $\mathrm{reg}(R_P) \geq n$.

To prove $(S_n''') \Rightarrow (S_n')$ we make induction on n. Since there is nothing to prove for $n = 0$, suppose $n > 0$ and assume for $n-1$. Let there be given any elements $x_1, \ldots, x_m$ in R with $n > m \in \mathbb{N}$ such that the ideal I_j in R generated by $x_1, \ldots, x_j$ is a nonunit ideal with $\mathrm{ht}_R I_j = j$ for $0 \leq j \leq m$. We are assuming (S_n''') and we want to show that the ideal I_m is unmixed. Obviously $(S_n''') \Rightarrow (S_{n-1}''')$ and hence by the induction hypothesis we have (S_{n-1}') and therefore by what we have proved above we get (S_{n-1}''). If $m < n-1$ then I_m is unmixed by (S_{n-1}'). So assume that $m = n-1$. If $m = 0$ then $n = 1$; clearly if $R = 0$ then S_n''' and S_n' hold for all $n \in \mathbb{N}$ (the first because the spec of the null ring is empty and the second because the null ring has no nonunit ideal), and if $R \neq 0$ then S_1''' and S_1' are both equivalent to saying that the zero ideal in R is an unmixed ideal of height zero. So let $m > 0$. Then by (S_{n-1}'') we see that I_{m-1} is an unmixed ideal of height $m - 1$ and the sequence $x_1, \ldots, x_{m-1}$ is R-regular. Hence the sequence $x_1, \ldots, x_m$ is R-regular by (†). Consequently by L4§7.1, for any associated prime P of I_m, upon letting $\phi : R \to R_P$ to be the canonical homomorphism, the sequence $\phi(x_1), \ldots, \phi(x_m)$ is (R_P)-regular and $M(R_P)$ is an associated prime of $\phi(I_m)R_P$. Suppose if possible that $\mathrm{ht}_R P > m$. We shall show that this leads to a contradiction and this will complete the proof of the unmixedness of I_m. Now $\min(n, \mathrm{ht}_R P) = m + 1$ and hence by (S_n''') we have $\mathrm{reg}(R_P) > m$. Therefore by (T66.4) and (T66.5) there exists $x_{m+1}' \in M(R_P)$ such that the sequence $\phi(x_1), \ldots, \phi(x_m), x_{m+1}'$ is (R_P)-regular and hence $M(R_P)$ is not an associated prime of $\phi(I_m)R_P$, which is a contradiction.

CONDITIONS FOR RNR (T87). For any given noetherian ring R we have: $(T) \Leftrightarrow (T') \Leftrightarrow (T'') \Leftrightarrow (T''') \Rightarrow (T^*)$ where $(T), (T'), (T''), (T''')$, and (T^*) refer to the following conditions.

(T). R is a reduced normal ring.

(T'). R is a finite direct sum of normal domains.

(T''). For every $P \in \mathrm{spec}(R)$ the local ring R_P is a normal domain.

(T'''). For every $P \in \mathrm{mspec}(R)$ the local ring R_P is a normal domain.

(T^*). For every $P \in \mathrm{nspec}(R)$ the ring R/P is a normal domain.

PROOF. By (Q18.2)(1) we get $(T) \Rightarrow (T')$ and $(T) \Rightarrow (T^*)$. By (Q18.2)(2) we get $(T') \Rightarrow (T'')$. Obviously $(T'') \Rightarrow (T''')$. By (Q18.2)(3) we get $(T''') \Rightarrow (T)$.

SERRE CRITERION (T88). A noetherian ring R is a reduced ring iff it satisfies conditions (R_0) and (S_1). A noetherian ring R is a reduced normal ring iff it satisfies conditions (R_1) and (S_2).

PROOF. We may suppose $R \neq 0$ because the null ring is obviously a reduced normal ring satisfying (R_n) and (S_n) for all $n \in \mathbb{N}$. Indeed, $R = 0 \Rightarrow \mathrm{QR}(R) = R$ and $\mathrm{spec}(R) = \emptyset$.

Now first assume (R_0) and (S_1). In view of (T86) we get $(S_1) \Rightarrow (S_1') \Rightarrow$ the zero ideal in R is unmixed $\Rightarrow$ every associated prime of zero in R has height zero. Moreover $(R_0) \Rightarrow$ for every height-zero prime P in R the local ring R_P is field and hence the ideal $[0 : P]_R$, which is the kernel of the natural homomorphism $R \to R_P$, is prime. Therefore by L4§7.1 we see that the zero ideal in R is the intersection of all the height-zero primes, and hence R is reduced.

Next assume that R is a reduced ring. Then in R we have $0 = P_1 \cap \cdots \cap P_n$ where $P_1, \ldots, P_n$ are all the distinct height-zero primes, and hence the zero ideal in R is unmixed which proves (S_1') and therefore by (T86) it proves (S_1). Moreover, for $1 \leq i \leq n$ we get $R_{P_i} = (R/P_i)_0 = $ a field, which proves (R_0).

Now assume that R is a reduced normal ring. Then, in view of what we proved above, we get (R_0) and (S_1). Hence to get (R_1), it suffices to show that for any height-one prime P in R the local ring R_P is regular and, in view of (T86), to get (S_2) it suffices to show that for any associated prime Q of any height-one principal ideal I in R we have $\operatorname{ht}_R Q = 1$. Now R_P is a normal domain by (T87), and hence R_P is regular by (T85.1). Again R_Q is a normal domain by (T87), and in view of L4§7.1 we see that $M(R_Q)$ is an associated prime of the nonzero principal ideal $\phi(I)R_Q$ where $\phi : R \to R_Q$ is the natural homomorphism; consequently $\dim(R_Q) = 1$ by (T82.1) to (T82.3), and hence $\operatorname{ht}_R Q = 1$ by L4§7.1.

Next, assuming (R_1) and (S_2), we want to show that R is a reduced normal ring. Since (R_1) and (S_2) clearly imply (R_0) and (S_1) respectively, by what we have proved above we see that R is reduced. To prove the normality of R, given any x, y in R such that $y \in S_R(R)$ and x/y is integral over R, we want to show that $x/y \in R$. This being obvious when $yR = R$, assume that $yR \neq R$. Then by (T24) we see that $\operatorname{ht}_R(yR) = 1$. Let

$$(x/y)^m + a_1(x/y)^{m-1} + \cdots + a_m = 0$$

with m in $\mathbb{N}_+$ and $a_1, \ldots, a_m$ in R be an equation of integral dependence for x/y. Given any associated prime Q of yR, upon letting $\phi : R \to R_Q$ be the natural homomorphism, by L4§7.1 we see that $\phi(y) \in S_{R_Q}(R_Q)$. Multiplying the above equation by y^m and then applying ϕ we get

$$\phi(x)^m + \phi(a_1)\phi(x)^{m-1}\phi(y) + \cdots + \phi(a_m)\phi(y)^m = 0.$$

Dividing this by $\phi(y)^m$ we get the equation

$$(\phi(x)/\phi(y))^m + \phi(a_1)(\phi(x)/\phi(y))^{m-1} + \cdots + \phi(a_m) = 0$$

which is an equation of integral dependence over R_Q for the element $\phi(x)/\phi(y)$ of $\operatorname{QR}(R_Q)$. By (T86) we have $(S_2) \Rightarrow (S_2')$, and hence $\operatorname{ht}_R Q = 1$; consequently R_Q is a one dimensional regular local domain by (R_1) and hence it is normal by (T85.1); therefore the above equation tells us that

$$\phi(x)/\phi(y) \in R_Q.$$

Thus

$$\phi(x) \in \phi(y)R_Q.$$

By L4§7.1 we have

$$\phi^{-1}(\phi(y)R_Q) = [yR : Q]_R$$

and hence $x \in [yR : Q]_R$. By L4§6 we see that

$$yR = \cap_{Q \in \operatorname{ass}_R(R/yR)} [yR : Q]_R$$

and hence $x \in yR$, i.e., $x/y \in R$.

COMMENT (C33). [**Local Analytic Geometry Book**]. Most of the material of (Q18) and (Q19) is taken from Chapter III of my Local Analytic Geometry book [A06] which was published in 1964 by the Academic Press and reprinted in 2001 by the World Scientific Publishing Company. Moreover, Chapter IV of that book contains the proof of regularity of a localization of a power series ring which was cited in (C32) of (Q17).

QUEST (Q20) Dedekind Domains and Chinese Remainder Theorem

We shall now discuss the concept of a DD = Dedekind Domain which was introduced by Dedekind for unifying the theories of algebraic numbers and algebraic curves. These domains are obtained by weakening the UFD property from unique factorization of elements in the ring of integers or one variable polynomials over a field to unique factorization of ideals in their integral closures in finite algebraic extensions of their quotient fields. Officially we define a DD to be a normal noetherian domain in which every nonzero prime ideal is maximal, i.e., a normal noetherian domain of dimension at most one. By a DFI (= Domain with Factorization of Ideals) we mean a domain in which every nonzero ideal is a product of nonzero prime ideals; the factorization is into a finite family with the usual convention that the product of the empty family is the unit ideal. By a UFI (= domain with Unique Factorization of Ideals) we mean a DFI in which the said factorization is unique (up to order).

Thus in a DFI every nonzero nonunit ideal I can be expressed as

$$I = P_1^{e_1} \ldots P_h^{e_h}$$

where $P_1, \ldots, P_h$ are pairwise distinct nonzero prime ideals with positive integers h and $e_1, \ldots, e_h$. A UFI is a DFI in which I determines the set $\{P_1, \ldots, P_h\}$ and the positive integers $e_1, \ldots, e_h$.

Referring to Comment (C25) of (Q16) and the preamble of (Q19) for the concept of fractional ideals and their inverses, for any domain R let $G(R)$ denote the set of all nonzero fractional R-ideals, let $H(R)$ denote the set of all nonzero ideals in R, let $T(R)$ denote the set of all invertible members of $G(R)$, let $P(R)$ denote the set of all nonzero prime ideals in R, and let $Q(R)$ denote the set of all nonzero maximal ideals in R. Note that then $G(R)$ is a (multiplicative) commutative monoid with R as its identity element, $H(R)$ is a submonoid of $G(R)$, and $Q(R) \subset P(R) \subset H(R)$. Observe that for any $I \in T(R)$ and $n \in \mathbb{N}$, the two meanings of I^{-n} coincide, i.e., $(I^n)^{-1} = (I^{-1})^n$. Also observe that $T(R) = \mathcal{U}(G(R))$ where for any monoid G we let $\mathcal{U}(G)$ stand for the set of all $A \in G$ having an inverse in G; note that when such an inverse exists, it is unique; this follows by noting that if $AB = 1 = AC$ in G then multiplying the equation $AB = 1$ by C we get $(AC)B = C$ and hence by the equation $1 = AC$ we get $B = C$; note that $\mathcal{U}(G)$ is a subgroup of G; also note that if R is any nonnull ring then $R^\times$ is a multiplicative monoid with $\mathcal{U}(R^\times) = U(R)$.

By a GFI (= domain with Group Factorization of Ideals) we mean a domain R which is a DFI such that $G(R)$ is a group. Since for any $B \in G(R)$ we can find an $A = tR \in H(R)$ with $t \in R^\times$ such that $AB \in H(R)$, by Lemma (T90.1) below it follows that a DFI is a GFI iff every nonzero ideal in it is invertible.

Referring to §1, recall the definition of the additive abelian group $\mathbb{Z}_\oplus^P$ where P is any indexing set; we call this the free additive abelian group on P; by $\mathbb{N}_\oplus^P$ we denote the submonoid of $\mathbb{Z}_\oplus^P$ consisting of all $\phi : P \to \mathbb{Z}$ in $\mathbb{Z}_\oplus^P$ with $\phi(P) \subset \mathbb{N}$, and we call this the free additive abelian monoid on P; for any $p \in P$ let $\phi_p \in \mathbb{N}_\oplus^P$ be defined by $\phi_p(q) = 1$ or 0 according as $q = p$ or $q \neq p$. Referring to L1§1, recall the definition of a module basis; this tells us what is a basis of an additive abelian group G by regarding G as a $\mathbb{Z}$-module; when G has a basis P we say that G is a free additive abelian group or that G is an internal free additive abelian group on P; note that for any $P \subset G$ there is a unique homomorphism $\mathbb{Z}_\oplus^P \to G$ which sends ϕ_p to p for all $p \in P$, and by the nonnegative portion of (G, P) we mean the image of $\mathbb{N}_\oplus^P$ under this homomorphism; note that P is a basis of G iff the said homomorphism is an isomorphism. All this carries over to a (multiplicative) commutative group G in an obvious manner; for instance $P \subset G$ is a basis of G iff every $I \in G$ has a unique expression

$$I = P_1^{e_1} \dots P_h^{e_h}$$

with pairwise distinct elements $P_1, \dots, P_h$ in P with h in $\mathbb{N}$ and $e_1, \dots, e_h$ in $\mathbb{Z}^\times$; now the nonnegative portion of (G, P) consists of those I for which $e_1, \dots, e_h$ are positive; this includes the case $I = R$ for which $h = 0$. If such an expression exists for every $I \in G$ (without necessarily being unique) then we say that P is a generating set of G. If G is a monoid and the expressions exist with $e_1, \dots, e_h$ in $\mathbb{N}_+$ then we say that P is a semigenerating set of G.

By a PFI (= domain with Prime Factorization of Ideals) we mean a domain R which is a GFI such that $P(R)$ is a basis of $G(R)$, and $H(R)$ is the nonnegative portion of $(G(R), P(R))$.

In (T89) we characterize DDs. In (T90) we prove a lemma on commutative monoids. In (T91) we prove some properties of DDs including the fact that a semilocal DD is a PID, where by a quasisemilocal ring we mean a ring having at least one and at most a finite number of maximal ideals, and by a semilocal ring we mean a noetherian quasisemilocal ring. In (T93) we show that the

CRT = Chinese Remainder Theorem

holds in every DD, where we say that the CRT hold in a ring R to mean that, for every positive integer n, the following condition $\mathrm{CR}(n)$ is satisfied:

$$\mathrm{CR}(n) \quad \begin{cases} \text{given any elements } (x_i)_{1 \le i \le n} \text{ and ideals } (A_i)_{1 \le i \le n} \text{ in } R \\ \text{with } x_i - x_j \in A_i + A_j \text{ for all } i \ne j, \\ \text{there exists } x \in R \text{ with } x - x_i \in A_i \text{ for all } i. \end{cases}$$

In (T92) we show that CRT holds in a ring R iff $+$ and $\cap$ are distributive for ideals in R, i.e., for any ideals A, B, C in R we have

(D1) $$A + (B \cap C) = (A + B) \cap (A + C)$$

and

(D2) $$A \cap (B + C) = (A \cap B) + (A \cap C).$$

CHARACTERIZATION OF DD (T89). DD $\Leftrightarrow$ DFI $\Leftrightarrow$ UFI $\Leftrightarrow$ GFI $\Leftrightarrow$ PFI.

PROOF. Clearly it suffices to show that DD $\Rightarrow$ DFI $\Rightarrow$ PFI $\Rightarrow$ DD.

To prove DD $\Rightarrow$ DFI let R be a DD. By L4§6(T5) every nonzero ideal I in R can be expressed as $I = Q_1 \cap \cdots \cap Q_n$ where $Q_1, \ldots, Q_n$ are a finite number of primary ideals in R whose radicals are pairwise coprime height-one prime ideals in R, and hence by L4§5(O20) we get $I = Q_1 \ldots Q_n$. Thus it suffices to show that for any height-one prime P in R, every P-primary ideal Q in R is of the form P^e with $e \in \mathbb{N}_+$. By L4§5(O8) and L4§7.1(T17) we see that QR_P is $M(R_P)$-primary with $Q = (QR_P) \cap R$ and for every $e \in \mathbb{N}_+$ we have $P^e = M(R_P)^e \cap R$. Consequently it suffices to show that every $M(R_P)$-primary ideal in R_P is of the form $M(R_P)^e$ for some $e \in \mathbb{N}_+$. In view of L4§7.1(T7.1), by (Q9)(T36) we see that R_P is a one-dimensional normal local domain, and hence we are done by (T85.1).

By (T90.4) and (T90.7) below we get DFI $\Rightarrow$ PFI.

By (T90.9) below we get PFI $\Rightarrow$ DD.

LEMMA ON COMMUTATIVE MONOIDS (T90). For any commutative monoid G we have the following.

(T90.1) If a finite product in G has an inverse then so do all its factors.

(T90.2) Let $Q \subset P \subset H$, where H is a submonoid of G, be such that:

(i) for all A in P and $B_1, \ldots, B_r$ in Q with $r \in \mathbb{N}$ and $A|_H (B_1 \ldots B_r)$ (where $A \mid_H B$ means $AA' = B$ for some $A' \in H$) we have $r > 0$ and $A = B_i$ for some i.

Then for any $P_1, \ldots, P_m$ in $P \cap \mathcal{U}(G)$ and $Q_1, \ldots, Q_n$ in Q with m, n in $\mathbb{N}$ and $P_1 \ldots P_m = Q_1 \ldots Q_n$ we have $m = n$ and after suitably relabelling $Q_1, \ldots, Q_n$ we have $P_i = Q_i$ for $1 \le i \le n$.

(T90.3) Let $P \subset H$, where H is a submonoid of G, be such that:

(ii) P is a semigenerating set of H,

(iii) for every $I \in G$ there exists $J \in H$ with $IJ \in H$, and

(iv) for every $I \in P$ there is $J \in G$ with $IJ \in \mathcal{U}(G)$.

Then G is a (multiplicative) group with generating set P.

(T90.4) Let $Q \subset P \subset H$, where H is a submonoid of G, be such that the above conditions (i) to (iv) hold and we have:

(v) $P \cap \mathcal{U}(G) \subset Q$.

Then G is a (multiplicative) group with basis P, and H is the nonnegative portion of (G, P).

(T90.5) If $G = G(R)$ with $P = P(R)$ where R is a domain, then for any $P_1, \ldots, P_m$ in $P \cap \mathcal{U}(G)$ and $Q_1, \ldots, Q_n$ in P with m, n in $\mathbb{N}$ and $P_1 \ldots P_m = Q_1 \ldots Q_n$ we have $m = n$ and after suitably relabelling $Q_1, \ldots, Q_n$ we have $P_i = Q_i$ for $1 \leq i \leq n$. [This is variation of (T90.2)].

(T90.6) If $G = G(R)$ where R is a DFI then the above condition (v) is satisfied with $Q = Q(R)$, $P = P(R)$, and $H = H(R)$.

(T90.7) If $G = G(R)$ where R is a DFI then the above five conditions (i) to (v) are satisfied with $Q = Q(R)$, $P = P(R)$, and $H = H(R)$.

(T90.8) If $G = G(R)$ where R is a domain then every member of $\mathcal{U}(G)$ is a finitely generated R-module.

(T90.9) If $G = G(R)$ where R is a domain such that G is a group with basis $P = P(R)$ and $H(R)$ is the nonnegative portion of (G, P), then R is a DD.

PROOF. By induction on the number of factors, (T90.1) follows by noting that if $C = AB$ in G with C having an inverse C' then $A(BC') = CC' = 1 = CC' = B(AC')$ and hence A and B have inverses.

To prove (T90.2) we make induction on m. The $m = 0$ case being easy, let $m > 0$ and assume for all smaller values of m. Then by (i) $n > 0$ and $P_1 = Q_i$ for some i. Relabelling $Q_1, \ldots, Q_n$ suitably we can arrange that $P_1 = Q_1$. Since $P_1 \in \mathcal{U}(G)$, upon multiplying both sides of the equation $P_1 \ldots P_m = Q_1 \ldots Q_n$ by P_1^{-1} we get $P_2 \ldots P_m = Q_2 \ldots Q_n$. Now by the induction hypothesis we get $m = n$ and after suitably relabelling $Q_2, \ldots, Q_n$ we have $P_i = Q_i$ for $2 \leq i \leq n$.

To prove (T90.3) note that, in view of (T90.1), by (iv) we get $P \subset \mathcal{U}(G)$ and hence by (ii) we get $H \subset \mathcal{U}(G)$ and therefore by (iii) we get $G \subset \mathcal{U}(G)$, i.e., G is a group. Now by (ii) and (iii) we see that P is a generating set of G.

To prove (T90.4), by (T90.3) we know that P is a generating set of the group G, and for the first assertion it suffices to show that, given any finite number of distinct elements $A_1, \ldots, A_r$ in P and any integers $b_1, \ldots, b_r, c_1, \ldots, c_r$ with $A_1^{b_1} \ldots A_r^{b_r} = A_1^{c_1} \ldots A_r^{c_r}$, we have $b_i = c_i$ for $1 \leq i \leq r$. Multiplying both sides by $A_1^{-c_1} \ldots A_r^{-c_r}$ we get $A_1^{d_1} \ldots A_r^{d_r} = 1$ where $d_i = b_i - c_i$ and it suffices to show that $d_i = 0$ for $1 \leq i \leq r$. Suitably relabelling $A_1, \ldots, A_r$ we can arrange that $d_i \geq 0$ for $1 \leq i \leq m$ and $d_i < 0$ for $m + 1 \leq i \leq m + n = r$. For $1 \leq i \leq m$ let $P_i = A_i$ and $e_i = d_i$, and for $1 \leq j \leq n$ let $Q_j = A_{m+j}$ and $f_j = -d_{m+j}$. Multiplying both sides of the equation $A_1^{d_1} \ldots A_r^{d_r} = 1$ by $Q_1^{f_1} \ldots Q_n^{f_n}$ we get $P_1^{e_1} \ldots P_m^{e_m} = Q_1^{f_1} \ldots Q_n^{f_n}$ and hence by (T90.2) we conclude that $e_1 = \cdots = e_m = f_1 = \cdots = f_n = 0$. This proves the first assertion. The second assertion is clear.

To prove (T90.5) we make induction on m. The $m = 0$ case being obvious, let $m > 0$ and assume for all smaller values of m. Suitably relabelling $P_1, \ldots, P_m$ we may suppose that P_1 is minimal, i.e., $P_i \not\subset P_1$ for those i for which $P_i \neq P_1$. Now $Q_1 \ldots Q_n \subset P_1$ and hence by (C7) we see that $n > 0$ and $Q_j \subset P_1$ for some j. Also $P_1 \ldots P_m \subset Q_j$ and hence by (C7) we get $P_i \subset Q_j$ for some i. Thus $P_i \subset Q_j \subset P_1$. Since P_1 is minimal, we obtain $P_i = Q_j = P_1$. Relabelling $Q_1, \ldots, Q_n$ suitably we can arrange that $P_1 = Q_1$. Since $P_1 \in \mathcal{U}(G)$, upon multiplying both sides of the equation $P_1 \ldots P_m = Q_1 \ldots Q_n$ by P_1^{-1} we get $P_2 \ldots P_m = Q_2 \ldots Q_n$. Now by the induction hypothesis we get $m = n$ and after suitably relabelling $Q_2, \ldots, Q_n$ we have $P_i = Q_i$ for $2 \leq i \leq n$.

To prove (T90.6) and (T90.7) assume that $G = G(R)$ where R is a DFI. Then (i) follows from (C7). (ii) is the definition of DFI. (iii) follows by the definition of a fractional R-ideal; indeed we can take $J = tR$ for some $t \in R^\times$. Suppose we have proved (v). Then to prove (iv) we can take a nonzero principal ideal L in R with $L \subset I$ and by (ii) write L as a finite product $L = P_1 \ldots P_m$ with $P_1, \ldots, P_m$ in P. Since $P_1 \ldots P_m \subset I$, by (C7) we get $m > 0$ and $P_i \subset I$ for some i. Clearly $L \in \mathcal{U}(G)$ and hence by (T90.1) we have $P_i \in \mathcal{U}(G)$ for all i. Therefore by (v) we get $P_i \in Q$ for all i. Consequently $P_i = I$ for some i. This proves (iv).

Thus we only have to prove (v). So let there be given any $A \in P \cap \mathcal{U}(R)$ and any $t \in R \setminus A$. Then we can write $A + tR = P_1 \ldots P_m$ and $A + t^2R = Q_1 \ldots Q_n$ for some m, n in $\mathbb{N}$ and nonzero prime ideals $P_1, \ldots, P_m, Q_1, \ldots, Q_n$ in R. Consider the residue class epimorphism $\phi : R \to S = R/A$ with $u = \phi(t)$ and let $\widehat{P}_i = \phi(P_i)$ and $\widehat{Q}_j = \phi(Q_j)$. Then $uS = \widehat{P}_1 \ldots \widehat{P}_m$ and $u^2S = \widehat{Q}_1 \ldots \widehat{Q}_n$ where $\widehat{P}_1, \ldots, \widehat{P}_m, \widehat{Q}_1, \ldots, \widehat{Q}_n$ are nonzero prime ideals in the domain S, and they are invertible by (T90.1). Clearly $\widehat{P}_1^2 \ldots \widehat{P}_m^2 = \widehat{Q}_1 \ldots \widehat{Q}_n$, and hence by (T90.5) we see that $n = 2m$ and, after suitably relabelling $Q_1, \ldots, Q_n$, for $1 \leq i \leq m$ we have $\widehat{Q}_{2i} = \widehat{Q}_{2i-1} = \widehat{P}_i$. Consequently for $1 \leq i \leq m$ we have $Q_{2i} = Q_{2i-1} = P_i$, and hence

$$A + t^2R = Q_1 \ldots Q_n = P_1^2 \ldots P_m^2 = (A + tR)^2.$$

Thus

$$A \subset A + t^2R = (A + tR)^2 \subset A^2 + tR$$

and hence every $x \in A$ can be written as $x = y + tz$ for some $y \in A^2$ and $z \in R$. Now $tz \in A$ with $t \notin A$ and hence $z \in A$. Thus $A = A^2 + tA$. Since A is invertible, multiplying the last equation by A^{-1} we get $A + tR = R$. This proves that $A \in Q$.

To prove (T90.8) assume that $G = G(R)$ where R is a domain, and let $A \in \mathcal{U}(G)$. Then we can express 1 as a finite sum

$$1 = \sum_{1 \leq i \leq n} x_i y_i$$

with $x_i \in A$ and $y_i \in A^{-1}$. Multiplying both sides by any $x \in A$ we get

$$x = \sum_{1 \le i \le n} x_i(xy_i)$$

with $xy_i \in R$. Thus $x_1, \ldots, x_n$ is a finite system of generators of A.

To prove (T90.9) note that R is noetherian by (T90.8), and R is of dimension ≤ 1 by (T90.6). To show that R is normal, in view of (T36) and (T85.1), it suffices to express R as an intersection of DVRs with quotient field $\mathrm{QF}(R)$. For every $t \in \mathrm{QF}(R)^\times$ we have a unique expression $tR = \prod_{I \in P} I^{V(I,t)}$ where $V(I,t)$ are integers at most a finite number of which are nonzero. Clearly $t \mapsto V(I,t)$ gives a real discrete valuation of $\mathrm{QF}(R)$ and upon letting $D(I)$ denote its valuation ring we have $R = \cap_{I \in P} D(I)$.

PROPERTIES OF DD (T91). Let R be any DD. Then we have the following.
(T91.1) If R is semilocal then R is a PID.
(T91.2) If I is any nonzero ideal in R then R/I is a PIR.
(T91.3) Every ideal in R is generated by two elements.
(T91.4) The localization of R at any multiplicative set in R is again a DD.

PROOF. To prove (T91.1) assume that R is a positive dimensional semilocal domain and let $P_1, \ldots, P_n$ be the distinct maximal ideals in R. If $P_i^2 = P_i$ for some i then extending these ideals to R_{P_i} we would get $M(R_{P_i})^2 = M(R_{P_i})$ in contradiction to Nakayama (T3). Therefore $P_i^2 \neq P_i$ and hence we can take $x_i \in P_i \setminus P_i^2$. In view of L4§5(O20), by comaximality we can find $y_i \in R$ with $y_i - x_i \in P_i^2$ such that $y_i - 1 \in P_j$ for all $j \neq i$. By (T89) we can uniquely express $y_i R$ as a product of nonnegative powers of $P_1, \ldots, P_n$, and hence we must have $y_i R = P_i$. Again by (T89) every nonzero ideal I in R can be expressed as $I = P_1^{e_1} \ldots P_n^{e_n}$ with $e_1, \ldots, e_n$ in $\mathbb{N}$. It follows that $y_1^{e_1} \ldots y_n^{e_n}$ is a generator of I.

To prove (T91.2) let I be any nonzero nonunit ideal in R. Then by (T89) we can write $I = P_1^{e_1} \ldots P_n^{e_n}$ where $n, e_1, \ldots, e_n$ are positive integers and $P_1, \ldots, P_n$ are distinct maximal ideals in R. In view of L4§5(O20), by (Q18.3)(2) we see that R/I is isomorphic to the finite direct sum $(R/P_1^{e_1}) \oplus \cdots \oplus (R/P_n^{e_n})$ and hence, in view of (Q18.4)(12), it suffices to show that $R/P_i^{e_i}$ is a PIR for every i. As above we can take $x_i \in P_i \setminus P_i^2$. Then $P_i^j = (x_i^j R) + P_i^{e_i}$ for $0 \le j \le e_i$, and hence $(P_i^j/P_i^{e_i})_{0 \le j \le e_i}$ are principal ideals in $R/P_i^{e_i}$. Since by (T89) these are the only ideals in $R/P_i^{e_i}$, we conclude that $R/P_i^{e_i}$ is a PIR.

To prove (T91.3) let J be any nonzero nonunit ideal in R. Take any $0 \neq t \in J$ and let $I = tR$. Then R/I is a PIR by (T91.2) and hence J/I is generated by the I-residue of some $s \in J$. It follows that $J = (s,t)R$.

In view of L4§7(T13), by (T36) we get (T91.4).

CHINESE REMAINDER LEMMA (T92). CRT holds in a ring R iff $+$ and $\cap$

are distributive for ideals in R. In greater detail, $CR(n \leq 2)$ is always true and we have $D1 \Rightarrow CRT \Rightarrow CR(n \leq 3) \Rightarrow D1 \Rightarrow D2$, where $CR(n \leq m)$ means condition $CR(n)$ holds for all $n \leq m$, and Di means condition (Di) holds for all ideals A, B, C in R.

PROOF. In $CR(n)$, for $n = 1$ we can take $x = x_1$, and for $n = 2$ we can write $x_1 - x_2 = y_1 - y_2$ with $y_i \in A_i$ for $1 \leq i \leq 2$ and take $x = x_1 - y_1 = x_2 - y_2$.

To prove $D1 \Rightarrow CRT$, we need to show that $D1 \Rightarrow CR(n)$ for all n, which we do by induction on n. In view of what we have said above, assuming $D1$ and $CR(n-1)$ for some $n \geq 2$, we want to prove $CR(n)$. So given any elements $(x_i)_{1 \leq i \leq n}$ and ideals $(A_i)_{1 \leq i \leq n}$ in R with $x_i - x_j \in A_i + A_j$ for all $i \neq j$, we have to find $x \in R$ with

$$x - x_i \in A_i \quad \text{for} \quad 1 \leq i \leq n.$$

By the induction hypothesis we can find $x' \in R$ with

$$x' - x_i \in A_i \quad \text{for} \quad 1 \leq i \leq n-1.$$

Now for $1 \leq i \leq n-1$ we have

$$x' - x_n = (x' - x_i) + (x_i - x_n) \in A_i + (A_i + A_n) = A_i + A_n$$

and hence

$$x' - x_n \in (A_1 + A_n) \cap \cdots \cap (A_{n-1} + A_n).$$

By repeated applications of $D1$ we get

$$A_n + (A_1 \cap \cdots \cap A_{n-1}) = (A_1 + A_n) \cap \cdots \cap (A_{n-1} + A_n).$$

The above two displays yield

$$x_n - x' \in A_n + (A_1 \cap \cdots \cap A_{n-1}).$$

Now the above display together with $CR(2)$ tells us that for some $x \in R$ we have $x - x_n \in A_n$ and $x - x' \in A_1 \cap \cdots \cap A_{n-1}$. Therefore, since $x' - x_i \in A_i$ for $1 \leq i \leq n-1$, we conclude that $x - x_i \in A_i$ for $1 \leq i \leq n$.

The implication $CRT \Rightarrow CR(n \leq 3)$ being obvious, it remains to show that $CR(n \leq 3) \Rightarrow D1 \Rightarrow D2$.

To prove $CR(n \leq 3) \Rightarrow D1$, assuming $CR(3)$ and noting that the LHS of $(D1)$ is obviously contained in its RHS, we want to establish the opposite inclusion. So given any ideals A, B, C in the ring R and any element y of $(A + B) \cap (A + C)$ we want to show that y belongs to $A + (B \cap C)$, i.e., we want to find $x \in A$ with $x - y \in B \cap C$. This is equivalent to finding $x \in R$ such that $x - 0 \in A$, $x - y \in B$, and $x - y \in C$. Since $0 - y \in A + B$, $0 - y \in A + C$, and $y - y \in B + C$, we are reduced to $CR(3)$.

To prove D1 $\Rightarrow$ D2, in view of what we have already proved, it suffices to show that CR(4) $\Rightarrow$ D2. So assuming CR(4) and noting that the RHS of (D2) is obviously contained in its LHS, we want to establish the opposite inclusion. So given any ideals A, B, C in the ring R and any element y of $A \cap (B + C)$ we want to show that y belongs to $(A \cap B) + (A \cap C)$, i.e., we want to find $x \in A \cap B$ with $x - y \in A \cap C$. This is equivalent to finding $x \in R$ such that $x - 0 \in A$, $x - 0 \in B$, $x - y \in A$, and $x - y \in C$. Since

$$\begin{cases} 0 - 0 \in A + B,\ 0 - y \in A + A,\ 0 - y \in A + C, \\ 0 - y \in B + A,\ 0 - y \in B + C,\ y - y \in A + C, \end{cases}$$

we are reduced to CR(4).

CHINESE REMAINDER THEOREM (T93). CRT holds in every DD.

PROOF. Let R be a DD. By (T89), for every nonzero ideal A in R we have a unique expression

$$A = \prod_{I \in P(R)} I^{V(I,A)}$$

where $V(I, A)$ are integers at most a finite number of which are nonzero. To prove (D1) and (D2), since they are obvious when one of the three ideals is zero, let A, B, C be any nonzero ideals in R. Then (D2) and (D1) are respectively equivalent to saying that for every $I \in P(R)$ we have the relations

$$\begin{cases} \max\{V(I,A), \min\{V(I,B), V(I,C)\}\} \\ = \min\{\max\{V(I,A), V(I,B)\}, \max\{V(I,A), V(I,C)\}\} \end{cases}$$

and

$$\begin{cases} \min\{V(I,A), \max\{V(I,B), V(I,C)\}\} \\ = \max\{\min\{V(I,A), V(I,B)\}, \min\{V(I,A), V(I,C)\}\}. \end{cases}$$

These relations follow from the fact that in $\mathbb{Z}$ the operations min and max are distributive relative to each other.

COMMENT (C34). [**Indian Kuttak Method and the CRT**]. The so called Chinese Remainder Theorem is nothing but the KUTTAK method of ancient Indian Mathematics as reproduced in Bhaskara's Beejganit. In India we used to learn it in grade school. Note that in the statement of CRT, the condition $x_i - x_j \in A_i + A_j$ for all $i \neq j$ is automatically satisfied if the ideals $A_1, \ldots, A_n$ are pairwise comaximal, and hence it is satisfied if the ideals $A_1, \ldots, A_n$ are distinct nonzero prime ideals in a DD.

QUEST (Q21) Real Ranks of Valuations and Segment Completions

In this Quest we shall prove the Real Valuation Characterization, which was stated in the preamble of Quest (Q19). It said that a valuation is real iff its valuation ring has no nonzero nonmaximal prime ideal. To put it in proper perspective, we introduce the real rank of a valuation as the order type (under reverse inclusion) of the set of all nonmaximal prime ideals in its valuation ring. This turns out to be the same as the order type (under inclusion) of the set of all nonzero isolated subgroups of the value group of the valuation, where by isolated subgroups we mean those subgroups which are also segments (= analogs of real intervals symmetric around zero). We give a complete characterization of the possible real ranks of valuation rings. To study these ranks we introduce the notion of the segment completion of a loset (= linearly ordered set) which is a slight variation of its Dedekind completion, i.e., the kind of completion which leads us from rationals to reals. We also generalize the idea of meromorphic series with exponents in an ordered abelian group to well ordered products and use them to construct ordered abelian groups. For the definitions of losets, segments, valuation rings of valuations, meromorphic series, and so on, see Lecture L2.

Recall that a segment of an ordered abelian group G is a nonempty subset H of G such that for all $h \in H$ and $g \in G$ with $|g| \leq |h|$ we have $g \in H$, and an isolated subgroup of G is a subgroup which is a segment. Let

$$\begin{cases} S(G) = \text{the set of all nonzero isolated subgroups of } G, \\ \overline{S}(G) = \text{the set of all isolated subgroups of } G, \text{ and} \\ \widehat{S}(G) = \text{the set of all segments of } G. \end{cases}$$

Note that then $S(G) \subset \overline{S}(G) \subset \widehat{S}(G)$ are losets ordered by inclusion; see L2§7(D5). More generally let

$$\begin{cases} \overline{S}'(G) = \text{the set of all isolated positive upper segments of } G, \text{ and} \\ \widehat{S}'(G) = \text{the set of all positive upper segments of } G \end{cases}$$

where by a positive upper segment of G we mean a subset H' of

$$G_+ = \{g \in G : g > 0\}$$

such that for all $h \in H'$ and $g \in G$ with $g \geq h$ we have $g \in H'$, and by an isolated positive upper segment of G we mean a positive upper segment H' of G such that for all g, h in $G_+ \setminus H'$ we have $g + h \notin H'$. For any subsets H, H' of G let

$$\begin{cases} p'(H') = \{g \in G : |g| < |h| \text{ for all } h \in H'\}, \text{ and} \\ q'(H) = \{g \in G : g > h \text{ for all } h \in H\} \end{cases}$$

and note that then

$$
(1) \quad
\begin{cases}
p' : \widehat{S}'(G) \to \widehat{S}(G) \text{ and } q' : \widehat{S}(G) \to \widehat{S}'(G) \text{ are} \\
\text{inclusion reversing bijections which are inverses of each other,} \\
\text{and they respectively induce inclusion reversing bijections} \\
\overline{S}'(G) \to \overline{S}(G) \text{ and } \overline{S}(G) \to \overline{S}'(G) \\
\text{which are inverses of each other.}
\end{cases}
$$

Recall that the real rank $\rho(G)$ of G is defined to be the order type of the loset $S(G)$, and we write $\rho(G) = d$ or $\rho(G) = \infty$ according as $S(G)$ is a finite set of cardinality $d \in \mathbb{N}$ or $S(G)$ is an infinite set.

Given any ring A let

$$
\mathrm{id}(A) = \text{the set of all ideals in } A \quad \text{and} \quad \mathrm{nid}(A) = \mathrm{id}(A) \setminus \{A\}
$$

and given any quasilocal ring R let

$$
\mathrm{spec}(R)^* = \mathrm{spec}(R) \setminus \{M(R)\}.
$$

Recall that for a valuation $v : K \to G \cup \{\infty\}$ of a field K, its value group $v(K^\times)$ is denoted by G_v and its valuation ring $\{x \in K : v(x) \geq 0\}$ is denoted by R_v. By L4§12(R7)(7.5) we know that id (R_v), nid(R_v), spec(R_v), and spec$(R_v)^*$, are losets under inclusion. By the negative $-E$ of a loset E we mean E with reverse order, i.e.,

$$
x \leq y \text{ in } -E \Leftrightarrow y \leq x \text{ in } E.
$$

We define the real rank $\rho(v)$ of v, also called the real rank $\rho(R_v)$ of R_v, to be the order type of the loset $-\mathrm{spec}(R_v)^*$, and we write $\rho(v) = \rho(R_v) = d$ or $\rho(v) = \rho(R_v) = \infty$ according as spec$(R_v)^*$ is a finite set of cardinality $d \in \mathbb{N}$ or spec$(R_v)^*$ is an infinite set. For any $J \subset R_v$ and $H \subset G_v$ let

$$
\begin{cases}
p''(J) = \{v(x) : 0 \neq x \in J\}, \text{ and} \\
q''(H) = \{x \in R_v : v(x) \in H \cup \{\infty\}\}
\end{cases}
$$

and note that then

$$
(2) \quad
\begin{cases}
p'' : \mathrm{nid}(R_v) \to \widehat{S}'(G_v) \text{ and } q'' : \widehat{S}'(G_v) \to \mathrm{nid}(R_v) \text{ are} \\
\text{inclusion preserving bijections which are inverses of each other,} \\
\text{and they respectively induce inclusion preserving bijections} \\
\mathrm{spec}(R_v) \to \overline{S}'(G_v) \text{ and } \overline{S}'(G_v) \to \mathrm{spec}(R_v) \\
\text{which are inverses of each other.}
\end{cases}
$$

Upon letting

$$
p_v = p'p'' \text{ and } q_v = q''q' \text{ where } p' \text{ and } q' \text{ are as in (1) with } G_v \text{ replacing } G
$$

by (1) and (2) we see that

(3)
$$\begin{cases} p_v : \mathrm{nid}(R_v) \to \widehat{S}(G_v) \text{ and } q_v : \widehat{S}(G_v) \to \mathrm{nid}(R_v) \text{ are} \\ \text{inclusion reversing bijections which are inverses of each other,} \\ \text{and they respectively induce inclusion reversing bijections} \\ \mathrm{spec}(R_v) \to \overline{S}(G_v) \text{ and } \overline{S}(G_v) \to \mathrm{spec}(R_v) \\ \text{which are inverses of each other, and} \\ \text{they also respectively induce inclusion reversing bijections} \\ \mathrm{spec}(R_v)^* \to S(G_v) \text{ and } S(G_v) \to \mathrm{spec}(R_v)^* \\ \text{which are inverses of each other.} \end{cases}$$

In connection with (3) we note that, as a direct description of the maps p_v and q_v, for all $J \in \mathrm{nid}(R_v)$ and $H \in \mathrm{seg}(G_v)$ we have

$$\begin{cases} p_v(J) = \{g \in G_v : |g| < v(x) \text{ for all } x \in J\}, \text{ and} \\ q_v(H) = \{x \in R_v : |g| < v(x) \text{ for all } g \in H\} \end{cases}$$

and we also note that

$$p_v(M(R_v)) = 0.$$

By (3) we see that

(4)
$$\rho(v) = \rho(R_v) = \rho(G_v)$$

and hence in particular:

(5)
$$v \text{ is real iff } \dim(R_v) \le 1.$$

By a segment cut (= nonempty lower segment) of a loset E we mean a nonempty subset L of E such that

$$\text{for all } (x, y) \in L \times (E \setminus L) \text{ we have } x < y.$$

It is a Dedekind cut means: if L has no max (= maximum) then $E \setminus L$ has no min (= minimum) and we exclude the case of $L = E$. Here max and min refer to max of L in L, and min of $E \setminus L$ in $E \setminus L$; see Comment (C35) below. By the segment completion of E we mean the set E^{sc} of all segment cuts of E which we (linearly) order by inclusion. By the Dedekind completion of E we mean the subset E^{dc} of all Dedekind cuts with the induced order. For every $x \in E$ we put

$$D_E(x) = \{z \in E : z \le x\} \quad \text{and} \quad D'_E(x) = \{z \in E : z < x\}$$

and this gives us order preserving injections

$$D_E : E \to E^{sc} \quad \text{and} \quad D'_E : E \to E^{sc}$$

which we call the closed Dedekind map of E and the open Dedekind map of E respectively. We say that E is segment-complete if the map D_E is surjective, i.e., if E becomes its own segment completion after we identify E with a subset of E^{sc} via D_E. Upon letting

$$\widetilde{E} = E \setminus \{E_\infty\} \text{ or } E$$

according as E does or does not have a maximum element E_∞, we respectively get

$$D_E(E) = D_E(\widetilde{E}) \coprod \{\infty\} \text{ or } D_E(\widetilde{E}) \quad \text{with} \quad \infty = E$$

and we say that E is Dedekind-complete if "it is its own Dedekind completion," i.e., if $D_E(E) = E^{dc}$.

Given any $F \subset E$, we get an order preserving injection

$$D_{F,E} : F^{sc} \to E^{sc}$$

where for all for all $H \in F^{sc}$ we put

$$D_{F,E}(H) = \begin{cases} D_E(\xi) & \text{if } H \text{ has a max } \xi \text{ in } E \text{ with } \xi \notin H \\ \cup_{x \in H} D_E(x) & \text{otherwise} \end{cases}$$

and we call $D_{F,E}$ the Dedekind map of (F, E). For the explanation of terms like "max of a set IN an overset" see Comment (C35) below. We write

$$F^{sc} \doteq E$$

to mean that $D_{F,E}(F^{sc}) = D_E(E)$ and read this by saying that F^{sc} is essentially equal to E. Note that then E is segment-complete iff $E^{sc} \doteq E$.

By the core of E we mean the set $C(E)$ of all $x \in E$ such that either $D'_E(x) = \emptyset$ or x has an immediate predecessor in E; note that, for any $x \in E$,

$$\begin{cases} \text{an immediate predecessor of } x \text{ in } E \text{ means} \\ \text{max of all elements smaller than } x, \end{cases}$$

i.e.,

$$\begin{cases} \text{an element } y \text{ in } E \text{ such that } y < x \\ \text{and there is no } z \in E \text{ with } y < z < x. \end{cases}$$

We call E segment-full if E is order isomorphic to T^{sc} for some loset T, i.e., if there exists an order preserving bijection of E onto T^{sc} for some loset T. We call E core-full if $C(E)^{sc} \doteq E$. We call E brim-full if every nonempty subset of E has a max in E, and every element of E is the max in E of some subset of $C(E)$. Using these notions, in Theorems (T95) and (T96) below, we characterize segment completions. In the proofs of these Theorems we shall tacitly use the OBSERVATION which says that if $F = C(F)$, and hence in particular if $F = C(E)$, then, in the first line in the above definition of $D_{F,E}$, the proviso "with $x \notin F$" may be dropped because now $\xi \in F \Rightarrow H = D_F(\xi) \Rightarrow \cup_{x \in H} D_E(x) = D_E(\xi)$.

Note that

$$(6) \qquad \text{for all } I \in E^{sc} \text{ we have } I = \bigcup_{x \in I} D_E(x)$$

and

$$(7) \qquad \begin{cases} \text{if } x \in E \text{ is such that } D'_E(x) \neq \emptyset \\ \text{then } D'_E(x) \text{ is an immediate predecessor of } D_E(x) \\ \text{and hence } D_E(x) \in C(E^{sc}) \end{cases}$$

and therefore

$$(8) \qquad \text{for all } x \in E \text{ we have } D_E(x) \in C(E^{sc}).$$

By a nonzero principal isolated subgroup of an ordered abelian group G we mean a member H of $S(G)$ for which there exists $0 < \tau \in G$ such that

$$H = \bigcap_{H' \in S(G) \text{ with } \tau \in H'} H'$$

and we put

$$P(G) = \text{the set of all nonzero principal isolated subgroups of } G.$$

Using $P(G)$, in Theorem (T96.1) below, we show that the loset $S(G)$ is segment-full. The rest of Theorem (T96) gives more examples of segment-fullness or otherwise.

As a companion to the above notion, for a quasilocal ring R, by a nonmaximal coprincipal prime ideal in R we mean a member J of $\text{spec}(R)^*$ for which there exists $0 \neq t \in M(R)$ such that

$$J = \bigcup_{J' \in \text{spec}(R)^* \text{ with } t \notin J'} J'$$

and we put

$$\text{coprin}(R)^* = \text{the set of all nonmaximal coprincipal prime ideals in } R.$$

Given any valuation $v : K \to G \cup \{\infty\}$ of a field K, by taking $v(t) = \tau$ we see that

$$(9) \qquad p_v(\text{coprin}(R_v)^*) = P(G_v)$$

which explains the terminology of nonzero principal isolated subgroup.

Turning to products, consider a family $(W_i)_{i \in E}$ of additive abelian groups (or more generally modules over a ring) indexed by a loset E; see §1. Just as the direct sum was defined to be the set of all ϕ in $\prod_{i \in E} W_i$ whose support $\text{supp}(\phi)$ is finite, we define the well ordered product $\prod_{i \in E}^{wo} W_i$ to be the set of all ϕ in $\prod_{i \in E} W_i$ whose support supp(ϕ) is well ordered. If $W_i = W$ for all $i \in E$ then $\prod_{i \in E}^{\text{wo}} W_i$ is denoted by

$$W_{\text{wo}}^{E}$$

and called the well ordered E-th power of W (note analogy with the restricted E-th power $W_{\oplus}^{E}$). Note that the sets $W_{\oplus}^{E}$ and W_{wo}^{E} respectively correspond to the sets $(W^E)_{\text{finite}}$ and $(W^E)_{\text{wellord}}$ considered in L2§2. Also note that if E is a finite set then the direct product, the well ordered product, and the direct sum, all coincide.

If each W_i is an ordered abelian group then by putting the lexicographic order on $\prod_{i \in E}^{\text{wo}} W_i$ this becomes a linearly ordered set which we denote by $\prod_{i \in E}^{\text{lex}} W_i$ and call it the lexicographic product; here lexicographic order means that for $\phi \neq \psi$ in $\prod_{i \in E}^{\text{wo}} W_i$ we have:

$$\phi < \psi \Leftrightarrow \phi(j) < \psi(j) \text{ where } j = \min\{i \in E : \phi(i) \neq \psi(i)\}.$$

In particular, if $W_i = W$ for all $i \in E$ then W_{wo}^{E} becomes linearly ordered and as such we denote it by

$$W_{\text{lex}}^{E} \text{ or } W^{[E]}$$

and call it the lexicographic E-th power of W. In case E is a finite set of cardinality $d \in \mathbb{N}$, say the set of integers $1, \ldots, d$, and we are using the tuple notation instead of the functional notation, we may write

$$W_1 \boxplus \cdots \boxplus W_d$$

in place of $\prod_{i \in E}^{\text{lex}} W_i$ and call it the lexicographic direct sum of $W_1, \ldots, W_d$, and we may write

$$W_{\text{lex}}^{d} \text{ or } W^{[d]}$$

in place of W_{lex}^{E} and call it the lexicographic d-th power of W.

If each W_i is a ring (instead of an ordered abelian group) and the indexing set is an ordered abelian group G then $\prod_{i \in G}^{\text{wo}} W_i$ becomes a ring by Cauchy multiplication. In particular if W is a ring then W_{wo}^{G} becomes a ring. When W is a field K this is a new incarnation of the meromorphic series field $K((X))_G$. Even when W is any ring, we may denote the ring W_{wo}^{G} by $W((X))_G$. [cf. L6§6(E45)].

By using well ordered products, in Theorem (T97.1) below, we shall construct an ordered abelian group whose set of all nonzero isolated subgroups is order isomorphic to the segment completion of any given loset. As a consequence, in Theorem (T97.2) below, we shall characterize the real ranks of ordered abelian groups, and hence of valuations.

CHARACTERIZATION THEOREM FOR SEGMENT COMPLETIONS
(T95). A loset is segment-full iff it is core-full iff it is brim-full.

PROOF. We establish this by a cyclical proof: segment-full $\Rightarrow$ brim-full $\Rightarrow$ core-full $\Rightarrow$ segment-full. To do the first implication, given any loset E we want to show that E^{sc} is brim-full; now, for any nonempty subset A of E^{sc}, clearly $\cup_{H \in A} H$ is a max of A in E^{sc}; likewise, every $I \in E^{sc}$ is clearly a max in E^{sc} of the subset B of $C(E^{sc})$ given by $B = \{J \in C(E^{sc}) : J \subset I\}$. To clarify the last "clearly," note that, upon letting $B^* = \{D_E(x) : x \in I\}$, by (6) and (7) we get $B^* \subset B$ with $I = \max(B^*) \leq \max(B) \leq I$ where max is taken in E^{sc}. To do the second implication, given any brim-full loset E we want to show that $C(E)^{sc} \doteq E$, i.e., we desire to show that for every $x \in E$ there exists $I \in C(E)^{sc}$ such that

$$D_E(x) = D_{C(E),E}(I)$$

and we desire to show that for every $J \in C(E)^{sc}$ there exists $\xi \in E$ such that

$$D_E(\xi) = D_{C(E),E}(J)$$

To satisfy the first desire, note that the brim-fullness of E tells us that x is the max in E of some $I' \subset C(E)$ and it suffices to take

$$I = \cup_{t \in I'} D_{C(E)}(t).$$

To satisfy the second desire, note that by the brim-fullness of E the nonempty subset J of E has a max in E and it suffices to take ξ to be this max. The third implication is obvious.

CONDITIONS FOR A LOSET TO BE SEGMENT FULL (T96). Here are some illustrations of segment completions and losets which are segment-full or not.

(T96.1) If G is any ordered abelian group then $C(S(G)) = P(G)$ and also $P(G)^{sc} \doteq S(G)$, and hence $S(G)$ is segment-full.

(T96.2) $\mathbb{R}$ is not segment-full and neither is $\mathbb{Q}$.

(T96.3) $\mathbb{Q}^{sc} = \mathbb{Q}^- \coprod \mathbb{R} \coprod \{\infty\}$ where $x \mapsto x^-$ gives a bijection $\mathbb{Q} \to \mathbb{Q}^-$ with x^- being an element smaller than x but bigger than every real number $z < x$, where $\mathbb{R} = \mathbb{Q}^{dc}$, and where $\infty = \max(\mathbb{Q}^{sc}) = \mathbb{Q}$ as a segment-cut of $\mathbb{Q}$. More generally, for any nonempty loset E, upon letting $F = E \setminus C(E)$ and $F^- = D_{F,E}(F)$, we have $E^{sc} = F^- \coprod E^{dc} \coprod \{\infty\}$ where $x \mapsto x^- = D'_E(x)$ gives a bijection $F \to F^-$ with x^- being an element smaller than $D_E(x) \in E^{dc}$ but bigger than every $z \in E^{dc}$ for which $z < D_E(x)$, where in a natural manner E^{dc} is a subset of E^{sc}, and where $\infty = \max(E^{sc}) = E$ as a segment-cut of E.

[Note that $C(\mathbb{Q}) = \emptyset = C(\mathbb{R})$].

[Also note that, informally speaking $\mathbb{Q}^{dc} = \mathbb{R} = \mathbb{R}^{dc}$, i.e, formally speaking $D_{\mathbb{Q},\mathbb{R}}(\mathbb{Q}^{dc}) = D_{\mathbb{R}}(\mathbb{R}) = \mathbb{R}^{dc}$].

(T96.4) An ordinal is segment-full iff it is not a limit ordinal, i.e., iff it has a "nonzero constant term" such as $\omega + 1$ or $\omega^\omega + 2\omega + 3$. More precisely, a woset is segment-full iff it has a max.

(T96.5) An ordinal is segment-complete iff it is finite. If it is infinite then its segment completion is obtained by "adding 1 to it," i.e., for example the segment completions of ω or $\omega + 1$ or $\omega^\omega + 2\omega$ or $\omega^\omega + 2\omega + 3$ are $\omega + 1$ or $\omega + 2$ or $\omega^\omega + 2\omega + 1$ or $\omega^\omega + 2\omega + 4$ respectively. More precisely, a woset is segment-complete iff it is finite; moreover, if it is infinite then its segment completion is order isomorphic to the woset obtained by augmenting it by a single element which is declared to be bigger than all the previous elements.

(T96.6) A loset E is segment-complete iff it is the negative of a woset, i.e., informally, iff it is the negative of an ordinal, like $-\omega = -\mathbb{N}_+ = \mathbb{Z}_-$.

PROOF. To prove (T96.1) let G be an ordered abelian group. Given any H in $S(G)$ for which there exists $0 < \tau \in G$ such that $H = \cap_{H' \in S(G) \text{ with } \tau \in H'} H'$, by L4§12(R7)(7.5) we see that $\cup_{H' \in S(G) \text{ with } \tau \notin H'} H'$ is empty or an immediate predecessor of H in $S(G)$ according as $D'_{S(G)}(H)$ is empty or not, and hence $H \in C(S(G))$; [cf. L6§6(E52)]. Thus $P(G) \subset C(S(G))$. Conversely, given any $H \in C(S(G))$, we can take $0 < \tau \in H$ such that τ does not belong to the immediate predecessor of H in $S(G)$ in case there is such a predecessor, and then by L4§12(R7)(7.5) we see that $H = \cap_{H' \in S(G) \text{ with } \tau \in H'} H'$ and hence $H \in P(G)$; [cf. L6§6(E52)]. Thus $P(G) \supset C(S(G))$. It follows that $P(G) = C(S(G))$.

To show that $P(G)^{sc} \doteq S(G)$, given any $H \in S(G)$, upon letting

$$I = \{I_\tau : 0 < \tau \in H\} \text{ with } I_\tau = \cap_{H' \in S(G) \text{ with } \tau \in H'} H'$$

we see that $I \in P(G)^{sc}$ with $D_{P(G),S(G)}(I) = D_{S(G)}(H)$. Conversely, given any $I \in P(G)^{sc}$, upon letting

$$H = \cup_{I' \in I} I'$$

we see that $H \in S(G)$ with $D_{P(G),S(G)}(I) = D_{S(G)}(H)$. This completes the proof of (T96.1).

To prove (T96.2), in view of (T95), we want to show that E in not brim-full where $E = \mathbb{R}$ or $\mathbb{Q}$. In both the cases, $C(E)$ is empty and hence the second condition of brim-fullness does not hold. Again in both the cases, E has no max in E and hence the first condition of brim-fullness is not satisfied. In the second case, for every irrational number x (i.e., an element of $\mathbb{R} \setminus \mathbb{Q}$ such as $\sqrt{2}$) the nonempty subset of $\mathbb{Q}$ consisting of all $y \in \mathbb{Q}$ with $y < x$ has no max in $\mathbb{Q}$, and hence E does not satisfy the first condition of brim-fullness. Thus (T96.2) can be proved in many different ways.

To prove (T96.3), given any nonempty loset E, upon letting the objects F, F^- and ∞ to be as in (T96.3), it follows that $x \mapsto x^- = D'_E(x)$ gives a bijection $F \to F^- = E^{sc} \setminus (E^{dc} \coprod \{\infty\})$ with x^- being an element smaller than $D_E(x) \in E^{dc}$ but bigger than every $z \in E^{dc}$ for which $z < D_E(x)$.

To prove (T96.4) and (T96.5), given any woset E, let J be the woset obtained by adjoining to E an extra element ∞ to E which is declared to be greater than all the previous elements, and let $A(E)$ denote the set of all $x \in E$ such that $D_E^*(x)$ is a finite set where we have put $D_E^*(x) = \{y \in E : x \leq y\}$. By the well order property we see that if $A(E)$ is nonempty then it has a smallest element.

Using the said smallest element and remembering that, according to L2§1, $|A(E)|$ and $||E||$ denote cardinal and ordinal respectively, we see that

(i) $|A(E)| \in \mathbb{N}$ (i.e, $A(E)$ is a finite set),

(ii) $|A(E)| = 0$ (i.e., $A(E)$ is empty) $\Leftrightarrow E$ has no max, and

(iii) $|A(J)| = |A(E)| + 1$ and hence $||J|| > ||E||$.

It only remains to find an order preserving bijective map $u : E^{sc} \to I$ where $I = E$ or J according as E is finite or infinite. This is done by letting $u(H) = \max(H)$ or $\min(E \setminus H)$ according as $H \in E^{sc}$ is finite or infinite.

To prove (96.6) let E be any loset. Given any nonempty subset F of E, upon letting

$$H = \{x \in E : x \leq y \text{ for some } y \in F\}$$

we get $H \in E^{sc}$ such that: F has a max iff H has a max (and when that is so the two max coincide). Also it is clear that $D_E : E \to E^{sc}$ is surjective iff every $H \in E^{sc}$ has a max (in H). It follows that D_E is surjective iff every nonempty subset of E has a max, i.e., iff $-E$ is a woset.

CHARACTERIZATION THEOREM FOR REAL RANKS (T97). Concerning real ranks we have the following.

(T97.1) For any loset E and any archimedean ordered nonzero abelian group $W \neq 0$, the loset $P(W^{[-E]})$ is order isomorphic to E and hence, by (T96.1), the loset $S(W^{[-E]})$ is order isomorphic to E^{sc}.

(T97.2) An order type is the real rank of some ordered abelian group iff it is the order type of the segment-completion of some loset, i.e., (by the definition of segment-fullness) iff it is the order type of a segment-full loset. Similarly, an order type is the real rank of some ordered abelian group iff it is the order type of a core-full loset. Likewise, an order type is the real rank of some ordered abelian group iff it is the order type of a brim-full loset.

PROOF. To prove (T97.1), given any loset E and any archimedean ordered abelian group $W \neq 0$, we can fix any $0 < \gamma \in W$. Now for every $x \in E$ let $\widehat{x} \in W^{[-E]}$ be defined by putting $\widehat{x}(y) = \gamma$ or 0 according as $y = x$ or $y \in E \setminus \{x\}$, and let

$$H_x = \bigcap_{H' \in S(W^{[-E]}) \text{ with } \widehat{x} \in H'} H'.$$

Clearly $x \mapsto H_x$ gives an order preserving injective map $E \to P(W^{[-E]})$. To prove that it is surjective, given any $H \in P(W^{[-E]})$, we have

$$H = \bigcap_{H' \in S(W^{[-E]}) \text{ with } \tau \in H'} H'$$

for some $0 < \tau \in W^{[-E]}$. Upon letting

$$x = \min\{z \in E : \tau(z) \neq 0\}$$

we clearly have

$$H = H_x.$$

The first sentence in (T97.2) follows from (T96.1) and (T97.1). The second and the third sentences follow from the first sentence and (T95).

COMMENT (C35). [**min, max, lub, glb**]. Extending the definitions made in L2§1, a smallest or least (resp: largest or greatest) element of a poset R is an element z of R such that for all $x \in R$ we have $z \leq x$ (resp: $z \geq x$); when z exists it is unique; we call it the min or the minimum (resp: the max or the maximum) of R and we denote it by some notation such as $\min(R)$ (resp: $\max(R)$). An upper (resp: lower) bound of R in an overposet S of R (i.e., a poset S with $R \subset S$ such that the partial order of R is induced by the partial order of S) is an element y of S such that $x \leq y$ (resp: $x \geq y$) for all $x \in R$. A least upper bound or lub (resp: greatest lower bound or glb) of R in S is a min (resp: max) of the set of all upper (resp: lower) bounds of R in S. By min (resp: max) of R in S we mean glb (resp: lub) of R in S and, without any flourish, min (resp: max) of R means min (resp: max) of R in R, or in an obvious overposet as in the definition of degree or order where the overposet is $\mathbb{Z} \cup \{-\infty\}$ or $\mathbb{Z} \cup \{\infty\}$.

Thus, in connection with the previous Quest Q(21), immediate predecessor of an element x of a loset E means max of $D'_E(x)$ in $D'_E(x)$.

To explicitly clarify the distinction between minimum and minimal (resp: maximum and maximal) by a minimal (resp: maximal) element of R we mean $z \in R$ such that there is no $x \in R$ with $x < z$ (resp: $x > z$). We do not abbreviate minimal (resp: maximal) to min (resp: max).

Some authors use supremum (resp: infimum) as a synonym for lub (resp: glb).

To summarize: we make do with min and max, with stress on "in" when taking these in an overset. So we avoid a plethora of terms by shunning lub, glb, sup, and inf.

QUEST (Q22) Specializations and Compositions of Valuations

In the last Quest we discussed the correspondence between prime ideals of a valuation ring and isolated subgroups of the value group. We shall now elucidate this further by introducing specializations and compositions of valuations. For related material also see L4§12(R7).

Given any valuation

$$v : K \to G \cup \{\infty\}$$

of a field K, in addition to the sets

$$\mathrm{spec}(R_v)^* \subset \mathrm{spec}(R_v) \subset \mathrm{nid}(R_v) \quad \text{and} \quad S(G) \subset \overline{S}(G) \subset \widehat{S}(G)$$

and the maps

$$p_v : \mathrm{nid}(R_v) \to \widehat{S}(G_v) \quad \text{and} \quad q_v : \widehat{S}(G_v) \to \mathrm{nid}(R_v)$$

introduced in Q(21), let

$$v^\times : K^\times \to G_v$$

be the group epimorphism from a multiplicative group to an additive group sending every x to $v(x)$, and note that

$$\ker(v^\times) = R_v \setminus M(R_v) = U(R_v) = \text{the group of all units in } R_v.$$

Also let

$$\pi_v : R_v \to \Delta_v = R_v/M_v \quad \text{with} \quad M_v = M(R_v)$$

be the residue class epimorphism, and let

$$\pi_v^\times : U(R_v) \to \Delta_v^\times$$

be the multiplicative group epimorphism sending every x to $\pi_v(x)$.

As explained in L4§12(R7), associated with any valuation ring R of K, we have the "canonical valuation" (= the divisibility valuation) γ_R of K with valuation ring R and value group $\Gamma(R) = K^\times/U(R)$. In the present situation, as abbreviation we may write γ_v instead of γ_{R_v}, and Γ_v instead of $\Gamma(R_v)$. Note that then

$$(0) \qquad \gamma_v^\times : K^\times \to \Gamma_v \quad \text{and} \quad \beta_v : \Gamma_v \to G_v \quad \text{with} \quad \beta_v \gamma_v^\times = v^\times$$

where the middle map is the unique order isomorphism satisfying the last equation.

Given any other valuation w of K with

$$(\bullet) \qquad\qquad\qquad R_v \subset R_w$$

we say that w specializes to v or that v generalizes to w and we write

$$w \searrow v \quad \text{or} \quad v \nearrow w.$$

Alternatively, we say that v is a specialization of w, or w is a generalization of v.

We claim that

$$(1) \qquad M_w \cap R_v = M_w \in \mathrm{spec}(R_v) \quad \text{with} \quad (R_v)_{M_w} = R_w \in \mathfrak{V}(R_v).$$

[This is surprising, since a prime ideal P in a domain A is usually much smaller than its extension $PR_P = M(R_P)$].

To prove (1) we first note that if R_w dominates R_v (i.e., if $M_w \cap R_v = M_v$) then $R_w = R_v$ because: $x \in K \setminus R_v \Rightarrow v(x) < 0 \Rightarrow v(1/x) > 0 \Rightarrow 1/x \in M_v \subset M_w \Rightarrow x \in K \setminus R_w$. It only remains to observe that, by L4§12(R7)(7.2), especially by the x or $1/x$ property L4§12(R7)(1), the localization $(R_v)_P$ of R_v at the prime ideal $P = M_w \cap R_v$ in it equals the valuation ring $R_{v'}$ of some valuation v' of K, and hence $R_w = (R_v)_P$ by the dominating case we just proved.

Let

$$\pi_{v,w} : R_v \to \pi_w(R_v) \quad \text{and} \quad \pi_{v,w}^{\times} : U(R_v) \to \pi_w(U(R_v))$$

be the surjective maps sending x to $\pi_w(x)$. By (1) we see that

(2) $\begin{cases} \pi_{v,w} : R_v \to \pi_w(R_v) \text{ is a ring epimorphism of quasilocal domains} \\ \text{whose kernel is the prime ideal } M_w \text{ in } R_v \end{cases}$

and hence

(3) $\begin{cases} \pi_{v,w}^{\times} : U(R_v) \to \pi_w(U(R_v)) = U(\pi_w(R_v)) \\ \text{is a multiplicative group epimorphism with kernel } M_w \cap U(R_v). \end{cases}$

Since $\ker(v^{\times}) = U(R_v) \subset U(R_w) = \ker(w^{\times})$, by (1) we also get a short exact sequence

(4) $$0 \longrightarrow p_v(M_w) \xrightarrow{p_{w,v}} G_v \xrightarrow{\alpha_{v,w}} G_w \longrightarrow 0$$

where $p_{v,w}$ is a subset injection, i.e., the natural injection of a subset into the set, and $\alpha_{v,w}$ is the unique order epimorphism such that

$$\alpha_{v,w} v^{\times} = w^{\times}.$$

In (4) we wrote the names of the maps on top of the corresponding arrows. Instead of that we could have written them on the bottom of the corresponding arrows. In case of vertical arrows we can write their names on their left side or the right side.

In view of L4§12(R7)(3) and L4§12(R7)(7.2), by (0) to (4) we see that

(5) $\begin{cases} v^{-1}(p_v(M_w)) = U(R_w) \\ \text{with } \pi_w(R_v) = \text{a divisibility ring of } \Delta_w \\ \text{and there exists a unique valuation} \\ v/w : \Delta_w \to p_v(M_w) \cup \{\infty\} \text{ of } \Delta_w \\ \text{with } R_{v/w} = \pi_w(R_v) \text{ and } G_{v/w} = p_v(M_w) \\ \text{such that } (v/w)(\pi_w(x)) = v(x) \text{ for all } x \in U(R_w) \\ \text{(the valuation } v/w \text{ is called } v \text{ divided by } w \text{).} \end{cases}$

The above construction of v/w may be illustrated by the commutative diagram

$$
\begin{array}{ccccccc}
U(R_v) & \xrightarrow{i_1} & U(R_w) & \xrightarrow{i_2} & K^\times & \xrightarrow{v^\times} & G_v \\
\downarrow{\scriptstyle\pi_{v,w}^\times} & & \downarrow{\scriptstyle\pi_w^\times} & & & & \uparrow{\scriptstyle p_{w,v}} \\
U(R_{v/w}) & \xrightarrow{i_3} & \Delta_w^\times & \xrightarrow{\gamma_{v/w}^\times} & \Gamma_{v/w} & \xrightarrow{\beta_{v/w}} & G_{v/w}
\end{array}
$$

where i_1, i_2, i_3 are subset injections. The commutativity of the left and the right rectangles, respectively, means that

$$i_3 \pi_{v,w}^\times = \pi_w^\times i_1$$

and

$$p_{w,v} \beta_{v/w} \gamma_{v/w}^\times \pi_w^\times = v^\times i_2.$$

In view of L4§12(R7)(1), L4§12(R7)(7.2), and L4§12(R7)(7.5), by (1) we see that

(6) $\begin{cases} \text{if } v \text{ is any valuation of a field } K \text{ then } P \mapsto (R_v)_P \\ \text{gives an inclusion reversing bijection of } \operatorname{spec}(R_v) \text{ onto} \\ \text{the set of all valuation rings of } K \text{ containing } R_v \text{ and} \\ \text{this set coincides with the set of all subrings of } K \text{ containing } R_v, \\ \text{and hence these sets are losets under inclusion.} \end{cases}$

In view of (L4§12(R7)(1) and L4§12(R7)(7.2), by (1) we see that

(7) $\begin{cases} \text{if } w \text{ is any valuation of a field } K \text{ then } S \mapsto R = \pi_w^{-1}(S) \\ \text{gives a bijection of the set of all valuation rings of } \Delta_w \text{ onto} \\ \text{the set of all valuation rings of } K \text{ contained in } R_w; \\ \text{in case } S = R_u \text{ for a valuation } u \text{ of } \Delta_w, \\ \text{we denote the valuation } \gamma_R \text{ of } K \text{ by } w \odot u \\ \text{and call it the composition of } w \text{ by } u \text{ [cf. L6§6(E56)].} \end{cases}$

In view of (7), in the situation of (6) we have

(8) $$\gamma_v = w \odot (v/w).$$

In view of (6) and (7), as an illustration of compositions of valuations, by L4§12(R7)(7.7) we see that

(9) $\begin{cases} \text{if } w \text{ is any valuation of a field } K \\ \text{and } A \text{ in any subring of } R_w \text{ with } M_w \cap A = P \subset Q \in \operatorname{spec}(A) \\ \text{then there exist valuations } u \text{ of } \Delta_w \text{ with } R_u \text{ dominating } \pi_w(A_Q) \\ \text{and for every such } u \text{ we have that } R_{w \odot u} \text{ dominates } A_Q. \end{cases}$

QUEST (Q23) UFD Property of Regular Local Domains

In UFD Theorem (T102) below we shall show that every regular local domain is a UFD. The proof will use the concept of pdim (= projective dimension) from (Q17). In UFD Lemma (T99) we prepare the groundwork by sharpening the UFD condition (6)(d) of (Q18.4) and proving the theorem for small dimensions. In Lemmas (T100) and (T101) we convert some material of (Q17) into a form more usable in the present context. Amongst the ideal (or module) theoretic colon operations $(:)_R$ and $[:]_R$ in a ring R, we gave preference to the bracket colon; so now in Lemma (T98) we give some properties of the parenthesis colon.

PARENTHETICAL COLON LEMMA (T98). For any ideals I, J in a noetherian ring R, upon letting $L = (I : J)_R$, we have the following.

(T98.1) $L = I \Leftrightarrow$ for every $P \in \mathrm{ass}_R(R/I)$ we have $J \not\subset P$.

(T98.2) If R is a local ring with $J = M(R) \neq 0$ and $L = I$, then $L = (I : xR)_R$ for some $x \in J \setminus J^2$.

(T98.3) If R is a local domain and I, J are nonzero principal ideals in R then: there exist nonzero principal ideals I', J' in R for which we have $L = (I' : J')_R$ and $I' \not\subset M(R)L$.

PROOF. To prove (T98.1) suppose $L = I$ and $P \in \mathrm{ass}_R(R/I)$. Then by L4§6 there exists $a \in R$ such that $P = (I : a)_R$. Since P is a prime ideal, $P \neq R$ and hence $a \notin I$. Thus $a \notin L$ and so there exists $x \in J$ such that $ax \notin I$, i.e., $x \notin (I : a)_R = P$. This proves that $J \not\subset P$. Conversely, suppose $J \not\subset P$ for every $P \in \mathrm{ass}_R(R/I)$. Then by L4§6 we can find $n \in \mathbb{N}$ and primary ideals $Q_1, \ldots, Q_n$ in R such that $I = \cap_{1 \leq i \leq n} Q_i$ and upon letting $\mathrm{rad}_R Q_i = P_i$ for $1 \leq i \leq n$ we have $\mathrm{ass}_R(R/I) = \{P_1, \ldots, P_n\}$. Now $J \not\subset P_i$ for $1 \leq i \leq n$ and hence by (22•) of L4§5(O24), there exists some $r \in J \backslash \cup_{1 \leq i \leq n} P_i$. Consequently, given any $s \in (I : J)_R$ we have $rs \in Q_i$ for $1 \leq i \leq n$. Since Q_i is primary and $r \notin P_i$ for every i, we obtain $s \in \cap_{1 \leq i \leq n} Q_i = I$. Thus $(I : J)_R \subset I$, and hence $L = I$.

To prove (T98.2) suppose R is local with $J = M(R) \neq 0$ and $L = I$. Then by Nakayama Lemma (Q2)(T3), $J \neq J^2$ and so there exists some $y \in J \setminus J^2$. Now by L4§6 we can find $n \in \mathbb{N}$ and prime ideals $P_1, \ldots, P_n$ such that $\mathrm{ass}_R(R/I) = \{P_1, \ldots, P_n\}$. Relabelling $P_1, \ldots, P_n$ if necessary, we can find r and m in $\mathbb{N}$ with $m \leq r \leq n$ such that: (i) $y \in P_i$ for $1 \leq i \leq m$ and $y \notin P_j$ for $m + 1 \leq j \leq r$, (ii) $P_j \not\subset P_i$ for $1 \leq i \leq m$ and $m + 1 \leq j \leq r$, and (iii) $\cup_{1 \leq i \leq r} P_i = \cup_{1 \leq i \leq n} P_i$. [cf. L6§6(E57) for a proof of this assertion and for an example showing that (i) and (ii) cannot always be arranged with $r = n$]. By (T98.1) we have $J \not\subset P_i$ and hence $J^2 \not\subset P_i$ for $1 \leq i \leq r$. Thus by (21•) of L4§5(O24), for $1 \leq i \leq m$ we have $\left(J^2 \cap P_{m+1} \cap \cdots \cap P_r\right) \not\subset P_i$ and hence by (22•) of L4§5(O24) we can find some $z \in J^2 \cap P_{m+1} \cap \cdots \cap P_r$ such that $z \notin \cup_{1 \leq i \leq m} P_i$. Let $x = y + z$. Clearly

$x \in J \setminus J^2$ and $x \notin \cup_{1 \le i \le r} P_i = \cup_{1 \le i \le n} P_i$. By L4§6 we can find primary ideals $Q_1, \ldots, Q_n$ in R such that $\mathrm{rad}_R Q_i = P_i$ for $1 \le i \le n$ and $I = \cap_{1 \le i \le n} Q_i$. Now if $a \in (I : xR)_R$ then $ax \in I$ and hence $ax \in Q_i$ with $x \notin P_i$ for $1 \le i \le n$. Consequently, $a \in \cap_{1 \le i \le n} Q_i = I$. Thus $(I : xR)_R \subset I$, and hence $L = (I : xR)_R$.

To prove (T98.3), first note that since $I \ne 0$ we have $L \ne 0$ and hence by Nakayama Lemma (Q2)(T3), $L \ne M(R)L$. Let $b \in L \setminus M(R)L$ and let c, d be nonzero elements of R such that $I = cR$ and $J = dR$. Now $b \in (cR : dR)_R$ and hence $bd = ac$ for some $a \in R$. Since R is a domain and b, d, c are nonzero, we have $a \ne 0$ and for any x, y in R we get: $xd = yc \Leftrightarrow xdb = ycb \Leftrightarrow xac = ybc \Leftrightarrow xa = yb$. Consequently $(cR : dR)_R = (bR : aR)_R$, and hence $I' = bR$ and $J' = aR$ have the desired properties.

UFD LEMMA (T99). For any local domain R we have the following.

(T99.1) R is a UFD iff for every ideal $V = aR + bR$ with $a \ne 0 \ne b$ in R, regarding V as an R-module, we have $\mathrm{pdim}_R V \le 1$.

[Note that, by (T80.4), (T80.10), and (T80.17), the condition $\mathrm{pdim}_R V \le 1$ for a finite module V over a local ring R is equivalent to the condition that V is isomorphic to F/E for a finite free submodule E of a finite free R-module F. Also note that if either $a = 0$ or $b = 0$ in a domain R then $\mathrm{pdim}_R(aR + bR) = 0$ because, for any nonzero element a in a domain R, multiplication by a gives an isomorphism of R onto aR. Conversely observe that if an ideal in a domain is a free module then it must be a principal ideal].

(T99.2) If R is regular of dimension ≤ 2 then R is a UFD.

(T99.3) If R is regular of dimension 3 then R is a UFD.

PROOF. To prove (T99.1), note that, for any $V = aR + bR$ with $a \ne 0 \ne b$ in R, we have the short exact sequence

$$(\dagger) \qquad 0 \longrightarrow E \xrightarrow{\ f_1\ } R^2 \xrightarrow{\ f_0\ } V \longrightarrow 0$$

of R-modules where $f_0(x, y) = ax - by$ for all $(x, y) \in R^2$ and $f_1 : E = \ker(f_0) \to R^2$ is the subset injection. Upon letting $g(x, y) = ax$ for all $(x, y) \in R^2$ we obtain an R-epimorphism with $g(E) = (aR) \cap (bR)$ and $E \cap \ker(g) = 0$, and hence

$$(\ddagger) \qquad E \approx (aR) \cap (bR).$$

If R is a UFD then, by (Q18.4)(6), the ideal $(aR) \cap (bR)$ is principal and hence it is a finite free R-module (indeed it is isomorphic to R^0 or R^1 according as it is zero or nonzero), and therefore by $(\dagger)$ and $(\ddagger)$ we get $V \approx R^2/E$ with finite free submodule E of the finite free R-module R^2, and so $\mathrm{pdim}_R V \le 1$. Conversely, if $\mathrm{pdim}_R V \le 1$ then, in view of (T80.10), by the short exact sequence $(\dagger)$ we see that $\mathrm{pdim}_R E = 0$, i.e. E is projective and hence free by (T80.4), and hence $(aR) \cap (bR)$ is principal by $(\ddagger)$; therefore R is a UFD by (Q18.4)(6).

In view of (T99.1), to prove (T99.2) and (T99.3) it suffices to show that for any ideal $V = aR + bR$ with $a \neq 0 \neq b$ in a regular local domain R of dimension ≤ 3, we have $\text{pdim}_R V \leq 1$. By Serre (T81.6) we know that $\text{pdim}_R V < \infty$. By (T69.4) and (T69.7) we see that $\text{reg}(R) = \dim(R)$, and since R is a domain, we also have $\text{reg}_R V \geq 1$ if $\dim(R) \geq 1$. Therefore by Auslander-Buchsbaum (T81.1) we conclude that $\text{pdim}_R V \leq \dim(R) - 1$ if $\dim(R) \geq 1$. This completes the proof of (T99.2). [Note that regular local domains of dimension zero and one are nothing but fields and DVRs respectively, and these are obviously UFDs; see (Q19)(T85)].

In case of (T99.3) we are assuming $\dim(R) = 3$ and we have shown $\text{pdim}_R V \leq 2$. We wish to show that $\text{pdim}_R V \leq 1$. In doing so we may assume that $\text{pdim}_R V \geq 1$ and we may use the set up of the above proof of (T99.1).

Since $1 \leq \text{pdim}_R V \leq 2$, in view of the short exact sequence (†), by (T80.10) we see that $\text{pdim}_R E \leq 1$, and also that if $\text{pdim}_R E = 0$ then $\text{pdim}_R V = 1$. Let $L = (bR : aR)_R$ and $J = M(R)$. For the projection map $h : R^2 \to R$ given by $h(x, y) = x$ we have $\ker(h) \cap E = 0$ and $h(E) = L$. Consequently h induces an isomorphism $E \to L$. Hence $\text{pdim}_R L \leq 1$ and it suffices to show that $\text{pdim}_R L = 0$. Using (T98.3) and replacing a and b by suitable nonzero elements of R we may assume that $b \notin JL$. Let $b_1 = b$. Clearly $b_1 \in L$. Let $\phi : L \to \overline{L} = L/(JL)$ be the residue class epimorphism, and let n be the dimension of $\overline{L}$ as a vector space over the field $\overline{R} = R/J$. Then $n \in \mathbb{N}_+$ and we can find elements $b_2, \ldots, b_n$ in L such that $\phi(b_1), \ldots, \phi(b_n)$ is a vector space basis of $\overline{L}$ over $\overline{R}$. By Nakayama (Q2)(T3), the ideal L is generated by $b_1, \ldots, b_n$. Hence there exists an R-epimorphism $t_0 : R^n \to L$ such that $t_0(e_i) = b_i$ for $1 \leq i \leq n$ where $e_1, \ldots, e_n$ is the basis of R^n given by $e_i = (0, \ldots, 1, 0, \ldots, 0) \in R^n$ with 1 in the i-th spot. Since $\text{pdim}_R L \leq 1$, by (T80.4) and (T80.10) we see that $\ker(t_0)$ is a finite free R-module and hence we can find $m \in \mathbb{N}$ together with an R-monomorphism $t_1 : R^m \to R^n$ such that $\text{im}(t_1) = \ker(t_0)$. Let $d_1, \ldots, d_m$ be the basis of R^m given by $d_i = (0, \ldots, 1, 0, \ldots, 0) \in R^m$ with 1 in the i-th spot. Now $b_n \in L = (bR : aR)_R$ with $b = b_1$ and hence for some $a' \in R$ we have $ab_n = a'b_1$, i.e., $ae_n - a'e_1 \in t_1(R^m)$. Also clearly $be_n - b'e_1 \in t_1(R^m)$ with $b' = b_n \in R$. Consequently

$$be_n - b'e_1 = \sum_{1 \leq i \leq m} c_i t_1(d_i) \quad \text{and} \quad ae_n - a'e_1 = \sum_{1 \leq i \leq m} c_i' t_1(d_i)$$

for some c_i, c_i' in R. Clearly $a(be_n - b'e_1) = b(ae_n - a'e_1)$, and therefore by the above display we see that for $1 \leq i \leq m$ we have $ac_i = bc_i'$ and hence $c_i \in L$. Since $\phi(b_1), \ldots, \phi(b_n)$ is a vector space basis of $\overline{L}$ over $\overline{R}$, we must have $\ker(t_0) \subset JR^n$; see Comment (C36) below. Therefore

$$t_1(d_i) = \sum_{1 \leq j \leq n} u_{ij} e_j \quad \text{with} \quad u_{ij} \in J \text{ for all } i, j.$$

So we have

$$be_n - b'e_1 = \sum_{1 \le i \le m} c_i \left(\sum_{1 \le j \le n} u_{ij} e_j \right).$$

If $n > 1$ then comparing the coefficients of e_n on the two sides of the above equation we get

$$b = \sum_{1 \le i \le m} c_i u_{in} \in JL$$

which is a contradiction. Therefore $n = 1$. Consequently $L = bR$, and hence $\mathrm{pdim}_R L = 0$.

SNAKE SUBLEMMA (T100) In the Snake Lemma (Q17)(T80.9) assume that the map $f_1^{(2)} : W_2' \to W_2$ is an isomorphism. Then the map $f_3^* : W_4^* \to W_3^*$ is an isomorphism. Moreover, if the map $f_1^{(0)} : W_0' \to W_0$ is an epimorphism then the map $f_1^{(1)} : W_1' \to W_1$ is an epimorphism.

[Let us depict the situation of the Snake Lemma (Q17)(T80.9) by the following commutative diagram.

$$
\begin{array}{ccccccccc}
& & 0 & & 0 & & 0 & & \\
& & \downarrow & & \downarrow & & \downarrow & & \\
0 & \longrightarrow & W_5^* & \xrightarrow{f_4^*} & W_4^* & \xrightarrow{f_3^*} & W_3^* & \xrightarrow{f_2^*} & W_2^* \\
& & \downarrow{\scriptstyle f_2^{(2)}} & & \downarrow{\scriptstyle f_2^{(1)}} & & \downarrow{\scriptstyle f_2^{(0)}} & & \\
0 & \longrightarrow & W_2' & \xrightarrow{f_1'} & W_1' & \xrightarrow{f_0'} & W_0' & \longrightarrow & 0 \\
& & \downarrow{\scriptstyle f_1^{(2)}} & & \downarrow{\scriptstyle f_1^{(1)}} & & \downarrow{\scriptstyle f_1^{(0)}} & & \\
0 & \longrightarrow & W_2 & \xrightarrow{f_1} & W_1 & \xrightarrow{f_0} & W_0 & \longrightarrow & 0 \\
& & \downarrow{\scriptstyle f_0^{(2)}} & & \downarrow{\scriptstyle f_0^{(1)}} & & \downarrow{\scriptstyle f_0^{(0)}} & & \\
& & W_2^* & \xrightarrow{f_1^*} & W_1^* & \xrightarrow{f_0^*} & W_0^* & \longrightarrow & 0 \\
& & \downarrow & & \downarrow & & \downarrow & & \\
& & 0 & & 0 & & 0 & &
\end{array}
$$

Here the three columns and the middle two rows are exact sequences of length two and one respectively. Also the top row starting with 0 followed by the bottom row ending with 0 is an exact sequence of length four which is called the SNAKE].

PROOF. Since $f_1^{(2)}$ is an isomorphism, the exactness of the length-two sequence $(f^{(2)}, W^{(2)}, 2)$ (i.e., the exactness of the first column of the above diagram) tells us that $W_5^* = 0 = W_2^*$, and hence the exactness of the length-four sequence $(f^*, W^*, 4)$ (i.e., the exactness of the Snake of the above diagram) tells us that f_3^* is an isomorphism. Moreover, if $f_1^{(0)}$ is an epimorphism then the exactness of the length-two sequence $(f^{(0)}, W^{(2)}, 2)$ (i.e., the exactness of the last column of the above diagram) tells us that $W_0^* = 0$, and hence (because of the equation $W_2^* = 0$) the exactness of the length-four sequence $(f^*, W^*, 4)$ (i.e., the exactness of the Snake of the above diagram) tells us that $W_1^* = 0$, and therefore the exactness of the length-two sequence $(f^{(1)}, W^{(2)}, 2)$ (i.e., the exactness of the middle column of the above diagram) tells us that $f_1^{(1)}$ is an epimorphism.

SUPPLEMENTAL PDIM LEMMA (T101). For any local ring R with $J = M(R)$ we have the following.

(T101.1) Let x be any element in $J \cap S_R(R)$, let $R' = R/(xR)$, and let U' be a finite R'-module with $\mathrm{pdim}_{R'} U' < \infty$. Then $\mathrm{pdim}_R U' \leq 1 + \mathrm{pdim}_{R'} U'$ where in the LHS we regard U' as an R-module.

(T101.2) Let V be a nonzero finite R-module with $\mathrm{pdim}_R V < \infty$, and let x be any element in $J \cap S_R(V)$. Then $\mathrm{pdim}_R(V/xV) = 1 + \mathrm{pdim}_R V$.

(T101.3) Let V be a nonzero submodule of a finite free R-module W_2 with $\mathrm{pdim}_R V < \infty$, and for any $x \in J \cap S_R(R) \cap S_R(W_2/V)$ let $W_1 = V + xW_2$ and $W_0 = V/xV$. Then either $\mathrm{pdim}_R W_1 = 0$ or $\mathrm{pdim}_R W_1 = 1 + \mathrm{pdim}_R V$. Moreover, if $\mathrm{pdim}_R W_1 = 0$ then $\mathrm{pdim}_R W_0 \leq 1$.

(T101.4) Assume that R is regular, and let U be any nonzero nonunit ideal in R with $\mathrm{pdim}_R U \leq \dim(R) - 2$. Then $(U : J)_R = U$.

PROOF. To prove (T101.1), we have the exact sequence

$$0 \longrightarrow R \longrightarrow R \longrightarrow R' \longrightarrow 0$$

of R-modules where the second map is given by $t \mapsto xt$ and the third map is the residue class epimorphism; moreover, since $x \in S_R(R) \cap \mathrm{ann}_R R'$, the R-module R' cannot be free and hence by (T80.4) we get $\mathrm{pdim}_R R' > 0 = \mathrm{pdim}_R R$; therefore by (T80.8) and the above exact sequence we get $\mathrm{pdim}_R R' = 1$. Consequently by (T80.11) we conclude that $\mathrm{pdim}_R R'^d = 1$ for all $d \in \mathbb{N}_+$. We now proceed by induction on $\mathrm{pdim}_{R'} U'$. If $\mathrm{pdim}_{R'} U' = 0$ then $U' \approx R'^d$ for some $d \in \mathbb{N}$ and hence $\mathrm{pdim}_R U' \leq 1$. So let $\mathrm{pdim}_{R'} U' > 0$ and assume for all smaller values of $\mathrm{pdim}_{R'} U'$. Clearly we can take an exact sequence

$$0 \longrightarrow U^* \longrightarrow R'^d \longrightarrow U' \longrightarrow 0$$

of R'-modules where d is some positive integer and U^* is some R'-module, and then by (T80.8) we get $\mathrm{pdim}_{R'} U^* = (\mathrm{pdim}_{R'} U') - 1$, and hence by the induction hypothesis we have $\mathrm{pdim}_R U^* \leq \mathrm{pdim}_{R'} U'$. Therefore, regarding the above sequence as an exact sequence of R-modules, by (T80.10) we see that $\mathrm{pdim}_R U' \leq 1 + \mathrm{pdim}_{R'} U'$.

To prove (T101.2), we have the exact sequence

$$0 \longrightarrow V \longrightarrow V \longrightarrow V/xV \longrightarrow 0$$

of R-modules where the second map is given by $t \mapsto xt$ and the third map is the residue class epimorphism; consequently, since $\mathrm{pdim}_R V < \infty$, by (T80.10) we see that $\mathrm{pdim}_R(V/xV) < \infty$. Since $x \in J \cap S_R(V)$, by (T69.2) we have $\mathrm{reg}_R(V/xV) = (\mathrm{reg}_R V) - 1$. Since V is nonzero with $x \in J$, by Nakayama Lemma (Q2)(T3), V/xV is nonzero. Thus by (T81.1) we can conclude that $\mathrm{pdim}_R(V/xV) = 1 + \mathrm{pdim}_R V$.

To prove (T101.3), we first observe that $x \in S_R(W_2/V)$ implies $(xW_2) \cap V = xV$, and hence there exists a unique R-epimorphism $f_0 : W_1 \to W_0$ such that $\alpha = f_0 \beta$ where $\alpha : V \to W_0$ is the residue class epimorphism and $\beta : V \to W_1$ is the subset injection. Clearly f_0 is an epimorphism whose kernel is xW_1. Thus upon letting the map $f_1 : W_2 \to W_1$ be given by $f_1(t) = xt$ for all $t \in W_2$, we get a short exact sequence $(f, W, 1)$ of R-modules. Next, we can take a finite free R-module W_0' together with an R-epimorphism $f_1^{(0)} : W_0' \to W_0$. Since W_0' is projective (because it is free), we can find an R-homomorphism $\phi : W_0' \to W_1$ such that $f_0 \phi = f_1^{(0)}$. Let

$$f_1^{(2)} : W_2' = W_2 \to W_2 \text{ be the identity map,}$$

let

$$f_1' : W_2' \to W_1' = W_2' \oplus W_0' \text{ be the natural injection,}$$

let

$$f_0' : W_1' \to W_0' \text{ be the natural projection,}$$

and let

$$\begin{cases} f_1^{(1)} : W_1' \to W_1 \text{ be the map defined by} \\ f_1^{(1)}(s, t) = xs + \phi(t) \text{ for all } (s, t) \in W_2' \oplus W_0'. \end{cases}$$

This gives us a short exact sequence $(f', W', 1)$ of R-modules such that for $0 \le i \le 1$ we have $f_1^{(i)} f_i' = f_i f_1^{(i+1)}$. For $0 \le j \le 2$ let $f_2^{(j)} : W_j^{(3)} = \ker(f_1^{(j)}) \to W_j' = W_j^{(2)}$ be the subset injection and let $f_0^{(j)} : W_j^{(1)} = W_j \to W_j/\mathrm{im}(f_1^{(j)}) = W_j^{(0)}$ be the residue class epimorphism. Now we are in the situation of Lemma (T100), and hence by that Lemma $f_1^{(1)} : W_1' \to W_1$ is an R-epimorphism and there exists an R-isomorphism $f_3^* : W_1^{(3)} = W_4^* \to W_3^* = W_0^{(3)}$.

Now (by partially depicting the last two columns of the diagram of (T100)) we have the following two vertical short exact sequences together with the horizontal isomorphism $f_3^* : W_4^* \to W_3^*$.

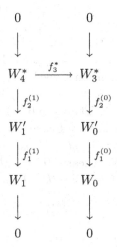

Notice that $W_0 = V/xV$ is nonzero by Nakayama, and hence it cannot be a free module because x is in its annihilator; consequently $\text{pdim}_R W_0 > 0$ by (T80.4). If $\text{pdim}_R W_1 = 0$ then applying (T80.8) to both the vertical exact sequences we get $\text{pdim}_R W_0 \le 1$. If $\text{pdim}_R W_1 > 0$ then again applying (T80.8) to both the vertical exact sequences we get $\text{pdim}_R W_1 = \text{pdim}_R W_0$, and hence (because $x \in S_R(R) = S_R(W_2) \subset S_R(V)$) by (T101.2) we conclude that $\text{pdim}_R W_1 = 1 + \text{pdim}_R V$.

To prove (T101.4), in view of (T98.1), it suffices to show that for every prime P in $\text{ass}_R(R/U)$ we have $J \not\subset P$, i.e., $\dim(R/P) > 0$, and hence in view of (T69.1)(2) it suffices to show that $\text{reg}_R(R/U) > 0$. By assumption we have $\text{pdim}_R U \le \dim(R) - 2$, and hence by (T80.10) we get $\text{pdim}_R(R/U) \le \dim(R) - 1$. Now we are done by noting that by the Auslander-Buchsbaum Theorem (T81.1) together with the Serre Theorem (T81.6) we have

$$\text{reg}_R(R/U) = \dim(R) - \text{pdim}_R(R/U).$$

UFD THEOREM (T102). Every regular local domain R is a UFD.

PROOF. In view of (T99.2), by induction on $\dim(R)$, it suffices to show that if $\dim(R) \ge 3$ then R is a UFD. In case $\dim(R) = 3$ we are reduced to (T99.3). So let $\dim(R) > 3$ and assume for all smaller values of $\dim(R)$. In view of (T99.1), given any ideal $V = aR + bR$ with $a \ne 0 \ne b$ in R, we want to show that $\text{pdim}_R V \le 1$. Let $J = M(R)$. If $a \notin J$ or $b \notin J$ then $V = R$ and hence $\text{pdim}_R V = 0$. So assume that $a \in J$ and $b \in J$. Since R is noetherian, the increasing sequence of ideals

$$V \subset (V : J)_R \subset (V : J^2)_R \subset \dots$$

must stop, i.e, for some $n \in \mathbb{N}$ we must have

$$(V : J^n)_R = (V : J^{n+1})_R.$$

Let $I = (V : J^n)_R$ and $L = (I : J)_R$. Then $L = I$, and hence by (T98.2) we can find $x \in J \setminus J^2$ with $L = (I : xR)_R$. Let $R' = R/(xR)$. By (Q15)(T71) we see that R' is a regular local domain whose dimension is one less than the dimension of R and hence, in view of (T99.1), upon letting $U = V + xR$ and $U' = U/(xR)$, by the induction hypothesis we get $\operatorname{pdim}_{R'} U' \leq 1$. Therefore by (T101.1) we see that $\operatorname{pdim}_R U' \leq 2$. Since $\operatorname{pdim}_R(xR) = 0$ (because xR is a free and hence a projective module over R), by (T80.10) we get $\operatorname{pdim}_R U \leq \operatorname{pdim}_R U'$. Consequently $\operatorname{pdim}_R U \leq 2$ and hence by (T101.4) we get $(U : J)_R = U$. Clearly $(U : J^{m+1})_R = ((U : J^m)_R : J)_R$ for all $m \in \mathbb{N}$, and hence $(U : J)_R = U$ implies $(U : J^n)_R = U$. Thus

$$I = (V : J^n)_R \subset (V + xR : J^n)_R = (U : J^n)_R = U = V + xR.$$

Consequently, every $u \in I$ can be written as $u = v + xc$ with $v \in V$ and $c \in R$; since $V \subset I$, we get $xc = u - v \in I$ and hence $c \in (I : xR)_R = I$. It follows that $I \subset V + xI$ and therefore $I = V + JI$. So by Nakayama Lemma (Q2)(T3) we get $I = V$ and hence $V = (V : xR)_R$. Consequently $x \in S_R(R/V)$.

   Therefore, since

$$U = V + xW_2 \quad \text{with} \quad W_2 = R,$$

by (T101.3) we see that

$$\begin{cases} \text{either } \operatorname{pdim}_R U = 0 \\ \text{or } \operatorname{pdim}_R U = 1 + \operatorname{pdim}_R V, \end{cases}$$

and moreover,

$$\begin{cases} \text{if } \operatorname{pdim}_R U = 0 \\ \text{then } \operatorname{pdim}_R(V/xV) \leq 1. \end{cases}$$

Since we also have $\operatorname{pdim}_R U \leq 2$, it follows that

$$\begin{cases} \text{either } \operatorname{pdim}_R(V/xV) \leq 1 \\ \text{or } \operatorname{pdim}_R V \leq 1. \end{cases}$$

Now $x \in S_R(R)$ implies $x \in S_R(V)$, and hence by (T101.2) we see that

$$\begin{cases} \text{if } \operatorname{pdim}_R(V/xV) \leq 1 \\ \text{then } \operatorname{pdim}_R V = 0. \end{cases}$$

Thus always $\operatorname{pdim}_R V \leq 1$.

COMMENT (C36). [**Finite Modules Over Local Rings**]. Let R be a local ring with $J = M(R)$, and let

$$\psi : R \to \overline{R} = R/J$$

be the residue class epimorphism. Given any finite R-module L, let

$$\phi : L \to \overline{L} = L/(JL)$$

be the residue class epimorphism. Let n be the dimension of $\overline{L}$ as a vector space over the field $\overline{R}$, and let $b_1, \ldots, b_n$ be elements in L such that $\phi(b_1), \ldots, \phi(b_n)$ is a basis of that vector space. Note that then by Nakayama (Q2)(T3), L is generated by $b_1, \ldots, b_n$ and hence we can take an R-epimorphism $t_0 : R^n \to L$ such that $t_0(e_i) = b_i$ for $1 \le i \le n$ where $e_1, \ldots, e_n$ is any given R-basis of R^n. We CLAIM that we must have $\ker(t_0) \subset JR^n$. To see this, taking summations over $1 \le i \le n$ we have

$$\begin{cases} \sum a_i e_i = e \in \ker(t_0) \text{ with } a_i \in R \text{ for all } i \\ \Rightarrow \sum a_i b_i = 0 \text{ by applying } t_0 \\ \Rightarrow \sum \psi(a_i)\phi(b_i) = 0 \text{ by applying } \phi \\ \Rightarrow \psi(a_i) = 0 \text{ for all } i \text{ because } \phi(b_1), \ldots, \phi(b_n) \text{ is an } \overline{R}\text{-basis of } \overline{L} \\ \Rightarrow a_i \in J \text{ for all } i \\ \Rightarrow e \in JR^n. \end{cases}$$

COMMENT (C37). [**Unique Factorization of Power Series**]. For the case when R is the ring of power series is a finite number of variables over a field, an alternative (concrete) proof of the UFD property of R, based on the Weierstrass Preparation Theorem, is given in my Local Analytic Geometry Book [A06].

QUEST (Q24) Graded Modules and Hilbert Polynomials

In §2 we talked about homogeneous ideals in a graded ring. We shall now generalize this to graded modules over a graded ring, and talk about their Hilbert functions. This will give rise to the Hilbert polynomial of a graded ring at a homogeneous ideal. In case of a polynomials ring, it turns out that the degree and the leading coefficient (= the coefficient of the highest degree term) of this polynomial correspond, respectively, to the dimension and the degree of the variety in projective space defined by the said homogeneous ideal. Likewise, as we shall see, the constant term of the Hilbert polynomial corresponds to the so-called arithmetic genus of the variety.

Referring to §2 for basic definitions, consider a graded ring

$$S = \sum_{i \in I} S_i \quad \text{with} \quad R = S_0 \quad \text{and} \quad \text{additive abelian monoid } I.$$

By a graded module over S (or a graded S-module) we mean an S-module

$$V = \sum_{i \in I} V_i \quad \text{(internal direct sum of } R\text{-modules)}$$

such that

for all $s_i \in S_i$ and $v_j \in v_j$ with i, j in I we have $s_i v_j \in V_{i+j}$.

The family $(V_i)_{i \in I}$ of R-submodules of V is called the gradation of V, and I is called the type of V. Elements of V_i are called homogeneous elements of degree i. Collectively, elements of $\cup_{i \in I} V_i$ are called homogeneous elements. The direct sum assumption tells us that every $v \in V$ has a unique expression

$$v = \sum_{i \in I} v_i$$

with $v_i \in V_i$; the element v_i is called the homogeneous component of v of degree i, and collectively the elements v_i are called the homogeneous components of v. Note that $v_i = 0$ for almost all i.

An S-submodule U of V is said to be a homogeneous submodule if it satisfies one and hence all of the following three mutually equivalent conditions:

(i) $U = \sum_{i \in I}(U \cap V_i)$ (as additive groups).
(ii) The homogeneous components of every element of U belong to U.
(iii) U is generated (as an S-module) by homogeneous elements.

Copying the proof of §2(1) we see that:

(1)   conditions (i), (ii), (iii) are mutually equivalent.

As in the case of an element, we call V_i the homogeneous component of V of degree i, and we note that for the homogeneous component U_i of U of degree i we have $U_i = U \cap V_i$. We use i-th homogeneous component as a synonym for homogeneous component of degree i. Also we may say component instead of homogeneous component.

Given any other graded S-module V', by a homogeneous S-homomorphism of degree $j \in I$ from V to V' we mean an S-homomorphism $\phi : V \to V'$ such that for all $i \in I$ we have $\phi(V_i) \subset V'_{i+j}$. By a homogeneous S-homomorphism we mean a homogeneous S-homomorphism of some degree. By a graded S-homomorphism of V to V' we mean a homogeneous S-homomorphism $\phi : V \to V'$ of degree 0.

Clearly:

(2)
$$\begin{cases} \text{if } \phi : V \to V' \text{ is any homogeneous } S\text{-homomorphism} \\ \text{then } \operatorname{im}(\phi) \text{ is a graded submodule of } V' \\ \text{and if either } I \text{ is cancellative or } \phi \text{ is of degree 0} \\ \text{then } \ker(\phi) \text{ is a graded submodule of } V. \end{cases}$$

Conversely, given any homogeneous submodule U of V, letting $\psi : V \to V/U$ be the canonical R-epimorphism and $(V/U)_i = \psi(V_i)$, we can convert V/U into a graded S-module with grading $(V/U)_{i \in I}$, and then clearly $\psi : V \to V/U$ becomes a graded S-epimorphism.

Thus:

(3)
$$\begin{cases} \text{if } U \text{ is any homogeneous submodule of } V \\ \text{then the canonical } R\text{-epimorphism } \psi : V \to V/U \\ \text{is a graded } S\text{-epimorphism.} \end{cases}$$

In view of (1) we see that:

(4)
$$\begin{cases} \text{if } (U_l)_{l \in L} \text{ is any family of} \\ \text{homogeneous submodules of } V \\ \text{then the submodules} \\ \sum_{l \in L} U_l \text{ and } \cap_{l \in L} U_l \text{ of } V \\ \text{are homogeneous.} \end{cases}$$

In view of (1) and (2) we see that:

(5)
$$\begin{cases} \text{if } \phi : V \to V' \text{ is any graded } S\text{-epimorphism} \\ \text{then } U \mapsto \phi(U) \text{ gives a bijection of} \\ \text{the set of all homogeneous submodules of } V \text{ containing } \ker(\phi) \\ \text{onto the set of all homogeneous submodules of } V'. \end{cases}$$

By a finite graded S-module we mean a graded S-module V which is finitely generated as an S-module. In view of (1), this is equivalent to saying that V is generated by a finite number of homogeneous elements.

Note that a homogeneous ideal in S is nothing but a homogeneous S-submodule of S where we regard S as a graded S-module in an obvious manner.

Now let S be a naturally graded ring, i.e., suppose that $I = \mathbb{N}$. Referring to (O21) and (O25) of L4§5 for the concepts of lengths of modules and artinian rings; assume that $R = S_0$ is an artinian ring. Given any graded module V over S, we put

$$\widehat{\mathfrak{h}}^{\mu}(S, V, n) = \ell_R(V_n)$$

and we call the function $\widehat{\mathfrak{h}}^{\mu}(S, V) : \mathbb{N} \to \mathbb{N} \cup \{\infty\}$, defined by $n \mapsto \ell_R(V_n)$, the modulized Hilbert function of S at V. In particular, given any homogeneous ideal Q in S, we note that S/Q is a graded module over S, and we put

$$\widehat{\mathfrak{h}}(S, Q, n) = \widehat{\mathfrak{h}}^{\mu}(S, S/Q, n) = \ell_R(S_n/Q_n)$$

and we call the function $\widehat{\mathfrak{h}}(S, Q) : \mathbb{N} \to \mathbb{N} \cup \{\infty\}$, defined by $n \mapsto \ell_R(S_n/Q_n)$, the Hilbert function of S at Q. Note that

$$\begin{cases} \text{if } R \text{ is a field} \\ \text{then } \ell_R(V_n) = [V_n : R] \\ \text{and } \ell_R(S_n/Q_n) = [S_n/Q_n : R]. \end{cases}$$

In a moment we shall prove the following:

HILBERT FUNCTION THEOREM (T103). Let $X_1, \ldots, X_N$ with $N \in \mathbb{N}$ be indeterminates over an artinian ring R. Let $S = R[X_1, \ldots, X_N]$ be the naturally graded polynomial ring with $R = S_0$. Let V be a finite graded module over S. Then for every nonnegative integer n we have

$$\ell_R(V_n) < \infty \quad \text{for all} \quad n \in \mathbb{N}$$

and there exists a unique univariate polynomial $\mathfrak{h}^{\mu}(S, V, Z)$ in an indeterminate Z with coefficients in $\mathbb{Q}$ such that

$$\mathfrak{h}^{\mu}(S, V, n) = \widehat{\mathfrak{h}}^{\mu}(S, V, n) \quad \text{for all} \quad n >> 0.$$

Moreover, upon letting

$$\mathfrak{t}^{\mu}(S, V) = \text{ the } Z\text{-degree of } \mathfrak{h}^{\mu}(S, V, Z)$$

we have

$$\mathfrak{t}^{\mu}(S, V) < N,$$

and if $d = \mathfrak{t}^{\mu}(S, V) \in \mathbb{N}$ then we have

$$\mathfrak{h}^{\mu}(S, V, Z) = \sum_{0 \le i \le d} \frac{\mathfrak{g}_i^{\mu}(S, V) Z(Z-1) \ldots (Z - i + 1)}{i!} \quad \text{with} \quad \mathfrak{g}_i^{\mu}(S, V) \in \mathbb{Z}$$

and upon letting

$$\mathfrak{g}^{\mu}(S, V) = \mathfrak{g}_d^{\mu}(S, V)$$

(and assuming $d = \mathfrak{t}^\mu(S, V) \in \mathbb{N}$) we have

$$\mathfrak{g}^\mu(S, V) > 0,$$

and if R is a field then (without assuming $\mathfrak{t}^\mu(S, V) \in \mathbb{N}$) we have

$$\mathfrak{t}^\mu(S, V) = \begin{cases} \dim(S/(0:V)_S) - 1 & \text{if } \dim(S/(0:V)S) > 0 \\ -\infty & \text{otherwise.} \end{cases}$$

We call $\mathfrak{h}^\mu(S, V, Z)$, $\mathfrak{t}^\mu(S, V)$, $\mathfrak{g}_i^\mu(S, V)$, $\mathfrak{g}^\mu(S, V)$, the modulized Hilbert polynomial, the modulized Hilbert transcendence, the modulized i-th Hilbert subdegree, and the modulized Hilbert degree of S at V, respectively.

[Note that, "for all $n >> 0$" means for all large (enough) n, i.e., for all integers n such that $n \geq n_0$ for some $n_0 \in \mathbb{N}$. Also note that if $V = S/Q$ where Q is a homogeneous ideal in S then $(0:V)_S = Q$].

Given any homogeneous ideal Q in S, with S and R as in (T103), we put

$$\begin{cases} \mathfrak{h}(S, Q, Z) = \mathfrak{h}^\mu(S, S/Q, Z) \\ \mathfrak{t}(S, Q) = \mathfrak{t}^\mu(S, S/Q) \\ \mathfrak{g}_i(S, Q) = \mathfrak{g}_i^\mu(S, S/Q) \\ \mathfrak{g}(S, Q) = \mathfrak{g}^\mu(S, S/Q) \end{cases}$$

and we call these the Hilbert polynomial, the Hilbert transcendence, the i-th Hilbert subdegree, and the Hilbert degree of S at Q, respectively.

Before turning to the following proof of (T103), in Quest (Q25) below, we shall illustrate the case when Q is the ideal of a hypersurface in a projective space over a field. Then in Quests (Q26) to (Q28) we shall include some auxiliary material to be used in the proof of (T103), which will be completed in parts (Q28.3) and (Q28.5) of (Q28).

QUEST (Q25) Hilbert Polynomial of a Hypersurface

In the naturally graded polynomial ring $S = k[X_1, \ldots, X_N]$ over a field k with integer $N > 1$, let Q be the homogeneous principal ideal in S generated by a nonzero homogeneous polynomial $f = f(X_1, \ldots, X_N)$ of degree $e > 0$. Then for every $n \geq e$ we have the k-isomorphism $S_{n-e} \to Q_n$ given by $\phi \mapsto f\phi$ and hence by (Q15)(C20) we get

$(\bullet)$ $\qquad [S_n/Q_n : k] = B_{N-1}(n + N - 1) - B_{N-1}(n + N - 1 - e)$

where for every $i \in \mathbb{N}$ we have put

$$B_i(Z) = \frac{Z(Z-1)\ldots(Z-i+1)}{i!} \in \mathbb{Q}[Z].$$

Given any i, j in $\mathbb{Z}$ with $i > 0$, we have

$$(i!)B_i(Z + j) = Z^i + \left(\sum_{j-i+1 \leq l \leq j} l\right) Z^{i-1} + \text{terms of degree} < i - 1$$

and hence

$$B_i(Z + j) = \frac{Z^i}{i!} + \frac{(2j - i + 1)Z^{i-1}}{(i-1)!2} + \text{terms of degree} < i - 1.$$

Therefore upon letting

$$H(Z) = B_{N-1}(Z + N - 1) - B_{N-1}(Z + N - 1 - e) \in \mathbb{Q}[Z]$$

we get

$$H(Z) = \sum_{0 \leq i \leq N-2} \frac{G_i Z(Z - 1) \ldots (Z - i + 1)}{i!} \quad \text{with} \quad G_i \in \mathbb{Q}$$

where

$$\deg_Z H(Z) = N - 2 \quad \text{with} \quad G_{N-2} = e$$

and by $(\bullet)$ we have

$$[S_n/Q_n : k] = H(n) \quad \text{for all} \quad n \geq e.$$

Note that $G_0 = H(0) = B_{N-1}(N - 1) - B_{N-1}(N - e - 1)$ and hence upon letting

$$p_a(S, Q) = (-1)^{N-2}(G_0 - 1)$$

we get

$$p_a(S, Q) = \frac{(e - 1)(e - 2) \ldots (e - N + 1)}{(N - 1)!}.$$

More generally, in the notation introduced just after (T103) and just before (Q25), for any homogeneous ideal Q in S, for which $d = \deg_Z \mathfrak{h}(S, Q, Z) \in \mathbb{N}$, we put

$$p_a(S, Q) = (-1)^d(\mathfrak{g}_0(S, Q) - 1)$$

and we call this the arithmetic genus of the variety in projective space defined by Q. More about this later. At present let us refer to pages 236-237 of my Resolution Book [A05] and pages 264-265 of my Engineering Book [A04].

Let us observe that if $N = 3$ and Q is the ideal generated by the e-degree nonzero homogeneous $f(X_1, X_2, X_3)$ then $p_a(S, Q) = \frac{(e-1)(e-2)}{2}$ which is the genus of the projective plane curve $f = 0$ assuming it to be nonsingular. Again more about this later. At present let us refer to pages 47-55 of the Engineering Book [A04] where it is shown that if $k = \mathbb{C}$ then the said curve is "topologically equivalent" to a sphere with

$$\frac{(e-1)(e-2)}{2}$$

handles.

QUEST (Q26) Homogeneous Submodules of Graded Modules

Again let I be an additive abelian monoid, and let S be a graded ring of type I over a ring R with gradation $(S_i)_{i \in I}$. Let us now generalize the material of §3 from the case of homogeneous ideals to homogeneous submodules of a graded S-module V.

Copying (1) and (2) of §3 we see that:

(1)
$$\begin{cases} \text{assuming } I \text{ to be cancellative,} \\ \text{for any homogeneous submodules } U \text{ and } T \text{ of } V, \\ \text{the ideal } (U:T)_S \text{ is homogeneous,} \\ \text{and for any homogeneous ideal } J \text{ in } S, \\ \text{the submodule } (U:J)_V \text{ is homogeneous,} \end{cases}$$

and:

(2)
$$\begin{cases} \text{assuming } I \text{ to be ordered,} \\ \text{for any homogeneous submodule } U \text{ of } V, \\ \text{the ideal } \mathrm{rad}_V U \text{ is homogeneous.} \end{cases}$$

Mimicking (3) and (4) of §3 we see that:

(3)
$$\begin{cases} \text{assuming } I \text{ to be ordered,} \\ \text{for any primary homogeneous submodule } U \text{ of } V, \\ \text{the prime ideal } \mathrm{rad}_V U \text{ in } S \text{ is homogeneous,} \end{cases}$$

and:

(4)
$$\begin{cases} \text{assuming } I \text{ to be ordered,} \\ \text{for any homogeneous submodule } U \neq V \text{ of } V, \text{ we have that:} \\ U \text{ is primary} \iff \text{ for all homogeneous elements } r \in S \text{ and } s \in V \\ \qquad\qquad \text{with } s \notin U \text{ and } rs \in U \text{ we have } r \in \mathrm{rad}_V U. \end{cases}$$

Copying (5) of §3 we see that:

(5)
$$\begin{cases} \text{assuming } I \text{ to be ordered,} \\ \text{and letting } (H)^* \text{ denote the submodule of } V \\ \text{generated by all homogeneous elements in } H \subset V, \\ \text{for any primary submodule } U \text{ of } V, \\ \text{the submodule } (U)^* \text{ is also primary.} \end{cases}$$

Recall that a primary decomposition of a submodule U of V is an expression of the form

$$U = Q_1 \cap \cdots \cap Q_n$$

where $Q_1, \ldots, Q_n$ are primary submodules of V with $n \in \mathbb{N}$, and this is an irredundant primary decomposition means the prime ideals $\mathrm{rad}_V Q_1, \ldots, \mathrm{rad}_V Q_n$ are pairwise distinct and for $1 \le i \le n$ we have

$$U \ne Q_1 \cap \cdots \cap Q_{i-1} \cap Q_{i+1} \cap \cdots \cap Q_n.$$

If the primary submodules $Q_1, \ldots, Q_n$ are all homogeneous then we call this a homogeneous primary decomposition or homogeneous irredundant primary decomposition respectively. Now copying (6) of §3 we see that:

(6)
$$\begin{cases} \text{assuming } I \text{ to be ordered,} \\ \text{if a homogeneous submodule } U \text{ of } V \text{ has a primary decomposition in } V \\ \text{then } U \text{ has a homogeneous irredundant primary decomposition in } V \\ \text{and all the associated primes of } U \text{ in } V \text{ are homogeneous.} \end{cases}$$

COMMENT (C38). [**Positive Portion and Irrelevant Submodules**]. In accordance with §3(C4), assuming I to be ordered, we define the positive portion $\Omega(V)$ of V by putting

$$\Omega(V) = \sum_{\{i \in I : i > 0\}} V_i$$

and we note that this is an R-submodule of V, and if S is nonnegatively graded then it is actually a homogeneous submodule of V. Assuming I to be ordered, a submodule U of V is said to be irrelevant or relevant according as $\Omega(S) \subset \mathrm{rad}_V U$ or $\Omega(S) \not\subset \mathrm{rad}_V U$.

COMMENT (C39). [**Relevant Portion and Isolated System of Prime Ideals**]. In accordance with §3(C5), assuming S to be a nonnegatively graded noetherian ring and V to be a finite graded S-module, given any submodule U of V, let $U = Q_1 \cap \cdots \cap Q_n$ be an irredundant primary decomposition of U in V with $\mathrm{rad}_V Q_i = P_i$. By a suitable labelling we may assume that $Q_1, \ldots, Q_m$ are relevant and $Q_{m+1}, \ldots, Q_n$ are irrelevant. By L4§5(O24)(22•) it follows that $P_1, \ldots, P_m$ is an "isolated system of prime ideals" of U in V, i.e., for $1 \le i \le m$ and $m+1 \le j \le n$ we have $P_j \not\subset P_i$, and hence by L4§6.1(T6′) we see that the intersection $Q_1 \cap \cdots \cap Q_m$ depends only on U and not on the particular irredundant primary decomposition. We define the relevant portion of U by putting

$$(U)^\sharp = Q_1 \cap \cdots \cap Q_m.$$

If U is homogeneous then in view of (6) we can arrange $Q_1, \ldots, Q_n$ to be homogeneous and hence $(U)^\sharp$ is homogeneous.

Mimicking (7) and (8) of §3 we get (7) and (8) stated below; for the definition of a semihomogeneous ring see §3(C6).

(7) $\begin{cases} \text{assuming } S \text{ to be a noetherian semihomogeneous ring over } R \\ \text{and } V \text{ to be a finite graded } S\text{-module,} \\ \text{for any homogeneous submodule } U \text{ of } V \text{ we have} \\ U \cap V_d = (U)^\sharp \cap V_d \text{ for all large enough } d \in \mathbb{N}_+, \\ \text{and moreover } (U)^\sharp \text{ is the largest such homogeneous submodule,} \\ \text{i.e., if } H \text{ is any homogeneous submodule of } V \text{ such that} \\ H \cap V_d = (U)^\sharp \cap V_d \text{ for all large enough } d \in \mathbb{N}_+ \\ \text{then } H \subset (U)^\sharp \text{ and } (H)^\sharp = (U)^\sharp, \end{cases}$

and:

(8) $\begin{cases} \text{assuming } S \text{ to be a noetherian semihomogeneous ring over } R \\ \text{and } V \text{ to be a finite graded } S\text{-module,} \\ \text{for any homogeneous submodule } U \text{ of } V \text{ we have} \\ (U)^\sharp = (U : \Omega(S)^d)_V \text{ for all large enough } d \in \mathbb{N}_+. \end{cases}$

QUEST (Q27) Homogeneous Normalization

In a moment we shall prove the following homogeneous version (T105) of the Normalization Theorem (T46) of (Q10); in Comment (C42) below we shall compare the two versions. As an aid to proving (T105), we shall first prove the following Lemma (T104); for the definition of an irrelevant ideal see §3(C4), and for the definition of a homogeneous ring see §3(C6); for further remarks on (T104) see Comment (C40) below.

IRRELEVANT IDEAL LEMMA (T104). Let S be a homogeneous ring over a ring R, and let $z_1, \ldots, z_m$ be a finite number of homogeneous elements in $S \setminus R$. Then S is integral over $R[z_1, \ldots, z_m]$ iff the ideal $(z_1, \ldots, z_m)S$ is irrelevant.

HOMOGENEOUS NORMALIZATION THEOREM (T105). In the naturally graded polynomial ring $S = k[X_1, \ldots, X_N]$ in indeterminates $X_1, \ldots, X_N$ over a field k with $N \in \mathbb{N}$ we have the following.

(T105.1) Given any homogeneous element Z_N in $S \setminus k$, there exist homogeneous elements $Z_1, \ldots, Z_{N-1}$ in $S \setminus k$ such that S is integral over $k[Z_1, \ldots, Z_N]$.

(T105.2) Given any nonunit homogeneous ideal J in S, there exists a positive integer c together with nonzero homogeneous polynomials $Y_1, \ldots, Y_N$ in S of degree c such that S is integral over $D = k[Y_1, \ldots, Y_N]$ and $J \cap D = (Y_{r+1}, \ldots, Y_N)D$ where $r = \dim(S/J)$.

(T105.3) Given any finite sequence $J_0 \subset \cdots \subset J_m$ of nonunit homogeneous ideals in S with $m \in \mathbb{N}$, there exists a positive integer c together with nonzero homogeneous polynomials $Y_1, \ldots, Y_N$ in S of degree c such that S is integral over $D = k[Y_1, \ldots, Y_N]$ and $J_i \cap D = (Y_{r_i+1}, \ldots, Y_N)D$ for $0 \le i \le m$ where we have $r_i = \dim(S/J_i)$. Moreover, for $0 \le i \le m$ and any such polynomials $Y_1, \ldots, Y_N$, upon letting $\phi_i : S \to S/J_i$ be the residue class epimorphism, the ring $\phi_i(S)$ is integral over the subring $\phi_i(D)$, and the latter ring is naturally isomorphic to the polynomial ring $T_i = k[X_1, \ldots, X_{r_i}]$ by the isomorphism $\psi_i : T_i \to \phi_i(D)$ such that $\psi_i(\kappa) = \phi_i(\kappa)$ for all $\kappa \in k$ and $\psi_i(X_j) = \phi_i(Y_j)$ for $1 \le j \le r_i$.

PROOF OF (T104). The assertion being obvious when R is null ring, let R be a nonnull ring. We are assuming that the R-module S_1 is generated by a finite number of elements $x_1, \ldots, x_n$. Upon replacing the elements $z_1, \ldots, z_m$ by their suitable powers, we may also assume that all the elements $z_1, \ldots, z_m$ belong to S_d for some common $d \in \mathbb{N}_+$. Then

$$z_i = V_i(x_1, \ldots, x_n) \quad \text{with} \quad V_i(x_1, \ldots, X_n) \in R[X_1, \ldots, X_n]_d$$

where $R[X_1, \ldots, X_n]$ is the naturally graded polynomial ring in indeterminates $X_1, \ldots, X_n$ over R.

First suppose that S is integral over $R[z_1, \ldots, z_m]$. For $1 \le i \le n$ let

$$x_i^{e_i} + \sum_{1 \le j \le e_i} u_{ij} x_i^{e_i - j} = 0 \quad \text{with} \quad e_i \in \mathbb{N}_+ \quad \text{and} \quad u_{ij} \in R[z_1, \ldots, z_m]$$

be an equation of integral dependence. Now

$$u_{ij} = \sum_{0 \le k \le \kappa} U_{ijk}(z_1, \ldots, z_m) \quad \text{with} \quad U_{ijk}(Z_1, \ldots, Z_m) \in R[Z_1, \ldots, Z_m]_k$$

where $R[Z_1, \ldots, Z_m]$ is the naturally graded polynomial ring in indeterminates $Z_1, \ldots, Z_m$ over R and κ is some positive integer. Upon letting $W_i(X_1, \ldots, X_n)$ in $R[X_1, \ldots, X_n]$ be defined by putting

$$W_i(X_1, \ldots, X_n) = X_i^{e_i} + \sum_{1 \le j \le e_i} \sum_{0 \le k \le \kappa} V_{ijk}(X_1, \ldots, X_n) X_i^{e_i - j},$$

where

$$\begin{cases} V_{ijk}(X_1,\ldots,X_n) \\ = U_{ijk}(V_1(X_1,\ldots,X_n),\ldots,V_m(X_1,\ldots,X_n)) \in R[X_1,\ldots,X_n]_{dk}, \end{cases}$$

the above equation of integral dependence tells us that $W_i(x_1,\ldots,x_n) = 0$. Upon letting

$$L'_j = \{l \in \mathbb{N}_+ : l \le \kappa \text{ with } ld = j\}$$

and

$$L''_j = \{l \in \mathbb{N}_+ : l \le \kappa \text{ with } ld \ne j\}$$

we get

$$W_i(X_1,\ldots,X_n) = W'_i(X_1,\ldots,X_n) + W''_i(X_1,\ldots,X_n)$$

with

$$\begin{cases} W'_i(X_1,\ldots,X_n) \\ = X_i^{e_i} + \sum_{1\le j\le e_i} \sum_{k\in L'_j} V_{ijk}(X_1,\ldots,X_n)X_i^{e_i-j} \in R[X_1,\ldots,X_n]_{e_i} \end{cases}$$

and

$$\begin{cases} W''_i(X_1,\ldots,X_n) \\ = \sum_{1\le j\le e_i} \sum_{k\in L''_j} V_{ijk}(X_1,\ldots,X_n)X_i^{e_i-j} \in R[X_1,\ldots,X_n]''_i \end{cases}$$

where

$$R[X_1,\ldots,X_n]''_i = \sum_{0\le p\le\kappa d \text{ with } p\ne e_i} R[X_1,\ldots,X_n]_p.$$

It follows that $W'_i(x_1,\ldots,x_n) = 0$ and hence (because $L'_{ij} \subset \mathbb{N}_+$) we see that $x_i^{e_i} \in (z_1,\ldots,z_m)S$. This being so for $1 \le i \le n$, we conclude that the ideal $(z_1,\ldots,z_m)S$ is irrelevant.

Next suppose that the ideal $(z_1,\ldots,z_m)S$ is irrelevant. In view of L4§10(E2), to prove that S is integral over $R[z_1,\ldots,z_m]$, it suffices to show that S is a finite $R[z_1,\ldots,z_m]$-module. The irrelevancy of $(z_1,\ldots,z_m)S$ tells us that all high enough powers of the elements $x_1,\ldots,x_n$ belong to $(z_1,\ldots,z_m)S$ and hence we can find an integer $k \ge d$ such that $P_i \subset (z_1,\ldots,z_m)S$ for all $i \ge k$ where P_i is the set of all monomials $x_1^{i_1}\ldots x_n^{i_n}$ with nonnegative integers $i_1,\ldots,i_n$ for which $i_1+\cdots+i_n = i$. Let A be the finite $R[z_1,\ldots,z_m]$-module generated by $\cup_{0\le i<k}P_i$. By induction on i we shall show that $P_i \subset A$ for all $i \in \mathbb{N}$, and this will establish the finite $R[z_1,\ldots,z_m]$-moduleness of S. Our assertion being obvious for $i < k$, let $i \ge k$ and assume for smaller i. Given any $r \in P_i$ we can write $r = st$ with $s \in P_k$ and $t \in P_{i-k}$. Now $s = y_1z_1+\cdots+y_mz_m$ with $y_1,\ldots,y_m$ in S_{k-d}, and hence $r = ty_1z_1+\cdots+ty_mz_m$

with $ty_1, \ldots, ty_m$ in S_j where $0 \le j = (i-k) + (k-d) = i-d < i$. Consequently, by the induction hypothesis, the elements $ty_1, \ldots, ty_m$ belong to A and hence $r \in A$.

PROOF OF (T105.1). Since Z_N in $S \setminus k$ is given, we must have $N > 0$. If k is infinite then we can argue as in the proof of the Special Case of (Q10)(T46). Namely, upon letting d to be the degree of Z_N, by L3§12(D4) there exists a k-linear automorphism θ of S such that

$$\theta(Z_N) = F_0 X_N^d + \sum_{1 \le i \le d} F_i X_N^{d-i}$$

with $0 \ne F_0 \in k$ and $F_i \in k[X_1, \ldots, X_{N-1}]$ for $1 \le i \le d$. Let $Z_j = \theta^{-1}(X_j)$ for $1 \le j \le N-1$. Then $Z_1, \ldots, Z_{N-1}$ are homogeneous elements in $S \setminus k$ such that S is integral over $k[Z_1, \ldots, Z_N]$.

The $N = 1$ case being obvious, here is a proof for the $N = 2$ case without assuming k to be infinite. Now $Z_2 = H(X_1, X_2)$ where $H(X_1, X_2)$ is a nonzero homogeneous polynomial of degree $e > 0$ with coefficients in k. We get $H(X_1, X_2) = X_1^e H(1, X_2/X_1)$ where $H(1, T)$ is a nonzero polynomial in an indeterminate T with coefficients in k. Upon letting e' be the degree of $H(1, T)$ we have $0 \le e' \le e$. Clearly we can take $G(T) \in k[T] \setminus k$, of some degree $d > 0$, such that $G(T)$ and $H(1, T)$ have no common factor in $k[T] \setminus k$. As in the proof of Basic Hensel's Lemma (T2) of L3§6, every $F(T) \in k[T]$ of degree $\le d + e' - 1$ can be written as

$$F(T) = H_F(T)G(T) + G_F(T)H(1, T)$$

with $H_F(T) \in k[T]$ of degree $< e' \le e$ and $G_F(T) \in k[T]$ of degree $< d$. In particular, for every nonnegative integer $q \le d + e - 1$ we have

$$T^q = H_q(T)G(T) + G_q(T)H(1, T)$$

with $H_q(T) \in k[T]$ of degree $< e$ and $G_q(T) \in k[T]$ of degree $< d$. Writing X_2/X_1 for T in this equation for T^q and multiplying both sides by X_1^{d+e-1} we get

$$X_1^{d+e-1-q} X_2^q = H_q'(X_1, X_2)Z_1 + G_q'(X_1, X_2)Z_2$$

where

$$H_q'(X_1, X_2) = X_1^{e-1} H_q(X_2/X_1) \in S_{e-1}$$

and

$$G_q'(X_1, X_2) = X_1^{d-1} G_q(X_2/X_1) \in S_{d-1}$$

with

$$Z_1 = X_1^d G(X_2/X_1) \in S_d.$$

Therefore $\Omega(S)^{d+e-1} \subset (Z_1, Z_2)S$ and hence by (T104) we see that S is integral over $k[Z_1, Z_2]$.

Since every finite field is of nonzero characteristic, it only remains to show that, assuming k to be of characteristic $p \neq 0$ and assuming the $N - 1$ case, we can do the case of any given $N \geq 3$. Now $Z_N = H(X_1, \ldots, X_N)$ where $H(X_1, \ldots, X_N)$ is a nonzero homogeneous polynomial of degree $e > 0$ with coefficients in k. Suitably relabelling $X_1, \ldots, X_N$ we can arrange that $H(X_1, \ldots, X_N) \notin k[X_1, \ldots, X_{N-1}]$. Thus

$$Z_N = H(X_1, \ldots, X_N) \in S_e \setminus S' \quad \text{with} \quad S' = k[X_1, \ldots, X_{N-1}].$$

Given any power $q > 0$ of p we have

$$H(X_1, \ldots, X_N)^q = H_q(X_1^q, \ldots, X_N^q) \quad \text{where} \quad 0 \neq H_q(X_1, \ldots, X_N) \in S_e$$

is obtained by raising every coefficient of H to its q-th power. We Claim that for some power $q > 0$ of p we have

$$\begin{cases} 0 \neq Z'_{N-1} \in S'_{qe} \\ \text{where } Z'_{N-1} = H_q(X_1^q, \ldots, X_{N-1}^q, G(X_1, \ldots, X_{N-1})) \\ \text{for some } G(X_1, \ldots, X_{N-1}) \in S'_q. \end{cases}$$

To complete the proof assuming this Claim, in view of (T104), by the $N-1$ case we can find homogeneous elements $Z_1, \ldots, Z_{N-2}$ in $S' \setminus k$ such that the ideal in S' generated by $Z_1, \ldots, Z_{N-2}, Z'_{N-1}$ is irrelevant, i.e., it contains S'_n for all $n \gg 0$. Let $Z_{N-1} = X_N^q - G(X_1, \ldots, X_{N-1})$. Then $Z_{N-1} \in S_q \setminus k$ and by the formula $x^i - y^i = (x - y)(x^{i-1} + x^{i-2}y + \cdots + y^{i-1})$ we see that $Z'_{N-1} - Z_N^q \in Z_{N-1}S$. Therefore the ideal in S generated by $Z_1, \ldots, Z_N$ is irrelevant and hence by (T104) we see that S is integral over $k[Z_1, \ldots, Z_N]$.

To prove the Claim, since $N - 1 \geq 2$, for every power $q > 0$ of p, there exists $G_q = G_q(X_1, \ldots, X_{N-1}) \in S'_q \setminus (S')^p$ where $(S')^p = \{G^p : G \in S'\}$, and we can take F_q in a fixed algebraic closure L of $K = k(X_1, \ldots, X_{N-1})$ with $F_q^q = G_q$. By an Obvious Fact included in Comment (C41) below, which discusses multiple roots, we have $[K(F_q) : K] = q$ and hence $F_q \neq F_{q'}$ for any two distinct powers $q > 0$ and $q' > 0$ of p. Since $H(X_1, \ldots, X_N)$, as a polynomial in X_N with coefficients in K, has only a finite number of distinct roots in L, we must have $H(X_1, \ldots, X_{N-1}, F_q) \neq 0$ for some q. It suffices to take $G(X_1, \ldots, X_{N-1}) = G_q(X_1, \ldots, X_{N-1})$ for this q.

PROOF OF (T105.2). For $N = 0$ there is nothing to show. So assume that $N > 0$. Let W be the set of all sequences $Y = (Y_1, \ldots, Y_N)$ in $S \setminus k$ such that $Y_1, \ldots, Y_N$ belong to $S_{c(Y)}$ for some (obviously unique) positive integer $c(Y)$ and S is integral over $D(Y) = k[Y_1, \ldots, Y_N]$. For every such sequence Y let $a(Y)$ be the smallest integer with $0 \leq a(Y) \leq N$ such that $Y_j \in J$ for $a(Y) + 1 \leq j \leq N$. Since W contains $X = (X_1, \ldots, X_N)$, it is nonempty, and therefore we may put $a(W) = \min\{a(Y) : Y \in W\}$ giving us an integer $a(W)$ with $0 \leq a(W) \leq N$, and we may fix some $Y \in W$ with $a(Y) = a(W)$. We shall show that then $a(Y) = \dim(S/J)$ with $J \cap D(Y) = (Y_{a(Y)+1}, \ldots, Y_N)D(Y)$ and this will complete the proof.

Since S is integral over $D(Y)$, the elements $Y_1, \ldots, Y_N$ must be algebraically independent over k. Hence, upon letting $S^* = k[X_1^*, \ldots, X_{a(Y)}^*]$ be the naturally graded polynomial ring in indeterminates $X_1^*, \ldots, X_{a(Y)}^*$, there is a unique k-isomorphism of rings $\alpha : S^* \to E(Y) = k[Y_1, \ldots, Y_{a(Y)}]$ such that $\alpha(X_i^*) = Y_i$ for $1 \leq i \leq a(Y)$.

Suppose if possible that $J \cap E(Y) \neq 0$. Then we must have $a(Y) > 0$ and we can find $0 \neq Z_{a(Y)} \in J \cap E(Y) \cap S_e$ for some $e \in \mathbb{N}$. Let $b(a(Y)) = e/c(Y)$. Then we must have $b(a(Y)) \in \mathbb{N}_+$ and upon letting $Z_{a(Y)}^* = \alpha^{-1}(Z_{a(Y)})$ we get $0 \neq Z_{a(Y)}^* \in S_{b(a(Y))}^*$. By (T105.1) we can find $0 \neq Z_i^* \in S_{b(i)}^*$ for $1 \leq i < a(Y)$ with $b(i) \in \mathbb{N}_+$ such that S^* is integral over $k[Z_1^*, \ldots, Z_{a(Y)}^*]$. Let $b = b(1) \ldots b(a(Y))$ and $Z_i = \alpha((Z_i^*)^{b/b(i)})$ for $1 \leq i \leq a(Y)$. Let $c(Z) = c(Y)b$ and $Z_i = Y_i^b$ for $a(Y) + 1 \leq i \leq N$. Then for $Z = (Z_1, \ldots, Z_N)$ we have $Z_i \in S_{c(Z)}$ for $1 \leq i \leq N$ with $c(Z) \in \mathbb{N}_+$. Clearly S is integral over $k[Z_1, \ldots, Z_N]$ and hence $Z \in W$ with $a(Z) \leq a(Y) - 1$ which is a contradiction. Therefore we must have $J \cap E(Y) = 0$.

Consequently $J \cap D(Y) = (Y_{a(Y)+1}, \ldots, Y_N)D(Y)$ and the ring $D(Y)/(J \cap D(Y))$ is isomorphic to the $a(Y)$-variable polynomial ring S^* over the field k. Since S is integral over $D(Y)$, it follows that S/J is integral over $D(Y)/(J \cap D(Y))$. Therefore either by (Q7)(T30.4) + (Q9)(T45.3) or by (Q10)(T47.7) we get $\dim(S/J) = a(Y)$. This completes the proof.

PROOF OF (T105.3). Some of the reasoning and references from the above proof of (T105.2) will now be used tacitly. To prove the first sentence we make induction on m. The $m = 0$ case follows from (T105.2). So let $m > 0$ and assume for all smaller values of m. Then by the induction hypothesis there exists a positive integer d together with nonzero homogeneous polynomials $Z_1, \ldots, Z_N$ in S of degree d such that S is integral over $E = k[Z_1, \ldots, Z_N]$ and $J_i \cap E = (Z_{r_i+1}, \ldots, Z_N)E$ for $0 \leq i \leq m - 1$ where $r_i = \dim(S/J_i)$. Let $M = r_{m-1}$, let $E' = k[Z_1, \ldots, Z_M]$, and let $S' = k[X_1', \ldots, X_M']$ be the naturally graded polynomial ring in indeterminates $X_1', \ldots, X_M'$. Then there is a unique k-isomorphism of rings $\alpha : S' \to E'$ such that $\alpha(X_i') = Z_i$ for $1 \leq i \leq M$. Let $J' = \alpha^{-1}(J_m \cap E')$. Then J' is a nonunit homogeneous ideal in S', and hence by (T105.2) there exists a positive integer d' together with homogeneous elements $Z_1', \ldots, Z_M'$ of degree d' in $S' \setminus k$ such that S' is integral over $D' = k[Z_1', \ldots, Z_M']$ and $J' \cap D' = (Z_{r'+1}', \ldots, Z_M')D'$ with $\dim(S'/J') = r'$. Now it suffices to take $c = dd'$ and $r_m = r'$ with $Y_i = \alpha(Z_i')$ for $1 \leq i \leq M$ and $Y_i = Z_i^{d'}$ for $M + 1 \leq i \leq N$.

The second sentence "Moreover ... " follows by using the "tacit" material.

COMMENT (C40). [**Irrelevant Ideals and Integral Dependence**]. In proving and applying (T104) we have tacitly used (and will continue to use) the following obvious facts. Let S be a homogeneous ring over a ring R, let $z_1, \ldots, z_m$ be a finite number of homogeneous elements in $S \setminus R$, and for $1 \leq i \leq m$ let $t_i = z_i^{d_i}$ with $d_i \in \mathbb{N}_+$. Then S is integral over $R[z_1, \ldots, z_m]$ iff S is integral over $R[t_1, \ldots, t_m]$. Likewise, the ideal $(z_1, \ldots, z_m)S$ is irrelevant iff the ideal $(t_1, \ldots, t_m)S$ is irrelevant. If $z_i \in S_{d_i}$ with $d_i \in \mathbb{N}_+$ for $1 \leq i \leq m$ then upon letting $d = d_1 \ldots d_m$ and $t_i = z_i^{d/d_i}$ we get $t_i \in S_d$. Thus, in checking the irrelevancy or the integralness of $S/R[z_1, \ldots, z_m]$ we may assume that the elements $z_1, \ldots, z_m$ are homogeneous of the same positive degree.

COMMENT (C41). [**Multiple Roots and Separable Polynomials**]. Before proving the Obvious Fact cited in the proof of (T105.1), recall that according to L1§8, given a nonconstant polynomial

$$f(Y) = a_0 Y^n + a_1 Y^{n-1} + \cdots + a_n = a_0 (Y - \alpha_1) \ldots (Y - \alpha_n)$$

of degree $n \in \mathbb{N}$ and coefficients $a_0, a_1, \ldots, a_n$ in a field K, we defined the polynomial f to be separable to mean that its roots $\alpha_1, \ldots, \alpha_n$ in an algebraic closure L of K are distinct, and we remarked that this is so iff f and its Y-derivative f_Y have no nonconstant common factor in $K[Y]$. This follows by noting that if $\alpha \in L$ and $m \in \mathbb{N}_+$ are such that

(†) $f(Y) = (Y - \alpha)^m g(Y)$ where $g(Y) \in L[Y]$ with $g(\alpha) \neq 0$

then differentiating both sides with respect to Y we get

$$f_Y(Y) = m(Y - \alpha)^{m-1} g(Y) + (Y - \alpha)^m g_Y(Y)$$

and hence upon letting p to be the characteristic of K (which may or may not be zero) we see that:

(1)   (†) $\Rightarrow f_Y(Y) = \begin{cases} (Y - \alpha)^{m-1} h(Y) \text{ where } h(Y) \in K[Y] \text{ with } h(\alpha) \neq 0 \\ \qquad\qquad\qquad\qquad\qquad\qquad \text{if } m \notin p\mathbb{Z} \\ (Y - \alpha)^m h(Y) \text{ where } h(Y) \in K[Y] \quad \text{if } m \in p\mathbb{Z}. \end{cases}$

Referring to L1§10(D3) for the definition of minimal polynomial, as a consequence of (1) we see that:

(2) $\begin{cases} \text{if } f(Y) \text{ is the minimal polynomial of } y \text{ over } K \text{ then:} \\ f(Y) \text{ is separable} \\ \Leftrightarrow f_Y(Y) \neq 0 \\ \Leftrightarrow f_Y(y) \neq 0 \\ \Leftrightarrow f(Y) \text{ and } f_Y(Y) \text{ have no nonconstant common factor in } K[Y]. \end{cases}$

Now the said Obvious Fact says that, given any $u \in K \setminus K^p$ where K is a field of characteristic $p \neq 0$ then for any $q = p^n$ with $n \in \mathbb{N}_+$ and any v in an overfield of K with $v^q = u$, we have $[K(v) : K] = q$. By induction on n, this follows from the Assertion saying that, for any y in an overfield of K with $y^p = x \in K \setminus K^p$ we have $[K(y) : K] = p$ and $y \notin K(y)^p$. To prove the first part of the Assertion, since y is a root of the polynomial $Y^p - x$, it suffices to show that $Y^p - x$ is the minimal polynomial of y over K. Otherwise we would have $Y^p - x = A(Y)B(Y)$ where $A(Y), B(Y)$ are monic polynomials of respective positive degrees a, b in Y with coefficients in K, and $A(Y)$ is the minimal polynomial of y over K. Since $Y^p - x = (Y - y)^p$, we must have $A(Y) = (Y - y)^a$ and therefore (because $0 < a < p$) by (2) we get $a = 1$ which contradicts the fact that $y \notin K^p$. To prove the second part of the Assertion, note that if $y \in K(y)^p$ then $y = (b_0 + b_1 y + \cdots + b_{p-1} y^{p-1})^p$ with $b_0, b_1, \ldots, b_{p-1}$ in K and hence $y = b_0^p + b_1^p x + \cdots + b_{p-1}^p x^{p-1} \in K$ which is a contradiction.

COMMENT (C42). [**Homogeneous and Nonhomogeneous Normalization Theorems**]. In the preamble to (T46) of (Q10) we noted that the length r of any normalization basis of an affine ring A equals the dimension of A. Similarly in (T105.2), the fact that the integer r equals $\dim(S/J)$ is not used in the rest of the proof but inserted at the end of it as a consequence of (Q7)(T30.4) + (Q9)(T45.7) or (Q10)(T47.7). Thus (T105.2) may be restated as saying that: given any nonunit homogeneous ideal J in S, there exists a positive integer c together with nonzero homogeneous polynomials $Y_1, \ldots, Y_N$ in S of degree c such that S is integral over $D = k[Y_1, \ldots, Y_N]$ and $J \cap D = (Y_{r+1}, \ldots, Y_N)D$ where r is some integer with $0 \leq r \leq N$; actually r equals $\dim(S/J)$ and is hence independent of the particular homogeneous polynomials $Y_1, \ldots, Y_N$ with the stated properties. Analogous remarks apply to (T105.3). Upon letting $A = S/J_0$ with $I_j = J_j/J_0$ for $1 \leq j \leq m$ in (T105.3) (and identifying k with its image in A), we get the following.

HOMOGENEOUS RING NORMALIZATION THEOREM (T106). Given any homogeneous ring over a field k, and given any finite sequence $I_1 \subset \cdots \subset I_m$ of nonunit homogeneous ideals in A with $m \in \mathbb{N}$, there exists a positive integer c together with nonzero homogeneous elements $Y_1, \ldots, Y_r$ in A of degree c with $\dim(A) = r$ such that the elements $Y_1, \ldots, Y_r$ are algebraically independent over k and such that the ring A is integral over the subring $D = k[Y_1, \ldots, Y_r]$ and $I_j \cap D = (Y_{r_j+1}, \ldots, Y_r)D$ for $1 \leq j \leq m$ where $r_j = \dim(A/I_j)$. Moreover, for $1 \leq j \leq m$ and any such elements $Y_1, \ldots, Y_r$, upon letting $\phi_j : A \to A/I_j$ be the residue class epimorphism, the ring $\phi_j(A)$ in integral over the subring $\phi_j(D)$, and the latter ring is naturally isomorphic to the polynomial ring $T_j = k[X_1, \ldots, X_{r_j}]$ by the isomorphism $\psi_j : T_j \to \phi_j(D)$ such that $\psi_j(\kappa) = \phi_j(\kappa)$ for all $\kappa \in k$ and $\psi_j(X_i) = \phi_j(Y_i)$ for $1 \leq i \leq r_j$.

QUEST (Q28) Alternating Sum of Lengths

As an analogue of assertion (11) of (Q17)(C30), in a moment we shall prove the following.

(Q28.1) Let (f, W, n) be an exact sequence of modules over a ring R such that $\ell_R(W_i) < \infty$ for $0 \leq i \leq n+1$. Then $\sum_{0 \leq i \leq n+1} (-1)^i \ell_R(W_i) = 0$.

We shall also prove the following.

(Q28.2) Let I be a cancellative additive abelian monoid, let S be a graded ring of type I over a ring $R = S_0$ such that S_j is a finite R-module for all $j \in I$, and let V be a finite graded S-module. Then V_i is a finite R-module for all $i \in I$. Moreover, if the ring R is artinian, then $\ell_R(V_i) < \infty$ for all $i \in I$.

Then, as a consequence of (Q28.1) and (Q28.2) we shall prove the following.

(Q28.3) For every $i \in \mathbb{N}$ consider the univariate polynomial $B_i(Z)$ of degree i with coefficients in $\mathbb{Q}$ given by $B_i(Z) = \frac{Z(Z-1)\ldots(Z-i+1)}{i!}$ and note that $B_0(Z) = 1$. Let $X_1, \ldots, X_N$ with $N \in \mathbb{N}$ be indeterminates over an artinian ring R, and let $S = R[X_1, \ldots, X_N]$ be the naturally graded polynomial ring with $R = S_0$. Let V be a finite graded module over S. Then $\ell_R(V_n) < \infty$ for all $n \in \mathbb{N}$ and there exists a unique $H(Z) \in \mathbb{Q}[Z]$ such that $H(n) = \ell_R(V_n)$ for all $n >> 0$. Moreover, upon letting d be the Z-degree of $H(Z)$ we have $d < N$, and if $d \in \mathbb{N}$ then we have $H(Z) = \sum_{0 \leq i \leq d} G_i B_i(Z)$ with $G_i \in \mathbb{Z}$ and $G_d > 0$.

As a supplement to (Q28.3) we shall prove the following.

(Q28.4) In the situation of (Q28.3), let $c \in \mathbb{N}_+$, let $S^* = R[X_1^*, \ldots, X_N^*]$ be the naturally graded polynomial ring in indeterminates $X_1^*, \ldots, X_N^*$ over R, and let $\alpha : S^* \to S$ be an R-monomorphism of rings such that $\alpha(X_1^*), \ldots, \alpha(X_N^*)$ belong to S_c, and S is integral over $\alpha(S^*)$. Then we have the following.

(I) Let V^* be a graded (S^*)-module together with an R-isomorphism $\beta : V^* \to V$ such that for all $s^* \in S^*$ and $v^* \in V^*$ we have $\beta(s^* v^*) = \alpha(s^*)\beta(v^*)$, and for all $n \in \mathbb{N}$ we have $\beta(V_n^*) = V_{cn} + V_{cn+1} + \cdots + V_{cn+c-1}$. Then V^* is a finite graded module over S^* and, upon letting $H^*(Z)$ be the unique member of $\mathbb{Q}[Z]$ with $\ell_R(V_n^*) = H^*(n)$ for all $n >> 0$, we have $d = d^*$ where $d^* = \deg_Z H^*(Z)$.

(II) Assume that R is a field and $(0 : V)_S \cap \alpha(S^*) = (\alpha(X_{r+1}^*)\ldots, \alpha(X_N^*))\alpha(S^*)$ for some $0 \leq r \leq N$. Then $d = r - 1$ or $d = -\infty$ according as $r > 0$ or $r = 0$.

Next, using (Q27)(T105.2) and (Q28.4) we shall prove the following.

(Q28.5) In the situation of (Q28.3) assume that R is a field. Then upon letting $r = \dim(S/(0 : V)_S)$ we have $d = r - 1$ or $d = -\infty$ according as $r > 0$ or $r \leq 0$.

PROOF OF (Q28.1). We prove this by induction on n. For $n = 0$ we have $W_1 \approx W_0$ and hence we have nothing to show. For $n = 1$, we are done by L4§5(O21)(17•). So let $n > 1$ and assume for $n - 1$. Let $W_i' = W_i$ or $\text{im}(f_{n-1})$ according as $0 \le i \le n-1$ or $i = n$. Let $f_i' = f_i$ for $0 \le i \le n-2$. Let $f_{n-1}' : W_n' \to W_{n-1}'$ be the subset injection. Let $W_i'' = W_n'$ or W_n or W_{n+1} according as $i = 0$ or 1 or 2. Let $f_1'' = f_n$, and let $f_0'' : W_1'' \to W_0''$ be the surjection given by putting $f_0''(x) = f_{n-1}(x)$ for all $x \in W_1''$. By L4§5(O21)(17•) it follows that $(f', W', n-1)$ and $(f'', W'', 1)$ are exact sequences of R-modules such that $\ell_R(W_i') < \infty$ for $0 \le i \le n$ and $\ell_R(W_i'') < \infty$ for $0 \le i \le 2$. Therefore by the induction hypothesis we have

$$\sum_{0 \le i \le n} (-1)^i \ell_R(W_i') = 0$$

and by the $n = 1$ case we have

$$\sum_{0 \le i \le 2} (-1)^i \ell_R(W_i'') = 0.$$

By adding or subtracting the second equation from the first according as n is odd or even we get

$$\sum_{0 \le i \le n+1} (-1)^i \ell_R(W_i) = 0$$

which completes the induction.

PROOF OF (Q28.2). By (Q24)(1), for the S-module V we can take a finite set of nonzero homogeneous generators $v_1, \ldots, v_m$ of degrees $d_1, \ldots, d_m$ respectively. Given any $i \in I$, by cancellativeness, upon relabelling $1, \ldots, m$ we can find $p \in \{0, 1, \ldots, m\}$ such that, for $1 \le j \le p$ there is a unique $j' \in I$ with $d_j + j' = i$, and for $p+1 \le j \le m$ there is no $j' \in I$ with $d_j + j' = i$. Since $S_{j'}$ is a finite R-module, we can find $q \in \mathbb{N}$ together with R-generators $u_{j1}, \ldots, u_{jq}$ of $S_{j'}$ for $1 \le j \le p$. Given any $v \in V_i$ we can write

$$v = \sum_{1 \le j \le m} s_j v_j \quad \text{with} \quad s_j \in S.$$

We can take a finite subset L of I with $\{j' : 1 \le j \le p\} \subset L$ such that, for every $1 \le j \le p$ we have $s_j = t_j + t_j'$ with $t_j \in S_{j'}$ and $t_j' \in \sum_{l \in L \setminus \{j'\}} S_l$, and for $p+1 \le j \le m$ we have $s_j \in \sum_{l \in L} S_l$. It follows that

$$v = \sum_{1 \le j \le p} t_j v_j.$$

For $1 \le j \le p$ we can write

$$t_j = \sum_{1 \le k \le q} r_{jk} u_{jk} \quad \text{with} \quad r_{jk} \in R.$$

Upon letting $v_{jk} = u_{jk}v_j$ we have $v_{jk} \in V_i$ and

$$v = \sum_{1 \leq j \leq p} \sum_{1 \leq k \leq q} r_{jk}v_{jk}.$$

Thus the R-module V_i is generated by the pq elements v_{jk}. If R is artinian then by L4§5(O25) it follows that $\ell_R(V_i) < \infty$.

PROOF OF (Q28.3). By (Q28.2) we have $\ell_R(V_n) < \infty$ for all $n \in \mathbb{N}$. The assertions about H will be proved by induction on N. By (Q24)(1), V has a finite set of generators in $\cup_{0 \leq i \leq e} V_i$ for some $e \in \mathbb{N}$, and hence if $N = 0$ then for all $n > e$ we have $V_n = 0$, i.e, $\ell_R(V_n) = 0$, and so $H(Z) = 0$ does the job. Now let $N > 0$ and assume for $N - 1$. Let $V' = \ker(\theta)$ and $U'' = V/\theta(V)$ where $\theta : V \to V$ is the degree-one homogeneous S-monomorphism given by $v \mapsto vX_N$. Then the graded S-modules V' and V'' are annihilated by X_N, and hence we may view them as finite graded modules over $R[X_1, \ldots, X_{N-1}]$. Therefore by the induction hypothesis there exist $H'(Z) \in \mathbb{Q}[Z]$ and $H''(Z) \in \mathbb{Q}[Z]$ with $d' = \deg_Z H'(Z) < N - 1$ and $d'' = \deg_Z H''(Z) < N - 1$ such that for some $m \in \mathbb{N}$ we have

$$H'(n) = \ell_R(V'_n) \quad \text{and} \quad H''(n) = \ell_R(V''_n) \quad \text{for all} \quad n \geq m$$

and such that if $d' \in \mathbb{N}$ then

$$H'(Z) = \sum_{0 \leq i \leq d'} G'_i B_i(Z) \quad \text{with} \quad G'_i \in \mathbb{Z}$$

and if $d'' \in \mathbb{N}$ then

$$H''(Z) = \sum_{0 \leq i \leq d''} G''_i B_i(Z) \quad \text{with} \quad G''_i \in \mathbb{Z}.$$

For every $n \in \mathbb{N}_+$ we have the exact sequence

$$0 \to V'_{n-1} \to V_{n-1} \to V_n \to V''_n \to 0$$

of R-modules where the second arrow is the subset injection, the third arrow is multiplication by X_N (i.e., induced by θ), and the fourth arrow is the residue class surjection. Applying (Q28.1) to this exact sequence we get

$$\ell_R(V_n) - \ell_R(V_{n-1}) = \ell_R(V''_n) - \ell_R(V'_{n-1})$$

and hence

(1) $\qquad \ell_R(V_n) - \ell_R(V_{n-1}) = H''(n) - H'(n-1) \quad \text{for all} \quad n > m.$

Clearly (this is the usual "binomial identity" used in proving the binomial theorem)

(2) $\qquad B_i(Z) = \begin{cases} B_i(Z-1) + B_{i-1}(Z-1) & \text{if } i \in \mathbb{N}_+ \\ B_i(Z) & \text{if } i = 0 \end{cases}$

where the second case is obvious and the first case follows by noting that

$$\begin{cases} \text{LHS} = \frac{Z(Z-1)...(Z-i+1)}{i!} \\ \quad = \frac{(Z-1)(Z-2)...(Z-i)}{i!} + \frac{(Z-1)(Z-2)...(Z-i+1)i}{i!} \\ \quad = \text{RHS.} \end{cases}$$

Upon letting

$$\begin{cases} H'''(Z) = H''(Z) - H'(Z-1) \in \mathbb{Q}[Z] \\ \text{with } d''' = \max(d', d'') \in \mathbb{N} \cup \{-\infty\} \end{cases}$$

by (2) we see that

$$(3) \qquad H'''(Z) = \sum_{0 \le i \le d'''} G_i''' B_i(Z-1) \quad \text{with} \quad G_i''' \in \mathbb{Z}$$

where, by the convention of an empty sum, if $d''' \notin \mathbb{N}$ then the RHS is 0. By (1) we have

$$\ell_R(V_n) - \ell_R(V_{n-1}) = H'''(n) \quad \text{for all} \quad n > m$$

and hence

$$\ell_R(V_n) = \ell_R(V_m) + \sum_{m < e \le n} H'''(e) \quad \text{for all} \quad n > m$$

and therefore by (3) we get

$$(4) \qquad \ell_R(V_n) = \ell_R(V_m) + \sum_{0 \le i \le d'''} G_i''' \sum_{m < e \le n} B_i(e-1) \quad \text{for all} \quad n > m.$$

By (2) we have

$$B_i(Z-1) = B_{i+1}(Z) - B_{i+1}(Z-1) \quad \text{for all} \quad i \ge 0$$

and hence

$$\sum_{m < e \le n} B_i(e-1) = B_{i+1}(n) - B_{i+1}(m) \quad \text{for all} \quad i \ge 0 \quad \text{and} \quad n > m$$

and therefore by (4) we see that

$$(5) \qquad \ell_R(V_n) = G_0 + \sum_{0 \le i \le d'''} G_{i+1} B_{i+1}(n) \quad \text{for all} \quad n > m.$$

where $G_j \in \mathbb{Z}$ is defined by putting

$$G_j = \begin{cases} \ell_R(V_m) - \sum_{0 \le i \le d'''} G_i''' B_{i+1}(m) & \text{if } j = 0 \\ G_{j-1}''' & \text{if } j \in \mathbb{N}_+ \text{ with } 0 \le j - 1 \le d'''. \end{cases}$$

Let $d \in \mathbb{N} \cup \{-\infty\}$ be defined by putting

$$d = \begin{cases} \max\{i+1 : 0 \le i \le d''' \text{ with } G_{i+1} \ne 0\} \\ \qquad \text{if } \{0 \le i \le d''' : G_{i+1} \ne 0\} \ne \emptyset, \\ 0 \qquad \text{if } \{0 \le i \le d''' : G_{i+1} \ne 0\} = \emptyset \text{ and } G_0 \ne 0 \\ -\infty \quad \text{if } \{0 \le i \le d''' : G_{i+1} \ne 0\} = \emptyset \text{ and } G_0 = 0. \end{cases}$$

Also let $H(Z) \in \mathbb{Q}[Z]$ be defined by putting

$$H(Z) = \sum_{0 \le i \le d} G_i B_i(Z).$$

Then

$$H(Z) = \sum_{0 \le i \le d} G_i B_i(Z) \quad \text{with} \quad G_i \in \mathbb{Z} \quad \text{and} \quad \deg_Z H(Z) = d$$

and by (5) we have

$$\ell_R(V_n) = H(n) \quad \text{for all} \quad n > m.$$

The uniqueness of $H(Z)$ follows from the fact that a nonzero univariate polynomial cannot have infinitely many roots, and the positivity of G_d follows from the fact that if $d \in \mathbb{N}$ then $\ell_R(V_n) > 0$ for infinitely many n.

PROOF OF (Q28.4)(I). Since S is integral over $\alpha(S^*)$, by L4§10(E2) we can find finite number of generators $(s_i)_{1 \le i \le p}$ for S as an $\alpha(S^*)$-module; since V is a finite S-module, it has a finite number of generators $(v_j)_{1 \le j \le q}$ as an S-module; the pq elements $s_i v_j$ clearly generate V as an $\alpha(S^*)$-module; therefore the pq elements $\beta^{-1}(s_i v_j)$ generate V^* as an (S^*)-module. If $H(Z) = 0$ then $\ell_R(V_n) = 0$ for all $n >> 0$ and hence $\ell_R(V_n^*) = 0$ for all $n >> 0$ and therefore $H^*(Z) = 0$. So assume that $H(Z) \ne 0$, i.e., $d \in \mathbb{N}$. Then (because $G_d > 0$), upon letting

$$\overline{H}(Z) = \sum_{0 \le j \le c-1} H(cZ + j)$$

we get $\overline{H}(Z) \in \mathbb{Q}[Z]$ with $\deg_Z \overline{H}(Z) = d$. Also clearly, say by L4§5(O21)(18•), we have $\ell_R(V_n^*) = \overline{H}(n)$ for all $n >> 0$. Therefore by the uniqueness part of (Q28.3) we get $H^*(Z) = \overline{H}(Z)$ and hence $d = d^*$.

PROOF OF (Q28.4)(II). Let $\beta : V^* = V \to V$ be the identity map and for all $n \in \mathbb{N}$ let

$$V_n^* = V_{cn} + V_{cn+1} + \cdots + V_{cn+c-1}.$$

Then V^* becomes a graded (S^*)-module by putting $s^* v^* = \alpha(s^*)\beta(v^*)$ for all $v^* \in V^*$ and $s^* \in S^*$. By (I) we know that V^* is a finite graded module over S^*. Also clearly

$$(0 : V^*)_{S^*} = (X^*_{r+1} \dots, X^*_N)S^*$$

and hence by identifying $S^*/(0 : V^*)_{S^*}$ with the naturally graded polynomial ring $T = k[X^*_1, \dots, X^*_r]$, we may view V^* as a faithful finite graded T-module, i.e., such that $(0 : V^*)_T = 0$. By (Q24)(1) we can take a finite number of nonzero homogeneous T-generators $v_1, \dots, v_m$ of V^* with $m \in \mathbb{N}_+$. If $(0 : v_i)_T \neq 0$ for $1 \leq i \leq m$ then by taking $0 \neq t_i \in (0 : v_i)_T$ and letting $t = t_1 \dots t_m$ we get $0 \neq t \in (0 : V^*)_T$ which is a contradiction. Therefore $(0 : v_i)_T = 0$ for some i and, upon letting e be the degree of v_i, we obtain a T-monomorphism $\gamma : T \to V^*$ of degree e given by $t \mapsto tv_i$. It follows that $[V^*_{n+e} : k] \geq [T_n : k]$ for all $n \in \mathbb{N}$. By (Q15)(C20) we also have $[T_n : k] = B_{r-1}(n + r - 1)$ for all $n \in \mathbb{N}$. Therefore by the uniqueness part of (Q28.3) we see that, if $r > 0$ then $d^* \geq r - 1$. But by (Q28.3) we have $d^* < r$. Consequently $d^* = r - 1$ or $d^* = -\infty$ according as $r > 0$ or $r = 0$. Therefore by (I) we get $d = r - 1$ or $d = -\infty$ according as $r > 0$ or $r = 0$.

PROOF OF (Q28.5). Upon letting

$$J = (0 : V)_S,$$

by (Q26)(1) we see that J is a homogeneous ideal in S. If $J = S$ then obviously $d = -\infty = \dim(S/J)$. So assume that $J \neq S$. Then, upon letting

$$r = \dim(S/J),$$

by (Q27)(T105.2) we can find a positive integer c together with nonzero homogeneous polynomials $Y_1, \dots, Y_N$ in S of degree c such that S is integral over $D = k[Y_1, \dots, Y_N]$ and $J \cap D = (Y_{r+1}, \dots, Y_N)D$. Let $S^* = k[X^*_1, \dots, X^*_N]$ be the naturally graded polynomial ring in indeterminates $X^*_1, \dots, X^*_N$ over k. Clearly we have a unique k-monomorphism $\alpha : S^* \to S$ of rings such that $\alpha(X^*_i) = Y_i$ for $1 \leq i \leq N$. Also clearly $\alpha(X^*_1), \dots, \alpha(X^*_N)$ belong to S_c, and S is integral over $\alpha(S^*)$. Thus we are reduced to (Q28.4)(II).

QUEST (Q29) Linear Disjointness and Intersection of Varieties

Geometrically speaking, in Theorem (T112) below we shall show that, if the intersection U of any two varieties V and W in the N-dimensional affine space over a field k is nonempty, then $\dim(U) \geq \dim(V) + \dim(W) - N$. According to L4§8, in algebraic terms, by taking the ideals of V and W in the polynomial ring $R = k[X_1, \dots, X_N]$, this says that for any ideals J and H in R, assuming $J + H$ to be a nonunit ideal, we have $\dim(R/(J + H)) \geq \dim(R/J) + \dim(R/H) - N$.

As a preparation for (T112), we shall first prove the following Theorem (T111) which geometrically says that, given any two irreducible affine varieties V and V' (in different affine spaces), the codimension of any irreducible component of their product $V \times V'$ equals the sum of their codimensions. In case k is algebraically closed, $V \times V'$ may be thought of as the set of all points $(a, a') \in k^N \times k^{N'}$ with $a \in V$ and $a' \in V'$. By the codimension of V in the affine space k^N we mean N minus the dimension of V. Algebraically speaking, (T111) says that given any prime ideal P in R and any prime ideal P' in $R' = k[X'_1, \ldots, X'_{N'}]$, letting $S = k[X_1, \ldots, X_N, X'_1, \ldots, X'_{N'}]$, for every associated prime Q of $PS + P'S$ in S we have $\operatorname{ht}_S Q = \operatorname{ht}_R P + \operatorname{ht}_{R'} P'$.

In turn, as a preparation for (T111), in (T110) we shall study extensions of ideals in a polynomial ring when the ground field is enlarged by adjoining any number of algebraically independent elements to it.

As further preparation for (T111), in Lemma (T109) we shall prove some properties of linear disjointness. Given any subrings R and R' of a ring S, such that $R \cap R'$ contains a subfield k of S, we say that R and R' are linearly disjoint over k to mean that: if $(x_i)_{1 \le i \le m}$ are any finite number of elements of R which are linearly independent over k and $(x'_j)_{1 \le j \le m'}$ are any finite number of elements of R' which are linearly independent over k, then the mm' elements $x_i x'_j$ of S are linearly independent over k. This linear disjointness is particularly significant for the ring-theoretic compositum of R and R'.

The ring-theoretic compositum of subrings R, R' of a ring S is defined to be the smallest subring of S which contains both R and R'. Equivalently it may defined to be the intersection of all subrings of S which contain both R and R'. For any subring k of the ring $R \cap R'$, the said compositum clearly coincides with $k[R, R']$, where we recall that for any subring k of a ring S, and any subset E of S, by $k[E]$ we denote the subring of S consisting of all finite sums $\sum a_{i_1 \ldots i_n} z_1^{i_1} \ldots z_n^{i_n}$ with $n, i_1, \ldots, i_n$ in $\mathbb{N}$, $a_{i_1 \ldots i_n}$ in k, and $z_1, \ldots, z_n$ in E. In all this, instead of two subrings R, R' of S, we may take any set (or family) of subrings of S.

In the above paragraph we may everywhere change "ring" to "field" with the understanding that $k[E]$ is replaced by $k(E)$ and the finite sums are replaced by quotients of finite sums with nonzero denominators. Usually the adjective "field-theoretic" is dropped from the phrase "field-theoretic compositum."

As an aid to (T109), in (T108) we shall prove some properties of a ring S which is a free module over a subring R. In particular we shall show that then R is ideally closed in S, by which we mean that for every ideal H in R we have $R \cap (HS) = H$. We shall give a second, more transparent, proof of this when S is a free R-algebra. An algebra over a ring R is said to be a free algebra if S has an R-basis of which $1 \in S$ is a member. Clearly a free algebra is a free module. Also clearly every algebra over a field is a free algebra.

In the above context, note that a module S over a ring R is a free R-module iff it has an R-basis $(x_i)_{i \in I}$ indexed by some indexing set I; when the set I is infinite, in a summation expression $z = \sum_{i \in I} z_i x_i$ with $z \in S$ and $z_i \in R$, it is understood that $z_i = 0$ for all except a finite number of i. At any rate, an R-algebra S is a free R-algebra iff it has an R-basis $(x_i)_{i \in I}$ such that for some $l \in I$ we have $x_l = 1 \in S$;

note that if R is a nonnull ring then this l must be unique.

As another aid to (T112), in (T107) given immediately below we shall obtain a corollary of the Generalized Principal Ideal Theorem (Q7)(T24) as applied to a catenarian domain defined in (Q12).

But the most noteworthy aspect of the proof of (T112) is the implicit use of a special set of generators of any given prime ideal P in the polynomial ring R which is a slight generalization of the set of generators of a maximal ideal in R given in L4§8(T22.3). We shall discuss this set of generators in Comment (C45) below where they will be used to give a very elementary proof of the regularity of the local ring R_P.

CATENARIAN DOMAIN COROLLARY (T107). Let $z_1, \ldots, z_n$ be any finite number of elements in a catenarian domain R, let H and H' be any ideals in R with $H' = H + (z_1, \ldots, z_n)R$, and let d be any nonnegative integer. Then we have the following.

(T107.1) Assume that for all $P \in \mathrm{nvspec}_R H$ we have $\mathrm{ht}_R P \le d$. Then for all $P' \in \mathrm{nvspec}_R H'$ we have $\mathrm{ht}_R P' \le d + n$.

(T107.2) Assume that for some $P \in \mathrm{nvspec}_R H$ we have $P + (z_1 \ldots, z_n)R \ne R$ and $\mathrm{ht}_R P \le d$. Then for some $P' \in \mathrm{nvspec}_R H'$ we have $\mathrm{ht}_R P' \le d + n$.

FREE MODULE LEMMA (T108). Let S be an overring of a ring R such that S is a free R-module. Then we have the following.

(a) Let $(x_i)_{i \in I}$ be any R-basis of S, and let H be any ideal in R. Then every $z \in HS$ can be expressed as a sum $z = \sum_i z_i x_i$ with $z_i \in H$. Moreover, for any sum $z = \sum_i z_i x_i$ with $z \in S$ and $z_i \in R$ we have that: $z \in HS \Leftrightarrow z_i \in H$ for all i.

(b) For any finite number of ideals $H_1, \ldots, H_n$ in R we have $(H_1 \cap \cdots \cap H_n)S = (H_1 S) \cap \cdots \cap (H_n S)$.

(c) R is ideally closed in S.

(d) If S is noetherian then so is R.

(e) For any ideal H in R, upon letting $\alpha : S \to S/(HS)$ be the residue class epimorphism, the ring $\alpha(S)$ is a free $\alpha(R)$-module. For any multiplicative set U in R, upon letting $\phi : R \to R_U$ and $\psi : S \to S_U$ be the canonical (localization) homomorphisms and upon letting $\iota : R \to S$ to be the subset injection, there is a unique homomorphism $\beta : R_U \to S_U$ with $\beta\phi = \psi\iota$ (β is actually a monomorphism), and moreover the ring S_U is a free $\beta(R_U)$-module.

(f) Let H be any ideal in R. Then for every $z \in R$ we have $(H : z)_R S = (HS : z)_S$ and hence in particular we have

$$R \cap Z_S(S/(HS)) \subset Z_R(R/H).$$

Similarly, for every multiplicative set U in R we have $[H : U]_R S = [HS : U]_S$.

(g) Assuming S to be noetherian, for every ideal H in R and every associated prime (resp: associated minimal prime) Q of HS in S we have that $Q \cap R$ is an associated prime (resp: associated minimal prime) of H in R.

LINEAR DISJOINTNESS LEMMA (T109). Let k, R, R' be subrings of a ring S such that k is a field with $k \subset R \cap R'$. Then we have the following.

(T109.1) The rings R, R' are linearly disjoint over the field k iff for every finite number of elements $x_1, \ldots, x_m$ of R which are linearly independent over k we have that the elements $x_1, \ldots, x_m$ are linearly independent over R'.

[This is the more checkable version of linear disjointness].

(T109.2) Assume that R, R', S are as in (T111). Then the rings R, R' are linearly disjoint over k and we have $S = k[R, R']$.

(T109.3) Assume that the rings R, R' are linearly disjoint over k and we have $S = k[R, R']$. Then S is a free R-algebra.

(T109.4) Assume that the rings R, R' are linearly disjoint over a subfield k, and $S = k[R, R']$. Then, given any nonunit ideals P and P' in R and R' respectively, upon letting $D = PS + P'S$ we have the following.

(i) $R \cap D = P$ and $R' \cap D = P'$. Hence in particular D is a nonunit ideal in S.

(ii) Upon letting $\phi : S \to S/D$ to be the residue class epimorphism, the subrings $\phi(R)$ and $\phi(R')$ of $\phi(S)$ are linearly disjoint over the subfield $\phi(k)$.

(iii) Assuming S to be noetherian, for every associated prime (resp: associated minimal prime) Q of D in S, the ideal $R \cap Q$ is an associated prime (resp: associated minimal prime) of P in R.

IDEALS UNDER TRANSCENDENTAL EXTENSION THEOREM (T110). Let $X_1, \ldots, X_N, (Y_i)_{i \in I}$ be indeterminates over a field k. Consider the field extensions $k \subset L = k((Y_i)_{i \in I})$ and consider the corresponding polynomial ring extensions $R = k[X_1, \ldots, X_N] \subset L[X_1, \ldots, X_N] = T$. For all ideals H and J in R and T respectively, let H^e and J^c be the extended and contracted ideals in T and R, i.e., the ideals given by $H^e = HT$ and $J^c = J \cap R$ respectively. Let C be the set of all ideals in R, and let E be the set of all extended ideals in T. Then we have the following; see L4§7(T13).

(I) R is ideally closed in T, i.e., C is the set of all contracted ideals in R.

(II) For every prime ideal P in R, the ideal P^e is a prime ideal in T. For every primary ideal Q in R, the ideal Q^e is a primary ideal in T. For every ideal H in R we have $(\mathrm{rad}_R H)^e = \mathrm{rad}_T H^e$. For any finite number of ideals $H_1, \ldots, H_n$ in R we have $(H_1 \cap \cdots \cap H_n)^e = H_1^e \cap \cdots \cap H_n^e$.

(III) Let $H = Q_1 \cap \cdots \cap Q_d$ be any irredundant primary decomposition of any ideal H in R with $\mathrm{rad}_R Q_j = P_j$, which is labelled so that $P_1, \ldots, P_{d'}$ are exactly all the minimal associated primes of H in R. Then $H^e = Q_1^e \cap \cdots \cap Q_d^e$ is an irredundant primary decomposition of H^e in T with $\mathrm{rad}_T Q_j^e = P_j^e$, and $P_1^e, \ldots, P_{d'}^e$ are exactly all the minimal associated primes of H^e in T.

(IV) For every ideal H in R we have $\mathrm{ht}_R H = \mathrm{ht}_T H^e$ and $\mathrm{dpt}_R H = \mathrm{dpt}_T H^e$.

CODIMENSION OF PRODUCT THEOREM (T111). Consider the polynomial ring $S = k[X_1, \ldots, X_N, X_1', \ldots, X_{N'}']$ in indeterminates $X_1, \ldots, X_N, X_1', \ldots, X_{N'}'$ over a field k with N, N' in $\mathbb{N}$. Let P and P' be any prime ideals in the polynomial subrings $R = k[X_1, \ldots, X_N]$ and $R' = k[X_1', \ldots, X_{N'}']$ of S respectively. Then $PS + P'S$ is a nonunit ideal in S, and for every associated prime Q of $PS + P'S$ in S we have $\mathrm{ht}_S Q = \mathrm{ht}_R P + \mathrm{ht}_{R'} P'$.

DIMENSION OF INTERSECTION THEOREM (T112). Considering the ring $R = k[X_1, \ldots, X_N]$ where $X_1, \ldots, X_N$ are indeterminates over a field k with $N \in \mathbb{N}$, for any prime ideals P and $\overline{P}$ in R and any associated minimal prime $\widehat{P}$ of $P + \overline{P}$ in R we have $\dim(R/\widehat{P}) \geq \dim(R/P) + \dim(R/\overline{P}) - N$, and for any ideals J and H in R with $J + H \neq R$ we have $\dim(R/(J + H)) \geq \dim(R/J) + \dim(R/H) - N$.

PROOF OF (T107.1). For any $P' \in \mathrm{nvspec}_R H'$ we have $H \subset P'$, and hence we can find $P \in \mathrm{nvspec}_R H$ with $P \subset P'$. Clearly $(P + (z_1, \ldots, z_n)R) \subset P'$, and hence $P' \in \mathrm{nvspec}_R(P + (z_1, \ldots, z_n)R)$, and therefore by (Q7)(T27) we have $\mathrm{ht}_{R/P}(P'/P) \leq n$. By assumption R is a catenarian domain, and $\mathrm{ht}_R P \leq d$. Therefore $\mathrm{ht}_R P' \leq d + n$.

PROOF OF (T107.2). By (T107.1), for some $P^* \in \mathrm{nvspec}_R(P + (z_1, \ldots, z_n)R)$ we have $\mathrm{ht}_R P^* \leq d + n$. Since $H' \subset P^*$, we can take $P' \in \mathrm{nvspec}_R H'$ with $P' \subset P^*$. Clearly $\mathrm{ht}_R P' \leq \mathrm{ht}_R P^*$, and hence $\mathrm{ht}_R P' \leq d + n$.

PROOF OF (T108)(a). Every $z \in HS$ can be written as a finite sum $z = \sum_j h_j s_j$ with $h_j \in H$ and $s_j \in S$. Also we can write each s_j as a finite sum $s_j = \sum_i r_{ji} x_i$ with $r_{ji} \in R$, and i running over a common finite subset of I. This gives us $z = \sum_i z_i x_i$ with $z_i = \sum_j h_j r_{ji} \in H$ for all i. The "Moreover" part follows by uniqueness (of linear expressions in terms of a basis).

PROOF OF (T108)(b). Follows from (T108)(a).

PROOF OF (T108)(c). By assumption S has an R-basis $(x_i)_{i \in I}$ where I is some indexing set. Given any ideal H in R and $y \in R \cap (HS)$ we want to show that $y \in H$.

Let us first prove this under the assumption $y = 1$. The said assumption implies that for every $j \in I$ we have $x_j = 1x_j \in HS$, and hence by (T108)(a) we can write $1x_j$ as a finite sum $1x_j = \sum_{i \in I'} u_i x_i$ with $u_i \in H$ where I' is a finite subset of I with $j \in I'$. By uniqueness, comparing coefficients of x_j we get $1 = u_j \in H$.

Now, without assuming $y = 1$, by (T108)(a) we can write y as a finite sum $y = \sum_i y_i x_i$ with $y_i \in H$. We can also write 1 as a finite sum $1 = \sum_i v_i x_i$ with $v_i \in R$. Applying the $y = 1$ case to the ideal in R generated by the elements v_i we see that 1 can be written as a finite sum $1 = \sum_i v_i r_i$ with $r_i \in R$. Multiplying the equation $1 = \sum_i v_i x_i$ by y we get $y = \sum_i y v_i x_i$ and comparing this with the equation $y = \sum_i y_i x_i$, by uniqueness we get $y_i = y v_i$ for all i. Multiplying the equation $1 = \sum_i v_i r_i$ by y we get $y = \sum_i y v_i r_i$ and substituting $y_i = y v_i$ in this we obtain $y = \sum_i y_i r_i$, where the RHS obviously belongs to H and hence $y \in H$.

PROOF OF (T108)(c) FOR FREE ALGEBRA. Here is a simpler proof assuming an R-basis $(x_i)_{i \in I}$ of S with $x_l = 1$ for some $l \in I$. Given any ideal H in R and $y \in R \cap (HS)$ we want to show that $y \in H$. By (T108)(a) we have $y = \sum_{i \in I'} y_i x_i$ with $y_i \in H$ where I' is a finite subset of I with $l \in I'$. By uniqueness $y = y_l \in H$.

PROOF OF (T108)(d). Follows from (T108)(c).

PROOF OF (T108)(e). By assumption S has an R-basis $(x_i)_{i \in I}$. For any $z \in S$ we have $z = \sum_{i \in I} z_i x_i$ with $z_i \in R$, and hence $\alpha(z) = \sum_{i \in I} \alpha(z_i) \alpha(x_i)$, and therefore $\alpha(x_i)_{i \in I}$ are $\alpha(R)$-generators of $\alpha(S)$. Moreover, if $\alpha(z) = 0$ then by (T108)(a) we get $\alpha(z_i) = 0$ for all i. Thus $\alpha(x_i)_{i \in I}$ is an $\alpha(R)$-basis of $\alpha(S)$, and hence $\alpha(S)$ is a free $\alpha(R)$-module. Turning to the multiplicative set U in R, and referring to L4§7, the unique existence and injectivity of β follows for any overring S of a ring R without assuming the freeness; all we need to do is, for every $y \in R$ and $u \in U$, let β send $\phi(y)/\phi(u) \in R_U$ to $\psi(y)/\psi(u) \in S_U$, and note that $\ker(\phi) = [0 : U]_R = R \cap [0 : U]_S$ with $\ker(\psi) = [0 : U]_S$.

Assuming the R-basis $(x_i)_{i \in I}$ of S, for every $u \in U$ we get $\psi(z)/\psi(u) = \sum_{i \in I} \beta(\phi(z_i)/\phi(u)) \psi(x_i)$ with $\beta(\phi(z_i)/\phi(u)) \in \beta(R_U)$, and hence $\psi(x_i)_{i \in I}$ are generators of S_U over $\beta(R_U)$. Moreover, if $\psi(z)/\psi(u) = 0$ in S_U then for some $v \in U$ we have $\sum_{i \in I} v z_i x_i = 0$ in S with $v z_i \in R$ for all i, and hence $\beta(\phi(z_i)/\phi(u)) = 0$ in $\beta(R_U)$ for all i. Thus $\psi(x_i)_{i \in I}$ is a basis of S_U over $\beta(R_U)$, and hence S_U is free $\beta(R_U)$-module.

PROOF OF (T108)(f). For any $z \in R$, the inclusion $(H : z)_R S \subset (HS : z)_S$ is obvious. To prove the reverse inclusion, given any $y \in (HS : z)_S$, by (T108)(a) we have $yz = \sum_i z_i x_i$ with $z_i \in H$ for all i. We can also write $y = \sum_i y_i x_i$ with $y_i \in R$ for all i, and multiplying this by z we get $zy = \sum_i z y_i x_i$ with $z y_i \in R$ for all i, and hence by uniqueness we obtain $z y_i = z_i \in H$ for all i. Consequently $y_i \in (H : z)_R$ for all i, and therefore $y \in (H : z)_R S$. Thus $(H : z)_R S = (HS : z)_S$. It follows that $\{z \in R : (HS : z)_S \neq HS\} \subset \{z \in R : (H : z)_R \neq H\}$, which is equivalent to saying that $R \cap Z_S(S/(HS)) \subset Z_R(R/H)$. Finally, for any multiplicative set U in R we get

$$[HS : U]_S = \bigcup_{u \in U} (HS : u)_S = \bigcup_{u \in U} (H : u)_R S = [H : U]_R S$$

where the middle equality follows from the equation $(H : z)_R S = (HS : z)_S$ which we just proved, and the last equality follows from the defining equation

$\cup_{u \in U}(H : u)_R = [H : U]_R$ together with the fact that $\cup_{u \in U}(H : u)_R S$ is an ideal in S because it equals $[HS : U]_S$.

PROOF OF (T108)(g). Assume that S (and hence by (d) also R) is noetherian. Let H be an ideal in R, let Q be an associated prime of HS in S, let $P = Q \cap R$, and let $U = R \setminus P$. Then $H \subset P \in \mathrm{spec}(R)$ and hence P contains an associated minimal prime of H in R. Therefore, in view of L4§6, we can find $1 \le m \le r \in \mathbb{N}_+$ and $1 \le n \le s \in \mathbb{N}_+$ together with irredundant primary decompositions

$$H = \bigcap_{1 \le j \le r} H_j \quad \text{and} \quad HS = \bigcap_{1 \le i \le s} J_i$$

of H in R and HS in S respectively, where the primes $P_j = \mathrm{rad}_R H_j$ are so arranged that $P_j \subset P$ for $1 \le j \le m$ and $P_j \not\subset P$ for $m + 1 \le j \le r$ and P_1 is an associated minimal prime of H in R, and the primes $Q_i = \mathrm{rad}_S J_i$ are so arranged that $Q_i \cap U = \emptyset$ for $1 \le i \le n$ and $Q_i \cap U \ne \emptyset$ for $n + 1 \le i \le s$ and $Q_1 = Q$. Let $\overline{H} = [H : U]_R$. Then by (T108)(f) we have $[HS : U]_S = \overline{H}S$ and therefore by L4§6 we get the irredundant primary decompositions

$$\overline{H} = \bigcap_{1 \le j \le m} H_j \quad \text{and} \quad \overline{H}S = \bigcap_{1 \le i \le n} J_i$$

of $\overline{H}$ in R and $\overline{H}S$ in S respectively. Hence by (Q1) we get

$$Z_R(R/\overline{H}) = \cup_{1 \le j \le m} P_j \quad \text{and} \quad Q \subset Z_S(S/(\overline{H}S)).$$

Therefore

$$P = R \cap Q \subset R \cap Z_S(S/(\overline{H}S)) \subset Z_R(R/\overline{H})$$

by (T108)(f). Thus $P_j \subset P$ for $1 \le j \le m$ and $P \subset \cup_{1 \le j \le m} P_j$, which by (Q17)(C31.1) implies that $P = P_j$ for some j, i.e., P is an associated prime of H in R.

Now suppose that Q is an associated minimal prime of HS in S. Then clearly $HS \subset P_1 S \subset Q$ implies that Q is an associated minimal prime of $P_1 S$ in S and hence $Q \subset Z_S(S/P_1 S)$ by (Q1). Consequently

$$P = R \cap Q \subset R \cap Z_S(S/P_1 S) \subset Z_R(R/P_1) = P_1$$

by (T108)(f). Hence $P = P_1$, and therefore P is an associated minimal prime of H in R.

PROOF OF (T109.1). First suppose that the rings R, R' are linearly disjoint over the field k. Let $x_1, \ldots, x_m$ be any finite number of elements of R which are linearly independent over k, and let $\sum_{1 \le j \le m} x_j y'_j = 0$ be any relation with $y'_1, \ldots, y'_m$ in R'. We want to show that then $y'_1 = \cdots = y'_m = 0$. Since k is a field, we can take a finite k-basis $x'_1, \ldots, x'_{m'}$ of the k-subspace of R' generated by $y'_1, \ldots, y'_m$, and we can write $y_i' = \sum_{1 \le j \le m'} a_{ij} x_j'$ with $a_{ij} \in k$. This gives us $\sum_{i,j} a_{ij} x_i x_j' = 0$, and hence by the linear disjointness of R, R' over k we get $a_{ij} = 0$ for all i, j, and therefore $y_1' = \cdots = y_m' = 0$.

Next suppose that every finite number of elements of R which are linearly independent over k are linearly independent over R'. Let $(x_i)_{1 \le i \le m}$ and $(x_j')_{1 \le j \le m'}$ be any finite number of k-linearly independent of R and R' respectively. Let $\sum_{i,j} a_{ij} x_i x_j' = 0$ be any relation with $a_{ij} \in k$. We want to show that then $a_{ij} = 0$ for all i, j. Now

$$\sum_{1 \le i \le m} x_i y_i' = 0 \quad \text{with} \quad y_i' = \sum_{1 \le j \le m'} a_{ij} x_j' \in R'$$

and hence by the linear independence of $x_1, \ldots, x_m$ over R' we get $y_1', \ldots, y_m' = 0$, and therefore by k-linear independence of $x_1', \ldots, x_{m'}'$ we get $a_{ij} = 0$ for all i, j.

PROOF OF (T109.2). Obviously $S = k[R, R']$. To prove linear disjointness, let $F_1 = F_1(X_1, \ldots, X_N), \ldots, F_m = F_m(X_1, \ldots, X_N)$ be any finite number of polynomials with coefficients in k, and assume that $F_1, \ldots, F_m$ are linearly independent over k. In view of (T109.1) it suffices to show that then $F_1, \ldots, F_m$ are linearly independent over R'. So let $\sum_{1 \le i \le m} F_i F_i' = 0$ be any relation with $F_i' = F_i'(X_1', \ldots, X_{N'}') \in k[X_1', \ldots, X_{N'}']$. Let $W_1, \ldots, W_n$ be all the monomials of degree $\le d$ in $X_1', \ldots, X_{N'}'$ where $d \in \mathbb{N}$ exceeds the degree of F_i' for $1 \le i \le m$. Then $F_i' = \sum_{1 \le j \le n} A_{ij} W_j$ with $A_{ij} \in k$, and substituting this in $\sum_{1 \le i \le m} F_i F_i' = 0$ we get the equation $\sum_{1 \le j \le n} B_j W_j = 0$ where $B_j = \sum_{1 \le i \le m} A_{ij} F_i \in R$. Regarding the last equation as a polynomial identity in $X_1', \ldots, X_{N'}'$ over R, we obtain the n equations $B_1 = 0, \ldots, B_n = 0$. Since $F_1, \ldots, F_m$ are linearly independent over k, these n equations tell us that $A_{ij} = 0$ for all i, j. Therefore $F_1' = 0, \ldots, F_m' = 0$. Thus $F_1, \ldots, F_m$ are linearly independent over R'.

PROOF OF (T109.3). Since k is a field, we can find an R-basis $(x_j')_{j \in I'}$ of R' with $x_{l'} = 1$ for some $l' \in I'$. Since R, R' are linearly disjoint over k, we get an R-basis $(x_j')_{j \in I'}$ of S with $x_{l'} = 1$.

PROOF OF (T109.4)(i) IN CASE OF (T111). Let R, R', S be as in (T111). Any $F \in R \cap D$ can be written as $F = \sum F_i P_i + \sum F_j' P_j'$ where the summations are finite and for the polynomials

$$\begin{cases} F = F(X_1, \ldots, X_N) \in k[X_1, \ldots, X_N] \\ F_i = F_i(X_1, \ldots, X_N, X_1', \ldots, X_{N'}') \in k[X_1, \ldots, X_N, X_1', \ldots, X_{N'}'] \\ F_j' = F_j'(X_1, \ldots, X_N, X_1', \ldots, X_{N'}') \in k[X_1, \ldots, X_N, X_1', \ldots, X_{N'}'] \\ P_i = P_i(X_1, \ldots, X_N) \in k[X_1, \ldots, X_{N'}] \\ P_j' = P_j'(X_1', \ldots, X_{N'}') \in k[X_1', \ldots, X_{N'}'] \end{cases}$$

we have $P_i \in P$ and $P_j' \in P'$. Take an algebraic closure $\widehat{k}$ of k and consider the ring extensions $\widehat{R} = \widehat{k}[X_1, \ldots, X_N]$ and $\widehat{R}' = \widehat{k}[X_1', \ldots, X_{N'}']$ of R and R' respectively. By (Q10)(C15) we see that $P'\widehat{R}'$ is a nonunit ideal in $\widehat{R}'$ and hence first by Zorn's Lemma we can take a maximal ideal M' in $\widehat{R}'$ containing $P'\widehat{R}'$ and then by the Nullstellensatz L4§8(T22.4) we can find elements $x_1', \ldots, x_{N'}'$ in $\widehat{k}$ such that for all $G'(X_1', \ldots, X_{N'}') \in M'$ we have $G'(x_1', \ldots, x_{N'}') = 0$. Putting $X_1' = x_1', \ldots, X_{N'}' = x_{N'}'$ in the polynomial identity $F = \sum F_i P_i + \sum F_j' P_j'$ we get $F = \sum \widehat{F}_i P_i$ where $\widehat{F}_i = F_i(X_1, \ldots, X_N, x_1', \ldots, x_{N'}')$. Thus $F \in R \cap (P\widehat{R}) = P$ by (Q10)(C15). Therefore $D \cap R = P$, and by symmetry $D \cap R' = P'$.

PROOF OF (T109.4)(i) IN GENERAL CASE. Mimicking the above proof, we can take a k-basis of R in three disjoint pieces $a = \{1\}, b = (b_i)_{i \in I}, c = (c_i)_{i \in J}$ where I, J are indexing sets, and where c is a k-basis of P. Similarly we can take a k-basis of R' in three disjoint pieces $a' = \{1\}, b' = (b_j')_{j \in I'}, c' = (c_j')_{j \in J'}$ where I', J' are indexing sets, and where c' is a k-basis of P'. By linear disjointness we get a k-basis of S in the nine disjoint pieces $\{1\}, b, c, b', c', bb', bc', b'c, cc'$ where $bb' = (b_i b_j')_{(i,j) \in I \times I'}$ and similarly for $bc', b'c, cc'$. Clearly $c, b'c, cc'$ and c', bc', cc' are k-bases of PS and $P'S$ respectively. Putting these together we get the k-basis $c, c', bc', b'c, cc'$ of D; the "intersection" of this with the k-basis $\{1\}, b, c$ of R gives the k-basis c of P. Therefore $D \cap R = P$. By symmetry $D \cap R' = P'$.

PROOF OF (T109.4)(ii). By (T109.4)(i) we have $R \cap D = P$ with $R' \cap D = P'$ and hence we have $k \cap D = 0$; we shall be using these three facts tacitly. Let $(x_i)_{1 \le i \le m}$ (resp: $(x_j')_{1 \le j \le m'}$) be any finite number of elements of R (resp: R') such that their images under ϕ are linearly independent over $\phi(k)$. We want to show that then the mm' elements $\phi(x_i x_j')$ are linearly independent over $\phi(k)$.

Clearly we can first find a finite number of elements $(x_i)_{m+1 \le i \le n}$ and $(\xi_i)_{1 \le i \le n}$ in R with $\xi_1 = 1$ such that the ϕ images of $(x_i)_{1 \le i \le n}$ are linearly independent over $\phi(k)$, the ϕ images of $(\xi_i)_{1 \le i \le n}$ are linearly independent over $\phi(k)$, and the ϕ images of $(x_i)_{1 \le i \le n}$ generate the same ϕ-vector subspace of $\phi(S)$ as the ϕ images of $(\xi_i)_{1 \le i \le n}$; then we can find a k-basis of R as in the above proof of (T109.4)(i) with the stipulation that the family $(\xi_i)_{2 \le i \le n}$ is contained in the family b.

Also we can first find a finite number of elements $(x_j')_{m'+1 \le i \le n'}$ and $(\xi_j')_{1 \le j \le n'}$ in R' with $\xi_1' = 1$ such that the ϕ images of $(x_j')_{1 \le j \le n'}$ are linearly independent over $\phi(k)$, the ϕ images of $(\xi_j')_{1 \le j \le n'}$ are linearly independent over $\phi(k)$, and the ϕ images of $(x_j')_{1 \le j \le n'}$ generate the same ϕ-vector subspace of $\phi(S)$ as the ϕ images of $(\xi_j')_{1 \le j \le n'}$; then we can find a k-basis of R' as in the above proof of (T109.4)(i) with the stipulation that the family $(\xi_j')_{2 \le j \le n'}$ is contained in the family b'.

By the above proof of (T109.4)(i) we now see that the nn' elements

$$\phi(\xi_i \xi_j')_{1 \le i \le n, 1 \le j \le n'}$$

are linearly independent over $\phi(k)$, and hence so are the nn' elements

$$\phi(x_i x'_j)_{1 \leq i \leq n, 1 \leq j \leq n'}.$$

PROOF OF (T109.4)(iii). In view of parts (i) and (ii) of (T109.4), we may replace S by S/D, i.e., without loss of generality, we may assume that $P = 0$ and $D = 0$. But then our assertion follows from (T108)(g) and (T109.3).

PROOF OF (T110). Part (I) follows from (Q10)(C15). We shall prove (II) in Comment (C44) below which itself will be based on Comment (C43) below. In view of (T108)(b) and (T109.3), part (III) follows from part (II). In view of part (III), the general case of part (IV) follows from its case when the ideal H is prime. By (I) and (II) we see that if $P_0 \subset P_1 \subset \cdots \subset P_n$ is any finite sequence of prime ideals in R with $P_0 \neq P_1 \neq \cdots \neq P_n$, then $P_0^e \subset P_1^e \subset \cdots \subset P_n^e$ is a finite sequence of prime ideals in T with $P_0^e \neq P_1^e \neq \cdots \neq P_n^e$. Therefore (IV) when H is a prime ideal follows by noting that in an N-variable polynomial ring over a field, every absolutely saturated chain of prime ideals has length N, and every prime ideal is part of some absolutely saturated chain of prime ideals; see (Q10)(T47) and the Terminology introduced in its preamble.

PROOF OF (T111). In Comment (C45) below we shall show how, using (T109) and (T110), together with special bases of the prime ideals P and P', the proof of (T111) can be reduced to the case when P and P' are maximal ideals. The proof of the maximal ideal case will be given in Comment (C46) below.

PROOF OF (T112). In Comment (C47) below we shall show how to deduce (T112) from (T107) and (T111), by using the ideal of a diagonal.

COMMENT (C43). [**Ideals in Polynomial Ring Extensions**]. Consider

(†) $\begin{cases} \text{the polynomial ring } S = R[Y] \\ \text{where } Y = (Y_i)_{i \in I} \text{ is a family of indeterminates} \\ \text{over a nonnull ring } R \end{cases}$

and let

(‡) $\begin{cases} W = (W_j)_{j \in J} \text{ be the set of all monomials in } Y \\ \text{and note that then } W_l = 1 \text{ for a unique } l \in J. \end{cases}$

Now clearly

(1) $\qquad W$ is an R-basis of S and hence S is a free R-algebra.

Indeed, allowing I to be an infinite set, a member of of $R[Y]$ may be defined to be a sum $\sum_{i \in I} a_i W_i$ where $a_i \in R$ is zero for all except a finite number of values of i. In other words, following the formal construction of polynomial and power series rings explained in L2§2, we put $R[Y] = (R^W)_{\text{finite}}$, i.e., we regard members of $R[Y]$ to be maps $W \to R$ whose support is finite. It is understood that any member of W_i involves only a finite number of Y_i and hence its degree is a nonnegative integer. For any nonzero element of $R[Y]$, its degree = its total degree = its Y-degree, is defined to be the maximum degree of the members of its support. The degree of $0 \in R[Y]$ is defined to be $-\infty$.

We claim that

(2) $\begin{cases} f(Y) \in R[Y]^\times \text{ and } g(Y) \in R[Y]^\times \text{ with } f(Y)g(Y) = 0 \\ \Rightarrow f(Y)h = 0 \text{ for some } h \in R^\times. \end{cases}$

To see this we make induction on the number n of the Y_i which occur in $f(Y)$ or $g(Y)$. If $n = 0$ then we can take $h = g(Y) \in R^\times$. So let $n > 0$ and assume for all smaller values of n. Let X be one of the Y_i which occurs in $f(Y)$ or $g(Y)$, and let Z be the remaining $n - 1$ of them. Let d and e be the X-degrees of $f(Y)$ and $g(Y)$ respectively. Now $f(Y) = \sum_{0 \le i \le d} f_i X^i$ where $f_i \in R[Z]$ with $f_d \neq 0$, and $g(Y) = \sum_{0 \le j \le e} g_j X^j$ where $g_j \in R[Z]$ with $g_e \neq 0$. Let us make a second induction on e. For a moment let $e = 0$; then $0 \neq \hat{g}(Z) = g(Y) \in R[Z]$ and hence we are done in case $n - 1 = 0$; so let $n - 1 > 0$ and let V be a monomial in Z of degree $>$ the Z-degree of f_i for $0 \le i \le d$; also let $\hat{f}(Z) = \sum_{0 \le i \le d} f_i(Z)V^i$; now $f(Y)g(Y) = 0 \Rightarrow f_i(Z)\hat{g}(Z) = 0$ for $0 \le i \le d \Rightarrow \hat{f}(Z)\hat{g}(Z) = 0$; therefore by the first induction hypothesis we get $\hat{f}(Z)h = 0$ for some $h \in R^\times$ and hence $f_i(Z)h = 0$ for $0 \le i \le d$ and therefore $f(Y)h = 0$. So let $e > 0$ and assume for all smaller values of e. If for some $u \in R[Z]$ we have $ug(Y) = 0$ then we must have $ug_e = 0$; hence if $f_i g(Y) = 0$ for $0 \le i \le d$, then $f_i g_e = 0$ for all $0 \le i \le d$, and therefore $f(Y)g_e = 0$ and we are done by the $e = 0$ case. So assume that $f_i g(Y) \neq 0$ for some i with $0 \le i \le d$ and let i be the largest such. Then $f_i g_e =$ coefficient of X^{i+e} in $f(Y)g(Y) = 0$ because $f(Y)g(Y) = 0$, and hence $\deg_X(f_i g(Y)) < e$; but $f(Y)(f_i g(Y)) = f_i(f(Y)g(Y)) = 0$, and therefore we are done by the second induction hypothesis.

Next we claim that

(3) R is a domain $\Rightarrow S$ is a domain

and

(4) $\begin{cases} \text{every zerodivisor in } R \text{ is nilpotent} \\ \Rightarrow \text{ every zerodivisor in } S \text{ is nilpotent} \end{cases}$

and

(5) R is reduced $\Rightarrow S$ is reduced.

Namely, (3) is obvious because the Y-degree of the product of any two nonzero members of $R[Y]$ equals the sum of their Y-degrees. To prove (4) assume that every zerodivisor in R is nilpotent, and let $f(Y) \in R[Y]^\times$ be a zerodivisor in S; then by (2) we have $f(Y)h = 0$ for some $h \in R^\times$; therefore every coefficient of $f(Y)$ is a zerodivisor in R and hence by assumption it is nilpotent; therefore $f(Y)$ is nilpotent because the set of all nilpotents in S form an ideal in S, namely $\mathrm{rad}_S 0$.

To prove (5) suppose that R is reduced and let $f(Y) \in R[Y]$ be such that $f(Y)^m = 0$ for some $m \in \mathbb{N}_+$. We will show by induction on the number n of the Y_i which occur in $f(Y)$ that $f(Y) = 0$. If $n = 0$ then $f(Y) \in R$ and hence $f(Y) = 0$ because R is reduced. So let $n > 0$ and assume for all smaller values of n. Let X be one of the Y_i which occurs in $f(Y)$ and let Z be the remaining $n - 1$ of them. If $f(Y) \neq 0$ then, upon letting d be the X-degree of $f(Y)$, we have $d \in \mathbb{N}$ and we can write $f(Y) = \sum_{0 \leq i \leq d} f_i X^i$ where $f_i \in R[Z]$ with $f_d \neq 0$. Now $f(Y)^m = \sum_{0 \leq i \leq md} F_i X^i$ where $F_i \in R[Z]$ with $F_{md} = f_d^m$. Since $f(Y)^m = 0$, we get $f_d^m = 0$ and hence $f_d = 0$ by the induction hypothesis. This is a contradiction, and so we are done.

Now for every ideal H in R let H^E be its extension to S, i.e.,

$$H^E = HS.$$

Slightly generalizing the substitution map described in L1§6 we see that:

(6)
$$\begin{cases} \text{if } \phi : R \to \overline{R} \text{ is any ring epimorphism with } \overline{R} \neq 0 \\ \text{then } \phi \text{ uniquely extends to a ring epimorphism} \\ \Phi : R[Y] \to \overline{R}[Y] \text{ such that} \\ \Phi(a) = \phi(a) \text{ for all } a \in R \text{ and } \Phi(Y_i) = Y_i \text{ for all } i \in I; \\ \text{moreover if } \ker(\phi) = H \text{ then } \ker(\Phi) = H^E. \\ \text{We call } \Phi \text{ the polynomial extension of } \phi. \end{cases}$$

We claim that for any ideal H in R we have:

(7) $$H \text{ is prime} \Rightarrow H^E \text{ is prime}$$

and

(8) $$H \text{ is primary} \Rightarrow H^E \text{ is primary}$$

and

(9) $$(\mathrm{rad}_R H)^E = \mathrm{rad}_S(H^E).$$

To see this first note that by (1) and (T108)(c) we have: $H \neq R \Rightarrow H^e \neq S$. Now upon letting $\phi : R \to \overline{R} = R/H$ be the residue class epimorphism and applying (6), by (3) and (4) we get (7) and (8) respectively. To prove (9) note that the inclusion $(\mathrm{rad}_R H)^E \subset \mathrm{rad}_S(H^E)$ is obvious; moreover, if $H \neq R$ then by (5) and (6) the ring

$S/(\mathrm{rad}_R H)^E$ is reduced and this implies the other inclusion $\mathrm{rad}_S(H^E) \subset (\mathrm{rad}_R H)^E$; finally, if $H = R$ then clearly $(\mathrm{rad}_R H)^E = \mathrm{rad}_S(H^E)$.

We shall conclude this Comment by giving two properties of S when R has a subfield k, i.e., when R is an overring of a field k. Namely, recalling that $Z_S(S)$ is the set of all zerodivisors in S, we claim that

(10)   if R has a subfield k then $k[Y]^\times \cap Z_S(S) = \emptyset$

and we also claim that

(11) $\begin{cases} \text{if } R \text{ has a subfield } k \text{ then} \\ \text{for any nonunit ideal } H \text{ in } R \text{ we have } k[Y]^\times \cap H^E = \emptyset. \end{cases}$

To prove (10) it suffices to note that, since every element of $k^\times$ is a unit in R, for any $f(Y) \in k[Y]^\times$ we have $f(Y)h \neq 0$ for all $h \in R^\times$ and hence by (2) we have $f(Y)g(Y) \neq 0$ for all $g(Y) \in R[Y]^\times$. Again since every element of $k^\times$ is a unit in R, in view of (T108)(a), by (1) we get (11).

COMMENT (C44). [**Ideals in Rational Function Ring Extensions**]. In L1§6 we discussed the rational function field in a family (or a set) of indeterminates (or algebraically independent elements) over a field. We shall now generalize this from a field to an overring of a field.

So

($\dagger$) $\begin{cases} \text{consider the polynomial ring } S = R[Y] \\ \text{where } Y = (Y_i)_{i \in I} \text{ is a family of indeterminates over a ring } R \end{cases}$

and

($\ddagger$) $\begin{cases} \text{consider the polynomial subring } k[Y] \text{ of } S \\ \text{where } k \text{ is a subfield of } R. \end{cases}$

In other words, as a ring, S is generated over the subring R by the family Y of elements which are algebraically independent over R, and $k[Y]$ is the subring of S generated by Y over the subfield k of R. In view of (C43)(10), we can form the localization $S_{k[Y]^\times}$ as a subring of the total quotient ring $\mathrm{QR}(S)$ of S. We call $S_{k[Y]^\times}$ the rational function ring in Y over R relative to k, and denote it by $R(Y)_k$.

Now for every ideal H in R let H^e be its extension to $T = R(Y)_k$, i.e.,

$$H^e = HT \quad \text{where} \quad T = R(Y)_k.$$

We claim that then:

(1) H is prime $\Rightarrow H^e$ is prime

and

(2) H is primary $\Rightarrow H^e$ is primary

and

(3) $$(\mathrm{rad}_R H)^e = \mathrm{rad}_S(H^e).$$

To see this first note that $H^e = (H^E)^{E'}$ where $H^E = HS$, and where for every ideal J in S we have put $J^{E'} = JT$. Therefore, in view of (C43)(11), by (C43)(7) (resp: (C43)(8)) we get (1) (resp: (2)) because, for any prime (resp: primary) ideal J in S with $J \cap k[y]^{\times} = \emptyset$, by L4§7 we know that $J^{E'}$ is a prime (resp: primary) ideal in T. Likewise, by (C43)(9) we get (3) because, for any ideal J in S, by L4§7 we know that $(\mathrm{rad}_S J)^{E'} = \mathrm{rad}_T(J^{E'})$.

Note that in the situation of (T110), i.e., if R is the finite variable polynomial ring $k[X_1, \ldots, X_N]$, then $T = L[X_1, \ldots, X_n]$ with $L = k(Y)$.

COMMENT (C45). [**Regularity of Localizations of Polynomial Rings**]. Let $R = k[X_1, \ldots, X_N]$ be the polynomial ring in a finite number of indeterminates $X_1, \ldots, X_N$ over a field k. Given any prime ideal P in R, letting $\mathrm{ht}_R P = h$ and $\mathrm{dpt}_R P = d$, by (Q10)(T47) we see that $d + h = N$ and $\mathrm{trdeg}_k \mathrm{QF}(R/P) = d$ where we have identified k with its image under the residue class epimorphism $\alpha : R \to R/P$, and hence by relabelling $X_1, \ldots, X_N$ suitably we can arrange matters so that the α-images of $X_1, \ldots, X_d$ form a transcendence basis of $\mathrm{QF}(R/P)$ over k. Let $Y = (Y_1, \ldots, Y_d)$ and $Z = (Z_1, \ldots, Z_h)$ where $Y_i = X_i$ and $Z_i = X_{d+i}$. Then $k[Y]^{\times} \cap P = \emptyset$ and hence, considering the localization

$$R^* = R_{k[Y]^{\times}} = k(Y)[Z]$$

and letting $P^* = PR^*$, by L4§7 we see that P^* is a maximal ideal in R^* with $\mathrm{ht}_{R^*} P^* = h$.

By L4§8(T22.3) we get a unique set of generators $f_i(Z_1, \ldots, Z_i)_{1 \le i \le h}$ of P^* such that

$$\begin{cases} f_i(Z_1, \ldots, Z_i) = Z_i^{n_i} + g_i(Z_1, \ldots, Z_i) \text{ and } n_i \in \mathbb{N}_+ \\ \text{with } g_i(Z_1, \ldots, Z_i) \in k(Y)[Z_1, \ldots, Z_i] \\ \text{and } \deg_{Z_j} g_i(Z_1, \ldots, Z_i) < n_j \text{ for } 1 \le j \le i \end{cases}$$

and such that $f_i(Z_1, \ldots, Z_i)$ is irreducible in $k[Z_1, \ldots, Z_i]$, for $1 \le i \le h$; for proof see the second half of (Q10) including (C16).

Multiplying the polynomial $f_i(Z_1, \ldots, Z_i)$ by a suitable element of $k[Y]^{\times}$, we get a set of generators $f_i^*(X_1, \ldots, X_{d+i})_{1 \le i \le h}$ of P^* such that

$$\begin{cases} f_i^*(X_1, \ldots, X_{d+i}) \\ = \widehat{g}_i(X_1, \ldots, X_d) X_{d+i}^{n_i} + g_i^*(X_1, \ldots, X_{d+i}) \text{ and } n_i \in \mathbb{N}_+ \\ \text{with } \widehat{g}_i(X_1, \ldots, X_d) \in k[X_1, \ldots, X_d]^{\times} \\ \text{and } g_i^*(X_1, \ldots, X_{d+i}) \in k[X_1, \ldots, X_{d+i}] \\ \text{and } \deg_{X_j} g_i^*(X_1, \ldots, X_{d+i}) < n_j \text{ for } d+1 \le j \le d+i \end{cases}$$

and such that $f_i^*(X_1, \ldots, X_{d+i})$ is irreducible in $k[X_1, \ldots, X_{d+i}]$, for $1 \le i \le h$; this time the polynomial $f_i^*(X_1, \ldots, X_{d+i})$ is unique only up to multiplication by nonzero elements of $k[X_1, \ldots, X_d]$.

At any rate, the height h prime ideal P^* is generated by h elements, and hence the local ring $R_P = R_{P^*}^*$ is an h dimensional regular local ring.

By (Q7)(T27) and (Q15)(T70) it follows that $(f_1^*, \ldots, f_h^*)R = Q_1 \cap \cdots \cap Q_r$ where $Q_1, \ldots, Q_r$ are a finite number of primary ideals in R whose radicals are pairwise distinct height h prime ideals in R with $Q_1 = P_1 = P$. Geometrically speaking, letting $V_1, \ldots, V_r$ be the irreducible varieties defined by $P_1, \ldots, P_r$ in the affine N-dimensional space over k, where $P_i = \mathrm{rad}_R Q_i$, what we have shown is that given an irreducible h-codimensional, i.e., $(N-h)$-dimensional, variety $V = V_1$, we can find h irreducible hypersurfaces $f_1^* = 0, \ldots, f_h^* = 0$ whose intersection consists of V_1 together with a finite number of other h-codimensional irreducible varieties $V_2, \ldots, V_r$ so that these hypersurfaces generate the prime ideal P of V "in the neighborhood of V," i.e., in the local ring R_P.

Let $R' = k[X_1', \ldots, X_{N'}']$ be the polynomial ring in another finite number of indeterminates $X_1', \ldots, X_{N'}'$ over k. Given any prime ideal P' in R', upon letting $\mathrm{ht}_{R'} P' = h'$ and $\mathrm{dpt}_{R'} P' = d'$, by (Q10)(T47) we see that $d' + h' = N'$ and $\mathrm{trdeg}_k \mathrm{QF}(R'/P') = d'$ where we have identified k with its image under the residue class epimorphism $\alpha' : R' \to R'/P'$, and hence by relabelling $X_1', \ldots, X_{N'}'$ suitably we can arrange matters so that the α'-images of $X_1', \ldots, X_{d'}'$ form a transcendence basis of $\mathrm{QF}(R'/P')$ over k. Let $Y' = (Y_1', \ldots, Y_{d'}')$ and $Z' = (Z_1', \ldots, Z_{h'}')$ where $Y_i' = X_i'$ and $Z_i' = X_{d'+i}'$. Let

$$R'^* = k(Y)[X_1', \ldots, X_{N'}'] = R'(Y)_k$$

and $P'^* = P' R'^*$. Then by (T110) we see that P'^* is a prime ideal in R'^* with $\mathrm{ht}_{R'^*} P'^* = h'$ and $\mathrm{dpt}_{R'^*} P'^* = d'$. Let

$$S = k[X_1, \ldots, X_N, X_1', \ldots, X_{N'}']$$

and let $D = PS + P'S$. Consider the localization

$$S^* = k(Y)[Z_1, \ldots, Z_h, X_1', \ldots, X_{N'}'] = S_{k[Y]^\times}$$

and let us put $D^* = P^* S^* + P'^* S^*$. Then clearly $D^* = DS^*$. Let $\widehat{P}_1, \ldots, \widehat{P}_m$ be all the distinct associated primes of D in S. Then by (T109.4)(iii) we see that for $1 \le i \le m$ we have $R \cap \widehat{P}_i = P$ and hence $k[Y]^\times \cap \widehat{P}_i = \emptyset$. Therefore, upon letting $\widehat{P}_i^* = \widehat{P}_i S^*$, by L4§7 we conclude that $\widehat{P}_1^*, \ldots, \widehat{P}_m^*$ are exactly all the distinct associated primes of D^* in S^*, and for $1 \le i \le m$ we have $\mathrm{ht}_{S^*} \widehat{P}_i^* = \mathrm{ht}_S \widehat{P}_i$.

Let $\alpha'^* : R'^* \to R'^*/P'^*$ be the residue class epimorphism. Then it is clear that $\alpha'^*(X_1'), \ldots, \alpha'^*(X_{d'}')$ is a transcendence basis of $\mathrm{QF}(R'^*/P'^*)$ over $\alpha'^*(k)$, and hence $k(Y)[Y']^\times \cap P'^* = \emptyset$.

So the above procedure of enlarging the "ground field" from k to $k(Y)$ can be repeated to enlarge it from $k(Y)$ to $k(Y, Y')$. Earlier we went from the unprimed to the primed direction. Now we go from the primed to the unprimed direction. Earlier we "converted" the prime ideal P into a maximal ideal. Now keeping P maximal, we also convert P' into a maximal ideal.

In greater detail, considering the localization

$$R'^{**} = R'^{*}_{k(Y)[Y']^{\times}} = k(Y, Y')[Z']$$

and letting $P'^{**} = P'^{*}R'^{**}$, by L4§7 we see that P'^{**} is a maximal ideal in R'^{**} with $\operatorname{ht}_{R'^{**}}P'^{**} = h'$. Let

$$R^{**} = k(Y, Y')[Z] = R^{*}(Y')_{k}$$

and $P^{**} = P^{*}R^{**}$. Then by (T110) we see that P^{**} is a maximal ideal in R^{**} with $\operatorname{ht}_{R^{**}}P^{**} = h$. Consider the localization

$$S^{**} = k(Y, Y')[Z_1, \ldots, Z_h, Z'_1, \ldots, Z'_{d'}] = S^{*}_{k(Y)[Y']^{\times}}$$

and let us put $D^{**} = P^{**}S^{**} + P'^{**}S^{**}$. Then clearly $D^{**} = D^{*}S^{**}$. By (T109.4)(iii) we see that for $1 \leq i \leq m$ we have $R'^{*} \cap \widehat{P}^{*}_{i} = P'^{*}$ and hence $k(Y)[Y']^{\times} \cap \widehat{P}^{*}_{i} = \emptyset$. Consequently, upon letting $\widehat{P}^{**}_{i} = \widehat{P}^{*}_{i}S^{**}$, by L4§7 we see that $\widehat{P}^{**}_1, \ldots, \widehat{P}^{**}_m$ are exactly all the distinct associated primes of D^{**} in S^{**}, and for $1 \leq i \leq m$ we have $\operatorname{ht}_{S^{**}}\widehat{P}^{**}_{i} = \operatorname{ht}_{S^{*}}\widehat{P}^{*}_{i} = \operatorname{ht}_{S}\widehat{P}_{i}$.

Therefore: $\operatorname{ht}_{S}Q = \operatorname{ht}_{R}P + \operatorname{ht}_{R'}P'$ for every associated prime Q of D in S $\Leftrightarrow$ $\operatorname{ht}_{S^{**}}Q^{**} = \operatorname{ht}_{R^{**}}P^{**} + \operatorname{ht}_{R'^{**}}P'^{**}$ for every associated prime Q^{**} of D^{**} in S^{**}. Consequently, since P^{**} and P'^{**} are maximal ideals in R^{**} and R'^{**} respectively, and $DS^{**} = D^{**}$, it suffices to prove (T111) when P and P' are maximal ideals in R and R' respectively. This we shall do in the next Comment (C46).

COMMENT (C46). [**Maximal Ideals in Polynomial Rings**]. Consider the polynomial ring $R = k[X_1, \ldots, X_N]$ in a finite number of indeterminates $X_1, \ldots, X_N$ over a field k. Given any maximal ideal P in R, by L4§8(T22.3) we get a unique set of generators $f_i(X_1, \ldots, X_i)_{1 \leq i \leq N}$ of P such that

$$\begin{cases} f_i(X_1, \ldots, X_i) = X_i^{n_i} + g_i(X_1, \ldots, X_i) \text{ and } n_i \in \mathbb{N}_{+} \\ \text{with } g_i(X_1, \ldots, X_i) \in k[X_1, \ldots, X_i] \\ \text{and } \deg_{X_j} g_i(X_1, \ldots, X_i) < n_j \text{ for } 1 \leq j \leq i \end{cases}$$

and such that $f_i(X_1, \ldots, X_i)$ is irreducible in $k[X_1, \ldots, X_i]$, for $1 \leq i \leq N$; for proof see the second half of (Q10) including (C16).

More generally, consider the ideal

$$J = (F_1, \ldots, F_N)R \quad \text{with} \quad F_i = F_i(X_1, \ldots, X_i) \in k[X_1, \ldots, X_i]$$

such that

$$\deg_{X_i}(F_i(X_1, \ldots, X_i) - X_i^{m_i}) < m_i \in \mathbb{N}_{+} \quad \text{for} \quad 1 \leq i \leq N.$$

Let

$$W = (X_1^{a_1} \dots X_N^{a_N} F_1^{b_1} \dots F_N^{b_N})_{0 \le a_1 < m_1, \dots, 0 \le a_N < m_N, 0 \le b_1 < \infty, \dots, 0 \le b_N < \infty}$$

and let V be the set of all members of W for which $b_1 + \cdots + b_N > 0$. By induction on N we shall show that then

(1) W is a k-basis of R, and V is a k-basis of J.

This being obvious for $N = 0$, let $N > 0$ and assume for $N - 1$. Let W^* (resp: V^*) be the set of all members of W (resp: V) for which $a_N = 0 = b_N$. Let $R^* = k[X_1, \dots, X_{N-1}]$ and consider the ideal

$$J^* = (F_1, \dots, F_{N-1})R^*.$$

Then by the induction hypothesis

(1′) W^* is a k-basis of R^*, and $V^*_{\cdot}$ is a k-basis of J^*,

and by repeated applications of the division algorithm we see that

(1″) $\begin{cases} (X_N^i F_N^j)_{0 \le i < m_N} \text{ with } 0 \le j < \infty \text{ is an } (R^*)\text{-basis of } R, \\ (X_N^i F_N^j)_{0 \le i < m_N} \text{ with } 0 < j < \infty \text{ is an } (R^*)\text{-basis of } F_N R. \end{cases}$

By (1′) and (1″) we get (1).

By (1) we see that

(2) $\begin{cases} J \cap k = 0 \text{ and letting } \alpha : R \to R/J \text{ be the canonical epimorphism} \\ \text{we have that } \alpha(R) \text{ is an } m\text{-dimensional vector space} \\ \text{over the field } \alpha(k) \text{ where } m \text{ is the product } m = m_1 \dots m_N. \end{cases}$

In geometric terms, (2) is some kind of Bézout Theorem saying that the hypersurfaces $F_1 = 0, \dots, F_N = 0$ meet in as many points as the product of the degrees $m_1, \dots, m_N$. At any rate, in view (O21) and (O25) of L4§5, by (2) we see that

(3) $\dim(R/J) = 0$

and hence by (Q10)(T47) we conclude that

(4) $\begin{cases} J \ne R \text{ and for every associated prime } Q \text{ of } J \text{ in } R \\ \text{we have } \operatorname{ht}_R Q = N. \end{cases}$

Let $R' = k[X_1', \dots, X_{N'}']$ be the polynomial ring in another finite number of indeterminates $X_1', \dots, X_{N'}'$ over k. Given any maximal ideal P' in R', by L4§8(T22.3) we get a unique set of generators $f_i'(X_1', \ldots, X_i')_{1 \le i \le N'}$ of P' such that

$$\begin{cases} f_i'(X_1', \ldots, X_i') = X_i'^{n_i'} + g_i'(X_1', \ldots, X_i') \text{ and } n_i' \in \mathbb{N}_+ \\ \text{with } g_i'(X_1', \ldots, X_i') \in k[X_1', \ldots, X_i'] \\ \text{and } \deg_{X_j'} g_i'(X_1', \ldots, X_i') < n_j' \text{ for } 1 \le j \le i \end{cases}$$

and such that $f_i'(X_1', \ldots, X_i')$ is irreducible in $k[X_1', \ldots, X_i']$, for $1 \le i \le h$; for proof see the second half of (Q10) including (C16).

By (Q10)(T47) we see that

(5) $$\mathrm{ht}_R P = N \text{ and } \mathrm{ht}_{R'} P' = N'.$$

Upon letting

$$S = k[X_1, \ldots, X_N, X_1', \ldots, X_{N'}'] \quad \text{and} \quad D = PS + P'S$$

by (4) we conclude that

(6) $$\begin{cases} D \ne S \text{ and for every associated prime } Q \text{ of } D \text{ in } S \\ \text{we have } \mathrm{ht}_S Q = N + N'. \end{cases}$$

COMMENT (C47). [**Diagonals of Product Spaces**]. We shall now use the idea of a diagonal to deduce (T112) from (T107) and (T111).

Consider the polynomial ring $R = k[X_1, \ldots, X_N]$ in indeterminates $X_1, \ldots, X_N$ over a field k with $N \in \mathbb{N}$, and let P and $\overline{P}$ be any prime ideals in R.

Let $R' = k[X_1', \ldots, X_N']$ be the polynomial ring in another set of indeterminates $X_1', \ldots, X_N'$ over k. Also let $S = k[X_1, \ldots, X_N, X_1' \ldots, X_N']$ be the polynomial ring in $2N$ indeterminates over k, and consider the ideal $\Delta = (X_1 - X_1', \ldots, X_N - X_N')S$. Let $\phi : R \to R'$ be the unique k-isomorphism of rings such that $\phi(X_i) = X_i'$ for $1 \le i \le N$. Let $P' = \phi(\overline{P})$ and $D = PS + P'S$. Then by (T111) we see that

$$\begin{cases} \text{for every associated prime } Q \text{ of } D \text{ in } S \\ \text{we have } \mathrm{ht}_S Q = \mathrm{ht}_R P + \mathrm{ht}_{R'} P'. \end{cases}$$

Therefore, by (Q10)(T47)+(Q12)(T56)+(T107) we see that for every associated minimal prime Q^* of $D + \Delta$ in S we have

$$\mathrm{ht}_S Q^* \le N + \mathrm{ht}_R P + \mathrm{ht}_{R'} P'$$

and hence we have

$$2N - \mathrm{ht}_S Q^* \ge (N - \mathrm{ht}_R P) + (N - \mathrm{ht}_R \overline{P}) - N$$

and therefore by (Q10)(T47) we have

$$\dim(S/Q^*) \ge \dim(R/P) + \dim(R/\overline{P}) - N.$$

Let $\psi : S \to R$ be the ring epimorphism such that $\psi(z) = z$ for all $z \in R$ and $\psi(X_i') = X_i$ for $1 \leq i \leq N$. Then clearly $\ker(\psi) = \Delta$ and $\psi(D) = P + \overline{P}$. Consequently the rings $S/(D + \Delta)$ and $R/(P + \overline{P})$ are isomorphic, and hence

(1) $\quad \begin{cases} \text{for every associated prime } \widehat{P} \text{ of } P + \overline{P} \text{ in } R \\ \text{we have } \dim(R/\widehat{P}) \geq \dim(R/P) + \dim(R/\overline{P}) - N. \end{cases}$

Given any ideals J and H in R with $J + H \neq R$, we can take prime ideals P and $\overline{P}$ in R with $J \subset P$ and $H \subset \overline{P}$ such that $\dim(R/P) = \dim(R/J)$ and $\dim(R/\overline{P}) = \dim(R/H)$. By (1) there exists a prime ideal $\widehat{P}$ in R with $P + \overline{P} \subset \widehat{P}$ such that

$$\dim(R/\widehat{P}) \geq \dim(R/P) + \dim(R/\overline{P}) - N.$$

Now $J + H \subset \widehat{P}$ and hence

$$\dim(R/(J + H)) \geq \dim(R/\widehat{P}).$$

Therefore

(2) $\qquad \dim(R/(J + H)) \geq \dim(R/P) + \dim(R/\overline{P}) - N.$

To explain the geometry behind the above proof, the variety (= the set of all common zeros) $V(\Delta)$ of the ideal Δ is the diagonal in the product $A \times A'$ of two affine N-spaces over k. Thus A is the set of all N-tuples with entries in k, and so is A'. The diagonal $V(\Delta)$ is the set of all $(a, a') \in A \times A'$ with $a = a'$. Let $V(P)$ and $V(\overline{P})$ be the varieties in A defined by the ideals P and $\overline{P}$ respectively. Let $V(P')$ be the "copy" of $V(\overline{P})$ in A'.

To indicate that the variety

$$V(P) \cap V(\overline{P}) \quad \text{in} \quad A$$

corresponds to the variety

$$V(\Delta) \cap (V(P) \times V(P')) \quad \text{in} \quad A \times A',$$

let A and A' be any sets together with a bijection $\psi : A \to A'$, let V and $\overline{V}$ be any subsets of A, let V' be the copy of $\overline{V}$ in A', and consider

$$\text{the diagonal } W = \{(a, a') \in A \times A' : a = a'\}.$$

Then for any $(a, a') \in A \times A'$, we clearly have:

$$(a, a') \in W \cap (V \times V') \Leftrightarrow a \in V \cap \overline{V}.$$

The reader may depict this by drawing A and A' as the coordinate axis in an affine plane.

COMMENT (C48). [**Intersection in Vector Space and Projective Space**]. The main result (T112) of (Q9) says that if the intersection U of any two varieties V and W in the N-dimensional affine space over a field k is nonempty, then we have

$$\dim(U) \geq \dim(V) + \dim(W) - N.$$

This is motivated by the easy to prove FACT that for any finite dimensional subspaces V and W of a k-vector space H, letting d denote vector space dimension, we always have

$$d(V \cap W) = d(V) + d(W) - d(V + W)$$

and hence if $d(H) = N \in \mathbb{N}$ then

$$d(V \cap W) \geq d(V) + d(W) - N.$$

Thus, unlike the case of affine varieties, we did NOT have to assume that $V \cap W$ is nonempty. As we shall soon see, the case of projective varieties is more like the vector space case.

Even in the affine case, the assumption that $V \cap W$ is nonempty means that either we are supposing the ground field k is algebraically closed, or we are dealing with the spectral affine space as described in L4§8.1.

For instance the circle $X^2 + Y^2 = 1$ and vertical line $X = a$ in the affine plane over the real number field, have a nonempty intersection iff $1 - a^2 \geq 0$. The mystery is solved by passing to the complex number field which is the algebraic closure of the reals.

But even after passing to the complex field, two parallel lines do not meet. As explained in L3§3, this and similar puzzles caused by "points of intersection going to infinity" are solved by going over to projective spaces by means of homogeneous coordinates. More about this later. As said in L3§3, for an elementary treatment of these matters you may browse in my Engineering Book [A04].

QUEST (Q30) Syzygies and Homogeneous Resolutions

In (Q17) we discussed projective resolutions of modules. In its original form, as was done in Hilbert's legendary 1890 paper [Hi1], it started as free resolutions of homogeneous ideals in polynomial rings. In this set-up, what we have called the pdim (= projective dimension) of a module corresponds to what we shall call the hdim (= homogeneous dimension) of a graded module as we proceed to explain.

All this gives rise to syzygies, originally an astronomical term to indicate the waning and waxing of the phases of the moon, adapted by Sylvester in his long 1853 paper [Sy2] to indicate a mathematical object.

Actually in (Q17) we have already worked with syzygies except that we did not give them a name. So what are they?

Given any noetherian ring R, any finite R-module V, and any $n \in \mathbb{N}$, by (Q17)(T80.17) we can find a preprojective R-resolution (f, W, n) of V of length n, and we call W_{n+1} an n-th R-syzygy of V. By (Q17)(T80.3) and (Q17)(T80.7) we see that if W'_{n+1} is any other n-th R-syzygy of V then the modules W_{n+1} and W'_{n+1} are R-equivalent, in symbols $W_{n+1} \sim_R W'_{n+1}$, by which we mean that for some finite projective R-modules U, U' we have an isomorphism $W_{n+1} \oplus U \approx W'_{n+1} \oplus U'$ of R-modules. By $\mathrm{syz}_R^n V$ we denote the collection of all n-th R-syzygies of V.

Before turning to homogeneous ideals in (and graded modules over) naturally graded polynomial rings, let us recall that $W_1 \oplus \cdots \oplus W_n$ denotes the (external) direct sum of a finite number of modules $W_1, \ldots, W_n$ over a ring R. In case all of these are graded modules $W_i = \sum_{\nu \in I} (W_i)_\nu$ over a graded ring $R = \sum_{\nu \in I} R_\nu$ of any type I, we convert $W_1 \oplus \cdots \oplus W_n$ into a graded R-module by putting

$$(W_1 \oplus \cdots \oplus W_n)_\nu = \{(x_1, \ldots, x_n) \in W_1 \oplus \cdots \oplus W_n : x_i \in (W_i)_\nu \text{ for } 1 \leq i \leq n\}.$$

We call this the graded (external) direct sum of $W_1, \ldots, W_n$ and continue to denote it by $W_1 \oplus \cdots \oplus W_n$. We define the homogeneous generating number of any graded R-module V by putting

$$\mathrm{hnb}_R V = \text{the smallest number of homogeneous generators of } V$$

and we note that this is ∞ iff V is not finitely generated, and it is 0 iff V is the zero module. In particular this defines the homogeneous generating number $\mathrm{hnb}_R I$ of any homogeneous ideal I (not to be confused with the type I) in R by putting

$$\mathrm{hnb}_R I = \text{the smallest number of homogeneous generators of } I.$$

Henceforth let us consider a naturally graded homogeneous ring $R = \sum_{\nu \in \mathbb{N}} R_\nu$ over a field $k = R_0$.

Given any graded R-module $V = \sum_{\nu \in \mathbb{N}} V_\nu$, we define the backward q-shift of V, with $q \in \mathbb{N}$, to be the graded R-module $^q V$ obtained by putting

$$^q V_\nu = \begin{cases} 0 & \text{if } 0 \leq \nu < q \\ V_{\nu - q} & \text{if } q \leq \nu < \infty. \end{cases}$$

As an R-module, $^q V$ is the same as V but with a different grading; so for any $r \in R$ and $v \in {}^q V$, the product rv is the same when we regard v as element of V. More generally we define the backward $(q_1, \ldots, q_n)$-shift of V, with $n, q_1, \ldots, q_n$ in $\mathbb{N}$, by putting

$$^{(q_1, \ldots, q_n)} V = {}^{q_1} V \oplus \cdots \oplus {}^{q_n} V \quad \text{(graded direct sum)}.$$

Note that, as ungraded modules over the ungraded ring R, the module $^{(q_1, \ldots, q_n)} V$ is isomorphic to the module V^n.

By a homogeneously finite free R-module we mean a graded R-module which is isomorphic (as a graded R-module) to $^{(q_1, \ldots, q_n)} R$ for some $n, q_1, \ldots, q_n$ in $\mathbb{N}$.

By putting $e_i = (0, \ldots, 0, 1, 0 \ldots, 0)$ with 1 in the i-th spot, clearly we get a homogeneous R-basis $(e_i \in {}^{(q_1, \ldots, q_n)} R_{q_i})_{1 \leq i \leq n}$ of ${}^{(q_1, \ldots, q_n)} R$; we may call this the canonical basis. Moreover, given any finite graded R-module V and any finite number of homogeneous elements (resp: generators) $(y_i \in V_{q_i})_{1 \leq i \leq n}$ of V, clearly we get a unique graded R-homomorphism (resp: R-epimorphism) ${}^{(q_1, \ldots, q_n)} R \to V$ which sends e_i to y_i for $1 \leq i \leq n$; we may call this the canonical homomorphism (resp: epimorphism) induced by the homogeneous elements (generators) $(y_i \in V_{q_i})_{1 \leq i \leq n}$.

Next we define a homogeneously finite projective R-module to be a finite graded R-module V having the property which says that if $\alpha : E \to F$ is any graded R-epimorphism of any finite graded R-modules and $\beta : V \to F$ is any graded R-homomorphism then $\beta = \alpha \gamma$ for some graded R-homomorphism $\gamma : V \to E$.

A graded R-module V is a graded direct summand of a homogeneously finite free module (resp: a homogeneously finite projective module) means there exists a graded R-module U such that the graded direct sum $V \oplus U$ is homogeneously finite free (resp: homogeneously finite projective).

By a graded exact sequence of R-modules we mean an exact sequence (f, W, n) of R-modules where the R-modules $W_0, \ldots, W_{n+1}$ as well as the R-homomorphisms $f_0, \ldots, f_n$ are graded. By a graded short exact sequence of R-modules we mean a graded exact sequence $(f, W, 1)$ of R-modules of length 1; this sequence splits homogeneously means there exists a graded R-homomorphism $g_0 : W_0 \to W_1$ such that $f_0 g_0$ is the identity map $W_0 \to W_0$.

By a graded R-resolution of a finite graded R-module V we mean a graded exact sequence (f, W, n) of R-modules such that $V = W_0$. The graded R-resolution (f, W, n) is called a homogeneously finite projective (resp: homogeneously finite free, homogeneously minimal free) graded R-resolution if for $1 \leq i \leq n + 1$ the module W_i is homogeneously finite projective (resp: homogeneously finite free, isomorphic to ${}^{(q_1, \ldots, q_{d(i)})} R$ with $d(i) = \mathrm{hnb}_R \mathrm{im}(f_{i-1}) \in \mathbb{N}$ and some $q_1, \ldots, q_{d(i)}$ in $\mathbb{N}$). The R-resolution (f, W, n) is called a homogeneously finite preprojective (resp: homogeneously finite prefree, homogeneously minimal prefree) R-resolution if for $1 \leq i \leq n$ the module W_i is homogeneously finite projective (resp: homogeneously finite free, isomorphic to ${}^{(q_1, \ldots, q_{d(i)})} R$ with $d(i) = \mathrm{hnb}_R \mathrm{im}(f_{i-1}) \in \mathbb{N}$ and some $q_1, \ldots, q_{d(i)}$ in $\mathbb{N}$).

By the homogeneous R-dimension of a finite graded R-module V, which we shall denote by $\mathrm{hdim}_R V$, we mean the smallest length of a homogeneously finite projective R-resolution of V if such exists; if there does not exist any homogeneously finite projective R-resolution of V then we put $\mathrm{hdim}_R V = \infty$. Note that V is homogeneously finite projective iff $\mathrm{hdim}_R V = 0$.

In Lemma (T113) and Theorem (T114) below we shall prove graded analogues of Lemma (Q17)(T80) and Theorem (Q17)(T81) respectively.

Given any finite graded R-module V and any nonnegative integer n, by (T113.9) below we can find a homogeneously finite preprojective R-resolution (f, W, n) of V of length n, and we call W_{n+1} an n-th homogeneous R-syzygy of V. By (T113.3) and (T113.7) below we see that if W'_{n+1} is any other n-th homogeneous R-syzygy of V then the modules W_{n+1} and W'_{n+1} are homogeneously R-equivalent, in symbols $W_{n+1} \sim_{[R]} W'_{n+1}$, by which we mean that for some homogeneously finite projective R-modules U, U' we have a graded R-isomorphism $W_{n+1} \oplus U \approx W'_{n+1} \oplus U'$ of graded direct sums. By $\mathrm{syz}^n_{[R]} V$ we denote the collection of all n-th homogeneous R-syzygies of V.

HOMOGENEOUS RESOLUTION LEMMA (T113). For any naturally graded homogeneous ring $R = \sum_{\nu \in \mathbb{N}} R_\nu$ over a field $k = R_0$ we have the following.

(T113.1) If a graded short exact sequence $(f, W, 1)$ of R-modules splits homogeneously then there exists a graded isomorphism $W_1 \approx W_2 \oplus W_0$. If a graded R-module V is a graded direct summand of a homogeneously finite free R-module then V is a finite module.

(T113.2) A finite graded R-module is homogeneously finite projective iff every length-one graded R-resolution of it splits homogeneously iff it is a graded direct summand of a homogeneously finite free module.

(T113.3) Every homogeneously finite free module is a homogeneously finite projective module. A graded direct sum of any finite number of homogeneously finite projective modules is a homogeneously finite projective module. A graded direct summand of a homogeneously finite projective module is a homogeneously finite projective module.

(T113.4) Given any finite graded R-module V, for every $i \in \mathbb{N}$ let $\lambda_i(V)$ be the k-vector-space dimension of $(V/(RV_0 + \cdots + RV_{i-1}))_i$. Then $\lambda_i \in \mathbb{N}$ with $\lambda_i = 0$ for all except a finite number of i, and we have $\mathrm{hnb}_R V = \sum_{0 \leq i < \infty} \lambda_i$. In other words, by taking elements $(f_{ij})_{1 \leq j \leq \lambda_i}$ in V_i which form a k-basis of V_i mod $(RV_0 + \cdots + RV_{i-1})_i$, we get a finite set of minimal homogeneous R-generators (f_{ij}) of V. Moreover, for the graded direct sum $V \oplus V'$, where V' is any other finite graded R-module, we have $\mathrm{hnb}_R(V \oplus V') = \mathrm{hnb}_R V + \mathrm{hnb}_R V'$. Finally, the module V is homogeneously finite projective iff it is homogeneously finite free.

(T113.5) Let there be given any two graded short exact sequences $(f, W, 1)$ and $(f', W', 1)$ of R-modules such that the modules W_1 and W'_1 are homogeneously finite projective, and assume that there exists a graded R-isomorphism $g : W_0 \to W'_0$. Then there exists a graded R-isomorphism $h : W_1 \oplus W'_1 \to W_1 \oplus W'_1$ such that $g f_0 l = f'_0 l' h$ where $l : W_1 \oplus W'_1 \to W_1$ and $l' : W_1 \oplus W'_1 \to W'_1$ are the natural projections. Moreover, for any such h we have $h(W_2 \oplus W'_1) = W_1 \oplus W'_2$ where we regard $W_2 \oplus W'_1$ and $W_1 \oplus W'_2$ as graded submodules of $W_1 \oplus W'_1$ via the injective maps f_1 and f'_1.

(T113.6) Conversely, let there be given any two graded short exact sequences $(f, W, 1)$ and $(f', W', 1)$ of R-modules such that the modules W_1 and W'_1 are homogeneously finite projective, and assume that there exists a graded R-isomorphism $h : W_1 \oplus W'_1 \to W_1 \oplus W'_1$ with $h(W_2 \oplus W'_1) = W_1 \oplus W'_2$ where we regard $W_2 \oplus W'_1$ and $W_1 \oplus W'_2$ as graded submodules of $W_1 \oplus W'_1$ via the injective maps f_1 and f'_1.

Then there exists a graded R-isomorphism $g : W_0 \to W_0'$ such that $gf_0l = f_0'l'h$ where $l : W_1 \oplus W_1' \to W_1$ and $l' : W_1 \oplus W_1' \to W_1'$ are the natural projections.

(T113.7) Generalizing (T113.5), let there be given any two graded exact sequences (f, W, n) and (f', W', n) of R-modules of equal length such that the modules W_i and W_i' are homogeneously finite projective for $1 \leq i \leq n$, and assume that there exists a graded R-isomorphism $g : W_0 \to W_0'$. Then we have graded R-isomorphisms

$$W_1 \oplus W_2' \oplus W_3 \oplus W_4' \oplus \cdots \oplus W_{n+1} \approx W_1' \oplus W_2 \oplus W_3' \oplus W_4 \oplus \cdots \oplus W_{n+1}'$$

or

$$W_1' \oplus W_2 \oplus W_3' \oplus W_4 \oplus \cdots \oplus W_{n+1} \approx W_1 \oplus W_2' \oplus W_3 \oplus W_4' \oplus \cdots \oplus W_{n+1}'$$

according as n is even or odd. Moreover: W_{n+1} is homogeneously finite projective iff W_{n+1}' is homogeneously finite projective.

(T113.8) (GRADED COMPARISON LEMMA) Let there be given any graded exact sequence (f, W, n) of R-modules such that the module W_i is homogeneously finite projective for $1 \leq i \leq n$. Then $\mathrm{hdim}_R W_0 \leq n$ iff W_{n+1} is homogeneously finite projective. More precisely, if the R-module $\mathrm{im}(f_m)$ is homogeneously finite projective for some integer $m \in \{0, 1, \ldots, n\}$ then for the smallest such integer m we have $\mathrm{hdim}_R W_0 = m$, and the R-module $\mathrm{im}(f_i)$ is homogeneously finite projective or not according as $m \leq i \leq n$ or $0 \leq i < m$. Moreover, if $\mathrm{hdim}_R W_0 \geq n$ then we have $n + \mathrm{hdim}_R W_{n+1} = \mathrm{hdim}_R W_0$ with the understanding that $n + \infty = \infty$.

(T113.9) Given any finite graded R-module V and any nonnegative integer n, there exists a homogeneously minimal prefree R-resolution (f, W, n) of V. Moreover, for any such R-resolution we have that (i) it is always homogeneously finite preprojective, and (ii) it is homogeneously finite projective $\Leftrightarrow \mathrm{hdim}_R V \leq n \Leftrightarrow W_{n+1}$ is a homogeneously finite projective R-module.

(T113.10) A graded R-module V is a homogeneously finite projective module iff (as an ungraded module over the ungraded ring R) it is a finite projective module.

(T113.11) For any finite graded R-module V we have $\mathrm{hdim}_R V = \mathrm{pdim}_R V$ where the pdim is taken in the sense of an ungraded module over an ungraded ring.

(T113.12) For any finite graded R-module V and any nonnegative integer n, every n-th homogeneous R-syzygy of V is an n-th R-syzygy of V as an ungraded module over an ungraded ring.

PROOF. The proofs of (T113.1) to (T113.3) are obtained by making some obvious small changes (such as inserting "graded" or "homogeneously" in appropriate places) in the proofs of (Q17)(T80.1) to (Q17)(T80.3). If the reader prefers to have things completely spelled out then here it is.

To prove (T113.1), first suppose that $(f, W, 1)$ is a graded short exact sequence of R-modules which splits homogeneously, i.e., there exists a graded R-homomorphism

$g_0 : W_0 \to W_1$ such that $f_0 g_0$ is the identity map $W_0 \to W_0$. Then clearly $(a, b) \mapsto$ $f_1(a) + g_0(b)$ gives an isomorphism of the graded direct sum $W_2 \oplus W_0$ onto W_1.

Next let V be a finite graded R-module such that for some graded R-module U and some $n, q_1, \ldots, q_n$ in $\mathbb{N}$ we have a graded isomorphism $g : {}^{(q_1,\ldots,q_n)}R \to V \oplus U$. Then upon letting $p : V \oplus U \to V$ be the natural projection, we get a graded R-epimorphism $pg : {}^{(q_1,\ldots,q_n)}R \to V$ and hence V is a finite graded R-module. This completes the proof of (T113.1).

To prove (T113.2), let V be a finite graded R-module. We shall cyclically prove that: V is a graded direct summand of a homogeneously finite free module $\Rightarrow$ V is homogeneously finite projective $\Rightarrow$ every length-one graded R-resolution of V splits homogeneously $\Rightarrow$ V is a graded direct summand of a homogeneously finite free module. First, assuming that for some graded R-module U and some $n, q_1, \ldots, q_n$ in $\mathbb{N}$ there is a graded isomorphism $g : {}^{(q_1,\ldots,q_n)}R \to V \oplus U$, let us show that V is homogeneously finite projective. So given any graded R-epimorphism $\alpha : E \to F$ of finite graded R-modules and any graded R-homomorphism $\beta : V \to F$, we want to find a graded R-homomorphism $\gamma : V \to E$ such that $\beta = \alpha\gamma$. Let $p : V \oplus U \to V$ be the natural projection, let $q : V \to V \oplus U$ be the natural injection, and let $(e_i)_{1 \leq i \leq n}$ be the canonical basis of ${}^{(q_1,\ldots,q_n)}R$. Since α is surjective, for $1 \leq i \leq n$ we can take $x_i \in E^{q_i}$ such that $\alpha(x_i) = \beta pg(e_i)$. Since $(e_i)_{1 \leq i \leq n}$ is a basis of ${}^{(q_1,\ldots,q_n)}R$, there is a unique graded R-homomorphism $\delta : {}^{(q_1,\ldots,q_n)}R \to E$ such that $\delta(e_i) = x_i$ for $1 \leq i \leq n$. Let $\gamma = \delta g^{-1} q$. Then γ is a graded R-homomorphism $V \to E$ with $\beta = \alpha\gamma$.

Next, assuming V is homogeneously finite projective, given any length-one graded R-resolution $(f, W, 1)$ of V, let us show that it splits homogeneously. Consider the identity map $i : W_0 \to W_0$. Since $f_0 : W_1 \to W_0$ is a graded R-epimorphism and W_0 is a homogeneously finite projective module, we can find a graded R-homomorphism $g_0 : W_0 \to W_1$ with $f_0 g_0 = i$.

Now, assuming every length-one graded R-resolution of V splits homogeneously, let us show that V is a graded direct summand of a homogeneously finite free module. Let $(y_i \in V_{q_i})_{1 \leq i \leq n}$ be a finite set of homogeneous generators of $W_0 = V$. Let $(e_i)_{1 \leq i \leq n}$ be the canonical basis of $W_1 = {}^{(q_1,\ldots,q_n)}R$, and let $f_0 : W_1 \to W_0$ be the unique graded R-epimorphism with $f_0(e_i) = y_i$ for $1 \leq i \leq n$. Let $\ker(f_0) = W_2$ and let $f_1 : W_2 \to W_1$ be the natural injection. Then $(f, W, 1)$ is a length-one graded R-resolution of V. Hence it splits homogeneously and therefore by (T113.1) there is a graded isomorphism $W_1 \approx W_2 \oplus W_0$. This completes the proof of (T113.2).

To prove (T113.3), first note that if in the definition of a homogeneously finite projective module we have $V = {}^{(q_1,\ldots,q_n)}R$ then, upon taking $(e_i)_{1 \leq i \leq n}$ to be the canonical basis of V and upon choosing any $y_i \in E_{q_i} \cap \alpha^{-1}(\beta(e_i))$, it suffices to let $\gamma : V \to E$ be the canonical R-homomorphism induced by the homogeneous elements $(y_i \in E_{q_i})_{1 \leq i \leq n}$; thus a homogeneously finite free module is homogeneously finite projective. Next, if $V_1, \ldots, V_m$ are any finite number of homogeneously finite projective R-modules then by (T113.2) we can find graded R-modules $U_1, \ldots, U_m$ such that the modules $V_1 \oplus U_1, \ldots, V_m \oplus U_m$ are homogeneously finite free, and now obviously the module $(V_1 \oplus \cdots \oplus V_m) \oplus (U_1 \oplus \cdots \oplus U_m)$ is homogeneously finite free and hence again by (T113.2) we conclude that the module $(V_1 \oplus \cdots \oplus V_m)$ is homogeneously finite projective. Finally, if V and U are graded R-modules such that $V \oplus U$ is homogeneously finite projective then by (T113.2) we can find a graded R-module T such that the module $(V \oplus U) \oplus T$ is projectively finite free, which is obviously isomorphic to the graded module $V \oplus (U \oplus T)$ and hence again by (T113.2) the module V is homogeneously finite projective. This completes the proof of (T113.3).

The proof of the third sentence "In other word ..." of (T113.4) is straightforward, and the second sentence "Then ..." follows from it. In turn, the fourth sentence "Moreover ..." follows from the second sentence. The if part of the fifth sentence "Finally ..." has been proved in the first sentence of (T113.3). Coming to the only if part of the fifth sentence, supposing V to be a homogeneously finite projective R-module, we want to show that V is isomorphic to the graded module $^{(q_1, \ldots, q_n)} R$ for some $n, q_1, \ldots, q_n$ in $\mathbb{N}$. So let $\mathrm{hnb}_R V = n$, let $(y_i \in V_{q_i})_{1 \le i \le n}$ be a minimal set of homogeneous generators of V, and let $^{(q_1, \ldots, q_n)} R \to V$ be the canonical epimorphism induced by them. Upon letting V' be the kernel of this epimorphism with $\mathrm{hnb}_R V' = n'$, by (T113.1) and (T113.2) we have $n' \in \mathbb{N}$ together with a graded R-isomorphism $^{(q_1, \ldots, q_n)} R \approx V \oplus V'$. Therefore by the earlier sentences of (T113.4) proved above we have $\mathrm{hnb}_R(V \oplus V') = n + n'$ and $\mathrm{hnb}_R^{(q_1, \ldots, q_n)} R = n$. Consequently $n = n + n'$ and hence $n' = 0$. Therefore $V' = 0$ and hence $^{(q_1, \ldots, q_n)} R \approx V$. This completes the proof of (T113.4).

The proofs of (T113.5) to (T113.8) are obtained by making the following changes in the proofs of (Q17)(T80.5) to (Q17)(T80.8) respectively:

$$
\begin{cases}
\text{T80} \to \text{T113} \\
\text{short exact} \to \text{graded short exact} \\
\text{exact} \to \text{graded exact} \\
\text{projective} \to \text{homogeneously finite projective} \\
\text{isomorphism} \to \text{graded isomorphism} \\
\text{submodule} \to \text{graded submodule} \\
\text{pdim} \to \text{hdim}.
\end{cases}
$$

To prove the existence of a homogeneously minimal prefree R-resolution (f, W, n) of the module V asserted in (T113.9), first let us put $\mathrm{hnb}_R W_0 = d(0)$ with $W_0 = V$. Let $(y_{0i} \in (W_0)_{q_{0i}})_{1 \le i \le d(0)}$ be a minimal set of homogeneous generators of W_0, and let $f_0 : W_1 = {}^{(q_{01},\dots,q_{0d(0)})}R \to W_0$ be the canonical R-epimorphism induced by them.

Now let us put $\mathrm{hnb}_R\ker(f_0) = d(1)$. Let $(y_{1i} \in (W_1)_{q_{1i}})_{1 \le i \le d(1)}$ be a minimal set of homogeneous generators of $\ker(f_0)$, and let $f_1 : W_2 = {}^{(q_{11},\dots,q_{1d(1)})}R \to W_1$ be the canonical R-homomorphism induced by them.

Next let us put $\mathrm{hnb}_R\ker(f_1) = d(2)$. Let $(y_{2i} \in (W_2)_{q_{2i}})_{1 \le i \le d(2)}$ be a minimal set of homogeneous generators of $\ker(f_1)$, and let $f_2 : W_3 = {}^{(q_{21},\dots,q_{2d(2)})}R \to W_2$ be the canonical R-homomorphism induced by them.

In this manner we construct the maps $f_i : W_{i+1} \to W_i$ for $0 \le i \le n-1$. Now it suffices to take $f_0 : W_1 = W_0 \to W_0 = V$ to be the identity map if $n = 0$, and $f_n : W_{n+1} = \ker(f_{n-1}) \to W_n$ to be the subset injection if $n > 0$.

The first part of the "Moreover" sentence follows from (T113.3) and the second part follows from (T113.8).

To prove (T113.10) let there be given any graded R-module V. If V is homogeneously finite projective, then by (T113.4) it is homogeneously finite free, and hence obviously it is finite free as an ungraded module, and therefore by (Q17)(T80.3) it is finite projective as an ungraded module. Conversely, assuming that V is finite projective as an ungraded module, we want to show that it is homogeneously finite projective. In other words, given any graded R-epimorphism $\alpha : E \to F$ of finite graded R-modules and given any graded R-homomorphism $\beta : V \to F$, we want to find a graded R-homomorphism $\gamma : V \to E$ such that $\beta = \alpha\gamma$. Since the module V is finite projective as an ungraded module, we can find an ungraded R-homomorphism $\widehat{\gamma} : V \to E$ such that $\beta = \alpha\widehat{\gamma}$. Let $(y_i \in V_{q_i})_{1 \le i \le n}$ be a finite set of homogeneous generators of V. We CLAIM that then there exists a unique graded R-homomorphism $\gamma : V \to E$ such that for $1 \le i \le n$ we have $\gamma(y_i) = (\widehat{\gamma}(y_i))_{q_i} \in E_{q_i}$. To prove this it suffices to show that all R-linear relations between $y_1,\dots,y_n$ are also R-linear relations between the proposed values of $\gamma(y_1),\dots,\gamma(y_n)$, i.e., for any $r_1,\dots,r_n$ in R we have:

$$\sum_{1 \le i \le n} r_i y_i = 0 \Rightarrow \sum_{1 \le i \le n} r_i \widehat{\gamma}(y_i)_{q_i} = 0.$$

We can write $r_i = \sum_{\mu \in \mathbb{Z}}(r_i)_\mu$ with $(r_i)_\mu \in R_\mu$ if $\mu \ge 0$ and $(r_i)_\mu = 0$ if $\mu < 0$. Equating homogeneous components we get:

$$\sum_{1 \le i \le n} r_i y_i = 0 \Rightarrow \sum_{1 \le i \le n} (r_i)_{\nu - q_i} y_i = 0 \text{ for all } \nu \in \mathbb{N}$$

Now for each $\nu \in \mathbb{N}$, first applying $\widehat{\gamma}$ and then equating homogeneous components of degree ν we get:

$$\sum_{1 \le i \le n} (r_i)_{\nu - q_i} y_i = 0 \Rightarrow \sum_{1 \le i \le n} (r_i)_{\nu - q_i} \sum_{j \in \mathbb{N}} \widehat{\gamma}(y_i)_j = 0 \Rightarrow \sum_{1 \le i \le n} (r_i)_{\nu - q_i} \widehat{\gamma}(y_i)_{q_i} = 0.$$

Summing the last equation over all $\nu \in \mathbb{N}$ we conclude that

$$\sum_{1 \leq i \leq n} (r_i)_{\nu-q_i} \widehat{\gamma}(y_i)_{q_i} = 0 \text{ for all } \nu \in \mathbb{N} \Rightarrow \sum_{1 \leq i \leq n} r_i \widehat{\gamma}(y_i)_{q_i} = 0.$$

This completes the proof of the CLAIM. Since α and β are graded, for $1 \leq i \leq n$ we have: $\beta(y_i) = \alpha(\widehat{\gamma}(y_i)) \Rightarrow \beta(y_i) = \alpha(\gamma(y_i))$. Therefore $\beta = \alpha\gamma$.

To prove (T113.11), given any finite graded R-module V, let us first show that $\mathrm{hdim}_n V \geq \mathrm{pdim}_R V$. If $\mathrm{hdim}_R V = \infty$ then we have nothing to do; so assume that $\mathrm{hdim}_R V = n \in \mathbb{N}$; by (T113.9) we can take a homogeneously finite projective R-resolution (f, W, n) of V; by (T113.10) this is a finite projective R-resolution of the ungraded module V, and hence by (Q17)(T80.8) we get $\mathrm{pdim}_R V \leq n$. Next let us show that $\mathrm{hdim}_n V \leq \mathrm{pdim}_R V$. If $\mathrm{pdim}_R V = \infty$ then we have nothing to do; so assume that $\mathrm{pdim}_R V = n \in \mathbb{N}$; by (T113.9) we can take a homogeneously finite preprojective R-resolution (f, W, n) of V; by (T113.10) this is a finite preprojective R-resolution of the ungraded module V, and hence by (Q17)(T80.8) it is a finite projective R-resolution of the ungraded module V, and therefore by (T113.10) it is a homogeneously finite projective R-resolution of (the graded) module V, and hence by (T113.8) we get $\mathrm{hdim}_R V \leq n$.

(T113.12) follows from (T113.9) and (T113.10).

HILBERT SYZYGY THEOREM (T114). For a naturally graded homogeneous ring $R = \sum_{\nu \in \mathbb{N}} R_\nu$ over a field $k = R_0$ we have the following.

(T114.1) For $n \in \mathbb{N}$ and $q_1, \ldots, q_n$ in $\mathbb{N}_+$, let $x_i \in R_{q_i}$ be such that for the homogeneous ideal $J_i = (x_1, \ldots, x_i)R$ in R we have $x_i \in S_R(R/J_{i-1})$ for $1 \leq i \leq n$. Then for the finite graded R-module R/J_i we have $\mathrm{hdim}_R(R/J_i) = i$ for $0 \leq i \leq n$.

(T114.2) If the ring R is regular with $\dim(R) = n \in \mathbb{N}$ then for every finite graded R-module V we have $\mathrm{hdim}_R V \leq n$.

(T114.3) If R is the n-variable polynomial ring over k with $n \in \mathbb{N}$ then the maximum of $\mathrm{hdim}_R V$ taken over all finite graded R-modules V equals n.

PROOF. In view of (T113.11), assertions (T114.1) and (T114.2) follow from (Q17)(T81.3) and (Q17)(T81.6) respectively. In (T114.3), R is an n-dimensional regular ring, say by (Q29)(C46), and hence it suffices to apply (T114.1) with $x_1, \ldots, x_n$ as the variables and then invoke (T114.2).

QUEST (Q31) Projective Modules Over Polynomial Rings

According to Lemmas (T113.4) and (T113.10) of the previous Quest (Q30), for any graded module V over a finite variable polynomial ring $R = k[X_1, \ldots, X_N]$, where k is a field, we have that: if V is a finite projective R-module then V is a finite free R-module. It was conjectured by Serre [Se2], and proved by Quillen [Qui] and Suslin [Su1], that this is true for any (ungraded) R-module V, and continues to be true when k is replaced by a Principal Ideal Domain K. The aim of this Quest is to give an exposition of Suslin's beautiful proof.

Recall that a module V over a ring R is projective means it has the "lifting property" which says that if $\alpha : E \to F$ is any R-epimorphism of any R-modules and $\beta : V \to F$ is any R-homomorphism then $\beta = \alpha\gamma$ for some R-homomorphism $\gamma : V \to E$; here γ is called a lifting homomorphism because it lifts the range of β from F to E; note that for any set-theoretic map $\beta : V \to F$, sometimes V is called the domain of (definition of) β and F is called the range of β. A free module clearly has the lifting property and so projectiveness is a generalization of freeness. The lifting homomorphism is drawn as a dotted arrow in the following diagram with commutative triangle and exact bottom row.

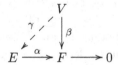

To reformulate the Serre Conjecture without using the concept of projective modules (and to give a motivation for that concept other than the lifting property), note that by (Q17)(T80.4) a finite module over a quasilocal ring is projective iff it is free, and by (Q17)(C30)(7′) a finite module over a noetherian ring is projective iff it is locally projective, i.e., iff its localization at any maximal ideal of that ring is projective over the localization of that ring at that maximal ideal.

So the Serre Conjecture says that a finite module over a polynomial ring is free iff it is locally free, i.e., iff its localization at any maximal ideal of that ring is free over the localization of that ring at that maximal ideal.

QUILLEN-SUSLIN THEOREM (T115). Let K be any ring. Consider the finite variable polynomial ring $R = K[X_1, \ldots, X_N]$ over K, and let there be given any finite projective R-module V.

(T115.1) Assume that $N = 1$. Given any polynomial $f = X_1^d + a_1 X_1^{d-1} + \cdots + a_d$ with $d \in \mathbb{N}$ and $a_1, \ldots, a_d$ in K, consider the multiplicative set $S(f) = \{f^i : i \in \mathbb{N}\}$ in R. Assume that $V_{S(f)}$ is a finite free $(R_{S(f)})$-module. Then V is a finite free R-module.

(T115.2) Assume that K is a PID. Then V is a finite free R-module.

PROOF OF (T115.1). We shall prove this in (T118.5) of Comment (C49) below.

PROOF OF (T115.2). We make induction on N. The case of $N = 0$ follows from (T119) of Comment (C50) below. So let $N > 0$ and assume for $N-1$. Let S be the multiplicative set in $K[X_1]$ consisting of all monic polynomials, i.e., expressions of the form $f = X_1^d + a_1 X_1^{d-1} + \cdots + a_d$ with $d \in \mathbb{N}$ and $a_1, \ldots, a_d$ in K. By (T120) of Comment (C50) below it follows that $K[X_1]_S$ is a PID, and clearly R_S is the $(N-1)$-variable polynomial ring $K[X_1]_S[X_2, \ldots, X_N]$. By (Q17)(C30)(4) we also know that V_S is a finite projective (R_S)-module. Therefore by the induction hypothesis we conclude that V_S is finite free (R_S)-module. For every $f \in S$ let $S(f) = \{f^i : i \in \mathbb{N}\}$. Upon letting $L = K[X_2, \ldots, X_N]$ we clearly have $R = L[X_1]$. In a moment we shall prove the implication:

$$V_S \text{ finite free over } R_S \Rightarrow V_{S(f)} \text{ finite free over } R_{S(f)} \text{ for some } f \in S.$$

In view of this implication, by taking L for K in (T115.1) we see that V is a finite free R-module.

To prove the above implication, let $\phi : R \to R_S$ and $\psi : V \to V_S$ be the canonical homomorphisms, and for any $f \in S$ let $\phi_f : R \to R_{S(f)}$ and $\psi_f : V \to V_{S(f)}$ be the canonical homomorphisms. We can first take a finite set of R-generators $x_1, \ldots, x_m$ of V together with a finite (R_S)-basis $y_1, \ldots, y_n$ of V_S, and then we can find g, h in S such that for $1 \leq j \leq n$ we have $y_j = \psi(z_j)/\phi(g)$ with $z_j \in V$, and for $1 \leq i \leq m$ we have

$$\psi(x_i) = \sum_{1 \leq j \leq n} \phi(r_{ij}) y_j / \phi(h)$$

with $r_{ij} \in R$ for $1 \leq j \leq n$. Let $f = gh \in S$ and for $1 \leq j \leq n$ let $\eta_j = \psi_f(z_j)$. We shall show that $\eta_1, \ldots, \eta_n$ are $(R_{S(f)})$-generators of $V_{S(f)}$ and they are linearly independent over $R_{S(f)}$; this will establish the fact that $\eta_1, \ldots, \eta_n$ is a finite $(R_{S(f)})$-basis of $V_{S(f)}$, and hence $V_{S(f)}$ is a finite free $(R_{S(f)})$-module.

Since R is a domain, the maps ϕ and ϕ_f are injective. Since V is a finite projective R-module, by (Q17)(T80.2) we see that V is a direct summand of a finite free R-module; consequently, since R is a domain, the maps ψ and ψ_f are also injective.

The above equations involving $\psi(x_i)$ together with the R-generation of V by $x_1, \ldots, x_m$ imply that for any $x \in V$ we can write

$$\psi(x) = \sum_{1 \leq j \leq n} \phi(r_j) \psi(z_j) / \phi(f)$$

with $r_j \in R$, and hence by the injectivity of ψ we get $xf = \sum_{1 \leq j \leq n} r_j z_j$, and therefore for every $e \in \mathbb{N}$ we have

$$\psi_f(x)/\phi_f(f^e) = \sum_{1 \leq j \leq n} \phi_f(r_j) \eta_j / \phi_f(f^{e+1}).$$

Thus the elements $\eta_1, \ldots, \eta_n$ are $(R_{S(f)})$-generators of $V_{S(f)}$.

Given any elements $u_1, \ldots, u_n$ in R with $u_1 z_1 + \cdots + u_n z_n = 0$, by applying ψ we get $\phi(gu_1)y_1 + \cdots + \phi(gu_n)y_n = 0$, and hence the (R_S)-linear-independence of $y_1, \ldots, y_n$ yields $\phi(gu_1) = \cdots = \phi(gu_n) = 0$, and therefore by the injectivity of ψ

and the domainness of R we get $u_1 = \cdots = u_n = 0$. Thus the elements $z_1, \ldots, z_n$ are linearly independent over R. Given any elements $v_1, \ldots, v_n$ in R and $d \in \mathbb{N}$ with

$$\phi_f(v_1)\eta_1/\phi_f(f^d) + \phi_f(v_n)\eta_n/\phi_f(f^d) = 0,$$

by the injectivity of ψ_f we get $v_1 z_1 + \cdots + v_n z_n = 0$ and hence by the R-linear-independence of $z_1, \ldots, z_n$ we get $v_1 = \cdots = v_n = 0$ and therefore we have $\phi_f(v_1)/\phi_f(f^d) = \cdots = \phi_f(v_n)/\phi_f(f^d) = 0$. Thus the elements $\eta_1, \ldots, \eta_n$ are linearly independent over $R_{S(f)}$.

COMMENT (C49). [**Modules Over Univariate Polynomial Rings**]. In this Comment, which is about thirty-five pages long, WE SHALL REPEATEDLY USE THE NOTATIONS AND DEFINITIONS AS WELL AS FORMULAS (1) TO (15) TO BE INTRODUCED IN THE NEXT THREE PAGES.

Let K be a ring and consider a finite variable polynomial ring $K[X_1, \ldots, X_N]$ over K. For any K module W we define the polynomial module $W[X_1, \ldots, X_N]$ in the following way. As an additive abelian group this consists of all finite expressions

$$B(X_1, \ldots, X_N) = \sum B_{j_1 \ldots j_N} X_1^{j_1} \ldots X_N^{j_N} \quad \text{with} \quad B_{j_1 \ldots j_N} \in W$$

(where $j_1, \ldots, j_N$ are nonnegative integers) which are added componentwise exactly as we add polynomials

$$A(X_1, \ldots, X_N) = \sum A_{i_1 \ldots i_N} X_1^{i_1} \ldots X_N^{i_N} \quad \text{with} \quad A_{i_1 \ldots i_N} \in K,$$

i.e., members of the ring $\overline{K} = K[X_1, \ldots, X_N]$; see L1§6. We regard W to be the subgroup of $\overline{W} = W[X_1, \ldots, X_N]$ consisting of those B whose degree is 0 or $-\infty$; here the degree of $B \neq 0$ is the maximum $j_1 + \cdots + j_N$ with $B_{j_1 \ldots j_N} \neq 0$, and the degree of $B = 0$ is $-\infty$. Again copying from L1§6, we convert $\overline{W}$ into a $\overline{K}$-module by the product rule

$$D(X_1, \ldots, X_N) = A(X_1, \ldots, X_N)B(X_1, \ldots, X_N) = \sum D_{i_1 \ldots i_N} X_1^{i_1} \ldots X_N^{i_N}$$

where

$$D_{k_1 \ldots k_N} = \sum_{i_1 + j_1 = k_1, \ldots, i_N + j_N = k_N} A_{i_1 \ldots i_N} B_{j_1 \ldots j_N}.$$

If U is another K-module and $\alpha : U \to W$ is a K-homomorphism, then its polynomial extension $\alpha[X_1, \ldots, X_N]$ is the unique $K[X_1, \ldots, X_N]$-homomorphism $U[X_1, \ldots, X_N] \to W[X_1, \ldots, X_N]$ given by $B \mapsto \sum \alpha(B_{j_1 \ldots j_N})X_1^{j_1} \ldots X_N^{j_N}$.

We are particularly interested in the univariate polynomial module $W[X]$ over the ring $K[X]$ where W is a finite projective module over K. Modules P, Q over $K[X]$ are almost isomorphic means, for some polynomial $f = X^d + a_1 X^{d-1} + \cdots + a_d$ with $d \in \mathbb{N}$ and $a_1, \ldots, a_d$ in K, the localized modules $P_{S(f)}, Q_{S(f)}$ over the ring $K[X]_{S(f)}$ are isomorphic, where $S(f)$ is the multiplicative set in $K[X]$ obtained by putting $S(f) = \{f^i : i \in \mathbb{N}\}$. In Theorem (T118.4) below we shall show that, if a finite projective $K[X]$-module V is almost isomorphic to $W[X]$ then it is actually isomorphic to $W[X]$.

As preparation for Theorem (T118) we shall prove Lemmas (T116) and (T117).

As notation to be used in these Lemmas, for modules P, Q over a ring R, an R-homomorphism $\sigma : P \to Q$ splits means the short exact $0 \to \sigma(P) \to Q \to Q/\sigma(P) \to 0$ splits where the second and third arrow is the natural injection and surjection respectively; we indicate this by saying that σ is a split R-homomorphism; we are particularly interested in the case when σ is also injective and we indicate this by saying that σ is a split R-monomorphism. Note that

$$(1) \qquad \begin{cases} \sigma : P \to Q \text{ is a split } R\text{-monomorphism} \\ \Leftrightarrow \eta\sigma = 1_P \text{ for some } R\text{-homomorphism } \eta : Q \to P \end{cases}$$

where 1_P is the identity map $P \to P$. To prove $\Leftarrow$, the existence of η tells us that σ is injective and Q is the internal direct sum $\sigma(P) \oplus \ker(\eta)$. The converse is even easier.

Given R-homomorphisms $\sigma : P \to Q$ and $\sigma' : P \to Q$, we write $\sigma \sim \sigma'$ to mean that $\sigma = \tau\sigma'$ for some R-automorphism $\tau : Q \to Q$; we verbalize this by saying that σ is equivalent to σ', or in greater detail, we may verbalize it by saying that σ is R-autoequivalent to σ'. Note that for R-homomorphisms $\sigma : P \to Q$ and $\sigma' : P \to Q$, the R-homomorphism $\sigma + \sigma' : P \to Q$ is defined by $p \mapsto \sigma(p) + \sigma'(p)$, and for $s \in R$ the R-homomorphism $s\sigma : P \to Q$ is defined by $p \mapsto s\sigma(p)$.

Given a pair (α, β) of R-homomorphisms $\alpha : P \to P$ and $\beta : P \to Q$ we can consider the diagonal direct sum $\alpha \oplus \beta : P \to P \oplus Q$. We introduce the set $I_R(\alpha, \beta)$ by putting

$$(2) \quad I_R(\alpha, \beta) = \{r \in R : (\alpha + s1_P) \oplus \beta \sim (\alpha + (s + r)1_P) \oplus \beta \text{ for all } s \in R\}$$

and we call this the R-invariance-set of the pair (α, β). We say that the pair (α, β) is admissible to mean that for some $m \in \mathbb{N}_+$ and R-homomorphisms $\gamma_i : Q \to P$ we have

$$(3) \qquad \sum_{0 \leq i \leq m} \gamma_i \beta \alpha^i = 1_P$$

where α^i stands for $\alpha \dots \alpha$ repeated i times.

Note that, given any R-homomorphism $\zeta_i : P \to R$ for $1 \leq i \leq n \in \mathbb{N}_+$, we have the diagonal direct sum

$$(4) \qquad \theta = \zeta_1 \oplus \cdots \oplus \zeta_n : P \to R^n$$

and given any R-homomorphism $\psi : R^n \to P$ we obtain an R-homomorphism

$$(5) \qquad \mu = \theta\psi : R^n \to R^n.$$

Let $(e_i)_{1 \le i \le n}$ be the canonical R-basis of R^n given by

(6) $e_i = (0, \ldots, 0, 1, 0 \ldots, 0) \in R^n$ with 1 in the i-th spot.

Then, given any R-homomorphism $\mu : R^n \to R^n$ we define the $n \times n$ matrix $H(\mu) = (H(\mu)_{ij})$ over R by writing

(7) $$\mu(e_i) = \sum_{1 \le j \le n} H(\mu)_{ji} e_j \quad \text{with} \quad H(\mu)_{ji} \in R$$

and we define $\det(\mu)$ by putting

(8) $$\det(\mu) = \det(H(\mu)).$$

Observe that in the notation of L3§12(D3) with $m = n$ we have $\phi_{H(\mu)} = \mu$. According to (D1) and (D2) of L3§12, the adjoint of $H(\mu)$ is the $n \times n$ matrix $H(\mu)^* = (H(\mu)^*_{ij})$ with

(9) $$H(\mu)^*_{ji} = \text{ the } (i,j)\text{-th cofactor of } H(\mu)$$

and by (E3) of L3§12 we have

(10) $$H(\mu)H(\mu)^* = H(\mu)^* H(\mu) = \det(\mu)1_n$$

where 1_n is the $n \times n$ identity matrix. We define the adjoint μ^* of μ to be the R-homomorphism $\mu^* : R^n \to R^n$ obtained by putting

(11) $$\mu^*(e_i) = \sum_{1 \le j \le n} H(\mu)^*_{ji} e_j \quad \text{for} \quad 1 \le i \le n.$$

In the situation of (5), considering the R-homomorphism

(12) $$\eta = \psi \mu^* : R^n \to P$$

by (13) we get

(13) $$\theta \eta = \det(\mu) 1_{R^n}.$$

Moreover, considering the R-homomorphism $\epsilon : R^n \to R^n$ such that

(14) $\epsilon(\rho_1, \ldots, \rho_n) = (0, \rho_1, \ldots, \rho_{n-1})$ for all $(\rho_1, \ldots, \rho_n) \in R^n$

and looking at the corresponding $n \times n$ matrix we see that

$$\epsilon^n = 0$$

i.e., ϵ^n is the zero R-homomorphism of R^n to R^n. Now writing

$$(\eta \epsilon \theta)^n = \eta \epsilon \theta \eta \epsilon \theta \ldots \eta \epsilon \theta$$

with $\eta \epsilon \theta$ repeated n times, and grouping the $n-1$ occurrences of $\theta \eta$, by (13) we get

$$(\eta \epsilon \theta)^n = \det(\mu)^{n-1} \eta \epsilon^n \theta$$

and hence by the equation $\epsilon^n = 0$ we conclude that

$$(15) \qquad\qquad\qquad (\eta\epsilon\theta)^n = 0.$$

INVARIANCE LEMMA (T116). Let P and Q be modules over a ring R, and let $\alpha : P \to P$ and $\beta : P \to Q$ be R-homomorphisms. Then we have the following.

(T116.1) $I_R(\alpha, \beta)$ is an additive subgroup of R.

(T116.2) For any R-homomorphism $\nu : Q \to P$ we have the set-theoretic equation of invariance $I_R(\alpha, \beta) = I_R(\alpha + \nu\beta, \beta)$.

(T116.3) Given any $n \in \mathbb{N}_+$ together with any R-homomorphisms $\phi : Q \to R$ and $\psi : R^n \to P$, let $\zeta_i : P \to R$ be the R-homomorphism given by

$$\zeta_i = \phi\beta\alpha^{i-1} \quad \text{for} \quad 1 \le i \le n,$$

let the notation be as in (4) to (15), and consider the following four conditions.

(i) P is a finite projective R-module such that for every maximal ideal J in R we have $\mathrm{gnb}_{R_J} P_J = n$.

(ii) $\psi\mu^*\theta = \det(\mu)1_P$.

(iii) $\alpha \oplus \beta \sim (\alpha + \det(\mu)1_P) \oplus \beta$.

(iv) $\det(\mu)R \subset I_R(\alpha, \beta)$.

Then we have (i) $\Rightarrow$ (ii) $\Rightarrow$ (iii), and we have (i) $\Rightarrow$ (iv).

PROOF OF (T116.1). Let there be given any $r, \widehat{r}$ in $I_R(\alpha, \beta)$ and any s in R. Then, since $r \in I_R(\alpha, \beta)$, by (2) we get

$$(\alpha + s1_P) \oplus \beta \sim (\alpha + (s+r)1_P) \oplus \beta$$

and, since $\widehat{r} \in I_R(\alpha, \beta)$, by taking $(\widehat{r}, s+r)$ for (r, s) in (2) we get

$$(\alpha + (s+r)1_P) \oplus \beta \sim (\alpha + (s+r+\widehat{r})1_P) \oplus \beta$$

and hence

$$(\alpha + s1_P) \oplus \beta \sim (\alpha + (s+r+\widehat{r})1_P) \oplus \beta$$

and therefore $r + \widehat{r}$ belongs to $I_R(\alpha, \beta)$. Since $r \in I_R(\alpha, \beta)$, by taking $(s-r, r)$ for (s, r) in (2) we get

$$(\alpha + (s-r)1_P) \oplus \beta \sim (\alpha + s1_P) \oplus \beta$$

and hence $-r$ belongs to $I_R(\alpha, \beta)$. Also obviously 0 belongs to $I_R(\alpha, \beta)$.

Thus $I_R(\alpha, \beta)$ is an additive subgroup of R.

PROOF OF (T116.2). For some helpful heuristics see the bracketed remark just after step (4') in the following proof of (T116.3).

Now let there be given any R-homomorphism $\nu : Q \to P$. Then we obtain an R-automorphism $\tilde{\nu} : P \oplus Q \to P \oplus Q$ given by $(p, q) \mapsto (p + \nu(q), q)$, and for every R-homomorphism $\lambda : P \to P$ we clearly have

$$(\bullet) \qquad\qquad \tilde{\nu}(\lambda \oplus \beta) = (\lambda + \nu\beta) \oplus \beta.$$

Moreover, given any $r \in I_R(\alpha, \beta)$ and $s \in R$, by (2) we have

$$(\alpha + s1_P) \oplus \beta = \tau((\alpha + (s + r)1_P) \oplus \beta)$$

for some R-automorphism $\tau : P \oplus Q \to P \oplus Q$. Multiplying both sides from the left by $\tilde{\nu}$, and writing $\tilde{\nu}\tau = \hat{\tau}\tilde{\nu}$ with $\hat{\tau} = \tilde{\nu}\tau\tilde{\nu}^{-1}$, we get

$$\tilde{\nu}((\alpha + s1_P) \oplus \beta) = \hat{\tau}\tilde{\nu}((\alpha + (s + r)1_P) \oplus \beta).$$

Now applying ($\bullet$) to both sides we get

$$(\alpha + \nu\beta + s1_P) \oplus \beta = \hat{\tau}(\alpha + \nu\beta + (s + r)1_P) \oplus \beta)$$

where $\hat{\tau} : P \oplus Q \to P \oplus Q$ is clearly an R-automorphism. Consequently by (2) we see that $r \in I_R(\alpha + \nu\beta, \beta)$. Thus $I_R(\alpha, \beta) \subset I_R(\alpha + \nu\beta, \beta)$. Replacing α by $\alpha + \nu\beta$ and then replacing ν by $-\nu$, we get the reverse inclusion $I_R(\alpha + \nu\beta, \beta) \subset I_R(\alpha, \beta)$. Therefore $I_R(\alpha, \beta) = I_R(\alpha + \nu\beta, \beta)$.

PROOF OF (T116.3). Before turning to (i) to (iv), we make some preliminary calculations. Some heuristics about the following steps $(1')$ to $(4')$ may be found in the bracketed remark just after step $(4')$. That remark tells us how to visualize the automorphism τ after step $(1')$ and its inverse δ written down between steps $(3')$ and $(4')$. Some heuristics and further details about the remaining part of the proof of (T116.3) may be found in the long bracketed remark at the end of that proof.

Now upon letting $\pi : R \to R^n$ be the R-homomorphism such that

$$\pi(\rho) = (\rho, 0 \ldots, 0) \quad \text{for all} \quad \rho \in R$$

we clearly have

$$(1') \qquad\qquad \epsilon\theta\alpha + \pi\phi\beta = \theta.$$

Let $\tau : P \oplus Q \to P \oplus Q$ be the R-homomorphism obtained by putting

$$\tau(p, q) = ((1_P + \eta\epsilon\theta)(p) + (\eta\pi\phi)(q), q) \quad \text{for all} \quad p \in P \quad \text{and} \quad q \in Q.$$

Then by $(1')$ we see that

$$(2') \qquad\qquad \tau(\alpha \oplus \beta) = (\alpha + \eta\theta) \oplus \beta.$$

Substituting $X = 1_P$ and $Y = \eta\epsilon\theta$ in the High-School factorization

$$X^{2n+1} + Y^{2n+1} = (X + Y) \sum_{0 \le i \le 2n} (-1)^i X^i Y^{2n-i}$$

and noting that $(\eta\epsilon\theta)^{2n+1} = 0$ by (15), we get

(3') $$(1_P + \eta\epsilon\theta)\gamma = \gamma(1_P + \eta\epsilon\theta) = 1_P$$

where the R-homomorphism $\gamma : P \to P$ is given by

$$\gamma = \sum_{0 \le i \le 2n} (-1)^i (\eta\epsilon\theta)^{2n-i}.$$

Let $\delta : P \oplus Q \to P \oplus Q$ be the R-homomorphism obtained by putting

$$\delta(p, q) = (\gamma(p) - (\gamma\eta\pi\phi)(q), q) \quad \text{for all} \quad p \in P \quad \text{and} \quad q \in Q.$$

Then by (3') we get

(4') $$\tau\delta = \delta\tau = 1_{P\oplus Q}$$

[The value of δ was suggested by thinking of τ as the 2×2 matrix $\left(\begin{smallmatrix} u & v \\ 0 & 1 \end{smallmatrix}\right)$ with $(u, v) = (1_P + \eta\epsilon\theta, \eta\pi\phi)$ and calculating its inverse to be $\left(\begin{smallmatrix} \gamma & -\gamma v \\ 0 & 1 \end{smallmatrix}\right)$ where γ is found by solving the equation $u\gamma = \gamma u = 1$. Thinking of $\alpha \oplus \beta$ as the column vector $\left(\begin{smallmatrix} \alpha \\ \beta \end{smallmatrix}\right)$, and multiplying it from the left by the matrix $\left(\begin{smallmatrix} u & v \\ 0 & 1 \end{smallmatrix}\right)$ gives the action of τ on $\alpha \oplus \beta$. Here and in the above proof of (T116.2), various maps from $P \oplus Q$ to $P \oplus Q$ may be thought of as 2×2 matrices, and various maps from P to $P \oplus Q$ may be thought of as 2×1 column vectors. Now the LHS of ($\bullet$) in the proof of (T116.2) reduces to multiplying such a column vector from the left by a 2×2 matrix].

By (4') we see that τ is an R-automorphism, and hence by (12) and (2') we get (ii) $\Rightarrow$ (iii).

Next assuming (i), let us show that (ii) holds. In view of (T80.4), (C29)(2), and (C30)(4) of (Q17), condition (i) says that for every $J \in \mathrm{mspec}(R)$ there is an (R_J)-isomorphism $f(J) : P_J \to (R_J)^n$. In view of (5) and (12), considering the R-homomorphism

(5') $$g = \psi(\theta\psi)^*\theta - \det(\theta\psi)1_P : P \to P,$$

condition (ii) says that $g = 0$. For any R-homomorphism $h : P \to P$, considering the induced (R_J)-homomorphism $h_J : P_J \to P_J$ we have

(6') $$h = 0 \Leftrightarrow h_J = 0 \text{ for all } J \in \mathrm{mspec}(R).$$

Considering the (R_J)-homomorphism $f(J)h_J f(J)^{-1} : (R_J)^n \to (R_J)^n$, for any $J \in \mathrm{mspec}(R)$ we also have

(7') $$h_J = 0 \Leftrightarrow f(J)h_J f(J)^{-1} = 0.$$

Given any $J \in \mathrm{mspec}(R)$, we shall show that

(8') $$f(J)g_J f(J)^{-1} = 0$$

and in view of (6') and (7') this will complete the proof of (i) $\Rightarrow$ (ii). Considering the (R_J)-homomorphisms

$$\widehat{\psi} = f(J)\psi_J : (R_J)^n \to (R_J)^n \quad \text{and} \quad \widehat{\theta} = \theta_J f(J)^{-1} : (R_J)^n \to (R_J)^n$$

and using (5') we get

(9') $$f(J)g_J f(J)^{-1} = C - D$$

where

$$C = \widehat{\psi}(\widehat{\theta}\widehat{\psi})^*\widehat{\theta}$$

and

$$D = f(J)\det(\theta\psi)1_{P_J} f(J)^{-1}.$$

Clearly

(10') $$C = \det(\widehat{\psi})\det(\widehat{\theta})1_{(R_J)^n} = \det(\widehat{\psi}\widehat{\theta})1_{(R_J)^n} = D$$

and hence by (9') we get (8').

Thus (i) $\Rightarrow$ (iii). Using this we deduce the stronger implication (i) $\Rightarrow$ (iii*), where (iii*) says that $\det(\mu) \in I_R(\alpha, \beta)$, which is equivalent to the condition saying that: (iii**) for every $s \in R$ we have $\alpha_s \oplus \beta \sim (\alpha_s + \det(\mu)1_P) \oplus \beta$ where the R-homomorphism $\alpha_s : P \to P$ is given by $\alpha_s = \alpha + s1_P$. Let us recall that the R-homomorphism $\theta : P \to R^n$ is the diagonal direct sum

$$\theta = \phi\beta \oplus \phi\beta\alpha \oplus \cdots \oplus \phi\beta\alpha^{n-1}$$

and let us consider the R-homomorphism $\theta_s : P \to R^n$ which is the diagonal direct sum

$$\theta_s = \phi\beta \oplus \phi\beta\alpha_s \oplus \cdots \oplus \phi\beta\alpha_s^{n-1}.$$

Since $\det(\mu) = \det(\theta\psi)$, to deduce the implication (i) $\Rightarrow$ (iii**) from the implication (i) $\Rightarrow$ (iii), it suffices to show that $\det(\theta_s\psi) = \det(\theta\psi)$. For $i \neq j$ and $\lambda \in R$ let $T_{ij}(n, \lambda) = (T_{ij}(n, \lambda)_{ab})$ be the $n \times n$ matrix over R given by

$$T_{ij}(n, \lambda)_{ab} = \begin{cases} 1 & \text{if } a = b \\ \lambda & \text{if } (a, b) = (i, j) \\ 0 & \text{if } a \neq b \text{ and } (a, b) \neq (i, j) \end{cases}$$

and note that clearly $\det(T_{ij}(n, \lambda)) = 1$. Moreover, for the R-homomorphism $t_{ij}(n, \lambda) : R^n \to R^n$ for which $H(t_{ij}(n, \lambda)) = T_{ij}(n, \lambda)$, we also have

$$\det(t_{ij}(n, \lambda)) = 1.$$

Now considering the R-homomorphism $\theta_{s,i} : P \to R^n$ which is the diagonal direct sum

$$\theta_{s,i} = \phi\beta \oplus \cdots \oplus \phi\beta\alpha^{i-1} \oplus \phi\beta\alpha_s^i \oplus \cdots \oplus \phi\beta\alpha_s^{n-1}$$

for $1 \leq i \leq n$, we clearly have $\theta_{s,1} = \theta_s$ and $\theta_{s,n} = \theta$. Moreover, considering the R-homomorphism $\theta_{s,i,j} : P \to R^n$ which is the diagonal direct sum

$$\theta_{s,i,j} = \phi\beta \oplus \cdots \oplus \phi\beta\alpha^{i-2} \oplus \left(\sum_{1 \leq l \leq j} \binom{i-1}{l-1} s^{l-1} \phi\beta\alpha^{i-l} \right) \oplus \phi\beta\alpha_s^i \oplus \cdots \oplus \phi\beta\alpha_s^{n-1}$$

for $1 \leq j \leq i \leq n$, we clearly have $\theta_{s,i,1} = \theta_{s,i}$ and $\theta_{s,i,i} = \theta_{s,i-1}$. For $2 \leq j \leq i \leq n$ we also have

(11′) $$t_{i,i-j+1}\left(n, \binom{i-1}{j-1} s^{j-1}\right) \theta_{s,i,j-1} = \theta_{s,i,j}$$

and hence

$$\det(\theta_{s,i,j-1}\psi) = \det(\theta_{s,i,j}\psi).$$

Therefore by double induction on i, j we get

$$\det(\theta_s \psi) = \det(\theta \psi).$$

Thus (i) $\Rightarrow$ (iii*) where (iii*) says that $\det(\mu) \in I_R(\alpha, \beta)$. Moreover, since $\det(\mu) = \det(\theta\psi)$, to prove (i) $\Rightarrow$ (iv), it only remains to show that for any $\lambda \in R$ we have $\det(\theta\psi_\lambda) = \det(\theta\psi)\lambda$ for some R-homomorphism $\psi_\lambda : R^n \to P$. For any integer i with $1 \leq i \leq n$, let $T_{ii}(n, \lambda) = (T_{ii}(n, \lambda)_{ab})$ be the $n \times n$ matrix over R given by

$$T_{ii}(n, \lambda)_{ab} = \begin{cases} 1 & \text{if } a = b \neq i \\ \lambda & \text{if } a = b = i \\ 0 & \text{if } a \neq b \end{cases}$$

and note that clearly $\det(T_{ii}(n, \lambda)) = \lambda$. Moreover, for the R-homomorphism $t_{ii}(n, \lambda) : R^n \to R^n$ for which $H(t_{ii}(n, \lambda)) = T_{ii}(n, \lambda)$, we also have

$$\det(t_{ii}(n, \lambda)) = \lambda.$$

Now it suffices to take $\psi_\lambda = \psi t_{ii}(n, \lambda)$.

[In Comment (C51) below we shall view the above matrices T_{ij} as elementary row and column operations. In that viewpoint we may interpret (11′) thus. Let $\theta_{s,i,j}$ "correspond" to the $n \times 1$ matrix $\Theta_{s,i,j} = ((\Theta_{s,i,j})_{a,b})$ where, for $1 \leq a \leq n$ and $b = 1$, the R-homomorphism $(\Theta_{s,i,j})_{a,b} : P \to R$ is the a-th term of the above displayed definition of $\theta_{s,i,j}$. Now capitalizing t, θ in (9′) amounts to saying that by adding $\binom{i-1}{j-1} s^{j-1}$ times the $(i - j + 1)$-th row of the matrix $\Theta_{s,i,j-1}$ to its i-th row we obtain the matrix $\Theta_{s,i,j}$. Similarly the various diagonal direct sums such as $\theta : P \to R^n$ correspond to row vectors of length n, i.e., $n \times 1$ matrices whose entries are maps $P \to R$.

Turning to the proof of (i) $\Rightarrow$ (ii) given between items (5′) to (10′), the following further details may be noted. Item (6′) follows from the property of localization which says that for an R-homomorphism $F : U \to V$ of R-modules, considering the localization $F_S : U_S \to V_S$ at a multiplicative set S is R, we have

$$(12')\qquad \begin{cases} F = 0 \Rightarrow F_S = 0 \text{ for every multiplicative set } S \text{ in } R, \\ \text{and: } F_J = 0 \text{ for every maximal ideal } J \text{ in } R \Rightarrow F = 0. \end{cases}$$

Here the first part of (12′) is obvious, and the second part follows by noting that if $u \in U$ is such that $F(u) = v \neq 0$ then $\mathrm{ann}_R(v)$ is a nonunit ideal in R and, taking a maximal ideal J in R with $\mathrm{ann}_R(v') \subset J$ and letting u_J and v_J denote the images of u and v under the natural homomorphisms $U \to U_J$ and $V \to V_J$ respectively, we obtain $F_J(u_J) = v_J \neq 0$.

Concerning the definitions of $\widehat{\psi}$ and $\widehat{\theta}$ displayed between items (8′) and (9'), we note that by a slight abuse of notation we have put $\psi_J = \overline{\psi}_J \omega$ and $\theta_J = \omega \overline{\theta}$ where $\omega : (R_J)^n \to (R^n)_J$ is the obvious isomorphism coming from (Q17)(C30)(3), and $\overline{\psi}_J : (R^n)_J \to P_J$ and $\overline{\theta}_J : P_J \to (R^n)_J$ are the localizations of ψ and θ respectively.

Item (10′) uses item (10) together with the fact that for the adjoints A^* and B^* of $n \times n$ matrices A and B we have

$$(13')\qquad\qquad\qquad (AB)^* = B^* A^*$$

which, after doing some fancy footwork, follows from the obvious fact that for the transposes A' and B' of $m \times n$ and $n \times r$ matrices A and B we have

$$(14')\qquad\qquad\qquad (AB)' = B'A'.$$

Namely, taking $A = H(\widehat{\theta})$ and $B = H(\widehat{\psi})$ in (13′) we get $(\widehat{\theta}\widehat{\psi})^* = \widehat{\psi}^*\widehat{\theta}^*$ and hence

$$C = \widehat{\psi}(\widehat{\theta}\widehat{\psi})^*\widehat{\theta} = (\widehat{\psi}\widehat{\psi}^*)(\widehat{\theta}^*\widehat{\theta})$$

and therefore by (10) we get

$$C = \det(\widehat{\psi})\det(\widehat{\theta})1_{(R_J)^n}.$$

At any rate, one reason why we need assumption (i) is that, unlike item (14′), item (13′) requires the matrices to be square matrices].

ADMISSIBILITY LEMMA (T117). Let P and Q be finite projective modules over a ring R, and let $\alpha : P \to P$ and $\beta : P \to Q$ be R-homomorphisms such that the pair (α, β) is admissible. Then we have the following.

(T117.1) $\alpha \oplus \beta : P \to P \oplus Q$ is a split R-monomorphism, and for every $s \in R$ and every R-homomorphism $\nu : Q \to P$, the pair $(\alpha + \nu\beta + s1_P, \beta)$ is admissible.

(T117.2) $I_R(\alpha, \beta)$ is an additive subgroup of R, and for any R-homomorphism $\nu : Q \to P$ we have $I_R(\alpha, \beta) = I_R(\alpha + \nu\beta, \beta)$.

(T117.3) Assume that for some $n \in \mathbb{N}_+$ and every maximal ideal J in R we have $\mathrm{gnb}_{R_J} P_J = n$. Then given any R-homomorphisms $\phi : Q \to R$ and $\psi : R^n \to P$, for the diagonal direct sum

$$\theta = (\phi\beta) \oplus (\phi\beta\alpha) \oplus \cdots \oplus (\phi\beta\alpha^{n-1}) : P \to R^n$$

we have $\det(\theta\psi)R \subset I_R(\alpha, \beta)$.

(T117.4) Assume that R is a field with $P \ne 0$, let $n \in \mathbb{N}_+$ be the vector-space dimension of P over R, and let $\phi : Q \to R$ be an R-homomorphism such that $\phi\beta \ne 0$. Then there exists an R-homomorphism $\nu : Q \to P$ such that the elements $(\phi\beta(\alpha + \nu\beta)^i)_{0 \le i \le n-1}$ of the dual space are linearly independent over R.

[If R is a field and P is a nonzero finite R-module then P is an R-vector-space whose dimension is a positive integer n, and the set of all R-homomorphisms $P \to R$ constitutes the "dual space" P' which is again an R-vector space whose dimension is n. Namely, taking a basis $(E_i)_{1 \le i \le n}$ of P, we get the "dual" basis $(E_i')_{1 \le i \le n}$ of P' by letting $E_i' : P \to R$ be defined by $E_i'(E_j) = 0$ or 1 according as $i \ne j$ or $i = j$].

(T117.5) For every $r \in R$ we have $\alpha \oplus \beta \sim (\alpha + r1_P) \oplus \beta$.

PROOF OF (T117.1). In the notation of (3), we get a unique R-homomorphism $\eta : P \oplus Q \to P$ such that for all $(p, q) \in P \oplus Q$ with $p \in P$ and $q \in Q$ we have

$$\eta(p, q) = \gamma_0(q) + \sum_{1 \le i \le m} \gamma_i \beta \alpha^{i-1}(p).$$

Clearly we have $\eta(\alpha \oplus \beta) = 1_P$ and hence $\alpha \oplus \beta$ is a split R-monomorphism by (1). Given any $s \in R$ and R-homomorphism $\nu : Q \to P$, upon letting

$$\widehat{\alpha} = \alpha + \nu\beta + s1_P$$

by induction on $i \in \mathbb{N}$ we see that

$$\alpha^i = \widehat{\alpha}^i + \sum_{0 \le j \le i-1} (\nu_{ij} \beta \widehat{\alpha}^j + s_{ij} \widehat{\alpha}^j)$$

with $s_{ij} \in R$ and R-homomorphism $\nu_{ij} : Q \to P$. Substituting in (3) we get

$$\sum_{0 \le i \le m} \widehat{\gamma}_i \beta \widehat{\alpha}^i = 1_P$$

where the R-homomorphism $\widehat{\gamma}_i : Q \to P$ is given by

$$\widehat{\gamma}_i = \gamma_i + \sum_{i+1 \le k \le m} (s_{ki} \gamma_k + \gamma_k \beta \nu_{ki}).$$

Therefore the pair $(\alpha + \nu\beta + s1_P, \beta)$ is admissible.

PROOF OF (T117.2). Follows from (T116.1) and (T116.2).

PROOF OF (T117.3). Follows from (T116.3).

PROOF OF (T117.4). By induction on $j \in \{0, \ldots, n-1\}$ we proceed to find an R-homomorphism $\nu_j : Q \to P$ such that the elements $(\phi\beta(\alpha + \nu_j\beta)^i)_{0 \leq i \leq j}$ are linearly independent (over R), and then by taking $\nu = \nu_{n-1}$ this will complete the proof. This is obvious for $j = 0$ because $\phi\beta \neq 0$ (and so we may take ν_0 to be any R-homomorphism $Q \to P$). So let

$$(1^*) \qquad\qquad 0 < j \leq n-1$$

and suppose that we have found $\nu_0, \ldots, \nu_{j-1}$. If

$$\phi\beta(\alpha + \nu_{j-1}\beta)^j \neq \sum_{0 \leq i \leq j-1} r_i\phi\beta(\alpha + \nu_{j-1}\beta)^i \quad \text{for all } r_0, \ldots, r_{j-1} \text{ in } R$$

then it suffices to take $\nu_j = \nu_{j-1}$. So now assume that

$$(2^*) \quad \phi\beta(\alpha + \nu_{j-1}\beta)^j = \sum_{0 \leq i \leq j-1} r_i\phi\beta(\alpha + \nu_{j-1}\beta)^i \quad \text{for some } r_0, \ldots, r_{j-1} \text{ in } R.$$

We claim that

$$(3^*) \quad \begin{cases} \text{if } R\text{-homomorphisms } \phi_1, \ldots, \phi_l \text{ from } R^n \text{ to } R \text{ are linearly independent} \\ \text{where } l \in \mathbb{N}, \text{ then } l \leq n \text{ and } \dim_R \cap_{1 \leq i \leq l} \ker(\phi_i) = n - l. \end{cases}$$

We can see this by induction on l; it being obvious for $l = 0$ (because by convention $\cap_{1 \leq i \leq 0}\ker(\phi_i) = R^n$), let $l > 0$ and assume for $l-1$; clearly $l \leq n$ and we can choose a basis $x_1, \ldots, x_n$ of R^n so that $x_1, \ldots, x_{n-l+1}$ is a basis of $\cap_{1 \leq i \leq l-1}\ker(\phi_i)$; suppose if possible that $\phi_l(x_j) = 0$ for $1 \leq j \leq n-l+1$; then $\phi_i(x_j) = 0$ for $1 \leq i \leq l$ and $1 \leq j \leq n-l+1$, i.e., $\ker(\phi_l) \supset \cap_{1 \leq i \leq l-1}\ker(\phi_i)$. Now by taking $(M, N) = (l-1, l)$ in L4§10(E5) we can find $(0, \ldots, 0) \neq (\rho_1, \ldots, \rho_l) \in R^l$ such that $\sum_{1 \leq i \leq l} \rho_i\phi_i(x_j) = 0$ for $1 \leq j \leq n$; this contradicts the linear independence of $\phi_1, \ldots, \phi_l$, and hence $\ker(\phi_l) \not\supset \cap_{1 \leq i \leq l-1}\ker(\phi_i)$ and therefore $\dim_R \cap_{1 \leq i \leq l} \ker(\phi_i) \leq n - l$. Since we know this dimension to be at least $n - l$, we get the desired conclusion. Let

$$P_1 = \cap_{0 \leq i \leq j-1}\ker(\phi\beta(\alpha + \nu_{j-1}\beta)^i)$$

and

$$P_2 = \cap_{0 \leq i \leq j-2}\ker(\phi\beta(\alpha + \nu_{j-1}\beta)^i).$$

Since the elements $(\phi\beta(\alpha + \nu_{j-1}\beta)^i)_{0 \leq i \leq j-1}$ are linearly independent and $P \approx R^n$, by (3^*) we see that

$$(4^*) \qquad\qquad \dim_R P_1 = n - j \geq 0$$

and

(5*) $$\dim_R P_2 = n - j + 1 \geq 0.$$

By (1*) and (4*) we also see that

(6*) $$P_1 \neq 0.$$

We claim that

(7*) $$(\alpha + \nu_{j-1}\beta)(P_1) \subset P_1.$$

To see this, given any $p \in P_1$, we want to show that $(\alpha + \nu_{j-1}\beta)(p) \in P_1$; now

$$\begin{cases} (\alpha + \nu_{j-1}\beta)(p) \in P_1 & \\ \Leftrightarrow \phi\beta(\alpha + \nu_{j-1}\beta)^i((\alpha + \nu_{j-1}\beta)(p)) = 0 \text{ for } 0 \leq i \leq j-1 & \text{by definition of } P_1 \\ \Leftrightarrow \phi\beta(\alpha + \nu_{j-1}\beta)^i(p) = 0 \text{ for } 1 \leq i \leq j & \text{obviously;} \end{cases}$$

for $1 \leq i \leq j-1$ the last equation holds because $p \in P_1$ and hence it holds also for $i = j$ by (2*).

Next we claim that

(8*) $$\beta(P_1) \neq 0.$$

To see this, note that $(\alpha + \nu_{j-1}\beta, \beta)$ is admissible by (T117.1) and hence by taking $\alpha + \nu_{j-1}\beta$ for α in (3) we get $P_1 \subset \overline{P}_1$ where $\overline{P}_1 = \sum_{0 \leq i \leq m} \gamma_i \beta(\alpha + \nu_{j-1}\beta)^i(P_1)$; by (7*) we see that $\overline{P}_1 \subset \widehat{P}_1$ where $\widehat{P}_1 = \sum_{0 \leq i \leq m} \gamma_i \beta(P_1)$; if $\beta(P_1) = 0$ then clearly $\widehat{P}_1 = 0$; hence if $\beta(P_1) = 0$ then $P_1 = 0$ in contradiction to (6*); therefore $\beta(P_1) \neq 0$.

By (8*) we can take $p \in P_1$ with $0 \neq \beta(p) \in Q$; in view of (4*) and (5*), by the definitions of P_1 and P_2 we have $P_1 \subset P_2 \subset P$ with $P_1 \neq P_2$; since P and Q are vector spaces over the field R, we can take an R-homomorphism $\nu' : Q \to P$ such that

(9*) $$\nu'(Q) \subset P_2 \quad \text{and} \quad \nu'(\beta(p)) \notin P_1.$$

Now considering the R-homomorphism

$$\nu_j = \nu_{j-1} + \nu' : Q \to P$$

and considering R-homomorphism

$$\Phi_i = \phi\beta(\alpha + \nu_j\beta)^i : P \to R \quad \text{for} \quad 0 \leq i \leq j$$

we shall first show that

(10*) $$\Phi_i = \phi\beta(\alpha + \nu_{j-1}\beta)^i \quad \text{for} \quad 0 \leq i \leq j-1$$

and then we shall show that

(11*)   the elements $(\Phi_i)_{0 \le i \le j}$ are linearly independent

and this will complete the proof.

To prove (10*) by induction on i, the assertion being obvious for $i = 0$, let $0 < i \le j - 1$ and assume for $i - 1$; now

$$\begin{cases} \Phi_i = \phi\beta(\alpha + \nu_{j-1}\beta)^{i-1}(\alpha + \nu_j\beta) & \text{by induction hypothesis} \\ \quad = \phi\beta(\alpha + \nu_{j-1}\beta)^{i-1}(\alpha + \nu_{j-1}\beta + \nu'\beta) & \text{by definition of } \nu_j \\ \quad = \phi\beta(\alpha + \nu_{j-1}\beta)^{i} + \phi\beta(\alpha + \nu_{j-1}\beta)^{i-1}\nu'\beta & \text{by expanding} \end{cases}$$

and hence it suffices to show that $\phi\beta(\alpha + \nu_{j-1}\beta)^{i-1}\nu'\beta = 0$; but this follows by noting that by (9*) we have $\mathrm{im}(\nu') \subset P_2$ and by the definition of P_2 we have $P_2 \subset \ker(\phi\beta(\alpha + \nu_{j-1}\beta)^{i-1})$.

To prove (11*) assume the contrary; then, because of (10*) and because (as said in the beginning of the proof) the elements $(\phi\beta(\alpha + \nu_{j-1}\beta)^i)_{0 \le i \le j-1}$ are linearly independent, we can find $(s_1, \ldots, s_j) \in R^j$ with $s_j \ne 0$ such that $\sum_{0 \le i \le j} s_i \Phi_i = 0$; evaluating the LHS at p we get $\sum_{0 \le i \le j} s_i \Phi_i(p) = 0$; since $p \in P_1$, by (10*) and the definition of P_1 we see that $\Phi_i(p) = 0$ for $0 \le i \le j - 1$; therefore $\Phi_j(p) = 0$; now

$$\begin{cases} \Phi_j = \phi\beta(\alpha + \nu_{j-1}\beta)^{j-1}(\alpha + \nu_j\beta) & \text{by (10*)} \\ \quad = \phi\beta(\alpha + \nu_{j-1}\beta)^{j-1}(\alpha + \nu_{j-1}\beta + \nu'\beta) & \text{by definition of } \nu_j \\ \quad = \phi\beta(\alpha + \nu_{j-1}\beta)^{j} + \phi\beta(\alpha + \nu_{j-1}\beta)^{j-1}\nu'\beta & \text{by expanding} \end{cases}$$

and hence, because $\Phi_j(p) = 0$, by evaluating the last line at p we get

$$\phi\beta(\alpha + \nu_{j-1}\beta)^{j}(p) + \phi\beta(\alpha + \nu_{j-1}\beta)^{j-1}\nu'\beta(p) = 0;$$

since $p \in P_1$, by (2*) and the definition of P_1 we have $\phi\beta(\alpha + \nu_{j-1}\beta)^{j}(p) = 0$, and hence by the above display we get

$$\phi\beta(\alpha + \nu_{j-1}\beta)^{j-1}\nu'\beta(p) = 0;$$

now by (9*) we have $\nu'\beta(p) \in P_2 \setminus P_1$, and hence by the definitions of P_1 and P_2 we get $\phi\beta(\alpha + \nu_{j-1}\beta)^{j-1}\nu'\beta(p) \ne 0$, contradicting the above display. Thus (11*) is established and the proof of (T117.4) is completed.

PROOF OF (T117.5). Let us start by proving our assertion in the case of constant gnb, i.e., when there exists $n \in \mathbb{N}$ such that for every $J \in \mathrm{mspec}(R)$ we have $\mathrm{gnb}_{R_J} P_J = n$. If $P = 0$ then our assertion is obvious. So assume that $P \ne 0$. Then by item (12') at the end of the proof of (T116.3) we must have $n \in \mathbb{N}_+$. Let Ψ be the set of all R-homomorphisms $\psi : R^n \to P$, and let Φ be the set of all R-homomorphisms $\phi : Q \to R$. Also let N be the set of all R-homomorphisms

$\nu : Q \to P$, for every $\nu \in N$ and $\phi \in \Phi$ consider the diagonal direct sum

$$\theta_{\nu,\phi} = \phi\beta \oplus \cdots \oplus \phi\beta(\alpha + \nu\beta)^{n-1} : P \to R^n,$$

and let $\widetilde{I}$ be the ideal in R obtained by putting

$$\widetilde{I} = \sum_{s \in R, \phi \in \Phi, \psi \in \Psi} \det(\theta_{\nu,\phi}\psi)R.$$

By (T117.1) to (T117.3) we see that $\widetilde{I} \subset I_R(\alpha, \beta)$. In a moment we shall show that $\widetilde{I} = R$ which will imply that $I_R(\alpha, \beta) = R$ and hence by taking $s = 0$ in (2) we will get our desired assertion.

Suppose if possible that $\widetilde{I} \neq R$. Then, because $\widetilde{I}$ is an ideal in R, by Zorn's Lemma $\widetilde{I} \subset J$ for some $J \in \mathrm{mspec}(R)$. Let $h : R \to \overline{R} = R/J$ be the residue class epimorphism. Then $\overline{P} = P/(JP)$ and $\overline{Q} = Q/(JQ)$ are finite dimensional vector spaces over the field $\overline{R}$, and the vector space dimension of $\overline{P}$ is n. Moreover, given any R-homomorphism $F : U \to V$ between any R-modules U and V, it "induces" an $\overline{R}$-homomorphism $\overline{F} : \overline{U} \to \overline{V}$ between the $\overline{R}$-modules $\overline{U} = U/(JU)$ and $\overline{V} = V/(JV)$. In this context we identify $\overline{R^n}$ with $\overline{R}^n$; then ψ, ϕ, ν as above induce $\overline{R}$-homomorphisms $\overline{\psi} : \overline{R}^n \to \overline{P}$, $\overline{\phi} : \overline{Q} \to \overline{R}$, $\overline{\nu} : \overline{Q} \to \overline{P}$ respectively. Also α and β induce $\overline{R}$-homomorphisms $\overline{\alpha} : \overline{P} \to \overline{P}$ and $\overline{\beta} : \overline{P} \to \overline{Q}$ respectively, and we have the diagonal direct sum

$$\overline{\alpha} \oplus \overline{\beta} : \overline{P} \to \overline{P} \oplus \overline{Q}.$$

Clearly the admissibility of the pair (α, β) implies the admissibility of the pair $(\overline{\alpha}, \overline{\beta})$. Also clearly

$$(12^*) \qquad\qquad h(\det(\theta_{\nu,\phi}\psi)) = \det(\theta_{\overline{\nu},\overline{\phi}}\overline{\psi})$$

for the diagonal direct sum

$$\theta_{\overline{\nu},\overline{\phi}} = \overline{\phi}\,\overline{\beta} \oplus \cdots \oplus \overline{\phi}\,\overline{\beta}(\overline{\alpha} + \overline{\nu}\overline{\beta})^{n-1} : \overline{P} \to \overline{R}^n.$$

We claim that:

$$(13^*) \qquad \begin{cases} \text{if } U \text{ is a projective } R\text{-module} \\ \text{then for every } \overline{R}\text{-homomorphism } \widetilde{F} : \overline{U} \to \overline{V} \\ \text{we have } \widetilde{F} = \overline{F} \text{ for some } R\text{-homomorphism } F : U \to V. \end{cases}$$

To see this, in the description of the "lifting property" at the beginning of this Quest (Q31), take α to be the residue class epimorphism $V \to \overline{V}$, and take β to be the composition $U \to \overline{U} \to \overline{V}$ where the first arrow is the residue class epimorphism and the second arrow is the given map $\widetilde{F}$; now it suffices to let F be the lifting homomorphism γ.

We also claim that:

$$(14^*) \qquad\qquad \text{for some } \phi \in \Phi \text{ we have } \overline{\phi}\,\overline{\beta} \neq 0.$$

To see this, note that $n > 0 \Rightarrow \overline{P} \neq 0$ and hence by the admissibility of (α, β), in view of (3) we get $\overline{\beta} \neq 0$. Therefore, since $\overline{Q}$ is a vector space over the field $\overline{R}$, we can find an $\overline{R}$-homomorphism $\overline{\phi} : \overline{Q} \to \overline{R}$ with $\overline{\phi}\, \overline{\beta} \neq 0$. Since Q is a projective R-module, by (13*) we have $\overline{\phi} = \overline{\phi}$ for some $\phi \in \Phi$. Now clearly $\overline{\phi}\, \overline{\beta} = \overline{\phi}\, \overline{\beta}$ and hence $\overline{\phi}\, \overline{\beta} \neq 0$.

In view of (14*), by (T117.4) we can find an $\overline{R}$-homomorphism $\widetilde{\nu} : \overline{Q} \to \overline{P}$ such that, upon letting

$$\zeta_i = \overline{\phi}\, \overline{\beta}(\overline{\alpha} + \widetilde{\nu}\, \overline{\beta})^{i-1},$$

the n elements $(\zeta_i)_{1 \leq i \leq n}$ of the dual space of $\overline{P}$ are linearly independent over $\overline{R}$. Since Q is a projective R-module, by (13*) we have $\widetilde{\nu} = \overline{\nu}$ for some $\nu \in N$. As an obvious property of vector spaces we see that

(15*) $\begin{cases} \text{if } \overline{P} \text{ is a vector space of finite positive dimension } n \text{ over field } \overline{R} \\ \text{and } \zeta_1, \ldots, \zeta_n \text{ are any linearly independent elements} \\ \text{in the dual space of } \overline{P}, \\ \text{then there exists an } \overline{R}\text{-homomorphism } \widetilde{\psi} : \overline{R}^n \to \overline{P} \\ \text{such that for the diagonal direct sum } \theta = \zeta_1 \oplus \cdots \oplus \zeta_n : \overline{P} \to \overline{R}^n \\ \text{we have } \det(\theta\widetilde{\psi}) \neq 0. \end{cases}$

Since R^n is a projective R-module, with $\widetilde{\psi}$ as in (15*), by (13*) we have $\widetilde{\psi} = \overline{\psi}$ for some $\psi \in \Psi$. Now by (12*) we get

$$\det(\theta_{\nu,\phi}\psi) \notin J$$

which is a contradiction because

$$\det(\theta_{\nu,\phi}\psi) \in \widetilde{I} \subset J.$$

Therefore we must have $\widetilde{I} = R$. This completes the proof in the constant gnb case.

In the general case, since P is a finite R-module, upon letting

$$n = \begin{cases} \max\{\mathrm{gnb}_{R_J} P_J : J \in \mathrm{mspec}(R)\} & \text{if } R \neq 0 \\ 0 & \text{if } R = 0 \end{cases}$$

we get $n \in \mathbb{N}$. We shall prove our assertion by induction on n. If $n = 0$ then we are done by the constant gnb case. So let $n > 0$ and assume for all smaller values of n.

In a moment we shall prove the following CLAIMS (16*) to (19*) in which we do NOT use the notation of the current Lemma (T117.5).

CLAIM (16*). Let P and Q be any modules over a ring R and let $\sigma : P \to Q$ and $\sigma' : P \to Q$ be any R-homomorphisms. Given elements u and v in R with $u + v = 1$ and $uv = 0$, consider the modules $P_u = P/(uP)$ and $Q_u = Q/(uQ)$ over the ring $R_u = R/(uR)$, and the modules $P_v = P/(vP)$ and $Q_v = Q/(vQ)$ over the ring $R_v = R/(vR)$. Also consider the (R_u)-homomorphisms $\sigma_u : P_u \to Q_u$ and $\sigma'_u : P_u \to Q_u$ and the (R_v)-homomorphisms $\sigma_v : P_v \to Q_v$ and $\sigma'_v : P_v \to Q_v$ induced by σ and σ'; the "induced homomorphism" is clarified, for instance, by noting that $\sigma_u : P_u \to Q_u$ is the unique (R_u)-homomorphism such that $\sigma_u \Lambda_{u,P} = \Lambda_{u,Q} \sigma$ where $\Lambda_{u,P} : P \to P_u$ and $\Lambda_{u,Q} : Q \to Q_u$ are the residue class epimorphisms. Then we have (16A*) and (16B*), where in (16A*) we are disregarding σ (with or without subscripts or superscripts):

(16A*) P is a finite projective module over $R \Leftrightarrow P_u$ and P_v are finite projective modules over R_u and R_v respectively. Similarly, Q is a finite projective module over $R \Leftrightarrow Q_u$ and Q_v are finite projective modules over R_u and R_v respectively.

(16B*) $\sigma \sim \sigma' \Leftrightarrow \sigma_u \sim \sigma'_u$ and $\sigma_v \sim \sigma'_v$.

CLAIM (17*). In the situation of (16*) let us write β for σ (with or without subscripts or superscripts), and given any R-homomorphisms $\alpha : P \to P$ and $\alpha' : P \to P$, let $\alpha_u : P_u \to P_u$ and $\alpha_v : P_v \to P_v$ be the R-homomorphisms induced by α, and let $\alpha'_u : P_u \to P_u$ and $\alpha'_v : P_v \to P_v$ be the R-homomorphisms induced by α'. Also consider the diagonal direct sums

$$\alpha \oplus \beta : P \to P \oplus Q, \quad \alpha_u \oplus \beta_u : P_u \to P_u \oplus Q_u, \quad \alpha_v \oplus \beta_v : P_v \to P_v \oplus Q_v,$$

and the diagonal direct sums

$$\alpha' \oplus \beta : P \to P \oplus Q, \quad \alpha'_u \oplus \beta_u : P_u \to P_u \oplus Q_u, \quad \alpha'_v \oplus \beta_v : P_v \to P_v \oplus Q_v.$$

Then we have (17A*) to (17C*):

(17A*) $\alpha \oplus \beta \sim \alpha' \oplus \beta \Leftrightarrow \alpha_u \oplus \beta_u \sim \alpha'_u \oplus \beta_u$ and $\alpha_v \oplus \beta_v \sim \alpha'_v \oplus \beta_v$.

(17B*) (α, β) is admissible $\Leftrightarrow (\alpha_u, \beta_u)$ and (α_v, β_v) are admissible.

(17C*) If $\alpha' = \alpha + r1_P$ with $r \in R$ then $\alpha'_u = \alpha_u + r_u 1_{P_u}$ and $\alpha'_v = \alpha_v + r_v 1_{P_v}$ where r_u and r_v are the images of r under the residue class epimorphisms $R \to R_u$ and $R \to R_v$ respectively.

CLAIM (18*). Let P be any finite module over a ring R, and let d be any nonnegative integer. Then there exists an ideal $L(P,d)$ in R such that for any $J \in \text{mspec}(R)$ we have: $\text{gnb}_{R_J} P_J > d \Leftrightarrow L(P,d) \subset J$.

CLAIM (19*). Let P be any finite projective module over a ring R. Assume that there exists a positive integer n such that $\text{gnb}_{R_J} P_J = n$ for some $J \in \text{mspec}(R)$, and $\text{gnb}_{R_J} P_J \leq n$ for all $J \in \text{mspec}(R)$. Then there exist elements u, v in R with $u + v = 1$ and $uv = 0$ such that in the notation of (16*) we have $\text{gnb}_{(R_u)_J}(P_u)_J = n$ for all $J \in \text{mspec}(R_u)$, and $\text{gnb}_{(R_v)_J}(P_v)_J < n$ for all $J \in \text{mspec}(R_v)$.

Assuming (16*) to (19*), let us continue with the proof of (T117.5). Indeed, in view of (16A*), (17*) and (19*), we are done by applying the constant gnb case to

$(P_u, Q_u, \alpha_u, \beta_u)$ and the induction hypothesis to $(P_v, Q_v, \alpha_v, \beta_v)$.

PROOF OF (16*). Recall that

(†) u, v are elements in R with $u + v = 1$ and $uv = 0$.

Multiplying the equation $u + v = 1$ first by u and then by v, in view of the equation $uv = 0$ we get

$$u^2 = u \quad \text{and} \quad v^2 = v$$

i.e., the elements u, v are idempotents in R. Consequently, as an internal direct sum of R-modules we have

$$P = (uP) \oplus (vP)$$

and therefore by (Q17)(T80.3) we see that:

$(1^\dagger)$ $\begin{cases} P \text{ is a finite projective } R\text{-module} \\ \Leftrightarrow uP \text{ and } vP \text{ are finite projective } R\text{-modules.} \end{cases}$

Similarly, as an internal direct sum of R-modules we have

$$Q = (uQ) \oplus (vQ)$$

and hence by (Q17)(T80.3) we see that:

$(2^\dagger)$ $\begin{cases} Q \text{ is a finite projective } R\text{-module} \\ \Leftrightarrow uQ \text{ and } vQ \text{ are finite projective } R\text{-modules.} \end{cases}$

By (†) it follows that:

$(3^\dagger)$ $\begin{cases} P_u \text{ is a finite projective } (R_u)\text{-module} \\ \Leftrightarrow vP \text{ is a finite projective } R\text{-module} \end{cases}$

and

$(4^\dagger)$ $\begin{cases} Q_u \text{ is a finite projective } (R_u)\text{-module} \\ \Leftrightarrow vQ \text{ is a finite projective } R\text{-module} \end{cases}$

and

$(5^\dagger)$ $\begin{cases} P_v \text{ is a finite projective } (R_v)\text{-module} \\ \Leftrightarrow uP \text{ is a finite projective } R\text{-module} \end{cases}$

and

$(6^\dagger)$ $\begin{cases} Q_v \text{ is a finite projective } (R_v)\text{-module} \\ \Leftrightarrow uQ \text{ is a finite projective } R\text{-module.} \end{cases}$

By $(1^\dagger)$ to $(6^\dagger)$ we get $(16A^*)$.

By $(\dagger)$ it also follows that:

$(7^\dagger)$ $\begin{cases} \text{if } \widetilde{u}\tau : uQ \to uQ \text{ and } \widetilde{v}\tau : vQ \to vQ \text{ are any } R\text{-automorphisms then} \\ \text{(the internal direct sum) } \tau = \widetilde{u}\tau \oplus \widetilde{v}\tau : Q \to Q \text{ is an } R\text{-automorphism,} \\ \text{and conversely every } R\text{-automorphism of } Q \text{ is of this form.} \end{cases}$

Clearly σ induces a unique R-homomorphism $\widehat{v}\sigma : vP \to vQ$ such that for all $p \in vP$ we have $(\widehat{v}\sigma)(p) = \sigma(p)$, and a unique R-homomorphism $\widehat{u}\sigma : uP \to uQ$ such that for all $p \in uP$ we have $(\widehat{u}\sigma)(p) = \sigma(p)$; similarly for σ'. Note that using the $\widehat{}$ notation for induced maps, in $(7^\dagger)$ we have $\widehat{u}\tau = \widetilde{u}\tau$ and $\widehat{v}\tau = \widetilde{v}\tau$. Moreover, in the notation of $(7^\dagger)$ we have

$$\sigma = \tau\sigma' \Leftrightarrow \widehat{u}\sigma = (\widetilde{u}\tau)(\widehat{u}\sigma') \text{ and } \widehat{v}\sigma = (\widetilde{v}\tau)(\widehat{v}\sigma')$$

and hence by $(7^\dagger)$ we see that

$(8^\dagger)$ $$\sigma \sim \sigma' \Leftrightarrow \widehat{u}\sigma \sim \widehat{u}\sigma' \text{ and } \widehat{v}\sigma \sim \widehat{v}\sigma'.$$

By $(\dagger)$ it follows that:

$(9^\dagger)$ $$\sigma_u \sim \sigma'_u \Leftrightarrow \widehat{v}\sigma \sim \widehat{v}\sigma'$$

and

$(10^\dagger)$ $$\sigma_v \sim \sigma'_v \Leftrightarrow \widehat{u}\sigma \sim \widehat{u}\sigma'.$$

By $(8^\dagger)$ to $(10^\dagger)$ we get $(16B^*)$.

More details of the implication $(\dagger) \Rightarrow (3^\dagger) + (7^\dagger) + (9^\dagger)$ will be given at the end of this Comment (C49). By symmetry, the implication $(\dagger) \Rightarrow (3^\dagger)$ yields the implication $(\dagger) \Rightarrow (4^\dagger) + (5^\dagger) + (6^\dagger)$, and the implication $(\dagger) \Rightarrow (9^\dagger)$ yields the implication $(\dagger) \Rightarrow (10^\dagger)$.

[NOTATION: $(3^\dagger)+(7^\dagger)+(9^\dagger) = $ TOTALITY OF THE THREE ASSERTIONS].

PROOF OF (17^*). Clearly there is a unique (R_u)-isomorphism

$$\Omega_u : (P \oplus Q)_u \to P_u \oplus Q_u$$

such that

$$\Omega_u(\Lambda_{u,P\oplus Q}(p,q)) = (\Lambda_{u,P}(p), \Lambda_{u,Q}(q)) \quad \text{for all} \quad (p,q) \in P \oplus Q$$

where

$$\Lambda_{u,P\oplus Q} : P \oplus Q \to (P \oplus Q)_u = (P \oplus Q)/u(P \oplus Q)$$

is the residue class epimorphism. Moreover, for the induced (R_u)-homomorphism

$$(\alpha \oplus \beta)_u : P_u \to (P \oplus Q)_u$$

we clearly have

$$\Omega_u(\alpha \oplus \beta)_u = \alpha_u \oplus \beta_u$$

and, similarly, for the induced (R_u)-homomorphism

$$(\alpha' \oplus \beta)_u : P_u \to (P \oplus Q)_u$$

we clearly have

$$\Omega_u(\alpha' \oplus \beta)_u = \alpha'_u \oplus \beta_u$$

and therefore

$$(11^\dagger) \qquad \alpha_u \oplus \beta_u \sim \alpha'_u \oplus \beta_u \Leftrightarrow (\alpha \oplus \beta)_u \sim (\alpha' \oplus \beta)_u.$$

Replacing u by v in $(11^\dagger)$ we get

$$(12^\dagger) \qquad \alpha_v \oplus \beta_v \sim \alpha'_v \oplus \beta_v \Leftrightarrow (\alpha \oplus \beta)_v \sim (\alpha' \oplus \beta)_v.$$

Taking $\alpha \oplus \beta$ and $\alpha' \oplus \beta$ for σ and σ' in (16B*), respectively, we obtain

$$(13^\dagger) \qquad \alpha \oplus \beta \sim \alpha' \oplus \beta \Leftrightarrow (\alpha \oplus \beta)_u \sim (\alpha' \oplus \beta)_u \text{ and } (\alpha \oplus \beta)_v \sim (\alpha' \oplus \beta)_v.$$

By $(11^\dagger)$ to $(13^\dagger)$ we get (17A*).

Turning to admissibility, given R-homomorphisms $\gamma_i : Q \to P$ for $0 \le i \le m$ with $m \in \mathbb{N}_+$, considering the R-homomorphism

$$\delta = \sum_{0 \le i \le m} \gamma_i \beta \alpha^i : P \to P,$$

for the induced R-homomorphisms $uP \to uP$ we have

$$\widehat{u}\delta = \sum_{0 \le i \le m} (\widehat{u}\gamma_i)(\widehat{u}\beta)(\widehat{u}\alpha)^i$$

and for the induced R-homomorphisms $vP \to vP$ we have

$$\widehat{v}\delta = \sum_{0 \le i \le m} (\widehat{v}\gamma_i)(\widehat{v}\beta)(\widehat{v}\alpha)^i.$$

Clearly for any $p \in P$ we have

$$\delta(p) = (\widehat{u}\delta)(up) + (\widehat{v}\delta)(vp)$$

and hence

$$\delta(p) = p \Leftrightarrow (\widehat{u}\delta)(up) = up \text{ and } (\widehat{v}\delta)(vp) = vp.$$

Consequently by (3) we see that

$$(14^\dagger) \qquad \text{if } (\alpha, \beta) \text{ is admissible then so are } (\widehat{u}\alpha, \widehat{u}\beta) \text{ and } (\widehat{v}\alpha, \widehat{v}\beta) .$$

Conversely, assuming $(\widehat{u}\alpha, \widehat{u}\beta)$ and $(\widehat{v}\alpha, \widehat{v}\beta)$ to be admissible, by (3) there exist R-homomorphisms $\gamma'_i : uQ \to uP$ for $0 \leq i \leq m'$ with $m' \in \mathbb{N}_+$ such that $\delta' = 1_{uP}$ where

$$\delta' = \sum_{0 \leq i \leq m'} \gamma'_i(\widehat{u}\beta)(\widehat{u}\alpha)^i$$

and there exit R-homomorphisms $\gamma''_i : vQ \to vP$ for $0 \leq i \leq m''$ with $m'' \in \mathbb{N}_+$ such that $\delta'' = 1_{uP}$ where

$$\delta'' = \sum_{0 \leq i \leq m''} \gamma''_i(\widehat{v}\beta)(\widehat{v}\alpha)^i.$$

Upon letting $m = m' + m''$ with $\gamma'_i = 0$ for $m' + 1 \leq i \leq m$ and $\gamma''_i = 0$ for $m'' + 1 \leq i \leq m$, we may assume that $m' = m'' = m$. Clearly there exists a unique R-homomorphism $\gamma_i : Q \to P$ with $\widehat{u}\gamma_i = \gamma'_i$ and $\widehat{v}\gamma_i = \gamma''_i$ for $0 \leq i \leq m$. Now considering the R-homomorphism $\delta = \sum_{0 \leq i \leq m} \gamma_i \beta \alpha^i : P \to P$, the equations $\delta' = 1_{uP}$ and $\delta'' = 1_{vP}$ yield the equation $\delta = 1_P$. Thus in view of (3) we see that

(15†) $\qquad$ if $(\widehat{u}\alpha, \widehat{u}\beta)$ and $(\widehat{v}\alpha, \widehat{v}\beta)$ are admissible then so is (α, β).

Clearly there exists a unique R-isomorphism $\overline{\Lambda}_{v,P} : uP \to P_v$ such that for all $p \in P$ we have $\overline{\Lambda}_{v,P}(up) = \Lambda_{v,P}(p)$, and there exists a unique R-isomorphism $\overline{\Lambda}_{v,Q} : uQ \to Q_v$ such that for all $q \in Q$ we have $\overline{\Lambda}_{v,Q}(uq) = \Lambda_{v,Q}(q)$. Assuming $(\widehat{u}\alpha, \widehat{u}\beta)$ to be admissible, by (3) there exist R-homomorphisms $\gamma'_i : uQ \to uP$ for $0 \leq i \leq m'$ with $m' \in \mathbb{N}_+$ such that

$$\sum_{0 \leq i \leq m'} \gamma'_i(\widehat{u}\beta)(\widehat{u}\alpha)^i = 1_{uP},$$

and clearly we get (R_u)-homomorphisms $\gamma^*_i = \overline{\Lambda}_{v,P}\gamma'_i(\overline{\Lambda}_{v,Q})^{-1} : Q_v \to P_v$ for $0 \leq i \leq m'$ such that

$$\sum_{0 \leq i \leq m'} \gamma^*_i \beta_v(\alpha_v)^i = 1_{P_v}$$

and hence (α_v, β_v) is admissible. Conversely assuming (α_v, β_v) to be admissible, by (3) there exist (R_v)-homomorphisms $\gamma^*_i : Q_v \to P_v$ for $0 \leq i \leq m'$ with $m' \in \mathbb{N}_+$ such that

$$\sum_{0 \leq i \leq m'} \gamma^*_i \beta_v(\alpha_v)^i = 1_{P_v},$$

and clearly we get R-homomorphisms $\gamma'_i = (\overline{\Lambda}_{v,P})^{-1}\gamma^*_i\overline{\Lambda}_{v,Q} : uQ \to uP$ for $0 \leq i \leq m'$ such that $\sum_{0 \leq i \leq m'} \gamma'_i(\widehat{u}\beta)(\widehat{u}\alpha)^i = 1_{uP}$ and hence $(\widehat{u}\alpha, \widehat{u}\beta)$ is admissible. Thus

(16†) $\qquad$ (α_v, β_v) is admissible $\Leftrightarrow$ $(\widehat{u}\alpha, \widehat{u}\beta)$ is admissible and hence by symmetry

$(17^\dagger)$ $\qquad\qquad (\alpha_u, \beta_u)$ is admissible $\Leftrightarrow (\widehat{v}\alpha, \widehat{v}\beta)$ is admissible.

By $(14^\dagger)$ to $(17^\dagger)$ we get (17B*).

The proof of (17C*) is straightforward.

PROOF OF (18*) AND (19*). Given any finite module P over a ring R and any nonnegative integer d, upon letting

$$L(P, d) = \sum_{(p_1, \ldots, p_d) \in P^d} \text{ann}(P/(Rp_1 + \cdots + Rp_d)),$$

we see that $L(P, d)$ is an ideal in R. For any $J \in \text{mspec}(R)$ let $\phi_J : R \to R_J$ and $\psi_J : P \to P_J$ be the natural localization homomorphisms. If $\text{gnb}_{R_J} P_J \leq d$ then we can find $(p_1, \ldots, p_d) \in P^d$ such that the (R_J)-module P_J is generated by $\psi_J(p_1), \ldots, \psi_J(p_d)$, and hence (because P is finitely generated) we can find $s \in R \backslash J$ such that $s \in \text{ann}(P/(Rp_1 + \cdots + Rp_d))$, and therefore $L(P, d) \not\subset J$. Conversely if $L(P, d) \not\subset J$ then we can find $(p_1, \ldots, p_d) \in P^d$ together with $s \in R \setminus J$ such that $s \in \text{ann}(P/(Rp_1 + \cdots + Rp_d))$, and hence the (R_J)-module P_J is generated by $\psi_J(p_1)/\phi_J(s), \ldots, \psi_J(p_d)/\phi_J(s)$, and therefore $\text{gnb}_{R_J} P_J \leq d$. This completes the proof of (18*).

Now assume that P is projective and there exists a positive integer n such that $\text{gnb}_{R_J} P_J = n$ for some $J \in \text{mspec}(R)$, and $\text{gnb}_{R_J} P_J \leq n$ for all $J \in \text{mspec}(R)$. By (T80.1) to (T80.3) of (Q17) we may assume that $P \oplus P' = R^m$ for some finite projective R-module P' and some integer $m \geq n$. By (C29)(2) and (C30)(3) of (Q17), for all $J \in \text{mspec}(R)$ we have $\text{gnb}_{R_J} P'_J = m - \text{gnb}_{R_J} P_J$. Therefore, upon letting $U' = L(P, n - 1)$ and $V' = L(P', m - n)$, by what we proved above we see that U' and V' are ideals in R such that for any $J \in \text{mspec}(R)$ we have

$$\text{gnb}_{R_J} P_J \begin{cases} = n \Leftrightarrow U' \subset J \\ < n \Leftrightarrow V' \subset J. \end{cases}$$

Let U (resp: V) be the intersection of all Q in $\text{nspec}(R)$ such that $Q \subset J$ for some J in $\text{mvspec}_R U'$) (resp: $\text{mvspec}_R V'$). Then clearly U and V are ideals in R with $U \cap V = \text{nrad}(R)$. Since P is finite projective, by (T80.4), C29(2), (C30)(3), and (C30)(4) of L4§17 we have $\text{gnb}_{R_Q} P_Q = \text{gnb}_{R_J} P_J$ for all $Q \subset J$ in $\text{spec}(R)$. Therefore for any $J \in \text{mspec}(R)$ we have

$$\text{gnb}_{R_J} P_J \begin{cases} = n \Leftrightarrow U \subset J \\ < n \Leftrightarrow V \subset J. \end{cases}$$

It follows that $U + V = R$ and hence we can find $u' \in U$ and $v' \in V$ with $u' + v' = 1$. Clearly $u'v' \in U \cap V$ and hence $(u'v')^e = 0$ for some $e \in \mathbb{N}_+$. Now

$$1 = (u' + v')^{2e} = a(u')^e + b(v')^e$$

with a, b in R and upon letting

$$u = a(u')^e \quad \text{and} \quad v = b(v')^e$$

we get u, v in R with $u \in U$ and $v \in V$ such that $u + v = 1$ and $uv = 0$.

Now let the notation be as in (16*), let $\Lambda_{u,R} : R \to R_u$ and $\Lambda_{v,R} : R \to R_v$ be the residue class epimorphisms, and note that then

$$\begin{cases} J \mapsto J_u = \Lambda_{u,R}(J) \text{ gives a bijection} \\ \mathrm{mvspec}_R U = \{J \in \mathrm{mspec}(R) : U \subset J\} \to \mathrm{mspec}(R_u) \end{cases}$$

and

$$\begin{cases} J \mapsto J_v = \Lambda_{v,R}(J) \text{ gives a bijection} \\ \mathrm{mvspec}_R V = \{J \in \mathrm{mspec}(R) : V \subset J\} \to \mathrm{mspec}(R_v). \end{cases}$$

Given any $J \in \mathrm{mvspec}_R U$, we can take elements $p_1, \ldots, p_n$ in P such that their images $\psi_J(p_1), \ldots, \psi_J(p_n)$ are generators of the (R_J)-module P_J and then, upon letting $\phi_{u,J} : R_u \to (R_u)_{J_u}$ and $\psi_{u,J} : P_u \to (P_u)_{J_u}$ be the natural localization homomorphisms, we see that the elements $\psi_{u,J}(\Lambda_{u,P}(p_1)), \ldots, \psi_{u,J}(\Lambda_{u,P}(p_n))$ are generators of the $((R_u)_{J_u})$-module $((P_u)_{J_u})$. Conversely, given any $\nu \in \mathbb{N}$ together with ν generators of the $((R_u)_{J_u})$-module $(P_u)_{J_u}$, upon multiplying them by a suitable element in $\phi_{u,J}(R_u \setminus J_u)$, we get ν generators of the $((R_u)_{J_u})$-module $(P_u)_{J_u}$ which are of the form $\psi_{u,J}(\Lambda_{u,P}(p_1)), \ldots, \psi_{u,J}(\Lambda_{u,P}(p_\nu))$ with $p_1, \ldots, p_\nu$ in R; it follows that then the elements $\psi_J(p_1), \ldots, \psi_J(p_\nu)$ are generators of the (R_J)-module P_J. Thus $\mathrm{gnb}_{(R_u)_{J_u}}(P_u)_{J_u} = n$.

Similarly for every $J \in \mathrm{mvspec}_R V$ we have $\mathrm{gnb}_{(R_v)_{J_v}}(P_v)_{J_v} = \mathrm{gnb}_{R_J} P_J$.

This completes the proof of (19*).

ALMOST ISOMORPHISM THEOREM (T118). For the univariate polynomial ring $R = K[X]$ over a ring K we have the following.

(T118.1) Let P, Q be R-modules such that $P = P_0[X]$ and $Q = Q_0[X]$ for some finite projective K-modules P_0 and Q_0. Let $\alpha_0 : P_0 \to P_0$ and $\beta_0 : P_0 \to Q_0$ be K-homomorphisms. Consider the R-homomorphisms $\alpha = \alpha_0[X] : P \to P$ and $\beta = \beta_0[X] : P \to Q$, and consider their diagonal direct sum $\alpha \oplus \beta : P \to P \oplus Q$. Assume that the diagonal direct sum $(\alpha + X 1_P) \oplus \beta : P \to P \oplus Q$ is a split R-monomorphism. Then the pair (α, β) is admissible and we have

$$\alpha \oplus \beta \sim (\alpha + X 1_P) \oplus \beta \sim (\alpha + 1_P) \oplus \beta.$$

(T118.2) Let P, Q be R-modules such that $P = P_0[X]$ and $Q = Q_0[X]$ for some finite projective K-modules P_0 and Q_0. Let $\alpha_0 : P_0 \to P_0$ and $\beta_0 : P_0 \to Q_0$ be K-homomorphisms such that the diagonal direct sum $\alpha_0 \oplus \beta_0 : P_0 \to P_0 \oplus Q_0$ is a split K-monomorphism. Consider the R-homomorphisms $\alpha = \alpha_0[X] : P \to P$ and $\beta = \beta_0[X] : P \to Q$, and consider their diagonal direct sum $\alpha \oplus \beta : P \to P \oplus Q$.

Assume that the diagonal direct sum $(X\alpha + 1_P) \oplus (X\beta) : P \to P \oplus Q$ is a split R-monomorphism. Then the pair (α, β) is admissible and we have

$$\alpha \oplus \beta \sim (\alpha + X1_P) \oplus \beta \sim (\alpha + 1_P) \oplus \beta.$$

(T118.3) Let P, Q be R-modules such that $P = P_0[X]$ and $Q = Q_0[X]$ for some finite projective K-modules P_0 and Q_0. Let $\sigma : P \to Q$ be an R-homomorphism such that $\sigma = \sigma_0[X]$ for some K-homomorphism $\sigma_0 : P_0 \to Q_0$. Let $\sigma' : P \to Q$ be an R-homomorphism such that $\sigma' = \sigma'_0[X]$ for some split K-monomorphism $\sigma'_0 : P_0 \to Q_0$. Assume that $\sigma + X\sigma' : P \to Q$ is a split R-monomorphism. Then $\sigma' + X\sigma : P \to Q$ is a split R-monomorphism and we have

$$\sigma \sim \sigma + X\sigma' \sim \sigma' + X\sigma \sim \sigma + \sigma' \sim \sigma'.$$

(T118.4) Let V be finite projective R-module and let W be an R-module such that $W = W_0[X]$ for some finite projective K-module W_0. Assume that V and W are almost R-isomorphic. Then they are R-isomorphic.

(T118.5) Let V be a finite projective R-module such that for some polynomial $f = X^d + a_1 X^{d-1} + \cdots + a_d$ with $d \in \mathbb{N}$ and $a_1, \ldots, a_d$ in K, upon considering the multiplicative set $S(f) = \{f^i : i \in \mathbb{N}\}$ in R, the $(R_{S(f)})$-module $V_{S(f)}$ is a finite free module. Then V is a finite free R-module.

PROOF OF (T118.1). The following FACT ($\sharp$), which we shall restate and prove at the end of this Comment (C49), is an easy consequence of characterization (1) of split monomorphisms.

FACT ($\sharp$). Let R be the univariate polynomial ring $K[X]$ over a ring K, and consider R-modules $P = P_0[X]$ and $Q = Q_0[X]$ where P_0 and Q_0 are finite modules over K. Also consider an R-homomorphism $a = \sum_{0 \le i \le n} X^i a_i[X] : P \to P$ and an R-homomorphism $b = \sum_{0 \le i \le n} X^i b_i[X] : P \to Q$ where $a_i : P_0 \to P_0$ and $b_i : P_0 \to Q_0$ for $0 \le i \le n \in \mathbb{N}$ are K-homomorphisms. Assume the diagonal direct sum $a \oplus b : P \to P \oplus Q$ is a split R-monomorphism. Then there exists $d \in \mathbb{N}$ together with K-homomorphisms $\eta_i : P_0 \to P_0$ and $\zeta_i : Q_0 \to P_0$ for $0 \le i \le d$ such that for the R-homomorphisms $\eta = \sum_{0 \le i \le d} X^i \eta_i[X] : P \to P$ and $\zeta = \sum_{0 \le i \le d} X^i \zeta_i[X] : Q \to P$ we have $\eta a + \zeta b = 1_P$.

In the case of (T118.1), by taking $(a_0, a_1) = (\alpha_0, 1_{P_0})$ and $(b_0, b_1) = (\beta_0, 0_{P_0})$ with $n = 1$ in ($\sharp$), where 0_{P_0} is the map $P_0 \to P_0$ which sends every element to 0, we get

$$\eta a + \zeta b = \sum_{0 \le i \le d+1} X^i \theta_i[X] = 1_P.$$

where the K-homomorphism $\theta_i : P_0 \to P_0$ is given by

$(\sharp')$
$$\theta_i = \begin{cases} \eta_0 \alpha_0 + \zeta_0 \beta_0 & \text{if } i = 0 \\ \eta_{i-1} + \eta_i \alpha_0 + \zeta_i \beta_0 & \text{if } 1 \le i \le d \\ \eta_d & \text{if } i = d+1 \end{cases}$$

and hence we get

$(\sharp_0')$
$$\theta_0 = 1_{P_0}$$

and for $1 \le i \le d+1$ we get

$(\sharp_i)$
$$\theta_i = 0_{P_0}.$$

Let H_i be the additive abelian group of all K-homomorphisms $P_0 \to P_0$ which can be expressed in the form $\sum_{0 \le j \le i} \gamma_j \beta_0 \alpha_0^j$ with suitable K-homomorphisms $\gamma_j : Q_0 \to P_0$, and note that then for all $i \in \mathbb{N}$ we have $H_i \subset H_{i+1}$. In view of $(\sharp')$ we see that

$$\begin{cases} (\sharp_{d+1}) \text{ implies } \eta_d = 0_{P_0}, \\ \text{and putting this in } (\sharp_d) \text{ implies } \eta_{d-1} \in H_1, \\ \text{and putting this in } (\sharp_{d-1}) \text{ implies } \eta_{d-2} \in H_2, \\ \text{and so on, until } (\sharp_1) \text{ implies } \eta_0 \in H_d, \end{cases}$$

and hence $\eta \in \widehat{H}_d$ where we are writing $\widehat{H}_i$ for the additive abelian group of all R-homomorphisms $P \to P$ which can be expressed in the form $\sum_{0 \le j \le i} \gamma_j \beta \alpha^j$ with suitable R-homomorphisms $\gamma_j : Q \to P$.

It follows that $\eta a + \zeta b \in \widehat{H}_m$ with $m = d+1$. Since $\eta a + \zeta b = 1_P$, by (3) we conclude that the pair (α, β) is admissible. Therefore by (T117.5) we get

$$\alpha \oplus \beta \sim (\alpha + X1_P) \oplus \beta \sim (\alpha + 1_P) \oplus \beta$$

(where we note that for any finite projective K-module P_0, the module $P_0[X]$ is obviously a finite projective module over $K[X]$).

PROOF OF (T118.2). By taking $(a_0, a_1) = (1_P, \alpha_0)$ and $(b_0, b_1) = (0_{P_0}, \beta_0)$ with $n = 1$ in $(\sharp)$, where 0_{P_0} is the map $P_0 \to P_0$ which sends every element to 0, we get

$$\eta \alpha + \zeta \beta = \sum_{0 \le i \le d+1} X^i \theta_i[X] = 1_P$$

where the K-homomorphism $\theta_i : P_0 \to P_0$ is given by

$(\flat')$
$$\theta_i = \begin{cases} \eta_0 & \text{if } i = 0 \\ \eta_{i-1} \alpha_0 + \eta_i + \zeta_{i-1} \beta_0 & \text{if } 1 \le i \le d \\ \eta_d \alpha_0 + \zeta_d \beta_0 & \text{if } i = d+1 \end{cases}$$

and hence we get

(♭₀′) $$\theta_0 = 1_{P_0}$$

and for $1 \le i \le d+1$ we get

(♭ᵢ) $$\theta_i = 0_{P_0}.$$

Let H_i and $\widehat{H}_i$ be as in the above proof of (T118.1). In view of (♭′) we see that

$$\begin{cases} (\flat_0') \text{ implies } \eta_0 = 1_{P_0}, \\ \text{and putting this in } (\flat_1) \text{ implies } \eta_1 + \alpha_0 \in H_1, \\ \text{and putting this in } (\flat_2) \text{ implies } \eta_2 - \alpha_0^2 \in H_2, \\ \text{and so on until } (\flat_d) \text{ implies } \eta_d - (-1)^d \alpha_0^d \in H_d, \\ \text{and then putting this in } (\flat_{d+1}) \text{ we get } \alpha_0^{d+1} \in H_{d+1}. \end{cases}$$

Thus $\alpha_0^e \in \cup_{0 \le i < \infty} H_i$ for some $e \in \mathbb{N}$; let e be the smallest such.

Since $\alpha_0 \oplus \beta_0$ is a split K-monomorphism, by taking $a_0 = \alpha_0$ and $b_0 = \beta_0$ with $n = 0$ in (♯) we can find a nonnegative integer c together with K-homomorphisms $y_i : P_0 \to P_0$ and $z_i : Q_0 \to P_0$ for $0 \le i \le c$ such that for the R-homomorphisms $y = \sum_{0 \le i \le c} X^i y_i[X] : P \to P$ and $z = \sum_{0 \le i \le c} X^i z_i[X] : Q \to P$ we have

$$y\alpha + z\beta = 1_P.$$

By "putting $X = 0$" in the last equation we get the new equation

$$y_0\alpha_0 + z_0\beta_0 = 1_{P_0}.$$

If $e > 0$ then multiplying both sides of the new equation from the right by α_0^{e-1} we get the newer equation

$$y_0\alpha_0^e + z_0\beta_0\alpha_0^{e-1} = \alpha_0^{e-1};$$

clearly the LHS of the above equation belongs to $\cup_{0 \le i < \infty} H_i$ and hence so does the RHS, which contradicts the minimality of e. Therefore we must have $e = 0$. Thus $1_{P_0} \in H_m$ for some $m \in \mathbb{N}_+$. Consequently $1_P \in \widehat{H}_m$, and hence by (3) we conclude that the pair (α, β) is admissible. Therefore by (T117.5) we get

$$\alpha \oplus \beta \sim (\alpha + X1_P) \oplus \beta \sim (\alpha + 1_P) \oplus \beta.$$

PROOF OF (T118.3). Since $\sigma_0' : P_0 \to Q_0$ is a split K-monomorphism, we can find a K-module M_0 together with a K-isomorphism $\xi_0 : Q_0 \to P_0 \oplus M_0$ such that, upon letting $\alpha_0' = 1_{P_0}$ and $\beta_0' = 0_{P_0,M_0}$ (i.e., the map $P_0 \to M_0$ which sends every element of P_0 to the 0 element of M_0), for the diagonal direct sum $\alpha_0' \oplus \beta_0' : P_0 \to P_0 \oplus M_0$ we have $\xi_0\sigma_0' = \alpha_0' \oplus \beta_0'$. Also clearly there are unique K-homomorphisms $\alpha_0 : P_0 \to P_0$ and $\beta_0 : P_0 \to M_0$ such that for the diagonal direct sum $\alpha_0 \oplus \beta_0 : P_0 \to P_0 \oplus M_0$ we have $\xi_0 \sigma_0 = \alpha_0 \oplus \beta_0$. Consider the R-homomorphisms $\alpha = \alpha_0[X] : P \to P$ and $\beta = \beta_0[X] : P \to M$ where $M = M_0[X]$. Since the map $\sigma + X\sigma' : P \to Q$ is a split R-monomorphisms, it follows that the diagonal direct sum $(\alpha + X1_P) \oplus \beta : P \to P \oplus M$ is a split R-monomorphism. Now by (T118.1) we conclude that the pair (α, β) is admissible and we have

$$\sigma \sim \sigma + X\sigma' \sim \sigma + \sigma'.$$

In view of (T117.1), the admissibility of (α, β) tells us that $\sigma = \sigma_0[X] : P \to Q$ is a split R-monomorphism, and hence $\sigma_0 : P_0 \to Q_0$ is a split K-monomorphism.

Since $\sigma_0 : P_0 \to Q_0$ is a split K-monomorphism, we can find a K-module $\widehat{M_0}$ together with a K-isomorphism $\widehat{\xi_0} : Q_0 \to P_0 \oplus \widehat{M_0}$ such that, upon letting $\widehat{\alpha_0'} = 1_{P_0}$ and $\widehat{\beta_0'} = 0_{P_0, \widehat{M_0}}$, for the diagonal direct sum $\widehat{\alpha_0'} \oplus \widehat{\beta_0'} : P_0 \to P_0 \oplus \widehat{M_0}$ we have $\widehat{\xi_0}\sigma_0 = \widehat{\alpha_0'} \oplus \widehat{\beta_0'}$. Also clearly there are unique K-homomorphisms $\widehat{\alpha_0} : P_0 \to P_0$ and $\widehat{\beta_0} : P_0 \to \widehat{M_0}$ such that the diagonal direct sum $\widehat{\alpha_0} \oplus \widehat{\beta_0} : P_0 \to P_0 \oplus \widehat{M_0}$ is a split K-monomorphism and we have $\widehat{\xi_0}\sigma_0' = \widehat{\alpha_0} \oplus \widehat{\beta_0}$. Consider the R-homomorphisms $\widehat{\alpha} = \widehat{\alpha_0}[X] : P \to P$ and $\widehat{\beta} = \widehat{\beta_0}[X] : P \to \widehat{M}$ where $\widehat{M} = \widehat{M_0}[X]$. Since the map $\sigma + X\sigma' : P \to Q$ is a split R-monomorphisms, it follows that the diagonal direct sum $(X\widehat{\alpha} + 1_P) \oplus (X\widehat{\beta}) : P \to P \oplus \widehat{M}$ is a split R-monomorphism. Now by (T118.2) we conclude that the pair $(\widehat{\alpha}, \widehat{\beta})$ is admissible and we have

$$\sigma' \sim \sigma' + X\sigma \sim \sigma + \sigma'.$$

By the above two displays it follows that

$$\sigma \sim \sigma + X\sigma' \sim \sigma' + X\sigma \sim \sigma + \sigma' \sim \sigma'.$$

In view of Lemma (T117.1), the admissibility of the pair $(\widehat{\alpha}, \widehat{\beta})$ implies that the pair $(\widehat{\alpha} + X1_P, \widehat{\beta})$ is admissible and hence the diagonal direct sum $(\widehat{\alpha} + X1_P) \oplus \widehat{\beta}$ is a split R-monomorphism. Consequently the map $\sigma' + X\sigma : P \to Q$ is a split R-monomorphism.

PROOF OF (T118.4). By assumption the localized modules $V_{S(f)}, W_{S(f)}$ over the ring $R_{S(f)}$ are isomorphic where $S(f)$ is the multiplicative set in R consisting of all nonnegative powers of a polynomial $f = X^d + a_1 X^{d-1} + \cdots + a_d$ with $d \in \mathbb{N}$ and $a_1, \ldots, a_d$ in K. Our assertion is obvious if either $K = 0$ or $d = 0$. So assume that $K \neq 0$ and $d > 0$. In view of the following FACT (♯♯), which we shall restate and prove at the end of this Comment (C49), V is R-isomorphic to a submodule $\widetilde{V}$ of W with $f^m W \subset \widetilde{V}$ for some $m \in \mathbb{N}_+$. Without loss of generality we may suppose that $V = \widetilde{V}$, i.e., $hW \subset V \subset W$ where upon letting $n = md \in \mathbb{N}_+$ we have $h = f^m = c_0 X^n + c_1 X^{n-1} + \cdots + c_n$ with $c_0, \ldots, c_n$ in K and $c_0 = 1$. We shall now complete the argument by using the following FACT (♯♯♯) which we shall restate and prove at the end of this Comment (C49).

FACT (♯♯). Let V be a finite projective module over a univariate polynomial ring $R = K[X]$ where K is a nonnull ring. Let W be an R-module such that $W = W_0[X]$ for a finite projective K-module W_0. Assume the localized modules $V_{S(f)}, W_{S(f)}$ over the ring $R_{S(f)}$ are isomorphic where $S(f)$ is the multiplicative set in R consisting of all nonnegative powers of a polynomial $f = X^d + a_1 X^{d-1} + \cdots + a_d$ with $d \in \mathbb{N}_+$ and $a_1, \ldots, a_d$ in K. Then V is R-isomorphic to a submodule $\widetilde{V}$ of W with $f^m W \subset \widetilde{V}$ for some $m \in \mathbb{N}_+$.

FACT (♯♯♯). Let V be a finite projective module over a univariate polynomial ring $R = K[X]$ where K is a nonnull ring. Let W be an R-module such that $W = W_0[X]$ for a finite projective K-module W_0. Assume V is a submodule of W with $hW \subset V$ for some polynomial $h = c_0 X^n + c_1 X^{n-1} + \cdots + c_n$ with $n \in \mathbb{N}_+$ where $c_1, \ldots, c_n$ are elements in K and c_0 is a unit in K. Viewing W (and every other R-module such as V) as a K-module, consider the K-homomorphism $\mu : W_0^n \to W$ given by $y = (y_1, \ldots, y_n) \mapsto y_1 + y_2 X + \cdots + y_n X^{n-1}$ for all $y \in W_0^n$ with $y_1, \ldots, y_n$ in W_0, and consider the R-module $U = U_0[X]$ where the K-module U_0 is given by $U_0 = \mu^{-1}(V \cap \mu(W_0^n))$. By a slight generalization of the usual division algorithm discussed in L1§10(R1) we see that, for any $y \in U_0$, there exists a unique $(\alpha_0(y), \beta_0(y)) \in U_0 \times W_0$ such that $X\mu(y) = \mu(\alpha_0(y)) + h\beta_0(y)$, and this gives K-homomorphisms $\alpha_0 : U_0 \to U_0$ and $\beta_0 : U_0 \to W_0$. Consider the diagonal direct sums $\sigma_0 = (-\alpha_0) \oplus \beta_0 : U_0 \to U_0 \oplus W_0$ and $\sigma_0' = 1_{U_0} \oplus 0_{U_0, W_0} : U_0 \to U_0 \oplus W_0$ where $0_{U_0, W_0} : U_0 \to W_0$ sends every element to 0. Also consider R-homomorphisms $\sigma = \sigma_0[X] : U \to U \oplus W$ and $\sigma' = \sigma_0'[X] : U \to U \oplus W$ where we have identified $(U_0 \oplus W_0)[X]$ with $U \oplus W$ in a straightforward manner. Let $\widehat{\mu} : U \to V$ be the R-homomorphism given by $\sum y^{(i)} X^i \mapsto \sum \mu(y^{(i)}) X^i$ for every finite sum $\sum y^{(i)} X^i$ with $y^{(i)} \in U_0$. Let $\widehat{h} : W \to V$ be the R-homomorphism given by $z \mapsto hz$ for every $z \in W$ (note the condition $hW \subset V$). Let $(\widehat{\mu}, -\widehat{h}) : U \oplus W \to V$ be the R-homomorphism given by $(p, q) \mapsto \widehat{\mu}(p) - \widehat{h}(q)$ for all $(p, q) \in U \times W$. Then:

(i) U_0 is a finite projective K-module,

(ii) there exists a K-homomorphism $\eta_0' : U_0 \oplus W_0 \to U_0$ with $\eta_0' \sigma_0' = 1_{U_0}$,

(iii) there exists a K-isomorphism $(U_0 \oplus W_0)/\sigma_0'(U_0) \approx W_0$, and

(iv)
$$0 \longrightarrow U \xrightarrow{\ \sigma + X\sigma'\ } U \oplus W \xrightarrow{\ (\widehat{\mu}, -\widehat{h})\ } V \longrightarrow 0$$

is a short exact sequence of R-homomorphisms which splits. So $U \oplus V \approx U \oplus W$.

Indeed, in view of (1) and (♯♯♯), our assertion follows by taking $P = U$ and $Q = U \oplus W$ in (T118.3) and noting that the equivalence

$$\sigma + X\sigma' \sim \sigma'$$

gives rise to an R-isomorphism

$$Q/(\sigma + X\sigma')(P) \approx Q/\sigma'(P)$$

which, by (iv) and (iii), is the same thing as an R-isomorphism $V \approx W$.

PROOF OF (T118.5). Note that now $V_{S(f)}$ is $(R_{S(f)})$-isomorphic to $W_{S(f)}$ where $W = W_0[X]$ with $W_0 = K^n$ for some $n \in \mathbb{N}$. Therefore by (T118.4) we see that V and W are R-isomorphic, and hence V is a finite free R-module.

PROOF OF $(\dagger) \Rightarrow (3^\dagger) + (7^\dagger) + (9^\dagger)$. We are assuming that R is a ring and

$(\dagger)$ $\qquad\qquad u, v$ are elements in R with $u + v = 1$ and $uv = 0$.

It follows that the elements u, v are idempotents, i.e.,

$$u^2 = u \quad \text{and} \quad v^2 = v,$$

and as an internal direct sum of R-modules we have

$$P = (uP) \oplus (vP).$$

Given an R-module P we are considering the module $P_u = P/(uP)$ over the ring $R_u = R/(uR)$, and we want to show that then:

$(3^\dagger)$ $\qquad \begin{cases} P_u \text{ is a finite projective } (R_u)\text{-module} \\ \Leftrightarrow vP \text{ is a finite projective } R\text{-module.} \end{cases}$

Namely, clearly there exists a unique R-isomorphism $\overline{\Lambda}_{u,P} : vP \to P_u$ such that for all $p \in P$ we have $\overline{\Lambda}_{u,P}(vp) = \Lambda_{u,P}(p)$ where $\Lambda_{u,P} : P \to P_u$ is the residue class epimorphism, and hence $(3^\dagger)$ is equivalent to saying that:

$(3_1^\dagger)$ $\qquad \begin{cases} P_u \text{ is a finite projective } (R_u)\text{-module} \\ \Leftrightarrow P_u \text{ is a finite projective } R\text{-module.} \end{cases}$

Now if P_u is a finite projective (R_u)-module then by (Q17)(T80.2) there is an (R_u)-isomorphism $R_u \oplus M \approx (R_u)^n$ for some (R_u)-module M and some nonnegative integer n; but as R-modules we have $R_u \oplus R_v \approx (vR) \oplus (uR) \approx R$, and hence we get the R-isomorphisms $P_u \oplus M \oplus (R_v)^n \approx (R_u)^n \oplus (R_v)^n \approx R^n$, and therefore again by (Q17)(T80.2) we see that P_u is a finite projective R-module. Since the R-module $N = R_u$ is annihilated by the ideal $J = uR$ in R, to prove the converse it suffices to note that:

$(3_2^\dagger)$ $\qquad \begin{cases} \text{if } J \text{ is an ideal in a ring } R \\ \text{and } N \text{ is a finite projective } R\text{-module annihilated by } J \\ \text{then } N \text{ is a finite projective } (R/J)\text{-module.} \end{cases}$

Before proving $(3_2^\dagger)$, let us prove that:

$(3_3^\dagger)$
$$\begin{cases} \text{if } J \text{ is an ideal in a ring } R \\ \text{and } (U_i)_{i \in I} \text{ is a family of } R\text{-modules} \\ \text{then } J(\oplus_{i \in I} U_i) = \oplus_{i \in I}(JU_i) \\ \text{and } (\oplus_{i \in I} U_i)/J(\oplus_{i \in I} U_i) \approx \oplus_{i \in I}(U_i/JU_i) \\ \text{which is an } R\text{-isomorphism as well as an } (R/J)\text{-isomorphism.} \end{cases}$$

To prove this note that

$$L = \{jm : j \in J \text{ and } m \in \oplus_{i \in I} U_i\}$$

is a set of R-generators of $J(\oplus_{i \in I} U_i)$. Also

$$\overline{L}_i = \{jm : j \in J \text{ and } m \in U_i\}$$

is a set of R-generators of JU_i, and hence upon letting

$$\widehat{L}_i = \{\mu \in \oplus_{i \in I} U_i : \mu(i) \in \overline{L}_i \text{ and } \mu(i') = 0 \text{ for all } i' \in I \setminus \{i\}\}$$

we have that $\cup_{i \in I}\widehat{L}_i$ is a set of R-generators of $\oplus_{i \in I}(JU_i)$. Clearly $L \subset \oplus_{i \in I}(JU_i)$, and for all $i \in I$ we have $\widehat{L}_i \subset J(\oplus_{i \in I} U_i)$. Therefore $J(\oplus_{i \in I} U_i) = \oplus_{i \in I}(JU_i)$. Now upon letting $\alpha_i : U_i \to D_i = U_i/JU_i$ be the residue class epimorphism and considering the R-homomorphism $\oplus_{i \in I}\alpha_i : \oplus_{i \in I} U_i \to \oplus_{i \in I} D_i$, by §1 we know that $\ker(\oplus_{i \in I}\alpha_i) = \oplus_{i \in I}\ker(\alpha_i)$ and $\mathrm{im}(\oplus_{i \in I}\alpha_i) = \oplus_{i \in I}\mathrm{im}(\alpha_i)$, and hence $\oplus_{i \in I}\alpha_i$ induces an R-isomorphism $(\oplus_{i \in I} U_i)/J(\oplus_{i \in I} U_i) \approx \oplus_{i \in I}(U_i/JU_i)$ which is also an (R/J)-isomorphism because both sides are annihilated by J.

To prove $(3_2^\dagger)$, by (Q17)(T80.2) there is an R-isomorphism $N \oplus M \approx R \oplus \cdots \oplus R$ (n times) where M is some R-module and n is some nonnegative integer, and now applying $(3_3^\dagger)$ twice we get an (R/J)-isomorphism $(N/JN) \oplus (M/JM) \approx (R/J)^n$. Clearly $N \approx (N/JN)$ and hence again by (Q17)(T80.2) we conclude that N is a finite projective (R/J)-module.

Given an R-module Q we want to show that:

$(7^\dagger)$
$$\begin{cases} \text{if } \widetilde{u}\tau : uQ \to uQ \text{ and } \widetilde{v}\tau : vQ \to vQ \text{ are any } R\text{-automorphisms then} \\ \text{(the internal direct sum) } \tau = \widetilde{u}\tau \oplus \widetilde{v}\tau : Q \to Q \text{ is an } R\text{-automorphism,} \\ \text{and conversely every } R\text{-automorphism of } Q \text{ is of this form.} \end{cases}$$

Namely, if $\widetilde{u}\tau : uQ \to uQ$ and $\widetilde{v}\tau : vQ \to vQ$ are any R-automorphisms then, because $Q = (uQ) \oplus (vQ)$ is an internal direct sum of R-modules, it follows that the internal direct sum $\tau = \widetilde{u}\tau \oplus \widetilde{v}\tau : Q \to Q$ is an R-automorphism. Conversely, given any R-automorphism $\tau : Q \to Q$, let $\widehat{u}\tau : uQ \to uQ$ and $\widehat{v}\tau : uQ \to vQ$ be the unique R-homomorphisms such that for all $q \in uQ$ we have $(\widehat{u}\tau)(q) = u\tau(q)$ and for all $q \in vQ$ we have $(\widehat{v}\tau)(q) = v\tau(q)$; then the injectivity of τ yields the injectivities of $\widehat{u}\tau$ and $\widehat{v}\tau$; moreover, any $q' \in uQ$ can be written as $q' = uq''$ with

$q'' \in Q$ and by the surjectivity of τ we get $q'' = \tau(q^*)$ for some $q^* \in Q$, and upon letting $q = uq^* \in uQ$, by the idempotency of u we obtain

$$u\tau(q) = u^2 q'' = uq'' = q';$$

thus $\widehat{u}\tau$ is surjective and by symmetry so is $\widehat{v}\tau$; clearly as internal direct sum we have

$$\tau = \widehat{u}\tau \oplus \widehat{v}\tau : Q \to Q.$$

Given R-homomorphisms $\sigma : P \to Q$ and $\sigma' : P \to Q$ let $\sigma_u : P_u \to Q_u$ and $\sigma'_u : P_u \to Q_u$, with $Q_u = Q/(uQ)$, be the unique (R_u)-homomorphisms such that $\sigma_u \Lambda_{u,P} = \Lambda_{u,Q}\,\sigma$ and $\sigma'_u \Lambda_{u,P} = \Lambda_{u,Q}\,\sigma'$, where $\Lambda_{u,P} : P \to P_u$ and $\Lambda_{u,Q} : Q \to Q_u$ are the residue class epimorphisms. Also let $\widehat{v}\sigma : vP \to vQ$ be the unique R-homomorphism such that for all $p \in vP$ we have $(\widehat{v}\sigma)(p) = \sigma(p)$, and let $\widehat{v}\sigma' : vP \to vQ$ be the unique R-homomorphism such that for all $p \in vP$ we have $(\widehat{v}\sigma')(p) = \sigma'(p)$. We want to show that then:

$(9^\dagger)$ $$\sigma_u \sim \sigma'_u \Leftrightarrow \widehat{v}\sigma \sim \widehat{v}\sigma'.$$

To prove this we note that

$$\sigma_u = \overline{\Lambda}_{u,Q}(\widehat{v}\sigma)(\overline{\Lambda}_{u,P})^{-1} \quad \text{and} \quad \sigma'_u = \overline{\Lambda}_{u,Q}(\widehat{v}\sigma')(\overline{\Lambda}_{u,P})^{-1}$$

where $\overline{\Lambda}_{u,P}$ is as above and $\overline{\Lambda}_{u,Q} : vQ \to Q_u$ is the unique R-isomorphism such that for all $q \in Q$ we have $\overline{\Lambda}_{u,Q}(vq) = \Lambda_{u,Q}(q)$ where $\Lambda_{u,Q} : Q \to Q_u$ is the residue class epimorphism. Moreover

$$T \mapsto T_u = \overline{\Lambda}_{u,Q}T(\overline{\Lambda}_{u,Q})^{-1}$$

gives a bijection of the set of all R-automorphisms of vQ onto the set of all (R_u)-automorphisms of Q_u. Also

$$\sigma_u = T_u\sigma'_u \Leftrightarrow \widehat{v}\sigma = T(\widehat{v}\sigma').$$

Therefore $\sigma_u \sim \sigma'_u \Leftrightarrow \widehat{v}\sigma \sim \widehat{v}\sigma'$.

FACT ($\natural$). Let R be the univariate polynomial ring $K[X]$ over a ring K, and consider R-modules $P = P_0[X]$ and $Q = Q_0[X]$ where P_0 and Q_0 are finite modules over K. Also consider an R-homomorphism $a = \sum_{0 \le i \le n} X^i a_i[X] : P \to P$ and an R-homomorphism $b = \sum_{0 \le i \le n} X^i b_i[X] : P \to Q$ where $a_i : P_0 \to P_0$ and $b_i : P_0 \to Q_0$ for $0 \le i \le n \in \mathbb{N}$ are K-homomorphisms. Assume the diagonal direct sum $a \oplus b : P \to P \oplus Q$ is a split R-monomorphism. Then there exists $d \in \mathbb{N}$ together with K-homomorphisms $\eta_i : P_0 \to P_0$ and $\zeta_i : Q_0 \to P_0$ for $0 \le i \le d$ such that for the R-homomorphisms $\eta = \sum_{0 \le i \le d} X^i \eta_i[X] : P \to P$ and $\zeta = \sum_{0 \le i \le d} X^i \zeta_i[X] : Q \to P$ we have $\eta a + \zeta b = 1_P$.

PROOF OF FACT ($\sharp$). In a moment we shall show that:

$$(1^{\bullet}) \quad \begin{cases} \text{given any } R\text{-homomorphism } \xi : U \to V \\ \text{and any nonnegative integer } d', \\ \text{where } R \text{ is the univariate polynomial ring } K[X] \text{ over a ring } K \\ \text{with } U = U_0[X] \text{ and } V = V_0[X] \\ \text{where } U_0 \text{ is finite } K\text{-module and } V_0 \text{ is some } K\text{-module,} \\ \text{we can write } \xi = \sum_{0 \le i \le d} X^i \xi_i[X] \\ \text{where } \xi_i : U_0 \to V_0 \text{ is some } K\text{-homomorphism} \\ \text{for } 0 \le i \le d \text{ with some integer } d \ge d'. \end{cases}$$

To prove ($\sharp$) using ($1^{\bullet}$), note that by item (1) we can find an R-homomorphism $c : P \oplus Q \to P$ such that $c(a \oplus b) = 1_P$. Upon letting $\eta = cc_P$ and $\zeta = cc_Q$ where $c_P : P \to P \oplus Q$ and $c_Q : Q \to P \oplus Q$ are the natural injections, we get R-homomorphisms $\eta : P \to P$ and $\zeta : Q \to P$ such that $\eta a + \zeta b = 1_P$. By applying ($1^{\bullet}$) twice, we find $d \in \mathbb{N}$ together with K-homomorphisms $\eta_i : P_0 \to P_0$ and $\zeta_i : Q_0 \to P_0$ for $0 \le i \le d$ such that $\eta = \sum_{0 \le i \le d} X^i \eta_i[X]$ and $\zeta = \sum_{0 \le i \le d} X^i \zeta_i[X]$.

To prove ($1^{\bullet}$), let $u_1, \dots, u_n$ be a finite number of K-generators of U_0; then they are also R-generators of U. We can write $\xi(u_j) = \sum_{0 \le i \le d} v_{ji} X^i$ for $1 \le j \le n$ with integer $d \ge d'$ and elements v_{ji} in V_0. We shall show that there exist unique K-homomorphisms $\xi_i : U_0 \to V_0$ for $0 \le i \le d$ such that $\xi_i(u_j) = v_{ji}$ for $1 \le j \le n$, and from this it will follow that $\xi = \sum_{0 \le i \le d} X^i \xi_i[X]$. The uniqueness is clear because $u_1, \dots, u_n$ are K-generators of U_0. To prove the existence, it suffices to show that

$$\sum_{1 \le j \le n} \lambda_j u_j = 0 \text{ with } \lambda_1, \dots, \lambda_n \text{ in } K \; \Rightarrow \; \sum_{1 \le j \le n} \lambda_j v_{ij} = 0 \text{ for } 0 \le i \le d.$$

Assuming the LHS we get $\sum_{1 \le j \le n} \lambda_j \xi(u_j) = 0$ and clearly

$$\sum_{1 \le j \le n} \lambda_j \xi(u_j) = \sum_{1 \le j \le n} \lambda_j \sum_{0 \le i \le d} v_{ji} X^i = \sum_{1 \le i \le d} X^i \sum_{1 \le j \le n} \lambda_j v_{ji}$$

and comparing coefficients of X^i we obtain $\sum_{1 \le j \le n} \lambda_j v_{ij} = 0$ for $0 \le i \le d$.

FACT ($\sharp\sharp$). Let V be a finite projective module over a univariate polynomial ring $R = K[X]$ where K is a nonnull ring. Let W be an R-module such that $W = W_0[X]$ for a finite projective K-module W_0. Assume the localized modules $V_{S(f)}, W_{S(f)}$ over the ring $R_{S(f)}$ are isomorphic where $S(f)$ is the multiplicative set in R consisting of all nonnegative powers of a polynomial $f = X^d + a_1 X^{d-1} + \dots + a_d$ with $d \in \mathbb{N}_+$ and $a_1, \dots, a_d$ in K. Then V is R-isomorphic to a submodule $\widetilde{V}$ of W with $f^m W \subset \widetilde{V}$ for some $m \in \mathbb{N}_+$.

PROOF OF FACT ($\sharp\sharp$). Every member of $S(f)$ is clearly a nonzerodivisor in R; consequently the localization map $R \to R_{S(f)}$ is injective and hence we may assume R to be a subring of $R_{S(f)}$. By (Q17)(T80.2) the localization maps $V \to V_{S(f)}$ and $W \to W_{S(f)}$ are also injective and hence we may assume V and W to be R-submodules of $V_{S(f)}$ and $W_{S(f)}$ respectively. Let $(v_i)_{1 \leq i \leq p}$ and $(w_j)_{1 \leq j \leq q}$ be finite sets of R-generators of V and W respectively, and let $\phi : V_{S(f)} \to W_{S(f)}$ be an $(R_{S(f)})$-isomorphism. Then we can find $n \in \mathbb{N}_+$ and $z_i \in W$ such that $\phi(v_i) = z_i/f^n$ for $1 \leq i \leq p$. Since ϕ is an isomorphism, $(z_i/f^n)_{1 \leq i \leq p}$ are $(R_{S(f)})$-generators of $W_{S(f)}$ and hence so are $(z_i)_{1 \leq i \leq p}$. Therefore we can find $m \in \mathbb{N}_+$ and $r_{ij} \in R$ such that $w_j = (1/f^m) \sum_{1 \leq i \leq p} r_{ij} z_i$ for $1 \leq j \leq q$. Let $\widetilde{V}$ be the R-submodule of W generated by $(z_i)_{1 \leq i \leq p}$ and consider the R-homomorphism $\psi : V \to \widetilde{V}$ given by $v \mapsto f^n \phi(v)$. Then ψ is injective because for any elements $r_1, \ldots, r_p$ in R we have: $f^n \phi(\sum_{1 \leq i \leq p} r_i v_i) = 0 \Rightarrow \phi(\sum_{1 \leq i \leq p} r_i v_i) = 0 \Rightarrow \sum_{1 \leq i \leq p} r_i v_i = 0$. It is surjective because for $1 \leq i \leq p$ we have: $f^n \phi(v_i) = f^n(z_i/f^n) = z_i$. Also $f^m W \subset \widetilde{V}$ because for $1 \leq j \leq q$ we have: $f^m w_j = \sum_{1 \leq i \leq p} r_{ij} z_i \in \widetilde{V}$.

ADDENDUM ($2^\bullet$) to FACT ($\sharp\sharp$). In ($\sharp\sharp$) we did not use the assumption that $W = W_0[X]$ with finite projective K-module W_0, but only its consequence that W is a finite projective R-module. Also we did not use the assumption that $R = K[X]$ and $f \in K[X]$ is monic of positive degree, but only its consequence that R is any ring and f is any nonzerodivisor in it.

FACT ($\sharp\sharp\sharp$). Let V be a finite projective module over a univariate polynomial ring $R = K[X]$ where K is a nonnull ring. Let W be an R-module such that $W = W_0[X]$ for a finite projective K-module W_0. Assume V is a submodule of W with $hW \subset V$ for some polynomial $h = c_0 X^n + c_1 X^{n-1} + \cdots + c_n$ with $n \in \mathbb{N}_+$ where $c_1, \ldots, c_n$ are elements in K and c_0 is a unit in K. Viewing W (and every other R-module such as V) as a K-module, consider the K-homomorphism $\mu : W_0^n \to W$ given by $y = (y_1, \ldots, y_n) \mapsto y_1 + y_2 X + \cdots + y_n X^{n-1}$ for all $y \in W_0^n$ with $y_1, \ldots, y_n$ in W_0, and consider the R-module $U = U_0[X]$ where the K-module U_0 is given by $U_0 = \mu^{-1}(V \cap \mu(W_0^n))$. By a slight generalization of the usual division algorithm discussed in L1§10(R1) we see that, for any $y \in U_0$, there exists a unique $(\alpha_0(y), \beta_0(y)) \in U_0 \times W_0$ such that $X\mu(y) = \mu(\alpha_0(y)) + h\beta_0(y)$, and this gives K-homomorphisms $\alpha_0 : U_0 \to U_0$ and $\beta_0 : U_0 \to W_0$. Consider the diagonal direct sums $\sigma_0 = (-\alpha_0) \oplus \beta_0 : U_0 \to U_0 \oplus W_0$ and $\sigma_0' = 1_{U_0} \oplus 0_{U_0, W_0} : U_0 \to U_0 \oplus W_0$ where $0_{U_0, W_0} : U_0 \to W_0$ sends every element to 0. Also consider R-homomorphisms $\sigma = \sigma_0[X] : U \to U \oplus W$ and $\sigma' = \sigma_0'[X] : U \to U \oplus W$ where we have identified $(U_0 \oplus W_0)[X]$ with $U \oplus W$ in a straightforward manner. Let $\widehat{\mu} : U \to V$ be the R-homomorphism given by $\sum y^{(i)} X^i \mapsto \sum \mu(y^{(i)}) X^i$ for every finite sum $\sum y^{(i)} X^i$ with $y^{(i)} \in U_0$. Let $\widehat{h} : W \to V$ be the R-homomorphism given by $z \mapsto hz$ for every $z \in W$ (note the condition $hW \subset V$). Let $(\widehat{\mu}, -\widehat{h}) : U \oplus W \to V$ be the R-homomorphism given by $(p, q) \mapsto \widehat{\mu}(p) - \widehat{h}(q)$ for all $(p, q) \in U \times W$. Then:

(i) U_0 is a finite projective K-module,

(ii) there exists a K-homomorphism $\eta'_0 : U_0 \oplus W_0 \to U_0$ with $\eta'_0 \sigma'_0 = 1_{U_0}$,

(iii) there exists a K-isomorphism $(U_0 \oplus W_0)/\sigma'_0(U_0) \approx W_0$, and

(iv)
$$0 \xrightarrow{\quad\quad} U \xrightarrow{\;\sigma + X\sigma'\;} U \oplus W \xrightarrow{\;(\widehat{\mu}, -\widehat{h})\;} V \xrightarrow{\quad\quad} 0$$

is a short exact sequence of R-homomorphisms which splits. So $U \oplus V \approx U \oplus W$.

PROOF OF FACT (♯♯♯). To show that U_0 is a projective K-module, first we note that since V is assumed projective as a $K[X]$-module and clearly $K[X]$ is a free K-module, it follows that V is projective as a K-module; namely by (Q17)(T80.2) we know that $V \oplus V' = V^*$ for some $K[X]$-module V' and some free $K[X]$-module V^*; now taking a $K[X]$-basis $(x_i)_{i \in I}$ of V^* and a K-basis $(y_j)_{j \in J}$ of $K[X]$ we get a K-basis $(x_i y_j)_{(i,j) \in I \times J}$ of V^* and hence again by (Q17)(T80.2) we see that V is projective as a K-module. Now $y \mapsto \mu(y)$ gives a K-monomorphism $U_0 \to V$ and, by the generalized division algorithm, every $v \in V$ can be uniquely expressed as

$(3^{\bullet})$ $\quad\quad\quad v = u + hw \quad$ with $\quad u \in \mu(U_0) \quad$ and $\quad w \in W$

and hence (as an internal direct sum of K-modules) we get $V = \mu(U_0) \oplus (hW)$, and therefore by (Q17)(T80.3) we conclude that U_0 is a projective K-module.

Note that, upon writing v, u, w as finite sums

$$v = \sum v^{(i)} X^i, \quad u = \sum u^{(i)} X^i, \quad w = \sum w^{(i)} X^i$$

with $v^{(i)}, u^{(i)}, w^{(i)}$ in W_0, the elements u and v in $(3^{\bullet})$ are the respective remainder and quotient when we divide w by h. Note that the summation for u can be taken over $0 \le i \le n-1$, i.e.,

$$u = \sum_{0 \le i \le n-1} u^{(i)} X^i.$$

Recall that $\deg_X w = \max\{i : w^{(i)} \ne 0\}$ if $w \ne 0$ and $\deg_X w = -\infty$ if $w = 0$; similarly for $\deg_X y$ for any $y = \sum y^{(i)} X^i \in U$ with $y^{(i)} \in U_0$. Since V is a finite $K[X]$-module, we can take a finite number of $K[X]$-generators $(v_j)_{1 \le j \le d}$ of V. By $(3^{\bullet})$, for $0 \le i \le n-1$ and $1 \le j \le d$ we can write

$$X^i v_j = u_{ij} + h w_{ij} \quad \text{with} \quad u_{ij} \in \mu(U_0) \quad \text{and} \quad w_{ij} \in W.$$

Any $v \in V$ can be expressed as $v = \sum_{1 \le j \le d} t_j v_j$ with $t_j \in K[X]$, and every t_j can be expressed as $t_j = r_j + h s_j$ where r_j, s_j are in $K[X]$ with the X-degree of r_j being $\le n-1$. This gives us

$$v = h\overline{w} + \sum_{0 \le i \le n-1, 1 \le j \le d} r_{ij} X^i v_j \quad \text{with} \quad \overline{w} \in W \quad \text{and} \quad r_{ij} \in K.$$

It follows that $(u_{ij})_{0 \le i \le n-1, 1 \le j \le d}$ are a finite number of K-generators of $\mu(U_0)$. Consequently $\mu(U_0)$ is a finite K-module and hence so is U_0.

This proves assertion (i). Assertions (ii) and (iii) are obvious.

To prove assertion (iv), note that because of the assumed finite projectiveness of the R-module V and in view of (Q17)(T80.2), the exactness of the sequence (iv) implies its splitting. Thus we only have to prove the exactness.

By $(3^\bullet)$ we get $(\widehat{\mu}, -\widehat{h})(U_0 \oplus W) = V$ and hence a fortiori $(\widehat{\mu}, -\widehat{h})(U \oplus W) = V$, i.e., $(\widehat{\mu}, -\widehat{h})$ is surjective.

Now by the definition of $\sigma + X\sigma'$ we see that

$$(4^\bullet) \quad \begin{cases} \text{if } Y' = sX^c \text{ with } s \in U_0 \text{ and } c \in \mathbb{N} \\ \text{then } (\sigma + X\sigma')(Y') = (p', q') \in U \times W \text{ where} \\ p' = sX^{c+1} - \alpha_0(s)X^c \text{ with } \alpha_0(s) \in U_0 \\ \text{and } q' = \beta_0(s)X^c \text{ with } \beta_0(s) \in W_0. \end{cases}$$

Applying $(4^\bullet)$ several times we conclude that

$$(5^\bullet) \quad \begin{cases} \text{if } Y = \sum_{0 \le i \le m} Y^{(i)}X^i \text{ with } Y^{(i)} \in U_0 \text{ for } 0 \le i \le m \in \mathbb{N}, \\ \text{then we have } (\sigma + X\sigma')(Y) = (p, q) \in U \times W \text{ where} \\ p = \sum_{0 \le i \le m+1}(Y^{(i-1)} - \alpha_0(Y^{(i)}))X^i \text{ with } Y^{(-1)} = Y^{(m+1)} = 0 \\ \text{and } q = \sum_{0 \le i \le m} \beta_0(Y^{(i)})X^i. \end{cases}$$

If $Y \ne 0$ then we may assume $Y^{(m)} \ne 0$ and this would make the X-degree of p to be $m + 1$ and therefore $p \ne 0$ and hence a fortiori $(p, q) \ne (0, 0)$, which yields the injectivity of $\sigma + X\sigma'$. Moreover, by taking $Y^{(i)}$ for y in the defining equation $X\mu(y) = \mu(\alpha_0(y)) + h\beta_0(y)$ we get $X\mu(Y^{(i)}) = \mu(\alpha_0(Y^{(i)})) + h\beta_0(Y^{(i)})$, and hence by the definition of $(\widehat{\mu}, -\widehat{h})$ applied to $(5^\bullet)$ we conclude that

$$(\widehat{\mu}, -\widehat{h})(p, q) = 0,$$

i.e.,

$$(\widehat{\mu}, -\widehat{h})(\sigma + X\sigma')(Y) = 0,$$

and therefore

$$(6^\bullet) \quad \operatorname{im}(\sigma + X\sigma') \subset \ker(\widehat{\mu}, -\widehat{h}).$$

It only remains to prove that $\ker(\widehat{\mu}, -\widehat{h}) \subset \operatorname{im}(\sigma + X\sigma')$, i.e.,

$$(7^\bullet) \quad \begin{cases} (p, q) \in U \times W \text{ with } \widehat{\mu}(p) = hq \\ \Rightarrow (\sigma + X\sigma')(Y) = (p, q) \text{ for some } Y \in U. \end{cases}$$

Now any $(p, q) \in U \times W$ can be uniquely written as

$$(8^\bullet) \quad \begin{cases} p = \sum_{0 \le i \le d} p^{(i)} X^i \text{ with } p^{(i)} \in U_0 \text{ for } 0 \le i \le d \in \mathbb{N} \\ \text{together with} \\ q = \sum_{0 \le i \le e} q^{(i)} X^i \text{ with } q^{(i)} \in W_0 \text{ for } 0 \le i \le e \in \mathbb{N}, \\ \text{and by the definition of } \widehat{\mu} \text{ we get} \\ \widehat{\mu}(p) = \sum_{0 \le i \le d} \mu(p^{(i)}) X^i. \end{cases}$$

Referring to $(8^\bullet)$, since $\deg_X \mu(p^{(i)}) \le n - 1$, the equation $\widehat{\mu}(p) = hq$ in $(7^\bullet)$ tells us that if $\deg_X p < 1$ then we must have $q = 0$ and hence $p = 0$ and therefore $(\sigma + X\sigma')(Y) = (p, q)$ with $Y = 0 \in U$. Thus

$$(9^\bullet) \qquad\qquad (7^\bullet) \text{ is true when } \deg_X p < 1.$$

Consequently by induction on $\deg_X p = d \ge 1$, the general case of $(7^\bullet)$ follows from $(6^\bullet)$ and the assertion which says that:

$$(10^\bullet) \quad \begin{cases} \text{given any } (p, q) \in U \times W \text{ with } \deg_X p \ge 1 \text{ and } \widehat{\mu}(p) = hq, \\ \text{there exists } (p', q') \in U \times W \\ \text{such that } (\sigma + X\sigma')(Y') = (p', q') \text{ for some } Y' \in U \\ \text{and } \deg_X (p - p') < \deg_X p. \end{cases}$$

In view of $(4^\bullet)$, by taking $Y' = p^{(d)} X^{d-1}$ we get $(10^\bullet)$.

ADDENDUM $(11^\bullet)$ to FACT (♯♯♯). $(7^\bullet)$ can be reduced to $(9^\bullet)$ by a closed formula instead of by induction. To obtain this formula heuristically, think of $\sigma + X\sigma'$ as $(X1_U - \alpha_0[X]) \oplus \beta_0[X]$ and divide p by $(X1_U - \alpha_0[X])$. To do the division, write $p = \sum_{0 \le i \le d} p^{(i)} X^i$ and $(X1_U - \alpha_0[X]) = X(1 - \alpha_0 X^{-1})$. Now expanding by the geometric series identity we get

$$p(X1_U - \alpha_0[X])^{-1} = \left(\sum_{0 \le i \le d} p^{(i)} X^i \right) \left(\sum_{0 \le i < \infty} \alpha_0^i X^{-i-1} \right)$$

and disregarding negative powers of X from this and calling the resulting expression Y' we obtain

$$Y' = \sum_{1 \le i \le d} \left[p^{(i)} X^{i-1} + \alpha_0(p^{(i)}) X^{(i-2)} + \cdots + \alpha_0^{i-2}(p^{(i)}) X + \alpha_0^{i-1}(p^{(i)}) \right].$$

At any rate, checking directly by $(4^\bullet)$ we see that $(\sigma + X\sigma')(Y') = (p', q') \in U \times W$ with $\deg_X (p - p') < 1$.

ADDENDUM (12•) to FACT (♯♯♯). Note that assertion (iv) is equivalent to saying that: (v) $\sigma + X\sigma'$ is a split R-monomorphism. Also note that:

$$\begin{cases} \text{(ii)}+\text{(v)} \\ \Rightarrow (-\alpha_0, \beta_0) \text{ is admissible} \\ \Rightarrow \sigma_0 \text{ is a split } K\text{-monomorphism} \\ \Rightarrow \sigma \text{ is a split } R\text{-monomorphism} \end{cases}$$

where the first implication uses (T118.1), the second implication uses (T117.1), and the third implication is obvious.

COMMENT (C50). [**Modules Over PIDs**]. We shall now prove the following Lemma (T119) (resp: Theorem (T120)) to take care of the first (resp: second) citation of (C50) made in the proof of Theorem (T115.2).

PID LEMMA (T119). Let K be a PID. Then, for any nonnegative integer n, every R-submodule of R^n is isomorphic to R^d for some nonnegative integer $d \leq n$. Moreover, every finite projective R-module is a finite free R-module.

PID THEOREM (T120). Let K be a PID. Let S be the multiplicative set in the the univariate polynomial ring $K[X]$ consisting of all monic polynomials. Then $K[X]_S$ is a PID.

PROOF OF (T119). In view of (Q17)(T80.2) the second assertion follows from the first assertion. To prove the first assertion, given any submodule P of R^n, by induction on $m \in \{0, \ldots, n\}$ we shall show that $P \cap R^m \approx R^{d(m)}$ for some nonnegative integer $d(m) \leq m$ where we have put

$$R^m = \{(r_1, \ldots, r_m, 0, \ldots, 0) : r_1, \ldots, r_m \text{ in } R\} \subset R^n.$$

This being obvious for $m = 0$, let $0 < m \leq n$ and assume that $P \cap R^{m-1} \approx R^{d(m-1)}$ for some nonnegative integer $d(m-1) \leq m - 1$. Clearly the set I of all $r_m \in R$ such that $(r_1, \ldots, r_m, 0 \ldots, 0) \in P$ for some $r_1, \ldots, r_{m-1}$ in R, is an ideal in R and hence, because R is a PID, we get $I = s_m R$ for some $s_m \in R$. If $s_m = 0$ then $P \cap R^m = P \cap R^{m-1} \approx R^{d(m-1)}$ and we are done. So assume $s_m \neq 0$ and take $s_1, \ldots, s_{m-1}$ in R such that $t \in P$ where $t = (s_1, \ldots, s_m, 0, \ldots, 0) \in R^n$. Now every $p \in P \cap R^m$ can be expressed as $p = q + at$ where $(q, a) \in (P \cap R^{m-1}) \times R$ is uniquely determined by p. Therefore $P \cap R^m$ equals the internal direct sum $(P \cap R^{m-1}) \oplus (tR)$ which is isomorphic to $R^{d(m-1)+1}$.

PROOF OF (T120). In Q(18.4)(7) we showed that every PID is a UFD, and in L3§8(J18) we showed that a univariate polynomial ring over a UFD is a UFD. Thus, in the present situation, since K is PID, we conclude that $K[X]$ is a UFD. Therefore our assertion that $K[X]_S$ is a PID, where S is the set of all monic polynomials in $K[X]$, will follow from the following Claims (1) to (3).

Claim (1). For every $P \in \mathrm{spec}(K[X]_S)$, the localization $(K[X]_S)_P$ is a PID.

Claim (2). If for a UFD T, the localization T_P is a PID for every $P \in \mathrm{mspec}(T)$, then T is a PID.

Claim (3). For any multiplicative set M of nonzero elements in any UFD R, the localization R_M is a UFD.

In these Claims, since we are dealing with domains, all localizations may be assumed to be subrings of the quotient fields of the domains.

To prove Claim (1), let $J = P \cap K[X]$ and $I = J \cap K$. Then $J \in \mathrm{spec}(K[X])$ with $J \cap S = \emptyset$ and $(K[X]_S)_P = K[X]_J$. By Q(18.4)(7) we know that K is of dimension at most 1 and hence either $I = 0$ or $I \in \mathrm{mspec}(K)$.

If $I = 0$ then upon letting $L = \mathrm{QF}(K)$ and $Q = JL[X]$ we have $Q \in \mathrm{spec}(L[X])$ and $K[X]_J = L[X]_Q$. Moreover, since $L[X]$ is a PID, so is $L[X]_Q$.

If $I \in \mathrm{mspec}(K)$ then let $\psi : K[X] \to \overline{K}[X]$ be the unique epimorphism with $\psi(X) = X$ such that for all $y \in K$ we have $\psi(y) = \phi(y)$ where ϕ is the residue class epimorphism of the domain K onto the field $\overline{K} = K/I$. Now $\ker(\psi) = IK[X] \subset J$ with $\psi(J) \in \mathrm{spec}(\overline{K}[X])$ and hence if $\psi(J) \neq 0$ then $\psi(J)$ contains a polynomial $g = X^d + b_1 X^{d-1} \cdots + b_d$ with $b_i \in \overline{K}$ for $1 \le i \le d \in \mathbb{N}_+$; but then taking $a_i \in K$ with $\phi(a_i) = b_i$ and letting $f = X^d + a_1 X^{d-1} \cdots + a_d$ we would get the contradiction $f \in J \cap S$. Therefore $\psi(J) = 0$, i.e., $J = IK[X]$ and hence, because I is principal, J is principal. Consequently, $K[X]_J$ is a PID (indeed it is a DVR).

To prove Claim (2), referring to (R2) and (R4) of L1§10, given any ideal N in T we can take a gcd y of N in T and then y would be a gcd of N in T_P for every $P \in \mathrm{mspec}(T)$ and hence because of the PIDness of T_P we would have $NT_P = yT_P$, and therefore by (Q15)(T65) we would get $N = yT$.

Before proving Claim (3) we shall prove the following auxiliary Claim (4). For the definition of irreducible elements and UFDs see L1§7, and for the definition of prime elements and principal prime ideals see L1§10. In analogy with a UFD, by a DFP (= Domain with Factorization into Prime elements) we mean a domain R in which every nonzero element z has a prime factorization, i.e., can be expressed as $z = uz_1 \ldots z_r$ where u is a unit in R and $z_1, \ldots, z_r$ are prime elements in R, and we call R a UFP (= domain with Unique Factorization into Prime elements) if (as in a UFD) this expression is unique upto order and associates. Note that then R is a DFP means every nonzero nonunit principal ideal I in R can be expressed as $I = P_1^{e_1} \ldots P_h^{e_h}$ where $P_1, \ldots, P_h$ are pairwise distinct nonzero principal prime ideals with positive integers $h, e_1, \ldots, e_h$, and R is a UFP means I determines the set $\{P_1, \ldots, P_h\}$ and the positive integers $e_1, \ldots, e_h$.

Claim (4). For any domain R we have the following.

(4.1) $0 \neq z \in R$ is prime $\Rightarrow z$ is irreducible.

(4.2) If R is a UFD then: $z \in R$ is irreducible $\Rightarrow z$ is prime.

(4.3) If R is a DFP then: $z \in R$ is irreducible $\Rightarrow z$ is prime.

(4.4) R is a UFD $\Leftrightarrow R$ is a UFP.

To prove Claim (4.1), let x, y be any elements in R with $z = xy$. Then, by the primeness of zR, upon relabelling x, y we may suppose that $x \in zR$, i.e., $x = zt$ for some $t \in R$. This gives $z = zty$ and hence $1 = ty$, i.e., y is a unit in R.

To prove Claim (4.2), let x, y be any elements in R with $xy \in zR$. Then by the existence of irreducible decompositions in R we can write $x = vx_1 \ldots x_s$ and $y = wy_1 \ldots y_t$ with units v, w in R and irreducible elements $x_1, \ldots, x_s, y_1, \ldots, y_t$ in R. Since irreducible z divides xy, by the uniqueness of irreducible decompositions in R we see that either z is an associate of x_i for some i or z is an associate of y_j for some j. Thus either $x \in zR$ or $y \in zR$.

To prove Claim (4.3), note that since R is a DFP, we can write $z = uz_1 \ldots z_r$ where u is a unit in R and $z_1, \ldots, z_r$ are prime elements in R. Since z is irreducible, we must have $r = 1$, and hence z is prime.

Claim (4.4) follows from Claims (4.1) to (4.3).

To prove Claim (3) note that by L4§7(T12), $P \mapsto Q = PR_M$ gives a product preserving bijection of the set of all nonzero principal prime ideals P in R with $\Gamma \cap M = \emptyset$ onto the set of all nonzero principal prime ideals Q in R_M and its inverse is given by $Q \mapsto P = Q \cap R$. Therefore: R is a UFP $\Rightarrow R_M$ is a UFP. Consequently Claim (3) follows from Claim (4.4).

COMMENT (C51). [**Elementary Row and Column Operations**]. While dealing with matrices, for instance as in the above proof of (C49)(T116.3), we frequently use elementary row and column operations.

To introduce these, referring to L3§1 and the first half of L3§12 for basic notation about matrices, consider a member $A = (A_{ij})$ of the set $\mathrm{MT}(m \times n, R)$ of all $m \times n$ matrices with entries A_{ij} in a ring R. Recall that 1_n is the $n \times n$ identity matrix. Let $T_{ij}(n, \lambda)$ be the $n \times n$ matrix over R obtained by changing the (i, j)-th entry of 1_n (from 0 or 1 according as $i \neq j$ or $i = j$) to $\lambda \in R$. For $i \neq j$ we call this a transvection matrix; more precisely, the (i, j)-th transvection matrix of order n at λ. They represent elementary row and column operations, in the sense that multiplying A from the left by $T_{ij}(m, \lambda)$ (resp: right by $T_{ij}(n, \lambda)$) amounts to adding the λ-th multiple of the j-th row to the i-th row (resp: i-th column to the j-th column) of A. Likewise we observe that multiplying A from the left by $T_{ii}(m, \lambda)$ (resp: right by $T_{ii}(n, \lambda)$) amounts to multiplying i-th row (resp: i-th column) of A by λ. We call $T_{ii}(n, \lambda)$ a dilatation matrix, or more precisely the i-th dilatation matrix of order n at λ. If λ is a unit in R then we call $T_{ii}(n, \lambda)$ a restricted dilatation matrix.

Note that, letting n to be a positive integer, for all λ in R we have

$$
(1) \qquad \det(T_{ij}(n, \lambda)) = \begin{cases} 1 & \text{if } i \neq j \\ \lambda & \text{if } i = j \end{cases}
$$

and for all λ, λ' in R we have

$$(2) \qquad\qquad T_{ij}(n,\lambda)T_{ij}(n,\lambda') = \begin{cases} T_{ij}(n,\lambda+\lambda') & \text{if } i \neq j \\ T_{ij}(n,\lambda\lambda') & \text{if } i = j. \end{cases}$$

By the subgroup of a group G generated by $H_1, H_2, \ldots, h_1, h_2, \ldots$, where $H_i \subset G$ and $h_j \in G$, we mean the set of all elements of G which can be expressed as finite products $x_1^{i_1} \ldots x_r^{i_r}$ with $i_1, \ldots, i_r$ in $\mathbb{Z}$ and $x_1, \ldots, x_n$ in $(\cup_i H_i) \cup (\cup_j \{h_j\})$. Note that this is the smallest subgroup of G containing $(\cup_i H_i) \cup (\cup_j \{h_j\})$; see L2§7(D3). Equivalently it is the intersection of all subgroups of G containing $(\cup_i H_i) \cup (\cup_j \{h_j\})$.

Recalling that $U(R)$ denotes the set of all units in R, we put

$$T_{ij}(n,R) = \{T_{ij}(n,\lambda) : \lambda \in R\} \quad \text{and} \quad T_{ii}^*(n,R) = \{T_{ii}(n,\lambda) : \lambda \in U(R)\}.$$

By (1) and (2) we see that

$$(3) \qquad \begin{cases} \text{for } i \neq j, \text{ "the mapping" } \lambda \mapsto T_{ij}(n,\lambda) \text{ gives a homomorphism} \\ \text{of the additive group } R^+ \text{ into the multiplicative group SL}(n,R) \\ \text{with image } T_{ij}(n,R) \end{cases}$$

and

$$(4) \qquad \begin{cases} \text{the mapping } \lambda \mapsto T_{ii}(n,\lambda) \text{ gives} \\ \text{a homomorphism } U(R) \to \text{GL}(n,R) \text{ of multiplicative groups} \\ \text{with image } T_{ii}^*(n,R). \end{cases}$$

We put

$$\text{SE}(n,R) = \begin{cases} \text{the subgroup of SL}(n,R) \text{ generated by} \\ \bigcup_{i \neq j \text{ in } \{1,\ldots,n\}} T_{ij}(n,R) \end{cases}$$

and

$$\text{GE}(n,R) = \begin{cases} \text{the subgroup of GL}(n,R) \text{ generated by} \\ \text{SE}(n,R) \cup \left(\bigcup_{1 \leq i \leq n} T_{ii}^*(n,R) \right) \end{cases}$$

and we respectively call these the n-dimensional special elementary group over R and the n-dimensional general elementary group over R.

In the following Theorem we shall show that if R is a field then $\text{SL}(n,R)$ is generated by transvection matrices, and $\text{GL}(n,R)$ is generated by transvection matrices together with restricted dilatation matrices. In Comment (C54) below we shall prove that the same thing is true if $n \geq 3$ and R is a finite variable polynomial ring over a field.

TRANSVECTION THEOREM (T121). For any field k and any positive integer n we have $\text{SL}(n,k) = \text{SE}(n,k)$ and $\text{GL}(n,k) = \text{GE}(n,k)$.

PROOF. The second assertion follows from the first assertion by noting that for any $D \in \mathrm{GL}(n,k)$ we have $T_{11}(\mu)D \in \mathrm{SL}(n,k)$ with $\mu = \det(D)^{-1}$. To prove the first assertion let any $E \in \mathrm{SL}(n,k)$ be given. By induction on l, with $0 \le l \le n$, we shall find products B and C of transvection matrices such that $BEC \in J(l)$ where

$$\begin{cases} J(l) = \{\widehat{E} = (\widehat{E}_{ij}) \in \mathrm{SL}(n,k) : \widehat{E}_{ii} - 1 = \widehat{E}_{ij} = 0 = \widehat{E}_{ji} \\ \qquad \text{for } 1 \le i \le l \text{ and } 1 \le j \le n \text{ with } j \ne i\}. \end{cases}$$

For $l = 0$ we can take $\widehat{E} = E$. Now let $1 \le l \le n$ and suppose we have found products B^* and C^* of transvection matrices such that for $E^* = (E^*_{ij}) = B^*EC^*$ we have $E^* \in J(l-1)$. Let us show that then there exist products B' and C' of transvection matrices such that $E' = (E'_{ij}) = B'EC' \in J(l-1)$ with $E'_{ll} = 1$. If $l = n$ then $E^*_{nn} = \det(E^*) = 1$ and hence $E^* =$ the identity matrix, and by taking $B' = B^*$ and $C' = C^*$ we get $E' = B'EC' \in J(l-1)$ with $E'_{ll} = 1$. If $l \ne n$ and $E^*_{jl} \ne 0$ for some $j > l$ then by taking $B' = T_{lj}(\lambda)B^*$ and $C' = C^*$, where $\lambda \in k$ with $E^*_{ll} + \lambda E^*_{jl} = 1$, we get $E' = B'EC' \in J(l-1)$ with $E'_{ll} = 1$. If $l \ne n$ and $E^*_{jl} = 0$ for all $j > l$ but $E^*_{li} \ne 0$ for some $i > l$ then by taking $B' = B^*$ and $C' = C^*T_{il}(\lambda)$, where $\lambda \in k$ with $E^*_{ll} + \lambda E^*_{li} = 1$, we get $E' = B'EC' \in J(l-1)$ with $E'_{ll} = 1$. If $l \ne n$ and $E^*_{jl} = 0$ for all $j > l$ and $E^*_{li} = 0$ for all $i > l$ then, since $\det(E^*) = 1$, we must have $E^*_{j'i'} \ne 0$ for some $j' > l$ and $i' > l$ and by taking $B' = T_{lj'}(1)B^*$ and $C' = C^*T_{i'l}(\lambda)$, where $\lambda \in k$ with $E^*_{ll} + \lambda E^*_{j'i'} = 1$, we get $E' = B'EC' \in J(l-1)$ with $E'_{ll} = 1$. Thus in all the cases we have found products B' and C' of transvection matrices such that $E' = B'EC' \in J(l-1)$ with $E'_{ll} = 1$. Now by letting

$$B = T_{l+1,l}(-E'_{l+1,l}) \ldots T_{n,l}(-E'_{n,l})B'$$

and

$$C = C'T_{l,l+1}(-E'_{l,l+1}) \ldots T_{l,n}(-E'_{l,n})$$

we get products B and C of transvection matrices such that $BEC \in J(l)$, which completes the induction on l. Taking $l = n$ we have found products B and C of transvection matrices such that $BEC \in J(n)$; clearly the inverse of a transvection matrix is a transvection matrix and $J(n)$ only contains the identity matrix; therefore E is a product of transvection matrices. Thus $\mathrm{SL}(n,k) = \mathrm{SE}(n,k)$.

COMMENT (C52). [**Completing Unimodular Rows**]. Elementary row and column operations can be used for completing unimodular tuples, by which we mean the following.

For

$$\begin{cases} \text{a nonnull ring } R \text{ and positive integer } n \text{ ,} \\ \text{an } n\text{-tuple } r = (r_1, \ldots, r_n) \in R^n \end{cases}$$

is said to be unimodular if the elements $r_1, \ldots, r_n$ generate the unit ideal in R. By $U_n(R)$ we denote the set of all such r, and we note that $r \mapsto r_1$ gives a bijection $U_1(R) \to U(R)$.

By B' we denote the transpose of a matrix B. In particular, thinking of r as a row vector, r' is a column vector, i.e., an $n \times 1$ matrix over R. By $U_n(R)'$ we denote the set of all r' with r varying in $U_n(R)$.

The n-tuple r is said to be H-completable, where H is a subset of $\mathrm{GL}(n, R)$, if r' is the first column of some member $A = (A_{ij})$ of H, i.e., if $A_{i1} = r_i$ for $1 \le i \le n$. Note that, this is equivalent to saying that for

the "row vectors" r and $e_1 = (1, 0, \ldots, 0)$ of length n

we have $r' = Ae_1'$, i.e., $\begin{pmatrix} r_1 \\ r_2 \\ \vdots \\ r_n \end{pmatrix} = A \begin{pmatrix} 1 \\ 0 \\ \vdots \\ 0 \end{pmatrix}$. More generally, for $1 \le i \le n$ we put

$$e_i = (0, \ldots, 0, 1, 0 \ldots, 0) \in R^n \text{ with 1 in the } i\text{-th spot}.$$

Note that $e_1, \ldots, e_n$ is an R-basis of $R^n = \mathrm{MT}(1 \times n, R)$, and $e_1', \ldots, e_n'$ is an R-basis of $\mathrm{MT}(n \times 1, R)$.

We say that r is completable if it is $\mathrm{GL}(n, R)$-completable. Given any $s \in R^n$ and any $H \subset \mathrm{GL}(n, R)$, we say that r is H-equivalent to s, in symbols $r \sim_H s$, to mean that $r' = As'$ for some $A \in H$. Thus r is H-completable means $r \sim_H e_1$. In particular, r is completable means $r \sim_{\mathrm{GL}(n,R)} e_1$.

If r' is the first column of some $n \times n$ matrix A over R and for some maximal ideal P in R we have $r_i \in P$ for $1 \le i \le n$, then applying the residue class epimorphism $\phi : R \to R/P$ we get $\phi(\det(A)) = \det((\phi(A_{ij})) = 0$ and hence $\det(A) \in P$ and therefore $A \notin \mathrm{GL}(n, R)$. Thus by Zorn's Lemma we see that

(1)   if r is completable then r is unimodular.

Let us observe that

(2) $\begin{cases} \text{for } n = 2 \text{ we have } \det\left(\begin{smallmatrix} r_1 & s_1 \\ r_2 & s_2 \end{smallmatrix}\right) = s_2 r_1 - s_1 r_2 \\ \text{and hence } r \text{ is } \mathrm{SL}(2, R)\text{-completable iff } r \in U_2(R). \end{cases}$

In case of $n \ge 2$ we get a monomorphism of $\mathrm{SL}(2, R)$ into $\mathrm{SL}(n, R)$ by sending every B in $\mathrm{SL}(2, R)$ to the block matrix $\left(\begin{smallmatrix} B & 0_{2 \times (n-2)} \\ 0_{(n-2) \times 2} & 1_{n-2} \end{smallmatrix}\right)$ with obvious notation; see L4§12(R6)(19). We call this monomorphism canonical and we denote its image by $\mathrm{SL}([2, n], R)$. In case of $n = 1$ we let $\mathrm{SL}([2, n], R)$ denote the identity subgroup of $\mathrm{SL}(n, R)$.

Reverting to n being any positive integer, we put

$$\mathrm{QSEP}(n, R) = \{AB : A \in \mathrm{SL}([2, n], R) \text{ and } B \in \mathrm{SE}(n, R)\}$$

and

$$QSE(n, R) = \text{the subgroup of } SL(n, R) \text{ generated by } QSEP(n, R)$$

and we respectively call these the n-dimensional quasielementary pregroup over R, and the n-dimensional quasielementary group over R. Note that, as defined in (C51), $SE(n, R)$ is the subgroup of $SL(n, R)$ generated by the transvection matrices $T_{ij}(n, \lambda)$ with $\lambda \in R$ and $i \neq j$ in $\{1, \ldots, n\}$.

In the rest of this Comment we may use the abbreviations

$(\bullet)$ $\qquad G(R) = GL(n, R)$ and $Q(R) = QSE(n, R)$ and $E(R) = SE(n, R)$

and we observe that

$(\bullet\bullet)$ $\qquad\qquad\qquad$ if $n \leq 2$ then $Q(R) = SL(n, R)$.

Note that if R is the univariate polynomial ring $K[X]$ over a nonnull ring K, then $r_i = r_i(X)$ is a polynomial in X with coefficients in K, and for any $a \in K$ we have

$$r(a) = (r_1(a), \ldots, r_n(a)) \in K^n \subset R^n$$

and more generally, for any K-algebra L and any $b \in L$, in a natural manner we have

$$r(b) = (r_1(b), \ldots, r_n(b)) \in L^n.$$

Similarly, if R is a finite variable polynomial ring $K[X_1, \ldots, X_N]$ over a nonnull ring K, then $r_i = r_i(X_1, \ldots, X_N)$ is a polynomial in $X_1, \ldots, X_N$ with coefficients in K, and for any $a \in K$ we have

$$\begin{cases} r(X_1, \ldots, X_{N-1}, a) \\ = (r_1(X_1, \ldots, X_{N-1}, a), \ldots, r_n(X_1, \ldots, X_{N-1}, a)) \\ \in K[X_1, \ldots, X_{N-1}]^n \subset R^n \end{cases}$$

and more generally, for any K-algebra L and any $b \in L$, in a natural manner we have

$$\begin{cases} r(X_1, \ldots, X_{N-1}, b) \\ = (r_1(X_1, \ldots, X_{N-1}, b), \ldots, r_n(X_1, \ldots, X_{N-1}, b)) \\ \in L[X_1, \ldots, X_{N-1}]^n \end{cases}$$

and so on.

When R is the univariate polynomial ring $K[X]$ over a nonnull ring K and L is a nonnull K-algebra, we consider the set

$$J_{L/K}(r) = \{a \in K : \text{ for all } b, c \text{ in } L \text{ with } b - c \in aL \text{ we have } r(b) \sim_{Q(L)} r(c)\}$$

where aL is the ideal in L given by $aL = \{al : l \in L\}$, and we call $J_{L/K}(r)$ the (L/K)-invariant-ideal of r.

Concerning the above concepts we shall now prove the following results (T122) to (T126). Recall that a submonic polynomial in the univariate polynomial ring $K[X]$ over a nonnull ring K is a nonzero polynomial with leading coefficient a unit in K, i.e., an expression of the form $a_0 X^d + a_1 X^{d-1} + \cdots + a_d$ where $d \in \mathbb{N}$ with $a_0 \in U(K)$ and elements $a_1, \ldots, a_d$ in K; note that d is the degree of the polynomial.

PID UNIMODULARITY LEMMA (T122). Assume that R is a PID with $n \geq 3$. Then we have the following.

(T122.1) If $r \in U_n(R)$ with $r_1 \neq 0$ then there exist elements $t_3, \ldots, t_n$ in R such that $(r_1, r_2 + t_3 r_3 + \cdots + t_n r_n) \in U_2(R)$.

(T122.2) The special elementary group $E(R)$ acts transitively on $U_n(R)'$, i.e., for all r, s in $U_n(R)$ we have $r \sim_{E(R)} s$.

INVARIANT IDEAL LEMMA (T123). Assume that the ring R is the univariate polynomial ring $K[X]$ over a nonnull ring K, let L be a nonnull K-algebra, and recall that let $r \in R^n$ with positive integer n. Then we have the following.

(T123.1) $J_{L/K}(r)$ is an ideal in K.

(T123.2) For every $A \in Q(L)$ and $s \in R^n$ with $r' = As'$ we have the equation $J_{L/K}(r) = J_{L/K}(s)$.

(T123.3) Assume that $n \geq 2$ and let $a \in K$ be such that $a = t_1 r_1 + t_2 r_2$ for some t_1, t_2 in R. Then $a \in J_{L/K}(r)$.

(T123.4) Assume that $n \geq 3$ and let $r \in U_n(R)$ be such that r_1 is submonic. Then for all b, c in L we have $r(b) \sim_{Q(L)} r(c)$.

COMPLETING UNIMODULAR ROWS LEMMA (T124). For any nonnull ring R and positive integer n, we have the following.

(T124.1) If $n \leq 2$ then every $r \in U_n(R)$ is $G(R)$-completable. If $n = 2$ then every $r \in U_n(R)$ is $Q(R)$-completable.

(T124.2) Assume that R is the univariate polynomial ring $K[X]$ over a nonnull ring K with $n \geq 3$, and let there be given any $r \in U_n(R)$ be such that r_1 is submonic. Then $r(X) \sim_{Q(R)} r(0)$.

(T124.3) Assume that R is the finite variable polynomial ring $K[X_1, \ldots, X_N]$ over a field K. Then $G(R)$ acts transitively on $U_n(R)'$, i.e., equivalently, every $r \in U_n(R)$ is $G(R)$-completable. Moreover, if $n \geq 2$, then $Q(R)$ acts transitively on $U_n(R)'$, i.e., equivalently, every $r \in U_n(R)$ is $Q(R)$-completable.

QUASIELEMENTARY GROUP LEMMA (T125). For any nonnullring R and any integer $n \geq 3$, $SE(n, R)$ is a normal subgroup of $GL(n, R)$ and hence we have $QSE(n, R) = QSEP(n, R)$.

COMPLETING UNIMODULAR ROWS THEOREM (T126). Assume that R is the finite variable polynomial ring $K[X_1, \ldots, X_N]$ over a field K, and we have $n \geq 3$. Then the special elementary group $E(R)$ acts transitively on $U_n(R)'$.

PROOF OF (T122.1). We can take elements $u, u_3, \ldots, u_n$ in R for which we have $(r_3, \ldots, r_n)R = uR$ and $u = u_3 r_3 + \cdots + u_n r_n$. Now $(r_1, r_2, u) \in U_3(R)$ and hence if $r_2 = 0$ then it suffices to take $t_i = u_i$ for $3 \leq i \leq n$. So assume that $r_2 \neq 0$. Referring to the proof of (C50)(T120), we can write

$$r_1 R = P_1^{e_1} \ldots P_h^{e_h} P_{h+1}^{e_{h+1}} \ldots P_k^{e_k} \quad \text{and} \quad r_2 R = P_1^{e_1'} \ldots P_h^{e_h'} P_{k+1}^{e_{k+1}} \ldots P_l^{e_l}$$

where $P_1, \ldots, P_l$ are pairwise distinct nonzero principal prime ideals in R and $e_1, \ldots, e_l, e_1', \ldots, e_h'$ are positive integers. We can also write $P_i = p_i R$ with $p_i \in R$ for $1 \leq i \leq l$. Since $(r_1, r_2, u) \in U_3(R)$, we must have $u \notin P_i$ for $1 \leq i \leq h$, and hence $(r_1, r_2 + up_{h+1} \ldots p_k) \in U_2(R)$. Thus, upon letting $t_i = u_i p_{h+1} \ldots p_k$ for $3 \leq i \leq n$, we get $(r_1, r_2 + t_3 r_3 + \cdots + t_n r_n) \in U_2(R)$.

PROOF OF (T122.2). If $r_1 = 0$ then $r_j \neq 0$ for some $j \in \{1, \ldots, n\}$ and we let $B = T_{1,j}(n, 1) \in E(R)$ and $u = (u_1, \ldots, u_n) \in U_n(R)$ where $u_1 = r_j$ with $u_i = r_i$ for $2 \leq i \leq n$; if $r_1 \neq 0$ then we let $B = 1_n \in E(R)$ and $u = r \in U_n(R)$. In both the cases we have $u' = Br'$ with $u_1 \neq 0$. By (T122.1) there exist elements $t_3, \ldots, t_n$ in R such that upon letting $v_1 = u_1$ and $v_2 = u_2 + t_3 u_3 + \cdots + t_n u_n$ with $v_i = u_i$ for $3 \leq i \leq n$ we have $v = (v_1, \ldots, v_n) \in R^n$ with $(v_1, v_2) \in U_2(R)$, i.e, $\lambda_1 v_1 + \lambda_2 v_2 = 1$ for some λ_1, λ_2 in R. Now upon letting

$$C = T_{23}(n, t_3) \ldots T_{2n}(n, t_n)$$

we get $C \in E(R)$ with $v' = Cu'$. Moreover, upon letting

$$D = T_{n1}(n, \lambda_1 - \lambda_1 v_n) T_{n2}(n, \lambda_2 - \lambda_2 v_n)$$

we get $D \in E(R)$ with $w' = Dv'$ where $w = (w_1, \ldots, w_n) \in R^n$ with $w_n = 1$ and $w_i = v_i$ for $1 \leq i \leq n - 1$. Finally, upon letting

$$E = T_{1,n}(n, -w_1) \ldots T_{n-1,n}(n, -w_{n-1})$$

we get $e_n' = Ew'$ where $e_n = (0, \ldots, 0, 1) \in R^n$. Thus upon letting $F = EDCB$ we have $F \in E(R)$ with $e_n' = Fr'$. Similarly $e_n' = \widehat{F}s'$ for some $\widehat{F} \in E(R)$. Now $r' = As'$ where $A = F^{-1}\widehat{F} \in E(R)$, and hence $r \sim_{E(R)} s$.

PROOF OF (T123.1). Clearly $0 \in J_{L/K}(r)$. Moreover, for all $a \in J_{L/K}(r)$ with $\kappa \in K$, and for all b, c in L with $b - c \in (\kappa a)L$, we have $b - c \in aL$ and hence $r(b) \sim_{Q(L)} r(c)$; thus $\kappa a \in J_{L/K}(r)$. Likewise, for all a_1, a_2 in $J_{L/K}(r)$ and for all b, c in L with $b - c$ in $(a_1 + a_2)L$, we have $b - \lambda a_1 = c + \lambda a_2$ for some $\lambda \in L$; since $a_1 \in J_{L/K}(r)$, we get $r'(b) = Br'(b - \lambda a_1)$ for some $B \in Q(R)$, and since $a_2 \in J_{L/K}(r)$, we get $r'(c + \lambda a_2) = Cr'(c)$ for some $C \in Q(R)$; this yields $r'(b) = (BC)r'(c)$ with $BC \in Q(L)$, and hence $a_1 + a_2 \in J_{L/K}(r)$.

PROOF OF (T123.2). For all $a \in J_{L/K}(r)$ and b, c in L with $b - c \in aL$ we have $r(b) = Br(c)$ for some $B \in Q(L)$. Clearly $s(b) = (A^{-1}BA)s(c)$ with $A^{-1}BA \in Q(L)$. Thus $a \in J_{L/K}(s)$. Consequently $J_{L/K}(r) \subset J_{L/K}(s)$. Similarly $J_{L/K}(s) \subset J_{L/K}(r)$. Therefore $J_{L/K}(r) = J_{L/K}(s)$.

PROOF OF (T123.3). Recall that

$$(3) \qquad\qquad a = t_1 r_1 + t_2 r_2.$$

Given any b, c in L with

$$(4) \qquad\qquad b = c + ad \quad \text{with} \quad d \in L$$

we want to show that $r'(b) = Ar'(c)$ for some $A \in Q(R)$. Using the grade-school identity

$$(5) \qquad X^{m+1} - Y^{m+1} = (X - Y)(X^m + X^{m-1}Y + \cdots + XY^{m-1} + Y^m)$$

and calculating in the trivariate polynomials ring $K[X, Y, Z]$ we find members $\rho_i(X, Y, Z), \tau_i(X, Y, Z)$ of that ring such that for $i = 1, 2$ we have

$$(6) \qquad \begin{cases} r_i(X + YZ) = r_i(X) + Z\rho_i(X, Y, Z) \\ \text{and} \\ t_i(X + YZ) = t_i(X) + Z\tau_i(X, Y, Z). \end{cases}$$

Consider the 2×2 matrix $\beta(X, Y, Z)$ over $K[X, Y, Z]$ given by

$$\beta(X, Y, Z) = \begin{pmatrix} 1 + \rho_1(X,Y,Z)t_1(X) + \tau_2(X,Y,Z)r_2(X) & \rho_1(X,Y,Z)t_2(X) - \tau_2(X,Y,Z)r_1(X) \\ \rho_2(X,Y,Z)t_1(X) - \tau_1(X,Y,Z)r_2(X) & 1 + \rho_2(X,Y,Z)t_2(X) + \tau_1(X,Y,Z)r_1(X) \end{pmatrix}$$

and let B be the 2×2 matrix over L obtained by putting $(X, Y, Z) = (c, d, a)$ in $\beta(X, Y, Z)$, i.e.,

$$B = \beta(c, d, a).$$

In a moment we shall establish the Claim saying that

$$\det(B) = 1 \quad \text{and} \quad B\begin{pmatrix} r_1(c) \\ r_2(c) \end{pmatrix} = \begin{pmatrix} r_1(b) \\ r_2(b) \end{pmatrix}.$$

Assuming this let us complete the proof. By the Claim we get $r(b) = \widehat{B}\widehat{r}$, where $\widehat{B}$ is the image of B under the canonical monomorphism of $SL(2, L)$ into $SL(n, L)$, and $\widehat{r} = (\widehat{r}_{i1})$ is the $n \times 1$ matrix over L given by $\widehat{r}_i = r_i(b)$ or $r_i(c)$ according as $i \leq 2$ or $i > 2$. By (4) and (5) we have

$$r_i(c) = r_i(b) + \lambda_i a \quad \text{with} \quad \lambda_i \in L \quad \text{for} \quad 3 \leq i \leq n.$$

Putting $X = c$ in the assumed identity $a = t_1r_1 + t_2r_2$ we get

$$a = t_1(c)r_1(c) + t_2(c)r_2(c).$$

By the above two displays we obtain

$$r_i(c) = r_i(b) + \lambda_i[t_1(c)r_1(c) + t_2(c)r_2(c)] \quad \text{for} \quad 3 \le i \le n.$$

Therefore, upon letting

$$D = T_{31}(-\lambda_3 t_1(c))T_{32}(-\lambda_3 t_2(c))\dots T_{n1}(-\lambda_n t_1(c))T_{n2}(-\lambda_n t_2(c)),$$

we see that

$$\widehat{r} = Dr'(c) \quad \text{with} \quad D \in \mathrm{SE}(n, L)$$

and hence upon letting $A = \widehat{B}D$ we conclude that

$$A \in Q(L) \quad \text{with} \quad r'(b) = Ar'(c).$$

Multiplying the 2×2 matrix $\beta(X, Y, Z)$ on the right by the row $\left(\begin{smallmatrix} r_1(X) \\ r_2(X) \end{smallmatrix} \right)$ and using (3) we obtain

$$\beta(X, Y, Z) \left(\begin{matrix} r_1(X) \\ r_2(X) \end{matrix} \right) = \left(\begin{matrix} r_1(X)+a\rho_1(X,Y,Z) \\ r_2(X)+a\rho_2(X,Y,Z) \end{matrix} \right)$$

and putting $(X, Y, Z) = (c, d, a)$ in this we get $B \left(\begin{smallmatrix} r_1(c) \\ r_2(c) \end{smallmatrix} \right) = \left(\begin{smallmatrix} r_1(b) \\ r_2(b) \end{smallmatrix} \right)$ which proves the second half of the Claim.

It only remains to show that $\det(B) = 1$. Letting

$$t_1, t_2, r_1, r_2, \tau_1, \tau_2, \rho_1, \rho_2, \beta$$

respectively stand for

$$t_1(X), \dots, r_2(X), \tau_1(X, Y, Z), \dots, \beta(X, Y, Z)$$

and calculating by the determinant rule $\det \left(\begin{smallmatrix} M_1 & N_1 \\ M_2 & N_2 \end{smallmatrix} \right) = M_1N_2 - M_2N_1$ we get

$$\det(\beta) = 1 + (t_1\rho_1 + r_1\tau_1 + t_2\rho_2 + r_2\tau_2) + (t_1r_1 + t_2r_2)(\tau_1\rho_1 + \tau_2\rho_2)$$

and hence in view of (3) we obtain

(7) $$\det(\beta) = 1 + \epsilon + a\delta$$

where

(8) $$\epsilon = t_1\rho_1 + r_1\tau_1 + t_2\rho_2 + r_2\tau_2 \quad \text{and} \quad \delta = \tau_1\rho_1 + \tau_2\rho_2.$$

By (6) and (8) we get

$$[t_1(X + YZ)r_1(X + YZ) + t_2(X + YZ)r_2(X + YZ)] - [t_1r_1 + t_2r_2] = Z(\epsilon + Z\delta)$$

where each of the bracketed quantity on the LHS is reduced to a by (3), and hence the LHS and therefore the RHS equals 0; dividing the resulting polynomial identity in $k[X, Y, Z]$ by Z conclude that

$$(9) \qquad\qquad \epsilon + Z\delta = 0.$$

Putting $(X, Y, Z) = (c, d, a)$ in (7) and (9) we get $\det(B) = 1$.

PROOF OF (T123.4). In view of (T123.1), given any $P \in \mathrm{mspec}(K)$, it suffices to find $a \in J_{L/K}(r)$ with $a \notin P$. Let ϕ be the residue class epimorphism of K onto the field $\overline{K} = K/P$. Let ψ be the unique epimorphism of $R = K[X]$ onto $\overline{R} = \overline{K}[X]$ with $\psi(X) = X$ such that $\psi(z) = \phi(z)$ for all $z \in K$. Then $\overline{R}$ is a PID with $(\psi(r_1), \ldots, \psi(r_n)) \in U_n(\overline{R})$, and hence by (T122.1) we can find $t_3, \ldots, t_n$ in R such that upon letting $s_1 = r_1$ and $s_2 = r_2 + t_3 r_3 + \cdots + t_n r_n$ with $s_i = r_i$ for $3 \leq i \leq n$ we have $s = (s_1, \ldots, s_n) \in R^n$ with $(\psi(s_1), \psi(s_2)) \in U_2(\overline{R})$, i.e, $\psi(\lambda_1 s_1 + \lambda_2 s_2) = 1$ for some λ_1, λ_2 in R. By (T123.2) we have $J_{L/K}(r) = J_{L/K}(s)$ and hence it suffices to find $a \in J_{L/K}(s)$ with $a \notin P$. Since $s_1 = r_1$, by assumption

$$s_1 = f = f(X) = a_0 X^d + a_1 X^{d-1} + \cdots + a_d$$

where $d \in \mathbb{N}$ with $a_0 \in U(K)$ and elements $a_1, \ldots, a_d$ in K. Also clearly

$$s_2 = g = g(X) = b_0 X^e + b_1 X^{e-1} + \cdots + b_e$$

where $e \in \mathbb{N}$ with elements $b_0, b_1, \ldots, b_e$ in K. Let

$$a = \mathrm{Res}_X(f, g) \in K.$$

Then by L4§1(T1), which was proved in L4§10(E7) and reproved in L4§12(R6)(34), we get $\phi(a) \neq 0$, i.e., $a \notin P$. By L4§12(R5)(5.1), also proved in L4§12(R6)(6.1), we have $a = t_1 s_1 + t_2 s_2$ for some t_1, t_2 in R, and hence by (T123.3) we get $a \in J_{L/K}(s)$.

PROOF OF (T124.1). For $n = 1$ the assertion is obvious, and for $n = 2$ it follows from (2).

PROOF OF (T124.2). In (T123.4) take $L = R$ and $(b, c) = (X, 0)$.

PROOF OF (T124.3). In view of (T124.1) we may suppose that $n \geq 3$. Let us make induction on N. For $N \leq 1$ our assertion follows from (T122.2). So let $N > 1$ and assume for $N - 1$. If $r_1 = 0$ then $r_j \neq 0$ for some $j \in \{1, \ldots, n\}$ and we let $B = T_{1,j}(n, 1) \in E(R)$ and $u = (u_1, \ldots, u_n) \in U_n(R)$ where $u_1 = r_j$ with $u_i = r_i$ for $2 \leq i \leq n$; if $r_1 \neq 0$ then we let $B = 1_n \in E(R)$ and $u = r \in U_n(R)$. In both the cases we have $u' = Br'$ with $u_1 \neq 0$. In view of of L3§12(4), by letting R undergo a K-automorphism, we can arrange matters so that u_1 is a submonic polynomial in X_N with coefficients in $\widehat{R} = K[X_1, \ldots, X_{N-1}]$. Let $\widehat{r}(X_N) = (\widehat{r}_1(X_N), \ldots, \widehat{r}_n(X_N))$ where $\widehat{r}_i(X_N)$ equals u_i considered as a member of $\widehat{R}[X_N]$. Then by (T124.2) we get

$\widehat{r}(X_N) \sim_{Q(R)} \widehat{r}(0)$. Clearly $\widehat{r}(0) \in U_n(\widehat{R})$ and hence by the induction hypothesis $\widehat{r}(0) \sim_{Q(\widehat{R})} e_1$ with $e_1 = (1, 0 \ldots, 0) \in R^n$. It follows that $r \sim_{Q(R)} e_1$, i.e., r is $Q(R)$-completable. This completes the induction.

PROOF OF (T125). We are given a nonnull ring R with positive integer n, and we want to show that

(†) $$n \geq 3 \Rightarrow \mathrm{SE}(n, R) \triangleleft \mathrm{GL}(n, R)$$

where we recall that $\leq$ and $\triangleleft$ denote subgroup and normal subgroup respectively. In view of (†), our second assertion that $n \geq 3 \Rightarrow \mathrm{QSE}(n, R) = \mathrm{QSEP}(n, R)$ follows by taking

$$(N, M, L) = (\mathrm{SL}([2, n], R), \mathrm{SE}(n, R), \mathrm{GL}(n, R))$$

in the Claim which says that:

(††) $$\begin{cases} N \leq L \text{ and } M \triangleleft L \text{ where } L \text{ is any group} \\ \Rightarrow \text{ every element in the subgroup of } L \text{ generated by } N \text{ and } M \\ \text{can be expressed as } \nu\mu \text{ with } \nu \in N \text{ and } \mu \in M. \end{cases}$$

To see this it suffices to note that $\mu\nu = \nu\widehat{\mu}$ with $\widehat{\mu} = \nu^{-1}\mu\nu \in M$.

We now proceed to prove (†) in several steps. First note that

(10) $$\begin{cases} \mathrm{SE}(n, R) \text{ contains every } n \times n \\ \text{uniupper triangular matrix } C = (C_{ij}) \text{ over } R, \\ \text{i.e., for which } C_{ii} = 1 \text{ for all } i \text{ and } C_{ij} = 0 \text{ for all } i > j \\ \text{and } \mathrm{SE}(n, R) \text{ also contains every } n \times n \\ \text{unilower triangular matrix } D = (D_{ij}) \text{ over } R, \\ \text{i.e., for which } D_{ii} = 1 \text{ for all } i \text{ and } D_{ij} = 0 \text{ for all } i < j. \end{cases}$$

Namely, assuming $n > 1$, we have

$$C = C_{n-1} \ldots C_2 C_1 \quad \text{with} \quad C_i = T_{i,i+1}(n, C_{i,i+1}) \ldots T_{i,n}(n, C_{i,n})$$

and

$$D = D_2 D_3 \ldots D_n \quad \text{with} \quad D_i = T_{i,1}(n, D_{i,1}) \ldots T_{i,i-1}(n, D_{i,i-1}).$$

Next we claim the following:

$$(11) \quad \begin{cases} \text{Let } A \in \mathrm{MT}(n \times m, R) \text{ and } B \in \mathrm{MT}(m \times n, R) \text{ with } m \in \mathbb{N}_+ \\ \text{be such that } 1_n + AB \in \mathrm{GL}(n, R). \\ \text{Then } 1_m + BA \in \mathrm{GL}(m, R) \\ \text{with } (1_m + BA)^{-1} = 1_m - B(1_n + AB)^{-1}A \\ \text{and we have } \begin{pmatrix} 1_n + AB & 0_{n,m} \\ 0_{m,n} & (1_m + BA)^{-1} \end{pmatrix} \in \mathrm{SE}(n+m, R). \end{cases}$$

To see this, first note that

$$\begin{cases} [1_m - B(1_n + AB)^{-1}A](1_m + BA) \\ = [1_m + BA] - B(1_n + AB)^{-1}(A + ABA) \\ = [1_m + BA] - B(1_n + AB)^{-1}(1_n + AB)A \\ = [1_m + BA] - BA \\ = 1_m \end{cases}$$

and hence

$$1_m + BA \in \mathrm{GL}(m, R) \quad \text{with} \quad (1_m + BA)^{-1} = 1_m - B(1_n + AB)^{-1}A.$$

Moreover, calculating with block matrices we have

$$\begin{cases} \begin{pmatrix} 1_n + AB & 0_{n,m} \\ 0_{m,n} & (1_m + BA)^{-1} \end{pmatrix} \\ = \begin{pmatrix} 1_n & 0_{n,m} \\ (1_m + BA)^{-1}B & 1_m \end{pmatrix} \begin{pmatrix} 1_n & -A \\ 0_{m,n} & 1_m \end{pmatrix} \begin{pmatrix} 1_n & 0_{n,m} \\ -B & 1_m \end{pmatrix} \begin{pmatrix} 1_n & (1_n + AB)^{-1}A \\ 0_{m,n} & 1_m \end{pmatrix} \end{cases}$$

and hence by (10) we see that the LHS belongs to $\mathrm{SE}(n+m, R)$.

Now we claim the following:

$$(12) \quad \begin{cases} \text{Let } V = (V_{i1}) \in \mathrm{MT}(n \times 1, R) \text{ and } W = (W_{1j}) \in \mathrm{MT}(1 \times n, R) \\ \text{be such that } \sum_{1 \leq i \leq n} W_{1i}V_{i1} = 0. \\ \text{Then } 1_n + VW \in \mathrm{GL}(n, R) \\ \text{and we have } \begin{pmatrix} 1_n + VW & 0_{n,1} \\ 0_{1,n} & 1_1 \end{pmatrix} \in \mathrm{SE}(n+1, R). \end{cases}$$

This follows by taking $(n, m, A, B) = (n, 1, V, W)$ in (11).

Next we claim the following:

$$(13) \quad \begin{cases} \text{In the situation of (12)} \\ \text{assume that } W_{1j} = 0 \text{ for some } j \in \{1, \dots, n\}. \\ \text{Then } 1_n + VW \in \mathrm{SE}(n, R). \end{cases}$$

To see this, note that the assertion is obvious for $n = 1$, and so suppose that $n > 1$. Let us first deal with the case when $j = n$. Now upon letting

$$\widetilde{V} = (\widetilde{V}_{i1}) \in \mathrm{MT}((n-1) \times 1, R) \quad \text{with} \quad \widetilde{V}_{i1} = V_{i1} \quad \text{for} \quad 1 \leq i \leq n-1$$

and

$$\widetilde{W} = (\widetilde{W}_{1j}) \in \mathrm{MT}(1 \times (n-1), R) \quad \text{with} \quad \widetilde{W}_{1j} = W_{1j} \quad \text{for} \quad 1 \leq j \leq n-1$$

we have

$$\sum_{1 \leq i \leq n-1} \widetilde{W}_{1i} \widetilde{V}_{i1} = 0$$

and hence by (12) we get

$$1_n + \widetilde{V}\widetilde{W} \in \mathrm{GL}(n-1, R) \quad \text{with} \quad \begin{pmatrix} 1_{n-1} + \widetilde{V}\widetilde{W} & 0_{n-1,1} \\ 0_{1,n-1} & 1_1 \end{pmatrix} \in \mathrm{SE}(n, R).$$

Clearly

$$1_n + VW = \begin{pmatrix} 1_{n-1} + \widetilde{V}\widetilde{W} & 0_{n-1,1} \\ U & 1_1 \end{pmatrix}$$

where

$$U = (U_{1j}) \in \mathrm{MT}(1 \times (n-1), R) \quad \text{with} \quad U_{1j} = V_{n1} W_{1j} \quad \text{for} \quad 1 \leq j \leq n-1$$

and hence

$$(1_n + VW)\widetilde{T} = \begin{pmatrix} 1_{n-1} + \widetilde{V}\widetilde{W} & 0_{n-1,1} \\ 0_{1,n-1} & 1_1 \end{pmatrix}$$

where

$$\widetilde{T} = T_{n,1}(n, -U_{1,1}) \ldots T_{n,n-1}(n, -U_{n-1,1}) \in \mathrm{SE}(n, R).$$

Therefore $1_n + VW \in \mathrm{SE}(n, R)$.

To reduce the $j < n$ case to the above case, upon letting

$$\widehat{T} = T_{nj}(n, 1) T_{jn}(n, -1) \in \mathrm{SE}(n, R)$$

together with

$$\widehat{V} = \widehat{T}^{-1} V \in \mathrm{MT}(1 \times n, R) \quad \text{and} \quad \widehat{W} = W\widehat{T}$$

we have

$$\widehat{T}^{-1}(1_n + VW)\widehat{T} = 1_n + \widehat{V}\widehat{W} \quad \text{with} \quad \sum_{1 \leq i \leq n} \widehat{W}_{1i} \widehat{V}_{i1} = 0 \quad \text{and} \quad \widehat{W}_{1n} = 0.$$

Indeed, $\widehat{W}_{1k} = W_{1k}$ or W_{in} or 0 according as $k \neq j, n$ or $k = j$ or $k = n$.

Now we claim the following:

(14)
$$\begin{cases}
\text{Let } V = (V_{i1}) \in \mathrm{MT}(n \times 1, R) \\
\text{be such that } (V_{11}, \ldots, V_{n1}) \in U_n(R), \\
\text{and let } \phi : R^n \to R \text{ be the } R\text{-homomorphism} \\
\text{with } \phi(e_i) = V_{i1} \text{ for } 1 \leq i \leq n \\
\text{where we recall that } e_i = (0, \ldots, 1, 0, \ldots, 0) \in R^n \\
\text{with 1 in the } i\text{-th spot.} \\
\text{Then } \ker(\phi) \text{ is the } R\text{-submodule of } R^n \\
\text{generated by the set } \{V_{1j}e_i - V_{1i}e_j : 1 \leq i < j \leq n\}.
\end{cases}$$

To see this note that, since $(V_{11}, \ldots, V_{n1}) \in U_n(R)$, we can write $1 = \sum_{1 \leq j \leq n} b_j V_{j1}$ with $b_j \in R$. Consider the R-homomorphism $\theta = 1_{R^n} - (\psi\phi) : R^n \to R^n$ where $\psi : R \to R^n$ is the unique R-homomorphism with $\psi(1) = \sum_{1 \leq j \leq n} b_j e_j$. Then $\phi\theta = 0$ and for all $x \in \ker(\phi)$ we clearly have $\theta(x) = x$, and hence $\theta(R^n) = \ker(\phi)$. Therefore, since the set $\{V_{j1}e_i - V_{i1}e_j : 1 \leq i < j \leq n\}$ is obviously contained in $\ker(\phi)$, it suffices to show that, for $1 \leq i \leq n$, the element $\theta(e_i)$ can be expressed as an R-linear combination of members of the said set. This follows by noting that

$$\begin{cases}
\theta(e_i) \\
= e_i - \psi(\phi(e_i)) \\
= e_i - \psi(V_{i1}) \\
= (\sum_{1 \leq j \leq n} b_j V_{j1} e_i) - (\sum_{1 \leq j \leq n} b_j e_j V_{i1}) \\
= \sum_{1 \leq j \leq n} b_j (V_{j1} e_i - V_{i1} e_j).
\end{cases}$$

Next we claim the following:

(15)
$$\begin{cases}
\text{Let } V = (V_{i1}) \in \mathrm{MT}(n \times 1, R) \text{ and } W = (W_{1j}) \in \mathrm{MT}(1 \times n, R) \\
\text{with } n \geq 3 \text{ be such that } (V_{11}, \ldots, V_{n1}) \in U_n(R) \text{ and } WV = 0_{1,1}. \\
\text{Then } 1_n + VW \in \mathrm{SE}(n, R).
\end{cases}$$

To prove this, for $1 \leq l \leq n$ let $E_l = ((E_l)_{1j}) \in \mathrm{MT}(1 \times n, R)$ be given by putting $(E_l)_{1j} = 1$ or 0 according as $j = l$ or $j \neq l$, and for $1 \leq l < k \leq n$ let $U_{lk} = ((U_{lk})_{1j}) \in \mathrm{MT}(1 \times n, R)$ be given by putting $U_{lk} = V_{k1}E_l - V_{l1}E_k$. Then, upon "identifying" R^n and R with $\mathrm{MT}(1 \times n, R)$ and $\mathrm{MT}(1 \times 1, R)$ respectively, in the notation of (14), for all $U \in \mathrm{MT}(1 \times n, R)$ we have $\phi(U) = UV$, and hence by (14) we can write

(15′) $W = \sum_{1 \leq l < k \leq n} \widehat{W}_{lk}$ where $\widehat{W}_{lk} = ((\widehat{W}_{lk})_{1j}) = a_{lk} U_{lk} \in \mathrm{MT}(1 \times n, R)$

with $a_{lk} \in R$. For $1 \leq l < k \leq n$ we clearly have: (15″) $\widehat{W}_{lk}V = 0$; moreover, since $n \geq 3$, we can take $j \in \{1, \ldots, n\} \setminus \{l, k\}$ and for any such j we clearly have

$(\widehat{W}_{lk})_{1j} = 0$; therefore by (13) we get $(15''')$ $1_n + V\widehat{W}_{lk} \in \mathrm{SE}(n, R)$. By items $(15')$ and $(15'')$ we obtain

$$(15'''') \qquad 1_n + VW = 1_n + \sum_{1 \le l < k \le n} V\widehat{W}_{lk} = \prod_{1 \le l < k \le n} (1_n + V\widehat{W}_{lk})$$

(with commuting factors) and hence by item $(15''')$ we get $1_n + VW \in \mathrm{SE}(n, R)$.

Although we shall not use it, for the sake of completeness we note the following consequence of (15) which follows from it by taking transposes.

(16) $\quad \begin{cases} \text{Let } V = (V_{i1}) \in \mathrm{MT}(n \times 1, R) \text{ and } W = (W_{1j}) \in \mathrm{MT}(1 \times n, R) \\ \text{with } n \ge 3 \text{ be such that } (W_{11}, \dots, W_{1n}) \in U_n(R) \text{ and } WV = 0_{1,1}. \\ \text{Then } 1_n + VW \in \mathrm{SE}(n, R). \end{cases}$

By using the basic property of left or right multiplication by a transvection matrix $T_{ij}(n, \lambda)$ discussed in the preamble of Comment (C51), we get the following assertion which generalizes both of them; to see that it does subsume both, take $A = 1_n$ or $B = 1_n$.

(17) $\quad \begin{cases} \text{Given any } n \times n \text{ matrices } A = (A_{ij}) \text{ and } B = (B_{ij}) \text{ over } R, \\ \text{let } (A_j = (A_{ij})_{1 \le i \le n})_{1 \le j \le n} \text{ be the columns of } A \\ \text{and let } (B_i = (B_{ij})_{1 \le j \le n})_{1 \le i \le n} \text{ be the rows of } B. \\ \text{Then for all } i \ne j \text{ in } \{1, \dots, n\} \text{ and } \lambda \in R \\ \text{we have } AT_{ij}(n, \lambda)B = AB + \lambda A_i B_j. \end{cases}$

Next we claim the following:

(18) $\quad \begin{cases} \text{Given any } n \times n \text{ matrices } A = (A_{ij}) \text{ and } B = (B_{ij}) \text{ over } R \\ \text{with } A \in \mathrm{GL}(n, R) \text{ and } B = A^{-1}, \\ \text{let } (A_j = (A_{ij})_{1 \le i \le n})_{1 \le j \le n} \text{ be the columns of } A \\ \text{and let } (B_i = (B_{ij})_{1 \le j \le n})_{1 \le i \le n} \text{ be the rows of } B. \\ \text{Then for all } i \ne j \text{ in } \{1, \dots, n\} \text{ we have } B_j A_i = 0_{1,1}. \end{cases}$

This follows by noting that the equality $BA = 1_n$ is equivalent to saying that $B_j A_i = 1_1$ or $0_{1,1}$ according as $j = i$ or $j \ne i$.

Now let us prove (†) as a consequence of (15), (17) and (18). Since $\mathrm{SE}(n, R)$ is generated by transvection matrices, it suffices to show that for all $i \ne j$ in $\{1, \dots, n\}$ with $\lambda \in R$, and for all $A \in \mathrm{GL}(n, R)$, we have $AT_{ij}(n, \lambda)A^{-1} \in \mathrm{SE}(n, R)$. Taking $B = A^{-1}$ in (17) we get $AT_{ij}(n, \lambda)A^{-1} = 1_n + \lambda A_i B_j$, and by (18) we conclude that $B_j A_i = 0_{1,1}$. Since $A \in \mathrm{GL}(n, R)$, by (1) we see that $(A_{1i}, \dots, A_{ni}) \in U_n(R)$. Therefore by taking $(V, W) = (A_i, B_j)$ in (15) we get $AT_{ij}(n, \lambda)A^{-1} \in \mathrm{SE}(n, R)$.

PROOF OF (T126). Given any $r \in U_n(R)$, by (T124.3) and (T125) we can find $A \in \mathrm{SL}([2, n], R)$ and $B \in \mathrm{SE}(n, R)$ such that $ABr' = e'_n$ where we recall that $e_n = (0, \ldots, 0, 1) \in R^n$. It follows that

$$r' = B^{-1}A^{-1}e'_n = B^{-1}e'_n$$

and hence $Br' = e'_n$. Similarly for any $s \in U_n(R)$ we have $Cs' = e'_n$ for some $C \in \mathrm{SE}(n, R)$. Thus $r' = (B^{-1}C)s'$ with $B^{-1}C \in \mathrm{SE}(n, R)$.

ADDENDUM (19). For proving (T124.2) and (T126), we only need the case of (T123.3) when the image of a is a nonzerodivisor in L. Here is the sketch of a shorter proof in that case. It suffices to discuss the Claim saying that for a suitable 2×2 matrix B over L we have $\det(B) = 1$ and $B\begin{pmatrix} r_1(c) \\ r_2(c) \end{pmatrix} = \begin{pmatrix} r_1(b) \\ r_2(b) \end{pmatrix}$. Take $B = (1/a)A$ with $A = \begin{pmatrix} r_1(b) & -t_2(b) \\ r_2(b) & t_1(b) \end{pmatrix} \begin{pmatrix} t_1(c) & t_2(c) \\ -r_2(c) & r_1(c) \end{pmatrix}$. Then B satisfies both the required equations except that its entries are in the total quotient ring of L. But the equation $b = c + ad$ tells us that every entry of A belongs to the ideal aL and hence every entry of B belongs to L.

ADDENDUM (20). For the special situation when R is the univariate polynomial ring $K[X]$ over a quasilocal ring K, here is a shorter direct proof of (T124.2) and also a proof of the assertion that then every $r \in U_n(R)$ with submonic r_i for some i is completable. In view of (T124.1) we may assume that $n \geq 3$. So let $r \in U_n(R)$ with $n \geq 3$ be such that for some $i \in \{1, \ldots, n\}$ and $d \in \mathbb{N}$ we have $r_i = \sum_{0 \leq m \leq d} a_m X^{d-m}$ where a_m are elements in K such that a_0 is a unit in K. We shall show that then $r(X) \sim_{E(R)} r(0)$ and r is $E(R)$-completable.

Let us first prove the following: (i) Assume that for some $j \in \{1, \ldots, n\} \setminus \{i\}$ and some t_i, t_j in R the polynomial $t_i r_i + t_j r_j$ is submonic of X-degree $e \in \mathbb{N}$; then for some $s \in U_n(R)$ and $A \in E(R)$ we have $r' = As'$ with s_1 monic of X-degree e and $\deg_X s_m < e$ for $2 \leq m \leq n$.

To see this, because of $n \geq 3$ we can take $k \in \{1, \ldots, n\} \setminus \{i, j\}$, and by the division algorithm we can find $\lambda \in R$ with $\deg_X(r_k - \lambda(t_i r_i + t_j r_j)) < e$. Let μ be the coefficient of X^e in $t_i r_i + t_j r_j$. Let $u \in R^n$ be given by $(u_i, u_j, u_k) = (r_i, r_j, r_k - (\lambda - \mu^{-1})(t_i r_i + t_j r_j))$ with $u_m = r_m$ for all $m \in \{1, \ldots, n\} \setminus \{i, j, k\}$. Then $u \in U_n(R)$ with u_k monic of X-degree e. Moreover we have $r' = Bu'$ with $B = T_{ki}(n, (\lambda - \mu^{-1})t_i)T_{kj}(n, (\lambda - \mu^{-1})t_j) \in E(R)$. If $k = 1$ then let $v = u$ with $C = 1_n$; if $k \neq 1$ then by the division algorithm we can find an element $\nu \in R$ such that $\deg_X(u_1 - \nu u_k) < e$, and we take $C = T_{1k}(n, \nu - 1)$ and $v \in R^n$ such that $v_1 = u_1 - (\nu - 1)u_k$ and $v_m = u_m$ for $2 \leq m \leq n$. In both the cases we have $v \in U_n(R)$ and $C \in E(R)$ with $u' = Cv'$ and v_1 is monic of X-degree e. By the division algorithm we find $\lambda_m \in R$ with $\deg_X(v_m - \lambda_m v_1) < e$ for $2 \leq m \leq n$ and letting $s \in R^n$ with $s_1 = v_1$ and $s_m = v_m - \lambda_m v_1$ for $2 \leq m \leq n$ we have $v' = Ds'$ where we have put $D = T_{21}(n, \lambda_2) \ldots T_{n1}(n, \lambda_n) \in E(R)$. Now we are done by putting $A = BCD$.

Next we prove the following: (ii) Assume $d > 0$ and for some $j \in \{1, \ldots, n\} \setminus \{i\}$ we have $r_j = \sum_{0 \le m \le d-1} b_m X^{d-1-m}$ where b_m are elements in K such that at least one of them is a unit in K. Then for some t_i, t_j in R the polynomial $t_i r_i + t_j r_j$ is monic of X-degree $d - 1$.

To see this let I be the set of all c in K such that for some t_i, t_j in R we have $t_i r_i + t_j r_j = \sum_{0 \le m \le d-1} c_m X^{d-1-m}$ with c_m in K such that $c_0 = c$. Then I is clearly an ideal in K. By induction on $l \in \{0, \ldots, d-1\}$ we shall show that for some t_{il}, t_{jl} in R we have $t_{il} r_i + t_{jl} r_j = \sum_{0 \le m \le d-1} c_{ml} X^{d-1-m}$ with c_{ml} in K such that $c_{ml} - b_{m+l} \in I$ for $0 \le m \le d - 1 - l$. For $l = 0$ we are done by taking $(t_{i0}, t_{j0}) = (0, 1)$. So let $0 < l \le d - 1$ and assume for $l - 1$. Then we are done by taking $(t_{il}, t_{jl}) = (Xt_{i,l-1} - (c_{0,l-1}/a_0), Xt_{j,l-1})$. This completes the induction. For $0 \le l \le d - 1$ we clearly have $c_{0l} \in I$ and hence $b_l \in I$. Consequently $I = R$ and hence $1 \in I$. Taking $c_0 = 1$ we see that $t_i r_i + t_j r_j$ is monic of X-degree $d - 1$.

Now by induction on d, let us prove that for some $F \in E(R)$ we have $r' = Fe'_1$ with $e_1 = (1, 0, \ldots, 0) \in R^n$. By (i) there exists $s \in U_n(R)$ and $A \in E(R)$ with $r' = As'$ such that s_1 is monic of X-degree d and $\deg_X s_m < d$ for $2 \le m \le n$. For $d - 0$ we are done by taking $F = A$. So let $d > 0$ and assume for $d - 1$. Since $s \in U_n(R)$, for some $j \in \{2, \ldots, n\}$ some coefficient of s_j must be a unit in K. Hence by (ii) we can find t_1, t_j in R such that $t_1 s_1 + t_j s_j$ is submonic of X-degree $d - 1$. Therefore by (i) we can find $u \in U_n(R)$ and $B \in E(R)$ with $s' = Bu'$ such that u_1 is submonic of X-degree $d - 1$. Consequently by the induction hypothesis we can find $C \in E(R)$ with $u' = Ce'_1$. The induction is completed by taking $F = ABC$.

Thus r is $E(R)$-completable. It follows that $r(X) \sim_{E(R)} r(0)$.

COMMENT (C53). [**Stably Free Modules**]. We shall now introduce stably free modules and use them, together with completing unimodular rows, to give an alternative proof of the Quillen-Suslin Theorem (T115.2) in the case of a finite variable polynomial ring over a field. This alternative proof, and its byproduct discussed in the next Comment (C54), are again due to Suslin [Su2].

A module V over a ring R is said to be stably free if for some nonnegative integers d, e we have an R-module isomorphism $V \oplus R^d \approx R^e$.

STABLY FREE MODULE THEOREM (T127). For any nonnull ring R we have the following.

(T127.1) For any finite R-module V and any surjective R-homomorphism $\phi : V \to V$ we have that ϕ is injective.

(T127.2) For any nonnegative integers d, e we have that: there is an R-module isomorphism $R^d \approx R^e \Leftrightarrow d = e$.

(T127.3) For any $r \in R^n$ with $n \in \mathbb{N}_+$ we have that: $r \in U_n(R) \Leftrightarrow \phi(r) = 1$ for some R-homomorphism $\phi : R^n \to R$.

(T127.4) For any $r \in R^n$ with $n \in \mathbb{N}_+$ we have that: r is completable $\Leftrightarrow r$

is completable to a basis of R^n, where the last phrase means there is an R-basis $(E_1, \ldots, E_n)$ of R^n with $r = E_1$.

(T127.5) Every stably free R-module is finite free $\Leftrightarrow$ for every $n \in \mathbb{N}_+$ and every $r \in U_n(R)$ we have that r is completable.

(T127.6) Assume that R is a domain and let V be a finite projective R-module. Then V is an R-submodule of a finite free R-module W so that $hW \subset V$ for some $0 \neq h \in R$.

(T127.7) Assume that R is the univariate polynomial ring $K[X]$ over a nonnull ring K such that every finite projective module over K is free. Let V be a finite projective R-module such that V is an R-submodule of a finite free R-module W so that $hW \subset V$ for some submonic $h \in K[X]$. Then V is a stably free R-module.

(T127.8) Assume that R is a finite variable polynomial ring $K[X_1, \ldots, X_N]$ over a field K. Then for any R-module we have:

$$\text{finite projective} \Leftrightarrow \text{stably free} \Leftrightarrow \text{finite free.}$$

PROOF OF (T127.1). Considering the univariate polynomial ring $R[T]$ over R, we extend the R-module structure on V to an $R[T]$-module structure by putting $g(T)v = \sum g_i \phi^i(v)$ for all $v \in V$ and $g(T) = \sum g_i T^i \in R[T]$ with $g_i \in R$; note that $\phi^0(v) = v$ and hence this indeed extends the given action ($=$ scalar multiplication) of R on V. Let $v_1, \ldots, v_m$ with $m \in \mathbb{N}_+$ be R-generators of V. Then because of the surjectivity of ϕ we can find $r_{ij} \in R$ such that for $1 \leq j \leq m$ we have $v_j = \sum_{1 \leq i \leq m} r_{ij}(Tv_i)$, i.e., $\sum_{1 \leq i \leq m}(\delta_{ij} - r_{ij}T)v_i = 0$ with $\delta_{ij} = 1$ or 0 according as $i = j$ or $i \neq j$. Considering the $m \times m$ matrix $A = (\delta_{ij} - r_{ij}T)$, and referring to (Q4)(C9) as well as L4§10(E4), by Cramer's Rule we get $\det(A)v_i = 0$ for $1 \leq i \leq m$. Clearly $\det(A) = 1 - h(T)T$ with $h(T) \in R[T]$ and so we have

$$v_i = h(T)Tv_i \quad \text{for} \quad 1 \leq i \leq m.$$

Given any $v = \sum_{1 \leq i \leq m} s_i v_i \in \ker(\phi)$ with $s_i \in R$, we have $Tv = \phi(v) = 0$ and hence $h(T)Tv = 0$ and therefore $\sum_{1 \leq i \leq m} s_i h(T)Tv_i = 0$ and so by the above display we get

$$0 = \sum_{1 \leq i \leq m}(s_i v_i - s_i h(T)Tv_i) = \sum_{1 \leq i \leq m} s_i v_i - \sum_{1 \leq i \leq m} s_i h(T)Tv_i = v.$$

PROOF OF (T127.2). If $d = e$ then obviously $R^d \approx R^e$. If $d \neq e$ and $R^d \approx R^e$ then we may suppose that $d > e$ and, because of the assumed isomorphism, this would give R-bases $(x_i)_{1 \leq i \leq d}$ and $(y_i)_{1 \leq i \leq e}$ which, by sending x_i to y_i or 0 according to $1 \leq i \leq e$ or $e+1 \leq i \leq d$, would provide a noninjective R-epimorphism $R^d \to R^d$ in contradiction to (T127.1).

PROOF OF (T127.3). If $r \in U_n(R)$ then $1 = \sum_{1 \leq i \leq n} b_i r_i$ with $b_i \in R$, and it suffices to take $\phi : R^n \to R$ to be the R-homomorphism defined by $\phi(e_i) = b_i$

for $1 \leq i \leq n$ where $e_i = (0, \ldots, 0, 1, 0, \ldots, 0) \in R^n$ with 1 in the i-th spot. If $\phi : R^n \to R$ is an R-homomorphism with $\phi(r) = 1$ then $\sum_{1 \leq i \leq n} b_i r_i = 1$ where $b_i = \phi(e_i)$ for $1 \leq i \leq n$.

PROOF OF (T127.4). If r is completable then r' is the first column of a member A of $\mathrm{GL}(n, R)$, and upon letting E_j be the transpose of the j-th column of A we clearly have $r = E_1$ and $(E_1, \ldots, E_n)$ is an R-basis of R^n. Conversely, if $r = E_1$ for an R-basis $(E_1, \ldots, E_n)$ of R^n, then, upon letting A be the $n \times n$ matrix over R whose j-th column is the transpose of E_j, we get a member A of $\mathrm{GL}(n, R)$ whose first column is r'.

PROOF OF (T127.5). Assuming every stably free R-module to be finite free, let there be given any $r \in U_n(R)$ with $n \in \mathbb{N}_+$. Then by (T127.3) there exists an R-homomorphism $\phi : R^n \to R$ with $\phi(r) = 1$. By (T80.1) to (T80.3) of (Q17) there is an R-isomorphism $\ker(\phi) \oplus R \approx R^n$, and hence by (T127.2) and our assumption we get $\ker(\phi) \approx R^{n-1}$. Therefore by letting $E_1 = r$ and taking an R-basis $(E_2, \ldots, E_n)$ of $\ker(\phi)$ we get an R-basis $(E_1, \ldots, E_n)$ of R^n. Consequently r is completable by (T127.4).

Conversely, assuming that every $r \in U_n(R)$ is completable for every $n \in \mathbb{N}_+$, we want to show that every stably free R-module is finite free. In view of (T127.2) and an obvious induction, it suffices to show that for any R-module V and any R-isomorphism $\alpha : R^n \to V \oplus R$ with $n \in \mathbb{N}_+$, we must have $V \approx R^{n-1}$. Let $\beta : V \oplus R \to R$ be the R-homomorphism given by $\beta(v, w) = w$ for all $v \in V$ and $w \in R$. Consider the R-homomorphism $\phi = \beta\alpha : R^n \to R$ and let

$$r = \alpha^{-1}(0, 1) \in R^n.$$

Then $\phi(r) = 1$ and hence by (T127.3) we get $r \in U_n(R)$. Therefore by (T127.4) and our assumption we can find an R-basis $(E_1, \ldots, E_n)$ of R^n with $E_1 = r$. Let $\gamma : V \oplus R \to V$ be the natural projection. Then clearly $\gamma\alpha(E_i)_{2 \leq i \leq n}$ is an R-basis of V and hence $V \approx R^{n-1}$.

PROOF OF (T127.6). We can take a finite number of R-generators $x_1, \ldots, x_m$ of V and by suitably relabelling them we can arrange that the elements $x_1, \ldots, x_n$ are linearly independent over R, but for $1 \leq i \leq m - n$ the elements $x_1, \ldots, x_n, x_{n+i}$ are linearly dependent over R. Let $\widetilde{W}$ be the R-submodule of V generated by $x_1, \ldots, x_n$. Then the above conditions on the elements $x_1, \ldots, x_m$ tell us that $\widetilde{W}$ is a finite free R-module and for some $0 \neq h \in R$ we have $hV \subset \widetilde{W}$. Thus $h\widetilde{W} \subset hV \subset \widetilde{W}$. By (Q17)(T80,2) $v \mapsto hv$ gives an R-isomorphism $\alpha : V \to hV$. Hence by "identification" we can find an overmodule W of V together with an R-isomorphism $\beta : \widetilde{W} \to W$ such that for all $u \in hV$ we have $\beta(u) = \alpha^{-1}(u)$. Now W is a finite free R-overmodule of V with $hW \subset V$.

PROOF OF (T127.7). Follows from Fact (♯♯♯) which was proved at the end of Comment (C49) and which was independent of the rest of that Comment.

PROOF OF (T127.8). In view of (Q17)(T80.3) we see that: (i) every finite free R-module is finite projective, and obviously: (ii) every finite free R-module is stably free. By (T124.3) and (T127.5) we also see that: (iii) every stably free R-module is finite free. By induction on N we shall now show that: (iv) every finite projective R-module is finite free, and this will complete the proof. This being obvious for $N = 0$, let $N > 1$ and assume for $N - 1$. Given any finite projective R-module V, by (T127.6) we can find a finite free R-overmodule W of V with $hW \subset V$ for some $0 \neq h \in R$. In view of of L3§12(4), by letting R undergo a K-automorphism, we can arrange matters so that h is a submonic polynomial in X_N with coefficients in $\widehat{R} = K[X_1, \ldots, X_{N-1}]$. Now, in view of (iii), by the induction hypothesis and (T127.7) we see that V is a finite free R-module. This completes the induction and hence the entire proof.

COMMENT (C54). [**Special Linear Groups Over Polynomial Rings**]. We shall now generalize (C51)(T121) by proving Suslin's Theorem (T130) stated below. In addition to (C52)(T126), the proof will also use properties of the Mennike Symbol $\mu(a, b)$ given in the following Lemma (T129).

Given any $(a, b) \in U_2(R)$ where R is any nonnull ring, we introduce the element $\mu(a, b) \in \mathrm{SL}(3, R)/\mathrm{SE}(3, R)$ thus; note that $\mathrm{SL}(3, R)/\mathrm{SE}(3, R)$ makes sense because $\mathrm{SE}(3, R)$ is a normal subgroup of $\mathrm{SL}(3, R)$ by (C52)(T125). Since $(a, b) \in U_2(R)$, we can find $(c, d) \in R^2$ with $ad - bc = 1$, and we let $\mu(a, b)$ be the image of the 3×3 matrix $\left(\begin{smallmatrix} a & b & 0 \\ c & d & 0 \\ 0 & 0 & 1 \end{smallmatrix} \right)$ under the residue class map $\mathrm{SL}(3, R) \to \mathrm{SL}(3, R)/\mathrm{SE}(3, R)$. To justify this we need to show that it is independent of the choice of (c, d). So let $(\widehat{c}, \widehat{d}) \in R^2$ be such that $a\widehat{d} - b\widehat{c} = 1$. Then

$$\begin{pmatrix} a & b \\ c & d \end{pmatrix} \begin{pmatrix} a & b \\ \widehat{c} & \widehat{d} \end{pmatrix}^{-1} = \begin{pmatrix} a & b \\ c & d \end{pmatrix} \begin{pmatrix} \widehat{d} & -b \\ -\widehat{c} & a \end{pmatrix} = \begin{pmatrix} 1 & 0 \\ c\widehat{d} - d\widehat{c} & 1 \end{pmatrix} \in \mathrm{SE}(2, R)$$

and hence

$$\begin{pmatrix} a & b & 0 \\ c & d & 0 \\ 0 & 0 & 1 \end{pmatrix} \begin{pmatrix} a & b & 0 \\ \widehat{c} & \widehat{d} & 0 \\ 0 & 0 & 1 \end{pmatrix}^{-1} \in \mathrm{SE}(3, R)$$

which shows that $\mu(a, b)$ is independent of the choice of (c, d).

We generalize the definitions of $\mathrm{SL}([2, n], R)$ and $\mathrm{QSE}(n, R)$ given in (C52) thus. For any positive integer $m < n$ we get the canonical monomorphism of $\mathrm{SL}(m, R)$ into $\mathrm{SL}(n, R)$ by sending every B in $\mathrm{SL}(m, R)$ to $\left(\begin{smallmatrix} B & 0_{m \times (n-m)} \\ 0_{(n-m) \times m} & 1_{n-m} \end{smallmatrix} \right)$ and we denote the image of this monomorphism by $\mathrm{SL}([m, n], R)$. We put

$$\mathrm{QSE}([m, n], R) = \text{subgroup of } \mathrm{SL}(n, R) \text{ generated by } \mathrm{SL}([m, n], R) \cup \mathrm{SE}(n, R)$$

and we call this the $[m, n]$-dimensional quasielementary group over R.

In proving (T129) we need Suslin's Lemma (T128) which is stated below and which deals with localization in the following sense. For the univariate polynomial ring $R = K[X]$ over a ring K and any given maximal ideal P in K, we shall consider the ring homomorphism

$$\psi_P : R \to R_P = K_P[X]$$

induced by the canonical homomorphism $\phi_P : K \to K_P$, i.e., given by $\psi_P(f) = \sum \phi_P(f_i) X^i$ for all $f = \sum f_i X^i$ with $f_i \in K$. For every $n \in \mathbb{N}_+$, the map ψ_P induces the ring homomorphism

$$\alpha_{P,n} : \text{MT}(n \times n, R) \to \text{MT}(n \times n, R_P) \quad \text{given by} \quad (A_{ij}) \mapsto (\psi_P(A_{ij})).$$

Under $\alpha_{P,n}$, the images of $\text{GL}(n, R)$, $\text{SL}(n, R)$, $\text{SE}(n, R)$ are clearly contained in $\text{GL}(n, R_P)$, $\text{SL}(n, R_P)$, $\text{SE}(n, R_P)$ respectively, and hence $\alpha_{P,n}$ induces a group homomorphism

$$\beta_{P,n} : \text{SL}(n, R)/\text{SE}(n, R) \to \text{SL}(n, R_P)/\text{SE}(n, R_P).$$

In a similar manner, given any element a in K, letting $S(a)$ be the multiplicative set $\{a^i : i \in \mathbb{N}\}$ in K, we shall consider the ring homomorphism

$$\psi_a : R \to R_a = K_{S(a)}[X]$$

which is induced by the canonical homomorphism $\phi_a : K \to K_{S(a)}$, i.e., which is given by $\psi_a(f) = \sum \phi_a(f_i) X^i$ for all $f = \sum f_i X^i$ with $f_i \in K$. For every $n \in \mathbb{N}_+$, the map ψ_a induces the ring homomorphism

$$\alpha_{a,n} : \text{MT}(n \times n, R) \to \text{MT}(n \times n, R_a) \quad \text{given by} \quad (A_{ij}) \mapsto (\psi_a(A_{ij})).$$

We put

$$\text{GE}(n, (K, R)) = \text{subgroup of } \text{GL}(n, R) \text{ generated by } \text{GL}(n, K) \cup \text{SE}(n, R)$$

and we call this the n-dimensional general elementary group over (K, R).

Note that R_P and R_a are only abbreviations for the displayed rings to be used in this Comment.

Given any ring R with positive integer n, and any ideal I in R, we put

$$\text{GL}(n, [R, I]) = \text{kernel of the map } \text{GL}(n, R) \to \text{GL}(n, R/I)$$

where $\text{GL}(n, R) \to \text{GL}(n, R/I)$ is the map obtained by sending every element of R to its image under the residue class epimorphism $R \to R/I$, and we put

$$\left\{ \begin{array}{l} \text{SE}(n, [R, I]) = \text{subgroup of } \text{SE}(n, R) \text{ generated by} \\ \qquad \text{all elements } \gamma T_{i,j}(n, \lambda) \gamma^{-1} \text{ where } i \neq j \\ \qquad \text{with } \lambda \in I \text{ and } \gamma \in \text{SE}(n, R). \end{array} \right.$$

In Cohn's Theorem (T131) stated below we shall give his [Coh] example showing that (T125) and (T130) are not true for $N = n = 2$.

SUSLIN'S LOCALIZATION LEMMA (T128). For the univariate polynomial ring $R = K[X]$ over a ring K, and any integer $n \geq 3$, we have the following.

[Note that for any $n \times n$ matrix $A = (A_{ij})$ over R and any $b \in K$ we let $A(b)$ stand for the $n \times n$ matrix over K given by $A(b) = (A_{ij}(b))$].

(T128.1) We have $\mathrm{GL}(n, [R, XR]) \cap \mathrm{SE}(n, R) = \mathrm{SE}(n, [R, XR]) =$ the subgroup of $\mathrm{SE}(n, R)$ generated by all elements $\gamma T_{ij}(n, \lambda)\gamma^{-1}$ where $i \neq j$ with $\lambda \in XR$ and $\gamma \in \mathrm{SE}(n, K)$.

(T128.2) Let there be given any $a \in K$ with $A \in \mathrm{SE}(n, [R_a, XR_a])$, and let there be given any positive integer s_0. Then there exists an integer $s \geq s_0$ together with $B \in \mathrm{SE}(n, [R, XR])$ such that $\alpha_{a,n}(B) = A(\phi_a(a^s)X)$.

(T128.3) Let there be given any $A \in \mathrm{GL}(n, [R, XR])$ with $a \in K$ such that $\alpha_{a,n}(A) \in \mathrm{SE}(n, [R_a, XR_a])$, and let there be given any positive integer s_0. Then there exists an integer $s \geq s_0$ such that $A(a^sX) \in \mathrm{SE}(n, [R, XR])$.

(T128.4) Let $A \in \mathrm{GL}(n, R)$ and $a \in K$ be such that $\alpha_{a,n}(A) \in \mathrm{SE}(n, R_a)$, and let there be given any positive integer s_0. Then there exists an integer $s \geq s_0$ such that for all c, d in K with $c - d \in a^sK$ we have $A(dX) \in \mathrm{GL}(n, R)$ and $A(cX)A(dX)^{-1} \in \mathrm{SE}(n, [R, XR])$.

(T128.5) Let $A \in \mathrm{GL}(n, [R, XR])$. Assume that there exist elements a, b in K with $aK + bK = K$ such that $\alpha_{a,n}(A) \in \mathrm{SE}(n, R_a)$ and $\alpha_{b,n}(A) \in \mathrm{SE}(n, R_b)$. Then $A \in \mathrm{SE}(n, R)$.

(T128.6) Let $A \in \mathrm{GL}(n, R)$ be such that $A(0) = 1_n$, and let I be the set of all $a \in K$ with $\alpha_{a,n}(A) \in \mathrm{SE}(n, R_a)$. Then I is an ideal in K.

(T128.7) Let $A \in \mathrm{GL}(n, R)$ be such that $A(0) = 1_n$ and for every maximal ideal P in K we have $\alpha_{P,n}(A) \in \mathrm{SE}(n, R_P)$. Then $A \in \mathrm{SE}(n, R)$.

(T128.8) Let $A \in \mathrm{GL}(n, R)$ be such that for every maximal ideal P in K we have $\alpha_{P,n}(A) \in \mathrm{GE}(n, (K_P, R_P))$. Then $A \in \mathrm{GE}(n, (K, R))$.

MENNIKE SYMBOL LEMMA (T129). For any nonnull ring R we have the following.

(T129.1) For any $a \in U(R)$ and $b \in R$ we have $\begin{pmatrix} a & b \\ 0 & a^{-1} \end{pmatrix} \in \mathrm{SE}(2, R)$.

(T129.2) For any $(a, b) \in U_2(R)$ we have the following.

(i) $a \in U(R) \Rightarrow \mu(a, b) = 1$.

(ii) $(\widehat{a}, b) \in U_2(R) \Rightarrow (a\widehat{a}, b) \in U_2(R)$ and $\mu(a\widehat{a}, b) = \mu(a, b)\mu(\widehat{a}, b)$.

(iii) $(b, a) \in U_2(R)$ and $\mu(a, b) = \mu(b, a)$.

(iv) $\lambda \in R \Rightarrow (a + \lambda b, b) \in U_2(R)$ and $\mu(a + \lambda b, b) = \mu(a, b)$.

(T129.3) Assume that R is the univariate polynomial ring $K[X]$ over a quasilocal ring K. Then for all $(f, g) \in U_2(R)$ with submonic f we have $\mu(f, g) = 1$.

(T129.4) Assume that R is the univariate polynomial ring $K[X]$ over a nonnull ring K. Let $(f, g) \in U_2(R)$ be such that for every maximal ideal P in K we have

$\beta_{P,3}(\mu(f,g)) = \mu(a,b)$ for some $(a,b) \in U_2(K_P)$. Then $\mu(f,g) = \mu(f(0),g(0))$.

(T129.5) Assume that R is the univariate polynomial ring $K[X]$ over a nonnull ring K. Let $(f,g) \in U_2(R)$ be such that the polynomial f is submonic. Then $\mu(f,g) = \mu(f(0),g(0))$.

(T129.6) Let m,n be positive integers with $2 \leq m < n$. Assume that $SE(p,R)$ acts transitively on $U_p(R)'$ whenever $m < p \leq n$. Then $SL(n,R) = QSE([m,n],R)$.

(T129.7) If R is a finite variable polynomial ring $K[X_1,\ldots,X_N]$ over a field K and n is an integer with $n \geq 3$ then $SL(n,R) = QSE([2,n],R)$.

GENERALIZED TRANSVECTION THEOREM (T130). Consider the finite variable polynomial ring $R = K[X_1,\ldots,X_N]$ over any field K with any integer $N \geq 0$, and let there be given any integer $n \geq 3$. Then we have $SL(n,R) = SE(n,R)$.

COHN'S TWO BY TWO THEOREM (T131). For any nonnull ring R we have (T131.1) to (T131.7) as stated after the Preamble.

PREAMBLE. For every $u \in R$ we put

$$E(u) = \begin{pmatrix} u & 1 \\ -1 & 0 \end{pmatrix}.$$

By $E(2,R)$ we denote the subgroup of $GL(2,R)$ generated by all $E(u)$ with $u \in R$. Recall that $\mathrm{diag}(a_1,\ldots,a_n)$ is the diagonal matrix whose (i,j)-th entry is a_i or 0 according as $i = j$ or $i \neq j$. By $D(n,R)$ we denote the group of all $\mathrm{diag}(a_1,\ldots,a_n)$ with $a_1,\ldots,a_n$ in $U(R)$. By $DE(n,R)$ we denote the subgroup of $GL(n,R)$ generated by $D(n,R) \cup SE(n,R)$. For all $a_1,\ldots,a_n$ in $U(R)$ with $\lambda \in R$ and $i \neq j$ we clearly have

$$\mathrm{diag}(a_1,\ldots,a_n)^{-1}T_{ij}(n,\lambda)\mathrm{diag}(a_1,\ldots,a_n) = T_{ij}(n,a_i^{-1}\lambda a_j)$$

and hence every element of $DE(n,R)$ can be expressed as $\alpha\beta$ with α in $D(n,R)$ and β in $SE(n,R)$. We consider Cohn's 2×2 matrix

$$C = \begin{pmatrix} 1+XY & X^2 \\ -Y^2 & 1-XY \end{pmatrix}$$

with entries in the bivariate polynomial ring $k[X,Y]$. Recall that the degree form of a nonzero polynomial is the sum of its highest degree terms. For instance the entries of $\begin{pmatrix} XY & X^2 \\ -Y^2 & -XY \end{pmatrix}$ are the degree forms of the entries of C.

(T131.1) $SE(2,R) = E(2,R)$.

(T131.2) For all u,v in R and a,b in $U(R)$ we have:

(i) $E(u)E(0)E(v) = -E(u+v)$,

(ii) $E(u)\mathrm{diag}(a,b) = \mathrm{diag}(b,a)E(b^{-1}ua)$, and (without reference to b)

(iii) $E(u)E(a)E(v) = E(u-a^{-1})\mathrm{diag}(a,a^{-1})E(v-a^{-1})$.

(T131.3) Every A in $DE(2,R)$ can be expressed as

$$A = \mathrm{diag}(a,b)E(u_1)\ldots E(u_m)$$

where a, b in $U(R)$ and $m \in \mathbb{N}$ with $u_1, \ldots, u_m$ in R are such that for $1 < i < m$ we have $u_i \notin U(R) \cup \{0\}$.

(T131.4) Let $R = k[X, Y]$ and

$$A = \begin{pmatrix} f & g \\ \widehat{f} & \widehat{g} \end{pmatrix} = \mathrm{diag}(a, b) E(u_1) \ldots E(u_m) \in \mathrm{DE}(2, R)$$

where $f, g, \widehat{f}, \widehat{g}$ in R with a, b in $k^\times$ and $m \in \mathbb{N}$ with $u_1, \ldots, u_m$ in R are such that for all $i \in \{j \in \mathbb{N} : 1 < j \le m \text{ or } 1 = j = m\}$ we have $u_i \notin k$. Then $\deg(f) > \deg(g)$.

(T131.5) Let $R = k[X, Y]$ and

$$B = \begin{pmatrix} F & G \\ \widehat{F} & \widehat{G} \end{pmatrix} \in \mathrm{GL}(2, R)$$

where $F, G, \widehat{F}, \widehat{G}$ are such that F, G are nonzero polynomials of equal degree and their degree forms are linearly independent over k. Then $B \notin \mathrm{DE}(2, R)$.

(T131.6) For $R = k[X, Y]$ we have $CE(0)C^{-1} = B$ with B as in (T131.5), and hence in particular $\mathrm{SE}(2, R)$ is not a normal subgroup of $\mathrm{SL}(2, R)$.

(T131.7) For $R = k[X, Y]$ we have $C \in \mathrm{SL}(2, R) \setminus \mathrm{SE}(2, R)$ and hence in particular $\mathrm{SL}(2, R) \ne \mathrm{SE}(2, R)$.

PROOF OF (T128.1). Given any $A \in \mathrm{GL}(n, [R, XR]) \cap \mathrm{SE}(n, R)$, we want to show that A can be expressed as a finite product of elements of the form $\gamma T_{ij}(n, \lambda) \gamma^{-1}$ where $i \ne j$ with $\lambda \in XR$ and $\gamma \in \mathrm{SE}(n, K)$. Since $A \in \mathrm{SE}(n, R)$, we can write

$$(1) \qquad A = T_{i_1 j_1}(n, f_1) \ldots T_{i_m j_m}(n, f_m)$$

where $m \in \mathbb{N}_+$ with $f_p = f_p(X) \in K[X]$ and $i_p \ne j_p$ in $\{1, \ldots, n\}$ for $1 \le p \le m$. Since $A \in \mathrm{GL}(n, [R, XR])$, we must have

$$(2) \qquad A(0) = \gamma_m = 1_n$$

where for $1 \le p \le m$ we have put

$$(3) \qquad \gamma_p = T_{i_1 j_1}(n, f_1(0)) \ldots T_{i_p j_p}(n, f_p(0)).$$

Now

$$f_p(X) - f_p(0) = X g_p(X) \quad \text{with} \quad g_p = g_p(X) \in K[X]$$

and we have

$$T_{i_p j_p}(n, f_p) = T_{i_p j_p}(n, f_p(0)) T_{i_p j_p}(n, X g_p)$$

and substituting this in (1), in view of (2) and (3), we see that

$$A = [\gamma_1 T_{i_1 j_1}(n, X g_1) \gamma_1^{-1}] \ldots [\gamma_m T_{i_m j_m}(n, X g_m) \gamma_m^{-1}].$$

PROOF OF (T128.2). If R_a is the null ring, i.e., equivalently, if $K_{S(a)}$ is the null ring, then A is the zero matrix and hence we are done by taking B to be the zero matrix and s to be any integer $\geq s_0$. So assume that R_a is a nonnull ring. By taking R_a for R in (T128.1) we can write

$$(4) \qquad A = A^{(1)} \ldots A^{(m)} \quad \text{with} \quad A^{(p)} = \gamma_p T_{i_p j_p}(n, X g_p) \gamma_p^{-1}$$

where $m \in \mathbb{N}_+$ with $\gamma_p \in \text{SE}(n, K_{S(a)})$ and $g_p = g_p(X) \in K_{S(a)}[X]$ and $i_p \neq j_p$ in $\{1, \ldots, n\}$ for $1 \leq p \leq m$. Let $(V_j^{(p)} = (V_{ij}^{(p)})_{1 \leq i \leq n})_{1 \leq j \leq n}$ be the columns of γ_p and let $(W_i^{(p)} = (W_{ij}^{(p)})_{1 \leq j \leq n})_{1 \leq i \leq n}$ be the rows of γ_p^{-1}. Then, for $1 \leq p \leq m$, by (C52)(17) we have

$$(5) \qquad A^{(p)} = 1_n + X g_p(X) V_{i_p}^{(p)} W_{j_p}^{(p)}$$

and by (C52)(18) we have

$$(6) \qquad W_{j_p}^{(p)} V_{i_p}^{(p)} = 0_{1,1}.$$

Now for $1 \leq l \leq n$ let $E_l = ((E_l)_{1j})$ be the $1 \times n$ matrix obtained by putting $(E_l)_{1j} = 1$ or 0 according as $j = l$ or $j \neq l$, and for $1 \leq l < k \leq n$ let

$$(7) \qquad U_{lkj_p}^{(p)} = ((U_{lkj_p}^{(p)})_{1j}) = V_{ki_p}^{(p)} E_l - V_{li_p}^{(p)} E_k \in \text{MT}(1 \times n, K_{S(a)}).$$

Then, in view of (6), by (C52)(15′) we can write

$$(8) \qquad W_{j_p}^{(p)} = \sum_{1 \leq l < k \leq n} a_{lkj_p}^{(p)} U_{lkj_p}^{(p)} \quad \text{with} \quad a_{lkj_p}^{(p)} \in K_{S(a)}.$$

In view of (5), (6) and (8), as in (15″) to (15⁗) of (C52) we see that

$$(9) \qquad U_{lkj_p}^{(p)} V_{i_p}^{(p)} = 0 \quad \text{with} \quad (U_{lkj_p}^{(p)})_{1j} = 0 \text{ for some } j \in \{1, \ldots, n\}$$

and

$$(10) \quad A^{(p)} = \prod_{1 \leq l < k \leq n} (1_n + X g_p(X) a_{lkj_p}^{(p)} V_{i_p}^{(p)} U_{lkj_p}^{(p)}) \quad \text{(with commuting factors).}$$

We can take integer $r \geq s_0$ together with polynomial $\tilde{g}_p = \tilde{g}_p(X) \in K[X]$ and element $\tilde{a}_{lkj_p}^{(p)} \in K$ as well as matrices $\tilde{V}^{(p)} = (\tilde{V}_{ij}^{(p)}) \in \text{MT}(n \times n, K)$ and $\tilde{U}_{lkj_p}^{(p)} = ((\tilde{U}_{lkj_p}^{(p)})_{1j}) \in \text{MT}(1 \times n, K)$ such that for all p, l, k, i, j we have

$$(11) \qquad g_p = \psi_a(\tilde{g}_p)/\phi_a(a^r) \quad \text{and} \quad a_{lkj_p}^{(p)} = \phi_a(\tilde{a}_{lkj_p}^{(p)})/\phi_a(a^r)$$

and

$$(12) \quad V_{ij}^{(p)} = \phi_a(\tilde{V}_{ij}^{(p)})/\phi_a(a^r) \quad \text{and} \quad (U_{lkj_p}^{(p)})_{1j} = \phi_a((\tilde{U}_{lkj_p}^{(p)})_{1j})/\phi_a(a^r).$$

Now for $1 \leq p \leq m$ let $(\widetilde{V}_j^{(p)} = (\widetilde{V}_{ij}^{(p)})_{1 \leq i \leq n})_{1 \leq j \leq n}$ be the columns of $\widetilde{V}^{(p)}$. Let $s = 4r$ and consider the $n \times n$ matrices over R given by

$$(13) \qquad B^{(p)} = \prod_{1 \leq l < k \leq n} (1_n + X\widetilde{g}_p(a^s X)\widetilde{a}_{lkj_p}^{(p)} \widetilde{V}_{i_p}^{(p)} \widetilde{U}_{lkj_p}^{(p)})$$

and

$$(14) \qquad B = B^{(1)} \ldots B^{(m)}.$$

Then by (10) to (13) we see that $\alpha_{a,n}(B^{(p)}) = A^{(p)}(\phi_a(a^s)X)$ for $1 \leq p \leq m$, and hence by (4) and (14) we get

$$(15) \qquad \alpha_{a,n}(B) = A(\phi_a(a^s)X).$$

In view of (9), (11), (12) and (13), by (C52)(13) we get $B^{(p)} \in \mathrm{SE}(n, R)$; this being so for $1 \leq p \leq m$, by (14) we obtain $B \in \mathrm{SE}(n, R)$. By (T128.1) we know that $A \in \mathrm{GL}(n, [R_a, XR_a])$ and therefore by (15) we get $B \in \mathrm{GL}(n, [R, XR])$. Now by (T128.1) we conclude that $B \in \mathrm{SE}(n, [R, XR])$.

PROOF OF (T128.3). By (T128.2) there exists integer $r \geq s_0$ together with $B \in \mathrm{SE}(n, [R, XR])$ such that $\alpha_{a,n}(B) = \alpha_{a,n}(\widehat{A})$ where $\widehat{A} = \widehat{A}(X) = A(a^r X)$. We can write $\widehat{A}$ and B as finite sums

$$\widehat{A} = 1_n + \sum_{p \in \mathbb{N}_+} X^p A^{(p)} \quad \text{and} \quad B = 1_n + \sum_{p \in \mathbb{N}_+} X^p B^{(p)}$$

with

$$A^{(p)} = (A_{ij}^{(p)}) \in \mathrm{MT}(n \times n, K) \quad \text{and} \quad B^{(p)} = (B_{ij}^{(p)}) \in \mathrm{MT}(n \times n, K).$$

Now for all p, i, j we clearly have $\phi_a(A_{ij}^{(p)}) = \phi_a(B_{ij}^{(p)})$ and hence there exists a positive integer t such that for all p, i, j we have $a^t A_{ij}^{(p)} = a^t B_{ij}^{(p)}$. It follows that

$$\widehat{A}(a^t X) = B(a^t X)$$

and hence upon letting $s = r + t$ we get

$$A(a^s X) = B(a^t X).$$

Since $B \in \mathrm{SE}(n, [R, XR])$, we also have $B(a^t X) \in \mathrm{SE}(n, [R, XR])$. Therefore

$$A(a^s X) \in \mathrm{SE}(n, [R, XR]).$$

PROOF OF (T128.4). The assertion being obvious when K is the null ring, assume the contrary. Consider the trivariate polynomial ring $\overline{R} = K[X, Y, Z]$ as an overring of R, and let $B = B(X, Y, Z)$ be the $n \times n$ matrix over $\overline{R}$ obtained by putting

$$B(X, Y, Z) = A(YX)A((Y + Z)X)^{-1}.$$

Then clearly $B \in \mathrm{GL}(n, [\overline{R}, Z\overline{R}])$. Since $\alpha_{a,n}(A) \in \mathrm{SE}(n, R_a)$, it follows that $\overline{\alpha}_{a,n}(B) \in \mathrm{SE}(n, [\overline{R}_a, Z\overline{R}_a])$, where $\overline{\alpha}_{a,n}$ is the obvious map from $\mathrm{MT}(n \times n, \overline{R})$ to $\mathrm{MT}(n \times n, \overline{R}_a)$ with $\overline{R}_a = K_{S(a)}[X, Y, Z]$, just like its unbarred version. Therefore by (T128.3) there exists an integer $s \geq s_0$ such that $B(X, Y, a^s Z) \in \mathrm{SE}(n, [\overline{R}, Z\overline{R}])$, i.e.,

$$A(YX)A((Y + a^s Z)X)^{-1} \in \mathrm{SE}(n, [\overline{R}, Z\overline{R}]).$$

Given any c, d, λ in K with $d = c + \lambda a^s$, by substituting $(Y, Z) = (c, \lambda)$ in the above display we see that $A(dX) \in \mathrm{GL}(n, R)$ and $A(cX)A(dX)^{-1} \in \mathrm{SE}(n, R)$ and hence $A(cX)A(dX)^{-1} \in \mathrm{SE}(n, [R, XR])$ by (T128.1). [cf. L6§6(E58)].

PROOF OF (T128.5). By (T128.4) there exists a positive integer s such that for all c, d in K we have that:

(16) $c - d \in a^s K \Rightarrow A(dX) \in \mathrm{GL}(n, R)$ and $A(cX)A(dX)^{-1} \in \mathrm{SE}(n, [R, XR])$

and

(17) $c - d \in b^s K \Rightarrow A(dX) \in \mathrm{GL}(n, R)$ and $A(cX)A(dX)^{-1} \in \mathrm{SE}(n, [R, XR])$.

Since $aK + bK = K$, either obviously or by L4§5(O26), we can find λ, λ^* in K such that $\lambda a^s + \lambda^* b^s = 1$. Taking $(c, d) = (\lambda a^s, 0)$ in (16) we get

(18) $A(0) \in \mathrm{GL}(n, R)$ and $A(\lambda a^s X)A(0)^{-1} \in \mathrm{SE}(n, [R, XR])$

and by taking $(c, d) = (1, \lambda a^s)$ in (17) we get

(19) $A(\lambda a^s X) \in \mathrm{GL}(n, R)$ and $A(X)A(\lambda a^s X)^{-1} \in \mathrm{SE}(n, [R, XR])$.

Multiplying (18) from the left by (19) we see that

(20) $A(0) \in \mathrm{GL}(n, R)$ and $A(X)A(0)^{-1} \in \mathrm{SE}(n, [R, XR])$.

Since $A \in \mathrm{GL}(n, [R, XR])$, we must have $A(0) = 1_n$. Therefore by (20) we get $A(X) \in \mathrm{SE}(n, [R, XR])$.

PROOF OF (T128.6). $A(0) = 1_n \Rightarrow 0 \in I$. For every $a \in I$ with $\lambda \in K$, we clearly have $\lambda a \in I$. Given any a, b in I, let $\widehat{R} = R_{(a+b)} = K_{S(a+b)}[X]$ and $\widehat{A} = \alpha_{(a+b),n}(A)$. By assumption $\alpha_{a,n}(A) \in \mathrm{SE}(n, R_a)$ and $\alpha_{b,n}(A) \in \mathrm{SE}(n, R_b)$; consequently $\widehat{\alpha}_{a,n}(\widehat{A}) \in \mathrm{SE}(n, \widehat{R}_a)$ and $\widehat{\alpha}_{b,n}(\widehat{A}) \in \mathrm{SE}(n, \widehat{R}_b)$ where the hatted maps $\widehat{\alpha}_{a,n} : \mathrm{MT}(n \times n, \widehat{R}) \to \mathrm{MT}(n \times n, \widehat{R}_a)$ and $\widehat{\alpha}_{b,n} : \mathrm{MT}(n \times n, \widehat{R}) \to \mathrm{MT}(n \times n, \widehat{R}_b)$

are the obvious incarnation of $\alpha_{a,n}$ and $\alpha_{b,n}$ respectively. Therefore by taking $\widehat{R}$ for R in (T128.5) we get $\widehat{A} \in \mathrm{SE}(n, \widehat{R})$.

PROOF OF (T128.7). Given any maximal ideal P in K, by assumption

$$\alpha_{P,n}(A) \in \mathrm{SE}(n, R_P)$$

and hence for some element $a \in K \setminus P$ we have $\alpha_{a,n}(A) \in \mathrm{SE}(n, R_a)$, and therefore a belongs to the ideal I defined in (T128.6), and so $I \not\subset P$. By Zorn's Lemma it follows that $I = K$. Therefore $1 \in I$, i.e., $A \in \mathrm{SE}(n, R)$.

PROOF OF (T128.8). The assertion being obvious when R is the null ring, we may assume the contrary. Then by (T125), $\mathrm{SE}(n, R)$ is a normal subgroup of $\mathrm{GL}(n, R)$, and hence by using item (††) at the beginning of the Proof of (T125) we can see that

(21) $$\mathrm{SE}(n, R) \supset \{ B \in \mathrm{GE}(n, (K, R)) : B(0) = 1_n \}.$$

Namely, given any B in the RHS, by (††) we can write $B = NM$ with $N \in \mathrm{GL}(n, K)$ and $M \in \mathrm{SE}(n, R)$; this yields $N = B(0)M^{-1}(0) \in \mathrm{SE}(n, K)$, and hence $B \in \mathrm{SE}(n, R)$.

Now let $\widehat{A} = A(0)^{-1}A$. Then $\widehat{A} \in \mathrm{GL}(n, R)$ with $\widehat{A}(0) = 1_n$, and by (21) we see that for every maximal ideal P in K we have $\alpha_{P,n}(\widehat{A}) \in \mathrm{SE}(n, R_P)$, and therefore by (T128.7) we get $\widehat{A} \in \mathrm{SE}(n, R)$. Since $A = A(0)\widehat{A}$ with $A(0) \in \mathrm{GL}(n, K)$, it follows that $A \in \mathrm{GE}(n, (K, R))$.

PROOF OF (T129.1). For any a_1, a_2 in $U(R)$, upon letting

$$A = \begin{pmatrix} 1 & a_2-1 \\ 0 & 1 \end{pmatrix} \begin{pmatrix} 1 & 0 \\ 1 & 1 \end{pmatrix} \begin{pmatrix} 1 & a_2^{-1}-1 \\ 0 & 1 \end{pmatrix} \begin{pmatrix} 1 & 0 \\ -a_2 & 1 \end{pmatrix}$$

we clearly have

(22) $$A \in \mathrm{SE}(2, R) \quad \text{and} \quad \begin{pmatrix} a_1 & 0 \\ 0 & a_2 \end{pmatrix} A = \begin{pmatrix} a_1 a_2 & 0 \\ 0 & 1 \end{pmatrix}.$$

Putting $a_1 = a$ and $a_2 = a^{-1}$ in (22) we see that

$$\begin{pmatrix} a & 0 \\ 0 & a^{-1} \end{pmatrix} \in \mathrm{SE}(2, R).$$

Now the the desired result follows by noting that

$$\begin{pmatrix} a & b \\ 0 & a^{-1} \end{pmatrix} = \begin{pmatrix} 1 & ab \\ 0 & 1 \end{pmatrix} \begin{pmatrix} a & 0 \\ 0 & a^{-1} \end{pmatrix} \quad \text{with} \quad \begin{pmatrix} 1 & ab \\ 0 & 1 \end{pmatrix} \in \mathrm{SE}(2, R).$$

PROOF OF (T129.2). If $a \in U(R)$ then $ad - bc = 1$ with $(c, d) = (0, a^{-1})$ and hence (i) follows by noting that by (T129.1) we have $\begin{pmatrix} a & b & 0 \\ c & d & 0 \\ 0 & 0 & 1 \end{pmatrix} \in \mathrm{SE}(3, R)$.

To prove (ii) assume that $(\widehat{a}, b) \in U_2(R)$, and let $c, d, \widehat{c}, \widehat{d}$ be elements in R such that $ad - bc = 1$ and $\widehat{a}\widehat{d} - b\widehat{c} = 1$. Considering the matrices

$$A = \begin{pmatrix} a & b & 0 \\ c & d & 0 \\ 0 & 0 & 1 \end{pmatrix} \quad \text{and} \quad \widehat{A} = \begin{pmatrix} \widehat{a} & b & 0 \\ \widehat{c} & d & 0 \\ 0 & 0 & 1 \end{pmatrix}$$

we want to show that

$$B_1 A B \widehat{A} B_2 = \begin{pmatrix} a\widehat{a} & b & 0 \\ c^* & d^* & 0 \\ 0 & 0 & 1 \end{pmatrix}$$

for some elements c^*, d^* in R and some members B_1, B, B_2 of SE$(3, R)$. To do this, by direct calculation we get

$$BA\widehat{B}\widehat{A}BT_{23}(3, a) = \begin{pmatrix} a\widehat{a} & b & 0 \\ \widehat{c} & 0 & -\widehat{d} \\ \widehat{a}c & d & 1 \end{pmatrix}$$

where

$$B = \begin{pmatrix} 1 & 0 & 0 \\ 0 & 0 & -1 \\ 0 & 1 & 0 \end{pmatrix} = T_{23}(3, -1)T_{32}(3, 1)T_{23}(3, -1) \in \text{SE}(3, R)$$

and (by taking transposes)

$$\widehat{B} = \begin{pmatrix} 1 & 0 & 0 \\ 0 & 0 & 1 \\ 0 & -1 & 0 \end{pmatrix} = T_{32}(3, -1)T_{23}(3, 1)T_{32}(3, -1) \in \text{SE}(3, R).$$

It only remains to note that clearly

$$\begin{pmatrix} a\widehat{a} & b & 0 \\ \widehat{c} & 0 & -\widehat{d} \\ \widehat{a}c & d & 1 \end{pmatrix} T_{31}(3, -\widehat{a}c)T_{32}(3, -d) = \begin{pmatrix} a\widehat{a} & b & 0 \\ c^* & d^* & -\widehat{d} \\ 0 & 0 & 1 \end{pmatrix}$$

for some elements c^*, d^* in R, and equally clearly

$$T_{23}(3, \widehat{d}) \begin{pmatrix} a\widehat{a} & b & 0 \\ c^* & d^* & -\widehat{d} \\ 0 & 0 & 1 \end{pmatrix} = \begin{pmatrix} a\widehat{a} & b & 0 \\ c^* & d^* & 0 \\ 0 & 0 & 1 \end{pmatrix}.$$

To prove (iii) let c, d be elements in R such that $ad - bc = 1$. Then

$$\begin{pmatrix} a & b & 0 \\ c & d & 0 \\ 0 & 0 & 1 \end{pmatrix} \widetilde{B} = \begin{pmatrix} -b & a & 0 \\ -d & c & 0 \\ 0 & 0 & 1 \end{pmatrix}$$

with

$$\widetilde{B} = \begin{pmatrix} 0 & 1 & 0 \\ -1 & 0 & 0 \\ 0 & 0 & 1 \end{pmatrix} = T_{21}(3, -1)T_{12}(3, 1)T_{21}(3, -1) \in \text{SE}(3, R)$$

and hence

$$\mu(a, b) = \mu(-b, a).$$

But

$$\mu(-b, a) = \mu(-1, a)\mu(b, a) = \mu(b, a)$$

where the first equation is by (ii) and the second equation is by (i). Therefore $\mu(a, b) = \mu(b, a)$.

To prove (iv) let c, d be elements in R such that $ad - bc = 1$, and note that then

$$\begin{pmatrix} a & b & 0 \\ c & d & 0 \\ 0 & 0 & 1 \end{pmatrix} T_{21}(3, \lambda) = \begin{pmatrix} a+\lambda b & b & 0 \\ c+\lambda d & d & 0 \\ 0 & 0 & 1 \end{pmatrix}.$$

PROOF OF (T129.3). We make induction on $\deg(f)$. In case of $\deg(f) = 0$, our assertion follows from (T129.2). So let $\deg(f) > 0$ and assume for all smaller values of $\deg(f)$. By the division algorithm we can write $g = qf + r$ where q, r are elements in R with $\deg(r) < \deg(f)$, and by parts (iii) and (iv) of (T129.2) we see that $(f, r) \in U_2(R)$ with $\mu(f, g) = \mu(f, r)$. Consequently, replacing g by r, we may assume that $\deg(g) < \deg(f)$. Now $(f, g) \in U_2(K[X])$ with quasilocal K implies that either $f(0)$ or $g(0)$ must belong to $U(K)$.

For a moment suppose that $g(0) \in U(K)$. Then letting $f^* = f - g(0)^{-1}f(0)g$ we have $(f^*, g) \in U_2(R)$ with $f^*(0) = 0$. By the last equation we can write $f^* = X\widehat{f}$ with submonic $\widehat{f} \in R$. Clearly (X, g) and $(\widehat{f}, g)$ both belong to $U_2(R)$ and by (T129.2)(ii) we have $\mu(f, g) = \mu(f^*, g) = \mu(X, g)\mu(\widehat{f}, g)$. Since $\deg(\widehat{f}) < \deg(f)$, by the induction hypothesis we get $\mu(\widehat{f}, g) = 1$. Clearly $g(0) = g - X\widehat{g}$ with $\widehat{g} \in R$ and hence by (T129.2) we also get $\mu(X, g) = \mu(X, g(0)) = 1$. Thus $\mu(f, g) = 1$.

Next suppose that $g(0) \notin U(K)$. If $g = 0$ then $\mu(f, g) = 1$ by (T129.2)(i). So assume that $g \neq 0$. Let $a = \operatorname{Res}_X(f, g) \in K$. Then, as in the proof of (C52)(T123.4), by L4§1(T1), which was proved in L4§10(E7) and reproved in L4§12(R6)(34), we get $a \in U(K)$. By L4§12(R5)(5.1), also proved in L4§12(R6)(6.1), we have $a = uf + vg$ for some u and v in R with $\deg(u) < \deg(g)$ and $\deg(v) < \deg(f)$. Multiplying u and v by a^{-1} and $-a^{-1}$ respectively, we may assume that $uf - vg = 1$. Since $g(0) \notin U(K)$, we must have $u(0) \in U(K)$. Now

$$T_{12}(3, 1)\begin{pmatrix} f & g & 0 \\ v & u & 0 \\ 0 & 0 & 1 \end{pmatrix} = \begin{pmatrix} f+v & g+u & 0 \\ v & u & 0 \\ 0 & 0 & 1 \end{pmatrix}.$$

and hence

$$(f + v, g + u) \in U_2(R) \quad \text{with} \quad \mu(f, g) = \mu(f + v, g + u).$$

Clearly $g(0) + u(0) \in U(K)$ and $f + v$ is submonic of the same degree as f, and hence by the above case of $g(0) \in U(K)$ we get $\mu(f+v, g+u) = 1$. Thus $\mu(f, g) = 1$. This completes the induction and establishes (T129.3).

PROOF OF (T129.4). Clearly we can take p, q in R with

$$A = \begin{pmatrix} f & g & 0 \\ p & q & 0 \\ 0 & 0 & 1 \end{pmatrix} \in \mathrm{SL}(3, R).$$

By hypothesis, for every maximal ideal P in K, the matrix $\alpha_{P,3}(A)$ belongs to the subgroup of $\mathrm{GL}(3, R_P)$ generated by $\mathrm{SL}(3, K_P)$ and $\mathrm{SE}(3, R_P)$, and hence it belongs to $\mathrm{GE}(3, (K_P, R_P))$. Therefore by (T128.8) we see that A belongs to $\mathrm{GE}(3, (K, R))$. Consequently, in view of (T125) and item (††) in the beginning of its proof, we can write $A = NM$ with N in $\mathrm{GL}(3, K)$ and M in $\mathrm{SE}(3, R)$. This yields $N = AM^{-1}$

and hence $N = A(0)M^{-1}(0)$. Therefore, upon letting $B = M^{-1}(0)M$, we get $A = A(0)B$ with $B \in \mathrm{SE}(3, R)$, and hence $\mu(f, g) = \mu(f(0), g(0))$.

PROOF OF (T129.5). By (T129.3) we see that for any maximal ideal P in K we have

$$\beta_{P,3}(\mu(f,g)) = 1 = \mu(1,0) \quad \text{with} \quad (1,0) \in U_2(K_P).$$

Therefore by (T129.4) we get $\mu(f, g) = \mu(f(0), g(0))$.

PROOF OF (T129.6). Given any $A = (A_{ij}) \in \mathrm{SL}(n, R)$, let $A_n = (A_{in})_{1 \le i \le n}$ be the last column of A. By the assumed transitivity we can find $B \in \mathrm{SE}(n, R)$ such that $BA_n = e'_n$ where e'_n is the transpose of the row vector $e_n = (0, \ldots, 0, 1) \in R^n$. Then upon letting $C = BA = (C_{ij})$ we have $C_{in} = 0$ or 1 according as $i \ne n$ or $i = n$. Therefore upon letting

$$D = BA\overline{B} \quad \text{with} \quad \overline{B} = T_{n,1}(n, -C_{n,1}) \ldots T_{n,n-1}(n, -C_{n,n-1}) \in \mathrm{SE}(n, R)$$

we get

$$D = \begin{pmatrix} \widehat{A} & 0_{n-1,1} \\ 0_{1,n-1} & 1_1 \end{pmatrix} \quad \text{with} \quad \widehat{A} \in \mathrm{SL}(n-1, R).$$

In view of (T125) we can write $D = \widehat{B}A$ with $\widehat{B} \in \mathrm{SE}(n, R)$. Repeating this argument $n - m$ times we get

$$\widetilde{B}A = \begin{pmatrix} \widetilde{A} & 0_{n-m,1} \\ 0_{1,n-m} & 1_{n-m} \end{pmatrix} \quad \text{with} \quad \widetilde{A} \in \mathrm{SL}(m, R).$$

Therefore $A \in \mathrm{QSE}([m, n], R)$. Thus $\mathrm{SL}(n, R) = \mathrm{QSE}([m, n], R)$.

PROOF OF (T129.7). In view of (T129.6), this follows from (C52)(T126).

PROOF OF (T130). In view of (T129.7), it suffices to show that: $(\bullet)$ for every $(f, g) \in U_2(R)$ we have $\mu(f, g) = 1$. To prove $(\bullet)$ we make induction on N. Since the $N = 0$ case follows from (C51)(T121), let $N > 0$ and assume for $N - 1$. Now clearly either $f \ne 0$ or $g \ne 0$, and hence in view of L3§12(4), by letting R undergo a K-automorphism, we can arrange matters so that either f or g is a submonic polynomial in X_N with coefficients in $\widehat{R} = K[X_1, \ldots, X_{N-1}]$. By (T129.2)(iii) and (T129.5) we see that $\mu(f, g) = \mu(f(0), g(0))$. Now by the induction hypothesis we get $\mu(f(0), g(0)) = 1$, and hence $\mu(f, g) = 1$. Thus the induction is complete and the theorem is established.

PROOF OF (T131.1). Follows by noting that for all $u \in R$ we have

$$E(u) = \begin{pmatrix} 1 & 1-u \\ 0 & 1 \end{pmatrix} \begin{pmatrix} 1 & 0 \\ -1 & 1 \end{pmatrix} \begin{pmatrix} 1 & 1 \\ 0 & 1 \end{pmatrix} \in \mathrm{SE}(2, R)$$

and

$$T_{12}(2, u) = \left(\begin{smallmatrix} 1 & u \\ 0 & 1 \end{smallmatrix}\right) = \left(\begin{smallmatrix} -u & 1 \\ -1 & 0 \end{smallmatrix}\right) \left(\begin{smallmatrix} 0 & 1 \\ -1 & 0 \end{smallmatrix}\right)^{-1} = E(-u)(E(0))^{-1}$$

and

$$T_{21}(2, u) = \left(\begin{smallmatrix} 1 & 0 \\ u & 1 \end{smallmatrix}\right) = \left(\begin{smallmatrix} 0 & 1 \\ -1 & 0 \end{smallmatrix}\right)^{-1} \left(\begin{smallmatrix} u & 1 \\ -1 & 0 \end{smallmatrix}\right) = (E(0))^{-1} E(u).$$

PROOF OF (T131.2). This follows by direct checking.

PROOF OF (T131.3). By (T131.2)(ii) we have $A = \text{diag}(a, b)E(u_1) \dots E(u_m)$ for some a, b in $U(R)$ and some $u_1, \dots, u_m$ in R with $m \in \mathbb{N}$. Let m be the smallest such. Then, in view of (T131.2), the minimality of m tells us that for $1 < i < m$ we must have $u_i \notin U(R) \cup \{0\}$.

PROOF OF (T131.4). We make induction on m. If $m = 0$ then $f = a \in k^\times$ with $g = 0$ and hence $\deg(f) > \deg(g)$. If $m = 1$ then $f = au_1 \in R \setminus k$ with $g = a \in k^\times$ and hence again $\deg(f) > \deg(g)$. If $m = 2$ then $f = a(u_1 u_2 - 1)$ with $g = au_1$; consequently in case of $u_1 = 0$ we have

$$\deg(f) = 0 > -\infty = \deg(g)$$

and in case of $u_1 \neq 0$ we have

$$\deg(f) = \deg(u_1) + \deg(u_2) > \deg(u_1) = \deg(g).$$

So let $m > 2$ and assume for $m - 1$. Then upon letting

$$A_1 = \left(\begin{smallmatrix} f_1 & g_1 \\ \widehat{f}_1 & \widehat{g}_1 \end{smallmatrix}\right) = \text{diag}(a, b)E(u_1) \dots E(u_{m-1})$$

by the induction hypothesis we have $\deg(f_1) > \deg(g_1)$. Also $A = A_1 E(u_m)$; consequently $f = f_1 u_m - g_1$ with $g = f_1$, and hence

$$\deg(f) = \deg(f_1) + \deg(u_m) > \deg(f_1) = \deg(g).$$

This completes the induction and establishes our assertion.

PROOF OF (T131.5). Suppose if possible that $B \in \text{DE}(2, R)$. Then by (T131.3) we can write $B = \text{diag}(a, b)E(u_1) \dots E(u_m)$ where a, b in $k^\times$ and $m \in \mathbb{N}$ with $u_1, \dots, u_m$ in R are such that for $1 < i < m$ we have $u_i \notin k$. Since F, G are nonzero polynomials of positive degree, we must have $m > 0$. Therefore, since $\deg(f) = \deg(g)$, by (T131.4) we must have $u_m \in k$. If $m = 1$ then $F = au_1 \in k$ in contradiction to $\deg(F) > 0$. If $m = 2$ then $F = a(u_1 u_2 - 1)$ and $G = au_1$ in contradiction to F and G being of equal positive degree with k-linearly-independent degree forms. So assume $m > 2$ and let

$$B_1 = \left(\begin{smallmatrix} F_1 & G_1 \\ \widehat{F}_1 & \widehat{G}_1 \end{smallmatrix}\right) = \text{diag}(a, b)E(u_1) \dots E(u_{m-1}).$$

Then by (T131.4) we have $\deg(F_1) > \deg(G_1)$. Also

$$B_1 = BE(u_m)^{-1} = \left(\begin{smallmatrix} F & G \\ \hat{F} & \hat{G} \end{smallmatrix}\right)\left(\begin{smallmatrix} 0 & -1 \\ 1 & u_m \end{smallmatrix}\right) = \left(\begin{smallmatrix} G & -F + Gu_m \\ \hat{G} & -\hat{F} + \hat{G}u_m \end{smallmatrix}\right)$$

which together with the assumption that the degree forms of the two polynomials F and G are k-linearly-independent tells us that $\deg(F_1) = \deg(G_1)$. Thus we have again reached a contradiction. Therefore we must have $B \notin \mathrm{DE}(2, R)$.

PROOF OF (T131.6). By direct calculation we see that

$$CE(0)C^{-1} = B = \left(\begin{smallmatrix} F & G \\ \hat{F} & \hat{G} \end{smallmatrix}\right)$$

with

$$F = -X^2(1 - XY) + Y^2(1 + XY) \quad \text{and} \quad X^4 + (1 + XY)^2.$$

Clearly F and G are polynomials of degree 4 and their degree forms are linearly independent over k. Therefore by (T131.1) and (T131.5) we see that $B \notin \mathrm{SE}(2, R)$ and hence $\mathrm{SE}(2, R)$ is not a normal subgroup of $\mathrm{SL}(2, R)$.

PROOF OF (T131.7). Clearly $\det(C) = 1$. Moreover, by (T131.1) and (T131.5) we get $C \notin \mathrm{SE}(2, R)$.

COMMENT (C55). [**Permutation Matrices**]. In addition to elementary row and column operations, and the operations of multiplying a row or column by a scalar, which we discussed in (C51), the other basic operations which can be performed on an $m \times n$ matrix $A = (A_{ij})$ over a ring R, are permuting rows or columns. To do these we respectively multiply A from the left by an $m \times m$ permutation matrix or from the right by an $n \times n$ permutation matrix.

Given any $\sigma \in S_n$ we let $T(n, \sigma) = (T(n, \sigma)_{ij})$ be the $n \times n$ matrix such that $T(n, \sigma)_{ij} = 1$ or 0 according as $i = \sigma(j)$ or $i \neq \sigma(j)$. Then for $B = AT(n, \sigma) = (B_{ij})$ we have $B_{ij} = A_{i\sigma(j)}$ for all i, j. Turning to rows, given any $\tau \in S_m$ we have $T(n, \tau^{-1})_{ij} = 1$ or 0 according as $j = \tau(i)$ or $j \neq \tau(i)$, and this time for $C = T(m, \tau^{-1})A = (V_{ij})$ we have $C_{ij} = A_{\tau(i)j}$ for all i, j.

By a permutation matrix we mean a matrix of the form $T(n, \sigma)$ for some $\sigma \in S_n$. Note that this belongs to $\mathrm{GL}(n, \mathbb{Z})$, but may also be regarded as a member of $\mathrm{GL}(n, R)$ for any ring R. Special cases of the following three easy to prove Lemmas about Matrices were used in (C54). The details of these three Lemmas may be left as exercises for the reader.

FIRST MATRIX LEMMA (T132). For any $n \in \mathbb{N}$ and any $\sigma \in S_n$ we have $\det(T(n, \sigma)) = \mathrm{sgn}(\sigma)$. For any $n \in \mathbb{N}_+$ and any nonnull ring R, the mapping $\sigma \mapsto T(n, \sigma)$ gives a monomorphism $S_n \to \mathrm{GL}(n, R)$. For any $n \in \mathbb{N}_+$ and any D in $\mathrm{GL}(n, R)$ where R is any nonnull ring, we have that: D is a permutation matrix iff each of its row has exactly one nonzero entry and the value of that entry is 1;

likewise: D is a permutation matrix iff each of its column has exactly one nonzero entry and the value of that entry is 1.

SECOND MATRIX LEMMA (T133). Let $\sigma \in S_n$ with $n \in \mathbb{N}_+$, let R be a nonnull ring, and let D be an $n \times n$ matrix over R which is obtained by changing an even or odd number of nonzero entries in $T(n, \sigma)$ from 1 to -1, according as σ is even or odd. Then $D \in \text{SE}(n, R)$.

THIRD MATRIX LEMMA (T134). Let $n \in \mathbb{N}_+$, let R be a nonnull ring, and let $D = (D_{ij})$ be an $n \times n$ matrix over R such that $D_{pq} \in U(R)$ for some p, q. Then there exist $E, \widehat{E}$ in $\text{SE}(n, R)$ such that upon letting $\widehat{D} = ED\widehat{E} = (\widehat{D}_{ij})$ we have: $\widehat{D}_{iq} = \widehat{D}_{pj} = 0$ for all $i \neq p$ and $j \neq q$, and we have: $\widehat{D}_{uv} = D_{uv}$ for all u, v with $D_{pv} = D_{uq} = 0$.

QUEST (Q32) Separable Extensions and Primitive Elements

Out of Theorems (T19) to (T30) stated in §§8-9 of L4, we have proved all except Theorem (T20) which we now proceed to prove. What we want to show is that, given any affine domain $T = k[x_1, \ldots, x_N]$ over an algebraically closed field k, upon letting $L = k(x_1, \ldots, x_N)$ and $r = $ the transcendence degree of L/k, there exist k-linear combinations $y_i = \sum_{1 \leq j \leq N} A_{ij} x_j$ with $A_{ij} \in k$ such that $y_1, \ldots, y_r$ is a transcendence basis of L/k and $L = k(y_1, \ldots, y_{r+1})$. In (T144) we shall obtain a somewhat stronger version of this result. We shall do this in two steps, by first showing that suitable k-linear combinations $y_1, \ldots, y_r$ of $x_1, \ldots, x_N$ constitute a separating transcendence basis of L/k, i.e., a transcendence basis so that L is separable algebraic over $K = k(y_1, \ldots, y_r)$ as defined below, and then showing that a suitable k-linear combination y_{r+1} of $x_1, \ldots, x_N$ provides a primitive element of L over K, i.e, an element for which $L = K(y_{r+1})$. These two steps will be given in (T143) and (T140) respectively. Actually, in (T143) we shall show that suitable k-linear combinations $y_1, \ldots, y_r$ of $x_1, \ldots, x_N$ provide a separating normalization basis of T/k, i.e., a separating transcendence basis of L/K such that the ring T is integral over the ring $R = k[y_1, \ldots, y_r]$. Using (T140), in (T141) we shall show that the integral closure of a normal noetherian domain in a finite separable algebraic extension of its quotient field is a finite module over that domain, and hence the integral closure S of R in L is a finite R-module; it follows that S, which is also the integral closure of T in L, is a finite T-module.

The proof of (T141) will mainly be based on the Vandermonde Determinant Theorem (T135) which is one of the oldest theorems about determinants dating back to 1771. The deduction of (T141) from (T135) will be via the consequence (T136) of (T135) which embeds the integral closure in a ring obtained by inverting the modified discriminant. Referring to L4§1 for the definition of the Y-discriminant $\text{Disc}_Y(f)$ of a univariate polynomial $f(Y) = a_0 Y^n + a_1 Y^{n-1} + \cdots + a_n$ with positive integer n we define its modified Y-discriminant $\mathrm{Disc}_Y^*(f)$ by putting

$$\mathrm{Disc}_Y^*(f) = (-1)^{n(n-1)/2}\mathrm{Disc}_Y(f).$$

Referring to L1§1 and L1§8, let us recall that a univariate polynomial $f = f(Y)$ over a field K is said to be separable if its roots can be separated, i.e., if it is devoid of multiple roots, i.e., it is not divisible by $(Y - \alpha)^2$ for any α in any overfield of K; if f is not separable then it is said to be inseparable; clearly the zero polynomial is inseparable. An element y in an overfield of K is said to be separable over K if $f(y) = 0$ for some separable f over K, i.e., equivalently, if y/K is algebraic and its minimal polynomial over K is separable; if y/K is algebraic but not separable then we say that y/K is inseparable; if $\mathrm{ch}(K) = p \neq 0$ (i.e., if the characteristic of K is a prime number p) and $y^{p^u} \in K$ for some $u \in \mathbb{N}_+$ then we say that y/K is purely inseparable; if $\mathrm{ch}(K) = 0$ then y/K is purely inseparable means $y \in K$. A field extension L/K is separable (resp: purely inseparable) means every element of L is separable (resp: purely inseparable) over K. A field extension L/K is inseparable means it is algebraic but not separable. A field is perfect means every algebraic extension of it is separable; otherwise it is imperfect. In (T139) we shall show that every characteristic zero field is perfect and so is every finite field.

As an aid to (T143), in Theorem (T142) we shall show that every algebraic function field over a perfect ground field is separably generated according to the following definitions. By an algebraic function field L (of r variables) over a ground field k we mean a finitely generated field extension of k (with $\mathrm{trdeg}_k L = r$), i.e., an overfield L of k of the form $L = k(x_1, \ldots, x_N)$ with $N \in \mathbb{N}$. The field extension L/k is separably generated (or the field L is separably generated over the subfield k) means there exists a transcendence basis $y_1, \ldots, y_r$ of L/k such that L is a separable algebraic field extension of $K = k(y_1, \ldots, y_r)$; we then call $y_1, \ldots, y_r$ a separating transcendence basis of L/k; in case of $r = 1$ we call y_1 a separating transcendental of L/k. For euphony we may say "separable algebraic" instead of "separable."

VANDERMONDE DETERMINANT THEOREM (T135). The Vandermonde Determinant

$$V(Y_1, \ldots, Y_n) = \det \begin{pmatrix} 1 & Y_1 & Y_1^2 & \ldots & Y_1^{n-1} \\ 1 & Y_2 & Y_2^2 & \ldots & Y_2^{n-1} \\ \cdots & \cdots & \cdots & \cdots & \cdots \\ \cdots & \cdots & \cdots & \cdots & \cdots \\ \cdots & \cdots & \cdots & \cdots & \cdots \\ 1 & Y_n & Y_n^2 & \ldots & Y_n^{n-1} \end{pmatrix}$$

has the factorization

$$V(Y_1, \ldots, Y_n) = \prod_{1 \leq i < j \leq n} (Y_j - Y_i).$$

This may be regarded as an identity in the n variable polynomial ring $\mathbb{Z}[Y_1, \ldots, Y_n]$ with $n \in \mathbb{N}_+$, and therefore it may also be regarded as an identity between two expressions involving any elements $Y_1, \ldots, Y_n$ in any ring.

PROOF. Let us write V_n for V, and let the above product be denoted by $W_n(Y_1, \ldots, Y_n)$. For $i < j$, by subtracting the i-th row of the above matrix from its j-th row we get a matrix whose every entry in the j-th row is divisible by $Y_j - Y_i$, and hence $V_n(Y_1, \ldots, Y_n)$ is divisible by $Y_j - Y_i$ in the polynomial ring $\mathbb{Z}[Y_1, \ldots, Y_n]$. Moreover, V_n and W_n are both homogeneous polynomials of the same degree $n(n-1)/2$. Therefore,

$$V_n(Y_1, \ldots, Y_n) = C_n W_n(Y_1, \ldots, Y_n)$$

with $C_n \in \mathbb{Z}$. Putting $Y_1 = 0$, the polynomial V_n is reduced to

$$Y_2 \ldots Y_n V_{n-1}(Y_2, \ldots, Y_n)$$

and the polynomial W_n is reduced to

$$Y_2 \ldots Y_n W_{n-1}(Y_2, \ldots, Y_n).$$

Also

$$V_1(Y_1) = 1 = W_1(Y_1).$$

So we are done by induction on n.

DISCRIMINANT INVERTING THEOREM (T136). Let $L = K(y)$ be a field extension where the element y is separable algebraic over K. Consider the minimal polynomial $f(Y) = Y^n + a_1 Y^{n-1} + \cdots + a_n$ of y over K, and let us factor it as $f(Y) = (Y - y_1) \ldots (Y - y_n)$ where $y_1, \ldots, y_n$ are elements in an overfield of L with $y = y_1$. Let Δ be the modified Y-discriminant of $f(Y)$, and let V be as in (T135). Then $0 \neq \Delta = V(y_1, \ldots, y_n)^2 \in K$. Now assume that R is a normal domain with quotient field K such that y is integral over R, and let S be the integral closure of R in L. Then $\Delta \in R$, and for the R-modules S and $(1/\Delta)R[y]$ we have $S \subset (1/\Delta)R[y]$. Hence, in particular, if R is noetherian then S is a finite R-module.

PROOF. Given any $z \in L$ we can write

(1) $$z = \sum_{1 \leq j \leq n} x_j y^{j-1} \quad \text{with} \quad x_j \in K.$$

Let

(2) $$z_i = \sum_{1 \leq j \leq n} a_{ij} x_j \quad \text{with} \quad a_{ij} = y_i^{j-1} \quad \text{for} \quad 1 \leq i \leq n.$$

Considering the "Vandermonde Matrix" $a = (a_{ij})$ and letting $b^{(j)}$ be the $n \times n$ matrix obtained by replacing the j-th column of a by the column vector $(z_1, \ldots, z_n)$, by Cramer's Rule L4§10(E4) we get

$$\det(a)x_j = \det(b^{(j)}) \quad \text{for} \quad 1 \le j \le n.$$

Clearly $\det(a) = V(y_1, \ldots, y_n)$ and hence, in view of L4§1(X1) proved in (28) to (30) of L4§12(R6), by (T135) we have

$$(3) \qquad\qquad 0 \ne \Delta = \det(a)^2 \in K$$

and hence by the previous equation we get

$$(4) \qquad\qquad \Delta x_j = \det(b^{(j)}) \det(a) \quad \text{for} \quad 1 \le j \le n.$$

Let $g(Y) = Y^m + b_1 Y^{m-1} + \cdots + b_m$ be the minimal polynomial of z over K. Then by (1) and (2) we see that $g(Y)$ is also the minimal polynomial of z_i over K for $1 \le i \le n$. Therefore if z is integral over R then so is z_i for $1 \le i \le n$.

Now assume that R is a normal domain with quotient field K such that y is integral over R, and let S be the integral closure of R in L. Also let $\overline{S}$ be the integral closure of R in $\overline{L} = K(y_1, \ldots, y_n)$ and assume that z is integral over R. Then the elements $y_1, \ldots, y_n, z_1, \ldots, z_n$ belong to $\overline{S}$ and hence so do the elements $\det(a), \det(b^{(1)}), \ldots, \det(b^{(n)})$. Therefore by (4) we see that

$$\Delta x_j \in \overline{S} \quad \text{for} \quad 1 \le j \le n.$$

and hence, because R is normal and Δ as well as x_j belong to K, we get

$$\Delta x_j \in R \quad \text{for} \quad 1 \le j \le n.$$

Consequently by (1) we obtain

$$z \in (1/\Delta)R[y].$$

Since z was a random element of S, we conclude that

$$S \subset (1/\Delta)R[y].$$

The rest follows from (O14)(5•) and (O25)(27•) of L4§5.

BASIC ROOTS OF UNITY THEOREM (T137). We have the following.

(T137.1) Let K be a field, and let n be a positive integer which is not divisible by the characteristic of K. Let L be an overfield of a splitting field of the polynomial $Y^n - 1$ over K, and let W be the set of all roots of $Y^n - 1$ in L. Then W is a cyclic group of order n, i.e., there exists $w \in L$ such that $w, w^2, \ldots, w^n = 1$ are exactly all the distinct n-th roots of 1 in L.

[Such an element w is called a primitive n-th root of 1 in L].

(T137.2) Let K be a field, and let V be a finite subgroup of $K^\times$. Then V is a cyclic group whose order m is not divisible by the characteristic of K. Moreover $Y^m - 1 = \prod_{v \in V}(Y - v)$.

PROOF OF (T137.1). Obviously $Y^n - 1$ has no common root with its derivative nY^{n-1}, and hence $|W| = n$; see (Q27)(C41). Let us prove the existence of w by induction on n. For $n = 1$ we have nothing to show. So let $n > 1$ and assume for all smaller values of n. If n is a power of a prime number π then we can take a root w of $Y^n - 1$ which is not a root of $Y^{n/\pi} - 1$ because the latter has only n/π distinct roots; since $w^n = 1 \neq w^{n/\pi}$, the order of w must be n. So suppose that n is not a power of a prime. Then we can write $n = n'n''$ where $n' < n$ and $n'' < n$ are positive integers with $\text{GCD}(n', n'') = 1$. By the induction hypothesis there exist elements w', w'' in W of orders n', n'' respectively. Let $w = w'w''$. Since $w^n = 1$, it follows that the order of w is n. So we are done.

PROOF OF (T137.2). In (T137.1) let L be a splitting field of $Y^n - 1$ over K where n is the LCM of the orders of the elements of V, and note that then V is a subgroup of W.

FACTS. In the above two proofs we used the following obvious facts:

(5) For any field K and positive integer n, the roots of $Y^n = 1$ belonging to K, i.e., the n-th roots of 1 in K, form a (finite commutative multiplicative) subgroup of $K^\times$, and the order of any element of this subgroup is nondivisible by $\text{ch}(K)$.

[The last statement follows by noting that if $\text{ch}(K) = p \neq 0$ then for all $q = p^u$ with $u \in \mathbb{N}_+$ and x, y in K we have $(x \pm y)^q = x^q \pm y^q$].

(6) For the order r of any element w in any group W, upon letting d to be the unique nonnegative generator of the principal ideal in $\mathbb{Z}$ consisting of all $e \in \mathbb{Z}$ with $w^e = 1$, we have $r = \infty$ or d according as $d = 0$ or $d > 0$.

(7) For any elements w' and w'' of finite coprime orders n' and n'' in any group W, with $w'w'' = w''w'$, we have that the order of $w'w''$ equals $n'n''$.

[Two integers are coprime means their GCD is 1].

(8) For any element w of finite order n in any group W and any integer m, the order of w^m equals $n/\text{GCD}(m, n)$.

(9) For any cyclic group W of finite order n, the mapping $V \mapsto |V| = m$ gives a bijection of the set of all subgroups of W onto the set of all positive integers which divide n, and the groups V and W/V are cyclic of order m and n/m respectively.

BASIC FINITE FIELD THEOREM (T138). We have the following.

(T138.1) For any finite field L, let $|L| = q$. Then the characteristic of L is a prime number p and q is a positive power of p. Moreover $Y^{q-1} - 1 = \prod_{w \in L^\times}(Y - w)$ and $Y^q - Y = \prod_{x \in L}(Y - x)$. Hence in particular $L^\times$ is a cyclic group of order $q - 1$, and $L^p = L$. Furthermore L is a splitting field of $Y^q - Y$ over every subfield (in particular over the prime subfield) of L.

[q is a positive power of p means $q = p^u$ for some $u \in \mathbb{N}_+$. For any field L of characteristic $p \neq 0$ and any positive power q of p we put $L^q = \{x^q : x \in L\}$. By (5) above we see that L^q is a subfield of L. The difference between this set L^q and the set of all q-tuples of elements of L should be clear from the context].

(T138.2) For any positive power q of any prime number p, there exists a field L with $|L| = q$, and any two fields of cardinality q are isomorphic.

PROOF OF (T138.1). Clearly the characteristic of L is a prime number p and, upon letting $[L : K] = u$ where K is the prime subfield ($=$ the prime subring) of L, we get $q = p^u$ with $u \in \mathbb{N}_+$; to see this, take a K-basis $z_1, \dots, z_u$ of L and note that then $(a_1, \dots, a_u) \mapsto a_1 z_1 + \cdots + a_u z_u$ gives a bijection $K^u \to L$. Now by (T137.2) we see that

$$Y^{q-1} - 1 = \prod_{w \in L^\times}(Y - w)$$

and $L^\times$ is a cyclic group of order $q - 1$. Therefore

$$Y^q - Y = \prod_{x \in L}(Y - x)$$

and hence $L^p = L$ where $L^p = \{x^p : x \in L\}$. It follows that L is a splitting field of $Y^q - Y$ over every subfield of L.

PROOF OF (T138.2). In view of the last sentence of (T138.1), the isomorphism follows from the uniqueness of splitting field discussed in L1§10(R8). To prove the existence let L be a splitting field of $Y^q - Y$ over $K = \mathbb{Z}/(p\mathbb{Z})$. Since the derivative of this polynomial equals -1, it has q distinct roots in L, and by (5) above they form a field, which must clearly coincide with L.

BASIC PERFECT FIELD THEOREM (T139). We have the following.

(T139.1) Every field K of characteristic zero is perfect.

(T139.2) A field K of characteristic $p \neq 0$ is perfect iff $K = K^p$.

(T139.3) Every finite field is perfect, and so is every algebraically closed field.

PROOF OF (T139.1). Given any element y in any overfield of K such that y is algebraic over K, let

$$f(Y) = Y^n + a_1 Y^{n-1} + \cdots + a_n$$

be the minimal polynomial of y over K. Its Y-derivative

$$f_Y(Y) = nY^{n-1} + \cdots$$

is a nonzero polynomial of degree $n-1$. Therefore $f(Y)$ is separable by (Q27)(C41). Hence y is separable over K. It follows that K is perfect.

PROOF OF (T139.2). First assume that $K = K^p$, and given any element y in any overfield of K such that y is algebraic over K, let $f(Y) = Y^n + a_1 Y^{n-1} + \cdots + a_n$ be the minimal polynomial of y over K. Suppose if possible that $f_Y(Y) = 0$. Then $n/p \in \mathbb{N}_+$, and for every integer $i \in \{1, \ldots, n\}$ with $a_i \neq 0$ we must have $i/p \in \mathbb{N}_+$; since by assumption $K = K^p$, we can write $a_i = b_i^p$ with $b_i \in K$. Let

$$g(Y) = Y^{n/p} + \sum_{i \in \{1, \ldots, n\} \text{ with } i \in p\mathbb{Z}} b_i Y^{(n-i)/p}.$$

Then $g(Y)$ is a nonzero member of $K[Y]$ of degree $n/p < n$, and $f(Y) = g(Y)^p$. This contradicts the irreducibility of $f(Y)$. Therefore $f_Y(Y) \neq 0$. Hence by (Q27)(C41) we see that y is separable over K. It follows that K is perfect.

Now conversely assume that $K \neq K^p$. Then we can find $x \in K \setminus K^p$ and we can take y in an overfield of K such that $y^p = x$. By (Q27)(C41) we see that $Y^p - x$ is the minimal polynomial of y over K. Clearly $f_Y(Y) = 0$ and hence again by (Q27)(C41) we conclude that y is inseparable over K. It follows that K is imperfect.

PROOF OF (T139.3). Follows from (T138.1), (T139.1), and (T139.2).

BASIC PRIMITIVE ELEMENT THEOREM (T140). Let $L = K(x_1, \ldots, x_N)$ be a finite algebraic field extension with $N \in \mathbb{N}$ such that, for every i with $2 \leq i \leq N$, the element x_i is separable over K. Then $L = K(y)$ for some $y \in L$. [Hence in particular every finite separable algebraic field extension has a primitive element]. Moreover, if K is infinite and $N > 0$ then y can be chosen to be a K-linear combination $A_1 x_1 + \cdots + A_N x_N$ of $x_1, \ldots, x_N$ with $A_1, \ldots, A_N$ in K.

PROOF. If K is finite then so is L and hence we are done by (T138.1). So suppose that K is infinite. If $N = 0$ then we are done by taking y to be any element of L. So also suppose that $N > 0$, and let us make induction on N. If $N = 1$ then we are done noting that $L = K(A_1 x_1)$ for all $A_1 \in K^\times$. So let $N > 1$ and assume that we have found $A_1, \ldots, A_{N-1}$ in K such that upon letting $u = A_1 x_1 + \cdots + A_{N-1} x_{N-1}$ and $K' = K(x_1, \ldots, x_{N-1})$ we have $K' = K(u)$. Let $v = x_N$. Let

$$f(Y) = Y^n + a_1 Y^{n-1} + \cdots + a_n \quad \text{and} \quad g(Y) = Y^m + b_1 Y^{m-1} + \cdots + b_m$$

be the minimal polynomials of u and v over K respectively. We can write

$$f(Y) = (Y - \alpha_1)\dots(Y - \alpha_n) \quad \text{and} \quad g(Y) = (Y - \beta_1)\dots(Y - \beta_m)$$

where $\alpha_1,\dots,\alpha_n,\beta_1\dots,\beta_m$ are elements in an overfield L' of L with

$$u = \alpha_1 \quad \text{and} \quad v = \beta_1.$$

By assumption the roots $\beta_1,\dots,\beta_m$ are pairwise distinct and hence for $2 \le i \le n$ and $2 \le j \le m$ we get $z_{ij} \in L'$ such that

$$z_{ij} = \frac{\alpha_1 - \alpha_i}{\beta_j - \beta_1}.$$

Since K is infinite, we can find $A_N \in K^\times$ such that $A_N \ne z_{ij}$ for $2 \le i \le n$ and $2 \le j \le m$. Let $y = u + A_N v$. We shall show that then $K(u,v) = K(y)$ and this will complete the induction and hence the proof.

Now upon letting

$$h(Y) = \left(\frac{-1}{A_N}\right)^n f(-A_N Y + y)$$

we have

$$h(Y) = Y^n + c_1 Y^{n-1} + \dots + c_n \in K(y)[Y]$$

with

$$h(Y) = (Y - v) \prod_{2 \le i \le n} \left(Y - \frac{y - \alpha_i}{A_N}\right).$$

Also

$$g(Y) = Y^m + b_1 Y^{m-1} + \dots + b_m \in K(y)[Y]$$

with

$$g(Y) = (Y - v) \prod_{2 \le j \le m} (Y - \beta_j).$$

By the choice of A_N we see that

$$A_N^{-1}(y - \alpha_i) \ne \beta_j \quad \text{for} \quad 2 \le i \le n \quad \text{and} \quad 2 \le j \le m.$$

Consequently $Y - v$ is a gcd of $g(Y)$ and $h(Y)$ in $L'[Y]$ and hence by L1§10(R2) on Long Division we can write $Y - v = P(Y)g(Y) + Q(Y)h(Y)$ with $P(Y), Q(Y)$ in $K(y)[Y]$. Therefore $v \in K(y)$ and hence by the equation $y = u + A_N v$ we get $u \in K(y)$. Thus $K(u,v) = K(y)$.

FINITENESS OF INTEGRAL CLOSURE THEOREM (T141). For any normal noetherian domain R with quotient field K, and any finite separable algebraic field extension L/K, the integral closure S of R in L is a finite R-module.

PROOF. By (T140) there exists an element z in L such that $L = K(z)$. Let $g(Y) = Y^n + \sum_{1 \le i \le n} b_i Y^{n-i}$ with $b_i \in K$ be the minimal polynomial of z over K. We can find $c \in R^\times$ such that $cb_i \in R$ for $1 \le i \le n$. Let $y = cz$ and $a_i = c^i b_i$. Then $L = K(y)$ and the minimal polynomial of y over K is given by $f(Y) = Y^n + \sum_{1 \le i \le n} a_i Y^{n-i}$ with $a_i \in R$ for $1 \le i \le n$. Consequently y is integral over R. Now we are done by (T136).

SEPARABLE GENERATION THEOREM (T142). For any finitely generated field extension $L = k(x_1, \ldots, x_N)$, with $N \in \mathbb{N}$ and $\mathrm{trdeg}_k L = r$, we have the following.

(T142.1) The extension L/k is separable algebraic (resp: purely inseparable) iff the element x_i is separable algebraic (resp: purely inseparable) over k for $1 \le i \le N$.

(T142.2) If $N = r + 1$ then $\ker(\phi)$ is a nonzero principal prime ideal in the polynomial ring $k[X_1, \ldots, X_N]$ where ϕ is the k-homomorphism of that ring into L which sends the indeterminate X_i to the element x_i for $1 \le i \le N$.

(T142.3) If L/k is separably generated then there exists a sequence of integers $1 \le i_1 < \cdots < i_r \le N$ such that $x_{i_1}, \ldots, x_{i_r}$ is a separating transcendence basis of L/k.

(T142.4) If L/k is not separably generated then $N > r$ and there exists a sequence of integers $1 \le i_1 < \cdots < i_{r+1} \le N$ such that $k(x_{i_1}, \ldots, x_{i_{r+1}})/k$ is not separably generated.

(T142.5) If k is perfect then L/k is separably generated.

FACTS. While proving (T142.1), we shall also prove the following related facts.

(10) If x is an element in a field K such that x is separable algebraic as well as purely inseparable over a subfield k of K, then x must belong to k.

(11) If x is an element in a field K such that x is separable algebraic (resp: purely inseparable) over a subfield k of K, then x is separable algebraic (resp: purely inseparable) over every subfield k' of K with $k \subset k'$.

(12) If K/k is any separable algebraic field extension for which $\mathrm{ch}(k) = p \ne 0$, then $k[K^p] = K$. Conversely, if K/k is any finite algebraic field extension for which $\mathrm{ch}(k) = p \ne 0$ and $k[K^p] = K$, then K/k is separable algebraic.

(13) For any element x in an overfield of a field k of $\mathrm{ch}(k) = p \ne 0$ we have that: x/k is separable $\Leftrightarrow k(x) = k(x^p) \Leftrightarrow k(x)/k$ is separable.

(14) If x is an element in a field K such that x is separable algebraic (resp: purely inseparable) over a subfield k' of K, and k'/k is a separable algebraic (resp: purely inseparable) field extension, then x is separable algebraic (resp: purely inseparable) over k.

(15) A field extension $K = k(\{y_j\}_{j \in J})$, where $\{y_j\}_{j \in J}$ is any family of elements, is separable algebraic (resp: purely inseparable) iff the element y_j is separable algebraic (resp: purely inseparable) over k for every $j \in J$.

PROOF OF (T142.1). To prove (10) let x be an element in a field K such that x is purely inseparable over a subfield k of K. If $\mathrm{ch}(k) = 0$ then by the definition of pure inseparability we get $x \in k$. If $\mathrm{ch}(k) = p \neq 0$ and $x \notin k$ then, upon letting u to be the smallest positive integer with $x^{p^u} \in k$, by (Q27)(C41) we see that $Y^{p^u} - x^{p^u}$ is the minimal polynomial of x over k, and since the Y-derivative of this polynomial obviously equals 0, again by (Q27)(C41) we see that x is not separable over k.

Since (11) is obvious, we proceed to prove (12). So let K/k be a field extension with $\mathrm{ch}(k) = p \neq 0$. Note that then $k[K^p] = K^p[k]$ and every element of k is algebraic over the field K^p and hence $K^p[k]$ is a field, i.e., $k[K^p]$ is a subfield of the field K. Therefore, if K/k is separable algebraic then $k[K^p] = K$ by (10) and (11). Now conversely suppose that K/k is finite algebraic with $k[K^p] = K$ and let $[K : k] = d \in \mathbb{N}_+$. Given any k-basis $w_1, \dots, w_d$ of K, every $x \in K$ can be written as $x = \sum_{1 \le i \le d} x_i w_i$ with $x_i \in k$ which yields

$$x^p = \sum_{1 \le i \le d} x_i^p w_i^p \quad \text{with} \quad x_i^p \in k$$

and hence, because of the assumption $k[K^p] = K$, it follows that $w_1^p, \dots, w_d^p$ is also a k-basis of K. Since every set of k-linearly-independent elements of K can be extended to a k-basis of K, it follows that for any k-linearly-independent elements $w_1, \dots, w_e$ of K, the elements $w_1^p, \dots, w_e^p$ are k-linearly-independent. Given any $y \in K$, let

$$f(Y) = Y^n + \sum_{1 \le i \le n} a_i Y^{n-i} \quad \text{with} \quad a_i \in k$$

be its minimal polynomial over k. We shall show that $f_Y(Y) \neq 0$ which by (Q27)(C41) will tell us that y is separable over k, and this will complete the proof of (12). Suppose if possible that $f_Y(Y) = 0$. Then $n/p \in \mathbb{N}_+$ and for every $i \in \{1, \dots, n\}$ with $a_i \neq 0$ we must have $i/p \in \mathbb{N}_+$. It follows that the elements $1, y^p, y^{2p}, \dots, y^{(n/p)p}$ are linearly dependent over k and hence, by what we just proved, so are the elements $1, y, y^2, \dots, y^{n/p}$. This is a contradiction. Therefore $f_Y(Y) \neq 0$.

To prove (13) note that if $k(x) = k(x^p)$ then by taking $K = k(x)$ in the converse part of (12) we see that $k(x)/k$ is separable. On the other hand, if x is separable over k then by (11) it is separable over $k(x^p)$ and obviously it is purely inseparable over $k(x^p)$, and therefore by (10) it belongs to $k(x^p)$, and hence $k(x) = k(x^p)$. If $k(x)/k$ is separable then obviously x/k is separable.

The purely inseparable part of (14) is obvious, and in case of characteristic zero, the separable part follows from (T139.1). So now suppose that $\mathrm{ch}(k) = p \neq 0$ and assume that x/k' and k'/k are separable. Let H be the subfield of k' obtained by adjoining to k all the coefficients of the minimal polynomial of x over k', i.e., $H = k(a_1, \ldots, a_n)$ where $a_1, \ldots, a_n$ are the said coefficients. Also let $J = H(x)$. Then obviously H/k is a finite separable algebraic field extension, and by (13) so is J/H. By the first part of (12) we get $J \subset H[J^p]$ as well as $H \subset k[H^p]$, and hence $J \subset k[J^p]$, and therefore $J = k[J^p]$. Consequently by the second part of (12) we conclude that J/k is separable and hence so is x/k.

Clearly (15) follows from (T142.1), and so it only remains to prove the latter which we proceed to do. The only if part is obvious. The characteristic zero case of the if part follows from (T139.1), and its purely inseparable case follows from (5). Supposing $\mathrm{ch}(k) = p \neq 0$ we shall establish the separable case of the if part by induction on N, and this completes the proof. For $N = 0$ we have nothing to show. So let $N > 0$ and assume for $N - 1$. Then upon letting $k' = k(x_1, \ldots, x_{N-1})$, by the induction hypothesis k'/k is separable. Since $K = k'(x_N)$, by (11) and (13) we see that K/k' is separable. Therefore K/k is separable by (14).

PROOF OF (T142.2). Since $N > r$, we see that the elements $x_1, \ldots, x_N$ are algebraically dependent over k and hence there exists an irreducible $F(X_1, \ldots, X_N)$ in $k[X_1, \ldots, X_N]$ belonging to $\ker(\phi)$. Since $N = r + 1$, there exists $j \in \{1, \ldots, N\}$ such that $x_1, \ldots, x_{j-1}, x_{j+1}, \ldots, x_N$ is a transcendence basis of L/k. For any such F and j, upon considering the ring $R = k[x_1, \ldots, x_{j-1}, x_{j+1}, \ldots, x_N]$, we must have $F(x_1, \ldots, x_{j-1}, Y, x_{j+1}, \ldots, x_N) =$ the minimal polynomial of x_j over $\mathrm{QF}(R)$ times some element of $R^\times$. It follows that $\ker(\phi)$ is generated by any such F. [Namely, let $S = k[X_1, \ldots, X_{j-1}, X_{j+1}, \ldots, X_N]$ and $G \in \ker(\phi)$. Then, for some $H \in S^\times$, F divides GH in $S[X_j]$. Hence F divides G in $S[X_N]$ because F is irreducible in $S[X_j]$ and $F \notin S$].

PROOF OF (T142.3). If $\mathrm{ch}(k) = 0$ then every algebraic extension of an overfield of k is separable by (T139.1), and hence every transcendence basis is a separating transcendence basis, and so our assertion follows from §§4-6 of L2. So assuming $\mathrm{ch}(k) = p \neq 0$, let us first consider the case $r = 1$. By assumption there exists a separating transcendental z of L/k. Since z/k is transcendental, we must have $z \notin k(z^p)$. Therefore $z/k(z^p)$ is inseparable by (Q27)(C41), and hence by (T142.1) we see that $x_j/k(z^p)$ is inseparable for some $j \in \{1, \ldots, N\}$.

We shall show that then x_j is a separating transcendence of L/k, and this will complete the proof of the $r = 1$ case. Let ϕ be the k-homomorphism of the polynomial ring $k[X, Z]$ into L which sends (X, Z) to (x_j, z). By (T142.2) the ideal $\ker(\phi)$ is generated by an irreducible polynomial $F(X, Z)$ in $k[X, Z]$. Let $g(X, Z) \in k(Z)[X]$ be such that $g(Y, z)$ is the minimal polynomial of x_j over $k(z)$. Since x_j is separable over $k(z)$, by (Q27)(C41) we get $g_X(x_j, z) \neq 0$ where subscripts denote partial derivatives. As in the above proof of (T142.2) we have $F(X, Z) = h(Z)g(X, Z)$ with $h(Z) \in k[Z]^\times$ and hence $F_X(X, Z) \neq 0$. If $F(X, Z) = f(X) \in k[X]$ then $f_X(x_j) \neq 0$ and hence x_j/k is separable by (Q27)(C41) and therefore $x_j/k(z^p)$ is separable by (11) in contradiction to our assumption.

Consequently $F(X, Z) \notin k[X]$, and hence $z/k(x_j)$ is algebraic and as in the above proof of (T142.2) we have $F(X, Z) = H(X)G(X, Z)$ where $H(X) \in k[X]^\times$ and $G(X, Z) \in k(X)[Z]$ are such that $G(x_j, Y)$ is the minimal polynomial of $z/k(x_j)$. If $z/k(x_j)$ is inseparable then by (Q27)(C41) we have $F(X, Z) = E(X, Z^p)$ with $E(X, Z) \in k[X, Z]$ and this implies $E_X(x_j, z^p) = F_X(x_j, z) \neq 0$ which, in view of (Q27)(C41), tells us that $x_j/k(z^p)$ is separable in contradiction to our assumption. Therefore $z/k(x_j)$ is separable, and hence x_j is a separating transcendence of L/k. This completes the proof in the $r = 1$ case.

In the general case we make induction on r. The assertion being trivial for $r = 0$, let $r > 1$ and assume for $r - 1$. By assumption there exists a separating transcendence basis $z_1, \ldots, z_r$ of L/k. Let $k_1 = k(z_1)$. Then by the induction hypothesis there exists a sequence $1 \leq j_2 < \cdots < j_r \leq N$ such that $x_{j_2}, \ldots, x_{j_r}$ is a separating transcendence basis of L/k_1. Let $k' = k(x_{j_2}, \ldots, x_{j_r})$. Then z_1 is a separating transcendence of L/k' and hence by the $r = 1$ case there exists $j_1 \in \{1, \ldots, N\} \setminus \{j_2, \ldots, j_r\}$ such that x_{j_1} is a separating transcendence of L/k'. Arranging the integers $j_1, \ldots, j_r$ as a sequence $1 \leq i_1 < \cdots < i_r \leq N$ we see that $x_{i_1}, \ldots, x_{i_r}$ is a separating transcendence basis of L/k. This completes the induction and hence the entire proof.

PROOF OF (T142.4). We make induction on N. If $N = 0$ then we have nothing to show. So let $N > 0$ and assume for all smaller values. Since we are supposing that L/k is not separably generated, we must have $N > r$. Now for some $j \in \{1, \ldots, N\}$, the element x_j is algebraic over the field

$$k' = k(x_1, \ldots, x_{j-1}, x_{j+1}, \ldots, x_N).$$

If k'/k is not separably generated then we are done by the induction hypothesis. So assume that k'/k is separably generated. Then by (T142.3) we can find a sequence $j_1 < \cdots < j_r$ in $\{1, \ldots, j-1, j+1, \ldots, N\}$ such that $x_{j_1}, \ldots, x_{j_r}$ is a separating transcendence basis of k'/k. We can write the integers $j, j_1, \ldots, j_r$ in the form of a sequence $1 \leq i_1 < \cdots < i_{r+1} \leq N$. By (11) the field extension $k(x_{i_1}, \ldots, x_{i_{r+1}})/k$ is not separably generated.

PROOF OF (T142.5). If $\mathrm{ch}(k) = 0$ then, in view of (T139.1), we are done by §§4-6 of L2. So assume that $\mathrm{ch}(k) = p \neq 0$. Now, in view of (T142.4), it suffices to prove our assertion when $N = r + 1$. In this case, upon letting ϕ to be the k-homomorphism of the polynomial ring $k[X_1, \ldots, X_N]$ into L which sends X_i to x_i for $1 \le i \le N$, by (T142.2) the ideal $\ker(\phi)$ is generated by an irreducible member $F(X_1, \ldots, X_N)$ of that ring.

Let there be given any $j \in \{1, \ldots, N\}$ such that $F(X_1, \ldots, X_N) \notin R_j$ where $R_j = k[X_1, \ldots, X_{j-1}, X_{j+1}, \ldots, X_N]$. Then the element x_j is algebraic over the field $k_j = k(x_1, \ldots, x_{j-1}, x_{j+1}, \ldots, x_N)$, and as in the above proof of (T142.2) we have

$$F(X_1, \ldots, X_N) = H_j(X_1, \ldots, X_{j-1}, X_{j+1}, \ldots, X_N) G_j(X_1, \ldots, X_N)$$

where

$$H_j(X_1, \ldots, X_{j-1}, X_{j+1}, \ldots, X_N) \in (R_j)^\times$$

and

$$G_j(X_1, \ldots, X_N) \in (\mathrm{QF}(R_j))[X_j]$$

are such that $G_j(x_1, \ldots, x_{j-1}, Y, x_{j+1}, \ldots, x_N)$ is the minimal polynomial of x_j over k_j. If $F(X_1, \ldots, X_N) \notin S_j$ where $S_j = k[X_1, \ldots, X_{j-1}, X_j^p, X_{j+1}, \ldots, X_N]$, then by (Q27)(C41) the element x_j would be separable over k_j, and so we would be done because $x_1, \ldots, x_{j-1}, x_{j+1}, \ldots, x_N$ would be a separating transcendence basis of L/k. So assume that $F(X_1, \ldots, X_N) \in S_j$ for every integer $j \in \{1, \ldots, N\}$ for which $F(X_1, \ldots, X_N) \notin R_j$. It follows that then $F(X_1, \ldots, X_N) \in k[X_1^p, \ldots, X_N^p]$ and hence by (T139.2) we get $F(X_1, \ldots, X_N) = J(X_1, \ldots, X_N)^p$ for some polynomial $J(X_1, \ldots, X_N) \in k[X_1, \ldots, X_N]$ which contradicts the irreducibility of F.

SEPARATING NORMALIZATION BASIS THEOREM (T143). Given an affine domain $T = k[x_1, \ldots, x_N]$ over a field k with $N \in \mathbb{N}$, let $L = k(x_1, \ldots, x_N)$ and let r be the transcendence degree of L/k. Assume that k is infinite and L/k is separably generated. Then T/k has a separating normalization basis $y_1, \ldots, y_r$. Moreover, we can choose the elements $y_1, \ldots, y_r$ to be k-linear combinations of $x_1, \ldots, x_N$, i.e., $y_i = \sum_{1 \le j \le N} A_{ij} x_j$ for $1 \le i \le r$ where $A = (A_{ij})$ is an $r \times N$ matrix over k of rank r.

[In view of (T139.3) and (T142.5), if k is algebraically closed then k is infinite and L/k is separably generated].

PROOF. In view of (T139.1), if $\mathrm{ch}(k) = 0$ then we are done by the Noether Normalization Theorem (Q10)(T46). Now assuming $\mathrm{ch}(k) = p \ne 0$, and slightly modifying the proof of (Q10)(T46), we prove our assertion by induction on N. If $N = 0$ then we have nothing to show. So let $N > 0$ and assume for all smaller values. If $N = r$ then we can take $y_i = x_i$ for $1 \le i \le r$. So let $N > r$. Then by (T142.3) we can find a permutation σ of the integers $(1, \ldots, N)$ such that $x_{\sigma(1)}, \ldots, x_{\sigma(r)}$ is a separating transcendence basis of L/k. Considering the polynomial rings

$$S = k[X_1, \ldots, X_r, X_N] \quad \text{and} \quad R = k[X_1, \ldots, X_r]$$

we can take

$$F(X_1, \ldots, X_r, X_N) \in S^\times \quad \text{and} \quad H(X_1, \ldots, X_r) \in R^\times$$

such that

$$\frac{F(x_{\sigma(1)}, \ldots, x_{\sigma(r)}, Y)}{H(x_{\sigma(1)}, \ldots, x_{\sigma(r)})}$$

is the minimal polynomial of $x_{\sigma(N)}$ over $k(x_{\sigma(1)}, \ldots, x_{\sigma(r)})$. Since $x_{\sigma(N)}$ is separable over $k(x_{\sigma(1)}, \ldots, x_{\sigma(r)})$, letting subscripts denote partial derivatives, by (Q27)(C41) we see that

$$(16) \qquad\qquad F_{X_N}(x_{\sigma(1)}, \ldots, x_{\sigma(r)}, x_{\sigma(N)}) \neq 0.$$

Let n be the (total) degree of $F(X_1, \ldots, X_r, X_N)$ and let $G(X_1, \ldots, X_r, X_N)$ be its degree form. Then $G(X_1, \ldots, X_r, X_N)$ is a nonzero homogeneous polynomial of degree $n \in \mathbb{N}_+$ and hence

$$(17) \qquad\qquad G(X_1, \ldots, X_r, 1) \in R^\times.$$

Given any

$$z = (z_1, \ldots, z_r) \in k^r$$

let $F^{(z)}$ in S be defined by putting

$$F^{(z)}(X_1, \ldots, X_r, X_N) = F(X_1 + z_1 X_N, \ldots, X_r + z_r X_N, X_N)$$

and note that then

$$F^{(z)}(X_1, \ldots, X_r, X_N) = G(z_1, \ldots, z_n, 1)X_N^n + \sum_{1 \leq i \leq n} G_i(X_1, \ldots, X_r)X_N^{n-i}$$

with

$$G_i(X_1, \ldots, X_r) \in R.$$

Consider the field

$$\widehat{k} = k(x_{\sigma(1)}, \ldots, x_{\sigma(r)}, x_{\sigma(N)})$$

and let $x_i^{(z)}$ and $x_N^{(z)}$ be the elements in it defined by putting

$$(18) \qquad x_i^{(z)} = x_{\sigma(i)} - z_i x_{\sigma(N)} \quad \text{for} \quad 1 \leq i \leq N - 1 \quad \text{with} \quad x_N^{(z)} = x_{\sigma(N)}$$

and note that then

$$\widehat{k} = k(x_1^{(z)}, \ldots, x_r^{(z)}, x_N^{(z)}).$$

Also

$$F^{(z)}(x_1^{(z)}, \ldots, x_r^{(z)}, x_N^{(z)}) = 0$$

and hence

(19)   if $G(z_1, \ldots, z_r, 1) \neq 0$ then $x_N^{(z)}$ is integral over $k[x_1^{(z)}, \ldots, x_r^{(z)}]$.

Considering the polynomial ring

$$\widehat{R} = \widehat{k}[Z_1, \ldots, Z_r]$$

and letting $\widehat{F}(Z_1, \ldots, Z_r) \in \widehat{R}$ be defined by putting

$$\begin{cases} \widehat{F}(Z_1, \ldots, Z_r) = F_{X_N}(x_{\sigma(1)}, \ldots, x_{\sigma(r)}, x_{\sigma(N)}) \\ \qquad + \sum_{1 \leq i \leq r} Z_i F_{X_i}(x_{\sigma(1)}, \ldots, x_{\sigma(r)}, x_{\sigma(N)}) \end{cases}$$

we have

$$F_{X_N}^{(z)}(x_1^{(z)}, \ldots, x_r^{(z)}, x_N^{(z)}) = \widehat{F}(z_1, \ldots, z_r)$$

and hence by (Q27)(C41) we see that

(20)   if $\widehat{F}(z_1, \ldots, z_r) \neq 0$ then $x_N^{(z)}$ is separable over $k(x_1^{(z)}, \ldots, x_r^{(z)})$.

By (16) we see that $\widehat{F}(0, \ldots, 0) \neq 0$, and hence $\widehat{F}(Z_1, \ldots, Z_r) \in \widehat{R}^{\times}$, and therefore upon letting

(21) $$\widehat{G}(Z_1, \ldots, Z_r) = G(Z_1, \ldots, Z_r, 1)\widehat{F}(Z_1, \ldots, Z_r)$$

by (17) we get

(22) $$\widehat{G}(Z_1, \ldots, Z_r) \in \widehat{k}[Z_1, \ldots, Z_r]^{\times}.$$

As a slight generalization of L3§12(E6) we see that:

(23) $$\begin{cases} \text{given any nonnegative integer } r \\ \text{and any nonzero polynomial } \widehat{G}(Z_1, \ldots, Z_r) \\ \text{in indeterminates } Z_1, \ldots, Z_r \text{ with coefficients in any field } \widehat{k} \\ \text{and given any infinite subset } k \text{ of } \widehat{k} \\ \text{there exist elements } z_1, \ldots, z_r \text{ in } k \text{ such that } \widehat{G}(z_1, \ldots, z_r) \neq 0. \end{cases}$$

By (21) to (23) we can take $z_1, \ldots, z_r$ in k such that

$$G(z_1, \ldots, z_r, 1) \neq 0 \neq \widehat{F}(z_1, \ldots, z_r)$$

and then by (19) and (20) we see that $x_1^{(z)}, \ldots, x_r^{(z)}$ is a separating transcendence basis of $k(x_{\sigma(1)}, \ldots, x_{\sigma(r)}, x_{\sigma(N)})$ over k, and the element $x_N^{(z)}$ is integral over the ring $k[x_{\sigma(1)}, \ldots, x_{\sigma(r)}]$. Let

$$T' = k[x_1^{(z)}, \ldots, x_{N-1}^{(z)}] \quad \text{and} \quad L' = k(x_1^{(z)}, \ldots, x_{N-1}^{(z)}).$$

Then T/T' is integral and L'/L is separable. Also L'/k is separably generated, and hence by the induction hypothesis there exist k-linear combinations $y_1, \ldots, y_r$ of $x_1^{(z)}, \ldots, x_{N-1}^{(z)}$ which form a separating normalization basis of T'/k. In view of (18) it follows that $y_1, \ldots, y_r$ are k-linear combinations of $x_1, \ldots, x_N$ and these k-linear combinations form a separating normalization basis of T/k. This completes the induction and hence the entire proof.

SUPPLEMENTED PRIMITIVE ELEMENT THEOREM (T144). Given an affine domain $T = k[x_1, \ldots, x_N]$ over a field k with $N \in \mathbb{N}$, let $L = k(x_1, \ldots, x_N)$ and let r be the transcendence degree of L/k. Assume that k is infinite and L/k is separably generated. Then there exist elements $y_1, \ldots, y_{r+1}$ in T such that $y_1, \ldots, y_r$ is a separating transcendence basis of L/k with $L = k(y_1, \ldots, y_{r+1})$, and T is integral over $k[y_1, \ldots, y_r]$. Moreover, we can choose the elements $y_1, \ldots, y_r$ to be k-linear combinations of $x_1, \ldots, x_N$, i.e., $y_i = \sum_{1 \le j \le N} A_{ij} x_j$ for $1 \le i \le r$ where $A = (A_{ij})$ is an $r \times N$ matrix of rank r over k. Likewise, if $N > r$ then the elements $y_1, \ldots, y_{r+1}$ can be chosen to be k-linear combinations of $x_1, \ldots, x_N$, i.e., $y_i = \sum_{1 \le j \le N} A_{ij} x_j$ for $1 \le i \le r+1$ where $A = (A_{ij})$ is an $(r+1) \times N$ matrix of rank $r+1$ over k.

[In view of (T139.3) and (T141.4), if k is algebraically closed then k is infinite and L/k is separably generated].

PROOF. By (T143) we can find an $r \times N$ matrix $A = (A_{ij})$ of rank r over k such that upon letting $y_i = \sum_{1 \le j \le N} A_{ij} x_j$ for $1 \le i \le r$ we have that $y_1, \ldots, y_r$ is a separating normalization basis of T/k. If $N = r$ then we are done by taking $y_{r+1} = 0$. So assume that $N > r$. Then we can find an $(r+1) \times N$ matrix $B = (B_{ij})$ of rank $r+1$ over k with $B_{ij} = A_{ij}$ for $1 \le i \le r$ and $1 \le j \le N$. Clearly if $L = k(y_1, \ldots, y_r)$ then it suffices to take $y_{r+1} = \sum_{1 \le j \le N} B_{r+1,j} x_j$. Likewise if $L \ne k(y_1, \ldots, y_r)$ then by (T140) there exist elements $C_{r+1,1}, \ldots, C_{r+1,N}$ in k such that upon letting $y_{r+1} = \sum_{1 \le i \le N} C_{r+1,j} x_j$ we have $L = k(y_1, \ldots, y_{r+1})$; now it suffices to note that, for the $(r+1) \times N$ matrix $D = (D_{ij})$ over k obtained by putting $D_{ij} = A_{ij}$ or C_{ij} according as $1 \le i \le r$ or $i = r+1$, we must have $\text{rk}(D) = r+1$ where we recall that rk denotes rank.

QUEST (Q33) Restricted Domains and Projective Normalization

By a restricted domain we mean a noetherian domain T such that the integral closure of T in any finite algebraic extension of the quotient field L of T is a finite module over T. By a strongly restricted domain we mean a restricted domain T such that every affine domain over T is again a restricted domain. By using

(T143) and some other material from the previous Quest (Q32), in (T145) we shall show that any affine domain over a field is a restricted domain, i.e., a field is a strongly restricted domain. In (T147) we shall show that, according to the following definitions, for a projective model E of K/A, where K is a function field over a strongly restricted domain A and $\overline{K}/K$ is a finite algebraic field extension, the normalization $\mathfrak{N}(E, \overline{K})$ of E in $\overline{K}$ is a projective model of $\overline{K}/A$. In (T146) we shall give an affine version of (T147).

Generalizing the concept of an algebraic function field over a field, by an algebraic function field K (of r variables) over a domain A we mean an algebraic function field K (of r variables) over the quotient field k of A; we also define the kroneckerian dimension of K over A by putting $\mathrm{krdim}_A K = r + \dim(A)$ with the usual convention that an integer $+ \infty = \infty$). Referring to L4§§8-9 for notation, given any quasilocal subring R of any domain $\overline{K}$ we put

$$\begin{cases} \mathfrak{N}(R, \overline{K}) = \{S \in \mathfrak{V}(\overline{R}) : S > R\} \\ \text{where } \overline{R} \text{ is the integral closure of } R \text{ in } \overline{K} \end{cases}$$

and we call this the normalization of R in $\overline{K}$. Likewise, given any set E of quasilocal subrings of a domain $\overline{K}$ we put

$$\mathfrak{N}(E, \overline{K}) = \bigcup_{R \in E} \mathfrak{N}(R, \overline{K})$$

and we call this the normalization of E in $\overline{K}$. Recall that $>$ denotes domination and hence, in view of L4§7.1(T17), by (Q9)(T38.2) we see that $\mathfrak{N}(R, \overline{K})$ equals the set of all the localizations of $\overline{R}$ at the various maximal ideals in it.

FINITE MODULE THEOREM (T145). Any affine domain over any field is a restricted domain. In other words, the integral closure of an affine domain T over a field k in any finite algebraic field extension of the quotient field L of T is a finite T-module.

PROOF. Now $T = k[x_1, \ldots, x_N]$ with $N \in \mathbb{N}$, and we want to show that the integral closure $\overline{T}$ of T in any finite algebraic field extension $\overline{L}$ of $L = k(x_1, \ldots, x_N)$ is a finite T-module.

Let us first consider the very special case when $\overline{L} = L$ and T/k has a separating normalization basis $y_1, \ldots, y_r$ where r is the transcendence degree of L/k. Now the ring $k[y_1, \ldots, y_r]$ is isomorphic to an r variable polynomial ring over the field k, and hence it is a normal noetherian domain; the noetherianness follows from the Hilbert Basis Theorem L4§3(T3), and the normality follows from L3§7(J3) together with L3§8(J20). Clearly $\overline{T}$ is the integral closure of $k[y_1, \ldots, y_r]$ in $\overline{L}$ and hence, by (T141), $\overline{T}$ is a finite module over $k[y_1, \ldots, y_r]$. Therefore, with stronger reason, $\overline{T}$ is a finite T-module.

Next let us consider the special case when $\overline{L} = L$. Referring to L2§§4-6, let $\widehat{L}$ be an algebraic closure of $\overline{L}$ and let $\widehat{k}$ be the (relative) algebraic closure of k in $\widehat{L}$; note that then $\widehat{k}$ is an (absolute) algebraic closure of k. By (T139.3) and (T142.5) we see that $\widehat{k}(x_1, \ldots, x_N)$ is separably generated over the infinite field $\widehat{k}$, and hence by (T143) we see that $\widehat{k}[x_1, \ldots, x_N]$ has a separating normalization basis $y_1, \ldots, y_r$ over $\widehat{k}$, where r is the transcendence degree of $\widehat{k}(x_1, \ldots, x_N)/\widehat{k}$, i.e., of L/k. In view of L3§8(J20) and L3§9(J26), for $1 \leq i \leq N$ we can find a monic polynomial

$$f_i(Y) = Y^{n_i} + \sum_{1 \leq j \leq n_i} a_{ij} Y^{n_i - j}$$

of degree $n_i \in \mathbb{N}_+$ with $a_{ij} \in \widehat{k}[y_1, \ldots, y_r]$ such that $f_i(Y)$ is the minimal polynomial of $x_i/\widehat{k}(y_1, \ldots, y_r)$. We can take a finite algebraic field extension k^*/k with $k^* \subset \widehat{k}$ such that $k^*[y_1, \ldots, y_r] \subset k^*[x_1, \ldots, x_N]$ and $a_{ij} \in k^*[y_1, \ldots, y_r]$ for all i, j. Let

$$T^* = k^*[x_1, \ldots, x_N] \quad \text{and} \quad L^* = k^*(x_1, \ldots, x_N).$$

In view of L4§10(E2)(J0), by (Q27)(C41) and (T142.1) we see that $y_1, \ldots, y_r$ is a separating normalization basis of T^*/k^*, and hence by the very special case we see that the integral closure $\overline{T}^*$ of T^* in L^* is a finite module over T^*. Consequently by L4§10(E2)STEP(E2.3) we see that $\overline{T}^*$ is a finite T-module. Therefore, since T is noetherian and $\overline{T}$ is a T-submodule of $\overline{T}^*$, by L4§5(O25)(27$^\bullet$) we conclude that $\overline{T}$ is a finite T-module.

Now let us consider the general case. Since $\overline{L}/L$ is finite algebraic, we can write

$$\overline{L} = L(x_{N+1}, \ldots, x_M)$$

with $N \leq M \in \mathbb{N}$ where the elements $x_{N+1}, \ldots, x_M$ are algebraic over L. As in the proof of (T141), upon multiplying these elements by suitable elements in $T^\times$ we may assume that they are integral over T. Let $T' = k[x_1, \ldots, x_M]$. Then T'/T is integral and $\overline{L} = k(x_1, \ldots, x_M)$. Clearly $\overline{T}$ is the integral closure of T' in $\overline{L}$ and hence by the special case $\overline{T}$ is a finite T'-module. By L4§10(E2)(J0) it follows that $\overline{T}$ is a finite T-module.

AFFINE NORMALIZATION THEOREM (T146). Let A be a strongly restricted domain, let K be an algebraic function field over A, let E be a model of K/A, let $\overline{K}/K$ be a finite algebraic field extension, and let $\overline{E} = \mathfrak{N}(E, \overline{K})$. Then $\overline{E}$ is a normal noetherian model of $\overline{K}/A$. Moreover, if E is an affine model (resp: complete model) of K/A then $\overline{E}$ is an affine model (resp: complete model) of $\overline{K}/A$.

In greater detail, if for the model E we have $E = \cup_{1 \leq i \leq n} \mathfrak{V}(B_i)$ where $n \in \mathbb{N}_+$ and B_i is an affine domain over A with quotient field K, then $\overline{E} = \cup_{1 \leq i \leq n} \mathfrak{V}(\overline{B}_i)$ where the normal noetherian affine domain $\overline{B}_i$ over A with quotient field $\overline{K}$ is obtained by putting $\overline{B}_i = $ the integral closure of B_i in $\overline{K}$.

PROOF. Since A is a strongly restricted domain, $\overline{B}_i$ is a finite module over B_i, and hence by L4§5(O25)(27$^\bullet$) we see that $\overline{B}_i$ is a normal noetherian affine domain over A with quotient field $\overline{K}$. In view of (T25) to (T28) of L4§9, which are proved in L4§12(R7), and in view of (Q9)(T36), the rest follows by considerations at the semimodel level. See (T148.1) and (T148.2) below.

PROJECTIVE NORMALIZATION THEOREM (T147). Let A be a strongly restricted domain, let K be an algebraic function field over A, let E be a projective model of K/A, let $\overline{K}/K$ be a finite algebraic field extension, and let $\overline{E} = \mathfrak{N}(E, \overline{K})$. Then $\overline{E}$ is a projective model of $\overline{K}/A$.

In greater detail, let $x_1, \ldots, x_n$ be any elements in $K^\times$ with $n \in \mathbb{N}_+$ such that $E = \mathfrak{W}(A; x_1, \ldots, x_n)$, i.e., $E = \cup_{1 \leq i \leq n} \mathfrak{V}(B_i)$ where $B_i = A[x_1/x_i, \ldots, x_n/x_i]$. Also let us put $\overline{B}_i =$ the integral closure of B_i in $\overline{K}$. Then $\overline{B}_i$ has a finite set of (B_i)-module generators $w_{i1}, \ldots, w_{im_i}$ with $m_i \in \mathbb{N}_+$, and for any such generators there exists $q_0 \in \mathbb{N}_+$ such that for every $q_0 \leq q \in \mathbb{N}_+$ we have $w_{ij}x_i^q/x_k^q \in \overline{B}_k$ for all i, j, k. Given any such q let us put $z_i = x_i^q$ for $1 \leq i \leq n$, and let all the remaining monomials $x_1^{i_1} \ldots x_n^{i_n}$ of degree q be labelled as $z_{n+1}, \ldots, z_\nu$ with $\nu = \binom{n+q-1}{q}$; also let us label the various products $w_{ij}x_i^q$ with $1 \leq i \leq n$ and $1 \leq j \leq m_i$ as $z_{\nu+1}, \ldots, z_m$ with $m - \nu = m_1 + \cdots + m_n$. Then $\overline{E} = \mathfrak{W}(A; z_1, \ldots, z_m)$, i.e., $\overline{E} = \cup_{1 \leq l \leq m} \mathfrak{V}(C_l)$ where $C_l = A[z_1/z_l, \ldots, z_m/z_l]$.

PROOF. Over and above what we have shown in (T146), all we need to prove is (i) the existence of $q_0 \in \mathbb{N}_+$ such that for every $q_0 \leq q \in \mathbb{N}_+$ we have $w_{ij}x_i^q/x_k^q \in \overline{B}_k$ for all i, j, k, and (ii) the equality

$$\cup_{1 \leq i \leq n} \mathfrak{V}(\overline{B}_i) = \cup_{1 \leq l \leq m} \mathfrak{V}(C_l).$$

To prove (i), since w_{ij} is integral over B_i, we have an equation $f_{ij}(w_{ij}) = 0$ where

$$f_{ij}(Y) = Y^{n_{ij}} + \sum_{1 \leq k \leq n_{ij}} a_{ijk} Y^{n_{ij} - k}$$

with $n_{ij} \in \mathbb{N}_+$ and $a_{ijk} \in B_i$. Clearly for every $a \in B_i$ we can find $d(a) \in \mathbb{N}_+$ such that $a = \theta_a(x_1, \ldots, x_n)/x_i^{d(a)}$ where $\theta_a(X_1, \ldots, X_n)$ is a homogeneous polynomial (which may be zero) of degree $d(a)$ in indeterminates $X_1, \ldots, X_n$ with coefficients in A. Take q_0 to be bigger than the values of $d(a_{ijk})$ for all i, j, k. Then for every $q \geq q_0$ we have $f_{ijk}(w_{ij}x_i^q/x_k^q) = 0$ where

$$f_{ijk}(Y) = Y^{n_{ij}} + \sum_{1 \leq k \leq n_{ij}} a_{ijk}(x_i^q/x_k^q)^k Y^{n_{ij} - k}$$

with $a_{ijk}(x_i^q/x_k^q)^k \in B_k$ and hence $w_{ij}x_i^q/x_k^q \in \overline{B}_k$.

To prove (ii), note that for $1 \leq i \leq n$ we clearly have

$$B_i = A[z_1/z_i, \ldots, z_\nu/z_i]$$

with

$$B_i[z_{\nu+1}/z_i, \ldots, z_m/z_i] \subset \overline{B}_i = B_i[(w_{ij})_{1 \leq j \leq m_i}] \subset B_i[z_{\nu+1}/z_i, \ldots, z_m/z_i]$$

and hence

$$C_i = \overline{B}_i$$

and therefore

(•) $$\cup_{1 \leq i \leq n} \mathfrak{V}(\overline{B}_i) \subset \cup_{1 \leq l \leq m} \mathfrak{V}(C_l).$$

By (T146) we know that $\cup_{1 \leq i \leq n} \mathfrak{V}(\overline{B}_i)$ is a complete model of $\overline{K}/A$, and by L4§9.1(T29), which was proved in L4§12(R8), we also know that $\cup_{1 \leq l \leq m} \mathfrak{V}(C_l)$ is a model of K'/A for some subfield K' of $\overline{K}$ with $A \subset K'$; consequently by (•) we see that $\cup_{1 \leq i \leq n} \mathfrak{V}(\overline{B}_i) = \cup_{1 \leq l \leq m} \mathfrak{V}(C_l)$. See (T148.3) and (T148.4) below.

In the proofs of (T146) and (T147) we used the following [cf. §6(E29)]:

AUXILIARY THEOREM (T148).

(T148.1) Let B be a domain with quotient field K, let A be a subdomain of B, and let $\overline{B}$ be the integral closure of B in a finite algebraic field extension $\overline{K}$ of K. Also let $E = \mathfrak{V}(B)$ and $\overline{E} = \mathfrak{V}(\overline{B})$. Then $\overline{E} = \mathfrak{N}(E, \overline{K})$ and $\overline{E}$ is a normal affine semimodel of $\overline{K}/A$.

(T148.2) Let A be a subdomain of a field K, let E be a semimodel of K/A given by $E = \cup_{l \in \Lambda} \mathfrak{V}(B_l)$ where $(B_l)_{l \in \Lambda}$ is a nonempty family of subdomains of K with quotient field K such that $A \subset B_l$, and let $\overline{E} = \cup_{l \in \Lambda} \mathfrak{V}(\overline{B}_l)$ where $\overline{B}_l = $ the integral closure of B_l in a finite algebraic field extension $\overline{K}$ of K. Then $\overline{E} = \mathfrak{N}(E, \overline{K})$ and $\overline{E}$ is a normal semimodel of $\overline{K}/A$. Moreover, if E is a complete semimodel of K/A then $\overline{E}$ is a complete semimodel of $\overline{K}/A$.

(T148.3) Let n, q be positive integers and let $x_1, \ldots, x_n$ be nonzero elements in an overfield K of a domain A. Then, upon letting $z_i = x_i^q$ for $1 \leq i \leq n$ and labelling the remaining monomials $x_1^{i_1} \ldots x_n^{i_n}$ of degree q as $z_{n+1}, \ldots, z_\nu$ with

$$\nu = \binom{n+q-1}{q}$$

for $1 \leq i \leq n$ we have $A[z_1/z_i, \ldots, z_\nu/z_i] = A[x_1/x_i, \ldots, x_n/x_i]$ and we have $\mathfrak{W}(A; z_1, \ldots, z_\nu) = \mathfrak{W}(A; x_1, \ldots, x_n)$.

(T148.4) Let E be a complete model of K/A where A is a subdomain of a field K, and let E' be a model of K'/A for some subfield K' of K with $A \subset K'$ such that $E \subset E'$. Then $E = E'$.

QUEST (Q34) Basic Projective Algebraic Geometry

Referring to L4§§7-9 for localization and affine varieties, let us now discuss their relation to homogeneous localization and projective varieties.

QUEST (Q34.1). [**Projective Spectrum**]. To start with, keeping in mind the definitions of graded rings and homogeneous ideals from §§2-3, for any subintegrally graded ring

$$A = \sum_{i \in I} A_i$$

where I is an additive submonoid of $\mathbb{Z}$, we let A_∞ denote the set of all homogeneous elements in A, i.e.,

$$A_\infty = \bigcup_{i \in I} A_i.$$

In analogy with the notation concerning spectral varieties in the affine case introduced in L4§5(O16) and L4§8, we introduce the corresponding notation in the projective case thus. We put

proj$(A) = $ the set of all relevant homogeneous prime ideals in A

and we call this the projective (or homogeneous) spectrum of A. We also put

mproj$(A) = $ the set of all maximal members of proj(A)

i.e., those $P \in \text{proj}(A)$ for which $P \subset Q \in \text{proj}(A) \Rightarrow P = Q$, and we call this the maximal prospectrum of A. For any $J \subset A$ we define the prospectral variety, the maximal prospectral variety, and the minimal prospectral variety of J in A by respectively putting

$$\text{vproj}_A J = \{P \in \text{proj}(A) : J \cap A_\infty \subset P\}$$

and

$$\text{mvproj}_A J = \text{mproj}(A) \cap \text{vproj}_A J$$

and

nvproj$_A J = $ the set of all minimal members of vproj$_A J$

i.e., those $P \in \text{vproj}_A J$ for which $P \supset Q \in \text{vproj}_A J \Rightarrow P = Q$. For any $J \subset A$ we also define the complementary prospectral variety, the maximal complementary prospectral variety, and the minimal complementary prospectral variety of J in A by respectively putting

$$\text{cproj}_A J = \text{proj}(A) \setminus \text{vproj}_A J$$

and

$$\mathrm{mcproj}_A J = \mathrm{mproj}(A) \setminus \mathrm{mvproj}_A J$$

and

$$\mathrm{ncproj}_A J = \mathrm{proj}(A) \setminus \mathrm{nvproj}_A J.$$

For any $U \subset \mathrm{proj}(A)$ we define the prospectral ideal of U in A by putting

$$\mathrm{iproj}_A U = \cap_{P \in U} P.$$

We also put

$$\mathrm{prd}(A) = \begin{cases} \text{the set of all homogeneous radical ideals in } A \\ \text{different from } \mathrm{rad}_A \Omega(A) \end{cases}$$

i.e., homogeneous ideals J with $J = \mathrm{rad}_A J \neq \mathrm{rad}_A \Omega(A)$. We put

$$\mathrm{psvt}(A) = \text{the set of all prospectral varieties in } A$$

i.e., the set of all subsets of $\mathrm{proj}(A)$ of the form $\mathrm{vproj}_A J$ with J varying over the set of all subsets of A; members of $\mathrm{psvt}(A)$ may be called varieties in $\mathrm{proj}(A)$. We also put

$$\mathrm{mpsvt}(A) = \text{the set of all maximal prospectral varieties in } A$$

i.e., the set of all subsets of $\mathrm{mproj}(A)$ of the form $\mathrm{mvproj}_A J$ with J varying over the set of all subsets of A; members of $\mathrm{mpsvt}(A)$ may be called varieties in $\mathrm{mproj}(A)$.

Given U, U' in $\mathrm{mpsvt}(A)$ or $\mathrm{psvt}(A)$, U' is a subvariety of U means $U' \subset U$, and U' is a proper subvariety of U means $U' \subset U$ with $U' \neq U$. Given U in $\mathrm{mpsvt}(A)$ or $\mathrm{psvt}(A)$, U is reducible means U is the union of two proper subvarieties; U is irreducible means it is nonempty and nonreducible. We put

$$\mathrm{impsvt}(A) = \text{the set of all irreducible members of } \mathrm{mpsvt}(A)$$

and

$$\mathrm{ipsvt}(A) = \text{the set of all irreducible members of } \mathrm{psvt}(A).$$

We define the projective dimension of A by putting

$$\mathrm{prodim}(A) = \dim(A) - 1$$

where we are following the usual convention according to which: if $\dim(A) = \infty$ then $\dim(A) - 1 = \infty$, and if $\dim(A) = -\infty$ then $\dim(A) - 1 = -\infty$. For any U in $\mathrm{mpsvt}(A)$ or $\mathrm{psvt}(A)$ we define the projective dimension of U by putting

$$\mathrm{prodim}(U) = \mathrm{prodim}(A/\mathrm{iproj}_A U).$$

QUEST (Q34.2). [**Homogeneous Localization**]. Now we define homogeneous localization, which will put L4§9 into proper perspective. So let

$$D = \sum_{n \in \mathbb{N}} D_n$$

be a semihomogeneous domain. According to the definitions made in §3(C6) we then have $D = D_0[D_1]$. We define the homogeneous quotient field $\mathfrak{K}(D)$ of D by putting

$$\mathfrak{K}(D) = \bigcup_{n \in \mathbb{N}} \{y_n/z_n : y_n \in D_n \text{ and } z_n \in D_n^{\times}\}$$

and we note that this is a subfield of $\mathrm{QF}(D)$. For any $P \in \mathrm{proj}(D)$ we define the homogeneous localization $D_{[P]}$ of D at P to be the subring of $\mathfrak{K}(D)$ given by

$$D_{[P]} = \bigcup_{n \in \mathbb{N}} \{y_n/z_n : y_n \in D_n \text{ and } z_n \in D_n \setminus P\}.$$

Note that this is a quasilocal domain with quotient field $\mathfrak{K}(D)$ and its maximal ideal is given by

$$M(D_{[P]}) = \bigcup_{n \in \mathbb{N}} \{y_n/z_n : y_n \in P \cap D_n \text{ and } z_n \in D_n \setminus P\}.$$

In analogy with the modelic spec $\mathfrak{V}(A)$ of a domain A introduced in L4§8.2, in case of $D_1 \neq 0$, we put

$$\mathfrak{W}(D) = \begin{cases} \text{the set of all homogeneous localizations } D_{[P]} \\ \text{with } P \text{ varying over } \mathrm{proj}(D) \end{cases}$$

and we call this the modelic proj of D.

Mimicking the homogenize-dehomogenize processes described in L3§4,

$$\text{for any } x \in D_1^{\times} \text{ we define } D_{/x} : D \to \mathrm{QF}(D)$$

to be the unique (ring) homomorphism such that

$$\text{for all } n \in \mathbb{N} \text{ and } f \in D_n \text{ we have } D_{/x}(f) = f/x^n$$

and we call this the dehomogenization map of D at x; moreover, by putting $F = D_{/x}(f)$, we call F the dehomogenization of f at x, and we call f a homogenization of F by x; furthermore, with F as above,

$$\begin{cases} \text{if } f \notin x^{n-m} D_m \text{ whenever } n > m \in \mathbb{N} \\ \text{then clearly } f \text{ is uniquely determined by } F \\ \text{and we call it the minimal homogenization of } F \text{ by } x. \end{cases}$$

Note that

(1)
$$D_{/x}(D) = D_0[D_1/x]$$

where

$$D_1/x = \{\xi/x : \xi \in D_1\}.$$

It follows that

(2) $\begin{cases} \text{if } (x_l)_{l\in\Lambda} \text{ is any family of generators of } D_1 \text{ as a module over } D_0 \\ \text{then } D_0[D_1/x] = D_0[(x_l/x)_{l\in\Lambda}]. \end{cases}$

Consequently

(3) $\begin{cases} \text{if } D \text{ is a homogeneous domain over a noetherian subdomain } D_0 \\ \text{then, for every } x \in D_1^\times, \text{ the domain } D_{/x}(D) \text{ is noetherian.} \end{cases}$

As evident properties of the dehomogenization map we note that

(4) $\begin{cases} \text{for any semihomogeneous domain } D \text{ and any } x \in D_1^\times \\ \text{we have } \ker(D_{/x}) = (x-1)D \text{ and, upon letting } D' = D_{/x}(D), \\ \text{the mappings } P \mapsto P' = D_{/x}(P) \mapsto P^* = P + (x-1)D = D_{/x}^{-1}(P') \\ \text{give inclusion preserving bijections} \\ \mathrm{cproj}_D(xD) \to \mathrm{spec}(D') \to \mathrm{vspec}_D(x-1)D \end{cases}$

and

(5) $\begin{cases} \text{in the situation of (4),} \\ \text{for any } P \in \mathrm{cproj}_D(xD) \text{ we have } D_{[P]} = D'_{P'} \\ \text{and there exists a unique epimorphism } D_{/x}^* : D_{P^*} \to D'_{P'} \\ \text{such that } D_{/x}^*(z) = D_{/x}(z) \text{ for all } z \in D \\ \text{and for this epimorphism we have } \ker(D_{/x}^*) = (x-1)D_{P^*}. \end{cases}$

Likewise, as an evident property of modelic proj we note that

(6) $\begin{cases} \text{for any semihomogeneous domain } D \text{ with } D_1 \neq 0 \\ \text{we have } \mathfrak{W}(D) = \mathfrak{W}(D_0, D_1), \\ \text{and if } (x_l)_{l\in\Lambda} \text{ is any family of generators of } D_1 \text{ as a module over } D_0 \\ \text{then we have } \mathfrak{W}(D) = \mathfrak{W}(D_0; (x_l)_{l\in\Lambda}) \end{cases}$

and hence in particular

(7)
$$\begin{cases} \text{if } D \text{ is a homogeneous domain with } D_1 \neq 0 \\ \text{then } D_1, \text{ as a module over } D_0, \text{ is generated by} \\ \text{a finite number of generators } x_1, \ldots, x_{N+1} \text{ in } D_1^\times \text{ with } N \in \mathbb{N} \\ \text{and for any such generators we have} \\ \mathfrak{W}(D) = \cup_{i=1}^{N+1} \mathfrak{V}(D_0[x_1/x_i, \ldots, x_{N+1}/x_i]). \end{cases}$$

In view of L4§9.1(T29), which was proved in L4§12(R8), by (6) and (7) we see that

(8)
$$\begin{cases} \text{if } D \text{ is a semihomogeneous domain with } D_1 \neq 0 \\ \text{then } \mathfrak{W}(D) \text{ is a semimodel of } \mathfrak{K}(D)/D_0 \end{cases}$$

and

(9)
$$\begin{cases} \text{if } D \text{ is a homogeneous domain with } D_1 \neq 0 \\ \text{then } \mathfrak{W}(D) \text{ is a complete model of } \mathfrak{K}(D)/D_0 \\ \text{which is also a projective model of } \mathfrak{K}(D)/D_0. \end{cases}$$

As a consequence of (3) to (5) we see that

(10)
$$\begin{cases} \text{if } D \text{ is a homogeneous domain over a noetherian subdomain } D_0 \\ \text{then the homogeneous localization } D_{[P]} \text{ of } D \\ \text{at any } P \in \mathrm{proj}(D) \text{ is a local ring.} \end{cases}$$

Concerning (6) to (9) we note that

(11)
$$\begin{cases} \text{if } D \text{ is any semihomogeneous domain then:} \\ D_1 \neq 0 \Leftrightarrow \Omega(D) \neq 0. \end{cases}$$

To relate this to L3§4 consider the polynomial ring

$$C = k[x_1, \ldots, x_{N+1}] = \sum_{d \in \mathbb{N}} C_d$$

in a finite number of variables $x_1, \ldots, x_{N+1}$ over a field k with $N \in \mathbb{N}$, which is regarded as a homogeneous domain over k in the usual manner. Given any

$$f(x_1, \ldots, x_{N+1}) \in C_d^\times$$

its dehomogenization at x_{N+1} is given by

$$F(X_1, \ldots, X_N) = \frac{f(x_1, \ldots, x_{N+1})}{x_{N+1}^d} \in B_{N,k} = k[X_1, \ldots, X_N]$$

where

$$C_{/x_{N+1}}(C) = B_{N,k} = k[X_1, \ldots, X_N]$$

with

$$X_i = \frac{x_i}{x_{N+1}} \quad \text{for} \quad 1 \leq i \leq N.$$

Operationally we simply have

$$F(X_1, \ldots, X_N) = f(X_1, \ldots, X_N, 1).$$

Homogenizing F by x_{N+1} we get back f by noting that

$$x_{N+1}^d F(x_1/x_{N+1}, \ldots, x_N/x_{N+1}) = f(x_1, \ldots, x_{N+1}).$$

This is the minimal homogenization of F iff f is not divisible by x_{N+1}; otherwise a smaller power of x_{N+1} will do the job of homogenizing.

Now (7) is partly captured by the display in L3§4 which was reproduced in L4§9.1 and which says that

$$(12) \qquad \mathbb{P}_k^N = \cup_{i=1}^{N+1}(\mathbb{P}_k^N \setminus H_i) = \cup_{i=1}^{N+1} \mathbb{A}_{k,i}^N$$

where the hyperplane H_i in $\mathbb{P}_k^N$ is given by

$$H_i : x_i = 0$$

and the affine N-space $\mathbb{A}_{k,i}^N$ is given by

$$\mathbb{A}_{k,i}^N = \{v = (v_1, \ldots, v_{N+1}) \in k^{N+1} : v_i = 1\}.$$

Let us note that such a point v is only a representative of the equivalence class of all the $(N+1)$-tuples $(cv_1, \ldots, cv_{N+1})$ with $cv_i = c \in k^\times$. Let us also note that, for $i \neq j$, the affine spaces $\mathbb{A}_{k,i}^N$ and $\mathbb{A}_{k,j}^N$ have a very large intersection. Namely

$$\mathbb{A}_{k,i}^N \cap \mathbb{A}_{k,j}^N = \mathbb{P}_k^N \setminus (H_i \cup H_j).$$

Thus, if a point $v = (v_1, \ldots, v_{N+1})$ belongs to the above described affine space

$$\mathbb{A}_{k,i}^N = \{v = (v_1, \ldots, v_{N+1}) \in k^{N+1} : v_i = 1\}$$

and is such that $v_j \neq 0$ then it corresponds to the point

$$\left(\frac{v_1}{v_j}, \ldots, \frac{v_{j-1}}{v_j}, 1, \frac{v_{j+1}}{v_j}, \ldots, \frac{v_{N+1}}{v_j} \right)$$

of the affine space $\mathbb{A}_{k,j}^N$.

In L4§9.1, as a slight change of notation, C is replaced by the $(N+1)$-variable polynomial ring

$$B = B_{N+1,k} = k[X_1, \ldots, X_{N+1}]$$

and for every positive integer $i \leq N+1$ we have put

$$\Gamma_i(B) = B_i' = B_{N,k,i} = k[Y_{1i}, \ldots, Y_{N+1,i}]$$

where

$$Y_{li} = X_l/X_i \quad \text{for} \quad 1 \le l \le N+1 \quad \text{with} \quad Y_{ll} = 1$$

and

$$\Gamma_i = B_{/X_i} : B \to k(X_1, \ldots, X_{N+1})$$

is the dehomogenization map obtained by putting

$$\Gamma_i(f(X_1, \ldots, X_{N+1})) = f(Y_{1,i}, \ldots, Y_{i-1,i}, 1, Y_{i+1,i}, \ldots, Y_{N+1,i})$$

for all $f(X_1, \ldots, X_{N+1}) \in k[X_1, \ldots, X_{N+1}]$. Note that now the hyperplane H_i in $\mathbb{P}_k^N$ is given by $H_i : X_i = 0$.

By taking B for D and $X_1, \ldots, X_{N+1}$ for $x_1, \ldots, x_{N+1}$, in this special case the equation in (7) becomes

(7′) $$\mathfrak{W}(B) = \cup_{i=1}^{N+1} \mathfrak{V}(B_i').$$

Likewise, given any positive integer $i \le N+1$, property (4) says that

(4*) $$\begin{cases} \ker(\Gamma_i) = (X_i - 1)B \\ \text{and the mappings} \\ P \mapsto P_i' = \Gamma_i(P) \mapsto P_i^* = P + (X_i - 1)B = \Gamma_i^{-1}(P_i') \\ \text{give inclusion preserving maps} \\ \text{cproj}_B(X_iB) \to \text{spec}(B_i') \to \text{vspec}_B(X_i - 1)B \end{cases}$$

and property (5) says that

(5*) $$\begin{cases} \text{for any } P \in \text{cproj}_B(X_iB) \text{ we have } B_{[P]} = (B_i')_{P_i'} \\ \text{and there exists a unique epimorphism } \Gamma_i^* : B_{P_i^*} \to (B_i')_{P_i'} \\ \text{such that } \Gamma_i^*(z) = \Gamma_i(z) \text{ for all } z \in B \\ \text{and for this epimorphism we have } \ker(\Gamma_i^*) = (X_i - 1)B_{P_i^*}. \end{cases}$$

The modelic projective space is defined in L4§9.1 by putting

$$(\mathbb{P}_k^N)^\delta = \mathfrak{W}(k; X_1, \ldots, X_{N+1})$$

which converts equation (7′) to the equation

(7*) $$(\mathbb{P}_k^N)^\delta = \cup_{i=1}^{N+1} \mathfrak{V}(B_i').$$

The fact that (7*) is partly recaptured by (12) comes down to the isomorphism portion of the display in L4§9.1 saying that

(12*) $$\mathbb{P}_k^N \approx (\mathbb{P}_k^N)^{\rho\delta} \subset (\mathbb{P}_k^N)^{\mu\delta} \subset (\mathbb{P}_k^N)^\delta.$$

To extend this isomorphism to the entire $(\mathbb{P}_k^N)^\delta$ we need to define varieties in $\mathbb{P}_k^N$ which we shall do in a moment. Note that (4*) and (5*) provide a more direct way of dealing with the injection θ of the last paragraph of L4§9.1 which led to the said isomorphism. In effect this was done by introducing the homogeneous localization $B_{[P]}$ which turns out to coincide with the ordinary localization $(B_i')_{P_i'}$ for every i for which P belongs to $\mathrm{cproj}_B(X_iB)$.

QUEST (Q34.3). [**Varieties in Projective Space**]. Let us reiterate that

$$B_{N+1,k} = k[X_1, \ldots, X_{N+1}] = \sum_{d \in \mathbb{N}} (B_{N+1,k})_d$$

is the $N+1$ variable polynomial ring over a field k with $N \in \mathbb{N}$, which we regard as a homogeneous domain over k. As in L4§8, for a moment let

$$k \subset \kappa \subset \lambda$$

be fields, i.e., let κ be an overfield of k, and let λ be an overfield of κ. Recall that

$$\mathbb{P}_\kappa^N = \begin{cases} \text{the set of all } [u] \text{ with} \\ u = (u_1, \ldots, u_{N+1}) \in \kappa^{N+1} \setminus \{(0, \ldots, 0)\} \end{cases}$$

where by $[u]$ we denote the equivalence class (under proportionality) containing u.

Given any

$$f = f(X_1, \ldots, X_{N+1}) \in (B_{N+1,\lambda})_\infty = \bigcup_{d \in \mathbb{N}} (B_{N+1,\lambda})_d$$

we may write $f([u]) = 0$ to mean that $f(u) = 0$ because this depends only on $[u]$ and not on u, i.e.,

$$\text{if } [u] = [u'] \text{ then: } f(u) = 0 \Leftrightarrow f(u') = 0.$$

Given any $J \subset B_{N+1,\lambda}$ we put

$$\mathbb{W}_\kappa(J) = \{[u] \in \mathbb{P}_\kappa^N : f([u]) = 0 \text{ for all } f \in J \cap (B_{N+1,\lambda})_\infty\}$$

and we call this the projective variety defined by J. For homogeneous polynomials $f, g, \ldots$ in $B_{N+1,\lambda}$ we may write $\mathbb{W}_\kappa(f, g, \ldots)$ in place of $\mathbb{W}_\kappa(\{f, g, \ldots\})$ and we may informally call this the projective variety $f = g = \cdots = 0$. By a variety in $\mathbb{P}_\kappa^N$ defined over k we mean a subset of $\mathbb{P}_\kappa^N$ which can be expressed in the form $\mathbb{W}_\kappa(J)$ for some $J \subset B_{N+1,k}$. We put

$$\mathrm{pvt}_k(\mathbb{P}_\kappa^N) = \text{ the set of all varieties in } \mathbb{P}_\kappa^N \text{ defined over } k.$$

Members of $\mathrm{pvt}_k(\mathbb{P}_k^N)$ may be called varieties in $\mathbb{P}_\kappa^N$. For any $U \subset \mathbb{P}_\kappa^N$ we put

$$\mathbb{J}_k(U) = \begin{cases} \text{the ideal in } B_{N+1,k} \text{ generated by} \\ \{f \in (B_{N+1,k})_\infty : f([u]) = 0 \text{ for all } [u] \in U\} \end{cases}$$

and we note that this is a homogeneous ideal in $B_{N+1,k}$; we call it the homogeneous ideal of U in $B_{N+1,k}$. For points $[u], [v], \ldots$ in $\mathbb{P}^N_\kappa$ we may write $\mathbb{J}_k([u], [v], \ldots)$ in place of $\mathbb{J}_k(\{[u], [v], \ldots\})$.

Given U, U' in $\text{pvt}_k(\mathbb{P}^N_\kappa)$, U' is a subvariety of U means $U' \subset U$, and U' is a proper subvariety of U means $U' \subset U$ with $U' \neq U$.

Given U in $\text{pvt}_k(\mathbb{P}^N_\kappa)$, U is reducible means U is the union of two proper subvarieties; U is irreducible means it is nonempty and nonreducible.

We put

$$\text{ipvt}_k(\mathbb{P}^N_\kappa) = \text{the set of all irreducible members of } \text{pvt}_k(\mathbb{P}^N_\kappa).$$

For any U in $\text{pvt}_k(\mathbb{P}^N_\kappa)$ we define the projective dimension of U by putting

$$\text{prodim}(U) = \text{prodim}(B_{N+1,k}/\mathbb{J}_k(U)).$$

We are denoting this, as well as the objects defined at the end of (Q34.1), by prodim so as to distinguish them from the pdim introduced in (Q17). Note that in (Q34.1) we defined the prodim of a subintegrally graded ring to be one less than its dim which itself was defined in L4§5(O22).

Given $U \in \text{pvt}_k(\mathbb{P}^N_\kappa)$, we denote the homogeneous ring $B_{N+1,k}/\mathbb{J}_k(U)$ by $k[U]^*$ and call it the homogeneous coordinate ring of U. In case $\mathbb{J}_k(U)$ is not the unit ideal in $B_{N+1,k}$, we may and we do identify k with a subfield of $k[U]^*$ and we note that then $k[U]^*$ becomes a homogeneous ring over k, and hence in particular it is a noetherian ring.

In case $U \in \text{pvt}_k(\mathbb{P}^N_\kappa)$ is such that $\mathbb{J}_k(U)$ is a homogeneous prime ideal in $B_{N+1,k}$, we denote the homogeneous quotient field of $k[U]^*$ by $k(U)^*$, and we call $k(U)^*$ the homogeneous function field of U (over k). We denote the homogeneous localization of $B_{N+1,k}$ at $\mathbb{J}_k(U)$ by $R_k(U)^*$, and call it the homogeneous local ring of U (over k); note that by (Q34.2)(10) this is indeed a local ring. We denote the residue field $R_k(U)^*/M(R_k(U)^*)$ of $R_k(U)^*$ by $k(U)^\sharp$, and we call this field $k(U)^\sharp$ the alternative homogeneous function field of U (over k); again we may and we do identify k with a subfield of $k(U)^\sharp$. Let $\Phi_U : B_{N+1,k} \to k[U]^*$ and $\Psi_U : R_k(U)^* \to k(U)^\sharp$ be the residue class epimorphisms. Now clearly there is a unique homomorphism $\psi_U : R_k(U)^* \to k(U)^*$ such that for all $d \in \mathbb{N}$ and for all $f_d \in (B_{N+1,k})_d$ and $g_d \in (B_{N+1,k})_d \setminus \mathbb{J}_k(U)$ we have $\psi_U(f_d/g_d) = \Phi_U(f_d)/\Phi_U(g_d)$. Obviously

$$\text{im}(\psi_U) = k(U)^* \quad \text{and} \quad \ker(\psi_U) = M(R_k(U)^*)$$

and hence there is a unique isomorphism $\phi_U : k(U)^\sharp \to k(U)^*$ for which we have $\phi_U \Psi_U = \psi_U$. As in the affine case, we call ψ_U and ϕ_U the natural epimorphism and the natural isomorphism respectively.

Note that $\mathbb{P}^N_k \subset \mathbb{P}^N_\kappa$, and the homogeneous ideal $\mathbb{J}_k([u])$ of any point $[u]$ in $\mathbb{P}^N_k$ is obviously the submaximal homogeneous ideal in $B_{N+1,k}$ generated by the N

elements

$$u_{N+1}X_1 - u_1X_{N+1}, \ldots, u_{N+1}X_N - u_NX_{N+1}.$$

Also clearly we have $\mathbb{W}_\kappa(\mathbb{J}_k([u])) = \{[u]\}$, and writing $[u]$ in place of $\{[u]\}$ we get the obvious relations $k = k([u])^* \subset R_k([u])^* =$ the homogeneous local ring of $[u]$. Also note that by a submaximal homogeneous ideal in $B_{N+1,k}$ we mean a member of $\mathrm{mproj}(B_{N+1,k})$, and note that the irrelevant ideal $\Omega(B_{N+1,k})$ is the unique maximal ideal in $B_{N+1,k}$ which is homogeneous.

To continue the discussion of (Q34.1), given any homogeneous ring A over k, for any U in $\mathrm{mpsvt}(A)$ or $\mathrm{psvt}(A)$ we denote $A/\mathrm{iproj}_A(U)$ by $k[U]^*$ and call it the homogeneous coordinate ring of U. In case $\mathrm{iproj}_A U$ is not the unit ideal in A, we may identify k with a subfield of $k[U]^*$ and we note that then $k[U]^*$ becomes a homogeneous ring over k, and hence it is a noetherian ring.

In case $\mathrm{iproj}_A U$ is a prime ideal in A, we denote the homogeneous quotient field of $k[U]^*$ by $k(U)^*$, and we call $k(U)^*$ the homogeneous function field of U (over k). We denote the homogeneous localization of A at $\mathrm{iproj}_A U$ by $R_k(U)^*$ and call it the homogeneous local ring of U (over k); note that by (Q34.2)(10) this is indeed a local ring. We denote the residue field $R_k(U)^*/M(R_k(U)^*)$ by $k(U)^\natural$ and we call it the alternative homogeneous function field of U (over k); we may and we do identify k with a subfield of $k(U)^\natural$. Let $\Phi_U : A \to k[U]^*$ and $\Psi_U : R_k(U)^* \to k(U)^\natural$ be the residue class epimorphisms. Clearly there is a unique homomorphism $\psi_U : R_k(U)^* \to k(U)^*$ such that for all $d \in \mathbb{N}$ and for all $f_d \in A_d$ and $g_d \in A_d \setminus \mathrm{ispec}_A U$ we have $\psi_U(f_d/g_d) = \Phi_U(f_d)/\Phi_U(g_d)$. Obviously $\mathrm{im}(\psi_U) = k(U)^*$ and $\ker(\psi_U) = M(R_k(U)^*)$ and hence it follows that there is a unique isomorphism $\phi_U : k(U)^\natural \to k(U)^*$ for which we have $\phi_U\Psi_U = \psi_U$. Again we call ψ_U and ϕ_U the natural epimorphism and the natural isomorphism respectively.

As straightforward consequences of the statements and proofs of L4§8(T19) to L4§8(T22) we get their respective projective versions (T149) to (T152) stated below [cf. §6(E31)]. Note that the proof of L4§8(T20) is given in (Q32), and the proofs of L4§8(T19), L4§8(T21), and L4§8(T22) are given in (Q10) and (Q11).

PROJECTIVE DIMENSION THEOREM (T149). Let $P \subset P'$ be any relevant homogeneous prime ideals in $B = B_{N+1,k}$, and let $\phi : B \to B/P$ be the residue class epimorphism. Then we have

$$1 + \mathrm{trdeg}_k\mathfrak{K}(B/P) = \mathrm{trdeg}_k\mathrm{QF}(B/P) = \dim(B/P) = \mathrm{dpt}_B P = N + 1 - \mathrm{ht}_B P$$

and

$$1 + \mathrm{trdeg}_k\mathfrak{K}(B/P) = \mathrm{trdeg}_k\mathrm{QF}(B/P) = \mathrm{ht}_{\phi(B)}\phi(P') + \mathrm{dpt}_{\phi(B)}\phi(P').$$

PROJECTIVE PRIMITIVE ELEMENT THEOREM (T150). Assume that the field k is algebraically closed. Given any relevant homogeneous prime ideal P in $B = B_{N+1,k}$, let $\phi : B \to B/P$ be the residue class epimorphism. Then after changing the variables by a homogeneous k-linear transformation, it can be arranged that $X_{N+1} \notin P$ and: $\phi(X_1)/\phi(X_{N+1}), \ldots, \phi(X_r)/\phi(X_{N+1})$ is a transcendence basis of $\mathfrak{K}(B/P)$ over k and either $N = r$ or $\phi(X_{r+1})/\phi(X_{N+1})$ is a primitive element of $\mathfrak{K}(B/P)$ over $\mathfrak{K}(\phi(B_{r,k}[X_{N+1}]))$, i.e., we have $\mathfrak{K}(B/P) = \mathfrak{K}(\phi(B_{r,k}[X_{N+1}]))(\phi(X_{r+1})/\phi(X_{N+1}))$. [In the latter case we have a homogeneous irreducible $f(X_1, \ldots, X_{r+1}, X_{N+1}) \in P \cap B_{r+1,k}[X_{N+1}]$ and the irreducible variety U in $\mathbb{P}_\kappa^N$ defined by P is "parametrized" by the irreducible hypersurface $f = 0$ in $\mathbb{P}_\kappa^{r+1}$; in the former case U is "parametrized" by $\mathbb{P}_\kappa^r$].

PROJECTIVE INCLUSION RELATIONS THEOREM (T151). Let A be any subintegrally graded ring. Let

$$
\begin{cases}
(R, S, V, V', V'', I) \text{ stand for} \\
(B_{N+1,k}, \mathbb{P}_\kappa^N, \mathbb{W}_\kappa, \mathrm{pvt}_k(\mathbb{P}_\kappa^N), \mathrm{ipvt}_k(\mathbb{P}_\kappa^N), \mathbb{J}_k) \\
\text{or } (A, \mathrm{mproj}(A), \mathrm{mvproj}_A, \mathrm{mpsvt}(A), \mathrm{impsvt}(A), \mathrm{iproj}_A) \\
\text{or } (A, \mathrm{proj}(A), \mathrm{vproj}_A, \mathrm{psvt}(A), \mathrm{ipsvt}(A), \mathrm{iproj}_A)
\end{cases}
$$

and let I' be the set of all ideals J in R such that $J = I(U)$ for some $U \subset S$. Then we have the following.

(T151.1) For any subsets J and J' of R we have: $J \subset J' \Rightarrow V(J') \subset V(J)$, and we have: $\mathrm{rad}_R(JR) = \mathrm{rad}_R(J'R) \Rightarrow V(J) = V(J')$.

(T151.2) For any subsets U and U' of S we have $U \subset U' \Rightarrow I(U') \subset I(U)$.

(T151.3) For any family of ideals $(J_l)_{l \in L}$ in R we have $V(\sum_{l \in L} J_l) = \cap_{l \in L} V(J_l)$, and if the family is finite then we have $V(\cap_{l \in L} J_l) = V(\prod_{l \in L} J_l) = \cup_{l \in L} V(J_l)$.

(T151.4) For any family of subsets $(U_l)_{l \in L}$ of S we have $I(\cup_{l \in L} U_l) = \cap_{l \in L} I(U_l)$.

(T151.5) For any $J \subset R$ we have $J \subset I(V(J))$ and: $J = I(V(J)) \Leftrightarrow J \in I' \Rightarrow J = \mathrm{rad}_R(JR)$.

(T151.6) For any $U \subset S$ we have $U \subset V(I(U))$ and: $U = V(I(U)) \Leftrightarrow U \in V'$.

(T151.7) $V'' = \{U \in V' : I(U) \in \mathrm{spec}(R)\}$.

(T151.8) Assuming R is noetherian (this is certainly so in case $R = B_{N+1,k}$), every $U \in V'$ can be expressed as a finite union $U = \cup_{1 \leq i \leq h} U_i$ with $U_i \in V''$, and this decomposition is unique up to order if it is irredundant, i.e., if no U_i can be deleted from it. [The irredundancy can be achieved by deleting some of the U_i, and then the remaining U_i are called the IRREDUCIBLE COMPONENTS of U].

PROJECTIVE NULLSTELLENSATZ (T152).

(T152.1) A field which is an affine ring over k is algebraic over k.

(T152.2) For any homogeneous submaximal ideal J in $B_{N+1,k}$, the homogeneous quotient field $\mathfrak{K}(B_{N+1,k}/J)$ is algebraic over k, i.e., over the image of k under the residue class map $B_{N+1,k} \to B_{N+1,k}/J$.

(T152.3) For any homogeneous submaximal ideal J in $B_{N+1,k}$ with $X_{N+1} \notin J$, upon letting $\gamma : B_{N+1,k} \to B_{N,k}$ be the unique $(B_{N,k})$-epimorphism with $\gamma(X_{N+1}) = 1$, the ideal $\gamma(J)$ in $B_{N,k}$ is maximal and has a unique set of N generators

$$(f_i(X_1, \ldots, X_i) \in B_{i,k})_{1 \le i \le N}$$

such that for $1 \le i \le N$ we have

$$f_i(X_1, \ldots, X_i) = X_i^{n_i} + g_i(X_1, \ldots, X_i)$$

where n_i is a positive integer and $g_i(X_1, \ldots, X_i) \in B_{i,k}$ with

$$\deg_{X_j} g_i(X_1, \ldots, X_i) < n_j \quad \text{for} \quad 1 \le j \le i.$$

(T152.4) If k is algebraically closed then the mapping $\beta \mapsto \mathbb{J}_k(\beta)$ gives a bijection $\mathbb{P}_k^N \to \mathrm{mproj}(B_{N+1,k})$.

(T152.5) If κ contains an algebraic closure of k then for any relevant homogeneous ideal J in $B_{N+1,k}$, we have $\mathbb{J}_k(\mathbb{W}_\kappa(J)) = \mathrm{rad}_{B_{N+1,k}} J$.

(T152.6) If κ contains an algebraic closure of k then the mapping $U \mapsto \mathbb{J}_k(U)$ gives inclusion reversing bijections

$$\mathrm{pvt}_k(\mathbb{P}_\kappa^N) \to \mathrm{prd}(B_{N+1,k}) \quad \text{and} \quad \mathrm{ipvt}_k(\mathbb{P}_\kappa^N) \to \mathrm{proj}(B_{N+1,k})$$

whose inverses are given by $J \mapsto \mathbb{W}_\kappa(J)$.

(T152.7) The mapping $U \mapsto \mathrm{iproj}_{B_{N+1,k}} U$ gives inclusion reversing bijections

$$\mathrm{mpsvt}(B_{N+1,k}) \to \mathrm{prd}(B_{N+1,k}) \quad \text{and} \quad \mathrm{impsvt}(B_{N+1,k}) \to \mathrm{proj}(B_{N+1,k})$$

whose inverses are given by $J \mapsto \mathrm{mvproj}_{B_{N+1,k}} J$.

(T152.8) The mapping $U \mapsto \mathrm{iproj}_{B_{N+1,k}} U$ gives inclusion reversing bijections

$$\mathrm{psvt}(B_{N+1,k}) \to \mathrm{prd}(B_{N+1,k}) \quad \text{and} \quad \mathrm{ipsvt}(B_{N+1,k}) \to \mathrm{proj}(B_{N+1,k})$$

whose inverses are given by $J \mapsto \mathrm{vproj}_{B_{N+1,k}} J$.

QUEST (Q34.4). [**Projective Decomposition of Ideals and Varieties**]. In (Q10)(C17) and (Q11)(C18) it was noticed that L4§8(T21) and L4§8(T22) lead to a connection between the primary decomposition of ideals given in L4§6 and the irreducible decomposition of affine varieties given in the bracketed remark appended to L4§8(T21). In a similar manner, the above Theorems (T151) and (T152) lead to the following connection between the primary decomposition of homogeneous ideals given in §3 of the current Lecture and the irreducible decomposition of projective varieties given in the bracketed remark appended to (T151).

In the geometric situation, let $k \subset \kappa$ be fields such that κ contains an algebraic closure of k. For any homogeneous ideal J in the finite variable polynomial ring

$B = k[X_1, \ldots, X_{N+1}]$ with $N \in \mathbb{N}$, take a primary decomposition

$$J = \cap_{1 \leq i \leq e} Q_i$$

where Q_i is a homogeneous ideal in B which is primary for the homogeneous prime ideal P_i in B, and the ideals Q_i are labelled so that $Q_1, \ldots, Q_d$ are relevant and $Q_{d+1}, \ldots, Q_e$ are irrelevant, i.e., the relevant portion of $(J)^\sharp$ of J is given by

$$(J)^\sharp = Q_1 \cap \cdots \cap Q_d.$$

Then $\mathbb{W}_\kappa(P_i) = \mathbb{W}_\kappa(Q_i) =$ an irreducible variety in $\mathrm{pvt}_k(\mathbb{P}_\kappa^N)$ for $1 \leq i \leq d$ and

$$\mathbb{W}_\kappa(J) = \cup_{1 \leq i \leq d} \mathbb{W}_\kappa(P_i)$$

is a decomposition of $\mathbb{W}_\kappa(J)$ into irreducible subvarieties. Moreover, in case the primary decomposition of J is irredundant and is so labelled that P_i is a minimal or embedded prime of J according as $1 \leq i \leq n$ or $n+1 \leq i \leq d$, then

$$\mathbb{W}_\kappa(J) = \cup_{1 \leq i \leq n} \mathbb{W}_\kappa(P_i)$$

is the irredundant decomposition of $\mathbb{W}_\kappa(J)$ into its irreducible components.

Conversely, for any variety U in $\mathrm{pvt}_k(\mathbb{P}_\kappa^N)$ let

$$U = \cup_{1 \leq i \leq n} U_i$$

be its decomposition into its distinct irreducible components. Then $\mathbb{J}_k(U_i)_{1 \leq i \leq n}$ are all the distinct associated (automatically minimal) primes of the ideal $\mathbb{J}_k(U)$ in B and we have

$$\mathrm{rad}_B \mathbb{J}_k(U) = \mathbb{J}_k(U) = \cap_{1 \leq i \leq n} \mathbb{J}_k(U_i).$$

Note that an irreducible component of U can be characterized by saying that it is an irreducible subvariety U' of U such that U has no irreducible subvariety $U'' \neq U'$ with $U' \subset U''$.

In the prospectral situation, given any subintegrally graded noetherian ring R, let us be in one of the last two cases of (T151). For any homogeneous ideal J in R take a primary decomposition

$$J = \cap_{1 \leq i \leq e} Q_i$$

where Q_i is a homogeneous ideal in R which is primary for the homogeneous prime ideal P_i in R, and the ideals Q_i are labelled so that $Q_1, \ldots, Q_d$ are relevant and $Q_{d+1}, \ldots, Q_e$ are irrelevant, i.e., the relevant portion of $(J)^\sharp$ of J is given by

$$(J)^\sharp = Q_1 \cap \cdots \cap Q_d.$$

Then $V(P_i) = V(Q_i) =$ an irreducible variety in S for $1 \leq i \leq d$ and

$$V(J) = \cup_{1 \leq i \leq d} V(P_i)$$

is a decomposition of $V(J)$ into irreducible subvarieties. Moreover, in case the primary decomposition of J is irredundant and is so labelled that P_i is a minimal or embedded prime of J according as $1 \le i \le n$ or $n + 1 \le i \le d$, then

$$V(J) = \cup_{1 \le i \le n} V(P_i)$$

is the irredundant decomposition of $V(J)$ into its irreducible components.

Conversely, for any variety U in S let

$$U = \cup_{1 \le i \le n} U_i$$

be its decomposition into its distinct irreducible components. Then $I(U_i)_{1 \le i \le n}$ are all the distinct associated (automatically minimal) primes of the ideal $I(U)$ in R and we have

$$\mathrm{rad}_R I(U) = I(U) = \cap_{1 \le i \le n} I(U_i).$$

Note that an irreducible component of U can be characterized by saying that it is an irreducible subvariety U' of U such that U has no irreducible subvariety $U'' \ne U'$ with $U' \subset U''$. All this is particularly significant when R contains a field k and there is a graded k-epimorphism $\phi : B \to R$ where $B = k[X_1, \ldots, X_{N+1}]$ is a finite variable polynomial ring over k, and $\ker(\phi)$ is the homogeneous ideal $\mathbb{J}_k(W)$ of some $W \in \mathrm{pvt}_k(\mathbb{P}^N_\kappa)$ for some overfield κ of k containing an algebraic closure of k; we may then regard R as the homogeneous coordinate ring of W.

QUEST (Q34.5). [**Modelic and Spectral Projective Spaces**]. Now we are ready to give the projective version of the affine enlargements described in L4§8.2. So consider the polynomial ring

$$B_{N+1,k} = k[X_1, \ldots, X_{N+1}]$$

in indeterminates $X_1, \ldots, X_{N+1}$ over a field k with $N \in \mathbb{N}$, and recall the display (Q34.2)(12*) which says that

$$\mathbb{P}^N_k \approx (\mathbb{P}^N_k)^{\rho\delta} \subset (\mathbb{P}^N_k)^{\mu\delta} \subset (\mathbb{P}^N_k)^\delta.$$

We want to enlarge the above isomorphism to the entire modelic projective space $(\mathbb{P}^N_k)^\delta = \mathfrak{W}(k; X_1, \ldots, X_{N+1})$. Note that after identifying k with a subfield of $R/M(R)$, as observed in L4§9.1, $(\mathbb{P}^N_k)^{\rho\delta}$ (resp: $(\mathbb{P}^N_k)^{\mu\delta}$) is the set of all $R \in (\mathbb{P}^N_k)^\delta$ for which $R/M(R) = k$ (resp: $R/M(R)$ is algebraic over k). Moreover, the above set-theoretic isomorphism of $\mathbb{P}^N_k$ with $\mathrm{im}(\theta) = (\mathbb{P}^N_k)^{\rho\delta}$ is given by the injective map $\theta : \mathbb{P}^N_k \to (\mathbb{P}^N_k)^\delta$ where for all $\beta = [u] \in \mathbb{P}^N_k$ we have $\theta(\beta) = R_k(\beta)^*$.

Before doing the projective modelic enlargement hinted above, let us first do the projective spectral enlargement corresponding to the affine spectral enlargement given in L4§8.1. As we have already noted, the homogeneous ideal $\mathbb{J}_k(\beta)$ of any point β in $\mathbb{P}^N_k$ is a submaximal homogeneous ideal in $B_{N+1,k}$. Also clearly $\beta \ne \beta'$ in $\mathbb{P}^N_k \Rightarrow \mathbb{J}_k(\beta) \ne \mathbb{J}_k(\beta')$. So via $\mathbb{J}_k$ we may ENLARGE $\mathbb{P}^N_k$ into $\mathrm{proj}(B_{N+1,k})$

which we may denote by $(\mathbb{P}_k^N)^\sigma$ and call it the spectral projective N-space over k. The mapping $\beta \mapsto \mathbb{J}_k(\beta)$ gives an injection $\mathbb{P}_k^N \to (\mathbb{P}_k^N)^\sigma$. The image of $\mathbb{P}_k^N$ under this injection may be denoted by $(\mathbb{P}_k^N)^{\rho\sigma}$ and called the rational spectral projective N-space over k. Also we may denote $\mathrm{mproj}(B_{N+1,k})$ by $(\mathbb{P}_k^N)^{\mu\sigma}$ and call it the maximal spectral projective N-space over k. Thus

$$\mathbb{P}_k^N \approx (\mathbb{P}_k^N)^{\rho\sigma} \subset (\mathbb{P}_k^N)^{\mu\sigma} \subset (\mathbb{P}_k^N)^\sigma$$

where $\approx$ is again denoting a set-theoretic isomorphism, i.e., a bijection.

Reverting to the projective modelic enlargement, we note that $P \mapsto (B_{N+1,k})_{[P]}$ gives an inclusion reversing bijection $(\mathbb{P}_k^N)^\sigma \to (\mathbb{P}_k^N)^\delta$. The images of $(\mathbb{P}_k^N)^{\rho\sigma}$ and $(\mathbb{P}_k^N)^{\mu\sigma}$ under this map are $(\mathbb{P}_k^N)^{\rho\delta}$ and $(\mathbb{P}_k^N)^{\mu\delta}$ which we have called the rational modelic projective N-space over k and the minimal modelic projective N-space over k respectively; note that because of the inclusion reversing property, $(\mathbb{P}_k^N)^{\mu\delta}$ is the set of all minimal members of $(\mathbb{P}_k^N)^\delta$. Now the isomorphisms

$$(\mathbb{P}_k^N)^{\rho\sigma} \approx (\mathbb{P}_k^N)^{\rho\delta} \quad \text{and} \quad (\mathbb{P}_k^N)^{\mu\sigma} \approx (\mathbb{P}_k^N)^{\mu\delta} \quad \text{and} \quad (\mathbb{P}_k^N)^\sigma \approx (\mathbb{P}_k^N)^\delta$$

give rise to the enlargements

$$\mathbb{P}_k^N \approx (\mathbb{P}_k^N)^{\rho\delta} \subset (\mathbb{P}_k^N)^{\mu\delta} \subset (\mathbb{P}_k^N)^\delta.$$

Given any $U \in (\mathbb{P}_k^N)^\sigma$, by (T152) we see that $k(U)^* = k \Leftrightarrow U \in (\mathbb{P}_k^N)^{\rho\sigma}$ whereas $k(U)^*/k$ is algebraic $\Leftrightarrow U \in (\mathbb{P}_k^N)^{\mu\sigma}$; consequently, points in $(\mathbb{P}_k^N)^{\rho\sigma}$ may be called rational and points in $(\mathbb{P}_k^N)^{\mu\sigma}$ may be called algebraic. Similarly, points in $(\mathbb{P}_k^N)^{\rho\delta}$ may be called rational and points in $(\mathbb{P}_k^N)^{\mu\delta}$ may be called algebraic. By (T152) we also see that if an overfield κ of k contains an algebraic closure of k then "points" of $(\mathbb{P}_k^N)^\sigma$ are the homogeneous prime ideals of "irreducible varieties in $\mathbb{P}_\kappa^N$ defined over k," and the corresponding "points" of $(\mathbb{P}_k^N)^\delta$ are their homogeneous local rings.

QUEST (Q34.6). [**Relation between Affine and Projective Varieties**]. To relate varieties in projective and affine spaces, consider the homogeneous ring

$$B = B_{N+1,k} = k[X_1, \ldots, X_{N+1}] = \sum_{d \in \mathbb{N}} B_d \quad \text{with} \quad B_\infty = \bigcup_{d \in \mathbb{N}} B_d$$

where $X_1, \ldots, X_{N+1}$ are indeterminates over a field k with $N \in \mathbb{N}$, and let

$$\gamma : B = \widehat{B}[X_{N+1}] \to \widehat{B} = B_{N,k} = k[X_1, \ldots, X_N]$$

be the unique $\widehat{B}$-epimorphism with

$$\gamma(X_{N+1}) = 1.$$

In the situation of (Q34.2) the dehomogenization map

$$\Gamma = B_{/X_{N+1}} : B \to k(X_1, \ldots, X_{N+1})$$

is the unique k-homomorphism such that

$$\Gamma(X_i) = Y_i = X_i/X_{N+1} \quad \text{for} \quad 1 \le i \le N+1 \quad \text{with} \quad Y_{N+1} = 1$$

and for it we have

$$\Gamma(B) = B' = k[Y_1, \ldots, Y_N]$$

[where we have abbreviated $\Gamma_{N+1}, Y_{i,N+1}, B'_{N+1}$ to Γ, Y_i, B' respectively]. Clearly

$$\Gamma = \widehat{\Gamma}\gamma$$

where

$$\widehat{\Gamma} : \widehat{B} \to k(X_1, \ldots, X_{N+1})$$

is the unique k-monomorphism such that

$$\widehat{\Gamma}(X_i) = Y_i \quad \text{for} \quad 1 \le i \le N.$$

So we may call γ the operational dehomogenization map of B at X_{N+1}.

As in (Q34.3), let κ be an overfield of k and recall that

$$\mathbb{P}^N_\kappa = \begin{cases} \text{the set of all } [u] \text{ with} \\ u = (u_1, \ldots, u_{N+1}) \in \kappa^{N+1} \setminus \{(0, \ldots, 0)\} \end{cases}$$

where by $[u]$ we denote the equivalence class (under proportionality) containing u. Also recall that

$$\mathbb{A}^N_\kappa = \{\alpha = (\alpha_1, \ldots, \alpha_N) \in \kappa^N\}.$$

As in (Q34.2)(12) we have

$$\mathbb{P}^N_\kappa = \cup_{i=1}^{N+1}(\mathbb{P}^N \setminus H_i) = \cup_{i=1}^{N+1} \mathbb{A}^N_{\kappa,i}$$

where the hyperplane H_i in $\mathbb{P}^N_\kappa$ is given by

$$H_i : X_i = 0$$

and the affine N-space $\mathbb{A}^N_{\kappa,i}$ is given by

$$\mathbb{A}^N_{\kappa,i} = \{v = (v_1, \ldots, v_{N+1}) \in \kappa^{N+1} : v_i = 1\}.$$

Let us put

$$H_\infty = H_{N+1} \quad \text{and} \quad \widehat{\mathbb{A}}^N_\kappa = \mathbb{A}^N_{\kappa,N+1}$$

and let us designate these as the hyperplane at infinity (of $\mathbb{P}^N_\kappa$) and the affine portion (of $\mathbb{P}^N_\kappa$) respectively; let us also refer to their points as points (of $\mathbb{P}^N_\kappa$) at infinity and at finite distance respectively. Clearly we obtain a unique bijection

$$\iota : \widehat{\mathbb{A}}^N_\kappa \to \mathbb{A}^N_\kappa$$

such that for all $v \in \widehat{\mathbb{A}}_\kappa^N$ we have

$$\iota(v) = \alpha \in \mathbb{A}_\kappa^N \text{ with } \alpha_i = v_i \text{ for } 1 \le i \le N.$$

For any $U \subset \mathbb{P}_\kappa^N$ let

$$\epsilon(U) = \iota(U \cap \widehat{\mathbb{A}}_\kappa^N)$$

and for any $\widehat{U} \subset \mathbb{A}_\kappa^N$ let

$$\pi(\widehat{U}) = \bigcap \{U \in \mathrm{pvt}_k(\mathbb{P}_\kappa^N) : \widehat{U} \subset \epsilon(U)\}$$

where, to avoid putting a complicated expression under the intersection sign, we have used the TYPOGRAPHICALLY CONVENIENT NOTATION according to which

$$\bigcap \{U \in \mathrm{pvt}_k(\mathbb{P}_\kappa^N) : \widehat{U} \subset \epsilon(U)\} = \bigcap_{U \in \Theta(\widehat{U})} U$$

with

$$\Theta(\widehat{U}) = \{U \in \mathrm{pvt}_k(\mathbb{P}_\kappa^N) : \widehat{U} \subset \epsilon(U)\}.$$

As temporary notation let us put

$$\widehat{T}_k(\mathbb{A}_\kappa^N) = \mathrm{avt}_k(\mathbb{A}_\kappa^N)$$

and

$$T_k(\mathbb{P}_\kappa^N) = \mathrm{pvt}_k(\mathbb{P}_\kappa^N)$$

and

$$T_k'(\mathbb{P}_\kappa^N) = \{U \in T_k(\mathbb{P}_\kappa^N) : \pi(\epsilon(U)) = U\}$$

and note that the last equation is equivalent to saying that $T_k'(\mathbb{P}_\kappa^N)$ is the set of all $U \in T_k(\mathbb{P}_\kappa^N)$ such that no irreducible component of U is contained in H_∞.

As temporary notation let us also put

$$\widehat{T}(\widehat{B}) = \text{the set of all ideals in } \widehat{B}$$

and

$$T(B) = \text{the set of all homogeneous ideals in } B$$

and

$$T'(B) = \{J \in T(B) : [J : X_{N+1}]_B = J\}$$

and note that the last equation is obviously equivalent to saying that

$$T'(B) = \{J \in T(B) : X_{N+1} \notin Z_B(B/J)\}$$

and hence by (Q1) it is equivalent to saying that $T'(B)$ is the set of all $J \in T(B)$ such that X_{N+1} does not belong to any associated prime of J in B.

To obtain a sort of inverse of the map γ,

$$\begin{cases} \text{for any } d \in \mathbb{N} \text{ and } f \in B_d \\ \text{such that } f \notin X_{N+1}^{d-e} B_e \text{ whenever } d > e \in \mathbb{N}, \\ \text{we put } \delta(\gamma(f)) = f \end{cases}$$

and we call $\delta(\gamma(f))$ the operational minimal homogenization of $\gamma(f)$ at X_{N+1}. Equivalently

$$\delta(0) = 0$$

and for any $0 \neq \widehat{f} = \widehat{f}(X_1, \ldots, X_N) \in \widehat{B}$ we have

$$\delta(\widehat{f}(X_1, \ldots, X_N)) = X_{N+1}^{\deg(\widehat{f})} \widehat{f}(X_1/X_{N+1}, \ldots, X_N/X_{N+1}).$$

Moreover

$$\text{for any } \widehat{J} \in \widehat{T}(\widehat{B}) \text{ we put } \delta(\widehat{J}) = \{\delta(\widehat{f}) : \widehat{f} \in \widehat{J}\}B$$

and we call $\delta(\widehat{J})$ the operational homogenization of $\widehat{J}$ at X_{N+1}.

Now, in view of the above discussion, we get Theorems (T153) and (T154) stated below.

IDEALS AND HOMOGENEOUS IDEALS RELATIONS THEOREM (T153).

In the above notation we have the following.

(T153.1) $J \mapsto \gamma(J)$ gives an inclusion preserving surjection $T(B) \to \widehat{T}(\widehat{B})$ which commutes with the ideal theoretic operations of radicals, intersections, sums, finite products, and quotients = parenthetical colons. [Here "commutes" means, for instance: $\text{rad}_{\widehat{B}}\gamma(J) = \gamma(\text{rad}_B J)$]. Moreover, if $J = \cap_{1 \leq i \leq e} Q_i$ is an irredundant primary decomposition of $J \in T(B)$ where the primary ideals $Q_i \in T(B)$ with radicals P_i are so labelled that $X_{N+1} \notin P_i$ or $X_{N+1} \in P_i$ according as $1 \leq i \leq d$ or $d + 1 \leq i \leq e$, then $\gamma(J) = \cap_{1 \leq i \leq d} \gamma(Q_i)$ is an irredundant primary decomposition of $\gamma(J) \in \widehat{T}(\widehat{B})$ and $\gamma(P_i)$ is the radical of the primary ideal $\gamma(Q_i) \in \widehat{T}(\widehat{B})$ for $1 \leq i \leq d$.

(T153.2) $J \mapsto \widehat{J} = \gamma(J)$ gives a bijection $T'(B) \to \widehat{T}(\widehat{B})$ whose inverse is given by $\widehat{J} \mapsto J = \delta(\widehat{J})$.

(T153.3) $J \mapsto \gamma(J)$ gives a bijection $T'(B) \cap \text{prd}(B) \to \text{rd}(\widehat{B})$, where we note that $T'(B) \cap \text{prd}(B)$ is the set of all those homogeneous radical ideals in B which are finite intersections of members of $\text{cproj}_B(X_{N+1}B)$.

(T153.4) $J \mapsto \gamma(J)$ gives a bijection $\text{cproj}_B(X_{N+1}B) \to \text{spec}(\widehat{B})$.

AFFINE AND PROJECTIVE VARIETIES RELATIONS THEOREM (T154). Assuming that κ contains an algebraic closure of k, in the above notation we have the following.

(T154.1) $U \mapsto \epsilon(U)$ gives an inclusion preserving surjection $T_k(\mathbb{P}^N_\kappa) \to \widehat{T}_k(\mathbb{A}^N_\kappa)$ which commutes with the set-theoretic operations of intersections and finite unions. Moreover, if $U = \cup_{1 \le i \le e} U_i$ is an irredundant decomposition of $U \in T_k(\mathbb{P}^N_\kappa)$ into irreducible varieties U_i in $T_k(\mathbb{P}^N_\kappa)$ which are labelled so that $U_i \not\subset H_\infty$ or $U_i \subset H_\infty$ according as $1 \le i \le d$ or $d + 1 \le i \le e$, then $\epsilon(U) = \cap_{1 \le i \le d} \epsilon(U_i)$ is an irredundant decomposition of $\epsilon(U) \in \widehat{T}_k(\mathbb{P}^N_\kappa)$ and $\epsilon(U_1), \ldots, \epsilon(U_d)$ are the irreducible components of $\epsilon(U)$.

(T154.2) $U \mapsto \widehat{U} = \epsilon(U)$ gives a bijection $T'_k(\mathbb{P}^N_\kappa) \to \widehat{T}_k(\mathbb{A}^N_\kappa)$ whose inverse is given by $\widehat{U} \mapsto U = \pi(\widehat{U})$.

(T154.3) $U \mapsto \epsilon(U)$ gives a bijection $T'_k(\mathbb{P}^N_\kappa) \cap \mathrm{ipvt}_k(\mathbb{P}^N_\kappa) \to \mathrm{iavt}_k(\mathbb{A}^N_\kappa)$ where we note that $T'_k(\mathbb{P}^N_\kappa) \cap \mathrm{ipvt}_k(\mathbb{P}^N_\kappa)$ is the set of all irreducible irreducible members of $\mathrm{ipvt}_k(\mathbb{P}^N_\kappa)$ which are not contained in H_∞.

QUEST (Q35) Simplifying Singularities by Blowups

We shall now say more about modelic blowups and their usage in simplifying singularities as initiated in L4§9.2 and L4§9.3. The material will be based on §1 of my Resolution Book [A05].

QUEST (Q35.1). [**Hypersurface Singularities**]. Geometrically speaking, the following Theorem, which corresponds to assertion (1.3.2) of §1 of [A05], says that the multiplicity of a hypersurface at any irreducible subvariety of it is at most equal to the degree of the hypersurface.

HYPERSURFACE SINGULARITY THEOREM (T155). Let $f(X_1, \ldots, X_n)$ be a nonzero polynomial of degree $\le d$ in indeterminates $X_1, \ldots, X_n$ over a field k, where d and n are nonnegative integers. Let R be the localization A_P of the polynomial ring $A = k[X_1, \ldots, X_n]$ at a prime ideal P in it. Then upon letting $e = \mathrm{ord}_R f(X_1, \ldots, X_n)$ we have $e \le d$.

PROOF. First we consider the special case when P is a maximal ideal in A. Now by L4§5(O8)(4•) and L4§7.1(T7) we have $M(R)^e \cap A = P^e$ and hence we see that $f(X_1, \ldots, X_n) \in P^e$. Let $\overline{k}$ be an algebraic closure of k and let $\overline{A} = \overline{k}[X_1, \ldots, X_n]$. Then $\overline{A}$ is integral over A and hence by (T38) and (T40) of (Q9) we can find a maximal ideal $\overline{P}$ in $\overline{A}$ with $\overline{P} \cap A = P$. Note that then $P^e \subset \overline{P}^e$ and hence $f(X_1, \ldots, X_n) \in \overline{P}^e$. By the Nullstellensatz L4§8(T22.4), which was proved in L5§5(Q10), there exist elements $r_1, \ldots, r_n$ in $\overline{k}$ such that

$$\overline{P} = (X_1 - r_1, \ldots, X_n - r_n)\overline{A}.$$

It follows that

$$f(X_1, \ldots, X_n) = \sum_{i_1 + \cdots + i_n = e} f_{i_1 \ldots i_n}(X_1, \ldots, X_n)(X_1 - r_1)^{i_1} \ldots (X_n - r_n)^{i_n}$$

where $f_{i_1 \ldots i_n}(X_1, \ldots, X_n)$ are polynomials in $X_1, \ldots, X_n$ with coefficients in $\overline{k}$. Let $g(X_1, \ldots, X_n)$ and $g_{i_1 \ldots i_n}(X_1, \ldots, X_n)$ be the polynomials in $X_1, \ldots, X_n$ with coefficients in $\overline{k}$ obtained by substituting $X_1 + r_1, \ldots, X_n + r_n$ for $X_1, \ldots, X_n$ in $f(X_1, \ldots, X_n)$ and $f_{i_1 \ldots i_n}(X_1, \ldots, X_n)$ respectively. Then $g(X_1, \ldots, X_n)$ is a nonzero polynomial of degree $\leq d$ in $X_1, \ldots, X_n$ with coefficients in $\overline{k}$. Clearly by substituting $X_1 + r_1, \ldots, X_n + r_n$ for $X_1, \ldots, X_n$ in the last displayed formula we get that

$$g(X_1, \ldots, X_n) = \sum_{i_1 + \cdots + i_n = e} g_{i_1 \ldots i_n}(X_1, \ldots, X_n) X_1^{i_1} \ldots X_n^{i_n}$$

and hence $g(X_1, \ldots, X_n)$ is either zero or is a polynomial of degree $\geq e$ in $X_1, \ldots, X_n$ with coefficients in $\overline{k}$. Therefore $e \leq d$.

Now we deduce the general case as a consequence of the special case considered above. So let $h : A \to A/P$ be the residue class epimorphism. By relabelling $X_1, \ldots, X_n$ suitably we may assume that $(h(X_{m_1}), \ldots, h(X_n))$ is a transcendence basis of $\mathrm{QF}(A/P)$ over $h(k)$. Let $B = k(X_{m+1}, \ldots, X_n)[X_1, \ldots, X_m]$ and $Q = PB$. Then, as in (Q29)(C45), we see that Q is a maximal ideal in B with $R = B_Q$, and clearly $f(X_1, \ldots, X_n)$ is a nonzero polynomial in $X_1, \ldots, X_m$ of degree $\leq d$ with coefficients in $k(X_{m+1}, \ldots, X_n)$. Therefore $e \leq d$ by the case considered above.

QUEST (Q35.2). [**Blowing-up Primary Ideals**]. In the following Theorem, which corresponds to assertion (1.3.3) of §1 of [A05], we shall study the effect of blowing up an ideal in an n-dimensional local domain R which is generated by a system of parameters for R, i.e., which is primary for $M(R)$ and is generated by n elements. Note that the following Theorem and its proof share some similarity with the arguments of (Q8).

PRIMARY IDEAL BLOWUP THEOREM (T156). Let R be an n-dimensional local domain with $n > 1$. Let $x_1, \ldots, x_n$ be elements in R which generate an ideal Q which is primary for $M(R)$. [Note that then we must have $x_i \neq 0$ for $1 \leq i \leq n$]. Let $A = R[x_2/x_1, \ldots, x_n/x_1]$, and let $h : A \to A/(M(R)A)$ be the residue class epimorphism. Then $M(R)A$ is a prime ideal in A such that $\dim(A_{M(R)A}) = 1$ and $M(R)A = \mathrm{rad}_A(QA)$ with $R \cap (M(R)A) = M(R)$, and the elements $h(x_2/x_1), \ldots, h(x_n/x_1)$ are algebraically independent over $h(R)$.

PROOF. Clearly $QA = x_1 A$. Since Q is primary for $M(R)$, there exists a positive integer e such that $M(R)^e \subset Q$, and then $(M(R)A)^e \subset x_1 A$. Let $X_1, \ldots, X_n$ be indeterminates.

Suppose if possible that $R \cap (M(R)A) \neq M(R)$. Then we must have $M(R)A = A$ and hence $x_1 A = A$. Consequently $x_1 y = 1$ for some nonzero $y \in A$. Since $y \in A$, there exists a nonzero polynomial $f(X_2, \ldots, X_n)$ of some degree d in $X_2, \ldots, X_n$ with coefficients in R such that

$$y = f(x_2/x_1, \ldots, x_n/x_1).$$

Now

$$x_1^d = x_1^{d+1} y = x_1 f'(x_1, \ldots, x_n)$$

where $f'(X_1, \ldots, X_n)$ is a nonzero homogeneous polynomial of degree d in indeterminates $X_1, \ldots, X_n$ with coefficients in R. In particular we get $x_1^d \in Q^{d+1}$ which is a contradiction by (T31.2) and (T34) of (Q8). Therefore $R \cap (M(R)A) = M(R)$.

Suppose if possible that the elements $h(x_2/x_1), \ldots, h(x_n/x_1)$ are algebraically dependent over the residue field $k = R/M(R)$. Then there exists a nonzero polynomial $F(X_2, \ldots, X_n)$ of some degree u in $X_2, \ldots, X_n$ with coefficients in R at least one of which is not in $M(R)$ such that $F(x_2/x_1, \ldots, x_n/x_1) \in M(R)A$. Since $F(x_2/x_1, \ldots, x_n/x_1) \in M(R)A$, there exists a polynomial $G(X_2, \ldots, X_n)$ in $X_2, \ldots, X_n$ with coefficients in $M(R)$ such that

$$F(x_2/x_1, \ldots, x_n/x_1) = G(x_2/x_1, \ldots, x_n/x_1).$$

Upon multiplying both sides of the above equation by x_1^v for a suitable integer $v \geq u$ we get that

$$F'(x_1, \ldots, x_n) = G'(x_1, \ldots, x_n)$$

where $F'(X_1, \ldots, X_n)$ is a nonzero homogeneous polynomial of degree v in the indeterminates $X_1, \ldots, X_n$ with coefficients in R at least one of which is not in $M(R)$, and $G'(X_1, \ldots, X_n)$ is either the zero polynomial or a nonzero homogeneous polynomial of degree v in the indeterminates $X_1, \ldots, X_n$ with coefficients in $M(R)$. In particular then

$$F'(x_1, \ldots, x_n) \in Q^{v+1}$$

which is a contradiction by (T31.2) and (T34) of (Q8). Therefore the elements $h(x_2/x_1), \ldots, h(x_n/x_1)$ are algebraically independent over k.

Since $h(A) = h(R)[h(x_2/x_1), \ldots, h(x_n/x_1)]$, we get that $h(A)$ is a domain and hence $M(R)A$ is a prime ideal in A. Since $(M(R)A)^e \subset x_1 A = QA$, by the Krull Principal Ideal Theorem (Q7)(T24) we conclude that

$$M(R)A = \mathrm{rad}_A(QA) \quad \text{and} \quad \dim(A_{M(R)A}) = 1.$$

QUEST (Q35.3). [**Residual Properties and Coefficient Sets**]. In (T156), instead of saying that the elements $h(x_2/x_1), \ldots, h(x_n/x_1)$ are algebraically independent over $h(R)$, we could have briefly said that the elements $x_2/x_1, \ldots, x_n/x_1$ are residually algebraically independent over R.

To generalize this, given any subring R of a ring A and any prime ideal N in A, let $h : A \to A/N$ be the residue class epimorphism. Then elements $y_1, \ldots, y_m$ in A are said to be residually algebraically independent (resp: dependent) over R relative to N if the elements $h(y_1), \ldots, h(y_m)$ are algebraically independent (resp: dependent) over $h(R)$, i.e., over $\mathrm{QF}(h(R))$. Moreover, by the residual transcendence degree of A over R relative to N we mean $\mathrm{trdeg}_{h(R)} h(A)$, i.e., $\mathrm{trdeg}_{\mathrm{QF}(h(R))} \mathrm{QF}(h(A))$, and we denote it by $\mathrm{restrdeg}_R A$. Likewise, A is said to be a residually algebraic (resp: residually finite algebraic, residually separable algebraic, residually finite separable algebraic, residually purely inseparable, residually finite purely inseparable) extension of R relative to N if $\mathrm{QF}(h(A))$ is an algebraic (resp: finite algebraic, separable algebraic, finite separable algebraic, purely inseparable, finite purely inseparable) extension of $\mathrm{QF}(h(R))$. Finally, A is said to be residually rational over R relative to N if $\mathrm{QF}(h(A)) = \mathrm{QF}(h(R))$.

If A is quasilocal with $N = M(A)$ then from the above definitions we may drop the phrase "relative to."

In a similar manner we may prefix the adjective "residually" to other properties without explicit definitions.

Note that the above definitions are consistent with the terminology introduced just before (Q10)(T47).

By a coefficient set of a quasilocal ring A we mean a subset κ of A which contains the elements 0 and 1 of A and which is mapped bijectively onto $A/M(A)$ by the residue class epimorphism $A \to A/M(A)$. By a coefficient field of a quasilocal ring A we mean a coefficient set of A which is a subring (and hence a subfield) of A. A quasilocal ring A is said to be equicharacteristic or nonequicharacteristic according as the characteristic of A coincides with the characteristic of $A/M(A)$ or not. Let us observe that any quasilocal ring has a coefficient set, but if it is nonequicharacteristic then it cannot have a coefficient field.

QUEST (Q35.4). [**Geometrically Blowing-up Simple Centers**]. In the two Theorems (T157) and (T158) below, which respectively correspond to assertions (1.4.1) and (1.4.2) of §1 of [A05], we shall see what happens when at a simple point P of an n-dimensional variety V we blow up an $(n - m)$-dimensional subvariety W having a simple point at P. Here m, n are integers with $0 < m \le n$.

Before dealing with these two Theorems in a formal algebraic set-up, as we shall do in (Q35.5), here in (Q35.4) we shall talk about them in a geometric fashion. In the algebraic set-up of (Q35.5) we shall

(1)
$$\begin{cases} \text{replace the pair } (V, P) \text{ by } (R, M(R)) \text{ where} \\ R \text{ is an } n\text{-dimensional regular local domain,} \\ \text{and the pair } (P, W) \text{ by a prime ideal } P \text{ in } R \text{ such that} \\ R/P \text{ is an } (n - m)\text{-dimensional regular local domain.} \end{cases}$$

In (Q35.5) we shall use some general properties of regular local rings which will be discussed in (Q35.9); note that by (Q14)(T64) such rings are domains.

To make the geometric discussion easy to visualize, let us focus our attention on the special case when P is the origin $(0, \ldots, 0)$ of the n-dimensional affine space V with coordinates $X_1, \ldots, X_n$ and W is the $(n - m)$-dimensional affine subspace $X_1 = \cdots = X_m = 0$. Let V' be the affine n-space with coordinates $X_1', \ldots, X_n'$. Then the blowup τ of V at P with "center" W (or at least one affine portion of it) is given by

$$\tau : \begin{cases} X_1' = X_1, \ X_2' = X_2/X_1, \ \ldots, \ X_m' = X_m/X_1, \\ X_{m+1}' = X_{m+1}, \ \ldots, \ X_n' = X_n. \end{cases}$$

In older times, for instance on page 1 of [A05], one could express this by writing $\tau : V \to V'$. But now-a-days, as in these Lectures, an arrow means a map. So we better say that the reverse τ' of τ is the map $\tau' : V' \to V$ given by

$$\tau' : \begin{cases} X_1 = X_1', \ X_2 = X_1' X_2', \ \ldots, \ X_m = X_1' X_m', \\ X_{m+1} = X_{m+1}', \ \ldots, \ X_n = X_n'. \end{cases}$$

By the equations of τ and τ' we see that

(2)
$$\begin{cases} \tau \text{ blows up } W \text{ into the exceptional hyperplane } E : X_1' = 0, \\ \text{and it blows up } P \text{ into the } (m - 1)\text{-dimensional} \\ \text{affine subspace } E_P : X_1' = X_{m+1}' = \cdots = X_n' = 0 \text{ of } E. \end{cases}$$

As an illustration let $m = n = 2$ and let us write (X, Y) and (X', Y') for (X_1, X_2) and (X_1', X_2') respectively. Now we may descriptively say that we are applying a QDT (= Quadratic Transformation) to the (X, Y)-plane with center at the origin $P = (0, 0)$. Generalizing the discussion of (X12) and (X13) of L4§9.3, consider a plane curve $C : f = 0$ where $f = f(X, Y)$ is a nonconstant bivariate polynomial; also see the references to my Engineering Book [A04] given in (X12) and (X13). Let the origin P be a d-fold point of C where d is some positive integer. Then

$$f(X, Y) = f_d(X, Y) + f_{d+1}(X, Y) + \ldots$$

where $f_d(X, Y)$ is a nonzero homogeneous polynomial of degree d, and for $i > 0$ the polynomial $f_{d+i}(X, Y)$ is either the zero polynomial or a nonzero homogeneous polynomial of degree $d + i$. Assuming the ground field k to be algebraically closed, by "rotating coordinates" we can arrange that X does not divide $f_d(X,Y)$, i.e., the line $X = 0$ is not tangent to C at P. Then

$$f_d(X,Y) = a_0 \prod_{1 \le i \le r} (Y - m_i X)^{e_i}$$

where $a_0 \in k^\times$, the constants $m_1, \ldots, m_r$ are pairwise distinct, and $e_1, \ldots, e_r$ are positive integers. Now the line $T_i : Y = m_i X$ of slope m_i passing through the origin is a tangent to C at P and its "tangential multiplicity" is e_i. Thus "counting properly" we get $e_1 + \cdots + e_r = d$ tangents to C at P. The "total transform" C' of C in V' is given by $C' : f' = 0$ where the polynomial $f'(X', Y')$ is obtained by substituting $X = X'$ and $Y = X'Y'$ in $f(X,Y)$. Now

$$\begin{cases} f'(X',Y') = f(X', X'Y') = f_d(X', X'Y') + f_{d+1}(X', X'Y') + \ldots \\ \qquad = X'^d [f_d(1, Y') + X' f_{d+1}(1, Y') + \ldots] \\ \qquad = X'^d f^*(X', Y') \end{cases}$$

where

$$f^*(X',Y') = f_d(1, Y') + X' f_{d+1}(1, Y') + X'^2 f_{d+2}(1, Y') + \ldots.$$

and we call $C^* : f^* = 0$ the "proper transform" of C in V'. Thus the total transform C' consists of the proper transform C^* together with d times the exceptional line $E = E_P$. By putting $X' = 0$ in in the above expression of $f^*(X', Y')$ we see that C^* meets E in the points $P_1' = (0, m_1), \ldots, P_r' = (0, m_r)$, and referring to L3§3 for the definitions of the multiplicity $\text{mult}_{P_i'} C^*$ of C^* at P_i' and the intersection multiplicity $\text{int}_{P_i'}(C^*, E)$ of C^* and E at P_i' we get

(3) $$\text{mult}_{P_i'} C^* \le \text{int}_{P_i'}(C^*, E) = e_i \text{ for all } i.$$

Since

(4) $$e_1 + \cdots + e_r = d = \text{mult}_P C$$

by (3) we see that

(5) $$\text{mult}_{P_i'} C^* \le \text{mult}_P C \text{ for all } i$$

and

(6) $$\begin{cases} \text{mult}_{P_i'} C^* = \text{mult}_P C \text{ for some } i \\ \Leftrightarrow r = 1 \text{ and } \text{mult}_{P_1'} C^* = \text{int}_{P_1'}(C^*, E) = e_1 = d = \text{mult}_P C. \end{cases}$$

As another illustration let $m = n > 2$. Again we say that we are applying a QDT to the $(X_1, \ldots, X_n)$-space with center at the origin $P = (0, \ldots, 0)$. Generalizing the discussion of (X14) of L4§9.3, let us consider a hypersurface $C : f = 0$ where

$f = f(X_1, \ldots, X_n)$ is a nonconstant n-variable polynomial. Let the origin P be a d-fold point of C where d is some positive integer. Then

$$f(X_1, \ldots, X_n) = f_d(X_1, \ldots, X_n) + f_{d+1}(X_1, \ldots, X_n) + \ldots$$

where for $i > 0$ the polynomial $f_{d+i}(X_1, \ldots, X_n)$ is either the zero polynomial or a nonzero homogeneous polynomial of degree $d + i$. The "total transform" C' of C in V' is given by $C' : f' = 0$ where the polynomial $f'(X_1', \ldots, X_n')$ is obtained by substituting $X_1 = X_1'$ and $X_j = X_1' X_j'$ for $2 \leq j \leq n$ in $f(X_1, \ldots, X_n)$. Now

$$\begin{cases} f'(X_1', \ldots, X_n') = \sum_{i \geq 0} f_{d+i}(X_1', X_1' X_2', \ldots, X_1' X_n') \\ \qquad\qquad = X_1'^d f^*(X_1', \ldots, X_n') \end{cases}$$

where

$$f^*(X_1', \ldots, X_n') = f_d(1, X_2', \ldots, X_n') + \sum_{i>0} X_1'^i f_{d+i}(1, X_2', \ldots, X_n')$$

and we call $C^* : f^* = 0$ the "proper transform" of C in V'. Thus the total transform C' consists of the proper transform C^* together with d times the exceptional hyperplane $E = E_P$. Given any point $P' = (0, a_2, \ldots, a_n)$ of $E \cap C^*$, by the last displayed equation we see that $\mathrm{mult}_{P'} C^* \leq \mathrm{mult}_Q D$ where the point $Q = (a_2, \ldots, a_n)$ is on the hypersurface $D : f_d(1, X_2, \ldots, X_n) = 0$ in the $(n-1)$-dimensional affine space with coordinates $X_2', \ldots, X_n'$, and by (T155) we get $\mathrm{mult}_Q D \leq d$. Therefore

$$(7) \qquad \begin{cases} \text{for every point } P' \text{ of } E \cap C^* \text{ we have} \\ \mathrm{mult}_{P'} C^* \leq d = \mathrm{mult}_P C. \end{cases}$$

As yet another illustration let $m < n$. Now we may descriptively say that we are applying an MDT (= Monoidal Transformation) to the $(X_1, \ldots, X_n)$-space V with center at the subspace $W : X_1 = \cdots = X_m = 0$. Generalizing the discussion of (X16) of L4§10, let us consider a hypersurface $C : f = 0$ where $f = f(X_1, \ldots, X_n)$ is a nonconstant n-variable polynomial. Let the origin P be a d-fold point of C where d is some positive integer. Then

$$f(X_1, \ldots, X_n) = f_d(X_1, \ldots, X_n) + f_{d+1}(X_1, \ldots, X_n) + \ldots$$

where for $i > 0$ the polynomial $f_{d+i}(X_1, \ldots, X_n)$ is either the zero polynomial or a nonzero homogeneous polynomial of degree $d + i$. The "total transform" C' of C in V' is given by $C' : f' = 0$ where the polynomial $f'(X_1', \ldots, X_n')$ is obtained by substituting $X_i = X_i'$ for $i \in \{1, m+1, \ldots, n\}$ and $X_j = X_1' X_j'$ for $2 \leq j \leq m$ in $f(X_1, \ldots, X_n)$. Let δ be the multiplicity $\mathrm{mult}_W C$ of C at W, i.e., δ is the largest integer such that for some polynomials $F_{i_1 \ldots i_m}(X_1, \ldots, X_n)$ we have

$$f(X_1, \ldots, X_n) = \sum_{i_1 + \cdots + i_m = \delta} F_{i_1 \ldots i_m}(X_1, \ldots, X_n) X_1^{i_1} \ldots X_m^{i_m}.$$

Now clearly $\delta \leq d$ and upon letting $F'_{i_1 \ldots i_m}(X'_1, \ldots, X'_n)$ be the polynomial obtained by substituting $X_i = X'_i$ for $i \in \{1, m+1, \ldots, n\}$ and $X_j = X'_1 X'_j$ for $2 \leq j \leq m$ in $F_{i_1 \ldots i_m}(X_1, \ldots, X_n)$ we have

$$f'(X'_1, \ldots, X'_n) = X'^{\delta}_1 f^*(X'_1, \ldots, X'_n)$$

where

$$f^*(X'_1, \ldots, X'_n) = \sum_{i_1 + \cdots + i_m = \delta} F'_{i_1 \ldots i_m}(X'_1, \ldots, X'_n) X'^{i_2}_2 \ldots X'^{i_m}_m$$

and we call $C^* : f^* = 0$ the "proper transform" of C in V'. Thus the total transform C^* consists of the proper transform C^* together with δ times the exceptional hyperplane E. As in (X16) we see that

(8)
$$\begin{cases} \text{if } C \text{ is equimultiple along } W \text{ at } P, \\ \text{i.e., if } \mathrm{mult}_W C = \delta = d = \mathrm{mult}_P C, \\ \text{then for every point } P' \text{ of } E_P \cap C^* \text{ we have} \\ \mathrm{mult}_{P'} C^* \leq d = \mathrm{mult}_P C. \end{cases}$$

In (X16) we have shown that in (8) the equimultiplicity condition is necessary.

QUEST (Q35.5). [**Algebraically Blowing-up Simple Centers**]. Let R be an n-dimensional regular local domain. As we shall see in (Q35.9)(1†): if $x_1, \ldots, x_n$ is a regular system of parameters for R, i.e., a sequence of generators of $M(R)$ of length n, then for every positive integer $m \leq n$ the ideal $P = (x_1, \ldots, x_m)R$ is a nonzero prime ideal in R for which the local ring R/P is regular, and conversely, if P is any nonzero prime ideal in R for which the local ring R/P is regular then $P = (x_1, \ldots, x_m)R$ for some regular system of parameters $x_1, \ldots, x_n$ for R and some positive integer $m \leq n$.

Now let $x_1, \ldots, x_n$ be a regular system of parameters for R. Then, by the above observation (Q35.9)(1†),

$$(0)R \subset (x_1)R \subset (x_1, x_2)R \subset \cdots \subset (x_1, \ldots, x_n)R$$

is an ascending chain of distinct prime ideals in R, and hence for any m with

$$0 < m \leq n$$

upon letting

$$P = (x_1, \ldots, x_m)R \quad \text{and} \quad S = R_P$$

we see that

$$\begin{cases} R/P \text{ and } S \text{ are regular local domains with} \\ \dim(R/P) = n - m \text{ and } \dim(S) = m. \end{cases}$$

Let

$$h : R \to k \quad \text{and} \quad h' : R \to T$$

be (ring) epimorphisms such that

$$\ker(h) = M(R) \quad \text{and} \quad \ker(h') = P.$$

Then clearly k is a field and there exists a unique

$$\text{epimorphism } h'' : T \to k \text{ with } h''h' = h.$$

Let

$$A = R[x_2/x_1, \ldots, x_m/x_1].$$

Let

$$B = T[X_2, \ldots, X_m] \quad \text{and} \quad A^* = k[X_2, \ldots, X_m]$$

where $X_2, \ldots, X_m$ are indeterminates. Let

$$S' = \text{the valuation ring of ord}_S.$$

Concerning this set-up we now prove the following Theorems (T157) and (T158).

PRIME IDEAL BLOWUP THEOREM (T157). For any $e \in \mathbb{N}$ we have

$$M(S')^e \cap A = P^e A = x_1^e A \quad \text{and} \quad P^e S' = M(S')^e = x_1^e S'$$

and

$$(P^e A) \cap R = M(S')^e \cap R = (M(S')^e \cap S) \cap R = M(S)^e \cap R = P^e.$$

PA is a prime ideal in A with

$$S' = A_{PA}.$$

For any $0 \neq w \in A$, upon letting

$$d = \text{ord}_S w$$

we have that d is the largest nonnegative integer such that $w \in P^d A$, which is equivalent to saying that d is the unique nonnegative integer such that

$$w/x_1^d \in A \quad \text{and} \quad w/x_1^{d+1} \notin A.$$

$M(R)A$ is a prime ideal in A and for it we have

$$M(R)A = (x_1, x_{m+1}, \ldots, x_n)A \quad \text{and} \quad (M(R)A) \cap R = M(R).$$

There exists a unique epimorphism $H : A \to A^*$ such that $H(x_i/x_1) = X_i$ for $2 \leq i \leq m$ and $H(u) = h(u)$ for all $u \in R$; there exists a unique epimorphism

$H' : A \to B$ such that $H'(x_i/x_1) = X_i$ for $2 \leq i \leq m$ and $H'(u) = h'(u)$ for all $u \in R$; and there exists a unique epimorphism $H'' : B \to A^*$ such that $H''(X_i) = X_i$ for $2 \leq i \leq m$ and $H''(u) = h''(u)$ for all $u \in T$. Moreover we have

$$H'' H' = H \quad \text{with} \quad \ker(H) = M(R)A \quad \text{and} \quad \ker(H') = PA.$$

[Geometrically speaking, as elucidated in assertion (2) of (Q35.4), the above epimorphisms H, H', H'' indicate that when at a simple point of an n-dimensional variety we blowup an $(n - m)$-dimensional subvariety having a simple point there, the subvariety blows up into an $(n - 1)$-dimensional "(nonsingular fibre) space" and the point blows up into an $(m - 1)$-dimensional "subspace (=fibre)" of it].

SIMPLE CENTER BLOWUP THEOREM (T158). Let H be as in (T157). Given any $R' \in \mathfrak{V}(A)$ such that R' dominates R, let

$$n' = \dim(R') \quad \text{with} \quad t = \mathrm{restrdeg}_R R'$$

and let

$$Q = A \cap M(R') \quad \text{with} \quad Q^* = \Pi(Q) \quad \text{and} \quad R^* = A^*_{Q^*}.$$

Then we have the following.

(1) For any $e \in \mathbb{N}$ we have

$$P^e R' = x_1^e R' \quad \text{with} \quad (P^e R') \cap A = P^e A \quad \text{and} \quad M(S')^e \cap R' = P^e R'.$$

PR' is a prime ideal in R' with

$$S' = R'_{PR'}.$$

For any $0 \neq w \in R'$, upon letting

$$d = \mathrm{ord}_S w$$

we have that d is the largest nonnegative integer such that $w \in P^d R'$, which is equivalent to saying that d is the unique nonnegative integer such that

$$w/x_1^d \in R' \quad \text{and} \quad w/x_1^{d+1} \notin R'.$$

$M(R)R'$ is a prime ideal in R' and for it we have

$$M(R)R' = (x_1, x_{m+1}, \ldots, x_n)R' \quad \text{and} \quad (M(R)R') \cap R = M(R).$$

(2) If $0 \neq w \in R$ is such that $\mathrm{ord}_S w = \mathrm{ord}_R w$, then upon letting $d = \mathrm{ord}_R w$ we have $\mathrm{ord}_{R'}(w/x_1^d) \leq d$.

[Geometrically this says that the multiplicity does not increase when we blowup an equimultiple simple center. See assertions (5), (7) and (8) of (Q35.4)].

(3) R' and R^* are regular local domains for which we have

$$\mathrm{restrdeg}_k R^* = t \quad \text{and} \quad \dim(R^*) = m - 1 - t \quad \text{with} \quad n \geq n' = n - t \geq n - m + 1.$$

If $D \subset A$ is such that $H(D)R^* = M(R^*)$ then

$$DR' + (x_1, x_{m+1}, \ldots, x_n)R' = M(R').$$

If m' is an integer with $1 \leq m' \leq m$ such that $x_i/x_1 \in M(R')$ for $2 \leq i \leq m'$ then there exist elements $y_1, \ldots, y_q$ with

$$q = n' - n + m - m'$$

such that

$$M(R') = (x_1, x_2/x_1, \ldots, x_{m'}/x_1, x_{m+1}, \ldots, x_n, y_1, \ldots, y_q)R'.$$

(4) The following six conditions are equivalent: (1*) R' is residually algebraic over R; (2*) $n' = n$; (3*) Q is a maximal ideal in A; (4*) Q^* is a maximal ideal in A^*; (5*) $\dim(R^*) = m - 1$; (6*) R^* is residually algebraic over k.

(5) R' is residually separable algebraic over R iff R^* is residually separable algebraic over k.

(6) R' is residually rational over R iff R^* is residually rational over k. If there exists $r_i \in R$ such that $(x_i/x_1) - r_i \in M(R')$ for $2 \leq i \leq m$, then

$$Q = (x_1, (x_2/x_1) - r_2, \ldots, (x_m/x_1) - r_m, x_{m+1}, \ldots, x_n)A$$

with

$$M(R') = (x_1, (x_2/x_1) - r_2, \ldots, (x_m/x_1) - r_m, x_{m+1}, \ldots, x_n)R'$$

and R' is residually rational over R. If R' is residually rational over R, then there exists $r_i \in R$ such that $(x_i/x_1) - r_i \in M(R')$ for $2 \leq i \leq m$. If R' is residually rational over R and κ is a coefficient set for R, then there exists $r_i \in \kappa$ such that $(x_i/x_1) - r_i \in M(R')$ for $2 \leq i \leq m$.

(7) There exists a unique epimorphism $H^* : R' \to R^*$ such that $H^*(u) = H(u)$ for all $u \in A$. Moreover,

$$\ker(H^*) = (\ker(H))R' = M(R)R' = (x_1, x_{m+1}, \ldots, x_n)R'.$$

PROOF OF (T157). For any $e \in \mathbb{N}$ we clearly have

$$P^e A = x_1^e A \quad \text{and} \quad P^e S' = M(S')^e = x_1^e S'$$

and by (Q35.9)(2[†]) we also have

$(\bullet)$ $$M(S')^e \cap R = (M(S')^e \cap S) \cap R = M(S)^e \cap R = P^e.$$

Now $\mathrm{ord}_S x_i = 1$ for $1 \leq i \leq m$ and hence

$$A \subset S'.$$

For any $0 \neq w \in A$ let $d = \text{ord}_S w$, i.e., $d = \text{ord}_{S'} w$; since $A \subset S'$, we must have $w/x_1^{d+1} \notin A$; since $w \in A$, there exists $c \in \mathbb{N}$ such that $wx_1^c \in R$; then $\text{ord}_S(wx_1^c) = d + c$ and hence by (Q35.9)($2^\dagger$) we get $wx_1^c \in P^{d+c}$; consequently $wx_1^c \in x_1^{d+c} A$ and hence $w/x_1^d \in A$; thus d is the largest nonnegative integer such that $w \in P^d A$. Therefore for every $e \in \mathbb{N}$ we have

$$M(S')^e \cap A = P^e A$$

and hence in view of (•) we get

$$(P^e A) \cap R = M(S')^e \cap R = (M(S')^e \cap S) \cap R = M(S)^e \cap R = P^e.$$

Since $PA = M(S') \cap A$, we see that PA is a prime ideal in A, and hence (because $PA = x_1 A =$ a principal ideal) A_{PA} is a one-dimensional regular local domain; clearly $A_{PA} \subset S'$ and these two local domains have the same quotient field; therefore $S' = A_{PA}$.

Let $A_1 = S[x_2/x_1, \ldots, x_m/x_1]$. Then $A \subset A_1 \subset S'$ and upon replacing (R, P) by $(S, M(S))$ in the above argument we get that $M(S') \cap A_1 = M(S)A_1$ and hence $(M(S)A_1) \cap A = PA$. Let $h_1 : A_1 \to A_1/M(S)A_1$ and $h_2 : A \to A/PA$ be the residue class epimorphisms; then by (T156) we know that the elements $h_1(x_2/x_1), \ldots, h_1(x_m/x_1)$ are algebraically independent over $h_1(S)$, and hence a fortiori the elements $h_2(x_2/x_1), \ldots, h_2(x_m/x_1)$ are algebraically independent over $h_2(S)$. Therefore there exists a unique epimorphism $H' : A \to B$ such that $H'(x_i/x_1) = X_i$ for $2 \leq i \leq m$ and $H'(u) = h'(u)$ for all $u \in R$; moreover, $\ker(H') = \ker(h_2) = PA = x_1 A$. Since $X_2, \ldots, X_m$ are indeterminates, there exists a unique epimorphism $H'' : B \to A^*$ such that $H''(X_i) = X_i$ for $2 \leq i \leq m$ and $H''(u) = h''(u)$ for all $u \in T$; clearly

$$\begin{cases} \ker(H'') = (\ker(h''))B = (h'(x_1), \ldots, h'(x_n))B \\ \qquad\qquad = (h'(x_{m+1}), \ldots, h'(x_n))B \end{cases}$$

and hence

$$H'^{-1}(\ker(H'')) = (x_1, x_{m+1}, \ldots, x_n)A.$$

Let $H(u) = H''(H'(u))$ for all $u \in A$. Then $H : A \to A^*$ is an epimorphism such that $H(x_i/x_1) = X_i$ for $2 \leq i \leq m$ and $H(u) = h(u)$ for all $u \in R$; clearly H is the only such epimorphism and

$$\ker(H) = H'^{-1}(\ker(H'')) = (x_1, x_{m+1}, \ldots, x_n)A = M(R)A$$

and hence in particular $M(R)A$ is a prime ideal in A. Since $\ker(H) = M(R)A$ and $\ker(h) = M(R)$ with $H(u) = h(u)$ for all $u \in R$, it follows that

$$(M(R)A) \cap R = M(R).$$

PROOF OF (T158). Now $R' = A_Q$ and $PA \subset M(R)A \subset Q$ and hence (1) follows from (T157). By (T157) we know that $\ker(H) = (x_1, x_{m+1}, \ldots, x_n)A = M(R)A$; therefore, since $M(R)A \subset Q$, we get (7).

Let $h^* : R^* \to R^*/M(R^*)$ be the residue class epimorphism, and for all $u \in R'$ let $h^{**}(u) = h^*(H^*(u))$. Then $h^{**} : R' \to R^*/M(R^*)$ is an epimorphism with $\ker(h^{**}) = M(R')$ and $h^{**}(R) = h^*(k)$. It follows that $\mathrm{restrdeg}_k R^* = t$ and R' is residually algebraic (resp: residually separable algebraic, residually rational) over R iff R^* is residually algebraic (resp: residually separable algebraic, residually rational) over k. Clearly Q is a maximal ideal in A iff Q^* is a maximal ideal in A^*; by the Nullstellensatz L4§8(T22), which was proved in (Q10), we also get that Q^* is a maximal ideal in A^* iff R^* is residually algebraic over k.

If there exists $r_i \in R$ such that $(x_i/x_1) - r_i \in M(R')$ for $2 \le i \le m$ then Q^* contains the maximal ideal

$$(X_2 - h(r_2), \ldots, X_m - h(r_m))A^*$$

in A^* and hence

$$Q^* = (X_2 - h(r_2), \ldots, X_m - h(r_m))A^*$$

and R^* is residually rational over k; since

$$Q = H^{-1}(Q^*) \quad \text{and} \quad \ker(H) = (x_1, x_{m+1}, \ldots, x_n)A$$

we deduce that if there exists $r_i \in R$ such that $(x_i/x_1) - r_i \in M(R')$ for $2 \le i \le m$ then

$$Q = (x_1, (x_2/x_1) - r_2, \ldots, (x_m/x_1) - r_m, x_{m+1}, \ldots, x_n)A$$

with

$$M(R') = (x_1, (x_2/x_1) - r_2, \ldots, (x_m/x_1) - r_m, x_{m+1}, \ldots, x_n)R'$$

and R' is residually rational over R. The last two statements in (6) are obvious. This completes the proof of (1), (5), (6), and (7); also in view of what we have shown so far, (4) would follow from (3).

Given any $0 \ne w \in R$ with $\mathrm{ord}_S w = \mathrm{ord}_R w$, let $d = \mathrm{ord}_R w$. Then $\mathrm{ord}_S w = d$ and hence $w = F(x_1, \ldots, x_m)$ where $F(X_1, \ldots, X_m)$ is a nonzero homogeneous polynomial of degree d in indeterminates $X_1, \ldots, X_m$ with coefficients in R at least one of which is not in P. Since $\mathrm{ord}_R w = d$, at least one of the coefficients of $F(X_1, \ldots, X_m)$ is not in $M(R)$. Let $G(X_2, \ldots, X_m)$ be the polynomial in $X_2, \ldots, X_m$ obtained by applying h to the coefficients of $F(1, X_2, \ldots, X_m)$. Then $w/x_1^d \in A$ with $H(w/x_1^d) = G(X_2, \ldots, X_m)$, and $G(X_2, \ldots, X_m)$ is a nonzero polynomial of degree $\le d$ in $X_2, \ldots, X_m$ with coefficients in k. Therefore by (T155) we get that $\mathrm{ord}_{R^*} H(w/x_1^d) \le d$, i.e., $\mathrm{ord}_{R^*} H^*(w/x_1^d) \le d$. Clearly

$$\mathrm{ord}_{R'}(w/x_1^d) \le \mathrm{ord}_{R^*} H^*(w/x_1^d)$$

and hence $\text{ord}_{R'}(w/x_1^d) \le d$. This proves (2).

It only remains to prove (3). Now by (Q10)(T47) and (Q29)(C45) R^* is a regular local domain with $\dim(R^*) = m - 1 - \text{restrdeg}_k R^*$. Since $\text{restrdeg}_k R^* = t$, we get $\dim(R^*) = m - 1 - t$. Let $D \subset A$ be such that $H(D)R^* = M(R^*)$. Then $H(D)A^* = Q^* \cap N_1 \cdots \cap N_s$ where N_j is a primary ideal in A^* with $N_j \not\subset Q^*$ for $1 \le j \le s$ with $s \in \mathbb{N}$. Since $\ker(H) = (x_1, x_{m+1}, \ldots, x_n)A$, we get

$$DA + (x_1, x_{m+1}, \ldots, x_n)A = Q \cap H^{-1}(N_1) \cap \cdots \cap H^{-1}(N_s).$$

Now $H^{-1}(N_j)$ is a primary ideal in A with $H^{-1}(N_j) \not\subset Q$ for $1 \le j \le s$. Therefore

$$DR' + (x_1, x_{m+1}, \ldots, x_n)R' = M(R').$$

Let m' be any integer with $1 \le m' \le m$ such that $x_i/x_1 \in M(R')$ for $2 \le i \le m'$ (for instance $m' = 1$). Let $A' = k[X_{m'+1}, \ldots, X_m]$. Since $X_2, \ldots, X_m$ are indeterminates, there exists a unique epimorphism $H_0 : A^* \to A'$ such that $H_0(X_i) = 0$ for $2 \le i \le m'$ and $H_0(u) = u$ for all $u \in A'$; note that then $\ker(H_0) = (X_2, \ldots, X_{m'})A^* \subset Q^*$. Let $Q' = H_0(Q^*)$. Then Q' is a prime ideal in A' and there exists a unique epimorphism $H_1 : R^* \to A'_{Q'}$ such that $H_1(u) = H_0(u)$ for all $u \in A^*$; note that then $\ker(H_1) = (X_2, \ldots, X_{m'})R^*$. Let $H_2 : A'_{Q'} \to A'_{Q'}/M(A'_{Q'})$ be the residue class epimorphism and let $H_3(u) = H_2(H_1(u))$ for all $u \in R^*$. Then $H_3 : R^* \to A'_{Q'}/M(A'_{Q'})$ is an epimorphism with $\ker(H_3) = M(R^*)$ and $H_3(k) = H_2(k)$.

It follows that $\text{restrdeg}_k A'_{Q'} = \text{restrdeg}_k R^*$ and hence

$$\text{restrdeg}_k A'_{Q'} = t.$$

Therefore upon letting

$$q = m - m' - t$$

by (Q10)(T47) and (Q29)(C45) we see that $A'_{Q'}$ is a regular local domain with $\dim(A'_{Q'}) = q$. Consequently there exist elements $y_1, \ldots, y_q$ in A such that

$$(H_0(H(y_1)), \ldots, H_0(H(y_q)))A'_{Q'} = M(A'_{Q'}).$$

Upon taking

$$D = \{x_2/x_1, \ldots, x_{m'}/x_1, y_1, \ldots, y_q\}$$

we get that $H(D)R^* = M(R^*)$ and hence

$$(x_1, x_2/x_1, \ldots, x_{m'}/x_1, x_{m+1}, \ldots, x_n, y_1, \ldots, y_q)R' = M(R').$$

Therefore if we show that $\dim(R') \ge n - t$ then it will follow that R' is a regular local domain with $\dim(R') = n - t$ and $q = n' - n + m - m'$ which will complete the proof.

Since $\dim(R^*) = m - t - 1$, there exist distinct prime ideals

$$Q_1 \subset Q_2 \subset \cdots \subset Q_{m-t}$$

in A^* such that $Q_1 = 0$ and $Q_{m-t} = Q^*$. Now

$$H^{-1}(Q_1) \subset H^{-1}(Q_2) \subset \cdots \subset H^{-1}(Q_{m-t})$$

are distinct prime ideals in A with

$$H^{-1}(Q_1) = \ker(H) \quad \text{and} \quad H^{-1}(Q_{m-t}) = Q.$$

We shall find distinct nonzero prime ideals

$$P_m'' \subset P_{m+1}'' \subset \cdots \subset P_n''$$

in A with $P_n'' = \ker(H)$ and this will prove that $\dim(R') \geq n - t$. Let H' and H'' be as in (T157). By (T157) we know that

$$H'^{-1}(\ker(H'')) = \ker(H) \quad \text{and} \quad \ker(H') \neq 0.$$

Consequently it suffices to find distinct prime ideals

$$P_m \subset P_{m+1} \subset \cdots \subset P_n$$

in B with $P_n = \ker(H'')$ because then we can take $P_j'' = H'^{-1}(P_j)$ for $m \leq j \leq n$. Let

$$P_j' = (h'(x_{m+1}), \ldots, h'(x_j))T$$

for $m \leq j \leq n$. Then

$$P_m' \subset P_{m+1}' \subset \cdots \subset P_n'$$

are distinct prime ideals in T with $P_n' = M(T)$. Let $h_j'' : T \to T/P_j'$ be the residue class epimorphism and let $P_j = P_j'B$. Since $X_2, \ldots, X_n$ are indeterminates, there exists a unique epimorphism $H_j'' : B \to h_j''(T)[X_2, \ldots, X_m]$ such that $H_j''(X_i) = X_i$ for $2 \leq i \leq m$ and $H_j''(u) = h_j''(u)$ for all $u \in T$; clearly $\ker(H_j'') = P_j$ and hence P_j is a prime ideal in B with $P_j \cap T = P_j'$. Therefore

$$P_m \subset P_{m+1} \subset \cdots \subset P_n$$

are distinct prime ideals in B with $P_n = M(T)B = \ker(H'')$.

QUEST (Q35.6). [**Dominating Modelic Blowup**]. For a nonzero ideal P in a quasilocal domain A we put

$$\mathfrak{W}(A, P)^\Delta = \{R' \in \mathfrak{W}(A, P) : R' > A\}$$

and we call this the dominating modelic blowup of P on A.

In (T159) below we shall give a supplement to Theorem (T30) of L4§9.2 which was proved in L4§12(R9). The six parts (T30.1) to (T30.6) of Theorem (T30) corresponded to the six parts (1.9.1) to (1.9.6) of assertion (1.9) of §1 of [A05]. The said supplement corresponds to the seventh part (1.9.7) of the said assertion.

SUPPLEMENTARY MODELIC BLOWUP THEOREM (T159). If A is any regular local domain and P is any nonzero prime ideal in A such that R/P is regular with $\dim(S) > 1$ where $S = A_P$ then, upon letting S' to be the valuation ring of ord_S, for every $R' \in \mathfrak{W}(A, P)^\Delta$ we have (1) $S' \in \mathfrak{V}(R')$ and $S \notin \mathfrak{V}(R')$. Moreover, (2) in the situation of the previous sentence, if P_1 is any prime ideal in A such that A/P_1 is regular and $R' \in \mathfrak{W}(A, P_1)$ then $P_1 = P$.

PROOF. By (T158)(1) we have $S' \in \mathfrak{V}(R')$. Since $\dim(S) > 1 = \dim(S')$, we see that $S \neq S'$. Since $S' \in \mathfrak{V}(R') \subset \mathfrak{W}(A, P)$ where the last set is an irredundant premodel of the quotient field of A, and S' dominates S, we get that $S \notin \mathfrak{W}(A, P)$. This proves (1). To prove (2) let P_1 be a nonzero prime ideal in A such that A/P_1 is regular and $R' \in \mathfrak{W}(A, P_1)$. Let $S_1 = A_{P_1}$. Since $\dim S > 1$, we get that P is not a principal ideal in A and hence $R' \neq A$ by L4§9.3(T30.6) because $R' \in \mathfrak{W}(A, P)$; since $R' \neq A$ and $R' \in \mathfrak{W}(A, P_1)$, again by L4§9.3(T30.6) we get that P_1 is not a principal ideal in A and hence $\dim S_1 > 1$. Now $S' \in \mathfrak{V}(R') \subset \mathfrak{W}(A, P_1)$, $\mathfrak{W}(A, P_1)$ is an irredundant premodel of the quotient field of A, $S' \neq S$, and S' dominates S; hence $S \notin \mathfrak{W}(A, P_1)$. Since $S \notin \mathfrak{W}(A, P_1)$ and $S \in \mathfrak{V}(A)$, by L4§9.3(T30.5) we get that $P_1 S \neq S$ and hence $S \subset S_1$; by symmetry we get that $S_1 \subset S$. Therefore $S_1 = S$ and hence $P_1 = P$.

QUEST (Q35.7). [**Normal Crossings, Equimultiple Locus, and Resolved Ideals**]. We have frequently spoken of simple and singular (= not simple) points of curves, surfaces, and varieties in general. A normal crossing is where two or more nonsingular (= devoid of singularities) varieties meet each other "nicely" or "transversally." The ideas of simple points and various types of normal crossings will now be modelized, i.e., put in the language of models, i.e., collections of local rings. This material corresponds to part (1.5) of §1 of the Resolution Book [A04].

Given a local domain R and $S \in \mathfrak{V}(R)$, we say that S has a simple point at R if the local ring $R/(R \cap M(S))$ is regular; by (Q35.9)(1†) this is equivalent to saying that $R \cap M(S)$ is generated by a subset of a system of regular parameters for R.

Let R be an n-dimensional regular local domain. Given $E \subset \mathfrak{V}(R)$, we say that E has a normal crossing at R if there exists a regular system of parameters $x_1, \ldots, x_n$ for R such that for every $S \in E$ we have $y_S R = R \cap M(S)$ for some subset y_S of $\{x_1, \ldots, x_n\}$. Given $E \subset \mathfrak{V}(R)$, we say that E has a strict normal crossing at R if E has a normal crossing at R and E contains at most two elements. Given a nonzero principal ideal I in R, we say that I has a normal crossing at R if $\{S' \in \mathfrak{V}(R) : \dim(S') = 1 \text{ and } IS' \neq S'\}$ has a normal crossing at R; note that this is equivalent to saying that there exists a regular system of parameters $(x_1, \ldots, x_n)$ for R together with nonnegative integers $a_1, \ldots, a_n$ such that $I = x_1^{a_1} \ldots x_n^{a_n} R$.

Again let R be an n-dimensional regular local domain. Given $E \subset \mathfrak{V}(R)$ and a nonzero principal ideal I in R, we say that (E, I) has a normal crossing at R if $E \cup \{S' \in \mathfrak{V}(R) : \dim(S') = 1 \text{ and } IS' \neq S'\}$ has a normal crossing at R; note that this is equivalent to saying that there exist regular parameters $(x_1, \ldots, x_n)$ for R together with nonnegative integers $a_1, \ldots, a_n$ and a subset y_S of $\{x_1, \ldots, x_n\}$ for each $S \in E$ such that $I = x_1^{a_1} \ldots x_n^{a_n} R$ and $R \cap M(S) = y_S R$ for each $S \in E$. Given $E \subset \mathfrak{V}(R)$ and a nonzero principal ideal I in R, we say that (E, I) has a strict normal crossing at R if (E, I) has a normal crossing at R and E contains at most two elements. Given $S \in \mathfrak{V}(R)$ and a nonzero principal ideal I in R, we say that (S, I) has a normal crossing at R if $(\{S\}, I)$ has a normal crossing at R. Given nonzero principal ideals J and I in R, we say that (J, I) has a quasinormal crossing at R if I has a normal crossing at R and for every nonzero principal prime ideal P in R with $J \subset P$ we have that PI has a normal crossing at R. Given a nonzero principal ideal I in R, we say that I has a quasinormal crossing at R if (I, R) has a quasinormal crossing at R. Note that for any nonzero principal ideal I in R the following four conditions are equivalent: (1) (I, I') has a quasinormal crossing at R for some nonzero principal ideal I' in R; (2) I has a quasinormal crossing at R; (3) for every nonzero principal prime ideal P in R with $I \subset P$ we have that R_P has a simple point at R; (4) $I = z_1 \ldots z_d R$ where $z_1, \ldots, z_d$ are elements in R with $\mathrm{ord}_R z_i = 1$ for $1 \leq i \leq d$ (we take $z_1 \ldots z_d R = R$ in case $d = 0$). Given $S \in \mathfrak{V}(R)$ and a nonzero principal ideal I in R, we say that (S, I) has a pseudonormal crossing at R if S has a simple point at R and for every nonzero principal prime ideal P in R with $I \subset P$ we have that $\{S, R_P\}$ has a normal crossing at R. Note that for any nonzero principal ideal I in R the following three conditions are equivalent: (1*) (S, I) has a pseudonormal crossing at R for some $S \in \mathfrak{V}(R)$; (2*) (R, I) has a pseudonormal crossing at R; (3*) I has a quasinormal crossing at R. Given $E \subset \mathfrak{V}(R)$ and a nonzero principal ideal I in R, we say that (E, I) has a pseudonormal crossing at R if I has a quasinormal crossing at R and for every $S \in E$ we have that (S, I) has a pseudonormal crossing at R.

For any ideal J in a regular local domain R, the set of all $S \in \mathfrak{V}(R)$ such that $\mathrm{ord}_S J = \mathrm{ord}_R J$ is called the equimultiple locus of (R, J) and is denoted by $\mathfrak{E}(R, J)$. For any nonnegative integer i, the set of all i-dimensional members of $\mathfrak{E}(R, J)$ is denoted by $\mathfrak{E}^i(R, J)$.

Let J be a nonzero principal ideal in a regular local domain R. We say that (R, J) is resolved if there exists a nonnegative integer d and a nonzero principal ideal J' in R with $\mathrm{ord}_R J' \leq 1$ such that $J = J'^d$. We say that (R, J) is unresolved if (R, J) is not resolved. Note that if either $\dim(R) \leq 1$ or $\mathrm{ord}_R J \leq 1$ then (R, J) is obviously resolved. Also note that if $\mathrm{ord}_R J \neq 0$ (i.e., if $J \neq R$) then the following six conditions are clearly equivalent: (1') (R, J) is resolved; (2') $\mathrm{ord}_R(\mathrm{rad}_R J) = 1$; (3') $R/\mathrm{rad}_R J$ is regular; (4') $J = (\mathrm{rad}_R J)^d$ where $d = \mathrm{ord}_R J$; (5') $\mathfrak{E}^1(R, J) \neq \emptyset$;

(6') $\mathrm{rad}_R J$ is a prime ideal in R and upon letting S' be the localization of R at $\mathrm{rad}_R J$ we have that

$$\mathfrak{E}(R, J) = \{S \in \mathfrak{V}(R) : S \subset S'\} = \{S \in \mathfrak{V}(R) : JS \neq S\}.$$

Note that if (R, J) is resolved and I is a nonzero principal ideal in R such that I has a quasinormal crossing at R then JI has a quasinormal crossing at R. Finally note that if (R, J) is resolved and I is a nonzero principal ideal in R such that (J, I) has a quasinormal crossing at R then JI has a normal crossing at R.

In Lemma T(160) below, which correspond to assertions (1.5.1) and (1.5.2) of §1 of the Resolution Book [A04], we shall prove some elementary results about the above concepts.

NORMAL CROSSING LEMMA (T160). For any regular local domain R we have the following.

(T160.1) Assuming $\dim(R) = n > 0$, let $(x_1, \ldots, x_n)$ be generators of $M(R)$, let $I = x_1^{a_1} \ldots x_n^{a_n} R$ where $a_1, \ldots, a_n$ are nonnegative integers, let m be an integer with $1 \leq m \leq n$, and let $z \in (x_1, \ldots, x_m)R$ with $\mathrm{ord}_R z = 1$ be such that (zR, I) has a quasinormal crossing at R. Then there exists an integer j with $1 \leq j \leq m$ such that upon letting $y_j = z$ and $y_i = x_i$ for all $i \neq j$ with $1 \leq i \leq n$ we have that $M(R) = (y_1, \ldots, y_n)R$ with $(x_1, \ldots, x_m)R = (y_1, \ldots, y_m)R$ and $I = y_1^{a_1} \ldots y_n^{a_n} R$

(T160.2) Let I be a nonzero principal ideal in R, and let $S \in \mathfrak{V}(R)$ be such that (S, I) has a normal crossing at R. Let $z \in R \cap M(S)$ with $\mathrm{ord}_R z = 1$ be such that (zR, I) has a quasinormal crossing at R. Then (zR, I) has a normal crossing at R.

PROOF. (T160.2) follows from (T160.1). To prove (T160.1) let A be the set of all integers i with $1 \leq i \leq m$ such that $a_i \neq 0$, and let B be the set of all integers i with $m < i \leq n$ such that $a_i \neq 0$. Now $z \in (x_1, \ldots, x_m)R$ and clearly $x_i \notin (x_1, \ldots, x_m)R$ whenever $m < i \leq n$; consequently, if $zR = x_q R$ for some $q \in A \cup B$ then we must have $q \in A$ and hence it suffices to take $j = q$. So now assume that $zR \neq x_i R$ whenever $i \in A \cup B$. Since $\mathrm{ord}_R z = 1$ and (zR, I) has a quasinormal crossing at R, there exist generators $(z_1, \ldots, z_n)$ of $M(R)$ such that $z_i R = x_i R$ whenever $i \in A \cup B$, and $z_e R = zR$ for some $e \notin A \cup B$ with $1 \leq e \leq n$. Let P be the ideal in R generated by the set of all x_i with $i \in A$, and let $Q = (z_1, \ldots, z_{e-1}, z_{e+1}, \ldots, z_n)R$. Now $Q \neq M(R)$ and hence by Nakayama (Q3)(T3) we get $M(R) \not\subset Q + M(R)^2$; consequently $z \notin Q + M(R)^2$; since $P \subset Q$, we get that $z \notin P + M(R)^2$. Since $z \in (x_1, \ldots, x_m)$, we get that $z = r_1 x_1 + \cdots + r_m x_m$ with $r_1, \ldots, r_m$ in R; since $z \notin P + M(R)^2$, we get that $r_p \notin M(R)$ for some $p \notin A$. It suffices to take $j = p$.

QUEST (Q35.8). [**Quadratic and Monoidal Transformations**]. We shall now formalize (= modelize) the concepts QDTs (= Quadratic Transformations) and MDTs (= Monoidal Transformations). This material corresponds to part (1.10) of §1 of the Resolution Book [A04].

Let R be a local domain, let $S \in \mathfrak{V}(R)$ with $\dim(S) > 0$, let J be an ideal in R, and let V be a valuation ring of $\mathrm{QF}(R)$ dominating R. By a monoidal transformation of (R, S) we mean an element of $\mathfrak{W}(R, R \cap M(S))^\Delta$. Since $\mathfrak{W}(R, R \cap M(S))$ is a projective model of $\mathrm{QF}(R)/R$, there exists a unique $R^* \in \mathfrak{W}(R, R \cap M(S))$ dominated by V; clearly R^* dominates R and hence R^* is a monoidal transform of (R, S); we call R^* the monoidal transform of (R, S) along V. Given a monoidal transform R' of (R, S), we define the (R, S, R')-transform of J to be the ideal in R' generated by the set of all $r \in R'$ such that $rx^d \in J$ for some $d \in \mathbb{N}$ and some $x \in R'$ for which $xR' = (R \cap M(S))R'$. By a monoidal transform of (R, J, S) we mean a pair (R', J') where R' is a monoidal transform of (R, S) and J' is the (R, S, R')-transform of J. By the monoidal transform of (R, J, S) along V we mean the pair (R^*, J^*) where R^* is a monoidal transform of (R, S) along V and J^* is the (R, S, R^*)-transform of J.

Given any positive dimensional local domain R, by a quadratic transform of R we mean a monoidal transform of (R, R), and given any valuation ring V of $\mathrm{QF}(R)$ dominating R, by the quadratic transform of R along V we mean a monoidal transform of (R, R) along V.

We are now going to study monoidal transforms R' of a regular local domain R "centered" at a positive dimensional $S \in \mathfrak{V}(R)$ having a simple point at R, and the (R, S, R')-transforms of nonzero principal ideals I, J in R. Note that in such a situation the considerations of Quest (Q35.5) are applicable. Concerning this matter we shall prove Theorems (T161) to (T172) below which correspond to assertions (1.10.1) to (1.10.12) of §1 of [A04]. Let us first introduce some more definitions.

Let R be a regular local domain, let S be a positive dimensional element of $\mathfrak{V}(R)$ having a simple point at R, and let J and I be nonzero principal ideals in R. Given a monoidal transform R' of (R, S), we define the (R, S, R')-transform of (J, I) to be the pair (J', I') where the ideal J' is the (R, S, R')-transform of the ideal J and $I' = (IR')((R \cap M(S))R')^d$ where $d = \mathrm{ord}_S J$; note that then I' is the unique principal ideal in R' such that $J'I' = (JI)R'$. By a monoidal transform of (R, J, I, S) we mean a triple (R', J', I') where R' is a monoidal transform of (R, S) and (J', I') is the (R, S, R')-transform of (J, I). Given a valuation ring V of $\mathrm{QF}(R)$ dominating R, by the monoidal transform of (R, J, I, S) along V we mean the triple (R^*, J^*, I^*) where R^* is the monoidal transform of (R, S) along V and (J^*, I^*) is the (R, S, R^*)-transform of (J, I).

Given a regular local domain R, by an iterated monoidal transform of R we mean a local domain R^* for which there exist finite sequences $(R_i)_{0 \le i \le m}$ and $(S_i)_{0 \le i < m}$ of local domains with $m \in \mathbb{N}$ such that: S_i is a positive dimensional element in $\mathfrak{V}(R_i)$ having a simple point at R_i for $0 \le i < m$, R_i is a monoidal transform of (R_{i-1}, S_{i-1}) for $0 < i \le m$, $R_0 = R$, and $R^* = R_m$. Note that for any iterated monoidal transform R^* of a regular local domain R we have that: R^* is regular, R^* and R have the same quotient field, $R^* \in \mathfrak{V}(A)$ for some affine domain A over R, R^* dominates R, $\dim(R^*) + \mathrm{restrdeg}_R R^* = \dim(R)$ and $h(R^*)$ is a function field over $h(R)$ (i.e., the quotient field of an affine domain over $h(R^*)$) where $h : R^* \to$ $R^*/M(R^*)$ is the residue class epimorphism; whence in particular the following three conditions are equivalent: (i) $\dim(R^*) = \dim(R)$; (ii) R^* is residually algebraic over R; (iii) R^* is residually finite algebraic over R. Also note that if R^* is an iterated monoidal transform of a regular local domain R such that $R^* \neq R$ then $0 < \dim(R^*) \leq \dim(R)$. Given a regular local domain R and a valuation ring V of $\mathrm{QF}(R)$ dominating R, by an iterated monoidal transform of R along V we mean an iterated monoidal transform R^* of R such that V dominates R^*.

FIRST MONOIDAL THEOREM (T161). Let R be a regular local domain and let S be a positive dimensional element of $\mathfrak{V}(R)$ having a simple point at R. Then we have the following.

(T161.1) $R \cap M(S)$ is a principal ideal in $R \Leftrightarrow \dim(S) = 1$.

(T161.2) $\dim(S) = 1 \Leftrightarrow R$ is a monoidal transform of $(R, S) \Leftrightarrow R$ is the only monoidal transform of (R, S).

PROOF. (T161.1) is obvious, and it implies (T161.2) by L4§9.2(T30.6).

SECOND MONOIDAL THEOREM (T162). Let R be a regular local domain, let S be a positive dimensional element of $\mathfrak{V}(R)$ having a simple point at R, and let R' be a monoidal transform of (R, S) such that $R' \neq R$. Then S is uniquely determined by the pair (R, R'), i.e., if S_1 is any positive dimensional element of $\mathfrak{V}(R)$ having a simple point at R such that R' is a monoidal transform of (R, S_1) then $S_1 = S$.

PROOF. Follows from (T159).

THIRD MONOIDAL THEOREM (T163). Let R be a regular local domain, let S be a positive dimensional element of $\mathfrak{V}(R)$ having a simple point at R, let J be a nonzero principal ideal in R, and let (R', J') be a monoidal transform of (R, J, S). Then we have the following.

(T163.1) There exists $w \in R$ together with $x \in R'$ such that $wR = J$ and $xR' = (R \cap M(S))R'$. Moreover, for any such w and x, upon letting $\mathrm{ord}_S J = d$, we have $w/x^d \in R'$ and $(w/x^d)R' = J'$.

(T163.2) If $S \in \mathfrak{E}(R, J)$ with $\dim(S) > 1$ then $\mathrm{ord}_{R'} J' \leq \mathrm{ord}_R J$.

(T163.3) If $S \in \mathfrak{E}(R, J)$ with $\dim(S) = 1$ then $R' = R$ and hence $J' = R'$, i.e., $\mathrm{ord}_{R'} J' = 0$.

PROOF. (T163.1) is obvious, and it implies (T163.2) by (T158)(2). (T163.3) is obvious.

FOURTH MONOIDAL THEOREM (T164). Let R be a regular local domain, let J be a nonzero principal ideal in R such that (R, J) is resolved, let S be a positive dimensional element of $\mathfrak{E}(R, J)$ having a simple point at R, and let (R', J') be a monoidal transform of (R, J, S). Then (R', J') is resolved.

PROOF. If $J = R$ then $J' = R'$ and we have nothing to show. So assume that $J \neq R$. Then $J = y^d R$ where $d = \text{ord}_R J \in \mathbb{N}_+$ and $y \in R$ with $\text{ord}_R y = 1$. Let $n = \dim(R)$ and $m = \dim(S)$. If $m = 1$ then $J' = R'$ by (T163) and we have nothing to show. So also assume that $m > 1$. Since S has a simple point at R, there exist generators $(y_1, \ldots, y_n)$ of $M(R)$ such that $R \cap M(S) = (y_1, \ldots, y_m)R$. Since $S \in \mathfrak{E}(R, J)$, we get that $y \in (y_1, \ldots, y_m)R$ and hence there exists $j' \in \{1, \ldots, m\}$ such that upon letting $(x_1, \ldots, x_n) = (y_1, \ldots, y_{j'-1}, y, y_{j'+1}, \ldots, y_n)$ we have $M(R) = (x_1, \ldots, x_n)R$ with $R \cap M(S) = (x_1, \ldots, x_m)R$. Upon relabelling $(x_1, \ldots, x_m)$ suitably we may assume that $x_i/x_1 \in R'$ for $2 \leq i \leq m$ and $J = x_j^d R$ for some $j \in \{1, \ldots, m\}$. Now $J' = (x_j/x_1)^d R'$. If $x_j/x_1 \notin M(R')$ then $J' = R'$ and we have nothing to show. So assume that $x_j/x_1 \in M(R')$. Then we must have $2 \leq j \leq m$. Let $n' = \dim(R')$. Then $n' \geq 2$ and there exist generators $(z_1, \ldots, z_{n'})$ of $M(R')$ such that $z_1 = x_1$ and $z_2 = x_j/x_1$. In particular $\text{ord}_{R'}(x_j/x_1) = 1$ and hence (R', J') is resolved.

FIFTH MONOIDAL THEOREM (T165). Let R be a regular local domain, let J be a nonzero principal ideal in R such that (R, J) is unresolved, let S be a positive dimensional element of $\mathfrak{E}(R, J)$ having a simple point at R, and let (R', J') be a monoidal transform of (R, J, S) such that $\text{ord}_{R'} J' = \text{ord}_R J$. Then $\dim(S) > 1$ and (R', J') is unresolved.

PROOF. Let $d = \text{ord}_R J$. Then $d > 0$ and hence by (T163) we get $\dim(S) > 1$. We can take $w \in R$ and $x \in R'$ such that $wR = J$ and $xR' = (R \cap M(S))R'$. Then $w/x^d \in R'$ and $(w/x^d)R' = J'$ with $w/x^d \notin xR'$ and $\text{ord}_{R'} x = 1$. Suppose if possible that (R', J') is resolved. Then $(w/x^d)R' = y^d R'$ with $y \in R'$ such that $\text{ord}_{R'} y = 1$. Let $R^* = R'_{yR'}$. Then R^* is a one-dimensional regular local domain and $\text{ord}_{R^*}(w/x^d) = d$. Also $x \notin yR'$ and hence $\text{ord}_{R^*} w = d$ and $(R \cap M(S))R^* = R^*$. Now $R^* \in \mathfrak{V}(R') \subset \mathfrak{W}(R, R \cap M(S))$ and $(R \cap M(S))R^* = R^*$, and hence by L4§9.2(T30.5) we see that $R^* \in \mathfrak{V}(R)$. Thus R^* is a one-dimensional regular local domain with $R^* \in \mathfrak{V}(R)$ and $\text{ord}_{R^*} J = d = \text{ord}_R J$; consequently $R \cap M(R^*)$ is a principal ideal in R with $\text{ord}_R(R \cap M(R^*)) = 1$ and $J = (R \cap M(R^*))^d$. This contradicts the assumption that (R, J) is unresolved.

SIXTH MONOIDAL THEOREM (T166). Let R be a regular local domain, let J and I be nonzero principal ideals in R, let S be a positive dimensional element of $\mathfrak{E}(R, J)$ such that (S, I) has normal crossing at R, and let (R', J', I') be a monoidal transform of (R, J, I, S). Then I' has a normal crossing at R'.

PROOF. Let $d = \mathrm{ord}_R J$ and $n = \dim(R)$ with $m = \dim(S)$ and $n' = \dim(R')$. Since (S, I) has a normal crossing, there exist generators $(x_1, \ldots, x_n)$ of $M(R)$ and nonnegative integers $a(1), \ldots, a(n)$ such that we have $R \cap M(S) = (x_1, \ldots, x_m)R$ and $I = x_1^{a(1)} \ldots x_n^{a(n)}$. Upon relabelling $x_1, \ldots, x_m$ suitably we may assume that $x_i/x_1 \in M(R')$ for $2 \le i \le p$ and $x_i/x_1 \in R' \setminus M(R')$ for $p < i \le m$, where p is an integer with $1 \le p \le m$. Let $q = n' - n + m - p$. Then $q \ge 0$ and there exist elements $y_1, \ldots, y_q$ in R' such that

$$M(R') = (x_1, x_2/x_1, \ldots, x_p/x_1, x_{m+1}, \ldots, x_n, y_1, \ldots, y_q)R'.$$

Now

$$I' = x_1^{d+a(1)+\cdots+a(m)}(x_2/x_1)^{a(2)} \ldots (x_p/x_1)^{a(p)} x_{m+1}^{a(m+1)} \ldots x_n^{a(n)} R'$$

and hence I' has a normal crossing at R'.

SEVENTH MONOIDAL THEOREM (T167). Let R be a regular local domain, let J and I be nonzero principal ideals in R such that (J, I) has a quasinormal crossing at R, let S be a positive dimensional element of $\mathfrak{E}(R, J)$ such that (S, I) has a normal crossing at R, and let (R', J', I') be a monoidal transform of (R, J, I, S). Then (J', I') has a quasinormal crossing at R'.

PROOF. We can take $x \in R$ such that $\mathrm{ord}_R x = 1$ and $(R \cap M(S))R' = xR'$. Let $d = \mathrm{ord}_R J$. By (T166) we know that I' has a normal crossing at R'. If $d = 0$ then $J' = R'$ and we have nothing more to show. So assume that $d \ne 0$. Now $J = z_1 \ldots z_d R$ with $z_i \in R \cap M(S)$ such that $\mathrm{ord}_R z_i = 1 = \mathrm{ord}_S z_i$ and $z_i I$ has a normal crossing at R for $1 \le i \le d$. By (T160.2) it follows that $(S, z_i I)$ has a normal crossing at R, and hence, upon letting (J_i', I_i') to be the (R, S, R')-transform of $(x^{d-1}R, z_i I)$, by (T166) we see that I_i' has a normal crossing at R' for $1 \le i \le d$. Now $J' = (z_1/x) \ldots (z_d/x)R'$, and for $1 \le i \le d$ we have that $z_i/x \in R'$ and $(z_i/x)I' = I_i'$. Therefore (J', I') has a quasinormal crossing at R'.

EIGHTH MONOIDAL THEOREM (T168). Let R be a regular local domain, let J and I be nonzero principal ideals in R, let S be a positive dimensional element of $\mathfrak{V}(R)$ such that (S, I) has a pseudonormal crossing at R, and let (R', J', I') be a monoidal transform of (R, J, I, S). Then I' has a quasinormal crossing at R'.

PROOF. We can take $x \in R$ such that $\mathrm{ord}_R x = 1$ and $(R \cap M(S))R' = xR'$. Let $d = \mathrm{ord}_R J$ and $e = \mathrm{ord}_R I$. Then $I' = x^d (IR')$, and $I = z_1 \ldots z_e R$ where $z_1, \ldots, z_e$ are elements in R such that $\mathrm{ord}_R z_i = 1$ and $(S, z_i R)$ has a normal crossing at R for $1 \le i \le e$ (we take $z_1 \ldots z_e R = R$ in case $e = 0$). Upon taking $(x^d R, R)$ for (J, I) in (T166) we see that $x^d R'$ has a normal crossing at R'. For $1 \le i \le e$, upon taking $(R, z_i R)$ for (J, I) in (T166) we see that $z_i R'$ has a normal crossing at R'. Since $I' = z_1 \ldots z_e x^d R'$, it follows that I' has a quasinormal crossing at R'.

NINTH MONOIDAL THEOREM (T169). Let R be an n-dimensional regular local domain, let $(x_1, \ldots, x_n)$ be generators of $M(R)$, let S be the localization of R at the prime ideal $(x_1, \ldots, x_m)$ for some m with $1 \le m \le n$, let R' be a monoidal transform of (R, S) such that $x_i/x_1 \in R'$ for $1 \le i \le m$, let $S' \in \mathfrak{V}(R) \cap \mathfrak{V}(R')$ be such that $S \subset S'$ with $S \ne S'$, and let $z \in R \cap M(S')$. Then $z/x_1 \in R' \cap M(S')$.

PROOF. Now

$$z \in R \cap M(S') \subset R \cap M(S) \not\subset M(S')$$

with

$$R' \subset S' \quad \text{and} \quad (R \cap M(S))R' = x_1 R'.$$

Therefore $z/x_1 \in R'$ and $x_1 \notin R' \cap M(S')$. Since $R' \cap M(S')$ is a prime ideal in R' and $z = (z/x_1)x_1 \in R' \cap M(S')$, we must have $z/x_1 \in R' \cap M(S')$.

TENTH MONOIDAL THEOREM (T170). Let R be an n-dimensional regular local domain, let $(x_1, \ldots, x_n)$ be generators of $M(R)$, let S and S' be the localization of R at the prime ideals $(x_1, \ldots, x_m)$ and $(x_2, \ldots, x_q)$ respectively where m and q are integers with $1 \le m \le q \le n$, and let R' be a monoidal transform of (R, S). Then: (1) $S' \in \mathfrak{V}(R')$ iff $x_i/x_1 \in M(R')$ for $2 \le i \le m$. Moreover: (2) If $S' \in \mathfrak{V}(R')$ then $\dim(R') = n$ with $M(R') = (x_1, x_2/x_1, \ldots, x_m/x_1, x_{m+1}, \ldots, x_n)R'$ and $R' \cap M(S') = (x_2/x_1, \ldots, x_m/x_1, x_{m+1}, \ldots, x_q)R'$.

PROOF. Clearly $x_1/x_i \notin S'$ for $2 \le i \le m$; consequently, if $S' \in \mathfrak{V}(R')$ then we have $x_1/x_i \notin R'$ for $2 \le i \le m$; hence, if $S' \in \mathfrak{V}(R')$ then $x_i/x_1 \in M(R')$ for $2 \le i \le m$. Now assume that $x_i/x_1 \in M(R')$ for $2 \le i \le m$. Then we have $\dim(R') = n$ with $M(R') = (x_1, x_2/x_1, \ldots, x_m/x_1, \ldots, x_n)R'$, and letting $A = R[x_2/x_1, \ldots, x_m/x_1]$ and $Q = A \cap M(R')$ we get that $A \subset R'$ and Q is a prime ideal in A with $R' = A_Q$ and

$$Q = (x_1, x_2/x_1, \ldots, x_m/x_1, x_{m+1}, \ldots, x_n)A.$$

Clearly $A \subset S'$ and hence upon letting $P = A \cap M(S')$ and

$$P' = (x_2/x_1, \ldots, x_m/x_1, x_{m+1}, \ldots, x_q)A$$

we get that $P' \subset P = $ a prime ideal in A and $S' = A_P$. Clearly $P' \subset Q$, and hence if we show that $P = P'$ then it will follow that $S' \in \mathfrak{V}(R')$ and

$$R' \cap M(S') = (x_2/x_1, \ldots, x_m/x_1, x_{m+1}, \ldots, x_q)R'.$$

To show that $P = P'$ let any $0 \ne z \in P$ be given. Since $0 \ne z \in A$, there exists a nonzero homogeneous polynomial $F(X_1, \ldots, X_m)$ of some degree e in indeterminates $X_1, \ldots, X_m$ with coefficients in R such that $zx_1^e = F(x_1, \ldots, x_m)$.

Now

$$zx_1^e \in P \cap R = M(S') \cap R = (x_2, \ldots, x_q)R$$

and hence $F(x_1, \ldots, x_m) \in (x_2, \ldots, x_m)R$. Clearly

$$F(x_1, \ldots, x_m) - x_1^e F(1, 0, \ldots, 0) \in (x_2, \ldots, x_m)R \subset (x_2, \ldots, x_q)R$$

and hence

$$x_1^e F(1, 0, \ldots, 0) \in (x_2, \ldots, x_q)R \subset P'.$$

Also $x_1^e \notin (x_2, \ldots, x_q)R$ and hence

$$F(1, 0, \ldots, 0) \in (x_2, \ldots, x_q)R \subset P'.$$

Clearly $z = F(1, x_2/x_1, \ldots, x_m/x_1)$ and

$$F(1, x_2/x_1, \ldots, x_m/x_1) - F(1, 0, \ldots, 0) \in (x_2/x_1, \ldots, x_m/x_1)A \subset P'.$$

Therefore $z \in P'$. Thus $P \subset P'$ and hence $P = P'$.

ELEVENTH MONOIDAL THEOREM (T171). Let R be an n-dimensional regular local domain, let R' be a quadratic transform of R, and let E be a set of $(n-1)$-dimensional elements of $\mathfrak{V}(R)$ such that every subset of E having at most two elements has a normal crossing at R. Then $E \cap \mathfrak{V}(R')$ contains at most one element, and $E \cap \mathfrak{V}(R')$ has a normal crossing at R'.

PROOF. If $E \cap \mathfrak{V}(R') = \emptyset$ then we have nothing to show. So assume that $E \cap \mathfrak{V}(R') \neq \emptyset$ and take $S' \in E \cap \mathfrak{V}(R')$. By assumption S' has a simple point at R and hence there exist generators $(x_1, \ldots, x_n)$ of $M(R)$ such that $R \cap M(S') = (x_2, \ldots, x_n)R$. Since $S' \in \mathfrak{V}(R')$, by (T170) we get that $\dim(R') = n$ with $M(R') = (x_1, x_2/x_1, \ldots, x_n/x_1)R'$ and $M(S') = (x_2/x_1, \ldots, x_n/x_1)R'$. Therefore S' has a simple point at R'. It now suffices to show that $E \cap \mathfrak{V}(R') = \{S'\}$. Suppose if possible that $E \cap \mathfrak{V}(R') \neq \{S'\}$ and take $S^* \in E \cap \mathfrak{V}(R')$ such that $S^* \neq S'$. By assumption $\{S', S^*\}$ has a normal crossing at R and hence there exist generators $(y_1, \ldots, y_n)$ of $M(R)$ such that $R \cap M(S') = (y_2, \ldots, y_n)R$ and $R \cap M(S^*) = (y_1, \ldots, y_{n-1})R$. Since $S' \in \mathfrak{V}(R')$ and $S^* \in \mathfrak{V}(R')$, by (T169) we get that $y_n/y_1 \in M(R')$ and $y_1/y_n \in M(R')$ which is a contradiction.

TWELFTH MONOIDAL THEOREM (T172). Let R be an n-dimensional regular local domain, let J and I be nonzero principal ideals in R, let S be a positive dimensional element of $\mathfrak{V}(R)$ such that (S, I) has a pseudonormal crossing at R, let (R', J', I') be a monoidal transform of (R, J, I, S), and let $S' \in \mathfrak{V}(R) \cap \mathfrak{V}(R')$ with $\dim(S') \geq n - 1$ be such that $\{S, S'\}$ has a normal crossing at R and (S', I) has a pseudonormal crossing at R. Then (S', I') has a pseudonormal crossing at R'.

PROOF. Let $d = \text{ord}_S J$ and $e = \text{ord}_R I$ with $m = \dim(S)$.

For a moment suppose that $m = 1$. Then $R' = R$. We can take $x \in R$ such that $R \cap M(S) = xR$. Then $\text{ord}_R x = 1$ and $I' = x^d I$. Since $\{S, S'\}$ has a normal crossing at R, we get that $\{S', x^d R'\}$ has a normal crossing at R'. Since $I' = x^d I$ and (S', I) has a pseudonormal crossing at R, we conclude that (S', I') has a pseudonormal crossing at R'.

Henceforth assume that $m > 1$. Then by (T159) we get $S \notin \mathfrak{V}(R')$. Since $S' \in \mathfrak{V}(R')$, we see that $S \notin \mathfrak{V}(S')$ and hence $R \cap M(S) \not\subset R \cap M(S')$. Therefore $\dim(S') = n-1$, and, since $\{S, S'\}$ has a normal crossing at R, there exist generators $(x_1, \ldots, x_n)$ of $M(R)$ such that

$$R \cap M(S) = (x_1, \ldots, x_m)R \quad \text{and} \quad R \cap M(S') = (x_2, \ldots, x_n)R.$$

Since $S' \in \mathfrak{V}(R')$, by (T170) we see that $\dim(R') = n$ with

$$M(R') = (x_1, x_2/x_1, \ldots, x_m/x_1, x_{m+1}, \ldots, x_n)R'$$

and

$$R' \cap M(S') = (x_2/x_1, \ldots, x_m/x_1, x_{m+1}, \ldots, x_n)R'.$$

Since (S, I) has a pseudonormal crossing at R and (S', I) has a pseudonormal crossing at R, we get $I = z_1 \ldots z_e R$ where $z_1, \ldots, z_e$ are elements in R with $\text{ord}_R z_i = 1$ such that for $1 \leq i \leq e$ we have that $(S, z_i R)$ has a normal crossing at R and $(S', z_i R)$ has a normal crossing at R (we take $z_1 \ldots z_e R = R$ in case $e = 0$). Now $I' = z_1 \ldots z_e x_1^d R'$ and clearly $(S', x_1^d R')$ has a normal crossing at R'. Therefore it suffices to show that $(S', z_i R')$ has a normal crossing at R' for $1 \leq i \leq e$. So let any i with $1 \leq i \leq e$ be given.

First suppose that $z_i \notin R \cap M(S')$. Since $z_i \in M(R)$, we can write $z_i = r_1 x_1 + \cdots + r_n z_n$ with $r_1, \ldots, r_n$ in R. Now $z_i \notin R \cap M(S')$ with $\text{ord}_R z_i = 1$ and $(S', z_i R)$ has a normal crossing at R; consequently there exist generators $(x_1', \ldots, x_n')$ of $M(R)$ such that $R \cap M(S') = (x_2', \ldots, x_n')R$ and $z_i R = x_1' R$. Now by Nakayama (Q3)(T3) we see that $z_i \notin (R \cap M(S')) + M(R)^2$. Since $R \cap M(S') = (x_2, \ldots, x_n)R$, we must have $r_1 \notin M(R)$ and hence $r_1 \notin M(R')$. Consequently $M(R') = (z_i, x_2/x_1, \ldots, x_m/x_1, x_{m+1}, \ldots, x_n)R'$. Therefore $(S', z_i R')$ has a normal crossing at R'.

Next suppose that $z_i \in R \cap M(S')$ and $z_i \notin R \cap M(S)$. Since $z_i \in R \cap M(S')$, we can write $z_i = s_2 x_2 + \cdots + s_n x_n$ with $s_2, \ldots, s_n$ in R. Since $z_i \notin R \cap M(S)$ with $\text{ord}_R z_i = 1$ and $(S, z_i R)$ has a normal crossing at R, there exist generators $(x_1^*, \ldots, x_n^*)$ of $M(R)$ such that $R \cap M(S) = (x_1^*, \ldots, x_m^*)R$ and $z_i R = x_q^* R$ for some q with $m < q \leq n$. Now by Nakayama (Q3)(T3) we get $z_i \notin (R \cap M(S)) + M(R)^2$. Since $R \cap M(S) = (x_1, \ldots, x_m)R$, we must have $s_p \notin M(R)$ for some p with $m \leq p \leq n$, and then $s_p \notin M(R')$. Consequently

$$M(R') = (x_1, x_2/x_1, \ldots, x_m/x_1, x_{m+1}, \ldots, x_{p-1}, z_i, x_{p+1}, \ldots, x_n)R$$

and

$$R' \cap M(S') = (x_2/x_1, \ldots, x_m/x_1, x_{m+1}, \ldots, x_{p-1}, z_i, x_{p+1}, \ldots, x_n)R'.$$

Therefore $(S', z_i R')$ has a normal crossing at R'.

Finally suppose that $z_i \in R \cap M(S')$ and $z_i \in R \cap M(S)$. Then we can write $z_i = t_1 x_1 + \cdots + t_n x_n$ and $z_i = t'_2 x_2 + \cdots + t'_n x_n$ with $t_1, \ldots, t_n, t'_2, \ldots, t'_n$ in R. From these two equations for z_i we get that $t_1 x_1 \in (x_2, \ldots, x_n)$. Now $x_1 \notin (x_2, \ldots, x_n)R$ and hence we must have $t_1 \in M(R)$. Since $\mathrm{ord}_R z_i = 1$, from the first equation for z_i we get $t_a \notin M(R)$ for some a with $2 \le a \le m$. From the above two equations for z_i we see that $(t_a - t'_a)x_a \in (x_1, \ldots, x_{a-1}, x_{a+1}, \ldots, x_n)R$. Now $x_a \notin (x_1, \ldots, x_{a-1}, x_{a+1}, \ldots, x_n)R$ and hence we must have $t_a - t'_a \in M(R)$; therefore $t'_a \notin M(R)$. Let $y_a = z_i$, and let $y_j = x_j$ for all $j \ne a$ with $1 \le j \le n$. Then from the above two equations for z_i we get $M(R) = (y_1, \ldots, y_n)R$ with $R \cap M(S) = (y_1, \ldots, y_m)R$ and $R \cap M(S') = (y_2, \ldots, y_n)R$. Since $S' \in \mathfrak{V}(R')$, by (T170) we get

$$M(R') = (y_1, y_2/y_1, \ldots, y_m/y_1, y_{m+1}, \ldots, y_n)R'$$

and

$$R' \cap M(S') = (y_2/y_1, \ldots, y_m/y_1, y_{m+1}, \ldots, y_n)R'.$$

Now $z_i R' = y_1(y_a/y_1)R'$ and hence $(S', z_i R')$ has a normal crossing at R'.

QUEST (Q35.9). [**Regular Local Rings**]. We shall now collect together some basic properties of regular local rings [cf. §6 (E34)].

$(1^\dagger)$ Let R be an n-dimensional regular local domain. Then by (Q7)(T29), (Q13)(T61), (Q14)(T64), (Q15)(T69.4), and (Q15)(T69.7) we see that: if $x_1, \ldots, x_n$ is a regular system of parameters for R, i.e., a sequence of generators of $M(R)$ of length n, then for every nonnegative integer $m \le n$ the ideal $P = (x_1, \ldots, x_m)R$ is a prime ideal in R for which the local ring R/P is regular, and conversely, if P is any prime ideal in R for which the local ring R/P is regular then $P = (x_1, \ldots, x_m)R$ for some regular system of parameters $x_1, \ldots, x_n$ for R and some nonnegative integer $m \le n$. Moreover, for any such P we have $\dim(R/P) = n - m$, and upon letting $S = R_P$ we have that S is an m-dimensional regular local domain.

$(2^\dagger)$ In the situation of $(1^\dagger)$, for any $0 \ne w \in R$ let d be the largest nonnegative integer such that $w \in P^d$. Then $w = F(x_1, \ldots, x_m)$ where $F(X_1, \ldots, X_m)$ is a nonzero homogeneous polynomial of degree d in indeterminates $X_1, \ldots, X_m$ with coefficients in S at least one of which is not in $M(S)$. Since $\dim(S) = m$ and $M(S) = (x_1, \ldots, x_m)S$, by (T31.2) and (T34) of (Q8) it follows that $w \in M(S)^d \setminus M(S)^{d+1}$, i.e., $\mathrm{ord}_S w = d$. Consequently, for every $e \in \mathbb{N}_+$ we have

$$M(S)^e \cap R = P^e$$

and hence P^e is primary for P by L4§5(O8) and L4§7.1(T17).

§6: DEFINITIONS AND EXERCISES

EXERCISE (E1). [**Isolated System of Prime Ideals**]. In §3(C5), given any family $(P_l)_{l \in L}$ of prime ideals in a ring R, with $I \subset L$, consider the conditions:

(i) For every $i \in I$ and $j \in L \setminus I$ we have $P_j \not\subset P_i$.

(ii) There exists a multiplicative set S in R such that for any $l \in L$ we have: $P_l \cap S = \emptyset \Leftrightarrow l \in I$.

(iii) There exists $T \subset R$ such that for any $l \in L$ we have: $T \not\subset P_l \Leftrightarrow l \in I$.

We call $(P_l)_{l \in L}$ with $I \subset L$ an isolated system to mean that condition (i) holds. Show that (ii) $\Rightarrow$ (i), and (iii) $\Rightarrow$ (i). Also show that if I is a finite set then (i) $\Rightarrow$ (ii), and if $L \setminus I$ is a finite set then (i) $\Rightarrow$ (iii). Moreover show that (ii) holds iff it holds with $S = \cap_{i \in I}(S \setminus P_i)$. Similarly show that (iii) holds iff it holds with $T = \cap_{l \in L \setminus I} P_l$. Establish limitations on these by showing that the implication (i) $\Rightarrow$ (ii), and the implication (i) $\Rightarrow$ (iii) are not true without the finiteness assumptions. Hint: Consider the infinite variable polynomial ring $R = k[X_1, X_2, \ldots]$ over a field k and let $L = \mathbb{N}$. In case of the first implication let $I = \mathbb{N}_+$ with $P_i = (X_1, X_2, \ldots, X_i)R$ for all $i \in \mathbb{N}_+$ and let $P_0 = (X_1, X_2, \ldots)R$. In case of the second implication let $I = \{0\}$ with $P_0 = 0$, and let $P_j = (X_j)R$ for all $j \in \mathbb{N}_+$.

EXERCISE (E2). [**Finite Generation of Ideals and Modules**]. In connection with the proof of §5(Q4)(T4), show that if S is any noetherian module over a ring A and $(H_i)_{i \in \mathbb{N}}$ is any family of subsets of S, then we can find $a \in \mathbb{N}$ and a finite number of elements $(y_{i,j})_{1 \le j \le d_i}$ in H_i for $0 \le i \le a$ with $d_i \in \mathbb{N}$ such that every element in $\cup_{e \in \mathbb{N}} H_e$ is an A-linear combination of the elements $(y_{i,j})_{0 \le i \le a, 1 \le j \le d_i}$. Hint: The noetherianness of S tells us that the A-submodule of S generated by $\cup_{e \in \mathbb{N}} H_e$ has a finite set of generators U; since U is finite, we can find $a \in \mathbb{N}$ and a finite number of elements $(y_{i,j})_{1 \le j \le d_i}$ in H_i for $0 \le i \le a$ with $d_i \in \mathbb{N}$ such that every element of U is an A-linear combination of the elements $(y_{i,j})_{0 \le i \le a, 1 \le j \le d_i}$; it follows that every element of U is an A-linear combination of the elements $(y_{i,j})_{0 \le i \le a, 1 \le j \le d_i}$.

EXERCISE (E3). [**Bracketed Colon Operation**]. In connection with the proof of §5(Q4)(T12), verify the following seven properties (1) to (7) of the bracketed colon operation $[U : S]_V$ on an R-submodule U of an R-module V on a ring R, with respect to a multiplicative set S in R defined in L4§5(O19); note that in (6) and (7) we have $n \in \mathbb{N}$ and by convention $\cap_{1 \le i \le 0} U_i = V$. Then, using these properties and also using (Q4)(T11), establish the following assertion ($\bullet$) which corresponds to the third sentence in the proof of (Q4)(T12) which reads "To prove (T12.2), first note..." Alternatively, in establishing ($\bullet$), instead of the said properties you may use localization as explained in L4§7. This is not surprising because of the correspondence between localization and bracketed colon given in L4§7(T12).

(1) $U \subset [U : S]_V$.

(2) U' is a submodule of $U \Rightarrow [U' : S]_V \subset [U : S]_V$.

(3) $[[U : S]_V : S]_V = [U : S]_V$.

(4) $S \cap \mathrm{rad}_V U \neq \emptyset \Rightarrow [U : S]_V = V$.

(5) U is a primary submodule of V and $S \cap \mathrm{rad}_V U = \emptyset \Rightarrow [U : S]_V = U$.

(6) $U = \cap_{1 \leq i \leq n} U_i$ with submodules U_i of $V \Rightarrow [U : S]_V = \cap_{1 \leq i \leq n}[U_i : S]_V$.

(7) Assume that $V = R = $ a noetherian ring and $U = [0 : P]_R$ for some prime ideal P in R. Let $U = \cap_{1 \leq i \leq n} U_i$ be an irredundant primary decomposition in R where U_i is a primary ideal in R with $\mathrm{rad}_R U_i = W_i$ for $1 \leq i \leq n$. Then for some $j \in \{1, \ldots, n\}$ we have $\mathrm{ht}_R W_j = 0$.

($\bullet$) If $J = Q_1 \cap \cdots \cap Q_d$ is an irredundant primary decomposition of an ideal J in R with $d \in \mathbb{N}_+$ such that the radicals $P_1, \ldots, P_d$ of $Q_1, \ldots, Q_d$ are prime ideals of positive heights in the ring R which is assumed to be noetherian, and if P is a prime ideal in R with $P \not\subset P_i$ for $1 \leq i \leq d$, then there exists a P-primary ideal Q in R with $J \not\subset Q$.

Hint for (3): (1) and (2) $\Rightarrow$ RHS of (3) $\subset$ LHS 0f (3).

Hint for (5): Use L4§5(O19)(11$^{\bullet}$).

Hint for (7): Use (T5) to (T7) of L4§6.

Hint for ($\bullet$): If $J \subset Q$ for every P-primary ideal Q then $J \subset [0 : P]_R$ by (Q4)(T11), and hence $[0 : P]_R \subset [J : P]_R \subset [[0 : P]_R : P]_R = [0 : P]_R$ where the inclusions are by (2) and the equality is by (3). This gives $[0 : P]_R = [J : P]_R$. Relabelling $Q_1, \ldots, Q_d$ we may assume that $Q_i \subset P$ for $1 \leq i \leq c$ and $Q_i \not\subset P$ for $c + 1 \leq i \leq d$ with $c \leq d$ in $\mathbb{N}$. Now by (4) and (5) we get the irredundant primary decomposition in R given by $[0 : P]_R = [J : P]_R = Q_1 \cap \cdots \cap Q_c$. Comparing this with (7), by the Prime Uniqueness Theorem L4§6(T6) we see that $c = n$ with $\{P_1, \ldots, P_c\} = \{W_1, \ldots, W_n\}$, and hence for some $j \in \{1, \ldots, c\}$ we must have $\mathrm{ht}_R P_j = 0$ which is a contradiction.

EXERCISE (E4). [**Principal Ideal Theorems**]. It may be tempting to assert a stronger version of §5(Q7)(T24) and §5(Q7)(T27) saying that: (T27*) if I and J are nonunit ideals in a noetherian ring R with $I = J + (xR)$ for some $x \in R$ then $\mathrm{ht}_R I \leq 1 + \mathrm{ht}_R J$. Use the example $R = A/(XZ, YZ)A$, where A is the trivariate power series ring $k[[X, Y, Z]]$ over a field k, to show that (T27*) is false. Also show that this ring R is a two dimensional reduced local ring.

Note that by (1) below, looking at the three dimensional (X, Y, Z)-space, the said ring R may be viewed as the "local affine coordinate ring at the origin" of the variety consisting of the horizontal (X, Y)-plane whose prime ideal is ZA together with the vertical Z-axis whose prime ideal is $(X, Y)A$. Also note that by (1) below, the kernel of the residue class epimorphism $\phi : A \to R$ is an intersection of the prime ideals ZA and $(X, Y)A$, neither of which contains the other, and hence $\phi(Z)R$ and $(\phi(X), \phi(Y))R$ are all the distinct height zero prime ideals in R and their intersection is the zero ideal. Quite generally, if $(P_l)_{l \in \Lambda}$ is any nonempty finite isolated system of prime ideals in any ring A, in the sense of (E1), then upon letting $\phi : A \to A/\cap_{l \in \Lambda} P_l$

be the residue class epimorphism we clearly have that $\phi(P_l)_{l \in \Lambda}$ are all the distinct height zero prime ideals in $\phi(A)$ and their intersection is the zero ideal which gives its unique irredundant primary decomposition.

Hint: Clearly

(1) $$(XZ, YZ)A = (ZA) \cap (X, Y)A$$

and we have the strictly increasing chains of prime ideals

(2) $(X, Y)A \subset (X, Y, Z)A$ and $(ZA) \subset (X, Z)A \subset (X, Y, Z)A.$

Therefore R is a two dimensional reduced local ring, and by letting J, I, x to be the images of $(X, Y)A, (X, Y, Z)A, Z$ under ϕ we see that $I = J + (xR)$ with $\operatorname{ht}_R J = 0$ and $\operatorname{ht}_R I = 2$.

EXERCISE (E5). [**Relative Independence of Parameters**]. Let $x_1 \ldots, x_n$ be a system of parameters of a local ring R, i.e., according to §5(Q16), a sequence of elements in R whose length n equals $\dim(R)$ and which generates a J-primary ideal I where $J = M(R)$. Let $Y_1, \ldots, Y_n$ be indeterminates over R. For any $A \subset R$ and $i \in \mathbb{N}$, let $A[Y_1, \ldots, Y_n]_i$ be the set of all homogeneous members of $R[Y_1, \ldots, Y_n]$ of degree i all of whose coefficients belong to A (including the zero polynomial in case $0 \in A$). As a consequence of (T31) and (T34) of §5(Q8) show that, if $G(Y_1, \ldots, Y_n) \in R[Y_1, \ldots, Y_n]_e$ with $e \in \mathbb{N}$ is such that $G(x_1, \ldots, x_n) \in JI^e$ then $G(Y_1, \ldots, Y_n) \in J[Y_1, \ldots, Y_n]_e$.

Hint: Since R is noetherian and I is J-primary, we have $J^d \subset I$ for some $d \in \mathbb{N}_+$. Let $F(Y_1, \ldots, Y_n) = G(Y_1, \ldots, Y_n)^d$ and $i = ed$. Then

$$F(Y_1, \ldots, Y_n) \in R[Y_1, \ldots, Y_n]_i \quad \text{with} \quad F(x_1, \ldots, x_n) \in I^{i+1}$$

and hence by (T31) and (T34) of §5(Q8) we get $F(Y_1, \ldots, Y_n) \in J[Y_1, \ldots, Y_n]_i$. Therefore $G(Y_1, \ldots, Y_n)^d \in J(R[Y_1, \ldots, Y_n])$, and by §5(Q7)(T30.1) we know that $J(R[Y_1, \ldots, Y_n])$ is a prime ideal in $R[Y_1, \ldots, Y_n]$. Consequently we must have $G(Y_1, \ldots, Y_n) \in J(R[Y_1, \ldots, Y_n])$. But clearly

$$J(R[Y_1, \ldots, Y_n]) \cap R[Y_1, \ldots, Y_n]_e = J[Y_1, \ldots, Y_n]_e$$

and hence $G(Y_1, \ldots, Y_n) \in J[Y_1, \ldots, Y_n]_e$.

EXERCISE (E6). [**Radical Description**]. Show that §5(Q9)(T43) remains valid if the rings R and S are not assumed to be domains. In other word, show that given any integral extension S of any ring R, and given any ideal H in we have $\operatorname{rad}_S(HS) = \{y \in S : y^n + \sum_{1 \le i \le n} x_i y^{n-i} = 0 \text{ with } n \in \mathbb{N}_+ \text{ and } x_i \in H\}$.

Hint: The proof of §5(Q9)(T43) verbatim works.

DEFINITION (D1). [**Minimal Polynomial**]. Given any element y of any overring L of a field K, we define the minimal polynomial of y/K, i.e., of y over

K, to be the unique monic generator of the kernel of the K-homomorphism of the univariate polynomial ring $K[Y]$ to L which sends Y to y in case the said kernel is nonzero, and to be the zero polynomial in case the said kernel is zero. Note that then the minimal polynomial is zero or nonzero according as y is transcendental or algebraic over K. Also note that in case L is an overfield of K, this definition agrees with Definition L1§10(D3).

EXERCISE (E7). [**Kronecker Divisibility**]. Let R be a normal domain with quotient field K, let y be an element in an overring L of K such that y is integral over R, and let $B(Y)$ be the minimal polynomial of y over K. Then $A(Y) \in R[Y]$. In greater detail, let $A(y) = 0$ be an equation of integral dependence of y over R, i.e., $A(Y) \in R[Y]$ is monic with $A(y) = 0$; then clearly $A(Y) = B(Y)C(Y)$ with monic $C(Y) \in K[Y]$, and hence by Kronecker's Theorem L3§8(J25) we get $B(Y) \in R[Y]$ and $C(Y) \in R[Y]$. As Exercise fill in the details.

EXERCISE (E8). [**Going Down**]. Let S be an integral extension of a ring R such that: (1) R is normal, (2) every nonzero element of R is a nonzerodivisor of S, and (3) R is a domain. Let $P^* \subset P$ be prime ideals in R, and let Q be a prime ideal in S lying above P. As a slight strengthening of the Going Down Theorem §5(Q9)(T44) show that then there exists a prime ideal Q^* in S lying above P^* such that $Q^* \subset Q$.

Hint: The total quotient ring L of S may be assumed to be an overring of the quotient field K of R, and hence the proof of §5(Q9)(T44) verbatim applies after changing the references to (T43) and Kronecker's Theorem L3§8(J25) by references to above Exercises (E6) and (E7) respectively.

EXERCISE (E9). [**Nongoing Down for Nonnormal Domains**]. Show that (E8) is not true if R is a nonnormal domain and S is a domain.

Hint: Consider the bivariate polynomial ring $S = k[Z, W]$ over a field k whose characteristic is different from 2, and consider the subring $R = k[Z, W^2 - 1, W^3 - W]$ of S. Then the domain S is an integral extension of the domain R. Now $(X, Y, Z) \mapsto (W^2 - 1, W^3 - W, Z)$ gives a k-epimorphism ϕ of the trivariate polynomial ring $B = k[X, Y, Z]$ onto R whose kernel is generated by $Y^2 - X^2 - X^3$ and hence we may regard R as the affine coordinate ring of the vertical cylinder C over the nodal cubic $Y^2 - X^2 - X^3 = 0$. It follows that the domain R is nonnormal. Consider the prime ideals $\widehat{P}^* = (XZ + Z + Y, Z^2 + 2Z - X)B \subset (X, Y, Z)B = \widehat{P}$ which are the ideals of an irreducible curve U^* and the point $U = (0, 0, 0)$ in the (X, Y, Z)-space respectively. This curve lies on C and hence upon letting $P^* = \phi(\widehat{P}^*)$ and $P = \phi(\widehat{P})$ we get prime ideals $P^* \subset P$ in R. The point U splits into the points $V = (0, 1)$ and $V_1 = (0, -1)$ of the (Z, W)-plane, i.e., $Q = (Z, W - 1)$ and $Q_1 = (Z, W + 1)$ are the prime ideals in S lying above P. Let Q^* be any nonzero principal prime ideal in S with $P^* \subset Q^*$; for the generators of $\widehat{P}^*$ we have $\phi(XZ + Z + Y) = (Z + W + 1)(W^2 - 1)$

and $\phi(Z^2 + 2Z - X) = (Z + W + 1)(Z + W - 1)$, and hence $Q^* = (Z + W + 1)S$. Thus $Q^* \subset Q_1$ but $Q^* \not\subset Q$. Therefore there is no prime ideal Q^* in S lying above P^* with $Q^* \subset Q$.

To further explain the geometry behind the above construction, referring to (Q35.4) for terminology and applying a QDT to the nodal cubic we get the parabola $(Y/X)^2 - 1 - X = 0$ which gives the parametrization $(X, Y) \mapsto (W^2 - 1, W^3 - 1)$. Correspondingly, applying an MDT to the cylinder we get the (Z, W)-plane. Now the $Z + W + 1 = 0$ in the (Z, W)-plane goes through the point V_1 but not through the point V. This indicates that the curve U^* "lies only on one branch of the cylinder near the point U."

EXERCISE (E10). [**Nongoing Down for Nonzerodivisors**]. Show that (E8) is not true if (1) and (3) hold but (2) does not.

Hint: Let R be the normal domain $\mathbb{Z}$. Let J be the ideal in the univariate polynomial ring $R[X]$ generated by $(X^2 - X, 2X)$. Since $J \cap R = 0$ we can take an overring S of R together with an R-epimorphism $\phi : R[X] \to S$ whose kernel is J. Upon letting $x = \phi(X)$ we have $S = R[x]$. Clearly $2x = 0$ and hence the nonzero element 2 of R is a zerodivisor in S. Also clearly $0 = P^* \subset P = 2R$ are prime ideals in R, and $Q = (2, x - 1)$ is a prime ideal in S lying above P. Let if possible Q^* be a prime ideal in S lying above P^* with $Q^* \subset Q$. Then $2x = 0 \in Q^*$ with $2 \notin Q^*$, and hence $x \in Q^*$. Therefore $x - 1 \notin Q^*$, which is a contradiction because $x - 1 \in Q$.

EXERCISE (E11). [**Nongoing Down for Nondomains**]. Show that (E8) is not true if (1) and (2) hold but (3) does not.

Hint: Consider the polynomial rings $A = k[X, Y] \subset k[X, Y, Z] = B$ over a field k. In A consider the prime ideals $XA = \widehat{P}^* \subset \widehat{P} = (X, Y)A$ and the $\widehat{P}$-primary ideal $\widehat{P}' = (X^2, Y)$. In B consider the prime ideals $(X, Z - 1)B = \widehat{Q}^\sharp \not\subset \widehat{Q} = (X, Y, Z)B$, and the $\widehat{Q}$-primary ideal $\widehat{Q}' = (X^2, Y, Z)$. Clearly $\widehat{P}^*, \widehat{P}, \widehat{P}'$ are the intersections of A with $\widehat{Q}^\sharp, \widehat{Q}, \widehat{Q}'$ respectively; this follows by noting that $k[X, Y, Z - 1] = k[X, Y, Z]$, and for the univariate polynomial ring $D = C[Z]$ over any ring C and any ideal H in C we have $C \cap (HD + ZD) = H$. Consequently, upon letting $J = \widehat{Q}^\sharp \cap \widehat{Q}'$ and $I = \widehat{P}^* \cap \widehat{P}'$, we have $A \cap J = I$. Clearly $k \cap \widehat{Q} = 0$, and hence $k \cap J = 0$. Therefore we can take affine rings $k[x, y] = R \subset S = k[x, y, z]$ over k together with a k-epimorphism $\phi : B \to S$ with kernel J such that $\phi(X, Y, Z) = (x, y, z)$ and $\phi(A) = R$. Clearly $X \notin I$ and $Y \notin I$ but $XY \in I$, and hence R is not a domain. Also clearly S is an integral extension of R; namely, $Z^2 - Z \in J$ and therefore z is integral over k and hence over R. Let $P^*, P, P', Q^\sharp, Q, Q'$ be the respective images of $\widehat{P}^*, \widehat{P}, \widehat{P}', \widehat{Q}^\sharp, \widehat{Q}, \widehat{Q}'$ under ϕ. Then $P^* \subset P$ are prime ideals in R, and P' is P-primary; likewise, $Q^\sharp \subset Q$ are prime ideals in S, and Q' is Q-primary; also Q lies above P. Clearly $\widehat{P}^* \neq \widehat{P}$, and hence $P^* \neq P$. Let if possible Q^* be a prime ideal in S lying above P^* with $Q^* \subset Q$; then we must have $Q^* \neq Q$ because $P^* \neq P$; therefore upon letting $\widehat{Q}^* = \phi^{-1}(Q^*)$ and noting that that $\widehat{Q} = \phi^{-1}(Q)$, it follows that $\widehat{Q}^* \subset \widehat{Q}$ are prime ideals in B with $\widehat{Q}^* \neq \widehat{Q}$; also we have $\widehat{Q}^\sharp \cap \widehat{Q}' = J \subset \widehat{Q}^* \subset \widehat{Q}$, and hence by L4§5(O24)(21$^\bullet$) we must have either $\widehat{Q}^\sharp \subset \widehat{Q}^* \subset \widehat{Q}$ or $\widehat{Q} \subset \widehat{Q}^* \subset \widehat{Q}$, which is a contradiction.

Clearly $J = \widehat{Q}^\sharp \cap \widehat{Q}'$ and $I = \widehat{P}^* \cap \widehat{P}'$ are irredundant primary decompositions in B and A respectively, and hence $0 = Q^\sharp \cap Q'$ and $0 = P^* \cap P'$ are irredundant primary decompositions of the zero ideal in S and R respectively. Therefore by L4§5(O24)(21$^\bullet$) we get $Z_S(S) = Q^\sharp \cup Q$ and $Z_R(R) = P^* \cup P$, and hence a nonzerodivisor in R stays a nonzerodivisor in S.

It only remains to show that R is normal. Clearly $I = (X^2, XY)A$. Hence every element of R can be written as $a + bx + c(y)y$ with a, b in k and $c(Y) \in k[Y]$, where a is uniquely determined by the element; moreover, the said element is a zerodivisor in R iff $a = 0$. Therefore every element of $\mathrm{QR}(R)$ can be written as $\frac{a' + b'x + c'(y)y}{a + bx + c(y)y}$ with $a \neq 0$. Multiplying the numerator and denominator by $a - bx$ we may assume that $b = 0$. Thus we can write any element t in $\mathrm{QR}(R)$ as $t = \frac{dx + e(y)}{f(y)}$ where $d \in k$ and $e(Y), f(Y)$ in $k[Y]$ with $0 \neq a = f(0) \in k$. Suppose t in integral over R. Since $xy = 0$, by cross multiplication we see that in $\mathrm{QR}(R)$ we have $\frac{dx}{f(y)} = \frac{dx}{a}$ and hence $\frac{dx}{f(y)} \in R$. Therefore $\frac{e(y)}{f(y)}$ is integral over $R = k[x, y]$, and hence it is integral over $k[y]$ because $x^2 = 0$. Clearly $I \cap k[Y] = 0$ and hence $k[y]$ is normal. Therefore $\frac{e(y)}{f(y)} \in k[y]$ and hence $t \in R$.

EXERCISE (E12). [**Jacobson Rings**]. In §5(Q11)(T50.3) we have noted that every affine domain over a field is a Jacobson ring. It is clear that a quasilocal domain R is Jacobson iff it is a field, i.e., iff it is zero dimensional, because in the positive dimensional case the zero prime ideal, being properly contained in the unique maximal ideal, cannot be an intersection of maximal ideals. Construct examples of noetherian domains which are not Jacobson but have infinitely many distinct maximal ideals.

Hint: Let $B = k[X_1, \ldots, X_N]$ where $X_1, \ldots, X_N$ are indeterminates over an algebraically closed field k with integer $N > 1$, and let W be the set of all maximal ideals in B containing the nonzero principal prime ideal $J = X_N B$. Geometrically, for the hyperplane $H : X_N = 0$ in the affine N-space $\mathbb{A}_k^N$ we have a bijection $H \to W$ given by $\alpha \mapsto \mathbb{I}(\alpha) = (X_1 - \alpha_1, \ldots, X_{N-1} - \alpha_{N-1}, X_N)$. Let S be the multiplicative set in B given by $S = \cap_{P \in W}(B \setminus P)$, and form the localization $R = B_S$. In view of the above cited result §5(Q11)(T50.3), by the Ideal Correspondence Theorem L4§7(T12) we see that: ($\bullet$) $P \mapsto PR$ gives a bijection of W onto $\mathrm{mspec}(R)$ and for the nonzero prime ideal JR in R we have $JR = \cap_{P \in W}(PR)$; consequently the zero prime ideal in R is not an intersection of maximal ideals and hence R is a noetherian domain which is not Jacobson. We are done by noting that clearly W is an infinite set. Also ($\bullet$) remains valid if we replace J by any nonzero prime ideal in B, and if J is not a maximal ideal then W is an infinite set. More generally, ($\bullet$) remains valid if we replace B by any noetherian Jacobson domain having a nonzero prime ideal J, and if $\mathrm{mvspec}_B J$ is an infinite set then so is $\mathrm{mspec}(R)$. Here the noetherianness of B is used only to get the noetherianness of R. Also everything works without assuming B to be a domain provided we take J to be a positive height prime ideal and replace PR by $\phi(P)R$ where $\phi : B \to B_S$ is the localization map, i.e., the canonical homomorphism. Thus if B is any positive dimensional Jacobson ring then a suitable localization of it is not Jacobson.

EXERCISE (E13). [**Krull's Struktursatz**]. Our §5(Q11)(T51) is the beautiful Struktursatz of Krull given on page 9 of his famous book [Kr2]. This was the main application of our §5(Q11)(T49.10) which may be paraphrased by saying that given any $U \in V'$: (i) for every $U'' \in V''$ with $U'' \subset U$ we have $U'' \in U^*$ for some irreducible component U^* of U, and (ii) U is the union of its irreducible components. The exercise is to give more details of the proof of this. Note that V' (resp: V'') is the set of all varieties (resp: irreducible varieties) in S where, for any ring R, we are letting S to be mspec(R) or jspec(R) or spec(R). Also note that the original version of (i) was stated only for $U = S$.

Hint: Let V be as in §5(Q11)(T49) and let us tacitly use §5(Q11)(T49.1) to §5(Q11)(T49.7). Then $U = V(J)$ and $U'' = V(P)$ for some ideal J in R and some $P \in W = \text{vspec}_R J$. Now every chain A in $\overline{A} = \{\overline{U} \in V'' : U'' \subset \overline{U} \subset U\}$ is of the form $V(Q)_{Q \in B}$ for some chain B in $\overline{B} = \{\overline{Q} \in W : \overline{Q} \subset P\}$. Clearly $\cap_{Q \in B} Q$ is a lower bound of B in $\overline{B}$, and hence $V(\cap_{Q \in B} Q)$ is an upper bound of A in $\overline{A}$. Therefore $\overline{A}$ has a maximal element U^* by Zorn's Lemma, and clearly such a U^* is an irreducible component of U with $U'' \subset U^*$. This proves (i), and (ii) follows from it by noting that $U = \cup_{P \in W} V(P)$.

DEFINITION (D2). **Domains, Ranges, Restrictions, and Conditions of Bijection**]. Given any (set theoretic) map $\phi : S \to T$, its domain S is denoted by dom(ϕ), and its range T is denoted by ran(ϕ). For any $A \subset S$ and $B \subset T$ with $\phi(A) \subset B$, by $\phi|_{(A,B)}$ we denote the map $A \to B$ which sends every $x \in A$ to $\phi(x)$ in B, and we call this the restriction of ϕ to (A, B); if $B = \phi(A)$ then we may denote this by $\phi|_A$ and call it the restriction of ϕ to A. By 1_S we denote the identity map of S, i.e., the map $1_S : S \to S$ given by $1_S(x) = x$ for all $x \in S$; this should not be confused with the identity element of a multiplicative monoid S which may also be denoted by 1_S, just as the zero element of an additive monoid S may be denoted by 0_S. As obvious conditions we note that the map ϕ is injective (resp: surjective, bijective) iff there exists a map $\psi : T \to S$ such that $\psi\phi = 1_S$ (resp: $\phi\psi = 1_T$, $\psi\phi = 1_S$ and $\phi\psi = 1_T$); moreover, if ϕ is bijective, then ϕ and ψ are inverses of each other.

EXERCISE (E14). [**Spectral Nullstellensatz**]. Give details of the proofs of Theorems (T50.1) and (T50.2) of §5(Q11), and show that the Jacobsonness of R is necessary for the $S = \text{mspec}(R)$ and $S = \text{jspec}(R)$ cases of these Theorems.

Hint for (T50.1): The spec case follows from (T48). If R is Jacobson then clearly

$\mathrm{jspec}(R) = \mathrm{spec}(R)$, and hence the jspec case is reduced to the spec case; moreover, the mspec case is also ok because

$$
\begin{cases}
\mathrm{rad}_R J = \bigcap_{P \in \mathrm{vspec}_R J} P & \text{by (T48)} \\
\quad = \bigcap_{P \in \mathrm{vspec}_R J} \bigcap_{Q \in \mathrm{mvspec}_R P} Q & \text{because } R \text{ is Jacobson} \\
\quad = \bigcap_{Q \in \mathrm{mvspec}_R J} Q \\
\quad = I(\mathrm{mvspec}_R J).
\end{cases}
$$

Hint for (T50.2): In view of (T50.1), $U \mapsto I(U)$ and $J \mapsto V(J)$ give maps $\phi : V' \to \mathrm{rd}(R)$ and $\psi : \mathrm{rd}(R) \to V'$ such that $\phi\psi = 1_{\mathrm{rd}(R)}$. By (T49.6) we also have $\psi\phi = 1_{V'}$. Therefore by the last sentence of (D2) we see that ϕ is a bijection whose inverse is ψ. In view of (T49.7), to get the assertion about $V'' \to \mathrm{spec}(R)$, replace ϕ by $\phi|_{V''}$.

Hint for necessity: If R is a positive dimensional quasilocal ring then in the mspec and jspec cases we have $S = \{M(R)\}$ and hence R is nonjacobson and (the conclusions of) (T50.1) and (T50.2) are false. Actually these conclusions are false whenever R is nonjacobson. To see this, it suffices to note that in the mspec and jspec cases, by substituting any $P \in \mathrm{spec}(R)$ for J in the equation $I(V(J)) = \mathrm{rad}_R J$ we can conclude that P is a Jacobson prime, i.e., a prime ideal which can be expressed as an intersection of maximal ideals.

EXERCISE (E15). [**Associated Graded Rings and Leading Ideals**]. Given a ring homomorphism $\phi : R \to R'$, and ideals I and I' in R and R' respectively, with $\phi(I) \subset I'$, let $J = \ker(\phi)$, and let $\mathrm{grad}_{(R,I)}(\phi) : \mathrm{grad}(R, I) \to \mathrm{grad}(R', I')$ be the induced homomorphism. In (1) and (2) of §5(Q13) we have noted that

(1) $\mathrm{ledi}_{(R,I)}(J) \subset \ker(\mathrm{grad}_{(R,I)}(\phi))$

and

(2)
$$
\begin{cases}
\text{if } \phi \text{ is surjective with } \phi(I) = I' \\
\text{then } \mathrm{grad}_{(R,I)}(\phi) \text{ is surjective} \\
\text{with } \mathrm{ledi}_{(R,I)}(J) = \ker(\mathrm{grad}_{(R,I)}(\phi)).
\end{cases}
$$

(i) Construct an example to show that in general the inclusion in (1) cannot be replaced by an equality, even when ϕ is injective. (ii) Assuming the hypothesis of (2), show that $\mathrm{grad}_{(R,I)}(\phi)$ is an isomorphism iff $\ker(\phi) \subset \cap_{n \in \mathbb{N}} I^n$.

Hint: (i) Let R, R' be local domains such that R' dominates R and for some $h \in R$ we have $\mathrm{ord}_{R'} h > \mathrm{ord}_R h = n \in \mathbb{N}_+$. Let I, I' be the maximal ideals in R, R' respectively, and let $\phi : R \to R'$ be the subset injection. Then $\phi(h) = h$ and

$$
\begin{cases}
0 \neq \mathrm{lefo}_R(h) \in \mathrm{grad}(R)_n \subset \mathrm{grad}(R) \\
\text{with } (\mathrm{grad}_R(\phi))(\mathrm{lefo}_R(h)) = \mathrm{grad}_{R'}^n(\phi(h)) = \mathrm{grad}_{R'}(h) = 0
\end{cases}
$$

and hence $\mathrm{grad}_R(\phi)$ is not injective. As a special case we can take R, R' to be the localizations of $k[X^2], k[X]$ at the prime ideals generated by X^2 and X respectively, where X is an indeterminate over a field k, and then upon letting $h = X^{2n}$ we get $\mathrm{ord}_R h = 2n > n = \mathrm{ord}_{R'} h$. Geometrically speaking, in the general case, $\mathrm{spec}(R')$ is a "ramified covering" of $\mathrm{spec}(R)$. In the special case, we have the covering $X \mapsto X^2$; its graph is the parabola $Y^2 = X$ and the covering map is the vertical projection; note that at every point other than the origin, the induced map of grads is injective.

DEFINITION (D3). **Generalized Associated Graded Rings**]. To generalize the matter of §5(Q13), given ideals $I \subset J$ in a ring R we define the associated graded ring $\mathrm{grad}(R, I, J)$ of (R, I, J) by putting

$$\mathrm{grad}(R, I, J) = \bigoplus_{n \in \mathbb{N}} (I^n/(JI^n)) \quad \text{(external direct sum of } \mathbb{Z}\text{-modules)}.$$

Now take over the material of (Q13), from the sentence "This is made into a ring..." to the sentence "The homomorphism $\Theta^x_{(R,I)}$...," after everywhere changing I to I, J, and I^{n+1} to JI^n. In particular the n-th graded component $\mathrm{grad}(R, I, J)_n$ corresponds to $\mathrm{grad}(R, I)_n$. Show that there is a unique graded ring homomorphism $\pi : \mathrm{grad}(R, I) \to \mathrm{grad}(R, I, J)$ such that for every n in $\mathbb{N}$ and $f \in I^n$ we have $\pi(\mathrm{grad}^n_{(R,I)}(f)) = \mathrm{grad}^n_{(R,I,J)}(f)$; we call π the natural homomorphism of $\mathrm{grad}(R, I)$ into $\mathrm{grad}(R, I, J)$.

EXERCISE (E16). [**Strong Relative Independence**]. Given any elements $x_1, \ldots, x_n$ in a ring R with $n \in \mathbb{N}$, let $I = (x_1, \ldots, x_n)R$, and let J be an ideal in R. To sharpen (E5), let $A[Y_1, \ldots, Y_n]_i$ be as defined there, and let us say that $x_1, \ldots, x_n$ are strongly independent over (R, J), or strongly independent relative to (R, J), if $I \subset J$ and the following condition is satisfied:

$$\begin{cases} G(Y_1, \ldots, Y_n) \in R[Y_1, \ldots, Y_n]_i \text{ with } i \in \mathbb{N} \text{ and } G(x_1, \ldots, x_n) \in JI^i \\ \Rightarrow G(Y_1, \ldots, Y_n) \in J[Y_1, \ldots, Y_n]_i. \end{cases}$$

Let us also say that $x_1, \ldots, x_n$ are strongly independent over R, or strongly independent relative to R, if $I \neq R$ and they are strongly independent over $(R, \mathrm{rad}_R I)$.

Note that obviously: (1) strong independence of $x_1, \ldots, x_n$ over (R, J) implies their independence over (R, J), and (2) if $I \subset J$ then $\pi \Theta^x_{(R,I)} = \Theta^x_{(R,I,J)}$.

Show that: (3) if $I \subset J \subset \mathrm{rad}_R I$ then $\ker(\pi) \subset \mathrm{nrad}(\mathrm{grad}(R, I))$, and therefore (4) independence of $x_1, \ldots, x_n$ over R implies their strong independence over R.

Hint: If $I \subset J$ then for any $e \in \mathbb{N}$ we clearly have

$$\ker(\pi) \cap \mathrm{grad}(R, I)_e = \{\mathrm{grad}^e_{(R,I)}\lambda_e(g) : g \in JI^e\}$$

where $\lambda_e : I^e \to I^e/I^{e+1}$ is the residue class epimorphism. Also clearly

$$\begin{cases} \mathrm{nrad}(\mathrm{grad}(R,I)) \cap \mathrm{grad}(R,I)_e \\ = \{\mathrm{grad}^e_{(R,I)}\lambda_e(g) : g \in I^e \text{ with } g^d \in I^{de+1} \text{ for some } d \in \mathbb{N}_+\}. \end{cases}$$

Therefore, if $I \subset J \subset \mathrm{rad}_R I$ then $\ker(\pi) \subset \mathrm{nrad}(\mathrm{grad}(R,I))$. This proves (3). To prove (4) assume that $x_1, \ldots, x_n$ are independent over R, i.e., they are independent over (R, J) with $J = \mathrm{rad}_R I \neq R$. Given any $G(Y_1, \ldots, Y_n) \in R[Y_1, \ldots, Y_n]_e$ with $e \in \mathbb{N}$ and $g \in JI^e$ where $g = G(x_1, \ldots, x_n)$, we want to show that $G(Y_1, \ldots, Y_n) \in J[Y_1, \ldots, Y_n]$. By (3) we find $d \in \mathbb{N}_+$ such that $F(x_1, \ldots, x_n) \in I^{de+1}$ where $F(Y_1, \ldots, Y_n) = G(Y_1, \ldots, Y_n)^d \in R[Y_1, \ldots, Y_n]_{de}$. Therefore by the independence of $x_1, \ldots, x_n$ over R we get $F(Y_1, \ldots, Y_n) \in J[Y_1, \ldots, Y_n]$. Consequently by the following Exercise (E17) we get $G(Y_1, \ldots, Y_n) \in J[Y_1, \ldots, Y_n]$.

EXERCISE (E17). [**Ideals in Polynomial Rings**]. Show that for a finite variable polynomial ring $S = R[Y_1, \ldots, Y_n]$ over any ring R, and any ideals I, J in R with $J = \mathrm{rad}_R I$, we have $JS = \mathrm{rad}_S(IS)$.

Hint: Note that $IS = I[Y_1, \ldots, Y_n]$ and $JS = J[Y_1, \ldots, Y_n]$. Make induction on degree and number of variables. Compare with (Q7)(T30) and (Q29)(C43).

EXERCISE (E18). [**Triviality of Associated Graded Ring**]. In connection with §5(Q13) show that for any ideal I in any noetherian ring R we have:

$$\begin{cases} I = I^2 \\ \Leftrightarrow I = I^n \text{ for all } n \in \mathbb{N}_+ \\ \Leftrightarrow \mathrm{grad}(R,I)_n = 0 \text{ for all } n \in \mathbb{N}_+ \\ \Leftrightarrow I = zR \text{ for some } z \in R \text{ with } z^2 = z. \end{cases}$$

Hint: The equivalence of the top three lines is obvious, and also it is obvious that the last line implies the first. To prove the first implies the last, assuming $I = I^2$ and taking $J = V = I$ in §5(Q3)(4) we find $z \in I$ such that $(1 - z)x = 0$ for all $x \in I$. This equation tells us that $x = xz$ for all $x \in I$ and hence $I = zR$. Taking $x = z$ we also get $z = z^2$.

EXERCISE (E19). [**Generating Number**]. In connection with the preamble of §5(Q15), show that for any ideal I in any noetherian ring R we have:

(1) $\mathrm{ht}_R I \leq \mathrm{gnb}_R(I/I^2) \leq \mathrm{gnb}_R I \leq 1 + \mathrm{gnb}_R(I/I^2)$, and

(2) $\mathrm{gnb}_R(I/I^2) > \dim(R) \Rightarrow \mathrm{gnb}_R I = \mathrm{gnb}_R(I/I^2)$.

Hint: The second inequality in (1) is obvious and the third can be proved as in the above Hint of (E18). Namely, upon letting $\mathrm{gnb}_R(I/I^2) = m$ we can take elements $a_1, \ldots, a_m$ in I which generate I mod I^2; upon letting $\pi : R \to R/(a_1, \ldots, a_m)R$ be the residue class epimorphism, we have $\pi(I) = \pi(I)^2$, and hence as in the above Hint $\pi(I) = \pi(zR)$ for some $z \in I$; clearly $I = (z, a_1, \ldots, a_m)R$ and hence $\text{gnb}_R I \leq 1 + n$.

For the first inequality in (1), we can take $P \in \text{vspec}_R I$ with $\text{ht}_R I = \text{ht}_S \phi(I)S$ where $\phi : R \to S = R_P$ is the localization map. Now

$$\text{ht}_S \phi(I)S \leq \text{gnb}_S \phi(I)S = \text{gnb}_S(\phi(I)S/\phi(I^2)S) \leq \text{gnb}_R(I/I^2)$$

where the first inequality is by Generalized Principal Ideal Theorem §5(Q7(T27)), the second equation is by Nakayama Lemma §5(Q3)(T3), and the third inequality is obvious. So we are done.

To prove (2) let $\text{gnb}_R(I/I^2) = m$. By the Prime Avoidance Corollary proved in §5(Q17)(C31,2) we can find elements $a_1, \ldots, a_m$ in I such that for every P in $\text{vspec}_R(a_1, \ldots, a_m)R \setminus \text{vspec}_R I$ we have $\text{ht}_R P \geq n$. Let $\pi : R \to R/(a_1, \ldots, a_m)R$ be the residue class epimorphism, and assume $n > \dim(R)$. Then we must have $\text{vspec}_R(a_1, \ldots, a_m)R \setminus \text{vspec}_R I = \emptyset$, and hence $\pi(I)^d = 0$ for some $d \in \mathbb{N}_+$. By the above Hint of (E18) we have $\pi(I) = \pi(zR)$ for some $z \in I$ with $\pi(z)^2 = 0$. Therefore $\pi(I) = 0$ and hence $\text{gnb}_R I = \text{gnb}_R(I/I^2)$.

EXERCISE (E20). [**Regularity and Inverse of the Maximal Ideal**]. In connection with the preamble of (Q15), show that for any local ring R with $\text{QR}(R) = K$ we have: $\text{reg}(R) > 1 \Rightarrow (R : M(R))_K = R$.

Hint: If $\text{reg}(R) > 1$ then we can take an R-regular sequence x, y, and for any such sequence we have: $R \subset (R : M(R))_K \subset (1/x)R \cap (1/y)R \subset R$. To prove the last inclusion, observe that $\text{QR}(R)$ is the set of all u/v with u in R and v in $S_R(R)$; moreover, for any u, u' in R and v, v' in $S_R(R)$ we have: $u/v = u'/v' \Leftrightarrow uv' = vu'$. In our case, by §5(Q15)(T66.3) we know that x, y are in $S_R(R)$ and hence the expression $(1/x)R \cap (1/y)R$ does make sense, and given any $z = p/x = q/y$ in $\text{QR}(R)$ with p, q in R we have $py = qx$ and therefore by the R-regularity of the sequence x, y we get $p = rx$ for some $r \in R$; thus $p/x = r \in R$.

DEFINITION (D4). [**Dimension and Subdimension Formulas**]. Theorems §5(Q12)(T58) and §5(Q15)(T70.12) tell us that the dimension formula of §5(Q12) holds in any CM ring R. In comparison with the dimension formula, we say that the subdimension formula holds in a local ring R if for every nonunit ideal I in R we have: $\text{ht}_R I = \dim(R) - \dim(R/I)$.

EXERCISE (E21). [**Subdimension Formula**]. Show that the subdimension formula holds in any CM local ring R.

Hint: Use §5(Q15)(T70.6).

EXERCISE (E22). [**Conditions for DVR**]. Strengthen Theorem (T85) of §5(Q19) by showing that if R is a local ring in which $M(R)^e$ with $e = 0, 1, 2, \ldots$ are exactly all the distinct nonzero ideals then R is a DVR.

Hint: In view of (T85), it suffices to show that R is a PID. Since all nonzero ideals in R are powers of $M(R)$, for any nonzero elements x, y in R we have $xR = M(R)^a$ and $y = M(R)^b$ with a, b in $\mathbb{N}$. Then $(xy)R = M(R)^a M(R)^b = M(R)^{a+b} \neq M(R)^{a+b+1}$ and hence $xy \neq 0$. Thus R is a domain. It only remains to show that for any $e \in \mathbb{N}$ we have $M(R)^e = zR$ for some $z \in R$. Since $M(R)^e \neq M(R)^{e+1}$, we can find $z \in M(R)^e \setminus M(R)^{e+1}$. Now $zR = M(R)^c$ for some $c \in \mathbb{N}$. Since $M(R)^c = zR \subset M(R)^e$, we must have $c \geq e$. Since $z \notin M(R)^{e+1}$, we must also have $c \leq e$. Therefore $c = e$.

EXERCISE (E23). [**Serre Conditions**]. In §5(Q19)(T86) it may be tempting to replace "Alternative Serre Conditions" (S'_n) and (S''_n) of §5(Q19) by their variations $(\widetilde{S}'_n)$ and $(\widetilde{S}''_n)$, or their variations $(\overline{S}'_n)$ and $(\overline{S}''_n)$, where for a noetherian ring R and nonnegative integer n these conditions say the following:

$(\widetilde{S}'_n)$. Nonunit ideal I in R with $0 \leq \mathrm{ht}_R I = m = n - 1$ and $I = (x_1, \ldots, x_m)R \Rightarrow I$ is unmixed.

$(\widetilde{S}''_n)$. Nonunit ideal I in R with $0 \leq \mathrm{ht}_R I = m = n - 1$ and $I = (x_1, \ldots, x_m)R \Rightarrow I$ is unmixed and the sequence $(x_1, \ldots, x_m)$ is R-regular.

$(\overline{S}'_n)$. Nonunit ideal I in R with $0 \leq \mathrm{ht}_R I = m < n$ and $I = (x_1, \ldots, x_m)R \Rightarrow I$ is unmixed.

$(\overline{S}''_n)$. Nonunit ideal I in R with $0 \leq \mathrm{ht}_R I = m < n$ and $I = (x_1, \ldots, x_m)R \Rightarrow I$ is unmixed and the sequence $(x_1, \ldots, x_m)$ is R-regular.

For the two dimensional reduced local ring R of (E4) show that R satisfies $(\widetilde{S}'_3)$ but it does not satisfy any one of the other three conditions $(\widetilde{S}''_n)$, $(\overline{S}'_n)$, $(\overline{S}''_n)$. Also show that for every $g \in R$ with $\mathrm{ht}_R(gR) = 1$ we have $M(R) \in \mathrm{ass}_R(R/gR)$ and therefore gR is not unmixed.

Hint: Since R is a two dimensional local ring, it obviously satisfies $(\widetilde{S}'_3)$. To show that it does not satisfy $(\widetilde{S}''_n)$, let $\phi : A = k[[X, Y, Z]] \to R = A/(XZ, YZ)A$ be the residue class epimorphism, and let $I = (\phi(Y), \phi(X - Z))R$. Clearly $\phi(Y)\phi(Z) = 0$ with $\phi(Z) \neq 0$; consequently $\phi(Y)$ is a zerodivisor in R, and hence the sequence $\phi(Y), \phi(X - Z)$ is not R-regular. Let $J = (Y, X - Z, XZ, YZ)A$. Then $J = \phi^{-1}(I)$. Also $Z^2 = XZ - (X - Z)Z \in J$ and hence $Z \in \mathrm{rad}_A J$, and therefore $M(A) = (X, Y, Z)A \subset \mathrm{rad}_A J$. Consequently $\mathrm{rad}_A J = M(R)$ and hence $\mathrm{rad}_R I = M(R)$. Therefore $\mathrm{ht}_R I = 2$.

Let $G = G(X, Y, Z) \in A$ be such that for $g = \phi(G)$ we have $\mathrm{ht}_R(gR) = 1$. By (E4), $\phi(ZA)$ and $\phi((X, Y)A)$ are exactly all the height zero primes in R and hence, in view of Krull's Principal Ideal Theorem §5(Q7)(T24), the equation $\mathrm{ht}_R(gR) = 1$ is equivalent to the condition $G \notin (ZA) \cup (X, Y)A$. Clearly $G \notin M(A)$ and hence upon letting $S = G(X, Y, 0) \in B = k[[X, Y]]$ and $T = G(0, 0, Z) \in C = k[[Z]]$ we get $G = S + T + UXZ + VYZ$ with $S \in M(B)$, $T \in M(C)$, $U \in A$, $V \in A$. Let $H = (G, XZ, YZ)A$. Then $\phi^{-1}(gR) = H$. We shall show that: (1) $TM(A) \subset H$ and (2) $T \notin H$. From this it will follow that $(gR : \phi(T))_R = M(R)$ and hence $M(R) \in \mathrm{ass}_R(R/xR)$ and therefore gR is not unmixed; consequently R does not satisfy condition $(\overline{S}'_n)$ and hence it does not satisfy condition $(\overline{S}''_n)$.

To prove (1) it suffices to note that TX, TY belong to H because $T \in M(C)$ and ZX, ZY belong to H, and moreover TZ belongs to H because $S \in M(B)$ and

$$(\bullet) \qquad\qquad T = G - S - UXZ - VYZ.$$

To prove (2), suppose if possible that $T \in H$. Then by $(\bullet)$ we get $S \in H$ and hence

$$(\bullet\bullet) \qquad G(X,Y,0) = \widehat{U}(X,Y,Z)XZ + \widehat{V}(X,Y,Z)YZ + \widehat{W}(X,Y,Z)G(X,Y,Z)$$

with $\widehat{U}(X,Y,Z), \widehat{V}(X,Y,Z), \widehat{W}(X,Y,Z)$ in A. By putting $Z = 0$ in $(\bullet\bullet)$ and remembering that $G \notin ZA$ we get $\widehat{W}(X,Y,0) = 1$. Therefore $\widehat{W}(X,Y,Z)$ is a unit in A and hence, remembering that $G(X,Y,0) \in M(B)$, by $(\bullet\bullet)$ we get $G(X,Y,Z) \in (X,Y)A$ which is a contradiction.

Remark: By §5(Q1) and §5(Q7)(T24), $S_R(R) \cap M(R) = \{g \in R : \text{ht}_R(gR) = 1\}$ and for any such g we have that: the R-regular sequence g cannot be extended to a longer R-regular sequence iff $M(R) \in \text{ass}_R(R/gR)$. By §5(Q15)(T66) we also know that any two maximal R-regular sequences have the same length. Therefore if g is nonunmixed for some $g \in R$ with $\text{ht}_R(gR) = 1$ then g is nonunmixed for every $g \in R$ with $\text{ht}_R(gR) = 1$.

EXERCISE (E24). [**Reduced Normality Quasicondition**]. In §5(Q19)(T87) we showed that Condition (T) implies "Quasicondition" (T^*) where these conditions on a noetherian ring R say the following:

(T). R is a reduced normal ring.

(T^*). For every $P \in \text{nspec}(R)$ the ring R/P is a normal domain.

By an example show that (T^*) does not imply (T).

Hint: Let R be the two dimensional reduced local ring as above. Upon letting $P_1 = ZA$ and $P_2 = (X,Y)A$ we have $\text{nspec}(R) = \{\phi(P_1), \phi(P_2)\}$ and $R_{\phi(P_i)} \approx A_{P_i}$ for $1 \le i \le 2$. Now A_{P_i} is a regular local domain by §5(Q7)(T30.4) and §5(Q17)(81.7), and hence it is normal by §5(Q14)(T64). Thus R satisfies (T^*). Note that the reference to §5(Q7)(T30.4) and §5(Q17)(81.7) for the regularity of A_{P_i} may be replaced by a reference to §5(Q17)(C32) and §5(Q19)(C33).

To show that R does not satisfy (T), in view of §5(Q19)(T86) and §5(Q19(T88), it suffices to show that R does not satisfy condition (S'_2) of §5(Q19)(T86) which in our case says that: for every $g \in R$ with $\text{ht}_R(gR) = 1$, the ideal (gR) is unmixed. So we are done by (E23).

EXERCISE (E25). [**Serre Quasicondition**]. In §5(Q19)(T86) we showed that Serre Condition (S_n) implies "Serre Quasicondition" (S_n^*), where for a noetherian ring R and nonnegative integer n these conditions say the following:

(S_n) $P \in \text{spec}(R)$ with $\text{reg}(R_P) < n \Rightarrow R_P$ is CM.

(S_n^*). $P \in \text{spec}(R)$ with $\text{ht}_R P \le n \Rightarrow R_P$ is CM.

By an example show that (S_n^*) does not imply (S_n).

Hint: Modify the example of (E23) and (E24) by taking A to be a four variable power series ring, i.e., let $\phi : A = k[[W,X,Y,Z]] \to R = A/(WZ,XZ,YZ)A$ be the residue class epimorphism, and let $P_1 = ZA$ and $P_2 = (W,X,Y)A$. Then as in (E23) and (E24), R is a three dimensional reduced local ring with $\mathrm{nspec}(R) = \{\phi(P_1), \phi(P_2)\}$ and for $1 \le i \le 2$ we have $R_{\phi(P_i)} \approx A_{P_i}$ where A_{P_i} where is a regular local domain, and hence a CM ring by §5(Q15)(T69.7). We claim that R satisfies (S_2^*) but not (S_2).

To prove that R does not satisfy (S_2), in view of §5(Q19)(T86), it suffices to show that R does not satisfy condition (S_2') of §5(Q19)(T86) which in our case says that: for every $g \in R$ with $\mathrm{ht}_R(gR) = 1$, the ideal (gR) is unmixed; the proof of this is very similar to the proof given in (E23).

To prove that R satisfies (S_2^*), given any prime ideal P in R with $\mathrm{ht}_R P \le 2$ we want to show that R_P is CM. In view of what we said above, we may suppose that $1 \le \mathrm{ht}_R P \le 2$. Let $Q = \phi^{-1}(P)$. Then we must have $(W,X,Y)A \not\subset Q$ and hence by L4§5(O22)(21•) we get $ZA \subset Q$. Therefore upon letting $\psi : A \to B = k[[W,X,Y]]$ to be the B-epimorphism which sends Z to 0, we see that $R_P \approx B_{\psi(Q)}$. It only remains to note that $B_{\psi(Q)}$ is regular as in (E24) and hence it is CM by §5(Q15)(T69.7).

EXERCISE (E26). [**Linear Disjointness**]. Let k, R, R' be subrings of a ring S such that k is a field with $k \subset R \cap R'$. Extend Lemma §5(Q29)(T109.1) by showing that the following five conditions are equivalent.

(1) The rings R, R' are linearly disjoint over the field k.

(2) For some k-basis $(x_i)_{i \in I}$ of R, the elements $(x_i)_{i \in I}$ are linearly independent over R'.

(3) Every k-basis of R is an (R')-basis of $k[R, R']$.

(4) For some k-bases $(x_i)_{i \in I}$ and $(x_j')_{j \in I'}$ of R and R' respectively, the elements $(x_i x_j')_{(i,j) \in I \times I'}$ are linearly independent over k.

(5) For all k-bases $(x_i)_{i \in I}$ and $(x_j')_{j \in I'}$ of R and R' respectively, the elements $(x_i x_j')_{(i,j) \in I \times I'}$ form a k-basis of $k[R, R']$.

As a consequence show that, in the situation of §5(Q29)(C44), (i) the rings $k[Y]$ and R are linearly disjoint over k and for their compositum we have $S = k[k[Y], R]$, and (ii) the rings $k(Y)$ and R are linearly disjoint over k and for their compositum we have $T = k[k(Y), R]$.

Hint: For (i) use the fact that the elements of the k-basis of $k[Y]$ consisting of all monomials in Y are linearly independent over R. Likewise, for (ii) use the fact that the elements of any k-basis of $k(Y)$ are linearly independent over R; to do this take common denominators in $k[Y]^\times$.

EXERCISE (E27). [**Split Monomorphisms**]. In connection with the material around display (1) of the preamble of §5(Q31)(C49), prove the following criterion.

$$\begin{cases} \text{For any split } R\text{-monomorphisms } \sigma : P \to Q \text{ and } \sigma' : P \to Q \\ \text{where } P \text{ and } Q \text{ are any modules over any ring } R, \text{ we have that:} \\ \sigma \sim \sigma' \Leftrightarrow \text{ there exists an } R\text{-isomorphism } Q/\sigma(P) \to Q/\sigma'(P). \end{cases}$$

EXERCISE (E28). [**Limitations on Normalization Theorem**]. Referring to the preamble of §5(Q33) for the definitions of restricted and strongly restricted domains, let us note the Finite Module Theorem §5(Q33)(T145) which says that every field is strongly restricted. A part of the Normalization Theorem §5(Q10)(T46) says that: given an affine domain $A = R[x_1, \ldots, x_n]$ over an infinite domain R, with $n \in \mathbb{N}_+$ and $\mathrm{trdeg}_R A = d$, assuming (1) R to be a field, (2) there exist d elements $y_1, \ldots, y_d$ in A such that A is integral over $R[y_1, \ldots, y_d]$, and moreover (3) $y_1, \ldots, y_d$ can be chosen to be R-linear combinations of $x_1, \ldots, x_n$. In the "Generalized Normalization Theorem" on page 124 of volume II of Zariski-Samuel's book [ZaS] it is "claimed" that (2) and (3) are true without assuming (1), and on the next two pages 125-126 this is used for "proving" a stronger version of §5(Q33)(T145) which says that every restricted domain is strongly restricted. The exercise is to show that without assuming (1), even (2) is untrue.

Hint: For any nonnegative integer d let $X_1, \ldots, X_d$ be indeterminates over $R = \mathbb{Z}$ and let $A = R[x_1, \ldots, x_n]$ where $n = d+1$ with $x_i = X_i$ for $1 \le i \le d$ and $x_n \in \mathbb{Q} \setminus \mathbb{Z}$ (for instance $x_n = 1/2$). Let if possible $y_1, \ldots, y_d$ be elements of A such that A is integral over $B = R[y_1, \ldots, y_d]$. Then $y_1, \ldots, y_d$ must be algebraically independent over $\mathbb{Q}$, and hence B is a normal domain. Therefore by (E7) the minimal polynomial $F(Y)$ of x_n over $\mathrm{QF}(B)$ belongs to $B[Y]$. But clearly $F(Y) = Y - x_n$. This is a contradiction.

EXERCISE (E29). [**Auxiliary Theorem**]. Prove Theorem §5(Q33)(T148).

Hint: To prove assertion (T148.3) about monomials, use Comment §5(Q15)(C20) about binomial coefficients.

To prove assertion (T148.1) and (T148.2), let A be a subdomain of a field K, let $\overline{K}$ be a finite algebraic field extension of K, let E be a semimodel of K/A given by $E = \cup_{l \in \Lambda} \mathfrak{V}(B_l)$ where $(B_l)_{l \in \Lambda}$ is a nonempty family of subdomains of K with quotient field K such that $A \subset B_l$, and let $\overline{E} = \cup_{l \in \Lambda} \mathfrak{V}(\overline{B}_l)$ where $\overline{B}_l =$ the integral closure of B_l in a finite algebraic field extension $\overline{K}$ of K. For any $l \in \Lambda$ and $P \in \mathrm{spec}(B_l)$, upon letting $S = B_l \setminus P$, by (T36) and (T37) of §5(Q9) we see that $(\overline{B}_l)_S$ is the integral closure of $(B_l)_P$ in $\overline{K}$, and hence by localization theory L4§7 we get $\mathfrak{V}((\overline{B}_l)_S) = \mathfrak{N}((B_l)_P, \overline{K})$. Consequently for any $l \in \Lambda$ we have $\mathfrak{V}(\overline{B}_l) = \mathfrak{N}(\mathfrak{V}(B_l), \overline{K})$. Therefore taking $B = B_l$ we get (T148.1), and by taking union over Λ we get $\overline{E} = \mathfrak{N}(E, \overline{K})$. By §5(Q9)(T36) we see that $\overline{E}$ is a normal semimodel of $\overline{K}/A$, and by valuation theory L4§12(R7) we see that if E is complete semimodel of K/A then $\overline{E}$ is complete semimodel of $\overline{K}/A$. This proves (T148.2).

The proof of assertion (T148.4) is straightforward.

EXERCISE (E30). [**Homogeneous Function Field**]. In the second paragraph before the statement of Theorem (T149) of §5(Q34.3) which starts with something like "In case iproj$_A U$ is the homogeneous prime ideal P in the homogeneous ring A over k," we discussed the homogeneous function field $k(U)^*$ and the alternative homogeneous function field $k(U)^\sharp$ of a projective variety U. This was the projective version of the third paragraph before the statement of Theorem (T19) of L4§8 which starts with something like "In case ispec$_A U$ is the prime ideal P in the affine ring A over k," and where we discussed the function field $k(U)^*$ and the alternative function field $k(U)^\sharp$ of an affine variety U. In both situations we had an epimorphism $\psi_U : R_k(U)^* \to k(U)^*$ of the local ring $R_k(U)^*$ of U over k which induced an isomorphism $\phi : k(U)^\sharp \to k(U)^*$. Recall that in the affine case $k(U)^*$ is the quotient field of A/P and $R_k(U)^*$ is the localization A_P, whereas in the projective case $k(U)^*$ is the homogeneous quotient field of A/P and $R_k(U)^*$ is the homogeneous localization $A_{[P]}$; in both cases $k(U)^\sharp = R_k(U)^*/M(R_k(U)^*)$.

Show that in the affine case, for the composition $\theta : A \to R_k(U)^* \to k(U)^\sharp$ (of the natural maps) we have $\ker(\theta) = P$ and $\mathrm{QF}(\mathrm{im}(\theta)) = k(U)^\sharp$, which gives a second definition of ϕ; also show that this does not work in the projective case.

Hint: See L4§7.1(T16).

EXERCISE (E31). [**Projective Theorems**]. Prove Theorems (T149) to (T152) stated in §5(Q34.3).

Hint: To prove (T149) and (T150), recall that P is a relevant homogeneous prime ideal in the $N + 1$ variable polynomial ring $B = k[X_1, \ldots, X_{N+1}]$ over a field k, let $\phi : B \to D = B/P$ be the graded residue class epimorphism, and take $x \in D_1^\times$. In view of (Q34.2)(4) and the paragraph before (T149), we have $\mathfrak{K}(D) = \mathrm{QF}(D_{/x}(D)) = \mathrm{QF}(D/(x-1)D)$. In view of §5(Q7)(T24) and the fact that both D and $D/(x-1)D$ are domains, we see that $(x-1)D$ is a prime ideal of height 1. Consequently (T149) and (T150) follow from L4§8(T19) and L4§8(T20) respectively. Similarly (T151) and (T152) follow from L4§8(T21) and L4§8(T22) respectively.

DEFINITION (D5). [**Homogeneous Localization**]. To put Theorems (T153) and (T154) of §5(Q34.6) in proper perspective, in the Hint to the next Exercise we shall generalize them by replacing the finite variable polynomial ring B over a field by any integrally graded ring $A = \sum_{i \in \mathbb{Z}} A_i$. First let us generalize the concept of homogeneous localization discussed in §5(Q34.2) to the homogeneous localization $A_{[S]}$ of any such A at any homogeneous multiplicative set S in A, i.e., at any multiplicative set S in A with $S \subset A_\infty = \bigcup_{i \in \mathbb{Z}} A_i$. We convert the localization A_S

into an integrally graded ring $A_S = \sum_{i \in \mathbb{Z}} (A_S)_i$ by letting

$$(A_S)_i = \bigcup_{j \in \mathbb{Z}} \{a/s : a \in A_{i+j} \text{ and } s \in S \cap A_j\}.$$

Note that the localization map $\phi : A \to A_S$ is now a graded ring homomorphism. We put $A_{[S]} = (A_S)_0$. For $P \in \text{proj}(A)$ we may write $A_{[P]} = $ instead of $A_{[A_\infty \setminus P]}$ and call it the homogeneous localization of A at P. Recall that in the good case, i.e, when $S \subset S_A(A)$, we may regard A_S, and hence also $A_{[S]}$, to be a subring of $\text{QR}(A)$. Thus in the situation of §5(Q34.2), by taking $A_i = D_i$ or 0 according as $i \geq 0$ or $i < 0$, the notation $D_{[P]}$ agrees with the present notation $A_{[P]}$.

EXERCISE (E32). [**Ideals And Homogeneous Ideals**]. Prove the relations between ideals and homogeneous ideals stated in §5(Q34.6)(T153) and the resulting relations between affine and projective varieties stated in §5(Q34.6)(T154).

Hint: To prove a generalization of §5(Q34.6)(T153) valid in the above situation of (D5), let

$$T(A) = \text{the set of all homogeneous ideals in } A$$

and

$$T(A_S) = \text{the set of all homogeneous ideals in } A_S.$$

Also let

$$T'_S(A) = \{J \in T(A) : [J : S]_A = J\}$$

and

$$\widehat{T}(A_{[S]}) = \text{the set of all ideals in } A_{[S]}.$$

Now let $\alpha : T(A) \to T(A_S)$ be the map given by $J \mapsto \phi(J)A_S$. By L4§7(T12) we see that α is an inclusion preserving surjective map which commutes with the ideal theoretic operations of radicals, intersections, sums, finite products, and quotients = parenthetical colons. Moreover, if $J = \cap_{1 \leq i \leq e} Q_i$ is an irredundant primary decomposition of $J \in T(A)$ where the primary ideals $Q_i \in T(A)$ with radicals P_i are so labelled that $S \cap P_i = \emptyset$ or $S \cap P_i \neq \emptyset$ according as $1 \leq i \leq d$ or $d + 1 \leq i \leq e$, then $\alpha(J) = \cap_{1 \leq i \leq d} \alpha(Q_i)$ is an irredundant primary decomposition of $\alpha(J) \in T(A_S)$ and $\alpha(P_i)$ is the radical of the primary ideal $\alpha(Q_i) \in T(A_S)$ for $1 \leq i \leq d$. In particular, α sends prime (resp; primary, radical) ideals to prime (resp: primary, radical) ideals.

Let $\widetilde{\alpha} : T'_S(A) \to T(A_S)$ be the map given by $J \mapsto \widetilde{J} = \alpha(J)$. Then clearly $\widetilde{\alpha}$ is a bijection whose inverse is given by $\widetilde{J} \mapsto J = \phi^{-1}(\widetilde{J})$. Moreover $\widetilde{\alpha}$ sends prime (resp: primary, radical) ideals to prime (resp: primary, radical) ideals.

Let $\beta : T(A_S) \to \widehat{T}(A_{[S]})$ be the map given by $J \mapsto J \cap A_{[S]}$. It is easy to check that β is an inclusion preserving surjective map which also commutes with the above mentioned ideal theoretic operations. β also send prime (resp: primary, radical) ideals to prime (resp: primary, radical) ideals. Furthermore, if $A_1 \cap S \neq \emptyset$, or more generally if $(A_S)_1 \cap U(A_S) \neq \emptyset$, then β is injective; namely, for any u in $(A_S)_1 \cap U(A_S)$ and any homogeneous ideal J in A_S we have $J \cap (A_S)_i = \beta(J)u^i$ for all $i \in \mathbb{Z}$ and hence $J = \sum_{i \in \mathbb{Z}} \beta(J)u^i$ is uniquely determined by $\beta(J)$. In general β need not be injective.

In the situation of (Q34.6) take $A_i = B_i$ or 0 according as $i \geq 0$ or $i < 0$. We get a (nongraded) ring isomorphism $\mu : B \to A$ by sending every finite sum $\sum_{i \in \mathbb{N}} x_i \in B$ with $x_i \in B_i$ to the finite sum $\sum_{i \in \mathbb{N}} x_i \in A$; this induces a bijection $\mu^* : T(B) \to T(A)$ given by $J \mapsto \mu(J)$. Taking $S = \{1, X_{N+1}, X_{N+1}^2, \dots\}$ we see that β is a bijection. We also get a bijection $\widetilde{\mu}^* : T'(B) \to T'_S(A)$ given by $J \mapsto \mu(J)$. We have a unique k-isomorphism of rings $\nu : \widehat{B} \to A_{[S]}$ which sends X_i to X_i/X_{N+1} for $1 \leq i \leq N$; this induces a bijection $\nu^* : \widehat{T}(\widehat{B}) \to \widehat{T}(A_{[S]})$ given by $J \mapsto \nu(J)$. Turning to the objects γ and δ defined in (Q34.6), we get a map $\gamma^* : T(B) \to \widehat{T}(\widehat{B})$ given by $J \mapsto \gamma(J)$, and a map $\delta^* : \widehat{T}(\widehat{B}) \to T'(B)$ given by $\widehat{J} \mapsto \delta(\widehat{J})$. Clearly

$$\beta \alpha \mu^* = \nu^* \gamma^* \quad \text{and} \quad \widetilde{\alpha}^{-1} \beta^{-1} \nu^* = \widetilde{\mu}^* \delta^*$$

and hence §5(Q34.6)(T153) follows from the above three paragraphs. Moreover, §5(Q34.6)(T154) follows from §5(Q34.3)(T151) and §5(Q34.6)(T153).

To see that β need not be injective, with A as in the above paragraph, take $S = k^\times$; then A_S may be identified with A, and $0A_S$ and $X_{N+1}A_S$ are distinct homogeneous ideals in A_S whose β images are the same, namely the zero ideal. To see that the condition $A_1 \cap S \neq \emptyset$ implies the condition $(A_S)_1 \cap U(A_S) \neq \emptyset$, note that $u \in A_1 \cap S \Rightarrow u/1 \in (A_S)_1 \cap U(A_S)$. To see that the converse is not true, in the above paragraph, taking $S = \{1, X_{N+1}^2, X_{N+1}^4, \dots\}$ we get $X_{N+1}/1 \in (A_S)_1 \cap U(A_S)$.

EXERCISE (E33). [**Blowup of a Line in Three Space**]. Verify Theorem §5(Q35.6)(T159) when A is the trivariate power series ring $k[[X, Y, Z]]$ over a field k, and P is the prime ideal in it generated by X and Y.

EXERCISE (E34). [**Regular Local Rings**]. Prove properties $(1^\dagger)$ and $(2^\dagger)$ of regular local rings stated in §5(Q35.9).

Hint: Let R be an n-dimensional regular local domain with $n \in \mathbb{N}$.

If $M(R) = (x_1, \dots, x_n)$ then upon letting $P = (x_1, \dots, x_m)R$ with $0 \leq m \leq n$, by §5(Q15)(T69.4) and §5(Q15)(T69.7) we get $\dim(R/P) = n - m$ and hence by (Q7)(T27) and (Q14)(T64) we see that R/P is a regular local domain, P is a prime ideal in R of height m, and R_P is an m-dimensional regular local domain.

Conversely, let P be a prime ideal in R such that R/P is an $(n-m)$-dimensional regular local domain for some integer m with $0 \leq m \leq n$. Let $\phi : R \to R' = R/P$ be the residue class epimorphism. Take a sequence $x = (x_1, \dots, x_n)$ of elements in R such that $(x_1, \dots, x_n)R = M(R)$, let $X = (X_1, \dots, X_n)$ be indeterminates over the field $k = R/M(R)$, and let $\Theta_R^x : k[X] \to \operatorname{grad}(R)$ be the graded ring isomorphism described in §5(Q13)(T61). Let $\phi(x) = (\phi(x_1), \ldots, \phi(x_n))$, let us "identify" $\phi(R)/M(\phi(R))$ with k, and let $\Theta_{\phi(R)}^{\phi(x)} : k[X] \to \operatorname{grad}(\phi(R))$ be the graded ring epimorphism described in §5(Q13)(T61). Clearly $\Theta_{\phi(R)}^{\phi(x)} = \operatorname{grad}_R(\phi)\Theta_R^x$. In view of §5(Q13)(T61.4), upon replacing $(x_1, \ldots, x_n)$ by suitable (homogeneous) R-linear combinations of them (whose determinant is a unit in R) we can arrange matters so that the kernel of $\Theta_{\phi(R)}^{\phi(x)}$ is generated by $X_1, \ldots, X_m$. By §5(Q13)(2) we know that $\operatorname{grad}_R(\phi) : \operatorname{grad}(R) \to \operatorname{grad}(\phi(R))$ is a graded ring epimorphism whose kernel is $\operatorname{ledi}_R(P)$. It follows that the leading ideal $\operatorname{ledi}_R(P)$ is generated by $\operatorname{grad}_R^1(x_1), \ldots, \operatorname{grad}_R^1(x_m)$. Therefore, again upon replacing $(x_1, \ldots, x_n)$ by suitable (homogeneous) R-linear combinations of them (whose determinant is a unit in R) we can arrange matters so that $Q \subset P$ where $Q = (x_1, \ldots, x_m)R$. By what we have proved above, Q is a prime ideal in R with $\dim(R/Q) = n - m$. Therefore we must have $P = Q$.

This proves $(1^\dagger)$. The proof of $(2^\dagger)$ follows from the references given there in a straight forward manner.

§7: NOTES

NOTE (N1). [**Noncancellative Quasiordered Monoids**]. As suggested in §2(C2), an ordered abelian group may be defined as an (additive) abelian group G with a linear order $\leq$ on it such that: (i) $i \leq i'$ and $j \leq j'$ in $G \Rightarrow i + j \leq i' + j'$, or equivalently such that: (ii) $i_1 = i_1', \ldots, i_a = i_a', j_1 < j_1', \ldots, j_b < j_b'$ in G with $a \in \mathbb{N}$ and $b \in \mathbb{N}_+ \Rightarrow i_1 + \cdots + i_a + j_1 + \cdots + j_b < i_1' + \cdots + i_a' + j_1' + \cdots + j_b'$. In defining an (additive) abelian monoid to be ordered we chose condition (ii), because it implies cancellativeness; what we get by replacing (ii) by (i) may be called a quasiordered abelian monoid. So we may ask whether every quasiordered abelian monoid is an ordered abelian monoid, i.e., whether the implication (i) $\Rightarrow$ (ii), and hence the implication (i) $\Rightarrow$ cancellative, is true also for an abelian monoid. That this is not so comes out of a general construction found in Bhaskara's ancient algebra book Beejganit [Bha] composed around Anno Domini 1150 in Ujjain which is the city in India where I was born. What Bhaskara is doing is to describe the properties of ∞ which he calls KHAHARA (in Sanskrit) and which is the entity you obtain when you divide an ordinary quantity by zero. So let there be given any abelian monoid G with a linear order $\leq$ satisfying (i); for instance G could be an ordered abelian group such as $\mathbb{Z}$ or $\mathbb{Q}$ or $\mathbb{R}$ or even the null group consisting of the single element 0. Let us adjoin an extra element ∞ to G to get a quasiordered abelian monoid $I = G \cup \{\infty\}$ by declaring $\infty + \infty = x + \infty$ and $x < \infty$ for all $x \in G$. The said declaration also shows that I is neither cancellative nor ordered.

NOTE (N2). [**Embedding Monoids Into Groups**]. In the above Note (N1), it is easily seen that for any abelian monoid G we have: (i) + cancellative $\Leftrightarrow$ (ii), i.e., a quasiordered abelian monoid is cancellative iff it is ordered. By mimicking the construction of the quotient field of a domain, it can also be seen that an abelian monoid is cancellative iff it can be embedded in an abelian group as a submonoid. This may be added to an appropriate list of Exercises.

§8: CONCLUDING NOTE

In my Historical Ramblings paper [A02] and in my Engineering Book [A04], I have divided Algebra into the High School Algebra of Polynomials and Power Series, the College Algebra of Rings and Ideals, and the University Algebra of Categories and Functors. There (see page 424 of [A02]) I declared Krull to be the college algebraist par excellence, listed Chevalley and Cohen as the follow-up leaders in that field, called Nagata [Nag] "a fitting successor to Krull," and on the previous page (page 423) I listed Dedekind and Emmy Noether as the predecessors of Krull. Zariski is the only person (see page 1 of [A02]) I included in the intersection of High School Algebra and College Algebra.

Lecture L6: Pause and Refresh

By reading the following summaries of the first five lectures, the rest of the book may become intelligible without studying the details of those lectures. In the first lecture we have introduced the basic structures of algebra such as groups, rings, fields, vector spaces, ideals, modules, polynomials, rational functions, euclidean domains, principal ideal domains, and unique factorization domains. In the second lecture, after introducing power series, meromorphic series, and valuations, we show the equivalence of well-ordering and Zorn's Lemma and use them to establish the existence of vector space basis, transcendence basis, algebraic closure, and maximal ideals. The third lecture deals with the power series theorems of Newton, Hensel, and Weierstrass. The fourth and fifth lectures deal with ideals, modules, varieties, and models which are the avatars of varieties full of local rings.

§1: SUMMARY OF LECTURE L1 ON QUADRATIC EQUATIONS

For **sets** (= collections of objects) S and T, a **map** $\phi : S \to T$ is an assignment which to every $x \in S$, i.e., to every **element** (= object) x of S, assigns $\phi(x) \in T$; this may be written $x \mapsto \phi(x)$; the element $\phi(x)$ is called the **image** of x under ϕ; we put $\mathrm{dom}(\phi) = S$ and $\mathrm{ran}(\phi) = T$ and call these the **domain** and **range** of ϕ respectively. The **composition** of maps $\phi : S \to T$ and $\psi : T \to U$ is the map $\psi\phi : S \to U$ given by $(\psi\phi)(x) = \psi(\phi(x))$. The map ϕ is **injective**, or is an **injection**, if $\phi(x) = \phi(y) \Rightarrow x = y$. A **subset** of S is a set R whose objects are amongst the objects of S; we write $R \subset S$; we may also write $S \supset R$ and call S an **overset** of R. We put $\phi(R) = \{\phi(x) : x \in R\}$ = the set of all $\phi(x)$ with x varying over R, and call this the **image** of R under ϕ. We put $\mathrm{im}(\phi) = \phi(S)$ and call this the **image** of ϕ; the map ϕ is **surjective**, or is a **surjection**, if $\phi(S) = T$. The map ϕ is **bijective**, or is a **bijection**, if it is injective as well as surjective. For a bijection $\phi : S \to T$ we have the **inverse** map $\phi^{-1} : T \to S$ given by $\phi^{-1}(y) = x \Leftrightarrow \phi(x) = y$. Without assuming the map $\phi : S \to T$ to be bijective, for any $y \in T$ we put $\phi^{-1}(y) = \{x \in S : \phi(x) = y\}$, and for any $U \subset T$ we put $\phi^{-1}(U) = \{x \in S : \phi(x) \in U\}$; moreover, for any $A \subset S$ and $B \subset T$ with $\phi(A) \subset B$, by $\phi|_{(A,B)}$ we denote the map $A \to B$ which sends every $x \in A$ to $\phi(x)$ in B, and we call this the **restriction** of ϕ to (A, B); if $B = \phi(A)$ then we may denote this by $\phi|_A$ and call it the **restriction** of ϕ to A.

The **empty set** is denoted by $\emptyset$. For subsets R_1 and R_2 of a set S, the **complement** of R_2 in R_1 is denoted by $R_1 \setminus R_2$, i.e., $R_1 \setminus R_2 = \{x \in R_1 : x \notin R_2\}$; needless to say that $x \notin R_2$ means x is not an element of R_2, just as $x \neq y$ means x is not equal to y, and so on. For subsets R_1 and R_2 of a set S, their **intersection** is denoted by $R_1 \cap R_2$ and their **union** is denoted by $R_1 \cup R_2$, i.e., $R_1 \cap R_2 = \{x \in S : x \in R_1 \text{ and } x \in R_2\}$ and $R_1 \cup R_2 = \{x \in S : x \in R_1 \text{ or } x \in R_2\}$. Similarly for more than two subsets $R_1, \ldots, R_m$ of a set S we have $\cap_{1 \leq i \leq m} R_i = \{x \in R : x \in R_i \text{ for all } i\}$ and

$\cup_{1 \le i \le m} R_i = \{x \in S : x \in R_i \text{ for some } i\}$. More generally we could have a family $(R_i)_{i \in I}$ of subsets R_i of a set S indexed by an **indexing set** I, i.e., $i \mapsto R_i$ gives a map of I into the set of all subsets of S, and then we put $\cap_{i \in I} R_i = \{x \in S : x \in R_i \text{ for all } i \in I\}$ and $\cup_{i \in I} R_i = \{x \in S : x \in R_i \text{ for some } i \in I\}$. A **partition** of a set S is a collection of nonempty subsets of S such that S is their union and any two of them have an empty intersection.

The set whose elements are $x_1, \ldots, x_e$ (which may or may not be distinct) is denoted by $\{x_1, \ldots, x_e\}$. By $\mathbb{N}_+ \subset \mathbb{N} \subset \mathbb{Z} \subset \mathbb{Q} \subset \mathbb{R} \subset \mathbb{C}$ we denote the sets of all positive integers, nonnegative integers, integers, rational numbers, real numbers, and complex numbers respectively. The **size** of any set S, i.e., the number of elements in it, is denoted by $|S|$; note that then $|S| \in \mathbb{N}$ or $|S| = \infty$ according as S is finite or infinite; later on we shall give a more precise meaning to different types of infinities, and then $|S|$ will denote the "cardinal number" of S. Clearly $|\emptyset| = 0$. For S_n defined below we have $|S_n| = n! = \prod_{1 \le i \le n} i$, and hence $|S_0| = 1$; convention: an empty product is 1 and an empty sum is 0.

A **group** G is a set with a binary operation which to every pair of elements x, y in G associates a product $xy \in G$ such that:

(i) $(xy)z = x(yz)$ for all x, y, z in G (associativity);

(ii) there is $1 \in G$ with $1x = x1 = x$ for all $x \in G$ (existence of identity);

(iii) for every $x \in G$ there is $x^{-1} \in G$ with $xx^{-1} = x^{-1}x = 1$ (existence of inverse).

A **subgroup** of a group G is a subset H which is a group under the same operation as G; we then write $H \le G$. If $H \le G$ with $H \ne G$ then we write $H < G$ and call H a **proper subgroup** of G. A normal subgroup of a group G is a subgroup H such that for all $x \in G$ we have $xHx^{-1} = H$ where $xHx^{-1} = \{xyx^{-1} : y \in H\}$; we then write $H \lhd G$. If $H \lhd G$ with $H \ne G$, then H is called a **proper normal subgroup** of G. A group G is **simple** if $G \ne 1$ and G has no proper normal subgroup $\ne 1$, where 1 denotes the **identity group** having only one element. The size $|G|$ of a group is its **order**; the order of $x \in G$ is the smallest $r \in \mathbb{N}_+$ with $x^r = 1$; if there is no such $r \in \mathbb{N}_+$ then the order of x is ∞. A group G is **cyclic** if it is generated by a single element x, i.e., if every element of G is a power of x; we then denote G by Z_r where r is the order of x which may be ∞; clearly Z_r is simple $\Leftrightarrow r$ is a prime number.

A **homomorphism** of a group G into a group J is a map $\phi : G \to J$ such that $\phi(1) = 1$ and $\phi(xy) = \phi(x)\phi(y)$ for all x, y in G; the **kernel** of ϕ is defined by $\ker(\phi) = \phi^{-1}(1)$, and we have $\text{im}(\phi) \le J$ and $\ker(\phi) \lhd G$; note that ϕ is injective iff (= if and only if) $\ker(\phi) = 1$ and when that is so we call ϕ a **monomorphism**; if ϕ is surjective, i.e., if $\text{im}(\phi) = J$, then we call ϕ an **epimorphism**; if ϕ is bijective then we call it an **isomorphism**; if $J = G$ and ϕ is an isomorphism then we call ϕ an **automorphism** of G.

For $H \le G$ we put $xH = \{xy : y \in H\}$ and call this a left **coset** of H in G (similar definition of a right coset), and by G/H we denote the set of all left cosets of H in G, and note that this is a partition of G; also we put $[G : H] = |G/H|$ and call this the **index** of H in G. If $H \lhd G$ then G/H becomes a group by defining $(xH)(yH) = (xy)H$, and we call G/H the **factor group** of G by H; now $x \mapsto xH$ gives an epimorphism $G \to G/H$ with kernel H; we call this the **canonical** epimorphism of G onto G/H.

A finite group G is **solvable** if there is a chain $1 = G_0 \lhd G_1 \lhd \cdots \lhd G_r = G$ such that G_i/G_{i-1} is cyclic of prime order for $1 \leq i \leq r$.

The set of all bijections of a set S onto itself forms a group under composition which we call the **symmetric group** on S and denote it by $\mathrm{Sym}(S)$. If $|S| = n \in \mathbb{N}$ then $\mathrm{Sym}(S) = S_n$. Any **permutation** σ on n letters, i.e., $\sigma \in S_n$, can be written as a product of a certain number ν of transpositions, and the **parity** of ν, i.e., its evenness or oddness, depends only on σ; we call the permutation σ **even** or **odd** according as ν is even or odd, and we define the **signature** $\mathrm{sgn}(\sigma)$ of σ to be 1 or -1 according as σ is even or odd; note that a **transposition** is an element τ in $S_n = \mathrm{Sym}(S)$ such that for some $i \neq j$ in S we have $\tau(i) = j$, $\tau(j) = i$, and $\tau(l) = l$ for all $l \in S \setminus \{i, j\}$. The set all even σ in S_n is a normal subgroup of S_n which we call the **alternating group** and denote it by A_n. We have $[S_n : A_n] = 1$ or 2 according as $n \leq 1$ or $n > 1$. For $n \geq 5$, A_n is simple and S_n is unsolvable. See (X1), (X2), and (E1) to (E6) of L1§11.

A group G is **commutative** or **abelian** if $xy = yx$ for all x, y in G. An **additive abelian group** is a commutative group in which the operation is written as a sum $x + y$ instead of a product, 0 is written for the identity, and $-x$ is written for the inverse of x. To contrast with this, a usual group, with the operation written as a product or multiplication, may be called **multiplicative**.

Deleting (iii) from the the definition of a group we get the definition of a **monoid**, and by deleting (ii) and (iii) we get the definition of a **semigroup**. The terms commutative or abelian semigroup, additive abelian semigroup, multiplicative semigroup, subsemigroup, and oversemigroup are obvious generalizations of the corresponding terms for groups; similarly for a monoid; note that the identity element of a submonoid is required to coincide with the identity group of the monoid. A semigroup H is **cancellative** means for any x, y, z in it we have the implication: $xy = xz$ and $yx = zx \Rightarrow y = z$. If H is a subsemigroup of a group G then clearly H is cancellative. As examples, $\mathbb{Z}$ is an additive abelian group, $\mathbb{N}$ is a submonoid of $\mathbb{Z}$, and $\mathbb{N}_+$ is a subsemigroup of $\mathbb{Z}$.

A **ring** R is an additive abelian group which is also a commutative multiplicative monoid such that the two operations are connected by the **distributive** laws saying that for all x, y, z in R we have $x(y + z) = xy + xz$ and $(y + z)x = yx + zx$; we call R a **null ring** if in it we have $1 = 0$, i.e., equivalently if $|R| = 1$. A **domain** is a nonnull ring whose nonzero elements form a cancellative multiplicative monoid. A **field** is a nonnull ring whose nonzero elements form a multiplicative group. If $R \subset S$ are rings under the same operations (and have the same zero and identity elements) then R is a subring of S or S is an overring of R. Similarly for subfields and overfields, and subdomains and overdomains.

As an additive abelian group, every subgroup I of a ring R is a normal subgroup and so we can form the factor group R/I; a typical member of R/I is a **residue class** (= additive coset) $a + I$ with $a \in R$. By an **ideal** in a ring R we mean an additive subgroup I of R such that for all $a \in R$ and $x \in I$ we have $ax \in I$, and when that is so we make R/I into a ring by defining $(a + I)(b + I) = (ab) + I$ for all a, b in R. We call R/I the **residue class ring** of R modulo I, and we call I a **maximal** ideal or a **prime** ideal in R according as R/I is a field or a domain; for more standard definitions of prime ideal and maximal ideal see (D1). Likewise, we call I a nonzero ideal or a nonunit ideal according as $I \neq \{0\}$ or $I \neq R$; note that every ideal contains the zero ideal $I = \{0\}$ and is contained in the **unit ideal** $I = R$; moreover, the zero ideal is prime $\Leftrightarrow R$ is a domain; likewise, I is the unit ideal $\Leftrightarrow R/I$ is the null ring. For any $a \in R$ we put $aR =$ the set of all multiples az of a with $z \in R$, and for any $W \subset R$ we put $WR =$ the set of all finite linear combinations $a_1 z_1 + \cdots + a_e z_e$ with $a_1, \ldots, a_e$ in W and $z_1, \ldots, z_e$ in R, and we note that these are ideals in R (by convention an empty sum is zero and hence $0 \in WR$); they are called ideals generated by a and W respectively; an ideal of the form aR is called a **principal ideal** in R and a called a generator of it; in case $W = \{w_1, w_2, \ldots\}$, we may denote WR by $(w_1, w_2, \ldots)R$ and call $w_1, w_2, \ldots$ its **generators**. The zero ideal, and more generally the (additive abelian) zero group may be denoted by 0 instead of $\{0\}$. It is easy to see that every ideal in $\mathbb{Z}$ is of the form $p\mathbb{Z}$ with $p \in \mathbb{N}$; moreover $p\mathbb{Z}$ is a maximal ideal in $\mathbb{Z}$ iff p is a prime number, and then $\mathbb{Z}/p\mathbb{Z}$ is the Galois field $\mathrm{GF}(p)$.

The set of all automorphisms of a group G is clearly a subgroup of $\mathrm{Sym}(G)$ and we denote it by $\mathrm{Aut}(G)$. For rings S and T, a (ring) homomorphism $\phi : S \to T$ is a homomorphism of additive abelian groups such that $\phi(1) = 1$ and $\phi(xy) = \phi(x)\phi(y)$ for all x, y in S; now $\ker(\phi)$ is an ideal in S, and the definitions of monomorphism, epimorphism, isomorphism, and automorphism carry over from the group case. Conversely, if $\phi : S \to S/I$ is the canonical ring epimorphism where I is an ideal in a ring S then $\ker(\phi) = I$. The set of all ring automorphisms of a ring S is clearly a subgroup of $\mathrm{Sym}(S)$ and we denote it by $\mathrm{Aut}(S)$; momentarily letting this Aut be written as Ring-Aut(S) and writing Group-Aut(S) for the automorphism group of the additive abelian group S we clearly have Ring-Aut$(S) \leq$ Group-Aut$(S) \leq \mathrm{Sym}(S)$; it is usually clear from the context which automorphism group is being considered. Finally for a subring R of a ring S, by an R-automorphism of S we mean an automorphism ϕ of S which leaves R elementwise fixed, i.e., $\phi(x) = x$ for all $x \in R$; these form a subgroup of $\mathrm{Aut}(S)$ and we denote it by $\mathrm{Aut}_R(S)$.

Dropping the commutativity of multiplication in a field (resp: ring, domain) we get the notion of a **skew-field** (resp: skew-ring, skew-domain). In the definitions of fields and rings we have spelled out two distributive laws, although they follow from each other, exactly because in case of skew-fields and skew-rings they do not. Moreover, in the definition of a field, by requiring the left-distributive law

$x(y+z) = xy+xz$, but not requiring the right-distributive law $(y+z)x = yx+zx$, we get the notion of a near-field. The concepts of subskew-field, overskew-field, and so on, are defined in an obvious manner. Also the above definitions of homomorphism, monomorphism, and so on, have obvious generalizations to semigroups, monoids, and so on.

See section L1§5 on **Modules and Vector Spaces** for the definitions of: a module V over a ring R, scalar multiplication in V, linearly independent and linearly dependent elements of V, submodule of V, submodule generated by a set of elements or a subset, generators of a submodule, **finitely generated** submodule, **basis** of a submodule, R-homomorphism (or R-linear map) $V \to V^*$ where V^* is another R-module (= module over R), R-monomorphism, R-isomorphism, left and right modules over a skew-ring, vector space over a field K (= module over K), subspace (= submodule of a vector space), vectors (= elements of a vector space), the dimension of a vector space V over a field K denoted by $\dim_K V$, and so on.

An overring of a ring R is an R-module in an obvious sense; thus R is an R-module, and an ideal in R is exactly an R-submodule of R. In particular, an overring L of a field K is a K-vector space, and we put $[L : K] = \dim_K L$. This applies to an overfield L of K, and then we call $[L : K]$ the **field degree** of L/K, i.e., of L over K. For a ring or skew-ring R, by R^+ we denote the **underlying additive abelian group** of R, and by $R^\times$ we denote the set of all **nonzero elements** of R; note that $K^\times$ is a multiplicative group in case of a field or skew-field K. Extending this notation to a vector space V we denote the set of all nonzero elements in it by $V^\times$. For an overfield L of a field K, the symbol $[L : K]$ usually denotes the field degree rather than the index of K^+ in L^+.

See section L1§6 on **Polynomials and Rational Functions** for the definitions of: **indeterminate** (= variable) over a ring, univariate polynomial ring $R[X]$ over a ring R, degree of a polynomial, monic polynomial, nonconstant polynomial, Cauchy multiplication rule, univariate rational function field $K(X)$ over a field K, derivative, quotient rule, multivariate polynomial ring $R[X_1, \ldots, X_m]$, degree and partial derivatives of a multivariate polynomial, multivariate rational function field $K(X_1, \ldots, X_m)$, substitution epimorphism $R[X_1, \ldots, X_m] \to R[x_1, \ldots, x_m] =$ the ring generated over R by elements $x_1, \ldots, x_m$ in an overring of R, **algebraic** or **transcendental** element over R, **algebraic set of elements over** R, **algebraically dependent** or **algebraically independent** sets of elements over R, field $K(x_1, \ldots, x_m)$ generated over a field K by elements $x_1, \ldots, x_m$ in an overfield of K, **transcendence basis** of an overfield L of K, and the number of elements in it called the **transcendence degree** of L/K (= L over K) denoted by $\mathrm{trdeg}_K L$.

See section L1§7 on **Euclidean Domains and Principal Ideal Domains** for the definition of: the **group** $U(R)$ **of all units** in a ring R (or two-sided units in a skew-ring R), **associates** and **irreducible elements** in a ring, UFD = **unique factorization domain**, PID = principal ideal domain, euclidean function, ED = euclidean domain, **special ED**, **quasispecial subset**, and **quasispecial** ED.

See section L1§8 on **Root Fields and Splitting Fields** for the definitions of: R-homomorphism (resp: R-monomorphism, R-epimorphism, R-isomorphism) between overrings of a ring R, root field of a univariate irreducible polynomial over a field K, splitting field of a nonconstant univariate polynomial f over K denoted by $\mathrm{SF}(f, K)$, **separable polynomial** f, the Galois group $\mathrm{Gal}(L^*, K)$ of the splitting field $L^* = \mathrm{SF}(f, K)$ of a nonconstant univariate separable polynomial f over K, the natural isomorphism $\mathrm{Gal}(L^*, K) \to \mathrm{Gal}(f, K) =$ the **Galois group** of f over K which is the group of all **relation preserving** permutations of the roots of f, and for any $H \leq \mathrm{Gal}(L^*, K)$ the **fixed field** $\mathrm{fix}_{L^*}(H)$ of H which is a field L between K and L^*.

In (T1) we state the theorem which includes the assertion that $H \mapsto L = \mathrm{fix}_{L^*}(H)$ gives an inclusion reversing bijection of the set of all subgroups H of $\mathrm{Gal}(L^*, K)$ onto the set of all fields L between K and L and the inverse bijection is given by $L \to \mathrm{Gal}(L^*, L)$. After stating other relevant theorems in (T2) to (T5), in (T6) we deduce Galois' Unsolvability Theorem which says that if a field k contains $n!$ distinct $n!$-th roots of 1 then, for $n \geq 5$, the generic n-th degree polynomial over $K = k(X_1, \ldots, X_n)$ cannot be solved by radicals.

In (D1) we define **divisibility** $x|y$ and prime elements in a ring. In (R2) to (R4) we define: gcd and GCD (= greatest common divisor), and coprime and relatively prime elements. In (D2) to (D4) we define: **characteristic** of a ring R denoted by $\mathrm{ch}(R)$, **minimal polynomial** of an algebraic element over a field, lcm and LCM (= least common multiple), and reduced form. In (R7) we define the (relative) **algebraic closure** of a field in an overfield. In (X2) we define the decomposition of a permutation into disjoint cycles. In (E1) we define the **modified discriminant**. In (N1) we define **derivations** and in (E11) and (E12) we describe their properties.

Proper Containment. By a **proper subset** A of a set B we mean $A \subset B$ with $A \neq B$. We indicate this by writing $A \subsetneqq B$ or $B \supsetneqq A$, and by saying that A is properly contained in B or B properly contains A. This agrees with the above definition of proper subgroup, and with the definitions in L4§8 and L5§5(Q16).

§2: SUMMARY OF LECTURE L2 ON CURVES AND SURFACES

In L2§1 we introduce the concepts of cartesian product, partial order, linear order, well order, cardinals, and ordinals. The **cartesian product** $S_1 \times \cdots \times S_m$ of sets $S_1, \ldots, S_m$ is the set of all m-tuples $(\gamma_1, \ldots, \gamma_m)$ with $\gamma_i \in S_i$. If $S_1 = \cdots = S_m = S$ then we write S^m for $S_1 \times \cdots \times S_m$. Also we define concepts of **algebraically closed** field, an (absolute) algebraic closure of a field, and the (relative) algebraic closure of a field in an overfield. To prove the existence of algebraic closures, we introduce the Axiom of Choice, Zorn's Lemma, and Well Ordering. These are based on the concepts of **posets** (= partially ordered sets), **losets** (= ordered sets = linearly ordered sets), and **wosets** (= well ordered sets). Cardinals are equivalence classes of sets under bijections, and ordinals are equivalence classes of well ordered sets under order preserving bijections. The **power set** $\widehat{\mathcal{P}}(T)$ is the set of all subsets of a set T, and the **restricted power set** $\widehat{\mathcal{P}}^\times(T) = \widehat{\mathcal{P}}(T) \setminus \{\emptyset\}$.

In L2§2 on **Power Series and Meromorphic Series** we introduce the notation $\mathrm{QF}(E)$ for the **quotient field** of a domain E. The quotient field of the power series ring $K[[Z_1, \ldots, Z_d]]$ over a field K is the meromorphic series field $K((Z_1, \ldots, Z_d))$. We define the **orders** and derivatives of meromorphic series. We use the geometric series identity $(1 - X)(1 + X + X^2 + \ldots) = 1$ to show that the group $U(S)$ of all units in the ring of power series $S = R[[Z_1, \ldots, Z_d]]$ over a ring R are exactly those power series $Q(Z_1, \ldots, Z_d)$ whose constant term $Q(0, \ldots, 0)$ belongs to $U(R)$.

In L2§3 on **Valuations** we generalize the concept of the order of a meromorphic series by defining a valuation of a field K to be a map $v : K \to G \cup \{\infty\}$ where G is an ordered abelian group G such that for all x, y in K we have:

    (1) $v(xy) = v(x) + v(y)$;
    (2) $v(x + y) \geq \min(v(x), v(y))$;
    (3) $v(x) = \infty \Leftrightarrow x = 0$.

We call G the assigned value group of v, and by the **value group** of v we mean the subgroup of G given by $G_v = \{v(x) : x \in K^\times\}$. By the **valuation ring** of v we mean the ring $R_v = \{x \in K : v(x) \geq 0\}$ and we note that this is clearly a subdomain (= subring which is a domain) of K which is its quotient field. We say that v is trivial over a subfield k of K, or that v is a valuation of K/k, to mean that $x \in k^\times \Rightarrow v(x) = 0$. For instance ord gives a valuation of the meromorphic series field $K((Z_1, \ldots, Z_d))$ over any field K which is trivial over K and whose value group is $\mathbb{Z}$. Here by an **ordered abelian group** we mean an additive abelian group G which is also an ordered set such that for all x, y, x', y' in G we have: $x \leq y$ and $x' \leq y' \Rightarrow x + x' \leq y + y'$. For instance G could be the set $\mathbb{R}^{[d]}$ of lexicographically ordered d-tuples of real numbers $r = (r_1, \ldots, r_d), s = (s_1, \ldots, s_d), \ldots$, where **lexicographic** order means: $r \leq s \Leftrightarrow$ either $r_i = s_i$ for $1 \leq i \leq d$, or for some j with $1 \leq j \leq d$ we have $r_i = s_i$ for $1 \leq i < j$ and $r_j < s_j$. Or G could be a subgroup of $\mathbb{R}^{[d]}$ such as $G = \mathbb{Z}$ or $\mathbb{Q}$ or $\mathbb{R}$.

The field $K((X))$ is generalized by taking any field K and any ordered abelian group G and considering the field $K((X))_G = \{A \in K^G : \mathrm{Supp}(A) \text{ is well ordered}\}$ where K^G is the set of all maps $G \to K$ and where the **support** of $A \in K^G$, denoted by $\mathrm{Supp}(A)$, is defined by putting $\mathrm{Supp}(A) = \{g \in G : A(g) \neq 0\}$. For any $A \in K((X))_G$ we define $\mathrm{ord}(A) = \min \mathrm{Supp}(A)$ if $A \neq 0$ and $\mathrm{ord}(A) = \infty$ if $A = 0$, and we note that now **ord** is a valuation of $K((X))_G/K$ with value group G.

In (T1) we state Newton's 1660 Theorem on the algebraic closure of $k((X))$ when k is an algebraically closed field with $\mathrm{ch}(k) = 0$, and in (T2) we formulate its generalization for $\mathrm{ch}(k) = p > 0$.

In L2§5 on **Zorn's Lemma and Well Ordering** we prove that these two are equivalent and they are equivalent to the Axiom of Choice. Then by using Zorn's Lemma we prove the: (R5) existence of maximal ideals, (R6) existence of algebraic closure, (R7) uniqueness of algebraic closure, existence of vector space bases as well as transcendence bases, (R9) linear order on cardinals, (R10) well order on ordinals. Let us state the contents of Remarks (R5) to (R8) in greater detail thus.

(R5) Any nonunit ideal in a ring R is contained in some maximal ideal in R.

(R6) Any field K has an algebraic closure, i.e., an algebraically closed field $\overline{K}$ which is algebraic over K.

(R7) The $\overline{K}$ of (R6) is unique in the sense that given any field isomorphism $\phi : K \to L$ and any algebraic closure $\overline{L}$ of L there exists an isomorphism $\overline{\phi} : \overline{K} \to \overline{L}$ such that $\overline{\phi}(x) = \phi(x)$ for all $x \in K$.

(R8) Let K be a field and let L be either a vector space over K or an overfield of K. Let W be a subset of L. In the vector space case (resp: overfield case): let us call W independent if every finite subset of W is linearly (resp: algebraically) independent over K, and let us call W generating if L coincides with KW (resp: if L is algebraic over $K(W)$); in both cases call W a basis if it is independent and generating; given any other subset U of L, let us say that U is dependent on W if every $u \in U$ belongs to KW (resp: is algebraic over $K(W)$). Then we prove the facts (W7) and (W8) stated below.

$$\begin{cases} W \text{ is a basis} & \Leftrightarrow \text{ it is a minimal generating set} \\ & \Leftrightarrow \text{ it is a maximal independent set} \end{cases} \quad \text{(W7)}$$

where the minimality means that W is a generating set but there is no generating set W' with $W' \subset W$ and $W' \neq W$, and the maximality means that W is an independent set but there is no independent set W' with $W \subset W'$ and $W \neq W'$.

$$W \text{ is generating } \Rightarrow \text{ there exists a basis } W' \text{ with } W' \subset W. \quad \text{(W8)}$$

In L2§7 on **Definitions and Exercises** we define: real and complex numbers, ordered fields, torsion subgroups and divisible groups, rational and real completions, Cauchy sequences and Dedekind cuts, rational and real ranks. In (D7) we define approximate roots of polynomials and in (E15) to (E17) list their basic properties.

More About Fields. In L3§12(E6) we show that if $G(Z_1, \ldots, Z_N)$ is any given polynomial over an infinite field k such that $G(a_1, \ldots, a_N) = 0$ for all $a_1, \ldots, a_N$ in k then $G(Z_1, \ldots, Z_N) = 0$. In L4§10(E1) we elucidate the above Remark (R6). In L5§5(Q27)(C41) we give a criterion for the minimal polynomial of an algebraic element to be separable. In L5§5(Q29)(T109) we discuss the linear disjointness of fields defined in the preamble of (Q29); we discuss this further in L5§6(E22). In L5§5(Q29)(C48) we give a dimension formula for subspaces of a vector spaces. In L5§5(Q32) we discuss separable and inseparable polynomials, separable and inseparable elements, purely inseparable elements, separable and inseparable field extensions, purely inseparable field extensions, finite fields, perfect fields, and the existence of primitive elements for finite separable algebraic field extensions. For any power q of any prime number p, in L5§5(Q32) we show that up to isomorphism there is a unique field of q elements; this is called the Galois field $\mathrm{GF}(q)$.

§3: SUMMARY OF LECTURE L3 ON TANGENTS AND POLARS

In L3§1 we introduce the concepts of rectangular matrices with entries in a ring, and groups of square matrices whose determinants are units in the ring of entries.

In L3§2 we study the **geometry** of quadrics and their pole-polar properties. In L3§3 to L3§5 this is expanded into the **geometry** of hypersurfaces, homogeneous coordinates, projective spaces, tangents, and singularities.

In L3§6 we prove Basic Hensel's Lemma which says that if a monic polynomial in Y whose coefficients are power series in X factors into coprime factors after putting $X = 0$ then it factors before putting $X = 0$. Using this and using the completing the square method of solving quadratic equations conceived by the fifth century Indian mathematician Shreedharacharya, in L3§7 we establish Newton's Theorem on fractional power series expansion which was stated in L2§3(T1) as a theorem on the algebraic closure of $k((X))$.

Newton's Theorem motivates the idea of **integral dependence** which we take up in L3§7 to L3§9. An element t in an overring S of a ring R is said to be integral over R if t satisfies a monic polynomial equation over R; a univariate polynomial over R is **monic** means its **leading coefficient**, i.e., the coefficient of the highest degree term in it, is 1. The set of all elements in S which are integral over R is a subring R' of S and it is called the **integral closure** of R in S; if $R' = R$ then we say that R is **integrally closed** in S; if R is integrally closed in its total quotient ring $QR(R)$, as defined in L4§7, then R is said to be **normal**. In particular a domain R is normal means it is integrally closed in its quotient field K. A domain R is **overnormal** means for every pair of monic polynomials in an indeterminate Y over K we have: $A(Y)B(Y) \in R[Y] \Rightarrow A(Y) \in R[Y]$ and $B(Y) \in R[Y]$. A domain R is **intval** means it is the intersection of a nonempty family of valuation rings of K, i.e., valuation rings of valuations of K. It turns out that for any domain we have: UFD $\Rightarrow$ intval $\Leftrightarrow$ overnormal $\Leftrightarrow$ normal. The GCD of all the coefficients of a nonzero polynomial over a UFD is its **content**. Gauss Lemma says that the content of a product is the product of their contents. This yields the fact that a finite variable polynomial ring over a UFD is a UFD. The proofs of some of the above assertions may be found in L4§10(E2), L4§10(E3), and L4§12(R7).

In L3§11 we deduce the multivariable version of Hensel's Lemma as a consequence of the Abstract WPT = Weierstrass Preparation Theorem and its companion the Abstract WDT = Weierstrass Division Theorem. These deal with the univariate power series ring $S = R[[Y]]$ over a complete Hausdorff quasilocal ring R. By a **quasilocal ring** we mean a ring R with a unique maximal ideal $M(R)$; note that a ring R is quasilocal iff nonunits in it form an ideal (which then equals $M(R)$); also note that the valuation ring R_v of any valuation v of a field K is a quasilocal ring with $M(R_v) = \{x \in K : v(x) > 0\}$. The **Weierstrass degree** of any $f \in S$ is defined by $\mathrm{wideg}(f) = \mathrm{ord}(F)$ where $F \in (R/M(R))[[Y]]$ is obtained by applying the residue class map $R \to R/M(R)$ to the coefficients of f. Now WPT says that every $f \in S$ with $\mathrm{wideg}(f) = d \in \mathbb{N}$ can uniquely be written as $f = \delta f^*$ where δ is a unit in S and f^* is a **distinguished polynomial** of degree d. i.e., $f^* = Y^d + f_1^* Y^{d-1} + \cdots + f_d^*$ with $f_1^*, \ldots, f_d^*$ in $M(R)$, and WDT says that any $g \in S$ can be uniquely written as $g = qf + r$ where $q \in S$ and $r \in R[Y]$ with $\deg(r) < d$.

The above abstract versions of WPT and WDT are converted into their concrete versions by taking R to be the m-variable power series ring $k[[X_1, \ldots, X_m]]$ over a field k. To apply these to the n-variable power series ring $S = k[[X_1, \ldots, X_n]]$ over a field k we need to ensure that wideg of a nonzero element f of S relative to X_n is finite so that we can take $Y = X_n$ and $m = n - 1$. In case the field k is infinite, we do this by rotating coordinates, i.e., by making a k-linear automorphism as explained in L3§12(D4). Alternatively, without assuming k to be infinite, again as explained in L3§12(D4), we achieve the same thing by making a more general polynomial automorphism.

About the polynomial ring $R[X_1, \ldots, X_m]$ and the power series ring $R[[X_1, \ldots, X_m]]$ in indeterminates $X_1, \ldots, X_m$ over a ring R with positive integer m let us note that, according the famous Hilbert Basis Theorem as proved in (T3) and (T4) of L4§3, if the ring R is noetherian then the rings $R[X_1, \ldots, X_m]$ and $R[[X_1, \ldots, X_m]]$ are also noetherian, where we note that in honor of Emmy Noether a ring R in which every ideal is finitely generated is said to be **noetherian**. It follows that the power series ring $k[[X_1, \ldots, X_m]]$ over a field k is a **local ring**, by which we mean a noetherian quasilocal ring. In L5§5(Q4) we prove the famous Krull Intersection Theorem which says that every local ring R is Hausdorff, i.e., in it we have $\cap_{n \in \mathbb{N}} M(R)^n = 0$.

Thus in a local ring R, for every $x \in R^\times =$ the set of all nonzero elements of R we have $\mathrm{ord}_R x \in \mathbb{N}$, where for any x in any quasilocal ring R we have put $\mathrm{ord}_R x = \max\{n \in \mathbb{N} : x \in M(R)^n\}$ with the understanding that if $x \in \cap_{n \in \mathbb{N}} M(R)^n$ then $\mathrm{ord}_R x = \infty$. Note that if R is the power series ring $k[[X_1, \ldots, X_m]]$ over a field k then for any

$$f = f(X_1, \ldots, X_m) = \sum_{i_1, \ldots, i_m} f_{i_1 \ldots i_m} X_1^{i_1} \ldots X_m^{i_m} \quad \text{with} \quad f_{i_1 \ldots i_m} \in k$$

we have

$$\mathrm{ord}_R f = \mathrm{ord}(f) = \min\{i_1 + \cdots + i_m : (i_1, \ldots, i_m) \in \mathrm{Supp}(f)\}$$

where $\mathrm{Supp}(f) = \{(i_1, \ldots, i_m) \in \mathbb{N}^m : f_{i_1 \ldots i_m} \neq 0\}$.

In (D1) to (D3) and (E1) to (E4) of L3§12 we discuss the theory of matrices including: **ranks, minors, cofactors, transposes, adjoints,** and **linear maps**.

In (D5) and (E12) to (E14) of L3§12 we generalize Hensel's Lemma by extending the notions of **ord, completeness, and Hausdorffness** from quasilocal rings to more general situations in the following manner.

Let I be an ideal in a ring R and let V be an R-module. V is I-**Hausdorff,** or

Hausdorff relative to I, means $\cap_{i\in\mathbb{N}}I^iV = 0$, where for any ideal J in R, by JV we denote the submodule of V given by $JV = \{\sum j_iv_i : j_i \in J \text{ and } v_i \in V\}$. For any $h \in V$ we define the (R,I)-**order** of h by putting $\text{ord}_{(R,I)}h = \max\{i \in \mathbb{N} : h \in I^iV\}$ and we note that then V is I-Hausdorff means for any $h \in V$ we have: $\text{ord}_{(R,I)}h = \infty \Leftrightarrow h = 0$. A sequence $x = (x_i)_{1\le i<\infty}$ in V is I-**Cauchy**, or Cauchy relative to I, means for every $E \in \mathbb{N}$ there exists $N_E \in \mathbb{N}$ such that for all $i > N_E$ and $j > N_E$ we have $x_i - x_j \in I^EV$. The sequence x is I-**convergent**, or convergent relative to I, means it converges to an I-limit $\xi \in V$, i.e., for every $E \in \mathbb{N}$ there exists $N_E \in \mathbb{N}$ such that for all $i > N_E$ we have $\xi - x_i \in I^EV$; we indicate this by some standard notation such as $x_i \to \xi$ as $i \to \infty$ or $\lim_{i\to\infty}x_i = \xi$. If V is I-Hausdorff then a limit when it exists is unique. V is I-**complete** means every I-Cauchy sequence in it has a limit in it. If R is quasilocal with $I = M(R)$ and $V = R$ then the adjective "relative to I" may be dropped from the above definitions.

§4: SUMMARY OF LECTURE L4 ON VARIETIES AND MODELS

In L4§1 we open up the lecture with Sylvester's 1840 theory of the **resultant** of two univariate polynomials f, g of degrees n, m which by eliminating the variable Y produces an $n+m$ by $n+m$ matrix $\text{Resmat}_Y(f, g)$ in their coefficients, the vanishing of whose determinant $\text{Res}_Y(f, g)$ gives a condition for them to have a common solution. In turn the vanishing of the **discriminant** $\text{Disc}_Y(f) = \text{Res}_Y(f, f_Y)$ gives a condition for f to have a multiple root.

Having already cited the Hilbert Basis Theorem proved in L4§3 which follows the general chit-chat about curves, surfaces, and varieties in L4§2, let us now describe L4§5 on ideals and modules consisting of Observations (O7) to (O27). Before coming to (O7), in the preamble of L4§5 we talk about intersections and sums of submodules of a module V over a ring, and make the definitions of the colon (or quotient) modules $(C : B)_V$ and $(C : a)_V$, where C is a submodule of V with $B \subset R$ and $a \in R$, by putting $(C : a)_V = \{v \in V : av \in C\}$ and $(C : B)_V = \cap_{b\in B}(C : a)_V$. In case of ideals we also consider their products, and we define the colon (or quotient) ideals $(C : B)_R$ and $(C : a)_R$, where C is a submodule of V with $B \subset V$ and $a \in V$, by putting $(C : a)_R = \{r \in R : rv \in C\}$ and $(C : B)_R = \cap_{b\in B}(C : a)_R$. Moreover, for a submodule U of V and an ideal I in R, we define their radicals by putting $\text{rad}_VU = \{r \in R : r^eV \subset U \text{ for some } e \in \mathbb{N}_+\}$ and $\text{rad}_RI = \{r \in R : r^e \in I \text{ for some } e \in \mathbb{N}_+\}$. Also we define zerodivisors and nilpotents. Finally we define **primary** and **irreducible** ideals as well as submodules.

In (O7), (O8), (O9), (O12), (O13), and (O18) we discuss primary ideals and submodules as well as **annihilators** and **colons** and relate them to **radicals**. In (O10) and (O11) we discuss basic isomorphism theorems.

In (O14) we define **noetherian modules** in terms of the three equivalent conditions of NNC (= noetherian condition saying that every submodule is finitely generated) and ACC (= ascending chain condition which says that every ascending chain of submodules stops) and MXC (= maximal condition which says that every nonempty family of submodules has a maximal element). In (O15) we show that every submodule U of a noetherian module V over a ring R has an **irredundant primary decomposition** $U = U_1 \cap \cdots \cap U_h$ where $U_1, \ldots, U_h$ are a finite number of primary submodules of V; here irredundant means none of the U_i can be deleted and the radicals $P_1, \ldots, P_h$ of $U_1, \ldots, U_h$ are distinct prime ideals in R.

In (O16) we introduce the **spectrum spec**(R) of a ring R as the set of all prime ideals in R. The set of all maximal ideals in R is denoted by **mspec**(R). The set of all members of spec(R) which contain an ideal J of R is called the **spectral variety** of J and is denoted by **vspec**$_R J$. The sets of all **maximal** and **minimal** members of vspec$_R J$ are denoted by **mvspec**$_R J$ and **nvspec**$_R J$ respectively. In L5§5(Q7)(C11) we also put **nspec**$(R) = $ nvspec$_R 0$. The **set of all spectral varieties** in R is denoted by **svt**(R). The **spectral ideal** of any subset W of spec(R) is defined by putting **ispec**$_R W = \cap_{P \in W} P$. The **set of all radical ideals** in R, i.e., the set of all ideals in R which are their own radicals, is denoted by **rd**(R). In L5§5(Q21) we also put **id**(R) (resp: **nid**(R)) = the set of all (resp: all nonunit) ideals in R. If R is a finite variable polynomial ring over a field then ispec$_R$ gives a bijection svt$(R) \to$ rd(R) whose inverse is given by vspec$_R$; this is part (T22.8) of the Hilbert Nullstellensatz L4§8(T22) proved in L5§5(Q11).

In (O16) and (O17) we define the **associator** ass$_R V$ and the **tight associator** tass$_R V$ of an R-module V as certain subsets of spec(R), and show that for the above irredundant primary decomposition $U = U_1 \cap \cdots \cap U_h$ of the submodule U of the noetherian R-module V we have ass$_R(V/U) = $ tass$_R(V/U) = \{P_1, \ldots, P_h\}$. Members of ass$_R(V/U)$ are called **associated primes** of U in V.

In (O19) we use **bracketed colon** to prove partial uniqueness of the primary components U_i in the above decomposition. A **multiplicative set** in a ring R is a subset S of R with $1 \in S$ such that S contains the product of every pair of elements in it. For any submodule U of an R-module V we put $[U : S]_V = \cup_{s \in S}(U : s)_V$ and call this the **isolated S-component** of U in V; the uniqueness follows by showing that $[U : S]_V = \cap_{P_i \cap S = \emptyset} U_i$. For any prime ideals $Q_1, \ldots, Q_n$ in R with $n \in \mathbb{N}_+$ we put $[U : (Q_1, \ldots, Q_n)]_V = [U : S]_V$ where $S = \cap_{1 \leq i \leq n}(R \setminus Q_i)$ and call this the isolated $(Q_1, \ldots, Q_n)$-component of U in V, and we note that now $\{1 \leq i \leq h : P_i \cap S = \emptyset\} = \{1 \leq i \leq h : P_i \subset Q_j$ for $1 \leq j \leq n\}$.

In (O20) we generalize the idea of primary decomposition to **quasiprimary** decomposition without assuming the noetherian condition. We also introduce the notation $Z_R(V) = $ **the set of all zerodivisors** of a module V over a ring R, and $S_R(V) = $ **the multiplicative set** $R \setminus Z_R(V)$ in R.

In (O21) we study the properties of the **length** $\ell_R(V)$ of a module V over a ring R which we define to be the maximum length $n \in \mathbb{N}$ of a finite sequence $0 = V_0 \subsetneqq V_1 \subset \cdots \subsetneqq V_n = V$ of submodules of V which we call a **normal series** of V of length n. When there does not exist a bound for the lengths of such sequences then we put $\ell_R(V) = \infty$. By a **simple module** we mean a module of length 1.

In (O22) we define **heights** and **depths** of ideals and **dimensions** of rings. By the dimension $\dim(R)$ of a ring R we mean the maximum length $n \in \mathbb{N}$ of a finite sequence (δ) $P_0 \subsetneq P_1 \subsetneq \cdots \subsetneq P_n$ of prime ideals in R. We call such a sequence a prime sequence in R of length n. When there does not exist a bound for such sequences then we put $\dim(R) = \infty$ provided R is not the null ring; if R is the null ring then we put $\dim(R) = -\infty$. By the height $\mathrm{ht}_R P$ (resp: depth $\mathrm{dpt}_R P$) of a prime ideal P in R we mean the maximum length of a prime chain (δ) in R with $P = P_n$ (resp: $P = P_0$); again the maximum can be ∞. We define the height $\mathrm{ht}_R J$ and the depth $\mathrm{dpt}_R J$ of a nonunit ideal J in R by putting $\mathrm{ht}_R J = \min\{\mathrm{ht}_R P : P \in \mathrm{vspec}_R J\} \in \mathbb{N} \cup \{\infty\}$ and $\mathrm{dpt}_R J = \max\{\mathrm{dpt}_R P : P \in \mathrm{vspec}_R J\} \in \mathbb{N} \cup \{\infty\}$. Let us complete the definitions of height and depth by putting $\mathrm{ht}_R R = -\infty = \mathrm{dpt}_R R$.

In (O25), analogous to noetherian modules, we define **artinian** modules in terms of the two equivalent conditions of DCC (= descending chain condition which says that every descending chain of submodules stops) and MNC (= minimal condition which says that every nonempty family of submodules has a minimal element). A ring is artinian if it is artinian as a module over itself. We show that a module has finite length iff it is artinian as well as noetherian. We show that a ring is artinian iff it is either the null ring or a zero-dimensional noetherian ring. We show that a finitely generated module over a noetherian (resp: artinian) ring is noetherian (resp: artinian).

In (O26) we define ideals A, B in a ring R to be **comaximal** to mean that $A + B = R$. We show that if $A_1, \ldots, A_n$ are a finite number of pairwise comaximal ideals in R then $\prod_{1 \le i \le n} A_i = \cap_{1 \le i \le n} A_i$, and given any $x_i \in R$ for $1 \le i \le n$ there exists $x \in R$ with $x \equiv x_i \bmod A_i$ for $1 \le i \le n$, i.e., $x - x_i \in A_i$ for $1 \le i \le n$.

In (O27) we define an **affine ring** over a ring R to be a finitely generated ring extension of R.

In L4§6 we recapitulate primary and quasiprimary decompositions.

In L4§7 we introduce the **localization** R_S of a ring R at a multiplicative set S in R, together with a ring homomorphism (called the canonical homomorphism or the localization map) $\phi : R \to R_S$ thus. We define R_S to be the set of all equivalence classes of pairs $(u, v) \in R \times S$ under the equivalence relation given by: $(u, v) \sim (u', v') \Leftrightarrow v''(uv' - u'v) = 0$ for some $v'' \in S$. The equivalence class containing (u, v) is denoted by u/v or $\frac{u}{v}$, and we add and multiply the equivalence classes by the rules $\frac{u_1}{v_1} + \frac{u_2}{v_2} = \frac{u_1 v_2 + u_2 v_1}{v_1 v_2}$ and $\frac{u_1}{v_1} \times \frac{u_2}{v_2} = \frac{u_1 u_2}{v_1 v_2}$. Also we "send" any $u \in R$ to $u/1 \in R_S$. This makes R_S into a ring and $u \mapsto u/1$ gives the canonical ring homomorphism $\phi : R \to R_S$. Clearly $\phi(S) \subset U(R_S)$ and $\ker(\phi) = [0 : S]_R$. The localization of R at $S_R(R)$ is the total quotient ring $\mathrm{QR}(R)$ of R; clearly $[0 : S_R(R)] = 0$ and hence we may and we do regard $\mathrm{QR}(R)$ to be an overring of R. Now: $\ker(\phi) = 0 \Leftrightarrow S \subset S_R(R)$. If $S \subset S_R(R)$ then there is a unique R-injection of R_S into $\mathrm{QR}(R)$; we call its image the localization of R at S in $\mathrm{QR}(R)$ and continue to denote it by R_S; let us call this the GOOD case; thus in the good case R_S is an overring of R. For any prime ideal P in R we let R_P stand for $R_{R \setminus P}$ and we call it the **localization** of R at P.

Some basic properties of localization are proved in Theorems (T9) to (T17) of L4§7 including L4§7.1. In particular, in (T12) we prove the salient fact which says that: $I \mapsto J = \phi(I)R_S$ gives a bijection of the set of all ideals I in R for which $[I : S]_S = I$ onto the set of of all ideals in R_S and its inverse is given by $J \mapsto \phi^{-1}(J)$; this bijection commutes with the ideal theoretic operations of radicals, intersections and quotients.

The **geometry** started in §§2-5 of L3 is picked up again in §§8-9 of L4. Common solutions of a finite number of polynomial equations in a finite number of variables with coefficients in a field form an **affine algebraic variety**. Some properties of these varieties are stated in Theorems (T19) to (T22) of L4§8 and proved in (Q10), (Q11), and (Q32) of L5§5. In L4§8.1 we relate these varieties to the spectral varieties discussed in the above item (O16) and then in L4§8.2, via localization, we relate them to **modelic specs** which are collections of local rings. Now a simple point is reincarnated as a **regular local ring** which means a local R whose **embedding dimension** emdim(R) equals its dimension, where emdim(R) is defined to be the smallest number of generators of $M(R)$. In L4§8.3, L4§9, L4§9.1, and L4§9.2 we formulate some properties of modelic specs as Theorems (T23) to (T30) which are proved in (R7) to (R9) of L4§12 together with (Q7), (Q14), and (Q17) of L5§5. In L4§9.2 we show how modelic specs give rise to **modelic blowups** and in L4§9.3 we use them for simplifying singularities.

In L4§11 we start a new column called Problems by doing which the student may get "mild satisfaction or Ph.D. thesis or fame."

In (R1) to (R5) of L4§12 we deal with Laplace development of determinants, block matrices, and Cramer's rule for solving linear equations.

In (R6) of L4§12 we concoct a resultant theory which makes it valid over rings with zerodivisors. This is done by converting the product rule for polynomials into matrix multiplication and expanding this into a product formula for the resultant matrix.

§5: SUMMARY OF LECTURE L5 ON PROJECTIVE VARIETIES

In L5§1 we define the direct sums of modules and related objects. A module V over a ring R is an **internal direct sum** of a family $(V_i)_{l \in I}$ of submodules if it is the sum $V = \sum_{i \in I} V_i$ and the sum is direct, i.e., for every finite subset I' of the indexing set I we have: $\sum_{i \in I'} v_i = 0$ with $v_i \in V_i$ for all $i \in I' \Rightarrow v_i = 0$ for all $i \in I'$. The fact that the sum is direct is expressed by writing $V = \sum_{i \in I} V_i$ (internal direct sum) or $V = \bigoplus_{i \in I} V_i$ (internal). In case I is a finite set, say $I = \{1, 2, \ldots, n\}$, we may write $V = V_1 \oplus V_2 \oplus \cdots \oplus V_n$ (internal).

The cartesian product $W = U_1 \times \cdots \times U_n$ of a finite number of R-module $U_1, \ldots, U_n$ is converted into an R-module by componentwise addition and scalar multiplication, and then we call W the **direct sum** of $U_1, \ldots, U_n$ and indicate this by writing $W = U_1 \oplus U_2 \oplus \cdots \oplus U_n$; to distinguish this from internal direct sum we may call it **external direct sum**. If $U_1 = \cdots = U_n = U$ then, as in the case of cartesian products, we put $W = U^n$ and call it the (module theoretic) direct sum of n copies of U. More generally, the cartesian product of a family $(U_i)_{i \in I}$ of sets U_i, where I is any indexing set (which need not be finite), is defined by putting $\prod_{i \in I} U_i =$ the set of all maps $\phi : I \to \bigcup_{i \in I} U_i$ such that $\phi(i) \in U_i$ for all $i \in I$. In case the U_i are R-modules, this becomes an R-module, called the **direct product**, by defining componentwise addition and scalar multiplication. The set of all $\phi \in \prod_{i \in I} U_i$ whose **support** $\text{supp}(\phi) = \{i \in I : \phi(i) \neq 0\}$ is finite is called the direct sum of the family and is denoted by $\bigoplus_{i \in I} U_i$; note that this is a submodule of the direct product $\prod_{i \in I} U_i$ and is hence an R-module; to distinguish this from the internal direct sum we may again call it the **external direct sum**.

For any $j \in I$ we get an obvious injective map $\mu_j : U_j \to \prod_{i \in I} U_i$ called the **natural injection**, and a surjective map $\nu_j : \prod_{i \in I} U_i \to U_j$ called the **natural projection**. Clearly $\nu_i \mu_i$ is the identity map of U_i. All this works for sets U_i. In the module case these maps are R-homomorphisms, and they give rise to an obvious R-monomorphism (again called the natural injection) $\mu_j : U_j \to \bigoplus_{i \in I} U_i$ and an obvious R-epimorphism (again called the natural projection) $\nu_j : \bigoplus_{i \in I} U_i \to U_j$. Letting $\overline{U}_i$ to be the image of μ_i we have $\bigoplus_{i \in I} U_i$ (external) $= \bigoplus_{i \in I} \overline{U}_i$ (internal) . If $U_i = $ a set U for all $i \in I$ then the cartesian product $\prod_{i \in I} U_i$ is denoted by U^I and is called the **set-theoretic I-th power** of U, and if U is an R-module then the module U^I is called the **module theoretic I-th power** of U. Finally, if the module $U_i = $ a module U for all $i \in I$ then the module $\bigoplus_{i \in I} U_i$ is denoted by $U_{\oplus}^I$ and is called the **module theoretic restricted I-th power** of U. Other related definitions such as **diagonal direct sum** may be found in L5§1.

In L5§2 we define a **graded ring** S over a ring R to be an overring of R such that $S = \sum_{i \in I} S_i$ as an internal direct sum of a family $(S_i)_{i \in I}$ of R-modules with $S_0 = R$ where the **type** I is any additive abelian monoid, and for all $f_i \in S_i$ and $f_j \in S_j$ with i, j in I we have $f_i f_j \in S_{i+j}$. An ideal in S is **homogeneous** if it is generated by homogeneous elements, i.e., elements of $\cup_{i \in I} S_i$. A ring homomorphism $\phi : S \to S'$ of graded rings of type I is **graded** if $\phi(S_i) \subset S_i'$ for all i. If $I = \mathbb{N}$ then we call S a **naturally graded ring**.

In L5§3 we define a **semihomogeneous ring** to be a naturally graded ring S such that $S = S_0[S_1]$; if S_1 is a finitely generated module over $R = S_0$ then we call S a **homogeneous ring**. Also we study the ideal theory of graded rings.

The very long L5§5 on **More About Ideals And Modules** is divided into Quests (Q1) ro (Q35) which we proceed to summarize.

Referring to (O14) to (O17) of the above Summary of L4, let $U = U_1 \cap \cdots \cap U_h$ be an irredundant primary decomposition of a submodule U of a finitely generated module V over a noetherian ring R, and let $P_1, \ldots, P_h$ be the prime ideals in R which are the respective radicals of $U_1, \ldots, U_h$. In the **Zerodivisor Theorem** (Q1)(T1) we show that then $Z_R(V/U) = P_1 \cup \cdots \cup P_h$.

In (Q2) we define a **faithful module** over a ring R to be an R-module D such that $\text{ann}_R D = 0$. Note that $\text{ann}_R D = \{r \in R : rd = 0 \text{ for all } d \in D\}$.

In (Q3) we define the **Jacobson radical** $\text{jrad}(R)$ of a ring R to be the intersection of all its maximal ideals, and we define a **Zariski ring** to be pair (R, J) where J is an ideal in a ring R which is contained in the Jacobson radical of R. In the very versatile **Nakayama Lemma** (Q3)(T3) we show that: if U is a submodule of a finitely generated module V over a ring R such that $V = U + JV$ for an ideal J in R with $J \subset \text{jrad}(R)$, then $V = U$. The particularly noteworthy case of this when R is a local ring with $J = M(R)$ was proved by the master algebraist Krull.

In (Q4) we prove: the **Krull Intersection Theorem** (T7) saying that in any local ring R we have $\cap_{n \in \mathbb{N}} M(R)^n = 0$, its Domainized Version (T10) saying that for any nonunit ideal J in any domain R we have $\cap_{n \in \mathbb{N}} J^n = 0$, and its variation which is called the **Artin Rees Lemma** (T4) and which says that, given any ideals V, W, J in a noetherian ring R, there exists $a \in \mathbb{N}$ such that for all $c \geq b \geq a$ in $\mathbb{N}$ we have $J^c W \cap V = J^{c-b}(J^b W \cap V)$.

In (Q5) we explain Nagata's **idealization principle** which converts module situations to ideal situations, and use it prove the module incarnations of the above results of (Q4). We also prove a Characterization Theorem (T20) for Zariski Rings.

In (Q6) we prove Cohen's Theorem saying that a ring is noetherian if all the prime ideals in it are finitely generated, and Eakin's Theorem saying that a ring is noetherian if some noetherian overring is a finitely generated module over it.

In (Q7) we prove Krull's **Generalized Principal Ideal Theorem** (T27) saying that: if an ideal I in a noetherian ring R is generated by n elements $x_1, \ldots, x_n$ with $n \in \mathbb{N}$ then $\text{ht}_R I \leq n$, and if x_i is a nonzerodivisor mod $x_1, \ldots, x_{i-1}$ for $1 \leq i \leq n$ then $\text{ht}_R P = n$ for every P in $\text{nvspec}_R I$. The Principal Ideal Theorem (T24) is the $n = 1$ case. We deduce the **Extended Dim-Emdim Theorem** (T29) saying that for any local ring R we have that $\text{emdim}(R) \geq \dim(R) =$ the smallest number of elements in R which generate an $M(R)$-primary ideal. We also deduce the **Dimension Lemma** (T30) about the polynomial ring $S = R[X_1, \ldots, X_m]$ as well as the power series ring $S = R[[X_1, \ldots, X_m]]$ in a finite number of variables $X_1, \ldots, X_m$ over a noetherian ring R, saying that: (1) S is noetherian, and for any nonunit ideal I in R and $j \in \{0, \ldots, m\}$, upon letting $I_j = IS + (X_1, \ldots, X_j)S$ we have $I_j \cap R = I$ with $\text{ht}_S I_j = j + \text{ht}_R I$, and if $I \in \text{spec}(R)$ then we also have $I_j \in \text{spec}(S)$; (2) $\dim(S) = m + \dim(R)$; (3) if R local and we are in the power series case then S is local with $M(S) = M(R)S + (X_1, \ldots, X_m)S$ and $\text{emdim}(S) = m + \text{emdim}(R)$, and hence: R is regular $\Leftrightarrow$ S is regular; (4) If the ring R is a field then $R[X_1, \ldots, X_m]$ is an m-dimensional noetherian ring and $R[[X_1, \ldots, X_m]]$ is an m-dimensional regular local ring; (5) in the polynomial case with $m = 1$, for any $I = J \cap R$ with $J \in \text{spec}(S)$ we have $I \in \text{spec}(R)$ and: (i) if $J = IS$ then $\text{ht}_S J = \text{ht}_R I$, whereas (ii) if $J \neq IS$ then $\text{ht}_S J = 1 + \text{ht}_R I$ and there exists $F(X_1) \in J$ of degree $E > 0$ such that the coefficient of X_1^E in F belongs to $R \setminus I$.

In the **Relative Independence Theorem** (Q8)(T34) we show that n elements in a noetherian ring generate an ideal of height n iff they are independent over that ring in the following sense. Considering the n variable polynomial ring $R[Y_1, \ldots, Y_n]$ in n variables $Y_1, \ldots, Y_n$ over a ring R with $n \in \mathbb{N}$, for any $A \subset R$ and $i \in \mathbb{N}$ we let $A[Y_1, \ldots, Y_n]_i = $ the set of all homogeneous members of $R[Y_1, \ldots, Y_n]$ of degree i all whose coefficients belong to A (including the zero polynomial in case $0 \in A$) and we put $A[Y_1, \ldots, Y_n]_\infty = \bigcup_{i \in \mathbb{N}} A[Y_1, \ldots, Y_n]_i$. Let J be an ideal in R. For elements $x_1, \ldots, x_n$ in R let $I = (x_1, \ldots, x_n)R$. The elements $x_1, \ldots, x_n$ are said to be **independent** over (R, J), or independent relative to (R, J), if $I \subset J$ and:

(†) $F(Y_1, \ldots, Y_n) \in R[Y_1, \ldots, Y_n]_\infty$ with $F(x_1, \ldots, x_n) = 0$
    $\Rightarrow F(Y_1, \ldots, Y_n) \in J[Y_1, \ldots, Y_n]_\infty$.

The elements $x_1, \ldots, x_n$ are said to be **independent** over R, or independent relative to R, if $I \neq R$ and the elements are independent over $(R, \mathrm{rad}_R I)$. Considering the R-epimorphism $\alpha : R[Y_1, \ldots, Y_n] \to R$ which sends Y_i to x_i for $1 \leq i \leq n$, we show that (♯) $\ker(\alpha) = (Y_1 - x_1, \ldots, Y_n - x_n)R[Y_1, \ldots, Y_n]$ and condition (†) is equivalent to the condition: (†′) $\ker(\alpha) \cap R[Y_1, \ldots, Y_n]_\infty \subset J[Y_1, \ldots, Y_n]_\infty$. Assuming $I \subset J$, we show that condition (†) is equivalent to the condition:

(‡) $F(Y_1, \ldots, Y_n) \in R[Y_1, \ldots, Y_n]_i$ with $i \in \mathbb{N}$ and $F(x_1, \ldots, x_n) \in I^{i+1}$
    $\Rightarrow F(Y_1, \ldots, Y_n) \in J[Y_1, \ldots, Y_n]_i$.

In case R is the power series ring $K[[X_1, \ldots, X_m]]$ in a finite number of variables over a field K, elements $x_1, \ldots, x_n$ in $M(R)$ are said to be analytically independent over K if there is no nonzero $f(Y_1, \ldots, Y_n) \in K[[Y_1, \ldots, Y_n]]$ with $f(x_1, \ldots, x_n) = 0$. In the **Analytic Independence Theorem** (Q8)(T35) we show that: (1) if the elements $x_1, \ldots, x_n$ are independent over $(R, M(R))$ then they are analytically independent over K, and (2) if $m = n$ and the ideal I is $M(R)$-primary then the elements $x_1, \ldots, x_n$ are analytically independent over K.

In (Q9) we show that the localization of any normal ring at a multiplicative set in it not containing any zerodivisor is again normal. A prime ideal Q in an overring S of a ring R **lies above** a prime ideal P in R means $P = Q \cap R$; we may also say that P **lies below** Q in R.

In (Q9) we mostly deal with an **integral extension** of a ring, i.e., an overring S of a ring R such that S is integral over R. We show that for any prime ideal P in R lying below a prime ideal Q in S we have: P is maximal $\Leftrightarrow Q$ maximal. In the **Lying Above Theorem** (T40) we show that for any prime ideal P in R, there exists a prime ideal Q in S lying above P. In the **Going Up Theorem** (T41) we show that given any prime ideals $P^* \subset P$ in R, and given any prime ideal Q^* in S lying above P^*, there exists a prime ideal Q in S lying above P for which we have $Q^* \subset Q$. In the **Going Down Theorem** (T44) we show that if R is a normal domain and S is a domain, then given any prime ideals $P^* \subset P$ in R, and any prime ideal Q in S lying above P, there exists a prime ideal Q^* in S lying above P^* with $Q^* \subset Q$. In the **Dimension Corollary** (T45) we show that: (1) above any ascending finite sequence of prime ideals in R there lies an ascending finite sequence of prime ideals in S; (2) any strictly ascending finite sequence of prime ideals in S lies above a strictly ascending finite sequence of prime ideals in R; (3) $\dim(S) = \dim(R)$; (4) for any prime ideal Q in S lying above any prime ideal P in R we have $\mathrm{dpt}_S Q = \mathrm{dpt}_R P$ and $\mathrm{ht}_S Q \le \mathrm{ht}_R P$; (5) if S is a domain and R is a normal domain, then for any prime ideal Q in S lying above any prime ideal P in R we have $\mathrm{ht}_S Q = \mathrm{ht}_R P$.

In (Q10) we prove the **Noether Normalization Theorem** (T46) which says that any affine ring A over a field k has a **normalization basis** over k, by which we mean a finite number of elements $Y_1, \dots, Y_r$ in A which are algebraically independent over k and which are such that the ring A is integral over the subring $k[Y_1, \dots, Y_r]$.

By a **saturated chain of prime ideals** in a ring R we mean a finite sequence $P_0 \subsetneq P_1 \subsetneq \cdots \subsetneq P_n$ of prime ideals in R such that there is no prime ideal P^* in R with $P_i \subsetneq P^* \subsetneq P_{i+1}$ for some $i \in \{0, \dots, n-1\}$; we call n the **length** of the chain; if $P \subset Q$ are prime ideals in R with $P = P_0$ and $Q = P_n$ then we say that $P_0 \subsetneq P_1 \subsetneq \cdots \subsetneq P_n$ is a saturated prime ideal chain between P and Q; if there is no prime P in R with $P \subsetneq P_0$ and there is no prime Q in R with $P_n \subsetneq Q$ then we say that $P_0 \subsetneq P_1 \subsetneq \cdots \subsetneq P_n$ is an **absolutely saturated prime ideal chain** in R. By an **infinite chain of prime ideals** in a ring R we mean a sequence of prime ideals $(P_i)_{i \in \mathbb{N}}$ in R such that for all $i \in \mathbb{N}$ we have $P_i \subsetneq P_{i+1}$; if $P \subset Q$ are prime ideals in R such that $P \subset P_i \subset Q$ for all $i \in \mathbb{N}$ then we say that this is an infinite prime ideal chain between P and Q.

For a subdomain R of a domain R', by $\mathrm{trdeg}_R R'$ we denote the transcendence degree of $\mathrm{QF}(R')$ over $\mathrm{QF}(R)$. More generally, for a prime ideal P' in a ring R' lying above a prime ideal P in a subring R, after identifying R/P with a subdomain of R'/P', by $\mathrm{trdeg}_{R/P} R'/P'$ we denote the transcendence degree of $\mathrm{QF}(R'/P')$ over $\mathrm{QF}(R/P)$; in case R is a subfield k of R', we write $\mathrm{trdeg}_k R'/P'$ in place of $\mathrm{trdeg}_{k/0} R'/P'$; in case P' and P are the respective maximal ideals in quasilocal rings R' and R, we may write $\mathrm{restrdeg}_R R'$ in place of $\mathrm{trdeg}_{R/P} R'/P'$ and call this the **residual transcendence degree** of R' over R.

Using (T46) we establish the **Extended Dimension Theorem** (T47) which subsumes L4§8(T19) and which says that for any affine ring A over a field k we have the following:

(T47.1) For any nonunit ideal J in A, letting $J_1, \dots, J_m$ be the distinct minimal primes of J in R we have $\dim(A/J) = \mathrm{dpt}_A J = \max_{1 \le i \le m} \mathrm{trdeg}_k (A/J_i) \in \mathbb{N}$, and for any prime ideal P in A we have $\dim(A/P) = \mathrm{dpt}_A P = \mathrm{trdeg}_k (A/P) \in \mathbb{N}$. Hence in particular, if A is a domain then $\dim(A) = \mathrm{trdeg}_k A \in \mathbb{N}$.

(T47.2) Every pair of prime ideals $P \subset Q$ in A has the property which says that: (*) there is no infinite chain of prime ideals between P and Q, and any two saturated chains of prime ideals between P and Q have the same length. Moreover: (**) this common length equals $\dim(A/P) - \dim(A/Q)$.

(T47.3) If A is a domain then the length of every absolutely saturated chain of prime ideals in R equals the dimension of R.

(T47.4) If A and H is a nonunit ideal H in A then $\mathrm{ht}_A H + \mathrm{dpt}_A H = \dim(A)$.

(T47.5) If A is a domain then, for any prime ideal P in A, the length of every saturated prime ideal chain in A between 0 and P equals $\mathrm{ht}_A P$, and the length of every saturated prime ideal chain in A between P and any member of $\mathrm{mvspec}_A P$ equals $\mathrm{dpt}_A P$.

(T47.6) If A is a domain and A' is an affine domain over A then for any prime ideal P' in A' lying above any prime ideal P in A, upon considering the localizations $R' = A'_{P'}$ and $R = A_P$, we have $\dim(R') + \mathrm{restrdeg}_R R' = \dim(R) + \mathrm{trdeg}_R R'$.

(T47.7) in the situation of (T46) we have $\dim(A) = r$.

In (Q10) and (Q11) we prove the **Inclusion Relations Theorem** L4§8(T21) and the **Hilbert Nullstellensatz** L4§8(T22).

In (Q11), the **Jacobson Spectrum And Nilradical** of a ring R are defined by the equation $\mathrm{jspec}(R) = \{P \in \mathrm{spec}(R) : P = \cap_{Q \in \mathrm{mvspec}_R P} Q\}$ and the equation $\mathrm{nrad}(R) = \cap_{P \in \mathrm{spec}(R)} P$. The ring R is said to be a **Jacobson ring** if $\mathrm{jspec}(R) = \mathrm{spec}(R)$. Using jspec we supplement L4§8(T21) in (Q11)(T50). Using nrad we prove the **Minimal Primes Theorem**(Q11)(T51) which says that, given any ideal J in any ring R, every member of $\mathrm{vspec}_R J$ contains some member of $\mathrm{nvspec}_R J$, and we have $\mathrm{rad}_R J = \cap_{P \in \mathrm{nvspec}_R J} P$.

A ring A is **catenarian** means every pair of prime ideals $P \subset Q$ in it satisfies the above condition $(*)$ of (T47.2). A ring A is **universally catenarian** means it is noetherian and the polynomial ring over it in any finite number of variables is catenarian. The **dimension formula** (resp: the **dimension inequality**) holds for a domain A relative to an overdomain A' means for every prime ideal P' in A', considering the localizations $R' = A'_{P'}$ and $R = A_P$ with $P = P' \cap A$, we have $\dim(R') + \mathrm{restrdeg}_R R' = \dim(R) + \mathrm{trdeg}_R R'$ (resp: $\dim(R') + \mathrm{restrdeg}_R R' \leq \dim(R) + \mathrm{trdeg}_R R'$). The dimension formula (resp: the dimension inequality) holds in a domain A means it holds for A relative to every affine domain over A. Thus (T47.6) says that the dimension formula holds in any affine domain over a field.

In (Q12) we sharpen Comment (Q7)(C13) on **Univariate Ideal Extensions**, which in the present summary corresponds to part (5) of the Dimension Lemma (Q7)(T30), by proving the **Multiple Ring Extension Lemma** (T55) which says that: the dimension inequality holds in any noetherian domains A relative to any affine domain A' over A, and if $A' = A[x_1, \ldots, x_m]$ for some $m \in \mathbb{N}$ and $x_1, \ldots, x_m$ in A' then for any prime ideal P' in A' lying above any prime ideal P in A, considering the localizations $R' = A'_{P'}$ and $R = A_P$ we have $\mathrm{restrdeg}_R R' \leq m$ with $\mathrm{trdeg}_R R' \leq m$, and if either $\mathrm{trdeg}_R R' = m$ or the polynomial ring $B = A[X_1, \ldots, X_m]$ in m variables is catenarian, then $\dim(R') + \mathrm{restrdeg}_R R' = \dim(R) + \mathrm{trdeg}_R R'$. The $m = 1$ case of (T55) is designated as the **Simple Ring Extension Lemma** (T54). In the **Catenarian Ring Theorem** (T58) we show that a ring A is universally catenarian iff A is a noetherian catenarian ring and the dimension formula holds in A/H for every prime ideal H in A.

In (Q13) on **Associated Graded Rings And Leading Ideals**, given any ideal I in a ring R, we define the **associated graded ring** of (R, I) to be the naturally graded ring $\text{grad}(R, I) = \bigoplus_{n \in \mathbb{N}} (I^n/I^{n+1})$ with external direct sum of $\mathbb{Z}$-modules. The multiplication is defined by $\text{grad}_{(R,I)}^m(f_m) \cdot \text{grad}_{(R,I)}^n(f_n) = \text{grad}_{(R,I)}^{m+n}(f_m f_n)$ for all $f_m \in I^m$ and $f_n \in I^n$ with m, n in $\mathbb{N}$. Here $\text{grad}_{(R,I)}^n : I^n \to \text{grad}(R, I)$ is the composition of the natural surjection $\lambda_n : I^n \to (I^n/I^{n+1})$ and the natural injection $\mu_n : (I^n/I^{n+1}) \to \text{grad}(R, I)$. We put $\text{grad}(R, I)_n = \text{grad}_{(R,I)}^n(I^n)$. For any $h \in R$ we put $\text{lefo}_{(R,I)}(h) = \text{grad}_{(R,I)}^n(h)$ or 0 according as $\text{ord}_{(R,I)}h = n \in \mathbb{N}$ or $\text{ord}_{(R,I)}h = \infty$; we call this the **leading form** of h relative to (R, I). For any ideal J in R we put $\text{ledi}_{(R,I)}(J) =$ the homogeneous ideal in $\text{grad}(R, I)$ generated by $(\text{lefo}_{(R,I)}(h))_{h \in J}$ and we call this the **leading ideal** of J relative to (R, I). Any family of generators $x = (x_l)_{l \in L}$ of I induces a unique graded ring epimorphism $\Theta_{(R,I)}^x : k[(X_l)_{l \in L}] \to \text{grad}(R, I)$ with $k = R/I$ such that $\Theta_{(R,I)}^x(z) = \mu_0(z)$ for all $z \in k$ and $\Theta_{(R,I)}^x(X_l) = y_l$ for all $l \in L$ where $(X_l)_{l \in L}$ are indeterminates over k and $y_l = \text{grad}_{(R,I)}^1(x_l)$; we say that $\Theta_{(R,I)}^x$ is **induced** by (R, I) and x. Note that $\text{grad}(R, I)$ is a semihomogeneous (homogeneous if L is finite) ring over the ring $\text{grad}(R, I)_0$ with $\text{grad}(R, I) = \text{grad}(R, I)_0[\text{grad}(R, I)_1] = \text{grad}(R, I)_0[(y_l)_{l \in L}]$.

In case R is a **quasilocal ring with** $I = M(R)$, in the above paragraph we write R in place of (R, I), except that we write $\text{grad}(R)$ and $\text{grad}(R)_n$ in place of $\text{grad}(R, I)$ and $\text{grad}(R, I)_n$ respectively.

A **valuation function** on a ring R is a map $v : R \to \mathbb{Z} \cup \{\infty\}$ such that for all x, y, z in R we have:

(i) $v(xy) = v(x) + v(y)$,

(ii) $v(x + y) \geq \min(v(x), v(y))$, and

(iii) $v(z) = \infty \Leftrightarrow z = 0$,

with the usual understanding about ∞. Thus in the notation of L3§11, a valuation function is like a valuation with R and $\mathbb{Z}$ playing the roles of the field K and the ordered abelian group G respectively. If v is a valuation function on a nonnull ring R then clearly R is a domain and we get a unique valuation $\overline{v} : \text{QF}(R) \to \mathbb{Z} \cup \{\infty\}$ such that for all x, y in $R^\times$ we have $\overline{v}(x/y) = v(x) - v(y)$; we say that the valuation $\overline{v}$ is induced by v; usually we denote $\overline{v}$ also by v.

Theorem (T60) says that for any given ideal I in a ring R we have the following: (1) $I \neq R$ and $\text{ord}_{(R,I)}$ is a valuation function on R iff R is I-Hausdorff and $\text{grad}(R, I)$ is a domain. (2) Upon letting $R^* = R/J^*$ and $I^* = I/J^*$ with $J^* = \cap_{n \in \mathbb{N}} I^n$, we have that R^* is I^*-Hausdorff, and if $\text{grad}(R, I)$ is a domain then R^* is a domain and $\text{ord}_{(R^*,I^*)}$ is a valuation function on R^*. (3) Let $\phi : R \to R'$ be a ring epimorphism, with $J = \ker(\phi)$, such that the leading ideal $\text{grad}_{(R,I)}(J)$ is a (homogeneous) prime ideal in $\text{grad}(R, I)$; then $\cap_{n \in \mathbb{N}}(J + I^n)$ is a prime ideal in R.

Given any local ring R with $\dim(R) = d$ and $\text{emdim}(R) = e$, let $x = (x_1, \ldots, x_m)$ be a finite number of generators of $M(R)$, let $X = (X_1, \ldots, X_m)$ be indeterminates over the field $k = R/M(R)$, and recall that $\Theta_R^x : k[X] \to \text{grad}(R)$ is a graded ring epimorphism where $k[X] = \sum_{n \in \mathbb{N}} k[X]_n$ is the naturally graded polynomial ring with $k[X]_n =$ the set of all homogeneous polynomials of degree n (including 0).

Theorem (T61) says that then we have: (1) $\ker(\Theta_R^x) \cap k[X]_1 = 0 \Leftrightarrow m = e$.

(2) $\ker(\Theta_R^x) = 0 \Leftrightarrow m = d$. (3) If $m = e$ then: R is regular $\Leftrightarrow \ker(\Theta_R^x) = 0$. (4) R is regular $\Leftrightarrow$ there is a permutation σ of $\{1, \ldots, m\}$ together with elements a_{ij} in k such that the ideal $\ker(\Theta_R^x)$ in $k[X]$ is generated by the $m - e$ elements $X_{\sigma(i)} - \sum_{1 \le j \le e} a_{ij} X_{\sigma(j)}$ for $e + 1 \le i \le m$.

In (Q14) we define a **Completely Normal Domain** to be a domain R such that: $x \in \mathrm{QF}(R)$ with $dx^n \in R$ for some $d \in R^\times$ and all $n \in \mathbb{N} \Rightarrow x \in R$.

Complete Normality Lemma (T63) says that for any given ring R we have the following: (1) If R is a completely normal domain then R is a normal domain. If R is a noetherian normal domain then R is a completely normal domain. (2) If I is an ideal in R such that for all $z \in R$ we have $\cap_{n \in \mathbb{N}}(zR + I^n) = zR$, and $\mathrm{grad}(R, I)$ is a completely normal domain, then R is a completely normal domain. (3) For any ideal I in R we have that: $I \ne R$ iff $\mathrm{grad}(R, I)$ is nonnull. (4) If R is noetherian and I is any ideal in R then $\mathrm{grad}(R, I)$ is noetherian. (5) If R is noetherian and has an ideal I with $I \subset \mathrm{jrad}(R)$ such that $\mathrm{grad}(R, I)$ is a normal domain, then R is a normal domain.

We proved L4§8(T24.1) as the **Krull Intersection Theorem** (Q4)(T7), we shall prove L4§8(T24.3) as **Another Ord Valuation Lemma** (Q15)(T71), and we prove L4§8(T24.2) as the **Ord Valuation Lemma** (Q14)(T64) which says that given any regular local ring R, for all nonzero elements x and y in R we have $\mathrm{ord}_R(xy) = \mathrm{ord}_R x + \mathrm{ord}_R y$, and hence R is a normal domain and we get a valuation $\mathrm{ord}_R : \mathrm{QF}(R) \to \mathbb{Z}$ of $\mathrm{QF}(R)$ by putting $\mathrm{ord}_R(x/y) = \mathrm{ord}_R x - \mathrm{ord}_R y$.

To summarize (Q15) on **Regular Sequences And CM Rings**, let R be a ring, let $x = (x_1, \ldots, x_n)$ be a sequence of elements in R of length $n \in \mathbb{N}$, for $0 \le m \le n$ let I_m be the ideal $(x_1, \ldots, x_m)R$, and let V be a module over R. We say that x is a V-**regular sequence**, if $I_n V \ne V$ and for $1 \le m \le n$ we have $x_m \in S_R(V/(I_{m-1}V))$. Moreover if J is an ideal in R with $I_n \subset J$ then we say that x is a V-**regular sequence in** J, and if for every $z \in J \cap S_R(V/(I_nV))$ we have $I_n V + zV = V$ then we say that x is a **maximal V-regular sequence in** J. We define the (R, J)-**regularity of** V by putting $\mathrm{reg}_{(R,J)} V =$ the maximum length of a V-regular sequence in J. If R is local with $J = M(R)$ then we call this the R-**regularity of** V and denote it by $\mathrm{reg}_R V$; in this case we write $\mathrm{reg}(R)$ in place of $\mathrm{reg}_R R$ and we call it the **regularity of** R. We define the **generating number** of V in R by putting $\mathrm{gnb}_R V =$ the smallest number of generators of V. An ideal I in R is said to be an **ideal-theoretic complete intersection** in R if $I \ne R$ and $\mathrm{ht}_R I = \mathrm{gnb}_R I$. A finitely generated R-module is **free** means it is isomorphic to R^d for some $d \in \mathbb{N}$. In case R is local and V is finitely generated, V is said to be a **CM (= Cohen-Macaulay)** module if $\mathrm{reg}_R V = \dim(R/(0 : V)_R)$. A local ring R is said to be a **CM ring** if it is a CM module over itself, i.e., if $\mathrm{reg}(R) = \dim(R)$. A noetherian ring R is said to be a **CM ring** if, for every maximal ideal P in R, the localization R_P is a CM ring. An ideal I in a noetherian ring R is said to be **unmixed** if for every prime P in $\mathrm{ass}_R(R/I)$ we have $\mathrm{ht}_R P = \mathrm{ht}_R I$. The **unmixedness theorem** is said to hold in a noetherian ring R if every ideal in R, which can be generated by as many elements as its height, is unmixed.

In **Cohen-Macaulay Theorem** (T70) we show that for any noetherian ring R we have the following:

(1) If R is CM and $x = (x_1, \ldots, x_n)$ is any R-regular sequence in R with $n \in \mathbb{N}$, then upon letting $I_n = xR$ we have $\operatorname{ht}_R I_n = n$ and for every $P \in \operatorname{ass}_R(R/I_n)$ we have $\operatorname{ht}_R P = n$, and hence in particular I_n is an ideal-theoretic complete intersection in R, and I_n is unmixed.

(2) If R is CM then for any nonunit ideal I in R we have $\operatorname{reg}_{(R,I)} R = \operatorname{ht}_R I$.

(3) If R is CM and I is a nonunit ideal I in R which is an ideal-theoretic complete intersection in R then I is generated by an R-regular sequence.

(4) If R is CM then its localization R_S at any multiplicative set S in it is CM, and hence in particular its localization R_P at any prime ideal P in it is CM.

(5) If R is CM then R is catenarian.

(6) If R is CM and local then the length of any absolutely saturated prime ideal chain in R equals $\dim(R)$ and for any nonunit ideal I in R we have $\operatorname{ht}_R I = \dim(R) - \dim(R/I)$.

(7) R is CM $\Leftrightarrow$ the unmixedness theorem holds in R.

(8) R is a finite variable polynomial ring over a field $\Rightarrow R$ is CM.

(9) R is a finite variable power series ring over a field $\Rightarrow R$ is CM.

(10) R is CM $\Rightarrow$ every finite variable polynomial ring over R is CM.

(11) R is CM $\Rightarrow$ every finite variable power series ring over R is CM.

(12) R is CM $\Rightarrow$ every homomorphic image of R is universally catenarian.

In **Another Ord Valuation Lemma** (T71) we show that given a local ring R, for any $x \in M(R) \setminus Z_R(R)$ we have that: (i) $R/(xR)$ is a local ring whose dimension is one less than the dimension of R, (ii) if $R/(xR)$ is regular then $\operatorname{ord}_R x = 1$, and (iii) if $\operatorname{ord}_R x = 1$ and R is regular then $R/(xR)$ is regular.

In (Q16) on **Complete Intersections And GST Rings** we start by making the following definitions. A local ring R is said to be a **complete intersection** if there exists a regular local ring T and an epimorphism $\phi : T \to R$ such that $\ker(\phi)$ is an ideal-theoretic complete intersection in T. By **parameters** for R we mean a sequence of elements in $M(R)$ whose length equals $\dim(R)$ and which generates an $M(R)$-primary ideal; by a **parameter ideal** in R we mean an ideal generated by parameters. By **regular parameters** for R we mean parameters which generate $M(R)$. The **socle** $\operatorname{soc}(R)$ of R is defined by putting $\operatorname{soc}(R) = (0 : M(R))_R$; this is an ideal in R with $M(R) \subset (0 : \operatorname{soc}(R))_R$; hence it is a finite dimensional vector space over $R/M(R)$; we define the **socle-size** $\operatorname{socz}(R)$ of R to be the vector space dimension of $\operatorname{soc}(R)$ over $R/M(R)$. We define the **CM type** $\operatorname{cmt}(R)$ of R to be the minimum of $\operatorname{socz}(R/Q)$ taken over all parameter ideals Q in R. The local ring R is said to be **GST (= Gorenstein)** if every parameter ideal in it is irreducible.

A noetherian ring R is said to be a **local complete intersection** if its localization R_P is a complete intersection for every maximal ideal P in it. A noetherian ring R is said to be **GST** if its localization R_P is GST for every maximal ideal P

in it. A noetherian ring R is said to be **regular** if its localization R_P is regular for every maximal ideal P in it. Given any ring R with total quotient ring K, for any R-submodule J of K we define its **inverse module** J^{-1} by putting $J^{-1} = (R : J)_K$ and we note that J^{-1} is again an R-submodule of K; by a **fractional** R-ideal J we mean an R-submodule J of K such that $tJ \subset R$ for some $t \in S_R(R)$. By the **conductor cond**$(R, \widetilde{R})$ of a ring R in an overring $\widetilde{R}$ of R we mean the ideal in R given by cond$(R, \widetilde{R}) = (R : \widetilde{R})_R$.

In **Socle Size Lemma** (T75) we show that for any local ring R we have the following: (1) If $\dim(R) = d > 0$ and there are parameters $(x_1, \ldots, x_d)$ for R such that upon letting $Q_e = (x_1^e, \ldots, x_d^e)R$ we have socz$(R/Q_e) = 1$ for all $e \in \mathbb{N}_+$ then reg$(R) > 0$. (2) If socz$(R/Q) = 1$ for every parameter ideal Q in R then R is CM. (3) If R is CM then for every parameter ideal Q in R we have cmt$(R) = $ socz(R/Q).

In **GST Ring Characterization Theorem** (T76) we show that a local ring is GST iff it is CM and some parameter ideal in it is irreducible.

In **Complete Intersection Theorem** (T77) we show that if a noetherian ring is a local complete intersection then it is GST; in particular, if a noetherian ring is regular then it is a complete intersection and hence it is GST.

In **One Dimensional GST Ring Characterization Theorem** (T78) we show that for any one-dimensional CM local ring R the following 7 conditions are equivalent, where $P = M(R)$.

(i) R is GST.

(ii) $\ell_R(P^{-1}/R) = 1$.

(iii) For every $t \in S_R(R) \cap P$, the ideal tR is irreducible.

(iv) For every $t \in S_R(R) \cap P$ we have $\ell_R(tR : P)_R/(tR)) = 1$.

(v) For every $t \in S_R(R) \cap P$ we have socz$(R/(tR)) = 1$.

(vi) For every fractional R-ideal J with $J \cap S_R(R) \neq \emptyset$ we have $J = (J^{-1})^{-1}$.

(vii) For every ideal I in R with $I \cap S_R(R) \neq \emptyset$ we have $\ell_R(I^{-1}/R) = \ell_R(R/I)$.

In **Something Is Twice Something Theorem** (T79) we show if R is a GST local ring of dimension one such that the integral closure R^* of R in the total quotient ring K of R is a finitely generated R-module, then upon letting C be the conductor of R in R^* we have $\ell_R(R^*/R) = \ell_R(R/C)$ and hence $\ell_R(R^*/C) = 2\ell_R(R/C)$ where all the lengths are nonnegative integers.

In **(Q17)** on **Projective Resolutions of Finite Modules** we prove (T24.4) of L4§8 which says that the localization of a regular local ring at any prime ideal in it is again regular. This is part (T81.7) of the **Dim-Pdim Theorem** (Q17)(T81) where the **projective dimension pdim**$_R V$ of a module V over a ring R is defined to be the smallest length of a projective R-resolution of V if such exists; otherwise we put pdim$_R V = \infty$. An **exact sequence** of R-modules is a sequence(f, W, n) of R-homomorphisms $f_i : W_{i+1} \to W_i$ for $0 \leq i \leq n \in \mathbb{N}$ such that im$(f_i) = \ker(f_{i-1})$ for $1 \leq i \leq n$ with $\ker(f_n) = 0$ and im$(f_0) = W_0$; the **length** of the sequence is n; by a **projective** R**-resolution of** V we mean such a sequence for which $W_0 = V$ and the module W_i is projective for $1 \leq i \leq n$. By a **projective** R**-module** we mean an R-module D having the property which says that if $\alpha : E \to F$ is any R-epimorphism of R-modules and $\beta : D \to F$ is any R-homomorphism then $\beta = \alpha\gamma$ for some R-homomorphism $\gamma : D \to E$. By a **finite module** we mean a finitely generated module. By the **global dimension gdim**(R) we mean the smallest nonnegative integer n, if it exists, such that $\text{pdim}_R V \leq n$ for every finite R-nodule V; otherwise we put $\text{gdim}(R) = \infty$. In part (T81.7) of the **Dim-Pdim Theorem** (Q17)(T81) we show that if R is a noetherian ring with $\dim(R) = d \in \mathbb{N}$ then: R is regular $\Leftrightarrow \text{gdim}(R) = d \Leftrightarrow \text{gdim}(R) < \infty \Leftrightarrow \text{pdim}_R V < \infty$ for every finite R-module V. What we proved previously in L4§5(O24)(22•) is now designated as the **Prime Avoidance Lemma** (C31.1) which says that if $J \subset P_1 \cup \cdots \cup P_n$ where J is an ideal in a ring R and $P_1, \ldots, P_n$ are prime ideals in R with $n \in \mathbb{N}_+$ then for some $i \in \{1, \ldots, n\}$ we have $J \subset P_i$.

In **(Q18) on Direct Sums of Algebras, Reduced Rings, And PIRs** we define an **R-algebra or algebra over a ring** R to be a ring S together with a ring homomorphism $\phi : R \to S$ which may be called the **underlying homomorphism**; S becomes an R-module by taking $rs = \phi(r)s$ for all $r \in R$ and $s \in S$. Let $\phi_1 : R \to R_1, \ldots, \phi_n : R \to R_n$ be R-algebras with $n \in \mathbb{N}$. Defining componentwise multiplication, the external direct sum module $R_1 \oplus \cdots \oplus R_n$ becomes a ring so that the diagonal direct sum $\phi_1 \oplus \cdots \oplus \phi_n : R \to R_1 \oplus \cdots \oplus R_n$ is a ring homomorphism; the R-algebra with this underlying homomorphism is called the **R-algebra-theoretic** or **multiplicative** direct sum of $R_1, \ldots, R_n$. Every ring is a $\mathbb{Z}$-algebra, and hence for any rings $R_1, \ldots, R_n$ we get the notion of the **ring-theoretic** or $\mathbb{Z}$-algebra theoretic direct sum $R_1 \oplus \cdots \oplus R_n$. We speak of a **finite direct sum of rings** instead of ring-isomorphic to a direct sum of rings.

A **PIR (= Principal Ideal Ring)** is a ring in which every ideal is principal, and a **SPIR (= Special PIR)** is a zero-dimensional local ring which is a PIR. A **reduced ring** is a ring having no nonzero nilpotent element.

In (Q18.1)(4) we prove the following: Let $T = R_1 \oplus \cdots \oplus R_n$ with $n \in \mathbb{N}_+$ be a direct sum of rings. For $1 \leq i \leq n$ let $f_i : T \to R_i$ be the natural projection. Let R be a nonnull reduced noetherian ring with height-zero primes $Q_1, \ldots, Q_n$ such that for $1 \leq i \leq n$ there is an epimorphism $\phi_i : R \to R_i$ with $\ker(\phi_i) = Q_i$. Then $\Phi = \phi_1 \oplus \cdots \oplus \phi_n : R \to T$ is a ring monomorphism with $f_i(\Phi(R)) = R_i$ for $1 \leq i \leq n$ and in a natural manner $\text{QR}(T)$ may be identified with $\text{QR}(\Phi(R))$.

In (Q18.4)(12) we prove the following: A finite direct sum of PIRs is itself a PIR. Every PIR is isomorphic to a finite direct sum of PIDs and SPIRs. A PIR whose zero ideal is primary is an SPIR or a PID according as its dimension is zero or positive. A nonnull PIR is an SPIR or a PID $\Leftrightarrow$ its zero ideal is primary $\Leftrightarrow$ it has exactly one height-zero prime ideal.

In **(Q19) on Invertible Ideals, Conditions for Normality, And DVRs**, we introduce **Serre Condition** (R_n) : $P \in \text{spec}(R)$ with $\text{ht}_R P \leq n \Rightarrow R_P$ is regular, and **Serre Condition** (S_n) : $P \in \text{spec}(R)$ with $\text{reg}(R_P) < n \Rightarrow R_P$ is CM. These are conditions on a ring R. An R-submodule I of the $\text{QR}(R)$ is **invertible** if

$IJ = R$ for some R-submodule J of $QR(R)$; this is so iff $II^{-1} = R$. A **DVR** is the valuation ring of a **real discrete valuation**, i.e, a valuation whose value group is order isomorphic to $\mathbb{Z}$. A valuation is **trivial** if its value group is 0.

In **Krull Normality Lemma** (T82.2) we show that if P is a nonzero maximal ideal in a normal noetherian domain with $P^{-1} \neq R$ then P is invertible.

In **Krull Normality Theorem** (T83) we show that a noetherian domain R is normal iff: the local ring R_P is normal for every height-one prime P in R and we have $R = \bigcap_{\{P \in \operatorname{spec}(R):\operatorname{ht}_R P=1\}} R_P$.

In **Normal Local Ring Lemma** (T84) we show that for any local ring R we have the following.

(1) If $M(R)$ is an associated prime of the zero ideal in R then $QR(R) = R$ and hence R is normal.

(2) If R is reduced and normal then R is a domain.

(3) If R is normal and $M(R)$ is not an associated prime of the zero ideal in R then R is a domain.

In **Conditions for DVR** (T85) we show that for any domain R we have the following.

(1) R is a DVR $\Leftrightarrow$ R is a one dimensional normal local domain $\Leftrightarrow$ R is a one dimensional regular local domain $\Leftrightarrow$ R is a positive dimensional local PID (where a local PID means a local ring which is also a principal ideal domain) $\Leftrightarrow$ R is a local PID in which $M(R)^e$ with $e = 0, 1, 2, \ldots$ are exactly all the distinct nonzero ideals.

(2) If R is the valuation ring of a nontrivial valuation v then: R is noetherian $\Leftrightarrow$ v is real discrete.

(3) If R is local then: R is the valuation ring of a nontrivial valuation $\Leftrightarrow$ R is regular of dimension one.

In **Serre Criterion** (T88) we show that a noetherian ring R is a reduced ring iff it satisfies conditions (R_0) and (S_1), whereas a noetherian ring R is a reduced normal ring iff it satisfies conditions (R_1) and (S_2).

In **(Q20) on Dedekind Domains And Chinese Remainder Theorem** we introduce several related terms as follows. A **DD** (= **Dedekind Domain**) is a normal noetherian domain in which every nonzero prime ideal is maximal. A **DFI** (= Domain with Factorization of Ideals) is a domain in which every nonzero ideal is a product of nonzero prime ideals. A **UFI** (= domain with Unique Factorization of Ideals) is a DFI in which the said factorization is unique (up to order). A **GFI** (= domain with Group Factorization of Ideals) is a domain R such that the set of all nonzero fractional R-ideals form a (multiplicative commutative) group.

In **Characterization of DD** (T89) we show that: DD $\Leftrightarrow$ DFI $\Leftrightarrow$ UFI $\Leftrightarrow$ GFI.

In **Properties of DD** (T91) we show that in a Dedekind Domain we have the following: (1) If R is semilocal then R is a PID. (2) If I is any nonzero ideal in R then R/I is a PIR. (3) Every ideal in R is generated by two elements. (4) The localization of R at any multiplicative set in R is again a DD.

In **Chinese Remainder Theorem** (T93) we show that given any elements

$(x_i)_{1\le i\le n}$ and ideals $(A_i)_{1\le i\le n}$ in a Dedekind Domain R with $x_i - x_j \in A_i + A_j$ for all $i \ne j$ where $n \in \mathbb{N}_+$, there exists $x \in R$ with $x - x_i \in A_i$ for all i.

In **(Q21) on Real Ranks of Valuations And Segment Completions** we introduce the **real rank** $\rho(v)$ **of a valuation** v as the order type (taken under reverse inclusion) of the set of all nonmaximal prime ideals in its valuation ring. By (Q21)(3) this equals the real rank $\rho(G_v)$ of the value group G_v which by L2§7(D5) is the order type (under inclusion) of the set of all nonzero isolated subgroups of G_v, where **isolated subgroup** is a subgroup which is also a **segment**, i.e., a subset H of G_v such that: $g \in G_v$ and $h \in H$ with $|g| \le |h| \Rightarrow g \in H$. It follows that a valuation is **real** (i.e., its value group is order isomorphic to a subgroup of $\mathbb{R}$) iff its valuation ring has no nonzero nonmaximal prime ideal.

By a **segment cut** of a loset E we mean a nonempty subset L of E such that for all $(x,y) \in L \times (E \setminus L)$ we have $x < y$. By the **segment completion** of E we mean the set E^{sc} of all segment cuts of E which we (linearly) order by inclusion. For every $x \in E$ we put $D_E(x) = \{z \in E : z \le x\}$ and $D'_E(x) = \{z \in E : z < x\}$ and this gives us order preserving injections $D_E : E \to E^{sc}$ and $D'_E : E \to E^{sc}$. We say that E is **segment-complete** if the map D_E is surjective. Given any $F \subset E$, by putting $D_{F,E}(H) = \{z \in E : z \le x \text{ for some } x \in H\}$ for all $H \in F^{sc}$, we get an order preserving injection $D_{F,E} : F^{sc} \to E^{sc}$; we write $F^{sc} \doteq E$ to mean that $D_{F,E}(F^{sc}) = D_E(E)$. By the **core** of E we mean the set $C(E)$ of all $x \in E$ such that either $D'_E(x) = \emptyset$ or x has an immediate predecessor in E; note that, for any $x \in E$, an **immediate predecessor** of x in E means an element y in E such that $y < x$ and there is no $z \in E$ with $y < z < x$. We call E **segment-full** if E is order isomorphic to T^{sc} for some loset T. We call E **core-full** if $C(E)^{sc} \doteq E$. We call E **brim-full** if every nonempty subset of E has a max in E, and every element of E is the max in E of some subset of $C(E)$. By L2§3 any ordered abelian group is the value group of some valuation, and hence to characterize the real ranks of valuations it suffices to characterize the real ranks of ordered abelian groups.

In **Characterization Theorem for Real Ranks** (T97.2) we prove that: an order type is the real rank of some ordered abelian group $\Leftrightarrow$ it is the order type of the segment-completion of some loset $\Leftrightarrow$ it is the order type of a segment-full loset $\Leftrightarrow$ it is the order type of a core-full loset $\Leftrightarrow$ it is the order type of a brim-full loset.

In **(Q22) on Specializations And Compositions of Valuations** we define these concepts. Given any valuations v and w of a field K with $R_v \subset R_w$, we say that v is a **specialization** of w or w is a **generalization** of v, and we indicate this by writing $w \searrow v$ or $v \nearrow w$. By L4§12(R7), associated with any **valuation ring** R of a field K (i.e., which means the valuation ring of some valuation of K) we have the **divisibility valuation** γ_R of K with valuation ring R and value group $\Gamma(R) = K^\times / U(R)$; if $R = R_v$ then we may write γ_v (resp: Γ_v) instead of γ_{R_v} (resp: $\Gamma(R_v)$). By L4§12(R7) we get the following assertions (6) to (8) of (Q22):

(6) If v is any valuation of a field K then $P \mapsto (R_v)_P$ gives an inclusion reversing bijection of $\mathrm{spec}(R_v)$ onto the set of all valuation rings of K containing R_v and this

set coincides with the set of all subrings of K containing R_v, and hence these sets are losets under inclusion.

(7) If w is any valuation of a field K then, letting $\pi_v : R_v \to \Delta_v = R_v/M(R_v)$ be the residue class epimorphism, we have that $S \mapsto R = \pi_w^{-1}(S)$ gives a bijection of the set of all valuation rings of Δ_w onto the set of all valuation rings of K contained in R_w; in case $S = R_u$ for a valuation u of Δ_w, we denote the valuation γ_R of K by $w \odot u$ and call it the **composition** of w by u.

(8) In the situation of (6) we have $\gamma_v = w \odot (v/w)$.

In (Q23) on UFD Property of Regular Local Domains, we prove the **UFD Theorem** (T102) which says that every regular local domain is a UFD. This we deduce from the **Parenthetical Colon Lemma** (T98) which says that for any ideals I, J in a noetherian ring R, upon letting $L = (I : J)_R$, we have the following:

(1) $L = I \Leftrightarrow$ for every $P \in \text{ass}_R(R/I)$ we have $J \not\subset P$.

(2) If R is a local ring with $J = M(R) \neq 0$ and $L = I$, then $L = (I : xR)_R$ for some $x \in J \setminus J^2$.

(3) If R is a local domain and I, J are nonzero principal ideals in R then: there exist nonzero principal ideals I', J' in R for which we have $L = (I' : J')_R$ and $I' \not\subset M(R)L$.

In (Q24) on Graded Modules And Hilbert Polynomials, we introduce these polynomials in terms of what are called Hilbert functions.

Considering a graded ring $S = \sum_{i \in I} S_i$ with $R = S_0$ and additive abelian monoid I, by a **graded module** over S (or a graded S-module) we mean an S-module $V = \sum_{i \in I} V_i$ with internal direct sum of R-modules such that for all $s_i \in S_i$ and $v_j \in v_j$ with i, j in I we have $s_i v_j \in V_{i+j}$. The family $(V_i)_{i \in I}$ of R-submodules of V is called the gradation of V, and I is called the type of V. Elements of V_i are called **homogeneous elements** of degree i. Collectively, elements of $\cup_{i \in I} V_i$ are called homogeneous elements. The direct sum assumption tells us that every $v \in V$ has a unique expression $v = \sum_{i \in I} v_i$ with $v_i \in V_i$; the element v_i is called the **homogeneous component** of v of degree i, and collectively the elements v_i are called the homogeneous components of v. Note that $v_i = 0$ for almost all i.

An S-submodule U of V is said to be a **homogeneous submodule** if it satisfies one and hence all of the following three mutually equivalent conditions:

(i) $U = \sum_{i \in I}(U \cap V_i)$ (as additive groups).

(ii) The homogeneous components of every element of U belong to U.

(iii) U is generated (as an S-module) by homogeneous elements.

As in the case of an element, we call V_i the **homogeneous component** of V of degree i, and we note that for the homogeneous component U_i of U of degree i we have $U_i = U \cap V_i$. We use i-th homogeneous component as a synonym for homogeneous component of degree i. Also we may say component instead of homogeneous component.

By a **finite** graded S-module we mean a graded S-module V which is finitely generated as an S-module. In view of (1), this is equivalent to saying that V is generated by a finite number of homogeneous elements.

Note that a homogeneous ideal in S is nothing but a homogeneous S-submodule of S where we regard S as a graded S-module in an obvious manner.

Now let S be a naturally graded ring, i.e., suppose that $I = \mathbb{N}$. Assume that $R = S_0$ is an artinian ring. Given any graded module V over S, we put

$$\widehat{\mathfrak{h}}^\mu(S, V, n) = \ell_R(V_n)$$

and we call the function $\widehat{\mathfrak{h}}^\mu(S, V) : \mathbb{N} \to \mathbb{N} \cup \{\infty\}$, defined by $n \mapsto \ell_R(V_n)$, the **modulized Hilbert function** of S at V. In particular, given any homogeneous ideal Q in S, we note that S/Q is a graded module over S, and we put

$$\widehat{\mathfrak{h}}(S, Q, n) = \widehat{\mathfrak{h}}^\mu(S, S/Q, n) = \ell_R(S_n/Q_n)$$

and we call the function $\widehat{\mathfrak{h}}(S, Q) : \mathbb{N} \to \mathbb{N} \cup \{\infty\}$, defined by $n \mapsto \ell_R(S_n/Q_n)$, the **Hilbert function** of S at Q. Note that if R is a field then $\ell_R(V_n) = [V_n : R]$ and $\ell_R(S_n/Q_n) = [S_n/Q_n : R]$. In (Q26) to (Q28) we prove the following:

HILBERT FUNCTION THEOREM (T103). Let $X_1, \ldots, X_N$ with $N \in \mathbb{N}$ be indeterminates over an artinian ring R. Let $S = R[X_1, \ldots, X_N]$ be the naturally graded polynomial ring with $R = S_0$. Let V be a finite graded module over S. Then for every nonnegative integer n we have

$$\ell_R(V_n) < \infty \quad \text{for all} \quad n \in \mathbb{N}$$

and there exists a unique univariate polynomial $\mathfrak{h}^\mu(S, V, Z)$ in an indeterminate Z with coefficients in $\mathbb{Q}$ such that

$$\mathfrak{h}^\mu(S, V, n) = \widehat{\mathfrak{h}}^\mu(S, V, n) \quad \text{for all} \quad n >> 0.$$

Moreover, upon letting

$$\mathfrak{t}^\mu(S, V) = \text{ the } Z\text{-degree of } \mathfrak{h}^\mu(S, V, Z)$$

we have

$$\mathfrak{t}^\mu(S, V) < N,$$

and if $d = \mathfrak{t}^\mu(S, V) \in \mathbb{N}$ then we have

$$\mathfrak{h}^\mu(S, V, Z) = \sum_{0 \le i \le d} \frac{\mathfrak{g}_i^\mu(S, V) Z(Z-1) \ldots (Z - i + 1)}{i!} \quad \text{with} \quad \mathfrak{g}_i^\mu(S, V) \in \mathbb{Z}$$

and upon letting

$$\mathfrak{g}^\mu(S, V) = \mathfrak{g}_d^\mu(S, V)$$

(and assuming $d = \mathfrak{t}^\mu(S, V) \in \mathbb{N}$) we have

$$\mathfrak{g}^\mu(S, V) > 0,$$

and if R is a field then (without assuming $\mathfrak{t}^\mu(S, V) \in \mathbb{N}$) we have

$$\mathfrak{t}^\mu(S, V) = \begin{cases} \dim(S/(0 : V)_S) - 1 & \text{if } \dim(S/(0 : V)S) > 0 \\ -\infty & \text{otherwise.} \end{cases}$$

We respectively call the quantities $\mathfrak{h}^\mu(S, V, Z)$, $\mathfrak{t}^\mu(S, V)$, $\mathfrak{g}_i^\mu(S, V)$, $\mathfrak{g}^\mu(S, V)$, the **modulized Hilbert polynomial**, the **modulized Hilbert transcendence**, the **modulized i-th Hilbert subdegree**, and the **modulized Hilbert degree** of S at V.

[Note that, "for all $n >> 0$" means for all large (enough) n, i.e., for all integers n such that $n \geq n_0$ for some $n_0 \in \mathbb{N}$. Also note that if $V = S/Q$ where Q is a homogeneous ideal in S then $(0 : V)_S = Q$].

Given any homogeneous ideal Q in S, with S and R as in (T103), we put

$$\begin{cases} \mathfrak{h}(S, Q, Z) = \mathfrak{h}^\mu(S, S/Q, Z) \\ \mathfrak{t}(S, Q) = \mathfrak{t}^\mu(S, S/Q) \\ \mathfrak{g}_i(S, Q) = \mathfrak{g}_i^\mu(S, S/Q) \\ \mathfrak{g}(S, Q) = \mathfrak{g}^\mu(S, S/Q) \end{cases}$$

and we again respectively call these quantities the **Hilbert polynomial**, the **Hilbert transcendence**, the **i-th Hilbert subdegree**, and the **Hilbert degree** of S at Q.

In (Q25) we illustrate the case when Q is the ideal of a hypersurface in some projective space. In (Q26) to (Q28) we include some material to be used in the proof of (T103), which is completed in parts (Q28.3) and (Q28.5) of (Q28).

In (Q27) on Homogeneous Normalization, we homogenize Normalization Theorem (Q10)(T46) by proving assertions (T104) to (T106) stated below. We define the **positive portion** $\Omega(V)$ of a graded module $V = \sum_{i \in I} V_i$ over a graded ring $S = \sum_{i \in I} S_i$ by putting $\Omega(V) = \sum_{\{i \in I : i > 0\}} V_i$, and we call a submodule U of V **irrelevant or relevant** according as $\Omega(S) \subset \mathrm{rad}_V U$ or $\Omega(S) \not\subset \mathrm{rad}_V U$. In particular these definition apply to ideals in homogeneous rings. In (C40) we make further comments about them.

IRRELEVANT IDEAL LEMMA (T104). Let S be a homogeneous ring over a ring R, and let $z_1, \ldots, z_m$ be a finite number of homogeneous elements in $S \setminus R$. Then S is integral over $R[z_1, \ldots, z_m]$ iff the ideal $(z_1, \ldots, z_m)S$ is irrelevant.

HOMOGENEOUS NORMALIZATION THEOREM (T105). In the naturally graded polynomial ring $S = k[X_1, \ldots, X_N]$ in indeterminates $X_1, \ldots, X_N$ over a field k with $N \in \mathbb{N}$ we have the following.

(T105.1) Given any homogeneous element Z_N in $S \setminus k$, there exist homogeneous elements $Z_1, \ldots, Z_{N-1}$ in $S \setminus k$ such that S is integral over $k[Z_1, \ldots, Z_N]$.

(T105.2) Given any nonunit homogeneous ideal J in S, there exists a positive integer c together with nonzero homogeneous polynomials $Y_1, \ldots, Y_N$ in S of degree

c such that S is integral over $D = k[Y_1, \ldots, Y_N]$ and $J \cap D = (Y_{r+1}, \ldots, Y_N)D$ where $r = \dim(S/J)$.

(T105.3) Given any finite sequence $J_0 \subset \cdots \subset J_m$ of nonunit homogeneous ideals in S with $m \in \mathbb{N}$, there exists a positive integer c together with nonzero homogeneous polynomials $Y_1, \ldots, Y_N$ in S of degree c such that S is integral over $D = k[Y_1, \ldots, Y_N]$ and $J_i \cap D = (Y_{r_i+1}, \ldots, Y_N)D$ for $0 \le i \le m$ where we have $r_i = \dim(S/J_i)$. Moreover, for $0 \le i \le m$ and any such polynomials $Y_1, \ldots, Y_N$, upon letting $\phi_i : S \to S/J_i$ be the residue class epimorphism, the ring $\phi_i(S)$ is integral over the subring $\phi_i(D)$, and the latter ring is naturally isomorphic to the polynomial ring $T_i = k[X_1, \ldots, X_{r_i}]$ by the isomorphism $\psi_i : T_i \to \phi_i(D)$ such that $\psi_i(\kappa) = \phi_i(\kappa)$ for all $\kappa \in k$ and $\psi_i(X_j) = \phi_i(Y_j)$ for $1 \le j \le r_i$.

HOMOGENEOUS RING NORMALIZATION THEOREM (T106). Given any homogeneous ring over a field k, and given any finite sequence $I_1 \subset \cdots \subset I_m$ of nonunit homogeneous ideals in A with $m \in \mathbb{N}$, there exists a positive integer c together with nonzero homogeneous elements $Y_1, \ldots, Y_r$ in A of degree c with $\dim(A) = r$ such that the elements $Y_1, \ldots, Y_r$ are algebraically independent over k and such that the ring A is integral over the subring $D = k[Y_1, \ldots, Y_r]$ and $I_j \cap D = (Y_{r_j+1}, \ldots, Y_r)D$ for $1 \le j \le m$ where $r_j = \dim(A/I_j)$. Moreover, for $1 \le j \le m$ and any such elements $Y_1, \ldots, Y_r$, upon letting $\phi_j : A \to A/I_j$ be the residue class epimorphism, the ring $\phi_j(A)$ is integral over the subring $\phi_j(D)$, and the latter ring is naturally isomorphic to the polynomial ring $T_j = k[X_1, \ldots, X_{r_j}]$ by the isomorphism $\psi_j : T_j \to \phi_j(D)$ such that $\psi_j(\kappa) = \phi_j(\kappa)$ for all $\kappa \in k$ and $\psi_j(X_i) = \phi_j(Y_i)$ for $1 \le i \le r_j$.

COMMENT (C40). [**Irrelevant Ideals and Integral Dependence**]. To apply (T104) we use the following facts. Let S be a homogeneous ring over a ring R, let $z_1, \ldots, z_m$ be a finite number of homogeneous elements in $S \setminus R$, and for $1 \le i \le m$ let $t_i = z_i^{d_i}$ with $d_i \in \mathbb{N}_+$. Then $S/R[z_1, \ldots, z_m]$ is integral iff $S/R[t_1, \ldots, t_m]$ is integral, and the ideal $(z_1, \ldots, z_m)S$ is irrelevant iff the ideal $(t_1, \ldots, t_m)S$ is irrelevant. If $z_i \in S_{d_i}$ with $d_i \in \mathbb{N}_+$ for $1 \le i \le m$ then upon letting $d = d_1 \ldots d_m$ and $t_i = z_i^{d/d_i}$ we get $t_i \in S_d$. So in checking irrelevancy or integralness we may assume $z_1, \ldots, z_m$ are homogeneous of the same positive degree.

In (Q29) on **Linear Disjointness And Intersection of Varieties** we prove **Dimension of Intersection Theorem** (T112) which says that, if the intersection U of any two varieties V and W in the N-dimensional affine space over a field k is nonempty, then $\dim(U) \ge \dim(V) + \dim(W) - N$. As preparation for (T112), we prove **Codimension of Product Theorem** (T111) which says that, given any two irreducible affine varieties V and V' (in different affine spaces), the codimension of any irreducible component of their product $V \times V'$ equals the sum of their codimensions; the **codimension** of a variety V in affine N-space means $N - \dim(V)$. As preparation for (T111), we prove some properties of linear disjointness; given any subrings R and R' of a ring S, such that $R \cap R'$ contains a subfield k of S, we say that R and R' are **linearly disjoint** over k to mean that: if $(x_i)_{1 \le i \le m}$ are any finite number of elements of R which are linearly independent over k and $(x'_j)_{1 \le j \le m'}$ are any finite number of elements of R' which are linearly independent over k, then the mm' elements $x_i x'_j$ of S are linearly independent over k.

This linear disjointness is particularly significant for the ring-theoretic compositum of R and R'. The ring-theoretic **compositum** of subrings R, R' of a ring S is defined to be the smallest subring of S which contains both R and R'. Equivalently it may be defined to be the intersection of all subrings of S which contain both R and R'. For any subring k of the ring $R \cap R'$, the said compositum clearly coincides with $k[R, R']$, where we recall that for any subring k of a ring S, and any subset E of S, by $k[E]$ we denote the subring of S consisting of all finite sums $\sum a_{i_1 \ldots i_n} z_1^{i_1} \ldots z_n^{i_n}$ with $n, i_1, \ldots, i_n$ in $\mathbb{N}$, $a_{i_1 \ldots i_n}$ in k, and $z_1, \ldots, z_n$ in E. In all this, instead of two subrings R, R' of S, we may take any set (or family) of subrings of S. Here we may everywhere change "ring" to "field" with the understanding that $k[E]$ is replaced by $k(E)$ and the finite sums are replaced by quotients of finite sums with nonzero denominators. Usually the adjective "field-theoretic" is dropped from the phrase "field-theoretic compositum."

In (Q30) on Syzygies And Homogeneous Resolutions we give Hilbert's original version of (Q17). In (Q17) we discussed projective resolutions of modules. In their original incarnation, as was conceived by David Hilbert, they started out as **free resolutions** of homogeneous ideals in polynomial rings; in that set-up, what we have called the pdim (= projective dimension) of a module corresponds to the **hdim (= homogeneous dimension)** of a graded module; for details see the text of (Q30).

All this gives rise to syzygies, originally an astronomical term to indicate the waning and waxing of the phases of the moon, adapted by Sylvester to indicate a mathematical object. Actually in (Q17) we have already worked with syzygies except that we did not give them a name. So what are they?

Now in the definition of a projective R-resolution (f, W, n) given in (Q17), by not requiring the last module W_{n+1} to be projective, we get hold of the definition of a **preprojective R-resolution**.

Given any noetherian ring R, any finite R-module V, and any $n \in \mathbb{N}$, we can find a preprojective R-resolution (f, W, n) of V of length n, and we call W_{n+1} an n-th R-**syzygy** of V. We can see that if W'_{n+1} is any other n-th R-syzygy of V then the modules W_{n+1} and W'_{n+1} are R-equivalent, in symbols $W_{n+1} \sim_R W'_{n+1}$, by which we mean that for some finite projective R-modules U, U' we have an isomorphism $W_{n+1} \oplus U \approx W'_{n+1} \oplus U'$ of R-modules. By $\mathrm{syz}_R^n V$ we denote the collection of all n-th R-syzygies of V.

In (Q31) on Projective Modules Over Polynomial Rings, we give two proofs (due to Suslin) of the freeness of any finite projective module over a finite variable polynomial ring over a ground ring K. The first applies when K is a PID, and the second works when K is a field. The main technique in the second proof is the idea of **completing unimodular rows**. A unimodular row over a ring R is an n-tuple $r = (r_1, \ldots, r_n) \in R^n$ which generates the unit ideal in R. This is completable means it is the first column of some matrix $A = (A_{ij})$ in $\mathrm{GL}(n, R)$, i.e., $A_{i1} = r_i$ for $1 \leq i \leq n$.

The **general linear group** $\mathrm{GL}(n, R)$ consists of all $n \times n$ matrices over R whose determinant is a unit in R. The **special linear group** $\mathrm{SL}(n, R)$ consists of all $n \times n$ matrices over R whose determinant is 1. The **special elementary group** $\mathrm{SE}(n, R)$ is the subgroup of $\mathrm{SL}(n, R)$ generated by all elementary matrices, i.e., matrices from the identity matrix by changing at most one zero entry to a nonzero entry. The **general elementary group** $\mathrm{GE}(n, R)$ is the subgroup of $\mathrm{GL}(n, R)$ generated by $\mathrm{SL}(n, R)$ and all restricted dilatation matrices.

As a byproduct of the second proof we establish **Suslin's Theorem** (T130) which says that for $n > 2$ and $R = K[X_1, \ldots, X_N]$ is a finite variable polynomial ring over a field K we have $\mathrm{SL}(n, R) = \mathrm{SE}(n, R)$. In (T131) we give a presentation of **Cohn's example** showing that (T130) is not true for $n = N = 2$. The proof of (T130) uses Theorem (T129) on properties of **Mennike Symbols** which are certain equivalence classes of $\mathrm{SL}(3, n)$ mod $\mathrm{SE}(3, n)$.

In (T132) to (T134) we discuss **permutation matrices**, i.e, $n \times n$ matrices of the form $T(n, \sigma) = (T(n, \sigma)_{ij})$ where $\sigma(i) = 1$ or 0 according as $i = \sigma(j)$ or $i \neq \sigma(j)$ for some permutation $\sigma \in S_n =$ the **symmetric group** on n letters.

In **(Q32) on Separable Extensions And Primitive Elements**, we establish Theorem L4§8(T20) by proving that, given any affine domain $T = k[x_1, \ldots, x_N]$ over an algebraically closed field k, upon letting $L = k(x_1, \ldots, x_N)$, there exist k-linear combinations $y_i = \sum_{1 \leq j \leq N} A_{ij} x_j$ with $A_{ij} \in k$ such that $y_1, \ldots, y_r$ is a transcendence basis of L/k and $L = k(y_1, \ldots, y_{r+1})$. In (T144) we give a stronger version of this result. Firstly, in (T143) we show that suitable k-linear combinations $y_1, \ldots, y_r$ of $x_1, \ldots, x_N$ constitute a **separating transcendence basis** of L/k, i.e., a transcendence basis so that L is separable algebraic over $K = k(y_1, \ldots, y_r)$ as defined below. Secondly, in (T140) we show that a suitable k-linear combination y_{r+1} of $x_1, \ldots, x_N$ provides a **primitive element** of L over K, i.e, an element for which $L = K(y_{r+1})$. Actually, in (T143) we show that suitable k-linear combinations $y_1, \ldots, y_r$ of $x_1, \ldots, x_N$ provide a **separating normalization basis** of T/k, i.e., a separating transcendence basis of L/K such that the ring T is integral over the ring $R = k[y_1, \ldots, y_r]$. Using (T140), in (T141) we show that the integral closure of a normal noetherian domain in a finite separable algebraic extension of its quotient field is a finite module over that domain, and hence the integral closure S of R in L is a finite R-module; it follows that S, which is also the integral closure of T in L, is a finite T-module. As an aid to (T141), in (T136) we embed the integral closure in a ring obtained by inverting the **modified discriminant** $\mathrm{Disc}_Y^*(f)$ of a univariate polynomial

$$f(Y) = a_0 Y^n + a_1 Y^{n-1} + \cdots + a_n$$

which is defined by putting

$$\mathrm{Disc}_Y^*(f) = (-1)^{n(n-1)/2}\mathrm{Disc}_Y(f).$$

According to L1§1 and L1§8, a univariate polynomial $f = f(Y)$ over a field K is **separable** if it is devoid of multiple roots; if f is not separable then it is **inseparable**. An element y in an overfield of K is **separable** over K if $f(y) = 0$ for some separable f over K, i.e., equivalently, if y/K is algebraic and its minimal polynomial over K is separable; if y/K is algebraic but not separable then y/K is **inseparable**; if $\mathrm{ch}(K) = p \neq 0$ (i.e., if the characteristic of K is a prime number p) and $y^{p^u} \in K$ for some $u \in \mathbb{N}_+$ then y/K is **purely inseparable**; if $\mathrm{ch}(K) = 0$ then y/K is purely inseparable means $y \in K$. A field extension L/K is separable (resp: purely inseparable) means every element of L is separable (resp: purely inseparable) over K. A field extension L/K is inseparable means it is algebraic but not separable. A field is **perfect** means every algebraic extension of it is separable; otherwise it is **imperfect**. In (T139) we show that every characteristic zero field is perfect and so is every finite field. In (T142) we show that every algebraic function field over a perfect ground field is separably generated according to the following definitions. By an **algebraic function field** L (of r variables) over a ground field k we mean a finitely generated field extension of k (with $\mathrm{trdeg}_k L = r$). The field extension L/k is **separably generated** means there exists a transcendence basis $y_1, \ldots, y_r$ of L/k such that L is a separable algebraic field extension of $K = k(y_1, \ldots, y_r)$; we then call $y_1, \ldots, y_r$ a **separating transcendence basis** of L/k; in case of $r = 1$ we call y_1 a **separating transcendental** of L/k.

In Quests **(Q33)** to **(Q35)** we continue the **geometrizing project** of L3§§2-5 and L4§§8-9. Out of these Quests, (Q33) is on Restricted Domains And Projective Normalization, (Q34) is on Homogeneous Localization, and (Q35) is on Simplifying Singularities by Blowups.

In **(Q33)** we define a **restricted domain** to be a noetherian domain T such that the integral closure of T in any finite algebraic extension of the quotient field L of T is a finite module over T, and we define a **strongly restricted domain** to be a restricted domain T such that every affine domain over T is again a restricted domain. In Theorem (T145) we show that any affine domain over a field is a restricted domain, i.e., a field is a strongly restricted domain. In Theorems (T146) and (T147) we respectively use this to discuss "normalizations" of affine and projective varieties in finite algebraic field extensions of their function fields.

In **(Q34.1)** we define the **Projective Spectrum proj**(A) of any subintegrally graded ring $A = \sum_{i \in I} A_i$, where I is an additive submonoid of $\mathbb{Z}$, by putting

$$\mathrm{proj}(A) = \text{ the set of all relevant homogeneous prime ideals in } A.$$

In **(Q34.2)** we define the **homogeneous quotient field** $\mathfrak{K}(D)$ of a semihomogeneous domain $D = \sum_{n \in \mathbb{N}} D_n$ by putting

$$\mathfrak{K}(D) = \bigcup_{n \in \mathbb{N}} \{y_n/z_n : y_n \in D_n \text{ and } z_n \in D_n^\times\}$$

and we note that this is a subfield of $\mathrm{QF}(D)$. For any $P \in \mathrm{proj}(D)$ we define the **homogeneous localization** $D_{[P]}$ of D at P to be the subring of $\mathfrak{K}(D)$ given by

$$D_{[P]} = \bigcup_{n \in \mathbb{N}} \{y_n/z_n : y_n \in D_n \text{ and } z_n \in D_n \setminus P\}$$

and we note that this is a quasilocal domain with quotient field $\mathfrak{K}(D)$ and

$$M(D_{[P]}) = \bigcup_{n \in \mathbb{N}} \{y_n/z_n : y_n \in P \cap D_n \text{ and } z_n \in D_n \setminus P\}.$$

Assuming $D \neq 0$, we define the **modelic proj** $\mathfrak{W}(D)$ of D by putting

$$\mathfrak{W}(D) = \begin{cases} \text{the set of all homogeneous localizations } D_{[P]} \\ \text{with } P \text{ varying over } \mathrm{proj}(A). \end{cases}$$

Note that then

$$\mathfrak{W}(D) = \mathfrak{W}(D_0, D_1)$$

and if $(x_l)_{l \in \Lambda}$ is any family of generators of D_1 as a module over D_0 then

$$\mathfrak{W}(D) = \mathfrak{W}(D_0; (x_l)_{l \in \Lambda})$$

and hence in particular if D_1, as a module over D_0, is generated by a finite number of generators $x_1, \ldots, x_{N+1}$ in $D_1^\times$ then

$$\mathfrak{W}(D) = \cup_{i=1}^{N+1} \mathfrak{V}(D_0[x_1/x_i, \ldots, x_{N+1}/x_i])$$

where according to the notation of L4§§8-9: we define the **modelic spec** $\mathfrak{V}(A)$ of any domain A by putting

$\mathfrak{V}(A) =$ the set of all localizations A_P with P varying over $\mathrm{spec}(A)$,

we define the **modelic blowup** $\mathfrak{W}(A, P)$ of any subdomain A of a field K and any nonzero A-submodule P of K by putting

$$\mathfrak{W}(A, P) = \bigcup_{0 \neq x \in P} \mathfrak{V}(A[Px^{-1}]),$$

and we define the **modelic proj** $\mathfrak{W}(A; (x_l)_{l \in \Lambda})$ of any family $(x_l)_{l \in \Lambda}$ of elements in K, with $x_j \neq 0$ for some $j \in \Lambda$, by putting

$$\mathfrak{W}(A; (x_l)_{l \in \Lambda}) = \bigcup_{j \in \Lambda \text{ with } x_j \neq 0} \mathfrak{V}(A[(x_l/x_j)_{l \in \Lambda}]);$$

if Λ is a finite set, say $\Lambda = \{1, \ldots, n\}$, then we may write $\mathfrak{W}(A; x_l, \ldots, x_n)$ instead of $\mathfrak{W}(A; (x_l)_{l \in \Lambda})$ and call it the **modelic proj** of $(x_1, \ldots, x_n)$ over A.

In (Q35) we define the **dominating modelic blowup** $\mathfrak{W}(R, P)^\Delta$ of a nonzero ideal P in a quasilocal domain R by putting

$$\mathfrak{W}(R, P)^\Delta = \{R' \in \mathfrak{W}(R, P) : R' > R\}$$

where a quasilocal ring R' **dominates** a quasilocal ring R, in notation $R' > R$ or $R < R'$, means R is a subring of R' with $M(R) = M(R') \cap R$. We are particularly interested in the case when R is a regular local domain and $P = R \cap M(S)$ for some $S \in \mathfrak{V}(R)$ such that S has a **simple point** at R by which we mean that R/P is a regular local domain. Any $R' \in \mathfrak{W}(R, P)^\Delta$ is now called a **monoidal transform** of (R, S), and by the (R, S, R')-**transform** of any nonzero principal ideal J in R we mean the ideal J' in R' given by $J'(M(R)R')^d = JR'$ where $d = \mathrm{ord}_R J$. Also we call (R', J') a **monoidal transform** of (R, J, S). We define $\mathfrak{E}(R, J)$ by putting $\mathfrak{E}(R, J) =$ the set of all $T \in \mathfrak{V}(R)$ with $\mathrm{ord}_T J = d$, and we call this the **equimultiple locus** of (R, J); for any $i \in \mathbb{N}$ we let $\mathfrak{E}^i(R, J)$ denote the set of all i-dimensional members of $\mathfrak{E}(R, J)$. Upon letting $d' = \mathrm{ord}_{R'} J'$, in (T158)(2) we show that if $S \in \mathfrak{E}(R, J)$ then $d' \leq d$; geometrically this says that the **multiplicity does not increase when we blowup an equimultiple simple center**. For any field K we put

$$\mathfrak{R}(K) = \text{the set of all valuation rings of } K$$

i.e., valuation rings with quotient field K, and for any subdomain A of K we put

$$\mathfrak{R}(K/A) = \text{the set of all valuation rings of } K/A$$

i.e., valuation rings of K which contains A; we respectively call these sets the **Riemann-Zariski space** of K and the Riemann-Zariski space of K/A. Now $\mathfrak{W}(R, P)$ is a **complete model** of K/R with $\mathrm{QF}(R) = K$, i.e., every $V \in \mathfrak{R}(K/P)$ dominates a unique $R' \in \mathfrak{W}(R, P)$ (which is called the **center** of V on $\mathfrak{W}(R, P)$); it follows that if V dominates R then $R' \in \mathfrak{W}(R, P)^\Delta$; in this case we call R' the **monoidal transform of** (R, S) **along** V; similarly we also call (R', J') the **monoidal transform of** (R, J, S) **along** V. In case of $S = R$ we may say **quadratic transform** of R or (R, J) instead of monoidal transform of (R, S) or (R, J, S) respectively.

To explain the above concept of "complete model" in greater detail we reproduce the definitions made in L4§9. For any field K we put

$$\mathfrak{R}'(K) = \text{the set of all quasilocal domains with quotient field } K$$

and for any subdomain A of K we put

$$\mathfrak{R}'(K/A) = \text{the set of all members of } \mathfrak{R}'(K) \text{ which contain } A$$

and we call these the **quasitotal Riemann-Zariski space** of K and the quasitotal Riemann-Zariski space of K/A respectively. We also put

$$\mathfrak{R}''(K) = \text{the set of all quasilocal domains which are subrings of } K$$

and

$$\mathfrak{R}''(K/A) = \text{the set of all members of } \mathfrak{R}''(K) \text{ which contain } A$$

and we call these the **total Riemann-Zariski space** of K and the total Riemann-Zariski space of K/A respectively. Domination converts $\mathfrak{R}(K), \ldots, \mathfrak{R}''(K/A)$ into posets (= partially ordered sets). A valuation v **dominates** a quasilocal ring R means R_v dominates R. By a **premodel** of K (resp: K/A) we mean a nonempty subset E of $\mathfrak{R}'(K)$ (resp: $\mathfrak{R}'(K/A)$). The premodel E is **irredundant** means any member of $\mathfrak{R}(K)$ (resp: $\mathfrak{R}(K/A)$) dominates at most one member of E. A quasilocal domain S can dominate at most one member R of an irredundant premodel E, and if R exists then we call it the **center** of S on E. By a **semimodel** (resp: **model**) of K/A we mean an irredundant premodel E of K/A which can be expressed as a union $E = \cup_{l \in \Lambda} \mathfrak{V}(B_l)$ for some family (resp: finite family) $(B_l)_{l \in \Lambda}$ of subrings B_l of K with quotient field K such that B_l is an overring of (resp: affine ring over) A. Note that if B is any subring of K with quotient field K such that B is an overring of (resp: affine ring over) A then $\mathfrak{V}(B)$ is a semimodel (resp: model) of K/A; we call it an **affine semimodel** (resp: **affine model**) of K/A. By a **projective model** of K/A we mean premodel E of K/A such that $E = \mathfrak{W}(A; x_1, \ldots, x_n)$ for some finite number of elements $x_1, \ldots, x_n$ in an overfield of K at least one of which is nonzero. A quasilocal ring S **dominates** a set E of quasilocal rings (or E is dominated by S) means S dominates some member of E; we may indicate this by writing $E < S$ (or $S > E$). A set E' of quasilocal rings **dominates** a set E of quasilocal rings (or E is dominated by E') means every member of E' dominates E; we may indicate this by writing $E < E'$ (or $E' > E$). A set E' of quasilocal rings **properly dominates** a set E of quasilocal rings (or E is properly dominated by E') means E' dominates E and every member of E is dominated by some member of E'. A semimodel or model of K/A is **complete** means it is dominated by $\mathfrak{R}(K/A)$.

Every projective model of K/A is easily seen to be a complete model of K/A. A model E (of K/A) is said to be **normal** (resp: **noetherian**) if every $R \in E$ is normal (resp: noetherian). A model E is said to be **nonsingular** if every $R \in E$ is a regular local ring. By the **dimension** $\dim(E)$ of a model E we mean $\max\{\dim(R) : R \in E\}$; note that then $\dim(E) \in \mathbb{N} \cup \{\infty\}$.

What we said about normalization in the above summary of (Q33) can now be said more precisely thus. Given any quasilocal subring R of any domain $\overline{K}$, by $\mathfrak{N}(R, \overline{K})$ we denote the set of all $S \in \mathfrak{V}(\overline{R})$ such that S dominates R where $\overline{R}$ is the integral closure of R in $\overline{K}$, and we call this the **normalization** of R in $\overline{K}$. Likewise, given any set E of quasilocal subrings of a domain $\overline{K}$, we introduce the **normalization** of E in $\overline{K}$ by putting $\mathfrak{N}(E, \overline{K}) = \cup_{R \in E} \mathfrak{N}(R, \overline{K})$.

In (T145) and (T147) we show that if E is a projective model of K/k where K is an algebraic function field over a field k, and $\overline{K}$ is a finite algebraic field extension of K, then $\mathfrak{N}(E, \overline{K})$ is a projective model of $\overline{K}/k$. Given any irreducible curve C in the projective plane over k, as an affine equation of C we can take a bivariate irreducible polynomial $f = f(X, Y)$ in $k[X, Y] \setminus k[X]$. Now we can take E to be the "modelic" projective line $\mathfrak{V}(k[X]) \cup \mathfrak{V}(k[1/X])$. We may then take K to be $k(X)$ and identify $\overline{K}$ with the function field of C given by $\mathrm{QF}(k[X, Y]/fk[X, Y])$. Now upon letting $\overline{C} = \mathfrak{N}(E, \overline{K})$, by Conditions (Q19)(T85) we see that $\overline{C}$ is a "modelic" nonsingular projective curve. This is Dedekind's 1882 method of desingularizing C. Riemann's 1851 method amounts to saying that, again by (Q19)(T85), $\mathfrak{N}(E, \overline{K})$ coincides with $\mathfrak{R}(\overline{K}/k)$. Max Noether's 1877 method, which is the only one so far capable of higher-dimensional generalization, uses monoidal transformations.

To give the definitions of some concepts which are relevant in Noether's method, let R be an n-dimensional regular local domain, let $E \subset \mathfrak{V}(R)$, and let I, J be nonzero principal ideals in R. We say that E has a **normal crossing** at R if there exists a regular system of parameters $x_1, \ldots, x_n$ for R such that for every $S \in E$ we have $y_S R = R \cap M(S)$ for some subset y_S of $\{x_1, \ldots, x_n\}$, and we say that E has a **strict normal crossing** at R if E has a normal crossing at R and E contains at most two elements. We say that I has a **normal crossing** at R if $\{S' \in \mathfrak{V}(R) : \dim(S') = 1 \text{ and } IS' \neq S'\}$ has a normal crossing at R. We say that (E, I) has a **normal crossing** at R if $E \cup \{S' \in \mathfrak{V}(R) : \dim(S') = 1 \text{ and } IS' \neq S'\}$ has a normal crossing at R, and we say that (E, I) has a strict normal crossing at R if (E, I) has a normal crossing at R and E contains at most two elements. Given $S \in \mathfrak{V}(R)$, we say that (S, I) has a normal crossing at R if $(\{S\}, I)$ has a normal crossing at R. We say that (J, I) has a **quasinormal crossing** at R if I has a normal crossing at R and for every nonzero principal prime ideal P in R with $J \subset P$ we have that PI has a normal crossing at R. We say that I has a **quasinormal crossing** at R if (I, R) has a quasinormal crossing at R. Given $S \in \mathfrak{V}(R)$, we say that (S, I) has a **pseudonormal crossing** at R if S has a simple point at R and for every nonzero principal prime ideal P in R with $I \subset P$ we have that $\{S, R_P\}$ has a normal crossing at R. We say that (E, I) has a **pseudonormal crossing** at R if I has a quasinormal crossing at R and for every $S \in E$ we have that (S, I) has a pseudonormal crossing at R. (R, J) is **resolved** means there exists a nonnegative integer d and a nonzero principal ideal J' in R with $\mathrm{ord}_R J' \leq 1$ such that $J = J'^d$.

Domination and Subgroups. In domination, $R < S$ indicates subset. In subgroups, $H \leq G$ indicates subset while $H < G$ indicates proper subset.

§6: DEFINITIONS AND EXERCISES

In Exercises (E1) to (E6) we shall give hints for completing the proofs of the six Galois Theory Theorems (T1) to (T6) of L1§8. In Exercises (E7) to (E50) we shall deal further with questions of Galois Theory, Group Theory, and Field Theory. In the remaining Exercises we shall deal with questions arising out of Lecture L5. For the basic definitions of separable and inseparable extensions etc., used in the following Exercises, see Lecture L1 and §5(Q32) of Lecture L5.

EXERCISE (E1). [**Fundamental Theorem of Galois Theory**]. Prove (T1) of L1§8. In other words, let L^*/K be a Galois extension, i.e., L^* is the splitting field of a nonconstant separable univariate polynomial $f(Y)$ over the field K. Let Λ be the set of all subfields L of L^* with $K \subset L$ and let us put $G(L) = \mathrm{Gal}(L^*, L)$, i.e, $G(L)$ is the Galois group of L^*/L which means $G(L)$ is the group of all automorphisms g of L^* such that $g(y) = y$ for all $y \in L$; note that L^*/L is a Galois extension because L^* is clearly the splitting field of $f(Y)$ over L. Let Γ be the set of all subgroups H of $G(L^*)$ and let us put $F(H) = \mathrm{fix}_{L^*}(H)$, i.e., $F(H)$ is the fixed field of H which means $F(H) = \{z \in L^* : g(z) = z \text{ for all } g \in H\}$. Prove that then (1) $L \mapsto G(L)$ gives a bijection $\Lambda \to \Gamma$ whose inverse is given by $H \mapsto F(H)$. Moreover show that (2) for every $L \in \Lambda$ we have $|G(L)| = [L^* : L]$.

HINT. Given any L and L' in Λ with $L \subset L'$, by L5§5(Q32)(T140) we can find $x \in L'$ with $L' = L(x)$. Let $e(Y)$ be the minimal polynomial of x over L. By L5§5(Q32)FACT(15) we know that $e(Y)$ is separable. Hence $e(Y) = \prod_{z \in W}(Y - z)$ where W is a subset of some overfield of L^* with $x \in W$, and upon letting m to be the Y-degree of $e(Y)$ we have $|W| = m \in \mathbb{N}_+$.

First for a moment suppose that $L' = L^*$. Then by (E2) below we have $W \subset L^*$ and $g \mapsto g(x)$ gives a bijection $G(L) \to W$, and hence $|G(L)| = [L^* : L]$. This proves (2). The said bijection also shows that if $H < G(L)$ then upon letting $W' = \{g(x) : g \in H\}$ and $m' = |W'|$ we get $m' \in \mathbb{N}_+$ with $m' < m$. Let $e'(Y) = \prod_{g \in H}(Y - g(x))$. Then $e'(Y) \in F(H)[Y]$ with $e'(x) = 0$; since $\deg_Y e'(Y) = m' < m$ and $e(Y)$ is the minimal polynomial of x/L, we must have $e'(Y) \notin L[Y]$; therefore $L \subsetneq F(H)$. This shows that (3) for every $L \in \Lambda$ and $H < G(L)$ we have $L \subsetneq F(H)$.

Next for a moment suppose that $L \neq L'$. Then we can find $x' \in W$ with $x' \neq x$. By (E7) below there exists $g \in G(L)$ with $g(x) = x'$, and hence $G(L') < G(L)$. This shows that (4) for any L, L' in Λ with $L \subsetneq L'$ we have $G(L') < G(L)$.

In view of L5§6(D2), to prove (1) it suffices to show that (1*) for every $L \in \Lambda$ we have $F(G(L)) = L$, and (2*) for every $H \in \Gamma$ we have $G(F(H)) = H$.

To prove (2*), given any $H \in \Gamma$, let $L = F(H)$; then $L \in \Lambda$ with $H \subset G(L)$, and hence by (3) we must have $H = G(L)$.

To prove (1*), given any $L \in \Lambda$, let $L' = F(G(L))$; then $L \subset L'$ with $G(L') = G(L)$, and hence by (4) we must have $L = L'$.

DEFINITION (D1). [**Normal Extensions**]. By a normal (field) extension of a field K we mean an algebraic extension L^* of K such that whenever a nonconstant irreducible univariate polynomial $e(Y) \in K[Y]$ has a root in L^* then it has all its roots in L^*, i.e., if $e(x) = 0$ for some $x \in L^*$ and $e(x') = 0$ for some x' in an overfield of L^* then $x' \in L^*$. It is clear that if L^* is a normal extension of K then L^* is a splitting field over K of a family J of nonzero univariate polynomials over K. It is also clear that if L^* is a finite normal extension of K then L^* is a splitting field of one single polynomial $f(Y)$; namely L^* is generated over K by a finite number of algebraic elements $x_1, \ldots, x_r$ and we can take $f(Y)$ to be the product of their minimal polynomials. The converse is proved in the following Exercise.

EXERCISE (E2). [**Splitting Fields**]. Let $f(Y)$ be a nonconstant univariate polynomial with coefficients in a field K, let L^* be a splitting field of f over K, let L be a field between K and L^*, let $e(Y)$ be a nonconstant irreducible polynomial in Y with coefficients in L, and let x, x' be roots of $e(Y)$ in an overfield of L^* with $x \in L^*$. Show that then $x' \in L^*$, and the L-isomorphism of root fields $\phi : L(x) \to L(x')$ given by $x \mapsto x'$ can be extended to some $g \in \mathrm{Gal}(L^*, L)$.

HINT. You are only being asked to give a more TRANSPARENT version of a part of (R6) to (R8) of L2§5 which was repeated in several sentences of L2§6 ending with the phrase "ϕ can be extended to an isomorphism of the splitting field L_J of $J = \{f\}$ over $L(x)$ onto the splitting field $L'_{J'}$ of $J' = \{f\}$ over $L(x')$." Note that since g is an L-isomorphism and L^* is a splitting field of f over L, we must have $g(L^*) = L^*$. Therefore $L^*(x') = L^*$ and $g \in \mathrm{Gal}(L^*, L)$.

Turning to the more TRANSPARENT version, clearly L^* and $L^*(x')$ are splitting fields of f over $L(x)$ and $L(x')$ respectively, and hence ϕ can be extended to an isomorphism $g : L^* \to L^*(x')$. Since L^* is a splitting field of f over K, its image $g(L^*)$ must be a splitting field of f over K. Therefore $x' \in g(L^*)$, and hence $L^*(x') = L^*$. It follows that $g \in \mathrm{Gal}(L^*, K)$ with $g(x) = x'$.

DEFINITION (D2). [**Lagrange Resolvent**]. Let $n \in \mathbb{N}_+$ and let K be a field which contains n distinct n-th roots of 1, and note that then by L5§5(Q27)(C41) and L5§5(Q32)(T137) n is not divisible by the characteristic of K, and K contains a primitive n-th root ζ of 1, i.e., there exists $\zeta \in K$ such that $\zeta^n = 1$ but $\zeta^m \neq 1$ for all $m < n$ in $\mathbb{N}_+$. Let L be any n-cyclic extension of K, i.e., a Galois extension of K such that $\mathrm{Gal}(L, K)$ is a cyclic group of order n, and let σ be a generator of the said Galois group. For any $i \in \mathbb{N}$ and $\theta \in L$, we define the Lagrange resolvent (ζ^i, θ) of ζ^i and θ relative to σ by putting

$$(\zeta^i, \theta) = \sum_{0 \le j \le n-1} \sigma^j(\theta) \zeta^{ij}$$

and we note that clearly

$$\sigma((\zeta^i, \theta)) = (\zeta^i, \theta) \zeta^{-i}$$

and hence

$$\sigma((\zeta^i, \theta)^n) = (\zeta^i, \theta)^n.$$

Now for every $i \in \{1, \ldots, n-1\}$ we have $\zeta^i - 1 \neq 0$ with

$$\sum_{0 \leq j \leq n-1} \zeta^{ij} = \frac{\zeta^{in} - 1}{\zeta^i - 1} = 0$$

and hence upon letting

$$A(\theta) = \{i \in \{0, \ldots, n-1\} : (\zeta^i, \theta) \neq 0\}$$

we get

$$\sum_{i \in A(\theta)} (\zeta^i, \theta) = n\theta.$$

By L5§5(Q32)(T140) we can find $u \in L$ with $L = K(u)$, and then by the above equation we get $L = K((\zeta^i, u)_{i \in A(u)})$. It follows that if p is any prime divisor of n and $v(p)$ is the largest integer such that $p^{v(p)}$ divides n then for some $i(p) \in A(u)$ we must have $\mathrm{GCD}(p^{v(p)}, i(p)) = 1$. Let $u_p = (\zeta^{i(p)}, u)^{n/p^{v(p)}}$. Then upon letting $\alpha - \prod u_p$, where the product is taken over all prime divisors p of n, we get $L - K(\alpha)$ with $0 \neq \alpha^n = a \in K$.

EXERCISE (E3). [**Cyclic Extensions of degree prime to characteristic**]. Prove (T2) of L1§8. In other words, let $n \in \mathbb{N}_+$ and let K be a field which contains n distinct n-th roots of 1. In (D2) above we have shown that any n-cyclic extension of K is obtained by adjoining an n-th root of some nonzero element a of K, i.e., it is of the form $K(\alpha)$ with $0 \neq \alpha^n = a \in K$. Show that conversely, for any $0 \neq a \in K$, the splitting field L of $Y^n - a$ over K is an m-cyclic extension of K where m is the largest divisor of n for which the polynomial $Y^m - a$ is irreducible in $K[Y]$.

[Note that L1§8(T3) follows from this Exercise and the above Exercise (E1)].

HINT. Now $Y^n - a = \prod_{\eta \in U}(Y - \alpha\eta)$ with $\alpha \in L$ and $U = $ the group of all n-th roots of 1 in K. Therefore $L = K(\alpha)$ and $g \mapsto g(\alpha)/\alpha$ gives a monomorphism $\mathrm{Gal}(L, K) \to U$. By (D2) above, U is cyclic of order n. So we are done by (E7) below.

DEFINITION (D3). [**Elementary Symmetric Functions**]. To relate the roots and the coefficients of a polynomial, consider the monic polynomial $F(Y)$ of degree $n \in \mathbb{N}_+$ in an indeterminate Y with coefficients in $L = k(Z_1, \ldots, Z_n)$, where $Z_1, \ldots, Z_n$ are indeterminates over a field k, given by

$$F(Y) = \prod_{1 \leq i \leq n}(Y - Z_i) = Y^n + \sum_{1 \leq i \leq n}(-1)^i E_i(Z_1, \ldots, Z_n)Y^{n-i}.$$

We call

$$E_i = E_i(Z_1, \ldots, Z_n) \in k[Z_1, \ldots, Z_n]$$

the i-th elementary symmetric function of $Z_1, \ldots, Z_n$ and we note that

$$
\begin{cases}
E_1 = Z_1 + \cdots + Z_n \\
E_2 = Z_1 Z_2 + \cdots + Z_1 Z_n + Z_2 Z_3 + \cdots + Z_{n-1} Z_n \\
E_3 = Z_1 Z_2 Z_3 + \cdots + Z_{n-2} Z_{n-1} Z_n \\
\vdots \\
E_{n-1} = Z_1 \ldots Z_{n-1} + \cdots + Z_2 \ldots Z_n \\
E_n = Z_1 \ldots Z_n
\end{cases}
$$

and more generally

$$
E_i = \sum_{B \in A(i)} \prod_{b \in B} Z_b
$$

where

$$
A(i) = \text{the set of all subsets of } \{1, \ldots, n\} \text{ of size } i
$$

with

$$
|A(i)| = \binom{n}{i} = \frac{n(n-1)\ldots(n-i+1)}{i!}.
$$

In particular 0 and 1 are the only elements of k which occur as coefficients in E_i. Thus E_i makes sense when k is replaced by any nonnull ring. In the expressions of $F(Y)$ and E_i we may substitute for $Z_1, \ldots, Z_n$ any values $z_1, \ldots, z_n$ in any overfield of k to get $f(Y)$ and e_i, and then e_i will be the i-th elementary symmetric function of the roots $z_1, \ldots, z_n$ of $f(Y)$ whose coefficients will be $\pm e_i$ depending on the parity of i. The functions E_i are obviously symmetric in the following sense.

DEFINITION (D4). [**Symmetric Functions**]. Continuing with (D3) above, let S_n be the group of all permutations of $\{1, \ldots, n\}$. Call a function $u(Z_1, \ldots, Z_n) \in L$ symmetric if for every $\sigma \in S_n$ we have $u(Z_{\sigma(1)}, \ldots, Z_{\sigma(n)}) = u(Z_1, \ldots, Z_n)$.

Let $K = k(E_1, \ldots, E_n)$. Then clearly L/K is a splitting field of the separable polynomial $F(Y)$ over K. Consequently $E_1, \ldots, E_n$ is a transcendence basis of L/K, and L/K is Galois with

$$
\mathrm{Gal}(L, K) \to \mathrm{Gal}(F, K) \subset \mathrm{Sym}(\{Z_1, \ldots, Z_n\}) \to S_n
$$

where the first arrow is the isomorphism which sends every $g \in \mathrm{Gal}(L, K)$ to its restriction $g|_{\{Z_1,\ldots,Z_n\}}$, and the second arrow is the isomorphism which sends every $\tau \in \mathrm{Sym}(\{Z_1, \ldots, Z_n\})$ to the unique $\mu(\tau) \in S_n$ such that $\tau(Z_i) = Z_{\mu(\tau)(i)}$ for $1 \le i \le n$. Given any $\sigma \in S_n$, we get a k-automorphism $\nu(\sigma)$ of L which sends every $u(Z_1, \ldots, Z_n) \in L$ to $u(Z_{\sigma(1)}, \ldots, Z_{\sigma(n)}) \in L$, and clearly $\nu(\sigma)$ is a K-automorphism with $\mu(\nu(\sigma)|_{\{Z_1,\ldots,Z_n\}}) = \sigma$. Therefore

$$
\mathrm{Gal}(F, K) = \mathrm{Sym}(\{Z_1, \ldots, Z_n\}).
$$

Since K is the fixed field of $\mathrm{Gal}(L, K)$, we see that K coincides with the set of all symmetric functions. Consequently, every symmetric function $u(Z_1, \ldots, Z_n)$ can be expressed uniquely as a rational function of the elementary symmetric functions, i.e, there exists a unique n-variable rational function

$$r(X_1, \ldots, X_n) \in k(X_1, \ldots, X_n)$$

such that

$$u(Z_1, \ldots, Z_n) = r(E_1, \ldots, E_n).$$

As polynomial rings, $k[Z_1, \ldots, Z_n]$ and $k[E_1, \ldots, E_n]$ are normal domains with quotients field L and K respectively. Since the elements $Z_1, \ldots, Z_n$ are obviously integral over $k[E_1, \ldots, E_n]$, the entire ring $k[Z_1, \ldots, Z_n]$ is integral over the subring $k[E_1, \ldots, E_n]$, and hence we get

$$K \cap k[Z_1, \ldots, Z_n] = k[E_1, \ldots, E_n].$$

Thus, assuming the symmetric rational function $u(Z_1, \ldots, Z_n)$ to be a polynomial, we must have

$$r(X_1, \ldots, X_n) \in k[X_1, \ldots, X_n].$$

Giving weight j to X_j for $1 \le j \le n$, we can write

$$r(X_1, \ldots, X_n) = \sum_{i \in \mathbb{N}} r_i(X_1, \ldots, X_n)$$

where r_i is isobaric [cf. L4§1(O3)] of weight i (or $r_i = 0$). Also let us write

$$u(X_1, \ldots, X_n) = \sum_{i \in \mathbb{N}} u_i(X_1, \ldots, X_n)$$

where u_i is homogeneous of degree i (or $u_i = 0$). Comparing terms of equal degree we must have

$$u_i(Z_1, \ldots, Z_n) = r_i(E_1, \ldots, E_n) \quad \text{for all} \quad i \in \mathbb{N}.$$

EXERCISE (E4). [**Newton's Symmetric Function Theorem**]. Prove (T4) of L1§8. In other words, show that for the generic n-th degree polynomial

$$F(Y) = Y^n + X_1 Y^{n-1} + \cdots + X_n$$

over $K = k(X_1, \ldots, X_n)$, where $X_1, \ldots, X_n$ are indeterminates over a field k, we have $\mathrm{Gal}(F, K) = S_n$.

HINT. See (D4) above.

EXERCISE (E5). [**More on Newton's Symmetric Function Theorem**]. Give a direct proof (without Galois theory) of the General Symmetric Function

Theorem which was established in (D4) above when k is a field, but you are to do this for any nonnull ring k. In other words, let $Z_1, \ldots, Z_n$ be indeterminates over a nonnull ring k with $n \in \mathbb{N}_+$, and show that every symmetric polynomial $u(Z_1, \ldots, Z_n) \in k[Z_1, \ldots, Z_n]$ can be uniquely expressed as

$$u(Z_1, \ldots, Z_n) = r(E_1, \ldots, E_n) \quad \text{with} \quad r(X_1, \ldots, X_n) \in k[X_1, \ldots, X_n]$$

where $X_1, \ldots, X_n$ are indeterminates and $E_1, \ldots, E_n$ are the elementary symmetric functions of $Z_1, \ldots, Z_n$. Also show that in the notation of (D4) we have

$$u_i(Z_1, \ldots, Z_n) = r_i(E_1, \ldots, E_n) \quad \text{for all} \quad i \in \mathbb{N}$$

and hence in particular the weight of $r(X_1, \ldots, X_n)$ equals the (total) degree of $u(Z_1, \ldots, Z_n)$; the weight of r is defined to be the largest weight of a monomial occurring in r if $r \neq 0$, and $-\infty$ if $r = 0$.

[Note that L1§8(T5) is proved in L1§11(E5) and, in view of this, L1§8(T6) follows from the above Exercises (E1), (E3), and (E4)].

HINT. The assertion about the u_i follows from the assertion about u, which we shall prove by double induction on n and the degree d of u. For $n = 1$ the assertion is obvious because then $E_1 = Z_1$. So let $n > 1$ and assume true for $n - 1$. Again the assertion is obvious for $d \leq 0$. So let $d > 0$ and assume for all smaller values of the degree. Let $F_1, \ldots, F_{n-1}$ be obtained by putting $Z_n = 0$ in $E_1, \ldots, E_{n-1}$ respectively. Clearly $u(Z_1, \ldots, Z_{n-1}, 0)$ is a symmetric polynomial in $Z_1, \ldots, Z_{n-1}$ of degree $\leq d$ and hence by the $n - 1$ case we get

$$u(Z_1, \ldots, Z_{n-1}, 0) = \overline{r}(F_1, \ldots, F_{n-1})$$

where $\overline{r}(X_1, \ldots, X_{n-1}) \in k[X_1, \ldots, X_{n-1}]$ has weight $\leq d$. Let

$$\overline{u}(Z_1, \ldots, Z_n) = u(Z_1, \ldots, Z_n) - \overline{r}(E_1, \ldots, E_{n-1}) \in k[Z_1, \ldots, Z_n]$$

and let $\overline{d}$ be the degree of $\overline{u}$. Then $\overline{d} \leq d$. Clearly $\overline{u}(Z_1, \ldots, Z_{n-1}, 0) = 0$ and hence $\overline{u}(Z_1, \ldots, Z_n)$ is divisible by Z_n. Also clearly $\overline{u}(Z_1, \ldots, Z_n)$ is symmetric and hence it is divisible by $Z_1 \ldots Z_n$. Thus

$$\overline{u}(Z_1, \ldots, Z_n) = v(Z_1, \ldots, Z_n) E_n$$

where $v(Z_1, \ldots, Z_n) \in k[Z_1, \ldots, Z_n]$ is symmetric of degree $e \leq d - n$. Therefore by the induction hypothesis on d, there is a unique $s(X_1, \ldots, X_n) \in k[X_1, \ldots, X_n]$ of weight e such that $v(Z_1, \ldots, Z_n) = s(E_1, \ldots, E_n)$. Now by letting

$$r(X_1, \ldots, X_n) = s(X_1, \ldots, X_n)X_n + \overline{r}(X_1, \ldots, X_{n-1})$$

we see that $r(X_1, \ldots, X_n) \in k[X_1, \ldots, X_n]$ has weight d and

$$u(Z_1, \ldots, Z_n) = r(E_1, \ldots, E_n).$$

EXERCISE (E6). [**Relations Preserving Permutations**]. Let there be given any nonconstant monic separable polynomial $f = f(Y) = \prod_{1 \leq i \leq n}(Y - \alpha_i) = Y^n + \sum_{1 \leq i \leq n} a_i Y^{n-1}$ with coefficients a_i in a field K. Show that then $\mathrm{Gal}(f, K)$ is the set of all relations preserving permutations of the roots, i.e., the set of all $\tau \in \mathrm{Sym}(\{\alpha_1, \ldots, \alpha_n\})$ such that for every n-variable polynomial $P(X_1, \ldots, X_n) \in K[X_1, \ldots X_n]$ with $P(\alpha_1, \ldots, \alpha_n) = 0$ we have $P(\tau(\alpha_1), \ldots, \tau(\alpha_n)) = 0$.

EXERCISE (E7). [**Cyclic Groups**]. Verify FACTS (5) to (9) of L5§5(Q32). In particular show that for any cyclic group W of finite order n, the mapping $V \mapsto |V| = m$ gives a bijection of the set of all subgroups of W onto the set of all positive integers which divide n, and the groups V and W/V are cyclic of order m and n/m respectively.

DEFINITION (D5). [**Artin-Schreier Extensions**]. Replacing the multiplicative group of roots of unity by the underlying additive group of the field $\mathrm{GF}(p)$, the cyclic extension of (E3) is converted to an Artin-Schreier extension, i.e., to the splitting field L of the Artin-Schreier polynomial $f(Y) = Y^p - Y - a$ over a field K of characteristic $p \neq 0$ for some $a \in K$. We claim that $\mathrm{Gal}(L, K)$ is a cyclic group of order p or 1 according as $f(Y)$ is irreducible in $K[Y]$ or not. In other words, if $\alpha^p - \alpha = a$ for some $\alpha \in K$ then $f(Y)$ factors into linear factors in $K[Y]$, and otherwise it is irreducible in $K[Y]$ with $\mathrm{Gal}(L, K)$ cyclic of order p. Note that if $\alpha^p - \alpha = a$ for some $\alpha \in L$ then for every $i \in \mathrm{GF}(p) \subset K$ we have $(\alpha + i)^p - (\alpha + i) = a$; it follows that if $\alpha \notin K$ then $g \mapsto g(\alpha) - \alpha$ gives an isomorphism of the multiplicative group $\mathrm{Gal}(L, K)$ onto the underlying additive group of $\mathrm{GF}(p)$. This proves the claim.

EXERCISE (E8). [**Cyclic Extensions of degree equal to characteristic**]. In (D5) above we have shown that the splitting field of the separable polynomial $f(Y) = Y^p - Y - a$ over a field of characteristic $p \neq 0$ with $a \in K$ is p-cyclic or 1-cyclic according as $f(Y)$ is irreducible in $K[Y]$ or not. Conversely show that any p-cyclic extension L of a field K of characteristic $p \neq 0$ is the splitting field of a separable irreducible polynomial of the form $f(Y) = Y^p - Y - a$ with $a \in K$.

HINT. Let σ be a generator of $\mathrm{Gal}(L, K)$. Since $[L : K]$ is a prime number, for any $u \in L \setminus K$ we have $L = K(u)$. Let $u_i = \sigma^i(u)$ and $s_l = \sum_{1 \leq i \leq p} u_i^l$. Also let $M = (M_{ij})$ be the $p \times p$ matrix with $M_{ij} = u_{i-1}^{j-1}$, and let $N = (N_{ij})$ be the $p \times p$ matrix with $N_{ij} = M_{ij}$ or s_{j-1} according as $i \neq 1$ or $i = 1$. Then N is obtained by adding the last $p - 1$ rows of M to its first row, and hence their determinants are equal. By L5§5(Q32)(T135) we see that the determinant of M is the Vandermonde Determinant $V(u_1, \ldots, u_p)$ which is nonzero. Therefore the determinant of N is nonzero. But $N_{11} = p = 0$, and hence for some $j \in \{2, \ldots, p\}$ we must have $N_{1j} \neq 0$. Let $l = j - 1$. Then $l \in \{1, \ldots, p - 1\}$ with $s_l = N_{1j}$, and hence $s_l \neq 0$.

But s_j is a symmetric function of $u_1, \ldots, u_p$, and hence $s_l \in K$. Let

$$\alpha = -s_l^{-1} \sum_{0 \leq i \leq p-1} iu_i^l.$$

Then

$$\sigma(\alpha) = -s_l^{-1} \sum_{0 \leq i \leq p-1} iu_{i+1}^l = -s_l^{-1} \sum_{0 \leq i \leq p-1} [(i+1)u_{i+1}^l - u_{i+1}^l] = \alpha + 1$$

and hence upon letting

$$f(Y) = Y^p - Y - a \quad \text{with} \quad a = \alpha^p - \alpha$$

we see that $f(Y)$ is a separable irreducible polynomial over K and its splitting field is L.

DEFINITION (D6). [**Field Polynomials and Norms and Traces**]. Out of the n elementary symmetric functions, the first and the last (i.e., the n-th), which are called the trace and the norm, are the easiest to use because they are respectively additive and multiplicative. We shall now introduce them formally. So let L/K be a finite algebraic field extension, and let n be the field degree, i.e., $[L : K] = n \in \mathbb{N}_+$. Given any $z \in L$, let $m = [K(z) : K]$ and let $e(Y)$ be the minimal polynomial of z over K. Then

$$e(Y) = \prod_{1 \leq i \leq m} (Y - y_i) = Y^m + \sum_{1 \leq i \leq m} b_i Y^{m-i}$$

with b_i in K and y_i in an overfield of L such that $y_j = z$ for some j. Now $n/m \in \mathbb{N}_+$ and we define the field polynomial of z relative to the field extension L/K to be the monic polynomial $f(Y)$ of degree n obtained by putting

$$f(Y) = e(Y)^{n/m} = \prod_{1 \leq i \leq n} (Y - z_i) = Y^n + \sum_{1 \leq i \leq n} a_i Y^{n-i}$$

with a_i in K and z_i in an overfield of L such that $z_j = z$ for some j. Note that $(-1)^i a_i$ is the i-th elementary symmetric function of $z_1, \ldots, z_n$. We define the norm $N_{L/K}(z)$ and the trace $T_{L/K}(z)$ of z relative to L/K by putting

$$N_{L/K}(z) = (-1)^n a_n \quad \text{and} \quad T_{L/K}(z) = -a_1$$

i.e.,

$$N_{L/K}(z) = z_1 \ldots z_n \quad \text{and} \quad T_{L/K}(z) = z_1 + \cdots + z_n.$$

DEFINITION (D7). [**Degrees of Separability and Inseparability**]. In the above situation, the roots $y_1, \ldots, y_m$ are distinct iff z is separable over K. If z is inseparable over K then we must have $\text{ch}(K) = p \neq 0$, and we may ask how many times does each root repeat. To answer this we define the exponent of inseparability of $e(Y)$, i.e., of any monic irreducible polynomial $e(Y)$ of degree $m \in \mathbb{N}_+$ over a field K of characteristic $p \neq 0$, to be the largest $\epsilon \in \mathbb{N}$ such that $e(Y) \in K[Y^{p^\epsilon}]$, and we note that then

$$m = \mu p^\epsilon \quad \text{with} \quad \mu \in \mathbb{N}_+.$$

We call μ and p^ϵ the degrees of separability and inseparability of $e(Y)$ respectively.

EXERCISE (E9). [**Exponent of Inseparability**]. Concerning the exponent of inseparability ϵ defined above, upon letting

$$x = z^{p^\epsilon} \in L$$

show that

$$e(Y) = d(Y^{p^\epsilon})$$

where

$$d(Y) = \prod_{1 \le i \le \mu} (Y - x_i) = Y^\mu + \sum_{1 \le i \le \mu} c_i Y^{\mu-i}$$

with $c_i = b_{ip^\epsilon}$ in K and x_i in an overfield of L such that $x_j = x$ for some j. Also show that $d(Y)$ is the minimal polynomial of the separable algebraic element x over K, and $Y^{p^\epsilon} - x$ is the minimal polynomial of the purely inseparable element z over $K(x)$. Thus the $(1/p^\epsilon)$-th powers of $x_1, \ldots, x_\mu$ are distinct, and each of them is repeated p^ϵ times amongst the roots $y_1, \ldots, y_m$ of $e(Y)$, and $(n/m)p^\epsilon$ times amongst the roots $z_1, \ldots, z_n$ of $f(Y)$.

HINT. See L5§5(Q27)(C41) and L5§5(Q32)(T142).

EXERCISE (E10). [**Behavior Under Finite Algebraic Field Extensions**]. In (D6) show that if L' is any field between K and L with $z \in L'$ then, upon letting $\delta = n/n'$ with $n' = [L' : K]$, for the field polynomial $f'(Y)$ of z relative to L'/K we have

$$f(Y) = f'(Y)^\delta$$

and for the norms and traces we have

$$N_{L/K}(z) = (N_{L'/K}(z))^\delta \quad \text{and} \quad T_{L/K}(z) = \delta T_{L'/K}(z).$$

EXERCISE (E11). [**Norm Giving Field Polynomial**]. In (D6) we defined the norm in terms of the field polynomial. Show that conversely the field polynomial $f(Y)$ of any element $z \in L$ relative to any finite algebraic field extension L/K can be expressed in terms of a norm by the formula

$$f(Y) = N_{L(Y)/K(Y)}(Y - z)$$

where we note that $L(Y)/K(Y)$ is clearly a field extension whose field degree is equal to the field degree n of L/K [cf.L5§6(E26)].

HINT. Taking an indeterminate T over $L(Y)$ and letting

$$\widehat{e}(Y) = (-1)^m e(T - Y)$$

we see that $\widehat{e}(Y)$ is the minimal polynomial of $T - z$ over $K(T)$ and hence

$$N_{K(T)(T-z)/K(T)}(T - z) = (-1)^m \widehat{e}(0) = e(T)$$

and therefore replacing T by Y we get

$$e(Y) = N_{K(z)(Y)/K(Y)}(Y - z).$$

In view of the norm portion of (E10), our assertion follows by raising both sides of the above equation to their δ-th powers.

DEFINITION (D8). [**Field Polynomial as Characteristic Polynomial**]. Noting that in (D6), L is a K-vector space of dimension n, multiplication by z gives the K-linear transformation (= map) $\tau_z : L \to L$ defined by $\tau_z(x) = zx$ for all $x \in L$. In (E12) below we shall show that the field polynomial $f(Y)$ equals the characteristic polynomial of any matrix of τ_z as defined in (D9) below.

DEFINITION (D9). [**Spur and Characteristic Matrix**]. Let K be a field and let L be a K-vector space of dimension $n \in \mathbb{N}_+$. Let $v = (v_1, \ldots, v_n)$ be a basis of L. The characteristic matrix $\mathrm{cmat}(A)$ and likewise the characteristic polynomial $\mathrm{cpol}(A)$ of an $n \times n$ matrix $A = (A_{ij})$ over K are defined by putting

$$\mathrm{cmat}(A) = YI_n - A \quad \text{and} \quad \mathrm{cpol}(A) = \det(YI_n - A)$$

where I_n is the $n \times n$ identity matrix and $\mathrm{cmat}(A)$ is an $n \times n$ matrix over $K[Y]$. Note that then

$$\mathrm{cpol}(A) = Y^n + \sum_{1 \le i \le n} \widehat{a}_i Y^{n-1} \quad \text{with} \quad \widehat{a}_i \in K.$$

Also we define the spur of A by putting

$$\mathrm{spur}(A) = \sum_{1 \le i \le n} A_{ii}.$$

We are particularly interested in these quantities when A is the matrix of a K-linear transformation (= map) $\tau : L \to L$ relative to v, i.e., when we have

$$\tau(v_j) = \sum_{1 \le i \le n} A_{ij} v_i \quad \text{for} \quad 1 \le j \le n.$$

Moreover we are also interested in the matrix B of τ relative to any other basis $w = (w_1, \ldots, w_n)$ of L.

EXERCISE (E12). [**Properties of Norms and Traces**]. In the situation of (D9) show that

(1) $$\mathrm{cpol}(B) = \mathrm{cpol}(A)$$

with

(2) $$\det(B) = \det(A) = (-1)^n \widehat{a}_n \quad \text{and} \quad \mathrm{spur}(B) = \mathrm{spur}(A) = -\widehat{a}_1.$$

Also show that if $\tau = \tau_z$ with τ_z as in (D8) then

(3) $$f(Y) = \mathrm{cpol}(A)$$

with

(4) $$N_{L/K}(z) = \det(A) \quad \text{and} \quad T_{L/K}(z) = \mathrm{spur}(A).$$

Moreover show that, in the situation of (D6), for any $z^* \in L$ we have

(5) $$N_{L/K}(zz^*) = N_{L/K}(z)N_{L/K}(z^*)$$

and

(6) $$T_{L/K}(z + z^*) = T_{L/K}(z) + T_{L/K}(z^*)$$

and for any $\kappa \in K$ we have

(7) $$N_{L/K}(\kappa) = \kappa^n \quad \text{and} \quad T_{L/K}(\kappa) = n\kappa$$

and

(8) $$N_{L/K}(\kappa z) = \kappa^n N_{L/K}(z) \quad \text{and} \quad T_{L/K}(\kappa z) = \kappa T_{L/K}(z).$$

Finally show that in the situation of (D6) we have

(9) $$N_{L/K}(z) = (y_1 \ldots y_m)^{n/m}$$

and

(10) $$T_{L/K}(z) = (n/m)(y_1 + \cdots + y_m)$$

and if $\mathrm{ch}(K) = p \neq 0$ then in the situation of (D7) and (E9) we have

(11) $$N_{L/K}(z) = (x_1 \ldots x_\mu)^{(n/m)p^\epsilon}$$

and

(12) $$T_{L/K}(z) = (n/m)p^\epsilon(x_1 + \cdots + x_\mu).$$

HINT. Since v and w are both bases of L, we have $B = CAC^{-1}$ for some $C \in \mathrm{GL}(n, K)$. Since the matrix $Y I_n$ commutes with every $n \times n$ matrix, we get $\mathrm{cmat}(B) = C\mathrm{cmat}(B)C^{-1}$, and hence by taking determinants we obtain (1).

Now $B = CAC^{-1}$ implies $\det(B) = \det(A)$ and $\mathrm{spur}(B) = \mathrm{spur}(A)$, and an easy calculation with the determinant $\mathrm{cpol}(A)$ yields the equations $\det(A) = (-1)^n \widehat{a}_n$ and $\mathrm{spur}(A) = -\widehat{a}_1$, which proves (2).

Now let the situation be as in (D8), and let $\tau = \tau_z$. Before turning to (3), let $L' = K(z)$ and $\delta = n/n'$ with $n' = m = [L' : K]$, let us consider the K-basis $u = (u_1, u_2, \ldots, u_m) = (1, z, \ldots, z^{m-1})$ of L', let $\tau' = \tau'_z : L' \to L'$ be the K-linear transformation defined by taking $\tau'_z(x) = zx$ for all $x \in L'$, and let $A' = (A'_{ij})$ be the $m \times m$ matrix of τ' relative to u. Then $A'_{im} = b_{m-i+1}$, and for $1 \leq j \leq m-1$ we have $A'_{ij} = 0$ or 1 according as $i \neq j+1$ or $i = j+1$. Therefore by expanding in terms of the last column we see that

$$(3') \qquad\qquad\qquad e(Y) = \mathrm{cpol}(A').$$

Let $t_1, t_2, \ldots, t_\delta$ be an L'-basis of L. For the alternative basis w we may assume that $w_{m(i-1)+j} = t_i u_j$ for $1 \leq i \leq \delta$ and $1 \leq j \leq m$. Now B is a diagonal block matrix with A' repeated δ times along the principal diagonal. Therefore $\mathrm{cpol}(B) = (\mathrm{cpol}(A'))^\delta$, and hence by (1) and (3') we get (3). By (2) and (3) we get (4).

For any $z^* \in L$ let A^* be the matrix of τ_{z^*} relative to v. Then clearly AA^* and $A + A^*$ are the respective matrices of τ_{zz^*} and τ_{z+z^*} relative to v. Therefore (5) and (6) follow from (4).

For any $\kappa \in K$, by (D6) we see that $N_{K/K}(\kappa) = T_{K/K}(\kappa) = \kappa$ and hence by (E10) we get (7). By (5) and (7) we get the first part of (8). The second part of (8) follows from (4) by noting that for the matrix $\overline{A}$ of $\tau_{\kappa z}$ we clearly have $\overline{A} = \kappa A$.

Finally (9) to (12) follow from the definition (D6).

EXERCISE (E13). [Condition For Inseparable Element]. In (D6) show that if z is inseparable over K then $T_{L/K}(z) = 0$.

HINT. See part (12) of (E12).

EXERCISE (E14). [Extending Derivations and Separable Extensions]. Referring to L1§12(N1) for the definition of derivations, and mimicking the calculus method of implicit differentiation discussed in L3§2, show that, given any (not necessarily finite) separable algebraic field extension L/K, every $D \in \mathrm{Der}(K, K)$ has a unique extension $E \in \mathrm{Der}(L, L)$, i.e, for any derivation D of K there is a unique derivation E of L such that for all $x \in K$ we have $E(x) = D(x)$.

HINT. For showing uniqueness, let $E \in \mathrm{Der}(L, L)$ be any extension of any given $D \in \mathrm{Der}(K, K)$. For any $y \in L$, let

$$f(Y) = Y^n + a_1 Y^{n-1} + \cdots + a_n$$

be its minimal polynomial over K. Then in view of L1§12(E12) we see that

$$E(f(y)) = f_Y(y)E(y) + f_D(y)$$

where

$$f_Y(Y) = nY^{n-1} + \sum_{1 \leq i \leq n-1} (n-i)a_i Y^{n-i-1}$$

and

$$f_D(Y) = \sum_{1 \leq i \leq n} D(a_i) Y^{n-i}.$$

Since $f(y) = 0$, we also have $E(f(y)) = 0$. By L5§5(Q27)(C41) we get $f_Y(y) \neq 0$, and hence

$$E(y) = -f_D(y)/f_Y(y)$$

which proves uniqueness.

Turning to existence, given any $D \in \text{Der}(K, K)$, for every $y \in L$ we define $E(y) \in K(y)$ by the above formula. Now clearly $E(x) = D(x)$ for all $x \in K$. To show that $E \in \text{Der}(L, L)$, in view of the primitive element theorem L5§5(Q32)(T140), it suffices to prove this in the case when $L = K(y)$ for some y. In view of the uniqueness proved above, this case is a consequence of the $|I| = 1$ case of the following Exercise (E15).

DEFINITION (D10). [**Polynomials in a Family and Pure Transcendental Extensions**]. As in L4§10(E17) we can consider the polynomial ring

$$R_I = K[Y]$$

in a family of (distinct) indeterminates

$$Y = (Y_i)_{i \in I}$$

over a field K where I is any indexing set which need not be finite. This makes sense for any ring K, but here we continue to assume K to be field. Given any family of elements

$$y = (y_i)_{i \in I}$$

in an overfield L of K, as in L1§6 we have the substitution map

$$\Phi_I : R_I \to L$$

which is the unique K-homomorphism of rings such that

$$\Phi_I(Y_i) = y_i \quad \text{for all} \quad i \in I.$$

Again L could be an overring of a ring K, but here we assume both to be fields. Let there be given any $D \in \text{Der}(K, L)$ and let

$$u = (u_i)_{i \in I}$$

be any other family of elements in L.

Note that y is a transcendence basis of L/K means $\ker(\Phi_I) = 0$ and $L/K(y)$ is algebraic. If $L = K(y)$ for some transcendence basis y of L/K then we say that L/K is pure transcendental. If $\ker(\Phi_I) = 0$ and $L/K(y)$ is separable algebraic then we call y a separating transcendence basis of L/K. If L/K has a separating transcendence basis then we say that L/K is separably generated. In accordance with L5§5(Q32) we say that L/K is finitely separably generated if L/K is finitely generated as well as separably generated.

Let Ω be the set of all finite subsets J of I, and let us label the elements of any such J as $J(i)_{1 \leq i \leq |J|}$ and let us put $m = |J|$ with $(Z_i, z_i, v_i) = (Y_{J(i)}, y_{J(i)}, u_{J(i)})$ for $1 \leq i \leq m$ where the reference to J is invisible in m, Z_i, z_i, v_i. Then clearly

$$\ker(\Phi_I) = \bigcup_{J \in \Omega} \ker(\Psi_J)$$

where

$$\Psi_J : S_J = K[Z_1, \ldots, Z_m] \to L$$

is the substitution map sending Z_i to z_i for $1 \leq i \leq m$.

For any $J \in \Omega$ clearly there is a unique homomorphism $\Psi_{D,J} : S_J \to L$ of underlying additive groups such that for all

$$f = f(Z_1, \ldots, Z_m) = \sum_{i_1, \ldots, i_m \text{ in } \mathbb{N}} a_{i_1 \ldots i_m} Z_1^{i_1} \ldots Z_m^{i_m} \in S_J$$

with $a_{i_1 \ldots i_m} \in K$ we have

$$\Psi_{D,J}(f) = f_D(z_1, \ldots, z_m) + \sum_{1 \leq i \leq m} f_{Z_i}(z_1, \ldots, z_m) v_i$$

where

$$f_D(Z_1, \ldots, Z_m) = \sum_{i_1, \ldots, i_m \text{ in } \mathbb{N}} D(a_{i_1 \ldots i_m}) Z_1^{i_1} \ldots Z_m^{i_m}$$

and

$$f_{Z_i}(Z_1, \ldots, Z_m) = \text{the } (Z_i)\text{-partial derivative of } f(Z_1, \ldots, Z_m).$$

Also clearly there is a unique homomorphism $\Phi_{D,I} : R_I \to L$ of underlying additive groups such that for all $J \in \Omega$ and $f \in S_J$, regarding S_J as a subring of R_I, we have

$$\Phi_{D,I}(f) = \Psi_{D,J}(f).$$

Moreover clearly

$$\ker(\Phi_{D,I}) = \bigcup_{J \in \Omega} \ker(\Psi_{D,J}).$$

EXERCISE (E15). [**Criterion for Extensions of Derivations**]. In the above situation of (D10) we have

$$\begin{cases} (1) \ \ker(\Phi_{D,I}) \text{ contains a set of generators of } \ker(\Phi_I) \\ \Rightarrow (2) \ \ker(\Phi_I) \subset \ker(\Phi_{D,I}) \\ \Rightarrow (3) \ D \text{ can be extended to some } E \in \mathrm{Der}(K(y), L) \\ \qquad \text{with } E(y_i) = u_i \text{ for all } i \in I \\ \Rightarrow (4) \ \ker(\Phi_I) \subset \ker(\Phi_{D,I}) \end{cases}$$

and moreover: $(2) \Rightarrow$ the E of (3) is unique.

HINT. By L1§12(E12) we see that $(3) \Rightarrow (4)$. By the product rule for derivations we have $(1) \Rightarrow (2)$. Now assuming (2) we get a unique homomorphism $F : K[y] \to L$ of underlying additive groups such that for all $f = f(Y) \in R_I$ we have

$$F(f(y)) = \Phi_{D,I}(f).$$

Clearly $F \in \mathrm{Der}(K[y], L)$ and F is an extension of D. By L1§12(E12) F can be uniquely extended to $E \in \mathrm{Der}(K(y), L)$.

EXERCISE (E16). [**Derivations and Purely Inseparable Extensions**]. Let L be an overfield of a field K of characteristic $p \neq 0$ and let $y \in L$ be such that $y^{p^e} \in K$ but $y^{p^{e-1}} \notin K$ for some $e \in \mathbb{N}_+$. Let there be given any $D \in \mathrm{Der}(K, L)$. Then D has an extension $E \in \mathrm{Der}(K(y), L)$ iff $D(y^{p^e}) = 0$. Moreover, if $D(y^{p^e}) = 0$ and u is any element of L then D has a unique extension $E \in \mathrm{Der}(K(y), L)$ with $E(y) = u$.

HINT. For any derivation E we have $E(y^{p^e}) = 0$, and by L5§5(Q27)(C41) we see that $Y^{p^e} - y^{p^e}$ is the minimal polynomial of y over K. Hence we are again reduced to the $|I| = 1$ case of the above Exercise (E15).

EXERCISE (E17). [**More About Purely Inseparable Extensions**]. Let L be an overfield of a field K of characteristic $p \neq 0$ such that $L^p \subset K$ (where $L^p = \{x^p : x \in L\}$), let $S \subset L$ be such that $L = K(S)$, and let $D \in \mathrm{Der}(K, L)$ be such that $D(x^p) = 0$ for all $x \in S$. Show that then D has an extension $E \in \mathrm{Der}(L, L)$.

HINT. Let W be the set of all pairs (L', E') where L' is a field between K and L, and $E' \in \mathrm{Der}(L', L')$ is an extension of D. In W define $(L', E') \leq (L'', E'')$ to mean that $L' \subset L''$ and E'' is an extension of E'. Then W is a partially ordered set having the Zorn property, and hence by Zorn's Lemma it has a maximal element (L', E'). By (E16) we must have $L' = L$.

EXERCISE (E18). [**Derivations and Separably Generated Extensions**]. Let L/K be a separably generated field extension, let $y = (y_i)_{i \in I}$ be a separating transcendence basis of L/K, and let $u = (u_i)_{i \in I}$ be a family of elements in L. Then D has a unique extension $E \in \mathrm{Der}(L, L)$ with $E(y_i) = u_i$ for all $i \in I$.

[The assumptions say that L is an overfield of a field K having a transcendence basis $y = (y_i)_{i \in I}$ such that L is a separable algebraic field extension of $K(y)$, and $u = (u_i)_{i \in I}$ is any family of elements in L; in particular we could have a pure transcendental field extension $L = K(y)$].

HINT. Follows from (E14) and (E16).

EXERCISE (E19). [**Criterion for Separable Algebraic Extension**]. Let L be a finitely generated field extension of a field K, i.e., $L = K(y_1, \ldots, y_m)$ where $y_1, \ldots, y_m$ are elements in L with $m \in \mathbb{N}$. Show that then L is separable algebraic over K iff $\mathrm{Der}_K(L, L) = 0$, i.e., iff for every $E \in \mathrm{Der}_K(L, L)$ we have $E(x) = 0$ for all $x \in L$.

HINT. The "only if" part follows from (E14). Conversely, suppose if possible that $\mathrm{Der}_K(L, L) = 0$ but y_j is not separable algebraic over $H = K(y_1, \ldots, y_{j-1})$ for some $j \in \{1, \ldots, m\}$, and let j be the largest such. Then y_j is either transcendental or inseparable over H. In the first case by (E18), and in the second case by (D7), (E9), and (E16), we can find $F \in \mathrm{Der}_H(H(x_j), H(x_j))$ with $F(y_j) = u \in H(y_j)^{\times}$ and then by (E14) we can extend F to some $E \in \mathrm{Der}(L, L)$, which is a contradiction.

DEFINITION (D11). [**Jacobian Matrix and Jacobian**]. For polynomials $f_1, \ldots, f_m$ in indeterminates $Y_1, \ldots, Y_m$ over a field K with $m \in \mathbb{N}_+$, we define the jacobian matrix of $f_1, \ldots, f_m$ relative to $Y_1, \ldots, Y_m$ by putting

$$\frac{\partial(f_1, \ldots, f_m)}{\partial(Y_1, \ldots, Y_m)} = \left(\frac{\partial f_i}{\partial Y_j} \right)$$

i.e., the $m \times m$ matrix whose (i, j)-th entry is the partial derivative

$$\frac{\partial f_i}{\partial Y_j} \in R = K[Y_1, \ldots, Y_m].$$

Its determinant is called the jacobian of $f_1, \ldots, f_m$ relative to $Y_1, \ldots, Y_m$ and we denote it by

$$\frac{J(f_1, \ldots, f_m)}{J(Y_1, \ldots, Y_m)} = \det \left(\frac{\partial f_i}{\partial Y_j} \right).$$

Given any K-homomorphism

$$\phi : R \to L$$

where L is any overfield of K, let

$$y_i = \phi(Y_i) \quad \text{for} \quad 1 \leq i \leq m.$$

We write

$$\frac{\partial(f_1, \ldots, f_m)}{\partial(Y_1, \ldots, Y_m)}(y_1, \ldots, y_m) \quad \text{or} \quad \left(\frac{\partial f_i}{\partial Y_j}(y_1, \ldots, y_m) \right)$$

for the $m \times m$ matrix obtained by substituting $(y_1, \ldots, y_m)$ for $(Y_1, \ldots, Y_m)$ in the above matrix, and we write

$$\frac{J(f_1, \ldots, f_m)}{J(Y_1, \ldots, Y_m)}(y_1, \ldots, y_m) \quad \text{or} \quad \det\left(\frac{\partial f_i}{\partial Y_j}(y_1, \ldots, y_m)\right)$$

for the corresponding determinant.

EXERCISE (E20). [**Jacobian Criterion of Separability**]. In (D11) assume that $L = K(y_1, \ldots, y_m)$. Show that then: L/K is separable algebraic iff

$$\det\left(\frac{\partial f_i}{\partial Y_j}(y_1, \ldots, y_m)\right) \neq 0$$

for some $f_1, \ldots, f_m$ in $\ker(\phi)$.

HINT. If the above determinant is nonzero then for any K-derivation D of L we have the m homogeneous linear equations $\sum_{1 \leq j \leq m} \frac{\partial f_i}{\partial Y_j}(y_1, \ldots, y_m)D(y_j) = 0$ for $1 \leq i \leq m$ whose determinant is nonzero, and hence $D(y_1) = \cdots = D(y_m) = 0$, and therefore $D = 0$; consequently by (E19) we see that L/K is separable algebraic. Conversely, assuming L/K to be separable algebraic, by (E19) we see that $\mathrm{Der}_K(L, L) = 0$; consequently by (E15) we see that the system of homogeneous linear equations $\sum_{1 \leq j \leq m} f_{Y_j}(y_1, \ldots, y_j)u_j = 0$ in $u_1, \ldots, u_m$, with f varying over $\ker(\phi)$, has $u_1 = \cdots = u_m = 0$ as the only solution in L^m, and hence the above determinant is nonzero for some $f_1, \ldots, f_m$ in $\ker(\phi)$.

EXERCISE (E21). [**Lagrange's Theorem**]. Show that: (1) if G, H, I are nonempty finite sets together with a surjection $\phi : G \to I$ as well as a bijection $\phi_x : H \to \phi^{-1}(\phi(x))$ for every $x \in G$, then $|G| = |H| \times |I|$. From (1) deduce that: (2) if H is any subgroup of any finite group G then for $I = |G/H|$ (= the set of all left cosets of H in G) we have $|G| = |H| \times |I|$ and hence in particular the order of H divides the order of G. From (2) deduce that: (3) the order of any element of a finite group divides the order of the group, and: (4) every group of prime order is a simple group as well as a cyclic group.

HINT. (1) is essentially the definition of multiplication in $\mathbb{N}_+$. To deduce (2) let ϕ and ϕ_x be defined by $\phi(x) = xH$ and $\phi_x(y) = xy$.

DEFINITION (D12). [**Action of a Group, Orbit, and Stabilizer**]. By an action of a group G on a set W we mean a (group) homomorphism $\theta : G \to \mathrm{Sym}(W)$, where we recall that $\mathrm{Sym}(W)$ is the group of all bijections $W \to W$; the elements of W may be called points of W. Note that if H is any subgroup of G then $\theta|_{(H,W)} : H \to \mathrm{sym}(W)$ is an action of H on W. For any $u \in W$, we may write $g(u)$ instead of $\theta(g)(u)$, i.e., we need not mention θ explicitly; we may simply say that G acts on W, or some such thing; this is similar to the nonmention of the underlying (ring) homomorphism $\phi : R \to S$ of an algebra in L5§5(Q18). For any $u \in W$ we define the G-orbit of u (in W or on W) by putting

$$\text{orb}_G(u) = \{g(u) : g \in G\}$$

and we define the G-stabilizer of u, or the stabilizer of u in G, by putting

$$\text{stab}_G(u) = \{g \in G : g(u) = u\}$$

and we note that this is a subgroup of G. For any $U \subset W$ we define the G-stabilizer of U, or the stabilizer of U in G, by putting

$$\text{stab}_G(U) = \{g \in G : g(U) = U\}$$

and we note that this is a subgroup of G, and we define the elementwise G-stabilizer of U, or the elementwise stabilizer of U in G, by putting

$$\text{estab}_G(U) = \{g \in G : g(u) = u \text{ for all } u \in U\}$$

and we note that this is a subgroup of G.

More precisely, we could call these the θ-orbit of u (in W), the θ-stabilizer of u (in G), the θ-stabilizer of U (in G), and the elementwise θ-stabilizer of U (in G), and denote them by $\text{orb}_\theta(u)$, $\text{stab}_\theta(u)$, $\text{stab}_\theta(U)$, and $\text{estab}_\theta(U)$ respectively.

If $G \leq \text{Sym}(W)$ then, unless otherwise stated, we take the action θ to be the subset injection.

Returning to any group G acting on any set W:

We generalize $\text{orb}_G(u)$ by putting $\text{orb}_H(u) = \{h(u) : h \in H\}$ for any $u \in W$ and $H \subset G$ and calling it the H-orbit of u (in W).

We say that $h \in G$ stabilizes $u \in W$ to mean that $h \in \text{stab}_G(u)$. We say that $H \subset G$ stabilizes $u \in W$ to mean that $H \subset \text{stab}_G(u)$.

We say that $h \in G$ stabilizes $U \subset W$ to mean that $h \in \text{stab}_G(U)$. We say that $H \subset G$ stabilizes $U \subset W$ to mean that $H \subset \text{stab}_G(U)$.

We say that $h \in G$ stabilizes $U \subset W$ elementwise to mean that $h \in \text{estab}_G(U)$. We say that $H \subset G$ stabilizes $U \subset W$ elementwise to mean that $H \subset \text{estab}_G(U)$.

EXERCISE (E22). [**Orbit-Stabilizer Lemma**]. Show that for any finite group G acting on a finite set W, and any $u \in W$, we have

$$|\text{orb}_G(u)| = |G|/|\text{stab}_G(u)|$$

i.e.,

$$|\text{orb}_G(u)| = [G : \text{stab}_G(u)]$$

or in words, the orbit size equals the index of the stabilizer.

HINT. For any u in W, and g, h in G, we have

$$g(u) = h(u) \Leftrightarrow (h^{-1}g)(u) = u \Leftrightarrow h^{-1}g \in \text{stab}_G(u)$$

and hence $g(u) \mapsto g\,\text{stab}_G(u)$ gives a bijection $\text{orb}_G(u) \to G/\text{stab}_G(u)$.

DEFINITION (D13). [**Orbit Set and Fixed Points**]. Let G be a group acting on a set W.

By an orbit of $H \subset G$ (in W) we mean a subset of W which is of the form $\mathrm{orb}_H(u)$ for some $u \in W$. We denote the set of all H-orbits in W by $\mathrm{orbset}_H(W)$ and call it the orbit set (or orbset) of H (in W).

For any $g \in G$ we denote the set of all fixed points of g (in W) by $\mathrm{fix}_W(g)$ (or $\mathrm{fix}(g)$), i.e., $\mathrm{fix}_W(g) = \{u \in W : g(u) = u\}$, and call it the fixed point set of g (in W). For any $H \subset G$ we denote the set of all fixed points of H (in W) by $\mathrm{fix}_W(H)$ (or $\mathrm{fix}(H)$), i.e., $\mathrm{fix}_W(H) = \cap_{g \in H}\mathrm{fix}_W(g)$, and call it the fixed point set of H (in W). Note that this agrees with (E1).

EXERCISE (E23). [**Orbit-Counting Lemma**]. Show that the number of orbits of a finite group G acting on a finite set W equals the average number of fixed points, i.e.,

$$|\mathrm{orbset}_G(W)| = \frac{\sum_{g \in G} |\mathrm{fix}_W(g)|}{|G|}.$$

HINT. Any orbit V of G on W is a nonempty finite subset of W, and for any nonempty finite set V we have $\sum_{u \in V}(1/|V|) = 1$, and hence in our case we get $\sum_{u \in V}(|G|/|V|) = |G|$, and therefore by E(21) we see that $\sum_{u \in V}|\mathrm{stab}_G(u)| = |G|$. Summing the last equation over $\mathrm{orbset}_G(W)$ we get

$$\sum_{u \in W} |\mathrm{stab}_G(u)| = \sum_{V \in \mathrm{orbset}_G(W)} \left[\sum_{u \in V} |\mathrm{stab}_G(u)| \right] = |\mathrm{orbset}_G(W)| \times |G|.$$

Considering the set P of all pairs (g, u) with $g \in G$ and $u \in W$ such that $g(u) = u$, and expressing it as a disjoint union in two different ways we see that $\coprod_{u \in W} \mathrm{stab}_G(u) \times \{u\} = P = \coprod_{g \in G}\{g\} \times \mathrm{fix}_W(g)$. Hence $\sum_{u \in W}|\mathrm{stab}_G(u)| = \sum_{g \in G}|\mathrm{fix}_W(g)|$ and therefore by the above displayed equation we get the desired equation.

DEFINITION (D14). [**Conjugation Action and Conjugacy Classes**]. Let G be any group. For any $g \in G$, by the g-conjugate of $h \in G$ (or the conjugate of h by g) we mean the element ghg^{-1}, and by the g-conjugate of $H \subset G$ (or the conjugate of H by g) we mean the set $gHg^{-1} = \{ghg^{-1} : h \in H\}$; note that if H is a subgroup of G then so is gHg^{-1}, and recall that H is a normal subgroup of G means H is a subgroup of G such that for all $g \in G$ we have $gHg^{-1} = H$. For any $g \in G$, clearly $h \mapsto ghg^{-1}$ gives an automorphism of G. By taking $g(h) = ghg^{-1}$ for all g, h in G, we get an action of G on G which we call the conjugation action. The orbits of G under this action are called conjugacy classes of G. These classes clearly form a partition of G, and hence if G is finite then their sizes add up to the size of G.

In other words, labelling the distinct conjugacy classes of a finite group G as $H_1, \ldots, H_s$ we have

$$G = \coprod_{1 \leq i \leq s} H_i$$

and hence $|G| = \sum_{1 \leq i \leq s} |H_i|$. Alternatively, letting $h_1, \ldots, h_s$ be representatives of the distinct conjugacy classes of G, i.e., picking $h_i \in H_i$, we have $\mathrm{orb}_G(h_i) = H_i$ for $1 \leq i \leq s$, and hence by (E22) we get

$$(1) \qquad |G| = \sum_{1 \leq i \leq s} [G : \mathrm{stab}_G(h_i)].$$

More generally let G be any group acting on any finite set W. Let $V_1, \ldots, V_b$ be the distinct orbits of G on W, and let $v_1, \ldots, v_b$ be their representatives, i.e, pick some $v_i \in V_i$ for $1 \leq i \leq b$. Then

$$W = \coprod_{1 \leq i \leq b} V_i \quad \text{with} \quad V_i = \mathrm{orb}_G(v_i)$$

and hence by (E22) we get

$$(2) \qquad |W| = \sum_{1 \leq i \leq b} [G : \mathrm{stab}_G(v_i)].$$

DEFINITION (D15). [**Normalizer, Centralizer, and Center of a Group**]. Let any group G act on itself by the conjugation action explained above. Now the G-stabilizer of $h \in G$ under this action is denoted by $C_G(h)$ and is called the G-centralizer of h or the centralizer of h (in G). Similarly the G-stabilizer and the elementwise G-stabilizer of $H \subset G$ under the said action are denoted by $N_G(H)$ and $C_G(H)$ and are called the G-normalizer and the G-centralizer of H or the normalizer and the centralizer of H (in G) respectively. We put $Z(G) = C_G(G)$ and we call $Z(G)$ the center of G.

We say that $k \in G$ centralizes $h \in G$ to mean that $k \in C_G(h)$; note that this is so iff k commutes with h, i.e., iff $kh = hk$. We say that $K \subset G$ centralizes $h \in G$ to mean that $K \subset C_G(h)$; note that this is so iff every $k \in K$ commutes with h.

We say that $k \in G$ centralizes $H \subset G$ to mean that $k \in C_G(H)$; note that this is so iff k commutes with every $h \in H$. We say that $K \subset G$ centralizes $H \subset G$ to mean that $K \subset C_G(H)$; note that this is so iff every $k \in K$ commutes with every $h \in H$.

We say that $k \in G$ normalizes $H \subset G$ to mean that $k \in N_G(H)$; note that this is so iff $kHk^{-1} = H$. We say that $K \subset G$ normalizes $H \subset G$ to mean that $K \subset N_G(H)$; note that this is so iff for every $k \in K$ we have $kHk^{-1} = H$.

The above three paragraphs give a direct characterization of normalizers and centralizers, without talking about actions of groups on sets. The resulting direct characterization of the center $Z(G)$ says that it is the set of all elements of G which commute with every element of G.

The following properties of the above concepts are easy to establish.

(1) For any $h \in G$ and $H \subset G$ we have $C_G(h) \leq G$ and $C_G(H) \leq G$.

(2) $Z(G) \triangleleft G$. Moreover: G is abelian $\Leftrightarrow G \subset Z(G) \Leftrightarrow G = Z(G)$.

(3) The normalizer of any subgroup H of G is the largest group between H and G in which H is normal, i.e., $H \triangleleft N_G(H)$ and for every $K \leq G$ with $H \triangleleft K$ we have $K \leq N_G(H)$.

EXERCISE (E24). [**Class Equation**]. Show that for any finite group G, upon letting $h_1, \ldots, h_s$ be representatives of the distinct conjugacy classes of G, we have

$$(1) \qquad\qquad |G| = \sum_{1 \leq i \leq s} [G : C_G(h_i)]$$

and upon letting $g_1, \ldots, g_r$ be representatives of the distinct conjugacy classes of G not contained in the center $Z(G)$ of G, we have

$$(2) \qquad\qquad |G| = |Z(G)| + \sum_{1 \leq i \leq r} [G : C_G(g_i)].$$

More generally show that for any finite group G acting on any finite set W, upon letting $v_1, \ldots, v_b$ be representatives of the distinct orbits of G on W, we have

$$(3) \qquad\qquad |W| = \sum_{1 \leq i \leq b} [G : \mathrm{stab}_G(v_i)]$$

and upon letting $u_1, \ldots, u_a$ be representatives of the distinct orbits of G on W of size > 1, we have

$$(4) \qquad\qquad |W| = |\mathrm{fix}_W(G)| + \sum_{1 \leq i \leq a} [G : \mathrm{stab}_G(u_i)].$$

HINT. (1) follows from (D14)(1), and (2) follows from (1) by picking $h_1, \ldots, h_s$ so that $h_i = g_i$ for $1 \leq i \leq r$ and $h_j \in Z(G)$ for $r + 1 \leq j \leq s$. (3) follows from (D14)(2), and (4) follows from (3) by picking $v_1, \ldots, v_b$ so that $v_i = u_i$ for $1 \leq i \leq a$ and $v_j \in \mathrm{fix}_W(G)$ for $a + 1 \leq j \leq b$.

DEFINITION (D16). [**Prime Power Group and Prime Power Subgroup**]. Let p be any prime. By a p-group we mean a finite group H such that $|H|$ is a power of p, i.e., $|H| = p^e$ for some $e \in \mathbb{N}$. By a p-subgroup of a finite group G we mean a subgroup H of G such that H is a p-group. By $\mathrm{Sub}_p(G)$ we denote the set of all p-subgroups of G.

EXERCISE (E25). [**Action of Prime Power Group**]. Let G be a finite group acting on a finite set W, let p be a prime number, and let H be a p-subgroup of G.

Show that then

$$|W| - |\text{fix}_W(H)| \in p\mathbb{Z}.$$

HINT. Obvious if $|H| = 1$, Otherwise apply (E24)(4) with $G = H$.

EXERCISE (E26). [**Cauchy's Theorem**]. Show that for any finite group $G \neq 1$ and any prime number p which divides the order of G we have the following.

(1) G has elements of order p.

(2) If G is a p-group then $Z(G) \neq 1$.

(3) G is a p-group iff every element of G is of p-power order.

HINT. Let h be a generator of a cyclic group H of order p, and consider the finite set $W = \{w = (w_1, \ldots, w_p) \in G^p : w_1 \ldots w_p = 1\}$. Then H acts on W by taking $h^z(w) = (w_{z+1}, \ldots, w_p, w_1, \ldots, w_z)$ for all $z \in \{0, \ldots, p-1\}$ and $w \in W$. Clearly $\text{fix}_W(H) = \{(g, \ldots, g) \in G^p : g^p = 1\}$ and hence $|\text{fix}_W(H)| \geq 1$ because $(1, \ldots, 1) \in \text{fix}_W(H)$. But by (E25) we have $|\text{fix}_W(H)| \in p\mathbb{Z}$, and hence $|\text{fix}_W(H)| > 1$. Therefore there exists $g \in G \setminus \{1\}$ with $g^p = 1$. It follows that the order of g is p. This proves (1). In (E24)(2) we clearly have $[G : C_G(g_i)] \in p\mathbb{Z}$ for $1 \leq i \leq r$, and hence $|Z(G)| \in p\mathbb{Z}$, and therefore $Z(G) \neq 1$, which proves (2). By (1) and (E21) we get (3).

EXERCISE (E27). [**Normalizer of Prime Power Subgroup**]. Let G be a finite group, let p be a prime number, and let H be a p-subgroup of G. Show that

$$(1) \qquad\qquad [G : H] - [N_G(H) : H] \in p\mathbb{Z}$$

and

$$(2) \qquad\qquad [G : H] \in p\mathbb{Z} \Rightarrow [N_G(H) : H] \in p\mathbb{Z} \Rightarrow N_G(H) \neq H.$$

HINT. Let G act on $W = G/H$ by taking $g(uH) = guH$ for all g, u in G. Then for any $u \in G$ we have

$$\begin{cases} uH \in \text{fix}_W(H) \\ \Leftrightarrow huH = uH \text{ for all } h \in H \\ \Leftrightarrow u^{-1}huH = H \text{ for all } h \in H \\ \Leftrightarrow u^{-1}hu \in H \text{ for all } h \in H \\ \Leftrightarrow u \in N_G(H). \end{cases}$$

Therefore $|\text{fix}_W(H)| = [N_G(H) : H]$ and hence by (E25) we get (1). By (1) we get (2).

DEFINITION (D17). [**Transitive Action and Transitive Group**]. Let a group G act on a set W. We say that the action of G on W is transitive, or G acts transitively on W, or G is transitive on W, if $W = \text{orb}_G(u)$ for some $u \in W$. Note that then $H \leq G$ acts on $V \subset W$ means $V = \cup_{u \in U} \mathrm{orb}_H(u)$ for some $U \subset V$, and $H \leq G$ acts transitively on $V \subset W$ means $V = \mathrm{orb}_H(u)$ for some $u \in V$.

By a transitive (permutation) group we mean a subgroup G of $\mathrm{Sym}(W)$, where W is some set, such that G acts transitively on W under the subset injection as action.

DEFINITION (D18). [**Sylow Subgroup and Prime Power Orbit**]. By $|G|_p$ we denote the highest power of a prime p which divides the size $|G|$ of a nonempty finite set G and, in case G is a finite group, by a p-Sylow subgroup of G we mean a subgroup P of G whose order equals $|G|_p$, i.e., $P \leq G$ such that $|P| = p^e = |G|_p$ and $|G| = \pi p^e$ with $e \in \mathbb{N}$ and $\pi \in \mathbb{N}_+ \setminus p\mathbb{Z}$; by $\mathrm{Syl}_p(G)$ we denote the set of all p-Sylow subgroups of G; note that then $\mathrm{Syl}_p(G) \subset \mathrm{Sub}_p(G)$. Assuming G to be a finite group acting on a finite set W, by a p-power G-orbit (in W) we mean a nonempty subset of W whose size is a power of p and which is of the form $\mathrm{orb}_G(u)$ for some $u \in W$.

EXERCISE (E28). [**Sylow Transitivity**]. Let G be a finite group acting on a finite set W. Using the notation of (D18) above, prove the following five assertions where in (1) to (4) u is any point of W and in (2) to (5) p is any prime.

(1) $|G|/|\mathrm{stab}_G(u)| = |\mathrm{orb}_G(u)|$.

(2) $|G|_p/|\mathrm{stab}_G(u)|_p = |\mathrm{orb}_G(u)|_p$.

(3) Let $H \leq G$ be such that $|H|_p = |G|_p$ (for instance H could be a p-Sylow subgroup of G). Then $|\mathrm{orb}_H(u)|_p \geq |\mathrm{orb}_G(u)|_p$.

(4) In the situation of (3) assume that $|\mathrm{orb}_G(u)| =$ a power of p. Then H acts transitively on $\mathrm{orb}_G(u)$, i.e., $\mathrm{orb}_G(u) = \mathrm{orb}_H(u)$.

(5) Every p-Sylow subgroup of G acts transitively on every p-power G-orbit.

HINT. (1) is a restatement of (E22). To prove (2) note that for any $H \leq G$ we clearly have $|G|_p/|H|_p =$ the highest power of p which divides $|G|/|H|$, and now apply (1) with $H = \mathrm{stab}_G(u)$. To prove (3) note that obviously we have $\mathrm{stab}_H(u) = H \cap \mathrm{stab}_G(u)$, and hence $|\mathrm{stab}_H(u)|$ divides $|\mathrm{stab}_G(u)|$, and therefore $|\mathrm{stab}_H(u)|_p$ divides $|\mathrm{stab}_G(u)|_p$, and hence our assertion follows from (2). To prove (4) note that

$$|\mathrm{orb}_H(u)| \geq |\mathrm{orb}_H(u)|_p \geq |\mathrm{orb}_G(u)|_p = |\mathrm{orb}_G(u)|$$

where the first inequality is obvious, the second inequality is (3), and the third equation is equivalent to the assumption of $|\mathrm{orb}_G(u)|$ being a power of p; therefore $|\mathrm{orb}_H(u)| \geq |\mathrm{orb}_G(u)|$; but obviously $|\mathrm{orb}_H(u)| \leq |\mathrm{orb}_G(u)|$; therefore we have $|\mathrm{orb}_H(u)| = |\mathrm{orb}_G(u)|$ and hence $\mathrm{orb}_H(u) = \mathrm{orb}_G(u)$. Finally (5) follows from (4) by taking H to be the given p-Sylow subgroup of G, and $\mathrm{orb}_G(u)$ to be the given p-power G-orbit.

DEFINITION (D19). [**Complete Set of Conjugates**]. For any group G, by a complete set of G-conjugates in G we mean a G-orbit under the conjugation action, i.e., a subset ω of G such that $\omega = \{ghg^{-1} : g \in G\}$ for some $h \in G$.

Likewise, by a complete set of G-conjugates in the set of subsets of G we mean a set Ω of subsets of G such that that $\Omega = \{gHg^{-1} : g \in G\}$ for some $H \subset G$.

In the above phrases the reference(s) to G may be dropped when it is clear from the context; for instance see (E29)(2) below.

EXERCISE (E29). [**Sylow's Theorem**]. Show that for any finite group G and prime p, with notation as in (D18), we have the following.

(1) Given any $H \in \mathrm{Sub}_p(G) \setminus \mathrm{Syl}_p(G)$ there exists $K \in \mathrm{Sub}_p(G)$ such that $H \lhd K$ with $[K : H] = p$. Hence given any $H \in \mathrm{Sub}_p(G)$ there exists $P \in \mathrm{Syl}_p(G)$ with $H \le P$. In particular $\mathrm{Syl}_p(G) \ne \emptyset$.

(2) Given any $H \in \mathrm{Sub}_p(G)$ and $P \in \mathrm{Syl}_p(G)$ there exists $g \in G$ such that $H \le gPg^{-1}$. In particular $\mathrm{Syl}_p(G)$ is a complete set of conjugate subgroups of G, i.e., $\mathrm{Syl}_p(G) = \{gPg^{-1} : g \in G\}$ for some $P \le G$.

(3) $|\mathrm{Syl}_p(G)| - 1 \in p\mathbb{Z}$ and $|\mathrm{Syl}_p(G)| = [G : N_G(P)]$ for every $P \in \mathrm{Syl}_p(G)$.

(4) If $P \in \mathrm{Syl}_p(G)$ then $N_G(N_G(P)) = N_G(P)$.

HINT. The first sentence of (1) clearly implies the other two; to prove the first, by (D15)(3) and (E27)(2) we have $H \lhd N_G(H)$ with $H \ne N_G(H)$, and hence by (E26)(1) we can find the desired K. The first sentence of (2) clearly implies the second; to prove the first we may assume that $H \ne 1$ and then letting G act on $W = G/P$ by taking $u(gP) = ugP$ for all u, g in G, by (E25) we can find $g \in G$ with $gP \in \mathrm{fix}_W(H)$, and now it suffices to note that for any $g \in G$ we have

$$
\begin{cases}
gP \in \mathrm{fix}_W(H) \\
\Leftrightarrow hgP = gP \text{ for all } h \in H \\
\Leftrightarrow g^{-1}hgP = P \text{ for all } h \in H \\
\Leftrightarrow g^{-1}Hg \le P \\
\Leftrightarrow H \le gPg^{-1}.
\end{cases}
$$

To prove (3) we can take $P \in \mathrm{Syl}_p(G)$; the assertion being obvious when $P = 1$, assume that $P \ne 1$; now P acts on $W = \mathrm{Syl}_p(G)$ by conjugation and $\mathrm{fix}_W(P) = \{P\}$ because for every $Q \in \mathrm{fix}_W(P)$ we have $P \in \mathrm{Syl}_P(N_G(Q)) = \{Q\}$ where the equality is by (2), and hence $|\mathrm{fix}_W(P)| = 1$; so by (E24)(4) we get $|\mathrm{Syl}_p(G)| - 1 \in p\mathbb{Z}$ because for $1 \le i \le a$ we clearly have $[P : \mathrm{stab}_P(u_i)] \in p\mathbb{Z}$; by (2) we know that, by conjugation, G acts transitively on W and by definition we have $\mathrm{stab}_G(P) = N_G(P)$, and hence by (E22) we get $|\mathrm{Syl}_p(G)| = [G : N_G(P)]$. To prove (4) let $Q = N_G(P)$;

since $P \triangleleft Q$, by (2) we get $\mathrm{Syl}_p(Q) = \{P\}$; now it suffices to note that

$$\begin{cases} g \in N_G(Q) \\ \Rightarrow gPg^{-1} \in \mathrm{Syl}_p(gQg^{-1}) \text{ and } gQg^{-1} = Q \\ \Rightarrow gPg^{-1} = P \\ \Rightarrow g \in Q. \end{cases}$$

EXERCISE (E30). [**Existence of Prime Power Subgroups**]. Reprove most of (E29)(1) along the following lines. For any finite group G and prime number p, let us write $|G| = mp^{d+e}$ with d, e in $\mathbb{N}$ and m in $\mathbb{N}_+ \setminus p\mathbb{Z}$. Let W be the set of all subsets of G of size p^e, and let U be the set of all members of W which are subgroups of G. Let G act on W by taking $g(u) = \{gx : x \in u\}$ for all $g \in G$ and $u \in W$. Let $V = \{v \in W : |\mathrm{orb}_G(v)| \notin p^{d+1}\mathbb{Z}\}$. Show that then

$$|W| \notin p^{d+1}\mathbb{Z}$$

and from this deduce that

$$\emptyset \neq U \subset V$$

and

$$v \mapsto \mathrm{stab}_G(v) \text{ gives a surjection } V \to U.$$

HINT. Now

$$|W| = \binom{mp^{d+e}}{p^e} = mp^d \prod_{1 \leq i < p^e} \frac{R_i = mp^{d+e} - i}{S_i = p^e - i}$$

and for $1 \leq i < p^e$, upon letting $i = m_i p^{e_i}$ with $e_i \in \mathbb{N}$ and $m_i \in \mathbb{N}_+ \setminus p\mathbb{Z}$, we get $R_i = A_i p^{e_i}$ and $S_i = B_i p^{e_i}$ with A_i, B_i in $\mathbb{N}_+ \setminus p\mathbb{Z}$, and hence $|W| \notin p^{d+1}\mathbb{Z}$. For every $u \in U$ we clearly have $\mathrm{orb}_G(u) = G/u$ and $\mathrm{stab}_G(u) = u$. Consequently $U \subset V$, and it suffices to show that for any $v \in V$ we have $|P| = p^e$ where $P = \mathrm{stab}_G(v)$. By (E22) we have $|P| = |G|/|\mathrm{orb}_G(v)|$ and hence $|P|/p^e \in \mathbb{N}_+$ and therefore $|P| \geq p^e$. For every $g \in P$ we have $g(v) = v$ and hence by choosing any $x \in v$ we see that $g \mapsto gx$ gives an injection $P \to v$ and hence $|P| \leq |v| = p^e$. Consequently $|P| = p^e$.

DEFINITION (D20). [**Exponential Notation and Subscript Notation**]. According to the exponential notation for action, given a group G acting on a set W: for any $g \in G$ and $u \in W$ we put $u^g = g(u)$ for $u \in W$, for any $g \in G$ and $U \subset W$ we put $U^g = g(U)$, and for any $H \subset G$ and $u \in W$ we put $u^H = \mathrm{orb}_H(u)$.

According to the exponential notation for conjugation, given a group G: for any $g \in G$ and $h \in G$ we put $h^g = ghg^{-1}$, and for any $g \in G$ and $H \subset G$ we put $H^g = gHg^{-1}$.

According to the subscript notation for action, given a group G acting on a set W: for any $u \in W$ we put $G_u = \text{stab}_G(u)$, and for any $U \subset W$ we put $G_U = \text{stab}_G(U)$ and $G_{[U]} = \text{estab}_G(U)$.

EXERCISE (E31). [**Conjugation Rule**]. Let $G \leq \text{Sym}(W)$ where W is a finite set of size n, and let us use the exponential notation for conjugation. Show that then, for all h, g in G, in the usual permutation notation we have

$$h = \begin{pmatrix} u_1 \ \cdots \ u_n \\ v_1 \ \cdots \ v_n \end{pmatrix} \Rightarrow h^g = \begin{pmatrix} g(u_1) \ \cdots \ g(u_n) \\ g(v_1) \ \cdots \ g(v_n) \end{pmatrix}.$$

i.e., we have

$$h^g(g(u)) = g(h(u)) \quad \text{for all} \quad u \in W.$$

DEFINITION (D21). [**Transportation and Permutation Isomorphisms**]. A bijection $f : W \to W^\flat$ of sets induces the isomorphism $f^\sharp : \text{Sym}(W) \to \text{Sym}(W^\flat)$ given by $f^\sharp(h)(f(u)) = f(h(u))$ for all $h \in \text{Sym}(W)$ and $u \in W$; this may be called transportation isomorphism since it transports the group structure from $\text{Sym}(W)$ to $\text{Sym}(W^\flat)$; in some sense, the above Conjugation Rule is also a transportation phenomenon; to bring out the analogy between these two, we note that in effect $f^\sharp(h) = fhf^{-1}$. In turn, for $G \leq \text{Sym}(W)$, $f^\sharp$ induces the (restriction) isomorphism $f^\sharp_G : G \to G^\flat$ where $G^\flat = f^\sharp(G)$; we call $f^\sharp_G$ a permutation-isomorphism, and we say that G and $G^\flat$ are permutation-isomorphic. If permutation groups G and $G^\flat$ are permutation-isomorphic then clearly: G is transitive $\Leftrightarrow$ $G^\flat$ is transitive, G is semi-regular $\Leftrightarrow$ $G^\flat$ is semi-regular, G is regular $\Leftrightarrow$ $G^\flat$ is regular; see (D22) below.

Upon taking $W^\flat = W$ and $f \in G$, by applying the above Conjugation Rule (with f replacing g) we get $f^\sharp(h) = h^f$ for all $h \in G$ and hence, in the subscript notation of (D20), for any $u \in W$ we have $f^\sharp(G_u) = G_{f(u)}$ and the induced isomorphism $G_u \to G_{f(u)}$ makes G_u permutation-isomorphic to $G_{f(u)}$. This provides a HINT for the following Exercise.

EXERCISE (E32). [**Conjugacy Lemma**]. Show that if $W = \{w_1, \ldots, w_n\}$ is a finite set of size n and $G \leq \text{Sym}(W)$ is transitive, then $G_{w_1}, \ldots, G_{w_n}$ are mutually permutation-isomorphic and they form a complete conjugacy class of subgroups of G. Also show that the said permutation-isomorphism $G_{w_i} \to G_{w_j}$ is induced by a bijection $W \to W$ which sends w_i to w_j.

DEFINITION (D22). [**Regular and Semi-regular Permutation Groups**]. Let G be a permutation group of degree n, i.e., $G \leq \text{Sym}(W)$ where W is a finite set of size n. G is semi-regular means that only the identity of G has a fixed point, i.e., $\text{fix}_W(g) = \emptyset$ for all $g \in G \setminus \{1\}$. G is regular means that G is both transitive and semi-regular. Notice that if G is regular then $|G| = n$; namely, taking any $u \in W$, we get a bijection $G \to W$ given by $g \mapsto g(u)$.

DEFINITION (D23). [**Left and Right Regular Representations**]. For any finite group X, the left and right multiplications in X correspond to the (group) monomorphisms

$$L_X : X \to \mathrm{Sym}(X) \quad \text{and} \quad R_X : X \to \mathrm{Sym}(X)$$

given by letting $L_X(x)(y) = xy$ and $R_X(x)(y) = yx^{-1}$ for all x, y in X. Their images $L_X(X)$ and $R_X(X)$ are clearly regular subgroups of $\mathrm{Sym}(X)$ and they may respectively be called the left and right regular permutation representations of X. Clearly $L_X(X)$ and $R_X(X)$ centralize each other.

EXERCISE (E33). [**Cayley's Theorem**]. As indicated in (D23), any finite group G is isomorphic to the regular permutation groups $L_X(X)$ and $R_X(X)$ which centralize each other. Show that they are actually the full centralizers of each other in $\mathrm{Sym}(X)$. Also show that they are permutation-isomorphic to each other.

NOTE. [**Hint to Cayley and Preamble to Burnside**]. Sometimes only the first sentence of (E33) is called Cayley's Theorem. The other two sentences are proved in (E34) below. Burnside's Theorem (E42) below may be viewed as further embellishment of Cayley's Theorem. From (E34) to (E42) our main aim is to reproduce the proof of Burnside's Theorem given in my 2002 paper [A07]. In the second 1911 edition of his famous book [Bur], Burnside gave two proofs of his theorem, one of which was taken from the first 1897 edition of [Bur]. Henceforth we shall tacitly use the notation of (D20). The HINTS (= Proofs) being rather long, we shall leave blank lines before them for better display.

EXERCISE (E34). [**Centralizer Lemma**]. Show that, given any $G \le \mathrm{Sym}(W)$ where W is any finite set, for the centralizer G' of G in $\mathrm{Sym}(W)$, we have the following:

(1) If G is transitive then G' is semi-regular.
[It follows that if a transitive subgroup G of $\mathrm{Sym}(W)$ is centralized by a subgroup G'' of $\mathrm{Sym}(W)$ then G'' is semi-regular].

(2) If G and G' are transitive then G and G' are regular, G is the centralizer of G' in $\mathrm{Sym}(W)$, and G' (resp: G) is the only transitive subgroup of $\mathrm{Sym}(W)$ which centralizes G (resp: G').
[It follows that if a transitive subgroup G of $\mathrm{Sym}(W)$ is centralized by a transitive subgroup G'' of $\mathrm{Sym}(W)$, then G and G'' are the full centralizers of each other in $\mathrm{Sym}(W)$, and G'' (resp: G) is the only transitive subgroup of $\mathrm{Sym}(W)$ which centralizes G (resp: G'')].

(3) If G is regular then G' is also regular, and by taking any $w \in W$ and letting $f : W \to G$ be the bijection whose inverse is given by $g \mapsto g(w)$, we

have $f^\sharp(G) = L_G(G)$ and $f^\sharp(G') = R_G(G)$ where $f^\sharp : \mathrm{Sym}(W) \to \mathrm{Sym}(G)$ is the isomorphism induced by f and $L_G(G)$ and $R_G(G)$ are the left and right permutation representations of G respectively, and hence G are G' are permutation-isomorphic.

[By (2) and (3) it follows that if G is regular then G' (resp: G) is the unique transitive subgroup of $\mathrm{Sym}(W)$ which centralizes G (resp: G')].

HINT. To prove (1) assume that G transitive. If G' is not semi-regular then we can find $u \neq v$ in W and $g \in G'$ with $u^g = u$ and $v^g \neq v$. Since G is transitive, we can find $h \in G$ with $h(u) = v$. Then $h^g(u^g) = v^g$ by the conjugation rule, and $h^g = h$ because h and g commute. Thus $v = h(u) = h^g(u^g) = v^g \neq v$ which is a contradiction. Therefore G' is semi-regular.

To prove (2) assume that G and G' are transitive, and let n be the size of W. Then by (1) they must be regular, and hence $|G| = n = |G'|$. If $G'' \leq \mathrm{Sym}(W)$ centralizes G then $G'' \leq G'$, and if G'' is also transitive then $|G''| \geq n$ and hence $G'' = G'$. Let $G^\dagger$ be the centralizer of G' in $\mathrm{Sym}(W)$. Since G is transitive and centralizes G', it follows that the $G \leq G^\dagger$ and $G^\dagger$ is transitive. Therefore by reversing the roles of G' and $G^\dagger$ in what we have already proved, it follows that $G = G^\dagger$ and G is the only transitive subgroup of $\mathrm{Sym}(W)$ which centralizes G'.

To prove (3) assume that G is regular. Take $w \in W$, let $f : W \to G$ be the bijection whose inverse is given by $g \mapsto g(w)$, let $f^\sharp : \mathrm{Sym}(W) \to \mathrm{Sym}(G)$ be the isomorphism induced by f, and let $L_G(G)$ and $R_G(G)$ be the left and right permutation representations of G respectively. Then $f^\sharp(G) = L_G(G)$, and hence upon letting $G^\ddagger = f^{\sharp-1}(R_G(G))$ we see that $G^\ddagger$ is a regular subgroup of $\mathrm{Sym}(W)$ which centralizes G and is permutation-isomorphic to it. By (2) we get $G^\ddagger = G'$.

DEFINITION (D24). [**Multitransitive and Antitransitive Groups**]. Let $G \leq \mathrm{Sym}(W)$ where W is a finite set of size n. Given $t \in \mathbb{N}$, we say that G is t-transitive if $t \leq n$ and any t points of W can be mapped to any other t points of W by an element of G (with prescribed order of the points), and we say that G is t-antitransitive if $t \leq n$ and the identity is the only element of G having t fixed points. G is (t, τ) means G is t-transitive and τ-antitransitive. Note that then: $(0, n)$ = every group, $(0, 1)$ = semi-regular, $(1, n)$ = transitive, and $(1, 1)$ = regular. By a **sharply t-transitive** group we mean a (t, t) group.

In the above type of definitions, when necessary, reference to W may be made explicit. For instance, instead of saying that G is t-transitive we may say that G is t-transitive on W.

Note that Cayley's Theorem (E33) says that every finite group is isomorphic to a regular permutation group, and the Centralizer Lemma (E34) says that in this manner we get all the regular permutation groups.

We call G **semi-transitive** if all the G-orbits have the same size, and we note:

(1) the **Obvious Lemma** which says that if G is semi-transitive then, upon letting m to be the common size of all the G-orbits and μ to their number, we

have $n = \mu m$ with $m = |u^G|$ for all $u \in W$.

If $n \geq 2$ then, for any $u \in W$, G_u may be viewed as a subgroup of $\mathrm{Sym}(W \setminus \{u\})$. We call G **sesqui-transitive** if G is transitive on W and, for every $u \in W$, G_u is semi-transitive on $W \setminus \{u\}$.

As a consequence of the Conjugacy Lemma (E32) we see that:

$$(2) \quad \begin{cases} G \text{ is two-transitive} \\ \Leftrightarrow G \text{ is transitive with } n \geq 2 \text{ and} \\ \quad G_u \text{ is transitive on } W \setminus \{u\} \text{ for some } u \in W \end{cases}$$

and

$$(3) \quad \begin{cases} G \text{ is regular} \\ \Leftrightarrow G \text{ is transitive and } G_u = 1 \text{ for some } u \in W \end{cases}$$

and so on.

As another obvious fact we note that: (4) if G is regular then, by fixing any $w \in W$, we get a bijection $f : W \to G$ whose inverse is given by $g \mapsto g(w)$.

The significance of this bijection $f : W \to G$ and the transportation isomorphism $f^\sharp : \mathrm{Sym}(W) \to \mathrm{Sym}(G)$ induced by it is exemplified by the Centralizer Lemma (E34) above and the Normality Lemma (E38) below.

DEFINITION (D25). [**Frobenius and Semi-Frobenius Groups**]. Again let $G \leq \mathrm{Sym}(W)$ where W is a finite set of size n. G is semi-Frobenius means that G is transitive, n is at least 2, and only the identity of G has 2 fixed points. G is Frobenius means that G is semi-Frobenius but is not regular. Note that then: $(1, 2) = $ semi-Frobenius, and $(1, 2) \setminus (1, 1) = $ Frobenius. A sharply 2-transitive group is also called **sharp-Frobenius**.

EXERCISE (E35). [**Order Lemma**].For any $G \leq \mathrm{Sym}(W)$ where W is a finite set of size n, we have the following:
  (1) G semi-transitive $\Rightarrow |G_u| = |G_v|$ for all u, v in W.
  (2) G semi-regular $\Rightarrow |u^G| = |G|$ for all $u \in W$.
  (3) G semi-regular $\Rightarrow G$ semi-transitive and $|G|$ divides n.
  (4) G transitive $\Rightarrow |G| = n|G_u|$ for all $u \in W$.
  (5) G regular $\Leftrightarrow G$ semi-regular and $|G| = n \Leftrightarrow G$ transitive and $|G| = n$.
  (6) G semi-Frobenius $\Rightarrow |G|$ divides $n(n-1)$.
  (7) G sesqui-transitive and $|G|$ divides $n^2 \Rightarrow |G| = n$.

HINT. Namely, (1) follows from (E22) by noting that if G is semi-transitive then for all u, v in W we have $|G_u| = |G|/m = |G_v|$ where m is the common size of all the G-orbits. (2) follows from (E22) by noting that if G is semi-regular then $G_u = 1$ for all $u \in W$. (3) follows from (D24)(1) and (2). (4) follows from (E22) by noting that if G is transitive then $|u^G| = n$ for all $u \in W$. (5) follows from (2) and (4) by noting that G is transitive $\Leftrightarrow |u^G| = n$ for all $u \in W$. (6) follows by noting that if G is semi-Frobenius then by (4) we get $|G| = n|G_u|$ for all $u \in W$, and by taking $(G_u, W \setminus \{u\})$ for (G, W) in (3) we see that $|G_u|$ divides $n - 1$. Finally, to prove (7) assume that G is sesqui-transitive and $|G|$ divides n^2; then fixing any $u \in W$, by (4) we have $|G| = n|G_u|$ and hence $|G_u|$ divides n and we want to show that $|G_u| = 1$, i.e., $G_u = 1$; now G_u is semi-transitive on $W \setminus \{u\}$, and hence by taking $(G_u, W \setminus \{u\})$ for (G, W) in (D24)(1) we get $n - 1 = \mu' m'$ where μ' is the number of orbits of G_u on $W \setminus \{u\}$ and m' is their common size; by taking $(G_u, W \setminus \{u\})$ for (G, W) in (E22) we also see that m' divides $|G_u|$ which we know divides n; thus m' divides n as well as $n - 1$ and hence we must have $m' = 1$, i.e., every orbit of G_u on $W \setminus \{u\}$ has size 1; therefore $G_u = 1$.

DEFINITION (D26). [**Invariant Partition and System of Imprimitivity**]. Let $G \leq \mathrm{Sym}(W)$ where W is a finite set of size n. By a G-**invariant partition** (of W) we mean a partition $T = (T_i)_{1 \leq i \leq r}$ of W into pairwise disjoint nonempty sets $T_1, \ldots, T_r$ (with $W = T_1 \cup \cdots \cup T_r$) such that for every $g \in G$ and $i \in \{1, \ldots, r\}$ we have $g(T_i) = T_j$ for some $j \in \{1, \ldots, r\}$ which may or may not be equal to i. The above partition T is trivial means either $r \leq 1$ or $|T_i| = 1$ for $1 \leq i \leq r$. A G-**system of imprimitivity** is a nontrivial G-invariant partition. By a G-**block** we mean a nonempty subset U of W such that for every $g \in G$ we have either $U^g = U$ or $U^g \cap U = \emptyset$; we call U a trivial block if either $U = W$ or $|U| = 1$. Note that if G is transitive then n and r are in $\mathbb{N}_+$ and there exists $s \in \mathbb{N}_+$ with $n = rs$ such that $|T_i| = s$ for $1 \leq i \leq r$; in this case we call (r, s) the type of T. The following Exercise relates blocks with orbits and partitions.

EXERCISE (E36). [**Block Lemma**]. Let $G \leq \mathrm{Sym}(W)$ where W is a finite set of size n. Show that we have the following:

(1) If $u \in W$ and $H \leq G$ then $(u^H)^g = (u^g)^{H^g}$ for all $g \in G$.

(2) If $u \in W$ and $G_u \leq H \leq G$ then u^H is a G-block with $|u^H| = [H : G_u]$.

(3) If U is a G-block and g, h are elements in G then U^g and U^h are G-blocks such that either $U^g = U^h$ or $U^g \cap U^h = \emptyset$.

(4) If $T = (T_i)_{1 \leq i \leq r}$ is a G-invariant partition then each T_i is a G-block.

(5) If G is transitive and U is a G-block then, upon letting $T = (T_i)_{1 \leq i \leq r}$ to be the family of all distinct sets of the form U^g as g varies over G (we then say that T is generated by U), we have that $r = n/|U|$ and T is a G-invariant partition of type $(r, |U|)$, and hence in particular $|U|$ divides n.

(6) Conversely, if G is transitive and $T = (T_i)_{1 \leq i \leq r}$ is a G-invariant partition then, for $1 \leq i \leq r$, T_i is a G-block which generates T. Moreover, if G is transitive then: a G-block is trivial $\Leftrightarrow$ the partition it generates is trivial.

(7) If G is transitive and $N \lhd G$ then the set of all N-orbits is a G-invariant partition, and hence N is semi-transitive and the size of any N-orbit divides n.

HINT. To prove (1) let $u \in W$ and $H \leq G$. Then for any $g \in G$, we have $h(u)^g = h^g(u^g)$ for every $h \in H$, and hence $(u^H)^g = (u^g)^{H^g}$.

To prove (2) let $u \in W$ and $G_u \leq H \leq G$. Then for any $g \in G$ we have: $(u^H)^g \cap u^H \neq \emptyset \Rightarrow g(f(u)) = h(u)$ for some $f, h \in H \Rightarrow h^{-1}gf \in G_u \subset H \Rightarrow g \in H \Rightarrow (u^H)^g = u^H$. Therefore u^H is a G-block. Moreover, for any $f, h \in H$ we have $f(u) = h(u) \Leftrightarrow h^{-1}f \in G_u$, and hence $|u^H| = [H : G_u]$.

To prove (3) let U be a G-block and let g, h be elements in G. Then for any $f \in G$ we have: $(U^g)^f \cap U^g \neq \emptyset \Rightarrow (fg)(u) = g(v)$ for some $u, v \in U \Rightarrow v \in U^e \cap U$ with $e = g^{-1}fg \in G \Rightarrow$ (because U is a block) $e(U) = U \Rightarrow$ (by applying g to both sides) $(fg)(U) = g(U) \Rightarrow (U^g)^f = U^g$. Therefore U^g is a block, and similarly so is U^h. Moreover: $U^g \cap U^h \neq \emptyset \Rightarrow g(u) = h(v)$ for some $u, v \in W \Rightarrow v \in U^e \cap U$ with $e = h^{-1}g \in G \Rightarrow$ (because U is a block) $e(U) = U \Rightarrow$ (by applying h to both sides) $g(U) = h(U) \Rightarrow U^g = U^h$.

(4) is obvious, and (5) follows from (3). Moreover, (6) follows from (4) and (5). Finally, to prove (7) assume that G is transitive and let $N \triangleleft G$. Then for any $g \in G$ we have $N^g = N$ and hence by taking N for H in (1) we get $(u^N)^g = (u^g)^N$ for all $u \in W$. Therefore by taking u^N for U in (5) we see that the set of all N-orbits is a G-invariant partition, and hence N is semi-transitive and the size of any N-orbit divides n.

DEFINITION (D27). [**Maximal and Minimal Normal Subgroups**]. Define a maximal subgroup of a group to be a proper subgroup which is not contained in any subgroup other than itself or the whole group; note that every nonidentity finite group has maximal subgroups. Define a minimal normal subgroup of a group to be a minimal element in the set of all nonidentity normal subgroups; note that every nonidentity finite group has minimal normal subgroups.

DEFINITION (D28). [**Primitive and Imprimitive Groups**]. Now we turn to the concept of primitivity which, like sesqui-transitivity, is between transitivity and two-transitivity. So let $G \leq \text{Sym}(W)$ where W is a finite set. Define G to be imprimitive if it is transitive and has a system of imprimitivity. Define G to be primitive if it is transitive but not imprimitive.

EXERCISE (E37). [**Primitivity Lemma**]. Let $G \leq \text{Sym}(W)$ where W is a finite set of size n. Show that we have the following:

(1) G primitive $\Leftrightarrow G$ transitive and has no nontrivial block $\Leftrightarrow G$ transitive and G_u maximal in G for every $u \in W$. [In view of (E32), the phrase "for every $u \in W$" may be replaced by the phrase "for some $u \in W$"].

(2) G two-transitive $\Rightarrow G$ primitive.

(3) G sesqui-transitive and imprimitive $\Rightarrow G$ semi-Frobenius.

HINT. The first implication in (1) follows from (E36)(5) and (E36)(6). To prove the second implication, first assume that G is transitive but has no nontrivial block, and let $G_u \leq H \leq G$ with $u \in W$; then by (E36)(2) we see that u^H is a G-block with $|u^H| = [H : G_u]$, and hence $[H : G_u] = 1$ or n; the transitivity of G tells us that $[G : G_u] = n$, and therefore we must have $H = G_u$ or G. Conversely, suppose that G is transitive and has a nontrivial block U; then we can find $u \neq v$ in U, and w in $W \setminus U$; now upon letting $H = \mathrm{stab}_G(U) = \{g \in G : U^g = U\}$ we have $H \leq G$, and by the blockness of U we get $G_u \leq H$; since $u \neq v$ in U, by the transitivity of G we find $h \in G$ with $h(u) = v$ and then $h \notin G_u$ but by the blockness of U we get $U^h = U$ and hence $h \in H$ and therefore $G_u \neq H$; since $u \neq w$ in W with $u \in U$ and $w \notin U$, by the transitivity of G we find $f \in G$ with $f(u) = w$ and then $f \notin H$ and hence $H \neq G$; thus G_u is not maximal in G.

To prove (2) assume that G is two-transitive. Let if possible, U be a nontrivial G-block. Then first by the nontriviality we can find $u \neq v$ in U and $w \in W \setminus U$, and then by the two-transitivity we can find $g \in G$ such that $g(u) = u$ and $g(v) = w$. Now $U^g \cap U \neq \emptyset$ because $u \in U^g \cap U$, and $U^g \neq U$ because $w \in U^g \setminus U$. This contradicts the blockness of U. Therefore G is primitive by (1).

Finally, to prove (3) assume that G is sesqui-transitive and imprimitive. Then there is a G-invariant partition $T = (T_i)_{1 \leq i \leq r}$ of W of type (r, s) with $r > 1$ and $s > 1$ and $T_i = \{w_{i1}, \ldots, w_{is}\}$ where $w_{i1}, \ldots, w_{is}$ are distinct points of W. Given any $u \neq v$ in W, upon letting $H = G_u$, we want to show that $H_v = 1$. By relabelling the T_i and the w_{ij} we may assume that $u = w_{11}$. Since G is sesqui-transitive, all the H-orbits on $W \setminus \{u\}$ have the same size, say m. Since $W \setminus \{u\}$ is a disjoint union of the H-orbits on it, m divides $|W \setminus \{u\}| = rs - 1$. Therefore $\mathrm{GCD}(m, s) = 1$. Fix any i with $1 < i \leq r$. Then clearly $u \notin (T_i)^H$ and $((T_i)^H)^h = (T_i)^H$ for all $h \in H$, and hence $(T_i)^H$ is a disjoint union of some H-orbits on $W \setminus \{u\}$ and therefore m divides $|(T_i)^H|$. Since T is a G-invariant partition, $(T_i)^H$ is also the disjoint union of some of the T_j, and hence $|(T_i)^H|$ is divisible by s. Since $\mathrm{GCD}(m, s) = 1$, it follows that ms divides $|(T_i)^H|$. Now we have the union $(T_i)^H = (w_{i1})^H \cup \cdots \cup (w_{is})^H$ with $|(w_{i1})^H| = \cdots = |(w_{is})^H| = m$ and hence $|(T_i)^H| \leq ms$; since ms divides $|(T_i)^H|$, we must have $|(T_i)^H| = ms$ and the said union must be disjoint. Since $T_i = \{w_{i1}, \ldots, w_{is}\}$ and the H-orbits $(w_{i1})^H, \ldots, (w_{is})^H$ are pairwise disjoint, we must have: (*) $(w_{ij})^H \cap T_i = \{w_{ij}\}$ for $1 \leq j \leq s$. If $v \in T_i$ and $h \in H_v$, then the G-blockness of T_i tells us that $(T_i)^h = T_i$ and, for $1 \leq j \leq s$, we clearly have $((w_{ij})^H)^h = (w_{ij})^H$, and hence by (*) we get $(w_{ij})^h = ((w_{ij})^H \cap T_i)^h = ((w_{ij})^H)^h \cap (T_i)^h = (w_{ij})^H \cap T_i = \{w_{ij}\}$; since i was any integer with $1 < i \leq r$ and h was any element of H_v, we conclude that: (**) $v \in T_i$ with $i \in \{2, \ldots, r\} \Rightarrow H_v \leq G_{[T_i]}$ where we recall that $G_{[T_i]} = \{g \in G : t^g = t$ for all $t \in T_i\}$. Clearly $H_v = (G_v)_u$, and hence by interchanging u and v in (**) we get: (1*) $v \in T_i$ with $i \in \{2, \ldots, r\} \Rightarrow H_v \leq G_{[T_1]}$. If $v \in T_i$ then, for $2 \leq j \leq s$, obviously $G_{[T_1]} \leq H_{w_{1j}}$ and hence by (1*) we get $H_v \leq H_{w_{1j}}$; again since i was any integer with $1 < i \leq r$, we conclude that: (2*) $v \in T_i$ with $i \in \{2, \ldots, r\} \Rightarrow H_v = H_{w_{12}} = \cdots = H_{w_{1s}}$. The only property of

v used in (2*) was that $v \notin T_1$; consequently: (3*) $H_{w_{12}} = \cdots = H_{w_{1s}} = H_w$ for all $w \in W \setminus T_1$. By (3*) we see that: (1') $H_{w_{1j}} = 1$ for $2 \leq j \leq s$. By (2*) and (1') we also see that: (2') $v \in T_i$ with $i \in \{2, \ldots, r\} \Rightarrow H_v = 1$. Finally by (1') and (2') we conclude that we always have $H_v = 1$.

DEFINITION (D29). [**Characteristic Subgroup**]. Define a characteristic subgroup of a group X to be a subgroup which is mapped onto itself by every automorphism of X. Clearly all "group theoretically well-defined" subgroups are characteristic subgroups. For instance:

(1) the center of X is a characteristic subgroup of X; similarly:

(2) if X is finite and has a unique p-Sylow subgroup for some prime divisor p of $|X|$ then it is also a characteristic subgroup of X; more generally:

(3) if X is finite and p is any prime then the subgroup of G generated by all of its p-Sylow subgroups (which is denoted by $p(G)$ and which may be defined as the subgroup of G generated by all of its elements of p-power order) is a characteristic subgroup of X; likewise:

(4) if X has a unique minimal normal subgroup then it is a characteristic subgroup of X; more generally:

(5) the subgroup of X generated by all of its minimal normal subgroups is a characteristic subgroup of X; moreover:

(6) if Y is a minimal normal subgroup of X, and Z is a nonidentity characteristic subgroup of Y, then (clearly Z is normal in X and hence by minimality) we must have $Y = Z$; finally:

(7) if Y is a minimal normal subgroup of a finite group X, and Y has a unique minimal normal subgroup Z, then (by (4) and (6) we get $Y = Z$ and by minimality Z is contained in every nonidentity normal subgroup of Y and hence) Y must be a simple group.

DEFINITION (D30). [**Elementary Abelian Group**]. By an additive elementary abelian group we mean an additive abelian group which is isomorphic to $\mathrm{GF}(q)^+$ for some prime power $q = p^e$ with prime p and positive integer e. Equivalently, it is the direct sum of a finite number of copies of an additive cyclic group of prime order. By an elementary abelian group we mean the multiplicative version of an additive elementary abelian group.

EXERCISE (E38). [**Normality Lemma**]. Let $G \leq \mathrm{Sym}(W)$ where W is a finite set of size n. Show that for any $1 \neq N \triangleleft G$ we have the following:

(1) G primitive $\Rightarrow N$ transitive.

(2) G two-transitive $\Rightarrow N$ sesqui-transitive.

(3) G two-transitive and N regular $\Rightarrow N$ elementary abelian.

HINT. To prove (1), note that if $N \triangleleft G$ with G primitive then, by (E36)(7), either all the N-orbits are singletons or W is the only N-orbit; if $N \neq 1$ then the first alternative is not possible and hence W is the only N-orbit, i.e., N is transitive.

To prove (2) assume that $1 \neq N \lhd G$ with G two-transitive. Then N is transitive by (E27)(2) and (1). For any $u \in W$ we have $N_u = N \cap G_u \lhd G_u$ with G_u transitive on $W \setminus \{u\}$, and hence by (E36)(7) we see that N_u is semi-transitive on $W \setminus \{u\}$. Therefore N is sesqui-transitive.

Finally, to prove (3), let $N \lhd G$ be such that N is regular. Fixing any $w \in W$, let $f : W \to N$ be the bijection whose inverse is given by $g \mapsto g(w)$, and let $f^\sharp : \mathrm{Sym}(W) \to \mathrm{Sym}(N)$ be the isomorphism induced by f. Then, for any $h \in G_w$ and $g \in N$, upon letting $u = g(w) \in W$ and $v = h(u) \in W$ and $g^* = f(v) \in N$, by the definition of f we have $f(u) = g$ and $g^*(w) = v$, and hence by the definition of $f^\sharp$ we get $f^\sharp(h)(g) \in N$ with $(f^\sharp(h)(g))(w) = v$, and by taking (g, h, w) for (h, g, u) in the Conjugation Rule (E31) we also get $g^h \in N$ with $g^h(w) = v$; since $g \mapsto g(w)$ gives a bijection $N \to W$, we must have $f^\sharp(h)(g) = g^h$. Thus, for every $h \in G_w$, the permutation $f^\sharp(h) : N \to N$ coincides with the conjugation automorphism given by $g \mapsto g^h$ and hence the order of any $g \in N$ equals the order of $f^\sharp(h)(g) \in N$. For any $g \neq 1 \neq g'$ in N we have $g(w) \neq w \neq g'(w)$ and, assuming G to be two-transitive, we can find $h \in G_w$ with $h(g(w)) = g'(w)$, and then clearly $f^\sharp(h)(g) = g'$, and hence the order of g equals the order of g'. Let $p > 1$ be the common order of all elements of $N \setminus \{1\}$. If $p = rs$ with $r > 1$ and $s > 1$ then taking $1 \neq g \in N$ we get $1 \neq g^r \in N$ and the order of g^r is s with $1 < s < p$ which is a contradiction. Therefore p must be prime. If the order of N were not a power of p then by Cauchy's Theorem (E26) N would contain an element whose order is a prime different from p which would be a contradiction. Therefore the order of N must be a power of p, and hence N has a nontrivial center M; obviously M is a characteristic subgroup of N and hence an automorphism of N cannot send an element in M to one not in M; but we have just shown that any two nonidentity elements of N are conjugate in G; therefore we must have $M = N$, i.e., N is abelian. Since the order of every nonidentity element of N is p, it follows that N is elementary abelian.

EXERCISE (E39). [**Fixed Point Lemma**]. Let $G \leq \mathrm{Sym}(W)$ where W is a finite set of size n. Assume that G is semi-Frobenius, and let $N = \{1\} \cup N'$ where N' is the set of all fixed point free elements of G. Show that we have the following:

(1) $|N| = n$, and for each prime divisor p of n we have that N has an element of order p, and every element of G whose order is a power of p belongs to N.

(2) If all elements of N' have equal order, then that order is a prime p, n is a power of p, N is the unique p-Sylow subgroup of G, N is a characteristic subgroup of G, and N is regular.

(3) If $G \lhd G^* \leq \mathrm{Sym}(W)$ with G^* two-transitive, then all elements of N' have equal order which is a prime p, n is a power of p, N is the unique p-Sylow subgroup of G, N is a characteristic subgroup of G, N is regular, and N is elementary abelian.

(4) If $G \lhd G^* \leq \mathrm{Sym}(W)$ with G^* two-transitive and G minimal normal in G^*,

then G is regular and $G = N =$ an elementary abelian group.

HINT. To prove this, first note that, assuming G to be semi-Frobenius, and letting $w_1, \ldots, w_n$ be the distinct elements of W, by (E35)(4) and (E35)(6) we have $|G| = mn$ where m is a divisor of $n - 1$ with $|G_{w_i}| = m$ for $1 \le i \le n$. Since G is semi-Frobenius, we have $G_{w_i} \cap G_{w_j} = \{1\}$ for all $i \ne j$, and hence $|\cup_{1 \le i \le n} G_{w_i}| = 1 + (m - 1)n$. Upon letting $N = \{1\} \cup N'$ where N' is the set of all fixed point free elements of G, we clearly have $N' = G \setminus \cup_{1 \le i \le n} G_{w_i}$, and hence $|N'| = mn - [1 + (m-1)n]$, and therefore $|N| = 1 + |N'| = n$. Since the order of G_{w_i} divides $n - 1$, the order of any nonidentity element of G_{w_i} must divide $n - 1$ and hence it cannot divide n; since $N' = G \setminus \cup_{1 \le i \le n} G_{w_i}$, every nonidentity element of G whose order divides n must belong to N'; therefore N contains every element of G whose order divides n. Since n divides $|G|$, by Sylow's (or Cauchy's) Theorem G does have elements of every prime order dividing n. Consequently, for each prime divisor p of n we have that N has an element of order p, and every element of G whose order is a power of p belongs to N. This completes the proof of (1).

To prove (2) assume that all elements of N' have the same order, say p. Then by (1), p must be prime and n must be a power of p. Since $|G| = mn$ with $\mathrm{GCD}(m, n) = 1$, n must be the highest power of p dividing $|G|$. By (1), $|N| = n$ and N contains every element of G whose order is a power of p. Therefore N must be the unique p-Sylow subgroup of G. Consequently N is a characteristic subgroup of G. Since G is transitive, W is an orbit of G; since $|W| =$ a power of p, and N is a p-Sylow subgroup of G, by Sylow Transitivity (E28)(5) we see that N is transitive on W; therefore, since $|N| = n$, by (E35)(5) we conclude that N is regular. This completes the proof of (2).

To prove (3) assume that $G \lhd G^* \le \mathrm{Sym}(W)$ with G^* two-transitive. Let $A = \{(h, u, v) \in N' \times W \times W : h(u) = v\}$. Then $(h, u, v) \mapsto (h, u)$ gives a bijection $A \to N' \times W$, and $|W| = n$ and by (1) we have $|N'| = n - 1$; therefore $|A| = (n-1)n$. Let us fix some $(h, u, v) \in A$. Upon letting $B = \{(u', v') \in W \times W : u' \ne v'\}$, given any $(u', v') \in B$, by the two-transitivity of G^* we can find $g \in G^*$ with $g(u) = u'$ and $g(v) = v'$, and now by the Conjugation Rule (5.4) and the normality of G we get $(h^g, u', v') \in A$; for each $(u', v') \in B$ we fix such $g \in G^*$ and then $(u', v') \mapsto (h^g, u', v')$ gives an injection $B \to A$; but $|B| = (n - 1)n = |A|$ and hence the said injection must be a bijection. Therefore $\{h^g : g \in G^*\} = N'$. Since conjugate elements have the same order, it follows that all elements of N' have equal order, say p. Now by (2) it follows that p is prime, n is a power of p, N is the unique p-Sylow subgroup of G, N is a characteristic subgroup of G, and N is regular. Since N is a regular normal subgroup of the two-transitive group G^*, by (E38)(3) we see that N is elementary abelian. This completes the proof of (3).

Finally, if $G \lhd G^* \le \mathrm{Sym}(W)$ with G^* two-transitive and G minimal normal in G^*, then by (3) we see that N is regular, elementary abelian, and a characteristic subgroup of G. Hence by (D29)(6) we get $G = N$. This completes the proof of (4).

DEFINITION (D31). [**Direct Product**]. The direct product of a finite number number of groups $X_1, \ldots, X_m$ is defined to be their cartesian product

$$X = X_1 \times \cdots \times X_m$$

equipped with componentwise multiplication, i.e., for any $y = (y_1, \ldots, y_m)$ and $z = (z_1, \ldots, z_m)$ in $X_1 \times \cdots \times X_m$ we have $yz = (y_1 z_1, \ldots, y_m z_m)$. To distinguished this from the mere cartesian product we may say something like "where the product is direct."

EXERCISE (E40). [**Double Normality Lemma**]. Let $G \leq \mathrm{Sym}(W)$ where W is a finite set of size n. Assume that G has two distinct minimal normal subgroups N and N'. Show that we have the following:

(1) $N \cap N' = 1$, N and N' centralize each other, and $NN' = N \times N' \lhd G$ where the product is direct.

(2) If G is primitive, then N and N' are the centralizers of each other in $\mathrm{Sym}(W)$, N and N' are regular, N and N' are nonabelian groups, G has no minimal normal subgroup other than N and N', and NN' is a characteristic subgroup of G with $|NN'| = n^2$.

[Now by (E34)(3) it follows that N and N' are permutation-isomorphic and, in the sense explained there, they are the left and right regular representations of each other].

(3) G cannot be two-transitive.

(4) If $G \lhd G^* \leq \mathrm{Sym}(W)$ with G primitive and minimal normal in G^*, then G^* cannot be two-transitive.

HINT. To prove this assume that G has two distinct minimal normal subgroups N and N'. Then obviously $N \cap N' \lhd G$ and hence by the minimality of N and N' we get $N \cap N' = 1$. Now the commutator group $[N, N']$ is the subgroup of G generated by all the commutators $ghg^{-1}h^{-1}$ with $g \in N$ and $h \in N'$; clearly $gh = hg \Leftrightarrow ghg^{-1}h^{-1} = 1$, and hence: $[N, N'] = 1 \Leftrightarrow N$ and N' centralize each other. For all $g \in N$ and $h \in N'$ by the normality of N' we have $ghg^{-1} \in N'$ and hence $ghg^{-1}h^{-1} \in N'$, and the normality of N gives $ghg^{-1}h^{-1} \in N$; therefore $[N, N'] \leq N \cap N'$. Consequently $[N, N'] = 1$. Hence N and N' centralize each other, and so $NN' = N \times N' \lhd G$ where the product is direct. This proves (1).

To prove (2) assume that G is primitive. Then by (E38)(1) we know that N and N' are transitive, and hence by (E34)(2) we see that N and N' are the centralizers of each other in $\mathrm{Sym}(W)$, N and N' are regular, and N' is the only transitive subgroup of $\mathrm{Sym}(W)$ which centralizes N. Since $N \cap N' = 1$ and N and N' are the centralizers of each other, they must be nonabelian. As we have just seen, by (E38)(1) and (E34)(2) it follows that if N'' is any minimal normal subgroup of G with $N'' \neq N$ such that N'' centralizes N then N'' is the only transitive subgroup of

Sym(W) which centralizes N, and since (as said above) N' is also the only transitive subgroup of Sym(W) which centralizes N, we must have $N'' = N$. Thus G has no minimal normal subgroups other than N and N', and hence NN' is a characteristic subgroup of G. Since N and N' are regular, by (E35)(5) we get $|N| = |N'| = n$; since NN' is the direct product of N and N', we must have $|NN'| = n^2$. This completes the proof of (2), and the bracketed remark at the end of it follows from (E34)(3).

If G were two-transitive, then by (E37)(2) G would be primitive, and hence by (2) N would be a nonabelian regular normal subgroup of G which would contradict (E38)(3). This proves (3).

To prove (4) let $G \triangleleft G^* \leq$ Sym(W) with G primitive and minimal normal in G^*. Then by (2) we know that NN' is a characteristic subgroup of G, and hence by (D29)(2) we get $NN' = G$. By (2) we have $|NN'| = n^2$, and hence by (E35)(7) we see that G cannot be sesqui-transitive, and therefore by (E38)(2) we conclude that G^* cannot be two-transitive.

EXERCISE (E41). [**Burnside's Lemma**]. Let $G \leq$ Sym(W) where W is a finite set of size n. Assuming G to be two-transitive, show that for any minimal normal subgroup N of G we have the following.

(1) N is primitive or semi-Frobenius.

(2) If N is semi-Frobenius then it is regular.

(3) If N is regular then it is elementary abelian.

(4) If N is primitive abelian simple then it is regular.

(5) If N is primitive then it is simple.

HINT. To prove this assume that G is two-transitive and let N be a minimal normal subgroup of G. Then by (E38)(2) N is sesqui-transitive, and hence by (E37)(3) it is primitive or semi-Frobenius, which proves (1). (2) follows by taking (N, G) for (G, G^*) in (E39)(4). (3) follows from (E38)(3). (4) follows by noting that if N is primitive then N is transitive and by (E35)(4) and (E37)(1) we see that for any $u \in W$ we have $N_u \neq N$ (because $n > 1$ by two-transitivity of G) and there is no group H with $N_u \leq H \leq N$ and $N_u \neq H \neq N$, and hence if N is also abelian simple (i.e., of prime order) then we must have $N_u = 1$, i.e., N is regular. Finally, (5) follows by noting that if N is primitive then by taking (N, G) for (G, G^*) in (E40)(4) we see that N has a unique minimal normal subgroup, and hence by (D29)(7) we conclude that N is simple.

EXERCISE (E42). [**Burnside's Theorem**]. Given any permutation group G (acting on a finite set), show that if G is two-transitive then it has a unique minimal normal subgroup N. Also show that if the said subgroup N is regular then it is elementary abelian, and if it is nonregular then it is primitive as well as nonabelian simple.

HINT. Namely, if G is two-transitive then by (E40)(3) we see that G has a unique minimal normal subgroup N. The rest follows from (E41).

NOTE. Let $N = \{1\} \cup N'$ where N' is the set of all fixed point free elements of G. Recall that G is semi-Frobenius means G is transitive and has no nonidentity element fixing two points, and G is Frobenius means G is semi-Frobenius but not regular. In the Fixed Point Lemma (E39)(4) we have proved that if G is semi-Frobenius and $G \triangleleft G^* \leq \mathrm{Sym}(W)$ with G^* two-transitive and G minimal normal in G^*, then G is regular and $G = N =$ an elementary abelian group. This motivates Frobenius' Theorem, proved in his 1901 paper, which says that if G is Frobenius then N is a regular normal subgroup of G. It is clear that Frobenius' Theorem implies the Fixed Point Lemma (E39)(4), and this gave rise to the second proof of Burnside's Theorem which, unlike his first proof, uses Frobenius' Theorem, and which he gave in the second edition of his book [Bur].

EXERCISE (E43). [**Simplicity of Alternating Groups**]. Show that the alternating group A_n is simple for $n \geq 5$.

HINT. Now $A_n \leq S_n = \mathrm{Sym}(W)$ with $W = \{1, \ldots, n\}$, and we want to show that, for any $1 \neq H \triangleleft A_n$ we have $H = A_n$. If H contains a 3-cycle then relabelling $1, \ldots, n$ we may assume that $(123) \in H$; now $\tau = (213) = (123)^2 \in H$ and for any $m \in \{4, \ldots, n\}$ we have $\sigma = (12)(3m) \in A_n$ and hence $(12m) = \sigma\tau\sigma^{-1} \in H$; consequently by L1§11(E3) we get $H = A_n$. Thus it suffices to show that H contains a 3-cycle. Let $d = \min\{|V(h)| : 1 \neq h \in H\}$ where $V(h) = \{u \in W : h(u) \neq u\}$, and take $h \in H$ with $|V(h)| = d$. We want to show that $d = 3$. Suppose if possible that $d \geq 4$. Then either (1) h is a product of disjoint transpositions which upon relabelling $1, \ldots, n$ looks like $h = (12)(34) \ldots$ with $V(h) = \{1, \ldots, d\}$, or (2) in the contrary case upon relabelling $1, \ldots, n$ we have $V(h) = \{1, \ldots, d\}$ with $d \geq 5$ and as a product of disjoint cycles of nonincreasing length $h = (123 \ldots) \ldots$. Now $g = (345) \in A_n$ and hence upon letting $h' = ghg^{-1}$ we have $h' \in H$. Referring to (E31), in case (1) we get $h' = (12)(45) \ldots$ and in case (2) we get $h' = (124 \ldots) \ldots$; consequently in both the cases, upon letting $h^* = h^{-1}h'$, we have $1 \neq h^* \in H$. But in case (1) we see that $V(h^*) \subset \{3, \ldots, d\} \cup \{5\}$ and in case (2) we see that $V(h^*) \subset \{2, \ldots, d\}$. This contradicts the minimality of d.

EXERCISE (E44). [**Unit Ideals in Polynomial Rings**]. Give a detailed proof of L2§5(T5). In other words, show that if $L = \coprod_{H \in \Omega} H$ is a partition of a nonempty set L (where Ω is a set of pairwise disjoint nonempty subsets of L), if $R_L = R[\{X_l : l \in L\}]$ is the polynomial ring in indeterminates $\{X_l : l \in L\}$ over a field R, if for each $H \in \Omega$ we are given a nonunit ideal J_H in the polynomial ring $R_H = R[\{X_h : h \in H\}]$, and if J_L is the ideal in R_L generated by all the ideals J_H

as H varies over Ω, then J_L is a nonunit ideal in R_L. Also show that this is not true if R is assumed to be a domain instead of a field.

HINT. For the negative assertion take $R_L = R[X, Y]$ with $R = \mathbb{Z}$, and take J_X and J_Y to be the ideals in $R_X = R[X]$ and $R_Y = R[Y]$ generated by $(2, X)$ and $(3, Y)$ respectively; then these are clearly nonunit ideals but the ideal generated by them in R_L is the unit ideal.

Turning to the positive assertion, first considering the case when L is finite, suppose if possible that J_L is the unit ideal. We can label the distinct members of Ω as $H(i)_{1 \leq i \leq n}$ with $n \in \mathbb{N}_+$ and we can label the distinct elements of $H(i)$ as $(ij)_{1 \leq j \leq m(i)}$ with $m(i) \in \mathbb{N}_+$. We can take an indeterminate Y over R_L, and let us do our work in finite algebraic field extensions of the field

$$K = R(Y, X_{11}, \ldots, X_{1m(1)}, \ldots, X_{n1}, \ldots, X_{nm(n)}).$$

Note that now J_L is an ideal in the subring R_L of K given by

$$R_L = R[X_{11}, \ldots, X_{1m(1)}, \ldots, X_{n1}, \ldots, X_{nm(n)}]$$

and $J_{H(i)}$ is a nonunit ideal in the subring $R_{H(i)}$ of R_L given by

$$R_{H(i)} = R[X_{i1}, \ldots, X_{im(i)}].$$

By our supposition we can write

$$1 = \sum_{1 \leq i \leq n} a_i b_i$$

where for $1 \leq i \leq n$ we have

$$a_i = a_i(X_{i1}, \ldots, X_{im(i)}) \in R_{H(i)} \setminus R^\times$$

and

$$b_i = b_i(X_{11}, \ldots, X_{1m(1)}, \ldots, X_{n1}, \ldots, X_{nm(n)}) \in R_L.$$

Let $S_0 = R(Y)$ and $K_0 = K$. For $1 \leq i \leq n$ we shall find elements $(y_{ij})_{1 \leq j \leq m(i)}$ in a finite algebraic field extension S_i of S_{i-1} inside an overfield K_i of K_{i-1} with $K_i = K_{i-1}(S_i)$ such that $a_i(y_{i1}, \ldots, y_{im(i)}) = 0$. In view of an obvious induction, it suffices to do this for $n = 1$. If $a_1 = 0$ then we can take $S_1 = S_0$ with $K_1 = K_0$ and $y_{1j} = 0$ for $1 \leq j \leq m(1)$. If $a_1 \neq 0$, then upon relabelling $X_{11} \ldots, X_{1m(1)}$ we may assume that $a_1 \notin R^* = R[X_{12}, \ldots, X_{1m(1)}]$. By applying L3§12(E6) to the product of the nonzero coefficients of a_1 as a polynomial in X_{11} over R^* we can find $(y_{1j})_{2 \leq j \leq m(1)}$ in S_0 such that for the polynomial $a^*(X_{11}) \in S_0[X_{11}]$ obtained by substituting $(y_{1j})_{2 \leq j \leq m(1)}$ for $(X_{1j})_{2 \leq j \leq m(1)}$ in a_1 we have $a^*(X_{11}) \notin S_0$. Now letting K_1 to be an overfield of K_0 such that $K_1 = K_0(S_1)$ where S_1 is a splitting field of a^* over S_0, we can find $y_{11} \in S_1$ with $a^*(y_{11}) = 0$.

By substituting $X_{ij} = y_{ij}$ for $1 \le i \le n$ and $1 \le j \le m(i)$ in the equation $1 = \sum_{1 \le i \le n} a_i b_i$ we get the contradiction $1 = 0$.

Now, without assuming L to be finite, suppose if possible that J_L is the unit ideal. Then we can find a finite number of distinct members $H(1), \ldots, H(n)$ of Ω with $n \in \mathbb{N}_+$ such that $1 = \sum_{1 \le i \le n} a_i b_i$ where for $1 \le i \le n$ we have $a_i \in J_{H(i)}$ and $b_i \in R_{H(1) \cup \cdots \cup H(n)}$. Clearly for $1 \le i \le n$ we can find a finite nonempty subset $L(i)$ of $H(i)$ such that for $1 \le i \le n$ we have $a_i \in R_{L(i)}$ and $b_i \in R_{L(1) \cup \cdots \cup L(n)}$. Thus we are reduced to the case when L is finite.

EXERCISE (E45). [**Generalized Power Series**]. Give details of the proof of L2§7(E13). In other words, given a field K and an ordered abelian group G, show that the set

$$K((X))_G = \{A \in K^G : \mathrm{Supp}(A) \text{ is well ordered}\}$$

is a field where addition is componentwise and multiplication is Cauchy multiplication. Concretely speaking, a typical member of $K((X))_G$ can be written as

$$A(X) = \sum_{i \in G} A_i X^i$$

where we are writing A_i for the previous $A(i)$ [cf. L2§3]. For proving that $K((X))_G$ is a field, the fact that multiplication makes sense is to be established by showing that

(V4)
$$\begin{cases} \text{for any } A, B \text{ in } K((X))_G \text{ and } g \in G, \\ \{(i,j) \in G^2 : i + j = g \text{ with } i \in \mathrm{Supp}(A) \text{ and } j \in \mathrm{Supp}(B)\} \\ \text{is a finite set} \end{cases}$$

and

(V5)
$$\begin{cases} \text{for any } A, B \text{ in } K((X))_G, \\ \mathrm{Supp}(AB) \text{ is well ordered.} \end{cases}$$

and the existence of inverse is to be established by showing that

(V6)
$$\begin{cases} 0 \ne A \in K((X))_G \\ \Rightarrow AA' = 1 \text{ for some } A' \in K((X))_G. \end{cases}$$

HINT. To prove (V4) let W be the set of all $(i,j) \in G^2$ such that $i + j = g$ for some $i \in \mathrm{Supp}(A)$ and $j \in \mathrm{Supp}(B)$. If W is empty then it is finite. So assume that W is nonempty. Then $U_1 = \{i \in \mathrm{Supp}(A) : i + j = g \text{ for some } j \in \mathrm{Supp}(B)\}$ is a nonempty subset of (the well ordered set) $\mathrm{Supp}(A)$ and hence it has a smallest element i_1. If $U_2 = \{i \in \mathrm{Supp}(A) \setminus \{i_1\} : i + j = g \text{ for some } j \in \mathrm{Supp}(B)\}$ is nonempty then (because it is a subset of $\mathrm{Supp}(A)$) it has a smallest element i_2 and clearly $i_1 < i_2$. If $U_3 = \{i \in \mathrm{Supp}(A) \setminus \{i_1, i_2\} : i + j = g \text{ for some } j \in \mathrm{Supp}(B)\}$ is nonempty then (because it is a subset of $\text{Supp}(A)$) it has a smallest element i_3 and clearly $i_1 < i_2 < i_3$. If this process continues indefinitely then we would have found a strictly decreasing infinite sequence $g - i_1 > g - i_2 > g - i_3 > \dots$ in $\text{Supp}(B)$ which would contradict the well orderedness of $\text{Supp}(B)$. Therefore U_{n+1} must be empty for some $n > 1$ in $\mathbb{N}$. It follows that W is the finite set

$$\{(i_1, g - i_1), \dots, (i_n, g - i_n)\}.$$

To prove (V5) let W be the set of all $g \in G$ such that $g = i + j$ for some $i \in \text{Supp}(A)$ and $j \in \text{Supp}(B)$. Clearly $\text{Supp}(AB) \subset W$ and hence it suffices to show that W is well ordered, because every subset of a well ordered set is well ordered. So given any nonempty subset V of W, we want show that V has a smallest element. Let U_1 be the set of all $i \in \text{Supp}(A)$ such that $i + j \in V$ for some $j \in \text{Supp}(B)$. For every $i \in U_1$ let $J(i)$ be the set of all $j \in \text{Supp}(B)$ such that $i + j \in V$. Then $J(i)$ is a nonempty subset of $\text{Supp}(B)$ and hence it has a smallest element $j(i)$. Now U_1 is a nonempty subset of $\text{Supp}(A)$ and hence it has a smallest element i_1. If $U_2 = \{i \in U_1 : j(i) < j(i_1)\}$ is nonempty then it has a smallest element i_2 because it is a subset of $\text{Supp}(A)$. If $U_3 = \{i \in U_1 : j(i) < j(i_2)\}$ is nonempty then it has a smallest element i_3 because it is subset of $\text{Supp}(A)$. If this process continues indefinitely then we would have found a strictly decreasing infinite sequence $j(i_1) > j(i_2) > j(i_3) > \dots$ in $\text{Supp}(B)$ which would contradict the well orderedness of $\text{Supp}(B)$. Therefore U_{n+1} must be empty for some $n > 1$ in $\mathbb{N}$. Now clearly $U_1 \supsetneqq U_2 \supsetneqq U_3 \cdots \supsetneqq U_n \supsetneqq U_{n+1} = \emptyset$. Also for any $m \in \{1, \dots, n\}$ and any $i \in U_m \setminus U_{m+1}$ we clearly have $i \geq i_m$ with $j(i) \geq j(i_m)$ and hence $i + j(i) \geq i_m + j(i_m)$. It follows that $\min\{i_1 + j(i_1), \dots, i_n + j(i_n)\}$ is the smallest element of V.

To prove (V6) we can write $A = \alpha X^f (1 - B)$ where $\alpha \in K^{\times}$ with $f \in G$ and $B \in K((X))_G$ with $\text{Supp}(B) \subset G_+$. Now it suffices to show that $1 + B + B^2 + B^3 + \dots$ makes sense as an element of $K((X))_G$ and there it is the inverse of $1 - B$. By taking $H = \text{Supp}(B)$, and noting that any subset of a finite (resp: well ordered) set is finite (resp: well ordered), this follows from the following claims (1) to (4) which we shall prove in a moment.

Let H be a well ordered subset of G_+. For any $m \in \mathbb{N}_+$ let (H, m) be the set of all $g \in G_+$ such that $g = h_1 + \cdots + h_m$ for some elements $h_1, \dots, h_m$ in H, and for any $g \in (H, m)$ let $g(H, m)$ be the set of all sequences $(h_1, \dots, h_m)$ of elements in H such that $g = h_1 + \cdots + h_m$. Also let $(H, \infty) = \cup_{m \in \mathbb{N}_+}(H, m)$, and for any $g \in (H, \infty)$ let $g(H)$ be the set of all $m \in \mathbb{N}_+$ such that $g \in (H, m)$. We claim that then

(1)   for any $m \in \mathbb{N}_+$ and $g \in (H, m)$ the set $g(H, m)$ is finite

and

(2)   for any $m \in \mathbb{N}_+$ the set (H, m) is a well ordered subset of G_+

and

(3)   the set (H, ∞) is a well ordered subset of G_+

and

(4)   for any $g \in (H, \infty)$ the set $g(H)$ is finite.

Namely, for any well ordered subset L of G_+, let (H, L) be the set of all $g \in G_+$ such that $g = h + l$ for some $h \in H$ and $l \in L$, and for any $g \in (H, L)$ let $g(H, L)$ be the set of all $(h, l) \in H \times L$ such that $g = h + l$. Then the above proof of (V4) subsumes a proof of (1*) which says that for any $g \in (H, L)$ the set $g(H, L)$ is finite and, the above proof of (V5) subsumes a proof of (2*) which says that (H, L) is a well ordered subset of G_+. Now by induction on m, (1) follows from (1*), and (2) follows from (2*).

In proving (3) and (4) we shall use the following easy to prove criterion of well orderedness which was implicitly used in the above proofs of (V4) and (V5) and which says that

(5)
$$
\begin{cases}
I \subset G \text{ is well ordered} \\
\Leftrightarrow \text{ every infinite sequence } g_1, g_2, \ldots \text{ in } I \text{ has a nondecreasing} \\
\quad \text{subsequence } g_{\mu(1)} \le g_{\mu(2)} \le \ldots \text{ with } 1 \le \mu(1) < \mu(2) < \ldots \\
\Leftrightarrow I \text{ has NO strictly decreasing infinite sequence } g_1 > g_2 > \ldots.
\end{cases}
$$

To give a cyclical proof of (5), first assume that I is well ordered, and let $g_1, g_2, \ldots$ be an infinite sequence in I. Then, by applying the well orderedness to the nonempty subset $\{g_1, g_2, \ldots\}$ of I, we find $\mu(1) \in \mathbb{N}_+$ such that $g_{\mu(1)} \le g_i$ for all $i > \mu(1)$, and by applying the well orderedness to the nonempty subset $\{g_{\mu(1)+1}, g_{\mu(1)+2}, \ldots\}$ of I, we find $\mu(1) < \mu(2) \in \mathbb{N}_+$ such that $g_{\mu(2)} \le g_i$ for all $i > \mu(2)$, and so on.

Next assume that every infinite sequence in I has a nondecreasing subsequence, and if possible let $g_1 > g_2 > \ldots$ be a strictly decreasing infinite sequence in I. But then clearly the infinite sequence $g_1, g_2, \ldots$ in I cannot have a nondecreasing subsequence.

Finally assume that I has no strictly decreasing infinite sequence, and if possible let J be a nonempty subset of I having no smallest element. Then taking any $g_1 \in J$, it is not the smallest element in J and hence $g_1 > g_2$ for some $g_2 \in I$. Again g_2 is not the smallest element in I and hence $g_2 > g_3$ for some $g_3 \in I$, and so on.

To prove (3) suppose if possible that (H, ∞) is not well ordered. Then by (5) it has a strictly decreasing infinite sequence $h_1 > h_2 > \ldots$. Now for every $i \in \mathbb{N}_+$ we can write $h_i = h_{i1} + \cdots + h_{im(i)}$ with $m(i) \in \mathbb{N}_+$ and $h_{ij} \in H$ for $1 \le j \le m(i)$. Upon relabelling $h_{i1}, h_{i2}, \ldots, h_{im(i)}$ we may assume that $h_{i1} \le h_{i2} \le \cdots \le h_{im(i)}$. By (5) we can find a nondecreasing subsequence $h_{\mu(1)m(\mu(1))} \le h_{\mu(2)m(\mu(2))} \le \cdots$. Upon replacing $h_1 > h_2 > \ldots$ by $h_{\mu(1)} > h_{\mu(2)} > \ldots$ we may also assume that

$h_{1m(1)} \leq h_{2m(2)} \leq \ldots$. Now letting $h = (h_i, h_{ij})_{1 \leq i < \infty, 1 \leq j \leq m(i)}$ we have $h \in \Gamma$ where Γ is the set of all systems $h' = (h'_i, h'_{ij})_{1 \leq i < \infty, 1 \leq j \leq m'(i)}$ of elements h'_i, h'_{ij} in H with $m'(i) \in \mathbb{N}_+$ such that $h'_i = h'_{i1} + \cdots + h'_{im'(i)}$ with $h'_{i1} \leq h'_{i2} \leq \cdots \leq h'_{im'(i)}$ and $h'_{1m'(1)} \leq h'_{2m'(2)} \leq \ldots$.

Since H is well ordered, there is $h^* \in \Gamma$ such that $h^*_{1m^*(1)} \leq h'_{1m'(1)}$ for all $h' \in \Gamma$. Let us call g', g^* in G_+ **archimedeanly comparable** to mean that $g^* \leq n'g'$ and $g' \leq n^*g^*$ for some n' and n^* in $\mathbb{N}_+$. Let Γ^* be the set of all $h' \in \Gamma$ such that $h'_{1m'(1)}$ and $h^*_{1m^*(1)}$ are archimedeanly comparable. Without loss of generality we may assume that $h \in \Gamma^*$. Since H is well ordered, there is $g^* \in H$ such that g^* is the smallest element in the set of all elements in H which are archimedeanly comparable to $h^*_{1,m^*(1)}$.

Now for every $h' \in \Gamma^*$ there is a unique $p(h') \in \mathbb{N}_+$ with $h'_1 \leq p(h')g^*$ such that $p(h') \leq q$ for all $q \in \mathbb{N}_+$ for which $h'_1 \leq qg^*$. Let $p \in \mathbb{N}_+$ be the smallest amongst $p(h')$ as h' varies over Γ^*. Let Γ' be the set of all $h' \in \Gamma^*$ with $p(h') = p$. Without loss of generality we may assume that $h \in \Gamma'$.

If $m(i) = 1$ for infinitely many values of i then by arranging these values in a strictly increasing sequence $i_1 < i_2 < \ldots$ of positive integers we get a strictly decreasing infinite sequence $h_{i_1} > h_{i_2} > \ldots$ of elements in H contradicting the well orderedness of H.

Therefore there is $n \in \mathbb{N}$ such that $m(i) > 1$ for all $i > n$ in $\mathbb{N}$. By (5) there exists a strictly increasing sequence of integers $n < l_1 < l_2 < \ldots$ such that $h_{l_1, m(l_1)-1} \leq h_{l_2, m(l_2)-1} \leq \ldots$. For all $i \in \mathbb{N}_+$ let $m^\dagger(i) = m(l_i) - 1$ with $h^\dagger_i = h^\dagger_{i1} + \cdots + h^\dagger_{im^\dagger(i)}$ where $h^\dagger_{ij} = h_{l_i, j}$ for $1 \leq j \leq m^\dagger(i)$. Also let $h^\dagger = (h^\dagger_i, h^\dagger_{ij})_{1 \leq i < \infty, 1 \leq j \leq m^\dagger(i)}$. Then clearly $h^\dagger \in \Gamma^*$ but $p(h^\dagger) < p(h)$ which is a contradiction. Thus (3) has been established.

To prove (4) let L be the set of all $g \in (H, \infty)$ for which $g(H)$ is infinite, and suppose if possible that L is nonempty. Then by (3) L has a smallest element h and, by (1), for all $i \in \mathbb{N}_+$ we can write $h = h_{i1} + \cdots + h_{im(i)}$ with $h_{ij} \in H$ and $1 < m(1) < m(2) < \ldots$ in $\mathbb{N}$. By (5), upon replacing $1, 2, \ldots$ by a suitable subsequence we may assume that $h_{11} \leq h_{21} \leq \ldots$. Let $g_i = h_{i2} + \cdots + h_{im(i)}$. Then $g_1 \geq g_2 \geq \ldots$ in (H, ∞). By (3) and (5), we can find $n \in \mathbb{N}$ and $g \in (H, \infty)$ such that $g_i = g$ for all $n < i \in \mathbb{N}$. Clearly $g \in L$ but $g < h$ which is a contradiction. Thus (4) has been established.

EXERCISE (E46). [**Simplicity of Projective Special Linear Group**]. Prove the simplicity of $\mathrm{PSL}(n, q)$ for $n \geq 2$ with $(n, q) \neq (2, 2), (2, 3)$ as asserted in L3§1.

HINT. In view of (E37)(2), this follows from (E47) to (E50) below.

EXERCISE (E47). [**Group Action and Iwasawa's Simplicity Criterion**]. Let a group G act primitively on a finite set W, i.e., let the action $\theta : G \to \mathrm{Sym}(W)$ be such that $\theta(G)$ is a primitive subgroup of $\mathrm{Sym}(W)$. Assume that G is generated by all the conjugates of an abelian normal subgroup H of G_u for some $u \in W$. Let G' be the commutator subgroup of G. Show that then, using the subscript notation introduced in (D20) above, we have the following:

(1) $N \lhd G \Rightarrow$ either $N \leq G_{[W]}$ or $G' \leq N$.

(2) $G_{[W]} \neq G = G' \Rightarrow G/G_{[W]}$ is simple.

NOTE. Note that for any action $\theta : G \to \mathrm{Sym}(W)$ of any group G on any set W we have $\ker(\theta) = G_{[W]}$. The action is **faithful** means $\ker(\theta) = 1$. Moreover, as above, if the group $\theta(G)$ has a certain **property** P then we may say that the group G has property P on W or that the action θ has property P. For instance, in case of a finite set W, we may that G is two-transitive (resp: regular, semi-regular, etc.) on W or G acts two-transitively (resp: regularly, semi-regularly, etc.) on W, to mean that $\theta(G)$ is two-transitive (resp: regular, semi-regular, etc.).

HINT. Clearly (2) follows from (1). To prove (1), for any $N \lhd G$ with $N \not\leq G_{[W]}$, by (E38)(1) we see that N is transitive on W. Consequently, given any $\gamma \in G$ there exists $\nu \in N$ such that $\gamma(u) = \nu(u)$; now $\nu^{-1}\gamma \in G_u$ and hence $\nu^{-1}\gamma H\gamma^{-1}\nu = H$ because $H \lhd G_u$; therefore

$$\gamma H\gamma^{-1} = \nu(\nu^{-1}\gamma H\gamma^{-1}\nu)\nu^{-1} = \nu H\nu^{-1} \leq G^*$$

where G^* is the subgroup of G generated by H and N. Since by assumption G is generated by the conjugates $\gamma H\gamma^{-1}$ as γ varies over G, we conclude that $G = G^*$. Now the residue class epimorphism $G^* \to G^*/N$ induces an isomorphism of G^*/N onto $H/(H \cap N)$ where the later is abelian because H is abelian. Therefore G^*/N, i.e., G/N is abelian, and hence $G' \leq N$.

EXERCISE (E48). [**Two-Transitive Action on Projective Space**]. Let $n > 1$ be an integer, and let k be a field. Let V be the vector space of all column vectors of size n over k, and let W be the associated projective space, i.e., the set of all one-dimensional subspaces of V. Consider the action $G = \mathrm{SL}(n, k)$ on W obtained by putting $A(kv) = k(Av)$ for all $A \in G$ and $0 \neq v \in V$. Also consider the induced action of $\overline{G} = \mathrm{PSL}(n, k)$ on W. Show that the action of G on W is two-transitive, and hence so is the action of $\overline{G}$.

NOTE. The definitions of t-transitive group, τ-antitransitive group, (t, τ) group, and sharp t-transitive group, introduced in (D24) are applicable without the set W being finite. For instance, $G \leq \mathrm{Sym}(W)$ is t-transitive means $|W| \geq t$ and any t points of W can be mapped to any other t points of W by an element of G. Similarly for t-transitive action, τ-antitransitive action, (t, τ) action, and sharp t-transitive action.

HINT. Given any nonzero elements v_1, v_2, w_1, w_2 in V such that $kv_1 \neq kv_2$ and

$kw_1 \neq kw_2$, we can find nonzero elements $v_3, \ldots, v_n, w_3, \ldots, w_n$ in V such that $v_1, \ldots, v_n$ and $w_1, \ldots, w_n$ are k-bases of V. Now we can find C in $\mathrm{GL}(n, k)$ such that $Cv_i = w_i$ for $1 \leq i \leq n$. Let $A \in \mathrm{GL}(n, k)$ be defined by $Av_1 = \mu w_1$ and $Av_i = w_i$ for $2 \leq i \leq n$ where $\mu = \det(C)^{-1}$. Then $A \in G$, and for $1 \leq i \leq n$ we have $A(kv_i) = kw_i$.

EXERCISE (E49). [**Transvection Matrix and Dilatation Matrix**]. Let the notation be as in (E48). In L5§5(Q31)(C51) we have defined the transvection matrix $T_{ij}(n, \lambda)$ and the dilatation matrix $T_{ii}(n, \lambda)$. In L5§5(Q31)(T121) we have shown that

$$(1) \qquad \begin{cases} G \text{ is generated by the matrices } T_{ij}(n, \lambda) \text{ as} \\ i \neq j \text{ vary over } \{1, \ldots, n\} \text{ and } \lambda \text{ varies over } k. \end{cases}$$

Let H denote the subgroup of G generated by the matrices $T_{1j}(n, \lambda)$ as j varies over $\{2, \ldots, n\}$ and λ varies over k. Let $e_i \in V$ be the column vector with 1 in the i-th spot and zeroes elsewhere. Let $u \in W$ consist of all multiples of e_1. Show that then H is an abelian normal subgroup of G_u, and G is generated by all the conjugates of H.

HINT. In L5§5(Q31)(C55) we have defined the permutation matrix $T(n, \sigma)$ for any $\sigma \in S_n$ and, in view of (1), the generation of G by all the conjugates of H follows by noting that

$$(2) \qquad T(n, \sigma) T_{ij}(n, \lambda) T(n, \sigma)^{-1} = T_{\sigma(i)\sigma(j)}(n, \lambda)$$

and hence

$$(3) \qquad S_{ii}(n, \sigma) T_{ij}(n, \lambda) S_{ii}(n, \sigma)^{-1} = T_{\sigma(i)\sigma(j)}(n, \mathrm{sgn}(\sigma)\lambda)$$

where

$$(4) \qquad S_{ii}(n, \sigma) = T(n, \sigma) T_{ii}(n, \mathrm{sgn}(\sigma)) \in G$$

and where $(2) \Rightarrow (3)$ because for all $l \in \{1, \ldots, n\}$ and $\mu \in k$ we have

$$(5) \qquad T_{ll}(n, \mu) T_{ij}(n, \lambda) T_{ll}(n, \mu)^{-1} = \begin{cases} T_{ij}(n, \lambda) & \text{if } i \neq l \neq j \\ T_{ij}(n, \lambda\mu) & \text{if } i = l \\ T_{ij}(n, \lambda\mu^{-1}) & \text{if } l = j. \end{cases}$$

For any $(\lambda_2, \ldots, \lambda_n) \in k^{n-1}$ consider the $n \times n$ matrix

$$S(\lambda_2, \ldots, \lambda_n) = (S(\lambda_2, \ldots, \lambda_n)_{ij})$$

where

$$S(\lambda_2,\ldots,\lambda_n)_{ij} = \begin{cases} 1 & \text{if } i = j \\ \lambda_j & \text{if } 1 = i \neq j \\ 0 & \text{if } 1 < i \neq j. \end{cases}$$

Clearly

$$(\lambda_2,\ldots,\lambda_n) \mapsto S(\lambda_2,\ldots,\lambda_n)$$

gives an isomorphism

$$(k^+)^{n-1} \to H^* = \{S(\lambda_2,\ldots,\lambda_n) : (\lambda_2,\ldots,\lambda_n) \in (k^+)^{n-1}\}$$

of an additive abelian group onto a multiplicative subgroup of G. Also clearly $H = H^* \leq G_u$. So it only remains to prove that for every $B \in G_u$ we have $BH^*B^{-1} = H^*$. But this follows by noting that for any $B = (B_{ij}) \in G$ we have: $B \in G_u$ iff $B_{i1} = 0$ for $2 \leq i \leq n$.

EXERCISE (E50). [**Perfect Group**]. By a perfect group we mean a group which equals its commutator subgroup. Let the notation be as in (E48) and (E49). Show that if either $n > 2$ or $|k| > 3$ then G and $\overline{G}$ are perfect groups.

HINT. In view of (E49)(1) it suffices to note that in case of $n > 2$ by taking any $l \in \{1,\ldots,n\} \setminus \{i,j\}$ we have

(1) $$T_{il}(n,1)T_{lj}(n,\lambda)T_{il}(n,1)^{-1}T_{lj}(n,\lambda)^{-1} = T_{ij}(n,\lambda)$$

and in case of $|k| > 3$, by taking any $\kappa \in k^\times$ with $\kappa^2 \neq 1$ and letting

$$S_{ij}(n,\kappa) = T_{ii}(n,\kappa)T_{jj}(n,\kappa^{-1}) \quad \text{and} \quad \delta = \lambda/(\kappa^2 - 1)$$

we have $S_{ij}(n,\kappa) \in G$ and, in view of (E49)(5),

(2) $$S_{ij}(n,\kappa)T_{ij}(n,\delta)S_{ij}(n,\kappa)^{-1}T_{ij}(n,\delta)^{-1} = T_{ij}(n,\lambda).$$

EXERCISE (E51). [**Socle-Size**]. Complete the proof of L5§5(Q16)(T75.3) by showing that if S is a 1-dimensional CM local ring, and yS and zS are parameter ideals in R then $\mathrm{socz}(S/(yS)) = \mathrm{socz}(S/(zS))$.

HINT. Now y, z belong to $M(S) \setminus Z_S(S)$ and hence so does yz. Let us consider the residue class epimorphisms $\alpha : S \to S/(yS)$ and $\beta : S \to S/(yzS)$, and let $\gamma : \alpha(S) \to \beta(S)$ be the unique R-module homomorphism such that for all $t \in S$ we have $\gamma(\alpha(t)) = \beta(zt)$. Clearly γ is injective and

$$\gamma((0 : M(\alpha(S)))_{\alpha(S)}) = (0 : M(\beta(S)))_{\beta(S)}.$$

Thus we get an R-module isomorphism $(0 : M(\alpha(S)))_{\alpha(S)} \to (0 : M(\beta(S)))_{\beta(S)}$. Consequently $\text{socz}(S/(yS)) = \text{socz}(S/(yz)S)$. Now by interchanging y and z we get $\text{socz}(S/(zS)) = \text{socz}(S/(yz)S)$. Therefore $\text{socz}(S/(yS)) = \text{socz}(S/(zS))$.

EXERCISE (E52). [**Coprincipal Primes and Isolated Subgroups**]. Citing L5§5(Q21) for the definitions of the above two concepts and the concepts of the **core** $C(E)$ of a loset (= linearly ordered set) E and the set $P(G)$ of all nonzero **principal isolated subgroups** of an ordered abelian group G, give more details of the proof of that part of L5§5(Q21)(T96.1) which says that $C(S(G)) = P(G)$ where $S(G)$ is the loset of all nonzero isolated subgroups of G. In other words: (i) given any $H \in P(G)$, i.e., given any H in $S(G)$ for which there exists $0 < \tau \in G$ such that $H = \cap_{H' \in S(G) \text{ with } \tau \in H'} H'$, show that the set $\cup_{H' \in S(G) \text{ with } \tau \notin H'} H'$ is empty or an immediate predecessor of H in $S(G)$ according as the set $\{L \in S(G) : L < H\}$ is empty or not, and hence $H \in C(S(G))$, and conversely: (ii) given any $H \in C(S(G))$, taking any $0 < \tau \in H$ such that τ does not belong to the immediate predecessor of H in $S(G)$ in case there is such a predecessor, show that $H = \cap_{H' \in S(G) \text{ with } \tau \in H'} H'$, and hence $H \in P(G)$.

HINT. Both (i) and (ii) can easily be deduced from the fact that $S(G)$ is a loset. This does not directly use L4§12(R7)(7.5). In the proof of (i) and (ii), the reference to L4§12(R7)(7.5) only signifies that, in view of (3) and (9) of L5§5(Q21), by realizing G as the value group G_v of a valuation $v : \widehat{K} \to G_v \cup \{\infty\}$ of a field $\widehat{K}$, the claim $C(S(G)) = P(G)$ becomes equivalent to the assertion which says that: ($\bullet$) $J \in \text{spec}(R_v)^* = \text{spec}(R_v) \setminus \{M(R_v)\}$ is coprincipal $\Leftrightarrow$ the set $W(J) = \{P \in \text{spec}(R_v)^* : J \subsetneq P\}$ is either empty or has a member Q for which the set $W(J, Q) = \{\overline{Q} \in \text{spec}(R_v)^* : J \subsetneq \overline{Q} \subsetneq Q\}$ is empty. Here J is coprincipal means for some $0 \neq t \in M(R_v)$ we have $J = \cup_{J' \in \text{spec}(R_v)^* \text{ with } t \notin J'} J'$. To establish ($\bullet$) we tacitly use L4§12(R7)(7.5) which states that $\text{spec}(R_v)$ is a loset under inclusion. To prove $\Rightarrow$, assuming $W(J) \neq \emptyset$, clearly $t \in P$ for all $P \in W(J)$, and it suffices to take $Q = \cap_{P \in W(J)} P$. To prove $\Leftarrow$, if $W(J) = \emptyset$ then take any $0 \neq t \in M(R_v) \setminus J$, whereas if there exists $Q \in W(J)$ with $W(J, Q) = \emptyset$ then take any $0 \neq t \in Q \setminus J$.

By (E45) we can take $\widehat{K} = K((X))_G$ with $v = \text{ord}$, or in view of (E53) below, we can take $\widehat{K}$ to be the quotient field of

$$K((X))_G^{\text{finite}} = \{A \in K^G : \text{Supp}(A) \text{ is finite}\}.$$

EXERCISE (E53). [**General Valuation Functions**]. In L5§5(Q13) we generalized the idea of a valuation on a field by defining a valuation function on a ring R to be a map $v : R \to \mathbb{Z} \cup \{\infty\}$ such that for all x, y, z in R we have:

(i) $v(xy) = v(x) + v(y)$,

(ii) $v(x + y) \geq \min(v(x), v(y))$, and

(iii) $v(z) = \infty \Leftrightarrow z = 0$,

with the usual understanding about ∞. We generalize this further by replacing $\mathbb{Z}$ by an ordered abelian group G, and calling v a generalized valuation function on the ring R. If v is a generalized valuation function on a nonnull ring R then clearly R is a domain and we get a unique valuation $\bar{v} : \mathrm{QF}(R) \to G \cup \{\infty\}$ such that for all x, y in $R^\times$ we have $\bar{v}(x/y) = v(x) - v(y)$; we say that the valuation $\bar{v}$ is induced by v; usually we denote $\bar{v}$ also by v.

In particular, for any field K, taking R to be the domain $K((X))_G^{\text{finite}}$ we get a generalized valuation function $\mathrm{ord} : R \to G \cup \{\infty\}$, and taking $\widehat{L}$ to be the quotient field of R we get a valuation $v : \widehat{L} \to G \cup \{\infty\}$ with $G_v = G$.

DEFINITION (D32). [**Lexicographic Products and Sums**]. From L5§1 we recall that the cartesian product $\prod_{i \in E} W_i$ of a family of sets $(W_i)_{i \in E}$ is the set of all maps $\phi : E \to \cup_{i \in E} W_i$ such that for all $i \in E$ we have $\phi(i) \in W_i$. Moreover, assuming each W_i to be an **additive abelian group**, the cartesian product is converted into an additive abelian group by taking componentwise sums, and then it is called the **direct product** which is again denoted by $\prod_{i \in E} W_i$. The **direct sum** $\oplus_{i \in E} W_i$ is the set of all ϕ in $\prod_{i \in E} W_i$ whose **support** $\mathrm{supp}(\phi) = \{i \in E : \phi(i) \neq 0\}$ is finite; this is a subgroup of the direct product.

Assuming E to be a **loset** (= linearly ordered set) and referring to L5§5(Q21), the **well ordered product** $\prod_{i \in E}^{\text{wo}} W_i$ is the set of all ϕ in $\prod_{i \in E} W_i$ whose support $\mathrm{supp}(\phi)$ is well ordered; this is a subgroup of the direct product containing the direct sum. If $W_i = W$ for all $i \in E$ then $\prod_{i \in E}^{\text{wo}} W_i$ and $\oplus_{i \in E} W_i$ are denoted by

$$W_{\text{wo}}^E \quad \text{and} \quad W_{\oplus}^E$$

and are called the **well ordered E-th power** and the **restricted E-th power** of W respectively. Note that these two sets respectively correspond to the sets $(W^E)_{\text{wellord}}$ and $(W^E)_{\text{finite}}$ considered in L2§2. Also note that if E is a finite set then the direct product, the well ordered product, and the direct sum, all coincide.

Assuming each set W_i to be an **ordered abelian group**, and by putting the lexicographic order on $\prod_{i \in E}^{\text{wo}} W_i$, this becomes a linearly ordered set which we denote by $\prod_{i \in E}^{\text{lex}} W_i$ and call it the **lexicographic product** (of the family); here **lexicographic order** means that for $\phi \neq \psi$ in $\prod_{i \in E}^{\text{wo}} W_i$ we have:

$$\phi < \psi \Leftrightarrow \phi(j) < \psi(j) \text{ where } j = \min\{i \in E : \phi(i) \neq \psi(i)\}.$$

Correspondingly, the subset $\oplus_{i \in E} W_i$ of $\prod_{i \in E}^{\text{wo}} W_i$ also becomes a loset and as such we denote it by $\boxplus_{i \in E} W_i$ and call it the **lexicographic direct sum** (of the family). If $W_i = W$ for all $i \in E$ then W_{wo}^E becomes a loset and as such we denote it by

$$W_{\text{lex}}^E \text{ or } W^{[E]}$$

and call it the **lexicographic E-th power** of W, and similarly, as a loset, $\boxplus_{i \in E} W_i$

is now denoted by

$$W_{\boxplus}^{E}$$

and called the **lexicographic restricted E-th power** of W.

In case E is a finite set of cardinality $d \in \mathbb{N}$, say the set of integers $1, \ldots, d$, and we are using the tuple notation instead of the functional notation, we may write $W_1 \boxplus \cdots \boxplus W_d$ in place of $\boxplus_{i \in E} W_i$ and call it the **lexicographic direct sum** of $W_1, \ldots, W_d$, and if $W_1 = \cdots = W_d = W$ then we may write

$$W_{\text{lex}}^{d} \text{ or } W^{[d]}$$

in place of W_{lex}^{E} and call it the **lexicographic d-th power** of W.

Given any family of sets $(T_i)_{i \in F}$, by $\widehat{\bigsqcup}_{i \in F} T_i$ we denote the set of all pairs (i, x) with $i \in F$ and $x \in T_i$, and we call this the **abstract disjoint union** of the family.

Given any family of **losets** $(S_i)_{i \in E}$ indexed by the **loset** E, we convert $\widehat{\bigsqcup}_{i \in E} S_i$ into a loset by defining

$$(i, x) \leq (i', x') \Leftrightarrow \text{ either (i) } i = i' \text{ with } x \leq x' \text{ or (ii) } i \leq i'$$

and we denote the new loset by $\bigsqcup_{i \in E} S_i$ and call it the **ordered disjoint union** of the family. Recalling that the real rank $\rho(W_i)$ of the ordered abelian group W_i is the order-type of the loset $S(W_i)$ consisting of all the nonzero isolated subgroups of W_i, we define the **lexicographic disjoint sum** $\uplus$ of real ranks by the equation $\uplus_{i \in E} \rho(W_i) = $ the order-type of $\bigsqcup_{i \in E} S(W_i)$.

In case E is a finite set of cardinality $d \in \mathbb{N}$, say the set of integers $1, \ldots, d$, and we are using the tuple notation instead of the functional notation, we may write $S_1 \sqcup \cdots \sqcup S_d$ and $\rho(W_1) \uplus \cdots \uplus \rho(W_d)$ in place of $\bigsqcup_{i \in E} S_i$ and $\uplus_{i \in E} \rho(W_i)$ and call these the **ordered disjoint union** of $S_1, \ldots, S_d$ and the **lexicographic disjoint sum** of $\rho(W_1), \ldots, \rho(W_d)$ respectively.

If $d \geq 1 = |S_d|$ or $d \geq 1 = |S_1|$ or $d > 1 = |S_1| = |S_d|$ then in place of $S_1 \sqcup \cdots \sqcup S_d$ we may respectively write $S_1 \sqcup \cdots \sqcup S_{d-1} \cup \{\infty\}$ or $\{-\infty\} \cup S_2 \sqcup \cdots \sqcup S_d$ or $\{-\infty\} \cup S_2 \sqcup \cdots \sqcup S_{d-1} \cup \{\infty\}$.

If $d \geq 1 = |S(W_d)|$ or $d \geq 1 = |S(W_1)|$ or $d > 1 = |S(W_1)| = |S(W_d)|$ then in place of $\rho(W_1) \uplus \cdots \uplus \rho(W_d)$ we may respectively write $\rho(W_1) \uplus \cdots \uplus \rho(W_{d-1}) + 1_\infty$ or $1_{-\infty} + \rho(W_2) \uplus \cdots \uplus \rho(W_d)$ or $1_{-\infty} + \rho(W_2) \uplus \cdots \uplus \rho(W_{d-1}) + 1_\infty$.

Recalling that $-E$ denotes the **negative of a loset** E, i.e., E with the reverse order, we claim that

(1) $\begin{cases} \text{if } (G_i)_{i \in E} \text{ is a family of nonzero ordered abelian groups} \\ \text{indexed by a woset } E \text{ and } G = \boxplus_{j \in E} G_j \text{ then} \\ \rho(G) = \uplus_{i \in -E} \rho(G_i). \end{cases}$

To prove (1) we find an order preserving bijection $\bigsqcup_{i \in -E} S(G_i) \to S(G)$ thus.

For any (i, H) with $i \in E$ and $H \in S(G_i)$ we clearly get $\widetilde{H} \in S(G)$ by putting

$$\widetilde{H} = \{\phi \in G : \phi(i) \in H \text{ and } \phi(j) = 0 \text{ for all } j < i\}.$$

Also clearly $H \mapsto \widetilde{H}$ gives an order preserving map $\sqcup_{i \in -E} S(G_i) \to S(G)$. Since E is a woset, for any $L \in S(G)$, the set $\{j \in E : \phi(j) \neq 0 \text{ for some } \phi \in L\}$ has a smallest element $j(L)$, and upon letting $J(L) = \{\phi(j(L)) : \phi \in L\}$ we clearly have $J(L) \in S(G_{j(L)})$ with $\widetilde{J(L)} = L$. Moreover, if $L = \widetilde{H}$ then we clearly have $j(L) = i$ with $J(L) = H$. Therefore by L5§6(D2) it follows that $H \mapsto \widetilde{H}$ gives a bijection $\sqcup_{i \in -E} S(G_i) \to S(G)$.

Now without using (1) we shall generalize it by proving the following claim (2). Note that $(2) \Rightarrow (1)$ by L5§5(Q21)(T96.6).

Recalling that E^{sc} is the **loset of all nonempty left segments** of a loset E, we claim that

(2) $\begin{cases} \text{if } (G_i)_{i \in E} \text{ is a family of nonzero ordered abelian groups} \\ \text{indexed by a loset } E \text{ and } \boxplus_{i \in -E} G_i \leq G \leq \prod_{i \in -E}^{\text{lex}} G_i \\ \text{then, regarding } G \text{ as an ordered abelian group} \\ \text{under the induced order, we have} \\ \rho(G) = \uplus_{i \in E^{sc}} \rho(G'_i) \\ \text{where for every } i \in E^* = \{\epsilon \in E^{sc} : \epsilon \text{ has a max } \delta(\epsilon) \text{ in } \epsilon\} \\ \text{we put } G'_i = G_{\delta(i)} \\ \text{and for every } i \in E^{sc} \setminus E^* \\ \text{we put } G'_i = \text{a nonzero archimedean ordered abelian group.} \end{cases}$

To prove (2) we find an order preserving bijection $\sqcup_{i \in E^{sc}} S(G'_i) \to S(G)$ thus. For any (i, H) with $i \in E^{sc}$ and $H \in S(G'_i)$ we clearly get $\widetilde{H} \in S(G)$ by putting

$$\widetilde{H} = \begin{cases} \{\phi \in G : \phi(\delta(i)) \in H \text{ and } \phi(j) = 0 \text{ for all } j \in E \setminus i\} & \text{if } i \in E^* \\ \{\phi \in G : \phi(j) = 0 \text{ for all } j \in E \setminus i\} & \text{if } i \in E^{sc} \setminus E^* \end{cases}$$

Also clearly $H \mapsto \widetilde{H}$ gives an order preserving map $\sqcup_{i \in E} S(G_i) \to S(G)$. For any $L \in S(G)$, upon letting $j(L) = \{l \in E : \phi(l) \neq 0 \text{ for some } \phi \in L\}$ we clearly have $j(L) \in E^{sc}$, and upon letting

$$J(L) = \begin{cases} \{\phi(\delta(j(L))) : \phi \in L\} & \text{if } j(L) \in E^* \\ G'_{j(L)} & \text{if } j(L) \in E^{sc} \setminus E^* \end{cases}$$

we clearly have $J(L) \in S(G'_{j(L)})$ with $\widetilde{J(L)} = L$. Moreover, if $L = \widetilde{H}$ then we clearly have $j(L) = i$ with $J(L) = H$. Therefore by L5§6(D2) it follows that $H \mapsto \widetilde{H}$ gives a bijection $\sqcup_{i \in -E} S(G'_i) \to S(G)$.

By (1) we see that

$$(3) \quad \begin{cases} \text{if } (G_i)_{1 \leq i \leq d} \text{ is a family of nonzero ordered abelian groups} \\ \text{with } d \in \mathbb{N} \text{ then} \\ \rho(G_1 \boxplus \cdots \boxplus G_d) = \rho(G_d) \uplus \cdots \uplus \rho(G_1). \end{cases}$$

Letting $\mathbb{N}_-$ be the loset of all negative integers, by (1) we also see that

$$(4) \quad \begin{cases} \text{if } (G_i)_{i \in \mathbb{N}_+} \text{ is a family of} \\ \text{nonzero archimedean ordered abelian groups} \\ \text{(for instance } 0 \neq G_i \leq \mathbb{R} \text{ such as } G_i = \mathbb{Z} \text{ or } \mathbb{Q} \text{ or } \mathbb{R}) \text{ then} \\ \rho(\boxplus_{i \in \mathbb{N}_+} G_i) = \text{order-type of } \mathbb{N}_- \end{cases}$$

and

$$(5) \quad \begin{cases} \text{if } (G_i)_{i \in \mathbb{N}_+ \cup \{\infty\}} \text{ is a family of} \\ \text{nonzero archimedean ordered abelian groups} \\ \text{(for instance } 0 \neq G_i \leq \mathbb{R} \text{ such as } G_i = \mathbb{Z} \text{ or } \mathbb{Q} \text{ or } \mathbb{R}) \text{ then} \\ \rho(\boxplus_{i \in \mathbb{N}_+ \cup \{\infty\}} G_i) = \text{order-type of } \{-\infty\} \cup \mathbb{N}_-. \end{cases}$$

By (2) we see that

$$(6) \quad \begin{cases} \text{if } (G_i)_{i \in \mathbb{N}} \text{ is a family of} \\ \text{nonzero archimedean ordered abelian groups} \\ \text{(for instance } 0 \neq G_i \leq \mathbb{R} \text{ such as } G_i = \mathbb{Z} \text{ or } \mathbb{Q} \text{ or } \mathbb{R}) \text{ then} \\ \rho(\boxplus_{i \in -\mathbb{N}} G_i) = \text{order-type of } \mathbb{N} \cup \{\infty\}. \end{cases}$$

In view of the $d = 2$ case of (3), by (4) and (6) we see that

$$(7) \quad \begin{cases} \text{if } (G_i)_{i \in \mathbb{Z}} \text{ is a family of} \\ \text{nonzero archimedean ordered abelian groups} \\ \text{(for instance } 0 \neq G_i \leq \mathbb{R} \text{ such as } G_i = \mathbb{Z} \text{ or } \mathbb{Q} \text{ or } \mathbb{R}) \text{ then} \\ \rho\left((\boxplus_{i \in -\mathbb{N}} G_{-i}) \boxplus (\boxplus_{i \in \mathbb{N}_+} G_i)\right) = \text{order-type of } \mathbb{Z} \cup \{\infty\} \end{cases}$$

and by (5) and (6) we see that

$$(8) \quad \begin{cases} \text{if } (G_i)_{i \in \mathbb{Z}} \text{ is a family of} \\ \text{nonzero archimedean ordered abelian groups} \\ \text{(for instance } 0 \neq G_i \leq \mathbb{R} \text{ such as } G_i = \mathbb{Z} \text{ or } \mathbb{Q} \text{ or } \mathbb{R}) \text{ then} \\ \rho\left((\boxplus_{i \in -\mathbb{N}} G_{-i}) \boxplus (\boxplus_{i \in \mathbb{N}_+ \cup \{\infty\}} G_i)\right) = \text{order-type of } \{-\infty\} \cup \mathbb{Z} \cup \{\infty\}. \end{cases}$$

EXERCISE (E54). [**Possible Real Ranks**]. Letting $\mathbb{N}_-$ be the loset of all negative integers, consider the losets given by:

$$E_3 = \{1, \ldots, d\} \text{ with } d \in \mathbb{N}, \quad E_4 = \mathbb{N}_-, \quad E_5 = \{-\infty\} \cup \mathbb{N}_-,$$

and

$$E_6 = \mathbb{N} \cup \{\infty\}, \quad E_7 = \mathbb{Z} \cup \{\infty\}, \quad E_8 = \{-\infty\} \cup \mathbb{Z} \cup \{\infty\}.$$

Given any field K, in view of (E45) and (53) above, by (D32) above we see that for $3 \le i \le 8$ there exists a K-valuation $v_i : K_i \to G_{v_i} \cup \{\infty\}$ of an overfield K_i of K whose real rank $\rho(v_i)$ is the order-type of E_i. The exercise is to give other proofs of this directly using the characterizations given in L5§5(Q21)(T97).

EXERCISE (E55). [**Impossible Real Ranks**]. Using the characterizations cited above show that there is no valuation (of any field) whose real rank is the order-type of any one of the losets [cf. L5§5(Q21)(T96.2)]:

$$\mathbb{N}, \quad \mathbb{Z}, \quad \mathbb{Q}, \quad \mathbb{R}, \quad \mathbb{Q} \cup \{\infty\}, \quad \mathbb{R} \cup \{\infty\},$$

and

$$\{-\infty\} \cup \mathbb{Q} \cup \{\infty\}, \quad \{-\infty\} \cup \mathbb{R} \cup \{\infty\}.$$

DEFINITION (D33). [**Principal Rank and Real Rank**]. In L5§5(Q21) we defined the real rank $\rho(G)$ of an ordered abelian group G to be the order-type of the loset $S(G)$ of all nonzero isolated subgroups of G. We also defined the real rank $\rho(v)$ of a valuation v, as well as the real rank $\rho(R_v)$ of its valuation ring, to be the same as the real rank $\rho(G_v)$ of its value group G_v.

Recalling that the set of all **nonzero principal isolated subgroups** of G is denoted by $P(G)$, let us introduce the **principal rank** $\rho'(G)$ of G by putting $\rho'(G) = $ the order-type of the loset $P(G)$. Let us also define the principal ranks $\rho'(v)$ and $\rho'(R_v)$ by putting

$$\rho'(v) = \rho'(R_v) = \rho'(G_v).$$

Recalling that $\mathrm{coprin}(R_v)^*$ is the set of all **nonmaximal coprincipal prime ideals** in R_v, we see that $\rho'(R_v)$ is the order-type of $\mathrm{coprin}(R_v)^*$ with reverse containment.

In L5§5(Q21)(T97.2) and in the above two Exercises we discussed the problem of characterizing real ranks. Turning to the problem of characterizing principal ranks, it turns out that $\rho'(G)$ can be the order-type of any loset E. Indeed, upon letting

$$G = \mathbb{R}^{[-E]}$$

by L5§5(Q21)(T97.1) we see that $\rho'(G)$ is the order-type of E. Moreover, by L5§5(Q21)(T96.1) we also see that $E = C(E^{sc})$ where the **core** $C(E)$ of E, and the **segment completion** E^{sc} of E, are as defined in L5§5(Q21).

All this may be compared with the material of the above Definition (D32).

EXERCISE (E56). [**Compositions of Valuations**]. In L5§5(Q22)(7) we have defined the composition of a valuation w by a valuation u which we have denoted by $w \odot u$. We have put a dot in the circle to distinguish it from $\psi \circ \phi$ which is how the basic composition of maps $\psi\phi$ defined in L1§2 is denoted by some authors. For clarity, we may call $w\odot u$ the valuation theoretic composition of w by u. The exercise is to give a concrete example of $w\odot u$. Moreover, referring to (D32) for the definition of lexicographic direct sum $G_1 \boxplus G_2 =$ the set of all pairs $(g_1, g_2) \in G_1 \times G_2$ with componentwise addition and lexicographic order, in the notation of L5§5(Q22)(7) show that there exists a unique order preserving isomorphism

$$\theta : G_{w\odot u} \to G_w \boxplus G_u$$

such that for all $x \in R_{w\odot u}$ we have

$$\theta(w \odot u)(x)) = (w(x), u(\pi_w(x))).$$

HINT. The "Moreover" part is straightforward. For the example, let X, Y be indeterminates over a field k, and consider the valuation v of $L = k(X, Y)$ with value group $\mathbb{Z}^{[2]} =$ the set of all lexicographically ordered pairs of integers (m, n), and which is defined by putting $v(aX^mY^n) = (m, n)$ for all a in $k^\times$ and m, n in $\mathbb{Z}$. Then $v = w \odot u$ with real discrete valutions w and u.

EXERCISE (E57). [**Relabelling Prime Ideals**]. Justify the relabelling claimed in the proof of L5§5(Q23)(T98.2). In other words let I, J be ideals in a local ring R such that $J = M(R)$ with $(I : J)_R = I$, let $y \in R$ be such that $y \in J \setminus J^2$, and let $P_1, \dots, P_n$ be the distinct members of $\mathrm{ass}_R(R/I)$. Show that $P_1, \dots, P_n$ can be relabelled to get r and m in $\mathbb{N}$ with $m \leq r \leq n$ such that
 (i) $y \in P_i$ for $1 \leq i \leq m$ and $y \notin P_j$ for $m + 1 \leq j \leq r$,
 (ii) $P_j \not\subset P_i$ for $1 \leq i \leq m$ and $m + 1 \leq j \leq r$, and
 (iii) $\cup_{1 \leq i \leq r} P_i = \cup_{1 \leq i \leq n} P_i$. By an example show that (i) and (ii) cannot always be arranged with $r = n$.

HINT. Relabelling $P_1, \dots, P_n$, we can find $n \geq r \in \mathbb{N}$ such that
 (1) for each j with $r + 1 \leq j \leq n$ there exists some i with $1 \leq i \leq r$ for which $P_j \subset P_i$, and such that
 (2) there are no inclusion relations amongst $P_1, \dots, P_r$, i.e., $P_i \not\subset P_j$ for all $i \neq j$ in $\{1, \dots, r\}$. Then $\cup_{1 \leq i \leq r} P_i = \cup_{1 \leq i \leq n} P_i$. By further relabelling $P_1, \dots, P_r$, we can find $r \geq m \in \mathbb{N}$ such that $y \in P_i$ for $1 \leq i \leq m$ and $y \notin P_j$ for $m + 1 \leq j \leq r$.
 For the desired example, in the trivariate power series ring $R = k[[t, y, z]]$ over a field k, take $J = (t, y, z)R$ and $I = (t^2, ty)R$. Then $(I : J)_R = I$, and the distinct members of $\mathrm{ass}_R(R/I)$ are $P_1 = (t, y)R$ and $P_2 = (t)R$ with $n = 2$. Now if we take $r = n$ then the only choice of relabelling and m satisfying (i) is the given labelling

and $m = 1$, but (ii) does not hold for this choice because $P_2 \subset P_1$.

EXERCISE (E58). [**Suslin Localization**].

(1) Note that in the last sentence of the proof of L5§5(Q31)(128.4) we have $A(cX)A(dX)^{-1} \in \mathrm{GL}(n, [R, XR])$ because by putting $X = 0$ in this matrix it is obviously reduced to the $n \times n$ identity matrix.

(2) Show that the proof of L5§5(Q31)(128.4) remains valid if the substitution $(Y, Z) = (c, \lambda)$ is changed to the substitution $(X, Y, Z) = (1, cX, \lambda X)$.

(3) Also show that the proof works if the phrase "Consider the trivariate polynomial ring $\overline{R} = K[X, Y, Z]$ as an overring of R," to the phrase "Consider the bivariate polynomial ring $\overline{R} = K[Y, Z]$," if the reference to X is dropped everywhere else, and if the substitution $(Y, Z) = (c, \lambda)$ is changed to the substitution $(Y, Z) = (cX, \lambda X)$.

BIBLIOGRAPHY

[A01] S. S. Abhyankar, *Ramification Theoretic Methods In Algebraic Geometry,* Princeton University Press, Princeton, 1959.

[A02] S. S. Abhyankar, *Historical ramblings in algebraic geometry and related algebra,* American Mathematical Monthly, vol. 83 (1976), pages 409-448. Kyoto 1977, pages 240-414.

[A03] S. S. Abhyankar, *On the semigroup of a meromorphic curve, Part I,* Proceedings of the International Symposium on Algebraic geometry, Kyoto 1977, pages 240-414.

[A04] S. S. Abhyankar, *Algebraic Geometry for Scientists and Engineers,* American Mathematical Society, 1990.

[A05] S. S. Abhyankar, *Resolution of Singularities of Embedded Algebraic Surfaces,* Springer, 1998.

[A06] S. S. Abhyankar, *Local Analytic Geometry,* World Scientific Publishing Company, Singapore, 2001.

[A07] S. S. Abhyankar, *Two step descent in modular Galois theory, theorems of Burnside and Cayley, and Hilbert's thirteenth problem,* Proceedings of the Saskatoon Valuation Theory Conference of August 1999, Fields Institute Communications, American Mathematical Society, vol. 32 (2002), pages 1-31.

[Ask] E. H. Askwith, *Pure Geometry,* Cambridge University Press, 1921.

[AuB] M. Auslander and D. Buchsbaum, *Homological dimension in local rings,* Transactions of the American Mathematical Society, vol. 85 (1957), pages 390-405. pages 2941-2969.

[Bel] R. J. T. Bell, *An Elementary Treatise on Coordinate Geometry of Three Dimensions,* Cambridge University Press, 1920.

[Bez] É. Bézout, *Théorie générale des équationes algébriques* Paris, 1770.

[Bha] Bhaskaracharya, *Beejganit,* Ujjain, India, 1150.

[Bur] W. Burnside, *Theory of Groups of Finite Order,* Cambridge University Press, First Edition 1897 and Second Edition 1911.

[Chr] G. Chrystal, *Textbook of Algebra, Parts I and II,* Originally Published by A. C. Black, Ltd, Edinburgh (1886-1889), Many Chelsea Reprints.

[Coh] P. M. Cohn, *On the structure of the GL_2 of a ring,* I. H. E. S., vol. 30 (1966), pages 5-53.

[Dic] L. E. Dickson, *Linear Groups,* Teubner, 1901.

[Hil] D. Hilbert *Über die Theorie der algebraischen Formen,* Mathematische Annalen, vol. 36 (1890), 473-534.

[Gal] E. Galois, *Oeuvres Mathématique,* Journal de Mathématiqes Pures et Appl., vol. 11 (1846), pages 381-444.

[Jac] N. Jacobson, *Basic Algebra I and II*, Freeman, 1980.

[Jor] C. Jordan, *Traité des Substitutions et des Équations Algébriques*, Gauthier-Villa, 1870.

[Kle] F. Klein, *Entwicklung der Mathematik im neunzehnten Jahrhundert*, Berlin, 1926.

[KLi] P. Kleidman and M. W. Liebeck, *The Subgroup Structure of the Finite Classical Groups*, Cambridge University Press, 1990.

[Kr1] W. Krull, *Elementare und klassische Algebra vom moderne Standpunkt*, Parts I and II, De Gruyter, Berlin, 1952-1959.

[Kr2] W. Krull, *Idealtheorie*, Springer-Verlag, Berlin, 1935.

[Mat] H. Matsumura, *Commutative Algebra*, Benjamin, 1989.

[Mo1] E. H. Moore, *A doubly-infinite system of simple groups*, Proceedings of the Chicago Worlds Fair, 1893, pages 208-242.

[Mo2] E. H. Moore, *A Two-fold generalization of Fermat's theorem*, Bulletin of the American Mathematical Society, vol. 2 (1896), pages 189-199.

[Nag] M. Nagata, *Local Rings*, John Wiley, 1962.

[PaR] K. H. Parshall and D. E. Rowe, *The Emergence of the American Mathematical Research Community*, American Mathematical Society, 1994.

[Qui] D. Quillen, *Projective modules over polynomial rings*, Inventiones Mathematicae, vol. 36 (1976).

[Sal] G. Salmon, *Higher Plane Curves*, Dublin, 1852.

[Se1] J.-P. Serre, *Alġbre Local–Multiplicité*, Springer Lecture Notes in Mathematics, vol. 7 (1958).

[Se2] J.-P. Serre, *Sue les modules projective*, Séminaires Dubreil-Pisot, (1960).

[Ste] E. Steinitz, *Algebraische Theorie der Köper*, Chelsea Publishing Company, 1950.

[Su1] A. Suslin, *Projective modules over polynomial rings*, Dokl. Akad. Nauk S.S.R., vol. 26 (1976).

[Su2] A. Suslin, *On the structure of the special linear group over polynomial rings*, Izv. Akad. Nauk S.S.R. Ser. Math., vol. 41 (1977), pages 235-252.

[Syl] J. J. Sylvester, *On a general method of determining by mere inspection the derivations from two equations of any degree*, Philosophical Magazine, vol. 16 (1840), pages 132–135.

[Sy2] J. J. Sylvester, *On a theory of syzygetic relations of two rational integral functions, comprising an application to the theory of Sturm's functions, and that of the greatest common measure*, Philosophical Transactions of the Royal Society of London, vol. 143 (1853), pages 407-458.

[VYo] O. Veblen and J. T. Young *Projective Geometry I–II*, Ginn and Company, 1910-1918.

[Wey] H. Weyl, *Classical Groups*, Princeton University Press, 1939.

[ZaS] O. Zariski and P. Samuel, *Commutative Algebra I and II*, Springer-Verlag, Berlin, 1986.

DETAILED CONTENTS

C = Comment, D = Definition, E = Exercise, MU = Mute = No Label, N = Note, O = Observation, P = Problem, Q= Quest, R = Remark, T = Theorem or Lemma, X = Example

NOTATION-SYMBOLS

multiplicative group to additive group; 382

$\{x_1, \ldots, x_e\}$: set of elements; 4, 599

$(\gamma_1, \ldots, \gamma_m) \in S_1 \times \cdots \times S_m$: with $\gamma_i \in S_i$ for $1 \le i \le m$; 32, 603

$S_1 \times \cdots \times S_m$: cartesian product; 32, 603

$S^m = S_1 \times \cdots \times S_m$: where for $1 \le i \le m$ we have $S_i = S$; 32, 603

$X = X_1 \times \cdots \times X_m$: direct product of groups; 670

S^T: set of all maps or functions $T \to S$; 37

$(S^T)_{\text{finite}}$: maps of finite support; 37

$(S^T)_{\text{wellord}}$: maps of well ordered support; 37

$G^{[T]}$: lexicographic power; 57

$\sum_{i \in I} V_i$ (direct or internal direct sum): internal direct sum; 202

$\bigoplus_{i \in I} V_i$ (internal): internal direct sum; 203

$V_1 \oplus V_2 \oplus \cdots \oplus V_n$ (internal): internal direct sum; 203

$U_1 \oplus U_2 \oplus \cdots \oplus U_n$: direct sum; 204

$U^n = U_1 \oplus U_2 \oplus \cdots \oplus U_n$ with $U = U_i$ for $1 \le i \le n$: module theoretic direct sum of n copies of U; 204

$W_1 \boxplus \cdots \boxplus W_d$: lexicographic direct sum; 377

$(W_1 \oplus W_2 \oplus \cdots \oplus W_n)_\nu$: ν-th homogeneous component of a graded direct sum; 434

$^q V$: backward shift; 434

$^{(q_1, \ldots)} V$: backward shift; 434

$^{(q_1, \ldots)} R \to V$: canonical homomorphism or epimorphism; 434

$-E$: negative of a loset; 373, 683

$\uplus_{i \in E} \rho(W_i)$: lexicographic disjoint sum; 683

$\rho(W_1) \uplus \cdots \uplus \rho(W_d)$: lexicographic disjoint sum; 683

$\widehat{\bigsqcup}_{i \in F} T_i$: abstract disjoint union; 683

$\sqcup_{i \in E} S_i$: ordered disjoint union; 683

$S_1 \sqcup \cdots \sqcup S_d$: ordered disjoint union; 683

$\coprod$: disjoint union; 71, 653

$\coprod_{i \in I} W_i$: partition of a set; 71, 664

$\prod_{i \in I} U_i$: cartesian product of sets; 203

$U^I = \prod_{i \in I} U_i$ with $U = U_i$ for all $i \in I$: set-theoretic I-th power; 203, 612

$\prod_{i \in I} U_i$: if the U_i are modules then their cartesian product is made into a module called the direct product; 203, 612

$U^I = \prod_{i \in I} U_i$ with $U = U_i$ for all $i \in I$: module theoretic I-th power; 204, 612

$\prod_{i \in I} \alpha_i$: direct product of maps; 204

$\bigoplus_{i \in I} U_i$: direct sum of modules; 204

$U_\oplus^I = \bigoplus_{i \in I} U_i$: module theoretic restricted I-th power; 204, 612

$W_\oplus^E$: restricted power; 377, 683

$W_\boxplus^E$: lexicographic restricted power; 683

$\bigoplus_{i \in I} \alpha_i$: direct sum of maps; 204

$\alpha_1 \oplus \cdots \oplus \alpha_n : U_1 \oplus \cdots \oplus U_n \to D_1 \oplus \cdots \oplus D_n$: direct sum of maps; 205

$\alpha_1 \oplus \cdots \oplus \alpha_n : U \to D_1 \oplus \cdots \oplus D_n$ (diagonal direct sum): diagonal direct sum of maps; 205

$\prod_{i \in E}^{\text{wo}} W_i$: well ordered product; 376, 682

$W_{\text{wo}}^E = \prod_{i \in E}^{\text{wo}} W_i$ with $W_i = W$ for all $i \in E$: well ordered power; 377, 682

$\prod_{i \in G}^{\text{wo}} W_i$: well ordered product as a ring; 377

$W_{\text{wo}}^G = W((X))_G$: meromorphic series ring; 377

$\prod_{i \in E}^{\text{lex}} W_i$: lexicographic product; 377, 682

$W_{\text{lex}}^E = W^{[E]} = \prod_{i \in E}^{\text{lex}} W_i$ with $W_i = W$ for all $i \in E$: lexicographic power; 377, 682

$W_{\text{lex}}^d = W^{[d]} = W_{\text{lex}}^E$ if E is the finite set $\{1, \ldots, d\}$: lexicographic power; 41-42, 57, 377, 683

$A = (A_{ij})$: matrix of (i, j)-th entry A_{ij}; 61

A^*: adjoint of a matrix; 90

A': transpose of a matrix; 90, 484

μ^*: adjoint of a homomorphism; 446

A_X: derivative; 27

A_{X_i}: partial derivative; 28

G_+: positive part of ordered abelian group G; 53

G_{0+}, G_-, G_{0-}: subsets of ordered abelian group G; 53

G/H: factor group; 5, 600

G_u: stabilizer of u in G; 660

G_U: stabilizer of U in G; 660

$G_{[U]}$: elementwise stabilizer of U in G; 660

u^g: g-image of $u \in W$ with $g \in G$ under action $G \to W$; 659

u^H: H-orbit of $u \in W$ with $H \le G$ under action $G \to W$; 659

h^g: g-conjugate of h, i.e., ghg^{-1}; 659

R/I: residue class ring; 7, 601

$/R$: (algebraic, integral, etc.) over R; 12, 78

NOTATION-WORDS

ACC: ascending chain condition; 113, 608

A_n: alternating group; 5, 25, 600

$\mathrm{ann}_R B$: annihilator; 112

$\mathrm{App}_D(F)$: approximate root; 58-59

$\mathrm{ass}_R V$: associator or assassinator; 116, 217

$\mathrm{Aut}(G)$: automorphisms group; 7-9, 599-601

$\mathrm{avt}_k(\mathbb{A}_\kappa^N)$: affine variety set; 146

C: Cohn's matrix $\begin{pmatrix} 1+XY & X^2 \\ -Y^2 & 1-XY \end{pmatrix}$; 503

$C_G(H)$: centralizer of H in G; 654

$C(E)$: core of E; 375, 623, 681

cd: common divisor; 18

$\mathrm{ch}(R)$: characteristic; 21, 603

cm: common multiple; 21

$\mathrm{cmat}(A)$: characteristic matrix of a matrix A; 644

CM: Cohen-Macaulay; 281, 618

CMR: Cohen-Macaulay Ring; 354, 360

$\mathrm{cond}(R, \widetilde{R})$: conductor; 310, 620

$\mathrm{coprin}(R)^*$: set of all nonmaximal coprincipal prime ideals in R; 376

$\mathrm{cpol}(A)$: characteristic polynomial of a matrix A; 644

$\mathrm{cproj}_A J$: complementary prospectral variety; 534

$\mathrm{crk}(A)$: column rank; 90

$\mathrm{CR}(n)$: condition characterizing Chinese Remainder Theorem; 366

CRT: Chinese Remainder Theorem; 365

CT: Classification Theorem; 5, 28

$\frac{dA(X)}{dX}$: X-derivative $A(X)$; 10

$\frac{\partial A(X_1,...,X_m)}{\partial X_j}$: partial derivative of the function $A(X_1, \ldots, X_m)$ relative to the variable X_j; 11

$\frac{\partial(f_1,...,f_m)}{\partial(Y_1,...,Y_m)}$: jacobian matrix; 650

(D1) and (D2): distributive laws for ideals; 366

$D_{/x}$: dehomogenization map at x; 536

$D_{/x}(f) = F$: means F is the dehomogenization of f while f is a homogenization of F ; 536

$D_{/x}^*$: epimorphism induced by the dehomogenization map at x; 537

$\mathrm{D}(n, R)$: group of all diagonal matrices; 503

$\mathrm{DE}(n, R)$: group generated by $\mathrm{D}(n, R)$ and $\mathrm{SE}(n, R)$; 503

D_E: closed Dedekind map; 374, 623

D_E': open Dedekind map; 374, 623

$D_{F,E}$: Dedekind map of a pair; 375, 623

^{dc}: Dedekind completion E^{dc}; 375

DCC: descending chain condition; 129, 610

DD: Dedekind Domain; 364, 622

defo: degree form; 93

$\det(A)$: determinant of a matrix; 61

$\det(\mu)$: determinant of a homomorphism; 446

DFI: Domain with Factorization of Ideals; 364, 622

$\dim_K V$: dimension of V over K; 9, 602

$\dim(E)$: dimension of a model; 156, 633

$\dim(R)$: dimension of a ring; 127, 610

$\dim(U)$: dimension of a variety; 147-148

$\mathrm{Disc}_Y(f)$: discriminant; 101-04, 60d

$\mathrm{Disc}_Y^*(f)$: modified discriminant; 26, 515, 516, 603, 630

$\mathrm{dom}(\phi)$: domain of ϕ where for any map ϕ we have $\phi : \mathrm{dom}(\phi) \to \mathrm{ran}(\phi)$; 584, 598

$\mathrm{dpt}_R J$: depth of an ideal; 127, 149, 610

INDEX

Numbers are page numbers where the item is first defined or further enhanced. If an item is claimed to be on page x but not found there, it may be found on page $x \pm n$ with small n. Symbols are listed under NOTATION-SYMBOLS and are separated from their meanings by a colon. Abbreviations of mathematical terms are alphabetized under NOTATION-WORDS and they are separated from their long forms by a colon; we follow the sequence: ordinary letters, blackboard bold letters, script letters, Greek letters, German letters. These two NOTATIONAL LISTS apeear at the end of the INDEX.